Gravity

This textbook explores approximate solutions to general relativity and their consequences. It offers a unique presentation of Einstein's theory by developing powerful methods that can be applied to astrophysical systems.

Beginning with a uniquely thorough treatment of Newtonian gravity, the book develops post-Newtonian and post-Minkowskian approximation methods to obtain weak-field solutions to the Einstein field equations. The book explores the motion of self-gravitating bodies, the physics of gravitational waves, and the impact of radiative losses on gravitating systems. It concludes with a brief overview of alternative theories of gravity.

Ideal for graduate courses on general relativity and relativistic astrophysics, the book examines real-life applications, such as planetary motion around the Sun, the timing of binary pulsars, and gravitational waves emitted by binary black holes. Text boxes explore related topics and provide historical context, and over 100 exercises present challenging tests of the material covered in the main text.

Eric Poisson is Professor of Physics at the University of Guelph. He is a Fellow of the American Physical Society and serves on the Editorial Boards of *Physical Review Letters* and *Classical and Quantum Gravity*.

Clifford M. Will is Distinguished Professor of Physics at the University of Florida and J. S. McDonnell Professor Emeritus at Washington University in St. Louis. He is a member of the US National Academy of Sciences, and Editor-in-Chief of *Classical and Quantum Gravity*. He is well known for his ability to bring science to broad audiences.

Gravity

Newtonian, Post-Newtonian, Relativistic

ERIC POISSON
University of Guelph

CLIFFORD M. WILL
University of Florida

CAMBRIDGE
UNIVERSITY PRESS

University Printing House, Cambridge CB2 8BS, United Kingdom

One Liberty Plaza, 20th Floor, New York, NY 10006, USA

477 Williamstown Road, Port Melbourne, VIC 3207, Australia

314-321, 3rd Floor, Plot 3, Splendor Forum, Jasola District Centre, New Delhi - 110025, India

79 Anson Road, #06-04/06, Singapore 079906

Cambridge University Press is part of the University of Cambridge.

It furthers the University's mission by disseminating knowledge in the pursuit of education, learning and research at the highest international levels of excellence.

www.cambridge.org
Information on this title: www.cambridge.org/9781107032866

First published 2014
Reprinted 2018

A catalogue record for this publication is available from the British Library

ISBN 978-1-107-03286-6 Hardback

Contents

Boxes

Preface

During the past forty years or so, spanning roughly our careers as teachers and research scientists, Einstein's theory of general relativity has made the transition from a largely mathematical curiosity with limited relevance to the real world to arguably the centerpiece of our effort to understand the universe on all scales.

At the largest scales, those of the universe as a whole, cosmology and general relativity are joined at the hip. You can't do one without the other. At the smallest scales, those of the Planck time, Planck length, and Planck energy, general relativity and particle physics are joined at the hip. String theory, loop quantum gravity, the multiverse, branes and bulk – these are arenas where the geometry of Einstein and the physics of the quantum may be inextricably linked. These days it seems that you can't do one without the other.

At the intermediate scales that interest astronomers, general relativity and astrophysics are becoming increasingly linked. You can still do one without the other, but it's becoming harder. One of us is old enough to remember a time when the majority of astronomers felt that black holes would never amount to much, and that it was a waste of time to worry about general relativity. Today black holes and neutron stars are everywhere in the astronomy literature, and gravitational lensing – the tool that relies on the relativistic bending of light – is used for everything from measuring dark energy to detecting exoplanets.

Given the surge of interest in general relativity, it is no surprise that the last several years have witnessed the publication of a multitude of new textbooks on Einstein's theory. Many of them are cut from a very similar cloth: they cover the fundamentals of the theory at an introductory level, including the spacetime formulation of special relativity, elements of differential geometry, the Einstein field equations, black holes, gravitational waves, and cosmology. This book is cut from a very different cloth. Here you will not (spoiler alert!) find any discussion of cosmology, and although black holes will appear in many places, you will not find anything about the joys and wonders of the Kerr metric.

This book is about *approximations* to Einstein's theory of general relativity, and their applications to planetary motion around the Sun, to the timing of binary pulsars, to gravitational waves emitted by binary black holes, and to many other real-life, astrophysical systems.

The first approximation to general relativity is, of course, Newton's gravity. Although the theories are conceptually very different, it must be admitted that the overwhelming majority of phenomena in the universe can be very adequately described by the laws of Newtonian gravity. To a high degree of accuracy, Newton rules the Sun, the Earth, the solar system, all normal stars, galaxies, and clusters of galaxies. Accordingly, almost a quarter of this book is devoted to Newton's theory. This choice reflects one of our (not so) hidden agendas. During our careers of teaching general relativity and advising graduate students, we have too often encountered students who are superbly motivated to study Einstein's theory, but who cannot say more than "inverse square law" and "elliptical orbits" when asked what they know about Newtonian gravity. In our view, general relativity is a theory of gravity, and if you wish to comprehend its importance for astrophysics, you must first master what Newton has to say about gravitating bodies, rotating bodies, tidally interacting bodies, perturbed Keplerian orbits, and so on. We therefore make it our mission, in Chapters 1, 2, and 3, to provide a thorough discussion of the wonders of Newtonian gravity.

In the following two chapters we quickly review special relativity, the foundations of general relativity as a metric theory of gravity, the mathematical formulation of the theory, and its most famous solution, the Schwarzschild metric. We emphasize that Chapters 4 and 5 are very much a minimal package. The coverage is sufficient for our intended purposes in the remainder of the book, but it is no substitute for a proper education in general relativity that can be acquired from the traditional textbooks.

We get to our main point by Chapter 6. This is the development of a set of systematic schemes, known as post-Minkowskian theory and post-Newtonian theory, for obtaining approximate solutions to the Einstein field equations. The idea is to go from the exact theory, which governs the behavior of arbitrarily strong fields, such as those near black holes, to a useful approximation that applies to weak fields, such as those inside and near the Sun, those inside and near white dwarfs, and those at a safe distance from neutron stars and black holes. The approximation, of course, reproduces the predictions of Newtonian theory, but we go beyond this and formulate a method of approximation that can be pushed systematically to higher and higher order, and generate increasingly accurate descriptions of a weak gravitational field. Along the way, we make the case that this approximation can also describe important situations involving compact objects such as neutron stars and black holes; not the up-close-and-personal geometry of a compact object, to be sure, but its motion around another body (compact or not), so long as the mutual gravitational attraction is weak.

This program occupies us through Chapters 6, 7, 8, and 9. In Chapter 10 we apply the approximation methods to the description of relativistic effects on the dynamics of the solar system, the measurement of time on the Earth's surface and in orbit, the bending of light by a massive body, and the dynamics of spinning bodies. In Chapter 11 we explore the rich physics of gravitational waves, and in Chapter 12 we investigate the impact of radiative losses on the dynamics of gravitating systems. We conclude the book in Chapter 13 with a brief overview of alternative theories of gravity.

The central theme of this book is therefore the physics of weak gravitational fields. The reader may object that we give up too much by eliminating strong fields from our discussion; after all, exact solutions to the Einstein field equations describe the full richness of curved spacetime, whether strong or weak. Unfortunately, there are *extremely few* exact solutions to Einstein's equations that are physically interesting. The Schwarzschild solution is obviously interesting and important, and so is the Kerr solution for rotating black holes (although the Kerr metric makes no appearance in this book). But no exact solution to Einstein's equations has ever been found that describes a simple double-star system in orbital motion. And no exact solution is known that describes any kind of bounded, physical system that radiates gravitational waves.

The problem is that Einstein's field equations are so complicated that it is almost always necessary to impose a high degree of symmetry (spherical symmetry, spatial homogeneity, stationarity, etc.) in order to make progress toward finding a solution. Furthermore, a solution to Einstein's equations is, by definition, a *spacetime*; it must encompass the entire past history and future fate of the system, everywhere in space. For a binary-star system, for example, the solution must, at least in principle, run from the distant past, when a tenuous cloud of gas coalesced to form the stars, all the way to the distant future, when the stars, having possibly collapsed to form neutron stars or black holes along the way, have merged into a single object (possibly a single black hole); it must also describe the gravitational waves that are generated during the entire time by the orbital motion and merger of the two stars, and by the relaxation of the merged object to a final stationary state. It should not come as a surprise that nobody has found a solution that describes such a wide range of phenomena. Ironically, a body of beautiful mathematical work has demonstrated conclusively that given suitable initial conditions, a solution to Einstein's equations *always exists*, at least within a specified part of the spacetime. Sadly, such existence theorems do not tell us how to find such solutions.

Often, when one talks about exact solutions to the Einstein field equations, one means analytic solutions, or solutions that can be expressed in terms of reasonably well-known mathematical functions. Perhaps this is too restrictive. What about numerical solutions? Given a sufficiently powerful computer, it should be possible to solve Einstein's equations numerically without imposing any symmetries. After all, the field equations of general relativity are partial differential equations, and these can readily be converted into the kind of difference equations that are suited to digital computing. This has turned out to be a very difficult challenge. Part of the difficulty is computational: simulation of the simplest spacetimes requires enormous computational power and memory. Part of the difficulty is mathematical: one must identify, from a broad spectrum of possibilities, a formulation of the field equations that is best suited for numerical work. There has been enormous progress on these fronts in the last 20 years, and spectacular breakthroughs have occurred in the last ten. Today (in 2013), numerical relativity is a major sub-branch of gravitational physics. It is now possible to simulate the final dozen orbits of two inspiralling and merging compact objects (black holes or neutron stars), the gravitational collapse of a dead stellar core on its way to form a supernova, the formation and evolution of accretion disks around black holes, the interaction of a binary neutron-star system with the strong magnetic fields it supports, and the generation of gravitational waves by such strongly gravitating systems.

As spectacular as this progress has been, at present it is still not possible to simulate the final thousand orbits of a compact binary inspiral. The limitations are both technical (a vast range of grid resolutions is required) and computational (insufficient memory and speed, even with the largest parallel processors). But approximately 990 of those orbits can be described by the weak-field methods that we develop in this book. It was found that there is a very good agreement between the approximation methods and those of numerical relativity when their domains of applicability overlap. So in addition to their obvious applications to the solar system, the weak-field methods have proved to be unreasonably effective in describing situations, such as the late stages of binary inspirals, where the fields are not so weak and the motions not so slow. And the combination of these methods with numerical relativity has proved to be a powerful tool for many important problems.

The vast majority of high-precision experiments that were carried out to test general relativity can be fully understood on the basis of the post-Newtonian methods that we develop in this book. And even though the departures from Newtonian gravity are very, very small on and around Earth, modern technology has made them not only detectable, but also *essentially important* in the precision measurement of time. A well-known example is the Global Positioning System, which simply would not work if relativistic corrections were not taken into account. Today every relativist proudly points to the GPS as an example – admittedly, perhaps, the only example – of a practical application of general relativity. We describe how this comes about in Chapter 10.

Finally, a central motivation for this book is the expectation that soon after its initial publication, gravitational waves will be measured directly and routinely, and that gravitational-wave astronomy, enabled by ground-based laser interferometers, by pulsar timing arrays, and possibly by a future space-based antenna, will become a new standard way of "listening" to the universe. The approximation methods that we develop in this book are *the* tools for understanding gravitational radiation, and it is our hope that students and researchers wishing to join this new scientific venture will turn to our book to learn and master these tools.

Acknowledgments

We would like to acknowledge colleagues and students who contributed important comments and corrections during the writing of this book: Emanuele Berti, Ryan Lang, Saeed Mirshekari, Laleh Sadeghian, Nico Yunes, and Ian Vega.

CMW is grateful to Washington University in St. Louis for its support during the early phase of writing, particularly during a sabbatical leave in 2010–2011. He also thanks the Institut d'Astrophysique de Paris for its hospitality during this sabbatical, and during extended stays in 2009, 2012, and 2013. Finally he is grateful to the US National Science Foundation for support under various grants.

EP thanks the University of Guelph for a sabbatical leave in 2008–2009, during which the writing of this book was initiated. He is grateful to the Canadian Institute for Theoretical Astrophysics at the University of Toronto for its generous hospitality during this sabbatical. Research support from the Natural Sciences and Engineering Research Council is also gratefully acknowledged. The writing of this book coincided with a stint as department chair during the years 2008–2013; this project did much to preserve the sanity of the co-author.

1 Foundations of Newtonian gravity

The central theme of this book is gravitation in its weak-field aspects, as described within the framework of Einstein's general theory of relativity. Because Newtonian gravity is recovered in the limit of very weak fields, it is an appropriate entry point into our discussion of weak-field gravitation. Newtonian gravity, therefore, will occupy us within this chapter, as well as the following two chapters.

There are, of course, many compelling reasons to begin a study of gravitation with a thorough review of the Newtonian theory; some of these are reviewed below in Sec. 1.1. The reason that compels us most of all is that although there is a vast literature on Newtonian gravity – a literature that has accumulated over more than 300 years – much of it is framed in old mathematical language that renders it virtually impenetrable to present-day students. This is quite unlike the situation encountered in current presentations of Maxwell's electrodynamics, which, thanks to books such as Jackson's influential text (1998), are thoroughly modern. One of our main goals, therefore, is to submit the classical literature on Newtonian gravity to a Jacksonian treatment, to modernize it so as to make it accessible to present-day students. And what a payoff is awaiting these students! As we shall see in Chapters 2 and 3, Newtonian gravity is most generous in its consequences, delivering a whole variety of fascinating phenomena.

Another reason that compels us to review the Newtonian formulation of the laws of gravitation is that much of this material will be recycled and put to good use in later chapters of this book, in which we examine relativistic aspects of gravitation. Newtonian gravity, in this context, is a necessary warm-up exercise on the path to general relativity.

In this chapter we describe the foundations of the Newtonian theory, and leave the exploration of consequences to Chapters 2 and 3. We begin in Sec. 1.1 with a discussion of the domain of validity of the Newtonian theory. The main equations are displayed in Sec. 1.2 and derived systematically in Secs. 1.3 and 1.4. The gravitational fields of spherical and nearly spherical bodies are described in Sec. 1.5, and in Sec. 1.6 we derive the equations that govern the center-of-mass motion of extended fluid bodies.

Gravitation rules the world, and before Einstein ruled gravitation, Newton was its king. In this chapter and the following two we pay tribute to the king.

1.1 Newtonian gravity

The gravitational theory of Newton is an extremely good representation of gravity for a host of situations of practical and astronomical interest. It accurately describes the structure

Table 1.1 Values of ε for representative gravitating systems.

System	ε
Earth's orbit around the Sun	10^{-8}
Solar system's orbit around the galaxy	10^{-6}
Surface of the Sun	10^{-5}
Surface of a white dwarf	10^{-4}
Surface of a neutron star	0.1
Event horizon of a black hole	~ 1

of the Earth and the tides raised on it by the Moon and Sun. It gives a detailed account of the orbital motion of the Moon around the Earth, and of the planets around the Sun. To be sure, it is now well established that the Newtonian theory is not an exact description of the laws of gravitation. As early as the middle of the 19th century, observations of the orbit of Mercury revealed a discrepancy with the prediction of Newtonian gravity. This famous discrepancy in the rate of advance of Mercury's perihelion was resolved by taking into account the relativistic corrections of Einstein's theory of gravity. The high precision of modern measuring devices has made it possible to detect relativistic effects in the lunar orbit, and has made it necessary to take relativity into account in precise tracking of planets and spacecraft, as well as in accurate measurements of the positions of stars using techniques such as Very Long Baseline Radio Interferometry (VLBI). Even such mundane daily activities as using the Global Positioning System (GPS) to navigate your car in a strange city require incorporation of special and general relativistic effects on the observed rates of the orbiting atomic clocks that regulate the GPS network. But apart from these specialized situations requiring very high precision, Newtonian gravity rules the solar system.

Newtonian gravity also rules for the overwhelming majority of stars in the universe. The structure and evolution of the Sun and other main-sequence stars can be completely and accurately treated using Newtonian gravity. Only for extremely compact stellar objects, such as neutron stars and, of course, black holes, is general relativity important. Newtonian gravity is also perfectly capable of handling the structure and evolution of galaxies and clusters of galaxies. Even the evolution of the largest structures in the universe, the great galactic clusters, sheets and voids, whose formation is dominated by the gravitational influence of dark matter, are frequently modelled using numerical simulations based on Newton's theory, albeit with the overall expansion of the universe playing a significant role.

Generally speaking, the criterion that we use to decide whether to employ Newtonian gravity or general relativity is the magnitude of a quantity called the "relativistic correction factor" ε:

$$\varepsilon \sim \frac{GM}{c^2 r} \sim \frac{v^2}{c^2}, \tag{1.1}$$

where G is the Newtonian gravitational constant, c is the speed of light, and where M, r, and v represent the characteristic mass, separation or size, and velocity of the system under consideration. The smaller this factor, the better is Newtonian gravity as an approximation. Table 1.1 shows representative values of ε for various systems.

Context is everything, of course. It is now accepted that general relativity, not Newtonian theory, is the "correct" classical theory of gravitation. But in the appropriate context, Newton's theory may be completely adequate to do the job at hand to the precision required. For example, Table 1.1 implies that a description of planetary motion around the Sun, at a level of accuracy limited to (say) one part in a million, can safely be based on the Newtonian laws. The Newtonian theory can also be exploited to calculate the internal structure of white dwarfs, provided that one is content with a level of accuracy limited to one part in one thousand. For more compact objects, such as neutron stars and black holes, Newtonian theory is wholly inadequate.

1.2 Equations of Newtonian gravity

Most undergraduate textbooks begin their treatment of Newtonian gravity with Newton's second law and the inverse-square law of gravitation:

$$m_I \boldsymbol{a} = \boldsymbol{F}\,, \tag{1.2a}$$

$$\boldsymbol{F} = -\frac{G m_G M}{r^2}\boldsymbol{n}\,. \tag{1.2b}$$

In the first equation, $\boldsymbol{F}$ is the force acting on a body of inertial mass m_I situated at position $\boldsymbol{r}(t)$, and $\boldsymbol{a} = d^2\boldsymbol{r}/dt^2$ is its acceleration. In the second equation, the force is assumed to be gravitational in nature, and to originate from a gravitating mass situated at the origin of the coordinate system. The force law involves m_G, the passive gravitational mass of the first body at $\boldsymbol{r}$, while M is the active gravitational mass of the second body. The quantity G is Newton's constant of gravitation, equal to $6.6738 \pm 0.0008 \times 10^{-11}\ \mathrm{m}^2\,\mathrm{kg}^{-1}\,\mathrm{s}^{-2}$. The force is attractive, it varies inversely with the square of the distance $r := |\boldsymbol{r}| = (x^2 + y^2 + z^2)^{1/2}$, and it points in the direction opposite to the unit vector $\boldsymbol{n} := \boldsymbol{r}/r$. An alternative form of the force law is obtained by writing it as the gradient of a potential $U = GM/r$, so that

$$\boldsymbol{F} = m_G \nabla U\,. \tag{1.3}$$

This *Newtonian potential* will play a central role in virtually all chapters of this book.

If the inertial and passive gravitational masses of the body are equal to each other, $m_I = m_G$, then the acceleration of the body is given by $\boldsymbol{a} = \nabla U$, and its magnitude is $a = GM/r^2$. Under this condition the acceleration is independent of the mass of the body. This statement is known as the *weak equivalence principle* (WEP), and it was a central element in Einstein's thinking on his way to the concepts of curved spacetime and general relativity. Although Newton did not explicitly use our formulation in terms of inertial and passive masses, he was well aware of the significance of their equality. In fact, he regarded this equality as so fundamental that he opened his treatise *Philosophiae Naturalis Principia Mathematica* with a discussion of it; he even alluded to his own experiments showing that the periods of pendulums were independent of the mass and type of material suspended, which establishes the equality of inertial and passive masses (he referred to them as the "quantity" and "weight" of bodies, respectively). Twentieth-century experiments have shown that the two types of mass are equal to parts in 10^{13} for a wide variety of materials (see Box 1.1).

Box 1.1

Tests of the weak equivalence principle

A useful way to discuss experimental tests of the weak equivalence principle is to parameterize the way it could be violated. In one parameterization, we imagine that a body is made up of atoms, and that the inertial mass m_I of an atom consists of the sum of all the mass and energy contributions of its constituents. But we suppose that the different forms of energy may contribute differently to the gravitational mass m_G than they do to m_I. One way to express this is to write

$$m_G = m_I(1+\eta),$$

where η is a dimensionless parameter that measures the difference. Because different forms of energy arising from the relevant subatomic interactions (such as electromagnetic and nuclear interactions) contribute different amounts to the total, depending on atomic structure, η could depend on the type of atom. For example, electrostatic energy of the nuclear protons contributes a much larger fraction of the total mass for high-Z atoms than for low-Z atoms.

Using this parameterization, we find from Eq. (1.2) that the acceleration of the body is given by

$$\boldsymbol{a} = -\frac{m_G}{m_I}\frac{GM}{r^2}\boldsymbol{n} = -(1+\eta)\frac{GM}{r^2}\boldsymbol{n}.$$

The difference in acceleration between two materials of different composition will then be given by

$$\Delta\boldsymbol{a} = \boldsymbol{a}_1 - \boldsymbol{a}_2 = -(\eta_1-\eta_2)\frac{GM}{r^2}\boldsymbol{n}.$$

One way to place a bound on $\eta_1-\eta_2$ is to drop two different objects in the Earth's gravitational field ($g = GM/r^2 \approx 9.8\,\mathrm{m\,s^{-2}}$), and compare their accelerations, or how long they take to fall. Although legend has it that Galileo Galilei verified the equivalence principle by dropping objects off the Leaning Tower of Pisa around 1590, in fact experiments like this had already been performed and were well known to Galileo; if he did indeed drop things off the Tower, he may simply have been performing a kind of classroom demonstration of an established fact for his students. Unfortunately, the "Galileo approach" is plagued by experimental errors, such as the difficulty of releasing the objects at exactly the same time, by the effects of air drag, and by the short time available for timing the drop.

A better approach is to balance the gravitational force (which depends on m_G) by a support force (which depends on m_I); the classic model is the pendulum experiments performed by Newton and reported in his *Principia*. The period of the pendulum depends on m_G/m_I, g, and the length of the pendulum. These experiments are also troubled by air drag, by errors in measuring or controlling the length of the pendulum, and by errors in timing the swing.

The best approach for laboratory tests was pioneered by Baron Roland von Eötvös, a Hungarian geophysicist working around the turn of the 20th century. He developed the *torsion balance*, schematically consisting of a rod suspended by a wire near its mid-point, with objects consisting of different materials attached at each end.

The point where the wire is attached to achieve a horizontal balance depends only on the gravitational masses of the two objects, so this configuration does not tell us anything. But if an additional gravitational force can be applied in a direction perpendicular to the supporting wire, and if there is a difference in m_G/m_I for the two bodies, then the rod will rotate in one direction or the other and the wire will twist until the restoring force of the twisted wire halts the rotation. There is no effect when m_G/m_I is the same for the two bodies. The additional force could be provided by a nearby massive body in the laboratory, a nearby mountain, the Sun, or the galaxy. Eötvös realized that, because of the centrifugal force produced by the rotation of the Earth, the wire hangs not exactly vertically, but is tilted slightly toward the south; at the latitude of Budapest, Hungary, the angle of tilt is about 0.1 degrees. Thus the gravitational acceleration of the Earth has a small component, about $g/400$, perpendicular to the wire, in a northerly direction. By slowly rotating the whole apparatus carefully about the vertical direction, Eötvös could compare the twist in two opposite orientations of the rod, and thereby eliminate a number of sources of error.

Eötvös found no measurable twist, within his experimental errors, for many different combinations of materials, and he was able to place an upper limit of $|\eta_1 - \eta_2| < 3 \times 10^{-9}$, corresponding to a limit on any difference in acceleration of the order of 7×10^{-11} m s^{-2}. Even though the driving acceleration is only a tiny fraction of g, there is an enormous gain in sensitivity to tiny accelerations, mainly because the apparatus is almost static and can be observed for long periods of time. Torsion balance experiments were improved by Robert Dicke in Princeton and Vladimir Braginsky in Moscow during the 1960s and 1970s, and again during the 1980s as part of a search for a hypothetical "fifth" force (no evidence for such a force was found). The most recent experiments, performed notably by the "Eöt-Wash" group at the University of Washington, Seattle, have reached precisions of a few parts in 10^{13}; these experiments used the Sun or the galaxy as the source of gravity.

All these experiments exploit only a tiny fraction of the available acceleration. The only way to make full use of g while maintaining high sensitivity to acceleration differences is to design a "perpetual" Galileo drop experiment, namely by putting the different bodies in orbit around the Earth. Various satellite tests of the equivalence principle are in preparation, with the goal of reaching sensitivities ranging from 10^{-15} to 10^{-18}. Such experiments come with a high monetary cost: compared to laboratory experiments, space experiments are extraordinarily expensive.

Another test of the equivalence principle was carried out using the Earth–Moon system. The two bodies have slightly different compositions, with the Earth dominated by its iron–nickel core, and the Moon dominated by silicates. If there were a violation of the equivalence principle, the two bodies would fall with different accelerations toward the Sun, and this would have an effect on the Earth–Moon orbit. Lunar laser ranging is a technique of bouncing laser beams off reflectors placed on the lunar surface during the American and Soviet lunar landing programs of the 1970s, and it has reached the capability of measuring the Earth–Moon distance at the sub-centimeter level. No evidence for such a perturbation in the Earth–Moon distance has been found, so that the Earth and the Moon obey the equivalence principle to a few parts in 10^{13}. We describe the laser ranging measurements of the Moon in more detail in Box 13.2.

The weak equivalence principle is one of the most important foundational elements of relativistic theories of gravity. We will return to it in Chapter 5, on our way to general relativity.

We shall assume that the weak equivalence principle holds perfectly, and make this an axiom of Newtonian gravity. We shall return to this principle in Chapter 5 and present it as an essential foundational element of general relativity, and we shall return to it again in Chapter 13 – in a different version known as the *strong equivalence principle* – and present it as a highly non-trivial property of massive, self-gravitating bodies in general relativity.

The weak equivalence principle allows us to rewrite Eqs. (1.2) in the form of an equation of motion for the body at $\boldsymbol{r}(t)$, and a field equation for the potential U:

$$\boldsymbol{a} = \nabla U\,, \tag{1.4a}$$

$$U = GM/r\,. \tag{1.4b}$$

These equations are limited in scope, and they do not yet form the final set of equations that will be adopted as the foundations of Newtonian gravity. Their limitation has to do with the fact that they apply to a point mass situated at $\boldsymbol{r}(t)$ being subjected to the gravitational force produced by another point mass situated at the origin of the coordinate system. We are interested in much more general situations. First, we wish to consider the motion of extended bodies made up of continuous matter (solid, fluid, or gas), allowing the bodies to be of arbitrary size, shape, and constitution, and possibly to evolve in time according to their own internal dynamics. Second, we wish to consider an arbitrary number of such bodies, and to put them all on an equal footing; each body will be subjected to the gravity of the remaining bodies, and each will move in response to this interaction.

These goals can be achieved by generalizing the primitive Eqs. (1.4) to a form that applies to a continuous distribution of matter. We shall perform this generalization in Secs. 1.3 and 1.4, but to complete the discussion of this section, we choose to immediately list and describe the resulting equations.

Our formulation of the fundamental equations of Newtonian gravity relies on a fluid description of matter, in which the matter distribution is characterized by a mass-density field $\rho(t, \boldsymbol{x})$, a pressure field $p(t, \boldsymbol{x})$, and a velocity field $\boldsymbol{v}(t, \boldsymbol{x})$; these quantities depend on time t and position $\boldsymbol{x}$ within the fluid. Our formulation relies also on the Newtonian potential $U(t, \boldsymbol{x})$, which also depends on time and position, and which provides a description of the gravitational field. The equations that govern the behavior of the matter are the *continuity equation*,

$$\frac{\partial \rho}{\partial t} + \nabla \cdot (\rho \boldsymbol{v}) = 0\,, \tag{1.5}$$

which expresses the conservation of mass, and *Euler's equation*,

$$\rho \frac{d\boldsymbol{v}}{dt} = \rho \nabla U - \nabla p\,, \tag{1.6}$$

which is the generalization of Eq. (1.4a) to continuous matter; here

$$\frac{d}{dt} := \frac{\partial}{\partial t} + \boldsymbol{v} \cdot \nabla \tag{1.7}$$

is the convective time derivative associated with the motion of fluid elements. The equation that governs the behavior of the gravitational field is *Poisson's equation*

$$\nabla^2 U = -4\pi G \rho\,, \tag{1.8}$$

where

$$\nabla^2 := \frac{\partial^2}{\partial x^2} + \frac{\partial^2}{\partial y^2} + \frac{\partial^2}{\partial z^2} \tag{1.9}$$

is the familiar Laplacian operator; Poisson's equation (known after its originator Siméon Denis Poisson, who unfortunately is not related to either author of this book) is the generalization of Eq. (1.4b) to continuous matter.

As was stated previously, these equations will be properly introduced in the following two sections. To complete the formulation of the theory we must impose a relationship between the pressure and the density of the fluid. This relationship, known as the *equation of state*, takes the general form of

$$p = p(\rho, T, \cdots), \tag{1.10}$$

in which the pressure is expressed as a function of the density, temperature, and possibly other relevant variables such as chemical composition. The equation of state encodes information about the microphysics that governs the fluid, and this information must be provided as an input in most applications of the theory.

A complete description of a physical situation involving gravity and a distribution of matter can be obtained by integrating Eqs. (1.5), (1.6), and (1.8) simultaneously and self-consistently. The solutions must be subjected to suitable boundary conditions, which will be part of the specification of the problem. All of Newtonian gravity is contained in these equations, and all associated phenomena follow as consequences of these equations.

1.3 Newtonian field equation

In this section we examine the equations that govern the behavior of the gravitational field, and show how Eq. (1.8) is an appropriate generalization of the more primitive form of Eq. (1.4b).

We recall that the relation $U = GM/r$ applies to a point body of active gravitational mass M situated at the origin of the coordinate system. Suppose that we are given an arbitrary number N of point bodies, and that we assign to each one a label $A = 1, 2, \cdots, N$. The mass and position of each body are then denoted M_A and $\boldsymbol{r}_A(t)$, respectively. If we *assume* that the total Newtonian potential U is a linear superposition of the individual potentials U_A created by each body, we have that the potential at position $\boldsymbol{x}$ is given by

$$U = \sum_A U_A = G \sum_A \frac{M_A}{|\boldsymbol{x} - \boldsymbol{r}_A|}. \tag{1.11}$$

The generalization of this relation to a continuous distribution of matter is straightforward. We convert the discrete sum $\sum_A M_A$ to a continuous integral $\int d^3x' \rho(t, \boldsymbol{x}')$, and we replace the discrete positions $\boldsymbol{r}_A$ with the continuous integration variable $\boldsymbol{x}'$. The result is

$$U(t, \boldsymbol{x}) = G \int \frac{\rho(t, \boldsymbol{x}')}{|\boldsymbol{x} - \boldsymbol{x}'|} d^3x', \tag{1.12}$$

one of the key defining equations for the Newtonian potential. The integral can be evaluated as soon as the density field $\rho(t, \boldsymbol{x}')$ is specified, regardless of whether ρ is a proper solution to the remaining fluid equations. As such, Eq. (1.12) gives U as a *functional* of an arbitrary function ρ. The potential, however, will be physically meaningful only when ρ itself is physically meaningful, which means that it must be a proper solution to the continuity and Euler equations.

The integral equation (1.12) can easily be transformed into a differential equation for the Newtonian potential U. The transformation relies on the identity

$$\nabla^2 \frac{1}{|\boldsymbol{x}-\boldsymbol{x}'|} = -4\pi\delta(\boldsymbol{x}-\boldsymbol{x}'), \tag{1.13}$$

in which $\delta(\boldsymbol{x}-\boldsymbol{x}') := \delta(x-x')\delta(y-y')\delta(z-z')$ is a three-dimensional delta function defined by the properties

$$\delta(\boldsymbol{x}-\boldsymbol{x}') = 0 \qquad \text{when } \boldsymbol{x} \neq \boldsymbol{x}', \tag{1.14a}$$

$$f(\boldsymbol{x})\delta(\boldsymbol{x}-\boldsymbol{x}') = f(\boldsymbol{x}')\delta(\boldsymbol{x}-\boldsymbol{x}') \qquad \text{for any smooth function } f(\boldsymbol{x}), \tag{1.14b}$$

$$\int \delta(\boldsymbol{x}-\boldsymbol{x}')\, d^3x' = 1 \qquad \text{for any domain of integration that encloses } \boldsymbol{x}. \tag{1.14c}$$

These properties further imply that $\delta(\boldsymbol{x}'-\boldsymbol{x}) = \delta(\boldsymbol{x}-\boldsymbol{x}')$. The identity of Eq. (1.13) is derived in Box 1.2. If we apply the Laplacian operator on both sides of Eq. (1.12) and exchange the operations of integration and differentiation on the right-hand side, we obtain

$$\begin{aligned}
\nabla^2 U &= G \int \rho(t, \boldsymbol{x}')\nabla^2 \frac{1}{|\boldsymbol{x}-\boldsymbol{x}'|}\, d^3x' \\
&= -4\pi G \int \rho(t, \boldsymbol{x}')\delta(\boldsymbol{x}-\boldsymbol{x}')\, d^3x' \\
&= -4\pi G\rho(t, \boldsymbol{x});
\end{aligned}$$

the identity was used in the second step, and the properties of the delta function displayed in Eq. (1.14) allowed us to evaluate the integral. The end result is Poisson's equation,

$$\nabla^2 U = -4\pi G\rho, \tag{1.15}$$

whose formulation was anticipated in Eq. (1.8).

It is possible to proceed in the opposite direction, and show that Eq. (1.12) provides a solution to Poisson's equation (1.15). A powerful tool in the integration of differential equations is the *Green's function* $G(\boldsymbol{x}, \boldsymbol{x}')$, a function of a field point $\boldsymbol{x}$ and a source point $\boldsymbol{x}'$. In the specific context of Poisson's equation, the Green's function is required to be a solution to

$$\nabla^2 G(\boldsymbol{x}, \boldsymbol{x}') = -4\pi\delta(\boldsymbol{x}-\boldsymbol{x}'), \tag{1.16}$$

which is recognized as a specific case of the general differential equation, corresponding to a point mass situated at $\boldsymbol{x}'$. Armed with such an object, a formal solution to Eq. (1.15) can be expressed as

$$U(t, \boldsymbol{x}) = G \int G(\boldsymbol{x}, \boldsymbol{x}')\rho(t, \boldsymbol{x}')\, d^3x'; \tag{1.17}$$

Box 1.2 **Proof that $\nabla^2|\boldsymbol{x}-\boldsymbol{x}'|^{-1} = -4\pi\delta(\boldsymbol{x}-\boldsymbol{x}')$**

To simplify the proof of Eq. (1.13) we set $\boldsymbol{x}' = \boldsymbol{0}$ without loss of generality; this can always be achieved by a translation of the coordinate system. This gives rise to the simpler equation

$$\nabla^2 r^{-1} = -4\pi\delta(\boldsymbol{x}), \tag{1}$$

with $r := |\boldsymbol{x}|$.

We first show that $\nabla^2 r^{-1} = 0$ whenever $\boldsymbol{x} \neq \boldsymbol{0}$. Derivatives of r^{-1} can be evaluated with the help of the identities

$$\frac{\partial r}{\partial x^j} = n_j, \qquad \frac{\partial n_j}{\partial x^k} = \frac{\partial n_k}{\partial x^j} = \frac{1}{r}\left(\delta_{jk} - n_j n_k\right),$$

where $x^j := (x, y, z)$ is a component notation for the vector $\boldsymbol{x}$, $n^j := x^j/r$, and δ_{jk} is the Kronecker delta, equal to one when $j = k$ and zero otherwise. These equations hold provided that $r \neq 0$. According to this we have that

$$\frac{\partial}{\partial x^j} r^{-1} = -\frac{1}{r^2} n_j$$

and

$$\frac{\partial^2}{\partial x^j \partial x^k} r^{-1} = \frac{1}{r^3}\left(3n_j n_k - \delta_{jk}\right).$$

Because $\boldsymbol{n}$ is a unit vector, it follows that $\nabla^2 r^{-1} = 0$ whenever $r \neq 0$.

To handle the special case $r = 0$ we introduce the vector $\boldsymbol{j} := \boldsymbol{\nabla} r^{-1}$ and write the left-hand side of Eq. (1) as $\boldsymbol{\nabla} \cdot \boldsymbol{j}$. Integrating this over a volume V bounded by a spherical surface S of radius η, we obtain

$$\int_V \boldsymbol{\nabla} \cdot \boldsymbol{j}\, d^3x = \oint_S \boldsymbol{j} \cdot d\boldsymbol{S}$$

by virtue of Gauss's theorem. Here $d\boldsymbol{S}$ is an outward-directed surface element on S, which can be expressed as $d\boldsymbol{S} = \boldsymbol{n}\eta^2\, d\Omega$, with $d\Omega$ denoting an element of solid angle centered at $\boldsymbol{n}$. The vector $\boldsymbol{j}$ is equal to $-\eta^{-2}\boldsymbol{n}$ on S, and evaluating the surface integral returns -4π.

Because $\nabla^2 r^{-1}$ vanishes when $\boldsymbol{x} \neq 0$ and integrates to -4π whenever the integration domain encloses $\boldsymbol{x} = \boldsymbol{0}$, we conclude that it is distributionally equal to $-4\pi\delta(\boldsymbol{x})$. The proof is complete.

the steps involved in establishing that this U is indeed a solution to Poisson's equation are identical to those that previously led us to Eq. (1.15) from Eq. (1.12). The difference is that in the earlier derivation the identity of the Green's function was already known. In the approach described here, the result follows simply by virtue of Eq. (1.16). It is not difficult, of course, to identify the Green's function: comparison with Eq. (1.13) allows us to write

$$G(\boldsymbol{x}, \boldsymbol{x}') = \frac{1}{|\boldsymbol{x} - \boldsymbol{x}'|}. \tag{1.18}$$

Not surprisingly, the Green's function represents the potential of a point mass situated at $\boldsymbol{x}'$.

1.4 Equations of hydrodynamics

In this section we develop the foundations for the equations of hydrodynamics, as displayed previously in Eqs. (1.5) and (1.6).

1.4.1 Motion of fluid elements

Definition of fluid element

We begin by describing any material body as being made up of *fluid elements*, volumes of matter that are very small compared to the size of the body, but very large compared to the inter-molecular distance, so that the element contains a macroscopic number of molecules. The fluid description of matter is a coarse-grained one in which the molecular fluctuations are smoothed over, and the fluid element is meant to represent a local average of the matter contained within. The coarse-graining could be described in great detail, for example, by introducing a microscopic density $\eta(t, \boldsymbol{x})$ that fluctuates wildly on the molecular scale, as well as smoothing function $w(|\boldsymbol{x} - \boldsymbol{x}'|)$ that varies over a much larger scale; the macroscopic density would then be defined as $\rho(t, \boldsymbol{x}) = \int \eta(t, \boldsymbol{x}')w(|\boldsymbol{x} - \boldsymbol{x}'|)\, d^3x'$. We will not go into such depth here, and keep the discussion at an intuitive, elementary level.

Each fluid element can be characterized by a mass density ρ (the mass of the element divided by its volume), a pressure p (the normal force per unit area acting on the surface of the element), and a velocity $\boldsymbol{v}$ (the average velocity of the molecules in the element). Other variables, such as viscosity, temperature, entropy, mean atomic weight, opacity, and so on, can also be introduced (some of these appear in Sec. 1.4.2). Apart from the velocity, all fluid variables are assumed to be measured by an observer who is momentarily at rest with respect to the fluid element. This description is adequate in a Newtonian setting, but it will have to be refined later, when we transition to the relativistic setting of Chapters 4 and 5.

Perhaps the most important aspect of a fluid element is that it keeps its contents intact as it moves within the fluid. During the motion the element may alter its shape and even its volume, but it will always contain the same collection of molecules; by definition no molecule is allowed to enter or leave the element. (It may be helpful to think of the molecules as being tagged, and of the fluid element as a bag that contains the tagged molecules.) A very important consequence of this property is that the total mass contained in a fluid element will never change; it is a constant of the element's motion.

Euler equation

We now apply Newton's laws to a selected fluid element of volume $\mathcal{V}$. The mass of the element is $\rho\mathcal{V}$, and from Newton's second law we have that

$$(\rho\mathcal{V})\boldsymbol{a} = \boldsymbol{F}, \tag{1.19}$$

where $\boldsymbol{F}$ is the net force acting on the element, and $\boldsymbol{a}$ is its acceleration. This can be expressed as $d\boldsymbol{v}/dt$, the rate of change of the element's velocity vector as it moves within the fluid. It is important to observe that this rate of change follows the motion of the fluid element, and that it does not keep the spatial position fixed; this observation gives rise to an important distinction between the *convective*, or *Lagrangian*, *derivative* d/dt, which follows the motion of the fluid, and the *partial*, or *Eulerian*, *derivative* $\partial/\partial t$, which keeps the spatial position fixed.

The Lagrangian time derivative d/dt takes into account both the intrinsic time evolution of fluid variables and the variations that result from the motion of each fluid element. The fluid changes its configuration in a time interval dt, and a selected fluid element moves from an old position $\boldsymbol{x}$ to a new position $\boldsymbol{x} + d\boldsymbol{x}$. A fluid quantity $f(t, \boldsymbol{x})$, such as the mass density or a component of the velocity vector, changes by $df = f(t + dt, \boldsymbol{x} + d\boldsymbol{x}) - f(t, \boldsymbol{x})$ when we follow the motion of the fluid element. To first order in the displacement this is $df = (\partial f/\partial t)dt + (\boldsymbol{\nabla} f) \cdot d\boldsymbol{x}$, or

$$\begin{aligned} \frac{d}{dt} f(t, \boldsymbol{x}) &= \frac{\partial}{\partial t} f(t, \boldsymbol{x}) + \frac{d\boldsymbol{x}}{dt} \cdot \boldsymbol{\nabla} f(t, \boldsymbol{x}) \\ &= \frac{\partial}{\partial t} f(t, \boldsymbol{x}) + \boldsymbol{v} \cdot \boldsymbol{\nabla} f(t, \boldsymbol{x}) . \end{aligned} \tag{1.20}$$

This equation provides a link between the Lagrangian and Eulerian time derivatives.

Returning to Eq. (1.19), we assume that the force $\boldsymbol{F}$ acting on the fluid element comes from gravity and pressure gradients. By analogy with the expression in Eq. (1.3), the gravitational force is written as

$$\boldsymbol{F}_{\text{gravity}} = (\rho \mathcal{V}) \boldsymbol{\nabla} U , \tag{1.21}$$

where we assume that the inertial mass density and passive gravitational mass density are equal, as dictated by the weak equivalence principle. To derive an expression for the pressure-gradient force, we consider a cubic fluid element, and for the moment we focus our attention on the x-component of the force. The normal force acting on the face at $x = x_1$ is $p(x_1)\mathcal{A}$, in which $\mathcal{A}$ is the cross-sectional area of the fluid element. Similarly, the normal force acting on the face at $x = x_2 = x_1 + dx$ is $-p(x_2)\mathcal{A}$, with the minus sign accounting for the different directions of the normal vector. It follows that the net force acting in the x-direction is $(p_1 - p_2)\mathcal{A} \approx -(dp/dx)\Delta x \mathcal{A} = -(dp/dx)\mathcal{V}$. Generalizing to three dimensions, we find that the pressure-gradient force is given by

$$\boldsymbol{F}_{\text{pressure}} = -\mathcal{V}\, \boldsymbol{\nabla} p . \tag{1.22}$$

Inserting Eqs. (1.21) and (1.22) within Eq. (1.19) and dropping the common factor of $\mathcal{V}$, we obtain *Euler's equation* of hydrodynamics in a gravitational field,

$$\rho \frac{d\boldsymbol{v}}{dt} = \rho \boldsymbol{\nabla} U - \boldsymbol{\nabla} p . \tag{1.23}$$

This equation (in spite of its name) is written in terms of the Lagrangian time derivative. An alternative formulation is

$$\rho\left[\frac{\partial \boldsymbol{v}}{\partial t} + (\boldsymbol{v}\cdot\boldsymbol{\nabla})\boldsymbol{v}\right] = \rho\boldsymbol{\nabla}U - \boldsymbol{\nabla}p\,, \tag{1.24}$$

and this involves the Eulerian time derivative.

Continuity equation

Conservation of the number of molecules in each fluid element implies that the mass of each element stays constant as it moves within the fluid. This is expressed mathematically as $d(\rho\mathcal{V})/dt = 0$, in terms of the Lagrangian time derivative. It is simple to show, however (see Box 1.3), that $\mathcal{V}^{-1}d\mathcal{V}/dt = \boldsymbol{\nabla}\cdot\boldsymbol{v}$, and the equation of mass conservation can be expressed as

$$\frac{d\rho}{dt} + \rho\boldsymbol{\nabla}\cdot\boldsymbol{v} = 0\,. \tag{1.25}$$

The Eulerian form of this equation is

$$\frac{\partial\rho}{\partial t} + \boldsymbol{\nabla}\cdot(\rho\boldsymbol{v}) = 0\,, \tag{1.26}$$

and in this guise it is known as the *continuity equation*.

Box 1.3 **Proof that $\mathcal{V}^{-1}d\mathcal{V}/dt = \boldsymbol{\nabla}\cdot\boldsymbol{v}$**

Consider a cubic fluid element of sides L, volume $\mathcal{V} = L^3$, moving with an averaged velocity $\boldsymbol{v}$. The face of the cube at $x + L/2$ moves with a velocity $\boldsymbol{v}(x + L/2, y, z)$, while the face at $x - L/2$ moves with a velocity $\boldsymbol{v}(x - L/2, y, z)$. In a time Δt the length of the cube in the x-direction changes by $[v_x(x + L/2, y, z) - v_x(x - L/2, y, z)]\Delta t \approx L(dv_x/dx)\Delta t$. Repeating this argument for the y- and z-directions, we find that the change in the cube's volume is

$$\Delta\mathcal{V} \approx L^3\left(1 + \frac{dv_x}{dx}\Delta t\right)\left(1 + \frac{dv_y}{dy}\Delta t\right)\left(1 + \frac{dv_z}{dz}\Delta t\right) - L^3 \approx \mathcal{V}\boldsymbol{\nabla}\cdot\boldsymbol{v}\Delta t\,.$$

Taking the limit $\Delta t \to 0$, we obtain the desired result.

1.4.2 Thermodynamics of fluid elements

We now focus our attention on a selected fluid element. We assume that the molecular mean free path (the average distance travelled by a molecule between collisions) as well as the photon mean free path (the average distance travelled by a photon before being scattered or absorbed by a molecule) are both very small compared to the size of the fluid element. Equivalently, we assume that the time required for the fluid element to change in a significant way is very long compared to the time scales that characterize interactions among molecules and photons within the fluid element. It follows from these assumptions that at any given moment of time, the fluid element can achieve a state of *local thermodynamic equilibrium* in which its matter content is characterized by a local temperature $T(t, \boldsymbol{x})$, and in which the photons are characterized by a black-body spectrum at the same temperature. We can also ascribe a local entropy $\mathcal{S}(t, \boldsymbol{x})$, a local internal energy $\mathcal{E}(t, \boldsymbol{x})$, and other thermodynamic variables to the fluid element. These quantities may vary from one fluid element to the next, and they may vary with time, but they do so on time and distance scales that are long compared to those associated with the molecular processes that drive each element toward equilibrium. We can therefore apply the laws of thermodynamics locally to each fluid element.

First and second laws

The first law of thermodynamics, applied to a fluid element, reads

$$d\mathcal{E} = \delta Q + \delta \mathcal{W}, \tag{1.27}$$

in which $\mathcal{E}$ is the internal energy of the fluid element, $\delta\mathcal{W} = -p\, d\mathcal{V}$ the work done on the fluid element, and δQ the heat absorbed. This can be expressed as

$$\delta Q = (\rho \mathcal{V}) q\, dt - \mathcal{V} \nabla \cdot \boldsymbol{H}\, dt, \tag{1.28}$$

where q is the energy per unit mass generated within the fluid element per unit time, and $\boldsymbol{H}$ is the heat-flux vector, defined in such a way that $\boldsymbol{H} \cdot d\boldsymbol{S}$ is the heat crossing an element of surface area (described by $d\boldsymbol{S}$) per unit time. In general, ρq represents heat that is generated internally (for example by chemical or nuclear reactions), and $\nabla \cdot \boldsymbol{H}$ represents heat imported from neighboring fluid elements (for example by heat conduction or radiation). The second term can be motivated by considering a cubic element, and examining the heat entering the element from the x-direction. The heat absorbed in a time dt is given by the flux $H_x(x)$ entering the face at x times the area $\mathcal{A}$ of that face, minus the flux $H_x(x + dx)$ leaving at $x + dx$ times the area $\mathcal{A}$. The net result is $-\partial H_x/dx(\mathcal{A}\, dx)$, and including the y- and z-directions gives $-\mathcal{V}\nabla \cdot \boldsymbol{H}\, dt$, as required.

Defining the energy density $\epsilon := \mathcal{E}/\mathcal{V}$, we can rewrite the first law of thermodynamics in the form

$$d\epsilon - \frac{\epsilon + p}{\rho} d\rho = (\rho q - \nabla \cdot \boldsymbol{H})\, dt. \tag{1.29}$$

A useful alternative variable is the internal energy per unit mass $\Pi := \epsilon/\rho$, for which the first law takes the form

$$d\Pi + p\, d\left(\frac{1}{\rho}\right) = \left(q - \frac{1}{\rho}\nabla \cdot \boldsymbol{H}\right) dt. \tag{1.30}$$

The second law of thermodynamics states that for any reversible process, $\delta Q = T\, d\mathcal{S}$, where $\mathcal{S}$ is the entropy of the fluid element. Introducing the entropy per unit mass $s := \mathcal{S}/(\rho\mathcal{V})$, we have that

$$T\, ds = \left(q - \frac{1}{\rho}\nabla \cdot \boldsymbol{H}\right) dt, \tag{1.31}$$

and the first law can now be expressed as

$$d\Pi + p\, d\left(\frac{1}{\rho}\right) = T\, ds\,. \tag{1.32}$$

As a special case of these laws, we consider a situation in which the fluid element, in addition to being in local thermodynamic equilibrium, is also in *thermal equilibrium* with neighboring elements. In such circumstances there is no net transfer of heat, and the element evolves adiabatically, with $ds = 0$. This requires that

$$\nabla \cdot \boldsymbol{H} = \rho q\,, \tag{1.33}$$

and the first law can then be expressed in the restricted form

$$d\Pi = -p\, d\left(\frac{1}{\rho}\right) = \frac{p}{\rho^2}\, d\rho\,. \tag{1.34}$$

Equation of state

Given a system of known composition, labelled symbolically by X, there exists a relation $p = p(\rho, T; X)$ between the pressure, density, and temperature called the *equation of state.* The equation of state is a necessary input into any application of the laws of thermodynamics, and to complete our discussion we provide a brief review of some of the equations of state that are relevant to the description of stellar configurations. We make no attempt to be complete here, as equations of state are the subject of a multitude of textbooks on statistical mechanics and thermodynamics.

The temperature inside most main-sequence stars is extremely high, and typically the kinetic energy of the atoms is very large compared to their interaction energy; the stellar matter can therefore be taken to be non-interacting, and to make up an ideal gas. Most stellar interiors are completely ionized, and the free electrons can also be treated as an ideal gas. The equation of state is then the familiar $p = nkT$, where n is the number density and k is Boltzmann's constant. The total pressure is the sum of the partial pressures, and the ionic contribution is

$$p_I = n_I kT = \frac{\rho}{\mu_I m_{\mathrm{H}}} kT\,, \tag{1.35}$$

where m_{H} is the atomic mass unit, and μ_I is the mean atomic number of the ions. This is defined by

$$\frac{1}{\mu_I} := \sum_i \frac{X_i}{\mathcal{A}_i}\,, \tag{1.36}$$

where X_i is the fraction by mass of the ith species ($\sum_i X_i = 1$), and $\mathcal{A}_i$ is its atomic mass number. For stars in which hydrogen and helium dominate over heavier elements (called *metals* by stellar astrophysicists), one often writes $\mu_I^{-1} := X + \frac{1}{4}Y + (1 - X - Y)\langle \mathcal{A}^{-1}\rangle$, in which X is the mass fraction of hydrogen, Y is the mass fraction of helium, and $\langle \mathcal{A}^{-1}\rangle$ is an average of $\mathcal{A}_i^{-1}$ over the metals.

For the electrons we have that

$$p_e = n_e kT = \frac{\rho}{\mu_e m_{\rm H}} kT \,, \tag{1.37}$$

where

$$\frac{1}{\mu_e} := \sum_i \frac{Z_i X_i}{A_i}, \tag{1.38}$$

with Z_i denoting the atomic number of the ith ionic species. Because $Z_i/A_i \approx 1/2$ for most elements except hydrogen (for which $Z/A = 1$), we can approximate μ_e^{-1} by $X + \frac{1}{2}(1 - X) = \frac{1}{2}(1 + X)$ for most stellar materials. The total gas pressure is then

$$p_{\rm gas} = \left(\frac{1}{\mu_I} + \frac{1}{\mu_e}\right) \frac{\rho}{m_{\rm H}} kT := \frac{\rho}{\mu m_{\rm H}} kT \,, \tag{1.39}$$

where $\mu^{-1} := \mu_I^{-1} + \mu_e^{-1}$. The energy density of such a classical ideal gas is given by $\epsilon_{\rm gas} = \frac{3}{2} p_{\rm gas}$.

Another important constituent of stars is radiation. As we have seen, under conditions of local thermodynamic equilibrium (which are upheld in stellar interiors) the radiation within each fluid element can be treated as a black body of the same temperature T as the fluid element. The equation of state and energy density for the radiation are given by

$$p_{\rm rad} = \frac{1}{3} a T^4 \,, \tag{1.40}$$

and $\epsilon_{\rm rad} = aT^4 = 3p_{\rm rad}$, where

$$a := \frac{8\pi^5 k^4}{15 h^3 c^3} \tag{1.41}$$

is the radiation constant ($\sigma := \frac{1}{4}ac$ is the Stefan–Boltzmann constant). The total pressure inside a star is then $p = p_{\rm gas} + p_{\rm rad}$, and the total energy density is $\epsilon = \epsilon_{\rm gas} + \epsilon_{\rm rad}$.

At the sufficiently high densities that characterize dead stars such as white dwarfs and neutron stars, matter becomes *degenerate*, and the equation of state changes dramatically. This occurs when the temperature T and number density n are such that the characteristic momentum (or uncertainty in the momentum) of a particle of mass m, $\Delta p \sim \sqrt{mkT}$, multiplied by the typical interparticle distance $\Delta x \sim n^{-1/3}$, starts running afoul of the Heisenberg uncertainty principle, which requires that $\Delta x \Delta p \geq \hbar$. This state of degeneracy occurs when $mkT \lesssim \hbar^2 n^{2/3}$, or when

$$T \leq T_{\rm F} \,, \qquad T_{\rm F} := \frac{\hbar^2}{2km}(3\pi^2 n)^{2/3} \tag{1.42}$$

after inserting the appropriate numerical coefficients. Here $T_{\rm F}$ is the *Fermi temperature* associated with a free fermion gas of number density n and constituent mass m. For a white dwarf, the electrons are degenerate, while the ions, being at least 2000 times more massive, are not. In a neutron star, as a consequence of the much higher density, the neutrons and the residual protons and electrons are degenerate. In laboratory situations involving low densities, the Fermi temperature is typically extremely low, and normal matter is rarely degenerate (an exception is the conduction electrons in metals, for which the Fermi temperature is much higher than room temperature). In a white dwarf, by contrast, the high densities involved imply that the Fermi temperature is of the order of 10^9 K, while the star's actual temperature typically ranges between 10^6 K and 10^7 K. For the even higher densities associated with neutron stars, the ratio T/T_F is even smaller; in this case the Fermi temperature is of order 10^{12} K, while the star's actual temperature is also comparable to 10^6 K.

To conclude our discussion we review the important case of the *polytropic* equation of state, in which p is related directly to the density, and in which T has been eliminated by assuming that each fluid element is in thermal equilibrium with neighboring elements. Under these conditions Eqs. (1.29) and (1.33) imply that $\rho\, d\epsilon - (\epsilon + p)\, d\rho = 0$, and we further assume that the fluid is such that the energy density is proportional to the pressure, so that

$$\epsilon = \eta\, p\,. \tag{1.43}$$

The dimensionless constant η (usually denoted n, which is avoided here because n has already been assigned the meaning of number density) is known as the *polytropic index*; we have seen that $\eta = \frac{3}{2}$ for an ideal gas, while $\eta = 3$ for a photon gas. Combining these relations we find that $\eta\rho\, dp - (\eta + 1)p\, d\rho = 0$, and this can be integrated to yield

$$p = K\rho^{\Gamma}\,, \qquad \Gamma := 1 + 1/\eta\,, \tag{1.44}$$

where K is an integration constant. This is the polytropic equation of state, which relates pressure and density during an adiabatic thermodynamic process; the exponent Γ is known as the *adiabatic index*.

1.4.3 Global conservation laws

The equations of hydrodynamics give rise to a number of important *global conservation laws*. These refer to global quantities, defined as integrals over the entire fluid system, that are constant in time whenever the system is *isolated*, that is, whenever the system is not affected by forces external to it. For fluids subjected to pressure forces and Newtonian gravity, the globally conserved quantities are total mass, momentum, energy, and angular momentum. Because these are fundamentally important in any physical context, we examine them in detail here, providing precise definitions and proofs of their conservation. For these derivations we introduce a number of mathematical tools that will prove helpful throughout this book.

Integral identities

The conserved quantities are all defined as integrals over a volume of space that contains the entire isolated system. The domain of integration V is largely arbitrary, and is constrained by only two essential conditions: it must be a fixed region of space that does not evolve in time, and it must contain all the matter. It is useful to think of this domain as extending beyond the matter; it could, in fact, extend all the way to infinity. An essential property of the boundary S of the region of integration is that all matter variables (such as the mass density ρ and the pressure p) vanish on S.

The global quantities are integrals of the form $\int_V f(t, \boldsymbol{x})\, d^3x$, in which $f(t, \boldsymbol{x})$ is a function of time and space that will typically involve the fluid variables. The integral itself is a function of time only; to simplify the notation we shall henceforth omit the label V on the integration symbol. For any such integral we have that

$$\frac{d}{dt}\int f(t, \boldsymbol{x})\, d^3x = \int \frac{\partial f}{\partial t}\, d^3x \,. \tag{1.45}$$

This property follows because V is independent of time, and because the variable of integration $\boldsymbol{x}$ also is independent of time.

We next consider integrals of the form $F(t) := \int \rho(t, \boldsymbol{x}) f(t, \boldsymbol{x})\, d^3x$, in which a factor of the mass density ρ was extracted from the original function f. As we shall prove below, such integrals obey the identity

$$\frac{d}{dt}\int \rho(t, \boldsymbol{x}) f(t, \boldsymbol{x})\, d^3x = \int \rho \frac{df}{dt}\, d^3x \,, \tag{1.46}$$

in which, as usual,

$$\frac{df}{dt} = \frac{\partial f}{\partial t} + \boldsymbol{v} \cdot \boldsymbol{\nabla} f \tag{1.47}$$

is the convective (or Lagrangian) time derivative.

We may generalize the result by allowing f to depend on two position vectors, $\boldsymbol{x}$ and $\boldsymbol{x}'$. We define the integral $F(t, \boldsymbol{x}) := \int \rho(t, \boldsymbol{x}') f(t, \boldsymbol{x}, \boldsymbol{x}')\, d^3x'$ and apply Eq. (1.46) to it. Because F depends on $\boldsymbol{x}$ in addition to t, the time derivative is correctly interpreted as a partial derivative that keeps the spatial variables fixed, and we obtain

$$\frac{\partial F}{\partial t} = \int \rho' \left(\frac{\partial f}{\partial t} + \boldsymbol{v}' \cdot \boldsymbol{\nabla}' f \right) d^3x' \,, \tag{1.48}$$

in which ρ' is the mass density expressed as a function of t and $\boldsymbol{x}'$, $\boldsymbol{v}'$ is the velocity field expressed in terms of the same variables, and $\boldsymbol{\nabla}'$ is the gradient operator associated with $\boldsymbol{x}'$. Now, the *Lagrangian time derivative* acting on F is $dF/dt = \partial F/\partial t + \boldsymbol{v} \cdot \boldsymbol{\nabla} F$, and from Eq. (1.48) and the definition of $F(t, \boldsymbol{x})$ we find that this can be expressed as

$$\frac{d}{dt}\int \rho(t, \boldsymbol{x}') f(t, \boldsymbol{x}, \boldsymbol{x}')\, d^3x' = \int \rho' \frac{df}{dt}\, d^3x' \,, \tag{1.49}$$

with

$$\frac{df}{dt} := \frac{\partial f}{\partial t} + \boldsymbol{v} \cdot \boldsymbol{\nabla} f + \boldsymbol{v}' \cdot \boldsymbol{\nabla}' f \tag{1.50}$$

denoting a generalized Lagrangian derivative.

For a final application of Eq. (1.46) we define $\mathcal{F}(t) := \int \rho(t, \boldsymbol{x}) F(t, \boldsymbol{x})\, d^3x = \int \rho\rho' f(t, \boldsymbol{x}, \boldsymbol{x}')\, d^3x' d^3x$. According to Eqs. (1.46) and (1.49) we find that the time derivative of this integral is given by

$$\frac{d}{dt}\int \rho(t, \boldsymbol{x}) \rho(t, \boldsymbol{x}') f(t, \boldsymbol{x}, \boldsymbol{x}')\, d^3x' d^3x = \int \rho\rho' \frac{df}{dt}\, d^3x' d^3x \,, \tag{1.51}$$

in which df/dt is once more given by Eq. (1.50).

We have yet to establish Eq. (1.46). The steps are straightforward, and they rely on the continuity equation (1.26), Gauss's theorem, and the fact that ρ vanishes on the boundary S of the domain of integration. We have

$$\begin{aligned}\frac{d}{dt}\int \rho(t,\boldsymbol{x})f(t,\boldsymbol{x})\,d^3x &= \int\left(\rho\frac{\partial f}{\partial t}+f\frac{\partial\rho}{\partial t}\right)d^3x\\ &= \int\left(\rho\frac{\partial f}{\partial t}-f\nabla\cdot(\rho\boldsymbol{v})\right)d^3x\\ &= \int\left(\rho\frac{\partial f}{\partial t}+\rho\boldsymbol{v}\cdot\nabla f\right)d^3x-\oint f\rho\boldsymbol{v}\cdot d\boldsymbol{S}\\ &= \int\rho\frac{df}{dt}\,d^3x\,. \end{aligned}\tag{1.52}$$

The continuity equation was used in the second step. In the third step the volume integral of the total divergence $\nabla\cdot(f\rho\boldsymbol{v})$ was expressed as a surface integral, which vanishes because $\rho=0$ on S. In the fourth step we recover Eq. (1.46), as required.

Mass, momentum, and center-of-mass

The *total mass* of the fluid system is

$$M := \int \rho(t,\boldsymbol{x})\,d^3x\,. \tag{1.53}$$

While the integral should in principle be a function of time, it is a direct consequence of Eq. (1.46) – applied with $f=1$ – that $dM/dt=0$. The total mass of the fluid system is a conserved quantity that does not change with time. This is an obvious consequence of the fact that mass is conserved within each fluid element.

The *total momentum* of the fluid system is

$$\boldsymbol{P} := \int \rho(t,\boldsymbol{x})\boldsymbol{v}(t,\boldsymbol{x})\,d^3x\,. \tag{1.54}$$

To verify that this is also a conserved quantity, we apply Eq. (1.46) with $f=\boldsymbol{v}$ and get

$$\frac{d\boldsymbol{P}}{dt}=\int\rho\frac{d\boldsymbol{v}}{dt}\,d^3x=\int\rho\nabla U\,d^3x-\int\nabla p\,d^3x \tag{1.55}$$

after inserting Euler's equation (1.23). The pressure integral is easy to dispose of: applying Gauss's theorem we find that it is equal to $\oint p\,d\boldsymbol{S}$, and this vanishes because $p=0$ everywhere on S. The integral involving the Newtonian potential requires more work, but we shall show presently that

$$\int\rho\nabla U\,d^3x=0\,, \tag{1.56}$$

a result that is fundamentally important in the Newtonian theory of gravity. With all this we find that $d\boldsymbol{P}/dt=0$, and conclude that total momentum is indeed conserved. This is a consequence of (or a statement of) Newton's third law, the equality of action and reaction.

The *center-of-mass* of the fluid system is situated at a position $\boldsymbol{R}(t)$ defined by

$$\boldsymbol{R}(t) := \frac{1}{M} \int \rho(t, \boldsymbol{x}) \boldsymbol{x} \, d^3x \,. \tag{1.57}$$

Because M is conserved, the center-of-mass velocity $\boldsymbol{V} := d\boldsymbol{R}/dt$ is given by

$$\boldsymbol{V} := \frac{1}{M} \int \rho(t, \boldsymbol{x}) \boldsymbol{v}(t, \boldsymbol{x}) \, d^3x \,, \tag{1.58}$$

as obtained from Eq. (1.46) by applying the identity $dx^j/dt = \partial x^j/\partial t + \boldsymbol{v} \cdot \boldsymbol{\nabla} x^j = v^j$ to each component of $\boldsymbol{x}$. The integral is recognized as the total momentum, and we find that $\boldsymbol{V}$ is a conserved quantity. It follows that the center-of-mass moves according to

$$\boldsymbol{R}(t) = \boldsymbol{R}(0) + \boldsymbol{V} t \,, \tag{1.59}$$

with $\boldsymbol{V} := \boldsymbol{P}/M$. It is always possible to choose a reference frame such that $\boldsymbol{R}(0) = \boldsymbol{0}$ and $\boldsymbol{V} = \boldsymbol{0}$, so that $\boldsymbol{R}(t) = \boldsymbol{0}$; this defines the *center-of-mass frame* of the fluid system.

To establish Eq. (1.56) we recall the expression of Eq. (1.12) for the gravitational potential, on which we apply the gradient operator. Focusing our attention on the x^j component of $\boldsymbol{\nabla} U$, we have that

$$\frac{\partial U}{\partial x^j} = G \int \rho' \frac{\partial}{\partial x^j} \frac{1}{|\boldsymbol{x} - \boldsymbol{x}'|} \, d^3x'. \tag{1.60}$$

The partial derivative can be evaluated explicitly, and can be seen to be equal to $-|\boldsymbol{x} - \boldsymbol{x}'|^{-3}(x^j - x'^j)$. Inserting $\partial U/\partial x^j$ within the integral of Eq. (1.56), we find that

$$\int \rho \frac{\partial U}{\partial x^j} \, d^3x = G \int \rho\rho' \frac{\partial}{\partial x^j} \frac{1}{|\boldsymbol{x} - \boldsymbol{x}'|} \, d^3x' d^3x \,. \tag{1.61}$$

To show that this vanishes we employ a clever trick that will recur frequently throughout this book. It consists of swapping the variables of integration ($\boldsymbol{x} \leftrightarrow \boldsymbol{x}'$), and of writing the integral in the alternative form

$$\begin{aligned} \int \rho \frac{\partial U}{\partial x^j} \, d^3x &= G \int \rho'\rho \frac{\partial}{\partial x'^j} \frac{1}{|\boldsymbol{x}' - \boldsymbol{x}|} \, d^3x d^3x' \\ &= G \int \rho\rho' \frac{\partial}{\partial x'^j} \frac{1}{|\boldsymbol{x} - \boldsymbol{x}'|} \, d^3x' d^3x \,. \end{aligned} \tag{1.62}$$

Explicitly we find that the partial derivative with respect to x'^j is equal to $+|\boldsymbol{x} - \boldsymbol{x}'|^{-3}(x^j - x'^j)$, which is equal and opposite to the derivative with respect to x^j. This property follows directly from the fact that $|\boldsymbol{x} - \boldsymbol{x}'|^{-1}$ depends on the *difference* between $\boldsymbol{x}$ and $\boldsymbol{x}'$. Taking this property into account in Eq. (1.62), we find that

$$\int \rho \frac{\partial U}{\partial x^j} \, d^3x = -G \int \rho\rho' \frac{\partial}{\partial x^j} \frac{1}{|\boldsymbol{x} - \boldsymbol{x}'|} \, d^3x' d^3x \,. \tag{1.63}$$

Comparing Eqs. (1.61) and (1.63), we find that the integral vanishes, as was already stated in Eq. (1.56).

Energy

The *total energy* of a fluid system comprises three components. The first is the kinetic energy

$$\mathcal{T}(t) := \frac{1}{2} \int \rho v^2 \, d^3x \,, \tag{1.64}$$

the second is the gravitational potential energy

$$\Omega(t) := -\frac{1}{2} \int \rho U \, d^3x = -\frac{1}{2} G \int \frac{\rho \rho'}{|\boldsymbol{x} - \boldsymbol{x}'|} \, d^3x' d^3x \,, \tag{1.65}$$

and the third is the internal thermodynamic energy

$$E_{\rm int}(t) := \int \epsilon \, d^3x = \int \rho \Pi \, d^3x \,. \tag{1.66}$$

In these expressions, ρ is the mass density expressed as a function of t and $\boldsymbol{x}$, ρ' is the mass density expressed in terms of t and $\boldsymbol{x}'$, $v^2 := \boldsymbol{v} \cdot \boldsymbol{v}$ is the square of the velocity vector, ϵ is the density of internal thermodynamic energy, and $\Pi := \epsilon/\rho$ is the specific internal energy. The total energy is

$$E := \mathcal{T}(t) + \Omega(t) + E_{\rm int}(t) \,, \tag{1.67}$$

and while $\mathcal{T}$, Ω, and $E_{\rm int}$ can each vary with time, we shall prove that E is a conserved quantity. The definition provided here for total kinetic energy is immediately plausible: we take the kinetic energy of each fluid element, $\frac{1}{2}(\rho \mathcal{V}) v^2$, and integrate over the entire fluid. The definition of total internal energy is also immediately plausible. The definition of total gravitational potential energy is more subtle, and its suitability is ultimately justified by the fact that the total energy turns out to be conserved. Nevertheless, we may observe that Ω is $-(\rho \mathcal{V}) U$, the potential energy of each fluid element in the field of all other elements, integrated over the entire fluid; the factor of $\frac{1}{2}$ is inserted to avoid a double counting of pairs of fluid elements.

To prove that E is conserved we calculate how each term in Eq. (1.67) changes with time. We begin with $\mathcal{T}$, and get

$$\frac{d\mathcal{T}}{dt} = \int \rho \boldsymbol{v} \cdot \frac{d\boldsymbol{v}}{dt} \, d^3x = \int \rho \boldsymbol{v} \cdot \boldsymbol{\nabla} U \, d^3x - \int \boldsymbol{v} \cdot \boldsymbol{\nabla} p \, d^3x \tag{1.68}$$

after involving Euler's equation. The first integral can be expressed as

$$\int \rho \boldsymbol{v} \cdot \boldsymbol{\nabla} U \, d^3x = G \int \rho \rho' \boldsymbol{v} \cdot \boldsymbol{\nabla} \frac{1}{|\boldsymbol{x} - \boldsymbol{x}'|} \, d^3x' d^3x \,, \tag{1.69}$$

or it can be expressed as

$$\int \rho \boldsymbol{v} \cdot \boldsymbol{\nabla} U \, d^3x = G \int \rho' \rho \, \boldsymbol{v}' \cdot \boldsymbol{\nabla}' \frac{1}{|\boldsymbol{x}' - \boldsymbol{x}|} \, d^3x d^3x' \tag{1.70}$$

by exploiting the "switch trick" introduced after Eq. (1.61). Adding the two expressions and dividing by 2, we obtain

$$\int \rho \boldsymbol{v} \cdot \boldsymbol{\nabla} U \, d^3x = \frac{1}{2} G \int \rho \rho' \big(\boldsymbol{v} \cdot \boldsymbol{\nabla} + \boldsymbol{v}' \cdot \boldsymbol{\nabla}' \big) \frac{1}{|\boldsymbol{x} - \boldsymbol{x}'|} \, d^3x' d^3x \,. \tag{1.71}$$

Because $|\boldsymbol{x}-\boldsymbol{x}'|^{-1}$ does not depend on time, we may re-express this as

$$\int \rho \boldsymbol{v} \cdot \boldsymbol{\nabla} U \, d^3x = \frac{1}{2} G \int \rho\rho' \frac{d}{dt} \frac{1}{|\boldsymbol{x}-\boldsymbol{x}'|} \, d^3x' d^3x \,, \tag{1.72}$$

in which d/dt is the generalized Lagrangian derivative of Eq. (1.50). Invoking now the integral identity of Eq. (1.51), applied with $f = |\boldsymbol{x}-\boldsymbol{x}'|^{-1}$, as well as the definition of Ω provided in Eq. (1.65), we finally arrive at

$$\int \rho \boldsymbol{v} \cdot \boldsymbol{\nabla} U \, d^3x = -\frac{d\Omega}{dt} \,, \tag{1.73}$$

which tells us how Ω changes with time.

Returning to Eq. (1.68), we next examine the second integral, in which we express $\boldsymbol{v} \cdot \boldsymbol{\nabla} p$ as $\boldsymbol{\nabla} \cdot (p\boldsymbol{v}) - p \boldsymbol{\nabla} \cdot \boldsymbol{v}$. The total divergence gives no contribution (because p vanishes on S), and Eq. (1.25) implies that $\boldsymbol{\nabla} \cdot \boldsymbol{v} = -\rho^{-1} d\rho/dt$. All this gives us

$$\int \boldsymbol{v} \cdot \boldsymbol{\nabla} p \, d^3x = \int \frac{p}{\rho} \frac{d\rho}{dt} \, d^3x \tag{1.74}$$

for the second integral. Inserting this and Eq. (1.73) within Eq. (1.68), we finally obtain

$$\frac{d\mathcal{T}}{dt} = -\frac{d\Omega}{dt} - \int \frac{p}{\rho} \frac{d\rho}{dt} \, d^3x \tag{1.75}$$

for the rate of change of the total kinetic energy.

The final step is to compute $dE_{\rm int}/dt$. Starting with Eq. (1.66) and involving Eq. (1.46) with $f = \Pi$, we find that

$$\frac{dE_{\rm int}}{dt} = \int \rho \frac{d\Pi}{dt} \, d^3x \,. \tag{1.76}$$

Assuming that each fluid element is at all times in thermal equilibrium with neighboring elements, we invoke the first law of thermodynamics as stated in Eq. (1.34): $\rho \, d\Pi = (p/\rho) \, d\rho$. This gives

$$\frac{dE_{\rm int}}{dt} = \int \frac{p}{\rho} \frac{d\rho}{dt} \, d^3x \tag{1.77}$$

for the rate of change of the total internal energy. Combining Eqs. (1.73), (1.75), and (1.77), we find that $dE/dt = 0$, and arrive at the conclusion that E is indeed conserved.

Angular momentum

The *total angular momentum* of a fluid system is defined by

$$\boldsymbol{J} := \int \rho \boldsymbol{x} \times \boldsymbol{v} \, d^3x \,. \tag{1.78}$$

The steps required to show that the angular momentum is conserved are now familiar. We use Eq. (1.46) with $f = \boldsymbol{x} \times \boldsymbol{v}$ to evaluate $d\boldsymbol{J}/dt$, and obtain

$$\frac{d\boldsymbol{J}}{dt} = \int \rho \boldsymbol{x} \times \boldsymbol{\nabla} U \, d^3x - \int \boldsymbol{x} \times \boldsymbol{\nabla} p \, d^3x \tag{1.79}$$

after inserting Euler's equation. The first integral is evaluated as

$$\begin{aligned}\int \rho \boldsymbol{x} \times \nabla U \, d^3x &= -G \int \rho\rho' \frac{\boldsymbol{x} \times (\boldsymbol{x} - \boldsymbol{x}')}{|\boldsymbol{x} - \boldsymbol{x}'|^3} \, d^3x' d^3x \\ &= G \int \rho\rho' \frac{\boldsymbol{x}' \times (\boldsymbol{x} - \boldsymbol{x}')}{|\boldsymbol{x} - \boldsymbol{x}'|^3} \, d^3x' d^3x \\ &= -\frac{1}{2} G \int \rho\rho' \frac{(\boldsymbol{x} - \boldsymbol{x}') \times (\boldsymbol{x} - \boldsymbol{x}')}{|\boldsymbol{x} - \boldsymbol{x}'|^3} \, d^3x' d^3x \,,\end{aligned}$$

so that

$$\int \rho \boldsymbol{x} \times \nabla U \, d^3x = 0 \,. \tag{1.80}$$

The "switch trick" was exploited in the second step, and in the third step we added the two expressions and divided by 2. For the second integral we make use of the vector-algebra identity $\boldsymbol{x} \times \nabla p = -\nabla \times (p\boldsymbol{x}) + p\nabla \times \boldsymbol{x}$; the second term vanishes identically, and the integral of the first term can be expressed as a vanishing surface integral. Thus

$$\int \boldsymbol{x} \times \nabla p \, d^3x = 0 \,, \tag{1.81}$$

and we have arrived at the conservation statement $d\boldsymbol{J}/dt = 0$.

Virial theorems

Another important set of global relations satisfied by an isolated fluid system is known as the *virial theorems*. They involve a number of new global quantities. The first is

$$I^{jk}(t) := \int \rho(t, \boldsymbol{x}) x^j x^k \, d^3x \,, \tag{1.82}$$

the *quadrupole moment tensor* of the mass distribution, an object that will accompany us throughout this book. The second is

$$\mathcal{T}^{jk}(t) := \frac{1}{2} \int \rho v^j v^k \, d^3x \,, \tag{1.83}$$

the *kinetic energy tensor* of the fluid system, a tensorial generalization of $\mathcal{T}$ defined by Eq. (1.64); it is easy to see that $\mathcal{T}$ is the trace of the kinetic energy tensor. The third is

$$\Omega^{jk}(t) := -\frac{1}{2} G \int \rho\rho' \frac{(x - x')^j (x - x')^k}{|\boldsymbol{x} - \boldsymbol{x}'|^3} \, d^3x' d^3x \,, \tag{1.84}$$

the *gravitational energy tensor* of the fluid system, a tensorial generalization of Ω defined by Eq. (1.65); once again it is easy to see that Ω is the trace of the gravitational energy tensor. And finally, the virial theorems involve

$$P(t) := \int p \, d^3x \,, \tag{1.85}$$

the *integrated pressure* of the fluid system.

The tensorial version of the virial theorem is a statement about the second time derivative of the quadrupole moment tensor. Applying these derivatives to Eq. (1.82) and making use of Eq. (1.46) yields

$$\frac{d^2 I^{jk}}{dt^2} = 2\int \rho v^j v^k \, d^3x + 2\int \rho x^{(j} \frac{dv^{k)}}{dt} \, d^3x \,, \tag{1.86}$$

where we inserted parentheses around the indices in the second integral to indicate symmetrization: $x^{(j} dv^{k)}/dt := \frac{1}{2}(x^j dv^k/dt + x^k dv^j/dt)$. (The operations of symmetrization and antisymmetrization of tensorial indices are described more fully in Box 1.4.) Inserting now the Euler equation within Eq. (1.86), we obtain

$$\begin{aligned} \frac{d^2 I^{jk}}{dt^2} &= 2\int \rho v^j v^k \, d^3x - 2G \int \rho\rho' \frac{x^{(j}(x - x')^{k)}}{|\boldsymbol{x} - \boldsymbol{x}'|^3} \, d^3x' d^3x \\ &\quad - 2\int x^{(j} \partial^{k)} p \, d^3x \,, \end{aligned} \tag{1.87}$$

in which $\partial^k p$ is a shorthand notation for $\partial p/\partial x^k$. To proceed we exploit the "switch trick" in the second integral, and integrate the third integral by parts. The end result is

$$\frac{1}{2}\frac{d^2 I^{jk}}{dt^2} = 2\mathcal{T}^{jk} + \Omega^{jk} + P\delta^{jk} \,, \tag{1.88}$$

the statement of the *tensor virial theorem*. Taking the trace of Eq. (1.88) returns

$$\frac{1}{2}\frac{d^2 I}{dt^2} = 2\mathcal{T} + \Omega + 3P \,, \tag{1.89}$$

the *scalar virial theorem*; here $I(t) := \int \rho r^2 \, d^3x$ is the trace of the quadrupole moment tensor.

Many applications of the virial theorems involve stationary systems, for which $d^2 I^{jk}/dt^2 = 0$. For such systems the virial theorems reduce to

$$2\mathcal{T}^{jk} + \Omega^{jk} + P\delta^{jk} = 0 \,, \tag{1.90a}$$

$$2\mathcal{T} + \Omega + 3P = 0 \,. \tag{1.90b}$$

Other applications involve periodic systems, for which we may integrate Eqs. (1.88) and (1.89) over a complete period of the system. In these circumstances the terms involving the quadrupole moment tensor disappear also, and Eqs. (1.90) continue to hold in a coarse-grained form; the equations now involve *averages* of $\mathcal{T}^{jk}$, Ω^{jk}, and P over a period of the system.

The virial theorems are powerful tools, and they can be exploited to great benefits in the study of stellar structure. In the context of this book we find them most useful in our study of post-Newtonian equations of motion (in Chapter 9) and gravitational waves (in Chapter 11).

Box 1.4 Symmetrized and antisymmetrized indices

We define symmetrized and antisymmetrized indices according to

$$A^{(j}B^{k)} := \frac{1}{2}\left(A^j B^k + A^k B^j\right),$$

$$A^{[j}B^{k]} := \frac{1}{2}\left(A^j B^k - A^k B^j\right).$$

These definitions apply equally well to tensors, for example $C^{(jk)} = \frac{1}{2}(C^{jk} + C^{kj})$ and $C^{[jk]} = \frac{1}{2}(C^{jk} - C^{kj})$. Tensors of higher ranks can also be symmetrized and antisymmetrized in an obvious way, so that a symmetrized rank-q tensor is defined by

$$C^{(k_1 k_2 \cdots k_q)} := \frac{1}{q!}\left(C^{k_1 k_2 \cdots k_q} + \cdots\right),$$

where the remaining terms consist of all possible permutations of the q indices. An antisymmetrized rank-q tensor is defined by

$$C^{[k_1 k_2 \cdots k_q]} := \frac{1}{q!}\left(C^{k_1 k_2 \cdots k_q} \pm \cdots\right),$$

where the sign of each term is positive when the index order is an even permutation of the original order, and negative when it is an odd permutation.

1.4.4 Mass–momentum tensor

An alternative formulation of the equations of hydrodynamics reveals a nice parallel with the relativistic equations to be introduced in Chapters 4 and 5. This formulation is based on the Eulerian (as opposed to Lagrangian) version of these equations, as given by Eqs. (1.24) and (1.26):

$$\frac{\partial \rho}{\partial t} + \nabla \cdot (\rho \boldsymbol{v}) = 0, \tag{1.91a}$$

$$\rho\left(\frac{\partial \boldsymbol{v}}{\partial t} + \boldsymbol{v} \cdot \nabla \boldsymbol{v}\right) = \rho \nabla U - \nabla p. \tag{1.91b}$$

It involves a repackaging of the main fluid variables, such as the mass density ρ and the momentum density $\rho\boldsymbol{v}$, into a *mass–momentum tensor* with components T^{tt}, T^{tj}, T^{jt}, and T^{jk}.

To elaborate the new formulation it is helpful to work in terms of vector components (such as x^j) instead of the vectors themselves. In this language, for example, we would write $\nabla \cdot \boldsymbol{v}$ as $\sum_j (\partial v^j / \partial x^j)$. To save a lot of unnecessary writing we adopt an important convention that was first introduced by Einstein in his papers on general relativity (he jokingly considered it to be a great mathematical discovery). According to the *Einstein summation convention*, we omit the summation symbol whenever two indices are repeated, and automatically sum the indices (such as j with j) over the full range of their values – in this case x, y, and z, or more generally, over all three spatial dimensions. Thus we write $\boldsymbol{\nabla} \cdot \boldsymbol{v}$ simply as $\partial v^j/\partial x^j$.

In the component language the continuity equation (1.91a) takes the form of

$$\frac{\partial \rho}{\partial t} + \frac{\partial}{\partial x^j}(\rho v^j) = 0\,, \tag{1.92}$$

and we can also show, by making use of the continuity equation, that the left-hand side of Eq. (1.91b) can be written as

$$\frac{\partial}{\partial t}(\rho v^j) + \frac{\partial}{\partial x^k}(\rho v^j v^k)\,.$$

To recast the right-hand side of Eq. (1.91b) we invoke Poisson's equation (1.15) and express ρ in terms of $\nabla^2 U$; this yields

$$\rho \frac{\partial U}{\partial x^j} = -\frac{1}{4\pi G}\nabla^2 U \frac{\partial U}{\partial x^j}\,, \tag{1.93}$$

which can be re-expressed as

$$\rho \frac{\partial U}{\partial x^j} = -\frac{1}{4\pi G}\frac{\partial}{\partial x^k}\left(\frac{\partial U}{\partial x^j}\frac{\partial U}{\partial x^k} - \frac{1}{2}\delta_{jk}\frac{\partial U}{\partial x^p}\frac{\partial U}{\partial x^p}\right)\,. \tag{1.94}$$

To verify the equality of these two expressions makes a worthy exercise in index manipulation. Recall that according to the summation convention, $(\partial U/\partial x^p)(\partial U/\partial x^p) = \boldsymbol{\nabla} U \cdot \boldsymbol{\nabla} U = |\boldsymbol{\nabla} U|^2$. Continuing our work on the right-hand side of Eq. (1.91b), we note that $\partial p/\partial x^j$ can be written as $\partial(p\delta^{jk})/\partial x^k$.

Collecting these results, we find that we may write the continuity and Euler equations in the compact form

$$\frac{\partial}{\partial t}T^{tt} + \frac{\partial}{\partial x^k}T^{tk} = 0\,, \tag{1.95a}$$

$$\frac{\partial}{\partial t}T^{jt} + \frac{\partial}{\partial x^k}T^{jk} = 0\,, \tag{1.95b}$$

where the components of the mass–momentum tensor are given by

$$T^{tt} := \rho\,, \tag{1.96a}$$

$$T^{tj} := \rho v^j\,, \tag{1.96b}$$

$$T^{jt} := \rho v^j\,, \tag{1.96c}$$

$$T^{jk} := \rho v^j v^k + p\,\delta^{jk} + \frac{1}{4\pi G}\left(\frac{\partial U}{\partial x^j}\frac{\partial U}{\partial x^k} - \frac{1}{2}\delta_{jk}\frac{\partial U}{\partial x^p}\frac{\partial U}{\partial x^p}\right)\,. \tag{1.96d}$$

Notice that the tensor is symmetric under an exchange of any pair of indices.

It is worthwhile to introduce yet more notation to simplify the appearance of Eqs. (1.95). We shall henceforth denote the partial derivatives of a function f by

$$\partial_t f := \frac{\partial f}{\partial t}\,, \qquad \partial_j f := \frac{\partial f}{\partial x^j}\,. \tag{1.97}$$

In addition, we shall use Greek indices (such as α and β) to denote all four space and time variables; Greek indices run over the values t, x, y, and z, while Latin indices continue to run over the three spatial values. In this notation, the partial derivatives of Eq. (1.97) are collectively denoted $\partial_\alpha f$. To reflect this new usage, we shall extend the Einstein summation convention so that it applies also to repeated Greek indices; here summation will run over the four spacetime values, while summation over repeated Latin indices will continue to run over the three spatial values.

With these new rules, the continuity and Euler equations (1.95) can be written in the wonderfully compact form

$$\partial_\beta T^{\alpha\beta} = 0 \,. \tag{1.98}$$

This is a set of four distinct equations, and we stress that summation of β extends over the four values t, x, y, and z. Some comments are in order. The first is an admission of humility: while we have managed to combine space and time indices into a seemingly unified expression, we hasten to point out that there is absolutely nothing relativistic about this. What we have here is a mere repackaging of the Newtonian equations of hydrodynamics, in a compact form that happens to anticipate the relativistic versions of Chapters 4 and 5. The second comment is a subtle one of interpretation. The original continuity and Euler equations, Eqs. (1.5) and (1.6), were derived from basic principles of Newtonian mechanics, and they would continue to hold even if Eq. (1.15), the equation governing the behavior of the gravitational potential, turned out to be invalid. By contrast, Poisson's equation was involved in the derivation of Eq. (1.98), and its validity therefore rests on the validity of the Newtonian field equation. This reveals that in the context of Newtonian mechanics, Eq. (1.98) is not quite as fundamental or general as the continuity and Euler equations. We shall see in Chapter 5 that the point of view is very different in a relativistic context; there it is Eq. (1.98) – or a suitable generalization thereof – that forms the fundamental starting point of a derivation of the relativistic continuity and Euler equations; in particular, we shall see that Eq. (1.98) is quite independent from, and indeed more fundamental than, the Einstein field equations.

Writing the equations of hydrodynamics in the form of Eqs. (1.95) or (1.98) suggests a very efficient way of deriving the conservation statements regarding mass, momentum, angular momentum, and center-of-mass motion for an isolated system. We first introduce new definitions for these quantities:

$$M := \int T^{tt} \, d^3x \,, \tag{1.99a}$$

$$P^j := \int T^{jt} \, d^3x \,, \tag{1.99b}$$

$$R^j := \frac{1}{M} \int T^{tt} x^j \, d^3x \,, \tag{1.99c}$$

$$J^j := \epsilon^{jpq} \int x^p T^{qt} \, d^3x \,, \tag{1.99d}$$

where ϵ^{jpq} is the completely anti-symmetric Levi–Civita symbol, whose value is 1 if jpq is an even permutation of 123 (or xyz), -1 if it is an odd permutation, and zero if any two indices take on the same value. The Levi–Civita symbol is a convenient tool for constructing cross products when using the index language; you may easily check that the components of $\boldsymbol{a} \times \boldsymbol{b}$ are $\epsilon^{jpq} a^p b^q$, if you keep in mind the implied summation of p and q. The new definitions of Eqs. (1.99) are fully equivalent to the old definitions of Eqs. (1.53), (1.54), (1.57), and (1.78).

It is then a simple matter to take the time derivative of each one of these quantities, to exploit the fact that, according to Eq. (1.95), a time derivative of T^{tt} or T^{jt} is equal to a spatial divergence, and to use Gauss's theorem to re-express the volume integral as a surface integral; because T^{tj} or T^{jk} vanishes on the surface, this integral vanishes, and the quantity is conserved. (This statement is true only when the domain of integration V extends over all space, and the boundary S is situated at infinity. Unlike our previous usage in Sec. 1.4.3, here it is not sufficient for S to merely enclose the fluid system. The reason is that while ρ and p can be trusted to vanish outside the matter, the Newtonian potential U cannot; instead it falls off as GM/r far away from the fluid system. To ensure that the surface integral associated with T^{jk} vanishes, it is necessary to place S at infinity, where $\partial_j U$ properly vanishes.) The end result of this efficient computation is the statement that M, $\boldsymbol{P}$, and $\boldsymbol{J}$ are constant, and that $d\boldsymbol{R}/dt = \boldsymbol{P}/M$. You will be asked to go through the steps of these computations in Exercise 1.5. It is interesting to observe that the proof of the constancy of $\boldsymbol{J}$ relies critically on the fact that T^{jk} is symmetric in its indices; the detailed expression for T^{jk} is never required.

Another result that follows by repeated use of Eq. (1.98) and Gauss's theorem is

$$\frac{1}{2}\frac{d^2}{dt^2}\int T^{tt} x^j x^k \, d^3x = \int T^{jk} \, d^3x \,. \tag{1.100}$$

Because $\int T^{tt} x^j x^k \, d^3x$ is the quadrupole moment tensor of the mass distribution, this is an alternative statement of the tensor virial theorem, which can be shown to be equivalent to the statement given previously in Eq. (1.88). You will be asked to go through the steps of this computation in Exercise 1.6.

It is important to appreciate that the formulation of the equations of motion using a mass–momentum tensor *does not* yield an expression for the conserved energy; for that we must follow our earlier derivation and deal directly with Euler's equation. In a relativistic formulation of the equations, however, the statement of energy conservation would also follow, and would in fact supersede the statement of mass conservation. The reason, of course, is that in a relativistic context, mass is but a form of energy, to be naturally included in a breakdown such as Eq. (1.67). In a relativistic context, therefore, mass conservation arises as a special case of energy conservation. In Newtonian mechanics conservation of mass is necessarily separated from conservation of energy.

1.5 Spherical and nearly spherical bodies

We next consider the problem of calculating the gravitational potential U for various kinds of bodies. The simplest case has already been dealt with: as we have seen, the potential of a single point mass M at the origin of the coordinate system is given by $U = GM/r$. We wish to go well beyond this simplest case, and to construct the Newtonian potential for a more realistic, finite-sized body with an arbitrary mass density ρ. A body can often be taken to be spherically symmetric as a first approximation, and we shall begin our presentation with a complete description of this idealized case. This is not sufficient, however, because deviations from spherical symmetry can often be very important. We shall therefore devote the remainder of this section to a description of non-spherical bodies. While our treatment will be exact, the formalism that we develop – based on a multipole expansion of the mass distribution and the gravitational potential – is most powerful when applied to bodies that deviate only modestly from spherical symmetry.

For our purposes in this section it is best to express the Laplacian operator in spherical polar coordinates (r, θ, ϕ), and to write Poisson's equation (1.15) as

$$\left(\frac{1}{r^2} \frac{\partial}{\partial r} r^2 \frac{\partial}{\partial r} + \frac{1}{r^2 \sin\theta} \frac{\partial}{\partial \theta} \sin\theta \frac{\partial}{\partial \theta} + \frac{1}{r^2 \sin^2\theta} \frac{\partial^2}{\partial \phi^2} \right) U = -4\pi G \rho \,, \tag{1.101}$$

in which ρ and U are functions of t, r, θ, and ϕ. We recall that the relation to Cartesian coordinates is given by $x = r \sin\theta \cos\phi$, $y = r \sin\theta \sin\phi$, and $z = r \cos\theta$.

1.5.1 Spherical bodies

The mass density ρ and gravitational potential U of a spherical body depend on t and r only, and in this case Eq. (1.101) reduces to

$$\frac{1}{r^2} \frac{\partial}{\partial r} \left(r^2 \frac{\partial U}{\partial r} \right) = -4\pi G \rho(t, r) \,. \tag{1.102}$$

Integrating once, we obtain

$$\frac{\partial U}{\partial r} = -\frac{4\pi G}{r^2} \int_0^r \rho(t, r') r'^2 \, dr' \,, \tag{1.103}$$

where the constant of integration (actually a function of t) was chosen so that the gravitational force at $r = 0$ vanishes, as it must by symmetry. To ensure this we make the physically reasonable requirement that the density ρ must be finite at $r = 0$; then the integral behaves as $\frac{1}{3}\rho(t, 0) r^3$ near $r = 0$, and $\partial U / \partial r$ properly vanishes at $r = 0$.

The integral in Eq. (1.103) defines the mass contained inside a sphere of radius r, which we write as

$$m(t, r) := \int_0^r 4\pi \rho(t, r') r'^2 \, dr' \,. \tag{1.104}$$

The mass density drops to zero at the surface of the body, and the body's total mass is

$$M := m(t, r = R) = \int_0^R 4\pi \rho(t, r') r'^2 \, dr' \,, \tag{1.105}$$

with R denoting the body's radius. It was already established in Sec. 1.4.3 that M does not depend on time, in spite of the fact that $m(t, r)$ itself may depend on time. For example, a pulsating star would display a time-dependent density ρ, a time-dependent internal mass function m, and a time-dependent stellar radius R, but its total mass M would still be constant.

Returning to Eq. (1.103), we find that the gravitational force acting on a fluid element of unit mass situated at radius r is given by

$$\frac{\partial U}{\partial r} = -\frac{Gm(t,r)}{r^2}. \tag{1.106}$$

We observe that in spherical symmetry, the force is completely determined by the mass contained inside the sphere of radius r, and that any spherical distribution of matter outside this sphere contributes nothing to the force. While this result is an almost trivial consequence of integrating Poisson's equation, it was far from obvious to Newton while he was struggling to prove it with the traditional geometrical methods that he adopted throughout the *Principia*. He describes this struggle in a commentary that follows Proposition 8 in Book 2:

> After I had found that the gravity toward a whole planet arises from and is compounded of the gravities toward the parts and that toward each of the individual parts it is inversely proportional to the squares of the distances from the parts, I was still not certain whether that proportion of the inverse square obtained exactly in a total force compounded of a number of forces, or only nearly so. For it could happen that a proportion which holds exactly enough at very great distances might be markedly in error near the surface of the planet, because there the distances of the particles may be unequal and their situations dissimilar. But at length, by means of Book 1, Propositions 75 and 76 and their corollaries, I discerned the truth of the proposition dealt with here.

In Propositions 75 and 76, Newton proves that spherical bodies attract each other with a force inversely proportional to the square of the distance between their centers; previously, in Proposition 71, he proved that the force on a particle outside a spherical distribution of matter is given by Eq. (1.106).

Outside the spherical body, $m(t,r) = M$ and Eq. (1.106) becomes

$$\frac{dU}{dr} = -\frac{GM}{r^2}. \tag{1.107}$$

The gravitational force is now time-independent and completely determined by the total mass of the body. These statements apply to the potential itself: the gravitational potential outside a spherical body is constant in time regardless of the time-dependence of the matter distribution, and determined by the body's total mass. As we shall see in Chapter 5, a similar statement can be made in general relativity, where it is known as *Birkhoff's theorem*.

The potential itself is determined by integrating Eqs. (1.106) and (1.107). We impose continuity of U at $r = R$ and the boundary condition that U should vanish at $r = \infty$. We obtain

$$U(t,r) = \frac{GM}{R} + G\int_r^R \frac{m(t,r')}{r'^2}\,dr' \tag{1.108}$$

inside the matter (for $r < R$), and

$$U(r) = \frac{GM}{r} \tag{1.109}$$

outside the matter (for $r > R$). Integration by parts reveals that an equivalent expression for the internal potential is

$$U(t,r) = \frac{Gm(t,r)}{r} + 4\pi G \int_r^R \rho(t,r')r'\,dr' \,. \tag{1.110}$$

This expression can be seen to apply outside the matter as well, where $\rho = 0$ and $m = M$. The potential at the center of the body is given by $U(t,0) = 4\pi G \int_0^R \rho(t,r')r'\,dr'$.

1.5.2 Non-spherical bodies

Multipole expansions

Spherical symmetry is a convenient simplification, but real bodies are seldom spherically symmetric. Rotation, interactions with other bodies, and stresses caused by solid materials (such as crusts in neutron stars) can lead to deviations from spherical symmetry. For bodies such as planets or stars, these deviations are usually small. This makes the method of *multipole expansions* a useful and powerful tool in modelling the gravitational field of such objects. If the deviations from spherical symmetry are small, the contributions to the gravitational potential of higher multipole moments of the mass distribution are progressively smaller, so that often a small number of moments is sufficient to give an accurate description of the gravitational field for most problems of interest.

Conversely, the determination of the multipole moments of a body by a precise measurement of its external field can supply important diagnostic information about its interior. Indeed, *geodesy* is the science of determining the Earth's gravitational field to high precision as a means of understanding the Earth's internal structure and dynamics. Modern geodetic measurements using precise tracking of Earth-orbiting satellites are determining the Earth's multipole moments up to $\ell = 360$, corresponding to variations in the Earth's surface gravity on a scale of 100 km. From these data it is possible to study such phenomena as the ongoing rebound of the continents following the disappearance of ice-age glaciers, seasonal variations of the amount of water in the Amazon basin, and the effects of volcanism and continental drift.

Some bodies, such as disk-shaped spiral galaxies, are not close to being spherical, and in such cases the method of multipole expansions must be relinquished in favor of other methods, mainly numerical, to solve for the gravitational potential. In the laboratory, for example, experiments to measure the gravitational constant G or to test the weak equivalence principle must carefully account for stray gravitational forces caused by nearby laboratory apparatus. This involves the determination of the gravitational field of strangely shaped objects, often including wires and knobs. The masses and shapes of all these objects must be measured precisely, and the gravitational field calculated numerically, often using techniques borrowed from engineering, such as "finite element" methods.

Spherical-harmonic decomposition

We shall focus our attention on a nearly spherical body, and provide a description of its gravitational field in terms of a multipole expansion. The key analytical tool to achieve this is a systematic expansion of our main variables, ρ and U, in terms of *spherical-harmonic functions* $Y_{\ell m}(\theta, \phi)$, in which ℓ is an integer that ranges from 0 to ∞, while m is a second integer that ranges from $-\ell$ to ℓ for each value of ℓ; the polar angles θ and ϕ were introduced previously in Eq. (1.101).

We recall that the spherical harmonics are solutions to the eigenvalue equation

$$\left(\frac{1}{\sin\theta}\frac{\partial}{\partial\theta}\sin\theta\frac{\partial}{\partial\theta} + \frac{1}{\sin^2\theta}\frac{\partial^2}{\partial\phi^2}\right) Y_{\ell m} = -\ell(\ell+1)Y_{\ell m}\,, \tag{1.111}$$

in which the left-hand side is recognized as the angular piece of the Laplacian operator in Eq. (1.101). For $m = 0$ they are given explicitly by

$$Y_{\ell 0}(\theta) = \sqrt{\frac{2\ell+1}{4\pi}}\,P_\ell(\cos\theta)\,, \tag{1.112}$$

where

$$P_\ell(\mu) := \frac{1}{2^\ell \ell!}\frac{d^\ell}{d\mu^\ell}(\mu^2-1)^\ell \tag{1.113}$$

are the well-known *Legendre polynomials*; in this case the spherical harmonics are independent of ϕ. For $m > 0$ they are given explicitly by

$$Y_{\ell m}(\theta, \phi) = \sqrt{\frac{2\ell+1}{4\pi}\frac{(\ell-m)!}{(\ell+m)!}}\,P_\ell^m(\cos\theta)e^{im\phi}\,, \tag{1.114}$$

where

$$P_\ell^m(\mu) := (-1)^m(1-\mu^2)^{m/2}\frac{d^m}{d\mu^m}P_\ell(\mu) \tag{1.115}$$

are the *associated Legendre functions*. For $m < 0$ we use the formula

$$Y_{\ell,-m}(\theta, \phi) = (-1)^m Y^*_{\ell m}(\theta, \phi)\,. \tag{1.116}$$

The spherical harmonics form a set of orthogonal functions, and they are normalized so that

$$\int Y_{\ell m}(\theta, \phi)Y^*_{\ell' m'}(\theta, \phi)\,d\Omega = \delta_{\ell\ell'}\delta_{mm'}\,, \tag{1.117}$$

where $d\Omega := \sin\theta\,d\theta d\phi$ is an element of solid angle in the direction specified by θ and ϕ; the integral extends over the entire two-sphere (any surface $r =$ constant), from $\phi = 0$ to $\phi = 2\pi$, and from $\theta = 0$ to $\theta = \pi$. The spherical harmonics also form a *complete set* of orthonormal functions, meaning that any square-integrable function on the two-sphere – any well-behaved function $f(\theta, \phi)$ – can be expanded as

$$f(\theta, \phi) = \sum_{\ell=0}^{\infty} \sum_{m=-\ell}^{\ell} f_{\ell m} Y_{\ell m}(\theta, \phi), \tag{1.118}$$

with coefficients given by

$$f_{\ell m} = \int f(\theta, \phi) Y^*_{\ell m}(\theta, \phi)\, d\Omega. \tag{1.119}$$

The spherical-harmonic functions of order $\ell = \{0, 1, 2, 3\}$ are listed in Box 1.5.

Box 1.5 **Spherical harmonics**

The spherical-harmonic functions of lowest order are given explicitly by

$$Y_{00} = \frac{1}{\sqrt{4\pi}},$$

$$Y_{10} = \sqrt{\frac{3}{4\pi}} \cos\theta,$$

$$Y_{11} = -\sqrt{\frac{3}{8\pi}} \sin\theta\, e^{i\phi},$$

$$Y_{20} = \sqrt{\frac{5}{16\pi}} (3\cos^2\theta - 1),$$

$$Y_{21} = -\sqrt{\frac{15}{8\pi}} \sin\theta \cos\theta\, e^{i\phi},$$

$$Y_{22} = \sqrt{\frac{15}{32\pi}} \sin^2\theta\, e^{2i\phi},$$

$$Y_{30} = \sqrt{\frac{7}{16\pi}} (5\cos^3\theta - 3\cos\theta),$$

$$Y_{31} = -\sqrt{\frac{21}{64\pi}} \sin\theta (5\cos^2\theta - 1)\, e^{i\phi},$$

$$Y_{32} = \sqrt{\frac{105}{32\pi}} \sin^2\theta \cos\theta\, e^{2i\phi},$$

$$Y_{33} = -\sqrt{\frac{35}{64\pi}} \sin^3\theta\, e^{3i\phi}.$$

As explained in the main text (in Sec. 1.5.3), the spherical harmonics can be expressed as an expansion in symmetric-tracefree (STF) tensors, according to

$$Y_{\ell m}(\theta, \phi) = \mathscr{Y}^{*\langle L\rangle}_{\ell m} n_{\langle L\rangle},$$

in which $\mathscr{Y}_{\ell m}^{\langle L\rangle}$ is a constant STF tensor, while $n_{\langle L\rangle}$ is a STF combination of unit radial vectors, with $\boldsymbol{n} := [\sin\theta\cos\phi, \sin\theta\sin\phi, \cos\theta]$. Specific examples are

$$\mathscr{Y}_{10}^{\langle z\rangle} = \sqrt{\frac{3}{4\pi}},$$

$$\mathscr{Y}_{11}^{\langle x\rangle} = -\sqrt{\frac{3}{8\pi}}, \quad \mathscr{Y}_{11}^{\langle y\rangle} = i\sqrt{\frac{3}{8\pi}},$$

$$\mathscr{Y}_{20}^{\langle xx\rangle} = -\sqrt{\frac{5}{16\pi}}, \quad \mathscr{Y}_{20}^{\langle yy\rangle} = -\sqrt{\frac{5}{16\pi}}, \quad \mathscr{Y}_{20}^{\langle zz\rangle} = 2\sqrt{\frac{5}{16\pi}},$$

$$\mathscr{Y}_{21}^{\langle xz\rangle} = -\frac{1}{2}\sqrt{\frac{15}{8\pi}}, \quad \mathscr{Y}_{21}^{\langle yz\rangle} = \frac{i}{2}\sqrt{\frac{15}{8\pi}},$$

$$\mathscr{Y}_{22}^{\langle xx\rangle} = \sqrt{\frac{15}{32\pi}}, \quad \mathscr{Y}_{22}^{\langle xy\rangle} = -i\sqrt{\frac{15}{32\pi}}, \quad \mathscr{Y}_{22}^{\langle yy\rangle} = -\sqrt{\frac{15}{32\pi}},$$

$$\mathscr{Y}_{30}^{\langle xxz\rangle} = -\sqrt{\frac{7}{16\pi}}, \quad \mathscr{Y}_{30}^{\langle yyz\rangle} = -\sqrt{\frac{7}{16\pi}}, \quad \mathscr{Y}_{30}^{\langle zzz\rangle} = 2\sqrt{\frac{7}{16\pi}},$$

$$\mathscr{Y}_{31}^{\langle xxx\rangle} = \sqrt{\frac{21}{64\pi}}, \quad \mathscr{Y}_{31}^{\langle xxy\rangle} = -\frac{i}{3}\sqrt{\frac{21}{64\pi}}, \quad \mathscr{Y}_{31}^{\langle xyy\rangle} = \frac{1}{3}\sqrt{\frac{21}{64\pi}},$$

$$\mathscr{Y}_{31}^{\langle xzz\rangle} = -\frac{4}{3}\sqrt{\frac{21}{64\pi}}, \quad \mathscr{Y}_{31}^{\langle yzz\rangle} = \frac{4i}{3}\sqrt{\frac{21}{64\pi}}, \quad \mathscr{Y}_{31}^{\langle yyy\rangle} = -i\sqrt{\frac{21}{64\pi}},$$

$$\mathscr{Y}_{32}^{\langle xxz\rangle} = \frac{1}{3}\sqrt{\frac{105}{32\pi}}, \quad \mathscr{Y}_{32}^{\langle xyz\rangle} = -\frac{i}{3}\sqrt{\frac{105}{32\pi}}, \quad \mathscr{Y}_{32}^{\langle yyz\rangle} = -\frac{1}{3}\sqrt{\frac{105}{32\pi}},$$

$$\mathscr{Y}_{33}^{\langle xxx\rangle} = -\sqrt{\frac{35}{64\pi}}, \quad \mathscr{Y}_{33}^{\langle xxy\rangle} = i\sqrt{\frac{35}{64\pi}}, \quad \mathscr{Y}_{33}^{\langle xyy\rangle} = \sqrt{\frac{35}{64\pi}}, \quad \mathscr{Y}_{33}^{\langle yyy\rangle} = -i\sqrt{\frac{35}{64\pi}}.$$

Only the independent, non-vanishing components are listed, and other components can be obtained by exploiting the index symmetries of $\mathscr{Y}_{\ell m}^{\langle L\rangle}$.

Reduced Poisson equation

To integrate Eq. (1.101) we decompose the mass density ρ and Newtonian potential U in spherical harmonics,

$$\rho(t,r,\theta,\phi) = \sum_{\ell m} \rho_{\ell m}(t,r) Y_{\ell m}(\theta,\phi), \tag{1.120a}$$

$$U(t,r,\theta,\phi) = \sum_{\ell m} U_{\ell m}(t,r) Y_{\ell m}(\theta,\phi), \tag{1.120b}$$

in which the coefficients

$$\rho_{\ell m}(t,r) = \int \rho(t,r,\theta,\phi) Y^*_{\ell m}(\theta,\phi)\, d\Omega\,, \tag{1.121a}$$

$$U_{\ell m}(t,r) = \int U(t,r,\theta,\phi) Y^*_{\ell m}(\theta,\phi)\, d\Omega \tag{1.121b}$$

are now allowed to depend on t and r; in Eq. (1.120) the summation sign is a shorthand notation for the double sum that appears in Eq. (1.118). Making the substitution in Eq. (1.101) produces the decoupled set of ordinary differential equations

$$\mathscr{L} U_{\ell m} = -4\pi G r^2 \rho_{\ell m}\,, \tag{1.122}$$

in which

$$\mathscr{L} := \frac{\partial}{\partial r} r^2 \frac{\partial}{\partial r} - \ell(\ell+1) \tag{1.123}$$

is an effective Laplacian operator (multiplied by r^2) that involves the radial coordinate only. Equation (1.122) is a dimensionally-reduced version of Poisson's equation, and we have simplified the problem of finding a solution to a partial differential equation in three variables to that of finding solutions to an infinite number of ordinary differential equations.

It is helpful here, as it was in Sec. 1.3, to integrate Eq. (1.122) with the help of a Green's function $g_\ell(r,r')$, which is required to be a solution to

$$\mathscr{L} g_\ell(r,r') = -4\pi \delta(r-r')\,. \tag{1.124}$$

It is easy to verify that once the Green's function is available, $U_{\ell m}$ can be obtained for any $\rho_{\ell m}$ by evaluating the integral

$$U_{\ell m}(t,r) = G \int g_\ell(r,r') \rho_{\ell m}(t,r') r'^2\, dr'\,. \tag{1.125}$$

The Green's function is not difficult to construct. First we observe that if $U_<(r)$ and $U_>(r)$ are independent solutions to the homogeneous equation $\mathscr{L}U = 0$, then

$$g(r,r') = U_<(r)\Theta(r'-r) + U_>(r)\Theta(r-r') \tag{1.126}$$

is a solution to Eq. (1.124) provided that $U_>(r') - U_<(r') = 0$ and $U'_>(r') - U'_<(r') = -4\pi/r'^2$. We omit the label ℓ to simplify the notation, and a prime on U indicates differentiation with respect to r; $\Theta(r-r')$ is the Heaviside step function, equal to one when $r - r' > 0$ and zero otherwise, and such that $\Theta'(r-r') = \delta(r-r')$. For $U_<(r)$ we choose a solution that is finite at $r = 0$, and this requirement forces $U_< \propto r^\ell$. For $U_>(r)$ we choose a solution that is finite at $r = \infty$, and this requirement forces $U_> \propto r^{-(\ell+1)}$. The junction conditions at $r = r'$ determine the constants of proportionality (which actually depend on r'), and we finally arrive at

$$g_\ell(r,r') = \frac{4\pi}{2\ell+1}\left[\frac{r^\ell}{r'^{\ell+1}}\Theta(r'-r) + \frac{r'^\ell}{r^{\ell+1}}\Theta(r-r')\right] = \frac{4\pi}{2\ell+1}\frac{r_<^\ell}{r_>^{\ell+1}}\,, \tag{1.127}$$

where $r_< := \min(r,r')$ and $r_> := \max(r,r')$.

Substituting Eq. (1.127) in Eq. (1.125) returns

$$U_{\ell m}(t,r) = \frac{4\pi G}{2\ell+1}\biggl[r^\ell \int_r^\infty \frac{\rho_{\ell m}(t,r')}{r'^{\ell+1}} r'^2\, dr' + \frac{1}{r^{\ell+1}} \int_0^r r'^\ell \rho_{\ell m}(t,r') r'^2\, dr' \biggr], \tag{1.128}$$

and this expression is ready to be inserted within Eq. (1.120). We express our final result as

$$U(t,r,\theta,\phi) = G \sum_{\ell m} \frac{4\pi}{2\ell+1} \biggl[q_{\ell m}(t,r) \frac{Y_{\ell m}(\theta,\phi)}{r^{\ell+1}} + p_{\ell m}(t,r) r^\ell Y_{\ell m}(\theta,\phi) \biggr], \tag{1.129}$$

where

$$q_{\ell m}(t,r) := \int_0^r r'^\ell \rho_{\ell m}(t,r') r'^2\, dr' \tag{1.130}$$

and

$$p_{\ell m}(t,r) := \int_r^R \frac{\rho_{\ell m}(t,r')}{r'^{\ell+1}} r'^2\, dr' . \tag{1.131}$$

These relations apply directly to the body's interior, in which $\rho_{\ell m} \neq 0$, and the integral defining $p_{\ell m}$ was truncated to a sphere of arbitrary radius R that surrounds the matter distribution. They apply also to the body's exterior, where $\rho_{\ell m} = 0$, but here they simplify to

$$U_{\rm ext}(t,r,\theta,\phi) = G \sum_{\ell m} \frac{4\pi}{2\ell+1} q_{\ell m}(t,R) \frac{Y_{\ell m}(\theta,\phi)}{r^{\ell+1}} ; \tag{1.132}$$

the term involving $p_{\ell m}$ vanishes, and $q_{\ell m}$ was evaluated at $r = R$, because the integrals of Eqs. (1.130) and (1.131) are now evaluated outside the matter distribution.

Integral solution

The results of Eqs. (1.129) and (1.132) can be reproduced by proceeding directly from the integral solution to Poisson's equation,

$$U(t,\boldsymbol{x}) = G \int \frac{\rho(t,\boldsymbol{x}')}{|\boldsymbol{x}-\boldsymbol{x}'|}\, d^3x' . \tag{1.133}$$

The strategy is to express $|\boldsymbol{x}-\boldsymbol{x}'|^{-1}$, the three-dimensional Green's function of Eq. (1.18), as the spherical-harmonic decomposition

$$\frac{1}{|\boldsymbol{x}-\boldsymbol{x}'|} = \sum_{\ell m} \frac{4\pi}{2\ell+1} \frac{r_<^\ell}{r_>^{\ell+1}} Y^*_{\ell m}(\theta',\phi') Y_{\ell m}(\theta,\phi) , \tag{1.134}$$

in which (r,θ,ϕ) are the spherical polar coordinates of the field point $\boldsymbol{x}$, while (r',θ',ϕ') are the coordinates of the source point $\boldsymbol{x}'$; we recall that $r_< := \min(r,r')$ and $r_> := \max(r,r')$. We recognize the radial Green's function $g_\ell(r,r')$ in the first factor within the sum in Eq. (1.134), and the product of spherical harmonics accounts for the angular dependence.

When we insert Eq. (1.134) within $U(t,\boldsymbol{x})$ and evaluate the integral outside the matter distribution, we find that $r_< = r'$ and $r_> = r$ because the variable of integration r' is always smaller than r; we end up once more with Eq. (1.132), with $q_{\ell m}(t, R)$ given by Eq. (1.130) and $\rho_{\ell m}$ given by Eq. (1.121). When instead we evaluate the integral inside the matter, the radial integration must be broken up into two pieces, the first ranging from $r' = 0$ to $r' = r$, for which $r_< = r'$ and $r_> = r$, and the second ranging from $r' = r$ to $r' = R$, for which $r_< = r$ and $r_> = r'$; we end up with Eq. (1.129), with $q_{\ell m}(t, r)$ and $p_{\ell m}(t, r)$ given by Eqs. (1.130) and (1.131), respectively.

To establish Eq. (1.134) we make use of two fundamental properties of the Legendre polynomials. The first is that the polynomials come with a *generating function*

$$\frac{1}{\sqrt{1 - 2\eta\mu + \eta^2}} = \sum_{\ell=0}^{\infty} \eta^\ell P_\ell(\mu), \tag{1.135}$$

in which η is an arbitrary number smaller than unity, and μ is the argument of the Legendre functions. In this representation, $P_\ell(\mu)$ is recognized as the set of coefficients in a Taylor expansion of the generating function in powers of η. In our particular application η is identified with $r_</r_>$, μ is identified with $\cos\gamma := \boldsymbol{x} \cdot \boldsymbol{x}'/(rr')$, and we identify the generating function with $r_> |\boldsymbol{x} - \boldsymbol{x}'|^{-1}$. Equation (1.135) yields

$$\frac{1}{|\boldsymbol{x} - \boldsymbol{x}'|} = \sum_{\ell=0}^{\infty} \frac{r_<^\ell}{r_>^{\ell+1}} P_\ell(\cos\gamma). \tag{1.136}$$

The second property is the *addition theorem*

$$P_\ell(\cos\gamma) = \frac{4\pi}{2\ell+1} \sum_{m=-\ell}^{\ell} Y^*_{\ell m}(\theta', \phi') Y_{\ell m}(\theta, \phi), \tag{1.137}$$

in which γ is related to the angles (θ, ϕ) and (θ', ϕ') by

$$\cos\gamma = \cos\theta\cos\theta' + \sin\theta\sin\theta'\cos(\phi - \phi'). \tag{1.138}$$

Inserting Eq. (1.137) in Eq. (1.136) returns Eq. (1.134).

Multipole moments

The quantities $I_{\ell m}(t) := q_{\ell m}(t, R)$, as defined by Eqs. (1.121) and (1.130), are the *multipole moments* of the mass distribution. Their full expression is

$$I_{\ell m}(t) := \int \rho(t, \boldsymbol{x}) r^\ell Y^*_{\ell m}(\theta, \phi)\, d^3x, \tag{1.139}$$

where the domain of integration extends over the volume occupied by the matter. (Because the domain is independent of the field point $\boldsymbol{x}$, it is no longer necessary to adorn the variables of integration with primes.) In terms of these the external potential is given by

$$U_{\text{ext}}(t, \boldsymbol{x}) = G \sum_{\ell m} \frac{4\pi}{2\ell+1} I_{\ell m}(t) \frac{Y_{\ell m}(\theta, \phi)}{r^{\ell+1}}; \tag{1.140}$$

this expression is copied directly from Eq. (1.132) – this is such an important result, we feel compelled to display it twice.

The moment corresponding to $\ell = m = 0$ is known as the *monopole moment*, and it is intimately related to the body's total mass:

$$I_{00} = \int \rho Y_{00}\, d^3x = \frac{M}{\sqrt{4\pi}}. \tag{1.141}$$

The moments corresponding to $\ell = 1$ (and $m = \{-1, 0, 1\}$) are known as the *dipole moments*. These all vanish when we place the origin of the coordinate system at the body's center-of-mass, so that $\int \rho \boldsymbol{x}\, d^3x = \boldsymbol{0}$. This conclusion follows from a simple computation:

$$I_{10} = \sqrt{\frac{3}{4\pi}} \int \rho r \cos\theta\, d^3x = \sqrt{\frac{3}{4\pi}} \int \rho z\, d^3x = 0, \tag{1.142a}$$

$$I_{1\pm1} = \mp\sqrt{\frac{3}{8\pi}} \int \rho r \sin\theta e^{\pm i\phi}\, d^3x = \mp\sqrt{\frac{3}{8\pi}} \int \rho(x \pm iy)\, d^3x = 0. \tag{1.142b}$$

Moments of higher degree are called quadrupole ($\ell = 2$), octopole ($\ell = 3$), hexadecapole ($\ell = 4$), and so on. When the body is spherically symmetric, only I_{00} is non-zero, and the potential U does not depend on the angles (θ, ϕ). When the body is axially symmetric about the z axis, only the moments with $m = 0$ are non-zero, and U is independent of the azimuthal angle ϕ. Notice that each $I_{\ell m}$ scales as MR^ℓ, with R denoting a characteristic length scale of the body. The mass multipole moments introduced here are in a close correspondence with the charge multipole moments defined in electromagnetism; in fact, our definitions are virtually identical to those adopted by J.D. Jackson (1998) in his famous textbook *Classical Electrodynamics*.

Axially symmetric bodies

A body is axially symmetric when its mass density is invariant under a rotation about a symmetry axis. The condition can apply exactly to an idealized body, or it can apply as a good approximation to a realistic body (the Sun, for example). Taking the z-direction to be aligned with the symmetry axis, we find that the mass density is independent of the azimuthal angle ϕ, and it follows that the only non-vanishing multipole moments are those with $m = 0$. It is conventional to express the moments in terms of dimensionless quantities J_ℓ defined by

$$J_\ell := -\sqrt{\frac{4\pi}{2\ell+1}} \frac{I_{\ell 0}}{MR^\ell}, \tag{1.143}$$

in which the characteristic length scale R is chosen to be the body's equatorial radius. This definition is adopted, for example, in *Allen's Astrophysical Quantities*, a repository of useful information about most areas of astrophysics.

The gravitational potential of an axially symmetric body can then be written in the form

$$U_{\rm ext}(t, \boldsymbol{x}) = \frac{GM}{r}\left[1 - \sum_{\ell=2}^{\infty} J_\ell \left(\frac{R}{r}\right)^\ell P_\ell(\cos\theta)\right]. \tag{1.144}$$

The dominant term in the sum is provided by the dimensionless quadrupole moment J_2, and this is frequently expressed in terms of the body's *principal moments of inertia*, defined by

$$I_1(t) \equiv A(t) := \int \rho(t, \boldsymbol{x})(y^2 + z^2)\, d^3x\,, \tag{1.145a}$$

$$I_2(t) \equiv B(t) := \int \rho(t, \boldsymbol{x})(x^2 + z^2)\, d^3x\,, \tag{1.145b}$$

$$I_3(t) \equiv C(t) := \int \rho(t, \boldsymbol{x})(x^2 + y^2)\, d^3x\,. \tag{1.145c}$$

We have

$$\begin{aligned}
J_2 &= -\frac{1}{MR^2}\int \rho r^2 \left[\frac{3\cos^2\theta - 1}{2}\right] d^3x \\
&= \frac{1}{MR^2}\int \rho \left[\frac{r^2 - 3z^2}{2}\right] d^3x \\
&= \frac{1}{MR^2}\int \rho \Big[x^2 + y^2 - \tfrac{1}{2}(x^2 + z^2) - \tfrac{1}{2}(y^2 + z^2)\Big] d^3x \\
&= \frac{C - \frac{1}{2}A - \frac{1}{2}B}{MR^2} \\
&= \frac{C - A}{MR^2}\,.
\end{aligned} \tag{1.146}$$

In the last step we used the fact that $A = B$ for an axially symmetric body.

1.5.3 Symmetric tracefree tensors

We next turn to an alternative decomposition of the gravitational potential that involves tensorial combinations of the unit vector $\boldsymbol{n} := \boldsymbol{x}/r$ instead of spherical harmonics. Each tensor that we shall construct from $\boldsymbol{n}$ will have the property of being symmetric under the exchange of any two of its indices, and of being tracefree in any pair of indices; these tensors are known as *symmetric tracefree tensors*, or *STF tensors*. The decompositions in STF tensors and spherical harmonics both involve building blocks that consist of irreducible representations of the rotation group labelled by a multipole index ℓ. The decomposition in spherical harmonics relies on spherical polar coordinates, and keeps the polar angles (θ, ϕ) segregated from the radial coordinate r. The decomposition in STF tensors relies on the original Cartesian coordinates, which are all put on an equal footing. In our experience we have found that it is helpful to be conversant in both languages; some applications are best handled with spherical harmonics, and some are most easily treated with STF tensors.

Taylor expansion of the external potential

We return to the integral representation of the gravitational potential U,

$$U(t, \boldsymbol{x}) = G\int \frac{\rho(t, \boldsymbol{x}')}{|\boldsymbol{x} - \boldsymbol{x}'|}\, d^3x'\,, \tag{1.147}$$

and consider a field point $\boldsymbol{x}$ that lies outside the matter distribution. With $|\boldsymbol{x}'| < |\boldsymbol{x}|$, we carry out a Taylor expansion of $|\boldsymbol{x} - \boldsymbol{x}'|^{-1}$ in powers of $\boldsymbol{x}'$:

$$\frac{1}{|\boldsymbol{x} - \boldsymbol{x}'|} = \frac{1}{r} - x'^j \partial_j \left(\frac{1}{r}\right) + \frac{1}{2} x'^j x'^k \partial_j \partial_k \left(\frac{1}{r}\right) - \cdots \tag{1.148a}$$

$$= \frac{1}{r} - x'^j \partial_j \left(\frac{1}{r}\right) + \frac{1}{2} x'^{jk} \partial_{jk} \left(\frac{1}{r}\right) - \cdots \tag{1.148b}$$

$$= \sum_{\ell=0}^{\infty} \frac{(-1)^\ell}{\ell!} x'^L \partial_L \left(\frac{1}{r}\right). \tag{1.148c}$$

Once more we adopt the Einstein summation convention and sum over repeated indices in an expression like $x'^j x'^k \partial_j \partial_k r^{-1}$. In the second line we introduce a condensed notation in which an expression like x^{jkn} stands for the product $x^j x^k x^n$, and ∂_{jkn} stands for $\partial_j \partial_k \partial_n$. In the third line we introduce an even more compact *multi-index* notation, in which an uppercase index such as L represents a collection of ℓ individual indices. Thus, x^L stands for $x^{j_1 j_2 \cdots j_\ell}$, ∂_L stands for $\partial_{j_1 j_2 \cdots j_\ell}$, and $x^L \partial_L$ involves a summation over all ℓ pairs of repeated indices.

Substituting Eq. (1.148) into Eq. (1.147) gives

$$U_{\text{ext}}(t, \boldsymbol{x}) = G \sum_{\ell=0}^{\infty} \frac{(-1)^\ell}{\ell!} I^{\langle L\rangle} \partial_{\langle L\rangle} \left(\frac{1}{r}\right), \tag{1.149}$$

with

$$I^{\langle L\rangle}(t) := \int \rho(t, \boldsymbol{x}') x'^{\langle L\rangle}\, d^3x' \tag{1.150}$$

defining a set of *STF multipole moments* for the mass distribution. The STF label, and the meaning of the angular brackets around the multi-index L, will be explained shortly. For the time being we may note that a comparison between Eq. (1.149) and Eq. (1.140) reveals a close relationship between $Y_{\ell m}/r^{\ell+1}$ and $\partial_{\langle L\rangle} r^{-1}$, and another between $I_{\ell m}$ and $I^{\langle L\rangle}$. The correspondence will be made precise below.

STF combinations

We next compute the derivatives of r^{-1} that appear in Eq. (1.149). We make repeated use of the identities

$$\partial_j r = n_j, \tag{1.151a}$$

$$\partial_j n_k = \partial_k n_j = \frac{1}{r}\left(\delta_{jk} - n_j n_k\right), \tag{1.151b}$$

where $\boldsymbol{n} := \boldsymbol{x}/r$ is a unit radial vector imagined to be expressed in terms of the Cartesian coordinates (x, y, z). We obtain

$$\partial_j r^{-1} = -n_j r^{-2}, \tag{1.152a}$$

$$\partial_{jk} r^{-1} = \big(3n_j n_k - \delta_{jk}\big) r^{-3}, \tag{1.152b}$$

$$\partial_{jkn} r^{-1} = -\Big[15 n_j n_k n_n - 3\big(n_j \delta_{kn} + n_k \delta_{jn} + n_n \delta_{jk}\big)\Big] r^{-4}, \tag{1.152c}$$

and so on; it is understood that $r \neq 0$ in these operations. We observe that the tensors on the right-hand side are all symmetric under an exchange of any two indices, and that they all vanish when a trace is taken over any pair of indices (which means that the indices within the pair are made equal and summed over); these are all examples of STF tensors. These properties are inherited from the definitions on the left-hand side: a tensor such as $\partial_{jkn} r^{-1}$ is necessarily symmetric because the partial derivatives commute with each other, and it is necessarily tracefree because, for example, $\delta^{jk}\partial_{jkn} r^{-1} = \nabla^2 \partial_n r^{-1} = \partial_n \nabla^2 r^{-1} = 0$. We conclude that each tensor in the collection $\partial_L r^{-1}$ is an STF tensor, a property that we can emphasize by enclosing L between angular brackets. More generally, any STF tensor will be distinguished with angular brackets; we shall write, for example, $A^{\langle jkn \rangle}$ for an STF tensor of rank 3, and $A^{\langle L \rangle}$ for an STF tensor of rank ℓ.

Conventionally, STF products of vectors such as n^j are obtained by beginning with the "raw" products $n^j n^k \cdots$ and then removing all traces, maintaining symmetry on all indices. Explicit examples are

$$n^{\langle jk \rangle} = n^j n^k - \frac{1}{3}\delta^{jk}, \tag{1.153a}$$

$$n^{\langle jkn \rangle} = n^j n^k n^n - \frac{1}{5}\big(\delta^{jk} n^n + \delta^{jn} n^k + \delta^{kn} n^j\big), \tag{1.153b}$$

$$\begin{aligned} n^{\langle jknp \rangle} = {}& n^j n^k n^n n^p - \frac{1}{7}\big(\delta^{jk} n^n n^p + \delta^{jn} n^k n^p + \delta^{jp} n^k n^n + \delta^{kn} n^j n^p \\ & + \delta^{kp} n^j n^n + \delta^{np} n^j n^k\big) + \frac{1}{35}\big(\delta^{jk}\delta^{np} + \delta^{jn}\delta^{kp} + \delta^{jp}\delta^{kn}\big). \end{aligned} \tag{1.153c}$$

For example $n^{\langle jkn \rangle}$ is tracefree because $\delta_{jk} n^{\langle jkn \rangle} = n^n - \frac{1}{5}(3n^n + n^n + n^n) = 0$, $\delta_{jn} n^{\langle jkn \rangle} = 0$, and $\delta_{kn} n^{\langle jkn \rangle} = 0$.

The general formula for such STF products is

$$\begin{aligned} n^{\langle j_1 j_2 \cdots j_\ell \rangle} = {}& \sum_{p=0}^{[\ell/2]} (-1)^p \frac{\ell!(2\ell - 2p - 1)!!}{(\ell - 2p)!(2\ell - 1)!!(2p)!!} \\ & \times \delta^{(j_1 j_2} \delta^{j_3 j_4} \ldots \delta^{j_{2p-1} j_{2p}} n^{j_{2p+1}} n^{j_{2p+2}} \ldots n^{j_\ell)}, \end{aligned} \tag{1.154}$$

in which $[\ell/2]$ is the largest integer not larger than $\ell/2$, equal to $\ell/2$ when ℓ is an even number and to $(\ell - 1)/2$ when ℓ is odd; all ℓ indices are enclosed within round brackets, which indicates the symmetrization operation defined in Box 1.4. In a more compact notation we have

$$n^{\langle L\rangle} = \sum_{p=0}^{[\ell/2]}(-1)^p\frac{(2\ell-2p-1)!!}{(2\ell-1)!!}\Big[\delta^{2P}n^{L-2P} + \mathrm{sym}(q)\Big], \tag{1.155}$$

where δ^{2P} stands for a product of p Kronecker deltas (with indices running from j_1 to j_{2p}), n^{L-2P} stands for a product of $\ell - 2p$ unit vectors (with indices running from j_{2p+1} to j_ℓ), and "sym(q)" denotes all distinct terms arising from permuting indices; the total number of terms within the square brackets is equal to $q := \ell!/[(\ell-2p)!(2p)!!]$.

Note that the tensor $n^{\langle jk\rangle}$ contains five independent components; the number would be six for a general symmetric tensor, but the vanishing trace removes one component from the total count. This number matches the five values of m that belong to $\ell = 2$. Similarly, the tensor $n^{\langle jkn\rangle}$ contains seven independent components, and this matches the seven values of m that belong to $\ell = 3$. It can be shown that, in general, $n^{\langle L\rangle}$ contains $2\ell + 1$ independent components, and this is also the number of integers in the interval between $-\ell$ and $+\ell$.

Comparing Eqs. (1.152) and (1.153) we find that $\partial_j r^{-1} = -n_j r^{-2}$, $\partial_{jk} r^{-1} = 3n_{\langle jk\rangle}r^{-3}$, and $\partial_{jkn} r^{-1} = -15n_{\langle jkn\rangle}r^{-4}$. The general rule can be obtained by induction:

$$\partial_L r^{-1} = \partial_{\langle L\rangle} r^{-1} = (-1)^\ell (2\ell-1)!!\,\frac{n_{\langle L\rangle}}{r^{\ell+1}}\,. \tag{1.156}$$

We may now return to Eq. (1.149), which can be expressed as

$$U_{\mathrm{ext}}(t,\boldsymbol{x}) = G\sum_{\ell=0}^{\infty}\frac{(2\ell-1)!!}{\ell!}\,I^{\langle L\rangle}\frac{n_{\langle L\rangle}}{r^{\ell+1}}\,, \tag{1.157}$$

and explain the reason for the angular brackets on $I^{\langle L\rangle}$. In the preceding step, displayed in Eq. (1.148c), we expressed $|\boldsymbol{x}-\boldsymbol{x}'|^{-1}$ as a sum of terms $x'^L\partial_L r^{-1}$, and substitution into Eq. (1.147) returned U as a sum of terms $I^L\partial_L r^{-1}$, with $I^L := \int \rho' x'^L\, d^3x'$ denoting the "raw" multipole moments. In view of Eq. (1.155), however, I^L differs from $I^{\langle L\rangle}$ by a sum of terms involving Kronecker deltas, and these automatically give zero when multiplied by the tracefree $\partial_L r^{-1}$. As a result, we find that $I^L\partial_L r^{-1} = I^{\langle L\rangle}\partial_{\langle L\rangle} r^{-1}$, and that U_{ext} can indeed be expressed as in Eq. (1.149).

It is worthwhile to display the main rule by which we were able to reach this conclusion: whenever an arbitrary tensor A^L multiplies an STF tensor $B_{\langle L\rangle}$, the outcome is

$$A^L B_{\langle L\rangle} = A^{\langle L\rangle}B_{\langle L\rangle}, \tag{1.158}$$

where $A^{\langle L\rangle}$ is the tensor obtained from A^L by complete symmetrization and removal of all traces.

STF identities

The STF tensors $n^{\langle L\rangle}$ satisfy a number of helpful identities, including

$$n_{\langle L\rangle}n^{\langle L\rangle} = \frac{\ell!}{(2\ell-1)!!}, \tag{1.159a}$$

$$n_j n^{\langle jL\rangle} = \frac{\ell+1}{2\ell+1}n^{\langle L\rangle}, \tag{1.159b}$$

$$n_{\langle L\rangle}n^{\langle jL\rangle} = \frac{(\ell+1)!}{(2\ell+1)!!}n^j. \tag{1.159c}$$

Other identities involve a second unit vector $\boldsymbol{n}'$:

$$n'_{\langle L\rangle}n^{\langle L\rangle} = \frac{\ell!}{(2\ell-1)!!}P_\ell(\mu), \tag{1.160a}$$

$$n'_{\langle L\rangle}n^{\langle jL\rangle} = \frac{\ell!}{(2\ell+1)!!}\left[\frac{dP_{\ell+1}}{d\mu}n^j - \frac{dP_\ell}{d\mu}n'^j\right], \tag{1.160b}$$

where $\mu := \boldsymbol{n}\cdot\boldsymbol{n}'$.

To establish these identities we begin with Eq. (1.160a), and write its left-hand side as $n'_L n^{\langle L\rangle}$ after invoking the rule of Eq. (1.158). We substitute Eq. (1.155) for $n^{\langle L\rangle}$ and perform the index contractions. For each value of p in the sum we find that $2p$ factors of $\boldsymbol{n}'$ multiply Kronecker deltas, returning unity, while the remaining $\ell - 2p$ vectors multiply an $\boldsymbol{n}$, returning $\mu^{\ell-2p}$. Because all q terms are equal to each other, we get

$$n'_{\langle L\rangle}n^{\langle L\rangle} = \sum_{p=0}^{[\ell/2]}(-1)^p\frac{\ell!(2\ell-2p-1)!!}{(\ell-2p)!(2\ell-1)!!(2p)!!}\mu^{\ell-2p}. \tag{1.161}$$

Making use of the identities $(2p)!! = 2^p p!$ and $(2\ell-2p-1)!! = (2\ell-2p)!/[2^{\ell-p}(\ell-p)!]$, we find that this is also

$$n'_{\langle L\rangle}n^{\langle L\rangle} = \frac{\ell!}{(2\ell-1)!!}\frac{1}{2^\ell}\sum_{p=0}^{[\ell/2]}(-1)^p\frac{(2\ell-2p)!}{p!(\ell-p)!(\ell-2p)!}\mu^{\ell-2p}, \tag{1.162}$$

and the sum (together with the prefactor of $2^{-\ell}$) is recognized as a representation of the Legendre polynomial $P_\ell(\mu)$. We have recovered Eq. (1.160a), and we notice that Eq. (1.159a) is a special case with $\boldsymbol{n}' = \boldsymbol{n}$, $\mu = 1$, and $P_\ell(\mu) = 1$.

To establish Eq. (1.159b) we observe that the product $n_j n^{\langle jL\rangle}$ is necessarily STF in the indices contained in L, and that it must therefore be proportional to $n^{\langle L\rangle}$. The constant of proportionality can be determined by making use of Eq. (1.159a) in the form of $n_{\langle jL\rangle}n^j n^L = (\ell+1)!/(2\ell+1)!!$; the end result is Eq. (1.159b). The identity of Eq. (1.159c) can be established by similar means.

To establish Eq. (1.160b) we observe that $n'_{\langle L\rangle}n^{\langle jL\rangle}$ must be a vector constructed from n^j and n'^j. We may write it as $(\ell+1)!(an^j + bn'^j)/(2\ell+1)!!$ and work to determine the coefficients a and b; the factor of $(\ell+1)!/(2\ell+1)!!$ is inserted for convenience.

Using Eqs. (1.159b) and (1.160a) it is easy to see that a and b must satisfy the equations $a + b\mu = P_\ell(\mu)$ and $a\mu + b = P_{\ell+1}(\mu)$. The solutions are $a = (\mu P_{\ell+1} - P_\ell)/(\mu^2 - 1)$ and $b = (\mu P_\ell - P_{\ell+1})/(\mu^2 - 1)$, and these can be re-expressed as $a = (\ell+1)^{-1} dP_{\ell+1}/d\mu$ and $b = -(\ell+1)^{-1} dP_\ell/d\mu$ by exploiting the recurrence relations satisfied by the Legendre polynomials. The end result is Eq. (1.160b).

Correspondence with spherical harmonics

We are now ready to reveal the correspondence between the STF tensors $n^{\langle L\rangle}$ and the spherical harmonics $Y_{\ell m}(\theta, \phi)$. It comes about when $\boldsymbol{n}$ is expressed as

$$\boldsymbol{n} = [\sin\theta\cos\phi, \sin\theta\sin\phi, \cos\theta], \tag{1.163}$$

and when it is recognized that $n^{\langle L\rangle}$ is a set of functions of θ and ϕ that can be decomposed in spherical harmonics, as in Eq. (1.118). The decomposition, however, involves a single value of ℓ (instead of a sum over all values), and the $2\ell+1$ values of m that belong to this ℓ. The reason is that $n^{\langle L\rangle}$ is not just any function; as we show in Box 1.6, it is an *eigenfunction* of the (angular piece of the) Laplacian operator: $r^2\nabla^2 n^{\langle L\rangle} = -\ell(\ell+1)n^{\langle L\rangle}$. Because $n^{\langle L\rangle}$ satisfies the same eigenvalue equation as $Y_{\ell m}(\theta,\phi)$, the expansion must be of the form

$$n^{\langle L\rangle} := N_\ell \sum_{m=-\ell}^{\ell} \mathscr{Y}^{\langle L\rangle}_{\ell m} Y_{\ell m}(\theta,\phi), \qquad N_\ell := \frac{4\pi\,\ell!}{(2\ell+1)!!}, \tag{1.164}$$

where $\mathscr{Y}^{\langle L\rangle}_{\ell m}$ is a constant STF tensor that satisfies $\mathscr{Y}^{\langle L\rangle}_{\ell,-m} = (-1)^m \mathscr{Y}^{*\langle L\rangle}_{\ell m}$, and N_ℓ is a normalization constant chosen for future convenience. In Box 1.5 we display a few members of $\mathscr{Y}^{\langle L\rangle}_{\ell m}$ for selected values of ℓ.

When we multiply Eq. (1.164) by $Y^*_{\ell m'}$ and integrate over the whole sphere, we obtain

$$\mathscr{Y}^{\langle L\rangle}_{\ell m} = \frac{1}{N_\ell}\int n^{\langle L\rangle} Y^*_{\ell m}(\theta,\phi)\, d\Omega. \tag{1.165}$$

When we next multiply this expression by $n'_{\langle L\rangle}$ and make use of Eq. (1.160a), we get

$$\mathscr{Y}^{\langle L\rangle}_{\ell m} n'_{\langle L\rangle} = \frac{1}{N_\ell}\int n^{\langle L\rangle} n'_{\langle L\rangle} Y^*_{\ell m}(\theta,\phi)\, d\Omega = \frac{2\ell+1}{4\pi}\int P_\ell(\mu) Y^*_{\ell m}(\theta,\phi)\, d\Omega. \tag{1.166}$$

When, finally, we insert the addition theorem of Eq. (1.137) with $\cos\gamma = \mu = \boldsymbol{n}\cdot\boldsymbol{n}'$ and perform the integration, we arrive at

$$Y_{\ell m}(\theta,\phi) = \mathscr{Y}^{*\langle L\rangle}_{\ell m} n_{\langle L\rangle}, \tag{1.167}$$

a decomposition of the spherical harmonics in STF tensors. Equation (1.167) is the inverse of Eq. (1.164), and N_ℓ was chosen so as to make the overall factor on the right-hand side of Eq. (1.167) equal to unity.

Box 1.6 **Proof that $r^2\nabla^2 n^{\langle L\rangle} = -\ell(\ell+1)n^{\langle L\rangle}$**

To prove that $n^{\langle L\rangle}$ satisfies the eigenvalue equation, it is convenient to work instead with the scalar field $\psi := A_{\langle L\rangle}n^{\langle L\rangle}$, in which $A_{\langle L\rangle}$ is an arbitrary STF tensor of constant elements. Because this tensor is arbitrary, it can be chosen so as to single out any particular element of $n^{\langle L\rangle}$, and it becomes sufficient to prove that ψ itself satisfies the eigenvalue equation.

We write the scalar field as $\psi = A_{\langle L\rangle}x^L/r^\ell$ and differentiate it once with respect to x^j. Because $A_{\langle L\rangle}$ is completely symmetric, we have that $A_{\langle L\rangle}\partial_j x^L = A_{\langle k_1k_2\cdots k_\ell\rangle}\partial_j x^{k_1k_2\cdots k_\ell} = \ell A_{\langle jk_2\cdots k_\ell\rangle}x^{k_2\cdots k_\ell} = \ell A_{\langle jL-1\rangle}x^{L-1}$. Combining this with $\partial_j r^{-\ell} = -\ell\, n_j r^{-(\ell+1)}$, we find that

$$\partial_j\psi = \ell A_{\langle jL-1\rangle}\frac{x^{L-1}}{r^\ell} - \ell A_{\langle L\rangle}\frac{x^L}{r^{\ell+1}}n_j\,.$$

Proceeding along the same lines to compute the second derivative, and noting that $A_{\langle jjL-2\rangle} = 0$ and $\partial_j n_j = 2/r$, we finally arrive at

$$r^2\nabla^2\psi = -\ell(\ell+1)\psi\,.$$

Because $\psi = A_{\langle L\rangle}n^{\langle L\rangle}$ and $A_{\langle L\rangle}$ is arbitrary, this proves that $n^{\langle L\rangle}$ itself satisfies the eigenvalue equation.

With the connection between STF tensors and spherical harmonics displayed in Eqs. (1.164) and (1.167), it is easy to show that the multipole moments of Eqs. (1.139) and (1.150) are related by

$$I_{\ell m} = \mathscr{Y}^{\langle L\rangle}_{\ell m} I_{\langle L\rangle}\,, \tag{1.168a}$$

$$I^{\langle L\rangle} = \frac{4\pi\ell!}{(2\ell+1)!!}\sum_{m=-\ell}^{\ell}\mathscr{Y}^{*\langle L\rangle}_{\ell m} I_{\ell m}\,. \tag{1.168b}$$

With these results, the equivalence of Eqs. (1.140) and (1.149) is immediately established.

The foregoing results give rise to another identity, which will be required in Chapter 6. We rewrite Eq. (1.164) in terms of a different ℓ' and different direction $\boldsymbol{n}'$ and get

$$n'^{\langle L'\rangle} = N_{\ell'}\sum_{m'=-\ell'}^{\ell'}\mathscr{Y}^{\langle L'\rangle}_{\ell' m'} Y_{\ell' m'}(\theta',\phi')\,. \tag{1.169}$$

Multiplying by $Y^*_{\ell m}(\theta',\phi')$, integrating over $d\Omega'$, and using the orthonormality of the spherical harmonics, we next obtain

$$\int Y^*_{\ell m}(\theta',\phi')n'^{\langle L'\rangle}\,d\Omega' = \delta_{\ell\ell'}N_\ell\,\mathscr{Y}^{\langle L\rangle}_{\ell m}\,. \tag{1.170}$$

If we now multiply each side by $Y_{\ell m}(\theta,\phi)$, sum over m, and insert Eq. (1.164), we finally obtain

$$\sum_{m=-\ell}^{\ell} Y^*_{\ell m}(\theta,\phi)\int Y_{\ell m}(\theta',\phi')n'^{\langle L'\rangle}\,d\Omega' = \delta_{\ell\ell'}\,n^{\langle L\rangle}\,; \tag{1.171}$$

this is the required identity.

Angular averages

There will be many occasions, in this book, when we need to calculate the average of a quantity $\psi(\theta, \phi)$ over the surface of a sphere:

$$\langle\langle \psi \rangle\rangle := \frac{1}{4\pi} \int \psi(\theta, \phi)\, d\Omega\,. \tag{1.172}$$

Of particular interest are the spherical average of products $n^j n^k n^n \cdots$ of radial vectors. These are easily computed using the fact that the average of the STF tensor $n^{\langle jkn\cdots\rangle}$ must be zero; this property follows directly from Eq. (1.164) and the identity $\int Y_{\ell m}(\theta, \phi)\, d\Omega = 0$ (unless $\ell = 0$). In this way we obtain

$$\langle\langle n^j \rangle\rangle = 0\,, \tag{1.173a}$$

$$\langle\langle n^j n^k \rangle\rangle = \frac{1}{3}\delta^{jk}\,, \tag{1.173b}$$

$$\langle\langle n^j n^k n^n \rangle\rangle = 0\,, \tag{1.173c}$$

$$\langle\langle n^j n^k n^n n^p \rangle\rangle = \frac{1}{15}\left(\delta^{jk}\delta^{np} + \delta^{jn}\delta^{kp} + \delta^{jp}\delta^{kn}\right), \tag{1.173d}$$

and so on. These results can also be established directly, by recognizing that the tensorial structure on the right-hand side is uniquely determined by the complete symmetry of the left-hand side and the fact that δ^{jk} is the only available geometrical object. The numerical coefficient can then be determined by taking traces; for example, $1 = \delta_{jk}\delta_{np}\langle\langle n^j n^k n^n n^p \rangle\rangle = \frac{1}{15}(9 + 3 + 3)$, and this confirms that the numerical coefficient must indeed be $\frac{1}{15}$.

The general expression for such angular averages can be shown to be given by

$$\langle\langle n^L \rangle\rangle = \frac{1}{(\ell + 1)!!}\Big[\delta^L + \text{sym}(q)\Big], \tag{1.174}$$

when ℓ is an even number, and $\langle\langle n^L \rangle\rangle = 0$ when ℓ is odd; we use the same notation as in Eq. (1.155), in which δ^L stands for a product of $\ell/2$ Kronecker deltas, and sym(q) denotes all distinct terms obtained by permuting indices; the total number of terms within the square brackets is equal to $q = (\ell - 1)!!$.

1.6 Motion of extended fluid bodies

In this final section of Chapter 1 we examine a specific kind of fluid system, one that consists of separated blobs of fluids surrounded by vacuum. Each blob is called a "body," and the bodies are imagined to be in orbital motion around one another, the motion governed by the mutual gravitational attractions. Examples of such systems abound in the universe: we may be speaking of a binary system of main-sequence stars, or of a solar system of (gaseous) planets orbiting a central sun. (Although the discussion below relies on the bodies being made up of a perfect fluid, the final results apply just as well to solid bodies such as Earth-like planets.)

1.6.1 From fluid configurations to isolated bodies

Consider, therefore, a fluid configuration that is broken up into a collection of separated bodies. The configuration is characterized by two length scales, the typical size R of each body, and the typical separation r between bodies. We assume that $R < r$, so that the bodies are indeed separated, and surrounded by vacuum regions of space; our discussion excludes contact binaries, in which two stars share a common envelope. Each body is assumed to be isolated, in the sense that no matter is ejected from, nor accreted by, the body. Our discussion therefore excludes stars with strong stellar winds, such as Wolf–Rayet stars, which can lose mass at a dynamically significant rate. It excludes also interacting binary-star systems, for which a transfer of mass from one star to the other can have important effects on the orbital motion. The assumption, however, is a good one for most binary stars, and also for our solar system, in which the effects of the solar wind and its associated mass loss on planetary motions can be safely neglected, at least over the time scales we might be interested in.

The formalism developed in this section applies to separated and isolated bodies, and it actually relies on the stronger inequality $R \ll r$, which states that the inter-body separation is very large compared with the extension of each body. The strong inequality comes with a number of important consequences that we now describe.

The external, inter-body dynamics is governed by mutual gravitational interactions, and it proceeds on an orbital time scale given approximately by $T_{\rm orb} \sim (r^3/Gm)^{1/2}$, where m is the mass of a typical body. The internal, intra-body dynamics is governed instead by hydrodynamical processes, and it proceeds on an internal time scale given approximately by $T_{\rm int} \sim (G\rho)^{-1/2} \sim (R^3/Gm)^{1/2}$. So $T_{\rm int} \ll T_{\rm orb}$ when $R \ll r$, and the internal and external dynamics take place over widely separated time scales. A consequence of this fact is that the internal and external dynamics are largely decoupled from each other. It is possible, for example, for an orbiting body to be in a state of (approximate) hydrodynamic equilibrium even when external gravitational forces are applied to it, and for the orbital motion to be (approximately) independent of the details of the internal state.

As we shall see in Chapter 2, the strong inequality $R \ll r$ also implies that the tidal interaction between bodies is small. When other sources of deformation (such as rotation) are also small, the bodies can be taken to be nearly spherical. In such circumstances the gravitational field of each body can be well approximated by a multipole expansion of the sort developed in Sec. 1.5.2.

The (approximate) decoupling of the internal and external dynamics, and the (approximate) near-spherical nature of the bodies, produce a substantial simplification of the mathematical description of the inter-body motion. Instead of the original description, which involved the fine-grained fluid variables $(\rho, p, \boldsymbol{v})$, the orbital motion can be described with a smaller set of coarse-grained variables that characterize each body as a whole; these include the body's mass, center-of-mass position, spin angular momentum, and a number of multipole moments which encapsulate the required details of the internal dynamics.

Our main goal in this section is to accomplish this coarse-grained description of the external dynamics. In Chapter 2 we shall return to the internal dynamics and describe the internal structure of self-gravitating bodies. It is important to bear in mind that while the internal and external problems are approximately decoupled from one another, they are not fully decoupled: Some aspects of the external dynamics (such as the tidal coupling between bodies) depend on internal processes, and aspects of the internal dynamics (such as tidal deformations) depend on the orbital motion; ultimately and fundamentally the internal and external problems are informed by each other. A simple example is provided by the Earth–Moon system. The Moon raises tides (both solid and oceanic) on the Earth, and these depend on the Moon's orbital position; the tidal deformation of the Earth then modifies its own gravitational potential, and this affects the orbit of the Moon.

1.6.2 Center-of-mass variables

We consider a fluid system that is broken up into a number N of separated and isolated bodies, in the sense provided in Sec. 1.6.1. Each body is assigned a label $A = 1, 2, \ldots, N$, and each body occupies a volume V_A bounded by a closed surface S_A. The mass density ρ is assumed to be equal to ρ_A inside V_A, and zero in the vacuum region between bodies. The fluid dynamics inside each body is governed by the Euler and continuity equations – Eqs. (1.23) and (1.25) – and the gravitational potential U is given everywhere by Eq. (1.12).

The total mass of body A is given by

$$m_A := \int_A \rho(t, \boldsymbol{x})\, d^3x\,, \tag{1.175}$$

where the domain of integration is a fixed region of space that extends slightly beyond the volume V_A; it is sufficiently small that it contains no other body, but sufficiently large that it continues to contain body A as it moves about in a small interval of time dt. It is easy to show, using the techniques developed in Sec. 1.4.3, that m_A is time-independent: $dm_A/dt = 0$.

We define the center-of-mass position of body A (see Box 1.7) by

$$\boldsymbol{r}_A(t) := \frac{1}{m_A}\int_A \rho(t, \boldsymbol{x})\boldsymbol{x}\, d^3x\,, \tag{1.176}$$

and we similarly define the center-of-mass velocity and acceleration by

$$\boldsymbol{v}_A(t) := \frac{1}{m_A}\int_A \rho(t, \boldsymbol{x})\boldsymbol{v}\, d^3x\,, \tag{1.177}$$

and

$$\boldsymbol{a}_A(t) := \frac{1}{m_A}\int_A \rho(t, \boldsymbol{x})\frac{d\boldsymbol{v}}{dt}\, d^3x\,. \tag{1.178}$$

The integration techniques of Sec. 1.4.3 imply that

$$\boldsymbol{v}_A = \frac{d\boldsymbol{r}_A}{dt}\,, \qquad \boldsymbol{a}_A = \frac{d\boldsymbol{v}_A}{dt}\,. \tag{1.179}$$

In addition to these variables we introduce

$$I_A^{\langle L\rangle}(t) := \int_A \rho(t, \boldsymbol{x})\,(\boldsymbol{x} - \boldsymbol{r}_A)^{\langle L\rangle}\, d^3x\,, \tag{1.180}$$

the STF multipole moments of body A, which refer to its center-of-mass position $\boldsymbol{r}_A(t)$; note that the dipole moment $I_A^j = \int_A \rho(x - r_A)^j\, d^3x$ vanishes by virtue of Eq. (1.176). These definitions, and the results of Eq. (1.179), form the core of the coarse-grained formulation of the external problem. Instead of the original fluid variables $(\rho, p, \boldsymbol{v})$, the equations of motion of each body will be written in terms of $\boldsymbol{r}_A(t)$ and $I_A^{\langle L\rangle}(t)$; instead of functions of time *and* space, the formulation involves functions of time only.

Box 1.7 Is the center-of-mass unique?

The definition of the center-of-mass position proposed in Eq. (1.176) is not unique. For example, we could equally well propose the alternative definition $\boldsymbol{r}_A := m_A^{-1}\int_A(\rho^2/\langle\rho\rangle)\boldsymbol{x}\,d^3x$, in which $\langle\rho\rangle$ is the mean density inside body A; this would in general produce a different position for the center-of-mass. The main requirements for a sensible definition of center-of-mass are that it be located somewhere inside the body (it should not wander too far off), that it be useful and convenient, and that it be used consistently in all developments. Once these requirements are satisfied, the freedom of choice is unlimited, and ultimately the most important aspect is the matter of convenience.

It is important to bear in mind that the choice carries no physical consequence: there is no measurable way to determine the true position of the center-of-mass. When astronomers track planets using telescopes, they track the geometrical center of the image, or the location of the edge as it occults a star or the Sun. When they bounce radar or laser beams off planets, they determine the distance between the beam emitter and a point on the surface. When they determine the motion of a planet by tracking a satellite orbiting around it, they perform a complicated reduction of the orbital data to determine what they call a "normal point," the effective center-of-mass of the planet that controls the spacecraft's orbit. Given the shape and orientation of the body, one only needs to be able to go back and forth between the center-of-mass, as conventionally defined, and the place on the surface that is actually being located, or to the trajectory of an orbiting satellite.

The choice of center-of-mass proposed in Eq. (1.176) is useful and convenient because $\boldsymbol{r}_A(t)$ remains at rest or moves uniformly when the body is not subjected to a net force, and because Eqs. (1.179) have a nice structure. As we shall see, the equations of motion for extended bodies based on Eq. (1.176) are about as simple as they can be (although we admit that simplicity is a subjective notion).

The lack of uniqueness becomes even more acute in special and general relativity, because there are now many different densities that could be involved in a definition of center-of-mass: density of rest mass alone, or density of rest mass plus other forms of internal energy, such as kinetic, thermodynamic, or even gravitational binding energy. One could even include contributions from the gravitational potential energy provided by other bodies in the system, so that the center-of-mass position of a body might depend on the location of its neighbors. In addition, the very act of integrating the vector $\boldsymbol{x}$ over the body is problematic in relativity, because of ambiguities associated with the choice of reference frame. The problem is most serious in general relativity, because of the additional ambiguities associated with the choice of coordinate system. And finally, deep and subtle issues arise in the definition of center-of-mass for spinning bodies in special and general relativity. There is a vast literature devoted to attempts to define *the* center-of-mass, most of it extremely formal, and little of it of practical use. In the relativistic part of this book we will accept the arbitrariness of the center-of-mass, and adopt definitions that are as useful and convenient as we can make them, even if they are not provided with complete relativistic rigor.

The equations of motion of body A are obtained by inserting Euler's equation (1.23) within Eq. (1.178). The term involving the pressure gradient is easily disposed of: it integrates to zero after invoking Gauss's theorem, because $p = 0$ on the boundary of the domain of integration. What remains is

$$m_A \boldsymbol{a}_A = \int_A \rho \nabla U \, d^3x \,. \tag{1.181}$$

Summing over all bodies and making use of Eq. (1.56), we find that

$$\sum_{A=1}^{N} m_A \boldsymbol{a}_A = \int_{\text{all space}} \rho \nabla U \, d^3x = 0 \,. \tag{1.182}$$

This is a statement of Newton's third law, and a confirmation that

$$\boldsymbol{R} := \frac{1}{m} \sum_A m_A \boldsymbol{r}_A \,, \tag{1.183}$$

the *barycenter* of the N-body system, moves uniformly with a constant velocity $\boldsymbol{V}$; here $m := \sum_A m_A$ is the total mass of the system.

1.6.3 Internal and external potential

The gravitational potential that appears in Eq. (1.181) is produced in part by body A, and in part by all the remaining bodies. To distinguish between these contributions we decompose U as

$$U = U_A + U_{\neg A} \,, \tag{1.184}$$

with

$$U_A(t, \boldsymbol{x}) = G \int_A \frac{\rho(t, \boldsymbol{x}')}{|\boldsymbol{x} - \boldsymbol{x}'|} \, d^3x' \tag{1.185}$$

denoting the piece produced by body A – the internal potential – and

$$U_{\neg A}(t, \boldsymbol{x}) = \sum_{B \neq A} G \int_B \frac{\rho(t, \boldsymbol{x}')}{|\boldsymbol{x} - \boldsymbol{x}'|} \, d^3x' \tag{1.186}$$

denoting the piece produced by the remaining bodies – the external potential.

When we insert Eq. (1.184) within Eq. (1.181) we find that the internal potential makes no contribution to the equations of motion. This comes as a consequence of the identity

$$\int_A \rho \nabla U_A \, d^3x = 0 \,, \tag{1.187}$$

which is the statement of Eq. (1.56) applied to body A instead of the entire fluid system; the identity is established by following the same sequence of steps that led to the derivation of Eq. (1.56). As a result of this simplification, the equations of motion become

$$m_A \boldsymbol{a}_A = \int_A \rho \nabla U_{\neg A} \, d^3x \,, \tag{1.188}$$

and they involve only the external potential of Eq. (1.186).

1.6.4 Taylor expansion of the external potential

At this stage we incorporate our assumption that the bodies are well separated, so that $R_A \ll r_{AB}$, with R_A denoting the characteristic size of body A, and $r_{AB} := |\boldsymbol{r}_A - \boldsymbol{r}_B|$ denoting the typical separation between bodies. The variable of integration $\boldsymbol{x}$ in Eq. (1.188) ranges over the small scale R_A, and because the external potential $U_{\neg A}$ varies over the much larger scale r_{AB}, it is appropriate to express it as the Taylor expansion

$$U_{\neg A}(t, \boldsymbol{x}) = U_{\neg A}(t, \boldsymbol{r}_A) + (x - r_A)^j \partial_j U_{\neg A}(t, \boldsymbol{r}_A) \\ + \frac{1}{2}(x - r_A)^{jk} \partial_{jk} U_{\neg A}(t, \boldsymbol{r}_A) + \cdots , \tag{1.189}$$

in which the potential is evaluated at $\boldsymbol{x} = \boldsymbol{r}_A$ after differentiation. In a compact multi-index notation, this is

$$U_{\neg A}(t, \boldsymbol{x}) = \sum_{\ell=0}^{\infty} \frac{1}{\ell!}(x - r_A)^L \partial_L U_{\neg A}(t, \boldsymbol{r}_A) . \tag{1.190}$$

Because the external potential satisfies Laplace's equation $\nabla^2 U_{\neg A} = 0$ within the volume occupied by body A, its partial derivatives form a STF tensor, and using the rule of Eq. (1.158) we can write

$$(x - r_A)^L \partial_L U_{\neg A} = (x - r_A)^L \partial_{\langle L \rangle} U_{\neg A} = (x - r_A)^{\langle L \rangle} \partial_{\langle L \rangle} U_{\neg A} . \tag{1.191}$$

This gives rise to our final expression for the external potential,

$$U_{\neg A}(t, \boldsymbol{x}) = \sum_{\ell=0}^{\infty} \frac{1}{\ell!}(x - r_A)^{\langle L \rangle} \partial_L U_{\neg A}(t, \boldsymbol{r}_A) , \tag{1.192}$$

from which we have removed the redundant angular brackets on $\partial_{\langle L \rangle}$.

The gradient of the external potential is then given by

$$\partial_j U_{\neg A}(t, \boldsymbol{x}) = \sum_{\ell=0}^{\infty} \frac{1}{\ell!}(x - r_A)^{\langle L \rangle} \partial_{jL} U_{\neg A}(t, \boldsymbol{r}_A) , \tag{1.193}$$

and substitution within Eq. (1.188) returns

$$m_A a_A^j = \sum_{\ell=0}^{\infty} \frac{1}{\ell!} I_A^{\langle L \rangle}(t)\, \partial_{jL} U_{\neg A}(t, \boldsymbol{r}_A) \tag{1.194}$$

after involving the definition of Eq. (1.180). At this stage we have the equations of motion expressed in terms of the multipole moments of body A and partial derivatives of the external potential $U_{\neg A}$ evaluated at $\boldsymbol{x} = \boldsymbol{r}_A$.

To proceed we must work on the external potential. We return to its definition of Eq. (1.186), and for each body B within the sum we express the variable of integration $\boldsymbol{x}'$ as

$$\boldsymbol{x}' = \boldsymbol{r}_B(t) + \bar{\boldsymbol{x}}' , \tag{1.195}$$

with $\bar{x}'$ describing a displacement from B's center-of-mass. Each term in the sum is of the form

$$G\int_B \frac{\rho(t, \boldsymbol{r}_B + \bar{\boldsymbol{x}}')}{|\boldsymbol{x} - \boldsymbol{r}_B - \bar{\boldsymbol{x}}'|}\, d^3\bar{x}', \tag{1.196}$$

in which the new integration variable $\bar{\boldsymbol{x}}'$ ranges over the small scale R_B. Because $\boldsymbol{x} - \boldsymbol{r}_B$ is of the order of the much larger scale r_{AB}, it is appropriate to express the denominator as the Taylor expansion

$$\begin{aligned}
|\boldsymbol{x} - \boldsymbol{r}_B - \bar{\boldsymbol{x}}'|^{-1} &= |\boldsymbol{x} - \boldsymbol{r}_B|^{-1} - \bar{x}'^p \partial_p |\boldsymbol{x} - \boldsymbol{r}_B|^{-1} + \frac{1}{2}\bar{x}'^{pq}\partial_{pq}|\boldsymbol{x} - \boldsymbol{r}_B|^{-1} + \cdots \\
&= \sum_{\ell'=0}^{\infty} \frac{(-1)^{\ell'}}{\ell'!} \bar{x}'^{L'} \partial_{L'} |\boldsymbol{x} - \boldsymbol{r}_B|^{-1} \\
&= \sum_{\ell'=0}^{\infty} \frac{(-1)^{\ell'}}{\ell'!} \bar{x}'^{\langle L' \rangle} \partial_{L'} |\boldsymbol{x} - \boldsymbol{r}_B|^{-1} .
\end{aligned} \tag{1.197}$$

Making the substitution in Eq. (1.196) and invoking once more the definition of Eq. (1.180), we arrive at

$$U_{\neg A}(t, \boldsymbol{x}) = G\sum_{B\neq A}\sum_{\ell'=0}^{\infty} \frac{(-1)^{\ell'}}{\ell'!} I_B^{\langle L' \rangle} \partial_{L'} |\boldsymbol{x} - \boldsymbol{r}_B|^{-1}, \tag{1.198}$$

an expression for the external potential that involves the multipole moments of each external body B.

We may now take the additional derivatives that are required in Eq. (1.194) and evaluate the result at $\boldsymbol{x} = \boldsymbol{r}_A$. The result is

$$\partial_{jL} U_{\neg A}(t, \boldsymbol{r}_A) = G\sum_{B\neq A}\sum_{\ell'=0}^{\infty} \frac{(-1)^{\ell'}}{\ell'!} I_B^{\langle L' \rangle} \partial_{jLL'} \left(\frac{1}{r_{AB}}\right), \tag{1.199}$$

with $r_{AB} := |\boldsymbol{r}_A - \boldsymbol{r}_B|$ denoting the inter-body distance. The notation requires some care in interpretation: by $\partial_{jLL'} r_{AB}^{-1}$ we mean "take $\ell' + \ell + 1$ derivatives of $|\boldsymbol{x} - \boldsymbol{r}_B|^{-1}$ with respect to $\boldsymbol{x}$ and evaluate the result at $\boldsymbol{x} = \boldsymbol{r}_A$." To simplify this and eliminate the risk of confusion, we choose to express the operation in the equivalent form

$$\partial^A_{jLL'}\left(\frac{1}{r_{AB}}\right),$$

which now means "take $\ell' + \ell + 1$ derivatives of r_{AB}^{-1} with respect to $\boldsymbol{r}_A$."

1.6.5 Equations of motion for isolated bodies

The hard work is over. Substitution of Eq. (1.199) into Eq. (1.194) returns our final expression for the center-of-mass acceleration of body A. We obtain

$$m_A a_A^j = G\sum_{B\neq A}\sum_{\ell=0}^{\infty}\sum_{\ell'=0}^{\infty} \frac{(-1)^{\ell'}}{\ell!\ell'!} I_A^{\langle L \rangle} I_B^{\langle L' \rangle} \partial^A_{jLL'}\left(\frac{1}{r_{AB}}\right). \tag{1.200}$$

This equation, as it stands, is exact. Because each multipole moment $I_A^{\langle L\rangle}$ scales as $m_A R_A^\ell$, each term in the sum scales as

$$\frac{Gm_A m_B}{r_{AB}^2}\left(\frac{R_A}{r_{AB}}\right)^\ell \left(\frac{R_B}{r_{AB}}\right)^{\ell'},$$

and the assumption that $R_A \ll r_{AB}$ ensures that each term gets progressively smaller; the equation is exact, but it is most useful as a starting point for an approximation scheme. For many applications involving a small ratio R_A/r_{AB}, the sums can be safely truncated after just a few terms. For other applications, however, a large number of terms may be required. An example is the motion of a satellite in a low Earth orbit, which is sensitive to many of Earth's multipole moments; in the satellite geodesy project GRACE (Gravity Recovery and Climate Experiment), multipole moments up to $\ell \sim 360$ have been measured.

To rewrite Eq. (1.200) in a friendlier form we first isolate the early terms in the sums over ℓ and ℓ', noting that the monopole moment of body A is simply its mass, $I_A = m_A$, and that its dipole moment vanishes, $I_A^j = 0$, by virtue of the definition of the center-of-mass. We next split the sums into a piece that is linear in the higher multipole moments (coming from the terms $\ell = 0$, $\ell' \geq 2$ or $\ell' = 0$, $\ell \geq 2$) and another piece that involves products of the moments. This yields

$$\begin{aligned} a_A^j = G\sum_{B\neq A}\Bigg\{ &-\frac{m_B}{r_{AB}^2} n_{AB}^j + \sum_{\ell=2}^{\infty}\frac{1}{\ell!}\left[(-1)^\ell I_B^{\langle L\rangle} + \frac{m_B}{m_A} I_A^{\langle L\rangle}\right]\partial_{jL}^A\left(\frac{1}{r_{AB}}\right) \\ &+ \frac{1}{m_A}\sum_{\ell=2}^{\infty}\sum_{\ell'=2}^{\infty}\frac{(-1)^{\ell'}}{\ell!\ell'!} I_A^{\langle L\rangle} I_B^{\langle L'\rangle}\partial_{jLL'}^A\left(\frac{1}{r_{AB}}\right)\Bigg\}, \end{aligned} \tag{1.201}$$

where $\boldsymbol{n}_{AB} := \boldsymbol{r}_{AB}/r_{AB}$ is a unit vector that points from body B to body A. This expression implies that $\sum_A m_A \boldsymbol{a}_A = \boldsymbol{0}$, a statement that was already established in Eq. (1.182).

The equations displayed in Eq. (1.201) form a complete set of equations of motion for the N bodies once their masses and multipole moments as functions of time are specified. They can be integrated once the initial position and velocity of each body are given. The equations, however, provide an *incomplete description* of the physical system, because they do not determine the time-evolution of the multipole moments; these will, in general, depend on the details of the internal structure of each body and the motion of the remaining bodies. The multipole moments capture the remaining coupling between the internal and external problems (as discussed in Sec. 1.6.1), and additional information must be supplied in order to turn Eq. (1.201) into a closed system of equations of motion. We shall return to this issue in Chapter 2.

The multipole moments of a perfectly spherical body vanish, $I_A^{\langle L\rangle} = 0$ for $\ell \neq 0$, and when all the bodies are spherical we find that Eq. (1.201) reduces to the familiar set of point-mass equations of motion,

$$a_A^j = -\sum_{B\neq A}\frac{Gm_B}{r_{AB}^2} n_{AB}^j\,. \tag{1.202}$$

When the bodies are not spherical, we observe that the motion of body A is affected by the distortion of the gravitational potential caused by the deformation of the other bodies; this influence is described by the terms in Eq. (1.201) that involve $I_B^{\langle L\rangle}$. It is affected also by the coupling of its own non-spherical mass distribution to gradients of the monopole field of each external body; this influence is described by the terms in Eq. (1.201) that are linear in $I_A^{\langle L\rangle}$. And finally, it is affected by couplings between its own multipole moments and those of the external bodies, as described by the last line in Eq. (1.201). This last effect is analogous to the dipole–dipole coupling in electrodynamics, except for the fact that there is no dipole moment in gravitation; the leading effect comes from a quadrupole–quadrupole interaction. The presence of terms involving $I_A^{\langle L\rangle}$ in the equations of motion implies that the motion of a body can depend on its internal structure, by virtue of its finite size and the non-spherical coupling of its mass distribution to the external gravitational field. This observation does not constitute a violation of the weak equivalence principle; a violation would imply a dependence on internal structure that remains even when the bodies have a negligible size.

1.6.6 Conserved quantities

In Sec. 1.4.3 we showed that the total mass, momentum, energy, and angular momentum of a fluid configuration are conserved as a consequence of the fluid's dynamics. We recall that the total momentum $\boldsymbol{P}$ is defined by Eq. (1.54), the total energy E is defined by Eq. (1.67), and the total angular momentum $\boldsymbol{J}$ is defined by Eq. (1.78). These quantities continue to be conserved when the fluid configuration describes a system of isolated bodies, and in this section we derive expressions for the total momentum, energy, and angular momentum of an N-body system.

We begin with the definition of total momentum, $\boldsymbol{P} = \int \rho \boldsymbol{v}\, d^3x$, in which the integral over all space may be written as a sum of integrals extending over each body. In each integral we decompose $\boldsymbol{v}$ as $(\boldsymbol{v} - \boldsymbol{v}_A) + \boldsymbol{v}_A$, with the first term describing a velocity relative to the center-of-mass of body A. The integral of $\rho(\boldsymbol{v} - \boldsymbol{v}_A)$ vanishes by virtue of the definition of the center-of-mass – refer back to Eq. (1.177) – and the second integral yields $m_A \boldsymbol{v}_A$. The final result is

$$\boldsymbol{P} = \sum_A m_A \boldsymbol{v}_A\,; \tag{1.203}$$

as expected, the total momentum is a simple sum of individual momenta.

A similar computation returns

$$\boldsymbol{J} = \sum_A \left(\boldsymbol{S}_A + m_A \boldsymbol{r}_A \times \boldsymbol{v}_A\right) \tag{1.204}$$

for the total angular momentum, where

$$\boldsymbol{S}_A := \int_A \rho \left(\boldsymbol{x} - \boldsymbol{r}_A\right) \times \left(\boldsymbol{v} - \boldsymbol{v}_A\right) d^3x \tag{1.205}$$

is the intrinsic angular momentum of body A – its spin. We see that the total angular momentum is a simple sum of individual spin and orbital angular momenta.

The calculation of the total energy involves a computation of the kinetic energy $\mathcal{T} = \frac{1}{2}\int \rho v^2\, d^3x$, the gravitational potential energy $\Omega = -\frac{1}{2}\int \rho U\, d^3x$, and the internal (thermodynamic) energy $E^{\rm int} = \int \epsilon\, d^3x$. For the kinetic and internal energies we immediately get

$$\mathcal{T} = \sum_A \Bigl(\mathcal{T}_A + \frac{1}{2} m_A v_A^2 \Bigr) \tag{1.206}$$

and

$$E^{\rm int} = \sum_A E_A^{\rm int}, \tag{1.207}$$

where

$$\mathcal{T}_A := \frac{1}{2} \int_A \rho \bigl| \boldsymbol{v} - \boldsymbol{v}_A \bigr|^2 d^3x \tag{1.208}$$

is the internal kinetic energy of body A, while

$$E_A^{\rm int} := \int_A \epsilon\, d^3x \tag{1.209}$$

is its own thermodynamic energy.

The computation of Ω requires more work. We first return to Eq. (1.184) and decompose the gravitational potential into internal and external pieces, $U = U_A + U_{\neg A}$. This gives rise to

$$\Omega = \sum_A \biggl(\Omega_A - \frac{1}{2} \int_A \rho U_{\neg A}\, d^3x \biggr), \tag{1.210}$$

in which

$$\Omega_A := -\frac{1}{2} \int_A \rho U_A\, d^3x \tag{1.211}$$

is the internal gravitational potential energy of body A. To evaluate the second term we follow the strategy of Sec. 1.6.4 and express $U_{\neg A}(t, \boldsymbol{x})$ as a Taylor expansion about $\boldsymbol{x} = \boldsymbol{r}_A$. Using the expression of Eq. (1.192), we obtain

$$-\frac{1}{2} \int_A \rho U_{\neg A}\, d^3x = -\frac{1}{2} m_A U_{\neg A}\bigl(t, \boldsymbol{r}_A\bigr) - \frac{1}{2} \sum_{\ell=2}^{\infty} \frac{1}{\ell!} I_A^{\langle L \rangle} \partial_L U_{\neg A}\bigl(t, \boldsymbol{r}_A\bigr), \tag{1.212}$$

in which $I_A^{\langle L \rangle}(t)$ are the multipole moments of body A, as defined by Eq. (1.180); there is no $\ell = 1$ term in the sum because $I_A^j = 0$ by virtue of the definition of the center-of-mass. In the remaining steps we express the external potential as an expansion in inverse powers of $r_{AB} := |\boldsymbol{r}_A - \boldsymbol{r}_B|$, as in Eq. (1.198). After some simplification we arrive at our final expression, which is recognized below as the collection of terms involving the nested sums over pairs of bodies.

Collecting results, we find that the total energy of a system of isolated bodies is given by

$$\begin{aligned} E = &\sum_A E_A + \sum_A \frac{1}{2} m_A v_A^2 - \frac{1}{2} \sum_A \sum_{B \neq A} \frac{G m_A m_B}{r_{AB}} \\ &- \frac{1}{2} \sum_A \sum_{B \neq A} \sum_{\ell=2}^{\infty} \frac{1}{\ell!} \Big[(-1)^\ell G m_A I_B^{\langle L \rangle} + G m_B I_A^{\langle L \rangle} \Big] \partial_L^A \bigg(\frac{1}{r_{AB}} \bigg) \\ &- \frac{1}{2} \sum_A \sum_{B \neq A} \sum_{\ell=2}^{\infty} \sum_{\ell'=2}^{\infty} \frac{(-1)^{\ell'}}{\ell! \ell'!} G I_A^{\langle L \rangle} I_B^{\langle L' \rangle} \partial_{LL'}^A \bigg(\frac{1}{r_{AB}} \bigg) , \end{aligned} \tag{1.213}$$

where $E_A := \mathcal{T}_A + \Omega_A + E_A^{\rm int}$ is the self-energy of body A. The manipulations following Eq. (1.67) can be adapted to each body, and the conclusion is that each self-energy is approximately conserved. Because the term $\sum_A E_A$ merely contributes an irrelevant constant to E, it can be safely removed from a conventional accounting of total energy, which holds that the energy should vanish when $v_A \to 0$ and $r_{AB} \to \infty$. In the final analysis we shall retain only the center-of-mass kinetic energies and the mutual interaction energies in Eq. (1.213).

When the bodies are spherical, so that $I_A^{\langle L \rangle} = 0$ for $\ell \neq 0$, the total energy reduces to

$$E = \sum_A \frac{1}{2} m_A v_A^2 - \frac{1}{2} \sum_A \sum_{B \neq A} \frac{G m_A m_B}{r_{AB}} , \tag{1.214}$$

the familiar expression for a system of point masses.

1.6.7 Equations of motion for binary systems

We next specialize the general discussion of this section to a system of two bodies. Our binary system consists of a first body of mass m_1 and multipole moments $I_1^{\langle L \rangle}$ at a position $\boldsymbol{r}_1$, and a second body of mass m_2 and multipole moments $I_2^{\langle L \rangle}$ at a position $\boldsymbol{r}_2$. The total mass of the binary system is $m := m_1 + m_2$. In place of $\boldsymbol{r}_1$ and $\boldsymbol{r}_2$ it is useful to work with the *barycenter position*

$$\boldsymbol{R} := \frac{m_1}{m} \boldsymbol{r}_1 + \frac{m_2}{m} \boldsymbol{r}_2 , \tag{1.215}$$

and the *relative separation*

$$\boldsymbol{r} := \boldsymbol{r}_1 - \boldsymbol{r}_2 . \tag{1.216}$$

This vector was denoted $\boldsymbol{r}_{12}$ in preceding subsections, and we simplify other notations in a similar way by defining

$$r := |\boldsymbol{r}| , \qquad \boldsymbol{n} := \boldsymbol{r}/r . \tag{1.217}$$

It is useful to note that $\boldsymbol{r}_{21} = -\boldsymbol{r}$, $\boldsymbol{n}_{21} = -\boldsymbol{n}$, and that $r_{21} = r$. In addition to the relative separation we also introduce the relative velocity $\boldsymbol{v} := d\boldsymbol{r}/dt = \boldsymbol{v}_1 - \boldsymbol{v}_2$ and the relative acceleration

$$\boldsymbol{a} := \frac{d^2 \boldsymbol{r}}{dt^2} = \boldsymbol{a}_1 - \boldsymbol{a}_2 . \tag{1.218}$$

Solving for $\boldsymbol{r}_1$ and $\boldsymbol{r}_2$, we find that

$$\boldsymbol{r}_1 = \boldsymbol{R} + \frac{m_2}{m}\boldsymbol{r}, \qquad \boldsymbol{r}_2 = \boldsymbol{R} - \frac{m_1}{m}\boldsymbol{r}. \tag{1.219}$$

The motion of the binary system is determined when $\boldsymbol{R}(t)$ and $\boldsymbol{r}(t)$ are both known as functions of time. The motion of the barycenter is uniform: as we saw at the end of Sec. 1.6.2, it is described by $\boldsymbol{R}(t) = \boldsymbol{R}(0) + \boldsymbol{V}t$, where $\boldsymbol{V} := \boldsymbol{P}/m$ is a constant velocity vector. The relative motion is governed by

$$a^j = -\frac{Gm}{r^2}n^j + Gm\sum_{\ell=2}^{\infty}\frac{1}{\ell!}\left[\frac{I_1^{\langle L\rangle}}{m_1} + (-1)^\ell\frac{I_2^{\langle L\rangle}}{m_2}\right]\partial_{jL}\left(\frac{1}{r}\right)$$
$$+ Gm\sum_{\ell=2}^{\infty}\sum_{\ell'=2}^{\infty}\frac{(-1)^{\ell'}}{\ell!\ell'!}\frac{I_1^{\langle L\rangle}}{m_1}\frac{I_2^{\langle L'\rangle}}{m_2}\partial_{jLL'}\left(\frac{1}{r}\right), \tag{1.220}$$

an effective one-body equation that can easily be obtained from Eq. (1.201). The derivation relies on the fact that $\partial_j^2 r_{21}^{-1} = -\partial_j^1 r_{12}^{-1} := -\partial_j r^{-1}$; in this notation ∂_j indicates partial differentiation with respect to the Cartesian coordinate r^j associated with the relative separation $\boldsymbol{r}$. From Eq. (1.213) we find that the total energy (excluding self-energies) of a two-body system is given by

$$E = \frac{1}{2}mV^2 + \frac{1}{2}\mu v^2 - \frac{G\mu m}{r}$$
$$- G\mu m\sum_{\ell=2}^{\infty}\frac{1}{\ell!}\left[\frac{I_1^{\langle L\rangle}}{m_1} + (-1)^\ell\frac{I_2^{\langle L\rangle}}{m_2}\right]\partial_L\left(\frac{1}{r}\right)$$
$$- G\mu m\sum_{\ell=2}^{\infty}\sum_{\ell'=2}^{\infty}\frac{(-1)^{\ell'}}{\ell!\ell'!}\frac{I_1^{\langle L\rangle}}{m_1}\frac{I_2^{\langle L'\rangle}}{m_2}\partial_{LL'}\left(\frac{1}{r}\right), \tag{1.221}$$

in which $\mu := m_1 m_2/m$ is the system's *reduced mass*.

As a specialization of these equations we assume that the multipole moments of one of the bodies, say body 1, are negligible. This simplification would apply, for example, to a planet orbiting the Sun (the planet has negligible moments), to a satellite orbiting the Earth (the satellite has negligible moments), or to a non-rotating black hole or neutron star orbiting a normal star (the compact object has negligible moments). In this case we find that the relative acceleration simplifies to

$$a^j = -\frac{Gm}{r^2}n^j + Gm\sum_{\ell=2}^{\infty}\frac{(-1)^\ell}{\ell!}\frac{I_2^{\langle L\rangle}}{m_2}\partial_{jL}\left(\frac{1}{r}\right), \tag{1.222}$$

and the total energy becomes

$$E = \frac{1}{2}mV^2 + \frac{1}{2}\mu v^2 - \frac{G\mu m}{r} - G\mu m\sum_{\ell=2}^{\infty}\frac{(-1)^\ell}{\ell!}\frac{I_2^{\langle L\rangle}}{m_2}\partial_L\left(\frac{1}{r}\right). \tag{1.223}$$

For a planet orbiting the Sun, or a spacecraft orbiting the Earth, m_1 is much smaller than m_2, and $m/m_2 \simeq 1$. In this case the equations describe the motion of a spherical body in the multipole field of a heavy, central object. In situations involving comparable masses, however, such as a black hole or neutron star orbiting a normal star, the ratio m/m_2 could be substantially larger than unity, reflecting the fact that the motion of both bodies can be strongly affected by the multipole moments of body 2.

Specializing even further, we now take body 2 to be symmetric about an axis aligned with the unit vector $\boldsymbol{e}$. The symmetry requires the body's multipole moments $I^{\langle L\rangle}$ to be proportional to the STF tensor $e^{\langle L\rangle}$, so that $I^{\langle L\rangle} = \alpha_\ell e^{\langle L\rangle}$ for some coefficient α_ℓ. We wish to relate this to the dimensionless multipole moments J_ℓ introduced in Eq. (1.143). To achieve this we align the z-direction with the vector $\boldsymbol{e}$ and invoke Eqs. (1.112), (1.167), and (1.168). After some algebra we obtain $\alpha_\ell = -mR^\ell J_\ell$, so that

$$I_2^{\langle L\rangle} = -m_2 R_2^\ell (J_\ell)_2\, e_2^{\langle L\rangle}\,; \tag{1.224}$$

to indicate that all quantities refer to body 2 we have inserted the label "2" on all quantities (such as mass, radius, symmetry axis, and multipole moments) that appear in Eq. (1.224).

Equation (1.222) can then be written as

$$a^j = -\frac{Gm}{r^2}\left[n^j - \sum_{\ell=2}^{\infty} \frac{(2\ell+1)!!}{\ell!}(J_\ell)_2 \left(\frac{R_2}{r}\right)^\ell e_2^{\langle L\rangle} n_{\langle jL\rangle}\right], \tag{1.225}$$

after making use of Eq. (1.156) to express the derivatives of r^{-1} in terms of STF products of the vector $\boldsymbol{n}$; the product $e_2^{\langle L\rangle} n_{\langle jL\rangle}$ could be further simplified by invoking Eq. (1.160b). With similar manipulations we can show that the total energy becomes

$$E = \frac{1}{2}mV^2 + \frac{1}{2}\mu v^2 - \frac{G\mu m}{r}\left[1 - \sum_{\ell=2}^{\infty}(J_\ell)_2 \left(\frac{R_2}{r}\right)^\ell P_\ell(\boldsymbol{e}_2\cdot\boldsymbol{n})\right]; \tag{1.226}$$

to arrive at this result we have made use of Eq. (1.159a) to express $e_2^{\langle L\rangle} n_{\langle L\rangle}$ in terms of Legendre polynomials.

For modestly deformed bodies, and for sufficiently large separations, the $\ell = 2$ term dominates in both Eq. (1.225) and Eq. (1.226). In this case the relative acceleration becomes

$$\boldsymbol{a} = -\frac{Gm}{r^2}\left\{\boldsymbol{n} - \frac{3}{2}(J_2)_2\left(\frac{R_2}{r}\right)^2 \left\{\left[5(\boldsymbol{e}_2\cdot\boldsymbol{n})^2 - 1\right]\boldsymbol{n} - 2(\boldsymbol{e}_2\cdot\boldsymbol{n})\boldsymbol{e}_2\right\}\right\}; \tag{1.227}$$

the expression involves the total mass $m := m_1 + m_2$, and it applies to a binary system of arbitrary mass ratio. This equation, the specialization of Eq. (1.220) to a spherical body moving in the monopole and quadrupole field of an axisymmetric body, is the foundation for the study of a number of important phenomena, including the effect of the solar quadrupole moment on the orbit of Mercury, and the precession of the planes of Earth-orbiting satellites. We shall return to these applications in Chapter 3.

1.6.8 Spin dynamics

As we saw back in Eq. (1.205), the *spin angular momentum* of body A is defined by

$$\boldsymbol{S}_A(t) := \int_A \rho(t,\boldsymbol{x})(\boldsymbol{x}-\boldsymbol{r}_A)\times(\boldsymbol{v}-\boldsymbol{v}_A)\,d^3x\,, \tag{1.228}$$

and it refers to its center-of-mass position $\boldsymbol{r}_A$ and velocity $\boldsymbol{v}_A$. In terms of components we use the permutation symbol ϵ^{jpq} to describe the vectorial product, and we write

$$S_A^j(t) := \epsilon^{jpq} \int_A \rho (x - r_A)^p (v - v_A)^q \, d^3x \, . \tag{1.229}$$

We wish to find an equation of motion for $\boldsymbol{S}_A(t)$, and we shall proceed by following the general method outlined in Secs. 1.6.2, 1.6.3, and 1.6.4.

We begin by differentiating Eq. (1.229) with respect to t. Exploiting once again the techniques developed in Sec. 1.4.3, we find that

$$\begin{aligned} \frac{dS_A^j}{dt} &= \epsilon^{jpq} \int_A \rho (v - v_A)^p (v - v_A)^q \, d^3x + \epsilon^{jpq} \int_A \rho (x - r_A)^p (dv/dt - a_A)^q \, d^3x \\ &= \epsilon^{jpq} \int_A \rho (x - r_A)^p \frac{dv^q}{dt} \, d^3x \, , \end{aligned} \tag{1.230}$$

where we have used the fact that $\int_A \rho (x - r_A)^p \, d^3x = 0$ by virtue of the definition of the center-of-mass position. In this we insert Euler's equation (1.23) and obtain

$$\frac{dS_A^j}{dt} = \epsilon^{jpq} \int_A \rho (x - r_A)^p \partial_q U \, d^3x - \epsilon^{jpq} \int_A (x - r_A)^p \partial_q p \, d^3x \, . \tag{1.231}$$

The second term, involving the pressure p, can be integrated by parts; after discarding the boundary term we are left with $\epsilon^{jpq} \delta_{pq} \int_A p \, d^3x$, which vanishes identically. We have obtained

$$\frac{dS_A^j}{dt} = \epsilon^{jpq} \int_A \rho (x - r_A)^p \partial_q U \, d^3x \, , \tag{1.232}$$

and in this we insert the decomposition of the gravitational potential in terms of internal and external pieces, as in Eq. (1.184). It is easy to show that the contribution from the internal potential,

$$\epsilon^{jpq} \int_A \rho \, x^p \partial_q U_A \, d^3x - \epsilon^{jpq} r_A^p \int_A \rho \, \partial_q U_A \, d^3x \, ,$$

is in fact zero. The first term vanishes by virtue of Eq. (1.80) (applied to body A instead of the entire N-body system), and the second term vanishes thanks to Eq. (1.187). The evolution of the spin is therefore governed by

$$\frac{dS_A^j}{dt} = \epsilon^{jpq} \int_A \rho (x - r_A)^p \partial_q U_{\neg A} \, d^3x \, , \tag{1.233}$$

which involves the gravitational potential $U_{\neg A}(t, \boldsymbol{x})$ produced by the bodies external to A.

At this stage we import Eq. (1.193), which provides an expression for $\partial_q U_{\neg A}(t, \boldsymbol{x})$ as a Taylor expansion in powers of $\boldsymbol{x} - \boldsymbol{r}_A$. Inserting this within Eq. (1.233), we obtain

$$\begin{aligned} \frac{dS_A^j}{dt} &= \epsilon^{jpq} \sum_{\ell=0}^{\infty} \frac{1}{\ell!} I_A^{pL} \partial_{qL} U_{\neg A}(t, \boldsymbol{r}_A) \\ &= \epsilon^{jpq} \sum_{\ell=0}^{\infty} \frac{1}{\ell!} I_A^{\langle pL \rangle} \partial_{qL} U_{\neg A}(t, \boldsymbol{r}_A) \, . \end{aligned} \tag{1.234}$$

In the second line we allowed ourselves to enclose the indices pL within angular brackets, recognizing that the difference between $I_A^{\langle pL\rangle}$ and I_A^{pL} involves a number of Kronecker deltas that either (i) force indices contained in ∂_{qL} to be equal, giving zero when acting on $U_{\neg A}$, or (ii) force the derivative operator to be of the form ∂_{pqL-1}, which vanishes when multiplied by ϵ^{jpq}. We next import Eq. (1.199) and obtain

$$\frac{dS_A^j}{dt} = G\epsilon^{jpq} \sum_{B\neq A} \sum_{\ell=0}^{\infty} \sum_{\ell'=0}^{\infty} \frac{(-1)^{\ell'}}{\ell!\ell'!} I_A^{\langle pL\rangle} I_B^{\langle L'\rangle} \partial^A_{\langle qLL'\rangle}\left(\frac{1}{r_{AB}}\right). \tag{1.235}$$

To write this in a friendlier form we observe that the terms with $\ell = 0$ make no contributions (because the dipole moment of body A vanishes), that the terms with $\ell' = 0$ involve m_B only, and that the terms with $\ell' = 1$ also make no contributions. We can therefore split the sum into two pieces, one linear in the moments of body A, and the other involving products of moments. We have

$$\begin{aligned}\frac{dS_A^j}{dt} &= G\epsilon^{jpq} \sum_{B\neq A} \sum_{\ell=1}^{\infty} \frac{1}{\ell!} m_B I_A^{\langle pL\rangle} \partial^A_{\langle qL\rangle}\left(\frac{1}{r_{AB}}\right) \\ &\quad + G\epsilon^{jpq} \sum_{B\neq A} \sum_{\ell=1}^{\infty} \sum_{\ell'=2}^{\infty} \frac{(-1)^{\ell'}}{\ell!\ell'!} I_A^{\langle pL\rangle} I_B^{\langle L'\rangle} \partial^A_{\langle qLL'\rangle}\left(\frac{1}{r_{AB}}\right),\end{aligned} \tag{1.236}$$

and this equation determines the behavior of each spin once the multipole moments and the center-of-mass motion of each body are specified.

We next specialize the discussion to an N-body system that consists of a spinning body A with non-vanishing multipole moments, and external bodies B with negligible multipole moments. In addition, we assume that body A is symmetric about an axis aligned with the unit vector $\boldsymbol{e}_A$. Under these conditions we have that

$$\boldsymbol{S}_A = S_A \boldsymbol{e}_A, \qquad S_A := |\boldsymbol{S}_A|, \tag{1.237}$$

and Eq. (1.224) implies that the body's multipole moments are given by

$$I_A^{\langle L\rangle} = -m_A R_A^{\ell} (J_\ell)_A e_A^{\langle L\rangle}. \tag{1.238}$$

This relation is inserted within Eq. (1.236), along with Eq. (1.156), and after some algebra we obtain

$$\frac{dS_A^j}{dt} = -\epsilon^{jpq} \sum_{B\neq A} \frac{Gm_A m_B}{r_{AB}} \sum_{\ell=1}^{\infty} (-1)^{\ell+1} \frac{(2\ell+1)!!}{\ell!} (J_{\ell+1})_A \left(\frac{R_A}{r_{AB}}\right)^{\ell+1} e_A^{\langle pL\rangle} n_{AB}^{\langle qL\rangle}. \tag{1.239}$$

This is simplified with the help of Eq. (1.160b), and we express the final result as

$$\frac{d\boldsymbol{S}_A}{dt} = -\sum_{B\neq A} \frac{Gm_A m_B}{r_{AB}} (\boldsymbol{e}_A \times \boldsymbol{n}_{AB}) \sum_{\ell=2}^{\infty} (-1)^{\ell} (J_\ell)_A \left(\frac{R_A}{r_{AB}}\right)^{\ell} \frac{dP_\ell}{d\mu_{AB}}, \tag{1.240}$$

in which $P_\ell(\mu_{AB})$ is a Legendre polynomial, and $\mu_{AB} := \boldsymbol{e}_A \cdot \boldsymbol{n}_{AB}$. This equation implies that the magnitude of the spin vector stays constant, because according to Eq. (1.237), $dS_A/dt = \boldsymbol{e}_A \cdot d\boldsymbol{S}_A/dt = 0$. And indeed, we observe that each term in the sum over B would give rise to a precession of $\boldsymbol{e}_A$ in the direction of $\boldsymbol{n}_{AB}$; after summation the precession is seen to take place in a direction given by a weighted average of all the vectors $\boldsymbol{n}_{AB}$.

As we shall see in Chapter 3, one notable consequence of Eq. (1.240) is the disturbance of the Earth's axis caused by the coupling of its equatorial bulge with the gravitational fields of the Sun and Moon. This leads to the famous precession of the equinoxes, with its cycle of approximately 26 000 years.

1.7 Bibliographical notes

The presentation of the basic equations of Newtonian gravity in Sec. 1.2 follows the standard treatment found in many undergraduate texts, including the venerable *Newtonian Mechanics* by French (1971). The theory, of course, was created in Newton's own *Principia*, which can be accessed in the superb English edition with extensive commentary by Cohen, Whitman, and Budenz (1999). The Eöt-Wash torsion balance experiment is described in Su *et al.* (1994) and Baessler *et al.* (1999).

The theory of Green's functions touched upon in Secs. 1.3 and 1.5 is developed systematically in many textbooks on mathematical methods, including the excellent *Mathematical Methods for Physicists* by Arfken, Weber, and Harris (2012).

The discussion of Sec. 1.4 relies on elements of fluid mechanics, thermodynamics, and statistical physics. Those are covered in many textbooks. An elegant and sophisticated development of fluid mechanics can be found in the classic *Fluid Mechanics* by Landau and Lifshitz (1987), and another useful resource is Kundu, Cohen, and Dowling (2011). A comprehensive presentation of thermodynamics and statistical physics can be found in Reif's *Fundamentals of Statistical and Thermal Physics*, now available in a new 2008 edition.

The development of multipole expansions to integrate Poisson's equation in Sec. 1.5 relies on the theory of spherical harmonics, a topic covered in most textbooks on mathematical methods. These developments are virtually identical to those related to the electrostatic potential, which are described in most textbooks on electromagnetism; the most comprehensive is the classic *Classical Electromagnetism* by Jackson (1998). The use of symmetric-tracefree tensors as substitutes for spherical harmonics was pioneered by Sachs (1961) and Pirani (1964); a systematic treatment can be found in Thorne (1980), and another useful resource is Damour and Iyer (1991). The citation from the *Principia* is taken from the Cohen, Whitman, and Budenz edition. The book *Allen's Astrophysical Quantities* is edited by Cox (2001).

An overview of the GRACE geodesy project, mentioned in Sec. 1.6, can be found at `www.csr.utexas.edu/grace/gravity/geodesy.html`.

1.8 Exercises

1.1 Show explicitly that, for a function $f(t, \boldsymbol{x}, \boldsymbol{x}')$,

$$\frac{\partial^2}{\partial t^2}\int \rho' f d^3x' = \int \rho'\left(\frac{\partial^2 f}{\partial t^2} + 2\boldsymbol{v}' \cdot \nabla' \frac{\partial f}{\partial t} + \frac{d\boldsymbol{v}'}{dt} \cdot \nabla' f + v'^j v'^k \partial_j \partial_k f\right) d^3x' .$$

1.2 Given the Newtonian potential $U(t, \boldsymbol{x})$, one can define a *superpotential* $X(t, \boldsymbol{x})$, a *superduperpotential* $Y(t, \boldsymbol{x})$, and another superlative potential $Z(t, \boldsymbol{x})$ that satisfy the equations

$$\nabla^2 X = 2U , \qquad \nabla^2 Y = 12X , \qquad \nabla^2 Z = 30Y .$$

Find explicit expressions for X, Y, and Z as integrals over the mass density $\rho(t, \boldsymbol{x}')$, assuming that ρ vanishes outside some finite region of space.

1.3 Using the expression for the superpotential X obtained in the preceding problem, show that

$$\frac{\partial^2}{\partial t^2} X(t, \boldsymbol{x}) = -\int \rho' \frac{d\boldsymbol{v}'}{dt} \cdot \frac{(\boldsymbol{x} - \boldsymbol{x}')}{|\boldsymbol{x} - \boldsymbol{x}'|} d^3x' + \int \frac{\rho'}{|\boldsymbol{x} - \boldsymbol{x}'|}\left\{v'^2 - \frac{[\boldsymbol{v}' \cdot (\boldsymbol{x} - \boldsymbol{x}')]^2}{|\boldsymbol{x} - \boldsymbol{x}'|^2}\right\} d^3x' .$$

1.4 Prove that

$$\int \rho(t, \boldsymbol{x}) x^j v^k \, d^3x = \frac{1}{2}\frac{dI^{jk}}{dt} + \frac{1}{2}\epsilon^{jkp} J^p ,$$

where I^{jk} is the quadrupole moment tensor of the mass distribution, and J^p is the total angular momentum.

1.5 Assuming that $T^{\alpha\beta} = 0$ far away from the system, use the equations of hydrodynamics in the form of $\partial_\beta T^{\alpha\beta} = 0$ to verify explicitly that the total mass M, momentum $\boldsymbol{P}$, and angular momentum $\boldsymbol{J}$ of an isolated system are all constant.

1.6 With the same assumptions as in the preceding problem, prove that a statement of the tensorial virial theorem is

$$\frac{d^2 I^{jk}}{dt^2} = \int T^{jk} \, d^3x ,$$

where $I^{jk} := \int T^{tt} x^j x^k \, d^3x$. Then show that with T^{jk} given by Eq. (1.96), the virial theorem takes the explicit form of Eq. (1.88).

1.7 Use the spherical-harmonic expansion of $|\boldsymbol{x} - \boldsymbol{x}'|^{-1}$ to verify that

$$U(t, r) = \frac{Gm(t, r)}{r} + 4\pi G \int_r^R \rho(t, r') r' dr'$$

for a spherical matter distribution.

1.8 Show explicitly that $\partial_{jknp} r^{-1} = 105 n^{\langle jknp \rangle} / r^5$.

1.9 Show that the forms of $n^{\langle jk\rangle}$, $n^{\langle jkn\rangle}$, $n^{\langle jknp\rangle}$ given by Eq. (1.153) satisfy the general formula of Eq. (1.155).

1.10 Find $n^{\langle jknpq\rangle}$ by explicit construction.

1.11 Show that the internal gravitational potential of Eq. (1.129) can be expressed as

$$U = G\sum_{\ell=0}^{\infty}\frac{(-1)^\ell}{\ell!}\Big[q^{\langle L\rangle}(t,r)\,\partial_L r^{-1} + p^{\langle L\rangle}(t,r)\,x^{\langle L\rangle}\Big],$$

where

$$q^{\langle L\rangle}(t,r) := \int_0^r \rho(t,\boldsymbol{x}')x'^{\langle L\rangle}\,d^3x', \qquad p^{\langle L\rangle}(t,r) := \int_r^R \rho(t,\boldsymbol{x}')\partial_L r'^{-1}\,d^3x'.$$

In the integral defining $q^{\langle L\rangle}(t,r)$, the domain of integration is the region of space bounded by a sphere of radius $r := |\boldsymbol{x}|$. In the integral defining $p^{\langle L\rangle}(t,r)$, the domain of integration is the region of space bounded inwardly by a sphere of radius r, and outwardly by a sphere of arbitrary radius R that lies outside the distribution of matter.

1.12 For $\ell = 2$, 3, and 4, show explicitly that $n'^{\langle L\rangle}n^{\langle L\rangle} = [\ell!/(2\ell-1)!!]P_\ell(\mu)$, where $\mu := \boldsymbol{n}'\cdot\boldsymbol{n}$.

1.13 Fill in all the steps that are required to establish the STF identities of Eqs. (1.159) and (1.160).

1.14 If $\boldsymbol{e}$ and $\boldsymbol{n}$ are unit vectors, show that

$$\begin{aligned} e^{\langle qL\rangle}n^{\langle pL\rangle} &= \frac{\ell!}{(2\ell+1)(2\ell+1)!!}\Bigg[\delta^{pq}\frac{dP_\ell}{d\mu} - (e^pe^q+n^pn^q)\frac{d^2P_{\ell+1}}{d\mu^2} \\ &\quad + e^{(p}n^{q)}\left(2\frac{d^2P_\ell}{d\mu^2} + (2\ell+1)\frac{dP_{\ell+1}}{d\mu}\right) + (2\ell+1)e^{[p}n^{q]}\frac{dP_{\ell+1}}{d\mu}\Bigg], \end{aligned}$$

where $\mu = \boldsymbol{e}\cdot\boldsymbol{n}$, and use this to verify Eq. (1.240). *Hint:* Exploit the fact that $e^{\langle qL\rangle} \propto \partial_{qL}(1/R)$ and $n^{\langle pL\rangle} \propto \partial_{pL}(1/r)$, where R and r are independent distance variables.

1.15 Determine the STF tensors $\mathscr{Y}^{\langle L\rangle}_{\ell m}$ for $\ell = 1$, $\ell = 2$, and $\ell = 3$, and thereby verify the results listed in Box 1.5.

1.16 Using the general equation of motion (1.201), show explicitly that $\sum_A m_A\boldsymbol{a}_A = \boldsymbol{0}$.

2 Structure of self-gravitating bodies

In Chapter 1 we introduced the foundations of Newtonian gravity, and presented the equations that govern the gravitational potential of spherical and nearly spherical bodies. We also examined the center-of-mass motion of extended bodies, and witnessed the remarkable near-decoupling of the external dynamics – the motion of each body as a whole – from the internal dynamics – the internal fluid motions within each body. As we saw in Chapter 1, the details of internal structure, encapsulated in multipole moments of the mass distribution, have a limited influence on the motion of the body as a whole. In this chapter we take the focus away from the external dynamics and examine the internal structure and dynamics of extended, self-gravitating bodies. We shall return to the theme of the near-decoupling of the external and internal dynamics, and reveal the limited influence of the center-of-mass motion and the external bodies on the structure of a selected body.

We begin in Sec. 2.1 with a review of the equations of fluid mechanics that are relevant to the internal dynamics; these are best formulated in the moving reference frame of a selected body A in an N-body system. In Sec. 2.2 we examine the simplest models of internal structure, involving spherical symmetry, assuming that the body is non-rotating and not influenced by external bodies. The simplicity permits a gentle acquisition of much insight into the structure of realistic bodies, and we shall introduce models of increasing complexity: incompressible fluids, polytropes, isothermal spheres, and degenerate fermion gases as models of white dwarfs. Rotation can only be ignored for so long, however, and in Sec. 2.3 we examine the physics of rotating, self-gravitating bodies; we first present elements of a general theory, and then construct models of incompressible, rigidly rotating bodies in hydrostatic equilibrium – the famous Maclaurin spheroids and Jacobi ellipsoids.

While our discussion in Sec. 2.3 is not restricted to slow rotations and small deviations away from spherical symmetry, it is severely limited by the assumption that the fluid is incompressible. In Sec. 2.4 we relax this assumption and formulate a general theory of deformed bodies that accommodates an arbitrary equation of state. Though limited to small deformations, the theory is sufficiently powerful that it can handle any perturbation that causes a deformation from spherical symmetry, including rotation and tidal fields created by external bodies. The theory, therefore, allows us to return to the main theme introduced previously – the near-decoupling of the external and internal dynamics – and to calculate the effects of a tidal field on the body's structure. The tidal dynamics of extended bodies is examined in some detail in Sec. 2.5, both in the context of static tides (which occur slowly, on a time scale that is long compared with the internal dynamical time scale) and dynamical tides (which occur rapidly).

2.1 Equations of internal structure

Our main goal in this chapter is to describe the internal structure of a body A in a system of N bodies, making some assumptions regarding its composition, and accounting for the influence of the external bodies. The motion of the body's center-of-mass position $\boldsymbol{r}_A(t)$ was examined in Sec. 1.6, and there we observed that this motion is largely (but not completely) insensitive to the details of internal structure, which are encapsulated in a number of multipole moments $I_A^{\langle L\rangle}(t)$. In this section we shall find that the internal structure is largely (though not completely) insensitive to the details of the center-of-mass motion and the presence of external bodies.

The foundations of our analysis are the same as in Chapter 1. We take the body to consist of a perfect fluid of mass density $\rho(t, \boldsymbol{x})$, pressure $p(t, \boldsymbol{x})$, and velocity field $\boldsymbol{v}(t, \boldsymbol{x})$. These quantities are governed by Euler's equation

$$\rho \frac{d\boldsymbol{v}}{dt} = \rho \boldsymbol{\nabla} U - \boldsymbol{\nabla} p \tag{2.1}$$

and the continuity equation

$$\frac{\partial \rho}{\partial t} + \boldsymbol{\nabla} \cdot (\rho \boldsymbol{v}) = 0 \,. \tag{2.2}$$

The gravitational potential $U(t, \boldsymbol{x})$ is produced by all the bodies in the system, and it is governed by Poisson's equation

$$\nabla^2 U = -4\pi G \rho \,. \tag{2.3}$$

The center-of-mass variables of body A were introduced back in Sec. 1.6.2. They comprise the body's total mass m_A, center-of-mass position $\boldsymbol{r}_A$, velocity $\boldsymbol{v}_A$, acceleration $\boldsymbol{a}_A$, and multipole moments $I_A^{\langle L\rangle}$. In Sec. 1.6.3 the gravitational potential was decomposed as $U = U_A + U_{\neg A}$, in terms of a piece U_A produced by body A alone – the internal potential – and a piece $U_{\neg A}$ produced by the remaining bodies – the external potential. In Sec. 1.6 we showed that the acceleration of body A is caused by the external potential, and its general expression was displayed in Eq. (1.200).

To determine the internal motions of body A we focus on the position $\bar{\boldsymbol{x}} := \boldsymbol{x} - \boldsymbol{r}_A$ of a fluid element relative to the center-of-mass position, and on its relative velocity $\bar{\boldsymbol{v}} := \boldsymbol{v} - \boldsymbol{v}_A$. The equation that governs the behavior of the relative velocity is easily obtained from Euler's equation, and we write it in the form

$$\rho \frac{d\bar{\boldsymbol{v}}}{dt} = \rho \bar{\boldsymbol{\nabla}} U_A - \bar{\boldsymbol{\nabla}} p + \rho \bar{\boldsymbol{\nabla}} \big(U_{\neg A} - \boldsymbol{a}_A \cdot \bar{\boldsymbol{x}} \big) \,, \tag{2.4}$$

in which $\bar{\boldsymbol{\nabla}}$ is the gradient operator in the relative coordinates $\bar{\boldsymbol{x}}$. The first two terms on the right-hand side account for the purely internal aspects of the body's dynamics, and the remaining terms account for the influence of the external bodies; the last term, involving the body's acceleration, is a fictitious force that arises because the body's center-of-mass frame is not inertial.

We recall from Sec. 1.6.3 that the internal potential is given explicitly by

$$U_A(t, \bar{\boldsymbol{x}}) = G \int_A \frac{\rho(t, \bar{\boldsymbol{x}}')}{|\bar{\boldsymbol{x}} - \bar{\boldsymbol{x}}'|} d^3\bar{x}' . \tag{2.5}$$

For the external potential we follow the strategy of Sec. 1.6.4 and express it as a Taylor expansion about the body's center-of-mass. From Eq. (1.192) we get

$$U_{\neg A}(t, \bar{\boldsymbol{x}}) = \sum_{\ell=0}^{\infty} \frac{1}{\ell!} \partial_L U_{\neg A}(t, \boldsymbol{0}) \bar{x}^L \tag{2.6a}$$

$$= U_{\neg A}(t, \boldsymbol{0}) + g_j(t)\bar{x}^j - \sum_{\ell=2}^{\infty} \frac{1}{\ell!} \mathcal{E}_L(t) \bar{x}^L , \tag{2.6b}$$

where

$$g_j(t) := \partial_j U_{\neg A}(t, \boldsymbol{0}) , \qquad \mathcal{E}_L(t) := -\partial_L U_{\neg A}(t, \boldsymbol{0}); \tag{2.7}$$

the derivatives of the external potential are evaluated at the center-of-mass $\bar{\boldsymbol{x}} = \boldsymbol{0}$. The multi-index notation, in which L stands for a collection of ℓ individual indices, was introduced back in Sec. 1.5.3. We observe that since the external potential satisfies Laplace's equation $\nabla^2 U_{\neg A} = 0$ within the volume occupied by the body, the tensors $\mathcal{E}_L$ are symmetric and tracefree (STF). We observe also that the term $\ell = 0$ in Eq. (2.6) is spatially constant, and that it plays no role whatever in Eq. (2.4); we may therefore remove it from the expansion. The body's acceleration can also be expressed in terms of the expanded external potential. From Eq. (1.194) we obtain

$$a_A^j = \sum_{\ell=0}^{\infty} \frac{1}{\ell!} \frac{I_A^{\langle L \rangle}(t)}{m_A} \partial_{jL} U_{\neg A}(t, \boldsymbol{0}) \tag{2.8a}$$

$$= g_j - \sum_{\ell=2}^{\infty} \frac{1}{\ell!} \frac{I_A^{\langle L \rangle}(t)}{m_A} \mathcal{E}_{jL}(t); \tag{2.8b}$$

the term with $\ell = 1$ vanishes because $I_A^j = 0$ by virtue of the definition of the center-of-mass.

With Eqs. (2.6) and (2.8) we find that the external terms in Euler's equation combine to give

$$U_{\neg A}^{\text{eff}} := U_{\neg A} - \boldsymbol{a}_A \cdot \bar{\boldsymbol{x}} = -\sum_{\ell=2}^{\infty} \frac{1}{\ell!} \left[\mathcal{E}_L(t) \bar{x}^L - \frac{I_A^{\langle L \rangle}(t)}{m_A} \mathcal{E}_{jL}(t) \bar{x}^j \right] . \tag{2.9}$$

The cancellation of g_j in this expression implies that the external terms in Eq. (2.4) are much smaller than the internal terms; this is the near-decoupling of the external dynamics from the internal dynamics. With $\bar{r}_c$ denoting a characteristic length scale within the body, we have that the internal potential scales as $U_A \sim G m_A / \bar{r}_c$. The (effective) external potential is dominated by the $\ell = 2$ term, and this scales as $G m_B \bar{r}_c^2 / r_{AB}^3$, with r_{AB} denoting a typical inter-body distance. The ratio of external and internal influences is therefore given by

$$\frac{U_{\neg A}^{\text{eff}}}{U_A} \sim \frac{m_B}{m_A} \left(\frac{\bar{r}_c}{r_{AB}} \right)^3 , \tag{2.10}$$

and this is indeed much smaller than unity when the bodies are well separated ($\bar{r}_c \ll r_{AB}$).

When the body is spherical, or when its deviations from spherical symmetry are sufficiently small, the coupling between the higher multipole moments $I_A^{\langle L \rangle}$ and $\mathcal{E}_{jL}$ can be neglected in the acceleration. In such circumstances the effective external potential simplifies to

$$U_{\neg A}^{\text{eff}} = -\sum_{\ell=2}^{\infty} \frac{1}{\ell!} \mathcal{E}_L(t) \bar{x}^L \,. \tag{2.11}$$

We shall return to this expression later on in the chapter.

In the next two sections we shall neglect the external terms and examine the equilibrium states of a completely isolated body. We shall incorporate the external terms in Secs. 2.4 and 2.5, and see how they affect the body's internal structure.

2.2 Equilibrium structure of a spherical body

We begin our exploration of the equilibrium structure of an isolated body with the simplest conceivable model, that of a non-rotating and spherically symmetric object.

2.2.1 Equations of body structure

Hydrostatic equilibrium

The equations that govern the equilibrium structure of an isolated body were already identified back in Sec. 1.5, and they can be recovered from Eq. (2.4) by setting $\bar{v}^j = 0$ and omitting the external terms. In spherical symmetry the equation of hydrostatic equilibrium is

$$\frac{dp}{dr} = \rho \frac{dU}{dr} \,, \tag{2.12}$$

in which we drop the bar over the radial variable and the label A on the gravitational potential to simplify the notation. Poisson's equation can be integrated to give

$$\frac{dU}{dr} = -\frac{Gm(r)}{r^2} \,, \tag{2.13}$$

in which $m(r)$ is the mass contained within a sphere of radius r; this is related to the density by

$$\frac{dm}{dr} = 4\pi r^2 \rho \,. \tag{2.14}$$

We recall from Sec. 1.5 that according to Eq. (2.13), the potential inside the body can be expressed as

$$U_{\text{in}} = \frac{Gm(r)}{r} + 4\pi G \int_r^R \rho(r') r' \, dr' \,, \tag{2.15}$$

in which R is the body's radius, at which $p = 0$. The potential outside the body is

$$U_{\text{out}} = \frac{GM}{r}, \tag{2.16}$$

in which $M := m(r = R)$ is the body's total mass. Again we adjust the notation employed in Sec. 2.1 and denote the total mass M instead of m_A; this switch of notation was also made in Chapter 1.

Equation of state; energy production and transport

These equations must be supplemented by an equation of state

$$p = p(\rho, T; X) \tag{2.17}$$

that relates the pressure to the density ρ, temperature T, and chemical composition X of the matter making up the body. In addition, equations must be provided to account for energy production and transport within the body. One such equation was already considered back in Sec. 1.4.2, where it was shown that in conditions of thermal equilibrium, the heat-flux vector $\boldsymbol{H}$ and the rate of energy production per unit mass q are related by the conservation equation $\nabla \cdot \boldsymbol{H} = \rho q$. In spherical symmetry this reduces to

$$\frac{1}{r^2}\frac{d}{dr}(r^2 H) = \rho q, \tag{2.18}$$

where H is the radial component of $\boldsymbol{H}$. This equation, in turn, must be supplemented by a relation $q = q(\rho, T; X)$ that links the rate of energy production to the local conditions within the fluid. Once $H(r)$ is known, the temperature profile is determined by the equation of radiative transport,

$$\frac{dT}{dr} = -\frac{3}{4ac}\kappa\rho\frac{H}{T^3}, \tag{2.19}$$

where κ is the mean opacity, which also depends on the local conditions within the fluid, and a is the radiation constant.

This set of equations is overly simplistic; realistic stellar models (like the standard solar model that describes the structure of our Sun) are far more complicated. Among many simplifying assumptions, we have taken the composition X to be uniform throughout the body; this assumption is violated in a highly evolved star, in which nuclear reactions have produced a large radial variation in the abundances of heavy elements. It is also violated for planets that might have a mantle or crust that cannot be modeled as a perfect fluid. We have also assumed that radiation is the only relevant mechanism of energy transport; this is violated in most main-sequence stars, which harbor large convection zones.

In our simplified picture, a stellar model is constructed by simultaneously integrating Eqs. (2.12), (2.13), (2.14), (2.18), and (2.19). The relevant boundary conditions at $r = 0$ are that $p = p_c$ (the central pressure), $m = 0$, $H = 0$, and $T = T_c$ (the central temperature). At the boundary $r = R$ we find that $p = 0$, $m = M$, $H = L/(4\pi R^2)$ with L denoting the stellar luminosity, and T achieves its surface value $T(R)$. This is a formidable set of equations; q and κ are typically provided in tabular form, and the equations must be integrated numerically.

Virial theorem and other integral properties

The equilibrium configuration must satisfy the virial theorem of Eq. (1.90), which we write in the restricted form

$$\Omega + 3P = 0\,, \tag{2.20}$$

in which $\Omega = -\frac{1}{2}\int \rho U\, d^3x$ is the body's gravitational potential energy, and $P = \int p\, d^3x$ is the integrated pressure. We have left the kinetic-energy term out of Eq. (2.20) because we are applying the theorem to a static configuration that has no kinetic energy.

The integrated pressure is $P = 4\pi \int_0^R pr^2\, dr$ for a spherical body, and integration by parts brings this to the form

$$P = -\frac{4\pi}{3}\int_0^R \frac{dp}{dr} r^3\, dr\,. \tag{2.21}$$

Substitution of Eqs. (2.12) and (2.13) produces

$$3P = 4\pi G \int_0^R \rho(r) m(r) r\, dr\,, \tag{2.22}$$

and we wish to show that the right-hand side is an expression for $-\Omega$. One way to establish this is to insert Eq. (2.15) within the definition of Ω; this yields

$$\Omega = -2\pi G \int_0^R \rho m r\, dr - 8\pi^2 G \int_0^R dr\, \rho(r) r^2 \int_r^R dr'\, \rho(r') r'\,, \tag{2.23}$$

which we re-express as

$$\Omega = -2\pi G \int_0^R \rho m r\, dr - 8\pi^2 G \int_0^R dr'\, \rho(r') r' \int_0^{r'} dr\, \rho(r) r^2 \tag{2.24}$$

by altering the order of integration in the second term. The right-most integral is recognized as $m(r')/(4\pi)$, and the second term becomes $-2\pi G \int_0^R \rho(r') m(r')\, dr'$, the same as the first term. Our final expression for Ω is therefore

$$\Omega = -4\pi G \int_0^R \rho(r) m(r) r\, dr\,, \tag{2.25}$$

and we confirm the validity of Eq. (2.20).

Other integral properties of the equilibrium configuration can be obtained in a similar way. Integration of dp/dr from $r = 0$ to $r = R$ produces $-p_c$, and substitution of Eqs. (2.12) and (2.13) within the integral reveals that

$$p_c = G \int_0^R \frac{\rho(r) m(r)}{r^2}\, dr\,. \tag{2.26}$$

We next integrate pr^3 from the center to the surface. Exploiting integration by parts as we did before, we find that

$$\int_0^R pr^3\,dr = \frac{1}{4}G\int_0^R \rho m r^2\,dr\,. \tag{2.27}$$

But $\rho r^2 = (4\pi)^{-1}dm/dr$, and the integrand can be expressed as the total derivative $(8\pi)^{-1}dm^2/dr$. Integration produces

$$M^2 = \frac{32\pi}{G}\int_0^R p(r)r^3\,dr\,. \tag{2.28}$$

When pressure depends only on density

When the pressure depends only on density, and does not depend on temperature, the main equations of hydrostatic equilibrium decouple from the energy equations, and they can be handled separately. These can be given either as a set of first-order differential equations,

$$\frac{dp}{dr} = -\rho\frac{Gm}{r^2}\,, \qquad \frac{dm}{dr} = 4\pi r^2\rho\,, \tag{2.29}$$

or they can be combined into a single second-order differential equation for the pressure,

$$\frac{1}{r^2}\frac{d}{dr}\left(\frac{r^2}{\rho}\frac{dp}{dr}\right) = -4\pi G\rho\,; \tag{2.30}$$

in both cases it is assumed that ρ is given as a function of the pressure. In most applications, especially those involving computational methods, the formulation of Eq. (2.29) is a more practical one. In some applications, however, Eq. (2.30) can be advantageous, as it sometimes leads to a differential equation that can be solved in closed form.

Box 2.1 **Newtonian gravity, neutrinos, and the Sun**

One of the most surprising successes of the standard solar model, with its foundation grounded in Newtonian gravity, was the role it played in the discovery of neutrino oscillations. The chain of nuclear reactions that convert hydrogen to helium in the Sun produces neutrinos as a by-product. In 1964, Raymond Davis Jr. and John Bahcall proposed an experiment to measure the flux of high-energy neutrinos from the decay of ^{8}B produced in a side chain of the solar nuclear reactions. These neutrinos are able to convert ^{37}Cl to radioactive ^{37}Ar. The experiment involved a swimming-pool sized container of ordinary cleaning fluid (C_2Cl_4), from which Davis could extract minute amounts of ^{37}Ar (a few atoms per month) using specially designed chemical techniques. Beginning in 1968, Davis reported that the flux of neutrinos was about one third of what was expected from standard solar models. While there was initial skepticism of his result, the "solar neutrino problem" survived numerous double-checks and refinements of his experiment.

Suspicion fell on the solar models used to predict the neutrino flux. To construct a solar model is an exceedingly complicated task. In addition to Newtonian gravity, one must provide the initial elemental abundances from which the Sun was formed, input all the relevant nuclear reactions with their measured rates, incorporate the correct heat transfer from core to surface (involving both radiative transport and convection), and evolve the Sun for 4.5 billion years. The resulting model must match the current solar luminosity and the surface elemental abundances. The reaction rates in the solar core are extremely sensitive to temperature; a 1% change in temperature can induce a 30% change in the neutrino flux from ^{8}B. Decades of re-analyses and refinements of the solar models failed to resolve the solar neutrino problem.

Meanwhile, developments in particle physics opened the possibility that neutrinos might not be strictly massless. As a consequence, they could undergo "neutrino oscillations," whereby an electron neutrino transmutes into a muon neutrino (and to a smaller extent into a tau neutrino) and back. Mikheev, Smirnov and Wolfenstein showed that this effect, while operative in vacuum, could be enhanced during passage through matter, such as the solar interior. The experimental results could be explained if a sufficient number of the initial ^{8}B electron neutrinos had converted to muon or tau neutrinos, because the conversion of ^{37}Cl to ^{37}Ar is induced only by electron neutrinos.

The solution to the solar neutrino problem was provided by new solar neutrino experiments, notably Kamiokande and super-Kamiokande in Japan, GALLEX in Germany, SAGE in Russia, and SNO in Canada. All the experiments confirmed the deficit of solar electron neutrinos. But the Japanese and Canadian experiments were sensitive to all three types of neutrinos, and SNO could actually distinguish between the different varieties of neutrinos; they ultimately verified that the *total* flux of neutrinos agreed completely with the predictions of the solar models based on Newtonian gravity.

2.2.2 Incompressible fluid

The simplest equilibrium structure is obtained when one assumes that the fluid is incompressible, that is, that the mass density is uniform throughout the body. We express this mathematically as

$$\rho(r) = \begin{cases} \rho_0 & r \leq R\,, \\ 0 & r > R\,, \end{cases} \tag{2.31}$$

where ρ_0 is a constant. Another way of expressing this is $\rho(r) = \rho_0 \Theta(R - r)$, with Θ denoting the Heaviside step function. The pressure of an incompressible fluid is unrelated to the density, and it must be determined by integrating the equation of hydrostatic equilibrium. The incompressible fluid is an exceedingly crude and entirely unphysical model – it leads, for example, to a formally infinite speed of sound within the body. Nevertheless, its simplicity makes it an attractive starting point for a study of equilibrium structures, and we shall have many occasions to return to it in this chapter.

From Eq. (2.14) we find that the mass function within the body is given by

$$m(r) = \frac{4\pi}{3}\rho_0 r^3 = M(r/R)^3\,, \tag{2.32}$$

with

$$M = \frac{4\pi}{3}\rho_0 R^3 \tag{2.33}$$

denoting the total mass. From Eqs. (2.12) and (2.13) we find that the pressure profile is given by

$$p(r) = p_c(1 - r^2/R^2)\,, \tag{2.34}$$

with

$$p_c = \frac{2\pi}{3} G\rho_0^2 R^2 = \frac{3}{8\pi}\frac{GM^2}{R^4} \tag{2.35}$$

denoting the central pressure $p(r = 0)$; the constant of integration was chosen so that $p(r)$ properly vanishes at the boundary $r = R$. And from Eq. (2.15) we find that the gravitational potential inside the body is given by

$$U_{\rm in} = \frac{GM}{2R}(3 - r^2/R^2)\,; \tag{2.36}$$

outside the body it takes the usual form $U_{\rm out} = GM/r$, and the potential is continuous (though not differentiable) at $r = R$.

Evaluation of Eq. (2.25) reveals that the gravitational potential energy of an incompressible body is

$$\Omega = -\frac{3}{5}\frac{GM^2}{R}\,, \tag{2.37}$$

and by virtue of the virial theorem, the integrated pressure is $P = -\frac{1}{3}\Omega = \frac{1}{5}GM^2/R$.

2.2.3 Polytropes and the Lane–Emden equation

Polytropic equation of state and polytropes

A *polytrope* is a body in hydrostatic equilibrium for which the matter satisfies the polytropic equation of state

$$p = K\rho^\Gamma\,, \qquad \Gamma := 1 + 1/n\,, \tag{2.38}$$

where K and Γ are constants; the related constant n is called the *polytropic index*. We first encountered this equation of state near the end of Sec. 1.4.2, where it was shown to result from thermal equilibrium when the energy density ϵ of a fluid element is related to the pressure p by $\epsilon = np$ (note that n was denoted η in Sec. 1.4.2). Equation (2.38) is an example of an equation of state that is independent of temperature, which implies that an equilibrium configuration can be constructed without having to consider the equations of energy production and transport.

Polytropes have the advantage of being more physical than models involving an incompressible fluid, but they are still a far cry from representing a realistic stellar structure. Nevertheless, they were studied extensively in the 19th and early 20th century in an effort to gain insight into stellar astrophysics at a time when almost nothing was known about how stars actually operate; the simplicity of Eq. (2.38) encouraged the development of a rich body of work that may not be directly applicable to real stars (although we shall see the connection with white dwarfs later in this section), but is nevertheless beautiful and worthy of study. This effort was initiated by Lord Kelvin, who noted that Eq. (2.38) should apply to a star in convective equilibrium. The study of stellar structure based on the polytropic equation of state was taken up by Lane and developed systematically by Ritter. The work, as it stood in 1907, was summarized in a treatise by Emden, and it was passed on (with further developments) to later generations by Chandrasekhar (1958) in his classic text *An Introduction to the Study of Stellar Structure.*

Scales

Our goal is to integrate the equations of hydrostatic equilibrium, either Eqs. (2.29) or Eq. (2.30), for a fluid with a polytropic equation of state. Before getting started with this task it is helpful to introduce a number of scaling quantities that are relevant to this problem. We have

$$\rho_c := \text{central density}, \tag{2.39a}$$

$$p_c := \text{central pressure}, \tag{2.39b}$$

$$r_0 := \text{length scale}, \tag{2.39c}$$

$$m_0 := \text{mass scale}. \tag{2.39d}$$

While ρ_c and p_c provide their own definitions, with the equation of state giving $p_c = K\rho_c^{1+1/n}$, r_0 and m_0 must still be determined. To define m_0 we simply note that it must scale as $\rho_c r_0^3$ and insert a convenient numerical factor; we set

$$m_0 := 4\pi\rho_c r_0^3 . \tag{2.40}$$

To define r_0 we appeal to the equation of hydrostatic equilibrium, $dp/dr = -G\rho m/r^2$, and note that the left-hand side scales as p_c/r_0 while the right-hand side scales as $G\rho_c m_0/r_0^2$; inserting a convenient numerical factor, we set

$$r_0^2 := \frac{(n+1)p_c}{4\pi G\rho_c^2} = \frac{(n+1)K}{4\pi G}\rho_c^{(1-n)/n} . \tag{2.41}$$

Combining this with Eq. (2.40), we find that the mass scale takes the explicit form

$$m_0 = \frac{(n+1)^{3/2}K^{3/2}}{(4\pi)^{1/2}G^{3/2}}\rho_c^{(3-n)/(2n)} . \tag{2.42}$$

From Eqs. (2.40) and (2.41) we also get $Gm_0/r_0 = (n+1)p_c/\rho_c$, a useful relation among the various scales. Other relations are

$$p_c = \frac{(4\pi)^{1/3}G}{n+1}\rho_c^{4/3}m_0^{2/3} \tag{2.43}$$

and

$$\frac{m_0^{n-1}}{r_0^{n-3}} = \frac{1}{4\pi}\left[\frac{(n+1)K}{G}\right]^n , \tag{2.44}$$

which reveals that m_0^{n-1}/r_0^{n-3} is actually independent of the central density ρ_c.

Dimensionless variables and Lane–Emden equation

Having introduced the relevant scales, the next step is to express the equations of hydrostatic equilibrium in dimensionless and scale-free form. We introduce

$$\xi := r/r_0 \tag{2.45}$$

as a dimensionless radial variable, and

$$\mu := m/m_0 \tag{2.46}$$

as a dimensionless mass function. We also write

$$\rho := \rho_c \theta^n , \tag{2.47}$$

and adopt θ as a dimensionless substitute for the density function. For the pressure we then have

$$p = p_c \theta^{n+1} , \tag{2.48}$$

in accordance with the equation of state of Eq. (2.38).

With these variables Eqs. (2.29) become

$$\frac{d\theta}{d\xi} = -\frac{\mu}{\xi^2} , \qquad \frac{d\mu}{d\xi} = \xi^2 \theta^n . \tag{2.49}$$

These equations are integrated outward from $\xi = 0$, where the boundary conditions

$$\theta(\xi = 0) = 1 , \qquad \mu(\xi = 0) = 0 \tag{2.50}$$

are imposed. Integration proceeds until $\theta = 0$ at $\xi = \xi_1$, which marks the body's boundary, where both the pressure and density vanish. The body's total mass is then

$$M = m_0 \mu_1 , \tag{2.51}$$

with $\mu_1 := \mu(\xi = \xi_1)$, while the body's radius is

$$R = r_0 \xi_1 . \tag{2.52}$$

From Eqs. (2.40), (2.51), and (2.52) it is easy to see that the mean density of a polytrope is given by

$$\bar{\rho} := \frac{3M}{4\pi R^3} = \rho_c \left(\frac{3\mu_1}{\xi_1^3} \right) . \tag{2.53}$$

It is useful to note that, according to Eq. (2.49), $\mu_1 = -\xi_1^2 \theta'(\xi_1)$, in which a prime indicates differentiation with respect to ξ.

As in Eq. (2.30), the equations displayed in Eq. (2.49) can be combined into a single, second-order differential equation for the density variable θ. This equation,

$$\frac{1}{\xi^2} \frac{d}{d\xi} \left(\xi^2 \frac{d\theta}{d\xi} \right) = -\theta^n , \tag{2.54}$$

Table 2.1 Numerical integration of the Lane–Emden equation for various polytropes. The first column lists the polytropic index n, and the second column lists $\Gamma = 1 + 1/n$. The third column lists ξ_1, the value of the radial variable at which $\theta = 0$. The fourth column lists μ_1, the value of the dimensionless mass function at $\xi = \xi_1$.

n	Γ	ξ_1	μ_1
1/2	3	2.752698054	3.788651185
2/3	5/2	2.871323871	3.538747902
1	2	3.141592654	3.141592654
3/2	5/3	3.653753736	2.714055120
2	3/2	4.352874596	2.411046012
3	4/3	6.896848619	2.018235951
4	5/4	14.97154635	1.797229914

is the famous *Lane–Emden equation*. For most applications the first-order formulation of Eq. (2.49) is more practical, but as we shall see below, for selected values of n Eq. (2.54) leads to a simple differential equation that can be integrated exactly.

Properties of polytropes

Because Eqs. (2.49) and (2.54) are independent of K and ρ_c, they can be integrated once and for all for any selected value of n; the solutions are scale-free and independent of both K and ρ_c. For a given n and K (that is, for a given equation of state), the solution gives rise to an entire family of stellar models parameterized by the central density ρ_c. As Eq. (2.42) reveals, the mass $M = m_0\mu_1$ increases with ρ_c when $n < 3$, but it decreases with increasing density when $n > 3$. From Eq. (2.41) we observe that the radius $R = r_0\xi_1$ increases with ρ_c when $n < 1$, but that it decreases with increasing density when $n > 1$. Combining these statements, we find that the mass increases with the radius when $n < 1$ and $n > 3$, while it decreases with increasing radius when $1 < n < 3$. Note that when $n = 3$ (or $\Gamma = 4/3$), the mass turns out to be *independent* of the central density; we shall come back to this observation in our study of white dwarfs. Note also that when $n = 1$ (or $\Gamma = 2$), it is the radius that becomes independent of the central density.

In Box 2.2 we describe how the Lane–Emden equations can be integrated numerically, and in Table 2.1 we display the results of such numerical integrations. In Fig. 2.1 we present plots of the density as a function of radius. The figure shows that as n increases and Γ decreases, so that the equation of state becomes increasingly soft, the polytropes become centrally dense, with a density profile that falls off increasingly rapidly away from $r = 0$. For stiffer equations of state (for n small and Γ large), the density becomes increasingly uniform.

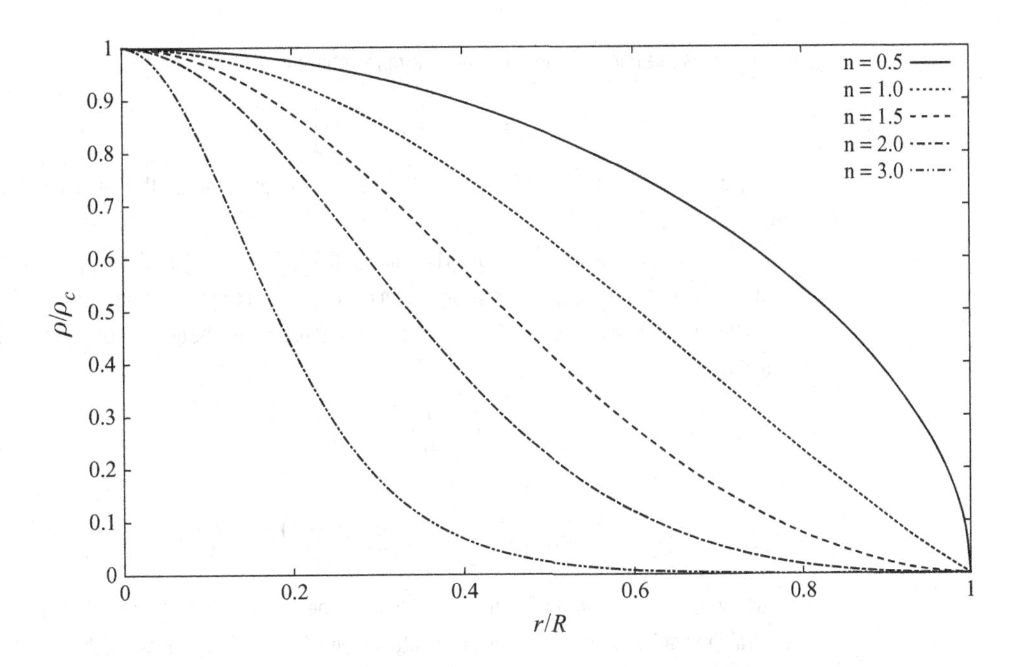

Fig. 2.1 Density versus radius for various polytropes. The density is normalized to the central density ρ_c, and the radius is normalized to the body radius R.

Box 2.2 Integration of the Lane–Emden equation

Equations (2.49) must be integrated outward from $\xi = 0$, and the boundary conditions are $\theta(0) = 1$ and $\mu(0) = 0$. A difficulty is immediately encountered with $d\theta/d\xi$, which evaluates to $0/0$ at $\xi = 0$; since μ goes to zero as $\frac{1}{3}\xi^3$, the equation is actually well-behaved at the center, but this formulation of the equations is ill suited to a direct numerical integration.

To avoid this difficulty it is helpful to adopt

$$\nu := \frac{\mu}{\xi^3}$$

as a substitute variable. The differential equations become

$$\frac{d\theta}{d\xi} = -\xi\nu, \qquad \frac{d\nu}{d\xi} = \frac{1}{\xi}(\theta^n - 3\nu),$$

and the equation for θ is now better behaved near $\xi = 0$. The boundary conditions are now $\theta(0) = 1$ and $\nu(0) = \frac{1}{3}$. The equation for ν is still presenting a problem at $\xi = 0$. To take care of this we perform the transformation

$$x := \ln \xi$$

and adopt x as a new independent variable. The differential equations become

$$\frac{d\theta}{dx} = -e^{2x}\nu\,, \qquad \frac{d\nu}{dx} = \theta^n - 3\nu\,,$$

and they are now well conditioned for numerical integration, except for the fact that integration must proceed from $x = -\infty$.

The solution to this last difficulty is to start the integration from $x = x_{\rm init}$, where $x_{\rm init}$ is large and negative. Starting values for the dependent variables can then be obtained by expanding θ and ν in powers of ξ, with coefficients determined by substituting the approximations within the differential equations. In this way we obtain

$$\theta = 1 - \frac{1}{6}\xi^2 + \frac{n}{120}\xi^4 + O(\xi^6)$$

and

$$\nu = \frac{1}{3} - \frac{n}{30}\xi^2 + \frac{n(8n-5)}{2520}\xi^4 + O(\xi^6)\,.$$

By choosing $\xi = e^{x_{\rm init}}$ sufficiently small, the errors can be adjusted to be below a chosen tolerance, and numerical integration of the equations can be attempted with standard methods such as the fourth-order Runge-Kutta algorithm.

Integration proceeds until $x = x_1 = \ln \xi_1$, which marks the first point at which θ changes sign. To determine the precise location requires finding the root of the equation $\theta(x) = 0$. The simplest way to achieve this is to perform a bisection search: One first identifies an x_0 such that $\theta(x_0) > 0$ and an x_2 such that $\theta(x_2) < 0$, so that x_1 must be between x_0 and x_2. One next evaluates $\theta(x)$ at the halfway point. If the sign is positive, then x_0 is moved to this new position while x_2 is left where it was; if the sign is negative, then x_0 is left alone while x_2 is brought to the new position. The cycle repeats until the interval containing x_1 has shrunk to a sufficiently small size. At this stage x_1 is known to sufficient numerical accuracy, and $\mu_1 = \nu(x_1)e^{3x_1}$ can be evaluated with the same degree of accuracy.

Gravitational potential energy

We wish to show that the gravitational potential energy of a polytrope is given by

$$\Omega = -\frac{3}{5-n}\frac{GM^2}{R}\,. \tag{2.55}$$

This result and the virial theorem imply that the integrated pressure is

$$P = -\frac{1}{3}\Omega = \frac{1}{5-n}\frac{GM^2}{R}\,. \tag{2.56}$$

These results indicate that models with $n \geq 5$ are peculiar; we shall examine the special case $n = 5$ below, and see what happens when $n > 5$.

Our starting point is Eq. (2.25), in which we substitute Eqs. (2.45), (2.46), and (2.47). After taking into account Eq. (2.40), we obtain

$$\Omega = -\frac{Gm_0^2}{r_0} \int_0^{\xi_1} \mu\,\theta^n \xi \, d\xi \,. \tag{2.57}$$

This integral can be written in a number of equivalent forms. We first use Eq. (2.49) to eliminate μ in favor of $-\xi^2\theta'$, in which a prime indicates differentiation with respect to ξ; after integration by parts we arrive at

$$\Omega = -\frac{Gm_0^2}{r_0} \frac{3}{n+1} \int_0^{\xi_1} \xi^2 \theta^{n+1} \, d\xi \,. \tag{2.58}$$

Proceeding from this expression, we next eliminate θ^n in favor of μ'/ξ^2 and write the integrand as $\theta\mu'$; integration by parts yields

$$\Omega = -\frac{Gm_0^2}{r_0} \frac{3}{n+1} \int_0^{\xi_1} \frac{\mu^2}{\xi^2} \, d\xi \,, \tag{2.59}$$

after expressing θ' as $-\mu/\xi^2$. If we next return to Eq. (2.57) and eliminate θ^n in the same way, we obtain the integrand $\mu\mu'/\xi$; integration by parts produces

$$\Omega = -\frac{Gm_0^2}{r_0} \biggl[\frac{\mu_1^2}{2\xi_1} + \frac{1}{2} \int_0^{\xi_1} \frac{\mu^2}{\xi^2} \, d\xi \biggr] . \tag{2.60}$$

The second term matches the expression displayed in Eq. (2.59), and combining these expressions, we finally arrive at

$$\Omega = -\frac{3}{5-n} \frac{Gm_0^2 \mu_1^2}{r_0 \xi_1} \,, \tag{2.61}$$

which is the same statement as Eq. (2.55).

Special cases with exact solutions

The Lane–Emden equation can be integrated exactly for special values of n. The first such case is $n = 0$, which is actually a singular limit of the formalism. While Eq. (2.54) is perfectly well behaved when $n = 0$, the polytropic equation of state is singular, and the density becomes unrelated to the pressure. Accordingly, the definition of θ provided by Eq. (2.47) breaks down, but we observe that Eq. (2.48) remains meaningful when $n = 0$; it becomes $p = p_c \theta$, and this relation supplies θ with a new definition. To keep the remaining equations meaningful, we introduce a new density scale ρ_0, a new length scale r_0, and a new mass scale m_0 defined by

$$r_0^2 = \frac{p_c}{4\pi G \rho_0^2} \,, \qquad m_0 = 4\pi \rho_0 r_0^3 \,; \tag{2.62}$$

these definitions replace Eqs. (2.40) and (2.41), and the density scale ρ_0 is unrelated to the central pressure p_c. The dimensionless variables then refer to the new scales: $\xi = r/r_0$, $\theta = p/p_c$, and $\mu = m/m_0$.

With these changes we find that Eqs. (2.49) or (2.54) apply just as well to the case $n = 0$. Integration is straightforward, and we find that the solutions are

$$\theta = 1 - \frac{1}{6}\xi^2, \qquad \mu = \frac{1}{3}\xi^3. \tag{2.63}$$

These results imply that the pressure vanishes at $\xi = \xi_1 = \sqrt{6}$, and that the body's dimensionless mass is $\mu_1 = 2\sqrt{6}$. Incorporating the scales, we have found that

$$R = \sqrt{6} r_0, \qquad M = 2\sqrt{6} m_0. \tag{2.64}$$

The fact that the mass function is exactly proportional to r^3 implies that the density is uniform within the body: at any radius r we have that $\rho = \rho_0$. The $n = 0$ limit of a polytrope, therefore, corresponds to an incompressible body, and the results obtained here are fully compatible with those described in Sec. 2.2.2. Note, in particular, that Eq. (2.55) reduces to Eq. (2.37) when $n = 0$.

Another case that admits an exact solution is $n = 1$, corresponding to $\Gamma = 2$. The general solution to Eq. (2.54) is $\theta = (c_1 \sin\xi + c_2 \cos\xi)/\xi$, where c_1 and c_2 are integration constants, and imposing the boundary conditions yields

$$\theta = \frac{\sin\xi}{\xi}, \qquad \mu = \sin\xi - \xi\cos\xi. \tag{2.65}$$

These results imply that the boundary is at $\xi = \xi_1 = \pi$, and the body's dimensionless mass is $\mu_1 = \pi$. Incorporating the scales of Eqs. (2.41) and (2.42), we find that

$$R = \sqrt{\frac{\pi K}{2G}}, \qquad M = \sqrt{\frac{2\pi K}{G}}\rho_c. \tag{2.66}$$

We note that R is independent of the central density, while M increases linearly with ρ_c.

A final case that gives rise to an exact solution is $n = 5$. To find this solution it is helpful to change the dependent variable from θ to $f := \theta^{-2}$. The Lane–Emden equation becomes

$$\frac{1}{2}ff'' - \frac{3}{4}(f')^2 + \frac{1}{\xi}ff' = 1, \tag{2.67}$$

in which a prime indicates differentiation with respect to ξ. This is a non-linear differential equation, but a solution can be found by substituting a trial solution of the form $f = 1 + \sum_{p=1}^{\infty} a_p \xi^{2p}$. This reveals that the exact solution is $f = 1 + \frac{1}{3}\xi^2$. In terms of the original variables, we have

$$\theta = \left(1 + \tfrac{1}{3}\xi^2\right)^{-1/2}, \qquad \mu = \tfrac{1}{3}\xi^3\left(1 + \tfrac{1}{3}\xi^2\right)^{-3/2}. \tag{2.68}$$

These results imply that the body does not have a well-defined surface: θ vanishes at $\xi = \xi_1 = \infty$. Nevertheless, the total mass is finite and equal to $\mu_1 = \sqrt{3}$. In this case Eq. (2.42) produces

$$M = \frac{18\sqrt{2}K^{3/2}}{(4\pi)^{1/2}G^{3/2}}\frac{1}{\rho_c^{1/5}}, \tag{2.69}$$

and we see that M decreases very slowly with an increasing central density ρ_c. We observed previously that Eq. (2.55) for the gravitational potential energy does not apply when $n = 5$;

in fact, the equation is meaningless when $n = 5$ and $R = \infty$, but it is nevertheless true that Ω is finite in this case. It can also be shown that all polytropes with $n \geq 5$ have an infinite radius.

2.2.4 Isothermal spheres

Another equation of state that leads to simple equilibrium configurations is

$$p = \frac{\rho k T}{\mu m_{\rm H}} + \frac{1}{3} a T^4 , \tag{2.70}$$

in which the temperature T is assumed to be uniform; μ is the mean atomic number (which is not to be confused with the dimensionless mass function), $m_{\rm H}$ is the atomic mass unit, and a is the radiation constant. The first term in the equation of state is the pressure exerted by an ideal gas, and the second term is the radiative pressure. This equation of state is relevant when convection brings large portions of the body into thermal equilibrium at a constant temperature; this is in contrast to radiative equilibrium, where the temperature varies because of radiation transport. The equation of state (without the radiation term) is also adopted in simple models of star clusters, where two-body scattering processes bring the stars into an approximate Maxwellian distribution of velocities, with a common "temperature" related to the velocity dispersion in the cluster.

Because T is constant, the radiation term does not give rise to a pressure gradient, and therefore it does not participate in the hydrostatic equilibrium. So for a given T, the effective equation of state is

$$p = K\rho , \tag{2.71}$$

where $K := kT/(\mu m_{\rm H})$; this is a special case of a polytropic equation of state with $\Gamma = 1$ and $n = \infty$. This limit is too singular to be handled by the Lane–Emden equation, and we must give it a separate treatment.

As usual we introduce the relevant scaling quantities: we have the central density ρ_c, the central pressure $p_c = K\rho_c$, and we introduce a length scale r_0 and a mass scale m_0 with the relations

$$r_0^2 := \frac{p_c}{4\pi G \rho_c^2} = \frac{K}{4\pi G \rho_c} \tag{2.72}$$

and

$$m_0 = 4\pi \rho_c r_0^3 . \tag{2.73}$$

A useful relation among these quantities is $Gm_0/r_0 = p_c/\rho_c = K$. The scale-free, dimensionless variables are

$$\xi := r/r_0 , \qquad e^{-\psi} := \rho/\rho_c = p/p_c , \qquad \mu := m/m_0 . \tag{2.74}$$

With these substitutions it is easy to show that the equations of hydrostatic equilibrium, displayed in Eq. (2.29), take the form of

$$\frac{d\psi}{d\xi} = \frac{\mu}{\xi^2} , \qquad \frac{d\mu}{d\xi} = \xi^2 e^{-\psi} . \tag{2.75}$$

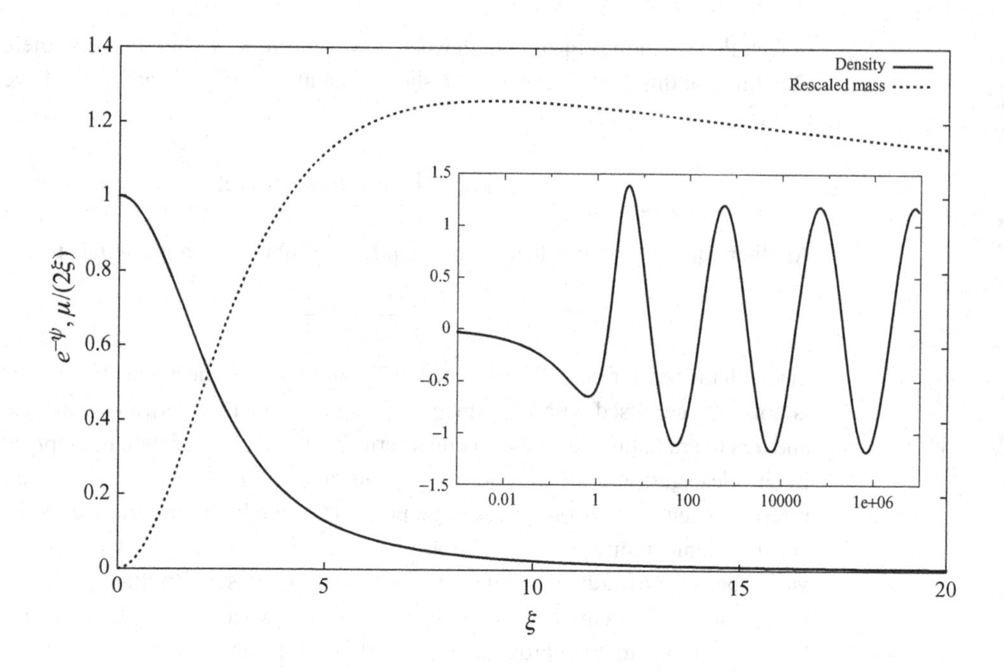

Fig. 2.2 Plots of the dimensionless density $e^{-\psi}$ and rescaled mass $\mu/(2\xi)$ as functions of the dimensionless radius ξ. While the density decreases monotonically with increasing ξ, it does so too slowly for the mass μ to converge to a finite limit; the figure reveals that $\mu/(2\xi)$ converges, but that the mass increases without bound. For reasons explained in the text, the inset shows a plot of $(\frac{1}{2}\xi^2 e^{-\psi} - 1)\xi^{1/2}$ as a function of $\log \xi$.

These can be combined into a single, second-order differential equation for ψ:

$$\frac{1}{\xi^2}\frac{d}{d\xi}\left(\xi^2\frac{d\psi}{d\xi}\right) = e^{-\psi}\,. \tag{2.76}$$

The equations are integrated outward from $\xi = 0$, with boundary conditions $\psi(0) = 0$ (so that $e^{-\psi} = 1$ at the center) and $\mu(0) = 0$; integration proceeds until $e^{-\psi} = 0$. The equations can be integrated once and for all for any K and central density ρ_c, and the solution describes an entire family of equilibrium structures parameterized by ρ_c. Numerical integration of the equations is facilitated by the tricks described in Box 2.2; it is useful to note that for $\xi \ll 1$, the functions are well approximated by the expansions $\psi = \frac{1}{6}\xi^2 - \frac{1}{120}\xi^4 + \frac{1}{1890}\xi^6 + O(\xi^8)$ and $\nu := \mu/\xi^3 = \frac{1}{3} - \frac{1}{30}\xi^2 + \frac{1}{315}\xi^4 + O(\xi^6)$.

In Fig. 2.2 we display the results of a numerical integration of Eqs. (2.75). The most important observation is that while the density decreases monotonically with increasing radius, it does so too slowly to give rise to a well-defined surface and a finite mass for the body. In fact, it can be shown that when $\xi \gg 1$, $e^{-\psi}$ behaves as

$$e^{-\psi} = \frac{2}{\xi^2}\left[1 + \frac{A}{\xi^{1/2}}\cos\left(\frac{\sqrt{7}}{2}\ln\xi + \delta\right) + O(\xi^{-1})\right], \tag{2.77}$$

where A and δ are constants that must be matched to the numerical results. The asymptotic behavior $e^{-\psi} \sim 2/\xi^2$ implies that the mass increases as $\mu \sim 2\xi$, and this is confirmed by the numerical results. The figure's inset shows a plot of $(\frac{1}{2}\xi^2 e^{-\psi} - 1)\xi^{1/2} \sim A\cos(\frac{\sqrt{7}}{2}\ln\xi + \delta)$, and the plot confirms the existence of logarithmic oscillations of constant amplitude.

These results make it clear that the isothermal equation of state cannot describe a complete star or star cluster. Nevertheless, it is often adopted to model the core of a star or star cluster, where the conditions are approximately isothermal. The model is cut off at some appropriate radius, and the solution is matched to another solution for the outer region of the body, where a different equation of state takes over.

2.2.5 White dwarfs

A white dwarf is a low-mass star that has come to the end state of its stellar evolution, following a long life on the main sequence. It is a dead star that is no longer able to support itself with radiative pressure produced by thermonuclear reactions in the core, because these have ceased after virtually all the hydrogen and helium have been converted to heavier elements. As the star starts to shrink (after it has ejected its outer layers), the density increases sufficiently for the electrons to become degenerate, and it is the degeneracy pressure of the electrons that will continue to support the star against further gravitational collapse. White dwarfs have masses typically ranging between $0.5M_\odot$ and $0.7M_\odot$, with the majority of them tightly clustered around $0.6M_\odot$; masses as low as $0.2M_\odot$ and as high as $1.3M_\odot$ have been observed. The typical radius of a white dwarf is of the order of 9×10^3 km, somewhat larger than the Earth; the typical density is of the order of $4 \times 10^8\ \mathrm{kg/m^3}$. A typical white dwarf has a composition dominated by ^{12}C and ^{16}O. In this section we construct the equation of state for a degenerate electron gas, and we integrate the equations of hydrostatic equilibrium to uncover the internal structure of a white dwarf.

Conditions for degeneracy and relativity

Electron degeneracy was reviewed briefly in Sec. 1.4.2, where it was shown to result when the temperature T becomes smaller than the Fermi temperature

$$T_\mathrm{F} = \frac{\hbar^2}{2m_e k}\left(3\pi^2 n_e\right)^{2/3}, \tag{2.78}$$

which was first introduced in Eq. (1.42); here m_e is the electron's mass, and n_e is its number density. This equation reveals that the Fermi temperature is high when the density is high, and these are the conditions that prevail inside a white dwarf.

To turn the criterion $T < T_\mathrm{F}$ into something more useful, we estimate the temperature of a white dwarf by relating it to its pressure and mass density; taking the ideal-gas equation of state as a rough guide, we have $kT \sim p/n_e$, where p is the pressure within the star. Because the matter is electrically neutral, and because most of the mass is contained in the ions (instead of the electrons), we write $n_e \sim \rho/m_\mathrm{H}$, in which ρ is the total mass density. With $p \sim GM^2/R^4$ and $\rho \sim M/R^3$, we find that $kT \sim Gm_\mathrm{H}M/R$. The Fermi temperature, on the other hand, is given by $kT_F = \hbar^2 M^{2/3}/(m_e m_H^{2/3} R^2)$, and the criterion for degeneracy is

$$M^{1/3} R < \frac{\hbar^2}{G m_e m_H^{5/2}}, \tag{2.79}$$

or

$$\left(\frac{M}{M_\odot}\right)^{1/3} \left(\frac{R}{R_\odot}\right) < 10^{-2} \tag{2.80}$$

after inserting the relevant numbers. For the typical white dwarf described previously, the left-hand side evaluates to 0.01, and the criterion is (marginally) satisfied.

The electrons are not only degenerate, they are also relativistic. (We are speaking of special relativity, not general relativity; as Table 1.1 indicates, the gravitational potential is still too modest for general relativity to play a significant role in the structure of white dwarfs.) To estimate the size of the relativistic corrections we examine the quantity $x := |\boldsymbol{p}|/(m_e c)$, the ratio of an electron's momentum to $m_e c$; relativistic effects are important when $x > 1$. For a degenerate gas the momentum is given by the Fermi momentum, which is linked to the density by the relation (derived below) $n_e = x^3/(3\pi^2 \lambda_e^3)$, in which $\lambda_e := \hbar/(m_e c)$ is the Compton wavelength of the electron. With this we find that

$$x \sim \frac{\lambda_e \rho^{1/3}}{m_H^{1/3}} \sim \frac{\lambda_e}{m_H^{1/3}} \frac{M^{1/3}}{R}, \tag{2.81}$$

or

$$x \sim 10^{-2} \left(\frac{M}{M_\odot}\right)^{1/3} \left(\frac{R_\odot}{R}\right) \tag{2.82}$$

after inserting the relevant numbers. For the typical white dwarf described previously, we have that $x \sim 1$, which indicates that relativistic effects are indeed important.

Equation of state

We wish to construct the equation of state of a degenerate, and relativistic, electron gas. To do this we rely on techniques borrowed from the kinetic theory of gases, which features the *distribution function* $f(\boldsymbol{x}, \boldsymbol{p})$ as a starting point of most calculations. This is defined so that

$$\frac{g}{(2\pi\hbar)^3} f(\boldsymbol{x}, \boldsymbol{p})\, d^3x\, d^3p$$

is the number of particles in a phase-space cell of volume $d^3x\, d^3p$ at position $\boldsymbol{x}$ and momentum $\boldsymbol{p}$. The factor g is the number of internal states associated with a particle of given momentum, and in the case of an electron (a spin-$\frac{1}{2}$ particle), this is equal to 2. It can be shown that the distribution function is a relativistic invariant, so that $\boldsymbol{x}$ and $\boldsymbol{p}$ can refer to any Lorentz frame.

The distribution function gives rise to the number density

$$n_e = \frac{2}{(2\pi\hbar)^3} \int f(\boldsymbol{x}, \boldsymbol{p})\, d^3p, \tag{2.83}$$

the energy density

$$\epsilon = \frac{2}{(2\pi\hbar)^3}\int Ef(\boldsymbol{x},\boldsymbol{p})\,d^3p\,, \tag{2.84}$$

and the pressure

$$p = \frac{1}{3}\frac{2}{(2\pi\hbar)^3}\int |\boldsymbol{p}|vf(\boldsymbol{x},\boldsymbol{p})\,d^3p\,, \tag{2.85}$$

where $E = \sqrt{|\boldsymbol{p}|^2c^2 + m_e^2c^4}$ is the relativistic energy, and $v = |\boldsymbol{p}|c^2/E$ the relativistic speed, of a particle with momentum $\boldsymbol{p}$. The equation for the pressure expresses the fact that pressure is a flux of momentum, and the factor of $\frac{1}{3}$ reflects the isotropy of the gas. Note that we must be cautious to distinguish the pressure p from the magnitude $|\boldsymbol{p}|$ of the momentum vector.

We are interested in the relation between the pressure p and the number density n_e. To work this out we need an expression for f, and in our case this is given by the famous Fermi–Dirac distribution. To simplify we assume that $T \ll T_{\rm F}$, and adopt the limiting form that applies at zero temperature. This is simple: $f = 1$ when $|\boldsymbol{p}|$ is smaller than the Fermi momentum $p_{\rm F}$, and $f = 0$ when $|\boldsymbol{p}| > p_{\rm F}$. Then the expression for the number density gives rise to $n_e = 2(2\pi\hbar)^{-3}\int_0^{p_{\rm F}} 4\pi|\boldsymbol{p}|^2\,d|\boldsymbol{p}|$, which integrates to

$$n_e = \frac{x^3}{3\pi^2\lambda_e^3}\,, \tag{2.86}$$

where $x := p_{\rm F}/(m_ec)$ is a dimensionless version of the Fermi momentum, and $\lambda_e := \hbar/(m_ec)$ is the electron's Compton wavelength. The electron gas is relativistic when $x > 1$, and Eq. (2.86) indicates that the average inter-particle distance $n_e^{-1/3}$ is then smaller than the Compton wavelength.

The integral for the pressure produces

$$p = \frac{m_ec^2}{3\pi^2\lambda_e^3}\phi(x)\,, \tag{2.87}$$

where

$$\phi(x) := \int_0^x \frac{y^4\,dy}{\sqrt{1+y^2}} = \frac{3}{8}\left[x\left(\tfrac{2}{3}x^2 - 1\right)\sqrt{1+x^2} + \ln\left(x + \sqrt{1+x^2}\right)\right]. \tag{2.88}$$

The function $\phi(x)$ admits the approximations

$$\phi(x) = \frac{1}{5}x^5 - \frac{1}{14}x^7 + O(x^9), \tag{2.89}$$

when $x \ll 1$, and

$$\phi(x) = \frac{1}{4}x^4 - \frac{1}{4}x^2 + O(1), \tag{2.90}$$

when $x \gg 1$. The equation of state is obtained by combining Eq. (2.86) for $n_e(x)$ with Eq. (2.87) for $p(x)$; in its exact formulation the equation of state is expressed in parametric form.

For our purposes we must relate the pressure to the mass density ρ instead of the number density n_e. To achieve this we return to the discussion of Sec. 1.4.2, and re-introduce the atomic mass unit $m_{\rm H}$ and the mean molecular weight per electron μ_e; this, we recall, is defined by the relation $\mu_e^{-1} := \sum_i Z_i X_i/\mathcal{A}_i$, in which Z_i is the atomic number of an ion of type i, X_i is the mass fraction of this ion, and $\mathcal{A}_i$ is its atomic mass number. In terms of these quantities we have that $\rho = \mu_e m_{\rm H} n_e$. Combining this with our previous results, we find that the equation of state of a white dwarf is given by

$$\rho = \frac{\mu_e m_{\rm H}}{3\pi^2\lambda_e^3} x^3, \qquad p = \frac{m_e c^2}{3\pi^2\lambda_e^3}\phi(x), \tag{2.91}$$

with $\phi(x)$ given by Eq. (2.88). For a typical white dwarf containing mostly ^{12}C ($Z = 6$, $\mathcal{A} = 12$) and ^{16}O ($Z = 8$, $\mathcal{A} = 16$), the mean molecular weight is $\mu_e = 2$.

In the non-relativistic regime $x \ll 1$ the approximation of Eq. (2.89) can be exploited, and the equation of state simplifies to

$$p = K\rho^{5/3}, \qquad K = \frac{(3\pi^2)^{2/3}\hbar^2}{5\mu_e^{5/3} m_e m_{\rm H}^{5/3}}; \tag{2.92}$$

this is a polytropic equation of state with $\Gamma = \frac{5}{3}$. In the extreme relativistic regime $x \gg 1$ the approximation of Eq. (2.89) takes over, and the equation of state simplifies to

$$p = K'\rho^{4/3}, \qquad K' = \frac{(3\pi^2)^{1/3}\hbar c}{4\mu_e^{4/3} m_{\rm H}^{4/3}}; \tag{2.93}$$

this is another polytropic equation of state with $\Gamma = \frac{4}{3}$.

Equations of structure

The scaling quantities relevant to a white-dwarf equilibrium are defined by

$$p_0 := \frac{m_e c^2}{3\pi^2\lambda_e^3}, \tag{2.94a}$$

$$\rho_0 := \frac{\mu_e m_{\rm H}}{3\pi^2\lambda_e^3}, \tag{2.94b}$$

$$r_0^2 := \frac{1}{f_c^2}\frac{p_0}{4\pi G\rho_0^2}, \tag{2.94c}$$

$$m_0 := 4\pi f_c^3 \rho_0 r_0^3, \tag{2.94d}$$

where the numerical factor f_c, which is defined precisely below, is introduced for convenience. A useful relation among these quantities is $Gm_0/r_0 = f_c p_0/\rho_0$. The length scale can be expressed as

$$r_0 = \frac{\sqrt{3\pi}}{2f_c\mu_e}\frac{m_{\rm Pl}}{m_{\rm H}}\lambda_e = 3.88466\times 10^6\,\frac{1}{f_c}\left(\frac{2}{\mu_e}\right)\,\text{m}, \tag{2.95}$$

where $m_{\rm Pl} := \sqrt{\hbar c/G}$ is the Planck mass, and the mass scale can be expressed as

$$m_0 = \frac{\sqrt{3\pi}}{2\mu_e^2} \frac{m_{\rm Pl}^3}{m_{\rm H}^2} = 0.721459 \left(\frac{2}{\mu_e}\right)^2 M_\odot \,. \tag{2.96}$$

The numerical value of the density scale is $\rho_0 = 1.94787 \times 10^9\, (\mu_e/2)\,\text{kg/m}^3$.

The scale-free, dimensionless variables are

$$\xi := r/r_0\,, \qquad x^3 := \rho/\rho_0\,, \qquad \phi(x) := p/p_0\,, \qquad \mu := m/m_0\,, \tag{2.97}$$

where $\phi(x)$ is the function defined in Eq. (2.88). In terms of these variables, the equations of hydrostatic equilibrium, Eqs. (2.29), become

$$\frac{dx}{d\xi} = -f_c \frac{\sqrt{1+x^2}}{x} \frac{\mu}{\xi^2}\,, \qquad \frac{d\mu}{d\xi} = \frac{1}{f_c^3}\xi^2 x^3\,. \tag{2.98}$$

The equations are integrated outward from $\xi = 0$, with boundary conditions $x(0) = x_c$ (related to the central density) and $\mu(0) = 0$. Integration proceeds until $x = 0$ at the stellar boundary $\xi = \xi_1$.

In place of x it is convenient to adopt a variable θ that is similar to the Lane–Emden variable introduced in Sec. 2.2.3. This is defined by

$$1 + f_c\theta := \sqrt{1+x^2}\,, \tag{2.99}$$

where

$$f_c := \sqrt{1+x_c^2} - 1 \tag{2.100}$$

is the precise definition for the numerical factor introduced previously. The new density variable θ begins at 1 when $\xi = 0$ and $x = x_c$, and it drops to zero when $\xi = \xi_1$ and $x = 0$. With θ substituting for x, the structure equations become

$$\frac{d\theta}{d\xi} = -\frac{\mu}{\xi^2}\,, \qquad \frac{d\mu}{d\xi} = \xi^2\theta^{3/2}\left(\theta + 2/f_c\right)^{3/2}\,. \tag{2.101}$$

As usual the equations can be combined into a single, second-order differential equation for θ:

$$\frac{1}{\xi^2}\frac{d}{d\xi}\left(\xi^2\frac{d\theta}{d\xi}\right) = -\theta^{3/2}\left(\theta + 2/f_c\right)^{3/2}\,. \tag{2.102}$$

The equations of white-dwarf structure bear a striking resemblance to the polytropic equations displayed in Eqs. (2.49) and (2.54). There is, however, a major difference: in the case of polytropes the structure equations were completely universal and independent of ρ_c; here the equations feature a direct dependence upon f_c, which is tied to the central density. For a selected value of f_c, a solution to Eqs. (2.101) or (2.102) yields a unique white-dwarf model with a radius and mass given by

$$R = \xi_1 r_0\,, \qquad M = \mu_1 m_0\,; \tag{2.103}$$

here $\mu_1 := \mu(\xi_1)$ is the dimensionless mass function evaluated at the stellar boundary.

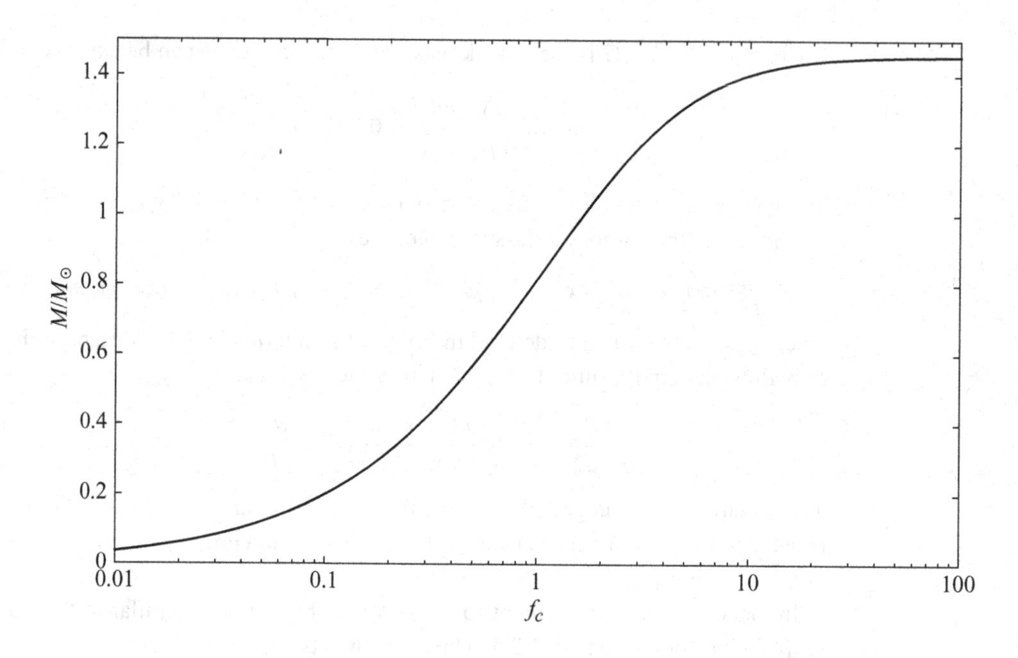

Fig. 2.3 Mass of a white dwarf, in units of the solar mass, as a function of the central-density parameter f_c. Models with $f_c \ll 1$ are non-relativistic, while models with $f_c \gg 1$ are extremely relativistic. The mass increases with f_c, and it asymptotes to the Chandrasekhar limit as $f_c \to \infty$.

Properties of white dwarfs

In Figs. 2.3 and 2.4 we display the results of a numerical integration of Eqs. (2.101) for many selected values of f_c. Models with $f_c \ll 1$ are non-relativistic, while models with $f_c \gg 1$ are extremely relativistic. In all cases the mean molecular weight per electron is set equal to $\mu_e = 2$. The main features revealed by the plots are that the mass increases with f_c, but never beyond a limit of approximately $1.46 M_\odot$, and that the radius decreases with increasing mass.

In the non-relativistic regime $f_c \ll 1$, Eqs. (2.101) simplify to $d\theta/d\xi = -\mu/\xi^2$ and $d\mu/d\xi = (2/f_c)^{3/2}\xi^2\theta^{3/2}$. With the rescalings

$$\xi = \left(\tfrac{1}{2}f_c\right)^{3/4}\bar{\xi}\,, \qquad \theta = \bar{\theta}\,, \qquad \mu = \left(\tfrac{1}{2}f_c\right)^{3/4}\bar{\mu}\,, \tag{2.104}$$

the equations become $d\bar{\theta}/d\bar{\xi} = -\bar{\mu}/\bar{\xi}^2$ and $d\bar{\mu}/d\bar{\xi} = \bar{\xi}^2\bar{\theta}^{3/2}$, and these are precisely the equations that govern the hydrostatic equilibrium of a polytrope with $n = \frac{3}{2}$. The essential properties of the solution are displayed in Table 2.1, which reveals that $\bar{\xi}_1 = 3.65375374$ and $\bar{\mu}_1 = 2.71405512$. These results, together with the approximation $f_c = \frac{1}{2}x_c^2 = \frac{1}{2}(\rho_c/\rho_0)^{2/3}$, imply that the mass and radius of a white dwarf are given by

$$M = 0.496028\left(\frac{2}{\mu_e}\right)^{5/2}\left(\frac{\rho_c}{10^9\ \mathrm{kg/m^3}}\right)^{1/2} M_\odot \tag{2.105}$$

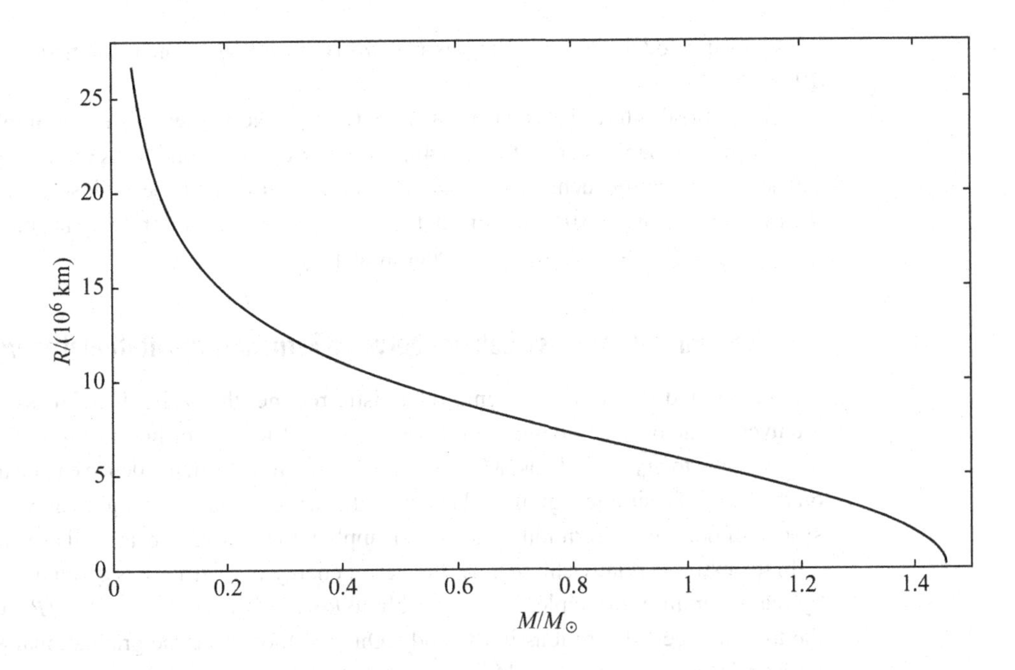

Fig. 2.4 Radius of a white dwarf, in units of 10^6 km, as a function of its mass, in units of the solar mass. The radius decreases with increasing mass, and goes to zero as the mass reaches the Chandrasekhar limit.

and

$$R = 1.12160 \times 10^7 \left(\frac{2}{\mu_e}\right)^{5/6} \left(\frac{10^9\ \text{kg/m}^3}{\rho_c}\right)^{1/6} \text{m} \tag{2.106}$$

in the non-relativistic regime. These results reveal that M increases with the central density as $\rho_c^{1/2}$, while R decreases as $\rho_c^{-1/6}$.

In the extreme relativistic regime $f_c \gg 1$, Eqs. (2.101) simplify to $d\theta/d\xi = -\mu/\xi^2$ and $d\mu/d\xi = \xi^2\theta^3$; in this limit the white dwarf is a polytrope with $n = 3$. In this case Table 2.1 reveals that $\xi_1 = 6.89684862$ and $\mu_1 = 2.01823595$. Together with the approximation $f_c = x_c = (\rho_c/\rho)^{1/3}$, these results imply that the mass and radius of a white dwarf are given by

$$M_{\text{Chandra}} = 1.45607 \left(\frac{2}{\mu_e}\right)^2 M_\odot \tag{2.107}$$

and

$$R = 3.34598 \times 10^7 \left(\frac{2}{\mu_e}\right)^{2/3} \left(\frac{10^9\ \text{kg/m}^3}{\rho_c}\right)^{1/3} \text{m} \tag{2.108}$$

in the extreme relativistic regime. These expressions reveal that R decreases as $\rho_c^{-1/3}$ with increasing central density, but that *the mass is independent of the central density*. The numerical results show that Eq. (2.107) is the maximum mass realized on the sequence of white-dwarf models; it is the famous *Chandrasekhar limit*, which was first discovered in 1930.

For a typical white dwarf of mass $M = 0.6M_\odot$, the numerical results imply that the central-density parameter is $f_c = 0.546$, revealing that the conditions are only mildly relativistic. The central density is $\rho_c = 3.19 \times 10^9\ \text{kg/m}^3$, and the radius of such a white dwarf is $R = 8.85 \times 10^6$ km. From these values we can infer that the mean density $\bar{\rho} = 3M/(4\pi R^3)$ is 0.129 times the central density.

Chandrasekhar mass: balance between Fermi and gravitational energies

We have found that in the extreme relativistic regime, the white-dwarf mass is equal to a universal factor multiplying the mass scale m_0, which, according to Eq. (2.96), is itself proportional to $m_{\rm Pl}^3/m_{\rm H}^2$. Thus, $M_{\rm Chandra} \propto m_{\rm Pl}^3/m_{\rm H}^2$, and this dependence can be understood on the basis of a simple argument that originates with Landau. The argument works for any system of degenerate fermions, and we can apply it to neutron stars as well as white dwarfs.

In the extreme relativistic regime, the Fermi energy of a fermion is given by $p_{\rm F}c$, which, by the uncertainty principle, is comparable to $\hbar c/\Delta x \sim \hbar c n^{1/3} \sim \hbar c N^{1/3}/R$, where N is the total number of fermions in the body. On the other hand, the gravitational energy per fermion is approximately $-GMm_{\rm H}/R$, with $M \sim Nm_{\rm H}$. (When the argument is applied to neutrons, the gravitational energy is dominated by the neutrons, and $m_{\rm H}$ is approximately the neutron's mass. When the argument is applied to electrons, the gravitational energy is dominated by the ions, and a factor of μ_e should be inserted to relate the number of fermions to the mass. In both cases the relevant mass is the atomic mass unit $m_{\rm H}$.) Collecting results, we find that the total energy of the system may be expressed as

$$E \sim \frac{\hbar c N^{4/3}}{R} - \frac{GN^2 m_{\rm H}^2}{R}. \tag{2.109}$$

The key point is that each term is inversely proportional to R. Now, the stability of the configuration is dictated by the overall sign of the total energy. When E is positive, a decrease in R produces an increase in energy, and this behavior indicates stability with respect to gravitational collapse. When E is negative, a decrease in R produces a decrease in energy, and this points to an instability.

Stability under gravitational collapse therefore places an upper limit on the number of fermions that can be relativistically degenerate. This limit is given by $\hbar c N^{4/3} > GN^2 m_{\rm H}^2$, or $N^{2/3} < (m_{\rm Pl}/m_{\rm H})^2$, recalling that $m_{\rm Pl} := \sqrt{\hbar c/G}$ is the Planck mass. We then find that the body's maximum mass is given by

$$M_{\rm max} \sim N_{\rm max} m_{\rm H} \sim \frac{m_{\rm Pl}^3}{m_{\rm H}^2}, \tag{2.110}$$

which is the same scaling as m_0, as defined by Eq. (2.96).

The value of the maximum mass does not depend on the identity of the degenerate fermion; apart from numerical factors, the result should be the same for electrons and for neutrons. So, if neutron stars were subjected to Newtonian gravity, and if their equation of state were indeed that of an ideal gas of degenerate fermions, then they would have a maximum mass also given by Eq. (2.107), but with μ_e replaced by μ_n, whose value depends on the ratio of free neutrons to neutrons still bound in nuclei. The catch, of course, is that neutron stars are *not* subjected to Newtonian gravity: because $\varepsilon \sim GM/(c^2 R) \sim 0.2$ for neutron stars, the gravitational fields are too strong for a Newtonian treatment, and they must be described by general relativity. The other catch is that neutron stars are not actually made up of an ideal gas of degenerate neutrons: the densities are so high (of the order of 10^{17} kg/m^3) that the neutrons are subjected to short-range strong interactions, beta decay and inverse beta decay, and the possible conversion to exotic particles such as pion condensates and strange hadrons. The neutrons are not free, and their interactions at such high densities are poorly understood; as a consequence, the maximum mass of neutron stars is somewhat uncertain, lying between about 2 and 3 solar masses.

2.3 Rotating self-gravitating bodies

The equilibrium structures examined in Sec. 2.2 were all spherically symmetric and non-rotating. Rotation, however, is everywhere: planets rotate, the Sun rotates, stars and galaxies rotate. Neutron stars observed as pulsars rotate; one of the fastest, PSR J1748-2446ad, spins at a rate of 716 times per second. Black holes are expected to have substantial rotation. Because astronomical objects are formed by gravitational collapse, the slightest amount of angular momentum in the progenitor material results in a rotational angular velocity that is magnified by the collapse, limited only by the shedding of material when the centrifugal forces exceed the gravitational forces.

Our previous models of non-rotating, spherically symmetric, self-gravitating bodies allowed us to explore many basic properties of stars and planets, but the idealization is not appropriate when applied to real astronomical bodies. We must consider rotation, and the related complication that rotating bodies are not spherical. And when we do so, the number of basic questions increases greatly, and their answers depend on the detailed nature of the body.

Do bodies rotate rigidly? That is, do they rotate with an angular velocity that is independent of position within the body? It turns out that this is impossible under some conditions, and the body must then undergo *differential rotation*, or must develop meridional currents, fluid flows in a direction parallel to the rotation axis. The Sun rotates differentially; the angular velocity in the polar regions is about 2/3 that at the equator, and the angular velocity at the center could be as much as 10 times higher than that at the surface.

What is the shape of a rotating fluid body? In a general situation the shape cannot be fixed *a priori*, and must be determined along with the rest of the body's structure. To keep the problem tractable, however, it is helpful to adopt a specific class of shapes for the body, and then determine whether these shapes are compatible with the equations of stellar structure. For example, the simplest shape one might expect for a rotating body is one for which one or more of the three cross sections has the form of an ellipse. If one of the cross sections is circular while the other two are elliptical, the shape is known as a *spheroid*;

if all three cross sections are elliptical, the shape is an *ellipsoid.* We shall see below that in the special case of a rigidly rotating body of uniform density, the equations of stellar structure are compatible with a spheroidal or ellipsoidal shape, provided that the angular velocity does not exceed a certain limit; beyond this maximum value, differential rotation is required to support the body. In the limit of slow rotation, for general density distributions and equations of state, we show in Sec. 2.4 that the shape is spheroidal to the first order of approximation. In general, however, the shape of a rotating body is neither a spheroid nor an ellipsoid, and it must be determined by solving the equations of stellar structure with no built-in assumptions – a most difficult task.

Are rotating bodies axially symmetric? Not necessarily: a rotating body can be in hydrostatic equilibrium and assume a shape that is not symmetric about the axis of rotation. In fact, we shall see that one can construct an equilibrium model of a rotating body that has two equally valid solutions, one an axisymmetric spheroid, and the other a non-axisymmetric ellipsoid; and one can show that some of the spheroids are subjected to instabilities that convert them into ellipsoids.

In this section we take advantage of simple models to address several of these questions.

2.3.1 Foundations of the theory of rotating bodies

Rotating frame

We again consider an isolated body, governed by Eq. (2.4), in which we continue to ignore the external terms. But now we assume that $\bar{\boldsymbol{v}}$, the velocity of a fluid element relative to the body's center-of-mass, can be decomposed as

$$\bar{\boldsymbol{v}} = \boldsymbol{\omega} \times \bar{\boldsymbol{x}} + \boldsymbol{u} . \tag{2.111}$$

The first term is a rotational velocity, which refers to a local frame of reference that rotates with an instantaneous angular velocity $\boldsymbol{\omega}(t, \boldsymbol{x})$ with respect to the center-of-mass frame; the angular velocity may in general depend on time and spatial position, but we assume that its direction is along a fixed axis described by the unit vector $\boldsymbol{e}$. The second term is the intrinsic velocity of the fluid element in the rotating frame of reference.

The acceleration of the fluid element is then given by

$$\begin{aligned}
\frac{d\bar{\boldsymbol{v}}}{dt} &= \boldsymbol{\omega} \times \bar{\boldsymbol{v}} + \frac{d\boldsymbol{\omega}}{dt} \times \bar{\boldsymbol{x}} + \frac{d\boldsymbol{u}}{dt} \\
&= \boldsymbol{\omega} \times (\boldsymbol{\omega} \times \bar{\boldsymbol{x}}) + \boldsymbol{\omega} \times \boldsymbol{u} + \frac{d\boldsymbol{\omega}}{dt} \times \bar{\boldsymbol{x}} + \frac{d\boldsymbol{u}}{dt} \\
&= -\omega^2 \boldsymbol{\varrho} + 2\boldsymbol{\omega} \times \boldsymbol{u} + \frac{d\boldsymbol{\omega}}{dt} \times \bar{\boldsymbol{x}} + \boldsymbol{a} ,
\end{aligned} \tag{2.112}$$

where we have used the fact that $d\boldsymbol{u}/dt = \boldsymbol{\omega} \times \boldsymbol{u} + \boldsymbol{a}$, with $\boldsymbol{a}$ denoting the acceleration in the rotating frame, and where

$$\boldsymbol{\varrho} := \bar{\boldsymbol{x}} - \boldsymbol{e}(\boldsymbol{e} \cdot \bar{\boldsymbol{x}}) \tag{2.113}$$

is the projection of the position vector to the plane perpendicular to the rotation axis. If we align the $\bar{z}$-direction with $\boldsymbol{e}$, then $\boldsymbol{\varrho} = [\bar{x}, \bar{y}, 0]$ and $\varrho := |\boldsymbol{\varrho}| = (\bar{x}^2 + \bar{y}^2)^{1/2}$ is the usual cylindrical radial coordinate of a $(\varrho, \bar{z}, \bar{\phi})$ coordinate system. The first term in Eq. (2.112) is the centrifugal acceleration, the second is the Coriolis acceleration, present if the fluid has intrinsic motion in the rotating frame; the third term is a rotational acceleration effect, and the final term is the intrinsic acceleration of the element in the rotating frame.

Steady rotation; equilibrium configuration

We now introduce some simplifying assumptions. First, the rotation is taken to be steady, in the sense that $\boldsymbol{\omega}$ is independent of time. Second, the fluid motion is taken to be purely rotational, in the sense that $\boldsymbol{u} = 0$. Third, the system is taken to be stationary when viewed by a non-rotating observer, so that $\partial\rho/\partial t = 0$. And fourth, the density is also taken to be constant when viewed by an observer following the fluid, so that $d\rho/dt = 0$.

The last two assumptions, combined with the equation of continuity, $\partial\rho/\partial t + \nabla\cdot(\rho\bar{\boldsymbol{v}}) = 0$ or $d\rho/dt + \rho\nabla\cdot\bar{\boldsymbol{v}} = 0$, imply that

$$\bar{\boldsymbol{v}}\cdot\nabla\rho = 0 \quad \text{and} \quad \nabla\cdot\bar{\boldsymbol{v}} = 0\,. \tag{2.114}$$

The second assumption implies that $\bar{\boldsymbol{v}} = \boldsymbol{\omega}\times\bar{\boldsymbol{x}}$, and it follows that

$$\begin{aligned} 0 &= (\boldsymbol{\omega}\times\bar{\boldsymbol{x}})\cdot\nabla\rho \\ &= \boldsymbol{\omega}\cdot(\bar{\boldsymbol{x}}\times\nabla)\rho \\ &= \omega\frac{\partial\rho}{\partial\bar{\phi}}\,. \end{aligned} \tag{2.115}$$

This is the statement that the mass density of a rotating body must be axially symmetric, independent of the azimuthal angle $\bar{\phi}$. We also have

$$\begin{aligned} 0 &= \nabla\cdot\bar{\boldsymbol{v}} \\ &= \nabla\cdot(\boldsymbol{\omega}\times\bar{\boldsymbol{x}}) \\ &= (\bar{\boldsymbol{x}}\times\nabla)\cdot\boldsymbol{\omega} \\ &= \frac{\partial\omega}{\partial\bar{\phi}}\,, \end{aligned} \tag{2.116}$$

the statement that the angular velocity also must be independent of $\bar{\phi}$.

With these assumptions, Euler's equation, Eq. (2.4), reduces to

$$\frac{1}{\rho}\nabla p = \nabla U + \omega(\bar{x})^2\boldsymbol{\varrho}\,. \tag{2.117}$$

This equation, together with Poisson's equation for the gravitational potential U, governs the equilibrium structure of our stationary rotating body.

Stratification

As we saw in Sec. 2.2, a feature of spherical bodies is that the surfaces of constant ρ, constant p, and constant U all coincide, because these surfaces are all spherical. In the case of a rotating body, is it still true that these level surfaces coincide? We wish to show that under some conditions on the angular velocity, surfaces of constant ρ and constant p continue to coincide, and there also exists a generalized potential Φ with coincident level surfaces. We shall find, however, that the surfaces of constant U do not in general coincide with the other level surfaces.

To explore this, we take the curl of Eq. (2.117) and obtain

$$-\frac{1}{\rho^2}\boldsymbol{\nabla}\rho \times \boldsymbol{\nabla}p = 2\omega\,\boldsymbol{\nabla}\omega \times \boldsymbol{\varrho}\,. \tag{2.118}$$

Now, surfaces of constant ρ and constant p will coincide when $\boldsymbol{\nabla}\rho \times \boldsymbol{\nabla}p = \boldsymbol{0}$, which implies (assuming that $\omega \neq 0$) that $\boldsymbol{\nabla}\omega \times \boldsymbol{\varrho} = \boldsymbol{0}$. Since we already know that $\partial\omega/\partial\bar{\phi} = 0$, this implies that $\partial\omega/\partial\bar{z} = 0$ and we conclude that $\omega = \omega(\varrho)$. We have found that when the body rotates uniformly ($\omega =$ constant), or when its angular velocity depends on ϱ only, then the surfaces of constant ρ and constant p coincide.

When ω is either constant or a function of ϱ only, we can further show that there exists a *centrifugal potential* $\Phi_{\rm C}$ associated with the centrifugal acceleration term in Eq. (2.117). This follows because the centrifugal acceleration has a vanishing curl,

$$\boldsymbol{\nabla} \times (\omega^2\boldsymbol{\varrho}) = 2\omega\boldsymbol{\nabla}\omega \times \boldsymbol{\varrho} = \boldsymbol{0}\,, \tag{2.119}$$

and is therefore given by the gradient of a scalar function. The potential is given by

$$\Phi_{\rm C}(\varrho) := \int^{\varrho} \omega(\varrho')^2\varrho'd\varrho'\,, \tag{2.120}$$

and a generalized potential

$$\Phi := U + \Phi_{\rm C} \tag{2.121}$$

can be introduced to simplify the form of the structure equations, which become

$$\boldsymbol{\nabla}p = \rho\boldsymbol{\nabla}\Phi\,. \tag{2.122}$$

Taking the curl of this equation, we conclude finally that with the stated assumptions, the surfaces of constant p, constant ρ, and constant Φ all coincide. Notice that this statement does not imply that surfaces of constant U and constant $\Phi_{\rm C}$ coincide with each other, nor with the level surfaces of the total potential Φ.

Global properties

Back in Sec. 1.4.3 we established the virial theorem of stationary fluid configurations,

$$2\mathcal{T} + \Omega + 3P = 0\,, \tag{2.123}$$

where $\mathcal{T}$ is the total kinetic energy, Ω the gravitational potential energy, and P the integrated pressure; these quantities are defined by Eqs. (1.64), (1.65), and (1.85), and the virial theorem was first stated in Eq. (1.90). Since $P \geq 0$ and Ω is negative, we can immediately derive a bound on the ratio of kinetic to potential energy for a rotating body:

$$\tau := \frac{\mathcal{T}}{|\Omega|} \leq \frac{1}{2}\,. \tag{2.124}$$

For a rotating body with $\bar{v} = \omega \times \bar{x}$, the kinetic energy is given by

$$\begin{aligned} \mathcal{T} &= \frac{1}{2}\int \rho \bar{v}^2 \, d^3\bar{x} \\ &= \frac{1}{2}\int \rho\omega^2\big[\bar{r}^2 - (e\cdot\bar{x})^2\big]\, d^3\bar{x} \\ &= \frac{1}{2}\int \rho\omega^2\varrho^2 \, d^3\bar{x} \,. \end{aligned} \tag{2.125}$$

The angular momentum is given by

$$\begin{aligned} J &= \int \rho\bar{x}\times(\omega\times\bar{x}) d^3\bar{x} \\ &= \int \rho\omega\big[e\bar{r}^2 - \bar{x}(e\cdot\bar{x})\big] d^3\bar{x} \\ &= e\int \rho\omega\varrho^2 d^3\bar{x} \,, \end{aligned} \tag{2.126}$$

where we use the axisymmetry of ρ and ω to obtain the final result. When the rotation is uniform, so that $\omega =$ constant, these equations simplify to

$$\mathcal{T} = \frac{1}{2}\omega^2 I \,, \qquad J = \omega I e \,, \tag{2.127}$$

where

$$I := \int \rho\varrho^2 \, d^3\bar{x} \tag{2.128}$$

is the body's moment of inertia about the axis of rotation.

We continue to assume that the rotation is uniform. In this case the centrifugal potential simplifies to

$$\Phi_C = \frac{1}{2}\omega^2\varrho^2 \,, \tag{2.129}$$

and from this we find that $\nabla^2\Phi_C = 2\omega^2$. For the total potential of Eq. (2.121) we then get $\nabla^2\Phi = -4\pi G\rho + 2\omega^2$. Integrating this equation over the body and using Gauss's theorem, we find that

$$\oint \nabla\Phi\cdot n dS = -4\pi GM + 2\omega^2 V \,, \tag{2.130}$$

where M and V are the mass and volume of the body, respectively, and n is an outward-pointing unit vector on the body's surface. But Eq. (2.122) implies $\nabla\Phi\cdot n = \rho^{-1}\nabla p\cdot n$, which must be negative everywhere on the surface, because p vanishes there and is positive everywhere inside. As a result, we obtain the Poincaré inequality

$$\omega^2 < 2\pi G\bar{\rho} \,, \tag{2.131}$$

where $\bar{\rho} = M/V$ is the body's mean density. Recall that the angular velocity required for a particle to be on a circular orbit at the body's equator is given approximately by $(d\phi/dt)^2 \sim GM/r_{\rm eq}^3 \sim G\bar{\rho}$, with $r_{\rm eq}$ denoting the equatorial radius. The condition

$\omega^2 \sim G\bar{\rho}$ marks the *mass-shedding limit*, because a body rotating faster than this will shed matter from its equatorial regions. The inequality of Eq. (2.131) is therefore the statement that a rotating body in equilibrium cannot exceed the mass-shedding limit. The Poincaré inequality provides a rather generous upper bound on a body's angular velocity, and we shall find that actual bounds are much tighter; the numerical coefficient on the right-hand side of Eq. (2.131) can be reduced by a factor approximately equal to 4.

Transformation to the rotating frame

We have developed the structure equations of a rotating body by referring to the rotating frame to measure the velocity $\boldsymbol{u}$ and acceleration $\boldsymbol{a}$ of a fluid element. We have, however, continued to measure the position $\bar{\boldsymbol{x}}$ relative to the non-rotating frame, and to conclude this section we wish to complete the transformation to the rotating frame. We shall now relax most of the assumptions introduced previously, except for the rather restrictive one that ω is taken to be a constant (independent of time and spatial position).

The original frame $\bar{\boldsymbol{x}} = (\bar{x}, \bar{y}, \bar{z})$ is attached to the body's center-of-mass, and is non-rotating. The new frame $\boldsymbol{x} = (x, y, z)$ that we introduce here is also attached to the center-of-mass, but is rotating rigidly with an angular velocity ω. We align the $\bar{z}$-direction with the rotation axis, and the transformation between the two frames is given by

$$\bar{x} = x\cos(\omega t) - y\sin(\omega t), \qquad \bar{y} = x\sin(\omega t) + y\cos(\omega t), \qquad \bar{z} = z. \tag{2.132}$$

The inverse transformation is obtained simply by reversing the sign of ω. For our developments it is useful to employ an index language in which $\bar{\boldsymbol{x}}$ is represented by the coordinates $x^{\bar{a}}$ while $\boldsymbol{x}$ is represented by x^j. The transformation can then be expressed as

$$x^{\bar{a}} = \Lambda^{\bar{a}}{}_j(t)x^j, \tag{2.133}$$

in which $\Lambda^{\bar{a}}{}_j(t)$ are the elements of the rotation matrix. The inverse transformation is

$$x^j = \Lambda^j{}_{\bar{a}}(t)x^{\bar{a}}, \tag{2.134}$$

with $\Lambda^j{}_{\bar{a}}(t)$ denoting the elements of the inverse matrix. The identities

$$\Lambda^{\bar{a}}{}_j\Lambda^j{}_{\bar{b}} = \delta^{\bar{a}}{}_{\bar{b}}, \qquad \Lambda^j{}_{\bar{a}}\Lambda^{\bar{a}}{}_k = \delta^j{}_k \tag{2.135}$$

express the fact that a transformation followed by its inverse is an identity transformation.

Suppose that we follow the motion of a fluid element. In the non-rotating frame the motion is described by $x^{\bar{a}}(t)$, while it is described by $x^j(t)$ in the rotating frame. The velocity of the fluid element is $v^{\bar{a}} = dx^{\bar{a}}/dt$ in the non-rotating frame, and $u^j = dx^j/dt$ in the rotating frame. The relation is provided by differentiating Eq. (2.133) with respect to time, and we obtain

$$v^{\bar{a}} = \Lambda^{\bar{a}}{}_j\left(u^j + \omega^j{}_k x^k\right), \tag{2.136}$$

where

$$\omega^j{}_k := \Lambda^j{}_{\bar{a}}\dot{\Lambda}^{\bar{a}}{}_k \tag{2.137}$$

is the angular-velocity tensor (the overdot on the right-hand side indicates differentiation with respect to time). A simple calculation reveals that

$$\omega_{jk} = -\epsilon_{jkn}\omega^n , \tag{2.138}$$

where ϵ_{jkn} is the permutation symbol and $\omega^n = \omega e^n$ are the components of the angular-velocity vector. Because $\omega^j{}_k x^k = (\boldsymbol{\omega} \times \boldsymbol{x})^j$, we see that Eq. (2.136) is essentially the same statement as Eq. (2.111). Differentiation of Eq. (2.136) then produces

$$\frac{dv^{\bar{a}}}{dt} = \Lambda^{\bar{a}}{}_j \left(a^j + 2\omega^j{}_k u^k - C^j{}_k x^k\right), \tag{2.139}$$

in which $a^j := du^j/dt$ is the acceleration of the fluid element in the rotating frame, and

$$C_{jk} := -\omega_{jn}\omega^n{}_k = \omega^2\left(\delta_{jk} - e_j e_k\right) \tag{2.140}$$

is the centrifugal tensor. The second term on the right-hand side of Eq. (2.139) is recognized as the Coriolis acceleration, while the third term is the centrifugal acceleration. Equation (2.139) is essentially the same statement as Eq. (2.112) when the angular velocity is restricted to be a constant.

We now wish to formulate the equations of fluid dynamics in the rotating frame. A subtle issue concerns the meaning of partial derivatives, because differentiating with respect to t while keeping $\bar{\boldsymbol{x}}$ constant (for a non-rotating observer) is very different from performing the operation while keeping $\boldsymbol{x}$ constant (for a rotating observer). The relation can be uncovered by expressing a function f of the variables $(t, x^{\bar{a}})$ as $f(t, \Lambda^{\bar{a}}{}_j x^j)$ and performing the differentiations. We obtain

$$\left(\frac{\partial f}{\partial t}\right)_{x^{\bar{a}}} = \left(\frac{\partial f}{\partial t}\right)_{x^j} - \left(\frac{\partial f}{\partial x^j}\right)\omega^j{}_k x^k , \tag{2.141a}$$

$$\left(\frac{\partial f}{\partial x^{\bar{a}}}\right) = \left(\frac{\partial f}{\partial x^j}\right)\Lambda^j{}_{\bar{a}} , \tag{2.141b}$$

in which the subscript on the time derivative indicates which variable is kept fixed.

With these rules it is easy to show that the continuity equation keeps its usual form in the rotating frame,

$$\left(\frac{\partial \rho}{\partial t}\right)_{x^j} + \frac{\partial}{\partial x^j}\left(\rho u^j\right) = 0 , \tag{2.142}$$

that Euler's equation becomes

$$\frac{du^j}{dt} = \partial_j U - \frac{1}{\rho}\partial_j p - 2\omega^j{}_k u^k + C^j{}_k x^k , \tag{2.143}$$

and that Poisson's equation also keeps its usual form,

$$\nabla^2 U = -4\pi G\rho , \tag{2.144}$$

in which the Laplacian operator refers to the rotating frame x^j.

When the configuration is stationary in the rotating frame ($u^j = 0$), Euler's equation reduces to

$$\frac{1}{\rho}\partial_j p = \partial_j \Phi\,, \tag{2.145}$$

where $\Phi = U + \Phi_C$ is the total potential introduced previously, with

$$\Phi_C = \frac{1}{2} C_{jk} x^j x^k = \frac{1}{2}\omega^2 \left(x^2 + y^2\right) \tag{2.146}$$

the centrifugal potential of Eq. (2.129), now written in terms of the rotating-frame coordinates x^j.

2.3.2 Rotating bodies of uniform density

Back in Sec. 2.2.2 we found that the mathematics of spherical bodies of uniform density were particularly simple, though physically unrealistic. In a similar way, rigidly rotating bodies of uniform density admit a relatively simple treatment in Newtonian theory, and the results can be expressed in terms of simple integrals that, unfortunately, cannot in general be evaluated in closed form. Even though the assumption of uniform density is no more realistic here than it was in the case of a spherical body, the resulting models capture many important properties of realistic rotating bodies. Because the mathematics are relatively simple, these models attracted the attention of many of the great mathematical physicists of the 18th and 19th centuries, and many of the resulting models bear their names: the spheroids of Maclaurin, the ellipsoids of Jacobi, Dirichlet, Dedekind, and Riemann. Although incomplete, this body of work was not much pursued in the 20th century; as Chandrasekhar famously expressed it, "the subject quietly went into a coma." It was, however, vigorously revived by Lebovitz and Chandrasekhar in the nineteen sixties, and the entire subject came to a beautiful close in Chandra's (1987) monumental *Ellipsoidal Figures of Equilibrium*.

In this section we assume that the figure of a rigidly rotating body of uniform density is an ellipsoid, and seek to determine the conditions that ensure consistency with the equations of hydrostatic equilibrium.

Gravitational potential inside an ellipsoid

The hardest part of the task ahead is to calculate the gravitational potential inside a body of uniform density, under the assumption that its surface is described by the equation

$$\frac{x^2}{a_1^2} + \frac{y^2}{a_2^2} + \frac{z^2}{a_3^2} = 1\,, \tag{2.147}$$

in which the constants a_1, a_2, and a_3 are the ellipsoid's semi-axes in the three principal directions. The description is given in the body's rotating frame, which is assumed to rotate rigidly about the z-direction with an angular velocity ω. When $a_1 = a_2$ the body is axially symmetric about the z-direction, and the surface is a spheroid.

We state the answer before getting on with the work:

$$U = \pi G\rho\big(A_0 - A_1x^2 - A_2y^2 - A_3z^2\big), \tag{2.148}$$

where

$$A_0 := a_1a_2a_3 \int_0^\infty \frac{du}{\Delta} \tag{2.149}$$

and

$$A_i := a_1a_2a_3 \int_0^\infty \frac{du}{\Delta\,(a_i^2 + u)}, \tag{2.150}$$

with

$$\Delta^2 := \big(a_1^2 + u\big)\big(a_2^2 + u\big)\big(a_3^2 + u\big). \tag{2.151}$$

The integrals must be evaluated numerically when $a_1 \neq a_2$, but they admit closed-form expressions when the body is axisymmetric; these will be revealed below. The quadratic form of Eq. (2.148) is remarkably simple, and it is easy to show that U satisfies Poisson's equation (for a constant density) for any set of A_is that satisfies the constraint $A_1 + A_2 + A_3 = 2$; as we shall see, the set of Eq. (2.150) does indeed possess this property.

To calculate U we follow the general strategy devised by Moulton in his classic text on celestial mechanics. We let $\boldsymbol{x} = (x, y, z)$ be the point at which U is evaluated (which is interior to the ellipsoid), and we let $\boldsymbol{x}' = (x', y', z')$ be a source point within the ellipsoid. To locate $\boldsymbol{x}'$ we employ spherical coordinates $(\tilde{r}, \tilde{\theta}, \tilde{\phi})$ centered upon $\boldsymbol{x}$, so that

$$x' = x + \tilde{r}\sin\tilde{\theta}\cos\tilde{\phi}, \qquad y' = y + \tilde{r}\sin\tilde{\theta}\sin\tilde{\phi}, \qquad z' = z + \tilde{r}\cos\tilde{\theta}. \tag{2.152}$$

The distance between the two points is evidently $\tilde{r}$. The source point is on the surface when $x' = x_s$, $y' = y_s$, $z' = z_s$, with (x_s, y_s, z_s) a solution to Eq. (2.147). In terms of the spherical coordinates, the surface is described by the equation $\tilde{r} = r_s(\tilde{\theta}, \tilde{\phi})$, and the function r_s is determined by inserting Eqs. (2.152) within Eq. (2.147). This gives rise to the quadratic equation

$$\alpha r_s^2 + 2\beta r_s + \gamma = 0, \tag{2.153}$$

with

$$\alpha := \frac{\sin^2\tilde{\theta}\cos^2\tilde{\phi}}{a_1^2} + \frac{\sin^2\tilde{\theta}\sin^2\tilde{\phi}}{a_2^2} + \frac{\cos^2\tilde{\theta}}{a_3^2}, \tag{2.154a}$$

$$\beta := \frac{x\,\sin\tilde{\theta}\cos\tilde{\phi}}{a_1^2} + \frac{y\,\sin\tilde{\theta}\sin\tilde{\phi}}{a_2^2} + \frac{z\,\cos\tilde{\theta}}{a_3^2}, \tag{2.154b}$$

$$\gamma := \frac{x^2}{a_1^2} + \frac{y^2}{a_2^2} + \frac{z^2}{a_3^2} - 1. \tag{2.154c}$$

We see that $\alpha > 0$, and since $\boldsymbol{x}$ is an interior point, we have that $\gamma < 0$; the sign of β depends on $\tilde{\theta}$ and $\tilde{\phi}$. We note that while α and γ are preserved under a reflection $(\tilde{\theta}, \tilde{\phi}) \rightarrow (\pi - \tilde{\theta}, \tilde{\phi} + \pi)$ across the origin at $\boldsymbol{x}$, β changes sign under the reflection. The appropriate solution to the quadratic equation is

$$r_s = \frac{-\beta + \sqrt{\beta^2 - \alpha\gamma}}{\alpha}; \tag{2.155}$$

the other solution would produce a negative radius.

The gravitational potential is given by

$$U(\boldsymbol{x}) = G \int \frac{\rho(\boldsymbol{x}')}{|\boldsymbol{x} - \boldsymbol{x}'|} d^3x' = G\rho \int \tilde{r}\, d\tilde{r} d\tilde{\Omega}, \tag{2.156}$$

where we make use of the fact that $d^3x' = \tilde{r}^2\, d\tilde{r} d\tilde{\Omega}$, with $d\tilde{\Omega} := \sin\tilde{\theta}\, d\tilde{\theta} d\tilde{\phi}$. Integration with respect to the radial variable yields

$$U = \frac{1}{2} G\rho \int r_s^2(\tilde{\theta}, \tilde{\phi})\, d\tilde{\Omega}, \tag{2.157}$$

and the remaining integration is over the usual range of the angular coordinates. Insertion of Eq. (2.155) gives

$$U = \frac{1}{2} G\rho \left[\int \frac{2\beta^2 - \alpha\gamma}{\alpha^2} d\tilde{\Omega} - 2 \int \frac{\beta\sqrt{\beta^2 - \alpha\gamma}}{\alpha^2} d\tilde{\Omega} \right], \tag{2.158}$$

and with the stated properties of α, β, and γ under a reflection across $\boldsymbol{x}$, we see that the second integral vanishes because contributions from one hemisphere cancel out the contributions from the opposite hemisphere. The potential simplifies to

$$U = \frac{1}{2} G\rho \int \frac{2\beta^2 - \alpha\gamma}{\alpha^2} d\tilde{\Omega}. \tag{2.159}$$

When we expand $2\beta^2 - \alpha\gamma$ in full, we find that it is a quadratic form that involves diagonal terms proportional to x^2, y^2, and z^2, and non-diagonal terms proportional to xy, xz, and yz; each term is multiplied by a specific function of $\tilde{\theta}$ and $\tilde{\phi}$. Examining the non-diagonal terms closely, we find that once again the integrals vanish by symmetry. What is left over is the expression of Eq. (2.148), with

$$A_0 = \frac{1}{2\pi} \int \frac{d\tilde{\Omega}}{\alpha}, \tag{2.160a}$$

$$A_1 = \frac{1}{2\pi a_1^2} \int \left(1 - \frac{2\sin^2\tilde{\theta}\cos^2\tilde{\phi}}{a_1^2 \alpha} \right) \frac{d\tilde{\Omega}}{\alpha}, \tag{2.160b}$$

$$A_2 = \frac{1}{2\pi a_2^2} \int \left(1 - \frac{2\sin^2\tilde{\theta}\sin^2\tilde{\phi}}{a_2^2 \alpha} \right) \frac{d\tilde{\Omega}}{\alpha}, \tag{2.160c}$$

$$A_3 = \frac{1}{2\pi a_3^2} \int \left(1 - \frac{2\cos^2\tilde{\theta}}{a_3^2 \alpha} \right) \frac{d\tilde{\Omega}}{\alpha}, \tag{2.160d}$$

where α continues to be given by Eq. (2.154).

The remaining task is to evaluate the integrals and bring them to the standard form displayed in Eqs. (2.149) and (2.150). The good news is that only A_0 requires evaluation, because all other integrals can be obtained by exploiting the identity

$$A_i = \frac{1}{a_i^2}\left(1 - a_i \frac{\partial}{\partial a_i}\right) A_0 , \tag{2.161}$$

which is easily established from the previous expressions; there is no summation over the index i in the second term. To evaluate A_0 we note first that the symmetries of the integrand allow us to restrict the range of integration to $0 < \tilde{\phi} < \frac{\pi}{2}$, at the cost of introducing a factor of 4; and in this range it is helpful to switch integration variables to $t := \tan\tilde{\phi}$, which varies from 0 to ∞. In terms of the new variable we have that $\alpha = (p + qt^2)/(1 + t^2)$ with

$$p := \frac{\sin^2\tilde{\theta}}{a_1^2} + \frac{\cos^2\tilde{\theta}}{a_3^2}, \qquad q := \frac{\sin^2\tilde{\theta}}{a_2^2} + \frac{\cos^2\tilde{\theta}}{a_3^2}. \tag{2.162}$$

With $d\tilde{\phi} = dt/(1 + t^2)$ the integration is immediate, and we obtain

$$A_0 = \frac{2}{\pi}\int_0^\pi \sin\tilde{\theta}\, d\tilde{\theta} \int_0^\infty \frac{dt}{p + qt^2} = \int_0^\pi \frac{\sin\tilde{\theta}\, d\tilde{\theta}}{\sqrt{pq}}. \tag{2.163}$$

To evaluate the remaining integral we note that the range of integration can be restricted to $0 < \tilde{\theta} < \frac{\pi}{2}$ (introducing a factor of 2), and in this range it is helpful to adopt $u := a_3^2 \tan^2\tilde{\theta}$ as a new integration variable, which varies from 0 to ∞. In terms of the new variable we have that

$$p = \frac{1}{a_1^2}\frac{a_1^2 + u}{a_3^2 + u}, \qquad q = \frac{1}{a_2^2}\frac{a_2^2 + u}{a_3^2 + u}, \tag{2.164}$$

and with $\sin\tilde{\theta}\, d\tilde{\theta} = \frac{1}{2}a_3(a_3^2 + u)^{-3/2}\, du$, we find that the integral for A_0 assumes the form originally given in Eq. (2.149). The expression for A_i is then recovered by exploiting the identity of Eq. (2.161).

The proof that $A_1 + A_2 + A_3 = 2$ begins with

$$A_1 + A_2 + A_3 = a_1 a_2 a_3 \int_0^\infty \frac{du}{\Delta}\left(\frac{1}{a_1^2 + u} + \frac{1}{a_2^2 + u} + \frac{1}{a_3^2 + u}\right), \tag{2.165}$$

and proceeds by recognizing that the expression within brackets is $2\Delta^{-1} d\Delta/du$. Integration can then be carried out, and since $\Delta^{-1} = 0$ at the upper limit while $\Delta^{-1} = (a_1 a_2 a_3)^{-1}$ at the lower limit, the result follows immediately.

Equilibrium conditions

The equilibrium conditions for a rigidly rotating ellipsoid of uniform density are derived from Eq. (2.145). When ρ is a constant the equation implies that $p/\rho = U + \Phi_C +$ constant, with U given by Eq. (2.148) and $\Phi_C = \frac{1}{2}\omega^2(x^2 + y^2)$. Making the substitutions, we find that the pressure is given by

$$\frac{p}{\pi G\rho} = \text{constant} - \left(A_1 - \frac{\omega^2}{2\pi G\rho}\right)x^2 - \left(A_2 - \frac{\omega^2}{2\pi G\rho}\right)y^2 - A_3 z^2 . \tag{2.166}$$

This, in particular, must apply to the surface of the ellipsoid (where $p = 0$), and because the surface is also described by Eq. (2.147), we find that the coefficients of the quadratic forms must be related by the constraints

$$a_1^2\left(A_1 - \frac{\omega^2}{2\pi G\rho}\right) = a_2^2\left(A_2 - \frac{\omega^2}{2\pi G\rho}\right) = a_3^2 A_3\,. \tag{2.167}$$

These equations give rise to the equilibrium conditions satisfied by the rotating ellipsoid.

The equations can be solved for the angular velocity, and the solution can be expressed in a number of ways:

$$\frac{\omega^2}{2\pi G\rho} = \frac{a_1^2 A_1 - a_3^2 A_3}{a_1^2} \tag{2.168a}$$

$$= \frac{a_2^2 A_2 - a_3^2 A_3}{a_2^2} \tag{2.168b}$$

$$= \frac{a_1^2 A_1 - a_2^2 A_2}{a_1^2 - a_2^2}\,. \tag{2.168c}$$

Only two of these relations are independent, and they are mutually compatible when

$$a_1^2 a_2^2 (A_1 - A_2) + (a_1^2 - a_2^2)\, a_3^2 A_3 = 0\,. \tag{2.169}$$

Inserting the integral expressions for A_i displayed in Eq. (2.150), we find that the angular velocity is given by

$$\frac{\omega^2}{2\pi G\rho} = \frac{a_1 a_3}{a_2}\left(a_2^2 - a_3^2\right) \int_0^\infty \frac{u}{\left(a_2^2 + u\right)\left(a_3^2 + u\right)} \frac{du}{\Delta} \tag{2.170a}$$

$$= \frac{a_2 a_3}{a_1}\left(a_1^2 - a_3^2\right) \int_0^\infty \frac{u}{\left(a_1^2 + u\right)\left(a_3^2 + u\right)} \frac{du}{\Delta} \tag{2.170b}$$

$$= a_1 a_2 a_3 \int_0^\infty \frac{u}{\left(a_1^2 + u\right)\left(a_2^2 + u\right)} \frac{du}{\Delta}\,, \tag{2.170c}$$

and that the compatibility condition becomes

$$\left(a_1^2 - a_2^2\right)\left[a_1^2 a_2^2 \int_0^\infty \frac{1}{\left(a_1^2 + u\right)\left(a_2^2 + u\right)} \frac{du}{\Delta} - a_3^2 \int_0^\infty \frac{1}{\left(a_3^2 + u\right)} \frac{du}{\Delta}\right] = 0\,. \tag{2.171}$$

The first two expressions for ω^2 imply that the principal axes of a rotating ellipsoid must be constrained by $a_1 > a_3$ and $a_2 > a_3$: the shortest axis must be the rotation axis, and the figure is oblate, not prolate. The compatibility condition implies either one of two statements: *either* $a_1 = a_2$ and the ellipsoid is axially symmetric, *or* the three axes must be adjusted to ensure that the integral constraint

$$a_1^2 a_2^2 \int_0^\infty \frac{1}{\left(a_1^2 + u\right)\left(a_2^2 + u\right)} \frac{du}{\Delta} = a_3^2 \int_0^\infty \frac{1}{\left(a_3^2 + u\right)} \frac{du}{\Delta} \tag{2.172}$$

is satisfied; for selected values of a_1 and a_2, this equation can be solved for a_3. The axisymmetric branch defines the *Maclaurin spheroids*, which will be described in detail below. The non-axisymmetric branch constrained by Eq. (2.172) defines the *Jacobi ellipsoids*, to which we turn next.

Table 2.2 Solutions to the equilibrium conditions for Jacobi ellipsoids. The first entry of the table, for $a_1 = a_2$, is the point of bifurcation from the Maclaurin sequence.

a_2/a_1	a_3/a_1	$\omega^2/(2\pi G\rho)$
1.00	0.5827241662	0.1871148374
0.95	0.5677381338	0.1869173572
0.90	0.5518726119	0.1862832772
0.85	0.5350575095	0.1851433886
0.80	0.5172160586	0.1834185183
0.75	0.4982641698	0.1810182506
0.70	0.4781097616	0.1778396039
0.65	0.4566520785	0.1737657518
0.60	0.4337810349	0.1686649517
0.55	0.4093766373	0.1623899695
0.50	0.3833085784	0.1547785068
0.45	0.3554361455	0.1456555096
0.40	0.3256086868	0.1348388928
0.35	0.2936670362	0.1221514037
0.30	0.2594465778	0.1074435407
0.25	0.2227831754	0.0906367278
0.20	0.1835242862	0.0718047786
0.15	0.1415500250	0.0513316734
0.10	0.09681524257	0.0302360801
0.05	0.04944350586	0.0109330058

Returning to Eq. (2.166), we insert the relations of Eq. (2.167) to simplify the expression for the pressure, and we adjust the constant to ensure that $p = 0$ on the surface of the ellipsoid. The end result of this exercise is

$$p = p_c\left(1 - \frac{x^2}{a_1^2} - \frac{y^2}{a_2^2} - \frac{z^2}{a_3^2}\right), \tag{2.173}$$

where $p_c := \pi G\rho^2 a_3^2 A_3$ is the central pressure.

Jacobi ellipsoids

Solutions to the compatibility condition of Eq. (2.172) are presented in Table 2.2, which also lists the corresponding values of ω^2 calculated according to Eq. (2.170c). Because a_1 can be adopted as the fundamental length scale of the rotating body, the values of a_2 and a_3 are given in units of a_1. The table is constructed by choosing a_2/a_1 and solving the compatibility condition for a_3/a_1; without loss of generality we take a_2 to be smaller than a_1. Two main observations can be made regarding the numerical results. The first is that as we proceed along the Jacobi sequence, with a_2/a_1 decreasing from unity, we find that a_3/a_1 decreases, giving rise to an increasingly elongated and flattened figure. The second is that *the angular velocity decreases along the sequence*, so that the most deformed figure is the one with the least angular velocity; the maximum value of ω^2 on the sequence is a factor of 0.187 smaller than the Poincaré bound of Eq. (2.131).

The mass of a Jacobi ellipsoid is given by

$$M = \int \rho\, d^3x = \frac{4\pi}{3}\rho a_1 a_2 a_3 , \tag{2.174}$$

its angular momentum is

$$J = \omega \int \rho(x^2 + y^2)\, d^3x = \frac{1}{5} M\omega\big(a_1^2 + a_2^2\big) , \tag{2.175}$$

and its total kinetic energy is

$$\mathcal{T} = \frac{1}{2}\omega J = \frac{1}{10} M\omega^2\big(a_1^2 + a_2^2\big) . \tag{2.176}$$

To perform these integrals it is helpful to adopt a system of ellipsoidal coordinates (s, θ, ϕ) related to the original system (x, y, z) by $x = a_1 s \sin\theta \cos\phi$, $y = a_2 s \sin\theta \sin\phi$, and $z = a_3 s \cos\theta$; the volume integral covers the interval $0 < s < 1$ and the usual range of the angular variables. From the results displayed in the table it is possible to show that J, when measured in units of $\sqrt{GM^3\bar{a}}$ with $\bar{a} := (a_1 a_2 a_3)^{1/3}$, increases along the Jacobi sequence.

Maclaurin spheroids

When $a_1 = a_2$ the rigidly rotating body becomes a Maclaurin spheroid. In this case A_0, A_i, and ω can all be expressed as simple functions of the *eccentricity*

$$e := \sqrt{1 - (a_3/a)^2} , \tag{2.177}$$

in which we have set $a := a_1 = a_2$. Inserting $a_3 = a\sqrt{1 - e^2}$ within Eqs. (2.149) and (2.150), we find that the integrals evaluate to

$$A_0 = 2a^2 \frac{\sqrt{1 - e^2}}{e} \arcsin e , \tag{2.178a}$$

$$A_1 = A_2 = \frac{\sqrt{1 - e^2}}{e^3} \arcsin e - \frac{1 - e^2}{e^2} , \tag{2.178b}$$

$$A_3 = -2\frac{\sqrt{1 - e^2}}{e^3} \arcsin e + \frac{2}{e^2} . \tag{2.178c}$$

On the other hand, the relation $\omega^2/(2\pi G\rho) = A_1 - (a_3/a)^2 A_3$ produces

$$\frac{\omega^2}{2\pi G\rho} = \frac{\sqrt{1 - e^2}}{e^3}(3 - 2e^2)\arcsin e - 3\frac{1 - e^2}{e^2} . \tag{2.179}$$

This expression for the angular velocity is plotted in Fig. 2.5. The figure reveals that the angular velocity first increases with the eccentricity, but that it reaches a maximum of $\omega^2/(2\pi G\rho) = 0.224666$ when $e = 0.929956$. It then decreases to zero as the eccentricity increases toward unity; in this limit we have an infinitely thin disk rotating with negligible angular velocity.

It is known, however, that the sequence of Maclaurin spheroids never comes close to these extremes. There exists on the sequence a *point of bifurcation* at which a Maclaurin

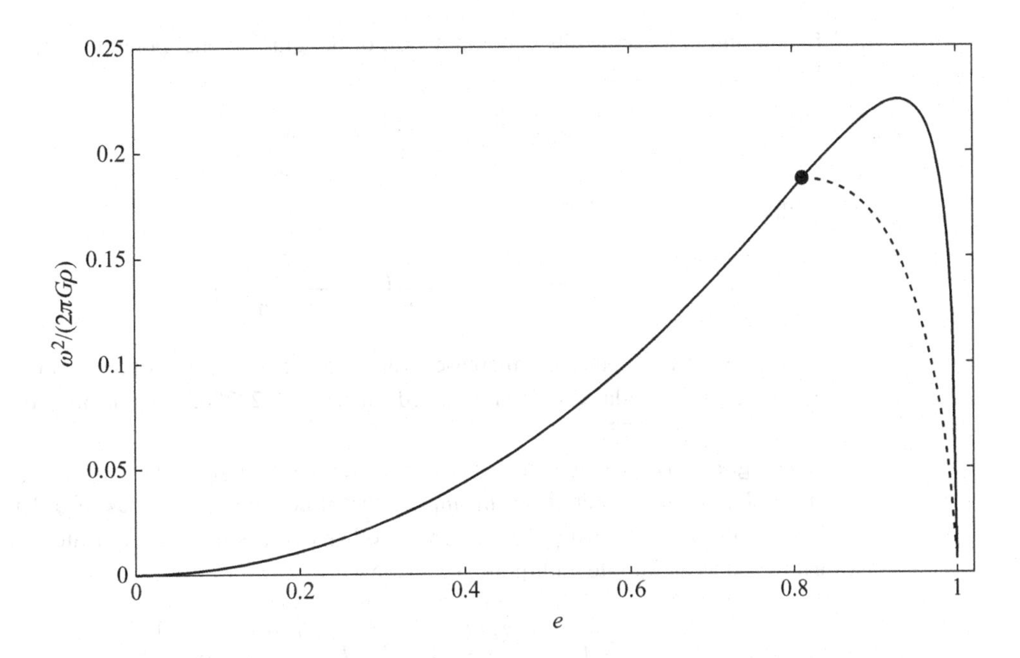

Fig. 2.5 Squared angular velocity $\omega^2/(2\pi G\rho)$ as a function of eccentricity e for Maclaurin spheroids. The plot reveals an ascending branch, along which the angular velocity increases with eccentricity, and a descending branch, along which the angular velocity decreases. The figure also shows the Jacobi sequence, which bifurcates from the Maclaurin sequence when $e = 0.812670$.

spheroid and a Jacobi ellipsoid with $a_1 = a_2$ are both valid solutions to the equilibrium equations; this point occurs at $e = 0.812670$, where $\omega^2/(2\pi G\rho) = 0.187115$. Going beyond the bifurcation point, it is observed that for a given eccentricity, the Jacobi ellipsoid has a smaller total energy than the corresponding Maclaurin spheroid, and this suggests that in the presence of a dissipative mechanism, such as viscosity or gravitational radiation damping, the Maclaurin spheroid is unstable to perturbations that eventually convert it to a Jacobi ellipsoid. This conclusion is confirmed by a stability analysis: the bifurcation point marks the onset of a secular instability that drives a Maclaurin spheroid toward a Jacobi ellipsoid. There is also, further along the sequence beyond $e = 0.952887$, a regime of dynamical instability that, even in the absence of dissipation, converts a Maclaurin spheroid to a Jacobi ellipsoid.

From the results obtained previously in the case of Jacobi ellipsoids, we find that the mass, angular momentum, and kinetic energy of a Maclaurin spheroid are given by

$$M = \frac{4\pi}{3}\rho a^3\sqrt{1-e^2}\,, \tag{2.180a}$$

$$J = \frac{2}{5}M\omega a^2\,, \tag{2.180b}$$

$$\mathcal{T} = \frac{1}{5}M\omega^2 a^2\,. \tag{2.180c}$$

In addition to these results, a computation of the gravitational potential energy – refer to Exercise 2.3 – reveals that

$$\Omega = -\frac{3}{5}\frac{GM^2}{a}\frac{\arcsin e}{e}, \tag{2.181}$$

so that

$$\tau := \frac{\mathcal{T}}{|\Omega|} = \frac{3}{2e^2}\left(1 - \frac{e\sqrt{1-e^2}}{\arcsin e}\right) - 1. \tag{2.182}$$

This relation implies that as e increases from 0 to 1, τ increases monotonically from 0 to $\frac{1}{2}$. The maximum angular velocity is reached when $\tau = 0.237902$, and the point of bifurcation occurs at $\tau = 0.137528$.

The gravitational potential inside a Maclaurin spheroid is given by Eq. (2.148) with $A_1 = A_2$, and the quadratic form implies that it contains multipoles of order $\ell = 0$ and $\ell = 2$ only. The external potential, on the other hand, is more complicated, and must be expressed as an infinite multipole expansion

$$U_{\rm ext} = \frac{GM}{r}\left[1 - \sum_{\ell=2}^{\infty} J_\ell \left(\frac{a}{r}\right)^\ell P_\ell(\cos\theta)\right], \tag{2.183}$$

where $\cos\theta := z/r$, and where the dimensionless multipole moments

$$J_\ell := -\frac{1}{Ma^\ell}\int \rho r^\ell P_\ell(\cos\theta)\, d^3x \tag{2.184}$$

were first introduced in Sec. 1.5.2 – refer to Eq. (1.143). The existence of an infinite number of multipole moments has to do with the fact that the internal and external potentials must match across a spheroidal surface; in this context, a simple internal potential can give rise to a complicated external potential. After performing the radial integration we find that the multipole moments become

$$J_\ell = -\frac{2\pi\rho}{(\ell+3)Ma^\ell}\int R^{\ell+3}(\theta) P_\ell(\cos\theta)\, d\cos\theta, \tag{2.185}$$

where $R(\theta)$, given explicitly by

$$\frac{1}{R^2} = \frac{\sin^2\theta}{a^2} + \frac{\cos^2\theta}{a_3^2}, \tag{2.186}$$

is the description of the spheroidal surface in the spherical coordinates (r, θ, ϕ). Mercifully these integrals can be evaluated, and after a short computation we obtain

$$J_\ell = \frac{3(-1)^{\ell/2+1}}{(\ell+1)(\ell+3)} e^\ell \tag{2.187}$$

for the multipole moments of a Maclaurin spheroid. You will be asked to go through the steps of this calculation in Exercise 2.4.

When the eccentricity is small, the various functions encountered previously can be expanded in powers of e, and they become

$$A_0 = 2a^2\left[1 - \frac{1}{3}e^2 - \frac{2}{15}e^4 + O(e^6)\right], \tag{2.188a}$$

$$A_1 = A_2 = \frac{2}{3}\left[1 - \frac{1}{5}e^2 - \frac{4}{35}e^4 + O(e^6)\right], \tag{2.188b}$$

$$A_3 = \frac{2}{3}\left[1 + \frac{2}{5}e^2 + \frac{8}{35}e^4 + O(e^6)\right], \tag{2.188c}$$

$$\frac{\omega^2}{2\pi G\rho} = \frac{4}{15}e^2\left[1 + \frac{1}{7}e^2 - \frac{8}{231}e^4 + O(e^6)\right], \tag{2.188d}$$

$$\Omega = -\frac{3}{5}\frac{GM^2}{a}\left[1 + \frac{1}{6}e^2 + \frac{3}{40}e^4 + O(e^6)\right], \tag{2.188e}$$

$$\tau = \frac{2}{15}e^2\left[1 + \frac{10}{21}e^2 + \frac{92}{315}e^4 + O(e^6)\right]. \tag{2.188f}$$

2.4 General theory of deformed bodies

When we relax the assumption of uniform density, the problem of finding the structure of a general rotating body becomes much more involved, even when we retain the assumption of rigid rotation. On the other hand, most planets and stars rotate sufficiently slowly that they deviate only slightly from spherical symmetry. For example, the Earth's dimensionless quadrupole moment J_2 is approximately 10^{-3}, and the higher J_ℓs are a thousand times smaller; for the Sun, $J_2 \simeq 2 \times 10^{-7}$. For these slowly-rotating bodies it is appropriate to take an approach in which the deviations from spherical symmetry are assumed to be small. This is the standard fare of *perturbation theory*, in which one starts with an unperturbed, spherical configuration and deforms it so slightly that all equations can be linearized with respect to the deviations from spherical symmetry.

The perturbative approach to deformed bodies is sufficiently general that we can consider bodies with arbitrary equations of state, and deformations that result not only from rotation, but also from tidal fields produced by external bodies. Our goal in this section is to develop this framework. We shall continue to restrict our attention to equilibrium configurations, but in Sec. 2.5.3 we will relax this assumption and consider fully dynamical perturbations of spherical bodies.

2.4.1 Fluid equations

We return to Eq. (2.4), the statement of Euler's equation in the moving (but non-rotating) frame of a given body A in a system of N separated bodies. We now keep the external terms associated with the remote bodies, and adopt the simplified form of Eq. (2.11), which neglects the influence of A's higher multipole moments on its motion in the external gravitational field – these vanish in spherical symmetry, and produce second-order corrections in perturbation theory. Assuming that A is rotating rigidly with angular velocity ω, we subject Euler's equation to a transformation to the rotating frame, in the way described near the end of Sec. 2.3.1 – refer to Eq. (2.139).

Collecting results, we find that Euler's equation takes the form of

$$\frac{du^j}{dt} = -\frac{1}{\rho}\partial_j p - 2\epsilon_{jkn}\omega^k u^n + \partial_j \Phi\,, \tag{2.189}$$

where $\boldsymbol{u}$ is the velocity vector in the rotating frame, $\boldsymbol{\omega} := [0, 0, \omega]$ is the angular-velocity vector, and

$$\Phi := U + \Phi_{\rm C} + U_{\rm tidal} \tag{2.190}$$

is a generalized potential that includes the body's gravitational potential U (previously denoted U_A), the centrifugal potential

$$\Phi_{\rm C} = \frac{1}{2}\omega^2(x^2 + y^2)\,, \tag{2.191}$$

and the tidal potential

$$U_{\rm tidal} = -\sum_{\ell=2}^{\infty} \frac{1}{\ell!}\mathcal{E}_L(t)x^L\,, \tag{2.192}$$

which was previously denoted $U^{\rm eff}_{\neg A}$. The STF tensors $\mathcal{E}_L(t) := -\partial_L U_{\neg A}(t, \boldsymbol{0})$ are the *tidal moments* associated with the external potential; they are obtained by differentiating $U_{\neg A}$ with respect to x^j and evaluating the result at the body's center-of-mass $\boldsymbol{x} = \boldsymbol{0}$. In our treatment the tidal moments are taken to be known functions of time; their specification depends on the particular application.

As usual the Euler equation must be supplemented by the continuity equation $\partial_t\rho + \partial_j(\rho u^j) = 0$ and Poisson's equation $\nabla^2 U = -4\pi G\rho$ for the body's potential; the expression of Eq. (2.192) implies that the tidal potential satisfies Laplace's equation $\nabla^2 U_{\rm tidal} = 0$, and we also have that $\nabla^2\Phi_{\rm C} = 2\omega^2$.

For our developments below it is useful to decompose the centrifugal potential $\Phi_{\rm C}$ into monopole and quadrupole pieces, according to

$$\Phi_{\rm C} = \Phi_{\rm C}^{\ell=0} + \Phi_{\rm C}^{\ell=2}\,, \tag{2.193}$$

with

$$\Phi_{\rm C}^{\ell=0} = \frac{1}{3}\omega^2(x^2 + y^2 + z^2) = \frac{1}{3}\omega^2 r^2 \tag{2.194}$$

and

$$\Phi_{\rm C}^{\ell=2} = \frac{1}{6}\omega^2(x^2 + y^2 - 2z^2) = -\frac{1}{3}\omega^2 r^2 P_2(\cos\theta)\,. \tag{2.195}$$

The monopole piece satisfies $\nabla^2\Phi_{\rm C}^{\ell=0} = 2\omega^2$, while $\nabla^2\Phi_{\rm C}^{\ell=2} = 0$.

The tidal potential depends on time through the tidal moments, which contain all the relevant information about the changing positions of the external bodies. This time dependence implies that there is an inherent contradiction between Euler's equation and the goal of finding equilibrium configurations. The contradiction would indeed be severe if the external bodies were in A's immediate vicinity, and in such circumstances no equilibrium solutions could ever be found. (We shall consider this situation in Sec. 2.5.3.) But the contradiction is very mild when the external bodies are remote and move on an orbital time scale $T_{\rm orb} \sim \sqrt{r_{AB}^3/GM}$ that is much longer than the internal, hydrodynamical time scale $T_{\rm int} \sim (G\rho)^{-1/2} \sim \sqrt{R^3/GM}$; here r_{AB} is a typical inter-body separation, R is the body's radius, and the time scales are indeed well separated when $r_{AB} \gg R$. In such circumstances, the time dependence contained in $\mathcal{E}_L(t)$ is too slow to take the body out of equilibrium; the solutions will still carry a dependence upon t inherited from the tidal moments, but this reflects a parametric dependence instead of a genuine dynamical dependence.

2.4.2 Unperturbed configuration

The unperturbed configuration of the fluid is spherically symmetric and in equilibrium; it corresponds to the absence of external bodies and rotational effects. The equations that govern such a configuration were presented in Sec. 2.2: we have the equation of hydrostatic equilibrium,

$$\frac{dp}{dr} = \rho\frac{d}{dr}\left(U + \Phi_{\rm C}^{\ell=0}\right), \tag{2.196}$$

and Poisson's equation

$$\frac{1}{r^2}\frac{d}{dr}\left(r^2\frac{dU}{dr}\right) = -4\pi G\rho \tag{2.197}$$

for the body potential. It is assumed that the density ρ and pressure p are related by an equation of state. For convenience we choose to insert the monopole piece of $\Phi_{\rm C}$ in the equation of hydrostatic equilibrium, in spite of the fact that it is part of the perturbation. This choice is motivated by the fact that unlike all other terms in the perturbing potential, $\Phi_{\rm C}^{\ell=0}$ is spherically symmetric, and thus naturally included as part of the unperturbed configuration.

To solve the unperturbed equations we introduce an effective potential

$$\tilde{U} := U + \Phi_{\rm C}^{\ell=0} \tag{2.198}$$

associated with an effective mass density

$$\tilde{\rho} := \rho - \frac{\omega^2}{2\pi G}, \tag{2.199}$$

so that $\nabla^2\tilde{U} = -4\pi G\tilde{\rho}$. If we also introduce an effective mass function $\tilde{m}(r)$ defined by

$$\frac{d\tilde{m}}{dr} = 4\pi r^2\tilde{\rho}, \tag{2.200}$$

then $d\tilde{U}/dr = -G\tilde{m}/r^2$ and the equation of hydrostatic equilibrium becomes

$$\frac{dp}{dr} = -\rho \frac{G\tilde{m}}{r^2} . \tag{2.201}$$

Note that this equation involves the actual mass density ρ together with the effective mass function $\tilde{m}$. We assume that these equations can be solved for a selected equation of state, and that $\rho(r)$, $p(r)$, $\tilde{m}(r)$, and $\tilde{U}(r)$ are known for the unperturbed configuration. The stellar radius R is such that $p(r = R) = 0$, and the total mass is $M = \tilde{m}(r = R) + \omega^2 R^3/(6\pi G)$.

2.4.3 Fluid perturbations

The introduction of $U_{\rm tidal}$ and $\Phi_{\rm C}^{\ell=2}$ in Euler's equation creates a perturbation of the fluid configuration. The density changes from ρ to $\rho + \delta\rho$, the pressure from p to $p + \delta p$, the body potential from U to $U + \delta U$, and the fluid velocity u^j is no longer zero if the equilibrium is disturbed. We suppose that the perturbation is small, and we work consistently to first order in all perturbation variables.

The perturbation of any fluid quantity Q can be described either in terms of its *Eulerian perturbation* δQ (as was done previously), or its *Lagrangian perturbation* ΔQ. The Eulerian point of view is macroscopic: Q is compared to its unperturbed value Q_0 at the same position in space, and

$$\delta Q := Q(t, \boldsymbol{x}) - Q_0(t, \boldsymbol{x}) . \tag{2.202}$$

The Lagrangian point of view is microscopic: Q is compared at the same fluid element, which is displaced by the vector $\boldsymbol{\xi}(t, \boldsymbol{x})$ relative to its position $\boldsymbol{x}$ in the unperturbed configuration. The Lagrangian perturbation is

$$\Delta Q := Q(t, \boldsymbol{x} + \boldsymbol{\xi}) - Q_0(t, \boldsymbol{x}) = \delta Q + \xi^j \partial_j Q_0 . \tag{2.203}$$

From the definition of the displacement vector it is obvious that

$$\Delta \boldsymbol{u} = \frac{d\boldsymbol{\xi}}{dt} , \tag{2.204}$$

where d/dt is the Lagrangian time derivative, which refers to the unperturbed flow. When the unperturbed fluid is in equilibrium, $d/dt = \partial/\partial t$.

The commutation rules,

$$\delta \partial_t = \partial_t \delta , \tag{2.205a}$$

$$\delta \partial_j = \partial_j \delta , \tag{2.205b}$$

$$\Delta \partial_t = \partial_t \Delta - (\partial_t \xi^j)\partial_j , \tag{2.205c}$$

$$\Delta \partial_j = \partial_j \Delta - (\partial_j \xi^k)\partial_k , \tag{2.205d}$$

$$\Delta \frac{d}{dt} = \frac{d}{dt} \Delta , \tag{2.205e}$$

are easy to establish, and they permit an efficient manipulation of the perturbed Euler equation. For example, $\delta(\partial_j Q) = \partial_j Q - \partial_j Q_0$ while $\partial_j(\delta Q) = \partial_j(Q - Q_0) = \partial_j Q - \partial_j Q_0$, and we see that the operations commute. As another example, $\Delta(\partial_j Q) = \delta(\partial_j Q) + \xi^k \partial_{kj} Q_0 = \partial_j(\delta Q) + \xi^k \partial_{jk} Q_0$ while $\partial_j(\Delta Q) = \partial_j(\delta Q) + \partial_j(\xi^k \partial_k Q_0) = \partial_j(\delta Q) + (\partial_j \xi^k)(\partial_k Q_0) + \xi^k \partial_{jk} Q_0$, and we see that these operations do not commute.

The mass of a selected portion of the fluid stays constant during a perturbation. In the unperturbed configuration the portion of the fluid occupies the volume V; in the perturbed configuration it occupies the volume $V + \Delta V$. The mass of the unperturbed configuration is $\int_V \rho\, d^3x$, the mass of the perturbed configuration is $\int_{V+\Delta V} (\rho + \delta\rho)\, d^3x$, and we insist that the two integrals must produce the same number. The first integral for the mass of the perturbed configuration can be expressed as $\int_V \rho\, d^3x + \oint_S \rho \boldsymbol{\xi} \cdot d\boldsymbol{S}$, where S is the closed surface bounding V, and $d\boldsymbol{S}$ is its (outward-directed) surface element. After converting the surface integral to a volume integral, we find that the statement of mass conservation is $\int_V [\delta\rho + \partial_j(\rho\xi^j)]\, d^3x = 0$. Because this statement must hold for any piece of the fluid, we have

$$\delta\rho = -\partial_j(\rho\xi^j), \tag{2.206}$$

or

$$\Delta\rho = -\rho\,\partial_j \xi^j. \tag{2.207}$$

These are time-integrated forms of the perturbed continuity equation.

By virtue of the equation of state satisfied by the fluid, the perturbation of the pressure is directly linked to the density perturbation. The link is provided by the quantity Γ_1, which is defined by the (Lagrangian) statement

$$\frac{\Delta p}{p} = \Gamma_1 \frac{\Delta\rho}{\rho}. \tag{2.208}$$

In general Γ_1 is a function of time and position within the fluid; for adiabatic changes it is given by c_P/c_V, the ratio of specific heats. According to this equation, the Eulerian perturbation of the pressure is given by

$$\delta p = -\Gamma_1 p\, \partial_j \xi^j - \xi^j \partial_j p. \tag{2.209}$$

The perturbation of Euler's equation can be formulated in either the Eulerian picture or the Lagrangian picture. We begin with the Lagrangian picture, and submit Eq. (2.189) to the operator Δ and use the commutation rules to simplify the results. For example, the perturbation of the left-hand side is $\Delta(du^j/dt)$, which can be written as $d(\Delta u^j)/dt$ by virtue of the commutation rules; invoking now Eq. (2.204), we write this in its final form of $d^2\xi^j/dt^2$. Proceeding in a similar way with the remaining terms, we find that the perturbed Euler equation can be expressed as

$$\frac{d^2\xi^j}{dt^2} = \frac{\Delta\rho}{\rho^2}\partial_j p - \frac{1}{\rho}\partial_j \Delta p - 2\epsilon_{jkn}\omega^k \frac{d\xi^n}{dt} + \partial_j \Delta\Phi + \left(\partial_j \xi^k\right)\left(\frac{1}{\rho}\partial_k p - \partial_k \Phi\right). \tag{2.210}$$

When the unperturbed state is an equilibrium configuration, the unperturbed velocity field vanishes, the factor multiplying $\partial_j \xi^k$ vanishes, and the equation simplifies to

$$\frac{\partial^2\xi^j}{\partial t^2} = \frac{\Delta\rho}{\rho^2}\partial_j p - \frac{1}{\rho}\partial_j \Delta p - 2\epsilon_{jkn}\omega^k \frac{\partial\xi^n}{\partial t} + \partial_j \Delta\Phi. \tag{2.211}$$

When the perturbed state is also an equilibrium configuration, $\partial\xi^j/\partial t = 0$ and the equation simplifies further.

Moving next to the Eulerian picture, we now submit Eq. (2.189) to the operator δ and use the commutation rules to simplify the result. For example, the perturbation of the first term on the right-hand side is $\delta(\rho^{-1}\partial_j p) = -\rho^{-2}(\delta\rho)\partial_j p + \rho^{-1}\delta(\partial_j p) = -\rho^{-2}(\delta\rho)\partial_j p + \rho^{-1}\partial_j \delta p$. Proceeding in a similar way with the other terms, and assuming once more that the unperturbed state is an equilibrium configuration, we find that the perturbed Euler equation can also be expressed as

$$\frac{\partial^2\xi^j}{\partial t^2} = \frac{\delta\rho}{\rho^2}\partial_j p - \frac{1}{\rho}\partial_j\delta p - 2\epsilon_{jkn}\omega^k\frac{\partial\xi^n}{\partial t} + \partial_j\delta\Phi \,. \tag{2.212}$$

The equation simplifies further when the perturbed configuration is also in equilibrium.

The perturbed Euler equation, either Eq. (2.211) or Eq. (2.212), can be solved once we incorporate the expressions obtained previously for $\delta\rho$ (or $\Delta\rho$) and δp (or Δp). A missing ingredient is the perturbation of the body's gravitational potential, which is best calculated in the Eulerian picture. It satisfies Poisson's equation

$$\nabla^2\delta U = -4\pi G\delta\rho \,, \tag{2.213}$$

and the solution can be fed into Euler's equation.

2.4.4 Perturbed equilibrium

Perturbation equations

When the perturbed fluid is also in equilibrium, $\partial\xi^j/\partial t = 0$ and Eq. (2.212) simplifies to

$$\frac{\delta\rho}{\rho^2}\partial_j p - \frac{1}{\rho}\partial_j\delta p + \partial_j\delta\Phi = 0 \,. \tag{2.214}$$

The perturbation of the generalized potential is decomposed as

$$\delta\Phi = \delta U + V \,, \tag{2.215}$$

where

$$V := U_{\rm tidal} + \Phi_{\rm C}^{\ell=2} \tag{2.216}$$

is the potential driving the perturbation; we recall that $U_{\rm tidal}$ is given by Eq. (2.192), and that $\Phi_{\rm C}^{\ell=2}$ is defined by Eq. (2.195).

Our task is to solve Eq. (2.214), assuming that V is specified. To achieve this we choose ξ^j to describe the displacement of a fluid element on a spherical surface $\rho =$ constant in the unperturbed configuration to the deformed surface $\rho =$ constant in the perturbed configuration (the mass density is numerically the same on both surfaces). This ensures that $\Delta\rho \equiv 0$, and Eq. (2.208) further ensures that $\Delta p = 0$. These results imply that

$$\partial_j\xi^j = 0 \tag{2.217}$$

and

$$\delta\rho = -\xi^j\partial_j\rho \,, \qquad \delta p = -\xi^j\partial_j p \,. \tag{2.218}$$

Because the unperturbed density ρ and pressure p depend on r only, these equations involve only ξ^r, the radial component of the displacement vector; we shall not need to solve for the angular components, which are constrained by the divergence-free condition.

Spherical-harmonic decomposition

To proceed it is helpful to decompose all perturbation quantities in spherical harmonics:

$$\xi^r = \sum_{\ell m} r f_{\ell m}(r) Y_{\ell m}(\theta, \phi), \tag{2.219a}$$

$$\delta\rho = \sum_{\ell m} \rho_{\ell m}(r) Y_{\ell m}(\theta, \phi), \tag{2.219b}$$

$$\delta p = \sum_{\ell m} p_{\ell m}(r) Y_{\ell m}(\theta, \phi), \tag{2.219c}$$

$$\delta U = \sum_{\ell m} U_{\ell m}(r) Y_{\ell m}(\theta, \phi), \tag{2.219d}$$

$$V = \sum_{\ell m} V_{\ell m}(r) Y_{\ell m}(\theta, \phi), \tag{2.219e}$$

in which (r, θ, ϕ) are spherical polar coordinates associated with the rotating frame x^j; note that a factor r has been inserted in the relation between ξ^r and its (dimensionless) coefficients $f_{\ell m}$. By virtue of Eqs. (2.192) and (2.195), the spherical-harmonic decompositions begin at $\ell = 2$, with m running from $-\ell$ to ℓ.

The equations for $\delta\rho$ and δp imply that

$$\rho_{\ell m} = -r\rho' f_{\ell m} \tag{2.220}$$

and

$$p_{\ell m} = -rp' f_{\ell m} = \frac{\rho G m}{r} f_{\ell m}, \tag{2.221}$$

where a prime indicates differentiation with respect to r. We invoked Eq. (2.201) to replace p' in the last equation. We no longer distinguish between the effective mass function $\tilde{m}(r)$ and the actual mass function $m(r)$, because the difference is of first order in the perturbation, and would therefore be of second order after multiplication by $f_{\ell m}$.

The spherical-harmonic components of δU have already been introduced and examined in Sec. 1.5.2 – refer to the discussion following Eq. (1.122). By virtue of Poisson's equation they satisfy the differential equation

$$r^2 U''_{\ell m} + 2r U'_{\ell m} - \ell(\ell+1) U_{\ell m} = -4\pi G r^2 \rho_{\ell m} \tag{2.222}$$

in the body's interior. Outside the body they are given by

$$U^{\rm out}_{\ell m}(r) = \frac{4\pi G}{2\ell+1} \frac{I_{\ell m}}{r^{\ell+1}}, \tag{2.223}$$

where $I_{\ell m}$ are the body's multipole moments.

Because $\nabla^2 V = 0$, the spherical-harmonic components of the driving potential satisfy the homogeneous version of Eq. (2.222),

$$r^2 V''_{\ell m} + 2r V'_{\ell m} - \ell(\ell+1) V_{\ell m} = 0\,. \tag{2.224}$$

This equation is satisfied both inside and outside the body, and the only admissible solution is

$$V_{\ell m}(r) = \frac{4\pi}{2\ell+1} d_{\ell m} r^\ell\,, \tag{2.225}$$

where $d_{\ell m}$ are the moments of the driving potential; these are readily computed once V is specified.

We now insert the spherical-harmonic decompositions within Eq. (2.214). It is easy to see that the radial component of the equation produces

$$p'_{\ell m} = -\frac{Gm}{r^2}\rho_{\ell m} + \rho\big(U'_{\ell m} + V'_{\ell m}\big)\,, \tag{2.226}$$

where we have used Eq. (2.201) to eliminate p'. The angular components imply

$$p_{\ell m} = \rho\big(U_{\ell m} + V_{\ell m}\big)\,. \tag{2.227}$$

When we differentiate this with respect to r and substitute back in the previous equation, we obtain

$$\frac{Gm}{r^2}\rho_{\ell m} = -\rho'\big(U_{\ell m} + V_{\ell m}\big)\,. \tag{2.228}$$

When combined with Eqs. (2.220) and (2.221), the last two equations yield

$$\frac{Gm}{r} f_{\ell m} = U_{\ell m} + V_{\ell m}\,. \tag{2.229}$$

This last equation summarizes the entire content of Euler's equation for a perturbed equilibrium.

It has become clear that the functions $f_{\ell m}(r)$ – or, as we shall see below, their substitutes $\eta_\ell(r)$ – determine the entire set of unknowns $\rho_{\ell m}$, $p_{\ell m}$, and $U_{\ell m}$. Our main goal, therefore, is to calculate them for a specified driving potential V. And from $f_{\ell m}(r)$ we wish to extract useful information about the deformation, such as the geometrical shape of the deformed surface, and the multipole moments $I_{\ell m}$ of the deformed mass distribution. Before we explain how $f_{\ell m}$ is computed and this information extracted, we summarize the results in Box 2.3.

Box 2.3 **Clairaut–Radau equation and Love numbers**

The potential driving the body to a perturbed equilibrium state from an unperturbed equilibrium is

$$V = U_{\rm tidal} + \Phi_{\rm C}^{\ell=2} = \sum_{\ell m} \frac{4\pi}{2\ell+1} d_{\ell m} r^\ell Y_{\ell m}(\theta,\phi)\,,$$

where the tidal potential is given by Eq. (2.192), and the quadrupole piece of the centrifugal potential is defined by Eq. (2.195). It is assumed that the unperturbed configuration is spherically symmetric, and the multipole expansion begins at $\ell = 2$.

The gravitational potential of the deformed body is

$$U = -\frac{GM}{r} + \sum_{\ell m} \frac{4\pi}{2\ell+1} \frac{GI_{\ell m}}{r^{\ell+1}} Y_{\ell m}(\theta, \phi),$$

where M is the body's total mass and $I_{\ell m}$ the multipole moments of the mass distribution. By virtue of the perturbation equations, these are related to the coefficients $d_{\ell m}$ of the driving potential by

$$GI_{\ell m} = 2k_\ell R^{2\ell+1} d_{\ell m},$$

where k_ℓ are the *gravitational Love numbers,* which depend on the details of the unperturbed configuration.

The shape of the deformed boundary is described by $r = R + \delta R$, with

$$\delta R = \sum_{\ell m} \frac{4\pi}{2\ell+1} R_{\ell m} Y_{\ell m}(\theta, \phi),$$

and the perturbation equations imply that the coefficients $R_{\ell m}$ are related to $d_{\ell m}$ by

$$R_{\ell m} = h_\ell \frac{R^{\ell+2}}{GM} d_{\ell m},$$

where h_ℓ are the *surficial Love numbers,* which also depend on the details of the unperturbed configuration.

To compute the Love numbers we require a solution to the *Clairaut–Radau equation,*

$$r\eta_\ell' + \eta_\ell(\eta_\ell - 1) + 6\mathcal{D}(\eta_\ell + 1) - \ell(\ell+1) = 0,$$

where a prime indicates differentiation with respect to r, and

$$\mathcal{D} := \frac{4\pi\rho(r)r^3}{3m(r)}$$

encodes the relevant information regarding the body's structure. The differential equation is integrated outward from $r = 0$, with the boundary condition $\eta_\ell(r = 0) = \ell - 2$. It proceeds to $r = R$, and $\eta_\ell(r = R)$ is computed. The Love numbers are then given by

$$k_\ell = \frac{\ell + 1 - \eta_\ell(R)}{2[\ell + \eta_\ell(R)]}, \qquad h_\ell = 1 + 2k_\ell.$$

Clairaut's equation

To determine $f_{\ell m}(r)$ we rely on Eq. (2.229), which we insert within Eq. (2.222), taking into account Eq. (2.224). We simplify the expression by making use of Eq. (2.220) and the equation $m' = 4\pi r^2 \rho$ for the unperturbed mass function. The end result is the second-order differential equation

$$r^2 f_{\ell m}'' + 6\mathcal{D}\big(rf_{\ell m}' + f_{\ell m}\big) - \ell(\ell+1)f_{\ell m} = 0, \tag{2.230}$$

where

$$\mathcal{D}(r) := \frac{4\pi\rho(r)r^3}{3m(r)} = \frac{\rho(r)}{\bar{\rho}(r)} \tag{2.231}$$

is a function that encodes the relevant details of the unperturbed configuration, contained in $\rho(r)$ and $m(r)$; $\bar{\rho} := 3m(r)/(4\pi r^3)$ is the mean density inside a sphere of radius r. This is *Clairaut's equation*, named after the French mathematician Alexis Clairaut (1713–1765). The equation is integrated outward from $r = 0$, and the boundary conditions can be determined by a local analysis near $r = 0$: because $\mathcal{D} \sim 1$ in the limit, we find that $f_{\ell m} \sim a_{\ell m} r^{\ell-2}$ with $a_{\ell m}$ an arbitrary constant, and from this it follows that $f'_{\ell m} \sim (\ell - 2)a_{\ell m} r^{\ell-3}$. These behaviors indicate that the Clairaut equation is ill-conditioned for numerical integration, and we cast it in a more practical form below. It should be noted that $f_{\ell m}(r)$ is determined up to a multiplicative factor $a_{\ell m}$, which is fixed by the junction conditions at $r = R$.

To extract useful information from $f_{\ell m}(r)$ we assume specifically that the unperturbed density $\rho(r)$ goes smoothly to zero at the boundary. We exclude discontinuities to ensure that $\rho_{\ell m}(r)$ is non-singular at $r = R$ – refer to Eq. (2.220) – and to ensure that $U_{\ell m}$ and its first derivative are continuous at $r = R$ – refer to Eq. (2.222). While discontinuities could be accommodated in a more complete treatment of the perturbation equations, we choose to exclude such complications from our discussion; we shall, however, examine the special case of an incompressible fluid below.

We first wish to relate the multipole moments $I_{\ell m}$ of the deformed mass distribution to the coefficients $d_{\ell m}$ of the driving potential; we expect a linear relationship mediated by the functions $f_{\ell m}$. To find this relation we evaluate Eq. (2.229) at $r = R$ and exploit the continuity of $U_{\ell m}$ at the boundary to equate it to its external expression of Eq. (2.223). This gives

$$\frac{GM}{R} f_{\ell m}(R) = \frac{4\pi}{2\ell + 1}\left[\frac{GI_{\ell m}}{R^{\ell+1}} + d_{\ell m} R^{\ell}\right]. \tag{2.232}$$

Repeating the procedure with the first derivative of Eq. (2.229), we also obtain

$$\frac{GM}{R}\Big[Rf'_{\ell m}(R) - f_{\ell m}(R)\Big] = \frac{4\pi}{2\ell + 1}\left[-(\ell + 1)\frac{GI_{\ell m}}{R^{\ell+1}} + \ell d_{\ell m} R^{\ell}\right]. \tag{2.233}$$

These equations can be solved for $d_{\ell m}$ and $I_{\ell m}$, and we get

$$d_{\ell m} = \frac{GM}{4\pi R^{\ell+1}}\Big[Rf'_{\ell m}(R) + \ell f_{\ell m}(R)\Big] \tag{2.234}$$

and

$$GI_{\ell m} = -\frac{GM}{4\pi} R^{\ell}\Big[Rf'_{\ell m}(R) - (\ell + 1) f_{\ell m}(R)\Big]. \tag{2.235}$$

These equations provide the desired relationship between $I_{\ell m}$ and $d_{\ell m}$. This is recast in a simpler form below.

We next determine the geometrical figure of the deformed boundary, at which the perturbed pressure $p + \delta p$ goes to zero. The deformed boundary is described mathematically by $r = R + \delta R(\theta, \phi)$, and the definition of the displacement vector ξ^j makes it clear that

$\delta R = \xi^r$. From this we find that

$$\delta R = \sum_{\ell m} \frac{4\pi}{2\ell+1} R_{\ell m} Y_{\ell m}(\theta, \phi), \tag{2.236}$$

with $4\pi(2\ell+1)^{-1} R_{\ell m} := R f_{\ell m}(R)$. We may express this in terms of $I_{\ell m}$ and $d_{\ell m}$ with the help of Eq. (2.232), and because $I_{\ell m}$ is linearly related to $d_{\ell m}$, we conclude that $R_{\ell m}$ is proportional to $d_{\ell m}$, with a coefficient that depends on $f_{\ell m}$ and its first derivative. This relation is given explicitly below.

Radau's equation and Love numbers

Clairaut's equation (2.230) is independent of the azimuthal number m, and the dependence of the solutions on m must be inherited from the junction conditions at $r = R$. And indeed, we have seen that $f_{\ell m}(r)$ is determined up to a multiplicative factor $a_{\ell m}$, and the dependence on m must be contained in this number. These considerations imply that the function

$$\eta_\ell := \frac{r f'_{\ell m}}{f_{\ell m}} = \frac{d \ln f_{\ell m}}{d \ln r}, \tag{2.237}$$

known as *Radau's function*, must be independent of m. And as it turns out, this combination of $f_{\ell m}$ and $f'_{\ell m}$ is precisely what is involved in the relation between $I_{\ell m}$ and $d_{\ell m}$. Returning to our previous expressions, we see that the relationship can be expressed as

$$G I_{\ell m} = 2 k_\ell R^{2\ell+1} d_{\ell m}, \tag{2.238}$$

where k_ℓ is a dimensionless quantity given by

$$k_\ell := \frac{\ell + 1 - \eta_\ell(R)}{2\big[\ell + \eta_\ell(R)\big]}. \tag{2.239}$$

The key point is that k_ℓ is independent of m, and determined by the Radau function, which is itself determined by Eq. (2.230). We see that the relation of Eq. (2.238) is valid for any perturbed equilibrium driven by a potential V characterized by its spherical-harmonic coefficients $d_{\ell m}$, and that the details of the body's composition are encapsulated in the constants k_ℓ. These all-important numbers are known as *gravitational Love numbers*, and they are named after the British geophysicist A.E.H. Love (1863–1940), who introduced them early in the 20th century. The numbers are also known in the astronomical and celestial mechanics literature as "apsidal constants," because they control the size of tidal and rotational deformations of stars in close binary systems, which lead to observable perturbations in the "line of apsides" (also known as the longitude of pericenter); we describe these perturbations in Sec. 3.4.4. The reader should be warned that our definition of k_ℓ is the one adopted by astronomers; geophysicists also use Love numbers to describe gravimeter measurements made on the surface of the Earth, but their definition of k_ℓ includes an additional factor of two.

Table 2.3 Gravitational Love numbers k_ℓ for various polytropes characterized by the polytropic index n and $\Gamma = 1 + 1/n$. The notation e-p at the end of each number stands for $\times\ 10^{-p}$.

ℓ	$n = 1/2$ $\Gamma = 3$	$n = 2/3$ $\Gamma = 5/2$	$n = 1$ $\Gamma = 2$	$n = 3/2$ $\Gamma = 5/3$	$n = 2$ $\Gamma = 3/2$	$n = 3$ $\Gamma = 4/3$	$n = 4$ $\Gamma = 5/4$
2	4.49154e-1	3.75966e-1	2.59909e-1	1.43279e-1	7.39384e-2	1.44430e-2	1.19488e-3
3	2.03384e-1	1.64696e-1	1.06454e-1	5.28485e-2	2.43940e-2	3.69989e-3	2.23093e-4
4	1.25063e-1	9.85460e-2	6.02413e-2	2.73931e-2	1.15077e-2	1.40970e-3	6.55706e-5
5	8.75838e-2	6.74240e-2	3.92925e-2	1.65688e-2	6.41997e-3	6.56211e-4	2.46927e-5
6	6.60100e-2	4.97911e-2	2.78270e-2	1.09837e-2	3.96610e-3	3.46899e-4	1.09340e-5
7	5.21751e-2	3.86488e-2	2.08099e-2	7.74547e-3	2.62762e-3	2.00562e-4	5.43272e-6

The relation between $R_{\ell m}$ and $d_{\ell m}$ can also be expressed in a similar form. From our previous results it is easy to see that

$$R_{\ell m} = h_\ell \frac{R^{\ell+2}}{GM} d_{\ell m} \,, \tag{2.240}$$

where h_ℓ is another dimensionless quantity given by

$$h_\ell = 1 + 2k_\ell \,. \tag{2.241}$$

These numbers also are independent of m and determined by the Radau function; they are known as *surficial Love numbers*, and are mostly of interest to geophysicists.

In principle, the Radau function can be obtained from $f_{\ell m}$ after integrating Eq. (2.230). It is more practical, however, to formulate a differential equation directly for η_ℓ. By inserting Eq. (2.237) within Eq. (2.230) it is easy to see that the equation becomes

$$r\eta_\ell' + \eta_\ell(\eta_\ell - 1) + 6\mathcal{D}(\eta_\ell + 1) - \ell(\ell + 1) = 0 \,, \tag{2.242}$$

which is known as *Radau's equation*. This is a first-order differential equation, but unlike Eq. (2.230) it is non-linear in the dependent variable. Radau's equation is integrated outward from $r = 0$ with the boundary condition

$$\eta_\ell(r = 0) = \ell - 2 \,, \tag{2.243}$$

and the integration proceeds to the boundary, where $\eta_\ell(r = R)$ is obtained. The computation produces the Love numbers k_ℓ and h_ℓ, and the deformed equilibrium of the body is completely determined. The function $\eta_\ell(r)$ and its differential equation are named after the Prussian-French astronomer and mathematician Rodolphe Radau (1835–1911).

In Table 2.3 we present the results of a computation for deformed polytropes. (Refer to Exercise 2.7 for helpful computational details.) We observe that k_ℓ decreases as n increases (Γ decreases, so that the equation of state becomes softer), and that this behavior is more pronounced as ℓ increases. This reflects the body's internal structure: when n is large and the equation of state is soft, the body is centrally dense with most of its mass confined to the core, where the factor r^ℓ involved in the definition of the multipole moments is small and getting smaller with increasing ℓ. Mathematically, this behavior comes about because

$\mathcal{D}$ decreases rapidly to zero when n is large; with $\mathcal{D}$ negligible near the surface, Eq. (2.230) implies that $f_{\ell m}(r) \propto r^{\ell+1}$ and $\eta_\ell \simeq \ell + 1$, which produces $k_\ell \simeq 0$.

We note that when $n = 1$ and $\rho = \rho_c \sin(\xi)/\xi$, where ξ is the dimensionless radial variable introduced in Sec. 2.2.3, Clairaut's equation can be integrated exactly to give

$$f_{\ell m} = \frac{\xi j_\ell(\xi)}{\xi \cos\xi - \sin\xi} \qquad (n = 1)\,, \tag{2.244}$$

where j_ℓ is a spherical Bessel function. This analytical result can be used as a check on the numerical computations.

It can be shown (see Exercise 2.9) that the magnitude of the gravitational acceleration of a test mass (for example a gravimeter) on the surface of a deformed body is perturbed from the nominal value GM/R^2 by a factor proportional to $[\ell + 2h_\ell - 2(\ell+1)k_\ell]d_{\ell m}$. The first term is the direct perturbation caused by the perturbing force, the second results from the displacement of the surface from its unperturbed location, and the third comes from the redistribution of the matter and the resulting perturbation of the body's gravitational field. Similarly, one can calculate the "tilt" of a gravimeter, the angle between the gravitational acceleration and a normal to the body's surface; this is controlled by the factor $(1 - h_\ell + 2k_\ell)d_{\ell m}$. For perfect-fluid bodies, the relation $h_\ell = 1 + 2k_\ell$ implies that the tilt vanishes – the surface is always perpendicular to a plumb line. But for a heterogeneous body like the Earth, with both solid and liquid components, the Love numbers are no longer linked, and gravimeter and tilt measurements can be used to shed light on the nature of the Earth's interior.

Uniform density

Our previous developments did not allow for a discontinuity of the unperturbed density ρ at the unperturbed boundary $r = R$. As an example of a model involving a discontinuous density and requiring a separate treatment, we examine the simplest case of a homogeneous body with density $\rho(r) = \rho_0 \Theta(R - r)$, in which ρ_0 is a constant and $\Theta(R - r)$ is the Heaviside step function. The body comes with a mass function $m(r) = (4\pi/3)\rho_0 r^3$ and a total mass $M = (4\pi/3)\rho_0 R^3$.

From Eq. (2.220) we find that

$$\rho_{\ell m} = \rho_0 R f_{\ell m}(R)\,\delta(r - R)\,, \tag{2.245}$$

indicating that the density perturbation is supported at the boundary only. The presence of a δ-function in $\rho_{\ell m}$ implies that the first derivative of $U_{\ell m}$ is discontinuous at the boundary, which invalidates many of the manipulations carried out previously. From Eq. (2.222) we indeed find that the junction conditions at $r = R$ are now given by

$$\left[U_{\ell m}\right] = 0\,, \qquad \left[U'_{\ell m}\right] = -\frac{3GM}{R^2} f_{\ell m}(R)\,, \tag{2.246}$$

where $[\psi] := \psi(r = R + \epsilon) - \psi(r = R - \epsilon)$ denotes the jump of a function ψ across $r = R$ (ϵ is a small, positive constant). Because $\rho_{\ell m}$ vanishes inside the body, the internal solution to Eq. (2.222) is $4\pi(2\ell+1)^{-1} c_{\ell m} r^\ell$ for some constants $c_{\ell m}$, and the external solution continues to be $4\pi(2\ell+1)^{-1} G I_{\ell m} r^{-\ell-1}$. Continuity of $U_{\ell m}$ at $r = R$ implies that

$c_{\ell m} = GI_{\ell m} R^{-2\ell-1}$, and the remaining junction condition yields

$$GI_{\ell m} = \frac{3}{4\pi} GMR^{\ell} f_{\ell m}(R), \tag{2.247}$$

which replaces Eq. (2.235).

The link to $d_{\ell m}$ is provided by Eq. (2.229), which requires no modification in this context because both $U_{\ell m}$ and $V_{\ell m}$ are continuous at the boundary. We find that

$$d_{\ell m} = \frac{2(\ell-1)}{3} \frac{GI_{\ell m}}{R^{2\ell+1}}, \tag{2.248}$$

and collecting results, we finally obtain the Love numbers of an incompressible fluid:

$$k_{\ell} = \frac{3}{4(\ell-1)}, \qquad h_{\ell} = \frac{2\ell+1}{2(\ell-1)}. \tag{2.249}$$

It is amusing to note that these results can be recovered on the basis of the recipe presented in Box 2.3, in spite of the fact that the discontinuity at $r = R$ invalidates its usage. Indeed, the uniform-density model implies that $\mathcal{D} = 1$ throughout the body, and in this case the solution to Clairaut's equation is $f_{\ell m} \propto r^{\ell-2}$, leading to $\eta_{\ell} = \ell - 2$. The Love numbers can then be computed by applying the general relations, and the results agree with our previous answers.

2.4.5 Rotational deformations

As a first application of the general theory we consider an isolated body subjected to a rigid rotation of angular velocity ω. We therefore leave $U_{\rm tidal}$ out of the perturbing potential V, but we include $\Phi_{\rm C}^{\ell=2}$ and seek to determine its effect on a body of arbitrary composition. In Sec. 2.5 we shall apply the general theory to a body subjected to tidal forces.

The perturbing potential is obtained from Eq. (2.195), which we write as

$$V = \Phi_{\rm C}^{\ell=2} = \frac{1}{6}\omega^2 r^2 (1 - 3\cos^2\theta) = -\frac{1}{3}\sqrt{\frac{4\pi}{3}}\omega^2 r^2 Y_{20}(\theta). \tag{2.250}$$

The potential is characterized by the single non-vanishing coefficient

$$d_{20} = -\frac{1}{3}\sqrt{\frac{5}{4\pi}}\omega^2, \tag{2.251}$$

and the perturbation is a pure axisymmetric quadrupole. To express our results below it is helpful to introduce a mean density $\bar{\rho}$ defined by $M := \frac{4\pi}{3}\bar{\rho}R^3$, as well as the parameter

$$\zeta := \frac{2\omega^2 R^3}{3GM} = \frac{\omega^2}{2\pi G\bar{\rho}}, \tag{2.252}$$

which is a dimensionless measure of the rotational deformation.

According to Eq. (2.238), the quadrupole moment of the rotating body is

$$GI_{20} = -\frac{2}{3}\sqrt{\frac{5}{4\pi}} k_2 \omega^2 R^5, \tag{2.253}$$

and this leads to a dimensionless quadrupole moment $J_2 := -\sqrt{4\pi/5} I_{20}/(MR^2)$ given by

$$J_2 = k_2 \zeta \,. \tag{2.254}$$

According now to Eqs. (2.236) and (2.240), the deformation of the boundary is described by $\delta R = (4\pi/5) h_2 (R^4/GM) d_{20} Y_{20}$, or

$$\delta R = \frac{1}{4} h_2 \zeta R (1 - 3\cos^2\theta) \,. \tag{2.255}$$

This allows us to define an equatorial radius

$$a := R + \delta R\left(\theta = \tfrac{\pi}{2}\right) = R\left(1 + \frac{1}{4} h_2 \zeta\right), \tag{2.256}$$

and a polar radius

$$a_3 := R + \delta R(\theta = 0) = R\left(1 - \frac{1}{2} h_2 \zeta\right), \tag{2.257}$$

and to express the equation describing the surface, to first order in ζ, as

$$\frac{1}{(R + \delta R)^2} = \frac{\sin^2\theta}{a^2} + \frac{\cos^2\theta}{a_3^2} \,. \tag{2.258}$$

Comparing with Eq. (2.186), we see that this is the equation of a spheroid. We conclude that to first order in perturbation theory, any rotationally deformed body assumes the shape of a spheroid.

The equatorial and polar radii give rise to an eccentricity given by

$$e^2 := 1 - \frac{a_3^2}{a^2} = \frac{3}{2} h_2 \zeta \,, \tag{2.259}$$

and the relation between the quadrupole moment and the eccentricity is then

$$J_2 = \frac{2k_2}{3h_2} e^2 \,. \tag{2.260}$$

For a body of uniform density, $k_2 = \frac{3}{4}$ and $h_2 = \frac{5}{2}$, and these relations become $\zeta = \frac{4}{15} e^2$ and $J_2 = \frac{1}{5} e^2$. Comparing with Eqs. (2.187) and (2.188d), we see that these results agree with those obtained previously for a Maclaurin spheroid of small eccentricity.

2.5 Tidally deformed bodies

Anybody who has spent time at the sea shore has experienced tides. In some parts of the world, such as the Bay of Fundy in Canada or the mouth of the River Severn in England, they are spectacular. In other areas the tides are barely noticeable. Tides are caused by inhomogeneities in the gravitational attraction of the Moon and Sun on the Earth. To lowest order, all parts of the Earth fall toward the Moon with the same acceleration; but the acceleration of a point on the side closest to the Moon is larger than that of a point at the center of the Earth, and this acceleration is itself larger than that of a point on the surface opposite to the Moon. Furthermore, points on the sides of the Earth perpendicular to the direction to the Moon have a small component of their acceleration toward the center of the Earth. As a result the tidal deformation has a characteristic quadrupolar shape, leading to the twice-per-day phenomenon of ocean tides. The solar tide is a little less than half as strong as the lunar tide. Depending on the relative position of the Sun and Moon in the sky, the tidal amplitudes vary from the large "spring tide" to the smaller "neap tide."

The ocean tides are very complex, because they depend not only on the gravitational attraction of the Sun and Moon, but also on the interaction between the water and the continental shelf where we landlubbers observe them. Less noticeable in daily life, but equally important, are the "solid" Earth tides, the tidal deformations of the full Earth. These are measured and analyzed to high precision using gravimeters, which measure time variations of g, the acceleration at the surface of the Earth. Measurements such as these are used to determine the Earth's gravitational Love numbers k_ℓ, which reveal information regarding its internal structure.

Tidal interactions are also important in many close binary-star systems. The modification to each star's gravitational potential caused by the tidal deformation leads to observable perturbations in the orbital motion, to be described in Chapter 3. These can be exploited to learn something about the stellar interiors through the inferred values of k_ℓ.

Tidal deformations can lead to heating and dissipation of angular momentum. In the Jovian system, tidal heating is responsible for the spectacular volcanic activity observed on Io's surface, which far exceeds in intensity the activity seen on Earth. In the Earth–Moon system, tidal torquing is responsible for the facts that the Moon is receding from the Earth and that the length of the day is slowly increasing. Interestingly, it was Immanuel Kant (1724–1804) who first suggested that this might be the case, long before the effect could be measured. He reasoned that the Moon presents the same face to the Earth because its rotation has slowed down by virtue of tidal dissipation, and that therefore the same must be happening to the Earth. Eventually, the two bodies will become tidally locked, rotating and revolving with the same angular velocity, and with the same side facing each other in perpetuity. If the Earth is still populated with sentient beings when this occurs, then half of them will never have the privilege of observing the *clair de lune*.

In this section we explore the physics of tides on a fluid body. We first examine *static tides* in Sec. 2.5.1, in which the external tidal field varies so slowly that it never takes the body out of hydrostatic equilibrium. The tools required to describe such tides were fashioned in Sec. 2.4, and this will be our second application of the general theory of deformed bodies. In Sec. 2.5.2 we incorporate viscous dissipation in our discussion of static tides, and reveal some of its most important consequences. We conclude with an exploration of *dynamical tides* in Sec. 2.5.3; here the tidal field varies so rapidly that it forces the body out of its equilibrium state. This is a vast domain of study, and our discussion will be limited to a very simplified model involving a body of uniform density.

2.5.1 Static tides

As a second application of the general theory of deformed bodies developed in Sec. 2.4, we consider a rotating or non-rotating body subjected to a tidal field created by one or more external bodies. We ignore the body's rotational deformation, which was treated separately in Sec. 2.4.5, and focus on the tidal deformation. We therefore leave $\Phi_{\rm C}^{\ell=2}$ out of the perturbing potential V, include $U_{\rm tidal}$, and seek to determine its effect on a body of arbitrary composition.

The tidal potential was first introduced in Sec. 2.1, where it was denoted $U_{\neg A}^{\rm eff}$, and a simplified expression valid for a body with small multipole moments $I_A^{\langle L\rangle}$ was first written down in Eq. (2.11) on page 66. Working in the body's rotating frame, we copy this here as

$$V = U_{\rm tidal} = -\sum_{\ell=2}^{\infty} \frac{1}{\ell!}\mathcal{E}_L(t)x^L\,, \tag{2.261}$$

in which the *tidal moments* $\mathcal{E}_L(t)$ are time-dependent STF tensors that serve to specify the tidal environment. They are defined by

$$\mathcal{E}_L(t) := -\partial_L U_{\neg A}(t,\mathbf{0})\,, \tag{2.262}$$

in which the gravitational potential $U_{\neg A}$ created by the external bodies is differentiated ℓ times with respect to x^j and the result evaluated at the body's center-of-mass $\boldsymbol{x} = \mathbf{0}$. The tidal moments depend on time, but as was discussed near the end of Sec. 2.4.1, this dependence is assumed to be sufficiently slow that the tidal field is never able to take the body out of hydrostatic equilibrium. This assumption will be relaxed in Sec. 2.5.3.

The tidal potential is naturally expressed as an expansion in STF tensors, but the general theory of Sec. 2.4 relies on an expansion in spherical harmonics. The translational tools were developed in Sec. 1.5.3, where we showed that the STF product $n^{\langle L\rangle}$, with $\boldsymbol{n} := \boldsymbol{x}/r$ denoting a unit vector in the direction of $\boldsymbol{x}$, can be expressed as

$$n^{\langle L\rangle} = \frac{4\pi\ell!}{(2\ell+1)!!}\sum_{m=-\ell}^{\ell} \mathscr{Y}_{\ell m}^{\langle L\rangle} Y_{\ell m}(\theta,\phi) \tag{2.263}$$

in terms of spherical harmonics, with $\mathscr{Y}_{\ell m}^{\langle L\rangle}$ the constant STF tensor defined by

$$Y_{\ell m}(\theta,\phi) = \mathscr{Y}_{\ell m}^{*\langle L\rangle} n_{\langle L\rangle}\,, \tag{2.264}$$

an expansion of the spherical-harmonic functions in STF tensors; the asterisk indicates complex conjugation. Inserting $x^L = r^\ell n^L$ within Eq. (2.261) and making use of the spherical-harmonic decomposition of $n^{\langle L\rangle}$, we quickly find that the coefficients $d_{\ell m}$ involved in the decomposition of the perturbing potential, as defined by Eq. (2.225), are given by

$$d_{\ell m} = -\frac{1}{(2\ell-1)!!}\mathscr{Y}_{\ell m}^{\langle L\rangle}\mathcal{E}_L\,. \tag{2.265}$$

A few steps of algebra next reveal that

$$\mathcal{E}_L = -\frac{4\pi\ell!}{2\ell+1}\sum_{m=-\ell}^{\ell} d_{\ell m}\mathscr{Y}_{\ell m}^{*\langle L\rangle} \tag{2.266}$$

is the inverse relation.

According to Eq. (2.238), the tidal field deforms the body so as to create the multipole moments $GI_{\ell m} = 2k_\ell R^{2\ell+1} d_{\ell m}$. We wish to express this in STF form, and the required translation was worked out in Sec. 1.5.3 – refer to Eq. (1.168). Collecting results, we find that the STF multipole moments of a body deformed by an external tidal field are given by

$$GI_{\langle L\rangle} = -\frac{2k_\ell}{(2\ell-1)!!}R^{2\ell+1}\mathcal{E}_L\,, \tag{2.267}$$

in which k_ℓ are the body's gravitational Love numbers. This expression can be inserted within the general multipole expansion of the body's potential, which is displayed in Eq. (1.157). We find

$$U = \frac{GM}{r} - \sum_{\ell=2}^{\infty}\frac{2k_\ell}{\ell!}\frac{R^{2\ell+1}}{r^{\ell+1}}\mathcal{E}_L n^L\,, \tag{2.268}$$

and the total gravitational potential is

$$U + U_{\rm tidal} = \frac{GM}{r} - \sum_{\ell=2}^{\infty}\frac{1}{\ell!}\Big[1 + 2k_\ell\big(R/r\big)^{2\ell+1}\Big]\mathcal{E}_L x^L\,. \tag{2.269}$$

This expression leaves out a term $g_j x^j$, where $g_j := \partial_j U_{\neg A}(\mathbf{0})$ is the body's acceleration in the field of the external bodies; as we saw back in Sec. 2.1, this term plays no role because it is cancelled out by a fictitious force originating from the translation to the body's moving frame. Focusing on the tidal terms, we see that the total perturbation at $r = R$ is given by a contribution proportional to $(1 + 2k_\ell)\mathcal{E}_L n^L$ from each tidal moment; the first unit of $\mathcal{E}_L n^L$ is the direct contribution from the tidal field, and the remaining $2k_\ell$ units come from the body's response to the applied tidal field.

The body's deformed surface is described by Eq. (2.236), in which we insert Eqs. (2.240) and (2.265). After conversion to an expansion in STF tensors, we find that the surface of a body deformed by tidal forces is described by $r = R + \delta R$ with

$$\delta R = -\sum_{\ell=2}^{\infty}\frac{1}{\ell!}h_\ell\frac{R^{\ell+2}}{GM}\mathcal{E}_L n^L\,, \tag{2.270}$$

where h_ℓ are the surficial Love numbers.

In many circumstances it is sufficient to keep only the leading, quadrupole term in the expansions of Eqs. (2.268) and (2.270). Truncating all expressions at $\ell = 2$, we find that the tidal potential simplifies to

$$U_{\rm tidal} = -\frac{1}{2}\mathcal{E}_{jk}(t)x^j x^k\,, \tag{2.271}$$

in which $\mathcal{E}_{jk}(t)$ is the tidal quadrupole moment. The body's quadrupole moment is then

$$GI_{\langle jk\rangle} = -\frac{2}{3}k_2 R^5\mathcal{E}_{jk}\,, \tag{2.272}$$

and its potential simplifies to

$$U = \frac{GM}{r} - k_2\frac{R^5}{r^3}\mathcal{E}_{jk}n^j n^k\,. \tag{2.273}$$

In this truncated description the surface deformation becomes

$$\delta R = -\frac{1}{2} h_2 \frac{R^4}{GM} \mathcal{E}_{jk} n^j n^k . \tag{2.274}$$

When the body is a member of an N-body system, the tidal field is supplied by the remaining $N-1$ bodies, and in this case the tidal quadrupole moment is given by

$$\begin{aligned}
\mathcal{E}_{jk} &= -\sum_{B \neq A} G m_B \partial^A_{jk} \frac{1}{r_{AB}} \\
&= -3 \sum_{B \neq A} \frac{G m_B}{r^3_{AB}} n^{\langle jk \rangle}_{AB} \\
&= -\sum_{B \neq A} \frac{G m_B}{r^3_{AB}} \left(3 n^j_{AB} n^k_{AB} - \delta^{jk} \right),
\end{aligned} \tag{2.275}$$

in which $r_{AB} := |\boldsymbol{r}_{AB}| = |\boldsymbol{r}_A - \boldsymbol{r}_B|$ is the distance between body A and body B, and $\boldsymbol{n}_{AB} := \boldsymbol{r}_{AB}/r_{AB}$ is a unit vector pointing from body B to body A. To arrive at this expression for $\mathcal{E}_{jk}$ we assume that the multipole moments $I^{\langle L \rangle}_B$ of the external bodies are small, so that they need not be included in the expression for the external potential. We have reverted to the notation of Sec. 1.6, in which the mass of body B is denoted m_B instead of M_B. Similarly, the mass of body A will be denoted m_A instead of M_A.

As a concrete example, imagine that our body is non-rotating, and that it is the first member of a two-body system on a circular orbit of radius $r := r_{12}$ and angular velocity $\Omega = \sqrt{Gm/r^3}$, in which $m = m_1 + m_2$. Taking the x-y plane to coincide with the orbital plane, we have that $\boldsymbol{n}_{12} = [\cos \Omega t, \sin \Omega t, 0]$, and we write $\boldsymbol{n} = [\sin\theta \cos\phi, \sin\theta \sin\phi, \cos\theta]$. With this we find that

$$\mathcal{E}_{jk} n^j n^k = -3 \frac{G m_2}{r^3} \left[\sin^2\theta \cos^2(\phi - \Omega t) - \frac{1}{3} \right]. \tag{2.276}$$

We see that the tidal field is proportional to Gm_2/r^3, and by virtue of its dependence on $\cos^2(\phi - \Omega t) = \frac{1}{2} + \frac{1}{2}\cos 2(\phi - \Omega t)$, we see that it oscillates at *twice* the orbital frequency Ω. The peak position of the tidal bulge is obtained when the argument of the cosine function vanishes, and we find that $\phi_{\rm bulge} = \Omega t$; the bulge is at all times aligned with the orbiting body.

2.5.2 Tidal dissipation

Our discussion of static tides was based on an assumption that the body is adequately modeled as a perfect fluid of mass density ρ and pressure p. For many applications this model is indeed adequate, but in some aspects it is deficient, and the deficiency is somewhat severe in the context of tidal dynamics. What is missing from a perfect-fluid model is a mechanism to dissipate energy, and in this section we attempt to incorporate this important effect in the physics of tides.

The simplest way to include a dissipative mechanism in fluid mechanics is to let the fluid be a viscous fluid, and to add the (kinematic) viscosity ν to the list of fluid variables, along with ρ and p. This is defined in such a way that the frictional force on a fluid element is given by the kinematic viscosity multiplied by the mass density multiplied by the element's cross-sectional area multiplied by the velocity gradient across the fluid element; the unit of ν is then m^2/s. The inclusion of viscosity in the fluid dynamics requires a modification of Euler's equation, which is generalized to the famous *Navier–Stokes equation*. It would take us much too far afield to formulate a perturbation theory based on the Navier–Stokes equation, and to find solutions that describe a tidally deformed body. Such developments are beyond the scope of this book, and we shall instead be satisfied with displaying the final outcome of this analysis. We provide no proof, but attempt to motivate the answer by appealing to an analogous physical system (see Box 2.4).

Box 2.4 **Driven harmonic oscillator**

As we shall see in Sec. 2.5.3, the physics of tidal deformations is closely analogous to the physics of a simple harmonic oscillator driven by an external force. In this analogy, the oscillator's position x from equilibrium plays the role of the tidal deformation (as measured by the mass quadrupole moment $I_{\langle jk \rangle}$), and the external force f plays the role of the applied tidal field (as measured by the tidal quadrupole moment $\mathcal{E}_{jk}$).

In the absence of damping, the oscillator's response to the external force is governed by

$$\ddot{x} + \omega^2 x = f,$$

in which an overdot indicates differentiation with respect to t, and ω is the oscillator's natural frequency. We assume that the internal time scale $T_{\rm int} := \omega^{-1}$ is very short compared with the external time scale $T_{\rm ext}$ associated with the behavior of the external force. The general solution to this equation is

$$x(t) = x(0)\cos\omega t + \frac{1}{\omega}\dot{x}(0)\sin\omega t + \frac{1}{\omega}\int_0^t f(t')\sin\omega(t-t')\,dt',$$

and repeated integration by parts turns this expression into

$$x(t) = \left[x(0) - \frac{1}{\omega^2}f(0) + \cdots\right]\cos\omega t + \frac{1}{\omega}\left[\dot{x}(0) - \frac{1}{\omega^2}\dot{f}(0) + \cdots\right]\sin\omega t \\ + \frac{1}{\omega^2}\big[f(t) + \cdots\big],$$

in which the neglected terms are smaller than the leading terms by a factor of order $(T_{\rm int}/T_{\rm ext})^2 \ll 1$. Averaging over the oscillations to keep track of the long-term motion, we arrive at

$$\langle x(t)\rangle = \frac{1}{\omega^2}\big[f(t) + \cdots\big],$$

and this result is analogous to Eq. (2.272).

We next incorporate dissipation in the oscillator's response by inserting a damping term in the differential equation, which now reads

$$\ddot{x} + 2\zeta\omega\dot{x} + \omega^2 x = f,$$

where ζ is a dimensionless parameter. We consider an overdamped situation with $\zeta \gg 1$, so that the damping time scale $T_{\rm damp} := (\zeta\omega)^{-1}$ is very short compared with the oscillation time scale $T_{\rm int}$. The general solution to the differential equation is

$$x(t) = x(0)e^{-\zeta\omega t}\left(\cosh\lambda\omega t + \frac{\zeta}{\lambda}\sinh\lambda\omega t\right) + \frac{1}{\lambda\omega}\dot{x}(0)e^{-\zeta\omega t}\sinh\lambda\omega t$$
$$+ \frac{1}{\lambda\omega}\int_0^t f(t')e^{-\zeta\omega(t-t')}\sinh\lambda\omega(t-t')\,dt',$$

where $\lambda := \sqrt{\zeta^2 - 1}$. Repeated integration by parts turns this into

$$x(t) = \text{transient terms} + \frac{1}{\omega^2}\big[f(t) - \tau\dot{f}(t) + \cdots\big],$$

where $\tau := 2\zeta/\omega$; the transient terms all decay exponentially, and the neglected terms are smaller than the leading term by a factor of order $(\tau/T_{\rm ext})^2 \ll 1$. This result is analogous to Eq. (2.277), and because $f(t) - \tau\dot{f}(t) = f(t-\tau) + \cdots$, we see that the effect of dissipation is to create a delay τ between the action of the force and the oscillator's response. We have assumed that the delay is short compared with the external time scale. Because $\tau = 2T_{\rm int}^2/T_{\rm damp} \gg T_{\rm int}$, this can be arranged when $T_{\rm int}/T_{\rm ext} \ll T_{\rm damp}/T_{\rm int} \ll 1$.

Equation (2.272) reveals that in the absence of viscosity, the body's quadrupole moment is related to the tidal quadrupole moment by $GI_{\langle jk\rangle}(t) = -\frac{2}{3}k_2R^5\mathcal{E}_{jk}(t)$, in which k_2 is the gravitational Love number and R is the body's radius; we see that the body responds instantaneously to the applied tidal field. In the presence of viscosity the relation becomes

$$GI_{\langle jk\rangle}(t) = -\frac{2}{3}k_2R^5\big[\mathcal{E}_{jk}(t) - \tau\dot{\mathcal{E}}_{jk}(t) + \cdots\big] \tag{2.277a}$$

$$= -\frac{2}{3}k_2R^5\big[\mathcal{E}_{jk}(t-\tau) + \cdots\big], \tag{2.277b}$$

in which an overdot indicates differentiation with respect to t; as for Eq. (2.272), this equation is formulated in the body's rotating frame. The new parameter τ has the dimension of time, and it represents a viscosity-induced delay between the action of the tidal field and the body's response. The viscous delay must be proportional to the fluid's kinematic viscosity, and it must be related to the body's radius R and its mass M; dimensional analysis reveals that it must be of the form $\tau \propto \bar{\nu}R/(GM)$, in which $\bar{\nu}$ is the averaged kinematic viscosity over the volume occupied by the fluid. The numerical coefficient will depend on the details of internal structure, and this dependence could be captured by introducing a third Love number, in addition to k_2 and h_2. For our purposes it is preferable to adopt τ itself as a dimensionful "Love quantity," a parameter that characterizes the body's internal structure. We assume that τ is much larger than the internal time scale $T_{\rm int} \sim (G\rho)^{-1/2} \sim \sqrt{R^3/GM}$ associated with the fluid's dynamics, and that it is also much smaller than the external time scale $T_{\rm ext} \sim \sqrt{r_{AB}^3/GM}$ associated with the orbital dynamics. The neglected terms in Eq. (2.277) are of order $(\tau/T_{\rm ext})^2 \ll 1$ relative to the leading term proportional to $\mathcal{E}_{jk}$.

The viscous delay also appears in a modified relation between the surface deformation δR and the applied tidal field. In this case it is possible to show that Eq. (2.274) becomes

$$\delta R = -\frac{1}{2} h_2 \frac{R^4}{GM} \left[\mathcal{E}_{jk}(t) - \tau \dot{\mathcal{E}}_{jk}(t) + \cdots \right] n^j n^k \tag{2.278a}$$

$$= -\frac{1}{2} h_2 \frac{R^4}{GM} \left[\mathcal{E}_{jk}(t - \tau) + \cdots \right] n^j n^k, \tag{2.278b}$$

and here also we see that the tidal deformation is delayed with respect to applied tidal field.

An important consequence of the inclusion of viscosity within the fluid dynamics is that it leads to energy dissipation, in the form of heat production, within the body. Another important consequence is that it gives rise to a transfer of angular momentum between the body and the remote bodies responsible for the tidal field. To calculate this effect we work initially in the non-rotating frame $x^{\bar{a}}$, and we recall that the tidal potential is given by $U_{\rm tidal} = -\frac{1}{2}\mathcal{E}_{\bar{a}\bar{b}} x^{\bar{a}} x^{\bar{b}}$, so that the density of tidal forces acting within the body is $f_{\bar{a}} = \rho \partial_{\bar{a}} U_{\rm tidal} = -\rho \mathcal{E}_{\bar{a}\bar{b}} x^{\bar{b}}$. These forces exert a torque, and integrating over the body, we find that the rate of change of the body's angular momentum is given by

$$\frac{dS_{\bar{a}}}{dt} = \epsilon_{\bar{a}\bar{b}\bar{c}} \int x^{\bar{b}} f^{\bar{c}} \, d^3\bar{x}. \tag{2.279}$$

We obtain

$$\frac{dS_{\bar{a}}}{dt} = -\epsilon_{\bar{a}\bar{b}\bar{c}} \, \mathcal{E}^{\bar{c}}_{\ \bar{p}} \int \rho x^{\bar{b}} x^{\bar{p}} \, d^3\bar{x} \tag{2.280}$$

after inserting our expression for the force density. The integral is the body's mass quadrupole moment, which can be decomposed as $I^{\bar{b}\bar{p}} = I^{\langle \bar{b}\bar{p} \rangle} + \frac{1}{3}\delta^{\bar{b}\bar{p}} I$. The trace term can be seen to give no contribution to the torque, and we find that

$$\frac{dS_{\bar{a}}}{dt} = -\epsilon_{\bar{a}\bar{b}\bar{c}} \, I^{\langle \bar{b}\bar{p} \rangle} \mathcal{E}^{\bar{c}}_{\ \bar{p}} \tag{2.281}$$

gives the rate at which the body's angular momentum changes as a result of the tidal interaction.

We next write the spin vector as $\boldsymbol{S} = S\boldsymbol{e}$, in terms of a magnitude S and a direction $\boldsymbol{e}$, and obtain an expression for dS/dt by projecting $d\boldsymbol{S}/dt$ in the direction of $\boldsymbol{e}$. Since we are now dealing with a scalar quantity, we may express each tensorial quantity in the body's rotating frame x^j. We obtain

$$\frac{dS}{dt} = -\epsilon_{jkn} e^j I^{\langle kp \rangle} \mathcal{E}_p^{\ n}, \tag{2.282}$$

in which we insert Eq. (2.277). We observe that the term proportional to $\mathcal{E}^{kp}$ makes no contribution to dS/dt, and our final expression is

$$\frac{dS}{dt} = \frac{2}{3} k_2 \tau R^5 \epsilon_{jkn} \mathcal{E}^k_{\ p} \dot{\mathcal{E}}^{pn}. \tag{2.283}$$

We note that the effect is proportional to τ; there is no transfer of angular momentum without viscosity.

As a concrete example we return to the discussion initiated at the end of Sec. 2.5.1. Once more we take our body to be the first member of a two-body system on a circular orbit of radius r and angular velocity $\Omega = \sqrt{Gm/r^3}$, in which $m = m_1 + m_2$. The body is rotating with an angular velocity ω, and we assume that the rotation axis $\boldsymbol{e}$ is perpendicular to the orbital plane. In the rotating frame the tidal quadrupole moment is given by

$$\mathcal{E}^{jk} = -\frac{Gm_2}{r^3}\left(3n_{12}^j n_{12}^k - \delta^{jk}\right), \tag{2.284}$$

in which $\boldsymbol{n}_{12} = [\cos(\Omega - \omega)t, \sin(\Omega - \omega)t, 0]$. With $\boldsymbol{n} = [\sin\theta\cos\phi, \sin\theta\sin\phi, \cos\theta]$ pointing toward a point on the surface, we have that the tidal deformation is proportional to

$$\mathcal{E}_{jk}(t-\tau)n^j n^k = -3\frac{Gm_2}{r^3}\left\{\sin^2\theta\cos^2\left[\phi - (\Omega - \omega)(t-\tau)\right] - \frac{1}{3}\right\}. \tag{2.285}$$

The peak position of the tidal bulge is now given by

$$\phi_{\text{bulge}} = \phi_2 - (\Omega - \omega)\tau, \tag{2.286}$$

in which $\phi_2 := (\Omega - \omega)t$ is the angular position of the second body as measured in the rotating frame of the first body. We observe that the viscous delay creates a misalignment between the tidal bulge and the direction of the orbiting body. When $\omega < \Omega$, that is, when the second body orbits faster than the first body rotates, we find that $\phi_{\text{bulge}} < \phi_2$; the tide *lags* behind the orbiting body. When $\omega > \Omega$, that is, when the first body rotates faster than the second body orbits, we find instead that $\phi_{\text{bulge}} > \phi_2$; the tide *leads* in front of the orbiting body.

The viscosity-produced misalignment between the tidal bulge and the orbiting body is directly implicated in the transfer of angular momentum. When the bulge lags behind the orbiting body, the tidal forces acting on the excess mass in the bulge create a positive torque that increases the body's angular momentum. When the bulge leads in front of the orbiting body, the torque is negative and produces a decrease in angular momentum. This intuition is confirmed with an explicit evaluation of dS/dt from Eq. (2.283), making use of Eq. (2.284). After making the substitutions and simplifying, we eventually arrive at

$$\frac{dS}{dt} = 6k_2\frac{Gm_2^2 R^5}{r^6}(\Omega - \omega)\tau. \tag{2.287}$$

This expression confirms that S increases when $\omega < \Omega$, and that it decreases when $\omega > \Omega$. A consequence of this transfer of angular momentum is that the body evolves toward an equilibrium state with $\omega = \Omega$; eventually the body will become *tidally locked*, and rotate at the same frequency as the orbital motion. This phenomenon is observed everywhere in the solar system, including in our own backyard: our Moon always shows the same face (ignoring librations), and this is the result of its tidal interaction with the Earth. In fact, the moons of all planets in the solar system tend to be tidally locked, unless their orbits are too large to permit a significant tidal interaction. The Earth, however, is not yet tidally locked to the Moon's orbital motion: the scaling of dS/dt with the square of the remote mass implies that the time scale for tidal locking is much longer for the larger body than for the smaller body.

2.5.3 Dynamical tides

The discussion of Secs. 2.5.1 and 2.5.2 was limited to tidal interactions that occur over an external time scale T_{ext} that is very long compared with the internal time scale T_{int} associated with the fluid dynamics of the deformed body. In this section we relax this condition, and consider a regime in which $T_{\rm ext}/T_{\rm int}$ may not be large; this is the realm of *dynamical tides*.

As we have done previously, the body is assumed to be spherical and in hydrostatic equilibrium in the absence of a tidal interaction. To keep the problem simple we further assume that the body is non-rotating and possesses a uniform density ρ_0. The body is perturbed by external bodies which generate a tidal potential

$$U_{\rm tidal} = -\frac{1}{2}\mathcal{E}_{jk}(t)\,x^j x^k\,, \tag{2.288}$$

in which $\mathcal{E}_{jk}(t)$ are time-dependent tidal moments. For an external object of mass M_2 at a distance r from the reference body, we have seen that $\mathcal{E}_{jk} \sim GM_2/r^3$.

Tidal response in the dynamical regime

The tidal forces created by $U_{\rm tidal}$ produce a (small) perturbation in the fluid configuration of the body. As we shall justify below, we can think of $\mathcal{E}_{jk}(t)$ as a driving force, and of the fluid configuration as a harmonic oscillator of natural frequency ω_2 that responds to this driving force. The response is measured by $I^{\langle jk\rangle}(t)$, the body's tracefree quadrupole moment, which vanishes in the unperturbed state. We shall find that

$$GI_{\langle jk\rangle}(t) = -\frac{2}{5}GMR^2\mathcal{F}_{\langle jk\rangle}(t)\,, \tag{2.289}$$

where

$$\mathcal{F}_{\langle jk\rangle}(t) := \frac{1}{\omega_2}\int_{-\infty}^{t} \mathcal{E}_{jk}(t')\sin\omega_2(t-t')\,dt' \tag{2.290}$$

is the typical response function of a driven oscillator – refer to Box 2.4. For a body of uniform density, the natural frequency associated with a quadrupolar ($\ell = 2$) driving force is given by

$$\omega_2 = \sqrt{\frac{4}{5}\frac{GM}{R^3}}\,. \tag{2.291}$$

This is comparable to the Keplerian angular velocity of an object orbiting just above the body's surface, and therefore larger than the orbital angular velocity Ω of any one of the external objects. For a single external object of mass M_2 at a distance r from the reference body, we have that $\Omega^2 \sim GM/r^3$ with $M := M_1 + M_2$, so that $\omega_2/\Omega \sim (M_1/M)^{1/2}(r/R)^{3/2} > 1$. A typical situation would have $r \gg R$ and $\omega_2/\Omega \gg 1$, but we are interested in close encounters with ω_2/Ω of order unity.

When the time scale over which $\mathcal{E}_{jk}(t)$ varies is very long compared with ω_2^{-1}, that is, when $\omega_2/\Omega \gg 1$, the integral of Eq. (2.290) can be evaluated by repeated integration by parts. This yields

$$\mathcal{F}_{\langle jk\rangle} = \omega_2^{-2}\,\mathcal{E}_{jk} - \omega_2^{-4}\,\ddot{\mathcal{E}}_{jk} + \omega_2^{-6}\,\overset{(4)}{\mathcal{E}}_{jk} + \cdots\,, \tag{2.292}$$

in which an overdot (or a number within brackets) indicates differentiation with respect to t. Inserting this and Eq. (2.291) within Eq. (2.289), we find that the quadrupole moment becomes

$$GI_{\langle jk \rangle} = -\frac{1}{2} R^5 \left(\mathcal{E}_{jk} - \omega_2^{-2} \ddot{\mathcal{E}}_{jk} + \omega_2^{-4} \overset{(4)}{\mathcal{E}}_{jk} + \cdots \right). \tag{2.293}$$

The leading term corresponds to the limit of static tides, and this result agrees with Eq. (2.272) when $k_2 = \frac{3}{4}$, the value of the gravitational Love number for an incompressible fluid.

Normal-mode analysis

To establish Eq. (2.289) we return to the formalism of fluid perturbations presented in Sec. 2.4.3 and relax the assumption that the perturbed configuration is an equilibrium state. We go back to Eq. (2.212), from which we eliminate the Coriolis term because the body is non-rotating. Recognizing that all perturbation variables depend linearly upon the displacement vector ξ^j, it is useful to write the perturbed Euler equation in the abstract form

$$\frac{\partial^2 \xi^j}{\partial t^2} + \mathscr{L}^j_k \xi^k = \partial^j U_{\text{tidal}} , \tag{2.294}$$

in which $\mathscr{L}^j_k$ is a linear differential operator that appears in implicit form in Eq. (2.212) – it is related to the terms involving $\delta\rho$, δp, and δU. We seek to integrate Eq. (2.294) with the boundary conditions that ξ^j be regular at $r = 0$ and continuous at $r = R$. We proceed via a normal-mode analysis of the equation.

The *normal modes* of oscillation of a fluid configuration are solutions to the homogeneous equation $\partial_{tt}\zeta^j + \mathscr{L}^j_k \zeta^k = 0$, which are taken to be in the form

$$\zeta^j(t, \boldsymbol{x}) = e^{-i\omega t} f^j(\boldsymbol{x}) , \tag{2.295}$$

where ω is the mode frequency and $f^j(\boldsymbol{x})$ its spatial profile, which is a solution to

$$\mathscr{L}^j_k f^k = \omega^2 f^j . \tag{2.296}$$

This is a second-order differential equation for f^j, and a general solution will contain two freely-specifiable constants of integration. Since the equation is homogeneous in f^j, one of these is an overall multiplicative constant, and the second constant is determined by one of the two boundary conditions that f^j must satisfy. The second boundary condition will not be satisfied unless ω, the only other tunable parameter, is chosen appropriately. Equation (2.296) therefore specifies an *eigenvalue problem* for the frequencies ω and mode functions f^j. The eigenfrequencies and eigenfunctions will be assigned an abstract label λ, which is typically discrete. (Later on λ will be identified with the spherical-harmonic indices ℓm.)

The differential operator $\mathscr{L}^j_k$ was shown by Chandrasekhar to be self-adjoint with respect to the integration measure $\rho\, d^3x$. This fact implies a number of useful properties for the eigenvalues and eigenfunctions. First, the eigenvalues ω^2_λ are real. Second, the eigenfunctions f^j_λ are orthogonal, in the sense that $\int \boldsymbol{f}^*_\lambda \cdot \boldsymbol{f}_{\lambda'} \rho\, d^3x = N_\lambda \delta_{\lambda,\lambda'}$, where an asterisk indicates complex conjugation and N_λ is a normalization constant. Third, the eigenfunctions are (believed to be) complete, in the sense that *any* vectorial function $\boldsymbol{\xi}$ admits an expansion of the form

$$\boldsymbol{\xi}(t,\boldsymbol{x}) = \sum_\lambda a_\lambda(t)\boldsymbol{f}_\lambda(\boldsymbol{x}), \tag{2.297}$$

in which the expansion coefficients are given by

$$a_\lambda(t) = \frac{1}{N_\lambda}\int \boldsymbol{\xi}\cdot\boldsymbol{f}^*_\lambda\,\rho\,d^3x. \tag{2.298}$$

These equations are the key to solving Eq. (2.294).

Assuming that the mode functions and frequencies have all been identified, we expand the displacement vector as in Eq. (2.297) and the driving force as

$$\partial^j U_{\rm tidal} = \sum_\lambda u_\lambda(t) f^j_\lambda(\boldsymbol{x}), \tag{2.299}$$

with

$$u_\lambda = \frac{1}{N_\lambda}\int \partial_j U^{\rm tidal} f^{*j}_\lambda\,\rho\,d^3x. \tag{2.300}$$

We substitute the expansions within Eq. (2.294), make use of Eq. (2.296), and take advantage of the orthogonality of the eigenfunctions. We obtain, after simplification, a sequence of ordinary differential equations for the mode coefficients a_λ:

$$\ddot{a}_\lambda + \omega^2_\lambda a_\lambda = u_\lambda. \tag{2.301}$$

Each equation describes a harmonic oscillator of natural frequency ω_λ subjected to a driving force $u_\lambda(t)$. We suppose that the driving force vanishes in the distant past, and assume that each mode is quiet before the action of the force; we therefore subject Eq. (2.301) to the initial conditions $a_\lambda(t=-\infty) = 0 = \dot{a}_\lambda(t=-\infty)$. It is easy to check that the solution is

$$a_\lambda(t) = \frac{1}{\omega_\lambda}\int_{-\infty}^t u_\lambda(t')\sin\omega_\lambda(t-t')\,dt'. \tag{2.302}$$

At this stage the problem is formally solved. With each frequency ω_λ and mode function $\boldsymbol{f}_\lambda$ previously identified, the displacement vector can be constructed as in Eq. (2.297). The density perturbation is then given by Eq. (2.206), the pressure perturbation by Eq. (2.209), and δU is obtained by integrating Eq. (2.213).

The normal modes of a fluid configuration depend on the unperturbed configuration. For a uniform-density model the eigenvalue problem can be shown to give rise to three separate classes of modes. In Cowling's terminology, we have the so-called *p-modes*, which are essentially acoustic waves driven by pressure fluctuations, the *g-modes*, which are essentially gravity waves driven by buoyancy, and the *f-modes* (also known as *Kelvin modes*), which are essentially gravity waves confined to the surface. (Note that the term "gravity wave" designates a fluid wave whose restoring force is gravity rather than fluid pressure. The terminology is standard in atmospheric physics, for example, but gravity waves should not be confused with the "gravitational waves" of general relativity.) In the case considered here, in which the perturbation is driven by a quadrupolar tidal force, it is known that the p-modes and the g-modes do not get excited by the tidal interaction;

the overlap integrals of Eq. (2.300) vanish for these modes. (This was actually established by one of us back in 1983, at a time when the other had not yet started his university education.) The only modes that matter in our problem are the f-modes, for which the eigenvalue equation can be integrated very easily.

f-modes

The defining property of these modes is the fact that they satisfy the divergence-free condition

$$\partial_j f^j = 0 . \tag{2.303}$$

The eigenvalue equation is obtained from Eq. (2.212), and we have that $\partial_j(\delta p/\rho - \delta U) = \omega^2 e^{-i\omega t} f_j$. The form of the equation implies that f_j must be the gradient of a scalar function ψ,

$$f_j = \partial_j \psi . \tag{2.304}$$

This, in turn, implies that the divergence-free condition becomes

$$\nabla^2 \psi = 0 , \tag{2.305}$$

and the mode equation simplifies to

$$\delta p/\rho - \delta U = \omega^2 e^{-i\omega t} \psi . \tag{2.306}$$

To solve these equations we expand ψ in spherical harmonics,

$$\psi(\boldsymbol{x}) = \sum_{\ell m} g_{\ell m}(r) Y_{\ell m}(\theta, \phi) , \tag{2.307}$$

and seek to determine the radial functions $g_{\ell m}(r)$. Laplace's equation immediately implies that they must be proportional to r^ℓ. The mode equation will then be used to determine the eigenfrequencies ω_ℓ.

From Eq. (2.206) we find that the mode produces a perturbation in density given by $\delta\rho = -e^{-i\omega t} f^j \partial_j \rho$, while Eq. (2.209) implies that $\delta p = -e^{-i\omega t} f^j \partial_j p$. For the uniform-density model under consideration, these equations become

$$\delta\rho = \rho_0 \delta(r - R) \sum_{\ell m} g'_{\ell m}(R) Y_{\ell m}(\theta, \phi) e^{-i\omega t} \tag{2.308}$$

and

$$\delta p/\rho = \frac{4\pi}{3} G\rho_0 r \sum_{\ell m} g'_{\ell m}(r) Y_{\ell m}(\theta, \phi) e^{-i\omega t} , \tag{2.309}$$

in which a prime indicates differentiation with respect to r. The perturbation in density gives rise to a change in the body's gravitational potential. Expanding in spherical harmonics as in Eq. (2.219d), we find that $U_{\ell m} \propto r^\ell$ inside the body, while $U_{\ell m} \propto r^{-\ell-1}$ outside the body. Demanding continuity of $U_{\ell m}$ at $r = R$, but imposing the proper discontinuity in $U'_{\ell m}$ to account for the delta function in $\delta\rho$, we arrive at

$$U_{\ell m} = \frac{4\pi}{2\ell+1} G\rho_0\, R g'_{\ell m}(R)\, (r/R)^\ell \tag{2.310}$$

inside the body. To get the solution outside the body we simply replace the last factor by $(R/r)^{\ell+1}$.

Collecting results, we find that the mode equation reduces to

$$\frac{4\pi}{3} G\rho_0\, r g'_{\ell m}(r) - \frac{4\pi}{2\ell+1} G\rho_0\, R g'_{\ell m}(R)\, (r/R)^\ell = \omega^2 g_{\ell m}(r)\,. \tag{2.311}$$

To simplify the notation we suppress the ℓm label and introduce the dimensionless quantities x, y, and ϵ such that $r := Rx$, $g := Rg'(R)y$, and $\omega^2 := (4\pi/3)G\rho_0\epsilon^2$. With these variables the mode equation becomes

$$x\frac{dy}{dx} - \epsilon^2 y = \frac{3}{2\ell+1} x^\ell\,, \tag{2.312}$$

which is to be integrated between $x = 0$ and $x = 1$. The solution must be regular at $x = 0$ and satisfy the boundary condition $dy/dx = 1$ at $x = 1$; this follows directly from its definition in terms of $g(r)$. It is already known that y must be proportional to x^ℓ, in order for ψ to satisfy Laplace's equation. The boundary condition at $x = 1$ determines the constant of proportionality: we find that $y = \ell^{-1}x^\ell$. It then follows from the mode equation that $\epsilon^2 = 2\ell(\ell-1)/(2\ell+1)$. The modes are now completely determined.

To summarize, we have found that the f-modes are labelled by the index $\lambda \equiv \ell m$, that the eigenfrequencies are independent of m and given by

$$\omega_\ell^2 = \frac{8\pi}{3} G\rho_0 \frac{\ell(\ell-1)}{2\ell+1}\,, \tag{2.313}$$

and that the eigenfunctions are

$$f^j_{\ell m} = \partial^j \psi_{\ell m}\,, \qquad \psi_{\ell m} = R^2 (r/R)^\ell Y_{\ell m}(\theta,\phi)\,. \tag{2.314}$$

The normalization is chosen so that $f^j_{\ell m}$ has the dimension of length. It is easy to check that $N_{\ell m} := \int \boldsymbol{f}^*_{\ell m} \cdot \boldsymbol{f}_{\ell m}\, \rho_0\, d^3x = \ell \rho_0 R^5$.

Tidal deformation

We may now involve the f-modes in the solution of our problem. We first return to the overlap integral of Eq. (2.300), which becomes

$$u_{\ell m} = \frac{1}{\ell R^{\ell+3}} \int \partial^j U_{\rm tidal}\, \partial_j \big(r^\ell \bar{Y}_{\ell m}\big)\, d^3x\,, \tag{2.315}$$

or

$$u_{\ell m} = \frac{1}{\ell R^{\ell+3}} \oint U_{\rm tidal}\, \partial_j \big(r^\ell \bar{Y}_{\ell m}\big)\, dS^j \tag{2.316}$$

after an integration by parts. Here $dS^j = R^2 n^j \, d\Omega$ is the surface element on the spherical boundary of the (unperturbed) body, and the integral simplifies to

$$u_{\ell m} = \frac{1}{R^2} \oint U_{\rm tidal} \bar{Y}_{\ell m} \, d\Omega \tag{2.317}$$

after evaluation of the radial derivative.

The tidal potential is integrated against spherical-harmonic functions on the surface of the body, and the operation is simplified by the fact that $U_{\rm tidal}$ is a quadrupolar potential. We write

$$U_{\rm tidal} = -\frac{1}{2} \mathcal{E}_{jk}(t) x^j x^k = -\frac{1}{2} R^2 \mathcal{E}_{jk}(t) n^{\langle jk \rangle}, \tag{2.318}$$

and to evaluate the integral we rely on the relationship between STF tensors and spherical harmonics that was uncovered back in Sec. 1.5.3. According to Eq. (1.164),

$$n^{\langle jk \rangle} = \frac{8\pi}{15} \sum_{m=-2}^{2} \mathscr{Y}^{\langle jk \rangle}_{2,m} Y_{2,m}(\theta, \phi), \tag{2.319}$$

where $\mathscr{Y}^{\langle L \rangle}_{\ell m}$ is the constant STF tensor defined by $Y_{\ell m} = \mathscr{Y}^{*\langle L \rangle}_{\ell m} n_{\langle L \rangle}$. Evaluation of the overlap integral yields

$$u_{2,m}(t) = -\frac{4\pi}{15} \mathscr{Y}^{\langle jk \rangle}_{2,m} \mathcal{E}_{jk}(t), \tag{2.320}$$

and all other coefficients $u_{\ell m}$ vanish. Substitution of this result into Eq. (2.302) produces

$$a_{2,m}(t) = -\frac{4\pi}{15} \mathscr{Y}^{\langle jk \rangle}_{2,m} \mathcal{F}_{jk}(t), \tag{2.321}$$

where

$$\mathcal{F}_{jk}(t) := \frac{1}{\omega_2} \int_{-\infty}^{t} \mathcal{E}_{jk}(t') \sin \omega_2 (t - t') \, dt' \tag{2.322}$$

is the response function of Eq. (2.290). The only relevant frequency is ω_2, and this is obtained by setting $\ell = 2$ within Eq. (2.313); the result is Eq. (2.291).

We may now construct the displacement vector. Substitution of $a_{\ell m}(t)$ into Eq. (2.297) produces

$$\begin{aligned} \xi^j &= \sum_m a_{2,m} f^j_{2,m} \\ &= \sum_m a_{2,m} \partial^j \left(r^2 Y_{2,m} \right) \\ &= \partial^j \left(-\frac{4\pi}{15} r^2 \mathcal{F}_{pq} \sum_m \mathscr{Y}^{\langle pq \rangle}_{2,m} Y_{2,m} \right) \\ &= \partial^j \left(-\frac{1}{2} r^2 \mathcal{F}_{pq} n^p n^q \right), \end{aligned} \tag{2.323}$$

and we finally arrive at

$$\xi^j = -\mathcal{F}^j_{\ k}(t) x^k. \tag{2.324}$$

The problem is now completely solved. From this we may obtain all other perturbation variables. For example, the velocity field is

$$v^j = -\dot{\mathcal{F}}^j_{\ k}(t) x^k , \tag{2.325}$$

the density perturbation is

$$\delta\rho = -R\rho_0 \delta(r-R) \mathcal{F}_{jk} n^j n^k , \tag{2.326}$$

the pressure perturbation is

$$\delta p = -\frac{4\pi}{3} G\rho_0^2 \mathcal{F}_{jk} x^j x^k , \tag{2.327}$$

and the perturbation in the body's gravitational potential is

$$\delta U = -\frac{4\pi}{5} G\rho_0 \mathcal{F}_{jk} x^j x^k . \tag{2.328}$$

Quadrupole moment

Our final task is to compute $I^{\langle jk \rangle}(t)$, the body's quadrupole moment tensor,

$$I^{\langle jk \rangle} = \int_{V+\Delta V} (\rho + \delta\rho) x^{\langle jk \rangle} \, d^3x , \tag{2.329}$$

in which $V + \Delta V$ is the region of space occupied by the deformed body (while V is the spherical region occupied by the unperturbed body). This can be written as

$$I^{\langle jk \rangle} = \int_V \rho\, x^{\langle jk \rangle} \, d^3x + \int_{\Delta V} \rho\, x^{\langle jk \rangle} \, d^3x + \int_V \delta\rho\, x^{\langle jk \rangle} \, d^3x . \tag{2.330}$$

The first integral gives the unperturbed quadrupole moment, which vanishes. The second integral can be expressed as $\oint_S \rho\, x^{\langle jk \rangle} \boldsymbol{\xi} \cdot d\boldsymbol{S}$, where S is the boundary of the unperturbed body, and $d\boldsymbol{S}$ is its surface element. The quadrupole moment is therefore

$$I^{\langle jk \rangle} = \oint_S \rho\, x^{\langle jk \rangle} \boldsymbol{\xi} \cdot d\boldsymbol{S} + \int_V \delta\rho\, x^{\langle jk \rangle} \, d^3x . \tag{2.331}$$

In this we insert Eq. (2.324) for $\boldsymbol{\xi}$, and Eq. (2.326) for $\delta\rho$.

Our expression for $I^{\langle jk \rangle}$ appears to be ill defined, because it is not clear what the value of ρ is on the surface, where it abruptly jumps from ρ_0 to zero. If, for example, the surface integral is evaluated just below $r = R$, then $\rho = \rho_0$ and we obtain a non-zero value. In this case the volume integral is zero, because $\delta\rho$ is proportional to $\delta(r - R)$ and the integral just misses the delta function. On the other hand, if the surface integral is evaluated just above $r = R$, then $\rho = 0$ and the integral is zero. In this case the volume integral does not vanish, and fortunately, the result turns out to be the same in either case. We arrive at the unambiguous expression

$$I^{\langle jk \rangle} = -\frac{8\pi}{15} \rho_0 R^5 \mathcal{F}^{\langle jk \rangle} . \tag{2.332}$$

This is the same statement as in Eq. (2.289). Our treatment of the tidal interaction is now complete.

2.6 Bibliographical notes

The physics of stellar structure described in Sec. 2.1 is covered in a lot more depth in a number of textbooks, which also discuss stellar evolution; among our favorites is the very accessible text by Hansen, Kawaler, and Trimble (2004). The phenomenon of neutrino oscillations, shown in Box 2.1 to have essential consequences on the physics of the Sun, was first proposed by Wolfenstein (1978) and explored further by Mikheev and Smirnov (1985 and 1986). Our discussion of polytropes in Sec. 2.2.3, isothermal spheres in Sec. 2.2.4, and white dwarfs in Sec. 2.2.5 is heavily inspired by Chandrasekhar's *An Introduction to the Study of Stellar Structure* (1958). Another useful reference on the astrophysics of white dwarfs and other compact bodies is Shapiro and Teukolsky (1983). The story of Chandrasekhar calculating the structure of white dwarfs and discovering the mass limit while traveling to England from India is famous; his discovery was published in Chandrasekhar (1931). Laudau's argument in favor of neutron stars was published in Landau (1932).

Our presentation of the theory of rotating bodies in Sec. 2.3 relies heavily on the excellent treatise by Tassoul (1978). The Maclaurin spheroids and Jacobi ellipsoids of Sec. 2.3.2 are explored in much greater detail in Chandrasekhar's *Ellipsoidal Figures of Equilibrium* (1987), a must-read for anyone interested in the structure and stability of rotating bodies. Our method to calculate the internal gravitational potential of an ellipsoid of uniform density is borrowed from Chapter IV of Moulton's text (1984).

The general theory of deformed fluid bodies developed in Sec. 2.4 can be pieced together from a number of useful sources, including Kopal (1959 and 1978) and Cox (1980). The formalism of fluid perturbations described in Sec. 2.4.3 is borrowed directly from Chandrashekhar's *Ellipsoidal Figures of Equilibrium*. The gravitational and surficial Love numbers were introduced by British geophysicist Augustus Edward Hough Love in his 1911 book. The Love numbers of polytropes were first computed by Brooker and Olle (1955).

The discussion of tides in Sec. 2.5 barely scratches the surface of a very rich field. A general introduction to the phenomenon can be found in Mccully (2006). Our presentation of the dynamical tides of a body of uniform density in Sec. 2.5.3 is based on Turner (1977) and Will (1983). The standard classification of fluid perturbation modes was introduced in Cowling (1941).

2.7 Exercises

2.1 Consider a power-series solution for the Lane–Emden equation, $\theta = \sum_p a_p \xi^p$.

(a) Show that the boundary conditions at $\xi = 0$ require that $a_0 = 1$ and $a_1 = 0$.

(b) Show that $a_p = 0$ for all odd values of p.

(c) Determine a_2, a_4, and a_6 for a general polytropic index n, and reproduce the expansions displayed in Box 2.2. Show also that your expansion agrees with the exact results for $n = 0$, $n = 1$, and $n = 5$.

2.2 Explain why Ω is ambiguous for a polytrope of index $n = 5$. Show by a direct calculation that, for $n = 5$, $\Omega = -\sqrt{2\pi}(81/16)K^{5/2}G^{-3/2}$.

2.3 Calculate the gravitational potential energy Ω for a Maclaurin spheroid, and verify the result of Eq. (2.181).

2.4 Fill in the gaps left in the calculation of the dimensionless multipole moments J_ℓ of a Maclaurin spheroid, and verify the result of Eq. (2.187).

2.5 Show that the surface of a rotationally deformed body can be described by

$$r = a(1 - \alpha + \alpha \cos 2L),$$

where a is the equatorial radius and L is the geographical latitude, which ranges from $\frac{\pi}{2}$ at the North Pole to $-\frac{\pi}{2}$ at the South Pole. Determine α in terms of the surficial Love number h_2 and the rotational-deformation parameter ζ. Evaluate α for the Earth and compare with the observed value of 0.001677.

2.6 Show that for a rotationally deformed body of arbitrary composition, the body's moment of inertia $I := \int \rho(x^2 + y^2)\, d^3x$ is given by $I = I_0 + \delta I$, with

$$I_0 = \frac{8\pi}{3} \int_0^R \rho r^4 \, dr$$

denoting the unperturbed piece, and

$$\delta I = \frac{2}{3} k_2 \zeta M R^2$$

the perturbation. Here k_2 is the gravitational Love number, and ζ is the deformation parameter. Assume that the unperturbed density goes smoothly to zero at $r = R$.

2.7 Show that for a body with a polytropic equation of state, the function $\mathcal{D}$ behaves as

$$\mathcal{D} = 1 - \frac{n}{15}\xi^2 + \frac{n(19n - 25)}{3150}\xi^4 + O(\xi^6)$$

near $\xi = 0$. Use this expression to derive the expansion

$$\begin{aligned}\eta_\ell = l - 2 + \frac{2n(\ell - 1)}{5(2\ell + 3)}\xi^2 - \frac{n(\ell - 1)}{525} \\ \times \frac{(76n - 100)\ell^2 + (144n - 300)\ell - (165n + 225)}{(2\ell + 3)^2(2\ell + 5)}\xi^4 \\ + O(\xi^6)\end{aligned}$$

for the Radau function.

2.8 The quadrupole moment J_2 of the Sun is known from helioseismology measurements to be 2.2×10^{-7}. Its rotation period is approximately 25 days. From these facts, estimate k_2 for the Sun and compare the result with the entry in Table 2.3 for the ideal gas equation of state ($\Gamma = 5/3$). From this comparison, what do you infer about the density distribution of the Sun as compared to a simple polytrope? What factors could lead to such a difference?

2.9 Consider a gravitating body with an $\ell = 2$ deformation, with $d_{2m} = d_2 Y^*_{2m}(\boldsymbol{e})$, where $\boldsymbol{e}$ is a given direction. A gravimeter measures the acceleration of a test body on the surface of the body. Assume that the surficial Love number h_2 is independent of k_2.

(a) Show that the variation in the magnitude of the acceleration over the surface is proportional to $1 + h_2 - 3k_2$.

(b) Show that the angle between the direction of the acceleration and the normal to the surface is proportional to $1 - h_2 + 2k_2$.

2.10 We consider two fluid configurations, one with a homogeneous mass density ρ_0, the other with a density $\rho(r)$ that deviates only slightly from ρ_0; we write $\rho = \rho_0 + \delta\rho$, and assume that $\delta\rho$ is small compared with ρ_0. We assume that the nearly homogeneous body has the same radius R and the same total mass M as the homogeneous body. We have seen that $\eta_\ell = \ell - 2$ for the homogeneous body, and selecting $\ell = 2$, we assume $\eta_2(r) \ll 1$ for the nearly homogeneous body.

(a) Prove that under the circumstances described here, Radau's equation for the nearly homogeneous body simplifies to

$$r\eta_2' + 5\eta_2 + 6\delta\mathcal{D} = 0\,,$$

where $\delta\mathcal{D} := \mathcal{D} - 1$ is also taken to be small.

(b) Calculate $\delta\mathcal{D}$ in terms of $\delta\rho$, and prove that

$$\eta_2(R) = -\frac{15}{\rho_0 R^5}\int_0^R \delta\rho(r) r^4\, dr\,.$$

(c) Calculate the moment of inertia $I := \int \rho(r)(x^2 + y^2)\, d^3x$ of the nearly homogeneous body, and relate it to I_0, the moment of inertia of the homogeneous body. Then show that

$$\eta_2(R) = -3\big(I/I_0 - 1\big)\,.$$

(d) Calculate k_2 to first order in $(I/I_0 - 1)$.

3 Newtonian orbital dynamics

In this chapter we apply the tools developed in the previous two chapters to an exploration of the orbital dynamics of bodies subjected to their mutual gravitational attractions. Many aspects of what we learned in Chapters 1 and 2 will be put to good use, and the end result will be considerable insight into the behavior of our own solar system. To be sure, the field of *celestial mechanics* has a rich literature that goes back centuries, and this relatively short chapter will only scratch the surface. We believe, however, that we have sampled the literature well, and selected a good collection of interesting topics. Some of the themes introduced here will be featured in later chapters, when we turn to relativistic aspects of celestial mechanics.

We begin in Sec. 3.1 with a very brief survey of celestial mechanics, from Newton to Einstein. In Sec. 3.2 we give a complete description of Kepler's problem, the specification of the motion of two spherical bodies subjected to their mutual gravity. In Sec. 3.3 we introduce a powerful formalism to treat Keplerian orbits perturbed by external bodies or deformations of the two primary bodies; in this framework of *osculating Keplerian orbits*, the motion is at all times described by a sequence of Keplerian orbits, with constants of the motion that evolve as a result of the perturbation. We shall apply this formalism to a number of different situations, and highlight a number of important processes that take place in the solar system and beyond. In Sec. 3.5 we examine the three-body problem and briefly touch upon the general case of N bodies. We conclude in Sec. 3.6 with a review of the Lagrangian formulation of Newtonian mechanics.

3.1 Celestial mechanics from Newton to Einstein

The triumph of Newton's mechanics and universal gravitation is largely contained in the confrontation with the observed motion of celestial bodies in our solar system, which was initiated by Kepler even before Newton's laws became available. The two-body dynamics of Newton's theory immediately accounted for Kepler's laws, which state that the orbits of bodies in the solar system are ellipses that trace out equal areas in equal times, with periods inversely proportional to the $3/2$ power of their diameters. Edmund Halley used Newton's equations to point out that the comets that had been observed in 1531, 1607, and 1682 were actually a single object that orbits the sun with a period ranging from 75 to 76 years. Refined orbital calculations carried out by Alexis Clairaut led to the successful prediction of the comet's return in 1759. The difficult problem of describing the detailed motion of the Moon, in particular the advance of its perigee, was successfully tackled by Clairaut using Newtonian theory; a first attempt was actually made by Newton himself, but with somewhat incorrect results that discouraged him and motivated him to start a new career at the Royal Mint. Finally, in a triumph of theoretical prediction, Urbain Jean Joseph Le Verrier in France, and independently John Couch Adams in England, pointed out that certain anomalies in the orbit of Uranus could best be explained by the existence of an additional planet, and each astronomer made a rough prediction of where such a planet might be found. In 1846, a day after receiving Le Verrier's prediction for its position, German astronomers discovered the new planet, which is now called Neptune.

This accumulation of successes built such confidence in Newton's theory that when the crisis occurred, the shock was almost palpable. The crisis was caused by Mercury. By the middle of the nineteenth century, astronomers had established that the perihelion of Mercury (the point of closest approach to the Sun) was advancing at a rate of 575 arcseconds per century relative to the fixed stars. Although the two-body solution of Newton's theory requires the perihelion to be fixed in direction, it seemed clear that the advance should be caused by the gravitational influences of the other planets (mostly Venus because of its proximity, and Jupiter because of its large mass) on Mercury's orbit. Fresh from his success with the prediction of Neptune, Le Verrier applied his methods to the problem of Mercury. He calculated the amount that each planet would contribute to Mercury's perihelion advance (see Table 3.1), but the total fell short of the measured value, by an amount comparable to 40 arcseconds per century. The modern value of the discrepancy is 42.98 ± 0.04 arcseconds per century, based upon improved measurements of Mercury's orbit using radar ranging, combined with improved data on the masses and orbits of the other planets, and accurate numerical ephemeris codes for calculating orbits.

The discrepancy could not be attributed to calculational errors or faulty observations, and no viable explanation could be found for the next 50 years. In the spirit of Neptune, Le Verrier and others supported the existence of another planet between Mercury and the Sun, which was given the provisional name Vulcan. But despite systematic astronomical searches, no credible evidence for such a planet was ever discovered. If changing the solar system would not do, perhaps a change of theory might fare better? Simon Newcomb proposed that all would be well with Mercury if the inverse square law of Newtonian gravity were changed to the inverse power of 2.0000001574. Such a change, however, would also contribute to the advance of the lunar perigee, and once improved data on the lunar orbit became available, Newcomb's proposal was shown not to be viable.

The resolution of the problem of Mercury is by now legendary. The place is Berlin, the time November 1915. Albert Einstein, using the new equations of general relativity, calculates the motion of Mercury and shows that the relativistic laws of orbital motion account for the notorious discrepancy. Einstein was overjoyed, and later wrote to a friend that this discovery gave him palpitations of the heart. The modern value for the shift predicted by general relativity, using the best data for all relevant quantities, is 42.98 arcseconds per century, in perfect agreement with the measured value.

Table 3.1 Planetary contributions to Mercury's perihelion advance (in arcseconds per century).

Planet	Advance
Venus	277.8
Earth	90.0
Mars	2.5
Jupiter	153.6
Saturn	7.3
Total	531.2
Discrepancy	42.9
Modern measured value	42.98 ± 0.04
General relativity prediction	42.98

3.2 Two bodies: Kepler's problem

Kepler's problem is to determine the motion of two bodies subjected to their mutual gravitational attraction, under the assumption that each body can be taken to be spherically symmetric. This is the simplest problem of celestial mechanics, but also one of the most relevant, because to a good first approximation, the motion of any planet around the Sun can be calculated while ignoring the effects of the other planets. It is also a problem that can be solved exactly and completely, in terms of simple functions.

3.2.1 Effective one-body description

The foundations for Kepler's problem were provided back in Sec. 1.6.7. We have a first body of mass m_1, position $\boldsymbol{r}_1$, velocity $\boldsymbol{v}_1 = d\boldsymbol{r}_1/dt$, and acceleration $\boldsymbol{a}_1 = d\boldsymbol{v}_1/dt$, and a second body of mass m_2, position $\boldsymbol{r}_2$, velocity $\boldsymbol{v}_2 = d\boldsymbol{r}_2/dt$, and acceleration $\boldsymbol{a}_2 = d\boldsymbol{v}_2/dt$. We place the origin of the coordinate system at the system's barycenter $\boldsymbol{R}$, so that $m_1\boldsymbol{r}_1 + m_2\boldsymbol{r}_2 = \boldsymbol{0}$. The position of each body is then given by

$$\boldsymbol{r}_1 = \frac{m_2}{m}\boldsymbol{r}, \qquad \boldsymbol{r}_2 = -\frac{m_1}{m}\boldsymbol{r}, \tag{3.1}$$

in which $m := m_1 + m_2$ is the total mass and $\boldsymbol{r} := \boldsymbol{r}_1 - \boldsymbol{r}_2$ the separation between bodies. Similar relations hold between $\boldsymbol{v}_1$, $\boldsymbol{v}_2$ and the relative velocity $\boldsymbol{v} := \boldsymbol{v}_1 - \boldsymbol{v}_2 = d\boldsymbol{r}/dt$. The relative acceleration $\boldsymbol{a} := \boldsymbol{a}_1 - \boldsymbol{a}_2 = d\boldsymbol{v}/dt$ is obtained from Eq. (1.220) by removing all terms that involve the multipole moments of each body (which vanish when the bodies are spherical); we have

$$\boldsymbol{a} = -\frac{Gm}{r^2}\boldsymbol{n}, \tag{3.2}$$

where $r := |\boldsymbol{r}|$ is the distance between bodies, and $\boldsymbol{n} := \boldsymbol{r}/r$ a unit vector that points from body 2 to body 1. This is the equation of motion for the relative orbit, and thanks to Eq. (3.1), its solution is sufficient to determine the individual motion of each body. Equation (3.2) can be interpreted as describing the motion of a fictitious particle at position $\boldsymbol{r}$ in the field of a gravitating center of mass m at $\boldsymbol{r} = \boldsymbol{0}$; as such our two-body problem has been reformulated as an effective one-body problem. This radical simplification of the original problem, which involved six independent degrees of freedom (the three components of each position vector) instead of the current three (the three components of the separation vector), is a consequence of the conservation of total momentum, which implies that the motion of the barycenter position $\boldsymbol{R}$ is uniform and therefore trivial.

According to Eq. (1.221), the total energy of the two-body system is given by

$$E = \frac{1}{2}\mu v^2 - \frac{G\mu m}{r}, \tag{3.3}$$

in which $\mu := m_1 m_2/m$ is the reduced mass of the system; to obtain this expression we once more dropped all terms involving multipole moments, and set $\boldsymbol{V} = d\boldsymbol{R}/dt = \boldsymbol{0}$ in the barycentric frame. The system's total angular momentum is given by Eq. (1.204), which reduces to

$$\boldsymbol{L} = m_1 \boldsymbol{r}_1 \times \boldsymbol{v}_1 + m_2 \boldsymbol{r}_2 \times \boldsymbol{v}_2 = \mu \boldsymbol{r} \times \boldsymbol{v} \tag{3.4}$$

for a two-body system. Because the bodies are assumed to have no spin, and because the angular momentum is contained entirely in the orbital motion, we use the standard notation $\boldsymbol{L}$ instead of $\boldsymbol{J}$ for the angular-momentum vector. It is a simple matter to verify that $dE/dt = 0$ by virtue of Eq. (3.2), and that $d\boldsymbol{L}/dt = \boldsymbol{0}$ by virtue of the sole fact that the acceleration $\boldsymbol{a}$ is directed along $\boldsymbol{r}$.

The constancy of the angular-momentum vector has far-reaching consequences on the motion of the bodies. Because $\boldsymbol{L}$ is a constant vector orthogonal to both $\boldsymbol{r}$ and $\boldsymbol{v}$, the motion must take place in a fixed plane that is at all times perpendicular to $\boldsymbol{L}$. We therefore have achieved another radical simplification of the problem: by confining the motion to a plane we have eliminated one degree of freedom from the original three associated with the effective one-body problem.

3.2.2 Orbital plane

We take the orbital plane to coincide with the x-y plane of the coordinate system, and we align $\boldsymbol{L}$ with the z-direction. To simplify the notation in subsequent developments, we write

$$\boldsymbol{L} = \mu \boldsymbol{h}, \qquad \boldsymbol{h} := \boldsymbol{r} \times \boldsymbol{v} = h\boldsymbol{e}_z, \tag{3.5}$$

with $h := |\boldsymbol{h}|$ denoting the magnitude of the constant vector $\boldsymbol{r} \times \boldsymbol{v}$.

It is helpful to describe the motion with the polar coordinates r and ϕ, defined such that the components of the separation vector are given by

$$\boldsymbol{r} = [r\cos\phi, r\sin\phi, 0]; \tag{3.6}$$

both r and ϕ depend on time. A vectorial basis in the orbital plane can be built from the constant unit vectors $\boldsymbol{e}_x$ and $\boldsymbol{e}_y$, but it is useful to introduce also the time-dependent unit vectors

$$\boldsymbol{n} := [\cos\phi, \sin\phi, 0], \qquad \boldsymbol{\lambda} := [-\sin\phi, \cos\phi, 0], \tag{3.7}$$

which are closely tied to the description of the orbital motion. The vector $\boldsymbol{n} := \boldsymbol{r}/r$ points from body 2 to body 1, while $\boldsymbol{\lambda}$ is orthogonal to it. The vectors satisfy

$$\frac{d\boldsymbol{n}}{d\phi} = \boldsymbol{\lambda}, \qquad \frac{d\boldsymbol{\lambda}}{d\phi} = -\boldsymbol{n}. \tag{3.8}$$

The basis is completed with $\boldsymbol{e}_z$, which is normal to the orbital plane and aligned with the angular-momentum vector.

The vectors $\boldsymbol{r}$, $\boldsymbol{v}$, and $\boldsymbol{a}$ can each be decomposed in the orbital basis. Simple computations involving Eqs. (3.8) produce

$$\boldsymbol{r} = r\,\boldsymbol{n}, \tag{3.9a}$$

$$\boldsymbol{v} = \dot{r}\,\boldsymbol{n} + r\dot{\phi}\,\boldsymbol{\lambda}, \tag{3.9b}$$

$$\boldsymbol{a} = \left(\ddot{r} - r\dot{\phi}^2\right)\boldsymbol{n} + \frac{1}{r}\frac{d}{dt}\left(r^2\dot{\phi}\right)\boldsymbol{\lambda}, \tag{3.9c}$$

in which an overdot indicates differentiation with respect to t.

3.2.3 First integrals

The acceleration of Eq. (3.9) may now be inserted within Eq. (3.2), and the absence of a component along $\boldsymbol{\lambda}$ immediately implies that $r^2\dot{\phi}$ is a constant of the motion. Because $r^2\dot{\phi}$ is equal to the z (and only non-vanishing) component of the vector $\boldsymbol{r} \times \boldsymbol{v}$, we have rediscovered the statement of angular-momentum conservation. Taking Eq. (3.5) into account, we have that

$$r^2\dot{\phi} = h. \tag{3.10}$$

In this form we can see that conservation of angular momentum gives rise to Kepler's second law: $r(r\,d\phi)$ is twice the area swept by the orbit as it advances by an angle $d\phi$, and $r^2\dot{\phi}$ is twice the area swept per unit time; conservation of angular momentum implies equal areas for equal times.

The radial component of the equation of motion yields $\ddot{r} - r\dot{\phi}^2 = -Gm/r^2$, or

$$\ddot{r} - \frac{h^2}{r^3} = -\frac{Gm}{r^2} \tag{3.11}$$

after involving Eq. (3.10). This second-order differential equation for $r(t)$ can be integrated once by applying the $\dot{r}$-trick: multiply the equation by $\dot{r}$ and recognize that each term is a total derivative with respect to time. Integration produces

$$\frac{1}{2}\dot{r}^2 + \frac{h^2}{2r^2} - \frac{Gm}{r} = \varepsilon, \tag{3.12}$$

in which ε is another constant of the motion. From Eq. (3.3) it is easy to see that $E = \mu\varepsilon$, and we have rediscovered the statement of energy conservation.

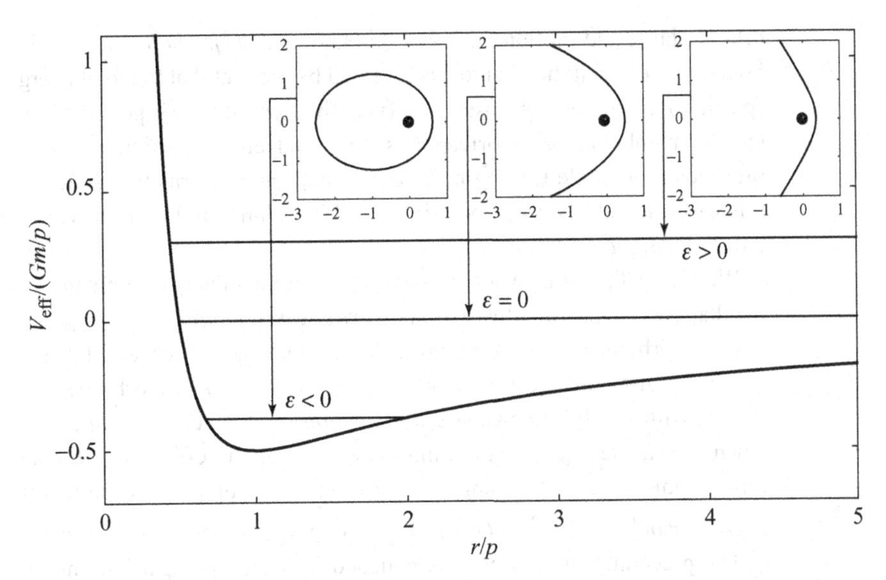

Fig. 3.1 Effective potential for Kepler's problem, together with lines of constant ε. The radial variable r is presented in units of $p := h^2/(Gm)$, and $V_{\rm eff}$ is presented in units of Gm/p. The regions of allowed motion correspond to $\varepsilon \geq V_{\rm eff}(r)$, and turning points occur when $\varepsilon = V_{\rm eff}(r)$. The different types of Keplerian motion (hyperbolic, parabolic, and elliptical) are shown.

It is instructive to rewrite Eq. (3.12) in the form

$$\frac{1}{2}\dot{r}^2 = \varepsilon - V_{\rm eff}(r), \tag{3.13}$$

in which the effective radial potential is defined as

$$V_{\rm eff}(r) := \frac{h^2}{2r^2} - \frac{Gm}{r}. \tag{3.14}$$

This new form allows us to explore the qualitative features of Keplerian motion without having to perform additional calculations. In Fig. 3.1 we display a plot of $V_{\rm eff}(r)$ for $h \neq 0$, with zero on the vertical axis denoting the limiting value of the effective potential as $r \to \infty$. The potential consists of an attractive (negative) gravitational well, and a repulsive (positive) centrifugal barrier rising to infinity as $r \to 0$. Motion is allowed when $\dot{r}^2 \geq 0$, that is, when $\varepsilon \geq V_{\rm eff}(r)$, and regions where this condition is met are easily identified in the figure. A *turning point* occurs when $\dot{r} = 0$ and $\varepsilon = V_{\rm eff}(r)$; at such points the radial velocity changes sign, and the motion changes from incoming to outgoing, or from outgoing to incoming.

We can easily imagine how a particle would move in this effective potential. If the particle has an energy $\varepsilon > 0$, then there is a single turning point at some innermost radius $r_{\rm min}$, and the motion takes place for $r \geq r_{\rm min}$. The particle starts at infinity with a negative radial velocity $\dot{r} = -\sqrt{2\varepsilon}$ and a vanishing angular velocity $\dot{\phi}$. As r decreases, $\dot{r}$ becomes increasingly negative, and $\dot{\phi}$ increases to obey conservation of angular momentum, until the particle reaches the turning point at $r = r_{\rm min}$. At this point $\dot{r}$ turns positive, and the particle begins its way back to infinity. As we shall see below, the particle follows a hyperbola in the orbital plane, and motion with $\varepsilon > 0$ is known as *hyperbolic motion*. Such an orbit is not bound to the gravitating center, as revealed by the fact that the total energy is dominated by (positive) kinetic energy instead of (negative) gravitational potential energy. The limiting case of an unbound orbit corresponds to $\varepsilon = 0$. Here the particle begins from rest at infinity, proceeds to a single turning point at $r = r_{\rm min}$, and returns to a state of rest at infinity. In this case the path is a parabola in the orbital plane, and motion with $\varepsilon = 0$ is known as *parabolic motion*.

When $\varepsilon < 0$, that is, when gravitational potential energy dominates over kinetic energy, the diagram reveals that there are now two turning points at $r = r_{\rm min}$ and $r = r_{\rm max}$. In this case the orbital motion is bound to the gravitating center, and takes place between the innermost and outermost radii. We shall see that the bound orbit describes an ellipse, and motion with $\varepsilon < 0$ is known as *elliptical motion*. A special case of elliptical motion occurs when ε is made equal to the minimum value of the effective potential. In this case the turning points merge to a single radius r_0, and motion proceeds at this fixed radius; this is a *circular orbit* with $h^2 = Gmr_0$, $\dot{\phi} = \sqrt{Gm/r_0^3}$, and $\varepsilon = -Gm/(2r_0)$.

The preceding discussion was couched in terms of a particle moving in the effective potential $V_{\rm eff}(r)$. It is important to understand, however, that this "particle" at position $\boldsymbol{r}(t)$ is in fact a fictitious representation of the relative orbit, which, as we saw back in Eq. (3.2), is subjected to a fictitious gravitating center-of-mass m at $\boldsymbol{r} = \boldsymbol{0}$. The motion described previously is therefore a description of the relative orbital motion. But thanks to Eqs. (3.1), the motion of each body is merely a scaled version of the relative motion, and it can be described in the same language; the bodies move about each other on opposite sides of the barycenter. In the limit of small mass ratios, $m_1/m_2 \ll 1$, it becomes increasingly true that $\boldsymbol{r}_1 \to \boldsymbol{r}$ and $\boldsymbol{r}_2 \to \boldsymbol{0}$; in this limiting case m_1 becomes a test mass in the field of $m_2 \to m$, its orbit coincides with the relative orbit, and m_2 stays put at the barycentric position.

3.2.4 Solution to Kepler's problem

Formal solution; integrable systems

Formally, a solution to Kepler's problem can be obtained by integrating Eq. (3.13) to get

$$t - t_i = \pm \int_{r_i}^{r} \frac{dr'}{\sqrt{2[\varepsilon - V_{\rm eff}(r')]}}, \tag{3.15}$$

inverting the result for $r(t)$, and integrating Eq. (3.10) to get

$$\phi - \phi_i = h \int_{t_i}^{t} \frac{dt'}{r(t')^2}, \tag{3.16}$$

which gives $\phi(t)$. In these equations, t_i is the time at which $r = r_i$, and ϕ_i is the orbital angle at time t_i. By following this procedure we convert the task of solving the second-order differential equations of (3.2) to the task of performing two integrations, or, to use an older but still popular terminology, evaluating two quadratures. When a dynamical problem such as this one can be reduced to doing quadratures, the problem is said to be *completely integrable*. A full discussion of integrable systems is beyond the scope of this book, but roughly speaking, a dynamical system is completely integrable when it possesses a sufficient number of conserved quantities. In the case of Kepler's problem, it is the constancy of total momentum (which allowed us to eliminate the motion of the barycenter), of total angular momentum (which allowed us to restrict the motion to a plane and to reduce the angular equation to $\dot{\phi} = h/r^2$), and of energy (which resulted in an equation for $\dot{r}^2$) that make the problem completely integrable.

Completely integrable systems are mathematically elegant, and are extremely convenient when they come along, but unfortunately they are rather rare in physics. Fortunately for us, however, the Kepler problem is one of them. More generally, because Eq. (3.15) is valid for any effective potential $V_{\rm eff}(r)$, motion in any spherically symmetric, static potential is always integrable.

The formal solutions of Eqs. (3.15) and (3.16) are not very useful from a practical point of view. We could try to carry out the integrals explicitly, but the results would not be illuminating. Alternatively, we could evaluate the integrals numerically, but the resulting tables would be of limited utility, and they would provide very little insight. We shall therefore proceed differently.

Spatial solution; conic sections

As a first step toward integrating the equations of motion, we eliminate t from the system of equations (3.10), (3.11), and adopt the orbital angle ϕ as independent variable. This strategy allows us to unravel the spatial aspects of the orbit – its shape in the orbital plane – and we shall return to the problem of describing the motion in time at a later stage. We also adopt $u := 1/r$ as a convenient substitute for r, and derive a differential equation for it by applying the chain rule of differential calculus. For example, we have that $\dot{r} = -u^{-2}\dot{u} = -u^{-2}\dot{\phi}u' = -hu'$ and $\ddot{r} = -h^2u^2u''$, in which a prime indicates differentiation with respect to ϕ. Making the substitutions within Eq. (3.11), we quickly arrive at

$$u'' + u = \frac{Gm}{h^2}. \tag{3.17}$$

The general solution to this simple equation is

$$u = \frac{Gm}{h^2}\big[1 + e\cos(\phi - \omega)\big], \tag{3.18}$$

in which e and ω are arbitrary constants of integration.

Returning to the original radial variable, the spatial solution to Kepler's problem is

$$r = \frac{p}{1 + e\cos(\phi - \omega)}, \tag{3.19}$$

in which

$$p := \frac{h^2}{Gm} \tag{3.20}$$

is a quantity known as the orbit's *semi-latus rectum*. We note that a solution with $e < 0$ is equivalent to one with $e > 0$, provided that ω is changed to $\omega + \pi$; we will adopt the convention that e is never negative.

The curve described by Eq. (3.19) is a *conic section*, an ellipse when $e < 1$, a hyperbola when $e > 1$, and a parabola when $e = 1$, with $r = 0$ situated at one of the curve's foci. The parameter e is called the *eccentricity* of the orbit. Note that r achieves a minimum when $\phi = \omega$. This is the point of closest approach in the orbit, called the *periapsis* or *pericenter*, and ω is known as the *longitude of pericenter*; its role is to fix the orientation of the orbit in the orbital plane. The term pericenter is usually adapted to reflect the identity of specific astronomical bodies. For example, we have perihelion for the Sun, perigee for the Earth, perijove for Jupiter, and periastron for binary star systems. No consensus has emerged to date for the closest approach to a black hole, but the word peribothron is gaining popularity; it was crafted by our colleague Sterl Phinney from the Greek root "bothros," which means hole or pit.

We examine first the elliptical orbits with $e < 1$. In this case the motion described by Eq. (3.19) is periodic, with period $\Delta\phi = 2\pi$, and we have recovered Kepler's first law, that planets move on elliptical paths around the Sun. We have already seen that $\phi = \omega$ describes the periapsis or pericenter, and we now see that $\phi = \omega + \pi$ describes the point of greatest separation, called *apoapsis* or *apocenter*. The pericenter and apocenter distances from the focus are given by

$$r_{\rm peri} = \frac{p}{1+e}, \qquad r_{\rm apo} = \frac{p}{1-e}. \tag{3.21}$$

The sum of these is the major axis of the ellipse, and we define the *semi-major axis* a to be

$$a := \frac{1}{2}(r_{\rm peri} + r_{\rm apo}) = \frac{p}{1-e^2}. \tag{3.22}$$

The semi-latus rectum can also be expressed in terms of these quantities:

$$p = \frac{2r_{\rm peri}r_{\rm apo}}{r_{\rm peri} + r_{\rm apo}}. \tag{3.23}$$

A special case of elliptical motion occurs when $e = 0$. In this case $r = p$ is constant, and the orbit is circular.

We examine next the hyperbolic and parabolic orbits with $e \geq 1$. In these cases the motion is no longer periodic, and the notion of apocenter ceases to be meaningful. Similarly, while we can still define a quantity a related to $p := h^2/(Gm)$ by $a = p/(1-e^2)$, this quantity also loses its usefulness, being negative for hyperbolic orbits and formally infinite for parabolic orbits. The pericenter still occurs at $\phi = \omega$, with $r_{\rm peri} = p/(1+e)$, and r approaches infinity as $\cos(\phi - \omega) \to -e^{-1}$. It is easy to show that the net change between the asymptotic angle $\phi_\infty^{\rm in}$ of the incoming part of the orbit and the asymptotic angle $\phi_\infty^{\rm out}$ of the outgoing orbit is given by

$$\Delta\phi := \phi_\infty^{\rm out} - \phi_\infty^{\rm in} = 2\arccos(-e^{-1}). \tag{3.24}$$

This reduces to $\Delta\phi = 2\pi$ when $e = 1$.

Returning to the general case with any e, we invoke Eqs. (3.10), (3.19), and (3.20) to derive the following useful expressions for the radial and angular velocities of a Keplerian orbit:

$$\dot{r} = \sqrt{\frac{Gm}{p}}\, e \sin(\phi - \omega), \tag{3.25a}$$

$$\dot{\phi} = \sqrt{\frac{Gm}{p^3}} \big[1 + e\cos(\phi - \omega)\big]^2. \tag{3.25b}$$

From these we obtain

$$v^2 = \frac{Gm}{p}\big[1 + 2e\cos(\phi - \omega) + e^2\big] = Gm\left(\frac{2}{r} - \frac{1 - e^2}{p}\right) \tag{3.26}$$

for the squared orbital velocity. From Eq. (3.12) we get an expression for the total energy $E := \mu\varepsilon$,

$$E = -G\mu m \frac{1 - e^2}{2p} = -\frac{G\mu m}{2a}. \tag{3.27}$$

We recall that $\mu := m_1 m_2/m$ is the system's reduced mass, and that a is related to p by Eq. (3.22). And finally, from Eq. (3.5) we have that

$$\boldsymbol{L} = \mu\sqrt{Gmp}\, \boldsymbol{e}_z \tag{3.28}$$

is the system's total angular momentum.

Motion in time; eccentric anomaly

So far we have determined the orbit as a function of ϕ, and the description involves three arbitrary constants, p, e, and ω, known as *orbital elements*. As we have seen, the semi-latus rectum p is a substitute for angular momentum, while the associated semi-major axis $a := p/(1 - e^2)$ is a substitute for orbital energy. The true nature of the longitude of pericenter ω will be revealed below.

The description of the orbit is completed by giving ϕ as a function of time. This can be accomplished by integrating Eq. (3.25b), in the form of

$$t - T = \sqrt{\frac{p^3}{Gm}} \int_{\omega}^{\phi} \frac{d\phi'}{\big[1 + e\cos(\phi' - \omega)\big]^2}, \tag{3.29}$$

where the constant T, called the *time of pericenter passage*, is a fourth orbital element. This equation can be inverted to give $\phi(t)$, and Kepler's problem is now solved.

There remains, however, the practical problem of evaluating the integral of Eq. (3.29). We henceforth specialize to elliptical motion, and introduce another running parameter on the orbit, known as the *eccentric anomaly* u. This is defined by the relations

$$\cos f = \frac{\cos u - e}{1 - e\cos u}, \qquad \sin f = \frac{\sqrt{1 - e^2}\,\sin u}{1 - e\cos u}, \tag{3.30}$$

in which $f := \phi - \omega$ is the *true anomaly*, the orbital angle as measured from the pericenter. The inverted relations are

$$\cos u = \frac{\cos f + e}{1 + e\cos f}, \qquad \sin u = \frac{\sqrt{1-e^2}\,\sin f}{1 + e\cos f}, \tag{3.31}$$

and these relationships are neatly summarized by the half-angle formula

$$\tan\frac{1}{2}f = \sqrt{\frac{1+e}{1-e}}\,\tan\frac{1}{2}u. \tag{3.32}$$

These equations reveal that $u = 0$ at pericenter (where $\phi = \omega$ and $f = 0$), that $u = \pi$ at apocenter (where $f = \pi$), and that $u = 2\pi$ at the end of a complete orbital cycle. Other useful relations are

$$\frac{df}{du} = \frac{\sqrt{1-e^2}}{1 - e\cos u}, \qquad \frac{du}{df} = \frac{\sqrt{1-e^2}}{1 + e\cos f}. \tag{3.33}$$

The eccentric anomaly gives rise to a useful alternative description of the orbit. Making the substitutions reveals that

$$r = a(1 - e\cos u), \tag{3.34a}$$

$$\dot{r} = \sqrt{\frac{Gm}{a}}\frac{e\sin u}{1 - e\cos u}, \tag{3.34b}$$

$$\dot{u} = \sqrt{\frac{Gm}{a^3}}\frac{1}{1 - e\cos u}, \tag{3.34c}$$

$$v^2 = \frac{Gm}{a}\frac{1 + e\cos u}{1 - e\cos u} = Gm\left(\frac{2}{r} - \frac{1}{a}\right). \tag{3.34d}$$

The main advantage of this description resides in Eq. (3.34c), which can be immediately integrated to give

$$t - T = \sqrt{\frac{a^3}{Gm}}\,(u - e\sin u). \tag{3.35}$$

This is known as *Kepler's equation*, which gives t as a simple function of the eccentric anomaly u. The equation can be inverted to yield $u(t)$, see Box 3.1, and this can finally be inserted within Eqs. (3.30) or (3.32) to express f as a function of time.

A description in terms of the eccentric anomaly is often a judicious choice when it is required to integrate over time. As a simple example, we can calculate the orbital period P of an elliptical orbit using Eq. (3.35). When u increases by 2π, t increases by P, and with very little effort we find that

$$P = 2\pi\sqrt{\frac{a^3}{Gm}}. \tag{3.36}$$

While the period can also be obtained directly from Eq. (3.29), the integration is far more laborious. That P is proportional to $a^{3/2}$ is Kepler's third law of planetary motion.

Box 3.1 **Solving Kepler's equation**

We wish to find the eccentric anomaly u that corresponds to a given time t. For this purpose it is useful to rewrite Kepler's equation as

$$u - e\sin u = M := \sqrt{\frac{Gm}{a^3}}(t - T) = 2\pi\frac{t - T}{P},$$

and to map the *mean anomaly* M to the interval $0 \le M < 2\pi$ by subtracting an appropriate multiple of P from $t - T$.

A time-honored method to solve Kepler's equation is based on Newton's root-finding method. It is an implementation of the iterative scheme

$$u_{n+1} = u_n + \frac{M - (u_n - e\sin u_n)}{1 - e\cos u_n}$$

until u converges to the desired accuracy. The iterations are seeded with $u_0 = M$, and the scheme typically requires a small number of iterations (fewer than six or so) for machine-precision accuracy.

Constancy of the pericenter; Runge–Lenz vector

We have seen that Keplerian motion within a fixed orbital plane is characterized by four constants of the motion, the orbital elements p, e, ω, and T. We have seen that constancy of p is tied to conservation of angular momentum, and that constancy of $a := p/(1-e^2)$ is tied to conservation of energy (so that constancy of e is also assured). In addition, the appearance of T as an integration constant was expected from the fact that the gravitational potential Gm/r does not depend explicitly on time. Constancy of ω, however, is not related to the spherical symmetry of the potential nor its time independence; we must seek a deeper cause. It is worth emphasizing that constancy of ω is a very important property of a Keplerian orbit: It ensures that the orientation of the orbit stays fixed, that the position of the pericenter does not move, and when the orbit is bound, that the orbit retraces itself after each orbital cycle.

The constancy of ω is the result of a hidden symmetry of Kepler's problem, associated with the specific $1/r$ nature of the gravitational potential; the symmetry does not exist for other potentials. The symmetry gives rise to a conservation statement for the *Runge–Lenz vector*, defined by

$$\boldsymbol{A} := \frac{\boldsymbol{v} \times \boldsymbol{h}}{Gm} - \boldsymbol{n}, \tag{3.37}$$

in which $\boldsymbol{h} := \boldsymbol{r} \times \boldsymbol{v}$ and $\boldsymbol{n} := \boldsymbol{r}/r$. A short computation using $\boldsymbol{a} = -Gm\boldsymbol{n}/r^2$ and $\dot{r} = \boldsymbol{n} \cdot \boldsymbol{v}$ shows that

$$\frac{d\boldsymbol{A}}{dt} = \boldsymbol{0}, \tag{3.38}$$

and the manipulations do indeed reveal that constancy of $\boldsymbol{A}$ relies on the specific form of the gravitational acceleration. The Runge–Lenz vector can be evaluated explicitly by making use of the Keplerian results displayed previously, including Eqs. (3.5), (3.7), (3.9), and (3.25). The result is

$$A = e(\cos\omega\, e_x + \sin\omega\, e_y), \tag{3.39}$$

and it reveals that the vector points in the fixed direction of the pericenter. The vector has a length e, and constancy of A as a vector implies that both e and ω are constants of the motion.

3.2.5 Keplerian orbits in space

The description of Keplerian orbits given previously achieved a remarkable degree of simplicity, thanks in part to conservation of angular momentum, which guarantees that the motion takes place in a fixed orbital plane. The description was entirely "orbit-centric," in that the reference frame was selected specifically to give a simple description of the orbital plane; and the simplicity was achieved largely by adopting the polar coordinates (r, ϕ) attached to the orbital plane. In many applications this description is entirely adequate, but in many others it is necessary to provide a fuller description that is less orbit-centric. For example, one might wish to describe the motion of two or more planets around the Sun (under the assumption that the Sun's motion and inter-planet interactions are negligible), with each planet moving in a different orbital plane; in such a case one would like to adopt the same reference frame for all the planets. Another example, which will be explored in some detail below, involves a two-body system perturbed by external bodies, or by a deformation of each body from a spherical configuration; these perturbations cause the orbital plane to move, and these motions must be described relative to a fixed reference frame. For such applications we require a fuller description of the orbital motion in space, relative to a reference frame that is not attached to the orbital plane.

We were already given an *orbital frame* with coordinates (x, y, z), such that the fixed orbital plane coincides with the x-y plane, and such that the angular-momentum vector is aligned with the z-direction. The orbital frame comes with the constant basis vectors e_x, e_y, and e_z, as well as the time-dependent basis n, λ, and e_z. We now introduce a *fundamental frame* with coordinates (X, Y, Z), and seek to describe the orbital motion in this new frame. We adopt the X-Y plane as a reference plane in the new frame, and the Z-axis as a reference direction. The fundamental frame comes with a constant vectorial basis e_X, e_Y, and e_Z. We assume that the orbital and fundamental frames share the same origin, so that $X = Y = Z = 0$ when $x = y = z = 0$. The choice of fundamental frame is arbitrary, and is often dictated by convention or convenience. For example, in the description of planetary motion in the solar system, the reference plane is chosen to coincide with Earth's own orbital plane (called the *ecliptic*). As another example, in the description of satellites orbiting the Earth, the fundamental plane is chosen to coincide with Earth's equatorial plane. In each case the direction of the X-axis is selected by convention.

The description of the orbital motion relative to the fundamental frame requires the introduction of additional orbital elements. The situation is represented in Fig. 3.2, which shows the orbital plane crossing the fundamental plane at an angle ι called the *inclination*; this is the angle between the z-direction of the orbital frame and the Z-direction of the fundamental frame. The line of intersection between the two planes is known as the *line of nodes*, and the point at which the orbit cuts the fundamental plane from below is the

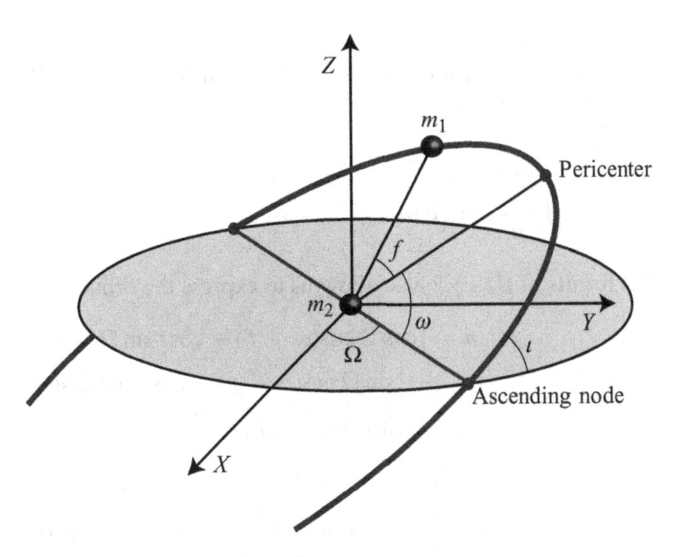

Fig. 3.2 Orbital motion viewed in the fundamental reference frame.

ascending node; the *descending node* is the point at which the orbit cuts the plane from above. The angle Ω between the X-direction and the line of nodes is the *longitude of the ascending node*. The diagram also shows ω, the *longitude of pericenter* introduced previously, which is now defined specifically as the angle between the line of nodes and the direction to the pericenter, as measured in the orbital plane. In this new description of the orbital motion, it is conventional to align the x-axis of the orbital plane with the direction to the pericenter, thereby deviating from our previous practice. And finally, the diagram shows the *true anomaly* f, the angle between the separation vector $\boldsymbol{r}$ and the direction to the pericenter, as measured in the orbital plane. The complete listing of orbital elements therefore consists of the principal elements p and e, the positional elements ι, Ω, and ω, and the time element T; the total number of elements is six, and it is no accident that this number corresponds to the number of initial conditions required to select a unique solution to Kepler's problem.

With these definitions and conventions, we show in Box 3.2 that the components of the separation vector $\boldsymbol{r}$ in the fundamental (X, Y, Z) frame are given by

$$r^X = r\big[\cos\Omega\cos(\omega + f) - \cos\iota\sin\Omega\sin(\omega + f)\big], \tag{3.40a}$$

$$r^Y = r\big[\sin\Omega\cos(\omega + f) + \cos\iota\cos\Omega\sin(\omega + f)\big], \tag{3.40b}$$

$$r^Z = r\sin\iota\sin(\omega + f), \tag{3.40c}$$

in which $r = p/(1 + e\cos f)$. Differentiation with respect to time and involvement of Eqs. (3.25) give us the components of the velocity vector,

$$v^X = -\sqrt{\frac{Gm}{p}}\Big\{\cos\Omega\big[\sin(\omega + f) + e\sin\omega\big] + \cos\iota\sin\Omega\big[\cos(\omega + f) + e\cos\omega\big]\Big\}, \tag{3.41a}$$

(continued overleaf)

$$v^Y = -\sqrt{\frac{Gm}{p}}\Big\{\sin\Omega\big[\sin(\omega+f)+e\sin\omega\big] - \cos\iota\cos\Omega\big[\cos(\omega+f)+e\cos\omega\big]\Big\}, \tag{3.41b}$$

$$v^Z = \sqrt{\frac{Gm}{p}}\sin\iota\big[\cos(\omega+f)+e\cos\omega\big]. \tag{3.41c}$$

The results of Box 3.2 also allow us to express the orbital basis vectors $\boldsymbol{n}$ and $\boldsymbol{\lambda}$ as

$$\begin{aligned}\boldsymbol{n} = {} & \big[\cos\Omega\cos(\omega+f) - \cos\iota\sin\Omega\sin(\omega+f)\big]\,\boldsymbol{e}_X \\ & + \big[\sin\Omega\cos(\omega+f) + \cos\iota\cos\Omega\sin(\omega+f)\big]\,\boldsymbol{e}_Y \\ & + \sin\iota\sin(\omega+f)\,\boldsymbol{e}_Z\end{aligned} \tag{3.42}$$

and

$$\begin{aligned}\boldsymbol{\lambda} = {} & \big[-\cos\Omega\sin(\omega+f) - \cos\iota\sin\Omega\cos(\omega+f)\big]\,\boldsymbol{e}_X \\ & + \big[-\sin\Omega\sin(\omega+f) + \cos\iota\cos\Omega\cos(\omega+f)\big]\,\boldsymbol{e}_Y \\ & + \sin\iota\cos(\omega+f)\,\boldsymbol{e}_Z.\end{aligned} \tag{3.43}$$

In addition, the direction to the pericenter is given by

$$\begin{aligned}\boldsymbol{e}_x = {} & \big(\cos\Omega\cos\omega - \cos\iota\sin\Omega\sin\omega\big)\,\boldsymbol{e}_X \\ & + \big(\sin\Omega\cos\omega + \cos\iota\cos\Omega\sin\omega\big)\,\boldsymbol{e}_Y \\ & + \sin\iota\sin\omega\,\boldsymbol{e}_Z,\end{aligned} \tag{3.44}$$

while the direction of the angular-momentum vector is

$$\boldsymbol{e}_z = \sin\iota\sin\Omega\,\boldsymbol{e}_X - \sin\iota\cos\Omega\,\boldsymbol{e}_Y + \cos\iota\,\boldsymbol{e}_Z. \tag{3.45}$$

In terms of these we have that $\boldsymbol{A} = e\boldsymbol{e}_x$ and $\boldsymbol{h} = h\boldsymbol{e}_z$, with $h := \sqrt{Gmp}$.

The description of the orbital motion in the fundamental frame allows us to provide simple definitions for the orbital elements. We have that

$$p := \frac{h^2}{Gm}, \tag{3.46a}$$

$$e := |\boldsymbol{A}|, \tag{3.46b}$$

$$\cos\iota := \boldsymbol{e}_z\cdot\boldsymbol{e}_Z = \frac{\boldsymbol{h}\cdot\boldsymbol{e}_Z}{h}, \tag{3.46c}$$

$$\sin\iota\sin\Omega := \boldsymbol{e}_z\cdot\boldsymbol{e}_X = \frac{\boldsymbol{h}\cdot\boldsymbol{e}_X}{h}, \tag{3.46d}$$

$$\sin\iota\sin\omega := \boldsymbol{e}_x\cdot\boldsymbol{e}_Z = \frac{\boldsymbol{A}\cdot\boldsymbol{e}_Z}{e}. \tag{3.46e}$$

To this we may add $a := p/(1-e^2)$ and the time element T, which is defined by Eq. (3.29). These definitions are elementary and fundamental, because they are formulated directly in terms of the fundamental vectors of the problem, the reduced angular-momentum vector $\boldsymbol{h} = \boldsymbol{r}\times\boldsymbol{v}$ and the Runge–Lenz vector $\boldsymbol{A}$ defined by Eq. (3.37). They will play a large role in the next section.

Box 3.2

Orbital and fundamental frames

A close examination of Fig. 3.2 and some thought reveal that the fundamental (X, Y, Z) frame can be obtained from the orbital (x, y, z) frame by a sequence of three rotations. The first is a rotation R_1 by an angle $-\omega$ around the *z*-axis, to align the rotated *x*-axis with the line of nodes. The second is a rotation R_2 by an angle $-\iota$ around the new *x*-axis, to align the rotated *z*-axis with the final *Z*-axis. The third is a rotation R_3 by an angle $-\Omega$ around the new *z*-axis, to align the rotated *x*-axis with the final *X*-axis. The rotation matrices are given by

$$\mathsf{R}_1 = \begin{pmatrix} \cos\omega & -\sin\omega & 0 \\ \sin\omega & \cos\omega & 0 \\ 0 & 0 & 1 \end{pmatrix}, \qquad \mathsf{R}_2 = \begin{pmatrix} 1 & 0 & 0 \\ 0 & \cos\iota & -\sin\iota \\ 0 & \sin\iota & \cos\iota \end{pmatrix},$$

$$\mathsf{R}_3 = \begin{pmatrix} \cos\Omega & -\sin\Omega & 0 \\ \sin\Omega & \cos\Omega & 0 \\ 0 & 0 & 1 \end{pmatrix},$$

and the overall transformation between the orbital and fundamental frames is $\mathsf{X} = \mathsf{R}_3\mathsf{R}_2\mathsf{R}_1\,\mathsf{x}$, with x representing a column vector with components (x, y, z), and X a column vector with components (X, Y, Z).

The end result of the transformation is

$$\begin{aligned} X &= (\cos\Omega\cos\omega - \cos\iota\sin\Omega\sin\omega)x \\ &\quad - (\cos\Omega\sin\omega + \cos\iota\sin\Omega\cos\omega)y \\ &\quad + (\sin\iota\sin\Omega)z, \\ Y &= (\sin\Omega\cos\omega + \cos\iota\cos\Omega\sin\omega)x \\ &\quad - (\sin\Omega\sin\omega - \cos\iota\cos\Omega\cos\omega)y \\ &\quad - (\sin\iota\cos\Omega)z, \\ Z &= (\sin\iota\sin\omega)x + (\sin\iota\cos\omega)y + (\cos\iota)z. \end{aligned}$$

The components of $\boldsymbol{r}$ displayed in Eqs. (3.40) follow from these results by inserting $x = r\cos f$, $y = r\sin f$, and $z = 0$.

From the coordinate transformation we can easily obtain the old basis vectors in terms of the new:

$$\begin{aligned} \boldsymbol{e}_x &= \frac{\partial \boldsymbol{x}}{\partial x} = (\cos\Omega\cos\omega - \cos\iota\sin\Omega\sin\omega)\boldsymbol{e}_X \\ &\quad + (\sin\Omega\cos\omega + \cos\iota\cos\Omega\sin\omega)\boldsymbol{e}_Y \\ &\quad + (\sin\iota\sin\omega)\boldsymbol{e}_Z, \\ \boldsymbol{e}_y &= \frac{\partial \boldsymbol{x}}{\partial y} = -(\cos\Omega\sin\omega + \cos\iota\sin\Omega\cos\omega)\boldsymbol{e}_X \\ &\quad - (\sin\Omega\sin\omega - \cos\iota\cos\Omega\cos\omega)\boldsymbol{e}_Y \\ &\quad + (\sin\iota\cos\omega)\boldsymbol{e}_Z, \\ \boldsymbol{e}_z &= \frac{\partial \boldsymbol{x}}{\partial z} = (\sin\iota\sin\Omega)\boldsymbol{e}_X - (\sin\iota\cos\Omega)\boldsymbol{e}_Y + (\cos\iota)\boldsymbol{e}_Z, \end{aligned}$$

in which the position vector $\boldsymbol{x}$ is expressed as $\boldsymbol{x} = X\,\boldsymbol{e}_X + Y\,\boldsymbol{e}_Y + Z\,\boldsymbol{e}_Z$.

The orbital vectors of Eqs. (3.42) and (3.43) are obtained by inserting these results within the relations $\boldsymbol{n} = \cos f\, \boldsymbol{e}_x + \sin f\, \boldsymbol{e}_y$ and $\boldsymbol{\lambda} = -\sin f\, \boldsymbol{e}_x + \cos f\, \boldsymbol{e}_y$, which are inherited from Eqs. (3.7); the new relations reflect the change of convention regarding the choice of x-direction.

3.3 Perturbed Kepler problem

We saw in the preceding section that the motion of two spherical bodies under their mutual gravitational attraction can be solved exactly and completely. As we shall see in Sec. 3.5, the same cannot be said of the three-body problem, and in general, the N-body problem does not admit an exact solution. Many situations of interest, however, involve more than two bodies. The Sun–Earth–Moon system is an extremely pertinent example, and the motion of any planet around the Sun is affected by the massive presence of Jupiter. While these systems cannot be given an exact description, we can nevertheless make progress by appealing to the fact that in many applications, the additional bodies have only a small effect on the orbital motion of a two-body system. In the Sun–Earth–Moon system, the dominant interaction is between the Sun and the Earth, and the gravitational effects of the Moon on Earth's orbital motion are small. Similarly, while Jupiter is indeed a massive body, it is sufficiently far from the other planets that its gravity has a small effect on them. The approximate analysis of small external influences on a system dominated by an internal interaction is the realm of *perturbation theory*, and in this section we formulate a perturbation theory for Kepler's problem.

3.3.1 Perturbing force

We return to the two-body problem of Sec. 3.2, but now suppose that the relative acceleration $\boldsymbol{a} := \boldsymbol{a}_1 - \boldsymbol{a}_2$ between bodies is given by

$$\boldsymbol{a} = -\frac{Gm}{r^2}\boldsymbol{n} + \boldsymbol{f}, \tag{3.47}$$

in which $m := m_1 + m_2$, $\boldsymbol{r} := \boldsymbol{r}_1 - \boldsymbol{r}_2$, $r := |\boldsymbol{r}|$, $\boldsymbol{n} := \boldsymbol{r}/r$, and where $\boldsymbol{f}$ is a perturbing force per unit mass, which may depend on $\boldsymbol{r}$, $\boldsymbol{v} := \boldsymbol{v}_1 - \boldsymbol{v}_2$, and time. The presence of the perturbing force implies that the Keplerian motion reviewed in Sec. 3.2 is no longer a solution to the equations of motion. As a consequence we can no longer expect $\boldsymbol{h} := \boldsymbol{r} \times \boldsymbol{v}$ to be a constant vector, the orbit to be elliptical, the pericenter to stay in a fixed position, and so on. But with $\boldsymbol{f}$ assumed to be small, we would expect the Keplerian description to remain approximately true. Our main goal is to find a useful way of describing the small deviations from Keplerian motion produced by the perturbing force. Below we shall follow the convention from celestial mechanics, and refer to $\boldsymbol{f}$ as a perturbing force, in spite of the fact that it is really a force per unit mass.

It is helpful to introduce, as we did previously, a vectorial basis adapted to the orbital motion of the two bodies. The first member of the basis will be $\boldsymbol{n}$, which is already singled out in Eq. (3.47). The third member will be $\boldsymbol{e}_z := \boldsymbol{h}/h$, which is necessarily orthogonal to $\boldsymbol{n}$; this vector can no longer be assumed to be a constant vector. The second member of the basis is then $\boldsymbol{\lambda}$, which is defined to be orthogonal to both $\boldsymbol{n}$ and $\boldsymbol{e}_z$. We take the triad of vectors $(\boldsymbol{n}, \boldsymbol{\lambda}, \boldsymbol{e}_z)$ to form a right-handed basis. When $\boldsymbol{f} = \boldsymbol{0}$ the vector basis reduces to the one introduced back in Sec. 3.2, but the bases are quite distinct in the presence of a perturbing force; for reasons that will become clear later, we perversely adopt the same notation in spite of this important distinction.

The vectorial basis can be used to decompose any vector relevant to the description of the perturbed orbital motion. Examples are

$$\boldsymbol{r} = r\,\boldsymbol{n}, \tag{3.48a}$$

$$\boldsymbol{v} = \dot{r}\,\boldsymbol{n} + v_\perp\,\boldsymbol{\lambda}, \tag{3.48b}$$

$$\boldsymbol{h} = h\,\boldsymbol{e}_z, \tag{3.48c}$$

in which $\dot{r} := \boldsymbol{v}\cdot\boldsymbol{n}$ and $v_\perp := \boldsymbol{v}\cdot\boldsymbol{\lambda}$; it is easy to verify that $h = rv_\perp$. Another example is the perturbing force, which is decomposed as

$$\boldsymbol{f} = \mathcal{R}\,\boldsymbol{n} + \mathcal{S}\,\boldsymbol{\lambda} + \mathcal{W}\,\boldsymbol{e}_z, \tag{3.49}$$

in terms of components $\mathcal{R}$, $\mathcal{S}$, and $\mathcal{W}$. Yet another example is the Runge–Lenz vector

$$\boldsymbol{A} := \frac{\boldsymbol{v}\times\boldsymbol{h}}{Gm} - \boldsymbol{n} = A\,\boldsymbol{e}_x, \qquad \boldsymbol{e}_x = \cos\alpha\,\boldsymbol{n} - \sin\alpha\,\boldsymbol{\lambda}, \tag{3.50}$$

which is given a component $A\cos\alpha$ along $\boldsymbol{n}$, and a component $-A\sin\alpha$ along $\boldsymbol{\lambda}$; there is no component along $\boldsymbol{e}_z$ because $\boldsymbol{A}$ is orthogonal to $\boldsymbol{h}$. Here we have defined the Runge–Lenz vector exactly as in Sec. 3.2, but we can no longer expect $\boldsymbol{A}$ to be a constant vector. We have given the vector a length A, and Eq. (3.50) serves as a definition for the unit vector $\boldsymbol{e}_x$, which is also not a constant vector. In an unperturbed situation A would be equal to e, α would be equal to the true anomaly f, and $\boldsymbol{e}_x$ would indeed be a constant vector; but none of these statements can be expected to hold when $\boldsymbol{f} \neq \boldsymbol{0}$.

The effect of the perturbing force $\boldsymbol{f}$ on the vectors $\boldsymbol{h}$ and $\boldsymbol{A}$ can be calculated by appealing directly to their definitions. We find that

$$\frac{d\boldsymbol{h}}{dt} = \boldsymbol{r}\times\boldsymbol{f} = -r\mathcal{W}\,\boldsymbol{\lambda} + r\mathcal{S}\,\boldsymbol{e}_z \tag{3.51}$$

and

$$Gm\frac{d\boldsymbol{A}}{dt} = \boldsymbol{f}\times\boldsymbol{h} + \boldsymbol{v}\times(\boldsymbol{r}\times\boldsymbol{f}) = 2h\mathcal{S}\,\boldsymbol{n} - \big(h\mathcal{R} + r\dot{r}\mathcal{S}\big)\,\boldsymbol{\lambda} - r\dot{r}\mathcal{W}\,\boldsymbol{e}_z. \tag{3.52}$$

These equations imply that

$$\frac{dh}{dt} = r\mathcal{S} \tag{3.53}$$

and

$$Gm\frac{dA}{dt} = h\sin\alpha\,\mathcal{R} + \big(2h\cos\alpha + r\dot{r}\sin\alpha\big)\mathcal{S}, \tag{3.54}$$

as well as

$$h\frac{d\boldsymbol{e}_z}{dt} = -r\mathcal{W}\boldsymbol{\lambda} \tag{3.55}$$

and

$$GmA\frac{d\boldsymbol{e}_x}{dt} = -\big[h\cos\alpha\,\mathcal{R} + (-2h\sin\alpha + r\dot{r}\cos\alpha)\mathcal{S}\big]\big(\sin\alpha\,\boldsymbol{n} + \cos\alpha\,\boldsymbol{\lambda}\big) - r\dot{r}\mathcal{W}\,\boldsymbol{e}_z. \tag{3.56}$$

These equations tell us, for example, that $\mathcal{S}$ produces a change in the magnitude of the angular-momentum vector, while $\mathcal{W}$ produces a change in its direction. Similarly, both $\mathcal{R}$ and $\mathcal{S}$ produce a change in $\boldsymbol{A}$'s magnitude, as well as a change of direction orthogonal to $\boldsymbol{e}_z$.

3.3.2 Osculating orbits

How are we to solve Eq. (3.47)? A direct approach, of course, is always possible: one inserts the given expression for the external force and integrates the second-order differential equations directly, either by analytical methods when the problem is sufficiently simple (a rare occasion), or by numerical methods. The approach provides an answer, but very little insight into the effects of the perturbation on various key aspects of the orbit. We shall favor instead a clever reformulation of the perturbed problem that was initially devised by Euler and Lagrange. This reformulation, known as the *method of osculating orbital elements*, is an application to orbital dynamics of the well-known method of variation of arbitrary constants to integrate differential equations (reviewed in Box 3.3).

Box 3.3 Variation of arbitrary constants

The method of variation of arbitrary constants to integrate differential equations is best introduced by examining a simple example. Consider a simple harmonic oscillator of unit frequency driven by an external force $f(t)$. The system is described by the differential equation

$$\ddot{x} + x = f, \tag{1}$$

in which x is the displacement from equilibrium, and an overdot indicates differentiation with respect to t. To integrate this equation we begin with the general solution to the homogeneous problem, $\ddot{x} + x = 0$. This solution can be expressed as

$$x(t) = x_0\cos t + v_0\sin t, \tag{2}$$

in which $x_0 := x(t=0)$ and $v_0 := \dot{x}(t=0)$ are the arbitrary constants of the solution. This solution also gives us

$$\dot{x}(t) = v_0\cos t - x_0\sin t. \tag{3}$$

Returning to the original differential equation with a driving force $f(t)$, we declare that Eqs. (2) and (3) shall also be a solution to Eq. (1), and evade the obvious contradiction by allowing x_0 and v_0 to become functions of time. Equation (3) will be compatible with Eq. (2) provided that $\dot{x}_0 \cos t + \dot{v}_0 \sin t = 0$, and substitution of Eqs. (2) and (3) within Eq. (1) yields $-\dot{x}_0 \sin t + \dot{v}_0 \cos t = f$. Solving for $\dot{x}_0$ and $\dot{v}_0$, we find that

$$\dot{x}_0 = -f \sin t, \qquad \dot{v}_0 = f \cos t, \tag{4}$$

and these equations now replace Eq. (1). The solutions are to be inserted within Eqs. (2) and (3) for a complete solution of the original problem.

These steps define the method of variation of arbitrary constants. One begins with the general solution to a homogeneous differential equation, with the appropriate number of arbitrary constants. One next solves the inhomogeneous problem by formally adopting the functional form of the homogeneous solution, and promoting the arbitrary constants to new variables. After some manipulations, the original problem acquires a new formulation as a set of first-order differential equations for the new variables.

The method

The starting point of the method of osculating orbital elements is to recall that the general solution to the unperturbed Kepler problem can be expressed as

$$\boldsymbol{r}(t) = \boldsymbol{r}_{\rm Kepler}(t, \mu^a), \qquad \boldsymbol{v}(t) = \boldsymbol{v}_{\rm Kepler}(t, \mu^a), \tag{3.57}$$

where $\boldsymbol{r}_{\rm Kepler}$ and $\boldsymbol{v}_{\rm Kepler}$ are the functional forms displayed in Eqs. (3.40) and (3.41), with μ^a denoting a collection of six constants of the motion, which we take to be the orbital elements $(p, e, \iota, \Omega, \omega, T)$. These vectors satisfy the differential equations

$$\frac{d\boldsymbol{r}_{\rm Kepler}}{dt} = \boldsymbol{v}_{\rm Kepler}, \qquad \frac{d\boldsymbol{v}_{\rm Kepler}}{dt} = -Gm\frac{\boldsymbol{r}_{\rm Kepler}}{r^3_{\rm Kepler}}, \tag{3.58}$$

which define the unperturbed Kepler problem.

The vectors $\boldsymbol{r}_{\rm Kepler}$ and $\boldsymbol{v}_{\rm Kepler}$ can also provide a solution to the perturbed problem if we let the orbital elements μ^a become functions of time: $\mu^a \to \mu^a(t)$. The original mathematical problem expressed by Eq. (3.47) can then be formulated as a set of first-order differential equations for μ^a, and the solutions to these are to be inserted within Eqs. (3.57) to obtain a complete solution to the perturbed problem. The meaning of the method is that at any time t_1, the orbit is taken to be a Keplerian orbit with orbital elements $\mu^a(t_1)$ that refer to this time only; at another time t_2 the orbit will still be Keplerian, but the orbital elements will have evolved to new values $\mu^a(t_2)$. Another way of describing this is to say that at any time t_1, there exists a Keplerian orbit with elements $\mu^a(t_1)$ that is *tangent* to the perturbed orbit at that time; this is the *osculating orbit*, the term originating in the Latin word *osculatio*, which means "to kiss."

We shall therefore express the solution to Eq. (3.47) as

$$\boldsymbol{r}(t) = \boldsymbol{r}_{\rm Kepler}\big(t, \mu^a(t)\big), \qquad \boldsymbol{v}(t) = \boldsymbol{v}_{\rm Kepler}\big(t, \mu^a(t)\big), \tag{3.59}$$

with $\boldsymbol{r}_{\rm Kepler}$ and $\boldsymbol{v}_{\rm Kepler}$ still standing for the functional forms of Eqs. (3.40) and (3.41), but with the orbital elements μ^a now allowed to be functions of time. Differentiating $\boldsymbol{r}$ with respect to time gives

$$\frac{d\boldsymbol{r}}{dt} = \frac{\partial \boldsymbol{r}_{\text{Kepler}}}{\partial t} + \sum_a \frac{\partial \boldsymbol{r}_{\text{Kepler}}}{\partial \mu^a}\frac{d\mu^a}{dt}. \tag{3.60}$$

The left-hand side is $\boldsymbol{v}$, and the first term on the right-hand side is $\boldsymbol{v}_{\text{Kepler}}$, because $\boldsymbol{r}_{\text{Kepler}}$ is differentiated with respect to t while keeping the orbital elements constant, thereby defining Keplerian motion. The equation is compatible with Eq. (3.59) provided that

$$\sum_a \frac{\partial \boldsymbol{r}_{\text{Kepler}}}{\partial \mu^a}\frac{d\mu^a}{dt} = \boldsymbol{0}, \tag{3.61}$$

and we have obtained our first osculating condition. Similarly, differentiating $\boldsymbol{v}$ with respect to t gives

$$\frac{d\boldsymbol{v}}{dt} = \frac{\partial \boldsymbol{v}_{\text{Kepler}}}{\partial t} + \sum_a \frac{\partial \boldsymbol{v}_{\text{Kepler}}}{\partial \mu^a}\frac{d\mu^a}{dt}. \tag{3.62}$$

The left-hand side is $\boldsymbol{a} = \boldsymbol{a}_{\text{Kepler}} + \boldsymbol{f}$, and the first term on the right-hand side is the Keplerian acceleration $\boldsymbol{a}_{\text{Kepler}}$. The second osculating condition is

$$\sum_a \frac{\partial \boldsymbol{v}_{\text{Kepler}}}{\partial \mu^a}\frac{d\mu^a}{dt} = \boldsymbol{f}, \tag{3.63}$$

and these equations can be solved for $d\mu^a/dt$.

Osculating equations

The purpose of the foregoing discussion was primarily to describe the conceptual aspects of the method of osculating orbital elements. While Eqs. (3.61) and (3.63) could in principle be solved for $d\mu^a/dt$ in terms of the perturbing force, it is easier in practice to obtain $d\mu^a/dt$ directly from the definitions of the orbital elements displayed in Eqs. (3.46). The orbital elements are given explicitly in terms of the fixed vectorial basis $(\boldsymbol{e}_X, \boldsymbol{e}_Y, \boldsymbol{e}_Z)$, as well as the angular-momentum vector $\boldsymbol{h}$ and the Runge–Lenz vector $\boldsymbol{A}$, which change according to Eqs. (3.51)–(3.56). The method authorizes us to substitute Keplerian relations on the right-hand side of these equations, so that h is at all times equal to $\sqrt{Gmp(t)}$, A is at all times equal to $e(t)$, α is at all times equal to the true anomaly f, r is at all times equal to $p/(1+e\cos f)$, and $\dot{r}$ is at all times equal to $(Gm/p)^{1/2} e \sin f$.

By following this approach we quickly arrive at

$$\frac{dp}{dt} = 2\sqrt{\frac{p^3}{Gm}}\frac{1}{1+e\cos f}\mathcal{S}, \tag{3.64a}$$

$$\frac{de}{dt} = \sqrt{\frac{p}{Gm}}\left[\sin f\,\mathcal{R} + \frac{2\cos f + e(1+\cos^2 f)}{1+e\cos f}\mathcal{S}\right], \tag{3.64b}$$

$$\frac{d\iota}{dt} = \sqrt{\frac{p}{Gm}}\frac{\cos(\omega+f)}{1+e\cos f}\mathcal{W}, \tag{3.64c}$$

$$\sin\iota\frac{d\Omega}{dt} = \sqrt{\frac{p}{Gm}}\frac{\sin(\omega+f)}{1+e\cos f}\mathcal{W}, \tag{3.64d}$$

(continued overleaf)

$$\frac{d\omega}{dt} = \frac{1}{e}\sqrt{\frac{p}{Gm}}\left[-\cos f\,\mathcal{R} + \frac{2+e\cos f}{1+e\cos f}\sin f\,\mathcal{S} - e\cot\iota\frac{\sin(\omega+f)}{1+e\cos f}\mathcal{W}\right], \tag{3.64e}$$

a listing of equations that govern the behavior of the osculating orbital elements. To these we can add

$$\frac{da}{dt} = 2\sqrt{\frac{a^3}{Gm}}(1-e^2)^{-1/2}\left[e\sin f\,\mathcal{R} + (1+e\cos f)\,\mathcal{S}\right], \tag{3.65}$$

in case $a := p/(1-e^2)$ is preferred over p in the selection of orbital elements. Note that p and e (or a and e) are affected only by components of $\boldsymbol{f}$ in the orbital plane, while Ω and ι are affected only by the component out of the plane. All components of the perturbing force affect ω.

We are missing an equation for the variation of T, the time of pericenter passage, which determines the true anomaly f via Eq. (3.29). It is more practical, however, to close the system of equations by providing an expression for df/dt, from which the true anomaly can be obtained directly. Because f is the angle between the (varying) pericenter and the position vector $\boldsymbol{r}$, we have that $\cos f = \boldsymbol{r}\cdot\boldsymbol{e}_x/r$, and this can immediately be differentiated with respect to time. In this we insert Eq. (3.56) for $d\boldsymbol{e}_x/dt$, the usual Keplerian relations for $d\boldsymbol{r}/dt$, and we obtain

$$\frac{df}{dt} = \sqrt{\frac{Gm}{p^3}}(1+e\cos f)^2 + \frac{1}{e}\sqrt{\frac{p}{Gm}}\left[\cos f\,\mathcal{R} - \frac{2+e\cos f}{1+e\cos f}\sin f\,\mathcal{S}\right] \tag{3.66}$$

after some simplification. The result can also be expressed as

$$\frac{df}{dt} = \left(\frac{df}{dt}\right)_{\text{Kepler}} - \left(\frac{d\omega}{dt} + \cos\iota\frac{d\Omega}{dt}\right), \tag{3.67}$$

which shows that df/dt differs from the usual Keplerian expression by a term $d\omega/dt + \cos\iota\, d\Omega/dt$ which possesses a direct geometrical meaning. We recall that ω is the angle from the (varying) pericenter to the (varying) line of nodes, as measured in the orbital plane, while Ω is the angle from the line of nodes to the (fixed) X-direction, as measured in the reference X-Y plane. The combination $d\omega + \cos\iota\, d\Omega$ can then be seen to describe the change in the direction to the pericenter relative to the X-direction, as measured entirely in the orbital plane. The non-Keplerian terms in Eqs. (3.66) and (3.67) therefore appear because the true anomaly f is measured relative to a varying set of directions.

The relevance of the combination $d\omega + \cos\iota\, d\Omega$ can also be inferred from Eq. (3.56), which gives the rate of change of the direction to the pericenter. After inserting the appropriate Keplerian relations in this expression, we find that

$$\frac{d\boldsymbol{e}_x}{dt} = \left(\frac{d\omega}{dt} + \cos\iota\frac{d\Omega}{dt}\right)(\sin f\,\boldsymbol{n} + \cos f\,\boldsymbol{\lambda}) + \left(\sin\omega\frac{d\iota}{dt} - \sin\iota\cos\omega\frac{d\Omega}{dt}\right)\boldsymbol{e}_z, \tag{3.68}$$

which reveals that $d\omega + \cos\iota\, d\Omega$ does indeed describe the change of $\boldsymbol{e}_x$ within the orbital plane, while $\sin\omega\, d\iota - \sin\iota\cos\omega\, d\Omega$ describes the change in the orthogonal direction.

First-order approximation

The formalism of osculating orbital elements, in the formulation of Eqs. (3.64) and (3.66), is *exactly equivalent* to the original formulation of the equations of motion in Eq. (3.47); no approximations have been introduced in the transcription. The usefulness of the formalism, however, is most immediate when the perturbing force is small, so that the changes in the orbital elements are small. In such a context one can achieve a very good approximation of the orbital dynamics by inserting the constant, zeroth-order values on the right-hand side of the equations, and integrating with respect to t to get the first-order changes. In such applications it can be convenient to use f as independent variable instead of t, and in this approximate context one can neglect the non-Keplerian terms on the right-hand side of Eq. (3.66). The system of osculating equations becomes

$$\frac{dp}{df} \simeq 2\frac{p^3}{Gm}\frac{1}{(1+e\cos f)^3}\mathcal{S}, \tag{3.69a}$$

$$\frac{de}{df} \simeq \frac{p^2}{Gm}\left[\frac{\sin f}{(1+e\cos f)^2}\mathcal{R} + \frac{2\cos f + e(1+\cos^2 f)}{(1+e\cos f)^3}\mathcal{S}\right], \tag{3.69b}$$

$$\frac{d\iota}{df} \simeq \frac{p^2}{Gm}\frac{\cos(\omega+f)}{(1+e\cos f)^3}\mathcal{W}, \tag{3.69c}$$

$$\sin\iota\frac{d\Omega}{df} \simeq \frac{p^2}{Gm}\frac{\sin(\omega+f)}{(1+e\cos f)^3}\mathcal{W}, \tag{3.69d}$$

$$\begin{aligned}\frac{d\omega}{df} \simeq \frac{1}{e}\frac{p^2}{Gm}\bigg[&-\frac{\cos f}{(1+e\cos f)^2}\mathcal{R} + \frac{2+e\cos f}{(1+e\cos f)^3}\sin f\,\mathcal{S} \\ &- e\cot\iota\frac{\sin(\omega+f)}{(1+e\cos f)^3}\mathcal{W}\bigg],\end{aligned} \tag{3.69e}$$

with

$$\begin{aligned}\frac{dt}{df} \simeq{}& \sqrt{\frac{p^3}{Gm}}\frac{1}{(1+e\cos f)^2} \\ &\times\left\{1 - \frac{1}{e}\frac{p^2}{Gm}\left[\frac{\cos f}{(1+e\cos f)^2}\mathcal{R} - \frac{2+e\cos f}{(1+e\cos f)^3}\sin f\,\mathcal{S}\right]\right\}\end{aligned} \tag{3.70}$$

providing the temporal information.

In most applications of the formalism it is found that the orbital elements undergo two types of changes. The first is an oscillation with a period equal to the orbital period P (or multiples of P), as given by Eq. (3.36); such changes are typically uninteresting. The second type is a steady drift that does not average out after a few orbital cycles; such changes, which are known as *secular changes*, are typically much more interesting, because they accumulate over time and eventually lead to large departures from the initial Keplerian orbit. The secular change of an orbital element μ^a over a complete orbit is given by

$$\Delta\mu^a = \int_0^P \frac{d\mu^a}{dt}\,dt = \int_0^{2\pi}\frac{d\mu^a}{df}\,df, \tag{3.71}$$

and division by the orbital period P gives a notion of average, or secular, time derivative:

$$\left(\frac{d\mu^a}{dt}\right)_{\rm sec} := \frac{\Delta\mu^a}{P}. \tag{3.72}$$

The integrations are best carried out in the form involving the true anomaly f, but the factors of $1 + e\cos f$ in the denominator can sometimes produce integrals that are difficult to evaluate. When such a situation is encountered, it is usually advantageous to make a change of variables from f to the eccentric anomaly u, as defined by Eqs. (3.30)–(3.33).

3.4 Case studies of perturbed Keplerian motion

Our purpose in this section is to explore some of the non-Keplerian aspects of celestial mechanics by working through a number of examples that are of real relevance to the solar system. We shall investigate these effects by exploiting the method of osculating orbital elements introduced in the preceding section.

3.4.1 Perturbations by a third body

We first examine a two-body system, such as the Sun–Mercury system, perturbed by a remote third body such as Jupiter. This is an example of a three-body problem, and we shall have occasion in Sec. 3.5 to give this problem a more complete treatment. Here we provide an approximate discussion, relying on the fact that the third body is remote, so that its gravitational influence on the two-body system is weak. This influence can be described as a perturbation, and its effects can be investigated by means of the method of osculating orbits.

To set up the problem we recall that the two-body system involves a mass m_1 at position $\boldsymbol{r}_1$ and a mass m_2 at position $\boldsymbol{r}_2$. The third body has a mass m_3 and a position $\boldsymbol{r}_3$. As usual we let $m := m_1 + m_2$, $\boldsymbol{r} := \boldsymbol{r}_1 - \boldsymbol{r}_2$, and we also introduce $\boldsymbol{R} := \boldsymbol{r}_3 - \boldsymbol{r}_2$; note that $\boldsymbol{R}$ no longer stands for the position of the two-body pericenter. We also have $r := |\boldsymbol{r}|$, $R := |\boldsymbol{R}|$, $\boldsymbol{n} := \boldsymbol{r}/r$, and $\boldsymbol{N} := \boldsymbol{R}/R$. We assume that $R \gg r$.

From Eq. (1.202) we find that the acceleration of the first body is given by

$$\boldsymbol{a}_1 = -Gm_2\frac{\boldsymbol{n}}{r^2} - Gm_3\frac{\boldsymbol{r}-\boldsymbol{R}}{|\boldsymbol{r}-\boldsymbol{R}|^3}, \tag{3.73}$$

while

$$\boldsymbol{a}_2 = Gm_1\frac{\boldsymbol{n}}{r^2} + Gm_3\frac{\boldsymbol{N}}{R^2} \tag{3.74}$$

is the acceleration of the second body. The relative acceleration is $\boldsymbol{a} := \boldsymbol{a}_1 - \boldsymbol{a}_2$, and removal of the Keplerian term gives us the perturbing force

$$\boldsymbol{f} = -Gm_3\left(\frac{\boldsymbol{r}-\boldsymbol{R}}{|\boldsymbol{r}-\boldsymbol{R}|^3} + \frac{\boldsymbol{N}}{R^2}\right). \tag{3.75}$$

Taking into account the assumption that R is large compared with r, we simplify this expression by expanding in powers of r/R, and obtain, to leading order,

$$f = -\frac{Gm_3 r}{R^3}\Big[\boldsymbol{n} - 3(\boldsymbol{n}\cdot\boldsymbol{N})\boldsymbol{N} + O(r/R)\Big]. \tag{3.76}$$

We see that the ratio of f to the Keplerian acceleration is of order $(m_3/m)(r/R)^3$, and therefore small by virtue of our assumption that $R \gg r$. The components of the perturbing force in the Keplerian basis $(\boldsymbol{n}, \boldsymbol{\lambda}, \boldsymbol{e}_z)$ are

$$\mathcal{R} := \boldsymbol{f}\cdot\boldsymbol{n} = -\frac{Gm_3 r}{R^3}\big[1 - 3(\boldsymbol{n}\cdot\boldsymbol{N})^2\big], \tag{3.77a}$$

$$\mathcal{S} := \boldsymbol{f}\cdot\boldsymbol{\lambda} = 3\frac{Gm_3 r}{R^3}(\boldsymbol{n}\cdot\boldsymbol{N})(\boldsymbol{\lambda}\cdot\boldsymbol{N}), \tag{3.77b}$$

$$\mathcal{W} := \boldsymbol{f}\cdot\boldsymbol{e}_z = 3\frac{Gm_3 r}{R^3}(\boldsymbol{n}\cdot\boldsymbol{N})(\boldsymbol{e}_z\cdot\boldsymbol{N}), \tag{3.77c}$$

and these can immediately be inserted within the equations that govern the evolution of the osculating orbital elements.

The expressions for $\mathcal{R}$, $\mathcal{S}$, and $\mathcal{W}$ refer to any two-body system perturbed by a remote third body, but to proceed we consider the specific situation mentioned previously: We examine the orbit of a planet like Mercury around the Sun, perturbed by an outer planet such as Jupiter. For simplicity we assume that Mercury and Jupiter move in the same orbital plane (when in reality they have a relative inclination of approximately 6 degrees), and we align the reference X-Y plane with this common orbital plane; this implies that $\iota = 0$, and since Ω is not defined, it may be set equal to zero. Again for simplicity, we assume that Jupiter moves on a circular orbit (a good approximation), with a constant radius R and angular frequency $\Omega_3 = \sqrt{G(m_2+m_3)/R^3}$. Its true anomaly is given by $F = \Omega_3 t$, and it changes very little in the course of a complete Mercury orbit; we have that

$$\Delta F := \Omega_3 P = 2\pi\left(\frac{m_2+m_3}{m_1+m_2}\right)^{1/2}\left(\frac{a}{R}\right)^{3/2} \ll 1, \tag{3.78}$$

with P denoting Mercury's orbital period, and $a := p/(1-e^2)$ its semi-major axis. The direction to Jupiter is

$$\boldsymbol{N} = \cos F\,\boldsymbol{e}_X + \sin F\,\boldsymbol{e}_Y, \tag{3.79}$$

and Eqs. (3.42), (3.43), and (3.45) reveal that

$$\boldsymbol{n} = \cos(f+\omega)\,\boldsymbol{e}_X + \sin(f+\omega)\,\boldsymbol{e}_Y, \tag{3.80a}$$

$$\boldsymbol{\lambda} = -\sin(f+\omega)\,\boldsymbol{e}_X + \cos(f+\omega)\,\boldsymbol{e}_Y, \tag{3.80b}$$

and $\boldsymbol{e}_z = \boldsymbol{e}_Z$. From these expressions we find that $\boldsymbol{n}\cdot\boldsymbol{N} = \cos(f+\omega-F)$, $\boldsymbol{\lambda}\cdot\boldsymbol{N} = -\sin(f+\omega-F)$, $\boldsymbol{e}_z\cdot\boldsymbol{N} = 0$, and the components of the perturbing force become

$$\mathcal{R} = -\frac{Gm_3 r}{R^3}\big[1 - 3\cos^2(f+\omega-F)\big], \tag{3.81a}$$

$$\mathcal{S} = -3\frac{Gm_3 r}{R^3}\sin(f+\omega-F)\cos(f+\omega-F), \tag{3.81b}$$

$$\mathcal{W} = 0, \tag{3.81c}$$

in which we may substitute $r = p/(1+e\cos f)$.

Mercury's orbital evolution in response to Jupiter's perturbation is governed by Eqs. (3.64). As we explained near the end of Sec. 3.3.2, we are primarily interested in the *secular variation* of the orbital elements p, e, and ω, as calculated according to Eq. (3.71) using the equations listed in Eq. (3.69). The integrals over f are easily evaluated by adopting the eccentric anomaly u as an integration variable, and by assuming that F is constant over Mercury's orbital cycle – refer to Eq. (3.78). After some manipulations we arrive at

$$\Delta p := -15\pi \frac{m_3 p^4}{m R^3} e^2 (1-e^2)^{-7/2} \sin 2(\omega - F), \tag{3.82a}$$

$$\Delta e := \frac{15\pi}{2} \frac{m_3 p^3}{m R^3} e (1-e^2)^{-5/2} \sin 2(\omega - F), \tag{3.82b}$$

$$\Delta \omega := \frac{3\pi}{2} \frac{m_3 p^3}{m R^3} (1-e^2)^{-5/2} \big[1 + 5\cos 2(\omega - F)\big] \tag{3.82c}$$

for the net change in the orbital elements after a complete Mercury orbit. From these results it can be inferred that $\Delta a = 0$: Mercury does not undergo a secular variation of its semi-major axis, and the change in $p = a(1-e^2)$ simply reflects a change in eccentricity.

These expressions reveal that depending on Jupiter's position on its orbit, Mercury's eccentricity e and longitude of perihelion ω either increase or decrease in the course of a complete orbit. After an extremely long time, much longer than Jupiter's own orbital period, the change in e will average out, but Mercury's perihelion will advance at an average rate of

$$\langle \Delta \omega \rangle = \frac{3\pi}{2} \frac{m_3}{m} \left(\frac{p}{R}\right)^3 (1-e^2)^{-5/2} = \frac{3\pi}{2} \frac{m_3}{m} \left(\frac{a}{R}\right)^3 (1-e^2)^{1/2} \tag{3.83}$$

per orbit.

Inserting the numerical values for Mercury and Jupiter from Table 3.2, we find that $\langle \Delta\omega \rangle = 1.79 \times 10^{-6}$ radians per orbit; the smallness of this effect provides ample justification for a calculation limited to first-order perturbation theory. It is customary to express the perihelion advance as a rate in units of arcseconds (as) per century. To achieve this conversion we use the fact that 1 rad $= 2.063 \times 10^5$ as, and take into account the fact that Mercury's orbital period is $P = 0.2408$ yr; we obtain $\langle (d\omega/dt)_{\rm sec} \rangle = 154$ as/century. This is very close to the accurately computed value of 153.6 as/century displayed in Table 3.1. The accuracy, however, would not be as good if we applied Eq. (3.83) to the perturbation produced by Earth's gravitational attraction; here we would get 69.2 as/century, quite a difference from the accurate value of 90 as/century. The reason for this lack of success is that $r/R \simeq 0.38$ for the Earth, so that an accurate computation of the perihelion advance requires higher-order terms in the expansion of the perturbing force in powers of r/R. In addition, the approximation formulated in Eq. (3.78) is not so good in the case of the Earth, because Earth's period is only four times longer than Mercury's (instead of forty times longer for Jupiter). Nevertheless, the method of osculating orbital elements would allow us to incorporate such details systematically in a straightforward manner, leading to the accurately calculated planetary perturbations listed in Table 3.1.

As we have seen in Sec. 3.1, the sum total of the contributions from all the planets to Mercury's perihelion advance does not account for the observed effect. The discrepancy,

Table 3.2 Orbital elements of selected planets. The astronomical unit (AU) is the Earth–Sun distance, equal to 149.60×10^6 km. The inclination is measured in degrees, minutes of arc, and seconds of arc. The inverse mass is measured in units of the inverse solar mass, with $M_\odot = 1.9889 \times 10^{30}$ kg.

Planet	Semi-major axis (AU)	Orbital period (yr)	Eccentricity	Inclination to ecliptic $\circ . \prime . \prime\prime$	Inverse mass $1/M_\odot = 1$
Mercury	0.387099	0.24085	0.205628	7.0.15	6010000
Venus	0.723332	0.61521	0.006787	3.23.40	408400
Earth	1.000000	1.00004	0.016722	0.0.0	328910
Mars	1.523691	1.88089	0.093377	1.51.0	3098500
Jupiter	5.202803	11.86223	0.04845	1.18.17	1047.39
Saturn	9.53884	29.4577	0.05565	2.29.22	3498.5

the famous 43 as/century, is accounted for by relativistic corrections to the equations of motion, which can be incorporated as additional contributions to the perturbing force. We will return to this question in Chapter 10.

3.4.2 The Kozai mechanism

In 1962, the Soviet dynamicist Michael Lidov and the Japanese astronomer Yoshihide Kozai independently discovered a remarkable phenomenon that occurs when an object orbits a massive body under the perturbing influence of a distant body, when the two orbits are inclined relative to each other. As a result of the perturbed dynamics, there is a periodic exchange between the orbital eccentricity and inclination, such that when one increases, the other decreases. This effect, known as the *Kozai mechanism*, has important implications for the behavior of asteroids, satellites of planets, extrasolar planets, and multiple-star systems.

To study this effect we re-examine the situation of the previous section, but now allow the orbiting body (m_1) to have an inclination ι relative to the orbit of the perturbing body (m_3), which is placed in the reference X-Y plane. For simplicity we take the perturbing body to move on a circular orbit, and we choose the line of nodes of the inclined orbit to be situated at $\Omega = 0$. The components of the perturbing force are still given by Eqs. (3.77), but the orbital basis vectors are now

$$\boldsymbol{n} = \cos(\omega + f)\,\boldsymbol{e}_X + \cos\iota\sin(\omega + f)\,\boldsymbol{e}_Y + \sin\iota\sin(\omega + f)\,\boldsymbol{e}_Z, \tag{3.84a}$$

$$\boldsymbol{\lambda} = -\sin(\omega + f)\,\boldsymbol{e}_X + \cos\iota\cos(\omega + f)\,\boldsymbol{e}_Y + \sin\iota\cos(\omega + f)\,\boldsymbol{e}_Z, \tag{3.84b}$$

$$\boldsymbol{e}_z = -\sin\iota\,\boldsymbol{e}_Y + \cos\iota\,\boldsymbol{e}_Z; \tag{3.84c}$$

the direction to the perturbed body is still given by Eq. (3.79). We calculate the secular variation of the orbital elements a, e, ω, and ι by inserting these expressions within Eqs. (3.69), integrating over a complete orbital cycle, and averaging over the position of the perturbing body. The final results are

$$\langle \Delta a \rangle = 0, \tag{3.85a}$$

$$\langle \Delta e \rangle = \frac{15\pi}{2} \frac{m_3}{m} \left(\frac{a}{R} \right)^3 e(1-e^2)^{1/2} \sin^2 \iota \sin \omega \cos \omega, \tag{3.85b}$$

$$\langle \Delta \omega \rangle = \frac{3\pi}{2} \frac{m_3}{m} \left(\frac{a}{R} \right)^3 (1-e^2)^{-1/2} \Big[5 \cos^2 \iota \sin^2 \omega + (1-e^2)(5\cos^2 \omega - 3) \Big], \tag{3.85c}$$

$$\langle \Delta \iota \rangle = -\frac{15\pi}{2} \frac{m_3}{m} \left(\frac{a}{R} \right)^3 e^2 (1-e^2)^{-1/2} \sin \iota \cos \iota \sin \omega \cos \omega. \tag{3.85d}$$

From Eqs. (3.85b) and (3.85d) we notice that $e(1-e^2)^{-1}\langle \Delta e \rangle + \tan \iota \langle \Delta \iota \rangle = 0$, which leads to the remarkable conservation law,

$$\sqrt{1-e^2} \cos \iota = \text{constant (over long time scales)}. \tag{3.86}$$

This is actually the consequence of angular-momentum conservation. The component of the total angular-momentum vector normal to the orbital plane of the perturbing body is $L_Z = L_Z^{1+2} + L_Z^3$, and if we assume that L_Z^3 is separately conserved (thereby ignoring the gravitational effects of m_1 and m_2 on the perturbing body), we find that $L_Z^{1+2} = \mu \boldsymbol{h} \cdot \boldsymbol{e}_Z = \mu \sqrt{Gmp} \cos \iota$ must be conserved. Because $a := p/(1-e^2)$ does not vary over long time scales, conservation of angular momentum implies that $(1-e^2)^{1/2} \cos \iota$ cannot vary, in agreement with our previous statement.

When the longitude of pericenter ω lies in the first or third quadrants, we see from Eqs. (3.85) that the eccentricity increases while the inclination decreases; when ω lies instead in the second or fourth quadrants, the eccentricity decreases while the inclination increases. The equations also reveal that as long as $\cos^2 \iota/(1-e^2) > 3/5$, ω increases monotonically, on the same time scale as the variations in e and ι. For orbits with sufficiently low inclinations, therefore, there is a periodic exchange between eccentricity and inclination as the pericenter keeps on advancing.

For high-inclination orbits, such that $\cos^2 \iota/(1-e^2) < 3/5$, the orbital evolution is very different. In this case, ω approaches a critical angle ω_c such that $\langle \Delta \omega \rangle = 0$, and the motion of the pericenter ceases; the critical angle is determined by

$$\cos^2 \omega_c = \frac{3(1-e^2) - 5\cos^2 \iota}{5(1-e^2-\cos^2 \iota)}. \tag{3.87}$$

Meanwhile, if e is initially in an increasing phase and ι in a decreasing phase, the ratio $\cos^2 \iota/(1-e^2)$ will be driven toward the limiting value of $3/5$, and ω will settle to a final value of $\omega_c = \frac{\pi}{2}$ or $\omega_c = \frac{3\pi}{2}$. The final outcome is a stationary solution of the perturbation equations with $\langle \Delta e \rangle = \langle \Delta \iota \rangle = \langle \Delta \omega \rangle = 0$, occurring at critical values e_c, ι_c, and ω_c constrained by

$$\omega_c = \frac{\pi}{2} \text{ or } \frac{3\pi}{2}, \qquad \frac{\cos^2 \iota_c}{1-e_c^2} = \frac{3}{5}. \tag{3.88}$$

This equilibrium is stable; it can be shown that a perturbation away from equilibrium leads to oscillations of e, ι, and ω about their critical values, with a period

$$P_{\rm Kozai} = \frac{2}{3\sqrt{15}} \frac{m}{m_3} \left(\frac{R}{a}\right)^3 \frac{1}{e_c \sin \iota_c} P, \tag{3.89}$$

where P is the period of the orbiting body. This stationary solution is called the *Kozai resonance*. You will be asked to verify these statements in Exercise 3.7.

The Kozai mechanism can have important consequences: as the orbital eccentricity increases, the orbiting body can pass sufficiently close to the inner regions of the system at pericenter to interact with other bodies there, or even be tidally captured or disrupted by the central body. Conversely, the body could find itself sufficiently far away at apocenter to interact with more distant objects and be ejected from the system. This mechanism accounts for many observed features of the asteroid belts, of minor satellites of Jupiter, and of objects within the Kuiper belt. As an example, many high-eccentricity comets are found to be in Kozai resonances, with their longitude of pericenter in the vicinity of $\frac{\pi}{2}$ or $\frac{3\pi}{2}$.

3.4.3 Effects of oblateness

As our next case study we examine a two-body system perturbed by the oblateness of one of its members. As we saw back in Sec. 2.3, the oblateness is most often due to rotational flattening, and here the deformed body is taken to have an axisymmetric shape; the deformation from spherical symmetry is measured by its quadrupole moment, and we neglect the gravitational influence of higher-order multipole moments. Examples of such systems abound. We could be dealing with Mercury orbiting an oblate, rotating Sun, or an artificial satellite orbiting the Earth, or else a neutron star orbiting a rotating, normal star.

The equations of motion for such a system were worked out back in Sec. 1.6.7, and according to Eq. (1.227), the perturbing force is given by

$$\boldsymbol{f} = \frac{3}{2} J_2 \frac{GmR^2}{r^4} \Big\{ \big[5(\boldsymbol{e}\cdot\boldsymbol{n})^2 - 1\big]\boldsymbol{n} - 2(\boldsymbol{e}\cdot\boldsymbol{n})\boldsymbol{e} \Big\}, \tag{3.90}$$

with J_2 denoting the dimensionless quadrupole moment of the deformed body, R its radius, and the unit vector $\boldsymbol{e}$ indicating the direction of the symmetry axis. As usual we have $m := m_1 + m_2$, $\boldsymbol{r} := \boldsymbol{r}_1 - \boldsymbol{r}_2$, $r := |\boldsymbol{r}|$, and $\boldsymbol{n} := \boldsymbol{r}/r$. In Eq. (3.90) we specifically identify the oblate body with m_2, and m_1 is assumed to be spherical; a swap of identities involves changing the sign of $\boldsymbol{n}$, and therefore the sign of the perturbing force.

We choose the reference X-Y plane to be orthogonal to the symmetry axis of m_2, so that $\boldsymbol{e}$ is aligned with the Z-direction. The orbital plane has an inclination ι relative to the reference plane, and the orbital basis is given by Eqs. (3.42)–(3.45). According to these equations, $\boldsymbol{e}\cdot\boldsymbol{n} = \sin\iota \sin(\omega + f)$, $\boldsymbol{e}\cdot\boldsymbol{\lambda} = \sin\iota\cos(\omega+f)$, $\boldsymbol{e}\cdot\boldsymbol{e}_z = \cos\iota$, and the components of the perturbing force in the orbital basis are

$$\mathcal{R} = \frac{3}{2} J_2 \frac{GmR^2}{r^4} \left[3 \sin^2 \iota \sin^2(\omega + f) - 1\right], \tag{3.91a}$$

$$\mathcal{S} = -3J_2 \frac{GmR^2}{r^4} \sin^2 \iota \sin(\omega + f) \cos(\omega + f), \tag{3.91b}$$

$$\mathcal{W} = -3J_2 \frac{GmR^2}{r^4} \sin \iota \cos \iota \sin(\omega + f), \tag{3.91c}$$

with $r = p/(1 + e \cos f)$.

We calculate the secular variation of the orbital elements p, e, ι, ω, and Ω by inserting these expressions within Eqs. (3.69), and integrating over a complete orbital cycle. (In this case there is no need to involve the eccentric anomaly u in the evaluation of the integrals; the factor of r^{-4} in the perturbing force ensures that there are no remaining factors of $1 + e \cos f$ in the denominators, and all integrations are elementary.) The end results are

$$\Delta p = 0, \tag{3.92a}$$

$$\Delta e = 0, \tag{3.92b}$$

$$\Delta \iota = 0, \tag{3.92c}$$

$$\Delta \omega = 6\pi J_2 \left(\frac{R}{p}\right)^2 \left(1 - \frac{5}{4} \sin^2 \iota\right), \tag{3.92d}$$

$$\Delta \Omega = -3\pi J_2 \left(\frac{R}{p}\right)^2 \cos \iota. \tag{3.92e}$$

The first two equations imply that $\Delta a = 0$; there is no secular change in the eccentricity, no secular change in the semi-major axis, and no secular change in the inclination. There are, however, secular changes in the line of nodes (measured by $\Delta\Omega$) and in the pericenter (measured relative to the line of nodes by $\Delta\omega$). The last two equations imply

$$\Delta \omega + \cos \iota \, \Delta \Omega = 3\pi J_2 \left(\frac{R}{p}\right)^2 \left(1 - \frac{3}{2} \sin^2 \iota\right); \tag{3.93}$$

as we saw back in Sec. 3.3.2, this is the pericenter advance relative to the reference X-direction, as measured in the orbital plane. At zero inclination, when $\boldsymbol{e} \cdot \boldsymbol{n} = 0$, the changes are produced by the r^{-4} modification to the gravitational acceleration, which is still directed along $\boldsymbol{n}$. For inclined orbits, the angular dependence of the perturbing force produces an additional contribution. An associated effect is a precession of the angular-momentum vector $\boldsymbol{h}$ around the symmetry axis $\boldsymbol{e}$, which produces a rotation of the line of nodes. The precession is predicted by Eq. (3.51), which becomes

$$\frac{d\boldsymbol{h}}{dt} = 3J_2 \frac{GmR^2}{r^3} (\boldsymbol{e} \cdot \boldsymbol{n})(\boldsymbol{e} \times \boldsymbol{n}) \tag{3.94}$$

in the case of the perturbing force of Eq. (3.90). This describes a precession because h does not change on a secular time scale, and because $\boldsymbol{e} \cdot \boldsymbol{h}$ stays constant during the orbital evolution.

For Mercury, the perihelion advance induced by the Sun's oblateness is negligible. Inserting the relevant orbital elements and the value $(J_2)_\odot = 2.2 \times 10^{-7}$, we obtain $(d\omega/dt)_{\rm sec} + \cos\iota (d\Omega/dt)_{\rm sec} = 0.03$ as/century, which is just below the observational uncertainties assigned to the measurement of the advance (refer to Table 3.1). The conclusion relies on a reliable determination of the Sun's oblateness, something that was difficult to come by. In fact, it is extremely difficult to measure the Sun's J_2 directly, because at planetary distances, its effects are just too small to be measured. The best way to obtain a reliable estimate would be to send a spacecraft to a low solar orbit, well inside the orbit of Mercury, and to measure precisely how J_2 affects its orbit; but despite a number of proposals for such a mission over the years, none has yet come to pass.

During the 1960s, Robert Dicke and Mark Goldenberg attempted to determine J_2 by measuring the Sun's visual shape. Because the surface of the Sun is an equipotential (recall the discussion in Sec. 2.3.1), its shape is affected by J_2 in a way that can be directly related to the deformation of the external gravitational field. The shape was measured by inserting a circular, opaque disk in front of a telescopic image of the Sun, leaving only a thin visible ring at the edge of the Sun, and measuring the difference in brightness of the visible ring between the equator and pole of the Sun. If the Sun were oblate, the ring at the equator should extend further beyond the occulting disk, and should therefore be brighter. But many factors had to be corrected for, including the effects of atmospheric distortion on the observed shape of the Sun, and the effects of possible temperature differences between the polar and equatorial regions of the Sun, which would lead to brightness differences not associated with the shape. Dicke and Goldenberg claimed to have measured a J_2 of the order of 2.5×10^{-5}, over 100 times larger than the currently accepted value. Dicke postulated that such a large oblateness would occur if the core of the Sun were rotating much faster than its outer layers, thereby generating more centrifugal flattening than would be expected on the basis of the observed surface rotation alone.

A value of J_2 this large would mean that solar oblateness contributes as much as 4 as/century to Mercury's perihelion advance, which would destroy the agreement of the measured advance with the prediction of general relativity. But it would have supported the scalar-tensor theory of gravity (to be presented in Chapter 13) that Dicke himself had developed with Carl Brans, whose prediction for the relativistic part of the advance could be fit to 39 as/century by tuning a coupling constant. (The flexibility to tune the Brans–Dicke parameter to obtain this low value was short lived; accurate observations of the relativistic light deflection eventually constrained the coupling constant to such an extent that the Brans–Dicke prediction for the perihelion advance of Mercury became only marginally different from Einstein's.) Later observations of the visible shape of the Sun by Henry Hill and others, along with observations to try to better understand the temperature differences, did not fully resolve this controversy.

The resolution came with the advance of helioseismology. This was the discovery that the Sun vibrates in a superposition of thousands of normal modes with an array of frequencies, as could be observed by measuring the frequency spectrum of Doppler-shifted solar spectral lines. The specific pattern of frequencies depends on the Sun's angular-velocity profile. Through a systematic program of ground-based and space-based observations of the Sun, it became possible to determine the Sun's rotational profile over much of its interior. The conclusion was that the core does not rotate much faster than the surface, and solar models consistent with this information produced the currently accepted value of $J_2 = 2.2 \times 10^{-7}$; this is approximately what one would infer from a Sun that rotates uniformly at its observed surface rate. The bottom line is that, as far as Mercury's motion and general relativity are concerned, the solar quadrupole moment does not play a significant role.

The quadrupole moment of the Earth does play an important role in the motion of artificial satellites. The effect on the line of nodes can be examined for a satellite placed on a circular orbit with radius a. Using $(J_2)_\oplus = 1.08 \times 10^{-3}$ and dividing $\Delta\Omega$ by the satellite's orbital period $P = 2\pi a^{3/2}/(Gm)^{1/2} = 83.91(a/R)^{3/2}$ minutes, we can show that

$$\left(\frac{d\Omega}{dt}\right)_{\rm sec} = -3639 \left(\frac{R}{a}\right)^{7/2} \cos\iota \quad \text{degrees/year}, \tag{3.95}$$

where R is the Earth's radius. For example, the orbit of the Laser Geodynamics Satellite (LAGEOS) I, with $a = 1.93R$ and $\iota = 109^\circ.8$, precesses at a rate of 120 degrees per year. A satellite with $a = 1.5R$ and $\iota = 65.9^\circ$ would precess at a rate of 360 degrees per year. Such an orbit is called *Sun synchronous* because it always presents the same face to the Sun, varying only in its inclination relative to the ecliptic because of the 23.5° tilt of the Earth's spin axis.

3.4.4 Tidally interacting bodies

In our next case study we examine a system of two bodies that deform each other by tidal interactions. The physics of tides was reviewed in some depth in Sec. 2.5, and here we consider a situation in which the orbital period $P = 2\pi a^{3/2}/(Gm)^{1/2}$ is long compared with the hydrodynamical time scale $T_{\rm int} \sim (G\rho)^{-1/2} \sim R^{3/2}/(Gm)^{1/2}$ that characterizes the internal dynamics of each body; this is the realm of *static tides*.

We first take the bodies to be non-rotating. According to the theory developed in Sec. 2.5.1, a body of radius R subjected to a tidal potential $U_{\rm tidal} = -\frac{1}{2}\mathcal{E}_{jk}x^j x^k$ acquires a mass quadrupole moment given by $GI_{\langle jk\rangle} = -\frac{2}{3}k_2 R^5 \mathcal{E}_{jk}$, with k_2 denoting the body's gravitational Love number, which depends on the details of its internal structure. In the case of a two-body system, we have a first body of mass m_1, radius R_1, Love number $(k_2)_1$, and quadrupole moment $I_1^{\langle jk\rangle}$, and a second body of mass m_2, radius R_2, Love number $(k_2)_2$, and quadrupole moment $I_2^{\langle jk\rangle}$. According to Eq. (2.275), the tidal quadrupole moment created by m_2 and acting on m_1 is given by $\mathcal{E}_1^{jk} = -(Gm_2/r^3)(3n^j n^k - \delta^{jk})$, while the tidal moment created by m_1 and acting on m_2 is $\mathcal{E}_2^{jk} = -(Gm_1/r^3)(3n^j n^k - \delta^{jk})$; as usual, we have that $m := m_1 + m_2$, $\boldsymbol{r} := \boldsymbol{r}_1 - \boldsymbol{r}_2$, $r := |\boldsymbol{r}|$, and $\boldsymbol{n} := \boldsymbol{r}/r$. From all this we find that the tidally-induced mass quadrupole moment of the first body is given by

$$I_1^{\langle jk\rangle} = \frac{2}{3}(k_2)_1 \frac{m_2 R_1^5}{r^3}\left(3n^j n^k - \delta^{jk}\right), \tag{3.96}$$

while

$$I_2^{\langle jk\rangle} = \frac{2}{3}(k_2)_2 \frac{m_1 R_2^5}{r^3}\left(3n^j n^k - \delta^{jk}\right) \tag{3.97}$$

is the mass quadrupole moment of the second body.

The equations of motion for two deformed bodies were obtained back in Sec. 1.6.7, and Eq. (1.220) provides an expression for the relative acceleration $\boldsymbol{a} := \boldsymbol{a}_1 - \boldsymbol{a}_2$. If we remove the Keplerian term, truncate the multipole expansions to the leading, quadrupole order, and neglect the quadrupole–quadrupole interaction (which is much smaller than the monopole–quadrupole interactions that we keep), we obtain

$$f^j = \frac{1}{2} G m \left(\frac{I_1^{\langle kn \rangle}}{m_1} + \frac{I_2^{\langle kn \rangle}}{m_2} \right) \partial_{jkn} \left(\frac{1}{r} \right) \tag{3.98}$$

for the perturbing force. Inserting the preceding expressions for the mass quadrupole moments, as well as Eq. (1.152) for $\partial_{jkn} r^{-1}$, we find that this becomes

$$\boldsymbol{f} = -6 \frac{G m}{r^7} \left((k_2)_1 \frac{m_2}{m_1} R_1^5 + (k_2)_2 \frac{m_1}{m_2} R_2^5 \right) \boldsymbol{n}. \tag{3.99}$$

The perturbing force is directed along $\boldsymbol{n}$, and its only non-vanishing component is $\mathcal{R}$. With $\mathcal{S} = \mathcal{W} = 0$, we see that the tidal interaction has no effect on p, ι, and Ω; the only orbital elements that can undergo a change are e and ω.

We insert our expression for $\mathcal{R}$ within Eqs. (3.69), integrate over a complete orbital cycle, and obtain the secular changes $\Delta e = 0$ and

$$\Delta \omega = 30\pi \left(1 + \frac{3}{2} e^2 + \frac{1}{8} e^4 \right) \left[(k_2)_1 \frac{m_2}{m_1} \left(\frac{R_1}{p} \right)^5 + (k_2)_2 \frac{m_1}{m_2} \left(\frac{R_2}{p} \right)^5 \right]. \tag{3.100}$$

These results imply that e (and therefore a) do not undergo a secular change, but the pericenter advances at a steady rate that depends on the orbital parameters, the body radii, and the gravitational Love numbers. Astronomers call this phenomenon the *apsidal advance*, and for a class of close binaries, this effect gives an important clue to the internal structure of each star, via the ability to infer k_2. These systems have masses of the order of a few solar masses, orbital periods of the order of 10 days, and modest eccentricities. The resulting apsidal advance can then be expressed as the rate

$$\begin{aligned} \left(\frac{d\omega}{dt} \right)_{\text{sec}} &= 0.06\, f(e) \left(\frac{M_\odot}{m} \right)^{5/3} \left(\frac{10 \text{ days}}{P} \right)^{13/3} \\ &\quad \times \left[\frac{(k_2)_1}{0.01} \frac{m_2}{m_1} \left(\frac{R_1}{R_\odot} \right)^5 + \frac{(k_2)_2}{0.01} \frac{m_1}{m_2} \left(\frac{R_2}{R_\odot} \right)^5 \right] \text{degrees/century}, \end{aligned} \tag{3.101}$$

where $f(e) = (1 + \frac{3}{2} e^2 + \frac{1}{8} e^4)/(1 - e^2)^5$. Many close binaries are eclipsing, and by combining light-curve data on the timing of eclipses with spectroscopic data on the orbital motion, it is possible to determine the masses and radii as well as the apsidal advance rate, and thereby estimate k_2 for each star. In most cases, there is good agreement between the values of k_2 obtained by implementing this method and those determined by stellar models for stars having the observed masses and spectra. As we shall see in Chapter 10, the general relativistic contribution to the apsidal advance is small, generally a few percent of the tidally induced advance.

Box 3.4 ***DI Herculis*: A tidal troublemaker**

There are a few systems for which the nice picture of a tidally induced apsidal advance does not work. The most famous is *DI Herculis*, a high eccentricity ($e \simeq 0.48$), high mass ($m \simeq 10\, M_\odot$) system with a 10.55 day period. In this case, the relativistic contribution to the apsidal advance is large, 2.34 degrees per century, while the tidal contribution is predicted to be 1.93 degrees per century, giving a total predicted advance of 4.27 degrees per century. Unfortunately, the *observed* apsidal advance is only $1.00 \pm .30$ degrees per century. While there were a few attempts to use this discrepancy as evidence against general relativity, more conventional explanations involve purely Newtonian mechanisms. First, there could be a large misalignment between the rotation axis of each star and the orbital plane, such that the contribution of the rotationally-induced quadrupole moments to $\Delta\omega$ would be negative, or retrograde. Second, there could be a third star in the vicinity of the system, whose perturbative effects on all the orbital elements would complicate how $\Delta\omega$ is inferred from the eclipse light-curve data, leading to a value in better agreement with the prediction. The book is not yet closed on *DI Herculis*.

Our discussion thus far has excluded the dissipative aspects of tidal dynamics that were reviewed in Sec. 2.5.2. To explore these effects we examine specifically the case of a moon orbiting a planet. We take the first body to be the planet, the second body to be the moon, and we place the bodies on a circular orbit of relative separation r. Both bodies are now rotating, and we take the moon to be tidally locked, so that it rotates with the same angular velocity as the orbit. The planet, on the other hand, rotates with an angular velocity ω_1 that differs, in general, from the orbital angular velocity Ω.

According to Eq. (2.277), the quadrupole moment of each body is given by

$$GI_{\langle jk\rangle} = -\frac{2}{3}k_2 R^5\left(\mathcal{E}_{jk} - \tau\dot{\mathcal{E}}_{jk}\right) \tag{3.102}$$

in the body's rotating frame. The effect of the first term was investigated previously, and we henceforth ignore it. We keep the second term, the one associated with the dissipative aspects of the tidal interaction; the parameter τ is the body's viscous delay. Because the moon is co-rotating with the orbit, the tidal field measured in its rotating frame is constant, and this implies that we can set $I_2^{\langle jk\rangle} = 0$; for our purposes here, the moon is not deformed by the tidal forces exerted by the planet. The planet, however, is deformed by the moon's tidal field, and we find that its quadrupole moment is given by

$$I_1^{\langle jk\rangle} = -2(k_2)_1\frac{m_2 R_1^5}{r^3}(\Omega - \omega_1)\tau_1\left(n^j\lambda^k + \lambda^j n^k\right) \tag{3.103}$$

in the rotating frame. Inserting this within the perturbing force of Eq. (3.98), we find that its dissipative piece is given by

$$\boldsymbol{f}_{\rm diss} = -6(k_2)_1 Gm\frac{m_2}{m_1}\frac{R_1^5}{r^7}(\Omega - \omega_1)\tau_1\,\boldsymbol{\lambda}. \tag{3.104}$$

Its only non-vanishing component is $\mathcal{S}_{\rm diss} := \boldsymbol{f}_{\rm diss}\cdot\boldsymbol{\lambda}$.

The dissipative piece of the perturbing force can now be inserted within the osculating equations (3.69). We find that r is the only orbital element that is affected by the perturbation, and that it changes by

$$\Delta r = -24\pi (k_2)_1 \frac{m_2}{m_1} \frac{R_1^5}{r^4} (\Omega - \omega_1)\tau_1 \tag{3.105}$$

in the course of a complete orbital cycle. The sign of the change is related to the sign of $\Omega - \omega_1$. When $\omega_1 < \Omega$, that is, when the moon orbits faster than the planet rotates, we find that $\Delta r < 0$, so that the moon slowly approaches the planet in a shrinking orbit; an example of such a situation is found in the Mars–Phobos system. When $\omega_1 > \Omega$, that is, when the planet rotates faster than the moon orbits, we find instead that $\Delta r > 0$, so that the moon recedes from the planet in an expanding orbit; the Earth–Moon system is a familiar example of such a situation. These conclusions are in accord with our discussion of angular-momentum transfer in Sec. 2.5.2. We saw that the planet gains spin angular momentum when $\omega_1 < \Omega$, and the shrinking orbit confirms that this gain comes at the expense of the orbital angular momentum. Similarly, the planet loses its spin when $\omega_1 > \Omega$, and this loss allows the orbit to gain angular momentum and to expand.

3.4.5 Luni-solar precession of the Earth

In this final case study we describe how the gravitational attraction of the Moon and the Sun on an oblate Earth causes a precession of its spin around the normal to the ecliptic plane. This is the origin of the famous *precession of the equinoxes*, with its period of approximately 26 000 years.

The motion of a body's spin vector $\boldsymbol{S}$ subjected to gravitational torques exerted by other bodies was determined back in Sec 1.6.8. According to Eq. (1.240), an axisymmetric body of mass m, radius R, and dimensionless quadrupole moment J_2, placed in the gravitational field of a first mass m_1 at a distance r_1, and of a second mass m_2 at a distance r_2, undergoes a spin precession described by

$$\frac{d\boldsymbol{S}}{dt} = -3GmJ_2R^2 \left[\frac{m_1}{r_1^3} (\boldsymbol{e}\cdot\boldsymbol{n}_1)(\boldsymbol{e}\times\boldsymbol{n}_1) + \frac{m_2}{r_2^3}(\boldsymbol{e}\cdot\boldsymbol{n}_2)(\boldsymbol{e}\times\boldsymbol{n}_2)\right]. \tag{3.106}$$

Here, $\boldsymbol{e}$ is the direction of the body's rotation axis, so that $\boldsymbol{S} = S\boldsymbol{e}$, $\boldsymbol{n}_1$ is the direction to m_1, and $\boldsymbol{n}_2$ the direction to m_2. Note that in this application, m stands for the body's mass instead of the total mass of the system. It is easy to see from Eq. (3.106) that S is a constant, so that $d\boldsymbol{S}/dt = Sd\boldsymbol{e}/dt$.

For simplicity, and as a rather good approximation, we take the Moon and the Sun to move on circular orbits in the ecliptic plane, which we identify with the reference X-Y plane. The direction to the Moon is then

$$\boldsymbol{n}_1 = \cos(\Omega_1 t)\,\boldsymbol{e}_X + \sin(\Omega_1 t)\,\boldsymbol{e}_Y, \tag{3.107}$$

with Ω_1 denoting the Moon's angular velocity, while the direction to the Sun is

$$\boldsymbol{n}_2 = \cos(\Omega_2 t + \psi)\,\boldsymbol{e}_X + \sin(\Omega_2 t + \psi)\,\boldsymbol{e}_Y, \tag{3.108}$$

with Ω_2 denoting the Sun's angular velocity, and ψ representing the Sun's initial phase on its orbit. The Earth's rotation axis is tilted by an angle $\alpha = 23.4^\circ$ relative to the ecliptic, and we express its direction as

$$\boldsymbol{e} = \sin\alpha \cos\beta\, \boldsymbol{e}_X + \sin\alpha \sin\beta\, \boldsymbol{e}_Y + \cos\alpha\, \boldsymbol{e}_Z, \tag{3.109}$$

in terms of a time-dependent precessional angle β. As we shall see, Eq. (3.106) implies that α remains constant on a secular time scale.

We insert these expressions within Eq. (3.106) and average the result over a large number of orbital cycles. We obtain

$$\left(\frac{d\boldsymbol{e}}{dt}\right)_{\rm sec} = -\frac{3GmJ_2R^2}{2S} \sin\alpha \cos\alpha \left(\frac{m_1}{r_1^3} + \frac{m_2}{r_2^3}\right)\left(-\sin\beta\, \boldsymbol{e}_X + \cos\beta\, \boldsymbol{e}_Y\right). \tag{3.110}$$

Substitution of Eq. (3.109) on the left-hand side confirms that $(d\alpha/dt)_{\rm sec} = 0$, and gives rise to a rate of change for the precessional angle,

$$\left(\frac{d\beta}{dt}\right)_{\rm sec} = -\frac{3GmJ_2R^2}{2S} \cos\alpha \left(\frac{m_1}{r_1^3} + \frac{m_2}{r_2^3}\right). \tag{3.111}$$

Inserting the relevant numbers, including $S = 5.86 \times 10^{33}$ kg m^2/s and $J_2 = 1.08 \times 10^{-3}$, we obtain $|(d\beta/dt)_{\rm sec}| = 7.74 \times 10^{-12}$ rad/s, or 5040 as/century, which is very close to the accepted value of 5029 as/century. The precessional period $P_{\rm prec} = 2\pi\, |(d\beta/dt)_{\rm sec}|^{-1}$ amounts to the previously quoted value of 26 000 years.

3.5 More bodies

We saw back in Sec. 3.2 that the problem of two spherical bodies moving under their mutual gravitational attraction can be solved completely and exactly. The same cannot be said, however, when the bodies have significant deformations from spherical symmetry, or when we go beyond two bodies; in such cases there are no exact solutions, except in very special situations. In this section we touch briefly on the three-body problem, and say a few words about the general N-body problem; throughout this section the bodies are assumed to be spherical.

3.5.1 The three-body problem

General problem

Adding just one body to a two-body system brings a remarkable complexity to the problem of motion. The equations of motion are simple enough to write down. In the familiar notation we have

$$a_1 = -Gm_2\frac{r_{12}}{r_{12}^3} - Gm_3\frac{r_{13}}{r_{13}^3}, \tag{3.112a}$$

$$a_2 = +Gm_1\frac{r_{12}}{r_{12}^3} - Gm_3\frac{r_{23}}{r_{23}^3}, \tag{3.112b}$$

$$a_3 = +Gm_1\frac{r_{13}}{r_{13}^3} + Gm_2\frac{r_{23}}{r_{23}^3}, \tag{3.112c}$$

where, for example, $\boldsymbol{r}_{12} := \boldsymbol{r}_1 - \boldsymbol{r}_2$. According to the discussion of Sec. 1.6.6, the equations of motion admit the conserved quantities

$$\boldsymbol{P} = m_1\boldsymbol{v}_1 + m_2\boldsymbol{v}_2 + m_3\boldsymbol{v}_3, \tag{3.113a}$$

$$\boldsymbol{L} = m_1\boldsymbol{r}_1 \times \boldsymbol{v}_1 + m_2\boldsymbol{r}_2 \times \boldsymbol{v}_2 + m_3\boldsymbol{r}_3 \times \boldsymbol{v}_3, \tag{3.113b}$$

$$E = \frac{1}{2}m_1v_1^2 + \frac{1}{2}m_2v_2^2 + \frac{1}{2}m_1v_2^2 - \frac{Gm_1m_2}{r_{12}} - \frac{Gm_1m_3}{r_{13}} - \frac{Gm_2m_3}{r_{23}}. \tag{3.113c}$$

Momentum conservation implies that the system's barycenter, situated at

$$\boldsymbol{R} := \frac{1}{M}\big(m_1\boldsymbol{r}_1 + m_2\boldsymbol{r}_2 + m_3\boldsymbol{r}_3\big), \tag{3.114}$$

with $M := m_1 + m_2 + m_3$, moves uniformly according to $\boldsymbol{R}(t) := \boldsymbol{R}(0) + \boldsymbol{P}t/M$. We shall describe the three-body system in the *barycentric frame*, and set $\boldsymbol{R} = \boldsymbol{0}$ at all times.

In the two-body problem it proved helpful to write the equations of motion in terms of the separation vector between the two bodies, which produces a reduction to an effective one-body problem. Here we follow Jacobi and accomplish a similar reduction to an effective two-body problem by introducing the alternative variables

$$\boldsymbol{r} := \boldsymbol{r}_1 - \boldsymbol{r}_2, \tag{3.115a}$$

$$\boldsymbol{\rho} := \boldsymbol{r}_3 - \frac{m_1}{m}\boldsymbol{r}_1 - \frac{m_2}{m}\boldsymbol{r}_2, \tag{3.115b}$$

in which $m := m_1 + m_2$; the vector $\boldsymbol{r}$ is the separation between body 1 and body 2, while $\boldsymbol{\rho}$ is the separation between body 3 and the barycenter of the 1–2 subsystem. The barycenter condition $\boldsymbol{R} = \boldsymbol{0}$ can be used to express each position vector in terms of the new variables; we find that

$$\boldsymbol{r}_1 = \frac{m_2}{m}\boldsymbol{r} - \frac{m_3}{M}\boldsymbol{\rho}, \tag{3.116a}$$

$$\boldsymbol{r}_2 = -\frac{m_1}{m}\boldsymbol{r} - \frac{m_3}{M}\boldsymbol{\rho}, \tag{3.116b}$$

$$\boldsymbol{r}_3 = \frac{m}{M}\boldsymbol{\rho}. \tag{3.116c}$$

Similar relations hold between the individual velocity vectors and the relative velocities $\boldsymbol{v} := d\boldsymbol{r}/dt$ and $\boldsymbol{V} := d\boldsymbol{\rho}/dt$. The separation vectors become $\boldsymbol{r}_{12} = \boldsymbol{r}$, $\boldsymbol{r}_{13} = -(\boldsymbol{\rho} - m_2\boldsymbol{r}/m)$, and $\boldsymbol{r}_{23} = -(\boldsymbol{\rho} + m_1\boldsymbol{r}/m)$ in the Jacobi variables.

Making the substitutions, we quickly find that the equations of motion of the effective two-body problem are

$$\frac{d^2\boldsymbol{r}}{dt^2} = -Gm\frac{\boldsymbol{r}}{r^3} + Gm_3\left(\frac{\boldsymbol{\rho} - m_2\boldsymbol{r}/m}{|\boldsymbol{\rho} - m_2\boldsymbol{r}/m|^3} - \frac{\boldsymbol{\rho} + m_1\boldsymbol{r}/m}{|\boldsymbol{\rho} + m_1\boldsymbol{r}/m|^3}\right), \tag{3.117a}$$

$$\frac{d^2\boldsymbol{\rho}}{dt^2} = -\frac{GM}{m}\left(m_1\frac{\boldsymbol{\rho} - m_2\boldsymbol{r}/m}{|\boldsymbol{\rho} - m_2\boldsymbol{r}/m|^3} + m_2\frac{\boldsymbol{\rho} + m_1\boldsymbol{r}/m}{|\boldsymbol{\rho} + m_1\boldsymbol{r}/m|^3}\right), \tag{3.117b}$$

and that the conserved quantities become

$$\boldsymbol{L} = \frac{m_1 m_2}{m}\boldsymbol{r}\times\boldsymbol{v} + \frac{mm_3}{M}\boldsymbol{\rho}\times\boldsymbol{V}, \tag{3.118a}$$

$$E = \frac{1}{2}\frac{m_1m_2}{m}v^2 + \frac{1}{2}\frac{mm_3}{M}V^2 - \frac{Gm_1m_2}{r} - \frac{Gm_1m_3}{|\boldsymbol{\rho} - m_2\boldsymbol{r}/m|} - \frac{Gm_2m_3}{|\boldsymbol{\rho} + m_1\boldsymbol{r}/m|} \tag{3.118b}$$

in the Jacobi variables.

The equations of motion (3.117) are extraordinarily complex, and in spite of massive efforts by the best physicists and mathematicians of past centuries, including Lagrange, Laplace, Jacobi, Hill, and especially Poincaré, the solution space is still poorly understood today. Special solutions are known. For example, when the three bodies are confined to a plane it is possible to find solutions with $r_{12} = r_{13} = r_{23}$, so that the bodies form an equilateral triangle. In this configuration, each body follows its own elliptical orbit around the barycenter, with the triangle expanding and shrinking as the bodies pass simultaneously through their apocenter and pericenter; in the special case of circular orbits, the three bodies rotate as a rigid equilateral triangle around the barycenter. Another example is the "figure-of-eight" solution, in which three equal-mass bodies chase each other on a closed orbit that takes the shape of the number eight; the solution has zero total angular momentum. Understanding the three-body problem is of considerable astrophysical interest, since over 500 stars in our galaxy have been identified as triple systems, including the north star Polaris and our nearest neighbor, Alpha Centauri.

Restricted problem

There is a special case of the three-body problem that offers sufficient simplicity to permit a detailed exploration. Known as the *restricted three-body problem*, this is a situation in which one of the three bodies has a negligible mass compared to the others, and the other two bodies are placed on a circular orbit, undisturbed by the third body. In this case the three-body problem reduces to an independent two-body problem with a specified solution (the circular orbit), and the decoupled problem of a test body moving in the gravitational field of the orbiting bodies.

We take the light body to be m_3, and let $m_3 \ll m_1$ and $m_3 \ll m_2$, so that $M \simeq m = m_1 + m_2$. In this limit the equations of motion become $d^2\bar{\boldsymbol{r}}/dt^2 = -Gm\bar{\boldsymbol{r}}/r^3$ for the two-body system, and

$$\frac{d^2\bar{\boldsymbol{\rho}}}{dt^2} = -Gm_1\frac{\bar{\boldsymbol{\rho}} - m_2\bar{\boldsymbol{r}}/m}{|\bar{\boldsymbol{\rho}} - m_2\bar{\boldsymbol{r}}/m|^3} - Gm_2\frac{\bar{\boldsymbol{\rho}} + m_1\bar{\boldsymbol{r}}/m}{|\bar{\boldsymbol{\rho}} + m_1\bar{\boldsymbol{r}}/m|^3} \tag{3.119}$$

for the third body; the reason for adorning each vector with an overbar will be revealed presently. In this limit we have $\bar{\boldsymbol{r}}_1 = (m_2/m)\bar{\boldsymbol{r}}$, $\bar{\boldsymbol{r}}_2 = -(m_1/m)\bar{\boldsymbol{r}}$, and $\boldsymbol{r}_3 = \bar{\boldsymbol{\rho}}$. The circular orbit of the two-body system is described by

$$\bar{\boldsymbol{r}} = r\cos(\Omega t)\,\bar{\boldsymbol{e}}_x + r\sin(\Omega t)\,\bar{\boldsymbol{e}}_y, \tag{3.120}$$

with $r = \text{constant}$ and $\Omega := \sqrt{Gm/r^3}$ denoting the orbital angular velocity.

The equations are presented in a frame $(\bar{x}, \bar{y}, \bar{z})$ that is attached to the barycenter of the three-body system. This frame is non-rotating, and accordingly, the motion of the two-body system is described by the circular orbit of Eq. (3.120). To proceed it is helpful to switch reference frames, and to choose instead a frame (x, y, z) that is co-rotating with the orbiting bodies. The tools to achieve this were developed near the end of Sec. 2.3.1, and the notation adopted here, with bars placed on the non-rotating coordinates and no decoration placed on rotating coordinates, follows the conventions adopted back in Chapter 2. The coordinate transformation is given by $\bar{x} = x\cos\Omega t - y\sin\Omega t$, $\bar{y} = x\sin\Omega t + y\cos\Omega t$, $\bar{z} = z$, and the two-body system appears stationary in the rotating frame; the separation vector is now given by

$$\boldsymbol{r} = r\,\boldsymbol{e}_x, \tag{3.121}$$

and body 1 is now fixed at coordinates $x_1 = (m_2/m)r$ and $y_1 = 0$, while body 2 is at coordinates $x_2 = -(m_1/m)r$ and $y_2 = 0$.

Simple manipulations reveal that the equations of motion of the test body become

$$\frac{d^2\boldsymbol{\rho}}{dt^2} + 2\boldsymbol{\Omega}\times\frac{d\boldsymbol{\rho}}{dt} = \boldsymbol{\nabla}\Phi \tag{3.122}$$

in the rotating frame, where $\boldsymbol{\Omega} := \Omega\,\boldsymbol{e}_z$ is the angular-velocity vector, and

$$\Phi = \frac{1}{2}\Omega^2\left[\rho^2 - (\boldsymbol{\rho}\cdot\boldsymbol{e}_z)^2\right] + \frac{Gm_1}{|\boldsymbol{\rho} - m_2\boldsymbol{r}/m|} + \frac{Gm_2}{|\boldsymbol{\rho} + m_1\boldsymbol{r}/m|}, \tag{3.123}$$

a generalized potential that includes both gravitational and centrifugal terms; in Eq. (3.122) the gradient operator $\boldsymbol{\nabla}$ refers to the three components of the vector $\boldsymbol{\rho}$ in the rotating frame. The equations admit a first integral, which is obtained by taking the scalar product of Eq. (3.122) with $d\boldsymbol{\rho}/dt$ and recognizing that each term is a total derivative with respect to t. Integration produces the *Jacobi integral*

$$\frac{1}{2}V^2 - \Phi = C, \tag{3.124}$$

where $V := |d\boldsymbol{\rho}/dt|$ is the speed of the test body in the rotating frame, and C is a constant of the motion, a generalized energy known as Jacobi's constant.

Equilibria of the restricted problem

The motions predicted by Eq. (3.122) are extremely diverse, and a complete exploration would require numerical integration of the equations. The question we shall investigate here is whether the equations admit stationary solutions with $d\boldsymbol{\rho}/dt = \boldsymbol{0} = d^2\boldsymbol{\rho}/dt^2$. If such solutions exist, they would occur at values of $\boldsymbol{\rho}$ such that $\boldsymbol{\nabla}\Phi = \boldsymbol{0}$, that is, at stationary points of the generalized potential. A little thought reveals that a stationary solution requires the test body to be in the same plane as the orbiting bodies.

To simplify the analysis we adopt r as a unit of length, and divide all distances by r so that they become dimensionless. Similarly, we adopt $m := m_1 + m_2$ as a unit of mass, and divide all masses by m so that they too become dimensionless. In those units, the first body is situated at $x_1 = m_2$, the second body at $x_2 = -m_1$, and the orbital angular velocity is $\sqrt{G}$. We take the test body to have coordinates $\boldsymbol{\rho} = (x_3, y_3, 0)$, and let

$$r_1 := r_{13} = |\boldsymbol{\rho} - m_2\boldsymbol{r}/m| = \sqrt{(x_3 - m_2)^2 + y_3^2}, \tag{3.125a}$$

$$r_2 := r_{23} = |\boldsymbol{\rho} + m_1\boldsymbol{r}/m| = \sqrt{(x_3 + m_1)^2 + y_3^2}. \tag{3.125b}$$

With $m_1 + m_2$ now restricted to be unity, the individual masses are determined by the mass ratio $q := m_1/m_2 \le 1$; we have that

$$m_1 = \frac{q}{1+q}, \qquad m_2 = \frac{1}{1+q}, \tag{3.126}$$

and we adopt the convention that the first body will be the less massive of the two main bodies.

The generalized potential is given by

$$G^{-1}\Phi = \frac{1}{2}(x_3^2 + y_3^2) + \frac{m_1}{r_1} + \frac{m_2}{r_2} \tag{3.127}$$

in the rescaled variables, and differentiation with respect to x_3 and y_3 produces the equilibrium conditions $x_3 - m_1(x_3 - m_2)/r_1^3 - m_2(x_3 + m_1)/r_2^3 = 0$ and $y_3 - m_1 y_3/r_1^3 - m_2 y_3/r_2^3 = 0$. These can be re-expressed as

$$0 = m_1(x_3 - m_2)(1 - r_1^{-3}) + m_2(x_3 + m_1)(1 - r_2^{-3}), \tag{3.128a}$$

$$0 = y_3\big[m_1(1 - r_1^{-3}) + m_2(1 - r_2^{-3})\big], \tag{3.128b}$$

by exploiting the identity $m_1 + m_2 = 1$. The existence of stationary solutions hinges on finding solutions to these algebraic equations. As we shall see below, there are five distinct solutions, which are known as the *Lagrange points* L_1, L_2, L_3, L_4, and L_5. The first three equilibria have $y_3 = 0$ and describe co-linear configurations, with all three bodies aligned on the x-axis; these equilibria are unstable. The last two equilibria have $r_1 = r_2 = 1$ and describe triangular configurations, with all three bodies equidistant from each other; these equilibria are stable for sufficiently small mass ratios. The five equilibria are represented in Fig. 3.3.

We first consider the solutions to Eqs. (3.128) with $y_3 = 0$. In this case $r_1 = |x_3 - m_2|$, $r_2 = |x_3 + m_1|$, and we must distinguish three subcases. For L_1 we take the test body to be between the other two bodies, so that $-m_1 < x_3 < m_2$; in this case we have that $r_1 = m_2 - x_3$, $r_2 = x_3 + m_1$, and Eq. (3.128a) can be expressed as

$$q = \frac{r_2 - r_2^{-2}}{r_1 - r_1^{-2}}, \qquad r_1 + r_2 = 1 \qquad (L_1). \tag{3.129}$$

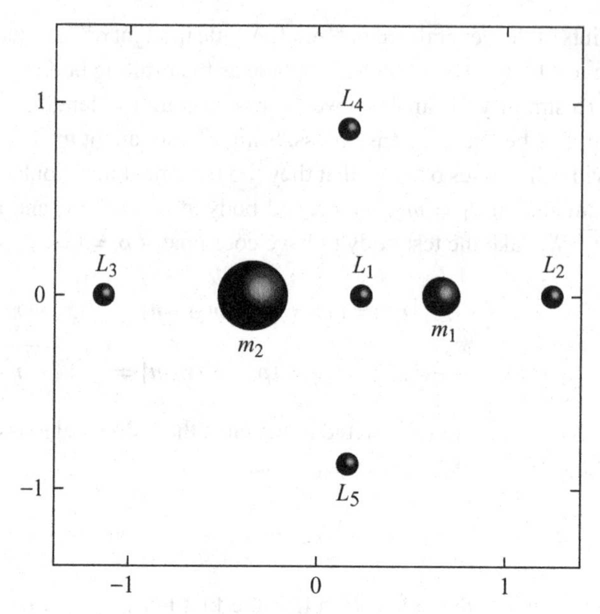

Fig. 3.3 The five Lagrange points of a restricted three-body system with mass ratio $q = \frac{1}{2}$.

It is easy to see that a solution exists for any value of $q \leq 1$, with $r_1 \leq \frac{1}{2}$. For example, with $q = \frac{1}{2}$ so that $m_1 = \frac{1}{3}$ and $m_2 = \frac{2}{3}$, the solution is $x_3 \simeq 0.23742$. For L_2 we take the test body to be beyond body 1, so that $x_3 > m_2$; in this case we have that $r_1 = x_3 - m_2$, $r_2 = x_3 + m_1$, and Eq. (3.128a) can now be expressed as

$$q = -\frac{r_2 - r_2^{-2}}{r_1 - r_1^{-2}}, \qquad r_2 - r_1 = 1 \qquad (L_2). \tag{3.130}$$

A solution exists for any value of $q \leq 1$ with $r_1 \leq 0.7$. For example, with $q = \frac{1}{2}$ the solution is $x_3 \simeq 1.24905$. Finally, for L_3 we take the test body to be beyond body 2, so that $x_3 < -m_1$; in this case we have that $r_1 = m_2 - x_3$, $r_2 = -(x_3 + m_1)$, and Eq. (3.128a) yields

$$q = -\frac{r_2 - r_2^{-2}}{r_1 - r_1^{-2}}, \qquad r_1 - r_2 = 1 \qquad (L_3). \tag{3.131}$$

A solution exists for any value of $q \leq 1$ with $1.7 \leq r_1 \leq 2$. For example, with $q = \frac{1}{2}$ the solution is $x_3 \simeq -1.13636$.

We next consider the solutions to Eqs. (3.128) with $y_3 \neq 0$. It is easy to see that a simultaneous solution to these equations requires $r_1 = r_2 = 1$, so that the configuration is an equilateral triangle. Inserting this constraint within Eq. (3.125) reveals that there are two distinct solutions described by $x_3 = \frac{1}{2} - m_1 = m_2 - \frac{1}{2}$ and $y_3 = \pm\frac{1}{2}\sqrt{3}$; the solution with $y_3 > 0$ is L_4, and the one with $y_3 < 0$ is L_5. For $q = \frac{1}{2}$, L_4 is situated at $x_3 = 0.16667$ and $y_3 = 0.86603$, while L_5 is situated at $x_3 = 0.16667$ and $y_3 = -0.86603$.

A stability analysis reveals that L_1, L_2, and L_3 are unstable equilibria: small displacements from these points produce motions that take the test body far away. The remaining Lagrange points, however, are stable provided that (refer to Exercise 3.12)

$$q < \frac{1 - \frac{1}{9}\sqrt{69}}{1 + \frac{1}{9}\sqrt{69}} \simeq 0.04; \tag{3.132}$$

in this case a small displacement produces a small oscillation around the equilibrium point.

In the solar system, the stable Lagrange points of the Sun–Jupiter system are occupied by large collections of asteroids. These orbit the Sun along Jupiter's orbital path, with one group, called the "Greek" asteroids, 60° ahead of it (at L_4), and the other group, called the "Trojan" asteroids, 60° behind it (at L_5). The asteroids remain in those positions because they are less disturbed by planetary perturbations, being protected by the weak potential well provided by the gravitational fields of the Sun and Jupiter combined with centrifugal forces. Similarly, there is a distribution of interplanetary dust at the Earth–Sun Lagrange points, and two small moons lie at the stable Lagrange points of the system formed by Saturn and Tethys, one of its moons.

It turns out that the L_1 and L_2 points of the Earth–Sun system are very useful places to park a spacecraft. Even though these equilibria are unstable, the instability is weak, and a relatively small amount of thrust is sufficient to keep the spacecraft in a small orbit around the Lagrange point. For example, the Solar and Heliospheric Observatory (SOHO) was placed at L_1, pointing continuously sunward. The Wilkinson Microwave Anisotropy Probe (WMAP) and the Planck satellite were both placed at L_2, so that they could point away from the Sun and Earth. The James Webb Space Telescope (the replacement for Hubble) is also destined for L_2.

3.5.2 The *N*-body problem

Very little can be said about the detailed motion of an N-body system, even when the equation of motion of each body is given by the simplest expression

$$\boldsymbol{a}_A = -\sum_{B \neq A} G m_B \frac{\boldsymbol{r}_{AB}}{r_{AB}^3}. \tag{3.133}$$

General statements can still be made, but they concern only the most global and coarse-grained aspects of the motion. For example, we know from Sec. 1.6.6 that the system's total momentum, energy, and angular momentum are conserved quantities. These, we recall, are given by $\boldsymbol{P} = \sum_A m_A \boldsymbol{v}_A$, $E = \mathcal{T} + \Omega$, and $\boldsymbol{L} = \sum_A m_A \boldsymbol{r}_A \times \boldsymbol{v}_A$, with

$$\mathcal{T} = \frac{1}{2} \sum_A m_A v_A^2 \tag{3.134}$$

denoting the total kinetic energy, and

$$\Omega = -\frac{1}{2} \sum_A \sum_{B \neq A} \frac{G m_A m_B}{r_{AB}} \tag{3.135}$$

the total gravitational potential energy. Another global statement is the virial theorem

$$\frac{1}{2}\frac{d^2 I}{dt^2} = 2\mathcal{T} + \Omega, \tag{3.136}$$

where $I = \sum_A m_A |\boldsymbol{r}_A|^2$ is the quadrupole-moment scalar of the N-body system. This version of the virial theorem is a specialization to isolated bodies of the general theorem of Sec. 1.4.3, which applies to any fluid configuration. Though they are limited in information content, the global statements are nevertheless useful because they rest upon the most fundamental aspects of Newtonian dynamics. They can, actually, be very powerful at probing the nature of complicated systems. For example, the virial theorem played an important role in deducing the existence of dark matter in the universe.

The virial theorem applies when an N-body system such as a star cluster can be regarded as being isolated and gravitationally bound, and it is useful when the system can also be regarded as being in a steady state, so that $\ddot{I} \approx 0$. In such situations the theorem implies that $\mathcal{T} \approx \frac{1}{2}|\Omega|$. The kinetic energy of the cluster can be expressed as $\frac{1}{2}M\langle v^2\rangle$, where M is the total mass of the cluster, and $\langle v^2\rangle$ is a suitable average of v^2 over all stars in the cluster. On the other hand, the gravitational potential energy is approximately equal to $-GM^2/(2R)$, where R is the cluster's radius. If the distribution of velocities is assumed to be isotropic, then $\langle v^2\rangle \approx 3\langle v_r^2\rangle$, with v_r denoting the component of the stellar velocity along the line of sight, which can be measured from the Doppler shift of the spectral lines (after removing the overall shift caused by the bulk motion of the cluster relative to the Earth). The radius R is determined from the cluster's angular size combined with an estimate of its distance. With these ingredients, the "virial mass" of the cluster is given by $GM_{\text{virial}} \simeq 6R\langle v_r^2\rangle$, and within a factor of order unity, we should expect this to be a reliable estimate of the cluster's total mass. The problem is that the virial mass determined in this way turns out to be as much as 10 times larger than the "visible mass" determined by measuring the brightness of the cluster and multiplying by the mass-to-light ratio known from stellar-structure theory for the kinds of stars contained in the cluster. When the virial method was applied to clusters of galaxies (with each galaxy viewed as a point mass), the discrepancy between the virial and visible masses was even larger.

One proposed explanation for the discrepancy held that star clusters are really not bound at all, so that the virial theorem does not apply to them. Another possible explanation is that the clusters are not in a steady state, so that the assumption $\ddot{I} \approx 0$ is not valid. But observations of many clusters and detailed analyses have suggested that these avenues are not promising. An alternative explanation is that there is additional mass in the system that contributes to the gravitational binding, but does not emit any light – this is the dark matter hypothesis. Eventually it was shown that the amount of dark matter implied by the virial-mass discrepancy in star clusters and galaxies is consistent with the amount required to explain the rotational properties of many spiral galaxies, the angular fluctuations in the cosmic microwave background radiation, and the growth of large-scale structure in the early universe. The dark matter hypothesis has now become a central feature of the standard cosmological model.

3.6 Lagrangian formulation of Newtonian dynamics

Our exploration of Newtonian gravity is coming to a close, but we would be remiss if we didn't touch upon the wonderful developments initiated by Maupertuis and Euler and completed by Lagrange and Hamilton. Their *principle of least action* has become a unifying principle for all of physics, being applicable to such disparate fields as mechanics, fluid dynamics, and field theory including general relativity. Our treatment here will be extremely brief; we expect the reader to have encountered this topic in greater depth elsewhere. We include it regardless because we shall have occasion to refer to Lagrangians and action functionals in later portions of the book, and wish to provide a quick reference manual to accompany these references.

3.6.1 Lagrangian and action principle

We consider a dynamical system with an arbitrary number of degrees of freedom. We describe the motion of the system with *generalized coordinates* $q^j(t)$, which may be chosen arbitrarily to give the simplest description; the generalized coordinates give rise to the *generalized velocities* $\dot{q}^j(t) := dq^j/dt$. The system possesses a kinetic energy $\mathcal{T}$ and a potential energy Ω which depend upon the generalized coordinates and velocities. The system's *Lagrangian function* is defined to be $L(\dot{q}^j, q^j) := \mathcal{T} - \Omega$. The system's *action functional* is

$$S[q] := \int_{t_1}^{t_2} L(\dot{q}^j, q^j)\, dt, \tag{3.137}$$

the integral of the Lagrangian function between two fixed times. The action can be evaluated for any path $q^j(t)$ that joins an initial configuration $q^j(t_1)$ to a final configuration $q^j(t_2)$, whether or not this path satisfies the equations of motion of the dynamical system.

The true path of the dynamical system – the one that satisfies the equations of motion – is identified as the one that produces an *extremum* of the action functional. This means the following. Suppose that the true path is $\bar{q}^j(t)$, and that we examine an arbitrary variation of this path described by the small displacements $\delta q^j(t)$, thereby constructing a trial path $q^j_{\rm trial}(t) = \bar{q}^j(t) + \delta q^j(t)$. We insist that the trial path should begin from the same configuration as the true path, and that it should also end in the same configuration as the true path, so that the displacements must obey $\delta q^j(t_1) = 0 = \delta q^j(t_2)$; apart from this the functions $\delta q^j(t)$ are arbitrary. The action functional can be evaluated on the true path, and it can be evaluated on the trial path; it will be an extremum when the difference $\delta S := S[q_{\rm trial}] - S[\bar{q}]$ vanishes to first order in the displacements. As we shall see, the condition $\delta S = 0$ gives us a means to identify $\bar{q}^j(t)$. It can be shown that when $t_2 - t_1$ is not too large, the extremum is in fact a minimum – the action is minimized when evaluated along the true path. When $t_2 - t_1$ increases beyond a certain threshold (called a kinetic focus), the extremum turns into a saddle point of the action functional.

To carry out the variation of the action we write

$$\begin{aligned}\delta S &= S[q_{\rm trial}] - S[\bar{q}] \\ &= \int_{t_1}^{t_2} \Big[L(\dot{q}^j_{\rm trial}, q^j_{\rm trial}) - L(\dot{\bar{q}}^j, \bar{q}^j) \Big] \, dt \end{aligned} \tag{3.138}$$

and express the trial Lagrangian as a Taylor expansion about the true path, obtaining

$$L(\dot{\bar{q}}^j, \bar{q}^j) + \frac{\partial L}{\partial \dot{q}^j} \delta \dot{q}^j + \frac{\partial L}{\partial q^j} \delta q^j$$

to first order in the displacements; summation over the repeated index j is understood. The variation of the action becomes

$$\delta S = \int_{t_1}^{t_2} \left(\frac{\partial L}{\partial \dot{q}^j} \frac{d}{dt} \delta q^j + \frac{\partial L}{\partial q^j} \delta q^j \right) dt, \tag{3.139}$$

and integration by parts of the first term produces

$$\delta S = \frac{\partial L}{\partial \dot{q}^j} \delta q^j \bigg|_{t_1}^{t_2} + \int_{t_1}^{t_2} \left(-\frac{d}{dt} \frac{\partial L}{\partial \dot{q}^j} + \frac{\partial L}{\partial q^j} \right) \delta q^j \, dt. \tag{3.140}$$

The boundary terms vanish because of the conditions $\delta q^j(t_1) = 0 = \delta q^j(t_2)$, and since the variations $\delta q^j(t)$ are otherwise arbitrary, we find that $\delta S = 0$ requires

$$\frac{d}{dt} \frac{\partial L}{\partial \dot{q}^j} - \frac{\partial L}{\partial q^j} = 0. \tag{3.141}$$

These are the *Euler–Lagrange equations* for the dynamical system, which give rise to second-order differential equations for the generalized coordinates $q^j(t)$. The Euler–Lagrange equations, therefore, provide the equations of motion to be satisfied by the dynamical system, as dictated by the principle of extremum action.

The Euler–Lagrange equations can be written in the form

$$\dot{p}^j = \frac{\partial L}{\partial q^j}, \tag{3.142}$$

in which $p^j := \partial L / \partial \dot{q}^j$ is the *generalized momentum* associated with the coordinate q^j. An immediate consequence of this equation is that when L does not depend explicitly on one (or more) of its generalized coordinates, say q^*, then the associated momentum p^* is necessarily a constant of the motion. In this way, momentum conservation is seen to arise as a consequence of a symmetry of the dynamical system.

3.6.2 Lagrangian mechanics of a two-body system

We apply these ideas to a system of two spherical bodies moving under their mutual gravitational attraction. We follow the description of Sec. 1.6.7 and adopt the variables $\boldsymbol{R} := (m_1/m)\boldsymbol{r}_1 + (m_2/m)\boldsymbol{r}_2$ and $\boldsymbol{r} := \boldsymbol{r}_1 - \boldsymbol{r}_2$ as generalized coordinates; here $m := m_1 + m_2$ is the total mass of the system, and $\mu := m_1 m_2 / m$ is the reduced mass. The corresponding generalized velocities are $\boldsymbol{V} := d\boldsymbol{R}/dt$ and $\boldsymbol{v} = d\boldsymbol{r}/dt$. The kinetic energy of the system is $\mathcal{T} = \frac{1}{2}mV^2 + \frac{1}{2}\mu v^2$, and its gravitational potential energy is $\Omega = -G\mu m/r$. The Lagrangian of the two-body system is therefore

$$L = \frac{1}{2}mV^2 + \frac{1}{2}\mu v^2 + \frac{G\mu m}{r}. \tag{3.143}$$

We see immediately that L does not depend explicitly on the coordinates $\boldsymbol{R}$, and this implies that the corresponding momentum $\boldsymbol{P} := \partial L/\partial \boldsymbol{V} = m\boldsymbol{V}$ is constant; we have recovered the fact that the total momentum of a two-body system is a conserved quantity. Taking this into account, we recognize that the first term in the Lagrangian is an irrelevant constant which plays no role in the dynamics of the relative orbit. We may remove it, and take the effective Lagrangian to be

$$L = \frac{1}{2}\mu v^2 + \frac{G\mu m}{r}. \tag{3.144}$$

The reduction to an effective one-body problem has been achieved at the level of the Lagrangian, instead of at the level of the equations of motion.

The gravitational potential energy Ω is spherically symmetric, and this motivates the use of spherical polar coordinates (r, θ, ϕ) instead of the original Cartesian coordinates (x, y, z) associated with the separation vector $\boldsymbol{r}$. The transformation, given by $x = r\sin\theta\cos\phi$, $y = r\sin\theta\sin\phi$, and $z = r\cos\theta$, can implemented directly within the Lagrangian, which becomes

$$L = \frac{1}{2}\mu\left(\dot{r}^2 + r^2\dot{\theta}^2 + r^2\sin^2\theta\,\dot{\phi}^2\right) + \frac{G\mu m}{r}. \tag{3.145}$$

Another symmetry is revealed: the Lagrangian does not depend explicitly on ϕ, and the generalized momentum $p_\phi = \partial L/\partial_\phi = \mu r^2 \sin^2\theta\,\dot{\phi}$ is a constant of the motion. Because ϕ is an angular coordinate, p_ϕ has the interpretation of an angular momentum, and it is in fact the component of the angular-momentum vector $\boldsymbol{L} = \mu\boldsymbol{r} \times \boldsymbol{v}$ in the z-direction. Because this direction was selected arbitrarily in the definition of the spherical coordinates, we can conclude that, in fact, all components of the angular-momentum vector are conserved. This statement implies that the motion takes place in a fixed orbital plane, and this plane can be identified with the equatorial plane $\theta = \frac{\pi}{2}$ without loss of generality.

Setting $\theta = \frac{\pi}{2}$ and $\dot{\theta} = 0$ in the Lagrangian, we finally obtain

$$L = \frac{1}{2}\mu\left(\dot{r}^2 + r^2\dot{\phi}^2\right) + \frac{G\mu m}{r} \tag{3.146}$$

for the fully reduced Lagrangian that governs the planar motion of a two-body system. At this stage it is easy to show that the Euler–Lagrange equations produce the orbital equations first obtained in Sec. 3.2.3.

3.6.3 Lagrangian mechanics of a test mass

To conclude this brief tour of Lagrangian mechanics, we examine the dynamics of a test body in the gravitational field of other massive bodies. In this case the dynamical system consists only of the test body, which is given a mass m, position $\boldsymbol{r}(t)$, and velocity $\boldsymbol{v}(t)$. The gravity of the external bodies is described by their gravitational potential $U(t, \boldsymbol{x})$, which is supposed to be known as a function of space and time. The particle's kinetic energy is $\frac{1}{2}mv^2$, and its potential energy – its interaction energy with the remaining bodies – is $-mU(\boldsymbol{r})$, in which the potential is evaluated at $\boldsymbol{x} = \boldsymbol{r}(t)$. The particle's Lagrangian is

$$L = \frac{1}{2}mv^2 + mU(\boldsymbol{r}). \tag{3.147}$$

In this case the Lagrangian possesses whatever symmetry is displayed by the gravitational potential U; in general there will be no particular symmetry.

The generalized momentum associated with $\boldsymbol{r}$ is $\boldsymbol{p} = \partial L/\partial \boldsymbol{v} = m\boldsymbol{v}$, which is recognized as the usual momentum vector. The Euler–Lagrange equations read

$$\frac{d\boldsymbol{p}}{dt} = m\nabla U(\boldsymbol{r}), \tag{3.148}$$

in which the potential is differentiated with respect to $\boldsymbol{r}$. The left-hand side is $m\boldsymbol{a} = md^2\boldsymbol{r}/dt^2$, and the right-hand side is recognized as the gravitational force exerted by the external bodies. We have recovered the familiar Newtonian equations of motion, $m\boldsymbol{a} = \boldsymbol{F}$ with $\boldsymbol{F} = m\nabla U$.

Thus ends our survey of Newtonian gravity. Now onward with relativity!

3.7 Bibliographical notes

Kepler's problem (Sec. 3.2) is a staple of Newtonian mechanics, and its solution is presented in most textbooks devoted to this topic, including French (1971). More complete presentations, including the three-dimensional aspects of Keplerian orbits (as reviewed in Sec. 3.2.5), are typically found in books on celestial mechanics; among our favorites are the old classic by Brouwer and Clemence (1961) and the more modern text by Murray and Dermott (2000).

The method of osculating orbital elements developed in Sec. 3.3 is also standard fare of books on celestial mechanics; these also contain many more applications of the formalism than the few presented in Sec. 3.4. The Kozai mechanism was discovered independently by Lidov (1962) and Kozai (1962). The first measurement of the solar oblateness by Dicke and Goldenberg is presented in their 1967 article. An accessible introduction to the rich field of helioseismology can be found in Narayanan (2012). The motion of the Moon is treated in great detail in Brown (1960). A broad overview of the gravitational N-body problem (Sec. 3.5), with applications to the solar system, stellar and galactic dynamics, and cosmology, is contained in the compilation of lectures edited by Steves and Maciejewski (2001).

Our presentation of Lagrangian mechanics in Sec. 3.6 does not do justice to this most elegant formulation of the laws of mechanics, and we add insult to injury by not even mentioning the Hamiltonian and Hamilton–Jacobi versions of the theory. These topics, fortunately, are thoroughly treated in many textbooks. A very terse but elegant presentation can be found in the classic *Mechanics* by Landau and Lifshitz (1976), and a more leisurely treatment in Goldstein, Poole, and Safko (2001).

3.8 Exercises

3.1 In this problem we examine the conic sections of Kepler's problem.

(a) For an orbit with $e \neq 1$ and $\omega = 0$, consider a coordinate transformation to new variables $\bar{x}$ and $\bar{y}$, of the form $\bar{x} = r\cos\phi + p\alpha$, $\bar{y} = r\sin\phi$, where $r = p/(1 + e\cos\phi)$ and α is a constant. Find the value of α that converts the curve to an ellipse or hyperbola, described by the equation

$$\frac{\bar{x}^2}{A^2} \pm \frac{\bar{y}^2}{B^2} = 1,$$

where the positive and negative signs correspond to an ellipse and hyperbola, respectively. Determine A and B in terms of the semi-latus rectum p and eccentricity e.

(b) For $e = 1$, find the transformation that converts the curve into a parabola described by $\bar{x} = C\bar{y}^2$. Determine C in terms of p.

3.2 Suppose that the solar system is filled with a uniform distribution of dark matter with constant mass density ρ. Taking this distribution into account, calculate the modified gravitational potential of the Sun, and find the perturbing force $\boldsymbol{f}$ acting on a planetary orbit. Find the relation between orbital period P and semi-major axis a for a circular orbit, and calculate the secular changes in the planet's orbital elements. Place a bound on ρ using suitable solar-system data.

3.3 In some relativistic theories of gravity, the "graviton" is not massless as in general relativity, but possesses a mass m_g. In the Newtonian limit, the graviton mass gives rise to a modified Poisson equation of the form

$$\left(\nabla^2 + \frac{1}{\lambda^2}\right) U = -4\pi G\rho,$$

in which $\lambda = h/(m_g c)$ is the Compton wavelength of the graviton. Show that the spherically symmetric potential of a body of mass m is given by $U = (Gm/r)e^{-r/\lambda}$. Applying this to the Sun as in the preceding problem, place a bound on λ using solar-system data.

3.4 A test body of mass μ orbits a body of mass m, radius R, and dimensionless quadrupole moment J_2 relative to a symmetry axis $\boldsymbol{e}$; all other J_ℓs are assumed to vanish. Prove that the following quantities are constants of the orbital motion:

(a) The total energy, given by

$$E = \frac{1}{2}\mu v^2 - \frac{G\mu m}{r} + \frac{1}{2}\frac{G\mu m J_2 R^2}{r^3}\left[3(\boldsymbol{n}\cdot\boldsymbol{e})^2 - 1\right].$$

(b) The angular momentum along $\boldsymbol{e}$, given by $L_e = \mu\boldsymbol{h}\cdot\boldsymbol{e}$, where $\boldsymbol{h} := \boldsymbol{r}\times\boldsymbol{v}$.

(c) A third quantity, constant to first order in J_2, given by

$$C = h^2 + J_2 R^2\left[(\boldsymbol{e}\cdot\boldsymbol{v})^2 - 2\frac{Gm}{r}(\boldsymbol{e}\cdot\boldsymbol{n})^2\right],$$

where $\boldsymbol{n} := \boldsymbol{r}/r$. This third constant is analogous to the "Carter constant" in the Kerr geometry of a rotating black hole.

3.5 Consider a spherical body on an inclined, circular orbit about an axisymmetric body of radius R and even multipole moments J_ℓ, with $\ell = 2, 4, 6$, and so on. To first order in perturbation theory, calculate the secular changes in the relevant orbital elements. In particular, show that:

(a) the inclination is constant, that is, $\Delta\iota = 0$;

(b) the line of nodes changes by an amount

$$\Delta\Omega = -3\pi \cos\iota \sum_{\ell=2}^{\infty} J_\ell \left(\frac{R}{p}\right)^\ell C_\ell,$$

where $C_2 = 1$, $C_4 = -\frac{5}{2}(1 - \frac{7}{4}\sin^2\iota)$, and $C_6 = \frac{35}{8}(1 - \frac{9}{2}\sin^2\iota + \frac{33}{8}\sin^4\iota)$.

3.6 The equations that govern the evolution of the osculating orbital elements have singularities when $e = 0$ and $\iota = 0$.

(a) Show that the bad behavior of the equations when $e = 0$ can be cured by implementing the transformation $\alpha = e\cos\omega$ and $\beta = e\sin\omega$. Obtain the osculating equations for the new elements α and β.

(b) Show that the bad behavior of the equations when $\iota = 0$ can be cured by implementing the transformation $\mu = \tan\iota\cos\Omega$ and $\nu = \tan\iota\sin\Omega$. Obtain the osculating equations for the new elements μ and ν.

3.7 Show that the period of oscillation of the Kozai resonance is given by Eq. (3.89), and that the orbital elements ω, ι, and e vary according to

$$\omega = \frac{\pi}{2} + A\sin(2\pi t/P_{\rm Kozai}),$$

$$\iota = \iota_c - \frac{1}{2}Ae_c\cos(2\pi t/P_{\rm Kozai}),$$

$$e = e_c + \frac{5}{6}A\sin\iota_c\cos\iota_c\cos(2\pi t/P_{\rm Kozai}),$$

on a secular time scale; here A is an arbitrary amplitude for the oscillations.

3.8 From the equations for Δe, $\Delta\omega$, and $\Delta\iota$ in the Kozai mechanism, show that

$$e^2\cos^2\omega\sin^2\iota - \frac{3}{5}e^2 + \cos^2\iota = \text{constant}.$$

3.9 (a) Consider a two-body system on a circular orbit with separation r and angular velocity $\Omega = (Gm/r^3)^{1/2}$, where $m = m_1 + m_2$ is the total mass. Body 2 may be treated as a point mass, but body 1 is tidally distorted by body 2. Because of tidal dissipation in body 1, r changes over one orbit by

$$\Delta r = -24\pi(k_2)_1 \frac{m_2}{m_1}\frac{R_1^5}{r^4}(\Omega - \omega_1)\tau_1,$$

where R_1, ω_1, $(k_2)_1$, and τ_1 are the radius, rotational angular velocity, Love number, and viscous delay of body 1, respectively – refer to Eq. (3.105). Find the rate of change of orbital angular momentum, $L = \mu r^2 \Omega$. Comparing your result with Eq. (2.287), what can you conclude about the total angular momentum of the system, $J = L + S_1$?

(b) The Earth–Moon distance is known to be increasing at a rate of about 3.8 cm/yr. Assuming a value $k_2 \approx 0.15$ for the Love number of the Earth, estimate the angle between the Earth's tidal bulge and the Earth–Moon direction. Does the bulge lie ahead of or behind that line?

(b) Show that when the Earth eventually becomes tidally locked with the Moon, the length of the day (and of the month) will be about 48 current days. You may ignore the rotational angular momentum of the Moon, and you may assume that, to a good approximation, the moment of inertia of the Earth about its rotation axis is given by $0.33 M_\oplus R_\oplus^2$.

3.10 Consider a point at a position $\bar{r}(t)$ on a circular orbit of radius r around a central body of mass m, orbiting with angular velocity Ω, with $\Omega^2 = Gm/r^3$. Consider also a test body moving on nearby orbit, at a position $\delta\bar{r}(t)$ relative to the point on the circular orbit. Assume that $\delta r \ll r$.

(a) In a coordinate system that rotates around the central body with angular velocity Ω, show that the equations of motion of the test body are given by

$$\frac{d^2}{dt^2}\delta\boldsymbol{r} + 2\boldsymbol{\Omega} \times \frac{d}{dt}\delta\boldsymbol{r} = \Omega^2\Big[3(\boldsymbol{n}\cdot\delta\boldsymbol{r})\boldsymbol{n} - (\boldsymbol{e}_z\cdot\delta\boldsymbol{r})\boldsymbol{e}_z\Big]$$

to first order in $\delta\boldsymbol{r}$; here $\boldsymbol{\Omega} := \Omega\boldsymbol{e}_z$ is the angular-velocity vector, and $\boldsymbol{n} := \boldsymbol{r}/r$.

(b) Prove that the general solution to the equations of motion takes the form of the linear superposition $\delta\boldsymbol{r} = c_1\delta\boldsymbol{r}_1 + c_2\delta\boldsymbol{r}_2 + c_3\delta\boldsymbol{r}_3 + c_4\delta\boldsymbol{r}_4$, where c_n are arbitrary constants, and

$$\begin{aligned}
\delta\boldsymbol{r}_1 &= \cos(\Omega t - \chi_1)\,\boldsymbol{n} - 2\sin(\Omega t - \chi_1)\,\boldsymbol{\lambda},\\
\delta\boldsymbol{r}_2 &= \boldsymbol{n} - \frac{3}{2}\Omega t\,\boldsymbol{\lambda},\\
\delta\boldsymbol{r}_3 &= \boldsymbol{\lambda},\\
\delta\boldsymbol{r}_4 &= \cos(\Omega t - \chi_4)\,\boldsymbol{e}_z,
\end{aligned}$$

are the four eigenmodes of the perturbed orbit; χ_n are arbitrary phases.

(c) Describe the motion that corresponds to each mode, and show that each mode is generated by a perturbation in the orbital elements (p, e, ι, Ω) relative to the unperturbed, circular orbit. Relate the constants c_n to the variations of the orbital elements.

(d) Find a solution with $c_2 = c_3 = 0$, but with $c_1 \neq 0$ and $c_4 \neq 0$, describing a relative orbit that is circular, with a constant radius δr. What is the angle between the plane of the relative orbit and that of the original, unperturbed orbit?

(e) Now find a solution describing three satellites that are moving on the same circular relative orbit, such that initially they are placed at the vertices of an equilateral triangle. Show that as each satellite follows its orbit, the constellation maintains the shape of an equilateral triangle. This configuration was adopted for the three satellites making up the Laser Interferometer Space Antenna (LISA), a proposed space-based gravitational-wave detector.

3.11 In this problem we explore further the general three-body problem introduced in Sec. 3.5.1.
(a) Show that there exist planar solutions for which $r_{12} = r_{13} = r_{23}$.
(b) Show that the general bound solution for $r := r_{AB}$ is given by

$$r = p/(1 + e\cos\phi),$$

$$r^2 \frac{d\phi}{dt} = (GMp)^{1/2},$$

$$\boldsymbol{n}_{AB} = \cos(\phi - \psi_{AB})\,\boldsymbol{e}_X + \sin(\phi - \psi_{AB})\,\boldsymbol{e}_Y,$$

with an orbital period given by $P = 2\pi a^{3/2}/(GM)^{1/2}$, where $a := p/(1 - e^2)$ and $M := m_1 + m_2 + m_3$, and where the three phases ψ_{AB} differ by $\frac{\pi}{3}$.
(c) Show that a_A, the semi-major axes of the individual orbits, are given by

$$a_1 = \frac{(m_2^2 + m_2 m_3 + m_3^2)^{1/2}}{M} a,$$

with the others obtained by suitable permutations.

3.12 Consider a small displacement $\delta\boldsymbol{\rho}$ about the Lagrange point L_4 or L_5 of the restricted three body problem. Expand $\boldsymbol{\nabla}\Phi$ to first order in the displacement, and include the Coriolis term in the equations of motion. Substitute $\delta\boldsymbol{\rho} = \delta\boldsymbol{\rho}_0 e^{-ipt}$ and show that the solutions for the frequency p are real, and thus that the Lagrange points are stable, if and only if the mass ratio q satisfies the criterion of Eq. (3.132).

3.13 Assuming that $q := m_1/m_2 \ll 1$, show that the unstable Lagrange points L_1 and L_2 lie on either side of body 1, at a distance d given by $d \simeq r(q/3)^{1/3}$, where r is the distance between bodies 1 and 2. Calculate d for the Earth–Sun system.

4 Minkowski spacetime

The preceding chapters were devoted to a Newtonian description of the gravitational interaction, and it is now time to embark on an exploration of its relativistic aspects. As we shall argue in the next chapter, a relativistic theory of gravity that respects the principle of equivalence reviewed in Sec. 1.2 must be a metric theory in which gravitation is a manifestation of the curvature of spacetime. The simplest metric theory of gravitation is Einstein's general relativity, and our task in this chapter and the next is to introduce its essential elements. Subsequent chapters will develop the weak-field limit of general relativity, and in these chapters we will return to notions (such as gravitational potentials and forces) that are familiar from Newtonian physics. But a proper grounding of the weak-field limit must rest on the exact theory, and we shall now work to acquire the required knowledge. It is, of course, unlikely that a mere two chapters will suffice to introduce all relevant aspects of general relativity. What we intend to cover here is a rather minimal package, the smallest required for the development of the weak-field limit.

This chapter is devoted to a description of physics in Minkowski spacetime (also known as flat spacetime), which codifies in a particularly elegant way the kinematical rules of special relativity. We begin in Sec. 4.1 with a description of spacetime itself, and we next present the (flat) spacetime formulation of some familiar laws of physics: In Sec. 4.2 we examine the special-relativistic formulation of hydrodynamics, in Sec. 4.3 we turn to electrodynamics, and in Sec. 4.4 we consider the dynamics of point particles. These spacetime formulations of the laws of physics exclude gravitation, and the inclusion of this important interaction can only be achieved by modifying the very structure of spacetime (from flat to curved); this is the topic of Chapter 5.

4.1 Spacetime

4.1.1 Lorentz transformation and spacetime

Einstein was the first to realize that a kinematical transformation between a reference frame S (believed to be "at rest") and another reference frame S' (moving with a uniform speed v relative to S) must be given by

$$t' = \gamma(t - vx/c^2), \qquad x' = \gamma(x - vt), \qquad y' = y, \qquad z' = z, \tag{4.1}$$

where

$$\gamma := \frac{1}{\sqrt{1-(v/c)^2}}, \tag{4.2}$$

if it is to observe the two fundamental axioms of relativistic physics. These are that the laws of physics must take the same form in all inertial frames, and that the speed of light c must be the same in all inertial frames. Einstein's *inertial frame* was the same as Galileo's: a frame in which a body freed from the action of external forces moves uniformly on a straight line. An inertial frame is a significant extrapolation from what we observe in nature, where bodies tend to slow down because of friction, and move on curved paths because of gravity. The first observation does not introduce an essential difficulty, because one can imagine reducing friction to negligible levels by various tricks (think of air hockey tables). But gravity is another matter, and Einstein soon realized that it would necessitate a radical rethinking of inertial frames and the nature of spacetime. We shall come to this in due course, but for the time being we restrict our attention to such ideal inertial frames, in the absence of gravity.

The transformation of Eq. (4.1) is known as a *Lorentz transformation*. It was first identified by Lorentz and Poincaré to give rise to an invariance of Maxwell's equations relative to a choice of inertial frame, and then recognized to have universal validity and to apply to actual measurements of space and time by Einstein. In Eq. (4.1) it was assumed that S' moves with respect to S in the x-direction; it is straightforward to generalize the Lorentz transformation to arbitrary translational directions (see Exercise 4.1).

The Lorentz transformations have a direct impact on the very structure of space and time, and they suggest the merging of each notion into a unified spacetime. To begin this discussion, let us define an *event* as any recognizable phenomenon that occurs at a specific position in space and at a specific moment in time as measured by a clock at that position; in a frame S an event is labeled by the three spatial coordinates x, y, and z, as well as a time coordinate t. Suppose that two events occur at different places but at the same time when observed in the frame S. These events are separated by a displacement Δx, and they are such that $\Delta t = 0$. When observed in S', however, we have that $\Delta x' = \gamma \Delta x$ and $\Delta t' = -\gamma v \Delta x/c^2 \neq 0$; the events are no longer simultaneous, and their spatial separation differs from Δx by a factor of γ. As another example, consider two other events that are seen in S to occur at the same spatial position (so that $\Delta x = 0$), but at two different times separated by the interval Δt. When observed in S' we find that $\Delta x' = -\gamma v \Delta t \neq 0$ and $\Delta t' = \gamma \Delta t$; the events are no longer in the same spatial position, and their separation in time differs from Δt by a factor of γ.

The message of the Lorentz transformations is that space by itself is not absolute (as it is in Newtonian mechanics), and time by itself also is not absolute. There is, nevertheless, a union of space and time – spacetime – that can be considered to be absolute. This comes about because the Lorentz transformation preserves the quadratic form $\Delta s^2 := -(c\Delta t)^2 + (\Delta x)^2 + (\Delta y)^2 + (\Delta z)^2$, which takes on the same value whether evaluated in S (as shown here), or evaluated in S' (by replacing all unprimed quantities with their primed version). This is conveniently written in the differential form

$$ds^2 = -(c dt)^2 + dx^2 + dy^2 + dz^2, \tag{4.3}$$

and Δs^2 or ds^2 gives us an absolute, or Lorentz-invariant, notion of (squared) distance between two events in spacetime. We shall refer to Δs^2 or ds^2 as the *spacetime interval*. Two inertial observers, one attached to S and another attached to S', will measure the same value of Δs^2 for the spacetime interval between two selected events.

Note that Δs^2 can be positive, negative, or zero. When two events are seen in S to be mostly separated in time, so that $(c\Delta t)^2 > (\Delta x)^2$, then $\Delta s^2 < 0$, and this conclusion is shared by any other inertial observer. Such events are said to have a *timelike* separation. There exists, in fact, a frame S' in which $\Delta x' = 0$, so that $\Delta s^2 = -(c\Delta t')^2$. In this frame, $\Delta t'$ is the interval of time between two events at the same spatial position, as measured by a clock that sits at that position, and we define this to be the *proper time interval* $\Delta\tau := \Delta t' = c^{-1}(-\Delta s^2)^{1/2}$ between the two events. It has the same value in any inertial frame.

When, on the other hand, two events are seen in S to be mostly separated in space, so that $(c\Delta t)^2 < (\Delta x)^2$, then $\Delta s^2 > 0$, and this conclusion also is shared by any other inertial observer. Such events have a *spacelike* separation. There exists a frame S' in which $\Delta t' = 0$, so that $\Delta s^2 = (\Delta x')^2$. In this case $\Delta x'$ gives the displacement between two simultaneous events, and this defines the *proper distance* $\Delta\ell := \Delta x' = (\Delta s^2)^{1/2}$ between the two events. This also is a Lorentz invariant.

When, finally, two events are linked by a signal propagating at the speed of light, so that $\Delta x = c\Delta t$, then $\Delta s^2 = 0$; the spacetime interval can be zero even when the events are widely separated both in time and in space. Such separations are called *null*, or *lightlike*. No Lorentz transformation can make two null-separated events occur simultaneously or at the same spatial position.

Box 4.1 Tests of special relativity

Special relativity, the physics built on Minkowski spacetime, has been so thoroughly integrated into the fabric of modern physics that its validity is rarely challenged, except by cranks and crackpots. But we should remember that it does rest on a strong empirical foundation, including some classic tests.

In addition to the famous Michelson–Morley experiment, which failed to find evidence of a variation of the speed of light with the Earth's velocity through a putative "aether," several classic experiments have been performed to verify that the speed of light is independent of the speed of the emitter. If the speed of light were given by $c + kv$, where v is the velocity of the emitter, and k is a parameter to be measured, then orbits of binary-star systems would appear to have an anomalous eccentricity unexplainable by normal Newtonian gravity. This test is not unambiguous at optical wavelengths, however, because light is absorbed and re-emitted by the intervening interstellar medium, thereby losing the memory of the speed of the source; this phenomenon is known to astronomers as extinction. But at X-ray wavelengths, the path length of extinction is tens of kiloparsecs, so nearby X-ray binary systems in our galaxy may be used to test the velocity dependence of light. Using data from three such systems, Kenneth Brecher in 1977 obtained a bound $|k| < 2 \times 10^{-9}$, for typical orbital velocities $v/c \sim 10^{-3}$.

At the other extreme, a 1964 experiment at CERN used ultrarelativistic neutral pions moving at $v/c \geq 0.99975$ as the source of light. Photons produced by the decay $\pi^0 \to \gamma + \gamma$ were collimated and timed over a flight path of 30 meters. The result for the speed was $2.9977 \pm 0.0004 \times 10^8$ m/s, in agreement with the laboratory value. This experiment thus set a bound $|k| < 10^{-4}$ for $v \approx c$.

The observational evidence for time dilation is overwhelming. In the 1930s, Ives and Stilwell measured the frequency shifts of radiation emitted in the forward and backward direction by moving ions of H_2 and H_3 molecules. The first-order Doppler shift cancels out from the sum of the forward and backward shifts, revealing the second-order time-dilation effect, which was found to agree with theory. (Ironically, Ives was a die-hard opponent of special relativity.)

A classic experiment published by Rossi and Hall in 1941 showed that the lifetime of μ-mesons was prolonged by the Lorentz factor $\gamma = (1 - v^2/c^2)^{-1/2}$. Muons are created in the upper atmosphere when cosmic-ray protons collide with nuclei of air, producing pions, which decay quickly to muons. With a rest half-life of 2.2×10^{-6} s, a muon travelling near the speed of light should travel only 2/3 of a kilometer on average before decaying to a harmless electron or positron and two neutrinos. Yet muons are the primary component of cosmic rays detected at sea level. With time dilation and a typical speed of $v/c \sim 0.994$, their lives as seen from Earth are prolonged by a factor of nine, enough for them to reach sea level. Rossi and Hall measured the distribution of muons as a function of altitude and also measured their energies, and confirmed the time-dilation formula. In fact, since collisions between cosmic-ray muons and DNA molecules are a non-negligible source of natural genetic mutations, one could argue that special relativity plays a role in evolution!

In an experiment performed in 1966 at CERN, muons produced by collisions at one of the targets in the accelerator were deflected by magnets so that they would move on circular paths in a storage ring. Their speeds were 99.7 percent of the velocity of light, and the observed twelve-fold increase in their lifetimes agreed with the prediction with 2 percent accuracy. Also, since the storage ring was 5 meters in diameter, the muons' accelerations were greater than the gravitational acceleration on the Earth's surface by a factor of 10^{19}; these accelerations had no effect on their decay rates.

The incorporation of Lorentz invariance into quantum mechanics provided further support for special relativity. The first attempt at integration was the discovery of the Dirac equation, the relativistic generalization of Schrödinger's equation, with its prediction of antiparticles and elementary particle spin. Complete integration came with the development of relativistic quantum field theory, which naturally embodies the Pauli exclusion principle by demanding that the creation and annihilation operators of spinor fields satisfy anti-commutation relations in order to satisfy Lorentz invariance. Because the Pauli exclusion principle explains the occupation of atomic energy levels by electrons, one could argue, with but a hint of chauvinism, that special relativity explains chemistry! The modern incarnations of quantum field theory, such as quantum electrodynamics, electroweak theory, quantum chromodynamics, and string theory, all have Lorentz invariance as foundations.

4.1.2 Metric tensor

It is convenient to combine the spatial coordinates (x, y, z) and the time t that label a spacetime event into a unified spacetime coordinate $x^\alpha = (ct, x, y, z)$. The Greek index α, and all other Greek indices that appear below, run over the values $\{0, 1, 2, 3\}$; we choose $x^0 = ct$ to represent time (rescaled by a factor of c so that all coordinates have dimensions of length), and $x^1 = x$, $x^2 = y$, $x^3 = z$ to represent the spatial coordinates. We shall often represent the spatial coordinates as the three-dimensional vector $\boldsymbol{x}$ with components x^j; the Latin index j (and all others like it) runs over the values $\{1, 2, 3\}$.

In this notation Eq. (4.3) can be expressed as

$$ds^2 = \eta_{\alpha\beta} dx^\alpha dx^\beta, \tag{4.4}$$

where $\eta_{\alpha\beta}$ is a diagonal matrix with entries $\eta_{00} = -1$, $\eta_{11} = \eta_{22} = \eta_{33} = 1$, and where (according to the Einstein summation convention introduced in Sec. 1.4.4) all repeated indices are summed over. The matrix has the purpose of converting coordinate intervals dx^α, which are affected by a change of reference frame, to the spacetime interval ds^2, which is a Lorentz invariant; it is called the *metric tensor* of Minkowski spacetime. Below we shall distinguish very clearly between vectors and tensors (such as dx^α) that are affected by a coordinate transformation, and scalar quantities (such as ds^2) that are spacetime invariants.

Equation (4.4) possesses the irresistible interpretation of expressing the inner product of the spacetime vector dx^α with itself. We shall adopt this geometric point of view, and use the metric $\eta_{\alpha\beta}$ to define the inner product between any two vectors. If A^α is a spacetime vector (an object that transforms as the coordinate increments dx^α under a Lorentz transformation) and B^α is another, then by definition $\eta_{\alpha\beta} A^\alpha B^\beta$ shall be their inner product; it may be verified that the inner product between two vectors is a Lorentz invariant. The inner product of a vector with itself, $\eta_{\alpha\beta} A^\alpha A^\beta$, is called the *norm* of the vector. Keep in mind that in Minkowski spacetime, norms are not necessarily positive: the norm of a timelike vector is always negative, and the norm of a lightlike vector is always zero.

We shall also use the Minkowski metric to raise and lower indices. Given a vector A^α, we define an associated quantity A_α (known as a dual vector) by the operation

$$A_\alpha = \eta_{\alpha\beta} A^\beta. \tag{4.5}$$

This operation is called "lowering the index," and it produces $A_0 = -A^0$, $A_1 = A^1$, $A_2 = A^2$, and $A_3 = A^3$. The operation can be inverted if we introduce the inverse Minkowski metric $\eta^{\alpha\beta}$, a diagonal matrix with entries $\eta^{00} = -1$ and $\eta^{11} = \eta^{22} = \eta^{33} = 1$. The inverse metric is defined by the statement

$$\eta^{\alpha\mu} \eta_{\mu\beta} = \delta^\alpha{}_\beta, \tag{4.6}$$

which is equivalent to the matrix equation $\eta^{-1}\eta = 1$. The inverse operation is

$$A^\alpha = \eta^{\alpha\beta} A_\beta, \tag{4.7}$$

and it is called "raising the index."

4.1.3 Kinematics of particles

The laws of physics must be formulated in spacetime. We begin this reformulation with an examination of the motion of particles. In Newtonian mechanics the motion of a particle is described mathematically by equations of the type $\boldsymbol{x} = \boldsymbol{r}(t)$, which assign to the particle, at any time t, the position determined by the vectorial function $\boldsymbol{r}(t)$. This gives rise to a parameterized curve in three-dimensional space, and absolute time t assumes the role of the parameter.

In spacetime we promote t to one of the coordinates, and resist the temptation of using it also as a parameter. Instead we look for a description of the motion of the general form $x^\alpha = r^\alpha(\lambda)$, which assigns, for any value of the parameter λ, the spacetime coordinates determined by the functions $r^\alpha(\lambda)$. We wish the parameter λ to be a Lorentz invariant, and for this we select *proper time* τ, the time as measured by a standard clock that is attached to the moving particle. This is a spacetime invariant because any inertial observer will agree, irrespective of her own motion, that a clock moving with the particle marks time at precisely the rate measured by this particular clock; she will not, of course, claim that her own clock marks time at this rate. The trajectory of the particle in spacetime, therefore, will be described by the parametric equations $x^\alpha = r^\alpha(\tau)$, with proper time τ assuming the role of the parameter; the trajectory is known as the particle's *world line* in spacetime.

The invariant $d\tau$ can be related to the spacetime interval ds^2 evaluated for two neighboring points on the world line. A central aspect of physics in flat spacetime is the statement, well supported by empirical evidence, that the rate at which a standard clock marks time may depend on its velocity, but it does not depend on its acceleration. The clock attached to the particle, therefore, marks time at precisely the same rate as a clock carried by an inertial observer, provided that the observer's velocity matches the particle's velocity when the measurement is carried out. To be sure, the velocities will be matched only momentarily if the particle is accelerated, but at that time the particle will be at rest relative to the observer. And at this moment, the two clocks will mark time at precisely the same rate. The observer's inertial frame will be referred to as the particle's *momentarily comoving Lorentz frame*, or MCLF. It is a most useful notion, and we will invoke it repeatedly to simplify subtle arguments.

As we have said, the particle is momentarily at rest in the comoving frame, and in this frame S', proper time τ advances at the same rate as observer time t'; we have $d\tau = dt'$. We also have that the (absent) spatial motion of the particle is described by $dx' = dy' = dz' = 0$ in the comoving frame. The spacetime interval along the world line is therefore $ds^2 = -(cdt')^2 = -(cd\tau)^2$. Because each side of the equation is a Lorentz invariant, we conclude that the relation

$$d\tau = c^{-1}\sqrt{-ds^2} \tag{4.8}$$

is valid in any reference frame. Proper time, therefore, measures the accumulation of spacetime interval along the particle's world line.

Having selected proper time as a natural parameter on the world line, we define the particle's velocity vector as

$$u^\alpha = \frac{dr^\alpha}{d\tau}. \tag{4.9}$$

Notice that this is a spacetime vector, with four components. Notice also that u^α, like $dr^\alpha = dx^\alpha$, does indeed transform as a vector under Lorentz transformations. The spacetime velocity vector can be related to $\boldsymbol{v} = d\boldsymbol{r}/dt$, the three-dimensional velocity vector of

Newtonian mechanics. If we factorize $dt/d\tau$ on the right-hand side of Eq. (4.9), we end up with $u^\alpha = (dt/d\tau)(dr^\alpha/dt)$, or

$$u^\alpha = \gamma(c, \boldsymbol{v}), \qquad \gamma := dt/d\tau, \tag{4.10}$$

if we break it down in terms of time and spatial components. The quantity γ, defined here as $dt/d\tau$, bears a close relationship with the Lorentz factor introduced in Eq. (4.2). The relation is revealed when we compute the spacetime norm of the velocity vector. We have $\eta_{\alpha\beta}u^\alpha u^\beta = (\eta_{\alpha\beta}dr^\alpha dr^\beta)/(d\tau)^2$, and in the numerator we recognize ds^2, the spacetime interval between two neighboring events on the world line; taking Eq. (4.8) into account, we conclude that

$$\eta_{\alpha\beta}u^\alpha u^\beta = -c^2. \tag{4.11}$$

Substituting Eq. (4.10) into this, we arrive quickly at

$$\gamma = \frac{1}{\sqrt{1-(v/c)^2}}, \tag{4.12}$$

an alternative expression for the factor γ. Here $v^2 := \boldsymbol{v}\cdot\boldsymbol{v}$ is the three-dimensional norm of the Newtonian velocity vector. It should be noted that this γ differs from the γ of Eq. (4.2) in one essential aspect: The velocity parameter that appears here is the particle's velocity vector $\boldsymbol{v}(t)$, which depends on time if the particle is accelerated; the velocity parameter that appears in Eq. (4.2) is the constant speed of the frame S' relative to S.

4.1.4 Momentum and energy

In Newtonian mechanics, the momentum $\boldsymbol{p}$ of a particle is obtained by multiplying the velocity vector $\boldsymbol{v}$ by the particle's mass m. In relativistic mechanics the closest analogue is the spacetime vector

$$p^\alpha = mu^\alpha, \tag{4.13}$$

with m denoting the particle's *rest-mass*, the mass as measured when the particle is momentarily at rest in some inertial frame S. It follows from Eqs. (4.11) and (4.13) that

$$\eta_{\alpha\beta}p^\alpha p^\beta = -m^2c^2. \tag{4.14}$$

From Eq. (4.10) we see that the spatial components of the momentum vector are $\boldsymbol{p} = \gamma m\boldsymbol{v}$, and the factor γ is a relativistic correction to the Newtonian expression. The time component is $p^0 = \gamma mc$, and to probe its significance we expand γ in powers of v/c; the result can be expressed as $cp^0 = mc^2 + \frac{1}{2}mv^2 + \frac{3}{8}mv^4/c^2 + \cdots$ In the first term we recognize the particle's rest-mass energy mc^2, in the second term we recognize the Newtonian kinetic energy $\frac{1}{2}mv^2$, and in the third and higher terms we have relativistic corrections. From this we conclude that cp^0 represents the relativistic energy of the particle. We can therefore express the momentum vector as

$$p^\alpha = (E/c, m\gamma\boldsymbol{v}), \qquad E = \gamma mc^2, \tag{4.15}$$

and Eq. (4.14) gives rise to the well-known formula $E^2 = p^2c^2 + m^2c^4$ for the relativistic energy of a particle; here $p^2 = \gamma^2 m^2 v^2$. The central message of Eq. (4.15) is that in relativistic mechanics, energy and momentum are no longer separate notions; they join together in a unified four-dimensional momentum vector.

Box 4.2

Relativistic mass

Because the rest mass of a particle is defined and measured in the particle's MCLF, it is a Lorentz invariant. It is a constant label that stays with the particle no matter what it is doing, and no matter what reference frame the observer uses to study the particle's motion. This definition of relativistic rest-mass is not universally accepted. For example, in an effort to make the momentum formula $\boldsymbol{p} = \gamma m \boldsymbol{v}$ look more "Newtonian," many authors have defined a "relativistic mass" γm that increases with velocity. In our view, this concept has sown more confusion than enlightenment, and fortunately, its usage has declined steadily over the years; we will *never* adopt it in this book. In our terminology, the quantity γm is the *energy* E of the particle divided by c^2, which does increase with velocity; m is the rest mass, which does not. The relativistic momentum is $\gamma m \boldsymbol{v}$. It is not like Newton's momentum; get over it.

The quantity E is the energy of the particle as measured by an observer at rest in the frame S. An interesting question is: What would its energy be if it were measured by a moving observer? We are given a particle moving with momentum p^α, an observer moving with velocity $u^\alpha_{\rm obs}$, and we wish to determine $E_{\rm obs}$, the particle's energy as measured by the moving observer. The observer, we note, may be accelerated; we make no requirement that either motion be inertial.

To answer this question we consider a Lorentz frame S' that is momentarily comoving with the observer at the moment that she encounters the particle and makes the measurement. Because the observer is momentarily at rest in this frame, the result of the measurement will be cp'^0, the time component of the momentum vector in S' multiplied by the speed of light. On the other hand, the observer's own velocity vector in S' possesses a time component only, and this is given by $u'^0_{\rm obs} = c$. We may therefore say that the measured energy is $p'^0 u'^0_{\rm obs}$, and in S' this is equal to $-\eta_{\alpha\beta} p'^\alpha u'^\beta_{\rm obs}$. Apart from the minus sign, this is the inner product between the vectors p^α and $u^\alpha_{\rm obs}$, and since the inner product is a spacetime invariant, it can be evaluated in the original frame S. We conclude that the particle's energy, as measured by the observer, can be expressed as

$$E_{\rm obs} = -p_\alpha u^\alpha_{\rm obs}. \tag{4.16}$$

Note that we have used the metric to lower the index on p^α. Substitution of Eqs. (4.10) and (4.15) into Eq. (4.16) reveals that

$$E_{\rm obs} = \frac{mc^2(1 - \boldsymbol{v} \cdot \boldsymbol{v}_{\rm obs}/c^2)}{\sqrt{1-(v/c)^2}\sqrt{1-(v_{\rm obs}/c)^2}}, \tag{4.17}$$

where $\boldsymbol{v}(t)$ is the Newtonian velocity of the particle, while $\boldsymbol{v}_{\rm obs}(t)$ is the observer's velocity; each velocity vector may depend on time.

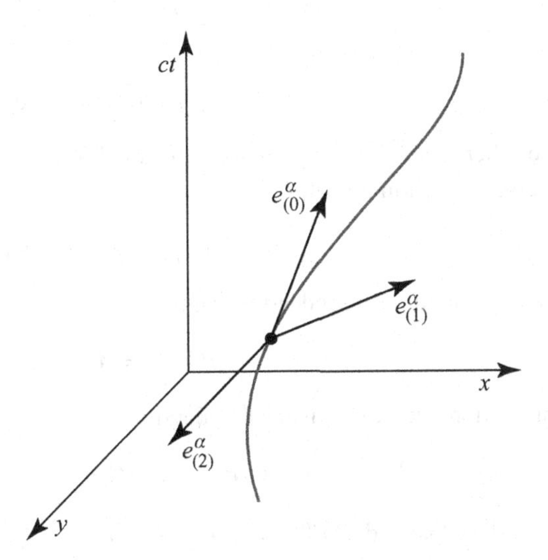

Fig. 4.1 Instantaneous rest frame of a particle moving on a timelike world line in spacetime.

4.1.5 Particle rest-frame

Suppose that we follow the motion of a particle with velocity vector u^α. At any point on the world line we may employ the unit vector $e^\alpha_{(0)} := u^\alpha/c$ to define a preferred (timelike) direction in spacetime; this is a unit vector because the norm of u^α is equal to $-c^2$. This direction and the three (spatial) directions orthogonal to it define the *particle's rest frame* at this moment (see Fig. 4.1). The spatial directions are spanned by three unit vectors $e^\alpha_{(1)}$, $e^\alpha_{(2)}$, and $e^\alpha_{(3)}$; these are chosen to be mutually orthogonal, and they are also orthogonal to $e^\alpha_{(0)}$.

Any vector field $A^\alpha(\tau)$ on the world line can be decomposed in the vectorial basis just defined. We have

$$A^\alpha = A^{(0)} e^\alpha_{(0)} + A^{(1)} e^\alpha_{(1)} + A^{(2)} e^\alpha_{(2)} + A^{(3)} e^\alpha_{(3)}, \tag{4.18}$$

and $A^{(0)} = -A_\alpha e^\alpha_{(0)}$, $A^{(j)} = A_\alpha e^\alpha_{(j)}$ are the projections of the vector onto the selected basis.

For concreteness, suppose that the particle is momentarily moving in the x-direction when viewed in a frame S. Then $e^\alpha_{(0)} = u^\alpha/c = \gamma(1, v/c, 0, 0)$, and an appropriate spatial basis is given by $e^\alpha_{(1)} = \gamma(v/c, 1, 0, 0)$, $e^\alpha_{(2)} = (0, 0, 1, 0)$, and $e^\alpha_{(3)} = (0, 0, 0, 1)$. It is easy to verify that these vectors are all of unit length and mutually orthogonal. We observe that when $v = 0$, the vector $e^\alpha_{(1)}$ points in the x-direction; in a frame S' momentarily comoving with the particle, the vector points in the x'-direction. (We shall make use of this observation in the following subsection.)

Given a vector field $A^\alpha(\tau)$ on the world line, its component along the timelike direction $e^\alpha_{(0)}$ can easily be extracted by acting with the longitudinal projection operator $-e^\alpha_{(0)} e_{(0)\beta}$;

we get

$$\left(-e^{\alpha}_{(0)}e_{(0)\beta}\right)A^{\beta} = A^{(0)}e^{\alpha}_{(0)}. \tag{4.19}$$

On the other hand, the components orthogonal to u^{α} can be extracted by acting with the transverse projection operator

$$P^{\alpha}_{\ \beta} := \delta^{\alpha}_{\ \beta} + e^{\alpha}_{(0)}e_{(0)\beta} = \delta^{\alpha}_{\ \beta} + u^{\alpha}u_{\beta}/c^2; \tag{4.20}$$

simple manipulations do indeed reveal that

$$P^{\alpha}_{\ \beta}A^{\beta} = A^{(j)}e^{\alpha}_{(j)}. \tag{4.21}$$

The projection operator satisfies the identities

$$P^{\alpha}_{\ \beta}u^{\beta} = 0 = u_{\alpha}P^{\alpha}_{\ \beta}, \qquad P^{\alpha}_{\ \mu}P^{\mu}_{\ \beta} = P^{\alpha}_{\ \beta}, \qquad P^{\alpha}_{\ \alpha} = 3, \tag{4.22}$$

and it can be expanded as $P^{\alpha\beta} = e^{\alpha}_{(1)}e^{\beta}_{(1)} + e^{\alpha}_{(2)}e^{\beta}_{(2)} + e^{\alpha}_{(3)}e^{\beta}_{(3)}$ in terms of the transverse basis $e^{\alpha}_{(j)}$.

4.1.6 Photons

A particle moving with the speed of light traces a world line along which ds^2 vanishes everywhere. According to Eq. (4.8), this implies that $d\tau = 0$ along the world line, and we conclude that the proper time of a photon is not defined. This, finally, implies that the photon's velocity vector u^{α} also is not defined.

Nevertheless, the momentum vector of a photon is well defined, in spite of the fact that $m = 0$ and u^{α} is not defined. It can be obtained by a limiting procedure in which the speed of the particle is taken to reach c, with $d\tau$ and m each taken to approach zero while keeping their ratio $d\tau/m$ fixed. In other words, the momentum vector is defined as the limit

$$p^{\alpha} = \lim \frac{dr^{\alpha}}{d(\tau/m)}, \tag{4.23}$$

in which the particle's speed approaches the speed of light. In the limit the rescaled world-line parameter τ/m becomes λ, the photon's world line is represented by the parametric relations $r^{\alpha}(\lambda)$, and we have $p^{\alpha} = dr^{\alpha}/d\lambda$. The momentum of a photon satisfies the null condition

$$p_{\alpha}p^{\alpha} = 0, \tag{4.24}$$

which states that the photon's rest-mass m is zero. The time component of the momentum vector (the photon's energy $\hbar\omega$ as measured by an observer attached to the frame S, divided by the speed of light) is not independent of the spatial components. If we write $p^{\alpha} = (\hbar\omega/c, \hbar\boldsymbol{k})$, with the spatial components expressed in terms of a wave vector $\boldsymbol{k}$, then Eq. (4.24) gives rise to the well-known dispersion relation $\omega^2 = c^2k^2$, where $k^2 = \boldsymbol{k}\cdot\boldsymbol{k}$. This allows us to express the momentum vector as

$$p^{\alpha} = \frac{\hbar\omega}{c}(1, \hat{\boldsymbol{k}}), \tag{4.25}$$

where $\hat{\boldsymbol{k}}$ is a unit vector (in the three-dimensional, Newtonian sense) that indicates the photon's direction of propagation, so that $\boldsymbol{k} = (\omega/c)\hat{\boldsymbol{k}}$.

Box 4.3 Photons: An alternative viewpoint

The reader might object to the limiting procedure used in Eq. (4.23), because it is not a physically realizable procedure. The deeper justification for Eqs. (4.24) and (4.25) comes from Maxwell's equations and quantum mechanics. In the limit where the wavelength of light is short compared to all other scales of variation in the problem (the geometrical optics limit), Maxwell's equations require that the spacetime wave vector $k_\alpha := \partial_\alpha S$ be null, where S is the phase of the wave. Quantization of Maxwell's equations then reveals that $\hbar k^\alpha = (\hbar\omega/c)(1, \hat{\boldsymbol{k}}) = p^\alpha$. The geometrical optics limit is discussed further in Box 5.6, in the context of curved spacetime.

The argument that leads to Eq. (4.16) applies just as well to photons, and we may immediately state that

$$\hbar\omega_{\rm obs} = -p_\alpha u^\alpha_{\rm obs} \tag{4.26}$$

is the photon's energy as measured by an observer moving with a velocity $u^\alpha_{\rm obs}$. With $u^\alpha_{\rm obs} = \gamma(c, \boldsymbol{v}_{\rm obs})$, this is

$$\hbar\omega_{\rm obs} = \hbar\omega \frac{1 - \hat{\boldsymbol{k}} \cdot \boldsymbol{v}_{\rm obs}/c}{\sqrt{1 - (v_{\rm obs}/c)^2}}, \tag{4.27}$$

and we have obtained the well-known formula for the Doppler effect applied to light and other forms of radiation. Although we derived Eq. (4.27) using relativistic methods, it is good to remember that it is really a mixture of the non-relativistic, first-order Doppler effect caused by the piling up or stretching out of the waves as seen by the moving observer – the $-\hat{\boldsymbol{k}} \cdot \boldsymbol{v}_{\rm obs}/c$ term, which can be applied to any type of wave by replacing c by the actual wave speed – and the relativistic time dilation of the observer's clock compared to the laboratory clock – the usual γ factor.

Another interesting phenomenon involving light is its *aberration*, the observed property that the apparent position of an astronomical body depends on the velocity of the observer. This phenomenon was first discovered by James Bradley in 1725, and a few years later he developed a theory of aberration based on Newton's corpuscular model for light. The explanation was simple: the apparent position shifts for the same reason that you have to tilt your umbrella forward when walking through an otherwise vertical rainfall. Bradley's theory did not survive the transition to a wave theory of light, but it is interesting to note that the wave theory could not account for the aberration until various *ad hoc* models of a dragged aether were concocted precisely to save the phenomenon. It was Einstein, finally, who showed how to reconcile aberration with Maxwell's theory in his famous 1905 paper on the electrodynamics of moving bodies.

Suppose that in a frame S, a photon with frequency ω is moving in the direction of the unit vector $\hat{\boldsymbol{k}}$. We align our coordinates so that $\hat{\boldsymbol{k}} = (\cos\alpha_{\rm rest}, \sin\alpha_{\rm rest}, 0)$; the photon moves in the x-y plane, and its trajectory makes an angle $\alpha_{\rm rest}$ with respect to the x-axis.

This is the angle that is measured by an observer at rest in the frame S, and we can derive a formula for $\cos\alpha_{\rm rest}$ that involves spacetime invariants only. For this purpose we re-introduce the photon's momentum vector $p^\alpha = c^{-1}\hbar\omega(1, \hat{\boldsymbol{k}})$, the observer's velocity vector $u^\alpha_{\rm rest} = (c, 0, 0, 0)$, and the spatial vector $e^\alpha_{\rm rest} = (0, 1, 0, 0)$ that points in the x-direction; this vector is orthogonal to u^α, and is one of the three spatial vectors that span the rest frame of our observer. It is easy to verify that

$$\cos\alpha_{\rm rest} = c\frac{p_\alpha e^\alpha_{\rm rest}}{-p_\beta u^\beta_{\rm rest}} \tag{4.28}$$

is the desired formula.

We next introduce another observer – let's call her Leslie – moving relative to S with a velocity u^α. We assume that at the time Leslie encounters the photon, she is momentarily moving in the x-direction, so that $u^\alpha = \gamma(c, v, 0, 0)$. At this moment Leslie is at rest relative to the comoving frame S', which moves uniformly with speed v with respect to S. For an observer (a third!) at rest in this frame, Eq. (4.28) applies, because it is expressed in terms of spacetime invariants. Because the angle α measured by Leslie is the same as the angle measured by the third observer (the one at rest in S'), we can immediately state that

$$\cos\alpha = c\frac{p_\alpha e^\alpha}{-p_\beta u^\beta}, \tag{4.29}$$

where e^α is a spatial vector that points in the direction of the x'-axis. An expression for this was worked out in the preceding subsection; we have that $e^\alpha = \gamma(v/c, 1, 0, 0)$. Making the substitutions for p^α, u^α, and e^α in Eq. (4.29) returns

$$\cos\alpha = \frac{\cos\alpha_{\rm rest} - v/c}{1 - (v/c)\cos\alpha_{\rm rest}}. \tag{4.30}$$

This gives the photon's angle α, as measured by an observer moving with speed v, in terms of the angle $\alpha_{\rm rest}$ measured by an observer at rest. Equation (4.30) is the mathematical description of the aberration of light. Like the Doppler effect, this is a mixture of non-relativistic and relativistic effects. To first order in v/c, the effect can be derived simply by calculating the angle between the vector $c\hat{\boldsymbol{k}} - \boldsymbol{v}$ (light's velocity as seen in S') and the vector $\boldsymbol{v}$. Relativistic effects occur at order $(v/c)^2$. The first-order effect applies just as well to raindrops, provided that you replace c by the appropriate speed of the projectile.

4.1.7 Particle dynamics

Our considerations so far have been entirely kinematical, and in the next three sections we will examine some dynamical aspects of physics in Minkowski spacetime. As a warm-up exercise we describe here what would be required of a dynamics of (massive) particles.

The acceleration vector of a particle with world line $r^\alpha(\tau)$ and velocity $u^\alpha(\tau) = dr^\alpha/d\tau$ is defined as

$$a^\alpha := \frac{du^\alpha}{d\tau}. \tag{4.31}$$

It is an important fact that in spacetime, the acceleration is everywhere orthogonal to the velocity:

$$a_\alpha u^\alpha = 0; \tag{4.32}$$

this is an immediate consequence of the normalization condition $\eta_{\alpha\beta}u^\alpha u^\beta = -c^2$. (To see this, differentiate with respect to τ, and use the symmetry of the metric tensor.) Equation (4.32) implies that the four components of the vector a^α are not all independent. If we write $u^\alpha = \gamma(c, \boldsymbol{v})$ and $a^\alpha = (a^0, \boldsymbol{a})$, then Eq. (4.32) implies that $a^0 = (\boldsymbol{a} \cdot \boldsymbol{v})/c$. Note that while $\boldsymbol{v} = d\boldsymbol{r}/dt$ is the Newtonian velocity vector, $\boldsymbol{a}$ is not equal to the Newtonian acceleration; we have $\boldsymbol{a} = d^2\boldsymbol{r}/d\tau^2$ instead of $d^2\boldsymbol{r}/dt^2$.

A dynamical law will relate the acceleration of a particle to the forces acting upon it. A relativistic version of Newton's second law must take a vectorial form in spacetime, and we therefore write

$$F^\alpha = ma^\alpha, \tag{4.33}$$

where m is the particle's rest mass and F^α is a force vector supplied by the dynamical theory. To be consistent with the basic kinematical constraint of Eq. (4.32), the force vector must be everywhere orthogonal to the velocity vector. If we write $F^\alpha = (F^0, \boldsymbol{F})$, then $F^0 = (\boldsymbol{F} \cdot \boldsymbol{v})/c$. Apart from the factor of c, we recognize on the right-hand side the relativistic generalization of the rate at which the spatial components of the force do work on the particle. The zeroth component of Eq. (4.33), therefore, is a relativistic statement of the work–energy theorem.

4.1.8 Free particle motion and maximum proper time

A freely-moving particle is one on which no forces act. For such a particle, $a^\alpha = 0$, $u^\alpha = u^\alpha_0 = \text{constant}$, $x^\alpha = x^\alpha_0 + u^\alpha_0 \tau$, and the particle moves on a straight line in spacetime. This motion, as trivial as it may seem, arises from an action principle, just as in Newtonian dynamics.

In Newtonian dynamics the action functional of a free particle is given by $S = \int_1^2 L\, dt$, with a Lagrangian function $L = \frac{1}{2}mv^2$. In relativistic dynamics the action must be a Lorentz invariant, and it must have the dimension of an energy multiplied by time. We adopt

$$S = -mc^2 \int_1^2 d\tau \tag{4.34}$$

as a suitable candidate with the required properties, and show that it leads to sensible results.

We saw back in Sec. 4.1.3 that the interval of proper time $d\tau$ along any world line is related to the spacetime interval ds^2 by $d\tau = c^{-1}\sqrt{-ds^2}$. Writing this in full, we have

$$\begin{aligned} d\tau &= \frac{1}{c}\sqrt{-\eta_{\alpha\beta}\, dr^\alpha dr^\beta} \\ &= \frac{1}{c}\sqrt{-\eta_{\alpha\beta}\frac{dr^\alpha}{dt}\frac{dr^\beta}{dt}}\, dt, \end{aligned} \tag{4.35}$$

where, in the second step, we divided the coordinate interval dr^α along the world line by the coordinate-time interval dt. These manipulations reveal that the action of Eq. (4.34) can be expressed in the standard form $S = \int_1^2 L\,dt$, with a relativistic Lagrangian

$$L = -mc\sqrt{-\eta_{\alpha\beta}\frac{dr^\alpha}{dt}\frac{dr^\beta}{dt}}. \tag{4.36}$$

A more explicit form is

$$L = -mc^2\sqrt{1 - v^2/c^2}, \tag{4.37}$$

where $v^2 := \boldsymbol{v}\cdot\boldsymbol{v}$. To see that the action of Eq. (4.34) is in fact a sensible choice, we expand the Lagrangian in powers of v/c and examine the non-relativistic limit. We get $L = -mc^2 + \frac{1}{2}mv^2 + \frac{1}{8}mv^4/c^2 + \cdots$, and we see that up to the irrelevant constant $-mc^2$, the Lagrangian is equal to the Newtonian kinetic energy $\frac{1}{2}mv^2$ to leading order; the additional terms are relativistic corrections to the kinetic energy.

The particle's dynamics is obtained by demanding that the action of Eq. (4.34) be stationary under arbitrary variations of the world line, with the usual provision that all world lines must link the same initial event 1 to the same final event 2. The calculus of variations implies that the Lagrangian of Eq. (4.36) must satisfy the Euler–Lagrange equations

$$\frac{d}{dt}\frac{\partial L}{\partial v^\alpha} - \frac{\partial L}{\partial r^\alpha} = 0, \tag{4.38}$$

where $v^\alpha := dr^\alpha/dt$. The Lagrangian is actually independent of r^α, and we obtain

$$\eta_{\alpha\beta}\frac{d}{dt}\left(\frac{v^\beta}{\sqrt{-\eta_{\mu\nu}v^\mu v^\nu}}\right) = 0. \tag{4.39}$$

This implies that

$$\frac{d}{dt}\left(\frac{dr^\alpha/dt}{d\tau/dt}\right) = \frac{d}{dt}\left(\frac{dr^\alpha}{d\tau}\right) = 0, \tag{4.40}$$

and converting the coordinate-time derivative to a proper-time derivative, we arrive at

$$a^\alpha = \frac{du^\alpha}{d\tau} = \frac{d^2r^\alpha}{d\tau^2} = 0, \tag{4.41}$$

which indeed corresponds to free particle motion.

The action functional of a relativistic particle is directly related to the elapsed proper time between the events 1 and 2. An extremum of the action is therefore an extremum of proper time, and it can be verified that the uniform motion of a free particle actually *maximizes* the elapsed proper time between the two events. This conclusion is a feature of timelike curves in spacetime; for spacelike curves we would find instead that a straight line minimizes the proper distance between two events.

4.2 Relativistic hydrodynamics

4.2.1 Fluid variables

In Chapter 1 we introduced a number of variables that describe the physical state of a Newtonian fluid. These were the velocity field $\boldsymbol{v}$, the mass density ρ, the pressure p, the internal energy density ϵ, the rate of heat generation q, and the heat flux $\boldsymbol{H}$; all of these are functions of time t and position $\boldsymbol{x}$ within the fluid. A spacetime formulation of the laws of hydrodynamics also involves such quantities, and it is straightforward, for example, to promote the Newtonian velocity field $\boldsymbol{v}$ to a relativistic velocity field $u^\alpha = \gamma(c, \boldsymbol{v})$. It is more delicate, however, to generalize the densities ρ and ϵ, because of issues of frame dependence that arise in special relativity. In which frame is the mass density to be defined? An answer is required, because Lorentz contraction implies that a density measured in one frame S will differ by a factor of γ from a density measured in another frame S'. The most useful answer turns out to be: Measure all densities at an event $x^\alpha = (ct, \boldsymbol{x})$ in spacetime in the Lorentz frame that is momentarily comoving with the fluid element at that event. Thus, the fluid's *proper mass density* $\rho(x^\alpha)$ shall be the mass per unit volume as measured in the MCLF of a fluid element at x^α; as such ρ is a spacetime invariant whose value does not depend on the Lorentz frame. Similarly, we define $\epsilon(x^\alpha)$ to be the proper density of internal (thermodynamic) energy contained in a fluid element at x^α, as measured in the comoving Lorentz frame. These quantities combine naturally into

$$\mu := \rho c^2 + \epsilon, \tag{4.42}$$

the total energy density (including rest-mass energy and internal energy) of the fluid element. Like ρ and ϵ, μ is a spacetime invariant.

We shall also agree that the fluid's pressure $p(x^\alpha)$ is to be measured in the momentarily comoving Lorentz frame at x^α, and like the densities considered previously, p also is a spacetime invariant. Our list of fluid variables is now complete; for our purposes in this book we shall not require relativistic generalizations of the heat variables q and $\boldsymbol{H}$.

4.2.2 Mass current

Having introduced the proper mass density $\rho(x^\alpha)$, which refers to an infinite number of momentarily comoving frames spread out throughout the fluid, we now wish to introduce quantities that describe the fluid's mass distribution as seen in a single (and global) Lorentz frame S; we shall think of this as the "laboratory frame." We define

$$c^{-1}j^0 := \text{fluid mass density, as measured in } S, \tag{4.43a}$$

$$j^k := \text{mass flux in the } x^k\text{-direction, as measured in } S; \tag{4.43b}$$

the mass flux $\boldsymbol{j}$ is defined so that $\boldsymbol{j} \cdot d\boldsymbol{S} = j^k dS_k$ is the mass crossing an element of surface dS_k per unit time. These quantities combine to form a spacetime vector j^α (which transforms appropriately under a Lorentz transformation), and the factor of c^{-1} was inserted to ensure that all components of j^α have a dimension of (mass)(velocity)/(volume). We shall call this vector the *mass current*.

The quantity $c^{-1}j^0$ is the fluid's mass density as measured in the laboratory frame S. This differs from the proper density ρ by a factor of γ, which takes into account the Lorentz contraction of a moving fluid element along its direction of motion. We have $c^{-1}j^0 = \gamma\rho$, and we see that $j^0 = \rho u^0$. The mass flux j^k is the mass density of a fluid element multiplied by its velocity (all measured in S), and we have $j^k = (\gamma\rho)v^k = \rho u^k$. We conclude that the mass-current vector can be related to the proper density ρ and the velocity field u^α by

$$j^\alpha = \rho u^\alpha. \tag{4.44}$$

This equation reveals very clearly that j^α is indeed a spacetime vector, because it is the product of a scalar ρ and a vector u^α.

Conservation of (rest-)mass is a very important property of a fluid, and we next work on a mathematical formulation of this property. We examine a three-dimensional volume V of the fluid, bounded by a two-dimensional surface Σ; the volume is arbitrarily large or small, and is taken to be in a fixed position in the laboratory frame S. The fluid flows freely in and out of the volume, and

$$-\frac{d}{dt}\int_V c^{-1}j^0\, d^3x = -\int_V \frac{\partial j^0}{\partial x^0}\, d^3x \tag{4.45}$$

is the amount of mass that leaves V per unit time. On the other hand,

$$\oint_\Sigma j^k dS_k = \int_V \frac{\partial j^k}{\partial x^k}\, d^3x \tag{4.46}$$

is the amount of mass that crosses the surface Σ per unit time; we use the divergence theorem to convert the surface integral into a volume integral. Conservation of mass dictates a strict equality between the two results, and the arbitrariness of V implies that

$$\partial_\alpha j^\alpha = 0. \tag{4.47}$$

This equation is the mathematical expression of local mass conservation; the mass-current vector is divergence-free in spacetime. We shall explore the consequences of this equation below.

4.2.3 Energy-momentum tensor

Having introduced the proper energy density μ in Eq. (4.42), we next turn to a description of energy density, energy flux, momentum density, and momentum flux as viewed in the laboratory frame S. But instead of defining a vector field, as we did for the mass current, we are forced here to define a tensor field. This can be seen most easily by considering the energy density of a fluid element that moves through the laboratory with speed v. The energy contained in the fluid element is increased by the relativistic factor γ, while its volume is decreased by the Lorentz contraction factor $1/\gamma$. Its energy density is therefore equal to $\mu\gamma^2 = \mu u^0 u^0/c^2$, and this suggests that we need a tensor to properly describe this quantity.

We define

$$T^{00} := \text{energy density}, \tag{4.48a}$$

$$cT^{0j} := \text{energy flux in the } x^j\text{-direction}, \tag{4.48b}$$

$$c^{-1}T^{j0} := \text{density of } j\text{-component of momentum}, \tag{4.48c}$$

$$T^{jk} := \text{flux of } j\text{-momentum in the } x^k\text{-direction}; \tag{4.48d}$$

these quantities are all measured in S. The fluxes are defined so that $cT^{0j}dS_j$ is the energy crossing an element of surface dS_j per unit time, while $T^{jk}dS_k$ is the j-component of the momentum vector crossing the element dS_k per unit time. The quantities defined above combine to form the energy-momentum tensor $T^{\alpha\beta}$, and each component of this tensor has a dimension of (energy)/(volume).

The argument that led to Eq. (4.47) can easily be adapted to describe the local conservation of energy and momentum. In this way we find that $\partial_\beta T^{0\beta} = 0$ is a statement of energy conservation, while $\partial_\beta T^{j\beta} = 0$ is a statement of momentum conservation (one for each spatial direction labeled by j). All together, these give rise to the tensorial statement

$$\partial_\beta T^{\alpha\beta} = 0 \tag{4.49}$$

of energy-momentum conservation. The energy-momentum tensor is divergence-free when energy and momentum are locally conserved.

Our considerations thus far have been general, and we next specialize them to the case at hand, that of a perfect fluid. We wish to relate $T^{\alpha\beta}$ to other fluid variables such as μ, p, and u^α. To begin we consider the special case of a fluid at rest, without pressure. In this case the fluid's energy density in S is the same as its proper density, and we have that $T^{00} = \mu$. In addition, the fluid's momentum is zero, and all fluxes vanish, so that T^{00} is the only non-vanishing component of the energy-momentum tensor. On the other hand, the fluid's velocity field is $u^\alpha = (c, 0, 0, 0)$, and we deduce the equality $T^{\alpha\beta} = \mu u^\alpha u^\beta/c^2$.

We next consider the case of a moving fluid, still without pressure. We select a fluid element and a moment of time, and we let $u^\alpha = \gamma(c, \boldsymbol{v})$ be the velocity vector of the fluid element at that time. The energy density of the fluid element as measured in S differs from μ by two factors of γ, as we discussed previously; we have $T^{00} = \mu\gamma^2$. The energy flux is this multiplied by the velocity of the fluid element: $cT^{0j} = (\mu\gamma^2)v^j$. The density of momentum can be expressed as (energy density)(momentum/energy), and since the momentum per energy of a fluid element is $p^j/(cp^0) = v^j/c^2$, this is $c^{-1}T^{j0} = (\mu\gamma^2)(v^j/c^2)$. And finally, the momentum flux is $T^{jk} = (\mu\gamma^2)(v^jv^k/c^2)$. Once more we find that these results are summarized by the equation $T^{\alpha\beta} = \mu u^\alpha u^\beta/c^2$.

This expression must be altered to account for the fluid's pressure. We return to the fluid at rest and re-examine the spatial components of the energy-momentum tensor. By definition we have that $T^{jk}dS_k$ is the rate of momentum transfer across an element of surface dS_k. This is also, by Newton's second law, the force acting on the surface element (directed out of the surface). In the case of a perfect fluid, for which there are no shear forces generated by viscosity, the only such force is created by pressure, and the pressure necessarily acts in the surface's normal direction. From this we conclude that for a perfect fluid at rest, T^{jk} must be equal to $p\,\delta^{jk}$. The result remains valid for a moving fluid, provided that the equality is stated in the MCLF of each fluid element. A more general statement must involve a tensor $P^{\alpha\beta}$ that replaces δ^{jk} and reduces to it in the comoving Lorentz frame. This tensor must be orthogonal to $u^\alpha u^\beta/c^2$, and it was identified back in Eq. (4.20) as the transverse projection operator $P^{\alpha\beta} = \eta^{\alpha\beta} + u^\alpha u^\beta/c^2$. We conclude, finally, that the pressure contribution to the energy-momentum tensor must be $pP^{\alpha\beta}$.

All in all we have obtained the following expression for the energy-momentum tensor of a perfect fluid:

$$T^{\alpha\beta} = \mu u^\alpha u^\beta/c^2 + p\left(\eta^{\alpha\beta} + u^\alpha u^\beta/c^2\right) = (\mu + p)u^\alpha u^\beta/c^2 + p\,\eta^{\alpha\beta}, \tag{4.50}$$

where, we recall, $\mu = \rho c^2 + \epsilon$ is the fluid's proper energy density.

We note that according to Eq. (4.50), the energy-momentum tensor is symmetric under an exchange of indices,

$$T^{\beta\alpha} = T^{\alpha\beta}. \tag{4.51}$$

This property is in fact very general, and not restricted to the specific case of a perfect fluid.

4.2.4 Fluid dynamics

All of fluid mechanics is contained in the conservation statements $\partial_\alpha j^\alpha = 0 = \partial_\beta T^{\alpha\beta}$, along with the assignments of Eqs. (4.44) and (4.50).

We first examine the consequences of Eq. (4.47), in which we substitute $j^\alpha = \rho u^\alpha$. If we also write $u^\alpha = \gamma(c, \boldsymbol{v})$, then the equation of mass conservation states that

$$\partial_t(\gamma\rho) + \partial_j(\gamma\rho v^j) = 0. \tag{4.52}$$

This is the relativistic generalization of Eq. (1.26), the Eulerian version of the continuity equation first encountered in Chapter 1; the relativistic equation reduces to the Newtonian version when $v/c \ll 1$ and $\gamma \simeq 1$. The equation of mass conservation can also be expressed in Lagrangian form, if we write it first in the form $u^\alpha \partial_\alpha \rho + \rho \partial_\alpha u^\alpha = 0$. The first term is the derivative of the proper density in the direction of the velocity field; if we focus our attention on a selected fluid element moving on a world line $r^\alpha(\tau)$, and write $u^\alpha = dr^\alpha/d\tau$ for its velocity vector, then $u^\alpha \partial_\alpha \rho$ is recognized as $d\rho/d\tau$, the change in proper density as we follow the world line of the fluid element. Equation (4.47) can therefore be expressed as

$$\frac{d\rho}{d\tau} + \rho\partial_\alpha u^\alpha = 0, \tag{4.53}$$

and this is the relativistic generalization of Eq. (1.25). This equation informs us that $\partial_\alpha u^\alpha = -\rho^{-1} d\rho/d\tau = \mathcal{V}^{-1} d\mathcal{V}/d\tau$, where $\mathcal{V} = \delta m/\rho$ is the volume of a fluid element of mass δm.

We next work on Eq. (4.49), in which we substitute Eq. (4.50). Simple manipulations first return

$$0 = u^\alpha\left(\frac{d\mu}{d\tau} + (\mu + p)\partial_\beta u^\beta\right) + (\mu + p)\frac{du^\alpha}{d\tau} + c^2 P^{\alpha\beta}\partial_\beta p, \tag{4.54}$$

in which we have written $d\mu/d\tau$ for $u^\beta \partial_\beta \mu$ and $du^\alpha/d\tau$ for $u^\beta \partial_\beta u^\alpha$, adopting the Lagrangian point of view that these derivatives follow the world line of a selected fluid element. The first set of terms in this expression is directed along u^α, while the remaining terms are orthogonal to u^α – refer to Eqs. (4.22) and (4.32). We set each orthogonal piece of this equation to zero, and obtain

$$\frac{d\mu}{d\tau} + (\mu + p)\partial_\beta u^\beta = 0 \tag{4.55}$$

and

$$(\mu + p)\frac{du^\alpha}{d\tau} + c^2 P^{\alpha\beta}\partial_\beta p = 0. \tag{4.56}$$

To elucidate the meaning of Eq. (4.55) we make the substitutions $\mu = \rho c^2 + \epsilon$ and $\partial_\alpha u^\alpha = -\rho^{-1} d\rho/d\tau$. After some cancellations we obtain

$$\frac{d\epsilon}{d\tau} - \frac{\epsilon + p}{\rho}\frac{d\rho}{d\tau} = 0. \tag{4.57}$$

This is *precisely* the same equation, Eq. (1.29), that was recognized in Chapter 1 as expressing the first law of thermodynamics for isentropic flows. To see this clearly, replace ρ by $\delta m/\mathcal{V}$ and write Eq. (4.57) as $d(\epsilon \mathcal{V}) = -p d\mathcal{V}$, which is a more recognizable form of the first law. It is a remarkable fact that the laws of thermodynamics do not require modifications in the transition from Newtonian physics to Minkowski spacetime, mainly because they are formulated in the momentarily comoving Lorentz frame of each fluid element.

Equation (4.56) is the relativistic generalization of Euler's equation, which was first displayed in Eq. (1.23); it involves the acceleration $du^\alpha/d\tau = u^\beta \partial_\beta u^\alpha$ of a fluid element and the (spatially projected) pressure gradient $\partial_\beta p$. Note that the presence of the projection operator $P^{\alpha\beta}$ in this equation guarantees that the acceleration $du^\alpha/d\tau$ is orthogonal to the velocity u^α.

A more explicit statement of Euler's equation is

$$(\rho + \epsilon/c^2 + p/c^2)\frac{du^\alpha}{d\tau} = -(\eta^{\alpha\beta} + u^\alpha u^\beta/c^2)\partial_\beta p. \tag{4.58}$$

An even more explicit version is obtained when we write $u^\alpha = \gamma(c, \boldsymbol{v})$ for the velocity field and focus on the spatial components of Eq. (4.58); this yields

$$\gamma(\rho + \epsilon/c^2 + p/c^2)(\partial_t + \boldsymbol{v}\cdot\boldsymbol{\nabla})(\gamma v^j) = -\partial_j p - \gamma^2 v^j(\partial_t + \boldsymbol{v}\cdot\boldsymbol{\nabla})p/c^2. \tag{4.59}$$

This differs from the non-relativistic version in a number of places. First, the quantity that multiplies the Lagrangian derivative $d(\gamma v^j)/dt$ on the left-hand side is $\rho + \epsilon/c^2 + p/c^2$ instead of just ρ; this tells us that in special relativity, the internal energy ϵ and the pressure p participate in the fluid's inertial response. Second, the Newtonian velocity v^j is multiplied by γ to convert it to a relativistic velocity, and the whole equation is multiplied by γ to convert the time derivative d/dt to a proper-time derivative $d/d\tau$. Third, the right-hand side is modified by a term involving dp/dt, the Lagrangian derivative of the pressure. All modifications become small when $v/c \ll 1$, and in the limit we recover $\rho(\partial_t + \boldsymbol{v}\cdot\boldsymbol{\nabla})\boldsymbol{v} = -\boldsymbol{\nabla} p$, the non-relativistic version of Euler's equation.

4.3 Electrodynamics

4.3.1 Maxwell's equations

The spacetime formulation of Maxwell's theory involves an electromagnetic field tensor $F^{\alpha\beta}$ whose components are defined, in a frame S, in terms of the electric field $\boldsymbol{E}$ and the magnetic field $\boldsymbol{B}$. The tensor is antisymmetric, $F^{\beta\alpha} = -F^{\alpha\beta}$, and it therefore contains six independent components. We make the assignments

$$F^{01} = c^{-1}E^x\,, \qquad F^{02} = c^{-1}E^y\,, \qquad F^{03} = c^{-1}E^z\,, \tag{4.60}$$

as well as

$$F^{12} = B^z\,, \qquad F^{23} = B^x\,, \qquad F^{31} = B^y\,. \tag{4.61}$$

These equations are summarized by $F^{0j} = c^{-1}E^j$ and $F^{ij} = \epsilon^{ijk}B_k$, where ϵ_{ijk} is the permutation symbol introduced in Sec. 1.4.4. The field $F^{\alpha\beta}$ transforms as a tensor under a Lorentz transformation, and this gives rise to the well-known transformation rules for $\boldsymbol{E}$ and $\boldsymbol{B}$ (see Exercise 4.8).

The spacetime formulation of electrodynamics must also involve a source for the electromagnetic field, and we take this to be a charged fluid. We introduce the proper charge density $\rho_e(x^\alpha)$ as the charge per unit volume that is measured by an observer momentarily comoving with the fluid element at x^α; as such ρ_e, like the proper mass density ρ introduced in Sec. 4.2, is a spacetime invariant whose value does not depend on the Lorentz frame. We introduce also a charge-current vector j_e^α whose definition is very close to that of the mass-current vector of Eq. (4.44). We have that $c^{-1}j_e^0$ is the charge density as measured in the laboratory frame S, while j_e^k is the charge flux (or current) in the x^k-direction. If u^α is the fluid's velocity field, then

$$j_e^\alpha = \rho_e u^\alpha\,. \tag{4.62}$$

The local statement of charge conservation is $\partial_\alpha j_e^\alpha = 0$.

The spacetime formulation of Maxwell's equations is

$$\partial_\beta F^{\alpha\beta} = \mu_0 j_e^\alpha\,, \qquad \partial_\alpha F_{\beta\gamma} + \partial_\gamma F_{\alpha\beta} + \partial_\beta F_{\gamma\alpha} = 0, \tag{4.63}$$

where μ_0 is a constant (known as the permeability of vacuum) that determines the units of the theory. It is connected to another constant ϵ_0 (known as the permittivity of vacuum, or the dielectric constant) by the relation $\epsilon_0\mu_0 = c^{-2}$. The first four equations are known as the inhomogeneous (or sourced) Maxwell equations; the remaining equations are known as the homogeneous (or source-free) Maxwell equations. It is easy to verify, for example, that when $\alpha = 0$, the first equation returns $\boldsymbol{\nabla}\cdot\boldsymbol{E} = (c^{-1}j_e^0)/\epsilon_0$, which is the usual statement of Gauss's law. As another example, setting $\alpha = 0$, $\beta = 1$, and $\gamma = 2$ in the second equation returns the z-component of $\partial_t\boldsymbol{B} + \boldsymbol{\nabla}\times\boldsymbol{E} = 0$, the usual statement of Faraday's law.

Note that by antisymmetry of the electromagnetic field tensor, $\partial_{\alpha\beta}F^{\alpha\beta}$ vanishes identically. This implies, together with the inhomogeneous Maxwell equations, that $\partial_\alpha j_e^\alpha$ is necessarily zero in Maxwell's theory; charge conservation is automatically enforced.

4.3.2 Vector potential

The scalar potential Φ and the vector potential $\boldsymbol{A}$ of the conventional formulation of electrodynamics can be combined into a spacetime vector $A^\alpha = (\Phi/c, \boldsymbol{A})$. The electromagnetic field can then be expressed in terms of the potential:

$$F_{\alpha\beta} = \partial_\alpha A_\beta - \partial_\beta A_\alpha. \tag{4.64}$$

These relations reproduce the familiar $\boldsymbol{E} = -\partial_t \boldsymbol{A} - \nabla\Phi$ and $\boldsymbol{B} = \nabla \times \boldsymbol{A}$. Their main virtue is that once $F_{\alpha\beta}$ is expressed in this way, the homogeneous Maxwell equations are automatically satisfied.

The remaining (inhomogeneous) equations can then be recast as differential equations for the potentials. These, however, cannot be uniquely determined, because Eq. (4.64) is unaffected by a *gauge transformation* of the form

$$A_\alpha \to A_\alpha + \partial_\alpha \chi, \tag{4.65}$$

in which χ is any function of the spacetime coordinates x^α. This gauge freedom can be exploited to simplify the equations to be satisfied by the potentials. A popular and useful choice for the gauge function χ enforces the *Lorenz-gauge condition*

$$\partial_\alpha A^\alpha = 0. \tag{4.66}$$

With this choice of gauge, the inhomogeneous Maxwell equations are readily shown to take the form of a wave equation for the vector potential:

$$\Box A^\alpha = -\mu_0 j_e^\alpha, \tag{4.67}$$

where

$$\Box := \eta^{\alpha\beta}\partial_\alpha\partial_\beta = -\frac{1}{c^2}\frac{\partial^2}{\partial t^2} + \nabla^2 \tag{4.68}$$

is the wave operator in Minkowski spacetime. The Lorenz gauge is named after Ludvig Lorenz (1829–1891), a Danish mathematician and physicist. He is often confused with his more famous Dutch colleague Hendrik Lorentz (1853–1928), and in the past the Lorenz gauge was almost universally known as the Lorentz gauge. This, however, is a historical slight that we do not wish to propagate.

Equations (4.67) and (4.68) indicate that the electromagnetic field propagates as a wave, and that the speed of propagation is c, the speed of light. This conclusion is not just an artifact of our choice of gauge: the electromagnetic field tensor itself can be shown to satisfy a wave equation. Simple manipulations starting from Eq. (4.63) do indeed reveal that $F^{\alpha\beta}$ must be a solution to $\Box F^{\alpha\beta} = -\mu_0(\partial^\alpha j_e^\beta - \partial^\beta j_e^\alpha)$.

4.3.3 Energy-momentum tensor

The electromagnetic field carries energy and momentum, and it exerts forces (and does work) on a charged fluid. These aspects of Maxwell's theory are encoded in the field's energy-momentum tensor

$$T^{\alpha\beta} = \frac{1}{\mu_0}\Big(F^{\alpha\mu}F^{\beta}_{\ \ \mu} - \frac{1}{4}\eta^{\alpha\beta}F_{\mu\nu}F^{\mu\nu}\Big). \tag{4.69}$$

In terms of $\boldsymbol{E}$ and $\boldsymbol{B}$ we have that $F_{\mu\nu}F^{\mu\nu} = -2(E^2/c^2 - B^2)$, $T^{00} = (E^2/c^2 + B^2)/(2\mu_0)$, $cT^{0j} = (\boldsymbol{E}\times\boldsymbol{B})^j/\mu_0$, and

$$T^{jk} = -\frac{1}{\mu_0 c^2}\Big(E^j E^k - \frac{1}{2}\delta^{jk}E^2\Big) - \frac{1}{\mu_0}\Big(B^j B^k - \frac{1}{2}\delta^{jk}B^2\Big), \tag{4.70}$$

where we write $E^2 := \boldsymbol{E}\cdot\boldsymbol{E}$ and $B^2 := \boldsymbol{B}\cdot\boldsymbol{B}$. We recognize in T^{00} the energy density of the electromagnetic field (as measured in the laboratory frame S), in cT^{0j} the j-component of the energy flux (also known as the Poynting vector), and T^{jk} is the Maxwell stress tensor. To illustrate its meaning we consider a field configuration such that $\boldsymbol{E} = 0$ and $\boldsymbol{B}$ points in the x-direction at some selected position in the laboratory frame. For this configuration $T^{xx} = -B^2/(2\mu_0)$, and this represents a tension (negative pressure) along the magnetic field lines. We also have $T^{yy} = T^{zz} = B^2/(2\mu_0)$, and this represents a pressure in the directions perpendicular to the field lines.

The field's energy and momentum are not conserved in general, because of the fact (pointed out previously) that the field exerts forces and does work. Straightforward manipulations starting from Eq. (4.69) reveal that

$$\partial_\beta T^{\alpha\beta} = \frac{1}{\mu_0}\Big[-F^{\alpha}_{\ \ \mu}(\partial_\beta F^{\mu\beta}) - \frac{1}{2}F^{\mu\nu}\big(\partial^\alpha F_{\mu\nu} + \partial_\nu F^{\alpha}_{\ \ \mu} + \partial_\mu F_{\nu}^{\ \ \alpha}\big)\Big], \tag{4.71}$$

and substitution of Maxwell's equations (4.63) simplifies this to

$$\partial_\beta T^{\alpha\beta} = -F^{\alpha}_{\ \ \mu}j^{\mu}_e. \tag{4.72}$$

This equation states that there is indeed an exchange of energy and momentum between the field and the charged fluid. The spatial components of the right-hand side of Eq. (4.72) become $-\gamma\rho_e(\boldsymbol{E} + \boldsymbol{v}\times\boldsymbol{B})$ when Eq. (4.62) is used and $F^{\alpha\beta}$ is expressed in terms of $\boldsymbol{E}$ and $\boldsymbol{B}$. We recognize this as (minus) the Lorentz-force density acting on a fluid with proper charge density ρ_e and (Newtonian) velocity field $\boldsymbol{v}$.

The total energy-momentum tensor of the system fluid + field includes a contribution $T^{\alpha\beta}_{\rm field}$ from the field and a contribution $T^{\alpha\beta}_{\rm fluid}$ from the fluid (as described in Sec. 4.2). The total energy and momentum are conserved when the system is isolated, and we then have

$$\partial_\beta\big(T^{\alpha\beta}_{\rm fluid} + T^{\alpha\beta}_{\rm field}\big) = 0. \tag{4.73}$$

This equation gives rise to a charged version of the relativistic Euler equation:

$$(\rho + \epsilon/c^2 + p/c^2)\frac{du^\alpha}{d\tau} = -\big(\eta^{\alpha\beta} + u^\alpha u^\beta/c^2\big)\partial_\beta p + \rho_e F^{\alpha}_{\ \ \beta}u^\beta. \tag{4.74}$$

The first term on the right-hand side describes the familiar pressure forces acting within the fluid; the second term represents the electromagnetic forces. The Euler equation is to be supplemented by the mass-conservation equation (4.53), the first law of thermodynamics (4.57), and Maxwell's equations (4.63).

4.4 Point particles in spacetime

Before moving on to general relativity and curved spacetimes, we return briefly to the description of point particles in Minkowski spacetime. Our goal here is to incorporate the point particle within the fluid language developed in the preceding sections.

A point particle can be viewed as a singular distribution of fluid, with non-vanishing mass m (and possibly charge q) but infinite mass density ρ (and possibly infinite charge density ρ_e). A point particle has no internal degrees of freedom, and therefore no internal energy density ϵ and no pressure p. In Newtonian mechanics the particle would move on a trajectory described by $\boldsymbol{x} = \boldsymbol{r}(t)$. In spacetime it moves on a world line $x^\alpha = r^\alpha(\tau)$, and its velocity vector is $u^\alpha = dr^\alpha/d\tau$.

In a frame S' that is momentarily comoving with the particle, the proper mass density ρ can be written as $m\delta_3(\boldsymbol{x}' - \boldsymbol{r}'(t'))$, in terms of a three-dimensional delta function evaluated on the world line. This expression, however, is not Lorentz invariant, because the delta function, like the volume element d^3x', is affected by a Lorentz transformation.

In an effort to replace the three-dimensional delta function with an invariant quantity, we examine the four-dimensional version

$$\delta\big(x^\mu - r^\mu(\tau)\big) := \delta\big(ct - r^0(\tau)\big)\delta_3\big(\boldsymbol{x} - \boldsymbol{r}(\tau)\big), \tag{4.75}$$

which is a spacetime invariant. Its defining property is that the integral

$$\int f(x^\alpha)\,\delta(x^\mu - r^\mu)\,d^4x$$

returns $f(r^\alpha)$ if the four-dimensional domain of integration includes the event r^α, and zero if the domain excludes it; here $f(x^\alpha)$ is an arbitrary test function of the spacetime coordinates. The spacetime volume element $d^4x = d(ct)d^3x$ is a Lorentz invariant, and this ensures that the delta function itself is a Lorentz invariant.

The four-dimensional delta function is a good starting point to define a proper mass density ρ, but the definition cannot simply be $m\delta(x^\mu - r^\mu)$. This is wrong for two reasons. First, the delta function has a dimension of inverse length raised to the fourth power, and this would not give rise to the expected dimension of (mass)/(volume) for the mass density. Second, the delta function is "on" for a single moment of time only, when $t = r^0(\tau)/c$; we would expect instead the mass density to be "on" at all times. The way out of these difficulties is to integrate the four-dimensional delta function with respect to proper time τ, and to define the particle's proper mass density as

$$\rho(x^\alpha) = mc \int \delta\big(x^\mu - r^\mu(\tau)\big)\,d\tau. \tag{4.76}$$

We shall see presently that this is indeed a well-motivated definition. For now we may observe that ρ is dimensionally correct, and that it is constructed entirely from scalar quantities (the rest-mass m, the delta function, and proper time) and operations (the integration over proper time).

To see what Eq. (4.76) implies, it is useful to change the variable of integration from τ to $r^0(\tau)$ and to rewrite the integral as

$$\rho = mc \int \frac{\delta(x^0 - r^0)}{u^0} \delta_3(\boldsymbol{x} - \boldsymbol{r})\, dr^0, \tag{4.77}$$

where $u^0 = dr^0/d\tau$. Integration is immediate, and we find

$$\rho(t, \boldsymbol{x}) = \frac{mc}{u^0} \delta_3\big(\boldsymbol{x} - \boldsymbol{r}(t)\big), \tag{4.78}$$

where the time t is in principle determined by $t = r^0(\tau)/c$. This expression for the mass density is valid in any Lorentz frame. In the comoving frame S' of the particle we have that $u'^0 = c$, and ρ reduces to $m\delta_3(\boldsymbol{x}' - \boldsymbol{r}')$, the expected expression. In another frame S, the factor $u^0/c = \gamma$ accounts for the Lorentz contraction of three-dimensional volume elements. We conclude that Eq. (4.76) is indeed a sensible, Lorentz-invariant definition of the proper mass density of a point particle.

The mass current j^α of a point particle is ρu^α, and this can be written either as $(mcu^\alpha/u^0)\delta_3(\boldsymbol{x} - \boldsymbol{r})$, or more elegantly as

$$j^\alpha = mc \int u^\alpha\, \delta\big(x^\mu - r^\mu(\tau)\big)\, d\tau, \tag{4.79}$$

in which the vector $u^\alpha(\tau)$ was inserted within the integral. Similarly, the energy-momentum tensor of a point particle can be shown to be given by

$$T^{\alpha\beta} = mc \int u^\alpha u^\beta\, \delta\big(x^\mu - r^\mu(\tau)\big)\, d\tau, \tag{4.80}$$

in which two factors of the velocity vector are inserted within the integral.

A charged particle possesses also a proper charge density ρ_e and a charge current $j_e^\alpha = \rho_e u^\alpha$. These can be expressed as

$$\rho_e = qc \int \delta\big(x^\mu - r^\mu(\tau)\big)\, d\tau \tag{4.81}$$

and

$$j_e^\alpha = qc \int u^\alpha\, \delta\big(x^\mu - r^\mu(\tau)\big)\, d\tau, \tag{4.82}$$

where q is the particle's electric charge.

To explore the consequences of these results we rely on the distributional identity

$$u^\alpha \partial_\alpha \delta\big(x^\mu - r^\mu(\tau)\big) = -\frac{d}{d\tau} \delta\big(x^\mu - r^\mu(\tau)\big), \tag{4.83}$$

which can be established by acting separately on the delta function with the differential operators $u^\alpha \partial_\alpha$ and $d/d\tau$, and comparing the results. Using this identity, we find that the conservation statements $\partial_\alpha j^\alpha = 0 = \partial_\alpha j_e^\alpha$ follow automatically. For example,

$$\partial_\alpha j^\alpha = mc \int u^\alpha \partial_\alpha \delta(x^\mu - r^\mu)\, d\tau = -mc \int \frac{d}{d\tau} \delta(x^\mu - r^\mu)\, d\tau = 0. \tag{4.84}$$

On the other hand, we have

$$\begin{aligned}\partial_\beta T^{\alpha\beta} &= mc\int u^\alpha u^\beta \partial_\beta \delta(x^\mu - r^\mu)\, d\tau \\ &= -mc\int u^\alpha \frac{d}{d\tau}\delta(x^\mu - r^\mu)\, d\tau,\end{aligned} \tag{4.85}$$

and this becomes

$$\partial_\beta T^{\alpha\beta} = mc\int \frac{du^\alpha}{d\tau}\delta(x^\mu - r^\mu)\, d\tau \tag{4.86}$$

after an integration by parts. The energy and momentum of a point particle are conserved when no forces are acting upon it; under these circumstances $\partial_\beta T^{\alpha\beta} = 0$ and Eq. (4.86) implies that $du^\alpha/d\tau = 0$. In general, however, the energy and momentum are not conserved, $\partial_\beta T^{\alpha\beta} \neq 0$, and the particle is accelerated.

As an example of a non-trivial dynamics, consider a situation in which a point particle carries a charge q and interacts with an electromagnetic field $F^{\alpha\beta}$. The total system is isolated, and in this case we have

$$\partial_\beta\big(T^{\alpha\beta}_{\text{particle}} + T^{\alpha\beta}_{\text{field}}\big) = 0. \tag{4.87}$$

We have seen in Eq. (4.72) that $\partial_\beta T^{\alpha\beta}_{\text{field}} = -F^\alpha_{\ \beta} j^\beta_e$, which can be written as

$$\partial_\beta T^{\alpha\beta}_{\text{field}} = -qc\int F^\alpha_{\ \beta}(\tau) u^\beta \delta(x^\mu - r^\mu)\, d\tau \tag{4.88}$$

if we make use of Eq. (4.82); here the field tensor $F^\alpha_{\ \beta}$ was inserted within the integral, where it is evaluated at $x^\alpha = r^\alpha(\tau)$. Combining this result with Eq. (4.86), we arrive at

$$m\frac{du^\alpha}{d\tau} = qF^\alpha_{\ \beta}u^\beta. \tag{4.89}$$

This is the relativistic expression of the Lorentz-force law. Equation (4.89) becomes $md(\gamma \boldsymbol{v})/d\tau = q\gamma(\boldsymbol{E} + \boldsymbol{v} \times \boldsymbol{B})$ when expressed in a frame S; the factor of γ converts the proper-time derivative to a coordinate-time derivative, and we recover the usual statement of the Lorentz-force equation.

The equations of motion of Eq. (4.89) can also be derived on the basis of an action principle. The action functional of a free particle can be generalized to include an interaction with an electromagnetic field. It is given by

$$S = -mc^2\int_1^2 d\tau + q\int_1^2 A_\alpha dx^\alpha, \tag{4.90}$$

and it is easy to show – refer to Exercise 4.10 – that the Euler–Lagrange equations derived from S reproduce Eq. (4.89).

4.5 Bibliographical notes

Our survey of special relativity and Minkowski spacetime is based on standard presentations found in a number of textbooks, including French (1968) and Rindler (1991), and the introductory texts on general relativity listed in the bibliographical notes of Chapter 5.

The experimental tests of special relativity reviewed in Box 4.1 are described in the following papers. The X-ray tests of the frame-independence of the speed of light are published in Brecher (1977), and the pion experiments were carried out by Alväger *et al.* (1964). The time-dilation experiments involving the motion of H_2 and H_3 molecules are described in Ives and Stilwell (1938), and the lifetime of muons was measured by Rossi and Hall (1941) and Farley *et al.* (1966). For a survey of classic and modern tests of special relativity and Lorentz invariance, see Will's (2006a) review, written on the occasion of the centenary of special relativity.

4.6 Exercises

4.1 The Lorentz transformation for an arbitrary velocity whose components in S are v^j is given by

$$x'^{\alpha} = \Lambda^{\alpha}_{\beta} x^{\beta} ,$$

where the components of Λ^{α}_{β} are given by

$$\begin{aligned} \Lambda^0_0 &= \gamma , \\ \Lambda^0_j &= -\gamma v_j / c , \\ \Lambda^j_0 &= -\gamma v^j / c , \\ \Lambda^j_k &= \delta^j_k + (\gamma - 1) n^j n_k , \end{aligned}$$

where $n^j = v^j / |\boldsymbol{v}|$.

(a) Show that this reduces to Eq. (4.1) when $\boldsymbol{v}$ is aligned with the x-direction.

(b) By considering the invariance of the interval ds^2, show that the Minkowski metric in S' is related to the Minkowski metric in S by

$$\eta_{\alpha\beta} = \eta'_{\gamma\delta} \Lambda^{\gamma}_{\alpha} \Lambda^{\delta}_{\beta} .$$

(c) Verify using the general Lorentz transformation that $\eta_{\alpha\beta}$ has the same diagonal form with entries $(-1, 1, 1, 1)$ as it had in S'. You can do this using matrix multiplication by recognizing that the transformation of the metric can be expressed as the matrix equation $\eta = \Lambda^{\mathrm{T}} \eta' \Lambda$.

4.2 Show that the composition of two Lorentz transformations $\Lambda^{\alpha}_{1\beta} \Lambda^{\beta}_{2\gamma}$ is also a Lorentz transformation, (a) by verifying that the Minkowski metric $\eta_{\alpha\beta}$ is unchanged under the combined transformation, and (b) by calculating the combined transformation explicitly for boosts in the x-direction with velocity v_1 and v_2 respectively.

4.3 Consider the three events in spacetime described in the inertial frame S by

$$A : (1, 1, 0, 0), \quad B : (2, 3, 0, 0), \quad C : (3, 2, 0, 0),$$

where the notation is (ct, x, y, z), all in some arbitrary units of length. For each pair of events AB, AC, and BC, determine if the interval is timelike, spacelike, or null. Find the proper distance or time between the events, as appropriate, and find the velocity of a moving frame S' in which the two events are simultaneous or at the same location, as appropriate.

4.4 Given a particle with momentum p^α and an observer with velocity u^α_{obs}, define the vector

$$V^\alpha_{\text{obs}} = c \frac{P_{\text{obs}}{}^\alpha{}_\beta \, p^\beta}{E_{\text{obs}}},$$

where $E_{\text{obs}} = -p_\alpha u^\alpha_{\text{obs}}$ and $P_{\text{obs}}{}^\alpha{}_\beta = \delta^\alpha_\beta + u^\alpha_{\text{obs}} u_{\text{obs}\,\beta}/c^2$.

(a) Show that V^α_{obs} is a genuine four-dimensional vector, and that it represents the three-dimensional velocity vector $\boldsymbol{v}$ of the particle as measured by the observer.

(b) Calculate the invariant quantity $\eta_{\alpha\beta} V^\alpha_{\text{obs}} V^\beta_{\text{obs}}$, and use it to show that for an observer and particle moving in the x-direction with velocities v_1 and v_2, respectively,

$$|\vec{V}_{\text{obs}}| = \frac{v_2 - v_1}{1 - v_1 v_2/c^2}.$$

This is the standard formula for the addition of velocities in special relativity.

4.5 Show that

$$\eta^{\alpha\beta} = -e^\alpha_{(0)} e^\beta_{(0)} + e^\alpha_{(1)} e^\beta_{(1)} + e^\alpha_{(2)} e^\beta_{(2)} + e^\alpha_{(3)} e^\beta_{(3)}$$

by checking its components (a) in a frame where the basis vectors are attached to a particle at rest, and (b) in a frame where the vectors are attached to a particle moving in the x-direction with velocity v.

4.6 A particle emits radiation isotropically in its own rest frame. As seen from a frame S' in which the particle moves with a speed $v \approx c$ such that $\gamma \gg 1$, show that the forward hemisphere of radiation in its rest frame is seen in S' to be beamed into a forward cone of opening angle $\theta \approx 1/\gamma$.

4.7 Three spaceships take off simultaneously from planet Earth, and head for the planet Romulus. An important meeting to discuss the borders of the Neutral Zone is scheduled to begin in exactly 15 years, Earth time. The planets are 12 light years apart. The ships travel according to precise flight plans issued by the dreaded Romulans:

- Ship #1, the USS Enterprise, captained by Jean-Luc Picard, travels on a straight line with a uniform speed $0.8\,c$ for the entire trip.

- Ship #2, the USS Voyager, captained by Kathryn Janeway, travels in a straight line, but with a varying speed, half the time (Earth time) at speed $0.7\,c$, and the other half of the time at speed $0.9\,c$.
- Ship #3, captained by the evil Romulan General Maldor, travels at constant speed $0.9\,c$, arriving early at Romulus, where the General hatches devious plots with his advisors while awaiting the start of the meeting.

Calculate the proper time elapsed between take-off and the start of the meeting according to the clocks carried by each of the three travelers, and point out the significance of the traveller with the largest elapsed proper time. What is the *minimum* possible proper time between take-off and the start of the meeting?

4.8 Using the general Lorentz transformation of Exercise 4.1, show that the electric and magnetic fields transform according to

$$\boldsymbol{E}' = \gamma \boldsymbol{E} + (1-\gamma)\boldsymbol{n}(\boldsymbol{n}\cdot\boldsymbol{E}) + \gamma \boldsymbol{v}\times\boldsymbol{B}\,,$$

$$\boldsymbol{B}' = \gamma \boldsymbol{B} + (1-\gamma)\boldsymbol{n}(\boldsymbol{n}\cdot\boldsymbol{B}) - \gamma\frac{\boldsymbol{v}}{c^2}\times\boldsymbol{E}\,.$$

4.9 Using the fact that a coordinate transformation induces a change in the volume element d^4x given by $d^4x' = J(x', x)d^4x$, where $J := \det(\partial x'^{\alpha}/\partial x^{\beta})$ is the Jacobian of the transformation, show that $d(ct)d^3x$ is a Lorentz invariant.

4.10 Verify that the Euler–Lagrange equations derived from the action

$$S = -mc^2\int_1^2 d\tau + q\int_1^2 A_\alpha dx^\alpha\,,$$

yield the relativistic form of the Lorentz-force law.

5 Curved spacetime

The relativistic formulation of the laws of physics developed in Chapter 4 excluded gravitation, and our task in this chapter is to complete the story by incorporating this all-important interaction (our personal favorite!). In Sec. 5.1 we explain why relativistic gravitation must be thought of as a theory of curved spacetime. In Sec. 5.2 we develop the elementary aspects of differential geometry that are required in a study of curved spacetime, and in Sec. 5.3 we show how the special-relativistic form of the laws of physics can be generalized to incorporate gravitation in a curved-spacetime formulation. We describe the Einstein field equations in Sec. 5.4, and in Sec. 5.5 we show how to solve them in the restricted context of small deviations from flat spacetime. We conclude in Sec. 5.6 with a description of spherical bodies in hydrostatic equilibrium, featuring the most famous (and historically the first) exact solution to the Einstein field equations; this is the Schwarzschild metric, which describes the vacuum exterior of any spherical distribution of matter (including a black hole).

5.1 Gravitation as curved spacetime

5.1.1 Principle of equivalence

Relativistic gravity

The relativistic Euler equation (4.59), unlike its Newtonian version of Eq. (1.23), does not contain a term that describes a gravitational force acting on the fluid. To insert such a term requires an understanding of how the Newtonian theory of gravitation can be generalized to a relativistic setting. It is tempting to attempt such a generalization by simply replacing the Poisson equation $\nabla^2 U = -4\pi G\rho$ with a Lorentz-invariant generalization such as $\Box U = 4\pi G T^{\mu}_{\ \mu}/c^2 = -4\pi G(\rho + \epsilon/c^2 - 3p/c^2)$, and replacing the term $\partial_j U$ in the Newtonian Euler equation by something like $P^{\alpha\beta}\partial_\beta U$. This attempt, however, would be unsuccessful. It would lead to inconsistencies – why, for example, is the gravitational piece of the stress tensor, $T_{jk} = (\partial_j U \partial_k U - \frac{1}{2}\delta_{jk}\nabla U \cdot \nabla U)/(4\pi G)$, absent from the right-hand side of the wave equation? And historically, such attempts have led to incorrect empirical consequences, such as no bending of light, or the wrong value for the perihelion advance of Mercury. Similar attempts to formulate a relativistic theory of gravitation based on a vector potential A^α have also failed.

What about a tensor theory? Our experience with Minkowski spacetime has revealed the important role of a tensor – the Minkowski metric $\eta_{\alpha\beta}$. In Minkowski spacetime, however, $\eta_{\alpha\beta}$ is a rather inert, somewhat boring object, merely providing the fixed arena in which special relativistic dynamics takes place. But if we open our minds to the possibility that $\eta_{\alpha\beta}$ could be an approximation to something more general – a spacetime metric $g_{\alpha\beta}$ – that could itself participate in the dynamics, then the possibility of a tensor theory of gravity emerges. We will therefore follow Einstein and argue in favor of a connection between gravitation and the curvature of spacetime.

Principles of equivalence

The key observation that leads to this connection is known as the *principle of equivalence*, which we state as:

> If a test body is placed at an initial event in spacetime and given an initial velocity there, and if the body subsequently moves freely, then its world line will be independent of its mass, internal structure, and composition.

Here, a "test body" is one that does not modify the gravitational field created by other (non-test) bodies, and a "freely-moving" body is one on which no forces are acting, except for the gravitational force; the test body, for example, is not allowed to possess an electric charge when an electromagnetic field is present. The principle of equivalence states that all test bodies move with the same acceleration in a gravitational field, irrespective of their mass or internal composition. This statement is known more precisely as the "weak equivalence principle," and in Newtonian theory it holds as a consequence of the equality between inertial mass and passive gravitational mass. This property is only accidental in Newtonian theory; it is a foundational axiom of relativistic gravitation. As we discussed in Box 1.1, there is ample, high-precision experimental support for the weak equivalence principle.

A stronger formulation of the principle of equivalence, known as "Einstein's equivalence principle," states that the weak version holds, and that, in addition:

> The outcome of any local, non-gravitational test experiment performed by a freely-moving apparatus is independent of the velocity of the apparatus and independent of when and where the experiment is carried out.

Here, a "local, non-gravitational test experiment" is any measurement that does not probe gravitational effects directly; this could be, for example, a measurement of the fine-structure constant, or a measurement of the critical temperature in a phase transition to a Bose–Einstein condensate. The measuring apparatus is assumed to be moving freely, in the sense provided previously: no forces other than gravity are acting on the apparatus.

An even stronger statement of the principle of equivalence, known as the "strong equivalence principle," states that the weak version holds even when the body is self-gravitating, and that the Einstein version holds even when local test experiments are allowed to probe gravitational effects. As we shall see, the Einstein equivalence principle implies that gravitation must be a manifestation of the curvature of spacetime, and that relativistic gravity must be formulated as a metric theory. The strong equivalence principle, however, is not valid for all metric theories of gravitation, but it is satisfied in Einstein's theory. We will see the strong equivalence principle in action in later chapters, particularly in Chapter 9, and will witness its failure when we turn to alternative theories of gravity in Chapter 13.

An important aspect of the principles of equivalence, which we have not yet stated, is that the spatial dimensions of the test body and the spatial dimensions of the freely-moving apparatus must be small compared with the length scale over which the gravitational field varies. The test body and the apparatus must not be allowed to probe the inhomogeneities of the gravitational field. Thus, all statements made with regard to the principles of equivalence are *local* in nature. With this important restriction in place, a succinct summary of the Einstein equivalence principle is:

> No measurement carried out in a suitably small laboratory moving freely in a gravitational field can reveal the existence of gravity locally, within the confines of the laboratory.

If all objects within the laboratory fall with the same acceleration, then all local measurements reveal a vanishing acceleration relative to a standard, freely-moving object, and no local measurement can reveal the presence of a gravitational field. In other words, all local aspects of gravity can be turned off by doing physics in a freely-moving frame of reference.

This statement is powerful. It implies, in particular, that the special-relativistic formulation of the laws of physics continues to hold locally in the freely-moving frames; gravity is not present in these frames, and there is no need to modify the laws of physics to account for it. And it follows that *the inertial frames of special relativity must be identified with the freely-moving frames of observers in a gravitational field*. In the old Newtonian language we would say that a freely-moving observer is accelerated (relative, say, to an observer at rest on the surface of the Earth), and that his reference frame is not inertial. In the new relativistic language we say instead that it is the freely-moving observer who is inertial, while the observer at rest on Earth is not; the freely-moving observer is not accelerated, while the observer at rest on the ground is. In everyday life we tend not to think in this way, but consider the following comparison: A person stands still in a long line for two hours waiting for admission to a rock concert, while an astronaut on the International Space Station works for two hours installing some equipment; which person ends up with sore feet? *In relativistic gravitation, acceleration measures a departure from free motion produced by non-gravitational forces. A freely-moving observer is unaccelerated, and his reference frame is inertial.*

The tale of Cliff and Eric

As we can now appreciate, all local aspects of gravitation are purely relative, and they can be turned off by adopting a freely-moving frame of reference. But as we shall now discuss, aspects of gravitation that probe inhomogeneities of the gravitational field are absolute; these can never be turned off.

Consider two remote observers in outer space, falling freely toward Earth. The first observer – let's call him Cliff – performs local measurements in his spacecraft, and infers the complete absence of gravity in his immediate vicinity. The second observer – Eric – also infers, on the basis of local measurements, the local absence of gravity. Cliff and Eric are both inertial observers (in the relativistic sense described previously). Their reference frames are distinct, however, and they are in fact accelerated with respect to each other. This relative acceleration is revealed by the fact that as Cliff and Eric both fall toward Earth, their spacecraft are slowly converging toward each other. The relative acceleration is caused by inhomogeneities in the gravitational field: the gravitational force is pulling the spacecraft in different directions. Cliff and Eric must agree that while gravity is absent locally, an interaction must be responsible for the relative acceleration of their inertial frames. This *is* gravity; absolute gravity is inhomogeneous gravity.

It is instructive to provide a mathematical description of the tale of Cliff and Eric. Cliff's trajectory toward Earth is described by $\boldsymbol{r}_{\mathrm{C}}(t)$, and this is a solution to

$$\frac{d^2\boldsymbol{r}_{\mathrm{C}}}{dt^2} = \boldsymbol{g}(\boldsymbol{r}_{\mathrm{C}}), \tag{5.1}$$

in which the (Newtonian, as opposed to relativistic) gravitational acceleration $\boldsymbol{g} = \boldsymbol{\nabla} U$ is evaluated at Cliff's position. Assuming that the spacecraft is sufficiently small that variations in $\boldsymbol{g}(\boldsymbol{x})$ are negligible within it, all objects in Cliff's spacecraft move with this acceleration. Suppose that a selected object is at a position $\boldsymbol{r}_{\mathrm{C}} + \boldsymbol{\xi}_{\mathrm{C}}$ within the spacecraft, where $\boldsymbol{\xi}_{\mathrm{C}}$ is the object's position relative to Cliff. Then $\boldsymbol{\xi}_{\mathrm{C}}(t)$ is a solution to $d^2\boldsymbol{\xi}_{\mathrm{C}}/dt^2 = \boldsymbol{g}(\boldsymbol{r}_{\mathrm{C}} + \boldsymbol{\xi}_{\mathrm{C}}) - \boldsymbol{g}(\boldsymbol{r}_{\mathrm{C}}) \simeq \boldsymbol{0}$. Relative to Cliff, the object moves on a straight line with a uniform velocity, and Cliff sees no local manifestation of gravity.

On the other hand, Eric's trajectory toward Earth is described by $\boldsymbol{r}_{\mathrm{E}}(t)$, a solution to

$$\frac{d^2\boldsymbol{r}_{\mathrm{E}}}{dt^2} = \boldsymbol{g}(\boldsymbol{r}_{\mathrm{E}}), \tag{5.2}$$

in which $\boldsymbol{g}$ is now evaluated at Eric's position. Assuming that Eric's spacecraft is sufficiently small, all objects within it move with this acceleration, and a selected object at a relative position $\boldsymbol{\xi}_{\mathrm{E}}$ moves on a straight line with a constant relative velocity $d\boldsymbol{\xi}_{\mathrm{E}}/dt$. Eric also sees no local manifestation of gravity.

Let now $\boldsymbol{\xi}_{\mathrm{EC}}(t) := \boldsymbol{r}_{\mathrm{E}}(t) - \boldsymbol{r}_{\mathrm{C}}(t)$ be Eric's position relative to Cliff. This is a solution to

$$\frac{d^2\boldsymbol{\xi}_{\mathrm{EC}}}{dt^2} = \boldsymbol{g}(\boldsymbol{r}_{\mathrm{E}}) - \boldsymbol{g}(\boldsymbol{r}_{\mathrm{C}}), \tag{5.3}$$

and because the displacement is now large, the right-hand side of this equation can no longer be approximated by zero. The gravitational accelerations at Eric's and Cliff's positions are different, and we can express this difference by means of a Taylor expansion:

$$g_j(\boldsymbol{r}_{\mathrm{E}}) = g_j(\boldsymbol{r}_{\mathrm{C}} + \boldsymbol{\xi}_{\mathrm{EC}}) = g_j(\boldsymbol{r}_{\mathrm{C}}) + \xi^k_{\mathrm{EC}}\partial_k g_j(\boldsymbol{r}_{\mathrm{C}}) + \cdots . \tag{5.4}$$

Combining this with $g_j = \partial_j U$, we arrive at

$$\frac{d^2\xi^j_{\mathrm{EC}}}{dt^2} = (\partial_{jk}U)\xi^k_{\mathrm{EC}}, \tag{5.5}$$

where the Newtonian potential $U(\boldsymbol{x})$ is evaluated at $\boldsymbol{x} = \boldsymbol{r}_{\mathrm{C}}(t)$ after differentiation, and where we have neglected quadratic and higher-order terms in the Taylor expansion of g_j. The left-hand side of the equation gives the relative acceleration between Cliff and Eric, and the right-hand side measures the inhomogeneities of the gravitational field, the failure of $\boldsymbol{g}$ to be a constant vector field. It is this failure that gives rise to the relative acceleration, and it is the homogeneities in $\boldsymbol{g}$ that produce an absolute manifestation of the gravitational interaction. The tidal distortions of the Earth by the Moon and the Sun are caused by precisely these relative accelerations, and so one frequently hears the statement that absolute gravity is *tidal* gravity.

We should be more honest and point out that even within Cliff's frame or Eric's frame, an object does not move precisely on a straight line. Had we expanded the acceleration of the object, $\boldsymbol{g}(\boldsymbol{r}_C + \boldsymbol{\xi}_C)$, to higher order in $\boldsymbol{\xi}_C$, we would have discovered that its acceleration relative to Cliff is actually given by

$$\frac{d^2\xi_C^j}{dt^2} = (\partial_{jk}U)\xi_C^k. \tag{5.6}$$

So the inhomogeneity of the gravitational field can manifest itself even within Cliff's or Eric's freely-moving frame. This is why we insist that the frame be *local* in order to apply the Einstein equivalence principle: It must be small enough that the effects of the relative acceleration described by Eq. (5.6) are negligible.

Box 5.1 **Uniform gravitational fields**

The principles of equivalence enunciated above are not what Einstein first named the principle of equivalence. Einstein's original formulation stated that *physics in a static and uniform gravitational field is equivalent to physics without gravity in a uniformly accelerated frame of reference*. This formulation played an essential role in guiding him toward general relativity. By considering the propagation of a light ray in an accelerating frame, and equating that to its behavior in a uniform gravitational field, he was led to the gravitational redshift effect in 1907 (his "happiest thought," as he later recounted) and to the deflection of light in 1911. These insights then led him to consider gravity as being linked to the geometry of curved spacetime, which was the critical step he needed to develop the full theory.

But his formulation is deeply flawed when taken literally. First, a uniform gravitational field does not exist in nature; in Newtonian gravity it would require as a source an infinite plane of matter in an otherwise empty universe. Second, static fields do not exist in the real world: stars rotate, planets orbit, supernovae explode, the universe expands. Third, the strict adoption of his principle has led to a pointless literature of apparent paradoxes, debates, and conundra.

The bottom line is that a uniform gravitational field is not a gravitational field at all. It is the "field" experienced by an observer undergoing constant acceleration in Minkowski spacetime, and nothing else. (See Misner, Thorne, and Wheeler's *Gravitation* or other books for a mathematical discussion of the accelerating frame, sometimes called a Rindler spacetime after Wolfgang Rindler, who elucidated many of its properties. See also Exercise 5.1.) From our point of view, it is not a gravitational field *precisely* because it is uniform. It possesses no inhomogeneities, and therefore is not gravity. Einstein's original formulation is not an equivalence, but a tautology.

It is possible, however, to restore respectability to Einstein's original version of the equivalence principle: simply insert the adjective "local" in suitable places. Any gravitational field is *approximately* uniform in a local region, and therefore Einstein's insights work equally well in a locally uniform field. But the use of "local" now permits the gravitational field to be a truly physical, inhomogeneous field on larger scales.

5.1.2 Metric theory of gravitation

As we have seen in the preceding subsection, the adoption of Einstein's equivalence principle produces a number of powerful statements. These include:

- gravity couples universally to matter, in a manner that is independent of mass, structure, and composition of small bodies;
- all local aspects of gravity are eliminated in freely-moving frames;
- freely-moving frames are inertial frames, and all special-relativistic formulations of the laws of physics continue to hold in freely-moving frames;
- freely-moving frames are unaccelerated; acceleration measures departures from free motion produced by non-gravitational forces;
- freely-moving frames extend over small regions only, and inhomogeneities in the gravitational field prevent their extension beyond these small regions;
- absolute aspects of gravity are revealed in the field inhomogeneities, which forbid a smooth meshing of inertial frames that are widely separated.

The first four statements imply that there must exist a tensor field in spacetime that couples universally to all forms of matter and reduces, in each local inertial frame attached to a freely-moving observer, to the Minkowski metric $\eta_{\mu\nu}$ of special relativity. This tensor field is denoted $g_{\alpha\beta}$ and named the *metric tensor* of the (now curved) spacetime. A relativistic theory of gravity shall be a theory of a dynamical metric tensor in spacetime. The last two statements imply that while the spacetime metric $g_{\alpha\beta}$ can be reduced locally to the Minkowski metric $\eta_{\mu\nu}$, it must differ from it in its global aspects.

Suppose that ξ^μ is a local Lorentzian coordinate system that covers a small inertial frame S attached to a freely-moving observer in a gravitational field. In these coordinates, the spacetime interval between two neighboring events within S is given by $ds^2 = \eta_{\mu\nu} d\xi^\mu d\xi^\nu$. The local coordinates ξ^μ are related to a global coordinate system x^α that covers a portion of spacetime that is much larger than the small neighborhood of the inertial observer. These global coordinates are a priori arbitrary, merely providing labels to spacetime events, and they will not be Lorentzian in the presence of field inhomogeneities. The relation can be expressed as $\xi^\mu = f^\mu(x^\alpha)$, where f^μ are functions of the global coordinates. By differentiation we have $d\xi^\mu = (\partial_\alpha f^\mu)\, dx^\alpha$, and the spacetime interval can be expressed as $ds^2 = (\eta_{\mu\nu} \partial_\alpha f^\mu \partial_\beta f^\nu) dx^\alpha dx^\beta$ in terms of the global coordinates. The quantity within brackets converts a coordinate increment dx^α to the spacetime interval ds^2, which is absolute, independent of the choice of reference frame and coordinate system. This conversion is precisely the role of a metric tensor, and we conclude that the spacetime interval is given by

$$ds^2 = g_{\alpha\beta}\, dx^\alpha dx^\beta \tag{5.7}$$

in the global coordinate system. This defines $g_{\alpha\beta}$, and the association $g_{\alpha\beta} = \eta_{\mu\nu} \partial_\alpha f^\mu \partial_\beta f^\nu$ is the statement that, locally, the spacetime metric can be obtained from the Minkowski metric by a transformation from local coordinates ξ^μ attached to an inertial frame to the global coordinates x^α. As we shall see in Sec. 5.2.5, the reverse statement is also true: It is *always* possible to reduce the global metric $g_{\alpha\beta}$ to a local Minkowski form by implementing a coordinate transformation $x^\alpha = F^\alpha(\xi^\mu)$ from the global coordinates x^α to the local Lorentzian coordinates ξ^μ.

5.1.3 Newtonian gravity as warped time

The connection between gravitation and curved spacetime can be revealed vividly through the following argument, which is limited in scope to Newtonian situations – weak fields and slow motions. We begin with a thought experiment.

Suppose that a particle of mass m is initially at rest at a position $\boldsymbol{x}_0$ in a gravitational field. Its total energy is initially equal to mc^2. The particle is released, and it falls freely toward a new position $\boldsymbol{x}$. In the course of its motion the particle acquires a kinetic energy $\frac{1}{2}mv^2$, which is equal to the difference between the gravitational potential energies at the initial and final positions: $\frac{1}{2}mv^2 = mU(\boldsymbol{x}) - mU(\boldsymbol{x}_0)$. Its total energy becomes $mc^2[1 + U(\boldsymbol{x})/c^2 - U(\boldsymbol{x}_0)/c^2]$. In this Newtonian context, v is small compared with c, and $U/c^2 \ll 1$.

At this stage of the experiment, all of the particle's energy is converted into a photon of energy $\hbar\omega(\boldsymbol{x}) = mc^2[1 + U(\boldsymbol{x})/c^2 - U(\boldsymbol{x}_0)/c^2]$. The photon climbs back toward $\boldsymbol{x}_0$, and it arrives there with an energy $\hbar\omega(\boldsymbol{x}_0)$. In the final step of the experiment, the photon is converted into a particle that will remain at rest at $\boldsymbol{x}_0$; its mass m' is such that $m'c^2 = \hbar\omega(\boldsymbol{x}_0)$.

Energy conservation demands that $m' = m$; a different outcome would imply that energy has been created or lost in a cyclic process that could be repeated any number of times. The photon must therefore lose energy to the gravitational field as it makes its way from $\boldsymbol{x}$ to $\boldsymbol{x}_0$, and we set $\hbar\omega(\boldsymbol{x}_0) = mc^2$. These results imply that

$$\frac{\omega(\boldsymbol{x})}{\omega(\boldsymbol{x}_0)} = 1 + U(\boldsymbol{x})/c^2 - U(\boldsymbol{x}_0)/c^2. \tag{5.8}$$

This simplifies when $\boldsymbol{x}_0$ is taken to be at infinity, where $U(\boldsymbol{x}_0) = 0$. We have then *Einstein's redshift formula*,

$$\frac{\omega(\boldsymbol{x})}{\omega_\infty} = 1 + U(\boldsymbol{x})/c^2. \tag{5.9}$$

As a photon climbs out of a gravitational field, it loses energy, and its frequency decreases, so that $\omega_\infty/\omega(\boldsymbol{x}) \approx 1 - U(\boldsymbol{x})/c^2 < 1$. The photon's wavelength increases, and its color is shifted toward the red end of the spectrum.

We next turn this thought experiment, and the redshift formula, into a conclusion about the very structure of spacetime in the presence of gravitation. We take the bold point of view that the frequency shift described by Eq. (5.9) is a manifestation of the fact that *time flows at a rate that depends on position within a gravitational field.* And we describe this phenomenon mathematically by the equation

$$\frac{dt(\boldsymbol{x})}{dt_\infty} = 1 - U(\boldsymbol{x})/c^2, \tag{5.10}$$

which is obtained from the redshift formula by letting $\omega(\boldsymbol{x})/\omega_\infty \equiv dt_\infty/dt(\boldsymbol{x})$. Here $dt(\boldsymbol{x})$ is the increment of time between a given number of oscillations of the photon's electromagnetic wave, as measured by a static observer at position $\boldsymbol{x}$ in the gravitational field (this observer is not moving freely, and therefore not inertial), while dt_∞ is the increment of time between the same number of oscillations as measured by a static observer at infinity (because there is no gravity there, this observer is inertial). The fact that $dt(\boldsymbol{x}) \neq dt_\infty$ implies that the spacetime metric at position $\boldsymbol{x}$ cannot be equal to the Minkowski metric. If it were, then $ds^2 = -d(ct)^2 + dx^2 + dy^2 + dz^2$ and a measurement of proper time by a static observer at $\boldsymbol{x}$ would be equal to $\sqrt{-ds^2}/c = dt \equiv dt_\infty$, in contradiction with Eq. (5.10). We must have, instead,

$$ds^2 = g_{00}\, d(ct)^2 + dx^2 + dy^2 + dz^2, \tag{5.11}$$

so that a measurement of proper time now gives $\sqrt{-g_{00}}\, dt$. If we identify this with $dt(\boldsymbol{x})$ and use Eq. (5.10), we conclude that $\sqrt{-g_{00}} = 1 - U(\boldsymbol{x})/c^2$.

The metric associated with a Newtonian gravitational field is therefore given approximately by

$$ds^2 = -\big[1 - 2U(\boldsymbol{x})/c^2\big]d(ct)^2 + dx^2 + dy^2 + dz^2. \tag{5.12}$$

This assignment encodes the fact that time flows at a rate that depends on position within a gravitational field, a postulated physical phenomenon that naturally explains the redshift of photons. We call attention to the fact that in Eq. (5.12), $dt \equiv dt_\infty$ represents the increment of time as measured by an observer at rest at infinity, where $U = 0$.

The metric of Eq. (5.12) does more than just explain the gravitational redshift of light. It also produces all phenomena associated with Newtonian gravity. To prove this point it will suffice to examine the motion of a test body in a spacetime with this metric.

As we saw back in Chapter 4, the free motion of a test body in Minkowski spacetime is determined by the vanishing of its acceleration, $a^\alpha = 0$, or equivalently by the extremum of its action $S := -mc^2 \int d\tau = \int L\, dt$, with associated Lagrangian

$$L = -mc\sqrt{-\eta_{\alpha\beta}\frac{dr^\alpha}{dt}\frac{dr^\beta}{dt}}. \tag{5.13}$$

But if we now use the metric of Eq. (5.12) instead of the Minkowski metric, we find that the Lagrangian becomes

$$L = -mc^2\sqrt{1 - 2U/c^2 - v^2/c^2}, \tag{5.14}$$

with $v^2 := \boldsymbol{v}\cdot\boldsymbol{v}$ and $\boldsymbol{v} = d\boldsymbol{r}/dt$. Expanding this to first order in the small quantities $(v/c)^2$ and U/c^2, the Lagrangian simplifies to

$$L = -mc^2 + \frac{1}{2}mv^2 + mU. \tag{5.15}$$

The first term is an irrelevant constant, the second term is the body's kinetic energy, and the third term is the gravitational potential energy. Substitution of this Lagrangian into the Euler–Lagrange equations gives rise to the familiar equations of motion of Newtonian gravity: $d^2\boldsymbol{r}/dt^2 = \boldsymbol{\nabla} U$, in which U is evaluated at $\boldsymbol{x} = \boldsymbol{r}(t)$ after differentiation. Newtonian gravity, therefore, is a manifestation of the fact that time flows at a rate that depends on position within a gravitational field.

5.2 Mathematics of curved spacetime

5.2.1 Metric

The lesson of the Einstein equivalence principle is that we must move from flat spacetime to curved spacetime to incorporate gravity into the laws of relativistic physics. Accordingly we must replace the Minkowski metric $\eta_{\alpha\beta}$ with a more general metric tensor $g_{\alpha\beta}$. This tensor is still symmetric under an exchange of its indices, but it is no longer diagonal; in general it possesses ten independent components, which are all functions of the spacetime coordinates x^α. The metric plays the same role as in Minkowski spacetime: it converts coordinate displacements dx^α, which evidently depend on the choice of coordinates, to an invariant spacetime interval ds^2. The formula is

$$ds^2 = g_{\alpha\beta}\, dx^\alpha dx^\beta. \tag{5.16}$$

The metric therefore achieves two purposes: it encodes geometrical information about the coordinate system, and it encodes physical information about the gravitational field. It is usually not easy to distinguish between these different aspects of the metric.

Under a change of coordinates described by $x^\alpha = f^\alpha(x'^\mu)$, where f^α are functions of the new coordinates x'^μ, the coordinate displacements change according to

$$dx^\alpha = \frac{\partial f^\alpha}{\partial x'^\mu}\, dx'^\mu \tag{5.17}$$

and the spacetime interval becomes

$$ds^2 = g_{\alpha\beta} \frac{\partial f^\alpha}{\partial x'^\mu} \frac{\partial f^\beta}{\partial x'^\nu}\, dx'^\mu dx'^\nu. \tag{5.18}$$

This expression shows that the metric is replaced by

$$g'_{\mu\nu} = g_{\alpha\beta} \frac{\partial f^\alpha}{\partial x'^\mu} \frac{\partial f^\beta}{\partial x'^\nu} \tag{5.19}$$

in the new coordinate system, so that $ds^2 = g'_{\mu\nu}\, dx'^\mu dx'^\nu$. The coordinate displacements change, and the metric also changes, but ds^2 remains the same during a coordinate transformation.

The coordinate systems considered here are completely general, and the transformations between them are not limited to linear transformations, as they were for Lorentz transformations. It would not do to restrict our attention to Lorentzian coordinates, because although these can always be installed locally in an inertial frame attached to a freely-moving observer, in general they cannot be extended globally to cover a large portion of the spacetime; and a system of locally Lorentzian coordinates installed in one inertial frame would not mesh smoothly with another system installed in another inertial frame. In this new setting, therefore, vectors (such as dx^α) and tensors (such as $g_{\alpha\beta}$) must be allowed to transform under a completely general class of coordinate transformations.

The metric $g_{\alpha\beta}$ is still used to define the inner product between two vectors. If A^α and B^α are vectors in a curved spacetime (so that they transform like dx^α under a general coordinate transformation), then their inner product is defined to be $g_{\alpha\beta} A^\alpha B^\beta$. It is easy to show that the inner product is invariant under a general coordinate transformation; the changes in the vectors are compensated for by the changes in the metric tensor.

We still use the metric to lower indices on vectors. Given a vector A^α, we define its dual A_α by

$$A_\alpha := g_{\alpha\beta} A^\beta. \tag{5.20}$$

It is important to appreciate that in curved spacetime, this operation involves more than a mere change of sign; in general the functions A_α will be very different from A^α. The raising operation can also be defined, if we first introduce the inverse metric $g^{\alpha\beta}$. This is the inverse to the matrix formed by all the components of the metric $g_{\alpha\beta}$; its defining relation is

$$g^{\alpha\gamma} g_{\gamma\beta} = \delta^\alpha{}_\beta, \tag{5.21}$$

which is equivalent to the matrix equation $g^{-1}g = 1$. If A_α is a dual vector, then

$$A^\alpha := g^{\alpha\beta} A_\beta \tag{5.22}$$

is its associated vector.

5.2.2 Tensor calculus

Vectors and components

In flat spacetime the preferred coordinate systems are Lorentzian, and those are characterized by straight coordinate lines. The basis vectors associated with these coordinate systems are constant, and under these conditions there is no need to distinguish between the derivative of *a vector* and the derivative of *its components*. In a curved spacetime the situation is very different: the coordinate lines are no longer straight, the basis vectors change in length and direction from place to place, and these changes produce an important distinction between the derivative of a vector and that of its components. (This situation is not limited to curved spacetimes. It is familiar also in three-dimensional flat space, when we choose to work in curvilinear coordinates instead of Cartesian coordinates.)

As we develop this idea we shall have to distinguish carefully between a geometric vector $\vec{A}$ and its components A^α in a selected coordinate system x^α. While the components change under a coordinate transformation, the geometric vector (which would be represented as an arrow in familiar vector calculus) does not. A geometric vector is related to its components via a set of basis vectors $\vec{e}_\alpha$ that also depend on the coordinate system. These are defined so that the vector $d\vec{x}$ that describes an infinitesimal displacement from one event in spacetime to another is given by

$$d\vec{x} = \vec{e}_\alpha \, dx^\alpha, \tag{5.23}$$

in terms of the basis vectors and the coordinate displacements. An equivalent way of stating this is that $\vec{e}_\alpha = \partial\vec{x}/\partial x^\alpha$. An arbitrary vector $\vec{A}$ can then be expressed as

$$\vec{A} = A^\alpha \vec{e}_\alpha \tag{5.24}$$

in terms of the basis vectors and its components A^α.

The inner product between $d\vec{x}$ and itself is the spacetime invariant ds^2. We have $ds^2 = d\vec{x} \cdot d\vec{x} = (\vec{e}_\alpha \cdot \vec{e}_\beta)\, dx^\alpha dx^\beta$, and comparing this with Eq. (5.16) reveals that

$$g_{\alpha\beta} = \vec{e}_\alpha \cdot \vec{e}_\beta; \tag{5.25}$$

the metric can be obtained by computing the inner products between all pairs of basis vectors. From this it follows that if $\vec{A} = A^\alpha \vec{e}_\alpha$ and $\vec{B} = B^\beta \vec{e}_\beta$ are vectors, then their inner product is $\vec{A} \cdot \vec{B} = A^\alpha B^\beta (\vec{e}_\alpha \cdot \vec{e}_\beta) = g_{\alpha\beta} A^\alpha B^\beta$, as expected.

Covariant differentiation of vectors

Suppose now that $\vec{A}$ is a vector field in spacetime. Its derivative with respect to coordinate x^β is

$$\partial_\beta \vec{A} = (\partial_\beta A^\alpha)\vec{e}_\alpha + A^\alpha (\partial_\beta \vec{e}_\alpha), \tag{5.26}$$

in which the first term accounts for the variation of the components, while the second accounts for the variation of the basis vectors. We write

$$\partial_\beta \vec{e}_\alpha = \Gamma^\mu_{\alpha\beta}\, \vec{e}_\mu, \tag{5.27}$$

which states the obvious fact that a change in basis vector induced by a coordinate displacement is itself a vector that can be decomposed in terms of basis vectors. Equation (5.27) provides a definition for the quantities $\Gamma^\mu_{\alpha\beta}$, which are known as *Christoffel symbols*. Because $\partial_\beta \vec{e}_\alpha = \partial^2 \vec{x}/\partial x^\alpha \partial x^\beta = \partial_\alpha \vec{e}_\beta$, we have that

$$\Gamma^\mu_{\beta\alpha} = \Gamma^\mu_{\alpha\beta}; \tag{5.28}$$

the Christoffel symbols are symmetric in their lower indices.

With Eq. (5.27) our previous expression for $\partial_\beta \vec{A}$ becomes $\partial_\beta \vec{A} = (\nabla_\beta A^\mu)\vec{e}_\mu$, where

$$\nabla_\beta A^\mu := \partial_\beta A^\mu + \Gamma^\mu_{\alpha\beta} A^\alpha \tag{5.29}$$

is known as the *covariant derivative* of the vector components A^μ. The covariant derivative accounts for the changes in the components as well as the changes in the basis vectors as the vector field is moved from one event in spacetime to the next. Note the distinction between the covariant derivative of a component, symbolized by ∇_β, and the partial derivative, symbolized by ∂_β.

The Christoffel symbols can be expressed neatly in terms of the metric. To produce this expression we differentiate the relation $g_{\alpha\beta} = \vec{e}_\alpha \cdot \vec{e}_\beta$ with respect to x^γ and substitute Eq. (5.27); this yields

$$\partial_\gamma g_{\alpha\beta} = g_{\alpha\mu}\Gamma^\mu_{\beta\gamma} + g_{\beta\mu}\Gamma^\mu_{\alpha\gamma}. \tag{5.30}$$

By permuting the indices we can produce two alternative versions of this equation,

$$\partial_\alpha g_{\gamma\beta} = g_{\gamma\mu}\Gamma^\mu_{\beta\alpha} + g_{\beta\mu}\Gamma^\mu_{\gamma\alpha} \tag{5.31}$$

and

$$\partial_\beta g_{\gamma\alpha} = g_{\gamma\mu}\Gamma^\mu_{\alpha\beta} + g_{\alpha\mu}\Gamma^\mu_{\gamma\beta}. \tag{5.32}$$

We next add the last two equations, subtract the first, and make use of Eq. (5.28). We arrive at

$$\partial_\alpha g_{\gamma\beta} + \partial_\beta g_{\gamma\alpha} - \partial_\gamma g_{\alpha\beta} = 2g_{\gamma\mu}\Gamma^\mu_{\alpha\beta}, \tag{5.33}$$

and solve for $\Gamma^\mu_{\alpha\beta}$ by multiplying both sides of this equation by the inverse metric $g^{\gamma\nu}$. The end result is

$$\Gamma^\mu_{\alpha\beta} = \frac{1}{2}g^{\mu\nu}\left(\partial_\alpha g_{\nu\beta} + \partial_\beta g_{\nu\alpha} - \partial_\nu g_{\alpha\beta}\right). \tag{5.34}$$

This formula gives rise to a practical method to compute the Christoffel symbols, starting from the metric tensor $g_{\alpha\beta}$.

Box 5.2 **Vector calculus in polar coordinates**

As a simple illustration of the formalism developed so far, we consider a two-dimensional flat plane charted by polar coordinates (r, ϕ). These are defined in terms of the original Cartesian coordinates (x, y) by $x = r\cos\phi$ and $y = r\sin\phi$.

The position vector of a point in the plane is given by $\vec{x} = (r\cos\phi)\vec{e}_x + (r\sin\phi)\vec{e}_y$, where the Cartesian basis vectors $\vec{e}_x$ and $\vec{e}_y$ are constant. The polar basis $\vec{e}_r$ and $\vec{e}_\phi$ is given by

$$\vec{e}_r = \frac{\partial\vec{x}}{\partial r} = \cos\phi\,\vec{e}_x + \sin\phi\,\vec{e}_y,$$

$$\vec{e}_\phi = \frac{\partial\vec{x}}{\partial\phi} = -r\sin\phi\,\vec{e}_x + r\cos\phi\,\vec{e}_y.$$

The metric on the plane is calculated either by substituting the relations $dx = \cos\phi\,dr - r\sin\phi\,d\phi$, $dy = \sin\phi\,dr + r\cos\phi\,d\phi$ into $ds^2 = dx^2 + dy^2$, or by computing the inner products between the basis vectors $\vec{e}_r$ and $\vec{e}_\phi$. In both cases we get the non-vanishing components $g_{rr} = 1$ and $g_{\phi\phi} = r^2$, so that

$$ds^2 = dr^2 + r^2\,d\phi^2.$$

The Christoffel symbols are calculated either by differentiating the basis vectors, or by employing Eq. (5.34). In both cases we get the non-vanishing components $\Gamma^r_{\phi\phi} = -r$ and $\Gamma^\phi_{r\phi} = \Gamma^\phi_{\phi r} = r^{-1}$.

Covariant differentiation of tensors

The action of the covariant-derivative operator ∇_β can be extended to tensors. We postulate that: (i) when ∇_β acts on a scalar field, it produces the same result as the partial-derivative operator ∂_β; and (ii) the covariant-derivative operator obeys the product rule of differential calculus, $\nabla_\beta(AB) = (\nabla_\beta A)B + A(\nabla_\beta B)$, in which A and B are any tensorial quantities (with indices suppressed).

The action of ∇_β on a tensor $A^{\mu\nu}$ can be determined if we examine the special case $A^{\mu\nu} = A^\mu B^\nu$. By invoking the product rule we quickly arrive at $\nabla_\beta(A^\mu B^\nu) = \partial_\beta(A^\mu B^\nu) + \Gamma^\mu_{\alpha\beta}(A^\alpha B^\nu) + \Gamma^\nu_{\alpha\beta}(A^\mu B^\alpha)$. The generalization to arbitrary tensors is immediate:

$$\nabla_\beta A^{\mu\nu} = \partial_\beta A^{\mu\nu} + \Gamma^\mu_{\alpha\beta} A^{\alpha\nu} + \Gamma^\nu_{\alpha\beta} A^{\mu\alpha}. \tag{5.35}$$

This rule can easily be extended to tensors with an arbitrary number of indices; there is one Christoffel symbol per tensorial index.

To determine the action of ∇_β on a dual vector A_μ we examine $\nabla_\beta(A_\mu B^\mu)$, in which B^μ is an arbitrary vector. Here the covariant derivative acts on a scalar quantity, and $\nabla_\beta(A_\mu B^\mu) = \partial_\beta(A_\mu B^\mu) = (\partial_\beta A_\mu)B^\mu + A_\mu(\partial_\beta B^\mu)$. On the other hand, the product rule implies $\nabla_\beta(A_\mu B^\mu) = (\nabla_\beta A_\mu)B^\mu + A_\mu(\nabla_\beta B^\mu)$. If we equate these results and use Eq. (5.29) to express $\nabla_\beta B^\mu$ in terms of partial derivatives and Christoffel symbols, we get $(\nabla_\beta A_\mu)B^\mu = (\partial_\beta A_\mu - \Gamma^\alpha_{\mu\beta} A_\alpha)B^\mu$. Since B^μ is arbitrary, we must have

$$\nabla_\beta A_\mu = \partial_\beta A_\mu - \Gamma^\alpha_{\mu\beta} A_\alpha. \tag{5.36}$$

From this we easily obtain the action of ∇_β on a tensor $A_{\mu\nu}$,

$$\nabla_\beta A_{\mu\nu} = \partial_\beta A_{\mu\nu} - \Gamma^\alpha_{\mu\beta} A_{\alpha\nu} - \Gamma^\alpha_{\nu\beta} A_{\mu\alpha}, \tag{5.37}$$

and the extension of this rule to tensors of arbitrary ranks is immediate.

Comparison of Eqs. (5.30) with (5.37) reveals that

$$\nabla_\gamma g_{\alpha\beta} = 0. \tag{5.38}$$

The metric tensor is covariantly constant, and this important fact is often described by the statement that the covariant-derivative operator ∇_β is *compatible* with the metric. Equation (5.38) implies that the covariant derivative of the inverse metric vanishes also: $\nabla_\gamma g^{\alpha\beta} = 0$. These results imply that the operations of index raising and lowering, and covariant differentiation, commute with each other. For example, $\nabla_\beta A_\mu = \nabla_\beta(g_{\mu\nu}A^\nu) = g_{\mu\nu}\nabla_\beta A^\nu$. This observation is very powerful and produces helpful simplifications in long calculations.

Metric determinant and volume elements

We conclude this subsection with some results involving the metric determinant $g := \det[g_{\alpha\beta}]$. First we mention the identity

$$\Gamma^\mu_{\mu\beta} = \frac{1}{2} g^{\mu\nu}\partial_\beta g_{\mu\nu} = \frac{1}{\sqrt{-g}}\partial_\beta\sqrt{-g}, \tag{5.39}$$

which involves the contraction of the Christoffel symbol over two of its indices. This is established by calculating the change in g that is induced by a change in each component of the metric tensor. It gives rise to a convenient expression for the *covariant divergence* of a vector field A^α:

$$\nabla_\alpha A^\alpha = \frac{1}{\sqrt{-g}}\partial_\alpha\big(\sqrt{-g}A^\alpha\big); \tag{5.40}$$

this result follows by direct computation.

Our final observation is that in a curved spacetime, the invariant four-dimensional volume element is

$$dV = \sqrt{-g}\, d^4x, \tag{5.41}$$

where $d^4x = dx^0 dx^1 dx^2 dx^3$ is the element of coordinate volume. This result is a consequence of two facts. The first is that the metric determinant changes according to $g' = gJ^2$ under a coordinate transformation $x^\alpha = f^\alpha(x'^\mu)$, where $J := \det[\partial f^\alpha/\partial x'^\mu]$ is the Jacobian of the transformation; this property follows directly from Eq. (5.19). The second is that d^4x changes also, but in a subtle way because it is not simply an algebraic string of differentials, but an *oriented* string. In fact, its proper definition is $d^4x := dx^0 \wedge dx^1 \wedge dx^2 \wedge dx^3$, where the wedge operation indicates that interchanging any pair of differentials produces a minus sign. The volume element becomes

$$d^4x = \frac{\partial f^0}{\partial x'^\mu}\frac{\partial f^1}{\partial x'^\nu}\frac{\partial f^2}{\partial x'^\rho}\frac{\partial f^3}{\partial x'^\omega} dx'^\mu \wedge dx'^\nu \wedge dx'^\rho \wedge dx'^\omega \tag{5.42}$$

under a coordinate transformation. The indices μ, ν, ρ and ω must all be different, and in the sum over all repeated indices, there will be a plus or minus sign depending on the number of permutations required to get the wedge product into the canonical order $(0, 1, 2, 3)$ corresponding to d^4x'. This yields

$$d^4x = J\, d^4x'. \tag{5.43}$$

Combining these facts, we find that $\sqrt{-g}\, d^4x = \sqrt{-g'}\, d^4x'$, and we verify the statement of Eq. (5.41).

An elementary example of Eq. (5.41) involves the polar coordinates (r, ϕ) introduced in Box 5.2. Here we are talking about an element of two-dimensional surface area, and $d^2x = dr d\phi$. The metric determinant is positive and given by $g = r^2$. The surface element is therefore $\sqrt{g}\, d^2x = r dr d\phi$, the familiar result from elementary calculus.

5.2.3 Parallel transport and geodesic equation

Covariant differentiation on a world line

We examine a timelike world line $x^\alpha = r^\alpha(\tau)$ in a curved spacetime, parameterized by proper time τ, the time measured by a clock moving on this world line; as before this is defined by $d\tau = \sqrt{-ds^2}/c$. The world line's tangent vector is $u^\alpha = dr^\alpha/d\tau$, and this satisfies the normalization condition

$$g_{\alpha\beta} u^\alpha u^\beta = -c^2, \tag{5.44}$$

which replaces Eq. (4.11). On this world line there exists a vector field $\vec{A}(\tau)$, and we wish to evaluate the derivative of this vector with respect to τ.

We proceed much as in the preceding subsection. We decompose $\vec{A}$ in terms of basis vectors $\vec{e}_\alpha$ and we differentiate:

$$\frac{d\vec{A}}{d\tau} = \frac{dA^\alpha}{d\tau}\vec{e}_\alpha + A^\alpha \frac{d\vec{e}_\alpha}{d\tau}. \tag{5.45}$$

Because $d\vec{e}_\alpha/d\tau = (\partial_\beta \vec{e}_\alpha)(dr^\beta/d\tau) = u^\beta \Gamma^\mu_{\alpha\beta}\vec{e}_\mu$, this is

$$\frac{d\vec{A}}{d\tau} = \left(\frac{dA^\mu}{d\tau} + \Gamma^\mu_{\alpha\beta}A^\alpha u^\beta\right)\vec{e}_\mu. \tag{5.46}$$

The quantity within brackets is the covariant derivative of A^μ along the world line. We denote this

$$\frac{DA^\mu}{d\tau} := \frac{dA^\mu}{d\tau} + \Gamma^\mu_{\alpha\beta}A^\alpha u^\beta, \tag{5.47}$$

so that $d\vec{A}/d\tau = (DA^\mu/d\tau)\vec{e}_\mu$. When the vector field $\vec{A}$ is defined also in the neighborhood of the world line (and not just directly on the world line), it becomes a function of all the spacetime coordinates x^α (instead of just proper time τ); then $dA^\mu/d\tau = u^\beta\partial_\beta A^\mu$ and the covariant derivative can expressed as $DA^\mu/d\tau = u^\beta\nabla_\beta A^\mu$.

Parallel transport and geodesics

The vector $\vec{A}$ is *parallel-transported* along the world line when it stays constant in both direction and magnitude. The mathematical statement of this is $d\vec{A}/d\tau = 0$, or

$$\frac{DA^\mu}{d\tau} = 0 \qquad \text{(parallel transport)}. \tag{5.48}$$

A timelike world line $r^\alpha(\tau)$ is a *geodesic* of the curved spacetime when its own tangent vector $\vec{u}$ is parallel-transported along the world line. A geodesic, defined in this way, is everywhere *locally straight*. The mathematical statement of the geodesic equation is

$$\frac{Du^\mu}{d\tau} = 0, \tag{5.49}$$

or

$$\frac{du^\mu}{d\tau} + \Gamma^\mu_{\alpha\beta}u^\alpha u^\beta = 0, \tag{5.50}$$

or else

$$\frac{d^2r^\mu}{d\tau^2} + \Gamma^\mu_{\alpha\beta}\frac{dr^\alpha}{d\tau}\frac{dr^\beta}{d\tau} = 0. \tag{5.51}$$

This last form is a system of second-order differential equations for the functions $r^\mu(\tau)$. These equations admit a unique solution given initial conditions $r^\mu(0)$ and $u^\mu(0)$ at some initial time $\tau = 0$.

The definition of a geodesic – a world line that parallel-transports its own tangent vector – is very fundamental and nicely geometrical. It applies just as well to spacelike and null geodesics. It applies to many kinds of spaces or manifolds. It applies to the two-dimensional surface of a sphere, where the geodesics are known as "great circles." It also applies to very abstract spaces, such as the space defined by the operations of the rotation group.

To illustrate the physical meaning of the geodesic equation we return to the Newtonian metric of Eq. (5.12) and calculate its geodesics. It is a simple exercise to show that for this spacetime, the non-vanishing Christoffel symbols are $\Gamma^0_{0j} = \Gamma^j_{00} = -\partial_j(U/c^2)$. It follows that in Newtonian situations, the spatial components of Eq. (5.51) reduce to

$d^2\boldsymbol{r}/dt^2 - \boldsymbol{\nabla}U = 0$. These are the Newtonian equations of motion, and this result provides a suggestion that a freely-moving body in curved spacetime moves on a geodesic.

This connection between the geodesic equation and the dynamics of a freely-moving body is confirmed when we recognize that Eq. (5.49) looks a lot like the vanishing of the acceleration of a free particle in Minkowski space. Indeed, in a local inertial frame in which the Christoffel symbols vanish (whose existence will be justified in Sec. 5.2.5), the equation is precisely $du^\mu/d\tau = a^\mu = 0$. There is therefore a direct link between the straight-line motion of a free particle in a local inertial frame and geodesic motion.

Another way to see the connection is to observe that the geodesic equation (5.51) can be obtained on the basis of a variational principle, in which the action functional is the elapsed proper time $\int_1^2 d\tau$ along a parameterized curve $r^\alpha(\tau)$ linking the fixed events 1 and 2. If, as we proposed in Sec. 5.1.3, the particle action introduced back in Sec. 4.1.8 is generalized to

$$S = -mc^2 \int_1^2 d\tau = -mc \int_1^2 \sqrt{-g_{\alpha\beta}\frac{dr^\alpha}{dt}\frac{dr^\beta}{dt}}\, dt \tag{5.52}$$

in curved spacetime, then its extremization with respect to world-line variations returns the geodesic equation. (You will be asked to prove this statement in Exercise 5.8.) These considerations, therefore, strongly point to the geodesic equation as a description of free particle motion in curved spacetime.

Another useful statement of the geodesic equation is

$$\frac{du_\mu}{d\tau} = \frac{1}{2} u^\alpha u^\beta \partial_\mu g_{\alpha\beta}, \tag{5.53}$$

which can be obtained by lowering the index on Eq. (5.50) and simplifying the result. This form of the geodesic equation reveals immediately that when the metric does not depend on one (or more) of its coordinates x^μ, then the corresponding component u_μ of the velocity (dual) vector is a constant of the motion.

Null geodesics

Our considerations thus far have been limited to timelike world lines. As we shall see below in Box 5.6, photons also move on geodesics of a curved spacetime. These geodesics, however, are such that $ds^2 = 0$, and these cannot be parameterized by proper time τ. A lightlike geodesic is instead parameterized by $\lambda = \lim(\tau/m)$, and its tangent vector is identified with the photon's momentum vector p^μ, which satisfies the null condition $g_{\mu\nu}p^\mu p^\nu = 0$. We have that $p^\mu = dr^\mu/d\lambda$, and this satisfies the geodesic equation

$$\frac{dp^\mu}{d\lambda} + \Gamma^\mu_{\alpha\beta} p^\alpha p^\beta = 0. \tag{5.54}$$

A variant of Eq. (5.53) is also valid: $dp_\mu/d\lambda = \frac{1}{2}p^\alpha p^\beta \partial_\mu g_{\alpha\beta}$, so that p_μ is a constant of the motion when the metric does not depend on coordinate x^μ. The geodesic equation for photons can also be derived from a variational principle, but with a twist. One uses the action

$$S = \int_1^2 \left(g_{\alpha\beta} \frac{dr^\alpha}{d\lambda} \frac{dr^\beta}{d\lambda} \right) d\lambda, \tag{5.55}$$

where we have dropped the mass (which is zero), some minus signs and factors of c (which are now irrelevant), and the square root, which has the desirable effect of avoiding the presence of zeros in the denominator when applying the Euler–Lagrange equations. We observe that the photon action is identically zero when the action is evaluated on a null geodesic. But it deviates from zero when evaluated on displaced paths, and its extremization is a well-defined procedure. It is interesting to note that the extremum giving a null geodesic between events 1 and 2 is neither a maximum nor a minimum, but a saddle point.

5.2.4 Curvature tensors

Riemann tensor

The symmetry of the Christoffel symbols in the lower indices implies that the action of two covariant derivatives on a scalar field f is independent of their order:

$$\nabla_\alpha \nabla_\beta f - \nabla_\beta \nabla_\alpha f = 0. \tag{5.56}$$

The same is not true, however, when the covariant derivatives act on a vector field A^μ; in this case

$$\nabla_\alpha \nabla_\beta A^\mu - \nabla_\beta \nabla_\alpha A^\mu = R^\mu{}_{\nu\alpha\beta} A^\nu, \tag{5.57}$$

and the operations do not commute. This equation defines the *Riemann curvature tensor* $R^\mu{}_{\nu\alpha\beta}$. A lengthy evaluation of the left-hand side of Eq. (5.57) shows that the Riemann tensor is given explicitly by

$$R^\alpha{}_{\beta\gamma\delta} = \partial_\gamma \Gamma^\alpha{}_{\beta\delta} - \partial_\delta \Gamma^\alpha{}_{\beta\gamma} + \Gamma^\alpha_{\mu\gamma} \Gamma^\mu{}_{\beta\delta} - \Gamma^\alpha_{\mu\delta} \Gamma^\mu{}_{\beta\gamma}. \tag{5.58}$$

The Riemann tensor is evidently antisymmetric in its last two indices. It also possesses additional symmetries that are not immediately revealed by Eqs. (5.57) and (5.58). These are

$$R_{\alpha\beta\delta\gamma} = -R_{\alpha\beta\gamma\delta}, \tag{5.59a}$$

$$R_{\beta\alpha\gamma\delta} = -R_{\alpha\beta\gamma\delta}, \tag{5.59b}$$

$$R_{\gamma\delta\alpha\beta} = +R_{\alpha\beta\gamma\delta}, \tag{5.59c}$$

$$R_{\mu\alpha\beta\gamma} + R_{\mu\gamma\alpha\beta} + R_{\mu\beta\gamma\alpha} = 0. \tag{5.59d}$$

By virtue of these symmetries, the Riemann tensor possesses 20 independent components in a four-dimensional spacetime.

Another important set of identities satisfied by the Riemann tensor is

$$\nabla_\alpha R_{\mu\nu\beta\gamma} + \nabla_\gamma R_{\mu\nu\alpha\beta} + \nabla_\beta R_{\mu\nu\gamma\alpha} = 0. \tag{5.60}$$

These are known as the *Bianchi identities*, and as we shall see, they play a fundamental role in Einstein's general relativity.

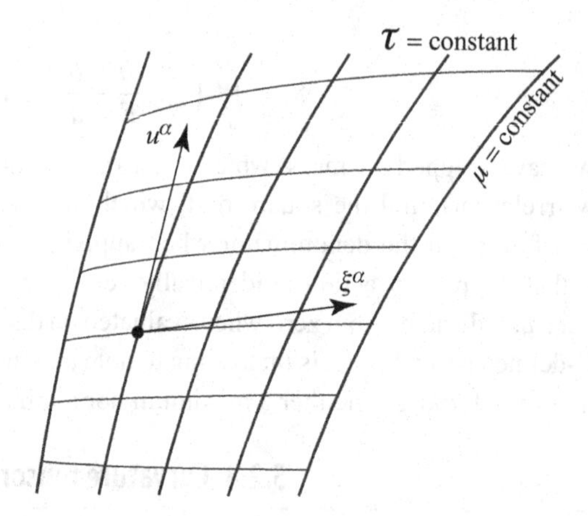

Fig. 5.1 Congruence of timelike geodesics. The vector u^α is tangent to each geodesic μ = constant. The deviation vector ξ^α is tangent to each curve τ = constant, and points from geodesic to geodesic.

Geodesic deviation

The geometrical meaning of the Riemann tensor is revealed most vividly by the *equation of geodesic deviation*, which governs the behavior of neighboring geodesics. We consider a continuous sequence of timelike geodesics parameterized by proper time τ, with each geodesic labelled by a parameter μ (refer to Fig. 5.1). This is sometimes called a *congruence* of timelike geodesics, and the entire congruence can be described by the parametric equations $x^\alpha = r^\alpha(\tau, \mu)$. When μ is kept fixed and τ varied in these equations, the displacement is along a selected geodesic within the congruence, and the geodesic's tangent vector is $u^\alpha = \partial r^\alpha / \partial \tau$. When, on the other hand, τ is kept fixed and μ varied, the displacement is across geodesics, and the vector $\xi^\alpha := \partial r^\alpha / \partial \mu$ is a *deviation vector* that points from geodesic to geodesic. We wish to derive an evolution equation for this deviation vector.

To begin we note that since u^α is defined as a vector field within the entire congruence, the geodesic equation can be expressed as $u^\beta \nabla_\beta u^\alpha = 0$. We note also that the definitions of u^α and ξ^α imply

$$\xi^\beta \partial_\beta u^\alpha - u^\beta \partial_\beta \xi^\alpha = \frac{\partial}{\partial \mu}\left(\frac{\partial r^\alpha}{\partial \tau}\right) - \frac{\partial}{\partial \tau}\left(\frac{\partial r^\alpha}{\partial \mu}\right) = 0, \tag{5.61}$$

and the equation can be re-expressed in covariant form as

$$\xi^\beta \nabla_\beta u^\alpha = u^\beta \nabla_\beta \xi^\alpha \tag{5.62}$$

by exploiting the symmetries of the Christoffel symbols. We next invoke Eq. (5.57) and write

$$\xi^\gamma u^\delta \left(\nabla_\gamma \nabla_\delta u^\alpha - \nabla_\delta \nabla_\gamma u^\alpha\right) = R^\alpha_{\ \beta\gamma\delta} u^\beta \xi^\gamma u^\delta. \tag{5.63}$$

We rewrite the first term on the left-hand side as

$$\xi^\gamma u^\delta \nabla_\gamma \nabla_\delta u^\alpha = \xi^\gamma \nabla_\gamma \left(u^\delta \nabla_\delta u^\alpha\right) - \xi^\gamma \left(\nabla_\gamma u^\delta\right)\left(\nabla_\delta u^\alpha\right), \tag{5.64}$$

and note that the first term on the right-hand side vanishes by virtue of the geodesic equation. Similarly, we rewrite the second term as

$$\xi^\gamma u^\delta \nabla_\delta \nabla_\gamma u^\alpha = u^\delta \nabla_\delta \bigl(\xi^\gamma \nabla_\gamma u^\alpha\bigr) - u^\delta \bigl(\nabla_\delta \xi^\gamma\bigr)\bigl(\nabla_\gamma u^\alpha\bigr), \tag{5.65}$$

and make use of Eq. (5.62) in the first term on the right-hand side. All of this produces

$$-R^\alpha_{\ \beta\gamma\delta} u^\beta \xi^\gamma u^\delta = u^\delta \nabla_\delta \bigl(u^\gamma \nabla_\gamma \xi^\alpha\bigr) - \bigl(u^\delta \nabla_\delta \xi^\gamma - \xi^\delta \nabla_\delta u^\gamma\bigr)\bigl(\nabla_\gamma u^\alpha\bigr), \tag{5.66}$$

and we see that the second term vanishes when we invoke Eq. (5.62) once more. The first term is recognized as $D^2\xi^\alpha/d\tau^2$, the second covariant derivative of the deviation vector along each geodesic within the congruence.

We have obtained the equation of geodesic deviation,

$$\frac{D^2\xi^\alpha}{d\tau^2} = -R^\alpha_{\ \beta\gamma\delta} u^\beta \xi^\gamma u^\delta, \tag{5.67}$$

which states that there is a relative acceleration between geodesics whenever the spacetime is curved, that is, whenever the Riemann curvature tensor is non-zero. To reflect on this we examine the situation in flat spacetime. Here the geodesics must be straight lines, and while these can diverge away from each other when they are not parallel, the rate at which they do so is necessarily constant. In this case the relative acceleration is zero, and this is compatible with Eq. (5.67) because the Riemann tensor vanishes in Minkowski spacetime. In a curved spacetime the situation is different: geodesics that start parallel to each other will eventually converge or diverge (think of two great circles on the surface of a sphere, starting parallel at the equator but converging together at both poles), and this effect is associated with a non-zero relative acceleration $D^2\xi^\alpha/d\tau^2$. Equation (5.67) then implies that the Riemann tensor cannot vanish when neighboring geodesics behave in this manner. We conclude that it is the Riemann tensor that measures the curvature of a four-dimensional spacetime.

Equation (5.67) is the precise statement of Eq. (5.6). The approximate version can be recovered by inserting $u^\alpha \simeq (c, \mathbf{0})$, and $\xi^\alpha \simeq (0, \boldsymbol{\xi})$ into Eq. (5.67), thus putting us in Cliff's momentary rest frame, and by using the Newtonian metric of Eq. (5.12) to show that $R_{0j0k} \simeq -\partial_{jk}(U/c^2)$. The spatial components of Eq. (5.67) reduce to

$$\frac{d^2\xi^j}{dt^2} \simeq -c^2 R^j_{\ 0k0}\xi^k \simeq (\partial_{jk}U)\xi^k, \tag{5.68}$$

and this equation describes the action of tidal gravitational forces in Cliff's laboratory.

The equation of geodesic deviation, in its exact formulation, carries the same interpretation: it describes the effects of tidal forces on neighboring observers that move freely in a gravitational field. The observers move on neighboring geodesics, their separation is described by the vector ξ^α, the tidal forces produce a relative acceleration $D^2\xi^\alpha/d\tau^2$, and since the tidal forces are caused by inhomogeneities in the gravitational field, we conclude that these are measured by the Riemann tensor. Note that a non-zero Riemann tensor unambiguously reveals the existence of a gravitational field; a non-trivial metric tensor does not, because the complexity of the metric could come entirely from the coordinates used to chart a flat spacetime. The Riemann tensor, therefore, is of fundamental importance in general relativity.

Ricci and Einstein tensors

Other curvature tensors can be obtained from the Riemann tensor. By contracting the first and third indices of the Riemann tensor we obtain the *Ricci tensor*

$$R_{\alpha\beta} := R^{\mu}{}_{\alpha\mu\beta}. \tag{5.69}$$

The symmetries of the Riemann tensor imply that $R_{\beta\alpha} = R_{\alpha\beta}$, and the Ricci tensor possesses ten independent components. By contracting its indices we obtain the *Ricci scalar*

$$R := R^{\mu}{}_{\mu} = g^{\alpha\beta} R_{\alpha\beta}. \tag{5.70}$$

Closely related to the Ricci tensor is the *Einstein tensor*

$$G_{\alpha\beta} := R_{\alpha\beta} - \frac{1}{2} g_{\alpha\beta} R, \tag{5.71}$$

which is also symmetric in its indices: $G_{\beta\alpha} = G_{\alpha\beta}$. The Einstein tensor possesses ten independent components, and its trace is given by $G := g^{\alpha\beta} G_{\alpha\beta} = -R$, because $g^{\alpha\beta} g_{\alpha\beta} = \delta^{\alpha}{}_{\alpha} = 4$.

The Bianchi identities of Eq. (5.60) give rise to

$$\nabla_{\beta} G^{\alpha\beta} = 0 \tag{5.72}$$

after two contractions of their indices. Equation (5.72) is known as the *contracted Bianchi identities*, and as we shall see, it plays a fundamental role in general relativity.

5.2.5 Curvature and the local inertial frame

Back in Sec. 5.1.2 we asserted that it is always possible to find a coordinate transformation $x^{\alpha} = F^{\alpha}(\xi^{\mu})$ that puts the metric $g_{\alpha\beta}$ into a local Minkowski form. We return to this issue here, and explicitly construct the coordinates that achieve this important goal. In fact, we shall construct two such coordinate systems. The first system, the *Riemann normal coordinates* ζ^{μ}, is such that at a selected event $\mathcal{O}$ in spacetime, the metric is equal to the Minkowski metric $\eta_{\mu\nu}$ and the Christoffel symbols $\Gamma^{\mu}_{\nu\lambda}$ vanish. The second system, the *Fermi normal coordinates* ξ^{μ}, is such that everywhere on a timelike geodesic γ, the metric is equal to the Minkowski form and the Christoffel symbols vanish. The Fermi coordinates are the mathematical embodiment of a local inertial frame in general relativity.

Riemann normal coordinates

We begin with the Riemann normal coordinates. The strategy we adopt to construct this coordinate system is very similar to what we might do in three-dimensional flat space to construct Cartesian coordinates. One way to proceed would involve the following steps. First, we select an origin $\mathcal{O}$. Second, we erect at $\mathcal{O}$ a set of three unit vectors $\boldsymbol{e}_{(j)}$ that point in mutually orthogonal directions; here the bracketed index serves to label each basis vector. Third, we select a point $\mathcal{P}$ and draw the straight line segment that links it to $\mathcal{O}$; the segment points in the direction of the unit vector $\boldsymbol{n}$, and $\mathcal{P}$ is at a distance r from the origin. Fourth, we decompose the vector $\boldsymbol{n}$ into the vector basis and express it as $\boldsymbol{n} = n^{j} \boldsymbol{e}_{(j)}$.

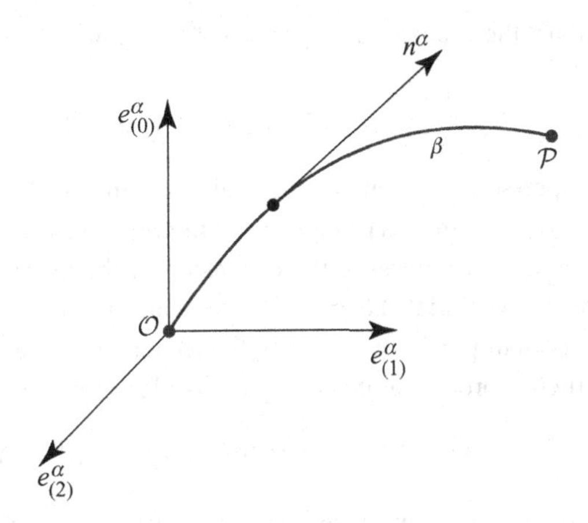

Fig. 5.2 Riemann normal coordinates about an event $\mathcal{O}$ in spacetime. The construction involves a geodesic segment β to which n^α is tangent.

And fifth, we assign to $\mathcal{P}$ the coordinates $x^j = rn^j$ and show that in these coordinates, the metric of three-dimensional flat space is given by δ_{jk}. The key features of this construction are that the coordinates are anchored to straight lines emanating from $\mathcal{O}$, and that they are defined directly in terms of the direction of each line and the distance measured along it. The generalization to curved spacetime suggests itself: use geodesics instead of straight lines, and define the coordinates in exactly the same way.

So let us now consider a curved spacetime with a metric $g_{\alpha\beta}$ initially expressed in an arbitrary coordinate system x^α. We select an event $\mathcal{O}$ in this spacetime, and at $\mathcal{O}$ we erect a set of four unit vectors $e^\alpha_{(\mu)}$ that point in mutually orthogonal directions (refer to Fig. 5.2); here the index α is the usual vector index that refers to the coordinates x^α, while the bracketed index (μ) is a label that allows us to distinguish each vector. The vector $e^\alpha_{(0)}$ is timelike, while the remaining vectors $e^\alpha_{(j)}$ are spacelike; the orthonormality condition can be compactly expressed by

$$g_{\alpha\beta} e^\alpha_{(\mu)} e^\beta_{(\nu)} = \eta_{\mu\nu}, \tag{5.73}$$

in which $\eta_{\mu\nu} := \text{diag}(-1, 1, 1, 1)$ is the Minkowski metric of flat spacetime.

We now select another event $\mathcal{P}$ in spacetime, and draw the geodesic segment β that links $\mathcal{P}$ to $\mathcal{O}$; this geodesic is assumed to be unique, and this usually requires $\mathcal{P}$ to be in a sufficiently close neighborhood of $\mathcal{O}$. The geodesic is parameterized by proper distance s if it is spacelike, so that $s = 0$ at $\mathcal{O}$ and $s = s_\mathcal{P}$ at $\mathcal{P}$, and by (rescaled) proper time $c\tau$ if it is timelike (with similar assignments at $\mathcal{O}$ and $\mathcal{P}$). Its tangent vector is the unit vector n^α, and at $\mathcal{O}$ this can be decomposed in the basis $e^\alpha_{(\mu)}$; we write

$$n^\alpha(\mathcal{O}) = n^\mu e^\alpha_{(\mu)}, \tag{5.74}$$

in which n^μ is a set of four coefficients that specify the direction of β relative to the vector basis. The *Riemann normal coordinates* of $\mathcal{P}$ are defined to be $\zeta^\mu := s_\mathcal{P} n^\mu$ when β is spacelike, and $\zeta^\mu := (c\tau_\mathcal{P}) n^\mu$ when β is timelike. As we show in Box 5.3, the construction implies that the spacetime metric near $\mathcal{O}$ becomes

$$g_{\mu\nu} = \eta_{\mu\nu} - \frac{1}{3}R_{\mu\lambda\nu\rho}(\mathcal{O})\zeta^\lambda\zeta^\rho + O(\zeta^3), \tag{5.75}$$

when expressed in Riemann normal coordinates. Because $\zeta^\mu = 0$ at $\mathcal{O}$, we see that $g_{\mu\nu}(\mathcal{O}) = \eta_{\mu\nu}$ so that as promised, the metric assumes the Minkowski form at the event $\mathcal{O}$. But we have more: because the expansion of the metric in powers of ζ^μ does not include linear terms, we find that $\partial_\lambda g_{\mu\nu}(\mathcal{O}) = 0$, which implies that *all Christoffel symbols vanish at* $\mathcal{O}$. This is an important property of the Riemann normal coordinates, and we have obtained a constructive proof that it is always possible to find coordinates that enforce the properties

$$g_{\mu\nu}(\mathcal{O}) = \eta_{\mu\nu}, \qquad \Gamma^\mu_{\nu\lambda}(\mathcal{O}) = 0, \tag{5.76}$$

at a selected event $\mathcal{O}$ in spacetime. The construction also shows that it is not possible to set the second derivatives of the metric tensor to zero at $\mathcal{O}$ unless the Riemann tensor vanishes there; a second-order deviation of the metric relative to the Minkowski values therefore signals the presence of curvature.

Box 5.3

Riemann normal coordinates

We wish to show that the construction of Riemann normal coordinates (RNC) detailed in the text leads to the metric of Eq. (5.75). The construction implies that the geodesic β that links $\mathcal{P}$ to $\mathcal{O}$ is described by

$$\zeta^\mu = sn^\mu \tag{1}$$

in RNC, and that $n^\mu = d\zeta^\mu/ds$ is a constant tangent vector on β. (To simplify the language we take β to be a spacelike geodesic.) When we vary the directions n^μ in these equations, we obtain other geodesics that also originate from $\mathcal{O}$, and

$$\eta^\mu_{(\nu)} := \frac{\partial\zeta^\mu}{\partial n^\nu} = s\delta^\mu_\nu \tag{2}$$

is a set of deviation vectors (one for each direction n^ν) that point from β to the displaced geodesics.

To show that $g_{\mu\nu} = \eta_{\mu\nu}$ at $\mathcal{O}$ we use the fact that n^μ is a unit vector. In RNC the mathematical statement of this is $g_{\mu\nu}n^\mu n^\nu = 1$, which is valid everywhere on β. In the original coordinates x^α we have instead $g_{\alpha\beta}n^\alpha n^\beta = 1$, in which we may insert Eqs. (5.73) and (5.74) to obtain $\eta_{\mu\nu}n^\mu n^\nu = 1$ at $\mathcal{O}$. Because the metric $g_{\mu\nu}$ at $\mathcal{O}$ must be independent of the directions n^μ, and because these are arbitrary, we conclude that $g_{\mu\nu}(\mathcal{O}) = \eta_{\mu\nu}$.

To show that the Christoffel symbols vanish at $\mathcal{O}$ we insert Eq. (1) within the geodesic equation and obtain $\Gamma^\mu_{\nu\lambda}n^\nu n^\lambda = 0$, which is valid everywhere on β. But because the Christoffel symbols at $\mathcal{O}$ must be independent of the directions n^μ, and because these are arbitrary, we conclude that $\Gamma^\mu_{\nu\lambda}(\mathcal{O}) = 0$. This, in turn, implies that $\partial_\lambda g_{\mu\nu}(\mathcal{O}) = 0$.

To relate the second derivatives of the metric at $\mathcal{O}$ to the Riemann tensor we apply the equation of geodesic deviation, Eq. (5.67), to each one of the deviation vectors of Eq. (2). Because the Christoffel symbols vanish at $\mathcal{O}$, we have that $\Gamma^\mu_{\nu\lambda} = \partial_\rho\Gamma^\mu_{\nu\lambda}(\mathcal{O})\zeta^\rho + \cdots$, and the covariant derivative of the deviation vectors along

β is given by

$$\frac{D\eta^{\mu}_{(\nu)}}{ds} = \delta^{\mu}_{\nu} + s^2 \partial_\rho \Gamma^{\mu}_{\nu\lambda}(\mathcal{O}) n^\lambda n^\rho + O(s^3).$$

The second covariant derivative is

$$\frac{D^2\eta^{\mu}_{(\nu)}}{ds^2} = 3s\, \partial_\rho \Gamma^{\mu}_{\nu\lambda}(\mathcal{O}) n^\lambda n^\rho + O(s^2),$$

and the equation of geodesic deviation becomes

$$\Big[3\partial_\rho \Gamma^{\mu}_{\nu\lambda}(\mathcal{O}) + R^{\mu}_{\ \lambda\nu\rho}(\mathcal{O})\Big] n^\lambda n^\rho + O(s) = 0.$$

Evaluating this at $\mathcal{O}$, and appealing once more to the arbitrariness of the directions n^μ, we arrive at

$$\partial_\lambda \Gamma^{\mu}_{\nu\rho} + \partial_\rho \Gamma^{\mu}_{\nu\lambda} = -\frac{1}{3}\Big(R^{\mu}_{\ \lambda\nu\rho} + R^{\mu}_{\ \rho\nu\lambda}\Big)$$

after properly symmetrizing the expression with respect to the indices λ and ρ. By permuting the indices we produce two alternative versions of this equation, which we add to and subtract from the original equation – refer to Eq. (5.30) for similar manipulations. This allows us to solve for the derivatives of the Christoffel symbols,

$$\partial_\rho \Gamma^{\mu}_{\nu\lambda}(\mathcal{O}) = -\frac{1}{3}\big(R^{\mu}_{\ \nu\lambda\rho} + R^{\mu}_{\ \lambda\nu\rho}\big),$$

in which the Riemann tensor is evaluated at $\mathcal{O}$. From this and Eq. (5.30) it is a simple matter to obtain

$$\partial_{\lambda\rho} g_{\mu\nu}(\mathcal{O}) = -\frac{1}{3}\big(R_{\mu\lambda\nu\rho} + R_{\mu\rho\nu\lambda}\big).$$

The metric of Eq. (5.75) is finally recovered by expanding $g_{\mu\nu}$ in powers of ζ^μ and inserting the expressions obtained here for the partial derivatives.

Fermi normal coordinates

We next turn to the Fermi normal coordinates. Instead of a single event $\mathcal{O}$ we now select an entire timelike geodesic γ in spacetime. The geodesic is parameterized by proper time τ, and everywhere on γ we erect a vector basis $e^{\alpha}_{(\mu)}$ that satisfies Eq. (5.73). We take $e^{\alpha}_{(0)}$ to be aligned with γ's tangent vector, and we assume that all vectors are parallel-transported along γ.

We next select an event $\mathcal{P}$ in spacetime, away from γ, and draw a spacelike geodesic β that passes through $\mathcal{P}$ and intersects γ orthogonally at $\mathcal{Q}$ (see Fig. 5.3); the requirement of orthogonality ensures that β is unique. The geodesic is parameterized by proper distance s, so that $s = 0$ at $\mathcal{Q}$ and $s = s_{\mathcal{P}}$ at $\mathcal{P}$, and its tangent vector is the unit vector n^α. At $\mathcal{Q}$ the vector can be decomposed as

$$n^\alpha(\mathcal{Q}) = n^j e^{\alpha}_{(j)}, \tag{5.77}$$

where $e^{\alpha}_{(j)}$ are the spatial members of the vector basis; the temporal member $e^{\alpha}_{(0)}$ is not involved because β is orthogonal to γ. The *Fermi normal coordinates* of $\mathcal{P}$ are defined to

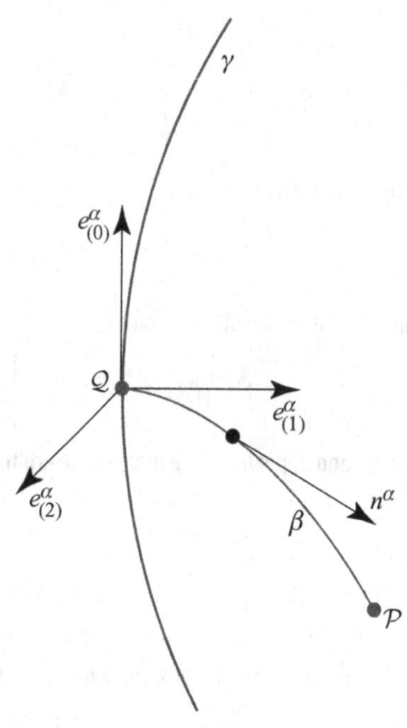

Fig. 5.3 Fermi normal coordinates about a timelike geodesic γ in spacetime. The construction involves a spacelike geodesic segment β which is orthogonal to γ, and to which n^α is tangent.

be $\xi^0 := c\tau_{\mathcal{Q}}$ and $\xi^j := s_{\mathcal{P}} n^j$. As we show in Box 5.4, the construction implies that the spacetime metric near γ becomes

$$g_{00} = -1 - R_{0p0q}(\gamma)\xi^p\xi^q + O(\xi^3), \tag{5.78a}$$

$$g_{0j} = \frac{2}{3} R_{jpq0}(\gamma)\xi^p\xi^q + O(\xi^3), \tag{5.78b}$$

$$g_{jk} = \delta_{jk} - \frac{1}{3} R_{jpkq}(\gamma)\xi^p\xi^q + O(\xi^3), \tag{5.78c}$$

when expressed in Fermi normal coordinates; here the Riemann tensor is evaluated on γ and expressed as a function of $\tau := \xi^0/c$. Because $\xi^j = 0$ everywhere on γ, we see that $g_{\mu\nu}(\gamma) = \eta_{\mu\nu}$, so that the metric assumes the Minkowski form on the entire timelike geodesic. Because the metric is constant on γ, and because its expansion away from γ does not include terms linear in ξ^j, we find that all first derivatives of the metric vanish on γ, which implies that the Christoffel symbols also vanish on γ. This amounts to a constructive proof that it is always possible to find coordinates that enforce the properties

$$g_{\mu\nu}(\gamma) = \eta_{\mu\nu}, \qquad \Gamma^\mu_{\nu\lambda}(\gamma) = 0, \tag{5.79}$$

everywhere along a selected timelike geodesic γ in spacetime. The Fermi normal coordinates provide the best mathematical realization of our notion of a freely-moving frame, in which gravity can be removed locally, up to effects associated with field inhomogeneities described by the Riemann tensor.

The significance of the Riemann tensor in Eq. (5.78) is clearly revealed when we work out the expression of the geodesic equation in Fermi normal coordinates. We examine a free particle moving on a world line $\xi^j = r^j(t)$ in a neighborhood of the reference geodesic γ; we adopt $t := \xi^0/c$ as the world-line parameter, and use the symbol t to emphasize the fact that while t is proper time on γ, it is not proper time on the particle's world line. The easiest way to obtain the geodesic equation is to insert the metric within $L = -mc\sqrt{-g_{\mu\nu}\dot{r}^\mu\dot{r}^\nu}$, the Lagrangian of Eq. (5.52). When we work consistently to second order in the spatial displacements r^j and velocities $\dot{r}^j := dr^j/dt$, we find that the Lagrangian simplifies to

$$L = -mc^2 + \frac{1}{2}mv^2 - \frac{1}{2}mc^2 R_{0j0k}r^jr^k + O(r^3), \tag{5.80}$$

where $v^2 := \delta_{jk}\dot{r}^j\dot{r}^k$. Substitution within the Euler–Lagrange equations produces

$$\frac{d^2r^j}{dt^2} = -c^2 R_{0j0k}r^k, \tag{5.81}$$

the statement of the geodesic equation in Fermi normal coordinates. Because $r^j(t)$ describes the displacement of the particle's world line relative to γ, it is not surprising that this takes the form of the equation of geodesic deviation, in the approximate version that was first displayed in Eq. (5.68). And this, of course, tells us once more that the Riemann tensor measures the inhomogeneities of the gravitational field, which prevent the metric of Eq. (5.78) from being constant beyond first order in the displacements ξ^j, and which prevent the particle's relative acceleration from vanishing in γ's inertial frame.

Box 5.4 Fermi normal coordinates

We wish to show that the construction of Fermi normal coordinates (FNC) detailed in the text leads to the metric of Eq. (5.78). We employ the Riemann normal coordinates (RNC), and rely on the metric of Eq. (5.75). Our strategy is to work out the transformation between the coordinate systems, and then to calculate how the metric changes under this transformation.

We suppose that the timelike geodesic γ passes through the origin $\mathcal{O}$ of the RNC, and that the basis vector $e^\alpha_{(0)}$ at $\mathcal{O}$ is aligned with γ's tangent vector. The vector basis at any other event on γ is then obtained by parallel transport, and our first task is to solve the equations of parallel transport for all four basis vectors. Setting $\tau = 0$ at $\mathcal{O}$, we expand the basis vectors in powers of τ,

$$e^\mu_{(\nu)}(\tau) = e^\mu_{(\nu)}(0) + \tau\,\dot{e}^\mu_{(\nu)}(0) + \frac{1}{2}\tau^2\,\ddot{e}^\mu_{(\nu)}(0) + O(\tau^3),$$

in which an overdot indicates differentiation with respect to τ. We invoke the definition of the RNC to set $e^\mu_{(\nu)}(0) = \delta^\mu_\nu$, insert the expansions within the equation of parallel transport,

$$\frac{de^\mu_{(\nu)}}{d\tau} + \Gamma^\mu_{\lambda\rho}e^\lambda_{(\nu)}\frac{d\zeta^\rho}{d\tau} = 0,$$

and solve order-by-order in τ. Because γ is described by $\zeta^0 = c\tau$ and $\zeta^j = 0$ in RNC, and because the Christoffel symbols are given by $\Gamma^\mu_{\nu\lambda} = -\frac{1}{3}(R^\mu_{\ \nu\lambda\rho} + R^\mu_{\ \lambda\nu\rho})\zeta^\rho + O(\zeta^2)$, we find that $\dot{e}^\mu_{(\nu)}(0) = 0$

and $\ddot{e}^{\mu}_{(\nu)}(0) = \frac{1}{3}c^2 R^{\mu}_{\ 0\nu 0}$, so that

$$e^{\mu}_{(\nu)}(\tau) = \delta^{\mu}_{\nu} + \frac{1}{6}(c\tau)^2 R^{\mu}_{\ 0\nu 0} + O(\tau^3).$$

This equation allows us to obtain the basis vectors at $\mathcal{Q}$.

Our next task is to launch a geodesic β that passes through the event $\mathcal{P}$ and intersects γ orthogonally at $\mathcal{Q}$. This geodesic is described by relations $\zeta^{\mu}(s)$ that can be obtained by integrating the geodesic equation,

$$\frac{d^2\zeta^{\mu}}{ds^2} + \Gamma^{\mu}_{\nu\lambda}\frac{d\zeta^{\nu}}{ds}\frac{d\zeta^{\lambda}}{ds} = 0.$$

Once more we express the solution as a Taylor expansion,

$$\zeta^{\mu}(s) = \zeta^{\mu}(0) + s\,\dot{\zeta}^{\mu}(0) + \frac{1}{2}s^2\,\ddot{\zeta}^{\mu}(0) + \frac{1}{6}s^3\,\dddot{\zeta}^{\mu}(0) + O(s^3),$$

and determine the various coefficients by inserting the expansion within the geodesic equation and solving order-by-order in s. The first two coefficients, actually, are determined by the boundary conditions at $\mathcal{Q}$: we must have that $\zeta^0(0) = c\tau_{\mathcal{Q}}$ and $\zeta^j(0) = 0$, and $\dot{\zeta}^{\mu}(0) \equiv n^{\mu}(\mathcal{Q})$ is given by Eq. (5.77) in terms of the direction coefficients n^j and the basis vectors $e^{\mu}_{(j)}$ obtained previously.

After some algebra we find that the solution to the geodesic equation can be expressed as

$$\zeta^0(s) = (c\tau_{\mathcal{Q}}) + \frac{1}{3}(c\tau_{\mathcal{Q}})R_{0p0q}(sn^p)(sn^q) + \cdots,$$

$$\zeta^j(s) = (sn^j) + \frac{1}{6}(c\tau_{\mathcal{Q}})^2 R^j_{\ 0p0}(sn^p) + \frac{1}{3}(c\tau_{\mathcal{Q}})R^j_{\ pq0}(sn^p)(sn^q) + \cdots.$$

This, without further ado, gives us the coordinate transformation, because the FNC of an event on β are given by $\xi^0 = c\tau_{\mathcal{Q}}$ and $\xi^j = sn^j$. We have

$$\zeta^0 = \xi^0 + \frac{1}{3}\xi^0 R_{0p0q}\xi^p\xi^q + \cdots,$$

$$\zeta^j = \xi^j + \frac{1}{6}(\xi^0)^2 R^j_{\ 0p0}\xi^p + \frac{1}{3}\xi^0 R^j_{\ pq0}\xi^p\xi^q + \cdots,$$

and the metric of Eq. (5.78) is obtained by inserting these expressions within

$$g_{\mu\nu}(\text{FNC}) = g_{\lambda\rho}(\text{RNC})\frac{\partial\zeta^{\lambda}}{\partial\xi^{\mu}}\frac{\partial\zeta^{\rho}}{\partial\xi^{\nu}},$$

the standard transformation law for the components of the metric tensor. The manipulations that lead to Eq. (5.78) are slightly tedious, but perfectly straightforward. A point of subtlety concerns the components of the Riemann tensor, which are evaluated at $\mathcal{O}$ in both the coordinate transformation and the RNC form of the metric. It is easy to see, however, that a displacement to $\mathcal{Q}$ does not change the metric components to second order in ζ^{μ}, nor the coordinate transformation to third order in ξ^{μ}. The Riemann tensor, therefore, can be safely evaluated at $\mathcal{Q}$ instead of $\mathcal{O}$, and because $\mathcal{Q}$ is an arbitrary event on γ, we conclude that the Riemann tensor is in fact a function of $\tau := \xi^0/c$ on the world line.

5.3 Physics in curved spacetime

5.3.1 From flat to curved spacetime

The rules to modify the laws of physics from their flat-spacetime formulation so that they hold in curved spacetime are exceedingly simple: start from a tensorial equation valid in any Lorentz frame, replace all occurrences of the Minkowski metric by $g_{\alpha\beta}$, and replace all partial derivatives by covariant derivatives; the result is a tensorial law valid in curved spacetime. The key to these rules is the important fact that according to the Einstein equivalence principle, the special-relativistic formulation of any law of physics is valid locally in any inertial frame attached to a freely-moving observer. All the hard work goes into rewriting this law in a form that is valid in any other reference frame, and in any coordinate system. And as we just saw, this work is really not that hard.

We prepare the way for this reformulation by recalling some useful mathematics from the preceding section. As we have seen, in the local (Fermi) coordinates ξ^μ, the metric on a timelike geodesic γ is equal to the Minkowski metric $\eta_{\mu\nu}$ and the Christoffel symbols $\Gamma^\lambda_{\mu\nu}$ vanish. There exists a transformation between the local coordinates ξ^μ and any system of global coordinates x^α; we write it as $\xi^\mu = f^\mu(x^\alpha)$, and these relations imply $d\xi^\mu = (\partial f^\mu/\partial x^\alpha)\,dx^\alpha$. The inverse transformation is $x^\alpha = F^\alpha(\xi^\mu)$, and these relations imply $dx^\alpha = (\partial F^\alpha/\partial \xi^\mu)\,d\xi^\mu$. It is easy to verify that

$$\frac{\partial F^\alpha}{\partial \xi^\mu}\frac{\partial f^\mu}{\partial x^\beta} = \delta^\alpha{}_\beta, \qquad \frac{\partial f^\mu}{\partial x^\alpha}\frac{\partial F^\alpha}{\partial \xi^\nu} = \delta^\mu{}_\nu, \tag{5.82}$$

follows from the preceding equations.

Up to terms involving the Riemann tensor, the spacetime interval in γ's immediate vicinity is given by $ds^2 = \eta_{\mu\nu}\,d\xi^\mu d\xi^\nu$ in the local coordinates. In the global coordinates it is $ds^2 = g_{\alpha\beta}\,dx^\alpha dx^\beta$, where

$$g_{\alpha\beta} = \eta_{\mu\nu}\frac{\partial f^\mu}{\partial x^\alpha}\frac{\partial f^\nu}{\partial x^\beta} \tag{5.83}$$

is an expression for the metric tensor. It is easy to verify that

$$g^{\alpha\beta} = \eta^{\mu\nu}\frac{\partial F^\alpha}{\partial \xi^\mu}\frac{\partial F^\beta}{\partial \xi^\nu} \tag{5.84}$$

is an appropriate expression for the inverse metric. We can compute the Christoffel symbols $\Gamma^\alpha_{\beta\gamma}$ by differentiating $g_{\alpha\beta}$ and inserting the result into Eq. (5.34). After simplification we obtain

$$\Gamma^\alpha_{\beta\gamma} = \frac{\partial F^\alpha}{\partial \xi^\mu}\frac{\partial^2 f^\mu}{\partial x^\beta \partial x^\gamma}. \tag{5.85}$$

To motivate the rules listed at the beginning of this section we examine the simple case of a particle moving freely in a curved spacetime. We already know the final outcome: this particle will move on a geodesic of the spacetime. We shall deduce this on the basis that in flat spacetime, a particle moving freely does so on a straight path. In special relativity, therefore, the particle's velocity vector u^μ satisfies $du^\mu/d\tau = 0$.

To generalize this law to curved spacetime we invoke the principle of equivalence and affirm that $du^\mu/d\tau = 0$ continues to apply locally in an inertial frame attached to any freely-moving observer. (This observer may or may not be moving together with the particle; we merely require that particle and observer come together at least once, at some selected event in spacetime.) In this frame $u^\mu = d\xi^\mu/d\tau$, and proper time τ continues to be defined in terms of the spacetime interval ds^2: $d\tau = \sqrt{-ds^2}/c$. We next carry out a transformation to the global coordinates x^α. We have that

$$u^\mu = \frac{\partial f^\mu}{\partial x^\alpha} u^\alpha, \tag{5.86}$$

where $u^\alpha = dx^\alpha/d\tau$ are the components of the velocity vector in the global coordinates. Differentiation with respect to τ gives

$$\begin{aligned} \frac{du^\mu}{d\tau} &= \frac{\partial f^\mu}{\partial x^\alpha}\frac{du^\alpha}{d\tau} + \frac{\partial^2 f^\mu}{\partial x^\beta \partial x^\gamma} u^\beta u^\gamma \\ &= \frac{\partial f^\mu}{\partial x^\alpha}\left(\frac{du^\alpha}{d\tau} + \frac{\partial F^\alpha}{\partial \xi^\nu}\frac{\partial^2 f^\nu}{\partial x^\beta \partial x^\gamma} u^\beta u^\gamma\right), \end{aligned} \tag{5.87}$$

where we used Eq. (5.82) in the second step. Invoking now Eq. (5.85), we recognize

$$\frac{du^\alpha}{d\tau} + \Gamma^\alpha_{\beta\gamma} u^\beta u^\gamma = \frac{Du^\alpha}{d\tau} \tag{5.88}$$

within the brackets, and we conclude that $du^\mu/d\tau = 0$ in the local coordinates becomes $Du^\alpha/d\tau = 0$ in the global coordinates. As expected, the particle moves on a geodesic of the curved spacetime.

The generalization of any other law of physics from flat to curved spacetime can always be accomplished by systematically running through the steps outlined in the preceding paragraph. The outcome is always the set of rules listed at the beginning of the section:

$$\eta_{\mu\nu} \to g_{\alpha\beta}, \qquad \partial_\mu \to \nabla_\alpha. \tag{5.89}$$

One can, in fact, obtain the rules without calculation. Instead of the route adopted previously, we could have started from the statement that $du^\mu/d\tau = 0$ holds in the local coordinates ξ^μ when the particle encounters the observer on the reference geodesic γ. We could have modified this immediately to $Du^\mu/d\tau = 0$, observing that the two forms differ by Christoffel symbols that vanish on γ. But this equation is in proper tensorial form in curved spacetime, and a transformation to any other coordinates system x^α will preserve its form; the equation becomes $Du^\alpha/d\tau = 0$ in the new coordinates, and once more we recover the geodesic equation.

As another application of the rules we may generalize Eq. (4.16) from flat to curved spacetime. The result is immediate, and we obtain the statement that

$$E_{\text{obs}} = -g_{\alpha\beta}\, p^\alpha u^\beta_{\text{obs}} \tag{5.90}$$

is the energy of a particle with momentum p^α as measured by an observer moving with velocity u^α_{obs}. This equation applies to massive particles as well as photons. In the case of a massive particle, the momentum vector is defined as in flat spacetime, $p^\alpha = mu^\alpha$, where

m is the particle's rest-mass and u^α is its own velocity vector. In the case of a photon, Eq. (5.90) generalizes Eq. (4.26).

It should be noted that the rule $\partial_\mu \to \nabla_\alpha$ always works when one generalizes a law of physics that involves first derivatives only. When second derivatives are involved, however, the rule can lead to ambiguities. This is because covariant derivatives do not commute when they act on vectors and tensors. Different orderings of $\partial_{\mu\nu}$ will therefore lead to different prescriptions that differ from each other by terms involving the Riemann tensor. Such ambiguities can usually be resolved by retreating to a more fundamental, first-order formulation of the law under consideration. When such a formulation does not exist, however, the ambiguity must be resolved differently, and this may require experimental input.

5.3.2 Hydrodynamics in curved spacetime

The curved-spacetime formulation of the laws of fluid mechanics involves the same variables as in flat spacetime, and most of these are defined in precisely the same way. Going through the list, we have the velocity field u^α, proper mass density ρ, proper density of internal (thermodynamic) energy ϵ, proper density of total energy $\mu = \rho c^2 + \epsilon$, pressure p, and mass current $j^\alpha = \rho u^\alpha$. The densities are all measured in an inertial frame that is momentarily comoving with a selected fluid element; this frame is attached to an observer moving freely in the gravitational field. We also have the fluid's energy-momentum tensor $T^{\alpha\beta}$, but its relation to the other fluid variables must be modified from Eq. (4.50) according to the rules of Eq. (5.89). We now have

$$T^{\alpha\beta} = (\mu + p)u^\alpha u^\beta/c^2 + pg^{\alpha\beta}, \tag{5.91}$$

which features the spacetime metric $g_{\alpha\beta}$ instead of $\eta_{\mu\nu}$.

The statement of mass conservation now takes the form

$$\nabla_\alpha j^\alpha = 0, \tag{5.92}$$

the curved-spacetime generalization of Eq. (4.47). Using Eq. (5.40) this can also be expressed as

$$\partial_\alpha\big(\sqrt{-g}\, j^\alpha\big) = 0, \tag{5.93}$$

and the factor of $\sqrt{-g}$ accounts for the dependence of volume elements on the metric determinant. Substitution of $j^\alpha = \rho u^\alpha$ into Eq. (5.92) yields

$$\frac{d\rho}{d\tau} + \rho\nabla_\alpha u^\alpha = 0, \tag{5.94}$$

which is the generalization of Eq. (4.53). Here, $d\rho/d\tau = u^\alpha\nabla_\alpha\rho$ is the (Lagrangian) derivative of the mass density along the world line of a selected fluid element.

The equation of energy-momentum conservation becomes

$$\nabla_\beta T^{\alpha\beta} = 0, \tag{5.95}$$

the generalization of the flat-spacetime version of Eq. (4.49). Insertion of Eq. (5.91) produces

$$\frac{d\mu}{d\tau} + (\mu + p)\nabla_\beta u^\beta = 0 \tag{5.96}$$

and

$$(\mu + p)\frac{Du^\alpha}{d\tau} + c^2\big(g^{\alpha\beta} + u^\alpha u^\beta/c^2\big)\nabla_\beta p = 0. \tag{5.97}$$

The first equation gives rise to

$$\rho\frac{d\epsilon}{d\tau} - (\epsilon + p)\frac{d\rho}{d\tau} = 0 \tag{5.98}$$

if we make use of Eq. (5.92), and we recognize this from Eq. (4.57) as a statement of the first law of thermodynamics for isentropic flows. The second equation is the curved-spacetime version of Euler's equation. In the first term we recognize $Du^\alpha/d\tau = u^\beta\nabla_\beta u^\alpha$ as the covariant acceleration of a selected fluid element, which would be zero if the fluid element were moving on a geodesic of the curved spacetime. This, however, is prevented by the pressure forces acting within the fluid; the fluid element is not moving freely in the gravitational field.

Box 5.5 Hydrostatic equilibrium

As a simple application of the laws of fluid mechanics in curved spacetime, we examine a static fluid configuration in a static spacetime. This fluid is in hydrostatic equilibrium, and we wish to find the relationship between the fluid's variables and the gravitational potential.

The metric of a static spacetime can always be put in the form

$$ds^2 = -e^{-2\Phi/c^2}\,d(ct)^2 + g_{jk}\,dx^j dx^k,$$

in terms of a gravitational potential $\Phi(x^j)$ and a spatial metric g_{jk} that also depends on the spatial coordinates x^j. The metric, and all the fluid variables, do not depend on the time coordinate t. In a weak-field situation we would have $\Phi/c^2 \ll 1$, and we would approximate g_{00} by $-1 + 2\Phi/c^2 + \cdots$; if we set $\Phi = U$ and $g_{jk} = \delta_{jk}$ we recover the Newtonian metric of Eq. (5.12).

In a static configuration the fluid elements do not move in the spatial directions, and the only non-vanishing component of the fluid's velocity field is u^0. The normalization condition $g_{\alpha\beta}u^\alpha u^\beta = -c^2$ implies that $u^0 = ce^{\Phi/c^2}$. This velocity field is not geodesic. It is easy to verify that the covariant acceleration is given by $Du^\alpha/d\tau = c^2e^{2\Phi/c^2}\Gamma^\alpha_{00}$, and a computation of the Christoffel symbols shows that $\Gamma^j_{00} = -c^{-2}e^{-2\Phi/c^2}g^{jk}\partial_k\Phi$ are the relevant, non-vanishing components. From this we obtain

$$\frac{Du^j}{d\tau} = -g^{jk}\partial_k\Phi$$

for the spatial components of the covariant acceleration.

The fluid elements are accelerated because they are are not moving freely in the gravitational field. Instead they are kept in place by the pressure forces acting within the fluid, which support it against gravity. The equation of hydrostatic equilibrium follows as a direct consequence of Euler's equation (5.97), which reduces, in our case, to $-(\mu + p)g^{jk}\partial_k\Phi + c^2g^{jk}\partial_k p = 0$. After lowering the j index, we arrive at

$$(\rho + p/c^2 + \epsilon/c^2)\partial_j\Phi = \partial_j p.$$

This is the (exact) relativistic statement of hydrostatic equilibrium. In Newtonian situations we may identify Φ with the Newtonian potential U and neglect the p/c^2 and ϵ/c^2 terms on the left-hand side; this gives rise to the familiar $\rho \nabla U = \nabla p$.

5.3.3 Electrodynamics in curved spacetime

The curved-spacetime formulation of Maxwell's theory involves an electromagnetic field tensor $F^{\alpha\beta}$ and a charge-current vector j_e^α. The statement of local charge conservation is $\nabla_\alpha j_e^\alpha = 0$, and Maxwell's equations take the form

$$\nabla_\beta F^{\alpha\beta} = \mu_0 j_e^\alpha, \qquad \nabla_\alpha F_{\beta\gamma} + \nabla_\gamma F_{\alpha\beta} + \nabla_\beta F_{\gamma\alpha} = 0. \tag{5.99}$$

The field tensor can be expressed in terms of a vector potential A^α. The relation is

$$F_{\alpha\beta} = \nabla_\alpha A_\beta - \nabla_\beta A_\alpha, \tag{5.100}$$

and this assignment ensures that the homogeneous Maxwell equations are automatically satisfied. The field tensor is left invariant under a gauge transformation of the form $A_\alpha \to A_\alpha + \nabla_\alpha \chi$, where χ is any scalar function of the spacetime coordinates x^α.

We may turn the inhomogeneous Maxwell equations into a set of wave equations for the vector potential. To achieve this we adopt the curved-spacetime version of the Lorenz gauge, in which the potential is required to satisfy the condition

$$\nabla_\alpha A^\alpha = 0. \tag{5.101}$$

We first insert Eq. (5.100) into the first of Eqs. (5.99) and get

$$\nabla_\beta \nabla^\alpha A^\beta - \nabla_\beta \nabla^\beta A^\alpha = \mu_0 j_e^\alpha. \tag{5.102}$$

The second term on the left-hand side involves the differential operator $\nabla_\beta \nabla^\beta = g^{\beta\gamma}\nabla_\beta \nabla_\gamma$, which reduces to $\eta^{\mu\nu}\partial_\mu \partial_\nu = -c^{-2}\partial^2/\partial t^2 + \nabla^2$ in flat spacetime; this is the curved-spacetime version of the wave operator, which we denote $\Box_g$. The first term can be simplified if we switch the order of the covariant derivatives, using the Riemann-tensor identity of Eq. (5.57). We have $\nabla_\beta \nabla_\alpha A^\beta = \nabla_\alpha \nabla_\beta A^\beta + R^\beta{}_{\nu\beta\alpha} A^\nu$, and the first term can be eliminated by virtue of the Lorenz-gauge condition. The second term involves the contraction of the Riemann tensor, which gives rise to the Ricci tensor $R_{\nu\alpha} = R_{\alpha\nu}$. All in all we have obtained

$$\Box_g A^\alpha - R^\alpha{}_\beta A^\beta = -\mu_0 j_e^\alpha, \tag{5.103}$$

which is the desired set of (coupled) wave equations for A^α. We recall that $\Box_g := g^{\beta\gamma}\nabla_\beta \nabla_\gamma$ is the covariant wave operator in curved spacetime.

We observe that Eq. (5.103) cannot be obtained directly from Eq. (4.67) by applying the rules of Eq. (5.89). The reason for this was already stated: the wave equation is a second-order differential equation, and the rules are ambiguous for such equations; different orderings of the derivatives lead to equations that differ from each other by Riemann-tensor terms. In the present case the ambiguity was resolved because the path leading to Eq. (5.103) originated from the first-order formulation of Maxwell's equations.

The energy-momentum tensor of the electromagnetic field is now

$$T^{\alpha\beta} = \frac{1}{\mu_0}\Big(F^{\alpha\mu}F^{\beta}_{\ \mu} - \frac{1}{4}g^{\alpha\beta}F_{\mu\nu}F^{\mu\nu}\Big), \tag{5.104}$$

and the exchange of energy and momentum between the field and the charge distribution is described by

$$\nabla_\beta T^{\alpha\beta} = -F^{\alpha}_{\ \beta} j^{\beta}_{e}. \tag{5.105}$$

This equation follows as a consequence of Maxwell's equations; the term on the right-hand side is (minus) the force density exerted on the charged fluid by the field.

Box 5.6 **Geometric optics**

Maxwell's equations imply that photons move on geodesics of a curved spacetime. In this context the term "photon" is employed in a classical sense, and designates a fictitious particle that follows the path of light rays. This is the domain of the *geometric-optics approximation* to Maxwell's theory, which is applicable when the characteristic wavelength associated with a field configuration is much smaller than any other scale of relevance, including the curvature scale set by the Riemann tensor. In our analysis we take these external scales to be of order unity, while the wavelength is taken to be much smaller than this.

The geometric-optics approximation is built into the following ansatz for the vector potential:

$$A^\alpha = \big[a^\alpha + i\epsilon b^\alpha + O(\epsilon^2)\big]e^{iS/\epsilon}.$$

The prefactor within square brackets is a slowly-varying, complex amplitude, while the exponential factor contains a rapidly-varying, real phase S/ϵ. The constant ϵ is a book-keeping parameter that we take to be small during our manipulations; at the end of our calculations we reset it to $\epsilon = 1$, so that S becomes the actual phase function. The amplitudes a^α and b^α, and the phase S, all depend on the spacetime coordinates x^α. It is understood that only the real part of A^α is physically significant.

In a local inertial frame attached to a freely-moving observer, the electromagnetic wave represented by A^α looks like a plane wave, and the phase function can be locally approximated by $S = -(\omega/c)\xi^0 + \boldsymbol{k}\cdot\boldsymbol{\xi} = k_\mu \xi^\mu$, where $k^\mu = (\omega/c, \boldsymbol{k}) = p^\mu/\hbar$ is the wave vector, related to the photon's momentum p^μ by a factor of $\hbar$. The wave vector may be defined by $k_\mu = \partial_\mu S$ in the local coordinates, and in the global coordinates x^α we shall have

$$k_\alpha = \partial_\alpha S.$$

The vector k^α is tangent to light rays, and we wish to show that it satisfies the null condition $k_\alpha k^\alpha = 0$ as well as the geodesic equation $k^\beta \nabla_\beta k^\alpha = 0$.

Differentiation of the vector potential gives

$$\nabla_\beta A^\alpha = \Big[\frac{i}{\epsilon}k_\beta a^\alpha - k_\beta b^\alpha + \nabla_\beta a^\alpha + O(\epsilon)\Big]e^{iS/\epsilon}.$$

At order ϵ^{-1} the Lorenz-gauge condition of Eq. (5.101) implies that $k_\alpha a^\alpha = 0$, and at order ϵ^0 we find that $k_\alpha b^\alpha = \nabla_\alpha a^\alpha$; the potential is orthogonal to the wave vector at leading-order only. A second differentiation reveals that

$$\Box_g A^\alpha = \left\{ -\frac{1}{\epsilon^2}(k_\beta k^\beta)a^\alpha + \frac{i}{\epsilon}\Big[(k_\beta k^\beta)b^\alpha + 2k^\beta \nabla_\beta a^\alpha + (\nabla_\beta k^\beta)a^\alpha\Big] + O(\epsilon^0) \right\} e^{iS/\epsilon}.$$

On the other hand, $R^\alpha_{\ \beta} A^\beta = O(\epsilon^0)$, and Maxwell's equations in the absence of charges imply that $\Box_g A^\alpha$ must vanish at orders ϵ^{-2} and ϵ^{-1}. At order ϵ^{-2} we obtain our null condition

$$k_\alpha k^\alpha = 0,$$

and at order ϵ^{-1} we get a differential equation for the amplitude a^α: $k^\beta \nabla_\beta a^\alpha = -\frac{1}{2}(\nabla_\beta k^\beta) a^\alpha$.

The geodesic equation follows immediately from the definition of k_α and the null condition. Differentiation of the null condition yields $0 = \nabla_\beta (k_\alpha k^\alpha) = 2k^\alpha \nabla_\beta k_\alpha = 2k^\alpha \nabla_\beta \nabla_\alpha S$. We next use the fact that two covariant derivatives commute when they act on a scalar function S, and rewrite our result as $0 = 2k^\alpha \nabla_\alpha \nabla_\beta S = 2k^\alpha \nabla_\alpha k_\beta$. This is

$$k^\alpha \nabla_\alpha k^\beta = 0,$$

the statement of the geodesic equation for light rays.

5.3.4 Point particles in curved spacetime

Very little needs to be changed when we generalize the description of point particles from Minkowski spacetime to curved spacetime. In the global coordinates x^α a particle moves on a world line $r^\alpha(\tau)$ parameterized by proper time τ, its velocity vector is $u^\alpha = dr^\alpha/d\tau$, and its mass density ρ continues to be represented by a Dirac delta function. This, however, must now come with an additional factor of $\sqrt{-g}$, to ensure that integrals such as

$$\int f(x^\alpha) \frac{\delta(x^\mu - r^\mu)}{\sqrt{-g}}\, dV$$

involve the correct, invariant volume element $dV := \sqrt{-g}\, d^4x$. The integral, we recall, returns $f(r^\alpha)$ if the event r^μ is contained within the domain of integration, and zero otherwise.

The mass density of a point particle of mass m is therefore given by

$$\rho(x^\alpha) = mc \int \frac{\delta\big(x^\mu - r^\mu(\tau)\big)}{\sqrt{-g}}\, d\tau, \tag{5.106}$$

its mass current by

$$j^\alpha = mc \int u^\alpha \frac{\delta\big(x^\mu - r^\mu(\tau)\big)}{\sqrt{-g}}\, d\tau, \tag{5.107}$$

and its energy-momentum tensor by

$$T^{\alpha\beta} = mc \int u^\alpha u^\beta \frac{\delta\big(x^\mu - r^\mu(\tau)\big)}{\sqrt{-g}}\, d\tau. \tag{5.108}$$

The delta function satisfies the distributional identity

$$u^\alpha \partial_\alpha \delta\big(x^\mu - r^\mu(\tau)\big) = -\frac{d}{d\tau}\delta\big(x^\mu - r^\mu(\tau)\big), \tag{5.109}$$

which was first established in Sec. 4.4.

With simple manipulations we can verify that the mass current has a vanishing divergence:

$$\begin{aligned}
\nabla_\alpha j^\alpha &= \frac{1}{\sqrt{-g}}\partial_\alpha\big(\sqrt{-g}\, j^\alpha\big) \\
&= \frac{mc}{\sqrt{-g}} \int u^\alpha \partial_\alpha \delta(x^\mu - r^\mu)\, d\tau \\
&= -\frac{mc}{\sqrt{-g}} \int \frac{d}{d\tau}\delta(x^\mu - r^\mu)\, d\tau \\
&= 0.
\end{aligned} \tag{5.110}$$

For the divergence of the energy-momentum tensor we get

$$\begin{aligned}
\nabla_\beta T^{\alpha\beta} &= \partial_\beta T^{\alpha\beta} + \Gamma^\beta_{\gamma\beta} T^{\alpha\gamma} + \Gamma^\alpha_{\gamma\beta} T^{\gamma\beta} \\
&= \frac{1}{\sqrt{-g}}\partial_\beta\big(\sqrt{-g}\, T^{\alpha\beta}\big) + \Gamma^\alpha_{\beta\gamma} T^{\beta\gamma} \\
&= \frac{mc}{\sqrt{-g}} \int u^\alpha u^\beta \partial_\beta \delta(x^\mu - r^\mu)\, d\tau + mc \int \Gamma^\alpha_{\beta\gamma} u^\beta u^\gamma \frac{\delta(x^\mu - r^\mu)}{\sqrt{-g}}\, d\tau;
\end{aligned}$$

this becomes

$$\nabla_\beta T^{\alpha\beta} = mc \int \left(\frac{du^\alpha}{d\tau} + \Gamma^\alpha_{\beta\gamma} u^\beta u^\gamma \right) \frac{\delta(x^\mu - r^\mu)}{\sqrt{-g}}\, d\tau \tag{5.111}$$

when we use the distributional identity within the first integral and integrate by parts. We recognize $Du^\alpha/d\tau$ within the integral, and deduce that $Du^\alpha/d\tau = 0$ when the particle does not exchange energy and momentum with an external agent (through the action of a force). In other words, the statement $\nabla_\beta T^{\alpha\beta} = 0$ implies that the particle is moving freely, and under these circumstances we recover the fact that the particle moves on a geodesic of the curved spacetime.

5.4 Einstein field equations

The Einstein field equations relate the curvature of spacetime to the distribution of matter within spacetime. They read

$$G^{\alpha\beta} = \frac{8\pi G}{c^4} T^{\alpha\beta}, \tag{5.112}$$

with the Einstein curvature tensor, Eq. (5.71), on the left-hand side, and the total energy-momentum tensor of all forms of matter (and fields) on the right-hand side. Taking into account the symmetries of the Einstein and energy-momentum tensors, the Einstein field equations are a set of ten second-order, partial differential equations for the metric tensor $g_{\alpha\beta}$. The equations are all coupled, and they are highly non-linear in the metric and its first derivatives; they are, however, linear in the second derivatives of the metric tensor.

A naive counting of the number of equations might suggest that, given suitable initial and boundary conditions for the metric, the solution to the Einstein field equations should be unique. This suggestion, however, is false, as the freedom to perform coordinate transformations must be retained; two metrics $g_{\alpha\beta}$ and $g'_{\mu\nu}$ related to each other by a coordinate transformation should both be valid solutions to the field equations. This freedom is guaranteed by the contracted Bianchi identities,

$$\nabla_\beta G^{\alpha\beta} = 0, \tag{5.113}$$

which reveal that of the ten field equations, only six are truly independent from each other. The Bianchi identities, together with the field equations, imply that

$$\nabla_\beta T^{\alpha\beta} = 0; \tag{5.114}$$

the energy-momentum tensor must have a vanishing divergence. This is a local statement of energy-momentum conservation for the matter (and field) distribution. This equation is of fundamental importance, because it determines (partially, if not fully) the equations of motion for the matter variables that make up the energy-momentum tensor. We witnessed this connection in the specific context of fluid mechanics in Sec. 5.3: we saw that the relativistic Euler equation (5.97) follows directly from Eq. (5.114).

The Einstein field equations have a fascinating structure. In the famous words of John Wheeler, the field equations (5.112) *tell spacetime how to curve*, given a description of the matter distribution through its energy-momentum tensor. At the same time, Eq. (5.114) *tells matter how to move* in a specified spacetime. But these equations are not independent, for they are intimately connected through the contracted Bianchi identities. Matter has no latitude: it must move in accordance to the spacetime that it produces.

This structure is unique to general relativity. In Maxwell's electrodynamics, for example, one can prescribe the motion of charges and determine the electromagnetic field that corresponds to this assumed distribution; no equations are violated when the charges do not respond to the field in a self-consistent manner, because external agents can always be introduced with the specific purpose of keeping the charges on their prescribed trajectories. In general relativity the motion of masses cannot be prescribed; it is necessarily dictated by the Einstein field equations. Any attempt to keep the masses on prescribed trajectories would involve external agents that are themselves a significant source of mass and energy; their contribution to the total $T^{\alpha\beta}$ would necessarily modify the spacetime in which the masses are moving.

In Einstein's theory the local statement of energy-momentum conservation, Eq. (5.114), is a consequence of the field equations, and the behavior of matter is intimately tied to the spacetime that the matter generates. Equation (5.114), however, should be expected to hold also in alternative theories of gravitation, in which the relationship between matter and curvature might involve additional mediators (such as a scalar field, in the case of scalar–tensor theories). The reason for this, as we shall elaborate below, resides in the fact that Eq. (5.114) is a direct consequence of the invariance of the matter action under general coordinate transformations. Because this principle of *general covariance* is a central aspect of any metric theory of gravitation (not just Einstein's), $\nabla_\beta T^{\alpha\beta} = 0$ is a fundamental equation that transcends its connection to the Einstein field equations and the Bianchi identities.

The equations of general relativity can be derived from an action principle. The total action S involves a gravitational piece given by the Hilbert action

$$S_{\rm grav} = \frac{c^3}{16\pi G} \int R\, dV, \tag{5.115}$$

where R is the Ricci scalar and $dV = \sqrt{-g}\, d^4x$ is the invariant volume element, as well as a matter piece given by

$$S_{\rm matter} = \int \mathscr{L}\, dV, \tag{5.116}$$

where $\mathscr{L}$ is the Lagrangian density for all matter (and field) variables. The Einstein tensor results from the functional variation of the gravitational action:

$$\frac{\delta S_{\rm grav}}{\delta g_{\alpha\beta}} = -\frac{c^3}{16\pi G} G^{\alpha\beta}. \tag{5.117}$$

The energy-momentum tensor, on the other hand, is defined by

$$\frac{\delta S_{\rm matter}}{\delta g_{\alpha\beta}} = \frac{1}{2c} T^{\alpha\beta}. \tag{5.118}$$

The Einstein field equations then follow from the requirement that $\delta(S_{\rm grav} + S_{\rm matter}) = 0$ under an arbitrary variation of the metric tensor. However, when we consider a specific variation $\delta g_{\alpha\beta}$ induced by a coordinate transformation, $\delta S_{\rm grav} = 0$ must follow as a matter of identity; it can be shown that the invariance of $S_{\rm grav}$ under such a variation gives rise to the Bianchi identities $\nabla_\beta G^{\alpha\beta} = 0$. And similarly, the invariance of $S_{\rm matter}$ under such a variation gives rise to Eq. (5.114). This result is the source of our earlier claim that $\nabla_\beta T^{\alpha\beta} = 0$ must hold in any metric theory of gravitation; it is a direct consequence of the principle of general covariance.

5.5 Linearized theory

The field equations of general relativity are exceedingly complicated, and it is a genuine miracle that so many exact solutions have been found over the years; but it is perhaps not a surprise that so few of these solutions possess an immediate physical significance. In the next section we shall explore the most famous of these exact solutions, the Schwarzschild metric, which describes the vacuum exterior to any spherical mass distribution. In this section we examine an approximate version of Einstein's theory, which applies when gravity is everywhere weak. As we shall see, there is more to weak-field gravity than was uncovered by Newton, and this exploration of the *linearized theory* will help us gain insight into the dynamical structure of general relativity. We warn the reader that our emphasis here is on conceptual matters; in the next chapters we develop more efficient ways of solving the Einstein field equations in weak-field situations, and we revisit topics introduced here with more depth and precision.

5.5.1 Metric and coordinate freedom

We consider a spacetime with a metric that can be expressed in the form

$$g_{\alpha\beta} = \eta_{\alpha\beta} + p_{\alpha\beta}, \tag{5.119}$$

where $\eta_{\alpha\beta}$ is the Minkowski metric, and where $p_{\alpha\beta}$ is a collection of ten independent fields that are each assumed to be small compared with unity. To keep track of the order of smallness we introduce the parameter $\epsilon \ll 1$, and we write $p_{\alpha\beta} = O(\epsilon)$. Equation (5.119) states that the spacetime metric deviates only slightly from the Minkowski metric, and this serves as our definition of a weakly-curved spacetime. This spacetime contains weak gravitational fields, and everywhere in this section we shall be working consistently to first order in ϵ, neglecting all terms of order ϵ^2 and higher. We shall refer to $\eta_{\alpha\beta}$ as the "background metric," and to $p_{\alpha\beta}$ as the "metric perturbation."

Equation (5.119) imposes a severe restriction on the coordinate freedom that is normally unlimited in the exact formulation of general relativity. We must restrict the coordinate transformations $x'^{\mu} = F^{\mu}(x^{\alpha})$ to those that preserve the decomposition of the metric into a dominant Minkowski piece and a small perturbation. The remaining coordinate freedom includes Lorentz transformations, which leave $\eta_{\alpha\beta}$ unchanged but alter $p_{\alpha\beta}$ as if it were a tensor field in flat spacetime, and small coordinate deformations of the form

$$x'^{\alpha} = x^{\alpha} + \zeta^{\alpha}(x^{\beta}), \tag{5.120}$$

where ζ^{α} is a vector field of order ϵ. Large coordinate transformations are excluded from our considerations.

To first order in ϵ the inverse to Eq. (5.120) is $x^{\alpha} = x'^{\alpha} - \zeta^{\alpha}(x'^{\beta})$, because the difference between $\zeta^{\alpha}(x'^{\beta})$ and $\zeta^{\alpha}(x^{\beta})$ is approximately $\zeta^{\beta}\partial_{\beta}\zeta^{\alpha}$, which is of order ϵ^2. This implies that $dx^{\alpha} = (\delta^{\alpha}{}_{\beta} - \partial_{\beta}\zeta^{\alpha})dx'^{\beta}$, and substitution into $ds^2 = g_{\alpha\beta}\, dx^{\alpha}dx^{\beta}$ reveals that the metric is given by

$$g'_{\alpha\beta} = \eta_{\alpha\beta} + p_{\alpha\beta} - \partial_{\alpha}\big(\eta_{\beta\mu}\zeta^{\mu}\big) - \partial_{\beta}\big(\eta_{\alpha\mu}\zeta^{\nu}\big) + O(\epsilon^2) \tag{5.121}$$

in the primed coordinates. This shows that the decomposition of Eq. (5.119) is indeed preserved, and that the perturbation becomes

$$p'_{\alpha\beta} = p_{\alpha\beta} - \partial_{\alpha}\zeta_{\beta} - \partial_{\beta}\zeta_{\alpha}, \qquad \zeta_{\alpha} := \eta_{\alpha\mu}\zeta^{\mu}, \tag{5.122}$$

in the new coordinates. Here we allow ourselves to lower the index on ζ^{μ} with the Minkowski metric $\eta_{\alpha\mu}$ instead of the full metric $g_{\alpha\mu}$, because the difference is of order ϵ^2; we shall make repeated use of this convention in what follows.

It is noteworthy that Eq. (5.122) looks like a natural tensorial generalization of $A_{\alpha} \to A'_{\alpha} = A_{\alpha} - \partial_{\alpha}\chi$, which is a gauge transformation of a vector potential A_{α} generated by a scalar field χ. For this reason the transformation of the metric perturbation under a small coordinate deformation is often referred to as a *gauge transformation* generated by the vector field ζ_α. We shall adopt this attractive terminology, but it is important to understand that the gauge freedom of the linearized theory is not a new symmetry of general relativity; it is simply the general covariance of the full theory restricted to the small coordinate deformations of Eq. (5.120).

5.5.2 Curvature and field equations

We next calculate the Christoffel symbols and curvature tensors associated with the metric of Eq. (5.119), working consistently to first order in ϵ. It is useful here to recall our previous convention that when an index is lowered on a quantity of order ϵ, the lowering can be accomplished with the Minkowski metric $\eta_{\alpha\beta}$; the same is true of index raising, which can be accomplished with $\eta^{\alpha\beta}$. This convention allows us to introduce

$$p^{\alpha\beta} := \eta^{\alpha\mu}\eta^{\beta\nu}p_{\mu\nu}, \qquad p := \eta^{\alpha\beta}p_{\alpha\beta}, \tag{5.123}$$

as convenient notations.

It is easy to verify that the inverse metric must be given by

$$g^{\alpha\beta} = \eta^{\alpha\beta} - p^{\alpha\beta}, \tag{5.124}$$

if $g^{\alpha\beta}$ is to respect its defining relation $g^{\alpha\mu}g_{\mu\beta} = \delta^\alpha{}_\beta$ up to terms of order ϵ^2. The Christoffel symbols are

$$\Gamma^\alpha_{\beta\gamma} = \frac{1}{2}\big(\partial_\beta p^\alpha{}_\gamma + \partial_\gamma p^\alpha{}_\beta - \partial^\alpha p_{\beta\gamma}\big), \tag{5.125}$$

and the Riemann tensor is

$$R_{\alpha\beta\gamma\delta} = \frac{1}{2}\big(\partial_{\beta\gamma}p_{\alpha\delta} - \partial_{\beta\delta}p_{\alpha\gamma} - \partial_{\alpha\gamma}p_{\beta\delta} + \partial_{\alpha\delta}p_{\beta\gamma}\big). \tag{5.126}$$

Under a gauge transformation we find that $\Gamma^\alpha{}_{\beta\gamma}$ changes by terms that depend on second derivatives of ζ_α, but the Riemann tensor is invariant:

$$R'_{\alpha\beta\gamma\delta} = R_{\alpha\beta\gamma\delta}. \tag{5.127}$$

This result illustrates the important fact that while the metric simultaneously encodes information about gravity and the adopted coordinate system, the Riemann tensor is all about gravitation.

From Eq. (5.126) we may form the Ricci tensor, Ricci scalar, and Einstein tensor, which are all gauge-invariant quantities. We obtain

$$R_{\alpha\beta} = -\frac{1}{2}\big(\Box p_{\alpha\beta} + \partial_{\alpha\beta}p - \partial_{\alpha\mu}p^\mu{}_\beta - \partial_{\beta\mu}p^\mu{}_\alpha\big), \tag{5.128}$$

$$R = -\Box p + \partial_{\mu\nu}p^{\mu\nu}, \tag{5.129}$$

and

$$G_{\alpha\beta} = -\frac{1}{2}\big(\Box p_{\alpha\beta} + \partial_{\alpha\beta}p - \partial_{\alpha\mu}p^\mu{}_\beta - \partial_{\beta\mu}p^\mu{}_\alpha\big) + \frac{1}{2}\eta_{\alpha\beta}\big(\Box p - \partial_{\mu\nu}p^{\mu\nu}\big). \tag{5.130}$$

Here we let $\Box := \eta^{\alpha\beta}\partial_{\alpha\beta}$ denote the flat-spacetime wave operator.

The form of the Einstein tensor can be simplified slightly if, instead of $p_{\alpha\beta}$, we express it in terms of the trace-reversed perturbation

$$\bar{p}_{\alpha\beta} := p_{\alpha\beta} - \frac{1}{2}\eta_{\alpha\beta}\, p. \tag{5.131}$$

The expression "trace-reversed" follows from the property that $\bar{p} = -p$, so that $p_{\alpha\beta} = \bar{p}_{\alpha\beta} - \frac{1}{2}\eta_{\alpha\beta}\,\bar{p}$. Making the substitution yields

$$G_{\alpha\beta} = -\frac{1}{2}\left(\Box\bar{p}_{\alpha\beta} - \partial_{\alpha\mu}\bar{p}^{\mu}{}_{\beta} - \partial_{\beta\mu}\bar{p}^{\mu}{}_{\alpha} + \eta_{\alpha\beta}\partial_{\mu\nu}\bar{p}^{\mu\nu}\right). \tag{5.132}$$

The Einstein field equations are

$$G^{\alpha\beta} = \frac{8\pi G}{c^4} T^{\alpha\beta}, \tag{5.133}$$

in which $T^{\alpha\beta}$, like $p_{\alpha\beta}$ and $G^{\alpha\beta}$, is imagined to be of order ϵ. The appropriate statement of the contracted Bianchi identities, to order ϵ, is

$$\partial_\beta G^{\alpha\beta} = 0, \tag{5.134}$$

and it is easy to verify that this does indeed follow automatically from Eqs. (5.130) or (5.132). Also to order ϵ, the appropriate statement of energy-momentum conservation is

$$\partial_\beta T^{\alpha\beta} = 0. \tag{5.135}$$

Physically, this equation implies that the matter fields that produce $T^{\alpha\beta}$ are allowed to exchange energy and momentum between themselves, *but not with the gravitational field*; such an exchange would be described by the exact equation $\nabla_\beta T^{\alpha\beta} = 0$, which differs from Eq. (5.135) by terms of order ϵ^2. Thus, in the linearized theory the dynamics of the matter fields cannot include gravity, and the theory cannot be applied to systems (like stars) that are bound together by gravitational forces. The dynamics must instead be dominated by non-gravitational forces. This is an important, and often unappreciated, limitation of the approximation.

5.5.3 Lorenz gauge

A substantial simplification of the Einstein tensor results when we exploit the coordinate freedom of Eq. (5.122) and impose the conditions

$$\partial^\beta\,\bar{p}_{\alpha\beta} = 0 \tag{5.136}$$

on the (trace-reversed) metric perturbation. This equation is a natural tensorial generalization of the Lorenz-gauge condition of electrodynamics, as expressed in Eq. (4.66); we therefore refer to Eq. (5.136) as the *Lorenz-gauge condition* for the gravitational potentials $p_{\alpha\beta}$. This choice of gauge can always be enforced. If we are presented with a $\bar{p}^{\rm old}_{\alpha\beta}$ that does not satisfy the Lorenz-gauge conditions, a gauge transformation will produce $\bar{p}^{\rm new}_{\alpha\beta} = \bar{p}^{\rm old}_{\alpha\beta} - \partial_\alpha\zeta_\beta - \partial_\beta\zeta_\alpha + \eta_{\alpha\beta}\partial_\mu\zeta^\mu$, and this will satisfy the gauge conditions when $0 = \partial^\beta\,\bar{p}^{\rm new}_{\alpha\beta} = \partial^\beta\,\bar{p}^{\rm old}_{\alpha\beta} - \Box\zeta_\alpha$. We see that the generator of the gauge transformation must satisfy the wave equation

$$\Box \zeta_\alpha = \partial^\beta \bar{p}^{\text{old}}_{\alpha\beta}, \tag{5.137}$$

and solutions to this equation can always be found. The solution, in fact, is not unique; to any solution ζ_α one can add the vector $\zeta_\alpha^{\text{hom}}$ provided that this satisfies the homogeneous version of the wave equation, $\Box\zeta_\alpha^{\text{hom}} = 0$. So while the Lorenz-gauge conditions can always be enforced, they do not completely specify the coordinate system. The conditions of Eq. (5.136) imply that of the ten original fields $p_{\alpha\beta}$, only six are independent.

With Eq. (5.136) the Einstein tensor becomes $G_{\alpha\beta} = -\frac{1}{2}\Box\bar{p}_{\alpha\beta}$, and the field equations take the form of a wave equation for the trace-reversed potentials:

$$\Box \bar{p}_{\alpha\beta} = -\frac{16\pi G}{c^4} T_{\alpha\beta}. \tag{5.138}$$

Solutions to this equation are easily obtained (the relevant techniques will be introduced in Chapter 6), and the metric is easily reconstructed from the potentials: $g_{\alpha\beta} = \eta_{\alpha\beta} + \bar{p}_{\alpha\beta} - \frac{1}{2}\eta_{\alpha\beta}\,\bar{p}$.

Because of its simplicity, Eq. (5.138) is an excellent starting point to an investigation of the nature of gravity in weak-field situations. In fact, in Chapter 6 we will recover the wave equation in an improved formulation of the theory, which involves a systematic expansion of the Einstein and energy-momentum tensors in powers of ϵ. So there is a lot that is good about Eq. (5.138), but one aspect that requires scrutiny is the suggested notion that all six independent degrees of freedom contained in $p_{\alpha\beta}$ are radiative degrees of freedom. If weak-field gravity is at all analogous to electrodynamics, we would expect instead that some degrees of freedom are non-radiative, and bound to the distribution of mass-energy (just as the Coulomb piece of the electric field is directly tied to the distribution of charge). A deeper analysis of the field equations will reveal that such is indeed the case: Of the six degrees of freedom, only two are radiative, while the remaining four are bound to the matter distribution. The suggestion from Eq. (5.138) that all degrees of freedom are radiative is merely an artifact of the choice of gauge.

5.5.4 Decomposition of the metric into irreducible pieces

We require a sophisticated decomposition of the metric perturbation $p_{\alpha\beta}$ that allows a clear identification of the degrees of freedom. To accomplish this we shall follow a nice treatment of the linearized theory that we learned from our friends Éanna Flanagan and Scott Hughes in their 2005 publication.

We first recall that under a Lorentz transformation, $p_{\alpha\beta}$ transforms as a tensor field in Minkowski spacetime. The general transformation includes a boost, in which the new frame moves with speed v relative to the old frame, and a rotation, in which the new spatial axes are rotated relative to the old frame. Here we are not interested in the boosts, and shall instead restrict our attention to pure rotations. The important observation for our purposes is that under a rotation of the spatial axes, p_{00} transforms as a scalar, p_{0j} transforms as a (Cartesian) vector, and p_{jk} transforms as a (Cartesian) tensor. We shall use this observation to decompose these quantities into their irreducible pieces (Box 5.7). Thus, p_{0j} will be decomposed into longitudinal and transverse pieces, while p_{jk} will be decomposed into trace, longitudinal-tracefree, longitudinal-transverse, and transverse-tracefree pieces. As we shall see, this decomposition clearly delineates the degrees of freedom contained in each component of $p_{\alpha\beta}$.

Box 5.7 **Decomposition of vectors and tensors into irreducible pieces**

We aim to show that a Cartesian vector field A^j can always be decomposed as

$$A^j = \partial^j A + A^j_{\rm T}, \qquad \partial_j A^j_{\rm T} = 0, \tag{1}$$

in terms of a *longitudinal piece* $\partial^j A$ and a *transverse piece* $A^j_{\rm T}$. Under some conditions that will be stated in a moment, this decomposition is unique. The origin of the terms "longitudinal" and "transverse" in this context will be clarified below. The longitudinal piece of A^j contains one independent component (the scalar A), and the transverse piece, by virtue of the condition $\partial_j A^j_{\rm T} = 0$, contains two independent components; these sum up to three independent components, the appropriate number for a vector field.

We shall show also that a *symmetric* tensor field B^{jk} can always be decomposed as

$$B^{jk} = \frac{1}{3}\delta^{jk} B + \left(\partial^{jk} - \frac{1}{3}\delta^{jk}\nabla^2\right)C + \partial^j C^k_{\rm T} + \partial^k C^j_{\rm T} + C^{jk}_{\rm TT}, \tag{2}$$

with the conditions

$$\partial_j C^j_{\rm T} = 0, \qquad \partial_k C^{jk}_{\rm TT} = 0 = \delta_{jk} C^{jk}_{\rm TT}. \tag{3}$$

The decomposition involves a trace piece $\frac{1}{3}\delta^{jk}B$, a longitudinal-tracefree piece $(\partial^{jk} - \frac{1}{3}\delta^{jk}\nabla^2)C$, a longitudinal-transverse piece $\partial^j C^k_{\rm T} + \partial^k C^j_{\rm T}$, and a transverse-tracefree piece $C^{jk}_{\rm TT}$. The six independent components of B^{jk} are contained in $B := \delta_{jk}B^{jk}$ (one component), C (one component), $C^j_{\rm T}$ (two components, by virtue of the divergence-free condition), and $C^{jk}_{\rm TT}$ (two components, by virtue of the three divergence-free conditions and the additional tracefree condition).

The uniqueness of the vector decomposition is established by noting that $\partial_j A^j = \nabla^2 A$. This is a Poisson equation for A, and the solution is unique, given A^j, provided that A goes to zero at infinity; this requires A^j to fall off sufficiently rapidly when $r := |\boldsymbol{x}| \to \infty$. Once A is found, Eq. (1) immediately gives $A^j_{\rm T}$.

The uniqueness of the tensor decomposition follows from similar considerations. We note first that $B = \delta_{jk}B^{jk}$ is obviously unique. Next we derive the equations $\partial_k B^{jk} = \frac{1}{3}\partial^j B + \frac{2}{3}\nabla^2\partial^j C + \nabla^2 C^j_{\rm T}$ and $\partial_{jk}B^{jk} = \frac{1}{3}\nabla^2 B + \frac{2}{3}\nabla^4 C$. With B determined, the second equation is a Poisson equation for $\nabla^2 C$, and the solution is unique provided that $\nabla^2 C \to 0$ when $r \to \infty$. With $\nabla^2 C$ known, C can be determined, and once more the solution is unique provided that $C \to 0$ when $r \to \infty$. Finally, the first equation is a Poisson equation for $C^j_{\rm T}$, and the solution is unique provided that $C^j_{\rm T} \to 0$ when $r \to \infty$. With B, C, and $C^j_{\rm T}$ determined, Eq. (2) gives $C^{jk}_{\rm TT}$.

To explain where Eqs. (1)–(3) and the associated terminology come from, we appeal to the Fourier transform. A vector field that is sufficiently well-behaved at infinity can always be decomposed as

$$A^j(\boldsymbol{x}) = \int \tilde{A}^j(\boldsymbol{k}) e^{i\boldsymbol{k}\cdot\boldsymbol{x}}\, d^3k,$$

in terms of plane waves $e^{i\boldsymbol{k}\cdot\boldsymbol{x}}$ with amplitudes $\tilde{A}^j(\boldsymbol{k})$. For each value of $\boldsymbol{k}$ in the domain of integration we can write $\tilde{A}^j = i\tilde{A}k^j + \tilde{A}^j_{\rm T}$, thereby decomposing the vector into a piece aligned with k^j – the longitudinal piece – and another piece orthogonal to k^j – the transverse piece; we demand that $ik_j\tilde{A}^j_{\rm T} = 0$. Substitution into the Fourier integral gives

$$\begin{aligned} A^j &= \int ik^j\tilde{A}e^{i\boldsymbol{k}\cdot\boldsymbol{x}}\,d^3k + \int \tilde{A}^j_{\rm T}e^{i\boldsymbol{k}\cdot\boldsymbol{x}}\,d^3k \\ &= \partial^j A + A^j_{\rm T}, \end{aligned}$$

where $A(\boldsymbol{x}) = \int \tilde{A}(\boldsymbol{k})e^{i\boldsymbol{k}\cdot\boldsymbol{x}}\,d^3k$. Note that the condition $ik_j\tilde{A}^j_{\rm T} = 0$ becomes $\partial_j A^j_{\rm T} = 0$ after evaluation of the Fourier integral. We have obtained the decomposition of Eq. (1) and clarified the meaning of the terms "longitudinal" and "transverse."

For a symmetric tensor field we begin with

$$B^{jk}(\boldsymbol{x}) = \int \tilde{B}^{jk}(\boldsymbol{k})e^{i\boldsymbol{k}\cdot\boldsymbol{x}}\,d^3k$$

and first decompose $\tilde{B}^{jk}$ into trace and tracefree pieces:

$$\tilde{B}^{jk} = \frac{1}{3}\delta^{jk}\tilde{B} + \tilde{C}^{jk},$$

where $\delta_{jk}\tilde{C}^{jk} = 0$. We next write $\tilde{C}^{jk} = -k^jk^k\tilde{C} + ik^j\tilde{C}^k_{\rm T} + ik^k\tilde{C}^j_{\rm T} + \tilde{C}^{jk}_{\rm T}$ and impose the conditions $ik_j\tilde{C}^j_{\rm T} = 0$ and $ik_k\tilde{C}^{jk}_{\rm T} = 0$; this provides a decomposition of $\tilde{C}^{jk}$ into a longitudinal piece $-k^jk^k\tilde{C}$, a longitudinal-transverse piece $ik^j\tilde{C}^k_{\rm T} + ik^k\tilde{C}^j_{\rm T}$, and a transverse piece $\tilde{C}^{jk}_{\rm T}$. This last tensor can still be decomposed into trace and tracefree pieces; we have $\tilde{C}^{jk}_{\rm T} = \frac{1}{3}\delta^{jk}\tilde{C}_{\rm T} + \tilde{C}^{jk}_{\rm TT}$, with $\delta_{jk}\tilde{C}^{jk}_{\rm TT} = 0$. Because $\tilde{C}^{jk}$ must itself be tracefree, we find that $\tilde{C}_{\rm T} = k^2\tilde{C}$, where $k^2 = \boldsymbol{k}\cdot\boldsymbol{k}$. All in all this gives us

$$\tilde{C}^{jk} = -\Big(k^jk^k - \frac{1}{3}\delta^{jk}k^2\Big)\tilde{C} + ik^j\tilde{C}^k_{\rm T} + ik^k\tilde{C}^j_{\rm T} + \tilde{C}^{jk}_{\rm TT}$$

for the tracefree piece of $\tilde{B}^{jk}$. Evaluating the Fourier integrals returns Eq. (2) for B^{jk} and the conditions of Eq. (3) for $C^j_{\rm T}$ and $C^{jk}_{\rm TT}$.

Following the rules of Box 5.7, the decomposition of the metric perturbation is accomplished by

$$p_{00} = 2U/c^2, \tag{5.139a}$$

$$p_{0j} = -4U_j/c^3 - \partial_j A/c, \tag{5.139b}$$

$$p_{jk} = 2\delta_{jk}V/c^2 + \Big(\partial_{jk} - \frac{1}{3}\delta_{jk}\nabla^2\Big)B + \big(\partial_j B_k + \partial_k B_j\big)/c^2 + h^{\rm TT}_{jk}, \tag{5.139c}$$

along with the conditions

$$\partial_j U^j = 0, \qquad \partial_j B^j = 0, \qquad \partial_k h^{jk}_{\rm TT} = 0 = \delta_{jk}h^{jk}_{\rm TT}. \tag{5.140}$$

The ten independent components of $p_{\alpha\beta}$ are contained in the potentials U (one component), U_j (two components), A (one component), V (one component), B (one component), B_j (two components), and $h^{\rm TT}_{jk}$ (two components). The various factors of 2, −4, and c that appear in Eq. (5.139) were inserted for later convenience.

We next figure out how the ten gravitational potentials that are contained in $p_{\alpha\beta}$ change under a gauge transformation. To achieve this we first decompose the generator ζ_α of the transformation into its own irreducible pieces. We write

$$\zeta_0 = \alpha/c, \tag{5.141a}$$

$$\zeta_j = 4\beta_j/c^2 + \partial_j\gamma, \qquad \partial^j\beta_j = 0, \tag{5.141b}$$

again inserting factors of c for convenience. Substitution of Eqs. (5.139) and (5.141) into Eq. (5.122) eventually returns

$$U' = U - \partial_t\alpha, \tag{5.142a}$$

$$U'_j = U_j + \partial_t\beta_j, \tag{5.142b}$$

$$V' = V - \frac{1}{3}c^2\nabla^2\gamma, \tag{5.142c}$$

$$h'^{\rm TT}_{jk} = h^{\rm TT}_{jk}, \tag{5.142d}$$

$$A' = A + \alpha + \partial_t\gamma, \tag{5.142e}$$

$$B' = B - 2\gamma, \tag{5.142f}$$

$$B'_j = B_j - 4\beta_j. \tag{5.142g}$$

We recall that $t := x^0/c$, and note that unlike all other gravitational potentials, the transverse-tracefree piece $h^{\rm TT}_{jk}$ is gauge-invariant.

5.5.5 Coulomb gauge and gauge-invariant potentials

Inspection of Eqs. (5.142) reveals that the variables

$$\Phi := U + \partial_t A + \frac{1}{2}\partial_{tt}B, \tag{5.143a}$$

$$\Phi_j := U_j + \frac{1}{4}\partial_t B_j, \tag{5.143b}$$

$$\Psi := V - \frac{1}{6}c^2\nabla^2 B, \tag{5.143c}$$

as well as $h^{\rm TT}_{jk}$, are all gauge-invariant. As such these potentials encode information that concerns the gravitational field only, and this information does not depend at all on the choice of coordinate system. The gauge-invariant potentials, therefore, represent the true degrees of freedom of the gravitational field. We have two scalar potentials Φ and Ψ, a vector potential Φ_j that contains two independent components (by virtue of the transverse condition $\partial^j\Phi_j = 0$), and a tensor potential $h^{\rm TT}_{jk}$ that also contains two independent components; the total number of degrees of freedom is six, and we have eliminated the four components of $p_{\alpha\beta}$ that encode coordinate information.

There exists a choice of gauge for which

$$A = B = B_j = 0 \qquad \text{(Coulomb gauge)}, \tag{5.144}$$

so that

$$p_{00} = 2U/c^2, \quad p_{0j} = -4U_j/c^3, \quad p_{jk} = 2\delta_{jk}V/c^2 + h^{\rm TT}_{jk} \qquad \text{(Coulomb gauge)}. \tag{5.145}$$

The potentials are constrained by the transverse condition $\partial_j U^j = 0$ and the transverse-tracefree conditions on $h^{\rm TT}_{jk}$. The condition $\partial_j U^j = 0$ is analogous to the choice of gauge defined by $\nabla \cdot \boldsymbol{A} = 0$ in electrodynamics, and for this reason we call the conditions of Eq. (5.144) the *Coulomb-gauge conditions*. We note the important property that in the Coulomb gauge, the gravitational potentials are equal to the gauge-invariant potentials:

$$\Phi = U, \qquad \Phi_j = U_j, \qquad \Psi = V \qquad \text{(Coulomb gauge)}. \tag{5.146}$$

This property makes the Coulomb gauge especially meaningful and convenient.

It is easy to see that the conditions of Eq. (5.144) can always be enforced. Suppose that we are presented with a $p^{\rm old}_{\alpha\beta}$, decomposed as in Eqs. (5.139), that does not satisfy the Coulomb-gauge conditions. A gauge transformation, decomposed as in Eqs. (5.141), produces a new perturbation $p^{\rm new}_{\alpha\beta}$, and the complete listing of changes is given by Eqs. (5.142). We have, in particular, $B^{\rm new}_j = B^{\rm old}_j - 4\beta_j$, and β_j can be chosen so as to set $B^{\rm new}_j = 0$. Similarly, $B^{\rm new} = B^{\rm old} - 2\gamma$, and γ can be chosen so that $B^{\rm new} = 0$. Finally, $A^{\rm new} = A^{\rm old} + \alpha + \partial_t \gamma$, and the condition $A^{\rm new} = 0$ determines α. We see that the Coulomb gauge can indeed be imposed, and moreover, we see that it completely specifies the coordinate system: ζ_α is completely determined by the gauge conditions.

5.5.6 Curvature and field equations (revisited)

It is a straightforward matter to insert Eqs. (5.145) into Eq. (5.126) and calculate the components of the Riemann tensor in the Coulomb gauge. But since the Riemann tensor is gauge-invariant, and the Coulomb-gauge potentials are equal to their gauge-invariant counterparts, one may simply make the substitutions $U \to \Phi$, $U_j \to \Phi_j$, and $V \to \Psi$ at the end of the calculation to obtain the general expression for the Riemann tensor, valid in any choice of gauge. This procedure yields

$$R_{0j0k} = -\frac{1}{c^2}\partial_{jk}\Phi - \frac{1}{c^4}\delta_{jk}\partial_{tt}\Psi - \frac{2}{c^4}\big(\partial_{tj}\Phi_k + \partial_{tk}\Phi_j\big) - \frac{1}{2c^2}\partial_{tt}h^{\rm TT}_{jk}, \tag{5.147a}$$

$$\begin{aligned} R_{0jkm} &= \frac{1}{c^3}\big(\delta_{jk}\partial_{tm}\Psi - \delta_{jm}\partial_{tk}\Psi\big) - \frac{2}{c^3}\big(\partial_{jk}\Phi_m - \partial_{jm}\Phi_k\big) \\ &\quad - \frac{1}{2c}\big(\partial_{tk}h^{\rm TT}_{jm} - \partial_{tm}h^{\rm TT}_{jk}\big), \end{aligned} \tag{5.147b}$$

$$\begin{aligned} R_{jkmn} &= \frac{1}{c^2}\big(\delta_{km}\partial_{jn}\Psi - \delta_{kn}\partial_{jm}\Psi - \delta_{jm}\partial_{kn}\Psi + \delta_{jn}\partial_{km}\Psi\big) \\ &\quad + \frac{1}{2}\big(\partial_{km}h^{\rm TT}_{jn} - \partial_{kn}h^{\rm TT}_{jm} - \partial_{jm}h^{\rm TT}_{kn} + \partial_{jn}h^{\rm TT}_{km}\big). \end{aligned} \tag{5.147c}$$

From this we next obtain the Einstein tensor, which is given by

$$G_{00} = -\frac{2}{c^2}\nabla^2\Psi, \tag{5.148a}$$

$$G_{0j} = -\frac{2}{c^3}\partial_{tj}\Psi + \frac{2}{c^3}\nabla^2\Phi_j, \tag{5.148b}$$

$$G_{jk} = -\frac{2}{3c^2}\delta_{jk}\nabla^2(\Phi - \Psi) - \frac{2}{c^4}\delta_{jk}\partial_{tt}\Psi + \frac{1}{c^2}\Big(\partial_{jk} - \frac{1}{3}\delta_{jk}\nabla^2\Big)(\Phi - \Psi)$$
$$+ \frac{2}{c^4}\big(\partial_{tj}\Phi_k + \partial_{tk}\Phi_j\big) - \frac{1}{2}\Box h^{\rm TT}_{jk}. \tag{5.148c}$$

We recall that $\Box := \eta^{\alpha\beta}\partial_{\alpha\beta}$ is the wave operator in flat spacetime. We recognize in Eq. (5.148) that the Einstein tensor is fully decomposed into its irreducible pieces, with G_{0j} containing longitudinal and transverse pieces, and G_{jk} containing trace, longitudinal-tracefree, longitudinal-transverse, and transverse-tracefree pieces.

Before we write down the Einstein field equations we should also decompose the energy-momentum tensor $T_{\alpha\beta}$ into its own irreducible pieces. We write

$$T^{00} = \varrho c^2, \tag{5.149a}$$

$$T^{0j} = (s^j + \partial^j s)c, \tag{5.149b}$$

$$T^{jk} = \tau\delta^{jk} + \Big(\partial^{jk} - \frac{1}{3}\delta^{jk}\nabla^2\Big)\sigma + \partial^j\sigma^k + \partial^k\sigma^j + \sigma^{jk}, \tag{5.149c}$$

and impose the conditions

$$\partial_j s^j = 0, \qquad \partial_j\sigma^j = 0, \qquad \partial_k\sigma^{jk} = 0 = \delta_{jk}\sigma^{jk}. \tag{5.150}$$

Here ϱ is the mass density of the matter distribution as measured by an observer at rest relative to the coordinate frame x^α; this should not be confused with ρ, the proper mass density of a perfect fluid. The vector $s^j + \partial^j s$ is the momentum density of the matter distribution, and T^{jk} is its stress tensor.

The decomposition of $T^{\alpha\beta}$ involves ten irreducible variables, but these cannot all be independent. A simple calculation reveals that $\partial_\beta T^{\alpha\beta} = 0$, the statement of energy-momentum conservation in the linearized theory, gives rise to the relations

$$\nabla^2 s = -\partial_t\varrho, \qquad \nabla^2\sigma^j = -\partial_t s^j, \qquad \nabla^2\sigma = -\frac{3}{2}(\partial_t s + \tau). \tag{5.151}$$

These imply that only ϱ, s^j, τ, and σ_{jk} – six independent variables in all – can be specified freely; the other four (s, σ_j, and σ) are determined by them.

Each irreducible piece of the Einstein tensor can be set equal to each irreducible piece of the energy-momentum tensor (multiplied by $8\pi G/c^4$). The independent pieces of the Einstein field equations are thus revealed to be

$$\nabla^2\Psi = -4\pi G\varrho, \tag{5.152a}$$

$$\nabla^2(\Phi - \Psi) = -\frac{12\pi G}{c^2}(\partial_t s + \tau), \tag{5.152b}$$

$$\nabla^2\Phi_j = -4\pi G s_j, \tag{5.152c}$$

$$\Box h^{\rm TT}_{jk} = -\frac{16\pi G}{c^4}\sigma_{jk}. \tag{5.152d}$$

We also find the additional equations $\partial_t \Psi = 4\pi G s$, $\Phi - \Psi = 8\pi G\sigma/c^2$, and $\partial_t \Phi_j = 4\pi G\sigma_j$, which are redundant by virtue of Eqs. (5.151).

Equations (5.152) are the culminating point of our considerations. They carry important insights into the nature of relativistic gravity. We note first that of the six degrees of freedom of the gravitational field, the four represented by Φ, Φ_j, and Ψ obey Poisson equations. Solutions for these potentials at a time t depend on the state of the matter variables at precisely the same time; the potentials march in step with the matter, much in the same way that the Newtonian potential U depends on the instantaneous profile of the mass density ρ. The remaining degrees of freedom, those represented by $h^{\rm TT}_{jk}$, behave very differently. These potentials obey wave equations, and their expression at a time t depends on the state of the matter variables at an earlier time t'; the delay allows for the propagation of a light signal from the source point $\boldsymbol{x}'$ to the field point $\boldsymbol{x}$. The main message is this: of the six degrees of freedom of the gravitational field, only two are radiative; the remaining four are not, and are directly tied to the matter distribution.

The gauge-invariant formulation of the linearized field equations is conceptually powerful because it cleanly separates the radiative from the non-radiative degrees of freedom. It does not, however, give rise to a practical method to integrate the field equations. Two difficulties arise. The first is that even if Eqs. (5.152) could be integrated explicitly for the gauge-invariant potentials Φ, Φ_j, Ψ, and $h^{\rm TT}_{jk}$, the problem would not be completely solved until *all* the gravitational potentials – U, U_j, V, $h^{\rm TT}_{jk}$, A, B, and B_j – are determined; this requires a choice of gauge and the integration of Eqs. (5.143). This first difficulty can be simply dealt with by adopting the Coulomb gauge and invoking Eqs. (5.146). The second difficulty is much more serious: to integrate Eqs. (5.152) one must first determine ϱ, s^j, τ, and σ^{jk}, the relevant irreducible pieces of the energy-momentum tensor. While this can always be done in principle, in most applications it is difficult and impractical. For this reason, the Lorenz-gauge formulation of the Einstein field equations, summarized by Eq. (5.138) on page 256, provides a much more user-friendly method of finding solutions.

5.5.7 Newtonian limit

The linearized theory developed in the preceding subsections possesses a rich dynamical structure. The theory features two scalar potentials Φ and Ψ, one transverse vector potential Φ_j, and a transverse-tracefree tensor potential $h^{\rm TT}_{jk}$. This is quite a bit more than in Newton's theory, and in this subsection we show how the Newtonian description (with its single scalar potential) emerges in an appropriate limit.

The Newtonian limit of linearized theory is defined by the statement that any speed v that characterizes the matter distribution must be small compared with the speed of light: $v/c \ll 1$. This inequality implies the existence of a hierarchy in the components of the energy-momentum tensor. We recall that T^{00} is a mass density multiplied by c^2, that T^{0j} is a mass flux multiplied by c, and that T^{jk} describes the stresses within the matter distribution. We expect the ratio T^{0j}/T^{00} to be of order v/c, and T^{jk}/T^{00} to be of order $(v/c)^2$. We therefore have $T^{jk} \ll T^{0j} \ll T^{00}$, and in the Newtonian limit we ignore T^{0j} and T^{jk} while we retain the services of $T^{00} = \varrho c^2$.

We formally define the Newtonian limit by setting $s^j = s = \tau = \sigma = \sigma^j = \sigma^{jk} = 0$. The field equations of Eqs. (5.152) imply that $\Psi = \Phi$ and $\Phi_j = 0$. The remaining field equations are

$$\nabla^2 \Phi = -4\pi G \rho \tag{5.153}$$

and $\Box h^{\rm TT}_{jk} = 0$. We recognize the first equation as Poisson's equation of Newton's theory, with Φ playing the role of the Newtonian potential U. The second equation describes the propagation of a free gravitational wave; this can be turned off by adopting a zero-wave initial condition.

If we next adopt the Coulomb gauge, then $U = V = \Phi$, and according to Eq. (5.145), the metric takes the form of

$$ds^2 = -(1 - 2U/c^2)\, d(ct)^2 + (1 + 2U/c^2)(dx^2 + dy^2 + dz^2), \tag{5.154}$$

with a potential U that satisfies $\nabla^2 U = -4\pi G \rho$. This is the Newtonian limit of general relativity.

We encountered a slightly different metric, given by $ds^2 = -(1 - 2U/c^2) d(ct^2) + dx^2 + dy^2 + dz^2$, back in Eq. (5.12). In the earlier context of Sec. 5.1, this metric was arrived at on the basis of a thought experiment involving a photon climbing up a Newtonian gravitational field. The argument was based entirely on the principle of equivalence, and did not refer at all to the Einstein field equations. As a result, the argument produced the correct expression for g_{00}, but it did not produce the corrections to the spatial portion of the metric. Nevertheless, the metric of Eq. (5.12) was seen to give rise to the correct equations of motion for a test body moving in a Newtonian gravitational field. Do the spatial terms in Eq. (5.154) spoil this earlier, successful result?

The answer is no. To see this, we construct the Lagrangian that governs geodesic motion in our improved version of the Newtonian metric. This is $L = -mc\sqrt{-g_{\alpha\beta}\dot{r}^\alpha \dot{r}^\beta}$ with $\dot{r}^\alpha := dr^\alpha/dt$, which evaluates to

$$\begin{aligned} L &= -mc^2 \sqrt{1 - 2U/c^2 - (1 + 2U/c^2)(v^2/c^2)} \\ &= -mc^2 \sqrt{1 - 2U/c^2 - v^2/c^2 + O(Uv^2/c^4)} \\ &= -mc^2 + \frac{1}{2} mv^2 + mU + O(mUv^2/c^2), \end{aligned} \tag{5.155}$$

where $v^2 := \delta_{jk}\dot{r}^j \dot{r}^k$. This computation reveals that the spatial terms in the metric are multiplied by $(v/c)^2$ in the Lagrangian, producing a contribution that is smaller than the Newtonian terms by a factor of $(v/c)^2$. This relativistic correction must be neglected in the Newtonian limit, and we conclude that the timelike geodesics of a spacetime with the metric of Eq. (5.154) are described by $\boldsymbol{a} = \boldsymbol{\nabla} U$. These, of course, are the Newtonian equations of motion.

The spatial terms in Eq. (5.154) do have a significant effect on the motion of light, because in this case a correction of order $(v/c)^2$ is actually a correction of order unity. We explore the relativistic deflection of light and its observable implications in Chapter 10.

5.6 Spherical bodies and Schwarzschild spacetime

We conclude this chapter with an exploration of spherical bodies in *exact* general relativity. While the internal description of such a body is necessarily complicated and requires detailed information about the matter content, the external description is exceptionally simple and provided by the most famous metric of all, the one associated with Karl Schwarzschild's name. The Schwarzschild metric is an exact solution to the Einstein field equations that describes the vacuum region exterior to any spherical distribution of matter. The matter distribution can be static (in the case, for example, of a star in hydrostatic equilibrium) or time-dependent (in the case, for example, of a star undergoing gravitational collapse). As long as the configuration is spherically symmetric, the external, vacuum gravitational field is static and described by the Schwarzschild metric. (There is actually a second solution discovered by Schwarzschild, which describes the gravitational field inside a spherical body in hydrostatic equilibrium, under the assumption that its mass density is constant.)

In the case of a complete gravitational collapse, there is no stellar interior, the vacuum region extends everywhere, and the vacuum Schwarzschild solution describes a non-rotating black hole. By virtue of *Israel's uniqueness theorem*, an isolated, stationary, non-rotating black hole must be spherical (regardless of the shape of its progenitor), and its metric must be the Schwarzschild vacuum metric.

Schwarzschild discovered his solutions in 1915, a few months before the formal publication by Einstein of the final form of the field equations. At the time Einstein had been able to obtain approximate solutions to his equations in weak-field situations, and it came as a surprise to him that an exact solution could be found, and that it could be expressed in such a simple form. Schwarzschild was a well-known German astronomer, and he was working at the Potsdam Observatory when World War I broke out. He carried out his famous work on general relativity while in hospital being treated for a rare auto-immune skin disease, which he had contracted while stationed at the Russian front; he died a few months later at the age of 42.

In this section we derive the Einstein field equations for a spherically-symmetric spacetime, and apply them to the vacuum exterior of any body, and to the interior of a static star. We will describe the exterior metric in a number of coordinate systems, work out the motion of a test mass, and consider the trajectory of a photon. We shall not be concerned here with the black-hole aspects of the spacetime, which are treated in detail in numerous textbooks.

5.6.1 Spherically symmetric spacetimes

Spherical symmetry encourages the use of spherical polar coordinates (r, θ, ϕ), in terms of which the metric of flat spacetime takes the form of $ds^2 = -d(ct)^2 + dr^2 + r^2(d\theta^2 + \sin^2\theta\, d\phi^2)$. Generalizing to curved spacetime, we assert that the metric of any spherically-symmetric spacetime can always be written in the form

$$ds^2 = -e^{-2\Phi/c^2}\, d(ct)^2 + e^{2\Lambda/c^2}\, dr^2 + r^2(d\theta^2 + \sin^2\theta\, d\phi^2), \tag{5.156}$$

in which $\Phi(t,r)$ and $\Lambda(t,r)$ are arbitrary functions of the coordinates t and r. While this may seem like a natural assumption, there are a number of subtle issues that must be addressed to justify it.

First, the assumed spherical symmetry implies that there exists in the spacetime a continuous sequence of concentric spheres, with the property that the spatial geometry on each sphere is the same everywhere. (By "sphere" we mean the two-dimensional surface of a sphere, not the three-dimensional interior.) The coordinate r is nothing but a label of each sphere, with the property that it increases monotonically outwards. On each sphere we lay down a standard grid based on polar coordinates θ and ϕ, in such a way that the proper distance between two neighboring points on the sphere is given by $d\ell^2 = r^2(d\theta^2 + \sin^2\theta\, d\phi^2)$. The line element informs us that r possesses two additional properties: $2\pi r$ is the circumference of a great circle on the sphere (take $\theta = \frac{\pi}{2}$ so that $d\ell = r\, d\phi$), and $4\pi r^2$ is the sphere's surface area (the surface element on the sphere is $dS = \sqrt{g}\, d\theta d\phi = r^2 \sin\theta\, d\theta d\phi$).

Second, the absence of $drd\theta$ and $drd\phi$ terms in Eq. (5.156) results from the fact that spherical symmetry allows us to orient successive spheres so that the radial direction – the direction connecting points with the same value of θ and ϕ on successive spheres – is everywhere orthogonal to the θ and ϕ directions. It also allows us to orient a sphere at one instant of time with the same sphere at a later instant of time, so that the time direction also is orthogonal to the θ and ϕ directions; this property explains the absence of $dtd\theta$ and $dtd\phi$ terms. A term that cannot be removed by appealing to spherical symmetry is the $d(ct)dr$ term, which is nevertheless absent from Eq. (5.156). This reflects a choice of time coordinate: one can always transform from a generic coordinate t' to a new coordinate $t = t(t', r)$ so as to eliminate an offending metric component g'_{0r}. The remaining metric components are g_{00} and g_{rr}, and for these we employ the functions $\Phi(t,r)$ and $\Lambda(t,r)$ as suitable substitutes (the exponential forms are chosen for convenience).

We assume that we are dealing with a single isolated body, so that the spacetime becomes asymptotically flat in the limit $r \to \infty$. This leads to the requirements that

$$\lim_{r\to\infty} \Phi(t,r) = 0, \qquad \lim_{r\to\infty} \Lambda(t,r) = 0, \tag{5.157}$$

and these boundary conditions ensure that the metric reduces to the Minkowski metric when $r \to \infty$. With these conditions we now recognize from Eq. (5.156) that the time coordinate t is proper time as measured by an observer at rest at infinity; for an observer at rest at position r in the gravitational field, proper time τ is related to coordinate time t by $\tau = \int e^{-\Phi(t,r)/c^2}\, dt$. The metric allows us to make another observation: while $2\pi r$ measures the circumference of great circles and $4\pi r^2$ measures the area of spheres, we see that r is *not* a measure of proper distance away from a center at $r = 0$; this is given instead by $\int_0^r e^{\Lambda(t,r')/c^2}\, dr'$.

In place of $\Lambda(t,r)$ it is helpful to employ instead a *relativistic mass-energy function* $m(t,r)$ defined by

$$e^{-2\Lambda/c^2} := f(t,r) := 1 - \frac{2Gm(t,r)}{c^2 r}. \tag{5.158}$$

As we shall see, this bizarre substitution produces a substantial simplification of the field equations, and the name "mass-energy function" will be motivated shortly. Notice the minus sign within the exponential: the mass function is related to g_{rr} by $1 - 2Gm/(c^2r) = (g_{rr})^{-1} = g^{rr}$.

It is straightforward, though tedious, to calculate the Christoffel symbols, Riemann tensor, Ricci tensor, and Einstein tensor for the metric of Eq. (5.156). (We cheat: we use computers to perform such computations.) The results for the relevant components of the Einstein tensor are

$$G^0{}_0 = -\frac{2G}{c^2r^2}\,\partial_r m, \tag{5.159a}$$

$$G^0{}_r = -\frac{2G}{c^3r^2}e^{2\Phi/c^2} f^{-1}\partial_t m, \tag{5.159b}$$

$$G^r{}_r = -\frac{2}{c^2r} f\partial_r \Phi - \frac{2Gm}{c^2r^3}. \tag{5.159c}$$

The remaining non-vanishing components of the Einstein tensor, $G^\theta{}_\theta$ and $G^\phi{}_\phi$, will not be required. They are in fact redundant, because they are related to those of Eq. (5.159) by the Bianchi identities $\nabla_\beta G^{\alpha\beta} = 0$.

The Einstein field equations $G^\alpha{}_\beta = (8\pi G/c^4)T^\alpha_\beta$ take the explicit form

$$\partial_r m = 4\pi r^2(-T^0_0/c^2), \tag{5.160a}$$

$$\partial_t m = 4\pi r^2 e^{-2\Phi/c^2} f(-T^0_r/c), \tag{5.160b}$$

$$\partial_r \Phi = -\frac{G}{r^2} f^{-1}\left[m + 4\pi r^3(T^r_r/c^2)\right], \tag{5.160c}$$

for a spherically-symmetric spacetime. The first two equations are first-order, partial differential equations for the mass function $m(t,r)$. The first equation bears a striking resemblance to Eq. (2.14), which determines the mass function in Newtonian gravity; it is this resemblance that motivates us to attach the name "mass" to this function. One difference, however, is that $-T^0_0$ must be interpreted as the total energy density of the matter distribution, with the factor of c^{-2} converting it into a mass density; the relativistic mass function, therefore, accounts for all forms of energy (including rest-mass energy, kinetic energy, and thermodynamic internal energy) within the spacetime. With this interpretation in place, the second equation also takes a suggestive form: $\partial_t(mc^2)$ is equal, up to relativistic corrections involving Φ and f, to the flux of energy crossing a sphere of constant r. The third member of Eq. (5.160) is an equation for the potential Φ involving the radial pressure $T^r{}_r$; note that there is no equation for $\partial_t \Phi$.

5.6.2 The vacuum Schwarzschild metric

Solution

In vacuum the energy-momentum tensor T^α_β vanishes, and the equations for the mass function immediately give

$$m(t,r) = M = \text{constant}. \tag{5.161}$$

With this assignment the equation for Φ integrates to $\Phi = -\frac{1}{2}c^2 \ln[1 - 2GM/(c^2 r)] + h(t)$, in which $h(t)$ is an arbitrary function of integration. The boundary conditions of Eq. (5.157) require that $h(t) = 0$, and we arrive at

$$e^{-2\Phi/c^2} = 1 - \frac{2GM}{c^2 r}. \tag{5.162}$$

The resulting metric,

$$ds^2 = -\left(1 - \frac{R}{r}\right) d(ct)^2 + \left(1 - \frac{R}{r}\right)^{-1} dr^2 + r^2(d\theta^2 + \sin^2\theta\, d\phi^2), \tag{5.163}$$

with

$$R := \frac{2GM}{c^2}, \tag{5.164}$$

is Schwarzschild's external metric, expressed in terms of the body's Schwarzschild radius R. So we discover, after a very short calculation, that a spherically-symmetric solution to Einstein's equations in vacuum is necessarily static, and given by the Schwarzschild metric; this statement is known as Birkhoff's theorem.

Box 5.8 Birkhoff's theorem in Newtonian gravity

There is an analogous statement of Birkhoff's theorem in Newtonian theory. For spherical symmetry, and in vacuum, Laplace's equation for the Newtonian potential is

$$\frac{1}{r^2}\frac{\partial}{\partial r}\left(r^2 \frac{\partial U(t,r)}{\partial r}\right) = 0.$$

This can be easily integrated to give $U(t, r) = GM/r + h(t)$, where GM is an integration constant, and $h(t)$ an arbitrary function of time. But since all physical manifestations of the potential depend on gradients, $h(t)$ is irrelevant, and the field is static.

As we have emphasized elsewhere, coordinates are merely labels of spacetime events; they have no physical significance whatever, and any coordinate system can be used to describe the geometry of spacetime, with equivalent physical results. Nevertheless, the choice of coordinates can have a significant impact on the ease of calculations. We remarked earlier that the Coulomb gauge of linearized theory is useful for illuminating certain aspects of the theory, but not for calculations that go beyond linearized theory (as will be seen abundantly in later portions of this book); for this the Lorenz gauge is far more useful.

Here too, our choice of coordinates was arbitrary – in this instance they are known as the Schwarzschild coordinates. Their main advantage is that they provide the simplest route to a solution of Einstein's equations and the simplest expression for the metric. One disadvantage is that a transformation from the polar coordinates (r, θ, ϕ) to quasi-Cartesian coordinates (x, y, z) produces a complicated metric with many off-diagonal terms.

Another disadvantage, which is shared by the isotropic and harmonic coordinates introduced below, is that the Schwarzschild coordinates are badly behaved at $r = R$. This, in the pure vacuum case – vacuum everywhere – marks the boundary of the black hole, known as the *event horizon*. Because we are not concerned here with the black-hole aspects of the spacetime, and because this infamous coordinate difficulty is resolved in all textbooks on general relativity, we shall have very little to say on this issue.

To illustrate further the arbitrariness of coordinates, we express the Schwarzschild metric in a number of alternative coordinate systems, each with its own advantages and disadvantages.

Isotropic coordinates

The transformation

$$r = r_{\rm iso}\left(1 + \frac{R}{4r_{\rm iso}}\right)^2 \tag{5.165}$$

brings the Schwarzschild metric to the new form

$$ds^2 = -\left(\frac{1 - R/4r_{\rm iso}}{1 + R/4r_{\rm iso}}\right)^2 d(ct)^2 + \left(1 + \frac{R}{4r_{\rm iso}}\right)^4 \left[dr_{\rm iso}^2 + r_{\rm iso}^2(d\theta^2 + \sin^2\theta\, d\phi^2)\right], \tag{5.166}$$

in which the spatial metric is proportional to the flat-space expression $dr_{\rm iso}^2 + r_{\rm iso}^2(d\theta^2 + \sin^2\theta\, d\phi^2)$. The coordinates t, θ, and ϕ are unchanged, and therefore they carry the same meaning as they do in Schwarzschild coordinates. The new radial coordinate $r_{\rm iso}$, however, does not possess a recognizable geometrical meaning: it does not measure the proper circumference of great circles, and it does not measure proper radial distance. A curious property of the isotropic coordinates is that the metric keeps its form under the inversion $r_{\rm iso} \to R^2/(4r_{\rm iso})$; this *isometry* is largely unphysical, but it has played a very useful role in the construction of initial-data sets for the numerical simulation of black-hole mergers.

The main advantage of this coordinate system is that the additional transformation $x = r_{\rm iso}\sin\theta\cos\phi$, $y = r_{\rm iso}\sin\theta\sin\phi$, $z = r_{\rm iso}\cos\theta$ produces a simple expression for the metric in Cartesian-like coordinates, with the components

$$g_{00} = -\left(\frac{1 - R/4r_{\rm iso}}{1 + R/4r_{\rm iso}}\right)^2, \tag{5.167a}$$

$$g_{jk} = \delta_{jk}\left(1 + \frac{R}{4r_{\rm iso}}\right)^4. \tag{5.167b}$$

When $r_{\rm iso}$ is large compared with the Schwarzschild radius R, these expressions become

$$g_{00} = -1 + R/r_{\rm iso} - \frac{1}{2}(R/r_{\rm iso})^2 + \cdots, \tag{5.168a}$$

$$g_{jk} = \delta_{jk}\left(1 + R/r_{\rm iso} + \frac{3}{8}(R/r_{\rm iso})^2 + \cdots\right), \tag{5.168b}$$

and they can be compared with the Newtonian metric of Eq. (5.154). In spherical symmetry, and in a vacuum exterior, the Newtonian potential is $U = GM/r_{\rm iso}$, so that $2U/c^2 = R/r_{\rm iso}$. The Newtonian metric has $g_{00} = -1 + 2U/c^2 = -1 + R/r_{\rm iso}$ and $g_{jk} = \delta_{jk}(1 + 2U/c^2) = \delta_{jk}(1 + R/r_{\rm iso})$. We see that the Schwarzschild metric, when it is expressed in isotropic coordinates, differs from this by terms of order $(R/r_{\rm iso})^2$ and higher; these corrections cannot be obtained in the linearized approximation that led to the Newtonian metric. The fact that the Schwarzschild metric with $R = 2GM/c^2$ reduces to the Newtonian metric when $r_{\rm iso}$ is large confirms the identification of M with the body's total mass.

Like the Schwarzschild coordinates, the isotropic coordinates become ill-behaved at the black-hole horizon, which is now situated at $r_{\rm iso} = \frac{1}{4}R$.

Harmonic coordinates

Another popular coordinate system for the Schwarzschild spacetime is obtained by the transformation

$$r = r_{\rm h} + \tfrac{1}{2}R. \tag{5.169}$$

For reasons that will be identified below, the new system is called *harmonic*, and in it the metric takes the form

$$ds^2 = -\left(\frac{1 - R/2r_{\rm h}}{1 + R/2r_{\rm h}}\right) d(ct)^2 + \left(\frac{1 + R/2r_{\rm h}}{1 - R/2r_{\rm h}}\right) dr_{\rm h}^2 + (r_{\rm h} + \tfrac{1}{2}R)^2(d\theta^2 + \sin^2\theta\, d\phi^2). \tag{5.170}$$

As before we find that t, θ, and ϕ carry the same geometrical meaning, but that $r_{\rm h}$ is linked neither to a proper circumference nor to a proper radial displacement. And the harmonic coordinates also are ill-behaved at the event horizon, which is now situated at $r_{\rm h} = \frac{1}{2}R$.

The additional transformation $x = r_{\rm h}\sin\theta\cos\phi$, $y = r_{\rm h}\sin\theta\sin\phi$, $z = r_{\rm h}\cos\theta$ produces a complicated metric, but we find it worthwhile to display it here:

$$g_{00} = -\frac{1 - R/2r_{\rm h}}{1 + R/2r_{\rm h}}, \tag{5.171a}$$

$$g_{jk} = \left(\frac{1 + R/2r_{\rm h}}{1 - R/2r_{\rm h}}\right) n_j n_k + \left(1 + R/2r_{\rm h}\right)^2 \left(\delta_{jk} - n_j n_k\right), \tag{5.171b}$$

where $n^j := x^j/r_{\rm h}$ is a radial unit vector, whose index is lowered with the Euclidean metric δ_{jk}, so that $n_j := \delta_{jk}n^k$. We note that in this Euclidean sense, the two terms in the spatial metric are orthogonal to each other, because $(\delta_{jk} - n_j n_k)n^k = 0$. This observation simplifies the computation of the inverse metric, which is given by

$$g^{00} = -\frac{1 + R/2r_{\rm h}}{1 - R/2r_{\rm h}}, \tag{5.172a}$$

$$g^{jk} = \left(\frac{1 - R/2r_{\rm h}}{1 + R/2r_{\rm h}}\right) n^j n^k + \left(1 + R/2r_{\rm h}\right)^{-2} \left(\delta^{jk} - n^j n^k\right). \tag{5.172b}$$

For future reference we also record the expression

$$\sqrt{-g} = \left(1 + R/2r_{\rm h}\right)^2 \tag{5.173}$$

for the metric determinant.

When $r_{\rm h}$ is large compared with R the metric becomes

$$g_{00} = -1 + R/r_{\rm h} - \frac{1}{2}(R/r_{\rm h})^2 + \cdots, \tag{5.174a}$$

$$g_{jk} = \left(1 + R/r_{\rm h}\right)\delta_{jk} + \frac{1}{4}(R/r_{\rm h})^2\left(\delta_{jk} + n_j n_k\right) + \cdots \tag{5.174b}$$

Here also we find that this agrees with the Newtonian metric of Eq. (5.154) to order $R/r_{\rm h}$, with deviations occurring at order $(R/r_{\rm h})^2$ and higher.

To explain the meaning of the phrase "harmonic coordinates" we examine the set of four scalar fields $X^{(0)} := ct$, $X^{(1)} := (r - \frac{1}{2}R)\sin\theta\cos\phi$, $X^{(2)} := (r - \frac{1}{2}R)\sin\theta\sin\phi$, and $X^{(3)} := (r - \frac{1}{2}R)\cos\theta$ in the Schwarzschild spacetime. We collectively denote the members of the set by $X^{(\mu)}$, and observe that while we have defined the scalar fields in terms of the original Schwarzschild coordinates (t, r, θ, ϕ), we have the simpler expressions $X^{(\mu)} = x^\mu$ in the Cartesian version (t, x, y, z) of the harmonic coordinates.

It is a matter of simple computation to show that each one of the four scalar fields satisfies the wave equation

$$\Box_g X^{(\mu)} = 0 \tag{5.175}$$

in Schwarzschild spacetime. Here $\Box_g := g^{\alpha\beta}\nabla_\alpha\nabla_\beta$ is the curved-spacetime wave operator, and we note the important fact that Eq. (5.175) is covariant and can therefore be expressed in any coordinate system; the equation is easiest to verify if we adopt the Schwarzschild coordinates (t, r, θ, ϕ). The term "harmonic" refers to solutions to Eq. (5.175), which is a generalization from three-dimensional Euclidean space to four-dimensional curved spacetime of Laplace's equation $\nabla^2 X = 0$; it has been a long tradition that solutions to Laplace's equation are called *harmonic functions*.

We can re-express Eq. (5.175) as $\nabla_\alpha X^{(\mu)\alpha} = 0$, where the vector fields $X^{(\mu)\alpha}$ are defined by $X^{(\mu)}_\alpha = \nabla_\alpha X^{(\mu)}$. With the divergence identity of Eq. (5.40), this is

$$\frac{1}{\sqrt{-g}}\partial_\alpha\Big(\sqrt{-g}g^{\alpha\beta}\partial_\beta X^{(\mu)}\Big) = 0, \tag{5.176}$$

and the wave equation continues to be covariant. We now, however, specialize the coordinate system to (t, x, y, z), the Cartesian version of our harmonic coordinates. We have already noted that $X^{(\mu)} = x^\mu$ in this coordinate system, and it follows that $\partial_\beta X^{(\mu)} = \delta^\mu_\beta$. Making the substitution reveals that the harmonic condition becomes

$$\partial_\beta\big(\sqrt{-g}g^{\alpha\beta}\big) = 0 \tag{5.177}$$

when it is expressed in the harmonic coordinates (t, x, y, z). An alternative statement of this is

$$g^{\mu\nu}\Gamma^\alpha_{\mu\nu} = 0. \tag{5.178}$$

It can be verified that these equations are indeed satisfied by the Cartesian-like metric of Eqs. (5.171); the computation relies on the results displayed in Eqs. (5.172) and (5.173). But they are *not* satisfied by the metric of Eqs. (5.170) expressed in spherical polar coordinates. Despite our use of covariant language in our discussion of Eq. (5.175), we are making a very specific coordinate choice in Eqs. (5.177) and (5.178), which are clearly not covariant equations. Another way of looking at this is to recognize that Eqs. (5.175) are four constraints on the coordinates $x^\mu \equiv X^{(\mu)}$.

Harmonic coordinates play a powerful role in our development of weak-field gravity in subsequent chapters, and our digression here serves mostly as a motivation for the coordinate conditions of Eq. (5.177). These, naturally enough, are known as the *harmonic coordinate conditions*.

Painlevé–Gullstrand coordinates

The problems of the preceding coordinate systems at the event horizon can be traced to the bad behavior of the time coordinate t, which becomes frozen at the event horizon. As we shall see in a moment, the coordinate time required for an observer to cross the horizon is always infinite, in spite of the fact that the required proper time τ is perfectly finite. To explore the features of the Schwarzschild spacetime near the horizon, it is necessary to define a new time coordinate T that is smoothly related to proper time as measured by an observer falling into the black hole.

A simple candidate for such a new time coordinate (though not the only one) is the one defined by

$$d(cT) := d(ct) + (1 - R/r)^{-1}\sqrt{R/r}\,dr; \tag{5.179}$$

the motivation behind this choice will be revealed shortly. Integration yields

$$cT = ct + 2R\left(\sqrt{r/R} + \frac{1}{2}\ln\frac{\sqrt{r/R} - 1}{\sqrt{r/R} + 1}\right), \tag{5.180}$$

and admitting for the moment that T is well-behaved across $r = R$, we see that t must diverge logarithmically as the horizon is approached. It is this behavior that signals the pathology of the Schwarzschild time coordinate.

Making the substitution of Eq. (5.179) into Eq. (5.163) brings the metric to the new form

$$ds^2 = -d(cT)^2 + \left[dr + \sqrt{R/r}\,d(cT)\right]^2 + r^2(d\theta^2 + \sin^2\theta\,d\phi^2), \tag{5.181}$$

and the coordinates (T, r, θ, ϕ) are known as the *Painlevé–Gullstrand coordinates*. The meaning of the time coordinate must still be elucidated, but the spatial coordinates (r, θ, ϕ) are the same as in the original system of Schwarzschild coordinates. The new coordinates possess the intriguing property that *the three-dimensional surfaces $T =$ constant are intrinsically flat*. This statement follows by inserting $dT = 0$ into Eq. (5.181) and observing that the spacetime interval becomes $ds^2 = dr^2 + r^2(d\theta^2 + \sin^2\theta\,d\phi^2)$; this is the flat-space metric expressed in spherical polar coordinates.

To identify the meaning of the Painlevé–Gullstrand time T we consider an observer – let's call her Meirong – moving radially in the Schwarzschild spacetime with constant angular coordinates (θ, ϕ); for such an observer the spacetime interval reduces to $ds^2 = -d(cT)^2 + [dr + \sqrt{R/r}\,d(cT)]^2$. If, in addition, Meirong moves on a world line described by the differential equation

$$\frac{dr}{dT} = -c\sqrt{R/r}, \tag{5.182}$$

then the spacetime interval simplifies further to $ds^2 = -d(cT)^2$. This shows that *T is proper time for our observer*, and Eq. (5.182) informs us that Meirong falls radially toward the black hole, having started from rest at $r = \infty$. Integration of Eq. (5.182) is immediate, and we find that

$$T = T_0 - \frac{2R}{3c}(r/R)^{3/2}, \tag{5.183}$$

where the constant of integration T_0 is the time at which Meirong reaches the black-hole singularity at $r = 0$; this expression confirms our expectation that $T(r)$ is perfectly well behaved at $r = R$. Finally, it is not hard to show that Meirong's world line, described by Eq. (5.183), is a timelike geodesic of the Schwarzschild spacetime; our observer is therefore falling freely toward the black hole.

5.6.3 Motion of a test mass

We examine the motion of a test mass in the equatorial plane ($\theta = \frac{\pi}{2}$) of the Schwarzschild spacetime. The particle is moving freely on a timelike geodesic of the spacetime, and our restriction to equatorial motion does not represent a loss of generality: It can be shown that in a spherically-symmetric spacetime, geodesic motion always proceeds within a fixed spatial plane, and this plane can always be chosen to be equatorial. For concreteness we assume that the particle follows a bounded trajectory, and moves between an innermost radius r_- (the pericenter) and an outermost radius r_+ (the apocenter); this *does* represent a loss of generality, because unbounded motion is also possible.

We begin with the form of the geodesic equation given by Eq. (5.53). This equation reveals that when the metric does not depend explicitly on one of its coordinates x^μ, the associated component u_μ of the velocity vector is a constant of the motion. Here we have two conserved quantities, u_0 and u_ϕ, and we express them as

$$u_0 =: -c\sqrt{1 + 2\varepsilon/c^2}, \qquad u_\phi =: h, \tag{5.184}$$

thereby defining the constants ε and h. Raising the index on the velocity vector gives us two equations of motion,

$$\dot{t} = \frac{\sqrt{1 + 2\varepsilon/c^2}}{1 - R/r} \tag{5.185}$$

and

$$\dot{\phi} = \frac{h}{r^2}. \tag{5.186}$$

We use an overdot to indicate differentiation with respect to proper time τ. Equation (5.186) informs us that h has the interpretation of a conserved angular momentum per unit mass; ε will presently be identified as a conserved orbital energy per unit mass.

The equation of motion for $r(\tau)$ can be obtained from the normalization condition $g_{\alpha\beta}u^\alpha u^\beta = -c^2$ for the velocity vector. A short calculation yields

$$\frac{1}{2}\dot{r}^2 + \nu(r) = \varepsilon, \tag{5.187}$$

where

$$\nu(r) = -\frac{GM}{r} + \frac{h^2}{2r^2}\left(1 - \frac{R}{r}\right) \tag{5.188}$$

is an effective potential for the radial component of the motion. Equation (5.187) takes the form of an energy equation, and it is this equation that reveals ε as an orbital energy per unit mass. A graph of $\nu(r)$ can be used to display the regions where motion is possible, given values for h and ε, and the energy diagram goes a long way toward a qualitative understanding of the possible motions (see Fig. 5.4). It is remarkable that Eqs. (5.186) and

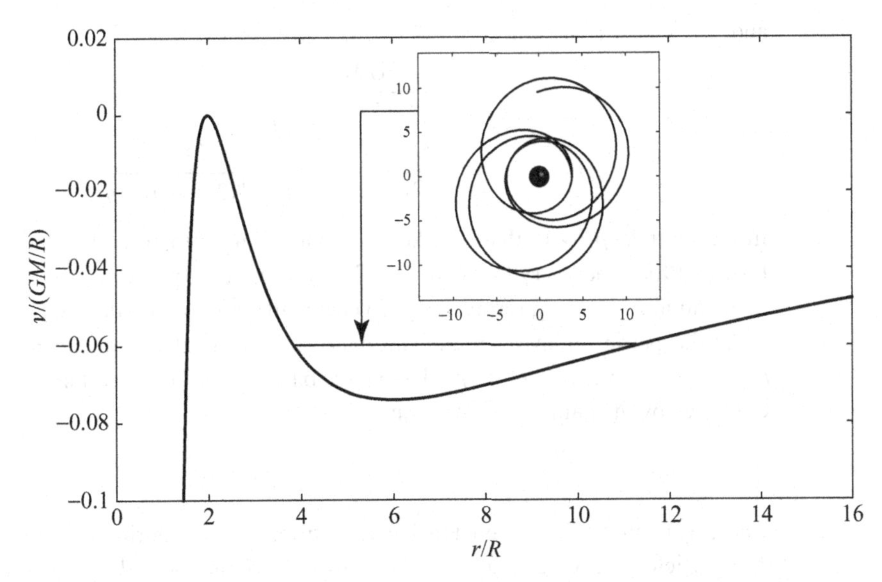

Fig. 5.4 Effective potential $\nu(r)$ for radial motion in the Schwarzschild spacetime, in units of GM/R. The angular-momentum parameter h is chosen to be larger than $\sqrt{6GMR}$, and set to such a value that a circular orbit occurs at $r = 6R$. The horizontal line is a curve of constant ε, and in this case the motion takes place between two turning points at $r_- = 3.8R$ and $r_+ = 11.4R$. The inset displays an orbit with parameters $p = 5.7R$ and $e = 0.5$.

(5.187) are so very similar to the Newtonian equations of motion in a central potential $U = GM/r$. There are two major points of departure: the overdot in the differential equations refers to proper time τ instead of coordinate time t, and the effective potential $\nu(r)$ contains a relativistic correction $-R/r$ in the centrifugal term. These differences are negligible when $r \gg R$, but they are very important when r is comparable to R; in particular, it is noteworthy that the centrifugal barrier is no longer infinite in the relativistic description, but instead goes to zero as $r \to R$.

As stated previously, we examine the bound motion of a particle between the turning points $r_\pm$; this situation requires that $h > \sqrt{6GMR}$. It is useful to introduce the orbital elements p and e, which are defined by

$$r_- =: \frac{p}{1+e}, \qquad r_+ =: \frac{p}{1-e}. \tag{5.189}$$

These are reminiscent of the definitions in Newtonian celestial mechanics; here p is a relativistic *semi-latus rectum*, while e is a relativistic *eccentricity*. The orbit is circular when $e = 0$.

The orbital elements are related to the constants of the motion h and ε, and the relationship can be worked out by factorizing $\varepsilon - \nu(r)$ as

$$k\left(\frac{1}{r} - \frac{1}{r_0}\right)\left(\frac{1}{r} - \frac{1}{r_-}\right)\left(\frac{1}{r} - \frac{1}{r_+}\right),$$

where k is a constant and r_0 is a third root of the equation $\varepsilon - \nu(r) = 0$. Comparing this form with Eq. (5.188) reveals the following relations: $k = \frac{1}{2}h^2 R$, $r_0 = R/(1 - 2R/p)$,

and

$$h^2 = \frac{GMp}{1 - \frac{1}{2}(3+e^2)R/p}, \tag{5.190a}$$

$$\varepsilon = -\frac{GM}{2p}(1-e^2)\frac{1-2R/p}{1-\frac{1}{2}(3+e^2)R/p}. \tag{5.190b}$$

In the limit $R/p \to 0$, that is, when p is much larger than the Schwarzschild radius R, Eqs. (5.190a) and (5.190b) reduce to the Newtonian expressions for the orbital angular momentum and energy, respectively. To ensure that $r_0 < r_-$ we demand that $p > (3+e)R$.

As the particle moves around the mass M, its radial coordinate proceeds from $r = r_- = p/(1+e)$ to $r = r_+ = p/(1-e)$ and back again. This radial motion is conveniently described by the parametric equation

$$r(\chi) = \frac{p}{1+e\cos\chi}, \tag{5.191}$$

where χ is an orbital parameter that runs from 0 to 2π during a complete radial cycle. The particle is at pericenter $p/(1+e)$ when $\chi = 0$, at apocenter $p/(1-e)$ when $\chi = \pi$, and back at pericenter when $\chi = 2\pi$. Equation (5.191) is very similar to the description of elliptical motion in Newtonian gravity, $r = p/(1+e\cos\phi)$, but it actually serves as a definition for χ, which is not equal to the azimuthal angle ϕ.

When Eqs. (5.190a), (5.190b), and (5.191) are inserted into Eq. (5.188), we obtain

$$\varepsilon - v(r) = \frac{GM}{2p}e^2\sin^2\chi\,\frac{1-(3+e\cos\chi)R/p}{1-\frac{1}{2}(3+e^2)R/p}, \tag{5.192}$$

and Eq. (5.187) can be turned into an equation of motion for χ. The end result is an expression for $\dot{\chi}$, whose reciprocal is

$$\frac{d\tau}{d\chi} = \sqrt{\frac{p^3}{GM}}(1+e\cos\chi)^{-2}\frac{[1-\frac{1}{2}(3+e^2)R/p]^{1/2}}{[1-(3+e\cos\chi)R/p]^{1/2}}. \tag{5.193}$$

This equation can be integrated to get proper time τ as a function of the orbital parameter χ. The azimuthal angle is obtained by integrating $d\phi/d\chi = \dot{\phi}(d\tau/d\chi)$; combining Eqs. (5.186), (5.190a), and (5.191) gives

$$\frac{d\phi}{d\chi} = \frac{1}{\sqrt{1-(3+e\cos\chi)R/p}}. \tag{5.194}$$

This equation shows that ϕ becomes equal to χ in the Newtonian limit $R/p \to 0$. Finally, similar manipulations produce

$$\frac{dt}{d\chi} = \sqrt{\frac{p^3}{GM}}(1+e\cos\chi)^{-2}\frac{[1-(1+e)R/p]^{1/2}[1-(1-e)R/p]^{1/2}}{[1-(3+e\cos\chi)R/p]^{1/2}[1-(1+e\cos\chi)R/p]}, \tag{5.195}$$

an equation that can be integrated to obtain $t(\chi)$.

Equations (5.193), (5.194), and (5.195), along with Eq. (5.191), form a complete set of orbital equations for the bounded motion of a test mass in the Schwarzschild spacetime.

Except for Eq. (5.194), which can be solved in terms of an elliptic function, the equations must be integrated numerically, and the formulation developed here is especially convenient for this purpose. The equations make the reduction to the Newtonian limit very easy to accomplish, and corrections of order R/p, $(R/p)^2$, and so on can be systematically worked out.

The orbital equations become especially simple when $e = 0$ and the orbit is circular. Then $r(\chi) = p$ and we find that the equations integrate to

$$\phi = \frac{1}{\sqrt{1-3R/p}}\chi, \tag{5.196a}$$

$$t = \sqrt{\frac{p^3}{GM}}\frac{1}{\sqrt{1-3R/p}}\chi, \tag{5.196b}$$

$$\tau = \sqrt{\frac{p^3}{GM}}\frac{\sqrt{1-3R/2p}}{\sqrt{1-3R/p}}\chi. \tag{5.196c}$$

These equations imply the simple relation $\phi = \Omega t$, where $\Omega := \sqrt{GM/p^3}$ is the orbital angular velocity as measured by an observer at rest at infinity; the proper angular velocity differs from this by a factor of $(1-3R/2p)^{-1/2}$. In this circular case χ loses its usefulness as an orbital parameter, and it is appropriate to replace it by ϕ.

Returning to the general case $e \neq 0$, we observe that in the course of a complete radial cycle, the orbital parameter χ increases by 2π, the azimuthal angle increases by $\Delta\phi$, the time coordinate increases by the orbital period P, and the particle's proper time increases by the proper period $\mathcal{P}$. These quantities are defined by

$$\Delta\phi = \int_0^{2\pi}\frac{d\phi}{d\chi}d\chi, \qquad P = \int_0^{2\pi}\frac{dt}{d\chi}d\chi, \qquad \mathcal{P} = \int_0^{2\pi}\frac{d\tau}{d\chi}d\chi. \tag{5.197}$$

The integrals can be evaluated numerically for each pair of orbital elements p and e, and analytical expressions can be obtained when R/p is small. We find

$$\Delta\phi = 2\pi + 6\pi\left(\frac{GM}{c^2p}\right) + \frac{3\pi}{2}(18+e^2)\left(\frac{GM}{c^2p}\right)^2 + \cdots, \tag{5.198}$$

$$P = 2\pi\sqrt{\frac{a^3}{GM}}\biggl[1 + 3(1-e^2)\left(\frac{GM}{c^2p}\right) + \frac{3}{2}(1-e^2)(4+5\sqrt{1-e^2})\left(\frac{GM}{c^2p}\right)^2 + \cdots\biggr], \tag{5.199}$$

$$\mathcal{P} = 2\pi\sqrt{\frac{a^3}{GM}}\biggl[1 + \frac{3}{2}(1-e^2)\left(\frac{GM}{c^2p}\right) + \frac{3}{8}(1-e^2)(17-e^2+4\sqrt{1-e^2})\left(\frac{GM}{c^2p}\right)^2 + \cdots\biggr], \tag{5.200}$$

where $a := p/(1-e^2)$ is the relativistic semi-major axis. To arrive at Eqs. (5.199) and (5.200) it is helpful to change the variable of integration from χ to u, defined by

$$\tan\frac{\chi}{2} = \sqrt{\frac{1+e}{1-e}}\tan\frac{u}{2}; \tag{5.201}$$

in Newtonian celestial mechanics u is known as the eccentric anomaly, refer to Eq. (3.32), and its range during a complete radial cycle is also from $u = 0$ to $u = 2\pi$.

The fact that $\Delta\phi > 2\pi$ implies that the orbit is not closed; for small values of R/p the orbit is approximately elliptical, but its major axis rotates by an angle $\Delta\phi - 2\pi$ in the course of each orbit. This is the *pericenter advance* of relativistic motion, an effect that was famously measured for Mercury and a number of binary pulsars. To first order in R/p we have that the advance is equal to $6\pi(GM/c^2p)$ per orbit. For Mercury this amounts to a tiny, but measurable 42.98 seconds of arc per century (refer to Table 3.1). For binary pulsars the effect is measured in degrees per year.

It is important to note that Eqs. (5.198), (5.199), and (5.200) do not apply when $e = 0$, because as we have seen, χ loses its meaning as orbital parameter in the circular case. The limit $e \to 0$ is therefore singular, and in this case we have the exact results

$$\Delta\phi = 2\pi, \qquad P = 2\pi\sqrt{\frac{p^3}{GM}}, \qquad \mathcal{P} = 2\pi\sqrt{\frac{p^3}{GM}}\sqrt{1-\frac{3R}{2p}}. \tag{5.202}$$

Equations (5.198), (5.199), and (5.200) are instances of *post-Newtonian expansions*, in which a quantity of interest is expressed as an expansion in powers of a post-Newtonian parameter $\varepsilon \sim (v/c)^2 \sim U/c^2$; here the adopted expansion parameter is $\varepsilon := GM/(c^2p)$. The leading term in these equations is the Newtonian answer, and this is labeled a 0PN term. The term of order ε is the first post-Newtonian correction, and it is labeled a 1PN term. And finally, the term of order ε^2 is a second post-Newtonian correction, or 2PN term. This kind of approximation to general relativity will be a central theme in the remainder of this book.

5.6.4 Motion of light

We next examine the motion of a photon in the equatorial plane of a Schwarzschild spacetime. The vector tangent to the photon's world line is p^α, and this satisfies the geodesic equation (5.54). A consequence of this equation is that both p_0 and p_ϕ are constants of the motion, and by a constant rescaling of the parameter λ on the world line, we can set $p_0 = -c$ and $p_\phi = h$. (This rescaling implies that p^α is not quite the photon's momentum vector; the unit of momentum has been changed by a factor $p_0^{\text{rescaled}}/p_0^{\text{conventional}} = c^2/(\hbar\omega_\infty)$, where ω_∞ is the photon's frequency as measured by an observer at rest at infinity. In these units, the parameter h represents the photon's angular momentum multiplied by c^2 and divided by $\hbar\omega_\infty$.)

We have already obtained two equations of motion,

$$\dot{t} = \frac{1}{1-R/r} \tag{5.203}$$

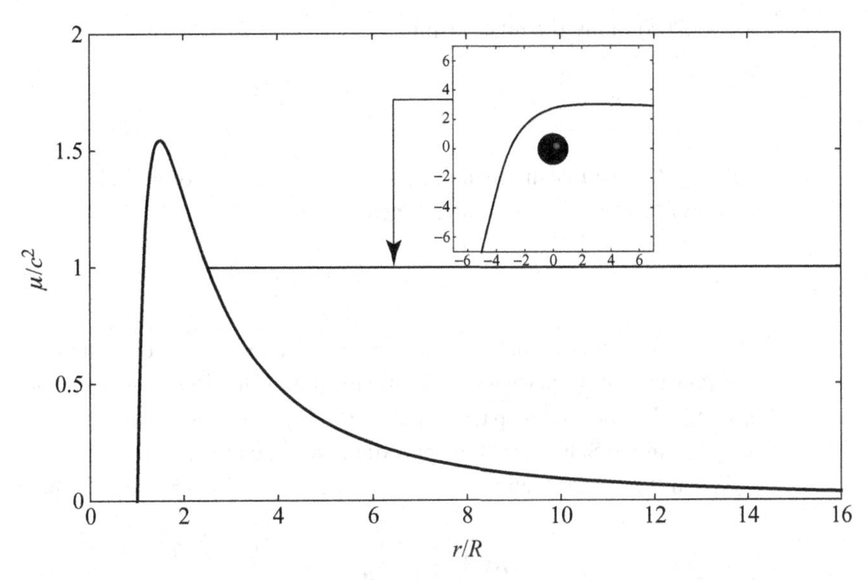

Fig. 5.5 Effective potential $\mu(r)$ for photon motion in the Schwarzschild spacetime, in units of c^2. The angular-momentum parameter h is chosen to be larger than h_c. The horizontal line marks the constant c^2, and in this case the photon meets a turning point at $r = p > \frac{3}{2}R$. The inset displays a photon orbit with $p = 2.5R$.

and

$$\dot{\phi} = \frac{h}{r^2}, \tag{5.204}$$

and the third follows as a consequence of the null condition $g_{\alpha\beta}p^\alpha p^\beta = 0$. We get

$$\dot{r}^2 + \mu(r) = c^2, \tag{5.205}$$

where

$$\mu(r) = \frac{h^2}{r^2}\left(1 - \frac{R}{r}\right) \tag{5.206}$$

is an effective potential for the radial component of the photon's motion. In these equations an overdot indicates differentiation with respect to the (rescaled) world-line parameter λ.

The effective potential of Eq. (5.206) contains only a centrifugal term – compare with Eq. (5.188) – and it represents a barrier. The potential is maximum at $r = \frac{3}{2}R = 3GM/c^2$, and a photon with $h = h_c := (3\sqrt{3}/2)Rc$ can move on a circular orbit at this radius, although the orbit is unstable (see Fig. 5.5). A photon with $h < h_c$ either moves in from $r = \infty$, or moves out to $r = \infty$, and this photon does not encounter a turning point of its radial motion. On the other hand, a photon with $h > h_c$ encounters a turning point at a radius $r = p > \frac{3}{2}R$. It moves in from $r = \infty$, turns around at $r = p$, and moves out to $r = \infty$; this type of behavior corresponds to the deflection of light by a massive body.

We consider this case in more detail, and examine the deflected motion of a photon with $h > h_c$. As we have seen, the motion has a radial turning point at $r = p$, and Eqs. (5.205) and (5.206) imply the relationship

$$h^2 = \frac{p^2c^2}{1 - R/p} \tag{5.207}$$

between the angular-momentum parameter h and the orbital element p. We describe the motion in terms of an orbital parameter η, and write

$$r(\eta) = \frac{p}{\sin\eta} \tag{5.208}$$

to represent a photon that starts at $r = \infty$ (when $\eta = 0$), moves in to $r = p$ (when $\eta = \frac{\pi}{2}$), and returns to $r = \infty$ (when $\eta = \pi$). This representation is analogous to the one adopted in Eq. (5.191), and in flat spacetime the orbital parameter η would be equal to the azimuthal angle ϕ; in the Schwarzschild spacetime these quantities are distinct.

When Eqs. (5.207) and (5.208) are inserted into Eq. (5.206) we obtain

$$c^2 - \mu(r) = \frac{c^2\cos^2\eta}{1 - R/p}\biggl(1 - \frac{1 + \sin\eta + \sin^2\eta}{1 + \sin\eta}\frac{R}{p}\biggr), \tag{5.209}$$

and Eq. (5.205) can be turned into an equation of motion for $\eta(\lambda)$. We get

$$\dot{\eta} = \frac{c\sin^2\eta}{p\sqrt{1 - R/p}}\biggl(1 - \frac{1 + \sin\eta + \sin^2\eta}{1 + \sin\eta}\frac{R}{p}\biggr)^{1/2}. \tag{5.210}$$

With this and Eq. (5.203) we obtain

$$\frac{dt}{d\eta} = \frac{p\sqrt{1 - R/p}}{c[1 - (\sin\eta)R/p]\sin^2\eta}\biggl(1 - \frac{1 + \sin\eta + \sin^2\eta}{1 + \sin\eta}\frac{R}{p}\biggr)^{-1/2}, \tag{5.211}$$

which allows computation of the time coordinate along the photon's world line. Then Eq. (5.204) yields

$$\frac{d\phi}{d\eta} = \biggl(1 - \frac{1 + \sin\eta + \sin^2\eta}{1 + \sin\eta}\frac{R}{p}\biggr)^{-1/2}, \tag{5.212}$$

which determines ϕ. Equations (5.208), (5.211), and (5.212) form a complete set of orbital equations for the motion of a deflected photon. These equations must be integrated numerically, and the formulation provided here is convenient for this purpose.

In the course of a complete orbit (from $\eta = 0$ to $\eta = \pi$) the azimuthal angle increases by an amount $\Delta\phi$ given by the integral of Eq. (5.212). The integration cannot be accomplished analytically in general, but an expansion in powers of R/p gives

$$\Delta\phi = \pi + 4\biggl(\frac{GM}{c^2p}\biggr) + \biggl(\frac{15\pi}{4} - 4\biggr)\biggl(\frac{GM}{c^2p}\biggr)^2 + \cdots. \tag{5.213}$$

In flat spacetime the answer would be the obvious $\Delta\phi = \pi$, and the difference between $\Delta\phi$ and π is the photon's *deflection angle* α. Equation (5.213) reveals that to leading order in a post-Newtonian expansion, $\alpha = 4GM/(c^2p)$. This is the famous deflection of light, to

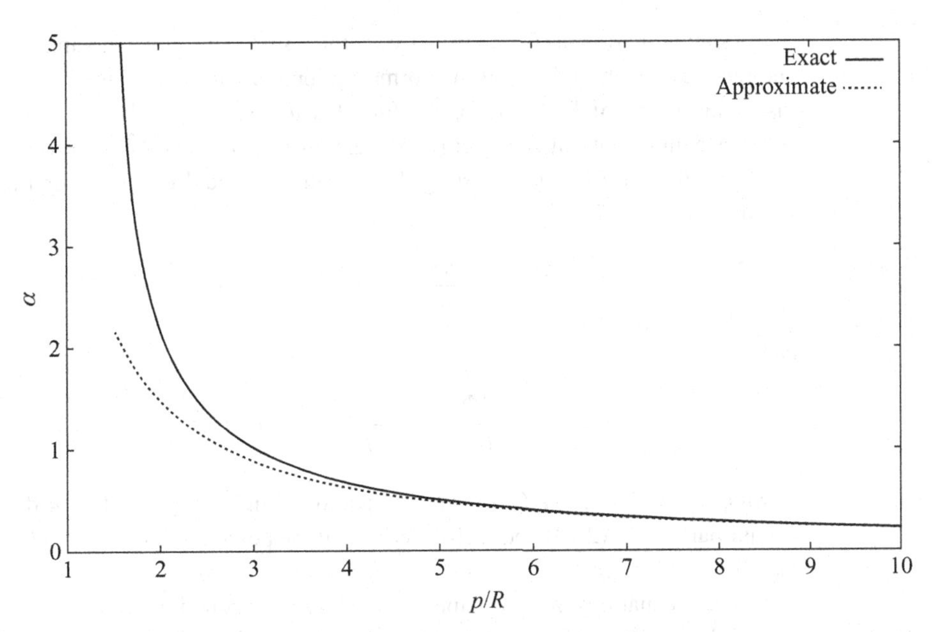

Fig. 5.6 Deflection of a photon in Schwarzschild spacetime. The deflection angle $\alpha := \Delta\phi - \pi$ is plotted as a function of the orbital parameter p/R. The higher curve is the exact result, computed by integrating Eq. (5.212) numerically. The lower curve is the approximate expression of Eq. (5.213). We observe that the curves coincide when $p/R \gg 1$. When $p/R = 3/2$, so that $p = 3GM/c^2$, the exact deflection angle is infinite: this is the photon's circular orbit.

which we will return in Chapter 10 in the context of post-Newtonian theory. The exact and approximate versions of the deflection angle are plotted in Fig. 5.6.

5.6.5 Spherical bodies in hydrostatic equilibrium

The previous subsections have dealt with the external Schwarzschild metric, which is a solution to the Einstein field equations in vacuum. We next turn to a description of the interior, assuming that the body is in hydrostatic equilibrium; this implies that the interior metric is time-independent. This discussion of stellar structure extends to general relativity the material that was developed in Sec. 2.2 in the context of Newton's theory.

Equations of stellar structure

We take the body to consist of a perfect fluid, and we adopt the form of Eq. (5.91) for its energy-momentum tensor:

$$T^{\alpha}_{\ \beta} = (\mu + p)u^{\alpha}u_{\beta}/c^2 + p\,\delta^{\alpha}_{\ \beta}. \tag{5.214}$$

Here μ is the fluid's proper energy density, p is the pressure, and u^{α} is the velocity field. The energy density is decomposed as $\mu = \rho c^2 + \epsilon$, in terms of a proper rest-mass density ρ and a proper density ϵ of internal (thermodynamic) energy. The first law of thermodynamics implies that these variables are related by $\rho\, d\epsilon = (\epsilon + p)\, d\rho$.

In a static situation the fluid's velocity vector has u^0 as its only non-vanishing component, and this can be determined by the normalization condition $g_{\alpha\beta}u^\alpha u^\beta = -c^2$, in which we insert the metric of Eq. (5.156). We find that $u^0 = ce^{\Phi/c^2}$ and $u_0 = -ce^{-\Phi/c^2}$, and the relevant components of the energy-momentum tensor are $-T^0_{\ 0}/c^2 = \mu/c^2 = \rho + \epsilon/c^2$, $-T^0_{\ r}/c = 0$, and $T^r_{\ r} = p$. Inserting these within the field equations of Eq. (5.160), we obtain

$$\frac{dm}{dr} = 4\pi r^2(\rho + \epsilon/c^2), \tag{5.215}$$

and

$$\frac{d\Phi}{dr} = -\frac{G}{r^2 f}\left(m + 4\pi r^3 p/c^2\right), \tag{5.216}$$

in which $f := 1 - 2Gm/(c^2 r)$. These equations can be compared with their Newtonian counterparts, Eqs. (2.13) and (2.14); recall that the potential Φ reduces to U in the Newtonian limit.

The field equations must be supplemented with the equation of hydrostatic equilibrium. This is obtained by working out the radial component of the conservation equation $\nabla_\beta T^{\alpha\beta} = 0$. Part of the work was already carried out in Sec. 5.3.2, and a good starting point for this task is Eq. (5.97), the curved-spacetime formulation of Euler's equation. An even better starting point is found in Box 5.5, which works out the condition of hydrostatic equilibrium for any static spacetime. Adapting this to the specific metric of Eq. (5.156), we find that the pressure of a spherical body in hydrostatic equilibrium is determined by

$$\frac{dp}{dr} = \left(\rho + \epsilon/c^2 + p/c^2\right)\frac{d\Phi}{dr}. \tag{5.217}$$

This equation can be compared with Eq. (2.13). Combining this with Eq. (5.216), we obtain

$$\frac{dp}{dr} = -\frac{G}{r^2 f}\left(\rho + \epsilon/c^2 + p/c^2\right)\left(m + 4\pi r^3 p/c^2\right), \tag{5.218}$$

an equation known as the *Tolman–Oppenheimer–Volkoff (TOV) equation*, after Richard C. Tolman (1881–1948), J. Robert Oppenheimer (1904–1967), and George Volkoff (1914–2000), who were among the first to study the structure of spherical bodies in general relativity.

The equations of relativistic stellar structure involve the fluid variables ρ, ϵ, p, and the metric variables m and Φ. There is one more variable (ϵ) than in the Newtonian theory, and the equations must be supplemented by the first law $\rho\, d\epsilon = (\epsilon + p)\, d\rho$ and an equation of state $p = p(\rho)$. The equations are integrated outward from $r = 0$ with the boundary conditions $\rho(0) = \rho_c$, $\epsilon(0) = \epsilon_c$, $p(0) = p_c$, $m(0) = 0$, and $\Phi(0) = \Phi_c$. Integration proceeds until the pressure vanishes at the boundary $r = R_0$, and $M := m(r = R_0)$ is the body's total mass. The constant Φ_c can be adjusted so that $e^{-2\Phi/c^2}$ matches the value $1 - 2GM/(c^2 R_0)$ at the boundary, and this ensures that the interior metric joins smoothly with the Schwarzschild exterior metric at $r = R_0$.

Incompressible fluid

The equations of relativistic stellar structure must typically be integrated numerically. There is, however, a simple situation that permits an analytical integration, the rather unphysical case of an incompressible fluid with $\rho = \rho_0 =$ constant and $\epsilon = 0$. This situation was the one considered by Karl Schwarzschild in 1915, and the resulting solution is known as the interior Schwarzschild solution. The corresponding Newtonian situation was examined back in Sec. 2.2.2.

The mass function of an incompressible fluid is obtained directly from Eq. (5.215), and is given by

$$m(r) = \frac{4\pi}{3}\rho_0 r^3 = M(r/R_0)^3, \tag{5.219}$$

in which M is the total mass and R_0 the stellar radius. With a little work it can then be shown that Eq. (5.218) integrates to

$$p = \rho_0 c^2 \frac{\sqrt{1 - Rr^2/R_0^3} - \sqrt{1 - R/R_0}}{3\sqrt{1 - R/R_0} - \sqrt{1 - Rr^2/R_0^3}}, \tag{5.220}$$

in which $R := 2GM/c^2 < R_0$ is the body's Schwarzschild radius. The pressure properly vanishes at $r = R_0$, and at $r = 0$ it is given by

$$p_c = \rho_0 c^2 \frac{1 - \sqrt{1 - R/R_0}}{3\sqrt{1 - R/R_0} - 1}. \tag{5.221}$$

This equation reveals that for a fixed density, the central pressure increases as the ratio R_0/R decreases, that is, as the body becomes increasingly compact. This is the expected behavior: more pressure is required to support a more compact body against its own weight. But the central pressure becomes *infinite* when R_0 reaches a critical value given by

$$R_{0,\text{crit}} = \frac{9}{8}R = \frac{9}{4}\frac{GM}{c^2}, \tag{5.222}$$

or when the mass reaches a critical value given by

$$M_{\text{crit}} = \left(\frac{3}{4\pi}\right)^{1/2}\left(\frac{4c^2}{9G}\right)^{3/2}\rho_0^{-1/2} = 5.69 M_\odot \left(\frac{4\times 10^{17}\,\text{kg/m}^3}{\rho_0}\right)^{1/2}; \tag{5.223}$$

the fiducial value adopted for ρ_0 is typical of densities found in neutron stars. We have found that the total mass of an incompressible body in general relativity cannot exceed M_{crit} if the body is to remain in hydrostatic equilibrium. This limiting mass is a purely relativistic phenomenon; there is no such limit in Newtonian gravity.

Relativistic polytropes

A maximum mass is a generic feature of relativistic stellar models, and the phenomenon is also witnessed in relativistic polytropes, the extension to general relativity of the Newtonian polytropes examined in Sec. 2.2.3. We consider a fluid with an equation of state

$$p = K\rho^{1+1/n}, \qquad \epsilon = np, \tag{5.224}$$

directly imported from Sec. 2.2.3. As in the Newtonian discussion, we introduce the scales ρ_c (central density), $p_c = K\rho_c^{1+1/n}$ (central pressure), $m_0 = 4\pi\rho_c r_0^3$ (mass scale), $r_0^2 = (n+1)p_c/(4\pi G\rho_c^2)$ (squared length scale), and work with scale-free variables θ, μ, and ξ, such that $\rho = \rho_c\theta^n$, $p = p_c\theta^{n+1}$, $m = m_0\mu$, and $r = r_0\xi$. We introduce also a relativistic parameter

$$b := \frac{p_c}{\rho_c c^2} = \frac{K\rho_c^{1/n}}{c^2}, \tag{5.225}$$

which is a dimensionless measure of how relativistic the central conditions are; when $b \ll 1$ the body is non-relativistic, and it is highly relativistic when $b > 1$.

It is a simple matter to convert Eqs. (5.215) and (5.218) to dimensionless forms involving the scale-free variables. We obtain

$$\frac{d\mu}{d\xi} = \xi^2\theta^n(1 + nb\theta) \tag{5.226}$$

and

$$\frac{d\theta}{d\xi} = -\left(\frac{\mu}{\xi^2} + b\xi\theta^{n+1}\right)\frac{1 + (n+1)b\theta}{1 - 2(n+1)b\mu/\xi}, \tag{5.227}$$

and these are relativistic versions of the Lane–Emden equations of Newtonian stellar structure; we see that they reduce to Eqs. (2.49) when $b \to 0$. The equations are integrated outward from $\xi = 0$ with the boundary conditions $\theta(0) = 1$ and $\mu(0) = 0$. Integration proceeds until $\theta = 0$ at $\xi = \xi_1$, which marks the body's boundary, where both the pressure and density vanish. The body's total mass is then $M = m_0\mu_1$ with $\mu_1 := \mu(\xi = \xi_1)$, while the body's radius is $R_0 = r_0\xi_1$.

For a selected equation of state, that is, for each choice of parameters K and n, integration of Eqs. (5.226) and (5.227) gives rise to a continuous sequence of stellar models parameterized by the central density ρ_c. Alternatively, because the central density is related to the relativistic parameter b by Eq. (5.225), it is convenient to adopt b as a parameter on the sequence. Because m_0 and r_0 depend on the central density (and therefore on b), it is necessary to rescale the mass and length units so as to eliminate this dependence before we produce plots of the mass M and radius R_0 as functions of b on the sequence. We therefore set

$$M = \bar{m}_0 b^{(3-n)/2}\mu_1, \qquad R_0 = \bar{r}_0 b^{(1-n)/2}\xi_1, \tag{5.228}$$

in which

$$\bar{m}_0 := m_0 b^{-(3-n)/2} = \frac{(n+1)^{3/2}K^{n/2}c^{3-n}}{(4\pi)^{1/2}G^{3/2}}, \tag{5.229a}$$

$$\bar{r}_0 := r_0 b^{-(1-n)/2} = \frac{(n+1)^{1/2}K^{n/2}c^{1-n}}{(4\pi)^{1/2}G^{1/2}}, \tag{5.229b}$$

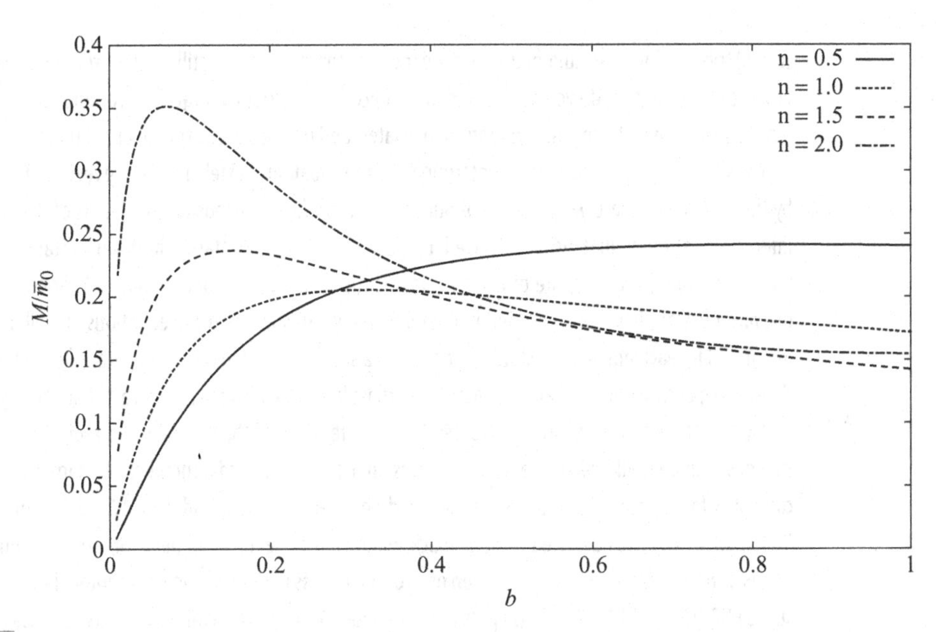

Fig. 5.7 Mass of a relativistic polytrope, in units of $\bar{m}_0$, as a function of the relativistic parameter b, for selected values of the polytropic index n. Models with $b \ll 1$ are non-relativistic, while models with b close to one are relativistic. In each case the mass reaches a maximum value at a critical value of b. A larger n means a stiffer equation of state, and a higher maximum mass.

are the rescaled mass and length units, respectively, which are independent of b and therefore constant on the sequence. Plots of $M/\bar{m}_0$ as a function of b for selected values of the polytropic index n are presented in Fig. 5.7.

Box 5.9 Neutron stars

Except for black holes, neutron stars are the most dense objects in the universe, and their gravitational fields are so strong that their description must be based upon general relativity. A typical neutron star has a mass around $1.4\,M_\odot$, a radius of the order of 10 km, a density comparable to $6.6 \times 10^{17}\,\mathrm{kg/m^3}$, and the strength of its gravitational field is measured by $GM/(c^2R) \simeq 0.21$. Most neutron stars harbour intense magnetic fields, ranging from 10^8 T to perhaps 10^{11} T, and many are active pulsars, emitting sharp pulses of radio waves or X-rays at very regular time intervals. Neutron stars are formed in supernovae events, during which the outer layers of a normal star at the end stage of its stellar evolution are ejected in a violent explosion, and the core undergoes gravitational collapse. The neutron-star state is one of the three possible forms of stellar death, along with the white-dwarf state for less massive stars (up to approximately 5 solar masses) and the black-hole state for more massive stars (from approximately 30 solar masses).

The internal composition of neutron stars is governed by the physics of nuclear matter at densities that far exceed those of ordinary nuclei. While this nuclear physics is fairly well understood at the relatively low densities found near the surface, it becomes less certain as the density increases toward the center. The outer layer of a neutron star is believed to be made of ordinary heavy nuclei, but this familiar form of matter gives way to a distribution of superfluid neutrons deeper inside the star. Deeper still the neutrons arrange themselves in lattice structures, and eventually the density becomes sufficiently large to accommodate exotic forms of matter such as pion, kaon, and hyperon condensates, and perhaps even strange quark matter.

The structure of a neutron star is determined by the equations of stellar structure reviewed in the text, and by the equation of state $p = p(\rho)$ of nuclear matter, which depends on the details of the strong nuclear interactions at the high densities involved. Because the equation of state is highly uncertain beyond normal nuclear densities, the structure of neutron stars is still poorly understood. In practice various models of the relevant nuclear physics are encoded in model equations of state, and the equations of stellar structure are integrated for each equation of state, giving rise to a sequence of stellar models parameterized by the central density ρ_c; each sequence features a maximum mass beyond which the models are dynamically unstable.

Models of neutron stars are constrained by measurements of their masses and radii. A measurement of the mass can be made when the neutron star is an active pulsar and a member of a binary system. In such cases it is found that most masses are clustered between $1.3\ M_\odot$ and $1.6\ M_\odot$, but a mass as high as $2.4 \pm 0.4\ M_\odot$ was deduced for the "black-widow pulsar," a 1.6 ms pulsar in orbit around a low-mass star, which has transferred a large fraction of its original mass to the neutron star. Another large-mass neutron star is PSR J1614-2230, a 3.15 ms pulsar in orbit around a $0.5\ M_\odot$ white dwarf; its mass was measured in 2010 to be $1.97 \pm 0.04\ M_\odot$ by exploiting the Shapiro time delay, a relativistic effect to be introduced in Sec. 10.2.5. These measurements imply that the maximum mass of a neutron star must exceed $2\ M_\odot$, and this observation rules out a number of model equations of state, including some associated with more exotic forms of matter.

The radius of a neutron star is much harder to estimate, but substantial progress has been achieved in the last few years by exploiting the measured light curves of X-ray bursters displaying photospheric radius expansion, and of transient low-mass X-ray binaries. These measurements permit the simultaneous estimation of both the mass and radius of a small sample of neutron stars (with fairly large uncertainties), and these estimates can be used to constrain the equation of state of nuclear matter. These studies reveal that a typical neutron star of $1.4\ M_\odot$ should have a radius between 11 km and 12 km, and confirm that the maximum mass should be larger than $2\ M_\odot$.

5.7 Bibliographical notes

Our survey of curved spacetime and general relativity is very incomplete, and the reader is invited to pursue an advanced education by consulting some of the excellent textbooks devoted to this subject. Among our favorites are the introductory texts by Schutz (2003 and 2009) and Hartle (2003), the venerable classics by Weinberg (1972) and Misner, Thorne, and Wheeler (1973), and the more modern treatments by Wald (1984) and Carroll (2003).

Our presentation of the linearized theory in Sec. 5.5 is modelled on Flanagan and Hughes (2005). Schwarzschild's original paper can be found in an English translation at `arXiv.org/abs/physics/9905030`. (*Warning*: The translators of the paper claim incorrectly that it proves the "science fiction" of black holes; Schwarzschild's solution uses slightly unusual coordinates, but it's still the spacetime of a black hole.) The Painlevé–Gullstrand coordinates to describe the Schwarzschild metric (Sec. 5.6.2) were discovered independently by Painlevé (1921) and Gullstrand (1922); their properties are described in more depth in Martel and Poisson (2001). Relativistic polytropes (Sec. 5.6.5) were first investigated by Tooper (1965).

The physics of neutron stars, summarized so briefly in Box 5.9, is reviewed much more thoroughly in Shapiro and Teukolsky (1983), Glendenning (2000), and Friedman and Stergioulas (2013). A survey of equations of state for nuclear matter is provided by Lattimer and Prakash (2001 and 2007), and Steiner *et al.* (2010) describe the state-of-the-art in mass and radius determinations. The recent discovery of a two solar-mass pulsar is reported in Demorest *et al.* (2010).

5.8 Exercises

5.1 (a) Transform the metric of Minkowski spacetime from the usual (ct, x, y, z) coordinates of an inertial frame, to new coordinates (ct', x', y', z') related by the coordinate transformation

$$\begin{aligned} ct &= \alpha^{-1}(1+\alpha x')\sinh(\alpha ct'), \\ x &= \alpha^{-1}(1+\alpha x')\cosh(\alpha ct'), \\ y &= y', \\ z &= z', \end{aligned}$$

where $\alpha := g/c^2$, with g a constant with the dimensions of an acceleration.

(b) For $\alpha ct' \ll 1$, show that the origin of the new frame ($x' = 0$) is uniformly accelerated, in the Newtonian sense.

(c) Show that a clock at rest at $x' = h$ runs fast compared to a clock at rest at $x' = 0$ by a factor $(1 + gh/c^2)$. How does this relate to the principle of equivalence?

(d) For an observer at the spatial origin of the new frame, calculate the components of the spacetime velocity $\vec{u}$ and acceleration $\vec{a}$ vectors. Verify that $\vec{u}\cdot\vec{u} = -c^2$, $\vec{u}\cdot\vec{a} = 0$, and that $\vec{a}\cdot\vec{a} = g^2$.

This is the Rindler spacetime of a uniformly accelerated observer.

5.2 In electrodynamics, a charge that accelerates relative to an inertial frame radiates electromagnetic radiation. A charge that is at rest or in uniform motion in an inertial frame does not radiate. Consider the following situations:

(a) a charge is at rest near the surface of the Earth;

(b) a charge is in free fall near the surface of the Earth.

Assume for simplicity that the Earth itself is at rest in an inertial frame. In which case does the charge radiate electromagnetic waves to infinity? Does this conflict with the Einstein equivalence principle? Why or why not?

5.3 Prove the following useful relations:

(a) $\nabla_\gamma g_{\alpha\beta} = 0$;

(b) $g_{\alpha\mu}\partial_\gamma g^{\mu\beta} = -g^{\mu\beta}\partial_\gamma g_{\alpha\mu}$;

(c) $\partial_\gamma g^{\alpha\beta} = -\Gamma^\alpha_{\mu\gamma} g^{\mu\beta} - \Gamma^\beta_{\mu\gamma} g^{\mu\alpha}$;

(d) $\Gamma^\alpha_{\alpha\gamma} = \partial_\gamma \ln(\sqrt{-g})$, where $g = \det(g_{\mu\nu})$;

(e) $\sqrt{-g} g^{\mu\nu}\Gamma^\alpha_{\mu\nu} = -\partial_\beta(\sqrt{-g} g^{\alpha\beta})$.

5.4 Prove the following results:

(a) Divergence of a vector:

$$\nabla_\alpha A^\alpha = \frac{1}{\sqrt{-g}}\partial_\alpha\left(\sqrt{-g}A^\alpha\right);$$

(b) Divergence of a second-rank tensor:

$$\nabla_\beta A_\alpha^{\ \beta} = \frac{1}{\sqrt{-g}}\partial_\beta\left(\sqrt{-g}A_\alpha^{\ \beta}\right) - \Gamma^\gamma_{\alpha\mu} A_\gamma^{\ \mu},$$

$$\nabla_\beta A^{\alpha\beta} = \frac{1}{\sqrt{-g}}\partial_\beta\left(\sqrt{-g}A^{\alpha\beta}\right) + \Gamma^\alpha_{\mu\nu} A^{\mu\nu};$$

(c) D'Alembertian of a scalar field:

$$\Box_g \phi := g^{\mu\nu}\nabla_\mu\nabla_\nu\phi = \frac{1}{\sqrt{-g}}\partial_\mu\left(\sqrt{-g}g^{\mu\nu}\partial_\nu\phi\right).$$

5.5 Show explicitly that if a vector $\vec{A}$ is parallel-transported along a world line, then $\vec{A}\cdot\vec{A}$ is constant along the world line.

5.6 (a) By transforming from x-y coordinates to r-ϕ coordinates, write down the metric of two-dimensional Euclidean space in polar coordinates, and calculate all the Christoffel symbols.

(b) Write down the geodesic equation, $d^2x^j/ds^2 + \Gamma^j_{nk}(dx^k/ds)(dx^n/ds) = 0$, for the two coordinates $x^1 = r$ and $x^2 = \phi$, where s is proper distance along the geodesic.

(c) Solve the geodesic equations explicitly for $r(s)$ and $\phi(s)$, and show that they are equivalent to the usual equations for straight-line motion, $x = As + B$, $y = Cs + D$, where A, B, C, and D are constants.

5.7 (a) Write down the metric of a two-dimensional sphere of unit radius, in standard θ–ϕ coordinates. Calculate the Christoffel symbols, and find the geodesic equations for $\theta(s)$ and $\phi(s)$.

(b) Show that the curve ϕ = constant, $\theta = As + B$ is a geodesic, and describe the curve in words.

(c) Show that the curve θ = constant, $\phi = Cs + D$ is a geodesic for only one specific value of θ. Find that value and describe the curve in words.

5.8 As we saw in the text, a timelike geodesic between events 1 and 2 in spacetime is a curve that maximizes the proper-time functional

$$\tau = \frac{1}{c}\int_1^2 L(r^\alpha, \dot{r}^\alpha)\, d\lambda, \qquad L(r^\alpha, \dot{r}^\alpha) := \sqrt{-g_{\alpha\beta}\dot{r}^\alpha \dot{r}^\beta},$$

where $r^\alpha(\lambda)$ is an arbitrary world line that links events 1 and 2, λ is an arbitrary parameter that ranges between λ_1 and λ_2, and $\dot{r}^\alpha = dx^\alpha/d\lambda$.

(a) Prove that the functional is invariant under a general reparameterization of the form $\lambda \to \lambda' = f(\lambda)$, where f is any monotonic function of its argument.

(b) Prove that the geodesic is described by a generalized form of the geodesic equation, $\ddot{r}^\alpha + \Gamma^\alpha_{\beta\gamma}\dot{r}^\beta \dot{r}^\gamma = \kappa \dot{r}^\alpha$. Express $\kappa(\lambda)$ in terms of L.

(c) Show that the geodesic equation assumes its usual form (with a zero right-hand side) when proper time τ is adopted as the parameter λ. (This can be done only after the variation has been performed.)

(d) Prove that the usual form of the geodesic equation is invariant under a reparameterization of the form $\tau \to \tau' = a\tau + b$, where a and b are constants; this defines the class of *affine parameters*.

(e) Show that the usual form of the geodesic equation can be derived on the basis of the alternative Lagrangian $L'(r^\alpha, \dot{r}^\alpha) = g_{\alpha\beta}\dot{r}^\alpha \dot{r}^\beta$, where $\dot{r}^\alpha = dx^\alpha/d\tau$.

5.9 (a) Show that the condition $R_{\alpha[\beta\gamma\delta]} = 0$ is equivalent to the cyclic condition of Eq. (5.59d).

(b) By counting components, show that the antisymmetry conditions $R_{\alpha\beta(\gamma\delta)} = 0$ and $R_{(\alpha\beta)\gamma\delta} = 0$, together with the cyclic condition $R_{\alpha[\beta\gamma\delta]} = 0$, are sufficient to leave 20 independent components of the Riemann tensor.

(c) Show that the antisymmetry conditions together with the symmetry $R_{\gamma\delta\alpha\beta} = R_{\alpha\beta\gamma\delta}$ are *not* sufficient, but must be supplemented by one additional constraint, given by $R_{\alpha[\beta\gamma\delta]} = 0$ with all four indices distinct.

5.10 (a) Verify that

$$R^\alpha{}_{\beta\gamma\delta} = \partial_\gamma \Gamma^\alpha{}_{\beta\delta} - \partial_\delta \Gamma^\alpha{}_{\beta\gamma} + \Gamma^\alpha_{\mu\gamma}\Gamma^\mu_{\beta\delta} - \Gamma^\alpha_{\mu\delta}\Gamma^\mu_{\beta\gamma}$$

by carrying out the explicit calculations of Eq. (5.57).

(b) Prove the Bianchi identity,

$$\nabla_\alpha R_{\mu\nu\beta\gamma} + \nabla_\gamma R_{\mu\nu\alpha\beta} + \nabla_\beta R_{\mu\nu\gamma\alpha} = 0.$$

Hint: Because it is a tensorial equation, the identity can be verified in any coordinate system. Choose a local inertial frame, in which the Christoffel symbols vanish (but not their derivatives!).

(c) Verify the contracted Bianchi identity, $\nabla_\beta G^{\alpha\beta} = 0$.

5.11 Show that the metric

$$ds^2 = -d(ct)^2 + (ct/a)^2\left[\frac{dr^2}{1 + r^2/a^2} + r^2\left(d\theta^2 + \sin^2\theta\, d\phi^2\right)\right],$$

where a is a constant with units of distance, is that of flat spacetime: (a) by finding a coordinate transformation that brings the line element to the Minkowski form everywhere in spacetime; and (b) by calculating at least five non-trivial components of its Riemann tensor.

5.12 (a) Calculate the Riemann curvature tensor for the metric of a two-dimensional sphere of unit radius.
(b) Repeat for an infinitely long, two-dimensional cylinder of unit radius. What do you make of your answer for the Riemann tensor?

5.13 (a) For a perfect fluid, show that the relativistic Euler equation can be written in the form

$$(\mu + p)\left(\frac{du_\alpha}{d\tau} - \frac{1}{2}u^\beta u^\gamma \partial_\alpha g_{\beta\gamma}\right) = -c^2 \partial_\alpha p - u_\alpha \frac{dp}{d\tau},$$

in which τ is proper time along the world line of a fluid element.
(b) If the fluid undergoes a *stationary* flow (pressure, density independent of t) in a *stationary* spacetime ($g_{\mu\nu}$ independent of t), prove that $(\mu + p)u_0\mathcal{V} = \text{constant}$ along the flow lines. This is the relativistic version of Bernoulli's equation.

5.14 Show that the mass density of a point particle can be written in the form

$$\rho(x^\alpha) = \frac{mc}{u^0\sqrt{-g}}\delta_3\big(\boldsymbol{x} - \boldsymbol{r}(t)\big).$$

5.15 By exploiting only spherical symmetry, we can always write the metric of a spherically-symmetric spacetime in the form

$$ds^2 = -e^{-2\Phi/c^2}\,d(ct)^2 - 2h\,d(ct)dr + e^{2\Lambda/c^2}\,dr^2 + r^2\big(d\theta^2 + \sin^2\theta\,d\phi^2\big),$$

in which Φ, Λ, and h are arbitrary functions of t and r. Show that it is always possible to find a transformation $t = F(T, r)$ so that the metric in $(T,\ r,\ \theta,\ \phi)$ coordinates has no off-diagonal $d(ct)dr$ term.

5.16 (a) Show that the isotropic coordinates

$$X^{(\mu)} := \big\{t,\ x_{\rm iso} = r_{\rm iso}\sin\theta\cos\phi,\ y_{\rm iso} = r_{\rm iso}\sin\theta\sin\phi,\ z_{\rm iso} = r_{\rm iso}\cos\theta\big\}$$

do not satisfy the harmonic coordinate condition, Eq. (5.175).
(b) Show that the spherical polar harmonic coordinates $(t, r_{\rm h}, \theta, \phi)$ do not satisfy the harmonic coordinate condition.

5.17 If $g_{\alpha\beta} := \eta_{\alpha\beta} + p_{\alpha\beta}$, show that to first order in $p_{\alpha\beta}$, the harmonic coordinate condition (5.177) is equivalent to the Lorenz-gauge condition of linearized theory, $\partial^\beta \bar{p}_{\alpha\beta} = 0$, where $\bar{p}_{\alpha\beta}$ is the trace-reversed perturbation, and indices are raised and lowered using $\eta_{\alpha\beta}$.

5.18 In the course of your study of general relativity you come across a vacuum solution to the Einstein field equations given by

$$ds^2 = -d(ct)^2 + \left(\frac{4\alpha ctx^j}{r(r^2-\alpha^2)}\right) d(ct)dx^j$$
$$+ \left[\frac{(r+\alpha)^4}{r^4}\delta_{jk} - \frac{4\alpha^2c^2t^2x^jx^k}{r^2(r^2-\alpha^2)^2}\right]dx^jdx^k,$$

in which α is a constant and $r^2 := \delta_{jk}x^jx^k$. You take it upon yourself to study the significance of this spacetime.

(a) Transform the metric from the Cartesian coordinates x^j to the standard spherical polar coordinates (r, θ, ϕ), and show that the metric is, in fact, spherically symmetric. *Hint:* What are $\delta_{jk}dx^jdx^k$ and $(x^j/r)dx^j$ in spherical coordinates?

(b) Calculate the acceleration of a body at rest at very large r, and use your result to relate the parameter α to the total mass M in the spacetime.

(c) Find a coordinate transformation that puts the metric in a static form, and confirm your result in part (b) by reading off the mass directly from the metric.

(d) Can you name this spacetime?

5.19 Calculate the potential $\Phi(r)$ for a spherical body of uniform density. Make sure to impose the proper boundary conditions at the boundary $r = R_0$.

5.20 Suppose that a free-falling observer moves radially inward in the Schwarzschild spacetime, having started from rest at infinity; denote the observer's four-velocity u^α. Furthermore, suppose that a photon is also moving radially inward, and that it is just about to catch up with the observer; denote the photon's four-momentum p^α and its energy-at-infinity $\hbar\omega_\infty$. On the basis of the general formula

$$\hbar\omega = -g_{\alpha\beta}p^\alpha u^\beta,$$

calculate the photon's frequency ω as measured by the observer when the photon catches up with her. Express your result in terms ω_∞ and r_0, the radius at which the photon and observer meet. Is this a redshift or a blueshift? Does the expression become singular at the event horizon? Should it?

6 Post-Minkowskian theory: Formulation

In this chapter we embark on a general program to specialize the formulation of general relativity to a description of weak gravitational fields. We will go from the exact theory, which governs the behavior of arbitrarily strong fields, such as those of neutron stars and black holes, to a useful approximation that applies to weak fields, such as those of planets, main-sequence stars, and white dwarfs. This approximation will reproduce the predictions of Newtonian theory, but we will formulate a method that can be pushed systematically to higher and higher order to produce an increasingly accurate description of a weak gravitational field. We shall find that the method is so successful that it can actually handle fields that are not so weak. For example, it provides a perfectly adequate description of gravity at a safe distance from a neutron star, and it can be used as a foundation to study the motion of a binary black-hole system, provided that the mutual gravity between bodies is weak.

The foundation for these methods is "post-Minkowskian theory," the topic of this chapter and the next. In post-Minkowskian theory the strength of the gravitational field is measured by the gravitational constant G, and the Einstein field equations are formally expanded in powers of G. At zeroth post-Minkowskian order there is no field, and one deals with Minkowski spacetime. At first post-Minkowskian order the gravitational field appears as a correction of order G to the Minkowski metric, and the (linearized) field equations are integrated to obtain this correction. The correction is refined by terms of order G^2 in the second post-Minkowskian approximation, and the process is continued until the desired degree of accuracy is achieved.

The formulation of the Einstein field equations that is best suited to this post-Minkowskian expansion was put forward by Landau and Lifshitz, and this framework is introduced in Sec. 6.1. In Sec. 6.2 we refine the Landau–Lifshitz formulation by imposing the harmonic coordinate conditions, and we show that the exact field equations can be expressed as a set of ten wave equations in Minkowski spacetime, with complicated and highly non-linear source terms. We explain how the metric can be systematically expanded in powers of the gravitational constant G and inserted within the wave equations; these are iterated a number of times, and each iteration increases the accuracy of the solution by one power of G.

In Sec. 6.3 we develop mathematical techniques to integrate the wave equation in flat spacetime. We begin by introducing the retarded Green's function for the wave equation, and we explain how the solution can be expressed as an integral over the past light cone of the spacetime point at which it is evaluated. Our methods involve a partition of three-dimensional space into near-zone and wave-zone regions, and we describe how the light-cone integral, decomposed into near-zone and wave-zone contributions, can be evaluated.

In Chapter 7 we implement the techniques developed here and construct the *second post-Minkowskian approximation* to the metric of a weakly curved spacetime. The post-Minkowskian approximation does not rely on an assumption that the matter distribution moves slowly. While this may be the typical context – in a gravitationally bound system, weak gravitational fields induce slow motions – we shall nevertheless divorce the weak-field assumption from a logically distinct slow-motion assumption, which is not required for the developments of this chapter. We shall eventually return to slow motions, however, and formulate an approximation method that incorporates both weak-field and slow-motion aspects. This is the domain of *post-Newtonian theory*, an approximation to general relativity that combines an expansion in powers of G (to measure the strength of the field) with an expansion in powers of c^{-2} (to measure the velocity of the matter distribution). Post-Newtonian theory is informally introduced in Chapter 7, but it is developed more systematically in Chapters 8, 9, and 10. The other main applications of post-Minkowskian theory, gravitational waves and radiation reaction, are the subject of Chapters 11 and 12.

6.1 Landau–Lifshitz formulation of general relativity

6.1.1 New formulation of the field equations

The post-Minkowskian approach to integrate the Einstein field equations is based on the Landau and Lifshitz formulation of these equations. In this framework the main variables are not the components of the metric tensor $g_{\alpha\beta}$ but those of the "gothic inverse metric"

$$\mathfrak{g}^{\alpha\beta} := \sqrt{-g}\,g^{\alpha\beta}, \tag{6.1}$$

where $g^{\alpha\beta}$ is the inverse metric and g the metric determinant. The factor of $\sqrt{-g}$ implies that $\mathfrak{g}^{\alpha\beta}$ is not a tensor; such objects, which differ from tensors by factors of the metric determinant, are known as *tensor densities*. Knowledge of the gothic metric is sufficient to determine the metric itself: note first that $\det[\mathfrak{g}^{\alpha\beta}] = g$, so that g can be directly obtained from the gothic metric; then Eq. (6.1) gives $g^{\alpha\beta}$, which can be inverted to obtain $g_{\alpha\beta}$.

In the Landau–Lifshitz formulation, the left-hand side of the field equations is built from

$$H^{\alpha\mu\beta\nu} := \mathfrak{g}^{\alpha\beta}\mathfrak{g}^{\mu\nu} - \mathfrak{g}^{\alpha\nu}\mathfrak{g}^{\beta\mu}. \tag{6.2}$$

This tensor density is readily seen to possess the same symmetries as the Riemann tensor,

$$H^{\mu\alpha\beta\nu} = -H^{\alpha\mu\beta\nu}, \qquad H^{\alpha\mu\nu\beta} = -H^{\alpha\mu\beta\nu}, \qquad H^{\beta\nu\alpha\mu} = H^{\alpha\mu\beta\nu}. \tag{6.3}$$

It also satisfies the remarkable identity

$$\partial_{\mu\nu}H^{\alpha\mu\beta\nu} = 2(-g)G^{\alpha\beta} + \frac{16\pi G}{c^4}(-g)t^{\alpha\beta}_{\text{LL}}, \tag{6.4}$$

in which $G^{\alpha\beta}$ is the Einstein tensor, and

$$(-g)t_{\rm LL}^{\alpha\beta} := \frac{c^4}{16\pi G}\biggl\{\partial_\lambda \mathfrak{g}^{\alpha\beta}\partial_\mu \mathfrak{g}^{\lambda\mu} - \partial_\lambda \mathfrak{g}^{\alpha\lambda}\partial_\mu \mathfrak{g}^{\beta\mu} + \frac{1}{2}g^{\alpha\beta}g_{\lambda\mu}\partial_\rho \mathfrak{g}^{\lambda\nu}\partial_\nu \mathfrak{g}^{\mu\rho}$$
$$- g^{\alpha\lambda}g_{\mu\nu}\partial_\rho \mathfrak{g}^{\beta\nu}\partial_\lambda \mathfrak{g}^{\mu\rho} - g^{\beta\lambda}g_{\mu\nu}\partial_\rho \mathfrak{g}^{\alpha\nu}\partial_\lambda \mathfrak{g}^{\mu\rho} + g_{\lambda\mu}g^{\nu\rho}\partial_\nu \mathfrak{g}^{\alpha\lambda}\partial_\rho \mathfrak{g}^{\beta\mu}$$
$$+ \frac{1}{8}\bigl(2g^{\alpha\lambda}g^{\beta\mu} - g^{\alpha\beta}g^{\lambda\mu}\bigr)\bigl(2g_{\nu\rho}g_{\sigma\tau} - g_{\rho\sigma}g_{\nu\tau}\bigr)\partial_\lambda \mathfrak{g}^{\nu\tau}\partial_\mu \mathfrak{g}^{\rho\sigma}\biggr\} \tag{6.5}$$

is the Landau–Lifshitz pseudotensor, so named because it does not transform as a tensor under a general coordinate transformation; the quantity $\partial_{\mu\nu}H^{\alpha\mu\beta\nu}$ is also a pseudotensor, and $(-g)G^{\alpha\beta}$ is a tensor density. Equation (6.4) is valid for any spacetime, whether or not its metric is a solution to the Einstein field equations.

The identity of Eq. (6.4) implies that the Einstein field equations, $G^{\alpha\beta} = (8\pi G/c^4)T^{\alpha\beta}$, can be expressed in the alternative, non-tensorial form

$$\partial_{\mu\nu}H^{\alpha\mu\beta\nu} = \frac{16\pi G}{c^4}(-g)\bigl(T^{\alpha\beta} + t_{\rm LL}^{\alpha\beta}\bigr). \tag{6.6}$$

As promised, the left-hand side involves $H^{\alpha\mu\beta\nu}$, and the right-hand side is built from $T^{\alpha\beta}$, the energy-momentum tensor of the matter distribution, and $t_{\rm LL}^{\alpha\beta}$. This form of the field equations provides the Landau–Lifshitz pseudotensor with a loose physical interpretation: it represents the distribution of gravitational-field energy in spacetime, which is added to the matter contribution on the right-hand side of the field equations.

By virtue of the antisymmetry of $H^{\alpha\mu\beta\nu}$ in the last pair of indices, we have that the equation

$$\partial_{\beta\mu\nu}H^{\alpha\mu\beta\nu} = 0 \tag{6.7}$$

holds as an identity. This, together with Eq. (6.6), implies that

$$\partial_\beta\Bigl[(-g)\bigl(T^{\alpha\beta} + t_{\rm LL}^{\alpha\beta}\bigr)\Bigr] = 0. \tag{6.8}$$

These are conservation equations for the total energy-momentum pseudotensor, expressed in terms of a partial-derivative operator. These equations are equivalent to the usual expression of energy-momentum conservation, $\nabla_\beta T^{\alpha\beta} = 0$, which involves only the matter's energy-momentum tensor and a covariant-derivative operator.

As we have seen, Eqs. (6.6) and (6.8) suggest that $t_{\rm LL}^{\alpha\beta}$ can be interpreted as an energy-momentum (pseudo)tensor for the gravitational field, and this interpretation is supported by the fact that the Landau–Lifshitz pseudotensor is quadratic in $\partial_\mu \mathfrak{g}^{\alpha\beta}$, just as the energy-momentum tensor of the electromagnetic field is quadratic in $\partial_\mu A^\alpha$. This interpretation, however, is not to be taken literally. It is, after all, based on a very specific reformulation of the Einstein field equations, and other reformulations would give rise to other candidates for the energy-momentum pseudotensor. And it is based on a non-tensorial quantity whose numerical value can change arbitrarily by performing a coordinate transformation; indeed, $t_{\rm LL}^{\alpha\beta}$ can be made to *vanish at any selected event in spacetime* by adopting Riemann normal coordinates in the neighborhood of this event (refer to Sec. 5.2.5). The literature abounds with attempts to introduce *the* energy-momentum tensor for the gravitational field. Such an object does not exist; do not fall prey to false prophets.

Box 6.1 **Two versions of energy-momentum conservation**

We state in the text that the two versions of energy-momentum conservation, $\nabla_\beta T^{\alpha\beta} = 0$ and $\partial_\beta[(-g)(T^{\alpha\beta} + t^{\alpha\beta}_{\rm LL})] = 0$, are equivalent. In fact, there is an important conceptual difference between these statements. The first equation is a direct consequence of the local conservation of energy-momentum, as observed in a local inertial frame; as such it is valid whether or not Einstein's equations are satisfied, or indeed, whether or not general relativity is the correct theory of gravity. The fact that it is compatible with the Bianchi identity, $\nabla_\beta G^{\alpha\beta} = 0$, is an added feature specific to Einstein's theory. There are alternative theories that lack this consistency, and yet $\nabla_\beta T^{\alpha\beta}$ is still zero.

By contrast, the second conservation equation follows only *after* using Einstein's equations to derive Eq. (6.6). Furthermore, the tedious calculations required to establish that the two versions are equivalent involve inserting the field equations (6.6) at various critical steps along the way.

The bottom line is that the conservation equation $\nabla_\beta T^{\alpha\beta} = 0$ is fundamental; the equation $\partial_\beta[(-g)(T^{\alpha\beta} + t^{\alpha\beta}_{\rm LL})] = 0$ is a consequence of Einstein's equations. If Einstein's equations are satisfied, then either equation may be adopted to express energy-momentum conservation, and the statements are equivalent in this sense.

Equations (6.1)–(6.8) form the core of the Landau–Lifshitz framework. It is out of the question to provide a derivation of these equations (the calculations are straightforward but extremely lengthy), but the following considerations, borrowed from Landau and Lifshitz in their influential book *The Classical Theory of Fields*, will provide at least a partial understanding of where they come from.

Let us write down the Einstein field equations, in their usual tensorial form

$$G^{\alpha\beta} = \frac{8\pi G}{c^4} T^{\alpha\beta}, \tag{6.9}$$

at an event P in spacetime, in a local coordinate system such that $\partial_\gamma g_{\alpha\beta}(P) \stackrel{*}{=} 0$. (We do not demand that $g_{\alpha\beta} \stackrel{*}{=} \eta_{\alpha\beta}$ at P; the special equality sign $\stackrel{*}{=}$ means "equals in the selected coordinate system.") In these coordinates the Riemann tensor at P involves only the metric and its second derivatives, and a short computation reveals that the Einstein tensor is given by

$$\begin{aligned} G^{\alpha\beta} \stackrel{*}{=} \frac{1}{2}\big(&g^{\alpha\lambda}g^{\beta\mu}g^{\nu\rho} + g^{\beta\lambda}g^{\alpha\mu}g^{\nu\rho} - g^{\alpha\lambda}g^{\beta\rho}g^{\mu\nu} - g^{\alpha\mu}g^{\beta\nu}g^{\lambda\rho} \\ &- g^{\alpha\beta}g^{\mu\lambda}g^{\nu\rho} + g^{\alpha\beta}g^{\mu\nu}g^{\lambda\rho}\big)\partial_{\mu\nu}g_{\lambda\rho}. \end{aligned} \tag{6.10}$$

If we now compute $\partial_{\mu\nu}H^{\alpha\mu\beta\nu}$, at the same point P and in the same coordinate system, we find after straightforward manipulations that it is given by

$$\begin{aligned} \partial_{\mu\nu}H^{\alpha\mu\beta\nu} \stackrel{*}{=} (-g)\big(&g^{\alpha\lambda}g^{\beta\mu}g^{\nu\rho} + g^{\beta\lambda}g^{\alpha\mu}g^{\nu\rho} - g^{\alpha\lambda}g^{\beta\rho}g^{\mu\nu} - g^{\alpha\mu}g^{\beta\nu}g^{\lambda\rho} \\ &- g^{\alpha\beta}g^{\mu\lambda}g^{\nu\rho} + g^{\alpha\beta}g^{\mu\nu}g^{\lambda\rho}\big)\partial_{\mu\nu}g_{\lambda\rho}. \end{aligned} \tag{6.11}$$

To arrive at this result we had to differentiate $(-g)$ using the rule $\partial_\mu(-g) = (-g)g^{\alpha\beta}\partial_\mu g_{\alpha\beta}$, which leads to $\partial_{\mu\nu}(-g) \stackrel{*}{=} (-g)g^{\alpha\beta}\partial_{\mu\nu}g_{\alpha\beta}$. We also had to relate derivatives of the inverse metric to derivatives of the metric itself; here we have used the rule $\partial_\mu g^{\alpha\beta} = -g^{\alpha\lambda}g^{\beta\rho}\partial_\mu g_{\lambda\rho}$, which leads to $\partial_{\mu\nu}g^{\alpha\beta} \stackrel{*}{=} -g^{\alpha\lambda}g^{\beta\rho}\partial_{\mu\nu}g_{\lambda\rho}$.

Our results imply that

$$\partial_{\mu\nu}H^{\alpha\mu\beta\nu} \stackrel{*}{=} 2(-g)G^{\alpha\beta} . \tag{6.12}$$

This is the same as Eq. (6.4), because $(-g)t^{\alpha\beta}_{\rm LL} \stackrel{*}{=} 0$ at P by virtue of the fact that each term in the Landau–Lifshitz pseudotensor is quadratic in $\partial_\mu \mathfrak{g}^{\alpha\beta}$, which vanishes at P in the selected coordinate system. It is therefore plausible that at any other event in spacetime, and in an arbitrary coordinate system, the identity (6.4) should hold, with a pseudotensor $(-g)t^{\alpha\beta}_{\rm LL}$ that restores all first-derivative terms that were made to vanish at P in the special coordinate system. To show that this pseudotensor takes the specific form of Eq. (6.5) requires a long computation.

6.1.2 Coordinate freedom

The Landau–Lifshitz formulation of general relativity is an *exact reformulation* of the standard form of the theory. No approximations are involved, and no restrictions are placed on the choice of coordinates. It has to be acknowledged, however, that the usefulness of the formalism is largely limited to situations in which (i) the coordinates $x^\alpha = (ct, x^j)$ are modest deformations of the Lorentzian coordinates of flat spacetime, and (ii) $\mathfrak{g}^{\alpha\beta}$ deviates only moderately from the Minkowski metric $\eta^{\alpha\beta}$. For these situations, which form the context of this book, the formalism is an excellent starting point for a systematic approximation method.

In other contexts the Landau–Lifshitz formulation can be a terrible approach. Even a simple problem such as finding the static, spherically symmetric, vacuum solution to the Einstein field equations, the Schwarzschild metric, which took us about six lines of mathematics back in Sec. 5.6, turns out to be a horrible undertaking in the Landau–Lifshitz approach. The lesson is that while the Landau–Lifshitz formulation of the field equations is mathematically equivalent to the tensorial formulation, it is not equivalent when it comes to the ease of performing calculations. In the post-Minkowskian context it is the preferred formulation; in other contexts it decidedly is not.

Given the practical restriction on the coordinate system, it is useful to observe that the Landau–Lifshitz formulation is manifestly invariant under Lorentz transformations, which we express in the general form

$$x^{\mu'} = \Lambda^{\mu'}_{\ \alpha} x^\alpha , \tag{6.13}$$

in which the transformation matrix $\Lambda^{\mu'}_{\ \alpha}$ is constant and possesses a unit determinant. (In fact, the formalism is invariant under all transformations that are linear in the coordinates, so long as the transformation matrix possesses a unit determinant; this ensures that g is not changed during the transformation. The transformation can also be generalized to include uniform translations, $x^\mu \to x^\mu + c^\mu$, where c^μ is a constant vector.) It is easy to show that $\mathfrak{g}^{\alpha\beta}$ and its partial derivatives transform as tensors under this class of transformations, and from this observation it follows immediately that all equations of the formalism are invariant under the transformation of Eq. (6.13).

6.1.3 Integral conservation identities

Because they involve a partial-derivative operator, the differential identities of Eq. (6.8) can immediately be turned into integral identities. We consider a three-dimensional region V, a fixed (time-independent) domain of the spatial coordinates x^j, bounded by a two-dimensional surface S. We assume that V contains at least some of the matter (so that $T^{\alpha\beta}$ is non-zero somewhere within V), but that S does not intersect any of the matter (so that $T^{\alpha\beta} = 0$ everywhere on S).

Total momentum and angular momentum: Volume integrals

We formally define a total momentum four-vector $P^\alpha[V]$ associated with the region V by the three-dimensional integral

$$P^\alpha[V] := \frac{1}{c}\int_V (-g)\big(T^{\alpha 0} + t^{\alpha 0}_{\rm LL}\big)\, d^3x. \tag{6.14}$$

This total momentum includes a contribution from the matter's momentum density $c^{-1}T^{\alpha 0}$, and a contribution from the gravitational field represented by $c^{-1}t^{\alpha 0}_{\rm LL}$; the factor of $(-g)$ is inserted so that we can take advantage of the conservation identities of Eq. (6.8). In flat spacetime and in Lorentzian coordinates, $P^\alpha[V]$ would have a firm interpretation as a total momentum vector associated with the energy-momentum tensor $T^{\alpha\beta}$. In curved spacetime, and in a coordinate system that cannot be assumed to be Lorentzian, the quantity defined by Eq. (6.14) does not have any direct physical meaning. It is, nevertheless, a useful quantity to introduce, as we shall have occasion to recognize.

The momentum four-vector can be decomposed into a time component $P^0[V]$ and a spatial three-vector $P^j[V]$. The time component can be used to define an energy $E[V] := cP^0[V]$ associated with the region V. Alternatively, we can define a total mass

$$M[V] := \frac{1}{c^2}\int_V (-g)\big(T^{00} + t^{00}_{\rm LL}\big)\, d^3x. \tag{6.15}$$

The three-momentum is given by

$$P^j[V] := \frac{1}{c}\int_V (-g)\big(T^{j0} + t^{j0}_{\rm LL}\big)\, d^3x. \tag{6.16}$$

In a similar way we introduce a total angular-momentum tensor $J^{\alpha\beta}[V]$ associated with the region V. This is defined by

$$J^{\alpha\beta}[V] := \frac{1}{c}\int_V \Big[x^\alpha\,(-g)\big(T^{\beta 0} + t^{\beta 0}_{\rm LL}\big) - x^\beta\,(-g)\big(T^{\alpha 0} + t^{\alpha 0}_{\rm LL}\big)\Big]\, d^3x, \tag{6.17}$$

and we note that the tensor is antisymmetric in its indices. The interpretation of $J^{\alpha\beta}[V]$ is easier to identify once it is decomposed into time and spatial components. The antisymmetry of the tensor implies that $J^{00}[V] = 0$. The time-space components can be expressed in the form

$$c^{-1} J^{0j}[V] = P^j[V]t - M[V]R^j[V], \tag{6.18}$$

where

$$R^j[V] := \frac{1}{M[V]c^2} \int_V (-g)(T^{00} + t^{00}_{\rm LL}) x^j \, d^3x \tag{6.19}$$

represents the position of the center-of-mass of the region V. Equation (6.18) reveals that when $c^{-1} J^{0j}[V]$ is a constant, it fixes the position of the center-of-mass at $t = 0$; when it is not a constant it measures the extent by which the center-of-mass fails to move with a total momentum $P^j[V]$. The spatial components of the angular-momentum tensor are

$$J^{jk}[V] = \frac{1}{c} \int_V \Big[x^j (-g)(T^{k0} + t^{k0}_{\rm LL}) - x^k (-g)(T^{j0} + t^{j0}_{\rm LL}) \Big] \, d^3x, \tag{6.20}$$

and this is best recognized in its equivalent vectorial form

$$J^j[V] := \frac{1}{2} \epsilon^j_{\ pq} J^{pq}[V] = \frac{1}{c} \int_V \epsilon^j_{\ pq} x^p (-g)(T^{q0} + t^{q0}_{\rm LL}) \, d^3x, \tag{6.21}$$

where ϵ_{jpq} is the completely antisymmetric permutation symbol. The integrand is the cross product between the position vector x^p and the momentum density $c^{-1}(-g)(T^{q0} + t^{q0}_{\rm LL})$ within V, and it is natural to interpret the integral as the total angular momentum contained in this region.

Total momentum and angular momentum: Surface integrals

The total momentum $P^\alpha[V]$ and angular momentum $J^{\alpha\beta}[V]$ were defined previously in terms of integrals over the three-dimensional region V. It is possible to provide alternative definitions in terms of surface integrals over the two-dimensional surface S that surrounds this region. This is advantageous when the volume integrals of Eq. (6.14) and (6.17) are ill-defined or difficult to compute.

Substituting Eq. (6.6) into Eq. (6.14) gives

$$P^\alpha[V] = \frac{c^3}{16\pi G} \int_V \partial_{\mu\nu} H^{\alpha\mu 0\nu} \, d^3x.$$

Summation over ν must exclude $\nu = 0$, because $H^{\alpha\mu 00} = 0$. We therefore have

$$P^\alpha[V] = \frac{c^3}{16\pi G} \int_V \partial_k \big(\partial_\mu H^{\alpha\mu 0k} \big) \, d^3x,$$

and this can be written as a surface integral by invoking Gauss's theorem. We have

$$P^\alpha[V] := \frac{c^3}{16\pi G} \oint_S \partial_\mu H^{\alpha\mu 0k} \, dS_k, \tag{6.22}$$

where dS_k is an outward-directed surface element on the two-dimensional surface S. Equation (6.22) can be adopted as an alternative definition for the total momentum enclosed by S; $H^{\alpha\mu 0k}$ must then be constructed from a solution to Einstein's equations for the given distribution of matter.

As before the momentum four-vector can be decomposed into time and spatial components. We have that the total mass $M[V]$ can be expressed as

$$M[V] := \frac{c^2}{16\pi G}\oint_S \partial_j H^{0j0k}\, dS_k, \tag{6.23}$$

and the total three-momentum is

$$P^j[V] := \frac{c^3}{16\pi G}\oint_S \partial_n H^{jn0k}\, dS_k - \frac{c^2}{16\pi G}\frac{d}{dt}\oint_S H^{0j0k}\, dS_k. \tag{6.24}$$

With similar manipulations we arrive at a surface-integral definition for the total angular momentum. One of the two terms that occur within the volume integral when we substitute Eq. (6.6) into Eq. (6.17) is $x^\alpha \partial_{k\mu} H^{\beta\mu 0k}$, which can be expressed as $\partial_k(x^\alpha \partial_\mu H^{\beta\mu 0k}) + \partial_\mu H^{\mu\beta 0\alpha}$. The first term gives rise to a surface integral, and the second term can be expanded as $\partial_0 H^{0\beta 0\alpha} + \partial_k H^{k\beta 0\alpha}$; in this, the first term can be ignored because it is symmetric in α and β, and the second term gives rise to another surface integral. Collecting results, we arrive at

$$J^{\alpha\beta}[V] := \frac{c^3}{16\pi G}\oint_S \bigl(x^\alpha \partial_\mu H^{\beta\mu 0k} - x^\beta \partial_\mu H^{\alpha\mu 0k} + H^{0\alpha k\beta} - H^{0\beta k\alpha}\bigr)\, dS_k, \tag{6.25}$$

and this can be adopted as an alternative definition for the total angular momentum enclosed by S.

The decomposition of $J^{\alpha\beta}[V]$ into time and spatial components first returns Eq. (6.18) together with the alternative expression

$$M[V]R^j[V] := \frac{c^2}{16\pi G}\oint_S \bigl(x^j \partial_n H^{0n0k} - H^{0j0k}\bigr)\, dS_k \tag{6.26}$$

for the position of the center-of-mass. It also returns

$$\begin{aligned} J^{jk}[V] :=\ & \frac{c^3}{16\pi G}\oint_S \bigl(x^j \partial_m H^{km0n} - x^k \partial_m H^{jm0n} + H^{0jnk} - H^{0knj}\bigr)\, dS_n \\ & - \frac{c^2}{16\pi G}\frac{d}{dt}\oint_S \bigl(x^j H^{0k0n} - x^k H^{0j0n}\bigr)\, dS_n \end{aligned} \tag{6.27}$$

as an alternative definition for the angular-momentum tensor.

Conservation statements

To obtain the conservation statements satisfied by $P^\alpha[V]$ and $J^{\alpha\beta}[V]$, we differentiate their defining expressions (in terms of volume integrals) with respect to x^0 and use the local conservation identity of Eq. (6.8). Starting with Eq. (6.14), we get

$$\begin{aligned} \frac{d}{dx^0}P^\alpha[V] &= \frac{1}{c}\int_V \partial_0\Bigl[(-g)\bigl(T^{\alpha 0} + t^{\alpha 0}_{\rm LL}\bigr)\Bigr]\, d^3x \\ &= -\frac{1}{c}\int_V \partial_k\Bigl[(-g)\bigl(T^{\alpha k} + t^{\alpha k}_{\rm LL}\bigr)\Bigr]\, d^3x. \end{aligned} \tag{6.28}$$

Converting this to a surface integral, and recalling our previous assumption that S does not intersect the matter distribution, so that $T^{\alpha\beta} = 0$ on S, we arrive at

$$\dot{P}^\alpha[V] = -\oint_S (-g)t^{\alpha k}_{\rm LL}\, dS_k, \tag{6.29}$$

in which an overdot indicates differentiation with respect to $t := x^0/c$. The rate of change of $P^\alpha[V]$ is therefore expressed as a flux integral over S, and the flux is measured by the Landau–Lifshitz pseudotensor (recall the definitions of fluxes provided back in Sec. 4.2). Equation (6.29) gives rise to the individual statements

$$\dot{M}[V] = -\frac{1}{c}\oint_S (-g)t^{0k}_{\rm LL}\, dS_k \tag{6.30}$$

and

$$\dot{P}^j[V] = -\oint_S (-g)t^{jk}_{\rm LL}\, dS_k \tag{6.31}$$

for the fluxes of mass and momentum three-vector across S.

Proceeding along similar lines for the angular-momentum tensor, we arrive at

$$\dot{J}^{\alpha\beta}[V] = -\oint_S \bigl[x^\alpha(-g)t^{\beta k}_{\rm LL} - x^\beta(-g)t^{\alpha k}_{\rm LL}\bigr]\, dS_k. \tag{6.32}$$

The symmetry of $t^{\alpha\beta}_{\rm LL}$ was essential in obtaining this result. When decomposed into time and spatial components, the statement becomes

$$c^{-1}\dot{J}^{0j}[V] = \dot{P}^j[V]t + \frac{1}{c}\oint_S x^j(-g)t^{0k}_{\rm LL}\, dS_k \tag{6.33}$$

and

$$\dot{J}^{jk}[V] = -\oint_S \bigl[x^j(-g)t^{kn}_{\rm LL} - x^k(-g)t^{jn}_{\rm LL}\bigr]\, dS_n. \tag{6.34}$$

Equation (6.33), when combined with Eq. (6.18), implies that

$$\frac{d}{dt}\bigl(M[V]R^j[V]\bigr) = P^j[V] - \frac{1}{c}\oint_S x^j(-g)t^{0k}_{\rm LL}\, dS_k. \tag{6.35}$$

6.1.4 Total mass, momentum, and angular momentum

The limit in which V is taken to include all of three-dimensional space is particularly interesting. In this limit $P^\alpha[V]$ is known to coincide with the Arnowitt–Deser–Misner four-momentum of an asymptotically-flat spacetime, and its physical interpretation as a measure of total momentum is robust. This statement is true whenever the coordinates x^α

coincide with a Lorentzian system at infinity; the coordinates do not have to be Lorentzian (and indeed, they could not be) at finite spatial distances.

Recalling the definitions of Eqs. (6.15) and (6.23), we define the total mass of the spacetime as

$$M := \frac{1}{c^2}\int_{\text{all space}} (-g)\big(T^{00} + t^{00}_{\text{LL}}\big)\, d^3x \tag{6.36a}$$

$$:= \frac{c^2}{16\pi G}\oint_\infty \partial_j H^{0j0k}\, dS_k. \tag{6.36b}$$

Recalling the definitions of Eqs. (6.16) and (6.24), we define the total three-momentum of the spacetime as

$$P^j := \frac{1}{c}\int_{\text{all space}} (-g)\big(T^{j0} + t^{j0}_{\text{LL}}\big)\, d^3x \tag{6.37a}$$

$$:= \frac{c^3}{16\pi G}\oint_\infty \partial_n H^{jn0k}\, dS_k - \frac{c^2}{16\pi G}\frac{d}{dt}\oint_\infty H^{0j0k}\, dS_k. \tag{6.37b}$$

Recalling the definitions of Eqs. (6.20) and (6.27), we define the total angular-momentum three-tensor of the spacetime as

$$J^{jk} := \frac{1}{c}\int_{\text{all space}} \Big[x^j(-g)\big(T^{k0} + t^{k0}_{\text{LL}}\big) - x^k(-g)\big(T^{j0} + t^{j0}_{\text{LL}}\big)\Big]\, d^3x \tag{6.38a}$$

$$:= \frac{c^3}{16\pi G}\oint_\infty \big(x^j\partial_m H^{km0n} - x^k\partial_m H^{jm0n} + H^{0jnk} - H^{0knj}\big)\, dS_n$$

$$- \frac{c^2}{16\pi G}\frac{d}{dt}\oint_\infty \big(x^j H^{0k0n} - x^k H^{0j0n}\big)\, dS_n. \tag{6.38b}$$

And finally, recalling Eqs. (6.19) and (6.26), we define

$$R^j := \frac{1}{Mc^2}\int_{\text{all space}} \big(T^{00} + t^{00}_{\text{LL}}\big)x^j\, d^3x \tag{6.39a}$$

$$:= \frac{c^2}{16\pi GM}\oint_\infty \big(x^j\partial_n H^{0n0k} - H^{0j0k}\big)\, dS_k \tag{6.39b}$$

as the position of the center-of-mass for the entire spacetime. The mass, momentum, angular momentum, and center-of-mass position of a spacetime can be defined either in terms of volume integrals over all space, or in terms of surface integrals at infinity. The surface integrals are especially powerful because they allow us to determine these quantities directly from the asymptotic behavior of the metric at large distances; an intimate knowledge of the material source is not required. This is reminiscent of the situation in electrodynamics: the total electric charge can be determined by integrating the normal component of the electric field over a surface enclosing the charge, and nothing need be known of the detailed distribution of charge within the surface.

Equations (6.30), (6.31), and (6.34) imply that the total mass M, total momentum P^j, and total angular momentum J^{jk} are constant in time whenever the surface integrals vanish in the limit $S \to \infty$. Under these circumstances, we have the conservation statements

$$M = \text{constant}, \qquad P^j = \text{constant}, \qquad J^{jk} = \text{constant}. \tag{6.40}$$

Furthermore, it can be shown that whenever the surface integrals vanish, the volume integrals of Eqs. (6.14) and (6.17) can be evaluated on any spacelike hypersurface and produce the same result. In particular, the momentum four-vector can be evaluated on a surface of simultaneity $t' =$ constant that is obtained from the original surface $t =$ constant by a Lorentz transformation; this observation can be used to show that P^α transforms as a four-vector under the transformation of Eq. (6.13).

In a similar way, Eq. (6.35) implies that $M\dot{R}^j = P^j$ whenever its surface integral vanishes, and whenever M itself is a constant. Assuming that $\boldsymbol{P}$ also is constant, we have

$$M\boldsymbol{R}(t) = M\boldsymbol{R}(0) + \boldsymbol{P}\,t, \tag{6.41}$$

where $\boldsymbol{R}(0)$ is the position of the center-of-mass at $t = 0$. This equation states that the center-of-mass moves uniformly with a velocity $\boldsymbol{P}/M$ (recall that $M = P^0/c$).

When Eq. (6.40) holds it is natural to adopt a reference frame in which $\boldsymbol{P}$ vanishes. This can always be achieved by performing a Lorentz transformation described by Eq. (6.13) and directing the boost in the direction of the momentum; the boost parameter must be set equal to $v = |\boldsymbol{P}|/M$. Once this is accomplished, it is also natural to place the origin of the spatial coordinates at the center-of-mass $\boldsymbol{R}$. This can always be achieved by translating the coordinates according to $\boldsymbol{x} \to \boldsymbol{x} - \boldsymbol{c}$, with $\boldsymbol{c}$ denoting a constant vector. It is easy to see that the translation changes the position of the center-of-mass according to $\boldsymbol{R} \to \boldsymbol{R} - \boldsymbol{c}$, and choosing $\boldsymbol{c} = \boldsymbol{R}$ places the center-of-mass at the origin of the spatial coordinates.

These choices define the *center-of-mass frame* of the spacetime:

$$\text{center-of-mass frame:} \qquad P^j = 0, \qquad R^j = 0. \tag{6.42}$$

As we have seen, this choice can be made whenever $\boldsymbol{P}$ is a constant vector, and whenever $M\dot{\boldsymbol{R}} = \boldsymbol{P}$. These conditions are fulfilled whenever the surface integrals of Eqs. (6.31) and (6.35) vanish when $S \to \infty$. This always happens when the spacetime is stationary. In the context of a radiating spacetime, however, the surface integrals cannot be assumed to vanish; in fact, the mass, momentum, and angular momentum of the spacetime are typically seen to change with time because the radiation transports energy, momentum, and angular momentum away from the source. Fortunately this effect can often be neglected in the context of approximate calculations.

We conclude this discussion with an illustration: We use the surface integrals to calculate the mass, momentum, and angular momentum of the Schwarzschild spacetime, first encountered back in Sec. 5.6. Expressing the metric of Eq. (5.163) in Cartesian coordinates, we find that

$$g_{00} = -\left(1 - \frac{R}{r}\right), \tag{6.43a}$$

$$g_{jk} = \delta_{jk} + \left(1 - \frac{R}{r}\right)^{-1} \frac{R}{r} n_j n_k, \tag{6.43b}$$

where $R := 2GM/c^2$ and $n^j := x^j/r$. It is then simple to show that $g = -1$ and

$$\mathfrak{g}^{00} = -\left(1 - \frac{R}{r}\right)^{-1}, \tag{6.44a}$$

$$\mathfrak{g}^{jk} = \delta^{jk} - \frac{R}{r} n^j n^k. \tag{6.44b}$$

We next compute $H^{\alpha\mu 0j}$ by substituting Eqs. (6.44) into Eq. (6.2), and insert the result within Eq. (6.22) to calculate $P^\alpha[r]$, the momentum vector associated with a surface S of constant r. The computations involve the surface element $dS_j = r^2 n_j\, d\Omega$ (where $d\Omega := \sin\theta\, d\theta d\phi$ is an element of solid angle), and they lead to $P^j[r] = 0$ and

$$M[r] = M\frac{r}{r - 2GM/c^2}. \tag{6.45}$$

The spatial momentum vanishes (as expected, since the coordinates are centered on the black hole), and in the limit $r \to \infty$ our previous result reduces to

$$M[\infty] = M. \tag{6.46}$$

The total energy is $cP^0[\infty] = Mc^2$, and M is recognized as the total gravitational mass of the Schwarzschild spacetime. A similar calculation reveals that the center-of-mass is situated at $R^j = 0$ and that the angular momentum vanishes.

6.2 Relaxed Einstein equations

6.2.1 Harmonic coordinates and a wave equation

It is advantageous at this stage to impose the four conditions

$$\partial_\beta \mathfrak{g}^{\alpha\beta} = 0 \tag{6.47}$$

on the gothic inverse metric. These are known as the *harmonic coordinate conditions*, and they were first encountered back in Sec. 5.6 (see Eq. (5.177)) in the context of the Schwarzschild solution. It is also useful to introduce the potentials

$$h^{\alpha\beta} := \eta^{\alpha\beta} - \mathfrak{g}^{\alpha\beta}, \tag{6.48}$$

where $\eta^{\alpha\beta} := \text{diag}(-1, 1, 1, 1)$ is the Minkowski metric expressed in Lorentzian coordinates $(x^0 := ct, x^j)$. In terms of these potentials the harmonic coordinate conditions read

$$\partial_\beta h^{\alpha\beta} = 0, \tag{6.49}$$

and in this context they are usually referred to as the *harmonic gauge conditions*. We observe that the harmonic conditions are preserved under the Lorentz transformations of

Eq. (6.13), and that the potentials $h^{\alpha\beta}$ transform as a tensor under this restricted class of coordinate transformations.

Box 6.2

Existence of harmonic coordinates

It seems plausible that the four harmonic coordinate conditions of Eq. (6.47) can always be imposed, given the four degrees of coordinate freedom inherent to general relativity, but it is worthwhile to see this explicitly. Given an initial coordinate system in which $\partial_\beta \mathfrak{g}^{\alpha\beta} \neq 0$, we make a coordinate transformation to $x'^\mu = f^\mu(x^\alpha)$. It is then straightforward to show that in the new coordinates

$$\partial_{\nu'} \mathfrak{g}^{\mu'\nu'} = \sqrt{-g'}\,\Box_g f^\mu(x^\alpha),$$

where $\Box_g := g^{\mu\nu}\nabla_\mu\nabla_\nu$ is the curved spacetime d'Alembertian operator acting on each one of the four functions f^μ, treated as a scalar function of x^α. Choosing each function to be *harmonic*, that is, a solution to $\Box_g f^\mu = 0$, ensures that the harmonic coordinate conditions will hold in the new coordinates.

The introduction of the potentials $h^{\alpha\beta}$ and the imposition of the harmonic gauge conditions simplify the appearance of the Einstein field equations. It is easy to verify that the left-hand side becomes

$$\partial_{\mu\nu} H^{\alpha\mu\beta\nu} = -\Box h^{\alpha\beta} + h^{\mu\nu}\partial_{\mu\nu}h^{\alpha\beta} - \partial_\mu h^{\alpha\nu}\partial_\nu h^{\beta\mu}, \tag{6.50}$$

where $\Box := \eta^{\mu\nu}\partial_{\mu\nu}$ is the flat-spacetime wave operator. The right-hand side of the field equations stays essentially unchanged, but the harmonic conditions do slightly simplify the form of the Landau–Lifshitz pseudotensor; as can be seen from Eq. (6.5), the first two terms of $(-g)t^{\alpha\beta}_{\rm LL}$ vanish in harmonic coordinates. Isolating the wave operator on the left-hand side, and putting everything else on the right-hand side, gives us the formal wave equation

$$\Box h^{\alpha\beta} = -\frac{16\pi G}{c^4}\tau^{\alpha\beta} \tag{6.51}$$

for the potentials $h^{\alpha\beta}$, where

$$\tau^{\alpha\beta} := (-g)\big(T^{\alpha\beta}[\mathsf{m}, g] + t^{\alpha\beta}_{\rm LL}[h] + t^{\alpha\beta}_{\rm H}[h]\big) \tag{6.52}$$

is the *effective energy-momentum pseudotensor* for the wave equation. We have introduced

$$(-g)t^{\alpha\beta}_{\rm H} := \frac{c^4}{16\pi G}\Big(\partial_\mu h^{\alpha\nu}\partial_\nu h^{\beta\mu} - h^{\mu\nu}\partial_{\mu\nu}h^{\alpha\beta}\Big) \tag{6.53}$$

as an additional (harmonic-gauge) contribution to the effective energy-momentum pseudotensor.

In our expression for $\tau^{\alpha\beta}$ we have indicated that the energy-momentum tensor $T^{\alpha\beta}$ is a functional of matter variables m, in addition to being a functional of the metric tensor $g_{\alpha\beta}$ (which is obtained from the gravitational potentials). As an example, when the matter consists of a perfect fluid, m collectively denotes variables such as the mass density ρ, pressure p, and velocity field u^α. We have also indicated that the Landau–Lifshitz and harmonic pseudotensors are functionals of $h^{\alpha\beta}$. As we shall see below, imposition of the gauge conditions (6.49) is equivalent to enforcing the conservation equations

$$\partial_\beta \tau^{\alpha\beta} = 0, \tag{6.54}$$

which can be compared with Eq. (6.8). It is easy to verify that $(-g)t_{\rm H}^{\alpha\beta}$ is separately conserved, in that it satisfies $\partial_\beta[(-g)t_{\rm H}^{\alpha\beta}] = 0$ as an identity.

The wave equation of Eq. (6.51) is the main starting point of post-Minkowskian theory. It is worth emphasizing the fact that this equation, together with Eq. (6.49) or (6.54), is an *exact formulation* of the Einstein field equations; no approximations have been introduced at this stage.

For a metric $g_{\alpha\beta}$ to satisfy the complete set of Einstein field equations, it is necessary for the potentials $h^{\alpha\beta}$ to satisfy *both* the wave equation *and* the gauge condition/conservation statement; it is the *union* of Eq. (6.51) and (6.49) or (6.54) that is equivalent to the original form of the Einstein field equations, $G^{\alpha\beta} = (8\pi G/c^4)T^{\alpha\beta}$. The two sets of equations play different roles. The wave equation (6.51) determines the gravitational potentials $h^{\alpha\beta}[\mathsf{m}]$ (and therefore the metric) as functions of the harmonic coordinates x^α, in terms of the matter variables m; these, however, remain undetermined until we also involve the conservation equation (6.54). It is this equation that determines the behavior of the matter variables in a curved spacetime whose metric is built from $h^{\alpha\beta}[\mathsf{m}]$. Solving both sets of equations therefore determines both the metric *and* the matter variables. This reminds us of John Wheeler's famous words: *matter tells spacetime how to curve, and spacetime tells matter how to move*; the decomposition of the field equations into a wave equation and a gauge condition/conservation statement provides a mathematical representation of this maxim.

We have just seen that when the complete set of Einstein field equations is integrated, one cannot solve for the metric independently of the matter variables, and one cannot solve for the matter variables independently of the metric. It is useful to observe, however, that when the equations are decomposed into the subsets [wave equation] and [gauge condition/conservation statement], one is entirely free to solve the wave equation (6.51) without also enforcing the gauge condition of Eq. (6.49) or the conservation statement of Eq. (6.54). Solving the wave equation independently of the gauge condition/conservation statement amounts to integrating only a subset of the Einstein field equations, and the procedure returns ten gravitational potentials $h^{\alpha\beta}[\mathsf{m}]$ expressed as functionals of undetermined matter variables m. The metric obtained from these potentials is also a functional of m, and it is not yet a solution to the Einstein field equations; it becomes a solution only when the gauge condition/conservation statement is imposed as an additional condition on the matter variables. The wave equation (6.51), taken by itself independently of Eqs. (6.49) or (6.54), is known as the *relaxed Einstein field equation*.

Box 6.3 Wave equation in flat and curved spacetimes

Because it involves second derivatives of the potentials, the term $h^{\mu\nu}\partial_{\mu\nu}h^{\alpha\beta}$ on the right-hand side of the field equations might have been more appropriately placed on the left-hand side, and joined together with the wave-operator term. In fact, there is a way of combining all second-order derivatives into a *curved-spacetime* wave operator. For this purpose we treat $h^{\alpha\beta}$ as a collection of ten scalar fields instead of as a tensor field.

The scalar wave operator associated with the metric $g_{\alpha\beta}$ (which is to be constructed from the potentials) is denoted $\Box_g$, and it has the following action on each of the ten potentials:

$$\begin{aligned}\Box_g h^{\alpha\beta} &= \frac{1}{\sqrt{-g}}\partial_\mu\big(\sqrt{-g}g^{\mu\nu}\partial_\nu h^{\alpha\beta}\big)\\ &= \frac{1}{\sqrt{-g}}\partial_\mu\Big[\big(\eta^{\mu\nu} - h^{\mu\nu}\big)\partial_\nu h^{\alpha\beta}\Big]\\ &= \frac{1}{\sqrt{-g}}\Big[\Box h^{\alpha\beta} - h^{\mu\nu}\partial_{\mu\nu}h^{\alpha\beta}\Big], \end{aligned} \tag{6.55}$$

where we have used the harmonic gauge conditions in the last step. This expression does indeed involve all second-derivative terms that appear in Eq. (6.51). The field equations could then be formulated in terms of $\Box_g$, and this was, in fact, the approach adopted by Kovacs and Thorne in their series of papers on the generation of gravitational waves. This approach, while conceptually compelling, is not as immediately useful for post-Minkowskian theory as the approach adopted here, which is based on the Minkowski wave operator. It is indeed much simpler to solve the wave equation in flat spacetime than it is to solve it in a curved spacetime with a complicated (and as yet unknown) metric.

6.2.2 Formal solution to the wave equation

The wave equation of Eq. (6.51) admits the formal solution

$$h^{\alpha\beta}(x) = \frac{4G}{c^4}\int G(x,x')\tau^{\alpha\beta}(x')\,d^4x', \tag{6.56}$$

where $x = (ct, \boldsymbol{x})$ is a field point and $x' = (ct', \boldsymbol{x}')$ a source point. The two-point function $G(x,x')$ is the *retarded Green's function* of the Minkowski wave operator, which satisfies

$$\Box G(x,x') = -4\pi\delta(x-x'), \tag{6.57}$$

and which is known to be a function of $x - x'$ only. (An explicit expression will be presented in Sec. 6.3.) This property is sufficient to prove that if the effective energy-momentum pseudotensor $\tau^{\alpha\beta}$ satisfies the conservation identities of Eq. (6.54), then the potentials $h^{\alpha\beta}$ will satisfy the harmonic gauge conditions of Eq. (6.49). The converse property, that $\partial_\beta\tau^{\alpha\beta} = 0$ when $\partial_\beta h^{\alpha\beta} = 0$, follows immediately from the wave equation (6.51).

To prove that $\partial_\beta h^{\alpha\beta} = 0$ when $\partial_\beta\tau^{\alpha\beta} = 0$, we begin by differentiating Eq. (6.56) with respect to x^β:

$$\partial_\beta h^{\alpha\beta} = \frac{4G}{c^4}\int \partial_\beta G(x,x')\tau^{\alpha\beta}(x')\,d^4x'. \tag{6.58}$$

Using the previously mentioned property that $G(x,x')$ depends on $x - x'$ only, we may write this as

$$\partial_\beta h^{\alpha\beta} = \frac{4G}{c^4}\int\big[-\partial_{\beta'}G(x,x')\big]\,\tau^{\alpha\beta}(x')\,d^4x', \tag{6.59}$$

in which the Green's function is now differentiated with respect to x'^{β}. Integrating by parts, we arrive at

$$\partial_\beta h^{\alpha\beta} = \frac{4G}{c^4} \int G(x, x') \partial_{\beta'} \tau^{\alpha\beta}(x')\, d^4x'. \tag{6.60}$$

This equation reveals directly that $h^{\alpha\beta}$ satisfies the harmonic gauge conditions when $\tau^{\alpha\beta}$ is conserved.

6.2.3 Iteration of the relaxed field equations

The question that concerns us now is this: Given the complexity of Eqs. (6.51)–(6.54), how can we construct solutions for a particular choice of matter variables m? Our answer will be: by successive approximations. We shall not attempt to find exact solutions to our equations; instead, we shall retreat to an approximate context in which our spacetime deviates only moderately from Minkowski spacetime. To construct the metric of this spacetime we consider a formal expansion of the form

$$h^{\alpha\beta} = G k_1^{\alpha\beta} + G^2 k_2^{\alpha\beta} + G^3 k_3^{\alpha\beta} + \cdots . \tag{6.61}$$

for the gravitational potentials. Such an expansion in powers of G is known as a *post-Minkowskian expansion*, and our hope is that the expansion – an asymptotic expansion that is not expected to converge – will give rise to an acceptable approximation to the true metric, at least in a useful portion of the spacetime. In the mathematical language of asymptotic expansions, our hope is that $g_{\alpha\beta}(x) - g^n_{\alpha\beta}(x) = O(G^{n+1})$ when x is within a wide domain $\mathscr{U}$ of the spacetime manifold; here $g^n_{\alpha\beta}$ is the metric obtained from Eq. (6.61) after truncating the asymptotic series to order G^n. Equation (6.61) gives rise to the successive approximations $h_0^{\alpha\beta} = 0$, $h_1^{\alpha\beta} = G k_1^{\alpha\beta}$, $h_2^{\alpha\beta} = G k_1^{\alpha\beta} + G^2 k_2^{\alpha\beta}$, and so on, for the gravitational potentials.

Box 6.4 The expansion parameter G

This development in powers of G is a formal device only. Because G has dimensions, its numerical value depends on the units in which it is evaluated, and it seems ridiculous to let it play the role of a "small" expansion parameter. For example, we were raised in geometrized units in which $G = 1$, and this does not look like a small quantity. The actual expansion parameter in a typical situation involving a characteristic mass m_c confined to a region of characteristic size r_c is $G m_c/(c^2 r_c)$, which is small in situations involving weak gravitational fields. Because the proper specification of the expansion parameter requires additional information that is specific to each situation considered, it is economical to stick with G as a formal expansion parameter, and let each physical situation dictate the translation to a meaningful, dimensionless parameter. The absence of a unique, dimensionless expansion parameter for the Einstein field equations is part of the reason why the expansions of post-Minkowskian and post-Newtonian theory are believed to be asymptotic sequences that may not converge.

In principle we might begin the process of solving the Einstein field equations by substituting Eq. (6.61) into Eq. (6.51) and plucking out terms that share the same power of G. In practice, however, it is more convenient to proceed by iterations, as we now explain.

In the *zeroth iteration* of the relaxed field equations we set $h_0^{\alpha\beta} = 0$ and immediately get $g_{\alpha\beta}^0 = \eta_{\alpha\beta}$, the metric of Minkowski spacetime. From this we construct $T^{\alpha\beta}[\mathsf{m}, g] = T^{\alpha\beta}[\mathsf{m}, \eta]$, $t_{\rm LL}^{\alpha\beta}[h] = t_{\rm LL}^{\alpha\beta}[h_0] = 0$, and $t_{\rm H}^{\alpha\beta}[h] = t_{\rm H}^{\alpha\beta}[h_0] = 0$. From all this we obtain $\tau_0^{\alpha\beta} = T^{\alpha\beta}[\mathsf{m}, \eta]$; this is the energy-momentum tensor of the matter variables m, and in the zeroth iteration these live in Minkowski spacetime.

In the *first iteration* of the relaxed field equations we solve the wave equation $\Box h^{\alpha\beta} = -(16\pi G/c^4)\tau_0^{\alpha\beta}$ for $h_1^{\alpha\beta} = Gk_1^{\alpha\beta}$. Because the source $\tau_0^{\alpha\beta}$ is known from the zeroth iteration, the wave equation can be integrated without difficulty (at least in principle), and this returns the potentials $h_1^{\alpha\beta}$ as functionals of the matter variables m, which have yet to be determined. From the potentials we form the metric $g_{\alpha\beta}^1$ and construct $\tau_1^{\alpha\beta}$, an improved version of the effective energy-momentum pseudotensor. This involves the material contribution $T^{\alpha\beta}[\mathsf{m}, g_1]$, as well as the field contributions $t_{\rm LL}^{\alpha\beta}[h_1]$ and $t_{\rm H}^{\alpha\beta}[h_1]$.

In the *second iteration* of the relaxed field equations we solve the wave equation $\Box h^{\alpha\beta} = -(16\pi G/c^4)\tau_1^{\alpha\beta}$ for $h_2^{\alpha\beta} = Gk_1^{\alpha\beta} + G^2k_2^{\alpha\beta}$, an improved version of the gravitational potentials. Because the source $\tau_1^{\alpha\beta}$ is known from the first iteration, the wave equation can once more be integrated, and $h_2^{\alpha\beta}$ are again functionals of the undetermined matter variables m. From the new potentials we form the metric $g_{\alpha\beta}^2$ and construct $\tau_2^{\alpha\beta}$, the latest version of the effective energy-momentum pseudotensor. The stage is ready for the next iteration.

After n iterations we obtain the potentials $h_n^{\alpha\beta} = Gk_1^{\alpha\beta} + G^2k_2^{\alpha\beta} + \cdots + G^nk_n^{\alpha\beta}$, the nth post-Minkowskian approximation to the true potentials $h^{\alpha\beta}$. These functions of the harmonic coordinates x^α are functionals of the matter variables m, which must now be determined. This is accomplished in the very last step of the procedure, the implementation of the gauge condition/conservation statement, which has not yet been invoked. We thus impose $\partial_\beta h_n^{\alpha\beta} = 0$ on our iterated solution to the relaxed field equations; this determines m and returns $g_{\alpha\beta}^n(x)$ as a proper tensor field in spacetime. Equivalently, we may enforce the conservation equation $\partial_\beta \tau_{n-1}^{\alpha\beta} = 0$, which (as we have seen) is formally equivalent to $\partial_\beta h_n^{\alpha\beta} = 0$. It is important to observe that while the gauge condition involves $h_n^{\alpha\beta}$, the conservation statement involves $\tau_{n-1}^{\alpha\beta}$; these quantities are linked by the iteration procedure described previously.

Let us illustrate the foregoing discussion by choosing the matter content of the spacetime to consist of N point masses labeled by an index $A = (1, 2, \ldots, N)$. In this case the collective matter variables m denote the set of vectors $\boldsymbol{r}_A(t)$, which give the position of each body in the harmonic system of coordinates. After n iterations of the relaxed field equations we obtain gravitational potentials of the form $h_n^{\alpha\beta}(x^\alpha; \boldsymbol{r}_A)$; these are functions of the spacetime coordinates x^α and functionals of the trajectories $\boldsymbol{r}_A(t)$. At this stage of the procedure the trajectories are not determined; the functions $\boldsymbol{r}_A(t)$ are completely arbitrary. In the final step we enforce the conservation equation $\partial_\beta \tau_{n-1}^{\alpha\beta} = 0$, and this produces equations of motion of the form

$$\frac{d^2\boldsymbol{r}_A}{dt^2} = O(G) + O(G^2) + \cdots + O(G^{n-1}). \tag{6.62}$$

These are used to determine $\boldsymbol{r}_A(t)$, and the task is completed: we have the metric and the motion of the individual bodies. These considerations indicate that two iterations of the relaxed field equations are required to obtain the Newtonian equations of motion – the $O(G)$ term on the right-hand side of Eq. (6.62).

It is important to understand that the iterations must be performed on the relaxed equations only, and not on the full set of Einstein field equations. In other words, one iterates the wave equation only, and leaves the gauge condition/conservation statement alone, until the final iteration is carried out; the gauge condition/conservation statement is enforced in the very last step of the procedure. It would indeed be misguided to enforce it at every iterative step. To see why, imagine that we choose to enforce $\partial_\beta \tau^{\alpha\beta} = 0$ immediately at the zeroth iteration. Because $\tau_0^{\alpha\beta} = T^{\alpha\beta}[\mathrm{m}, \eta]$, this is the conservation equation for matter fields in Minkowski spacetime, and it implies that the matter cannot be subjected to gravitational interactions. (In the illustrative case of point masses examined previously, the bodies would have to move on straight lines.) The next iteration would produce $h_1^{\alpha\beta}$ as sourced by this matter field, and the next version of the conservation statement, $\partial_\beta \tau_1^{\alpha\beta} = 0$, would imply that the matter is, after all, subjected to a gravitational interaction. (In our example, the point masses would now be allowed to move according to the Newtonian equations of motion, in a gravitational field determined as if the masses were moving on straight lines.) We have a contradiction, and this tension is best avoided by delaying the implementation of the gauge condition/conservation statement until the very last step of the iterative procedure.

As a small technical point, we might mention that the procedure does retain a limited amount of latitude. As described above, the penultimate step in the iterative procedure is to solve the wave equation $\Box h^{\alpha\beta} = -(16\pi G/c^4)\tau_{n-1}^{\alpha\beta}$ for $h_n^{\alpha\beta}$, given the known source $\tau_{n-1}^{\alpha\beta}$. The last step is to impose the additional conditions $\partial_\beta \tau_{n-1}^{\alpha\beta} = 0$. These steps can be switched: once $\tau_{n-1}^{\alpha\beta}$ is constructed from $h_{n-1}^{\alpha\beta}$ in the $(n-1)$th iteration, one can immediately enforce the conservation equation $\partial_\beta \tau_{n-1}^{\alpha\beta} = 0$. The final step is then to obtain $h_n^{\alpha\beta}$ by integrating the wave equation, and the gauge condition $\partial_\beta h_n^{\alpha\beta} = 0$ will automatically be satisfied by the solution.

We can be even more flexible. If we are interested *only* in the equations of motion that arise from the $(n-1)$th iteration, and *not* in the spacetime metric that is generated by that motion, then we do not actually have to complete the iterations to obtain $h_n^{\alpha\beta}$. The solutions $h_{n-1}^{\alpha\beta}$ are sufficient to insert into the conservation equations $\partial_\beta \tau_{n-1}^{\alpha\beta} = 0$, from which the motion of the system can be determined consistently to order G^{n-1}.

The iterative, post-Minkowskian method described in this section is technically demanding to carry out, and in the next chapter we shall develop a number of helpful techniques that permit its successful implementation. Before we start, however, we must learn how to solve a wave equation in flat spacetime. This is the topic of the following section.

6.3 Integration of the wave equation

At first sight the wave equation (6.51) appears to be highly non-linear, with the potentials $h^{\alpha\beta}$ present on both sides of the equation. In Sec. 6.2.3 we outlined an iterative procedure that ensures that in the course of each iteration, the wave equation is actually linear in $h^{\alpha\beta}$ and involves a known source term $\tau^{\alpha\beta}$. The task of solving the relaxed field equations therefore appears to be straightforward, and in this section we introduce a number of powerful techniques to integrate the wave equation.

For simplicity we shall eliminate all unnecessary tensorial indices on the wave equation, which we now write as

$$\Box\psi = -4\pi\mu. \tag{6.63}$$

The scalar potential $\psi(x)$ plays the role of $h^{\alpha\beta}$, and the source function $\mu(x)$ plays the role of $(4G/c^4)\tau^{\alpha\beta}$; the remaining factor of 4π is retained for later convenience. Here $x = (ct, \boldsymbol{x})$ labels a spacetime event, and we recall that

$$\Box := \eta^{\alpha\beta}\partial_{\alpha\beta} = -\frac{1}{c^2}\frac{\partial^2}{\partial t^2} + \nabla^2 \tag{6.64}$$

is the wave operator of Minkowski spacetime. The source function $\mu(x)$ is assumed to be known, but unlike the typical situations encountered in electrodynamics, for example, it cannot be assumed to be confined to a bounded region of three-dimensional space; it is instead taken to be distributed over all space. The reason originates from the post-Minkowskian context: as we have seen, during each iteration of the relaxed field equations, $\tau^{\alpha\beta}$ is built in part from $T^{\alpha\beta}$, which normally has compact support, and in part from $t^{\alpha\beta}_{\rm LL}$ and $t^{\alpha\beta}_{\rm H}$, which do not because they are constructed from $h^{\alpha\beta}$, which extends over all space. Our source term in Eq. (6.63) will therefore extend over all space, but μ is assumed to fall off sufficiently rapidly to ensure that ψ decays at least as fast as r^{-1} (where $r := |\boldsymbol{x}|$). Occasionally we shall find it useful to decompose μ into a piece $\mu_{\rm c}$ with compact support (analogous to $T^{\alpha\beta}$) and a piece $\mu_{\rm nc}$ with non-compact support.

A summary of our main results in this section is contained in Box 6.7.

6.3.1 Retarded Green's function

The central tool to integrate Eq. (6.63) is the *retarded Green's function* $G(x, x')$, a solution to

$$\Box G(x, x') = -4\pi\delta(x - x') = -4\pi\delta(ct - ct')\delta(\boldsymbol{x} - \boldsymbol{x}'), \tag{6.65}$$

with the property that $G(x, x')$ vanishes when x is in the past of x'. As we show in Box 6.5, the Green's function is given explicitly by

$$G(x, x') = \frac{\delta(ct - ct' - |\boldsymbol{x} - \boldsymbol{x}'|)}{|\boldsymbol{x} - \boldsymbol{x}'|}, \tag{6.66}$$

where

$$|\boldsymbol{x}-\boldsymbol{x}'| := \sqrt{(x-x')^2+(y-y')^2+(z-z')^2} \tag{6.67}$$

is the Euclidean distance between the field point $\boldsymbol{x}$ and the source point $\boldsymbol{x}'$. Alternatively, the Green's function can be expressed as

$$G(x,x') = 2\Theta(ct-ct')\,\delta\big[(ct-ct')^2-|\boldsymbol{x}-\boldsymbol{x}'|^2\big], \tag{6.68}$$

in terms of the flat spacetime interval Δs^2 between x and x'; here $\Theta(ct-ct')$ is the Heaviside step function, which is equal to one when $ct > ct'$ and zero when $ct < ct'$.

Box 6.5 Green's function for the wave equation

To construct a solution to Eq. (6.65) we write the Green's function as the Fourier transform

$$G(x,x') = \frac{1}{2\pi}\int \tilde{G}(k;\boldsymbol{x},\boldsymbol{x}')e^{-ik(ct-ct')}\,dk, \tag{1}$$

and we represent the time delta function as

$$\delta(ct-ct') = \frac{1}{2\pi}\int e^{-ik(ct-ct')}\,dk.$$

Substituting these expressions into Green's equation yields

$$\big(\nabla^2+k^2\big)\tilde{G}(k;\boldsymbol{x},\boldsymbol{x}') = -4\pi\,\delta(\boldsymbol{x}-\boldsymbol{x}'). \tag{2}$$

When $k=0$ this equation reduces to Green's equation for the Poisson equation, and from this comparison we learn that $\tilde{G}(0;\boldsymbol{x},\boldsymbol{x}') = |\boldsymbol{x}-\boldsymbol{x}'|^{-1}$.

We can anticipate that for $k\neq 0$, $\tilde{G}$ will be of the form

$$\tilde{G}(k;\boldsymbol{x},\boldsymbol{x}') = \frac{g(k,|\boldsymbol{x}-\boldsymbol{x}'|)}{|\boldsymbol{x}-\boldsymbol{x}'|}, \tag{3}$$

with g representing a function that stays non-singular when the second argument, $R := |\boldsymbol{x}-\boldsymbol{x}'|$, approaches zero. That $\tilde{G}$ should depend on the spatial variables through R only can be justified on the grounds that three-dimensional space is both homogeneous (so that $\tilde{G}$ can depend only on the vector $\boldsymbol{R} := \boldsymbol{x}-\boldsymbol{x}'$) and isotropic (so that only the length of the vector matters, and not its direction). That $\tilde{G}$ should behave as $1/R$ when R is small is justified by the following discussion.

We take Eq. (2) and integrate both sides over a sphere of small radius ε centered at $\boldsymbol{x}'$. Since $\nabla^2\tilde{G} = \boldsymbol{\nabla}\cdot\boldsymbol{\nabla}\tilde{G}$, we can use Gauss's theorem to get

$$\oint_{R=\varepsilon}\nabla\tilde{G}\cdot d\boldsymbol{S} + k^2\int_{R<\varepsilon}\tilde{G}\,d^3x = -4\pi,$$

where $d\boldsymbol{S}$ is the surface element on the sphere. In this equation, the volume integral is of order $\tilde{G}\varepsilon^3$ and it contributes nothing in the limit $\varepsilon\to 0$, unless $\tilde{G}$ happens to be as singular as $1/\varepsilon^3$. The surface integral, on the other hand, is equal to

$$4\pi\varepsilon^2\frac{d\tilde{G}}{dR}\bigg|_{R=\varepsilon}.$$

If $\tilde{G}$ were to behave as $1/\varepsilon^3$, then $d\tilde{G}/dR$ would be of order $1/\varepsilon^4$, the surface integral would contribute a term of order $1/\varepsilon^2$, and the left-hand side could never give rise to the required -4π. We conclude that $\tilde{G}$ cannot be so singular, and that the left-hand side is dominated by the surface integral. This implies that $\tilde{G}$ must be of order $1/\varepsilon$, as was anticipated in Eq. (3). Setting $\tilde{G} = g/R$ returns $-4\pi g(k, \varepsilon) + O(\varepsilon)$ for the surface integral, and this gives us the condition $g(k, 0) = 1$. We also recall that $g(0, R) = 1$.

We may now safely take $R \neq 0$ and substitute Eq. (3) into Eq. (2), taking its right-hand side to be zero. Since $\tilde{G}$ depends on $\boldsymbol{x}$ only through R, the Laplacian operator becomes

$$\nabla^2 \to \frac{1}{R^2}\frac{d}{dR}R^2\frac{d}{dR}.$$

Acting with this on $\tilde{G} = g/R$ yields g''/R and Eq. (2) becomes

$$g'' + k^2 g = 0,$$

with a prime indicating differentiation with respect to R. With the boundary condition at $R = 0$ specified previously, two linearly independent solutions to this equation are

$$g_\pm(k, R) = e^{\pm ikR}.$$

Substituting this into Eq. (3), and that into Eq. (1), we obtain

$$G_\pm(x, x') = \frac{1}{2\pi}\int \frac{e^{\pm ikR}}{R} e^{-ik(ct-ct')}\, dk = \frac{1}{2\pi R}\int e^{-ik(ct-ct'\mp R)}\, dk,$$

or

$$G_\pm(x, x') = \frac{\delta\big(ct - ct' \mp |\boldsymbol{x} - \boldsymbol{x}'|\big)}{|\boldsymbol{x} - \boldsymbol{x}'|}. \tag{4}$$

The function $G_+(x, x')$, which is non-zero when $ct - ct' = +R$, is known as the *retarded Green's function*; the function $G_-(x, x')$, which is non-zero when $ct - ct' = -R$, is known as the *advanced Green's function*.

The retarded Green's function can be expressed in the alternative form

$$G_+(x, x') = 2\Theta(ct - ct')\delta\big[(ct - ct')^2 - |\boldsymbol{x} - \boldsymbol{x}'|^2\big]. \tag{5}$$

The new argument of the delta function factorizes as $(ct - ct' - R)(ct - ct' + R)$, and when $c(t - t') > 0$ only the first factor may go through zero; the second factor is then equal to $2R$, and the delta function is distributionally equal to $\delta(ct - ct' - R)/(2R)$. At this stage the step function becomes redundant, because the delta function is active only when $c(t - t') > 0$, and we have reproduced Eq. (4).

Similarly, the advanced Green's function can be expressed as

$$G_-(x, x') = 2\Theta(ct' - ct)\delta\big[(ct - ct')^2 - |\boldsymbol{x} - \boldsymbol{x}'|^2\big].$$

In terms of the retarded Green's function $G(x, x')$, the solution to Eq. (6.63) is

$$\psi(x) = \int G(x, x')\mu(x')\, d^4x', \tag{6.69}$$

where $d^4x' = d(ct')d^3x'$. After substitution of Eq. (6.66) and integration over $d(ct')$, this becomes

$$\psi(t, \boldsymbol{x}) = \int \frac{\mu(t - |\boldsymbol{x} - \boldsymbol{x}'|/c, \boldsymbol{x}')}{|\boldsymbol{x} - \boldsymbol{x}'|} d^3x'. \tag{6.70}$$

This is the *retarded solution* to the wave equation, and the domain of integration extends over $\mathscr{C}(x)$, the *past light cone* of the field point $x = (ct, \boldsymbol{x})$.

6.3.2 Near zone and wave zone: slow-motion condition

In the following subsection the domain $\mathscr{C}(x)$ will be partitioned into a *near-zone domain* $\mathscr{N}$ and a *wave-zone domain* $\mathscr{W}$. Our task in this subsection is to introduce the important notions of near and wave zones in the general context of the wave equation (6.63).

To do so we introduce the following scaling quantities:

$$t_c := \text{characteristic time scale of the source}, \tag{6.71a}$$

$$\omega_c := \frac{2\pi}{t_c} = \text{characteristic frequency of the source}, \tag{6.71b}$$

$$\lambda_c := \frac{2\pi c}{\omega_c} = ct_c = \text{characteristic wavelength of the radiation}. \tag{6.71c}$$

The characteristic time scale t_c is the time required for noticeable changes to occur within the source; it is defined such that $\partial_t \mu$ is typically of order μ/t_c over the support of the source function. The characteristic frequency ω_c and wavelength λ_c are derived directly from t_c. If, for example, μ oscillates with a frequency ω, then $t_c \sim 2\pi/\omega$, $\omega_c \sim \omega$, and $\lambda_c \sim 2\pi c/\omega$.

The near zone and the wave zone are defined as

$$\text{near zone:} \quad r \ll \lambda_c = \frac{2\pi c}{\omega_c} = ct_c, \tag{6.72a}$$

$$\text{wave zone:} \quad r \gg \lambda_c = \frac{2\pi c}{\omega_c} = ct_c. \tag{6.72b}$$

Thus, the near zone is the region of three-dimensional space in which $r := |\boldsymbol{x}|$ is small compared with a characteristic wavelength λ_c, while the wave zone is the region in which r is large compared with this length scale. As we can see from the example of Box 6.6, the potential behaves very differently in the two zones: in the near zone the difference between $\tau := t - r/c$ and t is small (the field retardation is unimportant), and time derivatives are small compared with spatial derivatives; in the wave zone the difference between $\tau = t - r/c$ and t is large, and time derivatives are comparable to spatial derivatives. These properties are shared by all generic solutions to the wave equation.

Another important feature of the near zone concerns the quantity $(r/c)\partial_t \mu$. This is of order $(r/c)(\mu/t_c)$, or $(r/\lambda_c)\mu$, which is much smaller than μ. In the near zone, therefore,

$$\frac{r}{c}\frac{\partial \mu}{\partial t} = O\Big(\frac{r}{\lambda_c}\mu\Big) \ll \mu. \tag{6.73}$$

This states, simply, that the source retardation is unimportant within the near zone.

Thus far our considerations have been general, and our definitions of near and wave zones apply whether the source function μ is extended over all space or confined to a bounded region V. In addition, our definitions apply independently of the existence of a slow-motion condition, to which we turn next.

When the source function μ has a piece μ_c with compact support, we can introduce the additional scaling quantities

$$r_c := \text{characteristic length scale of the compact-support source}, \tag{6.74a}$$

$$v_c := \frac{r_c}{t_c} = \text{characteristic velocity within the source}. \tag{6.74b}$$

The characteristic radius r_c is defined such that μ_c vanishes outside a sphere of radius r_c; this part of μ has support only within this sphere. The characteristic velocity v_c is defined in terms of the scales r_c and t_c; it represents the speed with which changes in the source propagate across the region of space occupied by the source. In the case of a fluid, for example, v_c would be associated with the speed of sound within the fluid. In a binary-star system, v_c would be associated with the orbital velocities of the stars.

A *slow-motion condition* is in effect when the characteristic velocity v_c is small compared with the speed of light:

$$v_c \ll c \qquad \text{(slow-motion condition)}. \tag{6.75}$$

It then follows from Eq. (6.75) that

$$r_c \ll \lambda_c \qquad \text{(slow-motion condition)}; \tag{6.76}$$

this equation states that μ_c is necessarily situated deep within the near zone when a slow-motion condition is in effect.

Box 6.6 Dipole solution to the wave equation

We examine the solution to a specific version of Eq. (6.63),

$$\psi = (\boldsymbol{p}\cdot\boldsymbol{n})\bigg[\frac{\cos\omega(t-r/c)}{r^2} - \frac{\omega}{c}\frac{\sin\omega(t-r/c)}{r}\bigg],$$

which corresponds to $\mu = -\boldsymbol{p}\cdot\boldsymbol{\nabla}\delta(\boldsymbol{x})\cos\omega t$. Here $\boldsymbol{p}$ is a constant vector, $r := |\boldsymbol{x}|$, $\boldsymbol{n} := \boldsymbol{x}/r$ is the unit radial vector, and ω is an angular frequency. Physically speaking, this solution represents the scalar potential of a dipole of constant direction $\boldsymbol{p}$, oscillating in strength with a frequency $f = \omega/(2\pi)$; the wavelength of the radiation produced by the oscillating dipole is $\lambda = c/f = 2\pi c/\omega$.

Our first observation is that ψ behaves very differently depending on whether r is small or large compared with λ. When $r \ll \lambda = 2\pi c/\omega$, the trigonometric functions can be expanded in powers of $\omega r/c$, and the result is

$$\psi = (\boldsymbol{p}\cdot\boldsymbol{n})\frac{\cos\omega t}{r^2}\left[1 + O\left(\frac{\omega^2 r^2}{c^2}\right)\right] \qquad \text{(near zone)},$$

with a correction term that is quadratic in $r/\lambda \ll 1$. We observe also that in the *near zone* – the region $r \ll \lambda$ – the derivatives of ψ are related by

$$\frac{\partial_t \psi}{c|\nabla\psi|} = O\left(\frac{\omega r}{c}\right) \qquad \text{(near zone)}.$$

In the near zone, therefore, a time derivative is smaller than a spatial derivative (multiplied by c) by a factor of order $r/\lambda \ll 1$.

When, on the other hand, $r \gg \lambda = 2\pi c/\omega$, it is no longer appropriate to expand the trigonometric functions, and the potential must be expressed as

$$\psi = -(\boldsymbol{p}\cdot\boldsymbol{n})\frac{\omega}{c}\frac{\sin\omega\tau}{r}\left[1 + O\left(\frac{c}{\omega r}\right)\right] \qquad \text{(wave zone)},$$

in terms of the *retarded-time* variable $\tau := t - r/c$; here the difference between τ and t is large, and the correction term is linear in $\lambda/r \ll 1$. We observe also that in the *wave zone* – the region $r \gg \lambda$ – the derivatives of ψ are related by

$$\frac{\partial_t \psi}{c|\nabla\psi|} = O(1) \qquad \text{(wave zone)}.$$

To obtain this result we have used the fact that the spatial dependence contained in $\boldsymbol{n}$ and r^{-1} produces a spatial derivative of fractional order λ/r, while the spatial dependence contained in $\tau = t - r/c$ produces a spatial derivative of order unity. In the wave zone, therefore, a time derivative has the same order of magnitude as a spatial derivative (multiplied by c).

6.3.3 Integration domains

The integral of Eq. (6.70) extends over the past light cone $\mathscr{C}(x)$ of the field point x. To evaluate the integral we partition $\mathscr{C}(x)$ into two pieces, the *near-zone domain* $\mathscr{N}(x)$ and the *wave-zone domain* $\mathscr{W}(x)$. We place the boundary of the near and wave zones at an arbitrarily selected radius $\mathcal{R}$, with $\mathcal{R}$ imagined to be of the same order of magnitude as λ_c, the characteristic wavelength of the radiation emitted by μ. The near zone is then imagined as a three-dimensional ball of radius $\mathcal{R}$ that traces out a world tube $\mathscr{D}$ in spacetime. We let $\mathscr{N}(x)$ be the part of $\mathscr{C}(x)$ where $r' := |\boldsymbol{x}'| < \mathcal{R}$, and we let $\mathscr{W}(x)$ be the part of $\mathscr{C}(x)$ where $r' > \mathcal{R}$. The near-zone and wave-zone domains join together to form the complete light cone of the field point x: $\mathscr{N}(x) + \mathscr{W}(x) = \mathscr{C}(x)$. The domains are illustrated in Fig. 6.1.

We write Eq. (6.70) as

$$\psi(x) = \psi_{\mathscr{N}}(x) + \psi_{\mathscr{W}}(x), \tag{6.77}$$

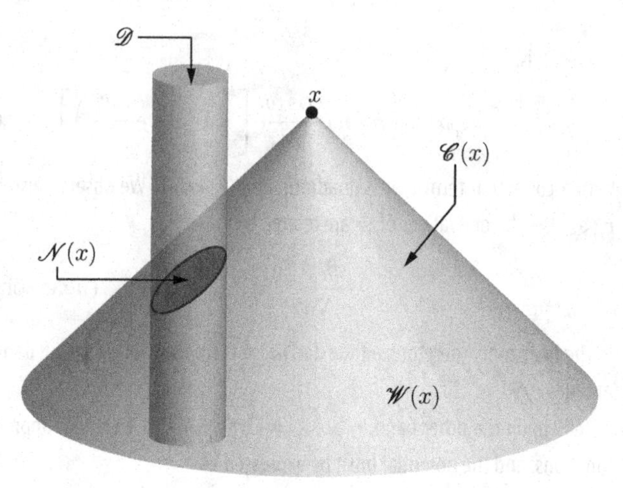

Fig. 6.1 Integration domains for the retarded solution of the wave equation: $\mathscr{C}(x)$ is the past light cone of the field point x; $\mathscr{D}$ is the world tube traced by a three-dimensional ball of radius $\mathcal{R}$, which contains the near-zone region of spacetime; $\mathscr{N}(x)$ is the intersection of $\mathscr{C}(x)$ with the near zone; and $\mathscr{W}(x)$ is the remaining piece of the light cone.

where

$$\psi_{\mathscr{N}}(x) = \int_{\mathscr{N}} G(x, x')\mu(x')\, d^4x' \tag{6.78}$$

is the near-zone portion of the light-cone integral, while

$$\psi_{\mathscr{W}}(x) = \int_{\mathscr{W}} G(x, x')\mu(x')\, d^4x' \tag{6.79}$$

is the wave-zone portion. Methods to evaluate $\psi_{\mathscr{N}}$ and $\psi_{\mathscr{W}}$ will be developed in the following two subsections. It is an important fact that while $\psi_{\mathscr{N}}$ and $\psi_{\mathscr{W}}$ will individually depend on the cutoff parameter $\mathcal{R}$, their sum $\psi = \psi_{\mathscr{N}} + \psi_{\mathscr{W}}$ will necessarily be independent of $\mathcal{R}$. The $\mathcal{R}$-dependence of $\psi_{\mathscr{N}}$ and $\psi_{\mathscr{W}}$ is therefore unimportant, and it can freely be ignored. This observation will serve as a helpful simplifying tool in many subsequent computations.

6.3.4 Integration over the near zone

In this subsection we develop methods to evaluate

$$\psi_{\mathscr{N}}(x) = \int_{\mathscr{N}} \frac{\mu(t - |\boldsymbol{x} - \boldsymbol{x}'|/c, \boldsymbol{x}')}{|\boldsymbol{x} - \boldsymbol{x}'|} d^3x', \tag{6.80}$$

the near-zone contribution to the complete solution $\psi = \psi_{\mathscr{N}} + \psi_{\mathscr{W}}$ to the wave equation. We recall that the domain of integration $\mathscr{N}$ is the intersection between $\mathscr{C}(x)$, the past light cone of the field point x, and the near zone $r' < \mathcal{R}$.

Wave-zone field point

We first evaluate Eq. (6.80) when x is situated in the wave zone, that is, when $r > \mathcal{R}$. For this purpose we introduce a modified integrand,

$$\frac{\mu(t - |\boldsymbol{x} - \boldsymbol{x}'|/c, \boldsymbol{x}')}{|\boldsymbol{x} - \boldsymbol{x}'|} = \int \frac{\mu(t - |\boldsymbol{x} - \boldsymbol{x}'|/c, \boldsymbol{y})}{|\boldsymbol{x} - \boldsymbol{x}'|} \delta(\boldsymbol{y} - \boldsymbol{x}')\, d^3y$$
$$=: \int g(\boldsymbol{x}, \boldsymbol{x}', \boldsymbol{y}) \delta(\boldsymbol{y} - \boldsymbol{x}')\, d^3y\,, \tag{6.81}$$

in which we can treat $\boldsymbol{x}'$ and $\boldsymbol{y}$ as independent variables. Knowing that $\boldsymbol{x}'$ lies within the near zone, we treat it as a small vector, and express g as a Taylor expansion about $\boldsymbol{x}' = \boldsymbol{0}$. Keeping just a few terms in this expansion, we have

$$g(\boldsymbol{x}, \boldsymbol{x}', \boldsymbol{y}) = g(\boldsymbol{x}, \boldsymbol{0}, \boldsymbol{y}) + \frac{\partial g}{\partial x'^j} x'^j + \frac{1}{2} \frac{\partial^2 g}{\partial x'^j x'^k} x'^j x'^k + \cdots\,, \tag{6.82}$$

in which all derivatives are evaluated at $\boldsymbol{x}' = \boldsymbol{0}$. But $\partial g / \partial x'^j = -\partial g / \partial x^j$ because g depends on $\boldsymbol{x}'$ only through the combination $|\boldsymbol{x} - \boldsymbol{x}'|$, and our Taylor expansion can be expressed as

$$g(\boldsymbol{x}, \boldsymbol{x}', \boldsymbol{y}) = g(\boldsymbol{x}, \boldsymbol{0}, \boldsymbol{y}) - \frac{\partial g}{\partial x^j} x'^j + \frac{1}{2} \frac{\partial^2 g}{\partial x^j x^k} x'^j x'^k + \cdots\,. \tag{6.83}$$

The derivatives of g are still evaluated at $\boldsymbol{x}' = \boldsymbol{0}$, but because the differentiation is now carried out with respect to $\boldsymbol{x}$, we can set $\boldsymbol{x}' = \boldsymbol{0}$ in g *before* taking the derivatives. Observing that g then becomes a function of $|\boldsymbol{x}| = r$ only, we have

$$g(\boldsymbol{x}, \boldsymbol{x}', \boldsymbol{y}) = g(r, 0, \boldsymbol{y}) - \frac{\partial g(r, 0, \boldsymbol{y})}{\partial x^j} x'^j + \frac{1}{2} \frac{\partial^2 g(r, 0, \boldsymbol{y})}{\partial x^j x^k} x'^j x'^k + \cdots\,. \tag{6.84}$$

Keeping all terms of the Taylor expansion, this is

$$g(\boldsymbol{x}, \boldsymbol{x}', \boldsymbol{y}) = \sum_{\ell=0}^{\infty} \frac{(-1)^\ell}{\ell!} x'^L \partial_L g(r, 0, \boldsymbol{y}), \tag{6.85}$$

where $L := j_1 j_2 \cdots j_\ell$ is a multi-index of the sort introduced back in Sec. 1.5.3. More explicitly, we have established the identity

$$\frac{\mu(t - |\boldsymbol{x} - \boldsymbol{x}'|/c, \boldsymbol{y})}{|\boldsymbol{x} - \boldsymbol{x}'|} = \sum_{\ell=0}^{\infty} \frac{(-1)^\ell}{\ell!} x'^L \partial_L \frac{\mu(t - r/c, \boldsymbol{y})}{r}. \tag{6.86}$$

The dependence of μ/r on the variables x^j is contained entirely within r.

Inserting this within Eq. (6.81) to restore $\boldsymbol{y} = \boldsymbol{x}'$, and substituting the result into Eq. (6.80), we arrive at

$$\psi_{\mathcal{N}}(t, \boldsymbol{x}) = \sum_{\ell=0}^{\infty} \frac{(-1)^\ell}{\ell!} \partial_L \left[\frac{1}{r} \int_{\mathcal{M}} \mu(\tau, \boldsymbol{x}') x'^L\, d^3x' \right], \tag{6.87}$$

where

$$\tau := t - r/c \tag{6.88}$$

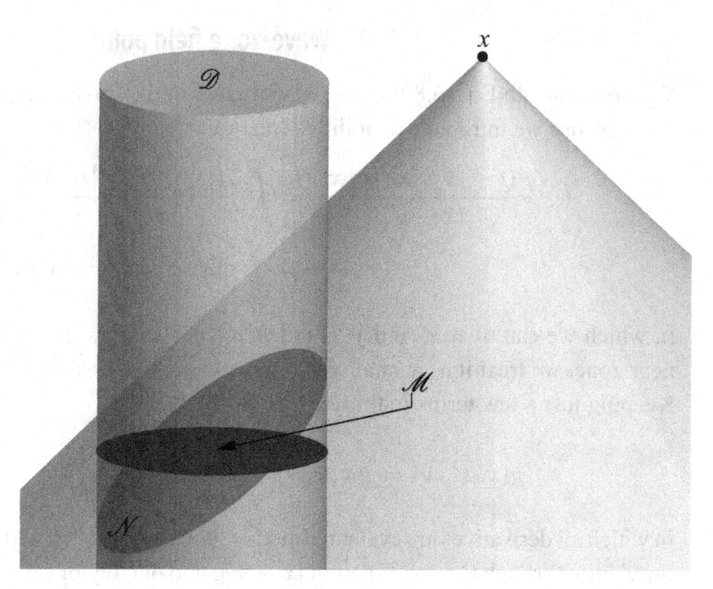

Fig. 6.2 Near-zone integration, wave-zone field point. The domain $\mathscr{M}$ is a surface of constant time bounded externally by the sphere $r' = \mathcal{R}$.

is a retarded-time variable. Note that the temporal dependence of the source function no longer involves $\boldsymbol{x}'$, the variable of integration. The domain of integration has therefore become a surface of constant time (the constant being equal to τ) bounded externally by the sphere $r' = \mathcal{R}$. This domain is denoted $\mathscr{M}$ in Eq. (6.87), and is illustrated in Fig. 6.2.

Equation (6.87) is valid everywhere within the wave zone. It simplifies when $r \to \infty$, that is, when $\psi_{\mathscr{N}}$ is evaluated in the *far-away wave zone*. In this limit we retain only the dominant r^{-1} term in $\psi_{\mathscr{N}}$, and we approximate Eq. (6.87) by

$$\psi_{\mathscr{N}} = \frac{1}{r}\sum_{\ell=0}^{\infty}\frac{(-1)^\ell}{\ell!}\int_{\mathscr{M}} \partial_L \mu(\tau, \boldsymbol{x}')x'^L\, d^3x' + O(r^{-2}). \tag{6.89}$$

The dependence of μ on x^j is contained in τ, so that $\partial_j \mu = -c^{-1}\mu^{(1)}\partial_j r = -c^{-1}\mu^{(1)} n_j$, in which $\mu^{(1)}$ denotes the first derivative of μ with respect to τ. We used the fact that

$$\partial_j r = n_j, \tag{6.90}$$

where $n^j = x^j/r$ is the unit radial vector. Invoking this result once more, we find that $\partial_{jk}\mu = c^{-2}\mu^{(2)} n_j n_k + O(r^{-1})$, and continuing along these lines reveals that in general, $\partial_L \mu = (-1)^\ell c^{-\ell}\mu^{(\ell)} n_L + O(r^{-1})$. Inserting this into our previous expression for $\psi_{\mathscr{N}}$, we find that Eq. (6.87) becomes

$$\psi_{\mathscr{N}}(t, \boldsymbol{x}) = \frac{1}{r}\sum_{\ell=0}^{\infty}\frac{1}{\ell! c^\ell} n_L \left(\frac{d}{d\tau}\right)^\ell \int_{\mathscr{M}} \mu(\tau, \boldsymbol{x}')x'^L\, d^3x' + O(r^{-2}) \tag{6.91}$$

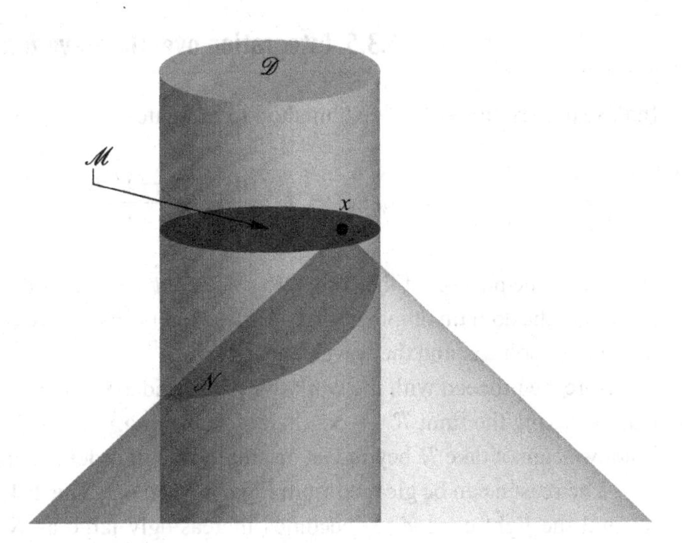

Fig. 6.3 Near-zone integration, near-zone field point.

in the far-away wave zone. This is a *multipole expansion* for the potential $\psi_{\mathcal{N}}$, in which each ℓ-pole moment $\int_{\mathcal{M}} \mu x^L \, d^3x$ is differentiated ℓ times with respect to τ. Note that $n_L x'^L = n_{j_1} n_{j_2} \cdots n_{j_\ell} x'^{j_1} x'^{j_2} \cdots x'^{j_\ell} = (\boldsymbol{n} \cdot \boldsymbol{x}')^\ell$.

Near-zone field point

We next evaluate Eq. (6.80) when x is situated in the near zone, that is, when $r = |\boldsymbol{x}| < \mathcal{R}$. In this situation, both $\boldsymbol{x}$ and $\boldsymbol{x}'$ lie within the near zone, and $|\boldsymbol{x} - \boldsymbol{x}'|$ can be treated as a small quantity. To evaluate the integral we simply Taylor-expand the time-dependence of the source function,

$$\mu(t - |\boldsymbol{x} - \boldsymbol{x}'|/c) = \mu(t) - \frac{1}{c}\frac{\partial \mu}{\partial t}|\boldsymbol{x} - \boldsymbol{x}'| + \frac{1}{2c^2}\frac{\partial^2 \mu}{\partial t^2}|\boldsymbol{x} - \boldsymbol{x}'|^2 + \cdots,$$

in which all derivatives are evaluated at time t. Substituting this expansion into Eq. (6.80) produces

$$\psi_{\mathcal{N}}(t, \boldsymbol{x}) = \sum_{\ell=0}^{\infty} \frac{(-1)^\ell}{\ell! c^\ell} \left(\frac{\partial}{\partial t}\right)^\ell \int_{\mathcal{M}} \mu(t, \boldsymbol{x}') |\boldsymbol{x} - \boldsymbol{x}'|^{\ell-1} \, d^3x', \tag{6.92}$$

which is valid everywhere within the near zone. Note that once more the domain of integration is $\mathcal{M}$, a surface of constant time bounded externally by the sphere $r' = \mathcal{R}$; here, however, the integral is evaluated at time t instead of at the retarded time τ. The geometry is illustrated in Fig. 6.3.

6.3.5 Integration over the wave zone

In this subsection we develop a method to evaluate

$$\psi_{\mathscr{W}}(x) = \int_{\mathscr{W}} \frac{\mu(t - |\boldsymbol{x} - \boldsymbol{x}'|/c, \boldsymbol{x}')}{|\boldsymbol{x} - \boldsymbol{x}'|} d^3x', \tag{6.93}$$

the wave-zone portion of the complete solution $\psi = \psi_{\mathscr{N}} + \psi_{\mathscr{W}}$ to the wave equation. We recall that the domain of integration $\mathscr{W}$ is the intersection between $\mathscr{C}(x)$, the past light cone of the field point x, and the wave zone $r' > \mathcal{R}$.

Before we proceed with the work, we pause and ask whether $\psi_{\mathscr{W}}(x)$ could be dispensed with by taking the limit $\mathcal{R} \to \infty$, thereby achieving $\psi_{\mathscr{N}} \to \psi$ and $\psi_{\mathscr{W}} \to 0$. The answer is no: we cannot take $\mathcal{R}$ beyond its original value of order λ_c, and we cannot dispense with $\psi_{\mathscr{W}}$. The reason can be gleaned from Figs. 6.2 and 6.3: The difference between the domain $\mathscr{M}$ and the light cone $\mathscr{C}(x)$ becomes increasingly large as $\mathcal{R}$ increases, and the Taylor expansion for $\mu(t - |\boldsymbol{x} - \boldsymbol{x}'|/c)$ becomes increasingly inaccurate; the resulting expression for $\psi_{\mathscr{N}}$ would then become increasingly unreliable as $\mathcal{R}$ increases beyond λ_c. This lesson was hard learned. Early attempts to integrate the wave equation of post-Minkowskian theory were indeed based on the limit $\mathcal{R} \to \infty$, with the expectation that $\psi_{\mathscr{N}}$ would make a good approximation to ψ. Such attempts led to a host of divergent integrals that had to be argued away or swept under the rug. While these methods could sometimes be teased to give correct physical results, their mathematical justification left a lot to be desired. The decomposition of ψ into near-zone and wave-zone pieces nicely overcomes all these difficulties.

Our method to integrate over $\mathscr{W}$ must reflect the nature of the integrand there, and the fact that we are integrating over a null cone instead of a surface of constant time. For the slow-motion systems that we will generally encounter, the compact-support piece of μ lies deep within the near zone, and therefore vanishes on $\mathscr{W}$. The extended piece survives, and it is built from potentials that are themselves solutions to the wave equation. This implies that for a given integration point $(ct', \boldsymbol{x}')$ on $\mathscr{W}$, μ_{nc} is predominantly a function of $t' - r'/c$. Integration over the light cone is therefore facilitated by adopting retarded time as a variable of integration. The strategy is therefore this: Express the integral of Eq. (6.93) in terms of the spherical coordinates (r', θ', ϕ'), and then switch variables from r' to $u' := ct' - r'$ in order to perform the integration.

The strategy lends itself to a nice geometrical representation (see Fig. 6.4). A surface $u' =$ constant is a future-directed null cone $\mathscr{F}$ that emanates from $r' = 0$. It intersects $\mathscr{C}(x)$ on a two-dimensional surface $\mathscr{S}(u')$ parameterized by the angular variables θ' and ϕ'. Integration on $\mathscr{C}(x)$ can therefore be achieved by integrating over $\mathscr{S}(u')$ and adding the contributions from each relevant $\mathscr{F}$. Integrating on $\mathscr{S}(u')$ amounts to varying θ' and ϕ' over their allowed range, and the integration over $\mathscr{C}(x)$ is completed by varying u', which ranges from $u' = -\infty$ to $u' = u := ct - r$; the final value of u' corresponds to a future null cone that is tangent to $\mathscr{C}(x)$, emanating from the spacetime event at which $r' = 0$ crosses $\mathscr{C}(x)$.

To make these ideas explicit, we first provide a mathematical expression for $\mathscr{S}(u')$. Because $ct' = ct - |\boldsymbol{x} - \boldsymbol{x}'|$ on $\mathscr{C}(x)$ and $ct' = u' + r'$ on $\mathscr{F}$, we find that it is described

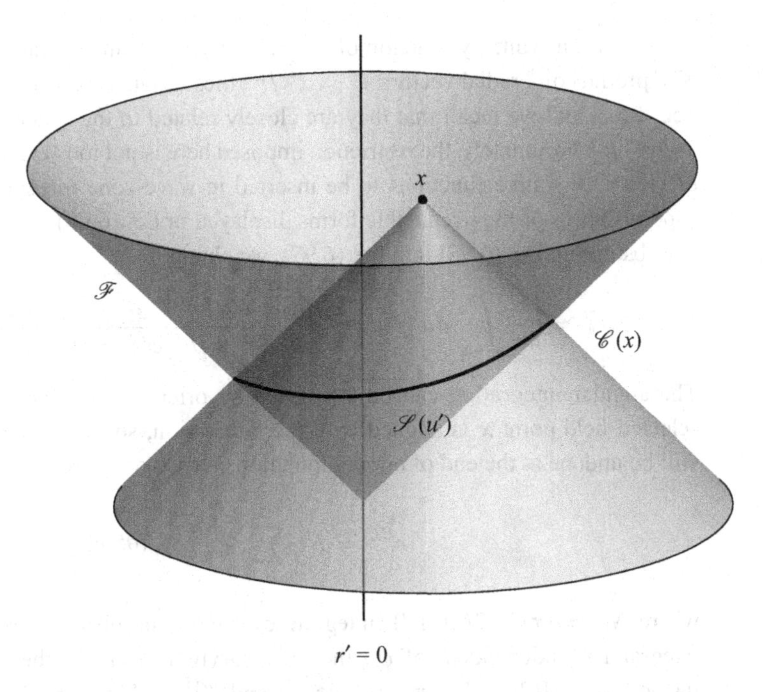

Fig. 6.4 Geometrical representation of the wave-zone integrations. $\mathscr{C}(x)$ is the past light cone of the field point x. $\mathscr{F}$ is the future light cone $u' = ct' - r' =$ constant with apex at $r' = 0$. $\mathscr{S}(u')$ is the two-dimensional surface of intersection between the past and future light cones.

by

$$u' = ct - r' - |\boldsymbol{x} - \boldsymbol{x}'|, \tag{6.94}$$

in which u' and t are constant. The equation can be solved for r' expressed as a function of θ' and ϕ':

$$r'(u', \theta', \phi') = \frac{(ct - u')^2 - r^2}{2(ct - u' - \boldsymbol{n}' \cdot \boldsymbol{x})}, \tag{6.95}$$

where $\boldsymbol{n}' := \boldsymbol{x}'/r'$. We next return to Eq. (6.93) and change variables from r' to u', using

$$\frac{\partial u'}{\partial r'} = \boldsymbol{n}' \cdot \boldsymbol{\nabla}' u' = \frac{u' - ct + \boldsymbol{n}' \cdot \boldsymbol{x}}{|\boldsymbol{x} - \boldsymbol{x}'|}. \tag{6.96}$$

This yields

$$\psi_{\mathscr{W}} = \int_{-\infty}^{u} du' \oint_{\mathscr{S}(u')} \frac{\mu((u' + r')/c, \boldsymbol{x}')}{ct - u' - \boldsymbol{n}' \cdot \boldsymbol{x}} r'(u', \theta', \phi')^2 \, d\Omega', \tag{6.97}$$

our new starting expression to calculate the wave-zone contribution to the potential $\psi(x)$.

To proceed it will be necessary to restrict our attention to source functions of the form

$$\mu(x') = \frac{1}{4\pi} \frac{f(\tau')}{r'^n} n'^{\langle L \rangle}, \tag{6.98}$$

where f is an arbitrary function of $\tau' = t' - r'/c$, n is an arbitrary integer, and $n'^{\langle L\rangle}$ is an STF product of ℓ radial vectors $n'^j = x'^j/r'$; these angular tensors were introduced back in Sec. 1.5.3, and we recall that they are closely related to the spherical-harmonic functions $Y_{lm}(\theta', \phi')$. Fortunately, the restriction imposed here is not too severe from a practical point of view: All source functions to be inserted in wave-zone integrals in this book will be superpositions of the irreducible forms displayed in Eq. (6.98).

Substituting Eq. (6.98) into Eq. (6.97), we obtain

$$\psi_{\mathscr{W}} = \frac{1}{4\pi} \int_{-\infty}^{u} du'\, f(u'/c) \oint_{\mathscr{S}(u')} \frac{n'^{\langle L\rangle}}{r'(u', \theta', \phi')^{n-2}} \frac{d\Omega'}{ct - u' - \boldsymbol{n}' \cdot \boldsymbol{x}}. \tag{6.99}$$

The angular integration can be simplified by orienting the coordinate axes so that the selected field point $\boldsymbol{x}$ is aligned with the z-direction, so that $\boldsymbol{n} = \boldsymbol{e}_z$; this specific choice will be undone at the end of our computation. We make use of Eq. (1.164),

$$n'^{\langle L\rangle} = N_\ell \sum_{m=-\ell}^{\ell} \mathscr{Y}^{\langle L\rangle}_{\ell m} Y_{\ell m}(\theta', \phi'), \tag{6.100}$$

where $N_\ell := 4\pi \ell!/(2\ell+1)!!$, integrate over $d\phi'$, and observe that since the rest of the integrand is independent of ϕ', the only surviving term in the sum is $m = 0$. Inserting now $Y_{\ell 0} = [(2\ell+1)/4\pi]^{1/2} P_\ell(\cos\theta')$ and $\mathscr{Y}^{\langle L\rangle}_{\ell 0} = [4\pi/(2\ell+1)]^{1/2} N_\ell^{-1} e_z^{\langle L\rangle}$ within the integral, we obtain

$$\psi_{\mathscr{W}} = \frac{1}{2} n^{\langle L\rangle} \int_{-\infty}^{u} du'\, f(u'/c) \int_{\mathscr{S}(u')} \frac{P_\ell(\xi)}{r'(u', \xi)^{n-2}(ct - u' - r\xi)} d\xi\,, \tag{6.101}$$

in which $\xi := \cos\theta'$ and

$$r'(u', \xi) := r'(u', \theta', 0) = \frac{(ct - u')^2 - r^2}{2(ct - u' - r\xi)}. \tag{6.102}$$

Switching integration variables from ξ back to r', using the fact that $\partial\xi/\partial r' = (ct - u' - r\xi)/rr'$, we recast $\psi_{\mathscr{W}}$ in the elegant form

$$\psi_{\mathscr{W}} = \frac{n^{\langle L\rangle}}{2r} \int_{-\infty}^{u} du'\, f(u'/c) \int_{\mathscr{S}(u')} \frac{P_\ell(\xi)}{r'^{(n-1)}} dr'\,, \tag{6.103}$$

in which ξ is now the function of r' determined by Eq. (6.102); an explicit expression will be provided below. We observe that the angular dependence of $\psi_{\mathscr{W}}$ is contained in the factor $n^{\langle L\rangle}$, with $\boldsymbol{n}$ previously chosen to be aligned with the z-direction. But since the remaining integral is now independent of all angles, the orientation of the coordinate axes has become irrelevant, and the special choice $\boldsymbol{n} = \boldsymbol{e}_z$ immaterial; we may now take $\boldsymbol{n}$ to point in the arbitrary direction specified by the polar angles θ and ϕ. The potential $\psi_{\mathscr{W}}$ has thus become a function of (t, r, θ, ϕ), with the dependence on t contained within $u = ct - r$.

To complete the wave-zone integration we must now give an explicit description of the closed surface $\mathscr{S}(u')$, and specify the limits of the integral over dr' so as to exclude the near zone from the domain of integration. The specific limits depend on whether the field point is in the near zone or in the wave zone.

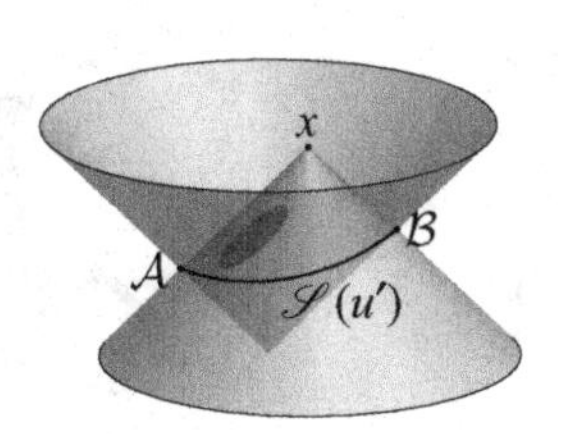

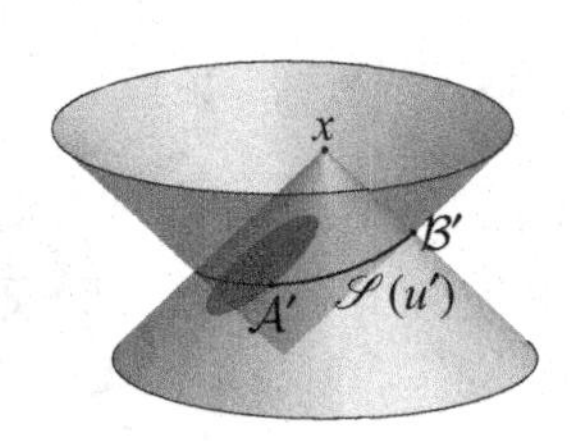

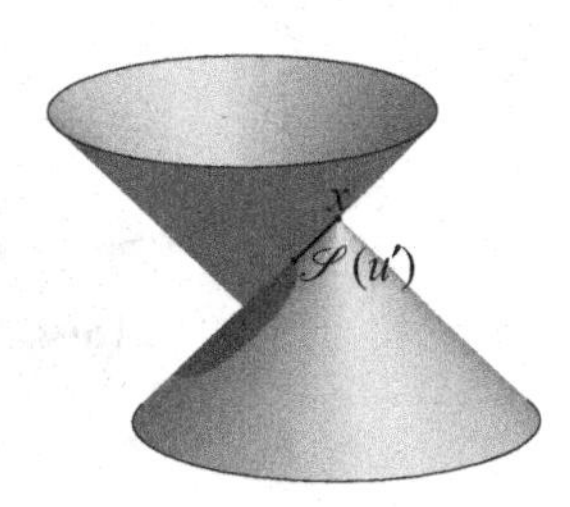

Fig. 6.5 Integration over the domain $\mathscr{W}(x)$, for a field point x in the wave zone, is carried out over each intersection surface $\mathscr{S}(u')$ in a sequence of future null cones $u' = \text{constant}$. The left panel corresponds to $u' < u - 2\mathcal{R}$; the integration runs from $\xi = -1$ (point $\mathcal{A}$) to $\xi = 1$ (point $\mathcal{B}$). The center panel corresponds to $u' > u - 2\mathcal{R}$; the intersection $\mathscr{S}(u')$ terminates at $\mathcal{A}'$, the boundary of the near zone $\mathscr{N}$. The right panel corresponds to $u' = u$; the cones are tangent, and $\mathscr{S}(u')$ runs from the edge of the near zone to x.

Wave-zone field point

To begin we assume that the field point x is situated in the wave zone, so that $r > \mathcal{R}$. We recall that $\mathscr{S}(u')$ is the intersection between the past null cone $\mathscr{C}(x)$ and the future null cone $u' = \text{constant}$. From Fig. 6.5 we see that when $u' < u - 2\mathcal{R}$, $\mathscr{S}(u')$ does not encounter the boundary of the near zone, and in this case ξ ranges from $\xi = -1$, at which $r' = \frac{1}{2}(ct - u' - r) = \frac{1}{2}(u - u')$, to $\xi = 1$, at which $r' = \frac{1}{2}(ct - u' + r) = \frac{1}{2}(u - u') + r$; these limits correspond to the events $\mathcal{A}$ and $\mathcal{B}$ in the left panel of Fig. 6.5. When $u - 2\mathcal{R} \le u' \le u$ we see that $\mathscr{S}(u')$ runs into the boundary of the near zone, and in this case the lower bound on r' must be $r' = \mathcal{R}$, with the corresponding value of $\xi > -1$ obtained from Eq. (6.102); the upper bound on r' is still $\frac{1}{2}(u - u') + r$, and these limits correspond to events $\mathcal{A}'$ and $\mathcal{B}'$ in the center panel of Fig. 6.5. The integration terminates when $u' = u$, as depicted on the right panel.

Defining $s := \frac{1}{2}(u - u')$ and the functions

$$A(s, r) := \int_{\mathcal{R}}^{r+s} \frac{P_\ell(\xi)}{r'^{(n-1)}}\, dr', \tag{6.104a}$$

$$B(s, r) := \int_{s}^{r+s} \frac{P_\ell(\xi)}{r'^{(n-1)}}\, dr', \tag{6.104b}$$

we obtain the final expression

$$\psi_{\mathscr{W}}(t, r, \theta, \phi) = \frac{n^{\langle L\rangle}}{r}\biggl\{\int_0^{\mathcal{R}} ds\, f(\tau - 2s/c)A(s, r) + \int_{\mathcal{R}}^{\infty} ds\, f(\tau - 2s/c)B(s, r)\biggr\} \tag{6.105}$$

for the wave-zone contribution to the potential $\psi(x)$, when x is situated in the wave zone. The quantity ξ that appears in A and B is determined by Eq. (6.102), in which we insert the definitions $u = ct - r$ and $s = \frac{1}{2}(u - u')$; this yields

$$\xi = \frac{r + 2s}{r} - \frac{2s(r + s)}{rr'}, \tag{6.106}$$

with $\xi = 1$ when $r' = r + s$ and $\xi = -1$ when $r' = s$.

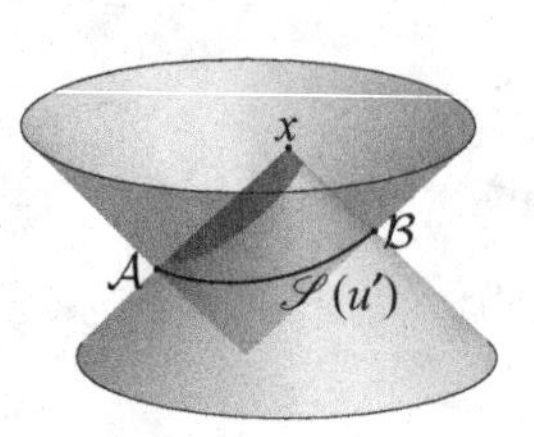

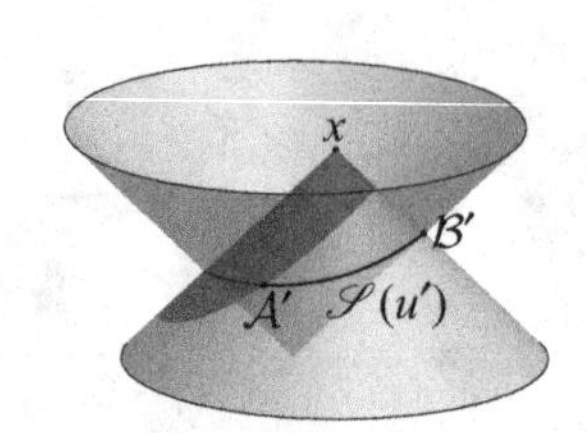

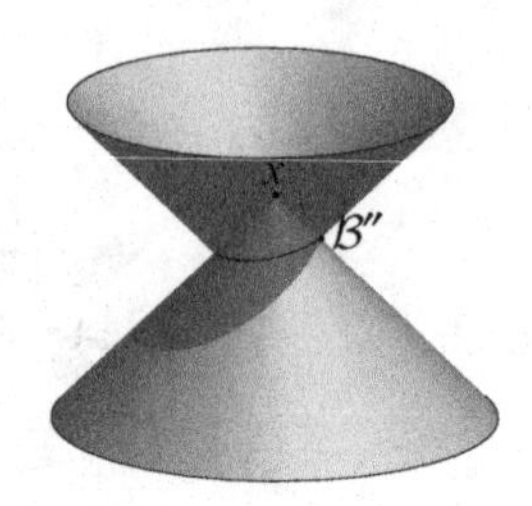

Fig. 6.6 Integration over the domain $\mathscr{W}(x)$, for a field point x in the near zone. The left panel corresponds to $u' < u - 2\mathcal{R}$; the integration runs from $\xi = -1$ (point $\mathcal{A}$) to $\xi = 1$ (point $\mathcal{B}$). The center panel corresponds to $u' > u - 2\mathcal{R}$; the intersection $\mathscr{S}(u')$ terminates at $\mathcal{A}'$, the boundary of the near zone $\mathscr{N}$. The right panel corresponds to $u' = u - 2\mathcal{R} + 2r$; the future cone intersects the past cone at $\xi = 1$ (point $\mathcal{B}''$) at the edge of the near zone.

Near-zone field point

We next take the field point x to be situated in the near zone, so that $r < \mathcal{R}$. In this case we find again that when $u' < u - 2\mathcal{R}$, $\mathscr{S}(u')$ does not encounter the near zone and ξ ranges from -1 to $+1$ (represented by the points $\mathcal{A}$ and $\mathcal{B}$ in the left panel of Fig. 6.6). When $u' > u - 2\mathcal{R}$, the integration runs from point $\mathcal{A}'$ in the center panel of Fig. 6.6, at which $r' = \mathcal{R}$, to point $\mathcal{B}'$, at which $\xi = 1$. But there is now a maximum value of u' at which the future null cone intersects $\mathscr{C}(x)$ at $\xi = 1$ (point $\mathcal{B}''$ in the right panel), corresponding to $u' = u - 2\mathcal{R} + 2r$; here the integration terminates. In this case, the minimum value of $s := \frac{1}{2}(u - u')$ is $\mathcal{R} - r$, and we obtain the expression

$$\psi_{\mathscr{W}}(t, r, \theta, \phi) = \frac{n^{\langle L \rangle}}{r} \left\{ \int_{\mathcal{R}-r}^{\mathcal{R}} ds f(\tau - 2s/c) A(s, r) + \int_{\mathcal{R}}^{\infty} ds f(\tau - 2s/c) B(s, r) \right\} \tag{6.107}$$

for the wave-zone contribution to the potential $\psi(x)$, when x is situated in the near zone. The functions $A(s, r)$ and $B(s, r)$ are again given by Eq. (6.104), and ξ is still given by Eq. (6.106).

Equation (6.105) is a concrete expression for the $\psi_{\mathscr{W}}(x)$ of Eq. (6.93) when the field point x is in the wave zone, and Eq. (6.107) is the corresponding expression when x is in the near zone. In both cases the source function $\mu(x')$ takes the form displayed in Eq. (6.98), with $f(\tau')$ describing its temporal behavior, r'^{-n} describing its radial profile, and $n'^{\langle L \rangle}$ describing its angular profile. Note that $\psi_{\mathscr{W}}(x)$ depends on the entire past history of the system, because f must be evaluated at retarded times $\tau - 2s/c$ all the way back to $-\infty$. This is a direct consequence of the fact that the source μ is not bounded by the near zone, and is generated by retarded fields that are themselves solutions to the wave equation. In post-Minkowskian theory, this feature is a consequence of the non-linearity of the Einstein field equations, which imply that the gravitational field itself generates gravity. While it may seem like a daunting task to evaluate the integrals of Eqs. (6.105) and (6.107), we shall find that they can be evaluated relatively easily for many interesting situations, with physically reasonable assumptions about the past behavior to ensure convergence.

Estimates

It is instructive to give crude estimates to the integrals of Eqs. (6.105) and (6.107). Suppose first that we wish to evaluate Eq. (6.105) in the far-away wave zone, and keep only its dominant r^{-1} part. Taking $P_\ell(\xi)$ to be of order unity, we approximate the functions defined by Eqs. (6.104) as $A \sim \int_{\mathcal{R}}^{\infty} p^{-(n-1)}\,dp \sim \mathcal{R}^{-(n-2)}$ and $B \sim \int_{s}^{\infty} p^{-(n-1)}\,dp \sim s^{-(n-2)}$; we ignore all numerical factors and exclude the special case $n = 2$. Inserting A into the first integral of Eq. (6.105) yields $\mathcal{R}^{-(n-2)} \int_0^{\mathcal{R}} f(\tau - 2s/c)\,ds$. Taking $\mathcal{R}$ to be small, we Taylor-expand $f(\tau - 2s/c)$ about $s = 0$ and integrate term by term. A typical term in the expansion is

$$\frac{\mathcal{R}^{q+1}}{c^q \mathcal{R}^{n-2}} f^{(q)}(\tau),$$

where the superscript (q) indicates the number of derivatives with respect to τ. As was motivated in the paragraph that follows Eq. (6.79), we are interested in the $\mathcal{R}$-independent part of $\psi_{\mathscr{W}}$. In order to extract this from our previous expansion, we retain the term $q = n - 3$ and discard all others. An estimate for the first integral is therefore $c^{-(n-3)} f^{(n-3)}(\tau)$. We next substitute B into the second integral of Eq. (6.105) and obtain $\int_{\mathcal{R}}^{\infty} s^{-(n-2)} f(\tau - 2s/c)\,ds$. Assuming that f and all its derivatives vanish in the infinite past, repeated integration by parts returns an expression of the form

$$\frac{f(\tau - 2\mathcal{R}/c)}{\mathcal{R}^{n-3}} + \frac{f^{(1)}(\tau - 2\mathcal{R}/c)}{c\mathcal{R}^{n-4}} + \frac{f^{(2)}(\tau - 2\mathcal{R}/c)}{c^2\mathcal{R}^{n-5}} + \cdots$$

The $\mathcal{R}$-independent part of this is easily seen to be of the form $c^{-(n-3)} f^{(n-3)}(\tau)$, as we had for the first integral. We conclude that a crude estimate for Eq. (6.105) is

$$\psi_{\mathscr{W}} \sim \frac{1}{c^{n-3}} \frac{n^{\langle L \rangle}}{r} f^{(n-3)}(\tau) \qquad \text{(far-away wave zone).} \tag{6.108}$$

The estimate ignores numerical factors, $\mathcal{R}$-dependent terms, and terms that decay faster than r^{-1}.

This estimate leads us to expect that the contribution from the wave-zone integral will be a small correction at any given iteration order of post-Minkowskian theory. First, the source function f is built from the pseudotensors $t_{\rm LL}^{\alpha\beta}$ and $t_{\rm H}^{\alpha\beta}$, which are quadratic in $h^{\alpha\beta}$ and therefore much smaller than the potentials themselves. Second, depending on n, the power with which the source falls off with r^{-1}, there will be additional time derivatives acting on f, generating additional powers of v_c/c. Accordingly, in many cases we will be able to ignore the contributions of the wave-zone integrals. But even when we are required to calculate those contributions, we will be able to do so using only the leading-order contributions to f. We will see a specific example of such a calculation in Sec. 7.4.

Suppose next that we wish to evaluate Eq. (6.107) deep within the near zone, for $r \ll \mathcal{R}$. Here the first integral of Eq. (6.107) is approximated as $\int_{\mathcal{R}-r}^{\mathcal{R}} ds f(\tau - 2s/c)A(s,r) \sim rf(\tau - 2\mathcal{R}/c)A(\mathcal{R},r)$, with $A(\mathcal{R},r) \sim \int_{\mathcal{R}}^{r+\mathcal{R}} p^{-(n-1)}\,dp \sim r\mathcal{R}^{-(n-1)}$. This produces the estimate

$$\frac{r^2}{\mathcal{R}^{n-1}} f(\tau - 2\mathcal{R}/c)$$

for the first integral, and the $\mathcal{R}$-independent part of this is $c^{-(n-1)}r^2 f^{(n-1)}(\tau)$. The second integral of Eq. (6.107) involves the domain of integration $\mathcal{R} < s < \infty$. Because s is large compared with r, we have the estimate $B \sim \int_s^{r+s} p^{-(n-1)}\,dp \sim rs^{-(n-1)}$. Inserting this inside the integral gives $r\int_{\mathcal{R}}^{\infty} s^{-(n-1)} f(\tau - 2s/c)\,ds$, and repeated integration by parts returns an expression of the form

$$\frac{rf(\tau - 2\mathcal{R}/c)}{\mathcal{R}^{n-2}} + \frac{rf^{(1)}(\tau - 2\mathcal{R}/c)}{c\mathcal{R}^{n-3}} + \frac{rf^{(2)}(\tau - 2\mathcal{R}/c)}{c^2\mathcal{R}^{n-4}} + \cdots .$$

The $\mathcal{R}$-independent part of this is of the form $c^{-(n-2)}rf^{(n-2)}(\tau)$. Collecting results, we conclude that a crude estimate for Eq. (6.107) is

$$\psi_{\mathcal{W}} \sim \frac{1}{c^{n-2}} n^{\langle L\rangle} \Big[f^{(n-2)}(\tau) + crf^{(n-1)}(\tau) \Big] \qquad \text{(near zone)}. \tag{6.109}$$

The estimate ignores numerical factors and all $\mathcal{R}$-dependent terms. In Sec. 7.3.4 we will learn that these contributions can be completely ignored for all our purposes in this book.

The case $n = 2$, for which μ falls off as r^{-2}, is special because the functions A and B are now logarithmic in $\mathcal{R}$ and s, and thus cannot be handled by our simple power-counting methods. We shall see that such terms are important in post-Minkowskian theory, and generate what are known as gravitational-wave "tails." We perform these computations, and describe these effects, in Chapter 11.

Box 6.7 Solution to the wave equation

The solution to the wave equation $\Box\psi = -4\pi\mu$ can be decomposed as

$$\psi = \psi_{\mathcal{N}} + \psi_{\mathcal{W}},$$

where $\psi_{\mathcal{N}}$ is the near-zone portion of the integral over the past light-cone $\mathscr{C}(x)$ of the field point x, while $\psi_{\mathcal{W}}$ is the wave-zone portion. The boundary between the near and wave zones is placed at $r' = \mathcal{R} = O(\lambda_c)$, where λ_c is a characteristic wavelength of the radiation.

When the field point $x = (ct, \boldsymbol{x})$ is in the wave zone,

$$\psi_{\mathcal{N}}(x) = \sum_{\ell=0}^{\infty} \frac{(-1)^\ell}{\ell!} \partial_L \left[\frac{1}{r} \int_{\mathcal{M}} \mu(\tau, \boldsymbol{x}') x'^L \, d^3x' \right],$$

$$\psi_{\mathcal{W}}(x) = \frac{n^{\langle L\rangle}}{r} \left\{ \int_0^{\mathcal{R}} ds f(\tau - 2s/c)A(s,r) + \int_{\mathcal{R}}^{\infty} ds f(\tau - 2s/c)B(s,r) \right\}.$$

And when x is in the near zone,

$$\psi_{\mathscr{N}}(x) = \sum_{\ell=0}^{\infty} \frac{(-1)^\ell}{\ell! c^\ell} \left(\frac{\partial}{\partial t}\right)^\ell \int_{\mathscr{M}} \mu(t, \boldsymbol{x}') |\boldsymbol{x} - \boldsymbol{x}'|^{\ell-1} \, d^3x',$$

$$\psi_{\mathscr{W}}(x) = \frac{n^{\langle L \rangle}}{r} \left\{ \int_{\mathcal{R}-r}^{\mathcal{R}} ds f(\tau - 2s/c) A(s,r) + \int_{\mathcal{R}}^{\infty} ds f(\tau - 2s/c) B(s,r) \right\}.$$

Here $\tau := t - r/c$ is retarded time, $\mathscr{M}$ is a surface of constant time bounded externally by the sphere $r' = \mathcal{R}$, and L is a multi-index that contains a number ℓ of individual spatial indices. For $\psi_{\mathscr{W}}$ we have assumed that the source function μ is of the specific form

$$\mu(x) = \frac{1}{4\pi} \frac{f(\tau)}{r^n} n^{\langle L \rangle},$$

in which $\boldsymbol{n} = \boldsymbol{x}/r$, and we have defined

$$A(s,r) := \int_{\mathcal{R}}^{r+s} \frac{P_\ell(\xi)}{r'^{(n-1)}} \, dr', \qquad B(s,r) := \int_{s}^{r+s} \frac{P_\ell(\xi)}{r'^{(n-1)}} \, dr',$$

where $\xi = (r + 2s)/r - 2s(r + s)/(rr')$.

6.4 Bibliographical notes

The formulation of the Einstein field equations detailed in Sec. 6.1 was first proposed by Landau and Lifshitz in their classic textbook *The Classical Theory of Fields*, now available in a fourth English edition (2000). Rigorous definitions for the total mass, momentum, and angular momentum of an asymptotically-flat spacetime were provided in a sequence of papers by Arnowitt, Deser, and Misner; their work is based on Hamiltonian methods, and is conveniently summarized in their 1962 review article.

The relaxation of the Einstein field equations described in Sec. 6.2 has become a standard tool of the field. The idea originated in Havas and Goldberg (1962), and it is beautifully summarized in Ehlers *et al.* (1976); another useful reference is Walker and Will (1980). The curved-spacetime formulation of the relaxed field equations in Box 6.3 was first proposed by Thorne and Kovacs (1975).

The mathematical methods introduced in Sec. 6.3 to integrate the wave equation when the source is extended over all space were first devised by Wiseman and Will (1991). They form the core of the DIRE approach (Direct Integration of the Relaxed Einstein equations) to post-Minkowskian theory, initiated by Will and Wiseman (1996) and developed systematically by Pati and Will (2000 and 2001). An alternative approach, based on a formal multipolar expansion of the potential outside the source, was pursued by Blanchet, Damour, Iyer, and their collaborators; this work is nicely summarized in Blanchet's *Living Reviews* article (2006).

6.5 Exercises

6.1 Show that $g_{\alpha\beta} = \sqrt{-\mathfrak{g}}\,\mathfrak{g}_{\alpha\beta}$, where $\mathfrak{g}_{\alpha\beta}$ is the matrix inverse to $\mathfrak{g}^{\alpha\beta}$, and $\mathfrak{g} = \det[\mathfrak{g}^{\alpha\beta}] = g$. If we define $\mathfrak{g}^{\alpha\beta} := \eta^{\alpha\beta} - h^{\alpha\beta}$, and $h^{\alpha\beta}$ is of order G, show that

$$(-g) = 1 - h + \frac{1}{2}h^2 - \frac{1}{2}h^{\mu\nu}h_{\mu\nu} + O(G^3),$$
$$g_{\alpha\beta} = \eta_{\alpha\beta} + h_{\alpha\beta} - \frac{1}{2}h\eta_{\alpha\beta} + h_{\alpha\mu}h^{\mu}{}_{\beta} - \frac{1}{2}hh_{\alpha\beta} + \left(\frac{1}{8}h^2 - \frac{1}{4}h^{\mu\nu}h_{\mu\nu}\right)\eta_{\alpha\beta} + O(G^3),$$

where indices on $h^{\alpha\beta}$ are lowered and contracted with the Minkowski metric.

6.2 Show that under the coordinate transformation $x'^{\mu} = f^{\mu}(x^{\alpha})$,

$$\mathfrak{g}^{\mu'\nu'} = J^{-1}\partial_{\alpha}f^{\mu}\partial_{\beta}f^{\nu}\mathfrak{g}^{\alpha\beta},$$
$$\partial_{\nu'}\mathfrak{g}^{\mu'\nu'} = \sqrt{-g'}\,\Box_g f^{\mu}(x^{\alpha}),$$

where $J := \det[\partial f^{\mu}/\partial x^{\alpha}]$ is the Jacobian of the transformation, and where for any scalar function f, $\Box_g f = (-g)^{-1/2}\partial_{\beta}(\mathfrak{g}^{\alpha\beta}\partial_{\alpha}f)$.

6.3 Consider the Schwarzschild metric in harmonic coordinates, given by Eqs. (5.171). Show explicitly that

$$\mathfrak{g}^{00} = -\frac{(1 + R/2r)^3}{1 - R/2r},$$
$$\mathfrak{g}^{jk} = \delta^{jk} - \left(\frac{R}{2r}\right)^2 n^j n^k,$$

where $R := 2GM/c^2$, and verify that the harmonic gauge condition $\partial_{\beta}\mathfrak{g}^{\alpha\beta} = 0$ is satisfied.

6.4 Consider the potentials $h^{\alpha\beta}$ for a stationary source ($\partial_0 h^{\alpha\beta} = 0$), in harmonic gauge. Show that the conserved quantities for the spacetime can be written in terms of the following surface integrals at infinity:

$$M = -\frac{c^2}{16\pi G}\oint_{\infty} r^2 \frac{\partial h^{00}}{\partial r}\,d\Omega,$$
$$P^j = -\frac{c^3}{16\pi G}\oint_{\infty} r^2 \frac{\partial h^{0j}}{\partial r}\,d\Omega,$$
$$J^{jk} = -\frac{c^3}{16\pi G}\oint_{\infty} r^2 \frac{\partial}{\partial r}\left(x^j h^{0k} - x^k h^{0j}\right)d\Omega,$$
$$R^j = -\frac{c^2}{16\pi GM}\oint_{\infty} r^4 \frac{\partial}{\partial r}\left(\frac{x^j h^{00}}{r^2}\right)d\Omega,$$

where $d\Omega = \sin\theta\, d\theta d\phi$ is the element of solid angle.

6.5 Consider the stationary metric given by

$$ds^2 = -\left(1 - \frac{R}{r}\right) d(ct)^2 - \frac{4GS}{c^2 r}\sin^2\theta\, d\phi dt + \left(1 + \frac{R}{r}\right)\left(dr^2 + r^2\, d\theta^2 + r^2 \sin^2\theta\, d\phi^2\right),$$

which is accurate to first order in G in a post-Minkowskian expansion; here $R = 2GM/c^2$ and S is a constant.

(a) Working to first order in G, find $\mathfrak{g}^{\alpha\beta}$ and verify that it is in the harmonic gauge.

(b) Using the surface integral formulation, find the mass, momentum, and angular momentum for this spacetime.

6.6 Using surface integrals, find the center-of-mass position of a spacetime for which

$$h^{00} = \frac{4GM}{c^2|\boldsymbol{x} - \boldsymbol{a}|},$$

where $\boldsymbol{a}$ is a constant vector.

6.7 Verify that the harmonic energy-momentum pseudotensor is conserved, so that $\partial_\beta[(-g)t_{\rm H}^{\alpha\beta}] = 0$.

6.8 Using the techniques of Sec. 6.3, find the solution to the wave equation $\Box\psi = -4\pi\mu$ when $\mu = -\boldsymbol{p}\cdot\boldsymbol{\nabla}\delta(\boldsymbol{x})\cos\omega t$, with $\boldsymbol{p}$ a constant vector. First take x to be in the wave zone, and find the solution there; then take x to be in the near zone. For the wave-zone expression, show that the sum over ℓ truncates. For the near-zone expression, show that the sum does not truncate. Compare your results with those of Box 6.6. Can you reconcile your results with the exact solution?

6.9 Using the techniques of Sec. 6.3, find the solution to the wave equation $\Box\psi = -4\pi\mu$ when μ is equal to $\mu_0(r/r_0)^4$ for $r < r_0$, and to $\mu_0(r_0/r)^4$ for $r > r_0$. You may take r_0 to be smaller than $\mathcal{R}$. You should find that

$$\psi = 4\pi\mu_0 r_0^2\left[\frac{2}{3} - \frac{1}{42}\left(\frac{r}{r_0}\right)^6\right]$$

for $r < r_0$, and

$$\psi = 4\pi\mu_0\frac{r_0^3}{r}\left(\frac{8}{7} - \frac{r_0}{2r}\right)$$

when $r > r_0$. Observe that while $\psi_{\mathcal{N}}$ and $\psi_{\mathcal{W}}$ both depend on $\mathcal{R}$, the final outcome for ψ is independent of $\mathcal{R}$.

7 Post-Minkowskian theory: Implementation

The theory was formulated in Chapter 6, and now we must get our hands dirty with its implementation. In this chapter we construct the *second post-Minkowskian approximation* to the metric of a curved spacetime produced by a bounded distribution of matter. For concreteness we choose the matter to consist of a perfect fluid. Our treatment allows the fluid to be of one piece (in the case of a single body), or broken up into a number of disconnected components (in the case of an N-body system).

Although the post-Minkowskian approximation does not *require* slow motion, we shall nevertheless assume that the fluid is subjected to a slow-motion condition of the sort described in Sec. 6.3.2: if v_c is a characteristic velocity within the fluid, we insist that $v_c/c \ll 1$. This amounts to incorporating a *post-Newtonian expansion* within the post-Minkowskian approximation. We do this for two reasons. First, our ultimate goal is to describe situations of astrophysical interest, and the virial theorem implies that $U \sim v^2$ for any gravitationally bound system; weak fields are naturally accompanied by slow motion. Second, any attempt to keep the velocities arbitrary in the post-Minkowskian expansion quickly leads to calculations that are unmanageable, and we prefer to avoid these complications here.

We begin in Sec. 7.1 by assembling the required tools and exploring the general structure of the gravitational potentials in the near and wave zones. In Sec. 7.2 we perform the first iteration of the relaxed field equations, and the outcome of this calculation is used as input in the second iteration, carried out in Sec. 7.3 for the near zone, and in Sec. 7.4 for the wave zone. Our main results are summarized in Boxes 7.5 and 7.7.

Before we proceed it is perhaps useful to recall the main results of the preceding chapter. We saw that in the Landau–Lifshitz formulation of general relativity, the Einstein field equations take the form of a wave equation for the gravitational potentials $h^{\alpha\beta} := \eta^{\alpha\beta} - \sqrt{-g}g^{\alpha\beta}$, together with the harmonic-gauge condition $\partial_\beta h^{\alpha\beta} = 0$; this is formally equivalent to the conservation equation $\partial_\beta \tau^{\alpha\beta} = 0$ for the effective energy-momentum pseudotensor, which acts as the source term in the wave equation. Each post-Minkowskian iteration of the wave equation gives rise to a new expression for the source, which is inserted back into the wave equation for the next iteration. After each iteration $h^{\alpha\beta}$ is expressed as an integral of the source over the past light cone of the field point $(t, \boldsymbol{x})$. Because the support of $\tau^{\alpha\beta}$ is not limited to the matter distribution, the domain of integration covers the entire light cone, and it is decomposed into a near-zone domain $\mathscr{N}$ and a wave-zone domain $\mathscr{W}$; the gravitational potentials are expressed as $h^{\alpha\beta} = h^{\alpha\beta}_{\mathscr{N}} + h^{\alpha\beta}_{\mathscr{W}}$. The boundary between the near and wave zones is placed at an arbitrary radius $r = \mathcal{R}$, with $\mathcal{R}$ chosen to be of the same order of magnitude as a characteristic wavelength of the gravitational radiation; while $h^{\alpha\beta}_{\mathcal{N}}$ and $h^{\alpha\beta}_{\mathcal{W}}$ individually depend on $\mathcal{R}$, their sum is guaranteed to be independent of the cutoff radius, and this dependence can therefore be ignored.

And now onward with an explicit implementation of these ideas.

7.1 Assembling the tools

We begin by gathering the various tools, formulae, and assumptions that are required in implementation of the post-Minkowskian expansion. Our discussion here will set the stage for the various applications to come, to post-Newtonian theory (Chapters 8 to 10), to gravitational waves (Chapter 11), and to gravitational radiation reaction (Chapter 12).

7.1.1 Fluid variables

A description of the laws of fluid mechanics in curved spacetime was presented in Sec. 5.3. There we saw that the matter variables m that are relevant to a perfect fluid are the proper mass density ρ, the proper internal energy density ϵ, the pressure p, and the velocity field u^α. The energy-momentum tensor of a perfect fluid is

$$T^{\alpha\beta} = (\rho + \epsilon/c^2 + p/c^2)u^\alpha u^\beta + p g^{\alpha\beta} \, . \tag{7.1}$$

The fluid dynamics is subjected to two conservation statements, a conservation of rest-mass expressed by $\nabla_\alpha(\rho u^\alpha) = 0$, and a conservation of energy-momentum expressed by $\nabla_\beta T^{\alpha\beta} = 0$.

For our purposes it is convenient to employ a slightly different set of matter variables. Noting that the components of u^α are not all independent (because of the normalization condition $g_{\alpha\beta}u^\alpha u^\beta = -c^2$), we express the four-velocity field as

$$u^\alpha = \gamma(c, \boldsymbol{v}) \, , \tag{7.2}$$

in terms of a three-velocity field $\boldsymbol{v}$ and a factor $\gamma := u^0/c$ that can be determined in terms of $\boldsymbol{v}$ by the normalization condition. Making the substitution within the equation of mass conservation, we find that it can be expressed in the form

$$\partial_t \rho^* + \partial_j(\rho^* v^j) = 0 \, , \tag{7.3}$$

in terms of a rescaled mass density defined by

$$\rho^* := \sqrt{-g}\gamma\rho = \sqrt{-g}\,\rho\, u^0/c \, . \tag{7.4}$$

To arrive at Eq. (7.3) we made use of the divergence identity of Eq. (5.40). Finally, we shall use $\Pi := \epsilon/\rho$ instead of ϵ; this is the fluid's internal energy per unit mass. Our final set of matter variables is therefore

$$\mathsf{m} := \{\rho^*, p, \Pi, \boldsymbol{v}\} \, , \tag{7.5}$$

and all other fluid variables can be determined in terms of this set.

The continuity equation (7.3) plays an important role in the description of a perfect fluid. We observe that unlike $\nabla_\beta T^{\alpha\beta} = 0$, which constrains the dynamics of the fluid, the statement of mass conservation is entirely kinematical in nature. Equation (7.3) states that the rest-mass of a fluid element does not change as we follow its motion within the fluid; this is tantamount to defining what one means by the phrase "fluid element," and the statement is indeed a piece of the kinematical description of the fluid. This is quite distinct, for example, from the statement of the first law of thermodynamics, $d\Pi - (p/\rho^2)\,d\rho = 0$ (refer to Sec. 1.4.2), which is dynamical in nature.

We assume that the fluid is subjected to a slow-motion condition. Recalling the scaling quantities introduced in Sec. 6.3.2, we have that r_c is the radius of a sphere that surrounds the matter distribution, t_c is a characteristic time scale associated with the fluid motions, $v_c = r_c/t_c$ is a characteristic velocity within the fluid, $\lambda_c = ct_c$ is a characteristic wavelength of the gravitational radiation produced by the moving fluid, and m_c is the characteristic mass of the matter distribution. We demand that $v_c/c \ll 1$, which implies that the fluid is situated deep within the near zone: $r_c \ll \lambda_c$.

The slow-motion condition gives rise to a hierarchy between the components of the energy-momentum tensor. From Eq. (7.1) we have the approximate relations $T^{00} \simeq \rho^* c^2$, $T^{0j} \simeq \rho^* v^j c$, and $T^{jk} \simeq \rho^* v^j v^k + p\,\delta^{jk}$, and these imply

$$T^{0j}/T^{00} \sim v_c/c\,, \qquad T^{jk}/T^{00} \sim (v_c/c)^2\,. \tag{7.6}$$

A glance at Eq. (6.51) then reveals that this hierarchy is inherited by the gravitational potentials:

$$h^{0j}/h^{00} \sim v_c/c\,, \qquad h^{jk}/h^{00} \sim (v_c/c)^2\,. \tag{7.7}$$

It is useful to express these relations more directly as

$$T^{00} = O(c^2)\,, \qquad T^{0j} = O(c)\,, \qquad T^{jk} = O(1)\,, \tag{7.8}$$

and (taking into account the factor c^{-4} in the field equations)

$$h^{00} = O(c^{-2}), \qquad h^{0j} = O(c^{-3}), \qquad h^{jk} = O(c^{-4}), \tag{7.9}$$

thereby introducing c^{-2} as a *post-Newtonian expansion parameter*. This notation serves as a powerful mnemonic to judge the importance of various terms in a post-Newtonian expansion. But it is a notational shortcut that must be used with care; it should be remembered, for example, that a relation such as $T^{jk} = O(1)$ really stands for something more meaningful, such as $T^{jk}/T^{00} \sim (v_c/c)^2$.

7.1.2 General structure of the potentials: Near zone

Having introduced the matter variables, the slow-motion condition, and the post-Newtonian hierarchy, we turn next to an examination of the general structure of the gravitational potentials $h^{\alpha\beta}$. These are determined by the relaxed field equations

$$\Box h^{\alpha\beta} = -\frac{16\pi G}{c^4}\tau^{\alpha\beta}, \tag{7.10}$$

in which

$$\tau^{\alpha\beta} = (-g)\big(T^{\alpha\beta} + t^{\alpha\beta}_{\rm LL} + t^{\alpha\beta}_{\rm H}\big) \tag{7.11}$$

is the effective energy-momentum pseudotensor of Eq. (6.52). We decompose the potentials as $h^{\alpha\beta} = h^{\alpha\beta}_{\mathscr{N}} + h^{\alpha\beta}_{\mathscr{W}}$, and first examine them when the field point $\boldsymbol{x}$ is in the near zone, where $r := |\boldsymbol{x}| < \mathcal{R}$.

Consulting Box 6.7, we see that $h^{\alpha\beta}_{\mathscr{N}}$ can be expressed as the expansion

$$h^{\alpha\beta}_{\mathscr{N}}(t,\boldsymbol{x}) = \frac{4G}{c^4}\sum_{\ell=0}^{\infty}\frac{(-1)^\ell}{\ell!c^\ell}\left(\frac{\partial}{\partial t}\right)^\ell \int_{\mathscr{M}} \tau^{\alpha\beta}(t,\boldsymbol{x}')|\boldsymbol{x}-\boldsymbol{x}'|^{\ell-1}\, d^3x', \tag{7.12}$$

in which $\mathscr{M}$ is a surface of constant time bounded externally by $r' := |\boldsymbol{x}'| = \mathcal{R}$. The first few terms are

$$\begin{aligned} h^{\alpha\beta}_{\mathscr{N}}(t,\boldsymbol{x}) = \frac{4G}{c^4}\bigg[&\int_{\mathscr{M}} \frac{\tau^{\alpha\beta}(t,\boldsymbol{x}')}{|\boldsymbol{x}-\boldsymbol{x}'|}\, d^3x' - \frac{1}{c}\frac{d}{dt}\int_{\mathscr{M}} \tau^{\alpha\beta}(t,\boldsymbol{x}')\, d^3x' \\ &+ \frac{1}{2c^2}\frac{\partial^2}{\partial t^2}\int_{\mathscr{M}} \tau^{\alpha\beta}(t,\boldsymbol{x}')|\boldsymbol{x}-\boldsymbol{x}'|\, d^3x' \\ &- \frac{1}{6c^3}\frac{\partial^3}{\partial t^3}\int_{\mathscr{M}} \tau^{\alpha\beta}(t,\boldsymbol{x}')(r^2 - 2\boldsymbol{x}\cdot\boldsymbol{x}' + r'^2)\, d^3x' + \cdots \bigg], \end{aligned} \tag{7.13}$$

and we see that each successive term comes with an additional factor of c^{-1}, signifying that it is smaller than the previous term by a factor of order $v_c/c \ll 1$. This is our first encounter with a post-Newtonian expansion in powers of c^{-2}, with fractional orders assigned to odd powers of c^{-1}.

The expansion of Eq. (7.13) is a direct consequence of the relaxed field equations. It simplifies when we take into account the conservation statement $\partial_\beta \tau^{\alpha\beta} = 0$ for the energy-momentum pseudotensor. When we examine the expansion for $h^{00}_{\mathscr{N}}$, for example, we note that the second term is given by $-\int_{\mathscr{M}} \partial_0 \tau^{00}\, d^3x'$. The conservation statement allows us to make the substitution $\partial_0\tau^{00} = -\partial_j \tau^{0j}$ inside the integral, which can then, by Gauss's theorem, be converted to a surface integral over $\partial\mathscr{M}$, the boundary of the region $\mathscr{M}$; this is the surface $r' = \mathcal{R}$. The surface integral would vanish if τ^{0j} were confined to the near zone, and in this case the expansion for $h^{00}_{\mathscr{N}}$ would skip the term at order c^{-1}. In general, however, τ^{0j} extends beyond the near zone, and the surface integral does not vanish. But since τ^{0j} is constructed from the potentials, the surface integral can be estimated and shown to be of a very high order in the post-Newtonian expansion, well beyond any order that we will encounter in this book. In practice, therefore, we can appeal to energy conservation and eliminate the second term in the expansion for $h^{00}_{\mathscr{N}}$.

In fact, the conservation equations $\partial_\beta \tau^{\alpha\beta} = 0$ can be put to good use to simplify and organize many terms in the expansion of $h^{\alpha\beta}_{\mathscr{N}}$. Particularly useful are a number of identities that follow from the conservation equations, namely,

$$\tau^{0j} = \partial_0\big(\tau^{00}x^j\big) + \partial_k\big(\tau^{0k}x^j\big), \tag{7.14a}$$

$$\tau^{jk} = \frac{1}{2}\partial_{00}\big(\tau^{00}x^jx^k\big) + \frac{1}{2}\partial_p\big(2\tau^{p(j}x^{k)} - \partial_q\tau^{pq}x^jx^k\big), \tag{7.14b}$$

(continued overleaf)

$$\tau^{0j}x^k = \frac{1}{2}\partial_0\big(\tau^{00}x^jx^k\big) + \tau^{0[j}x^{k]} + \frac{1}{2}\partial_p\big(\tau^{0p}x^jx^k\big), \tag{7.14c}$$

$$\tau^{jk}x^n = \frac{1}{2}\partial_0\big(2\tau^{0(j}x^{k)}x^n - \tau^{0n}x^jx^k\big) + \frac{1}{2}\partial_p\big(2\tau^{p(j}x^{k)}x^n - \tau^{np}x^jx^k\big), \tag{7.14d}$$

in which round and square brackets surrounding indices denote symmetrized and anti-symmetrized combinations, respectively. Exploiting these identities, we find after some manipulations that the various components of the gravitational potentials are now given by

$$\begin{aligned} h^{00}_{\mathcal{N}} = \frac{4G}{c^2}\bigg\{ &\int_{\mathcal{M}} \frac{c^{-2}\tau^{00}}{|\boldsymbol{x}-\boldsymbol{x}'|}\,d^3x' + \frac{1}{2c^2}\frac{\partial^2}{\partial t^2}\int_{\mathcal{M}} c^{-2}\tau^{00}|\boldsymbol{x}-\boldsymbol{x}'|\,d^3x' \\ &- \frac{1}{6c^3}\overset{(3)}{\mathcal{I}}{}^{kk}(t) + \frac{1}{24c^4}\frac{\partial^4}{\partial t^4}\int_{\mathcal{M}} c^{-2}\tau^{00}|\boldsymbol{x}-\boldsymbol{x}'|^3\,d^3x' \\ &- \frac{1}{120c^5}\Big[(4x^kx^l + 2r^2\delta^{kl})\overset{(5)}{\mathcal{I}}{}^{kl}(t) - 4x^k\overset{(5)}{\mathcal{I}}{}^{kll}(t) + \overset{(5)}{\mathcal{I}}{}^{kkll}(t)\Big] \\ &+ O(c^{-6})\bigg\} + h^{00}[\partial\mathcal{M}], \end{aligned} \tag{7.15a}$$

$$\begin{aligned} h^{0j}_{\mathcal{N}} = \frac{4G}{c^3}\bigg\{ &\int_{\mathcal{M}} \frac{c^{-1}\tau^{0j}}{|\boldsymbol{x}-\boldsymbol{x}'|}\,d^3x' + \frac{1}{2c^2}\frac{\partial^2}{\partial t^2}\int_{\mathcal{M}} c^{-1}\tau^{0j}|\boldsymbol{x}-\boldsymbol{x}'|\,d^3x' \\ &+ \frac{1}{18c^3}\Big[3x^k\overset{(4)}{\mathcal{I}}{}^{jk}(t) - \overset{(4)}{\mathcal{I}}{}^{jkk}(t) + 2\epsilon^{mjk}\overset{(3)}{\mathcal{J}}{}^{mk}(t)\Big] \\ &+ O(c^{-4})\bigg\} + h^{0j}[\partial\mathcal{M}], \end{aligned} \tag{7.15b}$$

$$\begin{aligned} h^{jk}_{\mathcal{N}} = \frac{4G}{c^4}\bigg\{ &\int_{\mathcal{M}} \frac{\tau^{jk}}{|\boldsymbol{x}-\boldsymbol{x}'|}\,d^3x' - \frac{1}{2c}\overset{(3)}{\mathcal{I}}{}^{jk}(t) + \frac{1}{2c^2}\frac{\partial^2}{\partial t^2}\int_{\mathcal{M}} \tau^{jk}|\boldsymbol{x}-\boldsymbol{x}'|\,d^3x' \\ &- \frac{1}{36c^3}\Big[3r^2\overset{(5)}{\mathcal{I}}{}^{jk}(t) - 2x^m\overset{(5)}{\mathcal{I}}{}^{jkm}(t) - 8x^n\epsilon^{mn(j}\overset{(4)}{\mathcal{J}}{}^{m|k)}(t) + 6\overset{(3)}{\mathcal{M}}{}^{jkmm}(t)\Big] \\ &+ O(c^{-4})\bigg\} + h^{jk}[\partial\mathcal{M}], \end{aligned} \tag{7.15c}$$

in which $\tau^{\alpha\beta}$ is expressed as a function of t and $\boldsymbol{x}'$ inside the integrals, a number within brackets placed above a symbol such as $\mathcal{I}^{jk}$ indicates the number of differentiations with respect to time, and $h^{\alpha\beta}[\partial\mathcal{M}]$ denotes the collected surface terms generated during our manipulations of the integrals (the details will not be displayed here). We have also introduced the following notation for the multipole moments of the source $\tau^{\alpha\beta}$:

$$\mathcal{I}^L(t) := \int_{\mathcal{M}} c^{-2}\tau^{00}(t,\boldsymbol{x})x^L\,d^3x\,, \tag{7.16a}$$

$$\mathcal{J}^{jL}(t) := \epsilon^{jab}\int_{\mathcal{M}} c^{-1}\tau^{0b}(t,\boldsymbol{x})x^{aL}\,d^3x\,, \tag{7.16b}$$

$$\mathcal{M}^{jkL} := \int_{\mathcal{M}} \tau^{jk}(t,\boldsymbol{x})x^L\,d^3x\,, \tag{7.16c}$$

in which L is a multi-index containing a number ℓ of individual indices, so that $A^L := A^{j_1 j_2 \cdots j_\ell}$ and $x^L := x^{j_1} x^{j_2} \ldots x^{j_\ell}$.

There is a lot to take in with the expansions of Eq. (7.15), and we shall now take the time to describe the structure of $h^{00}_{\mathcal{N}}$ in some detail. We begin with the first term on the right-hand side of Eq. (7.15a), and observe that it leads off at order c^{-2} with a Newtonian-like potential associated with the mass density $c^{-2}\tau^{00} \sim \rho^*$. Embedded within this term are corrections of order $(v_c/c)^2$ and higher that enter the detailed expression for τ^{00}, as well as corrections of order G and higher that arise in previous iterations of the relaxed field equations. But the leading contribution gives rise to Newtonian gravity.

The integral that appears in the second term in $h^{00}_{\mathcal{N}}$ is known as a superpotential, because the factor $|\boldsymbol{x} - \boldsymbol{x}'|$ appears in the numerator instead of the denominator; as we shall see, a superpotential is a potential sourced by another potential. Because of the time derivatives, this term leads off at order c^{-2} relative to the Newtonian term, or at overall order c^{-4} in $h^{00}_{\mathcal{N}}$; it is a "first post-Newtonian correction," or 1PN correction, to the gravitational potential. It also contains higher-order corrections coming from higher-order terms in $c^{-2}\tau^{00}$, just as we saw previously for the leading-order Newtonian term. It is instructive to note that the superpotential itself is of order $m_c r$, but since each time derivative produces a factor of $t_c^{-1} = v_c/r_c$, its contribution to $h^{00}_{\mathcal{N}}$ is a factor of order $(v_c/c)^2$ smaller than the Newtonian potential when r is comparable to r_c.

The third term in $h^{00}_{\mathcal{N}}$ involves three time derivatives of $\mathcal{I}^{kk}(t)$, the trace of the mass quadrupole moment. The factor of c^{-3} in front indicates that this term is a factor of order $(v_c/c)^3$ smaller than the leading Newtonian term, and therefore represents a 1.5PN contribution to the gravitational potential. We will show below that since this term depends on t only, it can always be absorbed into a redefinition of the time coordinate, and therefore be removed by a coordinate transformation. This observation suggests that the 1.5PN term does not play a physical role, and we shall have occasion to show that such is indeed the case. The expression for the 1.5PN term displayed in Eq. (7.15a) is derived in Box 7.1.

The integral that appears in the fourth term in $h^{00}_{\mathcal{N}}$ is sometimes called a superduperpotential because of the presence of $|\boldsymbol{x} - \boldsymbol{x}'|^3$ in the numerator; a superduperpotential is a potential sourced by a superpotential. Because of the time derivatives, this term leads off at order c^{-4} relative to the Newtonian term, and therefore represents a 2PN correction to the gravitational potential.

We now examine the fifth set of terms. The first member of the set involves the mass quadrupole moment differentiated five times with respect to time, and it scales as

$$\frac{r_c^2}{c^5}\frac{1}{t_c^5} m_c r_c^2 = (v_c/c)^5 \frac{m_c}{r_c}, \tag{7.17}$$

which is a correction of order $(v_c/c)^5$ relative to the Newtonian term. The other members have the same scaling, and this group of terms give rise to a 2.5PN correction to the gravitational potential. Unlike the 1.5PN term, this group depends on the spatial coordinates in addition to time, and it cannot be removed by a coordinate transformation. It gives rise to real, physical effects on the system. The nature of these effects can be anticipated from the fact that the 2.5PN terms involve an odd number of time derivatives, and are therefore antisymmetric under a time reflection $t \to -t$; this is in contrast with the 1PN and 2PN terms, which are symmetric under the time reflection. This property is associated with dissipative processes taking place within the system, representing a radiative loss of energy to gravitational waves. The 2.5PN contributions to the gravitational potentials are known as *radiation-reaction potentials*, and their effects will be explored in detail in Chapter 12.

Turning next to the other components of the gravitational potentials, we observe that they have a very similar structure. The component $h^{0j}_{\mathcal{N}}$ leads off at order c^{-3} with a Newtonian-like potential sourced by the mass-current density $c^{-1}\tau^{0j} \sim \rho^* v^j$. Comparing this with the leading term in $h^{00}_{\mathcal{N}}$, we see that it is smaller by a factor of order v_c/c, and it would be tempting to assign a 0.5PN label to this term. As we shall see below, however, all effects arising from $h^{0j}_{\mathcal{N}}$ will be the result of a coupling with other quantities that also scale as v_c/c; the result is a 1PN correction to the leading Newtonian effect. Keeping this context in mind, it is appropriate to reset the post-Newtonian counter and to declare that the leading term in $h^{0j}_{\mathcal{N}}$ makes a 1PN contribution to the gravitational potentials. The expansion of $h^{0j}_{\mathcal{N}}$ continues with a superpotential term at order c^{-5} which is assigned a 2PN label, and this is followed by 2.5PN contributions. The absence of a term at order c^{-4} is a consequence of momentum conservation; the manipulations that led to the disappearance of the c^{-3} term in $h^{00}_{\mathcal{N}}$ lead to the same conclusion here, and in both cases we see that these terms are absorbed in the surface integrals $h^{\alpha\beta}[\partial\mathcal{M}]$.

And finally, the components $h^{jk}_{\mathcal{N}}$ lead off at order c^{-4} with a Newtonian-like potential sourced by $\tau^{jk} \sim \rho^* v^j v^k$; this is smaller than the leading term in $h^{00}_{\mathcal{N}}$ by a factor of order $(v_c/c)^2$ and represents a 1PN contribution to the gravitational potentials. The next term, involving a single time derivative, does *not* vanish; use of Eqs. (7.14c) and (7.14d) converts it to three time derivatives of the mass quadrupole moment. This term represents a 1.5PN contribution, and it is followed by a superpotential term at 2PN order, and a set of 2.5PN contributions.

The potentials $h^{\alpha\beta}_{\mathcal{N}}$ provide the near-zone portion of the light-cone integral giving $h^{\alpha\beta}$ in terms of $\tau^{\alpha\beta}$, and we have yet to examine the wave-zone portion $h^{\alpha\beta}_{\mathcal{W}}$. We recall from Box 6.7 that this can be expressed as

$$h^{\alpha\beta}_{\mathcal{W}}(t, \boldsymbol{x}) = \frac{4G}{c^4}\frac{n^{\langle L\rangle}}{r}\left\{\int_{\mathcal{R}-r}^{\mathcal{R}} ds f^{\alpha\beta}(\tau - 2s/c)A(s, r) + \int_{\mathcal{R}}^{\infty} ds f^{\alpha\beta}(\tau - 2s/c)B(s, r)\right\}, \tag{7.18}$$

when $\tau^{\alpha\beta}$ can be put in the specific form

$$\tau^{\alpha\beta} = \frac{1}{4\pi}\frac{f^{\alpha\beta}(\tau)}{r^n} n^{\langle L\rangle}. \tag{7.19}$$

Here $\tau = t - r/c$ is retarded time, $n^{\langle L\rangle}$ is an angular STF tensor of the sort introduced back in Sec. 1.5.3, and the functions $A(s, r)$ and $B(s, r)$ are defined by Eq. (6.104). Although restrictive, the expression of Eq. (7.18) is nevertheless useful because the wave-zone sources

$\tau^{\alpha\beta}$ encountered below will always be decomposed in the elementary forms displayed in Eq. (7.19); the complete $h^{\alpha\beta}_{\mathcal{W}}$ can then be obtained by summing over these elementary contributions.

Little more can be said about the general structure of $h^{\alpha\beta}_{\mathcal{W}}$ in the near zone. The sources $f^{\alpha\beta}$ vanish in the first iteration of the relaxed field equations, because we are instructed to set $h^{\alpha\beta} = 0$ in $\tau^{\alpha\beta}$ and the material source is confined to the near zone. In the second and higher iterations, $h^{\alpha\beta}$ is no longer zero, and $\tau^{\alpha\beta}$ now extends into the wave zone; in these cases we have no choice but to plow through the detailed calculations to see what contribution $h^{\alpha\beta}_{\mathcal{W}}$ might make. We encounter some of these calculations later in this chapter, and then again in Chapter 11.

Box 7.1 Radiation-reaction terms in the potentials

To illustrate how the various radiation-reaction terms arise in the potentials, we examine the contribution

$$-\frac{1}{6c^3}\left(\frac{\partial}{\partial t}\right)^3 \int_{\mathcal{M}} \tau^{00}(t, \boldsymbol{x}')(r^2 - 2\boldsymbol{x}\cdot\boldsymbol{x}' + r'^2)\, d^3x'$$

to $h^{00}_{\mathcal{N}}$; this is the third line in Eq. (7.13). In the first term, r^2 can be brought outside the integral, giving $-\frac{1}{6}c^{-2}r^2\partial_t^2\int_{\mathcal{M}}\partial_0\tau^{00}\,d^3x' = \frac{1}{6}c^{-2}r^2\partial_t^2\int_{\mathcal{M}}\partial_j\tau^{0j}\,d^3x'$, which becomes a surface term, reflecting the fact that energy is conserved apart from a tiny flux of gravitational radiation. In the second term, $\boldsymbol{x}$ can be brought outside the integral, giving $\frac{1}{3}c^{-2}x^j\partial_t^2\int_{\mathcal{M}}\partial_0\tau^{00}x'^j\,d^3x' = \frac{1}{3}c^{-2}x^j\partial_t^2\int_{\mathcal{M}}\tau^{0j}\,d^3x'$ plus a surface term. This yields $-\frac{1}{3}c^{-1}x^j\partial_t\int_{\mathcal{M}}\partial_k\tau^{kj}\,d^3x'$, which gives another surface term. The elimination of this term reflects the conservation of momentum. The third term survives, giving $-\frac{1}{6}c^{-3}\dddot{I}^{kk}$ as shown in Eq. (7.15a).

The next term in $h^{00}_{\mathcal{N}}$ involving an odd number of time derivatives is

$$-\frac{1}{120c^5}\left(\frac{\partial}{\partial t}\right)^5 \int_{\mathcal{M}} \tau^{00}(t, \boldsymbol{x}')\Big[r^4 - 4r^2\boldsymbol{x}\cdot\boldsymbol{x}' + 4(\boldsymbol{x}\cdot\boldsymbol{x}')^2 + 2r^2r'^2 - 4r'^2\boldsymbol{x}\cdot\boldsymbol{x}' + r'^4\Big]\, d^3x'.$$

The first two terms can be shown to become surface integrals by appealing to the conservation identities of Eqs. (7.14), and the remaining four terms are displayed in Eq. (7.15a). Similar manipulations, albeit becoming progressively more complicated, yield the corresponding terms displayed in Eqs. (7.15) for $h^{0j}_{\mathcal{N}}$ and $h^{jk}_{\mathcal{N}}$.

7.1.3 Near-zone metric

We will need to construct the spacetime metric $g_{\alpha\beta}$ from the gravitational potentials $h^{\alpha\beta}$. The link is provided by the gothic inverse metric $\mathfrak{g}^{\alpha\beta} = \eta^{\alpha\beta} - h^{\alpha\beta}$, which is related to the inverse metric $g^{\alpha\beta}$ by $\mathfrak{g}^{\alpha\beta} = \sqrt{-g}g^{\alpha\beta}$. The inverse relation is $g_{\alpha\beta} = \sqrt{-\mathfrak{g}}\,\mathfrak{g}_{\alpha\beta}$, in which $\mathfrak{g}_{\alpha\beta}$ is the matrix inverse to $\mathfrak{g}^{\alpha\beta}$, and $\mathfrak{g} := \det[\mathfrak{g}^{\alpha\beta}]$. Given that $h^{\alpha\beta}$ is of order G, we can solve these equations and obtain the metric and its inverse as post-Minkowskian expansions in powers of G, and express the results in terms of the potentials $h^{\alpha\beta}$. We find

$$g_{\alpha\beta} = \eta_{\alpha\beta} + h_{\alpha\beta} - \frac{1}{2}h\eta_{\alpha\beta} + h_{\alpha\mu}h^{\mu}{}_{\beta} - \frac{1}{2}hh_{\alpha\beta} + \left(\frac{1}{8}h^2 - \frac{1}{4}h^{\mu\nu}h_{\mu\nu}\right)\eta_{\alpha\beta} + O(G^3), \tag{7.20a}$$

$$g^{\alpha\beta} = \eta^{\alpha\beta} - h^{\alpha\beta} + \frac{1}{2}h\eta^{\alpha\beta} - \frac{1}{2}hh^{\alpha\beta} + \left(\frac{1}{8}h^2 + \frac{1}{4}h^{\mu\nu}h_{\mu\nu}\right)\eta^{\alpha\beta} + O(G^3), \tag{7.20b}$$

$$(-g) = 1 - h + \frac{1}{2}h^2 - \frac{1}{2}h^{\mu\nu}h_{\mu\nu} + O(G^3), \tag{7.20c}$$

$$\sqrt{-g} = 1 - \frac{1}{2}h + \frac{1}{8}h^2 - \frac{1}{4}h^{\mu\nu}h_{\mu\nu} + O(G^3). \tag{7.20d}$$

It is understood that here, indices on $h^{\alpha\beta}$ are lowered with the Minkowski metric, so that $h_{\alpha\beta} := \eta_{\alpha\mu}\eta_{\beta\nu}h^{\mu\nu}$ and $h := \eta_{\mu\nu}h^{\mu\nu}$.

In practice, the construction of the metric from the potentials depends on the context, which dictates the degree of accuracy required of each metric component. Suppose that we are specifically interested in determining the geodesic motion of a slowly-moving particle in the near zone of a weakly-curved spacetime. As we saw back in Sec. 5.2.3, the motion is governed by a Lagrangian L given by

$$\begin{aligned} L &= -mc\sqrt{-g_{\alpha\beta}\frac{dr^\alpha}{dt}\frac{dr^\beta}{dt}} \\ &= -mc^2\sqrt{-g_{00} - 2g_{0j}v^j/c - g_{jk}v^jv^k/c^2}, \end{aligned} \tag{7.21}$$

where $r^\alpha = (ct, \boldsymbol{r})$ describes the particle's position in spacetime, and $v^j = dr^j/dt$ is its three-dimensional velocity vector. Newtonian gravity is reproduced by inserting the approximations $g_{00} = -1 + 2U/c^2 + O(c^{-4})$, $g_{0j} = O(c^{-3})$, and $g_{jk} = \delta_{jk} + O(c^{-2})$ within the Lagrangian, and expanding the square root to order c^{-2}; this yields

$$L = -mc^2 + \frac{1}{2}mv^2 + mU + O(c^{-2}), \tag{7.22}$$

in which U is the Newtonian potential. The first term is an irrelevant constant, and we indeed recognize $\frac{1}{2}mv^2 + mU$ as the Lagrangian of Newtonian gravity; the remaining terms of order c^{-2} are 1PN corrections. This simple exercise teaches us that a contribution of order c^{-2} to g_{00} is a Newtonian term, but that a term of order c^{-2} in g_{jk} is actually a post-Newtonian correction.

If we now want the post-Newtonian corrections to the motion, we must evaluate the Lagrangian to order c^{-2}, and this requires calculation of the metric to the following orders of approximation:

$$\begin{aligned} &O(c^{-4}) \quad \text{for} \quad g_{00}\,, \\ &O(c^{-3}) \quad \text{for} \quad g_{0j}\,, \\ &O(c^{-2}) \quad \text{for} \quad g_{jk}\,. \end{aligned}$$

In this case, a term of order c^{-4} in g_{00} gives rise to a post-Newtonian correction to the Lagrangian. The same is true of a term of order c^{-3} in g_{0j}, because it multiplies v^j/c in the Lagrangian, making the combination a term of order c^{-4}. And the same is also true of a term of order c^{-2} in g_{jk}, because it multiplies $v^j v^k/c^2$ in the Lagrangian. Generalizing the argument, we find that determination of the motion to nPN order requires calculation of the metric to the orders

$$\begin{aligned} O(c^{-2n-2}) &\quad \text{for} \quad g_{00}\,, \\ O(c^{-2n-1}) &\quad \text{for} \quad g_{0j}\,, \\ O(c^{-2n}) &\quad \text{for} \quad g_{jk}\,; \end{aligned}$$

as usual the orders in c^{-1} descend because of the additional factors of v^j/c in the Lagrangian.

Suppose next that we wish to determine the motion of a test body to 2.5PN order. The previous discussion indicates that we need g_{00} to order c^{-7}, g_{0j} to order c^{-6}, and g_{jk} to order c^{-5}. The metric is obtained from the potentials $h^{\alpha\beta}$, and recalling from Eqs. (7.15) that h^{00} leads off at order c^{-2}, h^{0j} at order c^{-3}, and h^{jk} at order c^{-4}, we find from Eq. (7.20) that the appropriate expression is

$$g_{00} = -1 + \frac{1}{2}h^{00} - \frac{3}{8}\left(h^{00}\right)^2 + \frac{5}{16}\left(h^{00}\right)^3 + \frac{1}{2}h^{kk}\left(1 - \frac{1}{2}h^{00}\right) + \frac{1}{2}h^{0j}h^{0j} + O(c^{-8}), \tag{7.23a}$$

$$g_{0j} = -h^{0j}\left(1 - \frac{1}{2}h^{00}\right) + O(c^{-7}), \tag{7.23b}$$

$$g_{jk} = \delta_{jk}\left[1 + \frac{1}{2}h^{00} - \frac{1}{8}\left(h^{00}\right)^2\right] + h^{jk} - \frac{1}{2}\delta_{jk}h^{mm} + O(c^{-6}), \tag{7.23c}$$

$$(-g) = 1 + h^{00} - h^{kk} + O(c^{-6}). \tag{7.23d}$$

To arrive at these results we actually had to carry the expansion of Eq. (7.20) to the third order in G, in order to capture the $(h^{00})^3$ term in g_{00}; this term is of order c^{-6}, and it is required for a complete expansion accurate through 2.5PN order.

Examining Eqs. (7.23), we begin to see how different orders in the post-Newtonian expansion of $h^{\alpha\beta}$ contribute to the metric. Beginning with g_{00}, we see from Eq. (7.15) that h^{00} contributes at all orders, from Newtonian order (c^{-2}) through 2.5PN order (c^{-7}), that h^{0j} contributes at 2PN order (c^{-6}) only, and that h^{jk} contributes at all orders beyond the Newtonian order (c^{-4}, c^{-5}, c^{-6}, and c^{-7}). With g_{0j} we find that h^{00} contributes at 2PN order (c^{-5}) only, while h^{0j} contributes at 1PN, 2PN, and 2.5PN orders (c^{-3}, c^{-5}, and c^{-6}). And finally, with g_{jk} we see that h^{00} contributes at 1PN, 2PN, and 2.5PN orders (c^{-2}, c^{-4}, and c^{-5}), while h^{jk} contributes at 2PN and 2.5PN orders (c^{-4} and c^{-5}).

We observe that each power of c^{-2} assigned to a contribution to $g_{\alpha\beta}$ translates to a specific post-Newtonian order. The translation, however, depends on the context. When the metric is examined in isolation, a term of order c^{-2} in g_{jk} could be declared to be of the same post-Newtonian order as a term of order c^{-2} in g_{00}. But when the metric is examined in the context of determining the motion of a slowly-moving particle, the c^{-2} term in g_{jk} is appropriately declared to be a 1PN term, while the c^{-2} term in g_{00} is labeled as a Newtonian contribution. The translation is again different when the motion is highly relativistic, with velocities v^j approaching the speed of light. In this case the coupling of the metric with powers of $v^j/c \simeq 1$ does not alter the post-Newtonian order, and a c^{-2} term in g_{jk} would again be declared to be a Newtonian contribution. Context is everything, and it must be specified before a meaningful post-Newtonian order can be assigned to a given expression.

Our considerations in this chapter, and the three chapters that follow, will be limited to post-Newtonian gravity, in which corrections of 2PN order and higher are neglected. In this 1PN context our expansion for the metric can be truncated to

$$g_{00} = -1 + \frac{1}{2}h^{00} - \frac{3}{8}(h^{00})^2 + \frac{1}{2}h^{kk} + O(c^{-6}), \tag{7.24a}$$

$$g_{0j} = -h^{0j} + O(c^{-5}), \tag{7.24b}$$

$$g_{jk} = \delta_{jk}\left(1 + \frac{1}{2}h^{00}\right) + O(c^{-4}), \tag{7.24c}$$

$$(-g) = 1 + h^{00} + O(c^{-4}). \tag{7.24d}$$

We return to the higher-order corrections in Chapter 12, when we study the effects of gravitational reaction in the near zone. There we shall be interested in all 2.5PN terms in the metric, those that scale as c^{-7} in g_{00}, as c^{-6} in g_{0j}, and as c^{-5} in g_{jk}. We shall see that with suitable care, we can study these radiative effects independently of the 1PN or 2PN influences.

7.1.4 General structure of the potentials: Wave zone

We proceed with an examination of the general structure of the gravitational potentials when the field point $\boldsymbol{x}$ is in the wave zone, where $r := |\boldsymbol{x}| > \mathcal{R}$. Consulting Box 6.7 once more, we see that we can express $h^{\alpha\beta}_{\mathcal{N}}$ as the multipole expansion

$$h^{\alpha\beta}_{\mathcal{N}}(t, \boldsymbol{x}) = \frac{4G}{c^4}\sum_{\ell=0}^{\infty}\frac{(-1)^\ell}{\ell!}\partial_L\left[\frac{1}{r}\int_{\mathcal{M}}\tau^{\alpha\beta}(\tau, \boldsymbol{x}')x'^L\,d^3x'\right], \tag{7.25}$$

in which $\tau := t - r/c$ is retarded time.

We first consider $h^{00}_{\mathcal{N}}$, and observe that the integral in Eq. (7.25) is just $c^2\mathcal{I}^L(\tau)$ as defined by Eqs. (7.16); the multipole moments are now evaluated at retarded time τ instead of time t. The first term ($\ell = 0$) in the series involves the monopole moment

$$M_0 := \mathcal{I}(\tau) = \int_{\mathcal{M}} c^{-2}\tau^{00}(\tau, \boldsymbol{x})\,d^3x, \tag{7.26}$$

and this represents the total mass contained within the near zone. Because of the conservation equations, we know that its time derivative can be converted to a surface integral on $\partial\mathcal{M}$, which can be shown to be small; the near-zone mass M_0 is therefore constant to a large degree of accuracy. The second term in the series involves

$$M_0 R_0^j := \mathcal{I}^j(\tau) = \int_{\mathcal{M}} c^{-2}\tau^{00}(\tau, \boldsymbol{x})x^j\,d^3x, \tag{7.27}$$

where R_0^j is the center-of-mass position associated with the domain $\mathscr{M}$. Its rate of change is related to the near-zone momentum

$$P_0^j := \int_{\mathscr{M}} c^{-1} \tau^{0j}(\tau, \boldsymbol{x})\, d^3x \tag{7.28}$$

by the conservation statement

$$\frac{d}{d\tau}\left(M_0 R_0^j\right) = P_0^j + \text{surface integral}\,, \tag{7.29}$$

and the momentum itself can be shown to satisfy

$$\frac{dP_0^j}{d\tau} = 0 + \text{surface integral}\,. \tag{7.30}$$

Because in each case the surface integral can be shown to be small, the total momentum is conserved to a large degree of accuracy, and the center-of-mass moves according to $d(M_0 R_0^j)/d\tau = P_0^j$. We may set $P_0^j = 0$ by working in the rest frame of the system, and set $R_0^j = 0$ by placing the center-of-mass at the spatial origin of the harmonic coordinates; the conservation equations ensure that R_0^j remains zero up to very small effects associated with the radiation of linear momentum. Thus, $h_{\mathscr{N}}^{00}$ consists of a static monopole piece plus time-dependent terms involving the quadrupole moment $\mathcal{I}^{jk}(\tau)$ and higher multipole moments.

Turning to $h_{\mathscr{N}}^{0j}$, and making use of the conservation identity of Eq. (7.14a), we can show that the $\ell = 0$ contribution to $h_{\mathscr{N}}^{0j}$ is of the form $(4G/c^3)r^{-1}\dot{\mathcal{I}}^j$ modulo surface terms; but since $\dot{\mathcal{I}}^j = P_0^j +$ surface integral, we find that this vanishes by virtue of our choice of reference frame. The $\ell = 1$ contribution involves $\int_{\mathscr{M}} \tau^{0j} x^k\, d^3x$, which according to Eq. (7.14c) can be converted to $\frac{1}{2}(\dot{\mathcal{I}}^{jk} - \epsilon^{mjk} J_0^m)$, where

$$J_0^m := \epsilon_{mjk} \int_{\mathscr{M}} x^j c^{-1} \tau^{0k}(\tau, \boldsymbol{x})\, d^3x \tag{7.31}$$

is the angular momentum contained within the near zone. The conservation identities can again be used to show that $dJ_0^m/d\tau$ vanishes up to a surface integral, so that $\boldsymbol{J}_0$ is constant except for a small radiative loss of angular momentum. Finally, looking at the $\ell = 0$ term in $h_{\mathscr{N}}^{jk}$ and using the identity of Eq. (7.14b), we find that we may convert it to $(2G/c^4)r^{-1}\ddot{\mathcal{I}}^{jk}$ modulo surface terms.

With these simplifications we obtain our final expression for $h_{\mathscr{N}}^{\alpha\beta}$ in the wave zone:

$$h_{\mathscr{N}}^{00} = \frac{4GM_0}{c^2 r} + \frac{4G}{c^2} \sum_{\ell=2}^{\infty} \frac{(-1)^\ell}{\ell!} \partial_L \left[\frac{\mathcal{I}^L(\tau)}{r}\right], \tag{7.32a}$$

$$h_{\mathscr{N}}^{0j} = -\frac{2G}{c^3} \frac{(\boldsymbol{n} \times \boldsymbol{J}_0)^j}{r^2} - \frac{2G}{c^3} \partial_k \left[\frac{\dot{\mathcal{I}}^{jk}(\tau)}{r}\right] + \frac{4G}{c^4} \sum_{\ell=2}^{\infty} \frac{(-1)^\ell}{\ell!} \partial_L \left[\frac{1}{r} \int_{\mathscr{M}} \tau^{0j}(\tau, \boldsymbol{x}') x'^L\, d^3x'\right], \tag{7.32b}$$

$$h_{\mathscr{N}}^{jk} = \frac{2G}{c^4} \frac{\ddot{\mathcal{I}}^{jk}(\tau)}{r} + \frac{4G}{c^4} \sum_{\ell=1}^{\infty} \frac{(-1)^\ell}{\ell!} \partial_L \left[\frac{1}{r} \int_{\mathscr{M}} \tau^{jk}(\tau, \boldsymbol{x}') x'^L\, d^3x'\right], \tag{7.32c}$$

in which overdots indicate differentiation with respect to $\tau = t - r/c$.

Still according to Box 6.7, we see that the wave-zone contribution $h^{\alpha\beta}_{\mathscr{W}}$ to the gravitational potentials is given by

$$h^{\alpha\beta}_{\mathscr{W}}(t,\boldsymbol{x}) = \frac{4G}{c^4}\frac{n^{\langle L\rangle}}{r}\left\{\int_0^{\mathcal{R}} ds f^{\alpha\beta}(\tau - 2s/c)A(s,r) + \int_{\mathcal{R}}^{\infty} ds f^{\alpha\beta}(\tau - 2s/c)B(s,r)\right\}, \tag{7.33}$$

when $\tau^{\alpha\beta}$ can be put in the specific form displayed in Eq. (7.19); the functions $A(s,r)$ and $B(s,r)$ are defined by Eq. (6.104). We shall learn how to evaluate these contributions below in Sec. 7.4, and then again in Chapter 11.

Box 7.2 Multipole structure of the wave-zone metric

By using extensions of the conservation identities (7.14), the wave-zone forms of the potentials $h^{\alpha\beta}_{\mathscr{N}}$ can be expressed elegantly in terms of a sequence of multipole moments. The general expressions are

$$h^{\alpha\beta}_{\mathscr{N}} = \frac{4G}{c^4}\sum_{\ell=0}^{\infty}\frac{(-1)^\ell}{\ell!}\partial_L\left[\frac{1}{r}\mathcal{M}^{\alpha\beta L}(\tau)\right],$$

where

$$\mathcal{M}^{00L} = c^2\mathcal{I}^L,$$

$$\mathcal{M}^{0jL} = \frac{c}{2(\ell+1)}\left(\dot{\mathcal{I}}^{jL} - \ell\epsilon^{mja_1}\mathcal{J}^{ma_2\cdots a_\ell}\right)(\mathrm{sym}\, a : L) + \frac{1}{(\ell+1)}\oint_{\partial\mathcal{M}}\tau^{0m}x^{jL}dS_m,$$

$$\begin{aligned}\mathcal{M}^{jkL} &= \frac{1}{(\ell+1)(\ell+2)}\ddot{\mathcal{I}}^{jkL} + \frac{2}{(\ell+2)}\epsilon^{ma_1(j}\dot{\mathcal{J}}^{m|k)a_2\cdots a_\ell}(\mathrm{sym}\, a : L)\\ &+ \frac{8(\ell-1)}{(\ell+1)}\mathcal{P}^{jk(a_1a_2\cdots a_\ell)}\\ &+ \frac{1}{(\ell+1)(\ell+2)}\oint_{\partial\mathcal{M}}\left[\tau^{mn}\partial_n(x^{jkL}) + \partial_\tau\tau^{0m}x^{jkL}\right]dS_m\\ &- \frac{2}{(\ell+2)}\oint_{\partial\mathcal{M}}\left[\tau^{n[a_1}x^{j]ka_2\cdots a_\ell} + (j \rightleftharpoons k)\right]dS_n\,(\mathrm{sym}\, a : L),\end{aligned}$$

where $\mathcal{I}^L$ and $\mathcal{J}^{jL}$ are defined in Eqs. (7.16), and

$$\mathcal{P}^{jkabL} := \int_{\mathcal{M}} x^{[a}\tau^{j][k}x^{b]L}\,d^3x.$$

The notation $(\mathrm{sym}\, a : L)$ means symmetrize on all ℓ a-indices.

7.1.5 Toward two iterations of the field equations

As we pointed out back in Sec. 6.2.3, to achieve the second post-Minkowskian approximation to the gravitational potentials $h^{\alpha\beta}$, we must carry out two iterations of the relaxed field equations and then impose the gauge condition/conservation statement. In other words, we must solve the wave equation $\Box h^{\alpha\beta} = -(16\pi G/c^4)\tau_1^{\alpha\beta}$ for the potentials $h_2^{\alpha\beta}$ and then impose the gauge condition $\partial_\beta h_2^{\alpha\beta} = 0$ or the conservation equation $\partial_\beta \tau_1^{\alpha\beta} = 0$. The starting point of these computations is the construction of the effective energy-momentum pseudotensor $\tau_1^{\alpha\beta}$, which depends on the fluid's energy-momentum tensor $T^{\alpha\beta}$ and the potentials generated during the *first iteration* of the relaxed field equations. Our very first task, therefore, is to perform the first iteration and obtain $\tau_1^{\alpha\beta}$.

7.2 First iteration

In this section we complete the first iteration of the relaxed field equations to obtain the gravitational potentials $h_1^{\alpha\beta}$. Our goal is to perform the computation to a degree of accuracy that is sufficient for the preparation of the second iteration, to be carried out in Secs. 7.3 and 7.4.

7.2.1 Energy-momentum tensor

In the first iteration of the field equations we replace $g_{\alpha\beta}$ by $\eta_{\alpha\beta}$ in the energy-momentum tensor of Eq. (7.1), and in the normalization condition for the velocity four-vector. Similarly, we set $\sqrt{-g} = 1$ in Eq. (7.4). We find that $\gamma = (1 - v^2/c^2)^{-1/2} = 1 + \frac{1}{2}(v/c)^2 + O(c^{-4})$, and Eq. (7.4) becomes

$$\rho = \left[1 - \frac{1}{2}(v/c)^2 + O(c^{-4})\right]\rho^*. \tag{7.34}$$

The components of the energy-momentum tensor are then

$$c^{-2}T_0^{00} = \rho^*\left[1 + \frac{1}{c^2}\left(\frac{1}{2}v^2 + \Pi\right) + O(c^{-4})\right], \tag{7.35a}$$

$$c^{-1}T_0^{0j} = \rho^* v^j\left[1 + \frac{1}{c^2}\left(\frac{1}{2}v^2 + \Pi + p/\rho^*\right) + O(c^{-4})\right], \tag{7.35b}$$

$$T_0^{jk} = \rho^* v^j v^k + p\,\delta^{jk} + O(c^{-2}). \tag{7.35c}$$

They are written as post-Newtonian expansions in flat spacetime, and these include both Newtonian and post-Newtonian contributions; terms occurring at 2PN order are neglected. Because they do not yet include 1PN terms involving the gravitational potentials, which will appear during the second iteration of the field equations, these post-Newtonian expansions are incomplete.

7.2.2 Near zone

We first take the field point $\boldsymbol{x}$ to be in the near zone, so that $r < \mathcal{R}$. To achieve the first iteration of the relaxed field equations, we set $\tau^{\alpha\beta} = T_0^{\alpha\beta}$ and make the substitution within Eqs. (7.13). Because the energy-momentum tensor is confined to the near zone, there is no need to truncate the integrals to the near-zone domain $\mathscr{M}$; they are naturally truncated to the volume occupied by the matter distribution. And because $T_0^{\alpha\beta}$ does not extend to the wave zone, the potentials $h_{\mathscr{W}}^{\alpha\beta}$ vanish, and $h^{\alpha\beta} = h_{\mathscr{N}}^{\alpha\beta}$.

As we shall see below in Sec. 7.3, for the purposes of preparing the second iteration of the field equations it is sufficient to compute h_1^{00} to order c^{-2}, h_1^{0j} to order c^{-3}, and to neglect h_1^{jk} because it is of order c^{-4}. This requirement implies that we can truncate Eqs. (7.35) to

$$c^{-2}T_0^{00} = \rho^* + O(c^{-2}), \tag{7.36a}$$

$$c^{-1}T_0^{0j} = \rho^* v^j + O(c^{-2}), \tag{7.36b}$$

$$T_0^{jk} = O(1). \tag{7.36c}$$

Making the substitution within Eq. (7.13) reveals that the potentials are given by

$$h_1^{00} = \frac{4}{c^2}U + O(c^{-4}), \tag{7.37a}$$

$$h_1^{0j} = \frac{4}{c^3}U^j + O(c^{-4}), \tag{7.37b}$$

$$h_1^{jk} = O(c^{-4}), \tag{7.37c}$$

in which U is a Newtonian potential defined by

$$U(t, \boldsymbol{x}) = G\int \frac{\rho^*(t, \boldsymbol{x}')}{|\boldsymbol{x} - \boldsymbol{x}'|}\, d^3x', \qquad \nabla^2 U = -4\pi G\rho^*, \tag{7.38}$$

in terms of the rescaled mass density ρ^*, and U^j is a vector potential defined by

$$U^j(t, \boldsymbol{x}) = G\int \frac{\rho^* v^j(t, \boldsymbol{x}')}{|\boldsymbol{x} - \boldsymbol{x}'|}\, d^3x', \qquad \nabla^2 U^j = -4\pi G\rho^* v^j, \tag{7.39}$$

in terms of the mass-current density $\rho^* v^j$. It is useful to note that by virtue of the continuity equation (7.3), the potentials satisfy

$$\partial_t U + \partial_j U^j = 0 \tag{7.40}$$

as a matter of identity.

Note that in Eq. (7.37), the corrections to h_1^{00} first occur at order c^{-4}. The expansion of Eq. (7.13), however, contains a term at order c^{-3} proportional to

$$\frac{d}{dt}\int \rho^*\, d^3x.$$

This vanishes because $m := \int \rho^*\, d^3x$; the total rest-mass within the fluid is conserved by virtue of Eq. (7.3). As was pointed out back in Sec. 7.1.1, the conservation of rest-mass is a basic kinematical requirement, quite divorced from any dynamical requirement based on energy-momentum conservation. This is an important point, because we recall that we are not at liberty to impose the conservation equations $\partial_\beta \tau^{\alpha\beta} = 0$ during the first iteration of the relaxed field equations; for this we must await the second iteration. With this in mind, you will notice that the corrections to h_1^{0j} first occur at order c^{-4}; this represents a term proportional to

$$\frac{d}{dt} \int \rho^* v^j \, d^3x$$

in the expansion of Eq. (7.13). The integral is the total momentum at Newtonian order, and it is tempting to declare that this term should vanish by virtue of momentum conservation. This temptation, however, must be resisted during the first iteration.

The gravitational potentials may be inserted within Eqs. (7.24) to obtain the near-zone metric. We obtain

$$g^1_{00} = -1 + \frac{2}{c^2} U + O(c^{-4}), \tag{7.41a}$$

$$g^1_{0j} = -\frac{4}{c^3} U^j + O(c^{-4}), \tag{7.41b}$$

$$g^1_{jk} = \left(1 + \frac{2}{c^2} U\right) \delta_{jk} + O(c^{-4}), \tag{7.41c}$$

and the metric determinant is $(-g_1) = 1 + 4U/c^2 + O(c^{-4})$. Recalling our discussion in Sec. 7.1.3, we see that this metric is not sufficiently accurate to obtain the motion of a test particle at post-Newtonian order, because it lacks the $O(c^{-4})$ contributions to g_{00}. It is sufficiently accurate, however, to serve as input in the second iteration of the relaxed field equations.

7.2.3 Wave zone

We next take the field point $\boldsymbol{x}$ to be in the wave zone, so that $r > \mathcal{R}$. To achieve the first iteration of the relaxed field equations, we could in principle set $\tau^{\alpha\beta} = T_0^{\alpha\beta}$, make the substitution within Eqs. (7.25), and evaluate the multipole moments explicitly. There is, however, no immediate need to proceed in this way. We can instead keep things simple by making direct use of Eqs. (7.25) and keeping the multipole moments unevaluated until we have completed the second iteration. An aspect of $\tau^{\alpha\beta}$ that we can incorporate is that it does not extend beyond the near zone; this implies that $h^{\alpha\beta}_{\mathcal{W}}$ vanishes, so that $h^{\alpha\beta} = h^{\alpha\beta}_{\mathcal{N}}$.

As we shall see below in Sec. 7.4, only h_1^{00} is required in the preparation of the second iteration. It is given by

$$h_1^{00} = \frac{4G}{c^2} \left\{ \frac{\mathcal{I}(\tau)}{r} - \partial_j \left[\frac{\mathcal{I}^j(\tau)}{r} \right] + \frac{1}{2} \partial_{jk} \left[\frac{\mathcal{I}^{jk}(\tau)}{r} \right] + \cdots \right\}, \tag{7.42}$$

in which $\mathcal{I}^L(\tau) := \int_{\mathcal{M}} c^{-2} \tau^{00}(\tau, \boldsymbol{x}) x^L \, d^3x$ are the multipole moments of the density $c^{-2}\tau^{00}$, expressed as functions of retarded time $\tau = t - r/c$. Note that we keep the dipole-moment term in the expansion, in spite of the fact that $\mathcal{I}^j$ will eventually be set equal to zero by a coordinate choice, as we indicated back in Sec. 7.1.4. The reason is that the ability to set $\mathcal{I}^j = 0$ relies on the conservation equation $\partial_\beta \tau^{\alpha\beta} = 0$, which we are not at liberty to impose during the first iteration.

Counting post-Newtonian orders is more subtle in the wave zone than it is in the near zone. The monopole term on the right-hand side of Eq. (7.42) is evidently of order $Gm_c/(c^2r)$, and we naturally assign a 0PN order to this term. To see about the dipole term, we perform the differentiation and express it as

$$-\frac{4G}{c^2}\partial_j\left[\frac{\mathcal{I}^j(\tau)}{r}\right] = \frac{4G}{c^2}\left(\frac{\dot{\mathcal{I}}^j}{cr} + \frac{\mathcal{I}^j}{r^2}\right)n_j, \tag{7.43}$$

in which $n^j := x^j/r$. Noting that $\mathcal{I}^j$ scales as $m_c r_c$, this term is of order

$$\frac{G}{c^2}m_c r_c\left(\frac{1}{ct_c r} + \frac{1}{r^2}\right) = \frac{Gm_c}{c^2r}\frac{r_c}{ct_c}\left(1 + \frac{ct_c}{r}\right). \tag{7.44}$$

This is smaller than $Gm_c/(c^2r)$ by a factor of order $(v_c/c)(1 + \lambda_c/r)$. The second factor is of order unity in the wave zone, and we conclude that the dipole term is smaller than the monopole term by a factor of order v_c/c. To this term we therefore assign a 0.5PN order. We do this in spite of the fact that the second term on the right of Eq. (7.43) is formally of Newtonian order. In the near zone, but outside the distribution of matter, this term would give the standard dipole contribution to the Newtonian potential, which normally would be set equal to zero by a suitable choice of coordinates. But because it falls off as r^{-2} and we are looking in the wave zone at distances $r > \lambda_c = r_c(c/v_c)$, it has decreased in size to such an extent that it is now comparable to (or even smaller than) the 0.5PN term produced by the time derivative of $\mathcal{I}^j$.

A simple extension of this argument reveals that the quadrupole term in h_1^{00} must be assigned a 1PN order. The octupole term, which would occur next in Eq. (7.42), gives a contribution at 1.5PN order, and the post-Newtonian counting becomes clear: an ℓ-pole term contributes at $(\ell/2)$PN order to the gravitational potential.

7.3 Second iteration: Near zone

In this section we face the challenging task of completing the second iteration of the relaxed field equations. Here we take the field point $\boldsymbol{x}$ to be in the near zone, so that $r < \mathcal{R}$. The wave zone will be considered next, in Sec. 7.4.

7.3.1 Effective energy momentum pseudotensor

Our first order of business is to use the potentials obtained in the first iteration to construct the effective energy-momentum pseudotensor of Eq. (6.52),

$$\tau^{\alpha\beta} = (-g)\left(T^{\alpha\beta} + t^{\alpha\beta}_{\rm LL} + t^{\alpha\beta}_{\rm H}\right), \tag{7.45}$$

with the Landau–Lifshitz contribution defined by Eq. (6.5), and the harmonic contribution defined by Eq. (6.53).

We begin by updating our expression for $T^{\alpha\beta}$, the fluid's energy-momentum tensor, which was given an incomplete post-Newtonian expansion back in Eq. (7.35). We return to Eq. (7.1) and substitute the near-zone metric displayed in Eq. (7.41). We also involve this metric in the normalization condition for u^α, and update our expression for γ to $1 + \frac{1}{2}(v/c)^2 + U/c^2 + O(c^{-4})$, which now incorporates the Newtonian potential U. Equation (7.34) becomes

$$\rho = \left[1 - \frac{1}{c^2}\left(\frac{1}{2}v^2 + 3U\right) + O(c^{-4})\right]\rho^*, \tag{7.46}$$

and the components of the energy-momentum tensor are now

$$c^{-2}(-g)T_1^{00} = \rho^*\left[1 + \frac{1}{c^2}\left(\frac{1}{2}v^2 + 3U + \Pi\right) + O(c^{-4})\right], \tag{7.47a}$$

$$c^{-1}(-g)T_1^{0j} = \rho^* v^j\left[1 + \frac{1}{c^2}\left(\frac{1}{2}v^2 + 3U + \Pi + p/\rho^*\right) + O(c^{-4})\right], \tag{7.47b}$$

$$(-g)T_1^{jk} = \rho^* v^j v^k + p\,\delta^{jk} + O(c^{-2}). \tag{7.47c}$$

We have multiplied $T_1^{\alpha\beta}$ by $(-g)$ because this is the combination that appears in $\tau^{\alpha\beta}$.

The hardest piece of the calculation by far (and this is always true) is the computation of $(-g)t_{\rm LL}^{\alpha\beta}$ to the appropriate degree of accuracy. To match the accuracy achieved in Eqs. (7.47) we need $c^{-2}(-g)t_{\rm LL}^{00}$ to orders $O(1)$ and $O(c^{-2})$, $c^{-1}(-g)t_{\rm LL}^{0j}$ to order $O(1)$ and $O(c^{-2})$, and $(-g)t_{\rm LL}^{jk}$ to order $O(1)$. To pluck out of Eq. (6.5) the terms of relevant orders, we use the facts recorded in Eq. (7.9), that the potentials scale as $h^{00} = O(c^{-2})$, $h^{0j} = O(c^{-3})$, and $h^{jk} = O(c^{-4})$. In addition, we use the property that $\partial_0 h^{00}$ is of order c^{-1} relative to $\partial_j h^{00}$. The dominant piece of $(-g)t_{\rm LL}^{\alpha\beta}$ will therefore come from $\partial_j h^{00} = 4\partial_j U/c^2$.

Armed with these observations, the reduction of $(-g)t_{\rm LL}^{\alpha\beta}$ to something manageable is well within reach. Let us, for example, examine the term

$$\frac{1}{4}(2g^{\alpha\lambda}g^{\beta\mu} - g^{\alpha\beta}g^{\lambda\mu})g_{\nu\rho}g_{\sigma\tau}\partial_\lambda h^{\nu\tau}\partial_\mu h^{\rho\sigma}$$

on the right-hand side of Eq. (6.5), in which we have replaced $\mathfrak{g}^{\alpha\beta}$ by $\eta^{\alpha\beta} - h^{\alpha\beta}$. A first source of simplification arises from the fact that each occurrence of $g_{\alpha\beta}$ can be replaced by $\eta_{\alpha\beta}$; this comes about because each factor of $h^{\alpha\beta}$ contributes a power of G, and we need to compute $(-g)t_{\rm LL}^{\alpha\beta}$ to order G^2 in the second post-Minkowskian approximation. A second source of simplification comes from the fact that at leading order, we can retain terms that involve $\partial_j h^{00}$ only. At this stage the previous expression becomes

$$\frac{1}{4}(2\eta^{\alpha j}\eta^{\beta k} - \eta^{\alpha\beta}\delta^{jk})\partial_j h^{00}\partial_k h^{00},$$

and it gives rise to a contribution $\frac{1}{4}\partial_j h^{00}\partial^j h^{00}$ to $(-g)t_{\rm LL}^{00}$, and a contribution $\frac{1}{2}\partial^j h^{00}\partial^k h^{00} - \frac{1}{4}\delta^{jk}\partial_n h^{00}\partial^n h^{00}$ to $(-g)t_{\rm LL}^{jk}$; there is no contribution to $(-g)t_{\rm LL}^{0j}$.

Keeping track of all terms that make up $(-g)t_{\rm LL}^{\alpha\beta}$, we eventually arrive at the expressions

$$\frac{16\pi G}{c^4}(-g)t_{\rm LL}^{00} = -\frac{7}{8}\partial_j h^{00}\partial^j h^{00} + O(c^{-6}), \tag{7.48a}$$

$$\frac{16\pi G}{c^4}(-g)t_{\rm LL}^{0j} = \frac{3}{4}\partial^j h^{00}\partial_0 h^{00} + (\partial^j h^{0k} - \partial^k h^{0j})\partial_k h^{00} + O(c^{-7}), \tag{7.48b}$$

$$\frac{16\pi G}{c^4}(-g)t_{\rm LL}^{jk} = \frac{1}{4}\partial^j h^{00}\partial^k h^{00} - \frac{1}{8}\delta^{jk}\partial_n h^{00}\partial^n h^{00} + O(c^{-6}). \tag{7.48c}$$

These results are sufficiently accurate for our immediate purposes. At a later stage, however, we shall need additional accuracy in our expression for $(-g)t^{jk}_{\rm LL}$, and we record this improved expression here:

$$\begin{aligned}\frac{16\pi G}{c^4}(-g)t^{jk}_{\rm LL} &= \frac{1}{4}(1-2h^{00})\partial^j h^{00}\partial^k h^{00} - \frac{1}{8}\delta^{jk}(1-2h^{00})\partial_n h^{00}\partial^n h^{00} \\ &\quad - \partial^j h^{0n}\partial^k h^0{}_n + \partial^j h^{0n}\partial_n h^{0k} + \partial^k h^{0n}\partial_n h^{0j} - \partial_n h^{0j}\partial^n h^{0k} \\ &\quad + \partial^j h^{00}\partial_0 h^{0k} + \partial^k h^{00}\partial_0 h^{0j} + \frac{1}{4}\partial^j h^{00}\partial^k h^n{}_n + \frac{1}{4}\partial^k h^{00}\partial^j h^n{}_n \\ &\quad + \delta^{jk}\biggl[-\frac{3}{8}(\partial_0 h^{00})^2 - \partial_n h^{00}\partial_0 h^{0n} - \frac{1}{4}\partial_n h^{00}\partial^n h^p{}_p \\ &\quad + \frac{1}{2}\partial_n h^0{}_p(\partial^n h^{0p} - \partial^p h^{0n})\biggr] + O(c^{-8}). \end{aligned} \tag{7.49}$$

It should be noted that this incorporates corrections of order c^{-2} relative to the leading-order expression of Eq. (7.48), and that to be consistent, we have terms such as $h^{00}\partial^j h^{00}\partial^k h^{00}$ that contain an additional power of the gravitational constant G.

With the substitutions of Eqs. (7.37), the Landau–Lifshitz pseudotensor becomes

$$c^{-2}(-g)t^{00}_{\rm LL} = -\frac{1}{4\pi Gc^2}\left(\frac{7}{2}\partial_j U\partial^j U\right) + O(c^{-4}), \tag{7.50a}$$

$$c^{-1}(-g)t^{0j}_{\rm LL} = \frac{1}{4\pi Gc^2}\Bigl[3\partial_t U\partial^j U + 4(\partial^j U^k - \partial^k U^j)\partial_k U\Bigr] + O(c^{-4}), \tag{7.50b}$$

$$(-g)t^{jk}_{\rm LL} = \frac{1}{4\pi G}\left(\partial^j U\partial^k U - \frac{1}{2}\delta^{jk}\partial_n U\partial^n U\right) + O(c^{-2}). \tag{7.50c}$$

To better understand the importance of these contributions to $\tau^{\alpha\beta}$, we estimate the order of magnitude of $c^{-2}(-g)t^{00}_{\rm LL}$ relative to ρ^*, the dominant contribution to $c^{-2}\tau^{00}$. We re-introduce the scaling quantities m_c, r_c, and v_c, and estimate the pseudotensor within the matter distribution. We have that $\rho^* \sim m_c/r_c^3$ and $U \sim Gm_c/r_c$. After differentiation we get $\partial_j U \sim Gm_c/r_c^2$, and all this produces

$$\frac{(-g)t^{00}_{\rm LL}}{\rho^* c^2} \sim \frac{Gm_c}{c^2 r_c}. \tag{7.51}$$

Since motion within the fluid is governed by gravity, we can rely on the virial theorem and claim that $Gm_c/r_c \sim v_c^2$. The end result is that $c^{-2}(-g)t^{00}_{\rm LL}$ is a quantity of order $(v_c/c)^2$ relative to ρ^*; it is comparable to the other 1PN terms that are displayed in Eq. (7.47).

The easiest piece of the calculation by far (and this is always true) is the computation of $(-g)t^{\alpha\beta}_{\rm H}$ to the required degree of accuracy. Using the information gathered previously, Eq. (6.53) returns

$$\frac{16\pi G}{c^4}(-g)t^{00}_{\rm H} = O(c^{-6}), \tag{7.52a}$$

$$\frac{16\pi G}{c^4}(-g)t^{0j}_{\rm H} = O(c^{-7}), \tag{7.52b}$$

$$\frac{16\pi G}{c^4}(-g)t^{jk}_{\rm H} = O(c^{-6}). \tag{7.52c}$$

These expressions should be compared with Eqs. (7.48); they imply that the harmonic pseudotensor makes no relevant contribution to $\tau^{\alpha\beta}$. For later reference we record the improved expression

$$\frac{16\pi G}{c^4}(-g)t_{\rm H}^{jk} = -h^{00}\partial_{00}h^{jk} + O(c^{-8}) \tag{7.53}$$

for the spatial components of the pseudotensor.

Collecting results, we have obtained

$$c^{-2}\tau_1^{00} = \rho^*\left[1 + \frac{1}{c^2}\left(\frac{1}{2}v^2 + 3U + \Pi\right)\right] - \frac{1}{4\pi Gc^2}\left(\frac{7}{2}\partial_j U\partial^j U\right) + O(c^{-4}), \tag{7.54a}$$

$$c^{-1}\tau_1^{0j} = \rho^* v^j\left[1 + \frac{1}{c^2}\left(\frac{1}{2}v^2 + 3U + \Pi + p/\rho^*\right)\right] + \frac{1}{4\pi Gc^2}\left[3\partial_t U\partial^j U + 4(\partial^j U^k - \partial^k U^j)\partial_k U\right] + O(c^{-4}), \tag{7.54b}$$

$$\tau_1^{jk} = \rho^* v^j v^k + p\delta^{jk} + \frac{1}{4\pi G}\left(\partial^j U\partial^k U - \frac{1}{2}\delta^{jk}\partial_n U\partial^n U\right) + O(c^{-2}), \tag{7.54c}$$

for the effective energy-momentum pseudotensor.

7.3.2 Energy-momentum conservation

At this stage of our development of the second post-Minkowskian approximation, we may impose the conservation equations

$$c^{-2}\partial_t\tau_1^{00} + c^{-1}\partial_j\tau_1^{0j} = 0, \qquad c^{-1}\partial_t\tau_1^{0j} + \partial_k\tau_1^{jk} = 0, \tag{7.55}$$

before calculating the second-iterated potentials $h_2^{\alpha\beta}$. At leading order the energy equation reproduces Eq. (7.3); not surprisingly, a statement of rest-mass conservation is included in the statement of energy conservation. The equation brings additional information at order c^{-2}, a statement of energy conservation for all relevant forms of fluid energy: kinetic, internal, and gravitational. We shall return to this theme below.

The momentum equation is equally informative. Using Eqs. (7.54), we have

$$\begin{aligned} c^{-1}\partial_t\tau_1^{0j} &= (\partial_t\rho^*)v^j + \rho^*\partial_t v^j + O(c^{-2}) \\ &= -v^j\partial_k(\rho^* v^k) + \rho^*\frac{dv^j}{dt} - \rho^* v^k\partial_k v^j + O(c^{-2}), \end{aligned} \tag{7.56}$$

where we have involved Eq. (7.3) and the definition of the total time derivative: $dv^j/dt = \partial_t v^j + v^k\partial_k v^j$. We also have

$$\partial_k\tau_1^{jk} = v^j\partial_k(\rho^* v^k) + \rho^* v^k\partial_k v^j + \partial^j p + \frac{1}{4\pi G}(\partial^j U)\nabla^2 U + O(c^{-2}). \tag{7.57}$$

Making the substitutions into Eq. (7.55), and replacing $\nabla^2 U$ by $-4\pi G\rho^*$, we arrive at

$$\rho^*\frac{dv^j}{dt} = \rho^*\partial^j U - \partial^j p + O(c^{-2}). \tag{7.58}$$

This is Euler's equation, which governs the dynamics of our perfect fluid at leading order in a post-Newtonian expansion. It was first obtained on the basis of Newtonian theory in

Chapter 1, and indeed, the foregoing computations have already been presented (in reverse order) in Sec. 1.4.4.

Recalling our discussion of the iteration procedure in Sec. 6.2.3, we observe that Euler's equation (i.e. Newtonian gravity) is the consequence of $\partial_\beta \tau_1^{\alpha\beta} = 0$, the conservation equation that goes along with the *second iteration* of the relaxed field equations. Performing a single iteration is not sufficient to produce this dynamics, because the equations of motion that are compatible with the first iteration, derived from the conservation equation $\partial_\beta \tau_0^{\alpha\beta} = 0$, *do not contain gravitational interactions*. This observation was also made in the context of the linearized approximation to general relativity, back in Sec. 5.5. So formally, a second iteration of the relaxed field equations is required to obtain the Newtonian equations of motion. Similarly, a third iteration is required to find the post-Newtonian equations of motion, and so on. But as we also discussed back in Sec. 6.2.3, the conservation equation compatible with the nth iteration requires ingredients that are collected during the $(n-1)$th iteration, and it can be formulated *before* completing the nth iteration to obtain the gravitational potentials. In practice, therefore, we may obtain the Newtonian equations of motion on the basis of the first-iterated potentials; the post-Newtonian equations on the basis of the second-iterated potentials, and so on.

As we saw back in Sec. 6.1.4, the local conservation equations (7.55) imply the existence of globally conserved quantities. From Eq. (6.36) we have the *total mass*

$$M := \frac{1}{c^2}\int (-g)\left(T^{00} + t_{\rm LL}^{00}\right) d^3x, \tag{7.59}$$

and from Eq. (6.37) we have the *total momentum*

$$P^j := \frac{1}{c}\int (-g)\left(T^{0j} + t_{\rm LL}^{0j}\right) d^3x. \tag{7.60}$$

In addition, it is useful to re-introduce the vector

$$R^j := \frac{1}{Mc^2}\int (-g)\left(T^{00} + t_{\rm LL}^{00}\right) x^j \, d^3x, \tag{7.61}$$

which denotes the position of the center-of-mass; this was first defined by Eq. (6.39). We recall that R^j is related to the total momentum by the equation $M dR^j/dt = P^j$, and that by adopting the center-of-mass frame of the spacetime, we can set both P^j and R^j to zero. It is worth pointing out that since $(-g)t_{\rm H}^{\alpha\beta}$ makes no relevant contribution to $\tau_1^{\alpha\beta}$ at this order, as we saw back in Eq. (7.52), the conserved quantities associated with $(-g)(T^{\alpha\beta} + t_{\rm LL}^{\alpha\beta})$ are the same as those associated with $\tau^{\alpha\beta}$.

The global quantities M, P^j, and R^j are defined in terms of integrals that extend over all space. We may still, however, evaluate them with the near-zone information available to us now, because their expressions turn out to be insensitive to the wave-zone aspects of the integrals. To evaluate M we insert our previous expression for $c^{-2}\tau_1^{00}$ within Eq. (7.59), which we truncate to the near-zone domain $\mathcal{M}$. The term involving U is handled as follows. We write

$$\partial_j U \partial^j U = \partial_j\left(U\partial^j U\right) - U\nabla^2 U = \partial_j\left(U\partial^j U\right) + 4\pi G\rho^* U \tag{7.62}$$

and observe that the first term gives rise to a surface integral that must be evaluated at $r = \mathcal{R}$; it makes an $\mathcal{R}$-dependent contribution to M that cancels out when the wave-zone portion of the integral is added to the near-zone portion. Collecting results, we arrive at

$$M = \int \rho^* \left[1 + \frac{1}{c^2} \left(\frac{1}{2} v^2 - \frac{1}{2} U + \Pi \right) \right] d^3x + O(c^{-4}) \tag{7.63}$$

for the total mass. The integral of ρ^* is m, the total rest-mass of the fluid, which is separately conserved. The integral of $\frac{1}{2}\rho^* v^2$ is $\mathcal{T}$, the fluid's total kinetic energy. The integral of $-\frac{1}{2}\rho^* U$ is Ω, the gravitational potential energy. And finally, the integral of $\rho^* \Pi = \epsilon + O(c^{-2})$ is $E_{\rm int}$, the total internal energy stored within the fluid. The sum of $\mathcal{T}$, Ω, and $E_{\rm int}$ is the total energy E, and this was shown to be constant (by virtue of Euler's equation and the first law of thermodynamics) back in Sec. 1.4.3. The total mass can therefore be expressed as $M = m + E/c^2 + O(c^{-4})$, and this equation possesses a clear interpretation: The total mass of the spacetime is a measure of all forms of energy, including rest-mass, kinetic, gravitational, and internal energies.

Similar manipulations reveal that R^j can be expressed as

$$R^j = \frac{1}{M} \int \rho^* x^j \left[1 + \frac{1}{c^2} \left(\frac{1}{2} v^2 - \frac{1}{2} U + \Pi \right) \right] d^3x + O(c^{-4}), \tag{7.64}$$

and Eq. (7.60) becomes

$$\begin{aligned} P^j = &\int \rho^* v^j \left[1 + \frac{1}{c^2} \left(\frac{1}{2} v^2 - \frac{1}{2} U + \Pi + p/\rho^* \right) \right] d^3x \\ &- \frac{1}{2c^2} \int \rho^* x^j \left(\partial_t U - v^k \partial_k U \right) d^3x + O(c^{-4}). \end{aligned} \tag{7.65}$$

The leading-order piece of the total momentum was shown to be constant (by virtue of Euler's equation) back in Sec. 1.4.3 of Chapter 1; with this improved expression the momentum is conserved to order c^{-2}.

It is instructive to examine the relationship between the total mass M, which is known to correspond to the ADM mass of the spacetime, and the near-zone mass M_0, defined by Eq. (7.26),

$$M_0 = \int_{\mathcal{M}} c^{-2} \tau^{00} \, d^3x, \tag{7.66}$$

which appears in the expression of Eq. (7.32) for h^{00} in the wave zone. It is easy to see that

$$M_0 = M + O(c^{-4}) \, . \tag{7.67}$$

This follows because the integrands for M and M_0 differ by $(-g)t_{\rm H}^{00}$, which makes no contribution at 1PN order, and because the wave-zone portion of the integral defining M makes no $\mathcal{R}$-independent contribution to the mass. Examining the relationship at higher post-Newtonian orders, we find that subtle differences between M_0 and M appear at order c^{-5}; these are explored in Exercise 7.8.

Similar manipulations reveal that

$$R_0^j = R^j + O(c^{-4}) \tag{7.68}$$

and

$$P_0^j = P^j + O(c^{-4}), \tag{7.69}$$

in which R_0^j is the position of the near-zone center-of-mass introduced in Eq. (7.27), and P_0^j is the near-zone momentum introduced in Eq. (7.28). These equalities imply that at 1PN order, a coordinate choice that enforces $R^j = 0 = P^j$ also enforces $R_0^j = 0 = P_0^j$.

7.3.3 Near-zone contribution to potentials

Armed with Eq. (7.54) for $\tau_1^{\alpha\beta}$, we are now ready to solve the relaxed field equations for the second-iterated potentials $h^{\alpha\beta} = h_{\mathcal{N}}^{\alpha\beta} + h_{\mathcal{W}}^{\alpha\beta}$. In this section we focus on the near-zone contribution $h_{\mathcal{N}}^{\alpha\beta}$, insert $\tau_1^{\alpha\beta}$ within Eqs. (7.15), and express the results in a convenient form. The spatial components h^{jk} require special care, because as we have observed in Sec. 7.1.3, the spatial trace h^{kk} contributes to the spacetime metric at 1PN order, while the remaining components contribute only at 2PN order. With this in mind, it is helpful to decompose the potentials into a "scalar class" comprising h^{00} and h^{kk}, a "vector class" comprising h^{0j}, and a "tensor class" comprising h^{jk}.

Scalar class

We begin with the computation of h^{00} and h^{kk}. Examining Eqs. (7.54), we observe that both τ_1^{00} and τ_1^{kk} contain a contribution proportional to $\partial_j U \partial^j U$, which does not have compact support. It is useful to re-express these terms by exploiting the identity

$$\nabla^2 U^2 = 2\partial_j U \partial^j U + 2U\nabla^2 U, \tag{7.70}$$

in which we may insert Poisson's equation $\nabla^2 U = -4\pi G\rho^*$. In this way we obtain

$$c^{-2}\tau_1^{00} = \rho^*\left[1 + \frac{1}{c^2}\left(\frac{1}{2}v^2 - \frac{1}{2}U + \Pi\right)\right] - \frac{7}{16\pi Gc^2}\nabla^2 U^2 + O(c^{-4}) \tag{7.71}$$

and

$$\tau_1^{kk} = \rho^*\left(v^2 - \frac{1}{2}U\right) + 3p - \frac{1}{16\pi G}\nabla^2 U^2 + O(c^{-2}) \tag{7.72}$$

for the relevant components of the energy-momentum pseudotensor.

Consulting Eq. (7.15), we see that the leading terms in both $h_{\mathcal{N}}^{00}$ and $h_{\mathcal{N}}^{kk}$ are Poisson integrals constructed from $c^{-2}\tau^{00}$ and τ^{kk}. To evaluate these we must distinguish between the pieces of the source functions that have compact support (those that are tied to the fluid variables), and those that depend on the Newtonian potential and extend beyond the matter distribution. To handle the compact-support pieces we introduce the potentials

$$\psi(t, \boldsymbol{x}) := G\int \frac{\rho^{*\prime}\left(\frac{3}{2}v'^2 - U' + \Pi'\right) + 3p'}{|\boldsymbol{x} - \boldsymbol{x}'|}\,d^3x', \tag{7.73a}$$

$$V(t, \boldsymbol{x}) := G\int \frac{\rho^{*\prime}\left(v'^2 - \frac{1}{2}U'\right) + 3p'}{|\boldsymbol{x} - \boldsymbol{x}'|}\,d^3x', \tag{7.73b}$$

in which primed quantities such as $\rho^{*\prime}$ indicate that the fluid variables are expressed as functions of t and $\boldsymbol{x}'$. These satisfy the Poisson equations

$$\nabla^2 \psi = -4\pi G\rho^*\left(\frac{3}{2}v^2 - U + \Pi + 3p/\rho^*\right), \tag{7.74a}$$

$$\nabla^2 V = -4\pi G\rho^*\left(v^2 - \frac{1}{2}U + 3p/\rho^*\right). \tag{7.74b}$$

With this notation we see that at leading order, the compact-support piece of $h^{00}_{\mathcal{N}}$ is given by $4U/c^2 + 4(\psi - V)/c^4$, while the compact-support piece of $h^{kk}_{\mathcal{N}}$ is $4V/c^4$; this choice of notation is motivated by the fact that once the potentials are inserted within the near-zone metric of Eq. (7.24), the leading-order, compact-support piece of g_{00} will involve only U and ψ.

Turning next to the Poisson integral involving $\nabla^2 U^2$, we evaluate it by making repeated use of integration by parts:

$$\begin{aligned}\frac{1}{4\pi}\int_{\mathcal{M}} \frac{\nabla'^2 U'^2}{|\boldsymbol{x}-\boldsymbol{x}'|} d^3x' &= \frac{1}{4\pi}\oint_{\partial\mathcal{M}} \frac{\partial'^j U'^2}{|\boldsymbol{x}-\boldsymbol{x}'|} dS'_j - \frac{1}{4\pi}\int_{\mathcal{M}} \partial'^j U'^2 \partial'_j \frac{1}{|\boldsymbol{x}-\boldsymbol{x}'|} d^3x' \\ &= \frac{1}{4\pi}\oint_{\partial\mathcal{M}} \left(\frac{\partial'^j U'^2}{|\boldsymbol{x}-\boldsymbol{x}'|} - U'^2 \partial'_j \frac{1}{|\boldsymbol{x}-\boldsymbol{x}'|}\right) dS'_j \\ &\quad + \frac{1}{4\pi}\int_{\mathcal{M}} U'^2 \nabla'^2 \frac{1}{|\boldsymbol{x}-\boldsymbol{x}'|} d^3x' \\ &= -U^2 + \frac{1}{4\pi}\oint_{\partial\mathcal{M}} \left(\frac{\partial'^j U'^2}{|\boldsymbol{x}-\boldsymbol{x}'|} - U'^2 \partial'_j \frac{1}{|\boldsymbol{x}-\boldsymbol{x}'|}\right) dS'_j. \end{aligned} \tag{7.75}$$

The surface term is evaluated at $r' = \mathcal{R}$, and because U' falls off as $(r')^{-1}$ at large distances from the matter distribution, it makes a contribution that scales as $\mathcal{R}^{-2}$. As with all $\mathcal{R}$-dependent terms in the potentials $h^{\alpha\beta}_{\mathcal{N}}$, we may discard it because it will eventually be cancelled by an equal and opposite term in $h^{\alpha\beta}_{\mathcal{W}}$.

It is interesting to note that if the Poisson integral of $\nabla^2 U^2$ were extended to infinity instead of being truncated to the domain $\mathcal{M}$, it would be exactly equal to $-U^2$. This may seem like a trivial observation, but we wish to call attention to the fact that the solution to the differential equation $\nabla^2 f = \nabla^2 g$ is not necessarily the obvious $f = g$. The actual solution may also include a solution to Laplace's equation $\nabla^2 f = 0$, and the correct mixture of particular and homogeneous solutions depends on the boundary conditions captured by the surface integral in Eq. (7.75). When the boundary conditions are such that the surface integral vanishes except for $\mathcal{R}$-dependent terms, the particular solution $f = g$ is justified. When, however, the surface integral returns contributions that are independent of $\mathcal{R}$, the relevant solution is no longer the simple $f = g$.

We have now taken care of the leading-order Poisson-integral terms in Eq. (7.15). Proceeding to the next order in $h^{00}_{\mathcal{N}}$, we examine the superpotential term, in which we may insert the leading-order expression $c^{-2}\tau^{00}_1 = \rho^* + O(c^{-2})$, because the correction at order c^{-2} would contribute to $h^{00}_{\mathcal{N}}$ at order c^{-6}. This gives rise to the post-Newtonian superpotential

$$X(t,\boldsymbol{x}) := G\int \rho^*(t,\boldsymbol{x}')|\boldsymbol{x}-\boldsymbol{x}'|\,d^3x'\,, \tag{7.76}$$

in which the integral over $\mathscr{M}$ is naturally truncated to the volume occupied by the matter distribution. With this notation we observe that the superpotential term in $h^{00}_{\mathscr{N}}$ is proportional to $\partial_t^2 X$. Since $\nabla^2|\boldsymbol{x}-\boldsymbol{x}'| = 2|\boldsymbol{x}-\boldsymbol{x}'|^{-1}$ when $\boldsymbol{x}\neq\boldsymbol{x}'$, we see that the superpotential satisfies the Poisson equation

$$\nabla^2 X = 2U\,, \tag{7.77}$$

and X is therefore sourced by the Newtonian potential. The connection between Eqs. (7.76) and (7.77) is further explored in Box 7.3.

Collecting results, we have obtained the following expressions for the scalar potentials $h^{00}_{\mathscr{N}}$ and $h^{kk}_{\mathscr{N}}$:

$$h^{00}_{2\mathscr{N}} = \frac{4}{c^2}U + \frac{1}{c^4}\biggl(7U^2 + 4\psi - 4V + 2\frac{\partial^2 X}{\partial t^2}\biggr) - \frac{2G}{3c^5}\dddot{\mathcal{I}}^{kk}(t) + O(c^{-6})\,, \tag{7.78a}$$

$$h^{kk}_{2\mathscr{N}} = \frac{1}{c^4}\bigl(U^2 + 4V\bigr) - 2\frac{G}{c^5}\dddot{\mathcal{I}}^{kk}(t) + O(c^{-6})\,. \tag{7.78b}$$

These expressions are accurate up to order c^{-6}, and they incorporate Newtonian, 1PN, and 1.5PN terms. Once we have obtained the spacetime metric from the potentials, the terms of order c^{-5} will be shown to represent coordinate artifacts that can be removed by a coordinate transformation.

Box 7.3 Definition of the superpotential

The post-Newtonian superpotential X is defined by Eq. (7.76), and this leads to the Poisson equation displayed in Eq. (7.77). Here we ask whether defining the superpotential through

$$\nabla^2 X = 2U$$

necessarily leads back to the integral representation of Eq. (7.76). We shall see that the answer to this question is subtle, and provides further illustration of the fact that boundary conditions and solutions to Laplace's equation sometimes play an important role in solving Poisson's equation.

The general solution to Poisson's equation for the superpotential is

$$X(t,\boldsymbol{x}) = -\frac{1}{2\pi}\int \frac{U(t,\boldsymbol{x}')}{|\boldsymbol{x}-\boldsymbol{x}'|}\,d^3x' + X_0(t,\boldsymbol{x}),$$

in which X_0 is a solution to $\nabla^2 X_0 = 0$. But the integral is ill defined; because U falls off as $(r')^{-1}$ at large distances, the integrand behaves as $(r')^{-2}$, and since it is multiplied by the integration measure $r'^2\,dr'$, the integral is linearly divergent. To provide a well-defined prescription for the Poisson integral, we truncate the domain of integration to $\mathscr{M}$. This amounts to modifying the Poisson equation to $\nabla^2 X = 2U\Theta(\mathcal{R}-r)$, in which Θ is the Heaviside step function; the modification produces no noticeable changes in the near zone.

Inserting the standard expression of Eq. (7.38) for the Newtonian potential, we find that the superpotential can be expressed as

$$X(t, \boldsymbol{x}) = G \int \rho^*(t, \boldsymbol{y}) K(\boldsymbol{x}; \boldsymbol{y})\, d^3y + X_0(t, \boldsymbol{x}),$$

in which the two-point function $K(\boldsymbol{x}; \boldsymbol{y})$ is defined by

$$K(\boldsymbol{x}; \boldsymbol{y}) := -\frac{1}{2\pi} \int_{\mathcal{M}} \frac{d^3x'}{|\boldsymbol{x} - \boldsymbol{x}'||\boldsymbol{x}' - \boldsymbol{y}|}.$$

To evaluate this we exploit the observation that K can depend on $\boldsymbol{x}$ and $\boldsymbol{y}$ only through the combination $\boldsymbol{x} - \boldsymbol{y}$, and thereby set $\boldsymbol{y} = \boldsymbol{0}$ to simplify the integral. Making use of the addition theorem for spherical harmonics, we find that $K(\boldsymbol{x}; \boldsymbol{0}) = -2 \int_0^{\mathcal{R}} (r_>)^{-1} r'\, dr'$, in which $r_>$ is the greater of r and r'. This returns $r - 2\mathcal{R}$, and we conclude that the two-point function is given by

$$K(\boldsymbol{x}; \boldsymbol{y}) = |\boldsymbol{x} - \boldsymbol{y}| - 2\mathcal{R}.$$

Inserting this within the integral for the superpotential, we obtain

$$X(t, \boldsymbol{x}) = G \int \rho^*(t, \boldsymbol{x}') |\boldsymbol{x} - \boldsymbol{x}'|\, d^3x' - 2Gm\mathcal{R} + X_0(t, \boldsymbol{x}),$$

with $m := \int \rho^*(t, \boldsymbol{x}')\, d^3x'$ denoting the total rest-mass of the matter distribution. Choosing $X_0 = 2Gm\mathcal{R}$, we reproduce the original definition of the superpotential.

It is interesting to note that since it is $\partial_t^2 X$ that appears in the gravitational potentials, the addition of $-2Gm\mathcal{R} + X_0$ to the integral is immaterial, so long as X_0 does not depend on time. The superpotential, therefore, is sufficiently robust to withstand the ambiguities associated with the choice of solution to $\nabla^2 X = 2U$.

Vector class

For our purposes it is necessary to evaluate the potential $h^{0j}_{\mathcal{N}}$ to order c^{-3} only. Our expression for $c^{-1}\tau_1^{0j}$ in Eq. (7.54b) is more accurate than we need, and we may truncate it to its leading term $\rho^* v^j + O(c^{-2})$. Consulting Eq. (7.15b), we see that the leading term in the potential is given by a Poisson integral constructed from $c^{-1}\tau_1^{0j}$, and we obtain

$$h^{0j}_{2\mathcal{N}} = \frac{4}{c^3} U^j + O(c^{-5}), \tag{7.79}$$

where U^j is the vector potential defined by Eq. (7.39). In principle we have enough information to calculate the correction terms at order c^{-5}, but these will not be needed in our future developments.

Tensor class

The computation of $h^{jk}_{\mathcal{N}}$ is more involved, because its source term contains a field contribution that is not as easy to deal with as it was with the scalar potentials. Returning to Eq. (7.54) and exploiting once more the identity of Eq. (7.70), we express τ_1^{jk} in the form

$$\tau_1^{jk} = \rho^*\left(v^j v^k - \frac{1}{2}U\delta^{jk}\right) + p\,\delta^{jk} - \frac{1}{16\pi G}\delta^{jk}\nabla^2 U^2 + \frac{1}{4\pi G}\partial^j U \partial^k U + O(c^{-2}). \tag{7.80}$$

Consulting Eq. (7.15c), we see that the leading term in the potential is a Poisson integral constructed from τ_1^{jk}. The first three terms have compact support, and they give rise to the tensorial potential

$$W^{jk}(t, \boldsymbol{x}) := G \int \frac{\rho^{*\prime}\left(v'^j v'^k - \frac{1}{2}U'\delta^{jk}\right) + p'\delta^{jk}}{|\boldsymbol{x} - \boldsymbol{x}'|}\, d^3x', \tag{7.81}$$

which satisfies the Poisson equation

$$\nabla^2 W^{jk} = -4\pi G\left(\rho^* v^j v^k - \frac{1}{2}\rho^* U\delta^{jk} + p\,\delta^{jk}\right). \tag{7.82}$$

The fourth term involves $\nabla^2 U^2$, which we know how to handle, and which produces a contribution proportional to $\delta^{jk}U^2$ to $h^{jk}_{\mathcal{N}}$. The fifth and final term is the hard one. To account for it we introduce another tensorial potential defined by

$$\chi^{jk}(t, \boldsymbol{x}) := \frac{1}{4\pi}\int_{\mathcal{M}} \frac{\partial^{j'} U' \partial^{k'} U'}{|\boldsymbol{x} - \boldsymbol{x}'|}\, d^3x', \tag{7.83}$$

which satisfies the Poisson equation

$$\nabla^2 \chi^{jk} = -\partial^j U \partial^k U. \tag{7.84}$$

Because the Poisson integral in Eq. (7.83) is truncated at $r' = \mathcal{R}$, the source term on the right-hand side of the Poisson equation should be multiplied by $\Theta(\mathcal{R} - r)$, as was discussed in Box 7.3. But since the truncation produces no noticeable changes within the near zone, we have kept it implicit in Eq. (7.84).

Armed with these tensorial potentials, we find that the gravitational potentials can be expressed as

$$h^{jk}_{2\mathcal{N}} = \frac{1}{c^4}\left(4W^{jk} + U^2\delta^{jk} + 4\chi^{jk}\right) - 2\frac{G}{c^5}\dddot{\mathcal{I}}^{jk}(t) + O(c^{-6}), \tag{7.85}$$

where we have included the $O(c^{-5})$ term for completeness.

Computation of χ^{jk}

We must now face the computation of χ^{jk}, as defined by Eq. (7.83). Returning to the standard expression of Eq. (7.38) for the Newtonian potential, we differentiate it and obtain

$$\begin{aligned}
\partial_{j'} U' &= G\int d^3y_1\, \rho_1^* \frac{\partial}{\partial x'^j}\frac{1}{|\boldsymbol{x}' - \boldsymbol{y}_1|} \\
&= -G\int d^3y_1\, \rho_1^* \frac{\partial}{\partial y_1^j}\frac{1}{|\boldsymbol{x}' - \boldsymbol{y}_1|},
\end{aligned} \tag{7.86}$$

in which $\boldsymbol{y}_1$ is an integration variable, and $\rho_1^* := \rho^*(t, \boldsymbol{y}_1)$. Expressing $\partial_{k'} U'$ in a similar way, in terms of an independent integration variable $\boldsymbol{y}_2$, and inserting these expressions in the Poisson integral for χ^{jk}, we arrive at

$$\chi^{jk} = G^2 \int d^3y_1 d^3y_2\, \rho_1^* \rho_2^* \frac{\partial^2}{\partial y_1^j \partial y_2^k} K(\boldsymbol{x}; \boldsymbol{y}_1, \boldsymbol{y}_2), \tag{7.87}$$

where

$$K(\boldsymbol{x}; \boldsymbol{y}_1, \boldsymbol{y}_2) := \frac{1}{4\pi} \int_{\mathscr{M}} \frac{d^3x'}{|\boldsymbol{x} - \boldsymbol{x}'||\boldsymbol{x}' - \boldsymbol{y}_1||\boldsymbol{x}' - \boldsymbol{y}_2|} \tag{7.88}$$

is a three-point function that must now be evaluated. This computation is presented in Box 7.4, and the end result is

$$K(\boldsymbol{x}; \boldsymbol{y}_1, \boldsymbol{y}_2) = 1 - \ln \frac{S}{2\mathcal{R}}, \tag{7.89}$$

where

$$S := r_1 + r_2 + r_{12}, \tag{7.90}$$

with the notations

$$r_1 := |\boldsymbol{x} - \boldsymbol{y}_1|, \qquad r_2 := |\boldsymbol{x} - \boldsymbol{y}_2|, \qquad r_{12} := |\boldsymbol{y}_1 - \boldsymbol{y}_2|. \tag{7.91}$$

We also introduce the corresponding separation vectors

$$\boldsymbol{r}_1 := \boldsymbol{x} - \boldsymbol{y}_1, \qquad \boldsymbol{r}_2 := \boldsymbol{x} - \boldsymbol{y}_2, \qquad \boldsymbol{r}_{12} := \boldsymbol{y}_1 - \boldsymbol{y}_2, \tag{7.92}$$

and the unit vectors

$$\boldsymbol{n}_1 := \frac{\boldsymbol{r}_1}{r_1}, \qquad \boldsymbol{n}_2 := \frac{\boldsymbol{r}_2}{r_2}, \qquad \boldsymbol{n}_{12} := \frac{\boldsymbol{r}_{12}}{r_{12}}. \tag{7.93}$$

The dependence of K on $\mathcal{R}$ comes from the fact that the domain of integration is truncated at $r' = \mathcal{R}$. This dependence plays no role, however, because K is differentiated as soon as it is inserted within Eq. (7.87). These derivatives are straightforward to compute, and we obtain

$$\frac{\partial^2 K}{\partial y_1^j \partial y_2^k} = \frac{(n_1^j - n_{12}^j)(n_2^k + n_{12}^k)}{S^2} - \frac{n_{12}^j n_{12}^k - \delta^{jk}}{S r_{12}}. \tag{7.94}$$

We then arrive at

$$\begin{aligned} \chi^{jk} &= G^2 \int \frac{\rho_1^* \rho_2^* (n_1^j - n_{12}^j)(n_2^k + n_{12}^k)}{S^2} d^3y_1 d^3y_2 \\ &\quad - G^2 \int \frac{\rho_1^* \rho_2^* (n_{12}^j n_{12}^k - \delta^{jk})}{S r_{12}} d^3y_1 d^3y_2. \end{aligned} \tag{7.95}$$

It is easy to check that each integral is symmetric in the jk indices; this property is evident in the second integral, and to establish it for the first it is necessary to swap the variables of integration, $\boldsymbol{y}_1 \leftrightarrow \boldsymbol{y}_2$, keeping in mind that $\boldsymbol{n}_{12} \to \boldsymbol{n}_{21} = -\boldsymbol{n}_{12}$ under this operation.

Note that the trace $\chi := \delta_{jk}\chi^{jk}$ is given by the Poisson potential of $\frac{1}{2}\partial_j U \partial^j U$. Using the identity of Eq. (7.70), it is easy to see that χ can be expressed as

$$\chi = -\frac{1}{2}U^2 + G\int_{\mathcal{M}} \frac{\rho'^* U'}{|\boldsymbol{x}-\boldsymbol{x}'|} d^3x' . \tag{7.96}$$

By inserting the Poisson integral for U, we can express this in the form

$$\chi = -\frac{1}{2}G^2 \int \frac{\rho_1^* \rho_2^* \, d^3y_1 d^3y_2}{|\boldsymbol{x}-\boldsymbol{y}_1||\boldsymbol{x}-\boldsymbol{y}_2|} + G^2 \int \frac{\rho_1^* \rho_2^* \, d^3y_1 d^3y_2}{|\boldsymbol{x}-\boldsymbol{y}_1||\boldsymbol{y}_1-\boldsymbol{y}_2|} . \tag{7.97}$$

The second integral can be written in the symmetric form

$$\frac{1}{2}\int \frac{\rho_1^* \rho_2^* \, d^3y_1 d^3y_2}{|\boldsymbol{x}-\boldsymbol{y}_1||\boldsymbol{y}_1-\boldsymbol{y}_2|} + \frac{1}{2}\int \frac{\rho_1^* \rho_2^* \, d^3y_1 d^3y_2}{|\boldsymbol{x}-\boldsymbol{y}_2||\boldsymbol{y}_1-\boldsymbol{y}_2|},$$

and this gives

$$\chi = \frac{1}{2}G^2 \int \rho_1^* \rho_2^* \left(-\frac{1}{r_1 r_2} + \frac{1}{r_1 r_{12}} + \frac{1}{r_2 r_{12}} \right) d^3y_1 d^3y_2 . \tag{7.98}$$

Our final expression is

$$\chi = \frac{1}{2}G^2 \int \rho_1^* \rho_2^* \frac{r_1 + r_2 - r_{12}}{r_1 r_2 r_{12}} d^3y_1 d^3y_2 , \tag{7.99}$$

and we may check that the trace of Eq. (7.95) reproduces this. The calculation is aided by the identities

$$\boldsymbol{n}_1 \cdot \boldsymbol{n}_2 = \frac{r_1^2 + r_2^2 - r_{12}^2}{2r_1 r_2}, \tag{7.100a}$$

$$\boldsymbol{n}_1 \cdot \boldsymbol{n}_{12} = \frac{r_2^2 - r_1^2 - r_{12}^2}{2r_1 r_{12}}, \tag{7.100b}$$

$$\boldsymbol{n}_2 \cdot \boldsymbol{n}_{12} = \frac{r_2^2 - r_1^2 + r_{12}^2}{2r_2 r_{12}}, \tag{7.100c}$$

involving the unit vectors defined by Eq. (7.93).

Box 7.4 Three-point function $K(x; y_1, y_2)$

The computation of the three-point function defined by Eq. (7.88) follows some of the same steps that were used to calculate the two-point function $K(\boldsymbol{x}; \boldsymbol{y})$ in Box 7.3.

We note first that $K(\boldsymbol{x}; \boldsymbol{y}_1, \boldsymbol{y}_2)$ satisfies

$$\nabla^2 K(\boldsymbol{x}; \boldsymbol{y}_1, \boldsymbol{y}_2) = -\frac{1}{|\boldsymbol{x}-\boldsymbol{y}_1||\boldsymbol{x}-\boldsymbol{y}_2|}, \tag{1}$$

and verify that $K_{\rm p} = -\ln S$ is a particular solution. The relation implies that

$$\nabla^2 K_{\rm p} = -\frac{1}{S^2}(S\nabla^2 S - \partial_j S \partial^j S),$$

and the various derivatives can be computed from the definition of S provided in Eq. (7.90). We have, for example, $\partial^j S = n_1^j + n_2^j$, from which it follows that

$$\partial^{jk} S = -\frac{n_1^j n_1^k - \delta^{jk}}{r_1} - \frac{n_2^j n_2^k - \delta^{jk}}{r_2}.$$

From this, and the helpful identities of Eqs. (7.100), we obtain

$$\nabla^2 S = 2\frac{r_1 + r_2}{r_1 r_2}, \qquad \partial_j S \partial^j S = \frac{(r_1 + r_2 - r_{12})S}{r_1 r_2}.$$

Collecting results, we confirm that $K_{\rm p}$ is a solution to $\nabla^2 K = -1/(r_1 r_2)$.

To this we must add a suitable solution $K_{\rm h}$ to Laplace's equation. The solution to the homogeneous equation must be non-singular in all three variables $\boldsymbol{x}$, $\boldsymbol{y}_1$, and $\boldsymbol{y}_2$, because the singularity structure required by Eq. (1) is already contained in $K_{\rm p}$. Furthermore, $K_{\rm h}$ must be dimensionless, and the only possibility is to make it equal to a constant. We are therefore looking for a solution of the form

$$K = K_0 - \ln(r_1 + r_2 + r_{12}),$$

where K_0 is a dimensionless constant. To determine this we carry out an independent computation of the special value $K(\boldsymbol{x}; \boldsymbol{0}, \boldsymbol{0})$, and compare our result to $K_0 - \ln(2r)$, which follows from the general expression.

From Eq. (7.88) we have

$$K(\boldsymbol{x}; \boldsymbol{0}, \boldsymbol{0}) = \frac{1}{4\pi}\int_{\mathcal{M}} \frac{d^3x'}{|\boldsymbol{x} - \boldsymbol{x}'||\boldsymbol{x}'|^2} = \frac{1}{4\pi}\int_0^{\mathcal{R}} \frac{dr'\, d\Omega'}{|\boldsymbol{x} - \boldsymbol{x}'|}.$$

Invoking the addition theorem for spherical harmonics, this is simply $\int_0^{\mathcal{R}} (r_>)^{-1}\, dr'$, and evaluating the integral gives $K(\boldsymbol{x}; \boldsymbol{0}, \boldsymbol{0}) = 1 + \ln(\mathcal{R}/r)$. This allows us to conclude that $K_0 = 1 + \ln(2\mathcal{R})$. Collecting results, we obtain the expression displayed in Eq. (7.89).

7.3.4 Wave-zone contribution to potentials

In this subsection we estimate $h^{\alpha\beta}_{\mathcal{W}}$, the wave-zone contribution to the second-iterated potentials, still assuming that the field point $\boldsymbol{x}$ is within the near zone. Techniques to carry out such a computation were described back in Sec. 6.3.5, and crude estimates were obtained toward the end of that section. These ignore numerical factors and terms that depend explicitly on $\mathcal{R}$, but they are sufficient to allow us to conclude that

$$h^{00}_{\mathcal{W}} = O(c^{-8}), \qquad h^{0j}_{\mathcal{W}} = O(c^{-8}), \qquad h^{jk}_{\mathcal{W}} = O(c^{-8}). \tag{7.101}$$

This is far beyond the 1PN accuracy of our calculations in this section, and we shall therefore ignore the wave-zone contribution to $h^{\alpha\beta}_2$.

To reach this conclusion we refer to Eq. (6.107), which applies to source terms of the form displayed in Eq. (6.98). In our current application, $\tau^{\alpha\beta}_1$ is built entirely from $(-g)t^{\alpha\beta}_{\rm LL}$

as displayed in Eqs. (7.48), by inserting the wave-zone potentials h_1^{00} and h_1^{0j} given by Eqs. (7.32). Focusing our attention on τ_1^{00} for concreteness, and ignoring all numerical and angle-dependent factors, we find that it has a structure given schematically by

$$\frac{G^2}{c^4}\left[\frac{M_0^2}{r^4} + \frac{M_0\mathcal{I}^{jk}}{r^6} + \frac{M_0\dot{\mathcal{I}}^{jk}}{cr^5} + \frac{M_0\ddot{\mathcal{I}}^{jk}}{c^2r^4} + \frac{M_0\dddot{\mathcal{I}}^{jk}}{c^3r^3} + \cdots\right], \tag{7.102}$$

in which the ellipsis designates terms of higher post-Newtonian order. Each term is of the form $f(\tau)/r^n$ required for the integration techniques of Sec. 6.3.5. Ignoring the overall factor of G^2/c^4, we see, for example, that for $n = 3$ we have $f = M_0\dddot{\mathcal{I}}^{jk}/c^3$, and that for $n = 4$ we have $f = M_0^2 + M_0\ddot{\mathcal{I}}^{jk}/c^2$. According to Eq. (6.109), an estimate of $h_{\mathcal{W}}^{00}$ for each contributing n is $c^{-(n-2)}f^{(n-2)} + c^{-(n-1)}rf^{(n-1)}$. The dominant term in a post-Newtonian expansion is $c^{-(n-2)}f^{(n-2)}$, and restoring the factor of G^2/c^4, we find that for each n, $h_{\mathcal{W}}^{00}$ is estimated as

$$\frac{G^2M_0}{c^8}\frac{d^4\mathcal{I}^{jk}}{d\tau^4}. \tag{7.103}$$

This is of order c^{-8}, and contributes to h_2^{00} at 3PN order. A similar result follows for the other components of $h_{\mathcal{W}}^{\alpha\beta}$, and we arrive at the statement of Eq. (7.101).

In fact, a detailed computation shows that these contributions are actually gauge artifacts that can be removed by a suitable coordinate transformation. The first instance in which $h_{\mathcal{W}}^{\alpha\beta}$ makes a non-trivial contribution to the near-zone potentials turns out to be at 4PN order. In any event, we see that $h_{\mathcal{W}}^{\alpha\beta}$ is far too small to contribute to our 1PN potentials, and for this reason we do not need to calculate it in detail.

7.3.5 Near-zone potentials: Final answer

We are now ready to collect our results and display the final expression for the second-iterated potentials $h_2^{\alpha\beta}$ in the near zone. Our results are summarized in Box 7.5.

Box 7.5 — Near-zone potentials

Combining Eqs. (7.78), (7.79), (7.85), and (7.101), we find that the near-zone gravitational potentials are given by

$$h_2^{00} = \frac{4}{c^2}U + \frac{1}{c^4}\left(7U^2 + 4\psi - 4V + 2\frac{\partial^2 X}{\partial t^2}\right) - \frac{2G}{3c^5}\dddot{\mathcal{I}}^{kk}(t) + O(c^{-6}),$$

$$h_2^{0j} = \frac{4}{c^3}U^j + O(c^{-5}),$$

$$h_2^{jk} = \frac{1}{c^4}\left(4W^{jk} + U^2\delta^{jk} + 4\chi^{jk}\right) - 2\frac{G}{c^5}\dddot{\mathcal{I}}^{jk}(t) + O(c^{-6}),$$

$$h_2^{kk} = \frac{1}{c^4}\left(U^2 + 4V\right) - \frac{2G}{c^5}\dddot{\mathcal{I}}^{kk}(t) + O(c^{-6}).$$

The potentials that make up $h^{\alpha\beta}$ satisfy the Poisson equations

$$\nabla^2 U = -4\pi G\rho^* ,$$

$$\nabla^2 \psi = -4\pi G\rho^* \left(\frac{3}{2}v^2 - U + \Pi + 3p/\rho^*\right),$$

$$\nabla^2 V = -4\pi G\rho^* \left(v^2 - \frac{1}{2}U + 3p/\rho^*\right),$$

$$\nabla^2 X = 2U ,$$

$$\nabla^2 U^j = -4\pi G\rho^* v^j ,$$

$$\nabla^2 W^{jk} = -4\pi G\left(\rho^* v^j v^k - \frac{1}{2}\rho^* U\delta^{jk} + p\delta^{jk}\right),$$

$$\nabla^2 \chi^{jk} = -\partial^j U \partial^k U .$$

The solutions are

$$U = G\int \frac{\rho^{*\prime}}{|\boldsymbol{x}-\boldsymbol{x}'|}\,d^3x' ,$$

$$\psi = G\int \frac{\rho^{*\prime}\left(\frac{3}{2}v'^2 - U' + \Pi'\right) + 3p'}{|\boldsymbol{x}-\boldsymbol{x}'|}\,d^3x' ,$$

$$V = G\int \frac{\rho^{*\prime}\left(v'^2 - \frac{1}{2}U'\right) + 3p'}{|\boldsymbol{x}-\boldsymbol{x}'|}\,d^3x' ,$$

$$X = G\int \rho^{*\prime}|\boldsymbol{x}-\boldsymbol{x}'|\,d^3x' ,$$

$$U^j = G\int \frac{\rho^{*\prime}v'^j}{|\boldsymbol{x}-\boldsymbol{x}'|}\,d^3x' ,$$

$$W^{jk} = G\int \frac{\rho^{*\prime}\left(v'^j v'^k - \frac{1}{2}U'\delta^{jk}\right) + p'\delta^{jk}}{|\boldsymbol{x}-\boldsymbol{x}'|}\,d^3x' ,$$

$$\chi^{jk} = G^2\int \frac{\rho_1^*\rho_2^*\left(n_1^j - n_{12}^j\right)\left(n_2^k + n_{12}^k\right)}{S^2}\,d^3y_1 d^3y_2 - G^2\int \frac{\rho_1^*\rho_2^*\left(n_{12}^j n_{12}^k - \delta^{jk}\right)}{Sr_{12}}\,d^3y_1 d^3y_2 .$$

The potentials are evaluated at time t and position $\boldsymbol{x}$; the sources are evaluated at the same time but at position $\boldsymbol{x}'$. We use the notation $\boldsymbol{r}_1 := \boldsymbol{x} - \boldsymbol{y}_1, r_1 := |\boldsymbol{r}_1|, \boldsymbol{n}_1 := \boldsymbol{r}_1/r_1$ (and similarly for $\boldsymbol{r}_2, r_2$, and $\boldsymbol{n}_2$), as well as $\boldsymbol{r}_{12} := \boldsymbol{y}_1 - \boldsymbol{y}_2, r_{12} := |\boldsymbol{r}_{12}|$, and $\boldsymbol{n}_{12} := \boldsymbol{r}_{12}/r_{12}$, in which $\boldsymbol{y}_1$ and $\boldsymbol{y}_2$ are integration variables. We also have $S := r_1 + r_2 + r_{12}$, and the trace of χ^{jk} is given by

$$\chi = -\frac{1}{2}U^2 + G\int_{\mathscr{M}} \frac{\rho^{*\prime}U'}{|\boldsymbol{x}-\boldsymbol{x}'|}\,d^3x' .$$

From Eq. (7.23) we find that the potentials give rise to the spacetime metric

$$g_{00} = -1 + \frac{2}{c^2}U + \frac{2}{c^4}\left(\psi - U^2 + \frac{1}{2}\frac{\partial^2 X}{\partial t^2}\right) - \frac{4G}{3c^5}\dddot{\mathcal{I}}^{kk}(t) + O(c^{-6}), \tag{7.104a}$$

$$g_{0j} = -\frac{4}{c^3}U_j + O(c^{-5}), \tag{7.104b}$$

$$g_{jk} = \delta_{jk}\left[1 + \frac{2}{c^2}U + \frac{2}{c^4}\left(\psi + U^2 - 2V + \frac{1}{2}\frac{\partial^2 X}{\partial t^2}\right)\right] + \frac{4}{c^4}\left(W^{jk} + \chi^{jk}\right) - 2\frac{G}{c^5}\dddot{\mathcal{I}}^{\langle jk\rangle}(t) + O(c^{-6}). \tag{7.104c}$$

This metric is too accurate for most of our purposes. As we indicated back in Sec. 7.1.3, in order to describe the slow motion of a weakly gravitating system accurately through 1PN order, we require g_{00} to order c^{-4}, g_{0j} to order c^{-3}, and g_{jk} to order c^{-2}. For this application our previous expressions can therefore be truncated to

$$g_{00} = -1 + \frac{2}{c^2}U + \frac{2}{c^4}\left(\psi - U^2 + \frac{1}{2}\frac{\partial^2 X}{\partial t^2}\right) + O(c^{-5}), \tag{7.105a}$$

$$g_{0j} = -\frac{4}{c^3}U_j + O(c^{-5}), \tag{7.105b}$$

$$g_{jk} = \delta_{jk}\left(1 + \frac{2}{c^2}U\right) + O(c^{-4}). \tag{7.105c}$$

This metric forms the basis of what is known as *post-Newtonian theory*. Chapters 8 through 10 will be devoted to the details and many applications of this approximation to general relativity.

We have previously indicated that the c^{-5} term in g_{00} is a coordinate artifact that has no impact on the physics of our gravitating system. Because it depends only on time, this term may in fact be removed by a transformation of the time coordinate given by

$$t = t' - \frac{2G}{3c^5}\ddot{\mathcal{I}}^{kk}(t') + O(c^{-7}). \tag{7.106}$$

It is a simple exercise to show that the time-time component of the transformed metric, expressed in terms of the new time t', no longer contains a term at order c^{-5}; the other components of the metric are not affected by the transformation. It should be noted that the transformed coordinates are no longer harmonic; the c^{-5} term must stay if we insist on using harmonic coordinates. A more careful calculation reveals that the transformation generates non-trivial terms in g_{00} at order c^{-7}, or at 2.5PN order; these must then be combined with other 2.5PN terms in order to give a correct description of radiation-reaction effects. We return to this theme in Chapter 12.

Box 7.6 Post-Minkowskian theory and the slow-motion approximation

The advantages of incorporating a slow-motion condition within post-Minkowskian theory should be pretty clear by now, quite apart from the physical relevance of slow motion within a weak-field context. Had we not expanded the various retarded potentials in powers of c^{-1} right from the start, we would have been faced with the need to evaluate fully retarded potentials such as

$$\int \frac{\rho^*(t - |\boldsymbol{x} - \boldsymbol{x}'|, \boldsymbol{x}')}{|\boldsymbol{x} - \boldsymbol{x}'|} d^3x',$$

$$\int \frac{\rho^*(t - |\boldsymbol{x} - \boldsymbol{x}'|, \boldsymbol{x}')}{|\boldsymbol{x} - \boldsymbol{x}'|} \frac{\rho^*(t - |\boldsymbol{x} - \boldsymbol{x}'| - |\boldsymbol{x}' - \boldsymbol{x}''|, \boldsymbol{x}'')}{|\boldsymbol{x}' - \boldsymbol{x}''|} d^3x' d^3x'',$$

$$\int_{\mathscr{M}} \frac{1}{|\boldsymbol{x} - \boldsymbol{x}'|} \partial'_j \frac{\rho^*(t - |\boldsymbol{x} - \boldsymbol{x}'| - |\boldsymbol{x}' - \boldsymbol{x}''|, \boldsymbol{x}'')}{|\boldsymbol{x}' - \boldsymbol{x}''|}$$
$$\times \partial'_k \frac{\rho^*(t - |\boldsymbol{x} - \boldsymbol{x}'| - |\boldsymbol{x}' - \boldsymbol{x}'''|, \boldsymbol{x}''')}{|\boldsymbol{x}' - \boldsymbol{x}'''|} d^3x' d^3x'' d^3x''';$$

these expressions are the fully retarded analogues of U, $\int \rho'^* U' |\boldsymbol{x} - \boldsymbol{x}'|^{-1} d^3x'$, and χ^{jk}, respectively.

Such potentials lead to hopeless complications. Even a relatively simple potential, such as the first one listed above, leads to difficult computations because of the need to account for the retardation condition. Examples of such complexities are known in Maxwell's theory, in which the evaluation of the retarded potential is difficult even for the simple case of a single point charge (remember the Liénard–Wiechert potentials?). The non-linear potentials are even more challenging, as they involve nested retardation conditions; such potentials do not occur in electromagnetism, because of the linearity of Maxwell's equations.

In the early 1960s, Peter Havas and Joshua Goldberg, together with their students and collaborators, worked on post-Minkowskian theory in order to study gravitational radiation, but they chose *not* to incorporate the slow-motion condition. Very quickly they ran into the difficulties noted above, and as a result, they were unable to go beyond the first iteration of the relaxed field equations. And even for the first-iterated potentials, they were able to evaluate quantities like the retarded Newtonian potential only for specific motions, such as circular orbits, where mathematical techniques from electrodynamics were available. In the 1970s, Havas's student Arnold Rosenblum worked on obtaining the second iteration, but progress was extremely slow, and his untimely death in 1991 essentially brought this program to an end without any definitive conclusion.

7.4 Second iteration: Wave zone

Our final task in this chapter is to obtain expressions for the second-iterated potentials when the field point $\boldsymbol{x}$ is in the wave zone, where $r := |\boldsymbol{x}| > \mathcal{R}$.

7.4.1 Near-zone contribution to potentials

Equations (7.32) give us formal expressions for the potentials $h^{\alpha\beta}_{\mathcal{N}}$ evaluated in the wave zone. Recalling our discussion of Sec. 7.2.3, in which we observed that each successive multipole moment brings an additional factor of v_c/c to the post-Newtonian ordering, we have that

$$h^{00}_{\mathcal{N}} = \underbrace{\frac{4GM}{c^2 r}}_{0+1\text{PN}} + \underbrace{\frac{2G}{c^2}\partial_{jk}\left[\frac{\mathcal{I}^{jk}(\tau)}{r}\right]}_{1\text{PN}} - \underbrace{\frac{2G}{3c^2}\partial_{jkn}\left[\frac{\mathcal{I}^{jkn}(\tau)}{r}\right]}_{1.5\text{PN}} + \cdots, \tag{7.107a}$$

$$h^{0j}_{\mathcal{N}} = -\underbrace{\frac{2G}{c^3}\frac{(\boldsymbol{n}\times\boldsymbol{J})^j}{r^2}}_{1\text{PN}} - \underbrace{\frac{2G}{c^3}\partial_k\left[\frac{\dot{\mathcal{I}}^{jk}(\tau)}{r}\right]}_{1\text{PN}}$$

$$\underbrace{-\frac{G}{3c^3}\partial_{kn}\left[\frac{\dot{\mathcal{I}}^{jkn}(\tau) - 2\epsilon^{mjk}\mathcal{J}^{mn}(\tau)}{r}\right]}_{1.5\text{PN}} + \cdots, \tag{7.107b}$$

$$h^{jk}_{\mathcal{N}} = \underbrace{\frac{2G}{c^4}\frac{\ddot{\mathcal{I}}^{jk}(\tau)}{r}}_{1\text{PN}} - \underbrace{\frac{2G}{3c^4}\partial_n\left[\frac{\ddot{\mathcal{I}}^{jkn}(\tau) + 4\epsilon^{mn(j}\dot{\mathcal{J}}^{m|k)}(\tau)}{r}\right]}_{1.5\text{PN}} + \cdots, \tag{7.107c}$$

is a post-Newtonian expansion of the potentials that is accurate through 1.5PN order. Recalling Eq. (7.67), we have replaced the monopole moment $\mathcal{I} = M_0$ – the near-zone mass – that originally appeared in Eq. (7.32) with the total mass M, since they agree to order c^{-4}. We recall that M is given by Eq. (7.63), so that it contains both a 0PN rest-mass contribution and 1PN corrections provided by the system's total energy. We have also replaced the near-zone angular momentum $\boldsymbol{J}_0$ by the total angular momentum $\boldsymbol{J}$, since these quantities agree to order c^{-2}.

The multipole moments that appear in Eqs. (7.107) are all functions of retarded time $\tau = t - r/c$. Formally they must be evaluated using the first-iterated forms $\tau^{\alpha\beta}_1$ for the energy-momentum pseudotensor, but since the multipole moments occur at 1PN and 1.5PN orders in the potentials, we may truncate $\tau^{\alpha\beta}_1$ to its leading-order expression $c^{-2}\tau^{00} = \rho^* + O(c^{-2})$ and $c^{-1}\tau^{0j} = \rho^* v^j + O(c^{-2})$. The multipole moments then take the explicit forms

$$\mathcal{I}^{jk}(\tau) = \int \rho^* x^j x^k \, d^3x + O(c^{-2}), \tag{7.108a}$$

$$\mathcal{I}^{jkn}(\tau) = \int \rho^* x^j x^k x^n \, d^3x + O(c^{-2}), \tag{7.108b}$$

$$\mathcal{J}^{jk}(\tau) = \epsilon^{jab} \int \rho^* v^a x^b x^k \, d^3x + O(c^{-2}). \tag{7.108c}$$

With these, our expressions for $h^{\alpha\beta}_{\mathcal{N}}$ are complete.

7.4.2 Wave-zone contribution to potentials

We turn next to the computation of $h^{\alpha\beta}_{\mathcal{W}}$ in the wave zone. To carry this out we insert the first-iterated potentials obtained in Sec. 7.2.3 within $\tau^{\alpha\beta}_1$, and solve the relaxed field equations for the second-iterated potentials. By virtue of Eq. (7.52), only the

Landau–Lifshitz pseudotensor of Eq. (7.48) makes a contribution to $\tau_1^{\alpha\beta}$. And by virtue of our requirement of 1.5PN overall accuracy for the potentials, we find that the only relevant piece of the first-iterated potentials is the Newtonian term in $h^{00}_{\mathcal{W}}$, given by

$$h^{00}_{\mathcal{W}} = \frac{4GM}{c^2 r} + O(c^{-4}). \tag{7.109}$$

Inserting this within Eq. (7.48), we find that the components of the energy-momentum pseudotensor are

$$\tau_1^{00} = -\frac{7GM^2}{8\pi r^4} + O(c^{-2}), \tag{7.110a}$$

$$\tau_1^{0j} = O(c^{-3}), \tag{7.110b}$$

$$\tau_1^{jk} = \frac{GM^2}{4\pi r^4}\left(n^j n^k - \frac{1}{2}\delta^{jk}\right) + O(c^{-2}), \tag{7.110c}$$

in which $n^j := x^j/r$.

To obtain $h^{\alpha\beta}_{\mathcal{W}}$ we rely on the methods of Sec. 6.3.5, which work for source terms of the form displayed in Eq. (6.98). Our first task is to decompose the effective stress tensor of Eq. (7.110c) in terms of STF angular tensors; refer to Sec. 1.5.3. We invoke the identity $n^j n^k = n^{\langle jk\rangle} + \frac{1}{3}\delta^{jk}$ and rewrite Eq. (7.110c) as

$$\tau_1^{jk} = \frac{G}{4\pi}\frac{M^2}{r^4}\left(n^{\langle jk\rangle} - \frac{1}{6}\delta^{jk}\right). \tag{7.111}$$

This and Eq. (7.110a) are now of the form of Eq. (7.19), and we identify $f^{00}_{\ell=0}$ with $-\frac{7}{2}GM^2$, $f^{jk}_{\ell=2}$ with GM^2, and $f^{jk}_{\ell=0}$ with $-\frac{1}{6}GM^2\delta^{jk}$. In each case we have that $n = 4$.

The contribution to $h^{\alpha\beta}_{\mathcal{W}}$ from each value of ℓ is given by Eq. (6.105), which we copy here as

$$h^{\alpha\beta}_{\mathcal{W}}(t, \boldsymbol{x}) = \frac{4G}{c^4}\frac{n^{\langle L\rangle}}{r}\left\{\int_0^{\mathcal{R}} ds f^{\alpha\beta}(\tau - 2s/c)A(s,r) + \int_{\mathcal{R}}^{\infty} ds f^{\alpha\beta}(\tau - 2s/c)B(s,r)\right\}, \tag{7.112}$$

in which $A(s,r) = \int_{\mathcal{R}}^{r+s} P_\ell(\xi)p^{-(n-1)}\,dp$, $B(s,r) = \int_s^{r+s} P_\ell(\xi)p^{-(n-1)}\,dp$, and $\xi = (r + 2s)/r - 2s(r+s)/(rp)$. Because $f^{\alpha\beta}$ is a constant, it can be taken outside of each integral, and the remaining computations are simple. For $\ell = 0$ we get

$$h^{00}_{\mathcal{W}} = 7\left(\frac{GM}{c^2 r}\right)^2\left(1 - 2\frac{r}{\mathcal{R}}\right), \tag{7.113a}$$

$$h^{jk}_{\mathcal{W}} = \frac{1}{3}\left(\frac{GM}{c^2 r}\right)^2\delta^{jk}\left(1 - 2\frac{r}{\mathcal{R}}\right), \tag{7.113b}$$

and for $\ell = 2$

$$h^{jk}_{\mathcal{W}} = \left(\frac{GM}{c^2 r}\right)^2 n^{\langle jk\rangle}\left(1 - \frac{4\mathcal{R}}{5r}\right). \tag{7.114}$$

Discarding all terms involving $\mathcal{R}$, as we are free to do, and adding the results, we arrive at

$$h^{00}_{\mathcal{W}} = 7\left(\frac{GM}{c^2 r}\right)^2, \tag{7.115a}$$

$$h^{jk}_{\mathcal{W}} = \left(\frac{GM}{c^2 r}\right)^2 n^j n^k. \tag{7.115b}$$

The post-Newtonian order of these contributions to h^{00} and h^{jk} is 1.5PN. To see this, we divide each of these expressions by $h^{00} \sim GM/(c^2 r)$ to obtain something proportional to $GM/(c^2 r)$. We next incorporate the fact that the Newtonian acceleration GM/r_c^2 is of order r_c/t_c^2, which makes GM of order r_c^3/t_c^2. Setting $r \sim \lambda_c = ct_c$, we finally get $h^{\alpha\beta}_{\mathcal{W}}/h^{00} \sim r_c^3/(c^3 t_c^3) = (v_c/c)^3$, and conclude that Eqs. (7.115) do indeed make contributions of 1.5PN order to the gravitational potentials.

We pull everything together and summarize our results in Box 7.7. It is instructive to note that in the limit of a static, spherically symmetric body, the results correspond precisely to the post-Newtonian expansion of the Schwarzschild metric. This statement is established in Exercise 7.7.

Box 7.7 **Wave-zone fields**

Combining Eqs. (7.107) and (7.115), we find that the wave-zone gravitational potentials are given by

$$h^{00} = \frac{4G}{c^2}\left[\frac{M}{r} + \frac{1}{2}\partial_{jk}\left(\frac{\mathcal{I}^{jk}}{r}\right) - \frac{1}{6}\partial_{jkn}\left(\frac{\mathcal{I}^{jkn}}{r}\right) + \frac{7}{4}\frac{GM^2}{c^2 r^2} + \cdots\right],$$

$$h^{0j} = \frac{4G}{c^3}\left[-\frac{1}{2}\frac{(\boldsymbol{n}\times\boldsymbol{J})^j}{r^2} - \frac{1}{2}\partial_k\left(\frac{\dot{\mathcal{I}}^{jk}}{r}\right) - \frac{1}{12}\partial_{kn}\left(\frac{\dot{\mathcal{I}}^{jkn} - 2\epsilon^{mjk}\mathcal{J}^{mn}}{r}\right) + \cdots\right],$$

$$h^{jk} = \frac{4G}{c^4}\left[\frac{1}{2}\frac{\ddot{\mathcal{I}}^{jk}}{r} - \frac{1}{6}\partial_n\left(\frac{\ddot{\mathcal{I}}^{jkn} + 2\epsilon^{mnj}\dot{\mathcal{J}}^{mk} + 2\epsilon^{mnk}\dot{\mathcal{J}}^{mj}}{r}\right) + \frac{GM^2}{4r^2}n^j n^k + \cdots\right].$$

The potentials are expressed in terms of $n^j = x^j/r$, and in terms of multipole moments that depend on retarded time $\tau = t - r/c$; overdots indicate differentiation with respect to τ. In h^{00} the mass term contains 0PN and 1PN contributions, the quadrupole term is a 1PN contribution, and the octupole and M^2 terms are 1.5PN contributions. The first two terms in h^{0j} are 1PN contributions, while the rest are 1.5PN. And finally, the quadrupole term in h^{jk} is a 1PN contribution, while the remaining terms are all 1.5PN contributions.

We have the total gravitational mass

$$M = \int \rho^*\left[1 + \frac{1}{c^2}\left(\frac{1}{2}v^2 - \frac{1}{2}U + \Pi\right)\right] d^3x + O(c^{-4}),$$

the total angular momentum

$$\boldsymbol{J} = \int \rho^* \boldsymbol{x}\times\boldsymbol{v}\, d^3x + O(c^{-2}),$$

and the mass and current multipole moments

$$\mathcal{I}^{jk}(\tau) = \int \rho^* x^j x^k \, d^3x + O(c^{-2}),$$

$$\mathcal{I}^{jkn}(\tau) = \int \rho^* x^j x^k x^n \, d^3x + O(c^{-2}),$$

$$\mathcal{J}^{jk}(\tau) = \epsilon^{jab} \int \rho^* x^a v^b x^k \, d^3x + O(c^{-2}).$$

We recall that M and $\boldsymbol{J}$ are conserved quantities. The gravitational potentials are evaluated in the center-of-mass frame, in which the total momentum $\boldsymbol{P}$ and center-of-mass position $\boldsymbol{R}$ are set equal to zero.

The multipole moments must be differentiated a number of times before they are inserted within the gravitational potentials. These operations are aided by the identity

$$\dot{F} = \int \rho^* \frac{df}{dt} \, d^3x,$$

where $F(t) := \int \rho^*(t, \boldsymbol{x}) f(t, \boldsymbol{x}) \, d^3x$ and $df/dt = \partial_t f + v^j \partial_j f$; this is established on the basis of the continuity equation $\partial_t \rho^* + \partial_j(\rho^* v^j) = 0$, as shown back in Sec. 1.4.3. The terms involving $d\boldsymbol{v}/dt$ are handled by invoking Euler's equation $\rho^*(dv^j/dt) = \rho^* \partial^j U - \partial^j p + O(c^{-2})$, which was shown in Sec. 7.3.2 to be a consequence of energy-momentum conservation.

In the far-away wave zone, where $r \gg \lambda_c$, the gravitational potentials reduce to

$$h^{00} = \frac{4G}{c^2 r}\left[M + \frac{1}{2c^2}\ddot{\mathcal{I}}^{jk} n_j n_k + \frac{1}{6c^3}\dddot{\mathcal{I}}^{jkn} n_j n_k n_n + \cdots\right],$$

$$h^{0j} = \frac{4G}{c^3 r}\left[\frac{1}{2c}\ddot{\mathcal{I}}^{jk} n_k + \frac{1}{12c^3}\left(\dddot{\mathcal{I}}^{jkn} - 2\epsilon^{mjk}\ddot{\mathcal{J}}^{mn}\right) n_k n_n + \cdots\right],$$

$$h^{jk} = \frac{4G}{c^4 r}\left[\frac{1}{2}\ddot{\mathcal{I}}^{jk} + \frac{1}{6c}\left(\dddot{\mathcal{I}}^{jkn} + 2\epsilon^{mnj}\ddot{\mathcal{J}}^{mk} + 2\epsilon^{mnk}\ddot{\mathcal{J}}^{mj}\right) n_n + \cdots\right].$$

The time-dependent piece of $h^{\alpha\beta}$ is dominated by the quadrupole moment of the mass distribution.

7.5 Bibliographical notes

The implementation of post-Minkowskian theory presented in this chapter is based on the DIRE approach (Direct Integration of the Relaxed Einstein equations) of Will and Wiseman (1996) and Pati and Will (2000 and 2001).

The fast-motion implementation of the theory reviewed in Box 7.6 was attempted by Goldberg, Havas, Rosenblum, and coworkers. Representative papers are Havas and Goldberg (1962), Smith and Havas (1965), and Rosenblum (1978).

7.6 Exercises

7.1 Show that in Eq. (7.13), the second term in the retarded expansion of $h^{00}_{\mathcal{N}}$ is given by the surface integral

$$\delta h^{00}_{\mathcal{N}} = \frac{4G}{c^4} \oint_{\partial\mathcal{M}} \tau^{0j} dS_j \,.$$

Using the first term of Eq. (7.48b) to estimate τ^{0j} in the wave zone, and taking the monopole and quadrupole contributions to h^{00} from Box 7.7, show that

$$\delta h^{00}_{\mathcal{N}} \sim \frac{G^2}{c^{10}} \overset{\cdots}{I}{}^{jk} \overset{\cdots}{I}{}^{jk}$$

after discarding terms that depend on the cutoff radius $\mathcal{R}$. Show that this makes a contribution to h^{00} at 4PN order.

7.2 Verify the identities of Eqs. (7.14). Using these, verify that the odd-order terms in Eq. (7.12) take the forms displayed in Eqs. (7.15), modulo surface terms.

7.3 In this problem we prove that at first post-Newtonian order, the integral of Eq. (7.59) defining the total mass M is insensitive to the wave-zone aspects of the integrand. To show this, decompose the integral into a near-zone portion $r < \mathcal{R}$ and a wave zone portion $r > \mathcal{R}$. Show that the $\partial_j(U\partial^j U)$ term in the energy-momentum pseudotensor makes a contribution

$$\Delta M_{\text{near}} = \frac{7G}{2c^2} \frac{M^2}{\mathcal{R}}$$

to the near-zone integral. Next, use the expression of Eq. (7.110) to show that the wave-zone contribution to the mass is given by

$$\Delta M_{\text{wave}} = -\frac{7G}{2c^2} \frac{M^2}{\mathcal{R}} .$$

Conclude that these contributions cancel out, and that the wave-zone portion of the integral makes no essential contribution to the mass.

7.4 As we saw in Sec. 7.3.3, the Poisson equation $\nabla^2 f = \nabla^2 g$ has the solution

$$f = g - \frac{1}{4\pi} \oint_{\partial\mathcal{M}} \left(\frac{\partial'^j g'}{|\boldsymbol{x} - \boldsymbol{x}'|} - g' \partial'_j \frac{1}{|\boldsymbol{x} - \boldsymbol{x}'|} \right) dS'_j \,.$$

Show that the surface term satisfies Laplace's equation for any point $\boldsymbol{x}$ within the near zone.

7.5 Consider the superduperpotential of ρ^*, defined by

$$Y(t, \boldsymbol{x}) := G \int \rho^*(t, \boldsymbol{x}') |\boldsymbol{x} - \boldsymbol{x}'|^3 \, d^3x' \,.$$

(a) Show that $\nabla^2 Y = 12X$, where X is the superpotential.

(b) Following the method of Box 7.3, show that the solution to $\nabla^2 Y = 12X$ can be expressed as

$$Y(t, \boldsymbol{x}) = G \int \rho^*(t, \boldsymbol{y}) K(\boldsymbol{x}; \boldsymbol{y})\, d^3y + Y_0(t, \boldsymbol{x}),$$

in terms of a two-point function K that satisfies $\nabla^2 K = 12|\boldsymbol{x} - \boldsymbol{y}|$; Y_0 is a solution to Laplace's equation.

(c) Calculate the two-point function, and determine Y_0 so that your answer for Y agrees with its starting definition.

7.6 Show that the quadrupole-moment piece of the wave-zone potential h^{00} in Box 7.7 is given explicitly by

$$\frac{2G}{c^2}\left(\frac{1}{c^2 r}\ddot{I}^{jk} + \frac{3}{cr^2}\dot{I}^{\langle jk\rangle} + \frac{3}{r^3} I^{\langle jk\rangle}\right) n_j n_k \,.$$

7.7 For a static, spherically-symmetric source, show that the wave-zone potentials given in Box 7.7 reduce to

$$\begin{aligned} h^{00} &= \frac{4GM}{c^2 r} + 7\left(\frac{GM}{c^2 r}\right)^2 + \cdots, \\ h^{0j} &= 0, \\ h^{jk} &= \left(\frac{GM}{c^2 r}\right)^2 n^j n^k + \cdots. \end{aligned}$$

Verify that this corresponds to the post-Newtonian expansion of the Schwarzschild metric in harmonic coordinates.

7.8 The total mass of a gravitating system is defined by the integral

$$M = \frac{1}{c^2}\int (-g)(T^{00} + t^{00}_{\rm LL})\, d^3x \,.$$

But the mass parameter that appears in the leading-order contribution to h^{00} in the wave zone is

$$M_0 = \frac{1}{c^2}\int_{\mathscr{M}} (-g)(T^{00} + t^{00}_{\rm LL} + t^{00}_{\rm H})\, d^3x \,.$$

Both masses satisfy a conservation law, because $\partial_\beta[(-g)t^{\alpha\beta}_{\rm H}] = 0$ identically. This problem explores whether $(-g)t^{00}_{\rm H}$ makes a contribution to the *value* of the mass.

(a) Defining $\tilde{t}^{\alpha\beta}_{\rm H} := (16\pi G/c^4)(-g)t^{\alpha\beta}_{\rm H} = \partial_\mu h^{\alpha\nu}\partial_\nu h^{\beta\mu} - h^{\mu\nu}\partial_{\mu\nu}h^{\alpha\beta}$, and using the harmonic gauge condition $\partial_\beta h^{\alpha\beta} = 0$, show that

$$\begin{aligned} \tilde{t}^{\alpha\beta}_{\rm H} &= 2\partial_0 h^{\alpha 0}\partial_0 h^{\beta 0} + 2h^{0(\alpha}\partial_0^2 h^{\beta)0} - h^{00}\partial_0^2 h^{\alpha\beta} \\ &\quad - 2\partial_0 h^{00}\partial_0 h^{\alpha\beta} - h^{\alpha\beta}\partial_0^2 h^{00} + \partial_j f^{j\alpha\beta}, \end{aligned}$$

where

$$f^{j\alpha\beta} := 2h^{0(\alpha}\partial_0 h^{\beta)j} + h^{k(\alpha}\partial_k h^{\beta)j} + h^{j(\alpha}\partial_0 h^{\beta)0}$$
$$- 2h^{0j}\partial_0 h^{\alpha\beta} - h^{jk}\partial_k h^{\alpha\beta} - \partial_0 h^{0j} h^{\alpha\beta} .$$

(b) Using this expression, show that the contribution of the harmonic energy-momentum pseudotensor to a near-zone momentum

$$P_0^\alpha := \frac{1}{c}\int_{\mathcal{M}} \tau^{\alpha 0}\, d^3x$$

and a near-zone angular momentum

$$J_0^{\alpha\beta} := \frac{2}{c}\int_{\mathcal{M}} x^{[\alpha}\tau^{\beta]0}\, d^3x$$

comes from integrals over the surface bounding the domain of integration.

(c) Show that $f^{j00} = \partial_k(h^{0j}h^{0k} - h^{00}h^{jk})$.

(d) Using the wave-zone form of the potentials from Box 7.7, and keeping only terms that are independent of the cutoff radius $\mathcal{R}$, show that M and M_0 are related by

$$M = M_0 - \frac{2}{3}\frac{GM_0}{c^5}\dddot{I}^{kk}(\tau) + O(c^{-7}) .$$

Show that the second term is a correction of order $(v_c/c)^5$ relative to the first term.

7.9 This problem explores how to solve the Landau–Lifshitz formulation of the Einstein field equations for the Schwarzschild geometry.

(a) Assuming static spherical symmetry, show that the general form of the gothic inverse metric in Cartesian coordinates can be written in the form

$$\mathfrak{g}^{00} = N(r) ,$$
$$\mathfrak{g}^{0j} = 0 ,$$
$$\mathfrak{g}^{jk} = \alpha(r)P^{jk} + \beta(r)n^j n^k ,$$

where N, α and β are arbitrary functions of r, n^j is a radial unit vector, and $P^{jk} := \delta^{jk} - n^j n^k$.

(b) Show that $\mathfrak{g}_{\alpha\beta}$ is given by $\mathfrak{g}_{00} = N^{-1}$, $\mathfrak{g}_{jk} = \alpha^{-1}P^{jk} + \beta^{-1}n^j n^k$, and that $\mathfrak{g} := \det[\mathfrak{g}^{\alpha\beta}] = N\alpha^2\beta$.

(c) Show that the imposition of the harmonic gauge condition leads to the constraint

$$\beta' = \frac{2}{r}(\alpha - \beta) ,$$

where a prime indicates differentiation with respect to r. Recall that $\partial^j F(r) = F'(r)n^j$, and $\partial^j n^k = r^{-1}P^{jk}$.

(d) Show that the three field equations that arise from the vacuum wave equation $\Box \mathfrak{g}^{\alpha\beta} = (16\pi G/c^4)\tau^{\alpha\beta}$ in harmonic coordinates have the form

$$X' + XY + \frac{1}{r}(2X - Y) = Q\,,$$
$$XY + \frac{1}{r}(2X + Y) = -Q\,,$$
$$Z' + YZ + \frac{2}{r}Z = Q\,,$$

where

$$X := \frac{\alpha'}{\alpha}\,, \quad Y := \frac{\beta'}{\beta}\,, \quad Z := \frac{N'}{N}\,,$$

and

$$Q := \frac{1}{8}\left(3Y^2 - Z^2 + 2YZ + 4XZ - 4XY\right).$$

Hint: One equation comes from the 00 component of the field equations; the other two come from splitting the jk components into a piece proportional to $n^j n^k$ and another piece proportional to P^{jk}. Use the gauge condition to simplify your expressions.

(e) By combining the first two field equations, obtain the solutions

$$X = 0 \quad \text{or} \quad r^4\beta^2 X = c\,,$$

where $c \neq 0$ is a constant.

(f) Choosing the solution $X = 0$, show that the solutions for α and β that satisfy appropriate asymptotic conditions at $r = \infty$ are

$$\alpha = 1\,, \quad \beta = 1 - \frac{a}{r^2}\,,$$

where a is an arbitrary constant. Find the solution for N, determine a, and verify that the result is the Schwarzschild metric in harmonic coordinates.

(g) What is your interpretation of the second class of solutions, represented by a non-zero value of c? Show that by combining the equation $r^4\beta^2 X = c$ with the gauge condition, you can eliminate α and obtain the following differential equation for β:

$$W'' - \frac{W'}{r} = c\frac{W'}{W^2}\,,$$

where $W := r^2\beta$. Spend some time (but not too much!) trying to find a closed form solution to this non-linear equation. (If you find one, please send it to us!)

7.10 Consider the harmonic gauge condition of Eq. (5.175), $\Box_g X^{(\mu)} = 0$, which is a scalar wave equation for the four scalar fields T, X, Y and Z. Using the metric in Schwarzschild coordinates to calculate the operator $\Box_g$, and defining $T := t$, $X := r_\mathrm{h}(r)\sin\theta\cos\phi$, $Y := r_\mathrm{h}(r)\sin\theta\sin\phi$, and $Z := r_\mathrm{h}(r)\cos\theta$, show that the harmonic condition reduces to a single differential equation for $r_\mathrm{h}(r)$, a Legendre equation of degree $\ell = 1$. Show that the solution that satisfies the condition that $r_{\rm h} \to r$ as $r \to \infty$ is given by

$$r_{\rm h} = r - \frac{1}{2}R + b\left[\left(r - \frac{1}{2}R\right)\ln\left(1 - \frac{R}{r}\right) + R\right],$$

where $R = 2GM/c^2$ and b is an arbitrary constant. What do you conclude about the uniqueness of harmonic coordinates? (We encounter this question again in Sec. 11.1.5, in the context of gravitational waves.) Is there a link between this and the second class of solutions in part (g) of the previous problem?

8 Post-Newtonian theory: Fundamentals

Post-Newtonian theory is the theory of weak-field gravity within the near zone, and of the slowly moving systems that generate it and respond to it. It was first encountered in Chapter 7, where it was embedded within the post-Minkowskian approximation; the idea relies on the slow-motion condition introduced in Sec. 6.3.2. But while post-Minkowskian theory deals with both the near and wave zone, here we focus exclusively on the near zone. In this chapter we develop the post-Newtonian theory systematically.

We begin in Sec. 8.1 by collecting the main ingredients obtained in Chapter 7, including the near-zone metric to 1PN order and the matter's energy-momentum tensor $T^{\alpha\beta}$. In Sec. 8.2 we present an alternative derivation of the post-Newtonian metric, based on the Einstein equations in their standard form; this is the "classic approach" to post-Newtonian theory, adopted by Einstein, Infeld, and Hoffmann in the 1930s, and by Fock, Chandrasekhar, and others in the 1960s. Although it produces the same results, we will see that the classic approach presents us with a number of ambiguities that are *not* present in the post-Minkowskian approach. In Sec. 8.3 we explore the coordinate freedom of post-Newtonian theory, and construct the most general transformation that preserves the post-Newtonian expansion of the metric. And in Sec. 8.4 we derive the laws of fluid dynamics in post-Newtonian theory; these will be applied to the motion of an N-body system in Chapter 9.

8.1 Equations of post-Newtonian theory

8.1.1 Post-Newtonian metric

We restrict our attention to a matter distribution that is subjected to a *slow-motion condition* of the sort first considered in Sec. 6.3.2. The distribution is characterized by a length scale r_c and a time scale t_c, and these give us the characteristic velocity $v_c := r_c/t_c$. We assume that this is much smaller than the speed of light,

$$v_c/c \ll 1, \tag{8.1}$$

and this defines what we mean by the slow-motion condition: all speeds within the matter distribution (such as the speed of sound within a body, or the speed of the body as a whole) shall be small compared with the speed of light. If $\lambda_c := ct_c$ is a characteristic wavelength of the gravitational radiation produced by the matter distribution, then Eq. (8.1) states that $r_c \ll \lambda_c$. The region of space occupied by the matter is therefore small compared with the characteristic wavelength; the matter is situated deep within the *near-zone region* of spacetime, defined by $r := |\boldsymbol{x}| \ll \lambda_c$.

Incorporating these assumptions, focusing on field points within the near zone, and carrying out two iterations of the relaxed field equations, we obtained the spacetime metric of a post-Newtonian system back in Sec. 7.3. The metric is displayed in Eq. (7.105), and we reproduce it here:

$$g_{00} = -1 + \frac{2}{c^2}U + \frac{2}{c^4}\left(\Psi - U^2\right) + O(c^{-5}), \tag{8.2a}$$

$$g_{0j} = -\frac{4}{c^3}U_j + O(c^{-5}), \tag{8.2b}$$

$$g_{jk} = \delta_{jk}\left(1 + \frac{2}{c^2}U\right) + O(c^{-4}), \tag{8.2c}$$

where

$$\Psi := \psi + \frac{1}{2}\partial_{tt}X. \tag{8.3}$$

The potentials that appear in the metric are defined by

$$U(t,\boldsymbol{x}) := G\int \frac{\rho^{*\prime}}{|\boldsymbol{x}-\boldsymbol{x}'|}\,d^3x', \tag{8.4a}$$

$$\psi(t,\boldsymbol{x}) := G\int \frac{\rho^{*\prime}\left(\frac{3}{2}v'^2 - U' + \Pi' + 3p'/\rho^{*\prime}\right)}{|\boldsymbol{x}-\boldsymbol{x}'|}\,d^3x', \tag{8.4b}$$

$$X(t,\boldsymbol{x}) := G\int \rho^{*\prime}|\boldsymbol{x}-\boldsymbol{x}'|\,d^3x', \tag{8.4c}$$

$$U^j(t,\boldsymbol{x}) := G\int \frac{\rho^{*\prime}v'^j}{|\boldsymbol{x}-\boldsymbol{x}'|}\,d^3x', \tag{8.4d}$$

in which the primed fluid variables are evaluated at time t and position $\boldsymbol{x}'$; these are determined by the equations of fluid dynamics to be derived in Sec. 8.4. As in the Newtonian theory, the dynamics of the fluid and the dynamics of the gravitational field are intimately coupled to each other. It should be noted that the potentials of Eqs. (8.4) are all *instantaneous potentials*: their profile at time t depends on the state of the system at the same time. The metric, however, does incorporate retardation effects that arise from solving the wave equation for the gravitational potentials $h^{\alpha\beta}$; these are captured by the superpotential term $\partial_{tt}X$ in g_{00}, which appears when h^{00} is expanded in powers of c^{-2} within the near zone.

The post-Newtonian metric makes a good approximation to the true spacetime metric in the near zone only; the approximation is not valid beyond $r = \lambda_c$. The reason for this limitation has already been invoked in Box 6.6. It has to do with the fact that while the behavior of the metric in the near zone is directly tied to the behavior of the matter, so that the metric varies slowly when the matter moves slowly, this is not so in the wave zone, where the radiative behavior of the metric asserts itself. Mathematically, the slow behavior of the metric in the near zone is expressed by the equation

$$\partial_0 g_{\alpha\beta} \sim \frac{v_c}{c}\,\partial_j g_{\alpha\beta}, \tag{8.5}$$

which states that derivatives with respect to $x^0 := ct$ are smaller than spatial derivatives by a factor of order $v_c/c \ll 1$. If we imagine, for example, a matter distribution that consists of N isolated bodies with positions $\boldsymbol{r}_A(t)$, then the metric will depend on time through the N position vectors, and temporal derivatives will be generated by spatial differentiation followed by differentiation of $\boldsymbol{r}_A(t)$ with respect to time; and we see that these operations do indeed bring out the additional factors of v_c/c. The situation is very different in the wave zone, because of the radiative nature of the metric when $r > \lambda_c$. Here the characteristic velocity of the field becomes the speed of light, and it is no longer related to the matter's velocity scale. As a result, $\partial_0 g_{\alpha\beta}$ is of the same order of magnitude as the spatial derivatives, and the slow-motion condition no longer has the same effect on the behavior of the metric.

8.1.2 Energy-momentum tensor

The metric of Eq. (8.2) was constructed from potentials $h_2^{\alpha\beta}$ obtained after two iterations of the relaxed Einstein equations. These potentials can then be involved in a computation of $\tau_2^{\alpha\beta}$, the effective energy-momentum pseudotensor, which can be substituted into the conservation statement $\partial_\beta \tau_2^{\alpha\beta} = 0$ to obtain the system's equations of motion. But since this conservation statement is formally equivalent to the covariant expression of energy-momentum conservation, $\nabla_\beta T^{\alpha\beta} = 0$, an alternative method to obtain the equations of motion is to compute $T^{\alpha\beta}$ to the required degree of accuracy, and to insert it within the covariant equation. This alternative method turns out to be simpler to implement than the original one involving $\tau_2^{\alpha\beta}$.

This program will be implemented in Sec. 8.4. In preparation for this discussion, we now compute $T^{\alpha\beta}$ to the required post-Newtonian order. We recall that

$$T^{\alpha\beta} = (\rho + \epsilon/c^2 + p/c^2)u^\alpha u^\beta + p g^{\alpha\beta}, \tag{8.6}$$

that $\rho = \rho^*(1 - v^2/2c^2 - 3U/c^2) + O(c^{-4})$ and that $u^\alpha = \gamma(c, \boldsymbol{v})$, with $\gamma := u^0/c = 1 + v^2/2c^2 + U/c^2 + O(c^{-4})$. These relations give us

$$c^{-2}T^{00} = \rho^*\left[1 + \frac{1}{c^2}\left(\frac{1}{2}v^2 - U + \Pi\right)\right] + O(c^{-4}), \tag{8.7a}$$

$$c^{-1}T^{0j} = \rho^* v^j\left[1 + \frac{1}{c^2}\left(\frac{1}{2}v^2 - U + \Pi + p/\rho^*\right)\right] + O(c^{-4}), \tag{8.7b}$$

$$T^{jk} = \rho^* v^j v^k\left[1 + \frac{1}{c^2}\left(\frac{1}{2}v^2 - U + \Pi + p/\rho^*\right)\right] + p\left(1 - \frac{2}{c^2}U\right)\delta^{jk} + O(c^{-4}), \tag{8.7c}$$

where $\Pi := \epsilon/\rho^*$. Note that T^{00} is expanded to order c^0, T^{0j} to order c^{-1}, and T^{jk} to order c^{-2}. The metric, on the other hand, is expanded to order c^{-4} for g_{00}, c^{-3} for g_{0j}, and c^{-2} for g_{jk}. As expressed here, the components of the energy-momentum tensor follow a reversed hierarchy of post-Newtonian orders compared to the components of the metric. The reason for this is that each component of the energy-momentum tensor must contain a leading-order piece and a post-Newtonian correction if it is to yield useful information at 1PN order. This is because the components will be inserted within the conservation equations, which take the schematic form $c^{-1}\partial_t T^{\alpha 0} \approx -\partial_k T^{\alpha k}$; while the leading-order pieces will deliver the system's Newtonian dynamics, it is the post-Newtonian corrections that will deliver the post-Newtonian dynamics.

8.1.3 Auxiliary potentials

For future reference we list here a number of post-Newtonian potentials that can also be associated with a perfect fluid; some appear in the potential ψ, while others will be used later in this chapter:

$$\Phi_1 := G \int \frac{\rho^{*\prime} v'^2}{|\boldsymbol{x} - \boldsymbol{x}'|} \, d^3x' , \tag{8.8a}$$

$$\Phi_2 := G \int \frac{\rho^{*\prime} U'}{|\boldsymbol{x} - \boldsymbol{x}'|} \, d^3x' , \tag{8.8b}$$

$$\Phi_3 := G \int \frac{\rho^{*\prime} \Pi'}{|\boldsymbol{x} - \boldsymbol{x}'|} \, d^3x' , \tag{8.8c}$$

$$\Phi_4 := G \int \frac{p'}{|\boldsymbol{x} - \boldsymbol{x}'|} \, d^3x' , \tag{8.8d}$$

$$\Phi_5 := G \int \rho^{*\prime} \partial_{j'} U' \frac{(x - x')^j}{|\boldsymbol{x} - \boldsymbol{x}'|} \, d^3x' , \tag{8.8e}$$

$$\Phi_6 := G \int \rho^{*\prime} v'_j v'_k \frac{(x - x')^j (x - x')^k}{|\boldsymbol{x} - \boldsymbol{x}'|^3} \, d^3x' , \tag{8.8f}$$

$$\Phi^j := G \int \rho^{*\prime} v'_k \frac{(x - x')^j (x - x')^k}{|\boldsymbol{x} - \boldsymbol{x}'|^3} \, d^3x' . \tag{8.8g}$$

Again a primed variable such as $\rho^{*\prime}$ stands for $\rho^*(t, \boldsymbol{x}')$, and $\partial_{j'} U'$ stands for the partial derivative of $U(t, \boldsymbol{x}')$ with respect to x'^j.

Referring to Eq. (8.4b), we see immediately that

$$\psi = \frac{3}{2}\Phi_1 - \Phi_2 + \Phi_3 + 3\Phi_4 . \tag{8.9}$$

In Sec. 8.4.4 we shall have occasion to prove that

$$\partial_{tj} X = \Phi_j - U_j , \tag{8.10a}$$

$$\partial_{tt} X = \Phi_1 + 2\Phi_4 - \Phi_5 - \Phi_6 . \tag{8.10b}$$

Another useful identity is

$$\partial_{jk} X = \delta_{jk} U - G \int \rho^{*\prime} \frac{(x - x')^j (x - x')^k}{|\boldsymbol{x} - \boldsymbol{x}'|^3} \, d^3x' . \tag{8.11}$$

This equation follows directly from the definition of the superpotential, and taking its trace confirms that $\nabla^2 X = 2U$. Combining Eqs. (8.3), (8.9), and (8.10b), we arrive at

$$\Psi = 2\Phi_1 - \Phi_2 + \Phi_3 + 4\Phi_4 - \frac{1}{2}\Phi_5 - \frac{1}{2}\Phi_6, \tag{8.12}$$

a useful decomposition of the post-Newtonian potential Ψ in terms of the auxiliary potentials.

8.1.4 Geodesic equations

To conclude this section we derive the form of the geodesic equation that governs the motion of a test particle in the post-Newtonian spacetime. We examine both the case of a test body that moves slowly ($v/c \ll 1$), and the case of a massless particle (such as a photon) that moves rapidly ($v/c \simeq 1$).

In either case the geodesic equation is

$$\frac{d^2r^\alpha}{d\lambda^2} + \Gamma^\alpha_{\beta\gamma}\frac{dr^\beta}{d\lambda}\frac{dr^\gamma}{d\lambda} = 0, \tag{8.13}$$

in which $r^\alpha(\lambda)$ describes the particle's world line in spacetime; the parameter λ is proper time τ in the case of a massive body, and an arbitrary affine parameter in the case of a photon. For our purposes it is useful to alter the parameterization of the world line and adopt the time coordinate $t := x^0/c$ instead of λ. There is a practical reason for this change: the motion of a planet or spacecraft, or the trajectory of a light ray in space, is generally tracked by an external observer who employs a clock that measures external time t instead of the planet's proper time λ; a description of the motion in terms of t is therefore much more useful to this observer. (We shall return to this theme in Chapter 10, and give a more precise description of the relation between t and the observer's clock time.) A straightforward application of the chain rule reveals that the geodesic equation becomes

$$\frac{dv^\alpha}{dt} = -\left(\Gamma^\alpha_{\beta\gamma} - \frac{v^\alpha}{c}\Gamma^0_{\beta\gamma}\right)v^\beta v^\gamma \tag{8.14}$$

when the world line is parameterized by t; here $v^\alpha := dr^\alpha/dt = (c, \boldsymbol{v})$. The time component of Eq. (8.14) returns $0 = 0$, and the motion of the particle is completely determined by the spatial components.

The Christoffel symbols required for the geodesic equation are obtained from Eq. (8.2), which we insert into Eq. (5.34). We get

$$\Gamma^0_{00} = -\frac{1}{c^3}\partial_t U + O(c^{-5}), \tag{8.15a}$$

$$\Gamma^0_{0j} = -\frac{1}{c^2}\partial_j U + O(c^{-4}), \tag{8.15b}$$

$$\Gamma^0_{jk} = \frac{2}{c^3}\big(\partial_j U_k + \partial_k U_j\big) + \frac{1}{c^3}\delta_{jk}\partial_t U + O(c^{-5}), \tag{8.15c}$$

$$\Gamma^j_{00} = -\frac{1}{c^2}\partial_j U - \frac{1}{c^4}\big(4\partial_t U_j + \partial_j\Psi - 4U\partial_j U\big) + O(c^{-6}), \tag{8.15d}$$

$$\Gamma^j_{0k} = \frac{1}{c^3}\delta_{jk}\partial_t U - \frac{2}{c^3}\big(\partial_k U_j - \partial_j U_k\big) + O(c^{-5}), \tag{8.15e}$$

$$\Gamma^j_{kn} = \frac{1}{c^2}\big(\delta_{jn}\partial_k U + \delta_{jk}\partial_n U - \delta_{kn}\partial_j U\big) + O(c^{-4}), \tag{8.15f}$$

and making the substitution within Eq. (8.14), we obtain

$$\frac{dv^j}{dt} = \partial_j U + \frac{1}{c^2}\Big[(v^2 - 4U)\partial_j U - \big(4v^k\partial_k U + 3\partial_t U\big)v^j - 4v^k\big(\partial_j U_k - \partial_k U_j\big) + 4\partial_t U_j + \partial_j\Psi\Big] + O(c^{-4}) \tag{8.16}$$

when the particle is a massive body that moves slowly, so that $v/c \ll 1$. In the first term we recognize the Newtonian acceleration field $\partial_j U$, and the remaining terms are post-Newtonian corrections. The 1PN terms are suppressed by factors of order $(v/c)^2$, vv_c/c^2, or $(v_c/c)^2$, where v is the magnitude of the particle's velocity, while v_c is the characteristic velocity scale of the matter distribution. For example $\partial_t U$ is of order v_c relative to $\partial_j U$, and this is multiplied by v^j/c^2 in the equations of motion; the contribution is therefore of order vv_c/c^2.

In the case of a photon we cannot take v to be much smaller than c, and the geodesic equation permits an expansion in powers of v_c/c only. The magnitude of $\boldsymbol{v}$ can be determined from the lightlike condition $g_{\alpha\beta}v^\alpha v^\beta = 0$. To leading order in the post-Newtonian expansion we find that $(v/c)^2 = 1 - 4U/c^2$, and this relation neglects terms of order $v_c U/c^3$. This implies that $\boldsymbol{v}$ can be expressed as

$$\boldsymbol{v} = c\left(1 - \frac{2}{c^2}U\right)\boldsymbol{n} + O(c^{-3}), \tag{8.17}$$

in terms of a unit vector $\boldsymbol{n}$. This equation reveals that the coordinate velocity of a photon deviates from c in curved spacetime. If we take v/c to be of order unity but continue to treat v_c/c as a small quantity, Eq. (8.14) produces

$$\frac{dv^j}{dt} = \left(1 + \frac{v^2}{c^2}\right)\partial_j U - \frac{4}{c^2}v^j v^k \partial_k U + O(c^{-3}). \tag{8.18}$$

The geodesic equation becomes

$$\frac{dn^j}{dt} = \frac{2}{c}(\delta^{jk} - n^j n^k)\partial_k U + O(c^{-2}) \tag{8.19}$$

after making the substitution of Eq. (8.17). We note that the right-hand side of Eq. (8.19) is orthogonal to n_j; this is as it should be, because $n_j dn^j/dt = \frac{1}{2}d(n_j n^j)/dt = 0$.

Box 8.1

Maxwell-like formulation of post-Newtonian theory

The main equations of post-Newtonian theory can be written in a form that displays a remarkable parallel with the equations of electrodynamics. These consist of Maxwell's equations,

$$\begin{aligned}
\nabla \cdot \boldsymbol{E} &= \frac{1}{\epsilon_0}\rho_e, \\
\nabla \cdot \boldsymbol{B} &= 0, \\
\nabla \times \boldsymbol{E} &= -\partial_t \boldsymbol{B}, \\
\nabla \times \boldsymbol{B} &= \frac{1}{c^2}\left(\frac{1}{\epsilon_0}\boldsymbol{j}_e + \partial_t \boldsymbol{E}\right),
\end{aligned}$$

which govern the behavior of the electric field $\boldsymbol{E}$ and magnetic field $\boldsymbol{B}$ in terms of the charge density $\rho_e = c^{-1}j_e^0$ and the current density $\boldsymbol{j}_e$, and the Lorentz-force law

$$m\frac{d(\gamma \boldsymbol{v})}{dt} = q\big(\boldsymbol{E} + \boldsymbol{v} \times \boldsymbol{B}\big),$$

with $\gamma = dt/d\tau$, which governs the behavior of a particle of mass m, charge q, and velocity $\boldsymbol{v}$ in the electromagnetic field.

From the post-Newtonian metric (8.2), we first define a gravito-electric potential $\Phi_g := -\frac{1}{2}c^2(1 + g_{00}) = -U - c^{-2}(\Phi - U^2) + O(c^{-4})$ and a gravito-magnetic potential $\boldsymbol{A}_g$ with components $cg_{0j} = -4c^{-2}U_j + O(c^{-4})$. We next define a gravito-electric field $\boldsymbol{E}_g := -\nabla\Phi_g - \partial_t \boldsymbol{A}_g = \nabla U + c^{-2}(\nabla\Psi - \nabla U^2 + 4\partial_t \boldsymbol{U})$ and a gravito-magnetic field $\boldsymbol{B}_g := \nabla \times \boldsymbol{A}_g = -4c^{-2}\nabla \times \boldsymbol{U}$; the relations between potentials and fields are the same as in electrodynamics. It is then a simple matter to show that the field equations of post-Newtonian theory can be put in the Maxwell-like form

$$\tilde{\nabla} \cdot \boldsymbol{E}_g = -4\pi G\rho^*\left[1 + \frac{1}{c^2}\left(\frac{3}{2}v^2 - 3U + \Pi + 3p/\rho^*\right)\right] - \frac{3}{c^2}\partial_{tt}U + O(c^{-4}),$$

$$\nabla \cdot \boldsymbol{B}_g = O(c^{-4}),$$

$$\nabla \times \boldsymbol{E}_g = -\partial_t \boldsymbol{B}_g,$$

$$\nabla \times \boldsymbol{B}_g = \frac{4}{c^2}\left(-4\pi G\rho^*\boldsymbol{v} + \partial_t \boldsymbol{E}_g\right) + O(c^{-4}),$$

where $\tilde{\nabla} \cdot \boldsymbol{E}_g$ denotes a "curved-space" divergence $(-g)^{-1/2}\nabla[(-g)^{1/2}\boldsymbol{E}_g]$, where $-g = 1 + 2U/c^2 + O(c^{-4})$ is the determinant of the post-Newtonian metric. It is also simple to show that the geodesic equation acquires a Lorentz-like form

$$\frac{d(g_s\gamma\boldsymbol{v})}{dt} = \gamma\left(\boldsymbol{E}_g + \boldsymbol{v} \times \boldsymbol{B}_g + v^2\nabla g_s\right) + O(c^{-4}), \qquad (1)$$

where $g_s := 1 + 2U/c^2$ is the coefficient of the spatial part of the PN metric, and $\gamma := dt/d\tau = 1 + c^{-2}(\frac{1}{2}v^2 + U) + O(c^{-4})$. This can be expressed in the more explicit form

$$\frac{d}{dt}\left\{\left[1 + \frac{1}{c^2}\left(\frac{1}{2}v^2 + 3U\right)\right]\boldsymbol{v}\right\} = \left[1 + \frac{1}{c^2}\left(\frac{3}{2}v^2 - U\right)\right]\boldsymbol{E}_g + \boldsymbol{v} \times \boldsymbol{B}_g + O(c^{-4}).$$

Apart from additional post-Newtonian terms, the equations are indeed remarkably similar to those of the Maxwell–Lorentz theory, with ρ^* playing the role of the charge density, $\rho^*\boldsymbol{v}$ that of the current density, and $-4\pi G$ playing the role of the coupling constant $1/\epsilon_0$.

There are, however, clear indications that gravity is different from electrodynamics. Apart from the additional post-Newtonian terms, the most important differences are seen in the sign of the coupling constant and the factor of 4 in the $\nabla \times \boldsymbol{B}_g$ equation. The gravitational coupling constant $-4\pi G$ is negative instead of positive, reflecting the fact that in gravity, like charges attract instead of repel. The factor of 4 reminds us that the gravitational potentials Φ_g and $\boldsymbol{A}_g$ originate from a tensor (the metric) instead of a vector; in quantum parlance we say that the graviton is a spin-2 particle, while the photon is a spin-1 particle. The Lorentz-like equation would be identical to that of electrodynamics through $O(c^{-2})$ were it not for the appearance of the spatial part of the metric, represented by the factor g_s and the extra factor of γ on the right-hand side of Eq. (1), reflecting its true origin in the geodesic equation.

Nevertheless, the Maxwell-like formulation of the post-Newtonian approximation to Einstein's equations and the geodesic equation can be very useful in specific situations, particularly when some of the additional post-Newtonian terms can be neglected. This occurs, for example, when the fields are stationary, when non-linear contributions (proportional to $\rho^* U$) can be ignored, or when the velocity field inside the source is particularly small. In such cases, solutions to the post-Newtonian equations can be imported with minimal modifications from electrodynamics, along with the attached intuition. The Maxwell-like formulation has been used to study everything from laboratory and space experiments to test general relativity to the behavior of matter around rotating black holes.

8.2 Classic approach to post-Newtonian theory

Before we proceed with our exploration of post-Newtonian theory, it is instructive to provide an alternative derivation of the metric based on the standard formulation of the Einstein field equations instead of the Landau–Lifshitz formulation reviewed in Chapter 6. We refer to this derivation as the *classic approach* to post-Newtonian theory, and our quick survey will reveal some of the ambiguities and conceptual difficulties associated with it. The *modern approach* to post-Newtonian theory, based on its post-Minkowskian foundation, is completely free of such ambiguities and conceptual difficulties.

We begin by *postulating* a form of the metric to 1PN order:

$$g_{00} = -1 + \frac{2}{c^2}U + \frac{2}{c^4}(\Psi - U^2) + O(c^{-6}), \tag{8.20a}$$

$$g_{0j} = -\frac{4}{c^3}U_j + O(c^{-5}), \tag{8.20b}$$

$$g_{jk} = \left(1 + \frac{2}{c^2}U\right)\delta_{jk} + O(c^{-4}), \tag{8.20c}$$

where U, U^j, and Ψ are gravitational potentials to be determined. The term of order c^{-2} in g_{00} is a Newtonian term. The terms of order c^{-4} in g_{00}, c^{-3} in g_{0j}, and c^{-2} in g_{jk} are post-Newtonian terms. The insertion of U^2 within g_{00} simplifies the form of the field equations. A blind post-Newtonian expansion of g_{jk} would introduce a general tensorial potential U_{jk} instead of the specific expression $U\delta_{jk}$ that involves the same potential U as in g_{00}. To keep the algebra simple, we anticipate the result of an integration of the Einstein field equations at lowest order, which reveals that indeed U_{jk} must be equal to $U\delta_{jk}$. (In fact, we have reached this conclusion already in Sec. 5.5, when we studied the linearized approximation of general relativity.) We impose a harmonic coordinate condition, as displayed in Eq. (6.47); this reduces to

$$\partial_t U + \partial_j U^j = 0 \tag{8.21}$$

at the required post-Newtonian order. Again we state that at the outset U, U^j, and Ψ are unknown functions to be determined by the field equations; apart from our assumption regarding the tensorial potential, there is no loss of generality in Eq. (8.20).

The standard formulation of the Einstein field equations is $G_{\alpha\beta} = (8\pi G/c^4)T_{\alpha\beta}$, and recalling the definition of the Einstein tensor from Eq. (5.71), this is

$$R_{\alpha\beta} - \frac{1}{2}Rg_{\alpha\beta} = \frac{8\pi G}{c^4}T_{\alpha\beta}, \tag{8.22}$$

in which $R_{\alpha\beta}$ is the Ricci tensor and $R := g^{\alpha\beta}R_{\alpha\beta}$ the Ricci scalar. Taking the trace yields $R = -(8\pi G/c^4)T$, in which $T := g^{\alpha\beta}R_{\alpha\beta}$, and making the substitution back in the field equations produces

$$R_{\alpha\beta} = \frac{8\pi G}{c^4}\bar{T}_{\alpha\beta}, \tag{8.23}$$

in which $\bar{T}_{\alpha\beta} := T_{\alpha\beta} - \frac{1}{2}Tg_{\alpha\beta}$ is the "trace-reversed" energy-momentum tensor. This form of the field equations is our starting point for the determination of the potentials U, U^j, and Ψ. It is advantageous because the computation of the Ricci tensor from the metric of Eq. (8.20) is relatively straightforward. The computation of the Einstein tensor would require additional steps and make the entire task more tedious.

A straightforward calculation using the Christoffel symbols of Eqs. (8.15) reveals that the components of the Ricci tensor are

$$R_{00} = -\frac{1}{c^2}\nabla^2 U + \frac{1}{c^4}\left(\partial_{tt}U + 4U\nabla^2 U - \nabla^2\Psi\right) + O(c^{-6}), \tag{8.24a}$$

$$R_{0j} = \frac{2}{c^3}\nabla^2 U_j + O(c^{-5}), \tag{8.24b}$$

$$R_{jk} = -\frac{1}{c^2}\nabla^2 U\delta_{jk} + O(c^{-4}). \tag{8.24c}$$

We have used Eq. (8.21) to eliminate terms involving $\partial_j U^j$ in favor of terms involving $\partial_t U$.

Importing the components of the energy-momentum tensor from Eq. (8.7), we have that

$$T_{00} = \rho^* c^2\left[1 + \frac{1}{c^2}\left(\frac{1}{2}v^2 - 5U + \Pi\right)\right] + O(c^{-2}), \tag{8.25a}$$

$$T_{0j} = -\rho^* v^j c + O(c^{-1}), \tag{8.25b}$$

$$T_{jk} = \rho^* v^j v^k + p\,\delta^{jk} + O(c^{-2}), \tag{8.25c}$$

to the required post-Newtonian order. To leading order this produces $\bar{T}_{00} = \frac{1}{2}\rho^* c^2 + O(1)$, and the c^{-2} piece of the 00 component of the field equations yields

$$\nabla^2 U = -4\pi G\rho^*. \tag{8.26}$$

We conclude that as expected, U is the standard Newtonian potential defined by Eq. (8.4). Also to leading order, $\bar{T}_{jk} = \frac{1}{2}\delta_{jk}\rho^* c^2 + O(1)$, and we see that the c^{-2} piece of the jk components of the field equations is automatically satisfied; this validates our assumed form for the spatial part of the metric. The c^{-3} piece of the $0j$ components of the field equations yields

$$\nabla^2 U_j = -4\pi G\rho^* v_j, \tag{8.27}$$

and U_j is the standard vector potential as defined by Eq. (8.4). Finally, the c^{-4} piece of the 00 component of the field equations yields

$$\partial_{tt}U + 4U\nabla^2 U - \nabla^2\Psi = 4\pi G\rho^*\left(\frac{3}{2}v^2 - 5U + \Pi + \frac{3p}{\rho^*}\right). \tag{8.28}$$

This is a Poisson equation for Ψ, and after substituting $\nabla^2 U = -4\pi G\rho^*$ on the left-hand side and rearranging, we see that Ψ is given by

$$\Psi = \psi + \frac{1}{2}\partial_{tt}F, \tag{8.29}$$

with ψ and F required to satisfy

$$\nabla^2\psi = -4\pi G\rho^*\left(\frac{3}{2}v^2 - U + \Pi + \frac{3p}{\rho^*}\right), \tag{8.30}$$

and

$$\nabla^2 F = 2U. \tag{8.31}$$

The solution to Eq. (8.30) is evidently the potential defined by Eq. (8.4b).

To identify the solution to Eq. (8.31) requires a more careful discussion, because the source term $2U$ is not limited to the region occupied by the matter distribution; it is distributed over all space. This discussion has already been provided in Box 7.3, where it was shown that the general solution is given by

$$F = X - 2Gm\mathcal{R} + F_0, \tag{8.32}$$

in which $X(t, \boldsymbol{x})$ is the standard superpotential as displayed in Eq. (8.4), $m := \int \rho^*\, d^3x$ is the total rest-mass of the fluid system, $\mathcal{R}$ is a constant length, and $F_0(t, \boldsymbol{x})$ is a solution to Laplace's equation. Demanding that F_0 does not depend on time implies that $\partial_{tt}F = \partial_{tt}X$, so that $\Psi = \psi + \partial_{tt}X$ as required by Eq. (8.3). Choosing $F_0 = 2Gm\mathcal{R}$ returns the stronger equality $F = X$, yielding the same expression for Ψ.

The preceding discussion indicates that when appropriate choices are made, the solution to Eq. (8.31) returns the correct expression for Ψ, as given by Eq. (8.3). The main question that arises is: Which guiding principle can be invoked to justify the choices made to specify this solution? The answer is simply that no such principle exists within the strict context of the classic approach to post-Newtonian theory; the ambiguities associated with F can only be resolved with ad hoc choices. For a more satisfying resolution, one must turn to the modern approach and its post-Minkowskian foundation.

In the classic approach, the superpotential arises as a particular solution to $\nabla^2 F = 2U$, and the choice of solution is ambiguous because U extends over all space. (The potentials U, U^j, and ψ do not share this ambiguity, because their source terms are tied to the matter distribution.) In the modern approach, the superpotential arises in an expansion of h^{00} in powers of c^{-1} in the near zone, and it is fundamentally defined as the integral $\int \rho^{*\prime}|\boldsymbol{x} - \boldsymbol{x}'|\, d^3x'$; the Poisson equation follows as a consequence of this definition. The advantages of the modern approach should be clear. First, the post-Minkowskian foundation provides a clear restriction of the post-Newtonian metric to the near zone, while no such restriction is immediately apparent in the classic approach. Second, while the classic approach features Poisson equations with ambiguous solutions, the modern approach defines all potentials in terms of near-zone integrals that are devoid of ambiguities.

A third advantage is concerned with the incorporation of retardation effects in the post-Newtonian metric. Our experience with post-Minkowskian theory allows us to locate the retardation in the $\partial_{tt}X$ term, but we would be hard pressed to provide this understanding if we were only familiar with the classic approach. Indeed, because the classic approach defines each potential in terms of a Poisson equation, each potential will necessarily be instantaneous, and the retardation effects will be implicit and hidden from view. There is actually a deeper problem that is revealed at higher post-Newtonian orders. A systematic development of the classic approach to higher orders would continue to introduce potentials that satisfy Poisson equations, and the ambiguities would pile up. In particular, it would quickly become unclear how to impose a condition that the metric should describe *outgoing* gravitational waves at infinity. Because the post-Newtonian expansion is necessarily limited to the near-zone region of space, the wave zone is inaccessible, and the boundary conditions cannot be formulated in a clean way. In the modern approach, the post-Minkowskian formulation of the problem is based on wave equations instead of Poisson equations, and the selection of retarded solutions ensures that the waves are properly outgoing in the wave zone. It is this specific choice of solution that provides the retardation and makes the near-zone metric completely unambiguous.

In his sequence of papers on post-Newtonian hydrodynamics written between 1964 and 1969, Chandrasekhar employed the classic approach outlined in this section to derive the metric and the equations of motion. Working with his student Yavuz Nutku, he continued to employ this method to obtain the 2PN equations of hydrodynamics. But when it came time in 1970 to move on to the 2.5PN equations of motion, an order at which the selection of outgoing-wave boundary conditions is essential, Chandra and his student Paul Esposito finally recognized the limitations described here. They converted to the modern approach.

8.3 Coordinate transformations

8.3.1 Introduction

In this section we explore the freedom that post-Newtonian theory possesses to perform coordinate transformations that preserve the post-Newtonian ordering of the metric. We wish to find the most general class of transformations

$$t = t(\bar{t}, \bar{x}^j), \qquad x^j = x^j(\bar{t}, \bar{x}^j), \tag{8.33}$$

that keeps the metric expressed as an expansion in powers of c^{-2}. We call these *post-Newtonian transformations*, and construct them step-by-step in Secs. 8.3.2 and 8.3.3.

We shall find that in general, the post-Newtonian transformations do not preserve the harmonic coordinate condition of Eq. (8.21). In Sec. 8.3.4 we specialize them to a class that keeps the coordinates harmonic; we call this restricted class the *harmonic transformations* of post-Newtonian theory. We describe a simple application of this formalism in Sec. 8.3.5, in which we examine the Newtonian potential of a moving body in its own (non-inertial) reference frame.

The post-Newtonian and harmonic transformations typically produce gravitational potentials that contain spatially-growing terms, even when the original potentials vanish in the formal limit $r \to \infty$. In Sec. 8.3.6 we specialize them further by demanding that the transformed potentials continue to vanish in the limit $\bar{r} \to \infty$; this property defines what is known as the *post-Galilean transformations* of post-Newtonian theory.

Within the post-Newtonian class of transformations there exists an interesting subclass that corresponds very closely to the ordinary gauge transformations of electrodynamics. We examine these in Sec. 8.3.7, and introduce the so-called *standard gauge* of post-Newtonian theory. This was the gauge that was adopted by Chandrasekhar in his pioneering work on the subject, and much of the older post-Newtonian literature is framed in this gauge. The standard gauge, however, has become less popular of late, and the more recent literature is uniformly cast in the harmonic gauge. We adhere to this choice in most of the book, but the standard post-Newtonian gauge is featured in Chapter 13, in which we examine alternative theories of gravity.

In this section we follow closely the treatment of post-Newtonian coordinate transformations developed by our friends Étienne Racine and Éanna Flanagan (2005). We recall that under the transformation of Eq. (8.33), the components of the metric tensor change according to

$$\bar{g}_{00} = \left(\frac{\partial t}{\partial \bar{t}}\right)^2 g_{00} + \frac{2}{c}\frac{\partial t}{\partial \bar{t}}\frac{\partial x^j}{\partial \bar{t}} g_{0j} + \frac{1}{c^2}\frac{\partial x^j}{\partial \bar{t}}\frac{\partial x^k}{\partial \bar{t}} g_{jk}, \tag{8.34a}$$

$$\bar{g}_{0j} = c\frac{\partial t}{\partial \bar{t}}\frac{\partial t}{\partial \bar{x}^j} g_{00} + \left(\frac{\partial t}{\partial \bar{t}}\frac{\partial x^k}{\partial \bar{x}^j} + \frac{\partial x^k}{\partial \bar{t}}\frac{\partial t}{\partial \bar{x}^j}\right) g_{0k} + \frac{1}{c}\frac{\partial x^k}{\partial \bar{t}}\frac{\partial x^n}{\partial \bar{x}^j} g_{kn}, \tag{8.34b}$$

$$\bar{g}_{jk} = c^2\frac{\partial t}{\partial \bar{x}^j}\frac{\partial t}{\partial \bar{x}^k} g_{00} + c\left(\frac{\partial t}{\partial \bar{x}^j}\frac{\partial x^n}{\partial \bar{x}^k} + \frac{\partial x^n}{\partial \bar{x}^j}\frac{\partial t}{\partial \bar{x}^k}\right) g_{0n} + \frac{\partial x^n}{\partial \bar{x}^j}\frac{\partial x^p}{\partial \bar{x}^k} g_{np}. \tag{8.34c}$$

Our most important results are summarized in Box 8.2. The reader is invited to peruse the summary before getting started with the details, so as to benefit from an overview of what is to come.

Box 8.2 Post-Newtonian transformations

The most general coordinate transformation that preserves the post-Newtonian ordering of the metric is given by

$$t = \bar{t} + \frac{1}{c^2}\alpha(\bar{t}, \bar{x}^j) + \frac{1}{c^4}\beta(\bar{t}, \bar{x}^j) + O(c^{-6}),$$

$$x^j = \bar{x}^j + r^j(\bar{t}) + \frac{1}{c^2}h^j(\bar{t}, \bar{x}^j) + O(c^{-4}),$$

where

$$\alpha = A(\bar{t}) + v_j \bar{x}^j,$$
$$h^j = H^j(\bar{t}) + H^j_{\ k}(\bar{t})\bar{x}^k + \frac{1}{2}H^j_{\ kn}(\bar{t})\bar{x}^k\bar{x}^n,$$

with

$$H_{jk} = \epsilon_{jkn}R^n(\bar{t}) + \frac{1}{2}v_j v_k - \delta_{jk}\Big(\dot{A} - \frac{1}{2}v^2\Big),$$
$$H_{jkn} = -\delta_{jk}a_n - \delta_{jn}a_k + \delta_{kn}a_j.$$

The functions A, r^j, H^j, and R^j are freely specifiable functions of time $\bar{t}$, while β is a free function of all the coordinates. The transformation is therefore characterized by ten arbitrary functions of time, and one arbitrary function of all the coordinates. An overdot indicates differentiation with respect to $\bar{t}$, and we have introduced $v^j := \dot{r}^j$ and $a^j := \dot{v}^j = \ddot{r}^j$. In addition, we let $v^2 = \delta_{jk}v^j v^k$.

The transformation preserves the post-Newtonian ordering of the metric, but it does not necessarily keep the coordinates harmonic. To preserve this also we must set

$$\beta = \frac{1}{6}\ddot{A}\delta_{jk}\bar{x}^j\bar{x}^k + \frac{1}{30}\big(\delta_{jk}\dot{a}_n + \delta_{jn}\dot{a}_k + \delta_{kn}\dot{a}_j\big)\bar{x}^j\bar{x}^k\bar{x}^n + \gamma(\bar{t}, \bar{x}^j),$$

and γ is required to satisfy Laplace's equation: $\bar{\nabla}^2\gamma = 0$, with $\bar{\nabla}^2$ denoting the Laplacian operator in the coordinates $\bar{x}^j$. The arbitrary function β has therefore been replaced by an arbitrary harmonic function γ.

Under a harmonic coordinate transformation the potentials become

$$\bar{U}(\bar{t}, \bar{x}^j) = \hat{U} - \dot{A} + \frac{1}{2}v^2 - a_j\bar{x}^j,$$
$$\bar{U}^j(\bar{t}, \bar{x}^j) = \hat{U}^j - v^j\hat{U} + \frac{1}{4}\Big(V^j + V^j_{\ k}\bar{x}^k + \frac{1}{2}V^j_{\ kn}\bar{x}^k\bar{x}^n + \partial_{\bar{j}}\gamma\Big),$$
$$\bar{\Psi}(\bar{t}, \bar{x}^j) = \hat{\Psi} - 4v^j\hat{U}_j + 2v^2\hat{U} + \big(A + v_k\bar{x}^k\big)\partial_{\bar{t}}\hat{U}$$
$$+ \Big(F^j + F^j_{\ k}\bar{x}^k + \frac{1}{2}F^j_{\ kn}\bar{x}^k\bar{x}^n\Big)\partial_{\bar{j}}\hat{U}$$
$$+ G + G_j\bar{x}^j + \frac{1}{2}G_{jk}\bar{x}^j\bar{x}^k + \frac{1}{6}G_{jkn}\bar{x}^j\bar{x}^k\bar{x}^n - \partial_{\bar{t}}\gamma,$$

where

$$V^j = (2\dot{A} - v^2)v^j - \dot{H}^j + \epsilon^j_{\ pq}v^p R^q,$$
$$V^j_{\ k} = \frac{3}{2}v^j a_k + \frac{1}{2}a^j v_k + \delta^j_{\ k}\Big(\frac{4}{3}\ddot{A} - 2v^n a_n\Big) - \epsilon^j_{\ kp}\dot{R}^p,$$
$$V^j_{\ kn} = \frac{6}{5}\big(\delta^j_{\ k}\dot{a}_n + \delta^j_{\ n}\dot{a}_k\big) - \frac{4}{5}\delta_{kn}\dot{a}^j,$$

(continued overleaf)

$$
\begin{aligned}
F^j &= H^j - Av^j, \\
F^j_k &= -\delta^j_k\Big(\dot{A} - \frac{1}{2}v^2\Big) - \frac{1}{2}v^j v_k + \epsilon^j_{\ kp} R^p, \\
F^j_{kn} &= -\big(\delta^j_k a_n + \delta^j_n a_k\big) + \delta_{kn} a^j, \\
G &= \frac{1}{2}\dot{A}^2 - \dot{A}v^2 + \frac{1}{4}v^4 + \dot{H}^j v_j, \\
G_j &= \Big(\dot{A} - \frac{1}{2}v^2\Big)a_j - \Big(\ddot{A} - \frac{3}{2}v^k a_k\Big)v_j - \epsilon_{jpq} v^p \dot{R}^q, \\
G_{jk} &= a_j a_k - v_j \dot{a}_k - \dot{a}_j v_k + \delta_{jk}(v_n \dot{a}^n) - \frac{1}{3}\delta_{jk}\dddot{A}, \\
G_{jkn} &= -\frac{1}{5}\big(\delta_{jk}\ddot{a}_n + \delta_{jn}\ddot{a}_k + \delta_{kn}\ddot{a}_j\big).
\end{aligned}
$$

The "hatted" potentials are equal to the original potentials evaluated at time $t = \bar{t}$ and position $x^j = \bar{x}^j + r^j(\bar{t})$. For example,

$$\hat{U}(\bar{t}, \bar{x}^j) := U\big(t = \bar{t}, x^j = \bar{x}^j + r^j(\bar{t})\big).$$

Because U now possesses, in addition to its original explicit time dependence, an implicit time dependence contained in $r^j(\bar{t})$, some care must be exercised when taking time derivatives. We have

$$\frac{\partial \hat{U}}{\partial \bar{t}} = \frac{\partial U}{\partial t} + v^j \frac{\partial U}{\partial x^j}, \qquad \frac{\partial \hat{U}}{\partial \bar{x}^j} = \frac{\partial U}{\partial x^j},$$

in which the right-hand sides are evaluated at $t = \bar{t}$ and $x^j = \bar{x}^j + r^j(\bar{t})$.

8.3.2 Newtonian transformations

We begin with a search for a coordinate transformation that preserves the form of the metric at Newtonian order:

$$g_{00} = -1 + \frac{2}{c^2}U + O(c^{-4}), \qquad g_{0j} = O(c^{-3}), \qquad g_{jk} = \delta_{jk} + O(c^{-2}). \tag{8.35}$$

Specifically, we demand that the new metric takes the form

$$\bar{g}_{00} = -1 + \frac{2}{c^2}\bar{U} + O(c^{-4}), \qquad \bar{g}_{0j} = O(c^{-3}), \qquad \bar{g}_{jk} = \delta_{jk} + O(c^{-2}), \tag{8.36}$$

with a new Newtonian potential $\bar{U}$ whose relation with the old one will be determined by the transformation.

Inspecting Eq. (8.34a) first, we note that the leading term in $\bar{g}_{00}$ will have the correct value of -1 if and only if $\partial t/\partial \bar{t} = 1 + O(c^{-2})$. Moving next to Eq. (8.34c), we see that $\bar{g}_{jk}$ will contain unwanted terms of order c^2 unless $\partial t/\partial \bar{x}^j = O(c^{-2})$. And the terms of order

c^0 will have the correct form if and only if

$$\delta_{jk} = \frac{\partial x^p}{\partial \bar{x}^j}\frac{\partial x^q}{\partial \bar{x}^k}\delta_{pq} + O(c^{-2}). \tag{8.37}$$

The Newtonian transformation must preserve the form of the spatial metric, and this means that it must be the combination of a translation of the spatial origin with a rotation of the coordinate axes: $x^j = r^j + R^j_{\ k}\bar{x}^k + O(c^{-2})$. Here $r^j(\bar{t})$ are arbitrary functions of time, and $R^j_{\ k}(\bar{t})$ are the components of a rotation matrix that satisfies $\delta_{pq}R^p_{\ j}R^q_{\ k} = \delta_{jk}$. At this stage of our considerations we conclude that the transformation must take the form

$$t = \bar{t} + \frac{1}{c^2}\alpha(\bar{t}, \bar{x}^j) + O(c^{-4}), \qquad x^j = r^j(\bar{t}) + R^j_{\ k}(\bar{t})\bar{x}^k + O(c^{-2}), \tag{8.38}$$

where α is (for now) an arbitrary function of the new coordinates, and r^j, $R^j_{\ k}$ are arbitrary functions of time.

We next examine Eq. (8.34b), in which we make the substitutions of Eq. (8.38). After simplification we notice that $\bar{g}_{0j}$ will contain unwanted terms at order c^{-1} unless

$$\partial_{\bar{j}}\alpha = \left(v^k + \dot{R}^k_{\ n}\bar{x}^n\right)R_{kj}; \tag{8.39}$$

an overdot indicates differentiation with respect to $\bar{t}$, and we introduced the notation $v^j := \dot{r}^j$. The previously stated condition on the rotation matrix implies that $\dot{R}^k_n R_{kj}$ is antisymmetric on n and j, and can therefore be expressed as $\dot{R}^k_n R_{kj} = \epsilon_{njm}\omega^m(t)$ for some vector ω^m. The equation for α is now

$$\partial_{\bar{j}}\alpha = v^k R_{kj} + \epsilon_{jmn}\omega^m\bar{x}^n, \tag{8.40}$$

and we see that the final term cannot be written as the gradient of any function. To eliminate the unwanted c^{-1} term in $\bar{g}_{0j}$ we must therefore set $\omega^m = 0$, so that $R^j_{\ k}$ describes a time-independent rotation of the coordinate axes. We choose to discard this uninteresting coordinate freedom by setting

$$R_{jk} = \delta_{jk}. \tag{8.41}$$

The general solution for α is then

$$\alpha = A(\bar{t}) + v_j(\bar{t})\bar{x}^j, \tag{8.42}$$

in which A is an arbitrary function of time $\bar{t}$. The Newtonian transformation of Eq. (8.38) is now fully characterized.

Box 8.3 **Rotating coordinates**

By forcing $\bar{g}_{0j}$ to vanish at order c^{-1}, we are consciously excluding rotating coordinate systems from the allowed class of post-Newtonian coordinates. To describe a purely rotating coordinate system, we would use the transformation

$$t = \bar{t} + O(c^{-4}), \qquad x^j = R^j_{\ k}(\bar{t})\bar{x}^k + O(c^{-2}),$$

which leads to a metric of the form

$$\bar{g}_{00} = -1 + \frac{2}{c^2}\bar{U} + \frac{1}{c^2}\left[\omega^2 - (\boldsymbol{\omega}\cdot\bar{\boldsymbol{n}})^2\right]\bar{r}^2 + O(c^{-4}),$$

$$\bar{g}_{0j} = \frac{1}{c}(\boldsymbol{\omega}\times\bar{\boldsymbol{x}})_j + O(c^{-3}),$$

$$\bar{g}_{jk} = \delta_{jk} + O(c^{-2}),$$

where $\omega_j(t) := \frac{1}{2}\epsilon_{jlm}\dot{R}^k_{\ l}R_{km}$ and $\bar{\boldsymbol{n}} = \bar{\boldsymbol{x}}/\bar{r}$. From a relativistic point of view, this metric presents a number of problems. One of these is that it does not reduce to the Minkowski metric when $\bar{r}\to\infty$. An even worse problem is that g_{00} vanishes when $\omega\bar{r}\sin\theta = c(1-U/c^2)$, where θ is the angle between $\boldsymbol{\omega}$ and $\bar{\boldsymbol{n}}$. This "light cylinder" is a place where the speed of a particle at rest in the rotating frame equals the local speed of light as measured in the global, non-rotating frame; particles at rest outside the light cylinder exceed the local speed of light.

Other issues connected with the rotating frame include the inability to synchronize clocks consistently around a circle at rest in this frame (the Sagnac effect, reviewed in Sec. 10.3.4), and the common misconception that the circumference of a rotating disk is shortened compared to 2π times its radius. These issues have generated so much misunderstanding that in the general relativity textbook by H.P. Robertson and T.W. Noonan (1968), there is a paragraph on this topic headed "That darned* rotating disk," with the asterisk indicating that the actual word selected by Robertson in his original lecture notes was much stronger!

As indicated in the text, we wish to preserve the post-Newtonian expansion of the metric, and therefore exclude rotating coordinate systems from our considerations. This doesn't mean, however, that rotating coordinates are never appropriate. They can indeed be very useful, provided that one stays well within the light cylinder. For example, a coordinate system that rotates with the Earth is an extremely powerful tool to describe post-Newtonian gravity in and around the Earth, including its effects on geocentric satellites, atomic timekeeping, and the Global Positioning System. These applications are discussed in detail in Chapter 10.

To determine how the Newtonian potential changes under the transformation, we return to Eq. (8.34a) and examine the terms of order c^{-2} after making the substitutions of Eq. (8.38). After simplification we find that the right-hand side is given by $-1 + 2c^{-2}(U - \partial_{\bar{t}}\alpha + \frac{1}{2}v^2) + O(c^{-4})$, where $v^2 := \delta_{jk}v^jv^k$. The new potential must therefore be $\bar{U} = U - \partial_{\bar{t}}\alpha + \frac{1}{2}v^2$. Here $\bar{U}$ is expressed in terms of the new coordinates $(\bar{t},\bar{x}^j)$, but U is still written in terms of the old coordinates (t, x^j). To make the equation more useful we should express U as a function of the new coordinates. To achieve this we write

$$U(t, x^j) = U(\bar{t} + c^{-2}\alpha + \cdots, \bar{x}^j + r^j + \cdots) \tag{8.43}$$

and perform a Taylor expansion of the right-hand side about the point $(\bar{t}, \bar{x}^j + r^j)$. This gives $U(t, x^j) = U(\bar{t}, \bar{x}^j + r^j) + O(c^{-2})$, and we find that the terms of order c^{-2} play no role in the transformation of the Newtonian potential. (They do, however, appear in the post-Newtonian transformation of the following subsection.)

To distinguish clearly between the sets of arguments $(\bar{t}, \bar{x}^j)$, $(\bar{t}, \bar{x}^j + r^j)$, and (t, x^j) we introduce the "hatted" potential

$$\hat{U}(\bar{t}, \bar{x}^j) := U(\bar{t}, \bar{x}^j + r^j). \tag{8.44}$$

This is the original potential U evaluated at time $t = \bar{t}$ and position $x^j = \bar{x}^j + r^j(\bar{t})$. In terms of this we have

$$U(t, x^j) = \hat{U}(\bar{t}, \bar{x}^j) + O(c^{-2}) \tag{8.45}$$

and the transformed potential is $\bar{U} = \hat{U} - \partial_{\bar{t}}\alpha + \frac{1}{2}v^2$. Using Eq. (8.42), this is

$$\bar{U} = \hat{U} - \dot{A} + \frac{1}{2}v^2 - a_j\bar{x}^j, \tag{8.46}$$

where $a^j := \dot{v}^j = \ddot{r}^j$. All members of this equation are functions of the new coordinates $(\bar{t}, \bar{x}^j)$.

When it is expressed in terms of U as in Eq. (8.44), the hatted potential $\hat{U}$ possesses both an explicit and an implicit dependence upon the time coordinate $\bar{t}$. The explicit dependence is contained in U's temporal argument, while the implicit dependence appears via $r^j(\bar{t})$ in the spatial arguments. Some care must therefore be exercised when computing partial derivatives. We have, for example,

$$\frac{\partial \hat{U}}{\partial \bar{t}} = \left(\frac{\partial U}{\partial t} + v^j \frac{\partial U}{\partial x^j} \right)_{t=\bar{t},\, x=\bar{x}+r}, \tag{8.47}$$

in which the substitutions $t = \bar{t}$, $x^j = \bar{x}^j + r^j(\bar{t})$ are made after differentiating U with respect to its original variables t and x^j. Spatial derivatives, on the other hand, are given simply by

$$\frac{\partial \hat{U}}{\partial \bar{x}^j} = \left(\frac{\partial U}{\partial x^j} \right)_{t=\bar{t},\, x=\bar{x}+r}. \tag{8.48}$$

8.3.3 Post-Newtonian transformations

To proceed to the next order we write the coordinate transformation as

$$t = \bar{t} + \frac{1}{c^2}\alpha(\bar{t}, \bar{x}^j) + \frac{1}{c^4}\beta(\bar{t}, \bar{x}^j) + O(c^{-6}), \tag{8.49a}$$

$$x^j = \bar{x}^j + r^j(\bar{t}) + \frac{1}{c^2}h^j(\bar{t}, \bar{x}^j) + O(c^{-4}), \tag{8.49b}$$

where α is given by Eq. (8.42), while β and h^j represent the additional coordinate freedom that appears at 1PN order.

The function β will remain arbitrary. To constrain h^j we examine the $O(c^{-2})$ terms on the right-hand side of Eq. (8.34c) and demand that, in accordance with Eq. (8.20), they be equal to $2\bar{U}\delta_{jk}$. With Eqs. (8.45) and (8.46), we find after simplification that h^j must be a solution to the differential equation

$$\partial_{\bar{j}} h_k + \partial_{\bar{k}} h_j = -2\delta_{jk}\left(\dot{A} - \frac{1}{2}v^2 + a_n \bar{x}^n\right) + v_j v_k. \tag{8.50}$$

The general solution to this equation is the sum of a particular solution and the general solution to the homogeneous equation $\partial_{\bar{j}} h_k + \partial_{\bar{k}} h_j = 0$. (The expert will recognize this as Killing's equation in a flat, three-dimensional space.)

The form of the homogeneous equation reveals that h^j_{hom} must be linear in the coordinates: $h^j_{\text{hom}} = H^j + A^j{}_k \bar{x}^k$, where H^j and $A^j{}_k$ are functions of $\bar{t}$. Substitution within the differential equation reveals that $H^j(\bar{t})$ is arbitrary, but that A_{jk} must be an antisymmetric tensor. Such a tensor contains three independent components, and we can always express it as $A_{jk} = \epsilon_{jkn} R^n$, in terms of a vector R^n that also contains three independent components. We have obtained $h^j_{\text{hom}} = H^j(\bar{t}) + \epsilon^j{}_{kn} \bar{x}^k R^n(\bar{t})$. The first term represents a translational component to the coordinate transformation, which combines with the Newtonian translation to form the total translation $r^j + c^{-2} H^j$. The second term represents a rotation of the coordinate axes, and the rotation matrix is $R^j{}_k = c^{-2} \epsilon^j{}_{kn} R^n$.

The form of the inhomogeneous equation reveals that a particular solution h^j_{part} will be quadratic in the coordinates. We adopt $B_{jk}\bar{x}^k + \frac{1}{2} B_{jkn} \bar{x}^k \bar{x}^n$ as a trial solution, and observe that B_{jkn} can be chosen to be symmetric in the last pair of indices. Substitution within the differential equation shows that B_{jk} is constrained by $B_{jk} + B_{kj} = -2\delta_{jk}(\dot{A} - \frac{1}{2}v^2) + v_j v_k$, while B_{jkn} is constrained by $B_{jkn} + B_{kjn} = -2\delta_{jk} a_n$. The solutions are readily identified as $B_{jk} = -\delta_{jk}(\dot{A} - \frac{1}{2}v^2) + \frac{1}{2} v_j v_k$ and $B_{jkn} = -\delta_{jk} a_n - \delta_{jn} a_k + \delta_{kn} a_j$.

Collecting results, we write our final expression for $h^j = h^j_{\text{hom}} + h^j_{\text{part}}$ as

$$h^j = H^j(\bar{t}) + H^j{}_k(\bar{t})\bar{x}^k + \frac{1}{2} H^j{}_{kn}(\bar{t})\bar{x}^k \bar{x}^n, \tag{8.51}$$

with

$$H_{jk} = \epsilon_{jkn} R^n(\bar{t}) + \frac{1}{2} v_j v_k - \delta_{jk}\left(\dot{A} - \frac{1}{2}v^2\right), \tag{8.52a}$$

$$H_{jkn} = -\delta_{jk} a_n - \delta_{jn} a_k + \delta_{kn} a_j. \tag{8.52b}$$

This piece of the coordinate transformation involves the six arbitrary functions of time that are contained in $H^j(\bar{t})$ and $R^j(\bar{t})$.

To determine how the vector potential U^j transforms under the post-Newtonian transformation of Eqs. (8.49), we make the substitutions in Eq. (8.34) and demand that g_{0j} keeps its post-Newtonian form of $-4c^{-3} U_j + O(c^{-5})$. A careful evaluation of Eq. (8.34b) reveals that the new vector potential $\bar{U}_j$ is given by

$$4\bar{U}_j = 4(\hat{U}_j - v_j \hat{U}) + \partial_{\bar{j}} \beta + v_j \partial_{\bar{t}} \alpha - v^k \partial_{\bar{j}} h_k - \partial_{\bar{t}} h_j, \tag{8.53}$$

in which $\hat{U}_j$ is defined by analogy with Eq. (8.44). Taking into account Eqs. (8.42) and (8.51), we finally arrive at

$$\bar{U}_j = \hat{U}_j - v_j \hat{U} + \frac{1}{4}\left(V_j + V_{jk}\bar{x}^k + \frac{1}{2} V_{jkn} \bar{x}^k \bar{x}^n + \partial_{\bar{j}} \beta\right), \tag{8.54}$$

with

$$V_j = (2\dot{A} - v^2)v_j - \dot{H}_j + \epsilon_{jpq}v^p R^q, \tag{8.55a}$$

$$V_{jk} = \frac{3}{2}v_j a_k + \frac{1}{2}v_k a_j + \delta_{jk}\big(\ddot{A} - 2v^n a_n\big) - \epsilon_{jkp}\dot{R}^p, \tag{8.55b}$$

$$V_{jkn} = \delta_{jk}\dot{a}_n + \delta_{jn}\dot{a}_k - \delta_{kn}\dot{a}_j. \tag{8.55c}$$

We are now ready to derive the transformation equation for the post-Newtonian potential Ψ. We proceed along the same lines as for the vector potential, but before we begin we must recover the $O(c^{-2})$ terms that were discarded back in Eq. (8.45); these were not needed previously, but they appear in the transformed version of g_{00} at order c^{-4}. If we write $U(t, x^j)$ as $U(\bar{t} + c^{-2}\alpha + \cdots, \bar{x}^j + r^j + c^{-2}h^j + \cdots)$ and expand to first order in c^{-2}, we obtain

$$U(t, x^j) = \hat{U} + \frac{1}{c^2}\alpha\Big(\partial_{\bar{t}}\hat{U} - v^j\partial_{\bar{j}}\hat{U}\Big) + \frac{1}{c^2}h^j\partial_{\bar{j}}\hat{U} + O(c^{-4}); \tag{8.56}$$

the right-hand side is expressed as a function of $(\bar{t}, \bar{x}^j)$, and we have used Eqs. (8.47) and (8.48) to relate the partial derivatives of U to those of $\hat{U}$. After substitution of Eq. (8.49) into Eq. (8.34a) we find that the new post-Newtonian potential must be given by

$$\begin{aligned}\bar{\Psi} = {} & \hat{\Psi} + 2v^2\hat{U} + \alpha\partial_{\bar{t}}\hat{U} + (h^j - \alpha v^j)\partial_{\bar{j}}\hat{U} - 4v_j\hat{U}^j \\ & + \frac{1}{2}(\partial_{\bar{t}}\alpha)^2 - v^2\partial_{\bar{t}}\alpha + \frac{1}{4}v^4 - \partial_{\bar{t}}\beta + v_j\partial_{\bar{t}}h^j.\end{aligned} \tag{8.57}$$

After taking into account Eqs. (8.42) and (8.51), we finally arrive at

$$\begin{aligned}\bar{\Psi} = {} & \hat{\Psi} - 4v_j\hat{U}^j + 2v^2\hat{U} + \big(A + v_k\bar{x}^k\big)\partial_{\bar{t}}\hat{U} + \Big(F^j + F^j_{\ k}\bar{x}^k + \frac{1}{2}F^j_{\ kn}\bar{x}^k\bar{x}^n\Big)\partial_{\bar{j}}\hat{U} \\ & + G + G_j\bar{x}^j + \frac{1}{2}G_{jk}\bar{x}^j\bar{x}^k - \partial_{\bar{t}}\beta,\end{aligned} \tag{8.58}$$

with

$$F^j = H^j - Av^j, \tag{8.59a}$$

$$F^j_{\ k} = -\delta^j_{\ k}\Big(\dot{A} - \frac{1}{2}v^2\Big) - \frac{1}{2}v^j v_k + \epsilon^j_{\ kp}R^p, \tag{8.59b}$$

$$F^j_{\ kn} = -\big(\delta^j_{\ k}a_n + \delta^j_{\ n}a_k\big) + \delta_{kn}a^j, \tag{8.59c}$$

$$G = \frac{1}{2}\dot{A}^2 - \dot{A}v^2 + \frac{1}{4}v^4 + \dot{H}^j v_j, \tag{8.59d}$$

$$G_j = \Big(\dot{A} - \frac{1}{2}v^2\Big)a_j - \Big(\ddot{A} - \frac{3}{2}v^k a_k\Big)v_j - \epsilon_{jpq}v^p\dot{R}^q, \tag{8.59e}$$

$$G_{jk} = a_j a_k - v_j\dot{a}_k - \dot{a}_j v_k + \delta_{jk}(v_n\dot{a}^n). \tag{8.59f}$$

The hatted potential $\hat{\Psi}$ is defined by analogy with Eq. (8.44).

Our task of constructing the most general coordinate transformation that preserves the post-Newtonian form of the metric is now complete. At Newtonian order the transformation is characterized by an arbitrary translation $r^j(\bar{t})$ and a shift $\alpha = A(\bar{t}) + v_j(\bar{t})\bar{x}^j$ of the time coordinate at order c^{-2}. At post-Newtonian order the transformation involves an additional component $H^j(\bar{t})$ to the translation, as well as a rotation governed by the vector $R^j(\bar{t})$. In addition, the transformation involves an arbitrary shift $\beta(\bar{t}, \bar{x}^j)$ of the time coordinate at order c^{-4}. All in all we have ten arbitrary functions of time, and one free function β of all the coordinates. The transformed potentials $\bar{U}$, $\bar{U}^j$, and $\bar{\Psi}$ are obtained from the old ones by employing Eqs. (8.46), (8.54), and (8.58), respectively.

8.3.4 Harmonic transformations

The general transformation of the preceding subsection does not, in general, preserve the harmonic condition of Eq. (8.21). It is possible, however, to specialize $\beta(\bar{t}, \bar{x}^j)$ so that we also have

$$\partial_{\bar{t}}\bar{U} + \partial_{\bar{j}}\bar{U}^j = 0 \tag{8.60}$$

in the new coordinates. This restriction of the coordinate freedom defines what we shall call the class of *harmonic coordinate transformations*.

In view of Eq. (8.47) we find that the harmonic condition is

$$\partial_{\bar{t}}\hat{U} - v^j\partial_{\bar{j}}\hat{U} + \partial_{\bar{j}}\hat{U}^j = 0 \tag{8.61}$$

when it is expressed in terms of the hatted potentials. If we substitute Eqs. (8.46) and (8.54) into Eq. (8.60) and make use of Eq. (8.61), we find that the harmonic condition is preserved when β satisfies the Poisson equation

$$\bar{\nabla}^2\beta = \ddot{A} + \dot{a}_j\bar{x}^j. \tag{8.62}$$

Here $\bar{\nabla}^2$ is the Laplacian operator in the coordinates $\bar{x}^j$. The general solution to this equation is the sum of a particular solution and the general solution to Laplace's equation. The particular solution must be cubic in the coordinates, and we construct it with the help of the ansatz $\frac{1}{2}C_{jk}\bar{x}^j\bar{x}^k + \frac{1}{6}C_{jkn}\bar{x}^j\bar{x}^k\bar{x}^n$, in which C_{jk} and C_{jkm} depend on $\bar{t}$ and are completely symmetric tensors. This property and the differential equation imply that $C_{jk} = \frac{1}{3}\delta_{jk}\ddot{A}$ and $C_{jkn} = \frac{1}{5}(\delta_{jk}\dot{a}_n + \delta_{jn}\dot{a}_k + \delta_{kn}\dot{a}_j)$. We have obtained

$$\beta = \frac{1}{6}\ddot{A}\delta_{jk}\bar{x}^j\bar{x}^k + \frac{1}{30}(\delta_{jk}\dot{a}_n + \delta_{jn}\dot{a}_k + \delta_{kn}\dot{a}_j)\bar{x}^j\bar{x}^k\bar{x}^n + \gamma(\bar{t}, \bar{x}^j), \tag{8.63}$$

where γ is any harmonic function that satisfies $\bar{\nabla}^2\gamma = 0$.

Making the substitution in Eqs. (8.46), (8.54), and (8.58), we obtain the results listed in Box 8.2. [Note that the expressions for V_{jk}, V_{jkn}, G_{jk}, and G_{jkn} that appear in the Box are different from those given by Eqs. (8.55) and (8.59). The differences are accounted for by the terms generated by $\partial_{\bar{j}}\beta$ and $\partial_{\bar{t}}\beta$.] The transformation is still characterized by the ten arbitrary functions of time that are contained in $A(\bar{t})$, $r^j(\bar{t})$, $H^j(\bar{t})$, and $R^j(\bar{t})$, but the remaining freedom is now restricted to a harmonic function $\gamma(\bar{t}, \bar{x}^j)$.

8.3.5 Comoving frame of a moving body

The general post-Newtonian transformations, and the restricted class of harmonic transformations, contain an enormous amount of freedom, and the transformations introduce spatially-growing terms in the potentials. For example, the transformation of the Newtonian potential is

$$\bar{U} = \hat{U} - \dot{A} + \frac{1}{2}v^2 - a_j \bar{x}^j, \tag{8.64}$$

and the last term grows linearly with $\bar{r}$. Similarly, $\bar{U}^j$ contains terms that grow like $\bar{r}^2$, and in the harmonic case, $\bar{\Psi}$ grows like $\bar{r}^3$. In view of this situation, a natural question to ponder is: What purpose is there in all this coordinate freedom?

We shall have occasion to give a more complete answer to this question in Sec. 9.4, but here we consider a simple application of the formalism that should illustrate its usefulness. We consider a spherical body of mass m whose center-of-mass is situated at $x^j = r^j(t)$ in an inertial frame of reference. The body creates a gravitational potential U_{body}, and it is surrounded by an external matter distribution that creates a potential U_{ext}. The total potential is $U = U_{\text{body}} + U_{\text{ext}}$, and we wish to examine its form in the non-inertial frame attached to the moving body. The coordinate transformation is given by $t = \bar{t} + c^{-2}\alpha + O(c^{-4})$ and $x^j = \bar{x}^j + r^j(\bar{t}) + O(c^{-2})$, with $\alpha = A(\bar{t}) + v_j \bar{x}^j$.

In the original (inertial) coordinates we have that the potential outside the body is given by

$$U_{\text{body}} = \frac{Gm}{|\boldsymbol{x} - \boldsymbol{r}(t)|}. \tag{8.65}$$

To simplify its expression we expand the external potential in a Taylor series about $x^j = r^j(t)$:

$$U_{\text{ext}} = U(t, r^j) + (x-r)^j \partial_j U_{\text{ext}}(t, r^j) + \frac{1}{2}(x-r)^j (x-r)^k \partial_{jk} U_{\text{ext}}(t, r^j) + \cdots. \tag{8.66}$$

The hatted potentials are

$$\hat{U}_{\text{body}} = \frac{Gm}{\bar{r}} \tag{8.67}$$

and

$$\hat{U}_{\text{ext}} = U_{\text{ext}}(\bar{t}, r^j) + \bar{x}^j \partial_j U_{\text{ext}}(\bar{t}, r^j) + \frac{1}{2}\bar{x}^j \bar{x}^k \partial_{jk} U_{\text{ext}}(\bar{t}, r^j) + \cdots. \tag{8.68}$$

We see that the total potential $\hat{U} = \hat{U}_{\text{body}} + \hat{U}_{\text{ext}}$ naturally contains growing terms that are associated with the external matter, in addition to the decaying term that is associated with the reference body. Note that after each differentiation, the external potential is evaluated at $t = \bar{t}$ and $x^j = r^j(\bar{t})$.

The transformed potential in the comoving frame of the body is $\bar{U} = \bar{U}_{\text{body}} + \bar{U}_{\text{ext}}$, with

$$\bar{U}_{\text{body}} = \frac{Gm}{\bar{r}} \tag{8.69}$$

and

$$\bar{U}_{\rm ext} = \left[U_{\rm ext}(\bar{t}, r^j) - \dot{A} + \frac{1}{2}v^2\right] + \bar{x}^j\left[a_j - \partial_j U_{\rm ext}(\bar{t}, r^j)\right] + \frac{1}{2}\bar{x}^j\bar{x}^k\partial_{jk}U_{\rm ext}(\bar{t}, r^j) + \cdots. \tag{8.70}$$

We simplify this by first exploiting the coordinate freedom, which allows us to set

$$\dot{A} = \frac{1}{2}v^2 + U_{\rm ext}(\bar{t}, r^j); \tag{8.71}$$

this is a differential equation that determines $A(\bar{t})$ up to an uninteresting constant of integration. We also make use of the fact that our body moves according to the Newtonian equations of motion, so that

$$a_j = \partial_j U_{\rm ext}(\bar{t}, r^j). \tag{8.72}$$

We recall that a^j stands for $d^2r^j/d\bar{t}^2$.

Our end result for the comoving-frame gravitational potential is

$$\bar{U} = \frac{Gm}{\bar{r}} + \bar{U}_{\rm tidal}, \tag{8.73}$$

where

$$\bar{U}_{\rm tidal} = \frac{1}{2}\bar{x}^j\bar{x}^k\partial_{jk}U_{\rm ext}(\bar{t}, r^j) + \cdots \tag{8.74}$$

is what remains of the external potential. As its label indicates, it is this potential that is responsible for the tidal interaction between the moving body and the external matter distribution. The Newtonian physics of tidally deformed bodies was explored in some detail in Sec. 2.5.

The coordinate transformation that takes us from the inertial frame to the moving frame is

$$t = \bar{t} + \frac{1}{c^2}\int\left[\frac{1}{2}v^2 + U_{\rm ext}(\bar{t}, r^j)\right]d\bar{t} + \frac{1}{c^2}v_j(\bar{t})\bar{x}^j + O(c^{-4}) \tag{8.75}$$

and

$$x^j = \bar{x}^j + r^j(\bar{t}) + O(c^{-2}). \tag{8.76}$$

The transformation can of course be generalized to post-Newtonian order, and we go through this exercise in Sec. 9.4.

8.3.6 Post-Galilean transformations

As we have seen, it can prove useful to exploit the full freedom contained in the general post-Newtonian transformations, or the restricted class of harmonic transformations, when one considers a bounded domain of spacetime such as the neighborhood of a moving body. When the considerations are global, however, the general freedom is too vast, and one would like to constrain it so as to eliminate the spatially growing terms in the potentials. In this section we assume that the original potentials U, U^j, and Ψ vanish in the formal limit

$r \to \infty$, and we specialize the post-Newtonian transformations so that the new potentials $\bar{U}$, $\bar{U}^j$, and $\bar{\Psi}$ share this property. This restricted class of coordinate transformations is known as the *post-Galilean transformations* of post-Newtonian theory. The name was coined by Chandrasekhar and Contopoulos in their classic 1967 paper.

Construction

Inspection of Eq. (8.46) reveals that the Newtonian potential will grow linearly with $\bar{r}$ unless $a^j = 0$. Discarding an uninteresting constant translation of the coordinates, this means that $r^j(\bar{t})$ must be of the form

$$r^j = V^j \bar{t}, \tag{8.77}$$

with V^j a constant vector. To eliminate the spatially-constant term in $\bar{U}$ we must also set $\dot{A} = \frac{1}{2}V^2$, so that

$$A = \frac{1}{2}V^2 \bar{t}. \tag{8.78}$$

Here $V^2 := \delta_{jk} V^j V^k$, and we again discard an uninteresting integration constant. These results imply that the post-Galilean transformation leaves the Newtonian potential invariant:

$$\bar{U} = \hat{U}, \tag{8.79}$$

where $\hat{U} = U(\bar{t}, \bar{x}^j + V^j \bar{t})$.

Our results can be used to simplify the general expression for $\bar{U}^j$, as it appears in Eq. (8.54). To keep $\bar{U}^j$ from growing we must set $\dot{R}^j = 0$. The rotation of the coordinate axes described by $R^j(\bar{t})$ must therefore be constant in time, and we choose to eliminate this uninteresting freedom by setting

$$R^j = 0. \tag{8.80}$$

To eliminate the spatially-constant term in $\bar{U}^j$ we set $\partial_{\bar{j}} \beta = \dot{H}^j$, which integrates to $\beta = \beta_0(\bar{t}) + \dot{H}_j \bar{x}^j$, where β_0 and H^j are arbitrary functions of time. We observe that β is a harmonic function, and that its expression is compatible with Eq. (8.63); the transformation is therefore within the class of harmonic transformations. With this result we find that the vector potential transforms as

$$\bar{U}^j = \hat{U}^j - V^j \hat{U} \tag{8.81}$$

under a post-Galilean transformation.

Moving on to $\bar{\Psi}$, as it appears in Eq. (8.58), we find that the removal of the growing term requires $\ddot{H}^j = 0$, so that $\dot{H}^j$ must be a constant vector. This vector is in principle arbitrary, but we choose to restrict the coordinate freedom by making it proportional to V^j. We write it as $\dot{H}^j = \frac{1}{2}V^2 V^j$, inserting an arbitrary numerical coefficient of $\frac{1}{2}$ for reasons that will be made clear below, and the factor of V^2 for proper dimensionality. Our choice for $H^j(\bar{t})$ is therefore

$$H^j = \frac{1}{2}V^2 V^j \bar{t}. \tag{8.82}$$

To eliminate the spatially-constant term in $\bar{\Psi}$ we must set $\dot{\beta}_0 = -\frac{1}{8}V^4 + \dot{H}_j V^j$. With our previous choice for H^j this is $\dot{\beta}_0 = \frac{3}{8}V^4$, and our final expression for β is

$$\beta = \frac{3}{8}V^4\bar{t} + \frac{1}{2}V^2 V_j \bar{x}^j. \tag{8.83}$$

With all this we find that the post-Newtonian potential transforms as

$$\bar{\Psi} = \hat{\Psi} - 4V_j\hat{U}^j + 2V^2\hat{U} + \left(\frac{1}{2}V^2\bar{t} + V_j\bar{x}^j\right)\partial_{\bar{t}}\hat{U} - \left(\frac{1}{2}V^j V_k \bar{x}^k\right)\partial_{\bar{j}}\hat{U} \tag{8.84}$$

under a post-Galilean transformation. We note that the terms involving $\partial_{\bar{t}}\hat{U}$ and $\partial_{\bar{j}}\hat{U}$ are multiplied by quantities that grow linearly with $\bar{r}$. Because $\hat{U}$ decays as $\bar{r}^{-1}$, and its derivatives as $\bar{r}^{-2}$, we see that $\bar{\Psi}$ properly vanishes in the formal limit $\bar{r} \to \infty$.

Collecting results, we find that the post-Galilean transformation is a three-parameter family described by

$$t = \left(1 + \frac{1}{2}\frac{V^2}{c^2} + \frac{3}{8}\frac{V^4}{c^4}\right)\bar{t} + \frac{1}{c^2}\left(1 + \frac{1}{2}\frac{V^2}{c^2}\right)V_j\bar{x}^j + O(c^{-6}), \tag{8.85a}$$

$$x^j = \left(\delta^j_{\ k} + \frac{1}{2}\frac{V^j V_k}{c^2}\right)\bar{x}^k + \left(1 + \frac{1}{2}\frac{V^2}{c^2}\right)V^j\bar{t} + O(c^{-4}); \tag{8.85b}$$

the parameters are the three components of the vector V^j. This is nothing but a Lorentz transformation expanded in powers of $(V/c)^2$. The coordinates $(\bar{t}, \bar{x}^j)$ define a frame $\bar{S}$ that is boosted with respect to the original frame S; the boost takes place in the direction of the velocity vector V^j.

Boosted potentials

In the foregoing discussion the boosted potentials $\bar{U}$, $\bar{U}^j$, and $\bar{\Psi}$ were expressed in terms of the "hatted potentials" $\hat{U}$, $\hat{U}^j$, and $\hat{\Psi}$; these, we recall, are the original potentials evaluated at time $t = \bar{t}$ and position $x^j = \bar{x}^j + V^j\bar{t}$. This representation of the transformed potentials was optimal in the context of the general theory of post-Newtonian transformations, as developed in the previous sections. It is not optimal in the restricted context of post-Galilean transformations, because of the schizophrenic nature of the hatted potentials, which live partly in the frame S and partly in the frame $\bar{S}$. An indication that the representation is indeed not optimal comes from our previous expression for $\bar{\Psi}$, which displays a curious and unwanted explicit dependence upon $\bar{t}$ and $\bar{x}^j$.

We therefore proceed differently. We shall (i) postulate plausible expressions for the transformed potentials $\bar{U}$, $\bar{U}^j$, and $\bar{\Psi}$; (ii) relate these to the original potentials U, U^j, and Ψ; and (iii) show that under the transformation of Eqs. (8.85), the transformed metric $\bar{g}_{\alpha\beta}$ keeps the standard post-Newtonian form of Eq. (8.20), with the understanding that the new metric is expressed in terms of the new potentials.

Our proposed expressions for the transformed potentials are

$$\bar{U}(\bar{t}, \bar{\boldsymbol{x}}) = G \int \frac{\bar{\rho}^*(\bar{t}, \bar{\boldsymbol{x}}')}{|\bar{\boldsymbol{x}} - \bar{\boldsymbol{x}}'|} d^3\bar{x}', \tag{8.86a}$$

$$\bar{U}^j(\bar{t}, \bar{\boldsymbol{x}}) = G \int \frac{\bar{\rho}^* \bar{v}^j(\bar{t}, \bar{\boldsymbol{x}}')}{|\bar{\boldsymbol{x}} - \bar{\boldsymbol{x}}'|} d^3\bar{x}', \tag{8.86b}$$

$$\bar{\psi}(\bar{t}, \bar{\boldsymbol{x}}) = G \int \frac{\bar{\rho}^*(\frac{3}{2}\bar{v}^2 - \bar{U} + \bar{\Pi} + 3\bar{p}/\bar{\rho}^*)(\bar{t}, \bar{\boldsymbol{x}}')}{|\bar{\boldsymbol{x}} - \bar{\boldsymbol{x}}'|} d^3\bar{x}', \tag{8.86c}$$

$$\bar{X}(\bar{t}, \bar{\boldsymbol{x}}) = G \int \rho^*(\bar{t}, \bar{\boldsymbol{x}}')|\bar{\boldsymbol{x}} - \bar{\boldsymbol{x}}'|\, d^3\bar{x}', \tag{8.86d}$$

and the transformed post-Newtonian potential is $\bar{\Psi} = \bar{\psi} + \frac{1}{2}\partial_{\bar{t}\bar{t}}\bar{X}$. These expressions are natural: the new potentials are defined just as the old potentials in terms of the boosted coordinates and the fluid variables $\bar{\rho}^*$, $\bar{p}$, $\bar{\Pi}$, and $\bar{\boldsymbol{v}}$ that would be measured in the frame $\bar{S}$ instead of the original frame S. It is useful to introduce transformed versions of the auxiliary potentials listed in Eqs. (8.8); in terms of these we have $\bar{\Psi} = 2\bar{\Phi}_1 - \bar{\Phi}_2 + \bar{\Phi}_3 + 4\bar{\Phi}_4 - \frac{1}{2}\bar{\Phi}_5 - \frac{1}{2}\bar{\Phi}_6$.

Our first task is to express the old Newtonian potential $U(t, \boldsymbol{x})$ in terms of the new potentials. This is not entirely straightforward. A major source of subtlety is the important fact that in Eqs. (8.86), the integration variables $\bar{\boldsymbol{x}}'$ describe the position of a fluid element at time $\bar{t}$, the same time at which the potentials are being evaluated. The spacetime events P and P', respectively labeled by the coordinates $(\bar{t}, \bar{\boldsymbol{x}})$ and $(\bar{t}, \bar{\boldsymbol{x}}')$, are simultaneous in the frame $\bar{S}$. But they are not simultaneous in the original frame S, and we must take this property carefully into account.

We examine the situation in the original frame S (see Fig. 8.1). The figure shows a spacetime diagram in which we display the field point P as well as the world line of a selected fluid element. Two events are shown on this world line: the source point Q' is simultaneous with P in the frame S, while P' is simultaneous with P in the frame $\bar{S}$. In the frame S the coordinates of P are $(t, \boldsymbol{x})$, the coordinates of Q' are $(t, \boldsymbol{x}')$, and the coordinates of P' are $(\tau, \boldsymbol{\xi})$. In the frame $\bar{S}$ the coordinates of P are $(\bar{t}, \bar{\boldsymbol{x}})$, the coordinates of Q' are $(\bar{\tau}, \bar{\boldsymbol{\xi}})$, and the coordinates of P' are $(\bar{t}, \bar{\boldsymbol{x}}')$. Note that the coordinates $(\tau, \boldsymbol{\xi})$ and $(\bar{\tau}, \bar{\boldsymbol{\xi}})$ refer to different events in spacetime. In the frame S the world line is described by the time-dependent position vector $\boldsymbol{r}$; we have that $\boldsymbol{x}' := \boldsymbol{r}(t)$ and $\boldsymbol{\xi} := \boldsymbol{r}(\tau)$. In the frame $\bar{S}$ the world line is described by $\bar{\boldsymbol{r}}$, and we have that $\bar{\boldsymbol{x}}' := \bar{\boldsymbol{r}}(\bar{t})$ and $\bar{\boldsymbol{\xi}} := \bar{\boldsymbol{r}}(\bar{\tau})$. In the frame S the velocity of the fluid element at Q' is $\boldsymbol{v}' := \dot{\boldsymbol{r}}(t)$, while in $\bar{S}$ the velocity of the fluid element at P' is $\bar{\boldsymbol{v}}' := \dot{\bar{\boldsymbol{r}}}(\bar{t})$; the overdots indicate differentiation with respect to the relevant time variable.

The coordinates of the field point P transform as in Eq. (8.85). The coordinates of the source point Q' transform as

$$t = \left(1 + \frac{1}{2}\frac{V^2}{c^2} + \frac{3}{8}\frac{V^4}{c^4}\right)\bar{\tau} + \frac{1}{c^2}\left(1 + \frac{1}{2}\frac{V^2}{c^2}\right)V_j\bar{\xi}^j + O(c^{-6}), \tag{8.87a}$$

$$x'^j = \left(\delta^j{}_k + \frac{1}{2}\frac{V^j V_k}{c^2}\right)\bar{\xi}^k + \left(1 + \frac{1}{2}\frac{V^2}{c^2}\right)V^j\bar{\tau} + O(c^{-4}). \tag{8.87b}$$

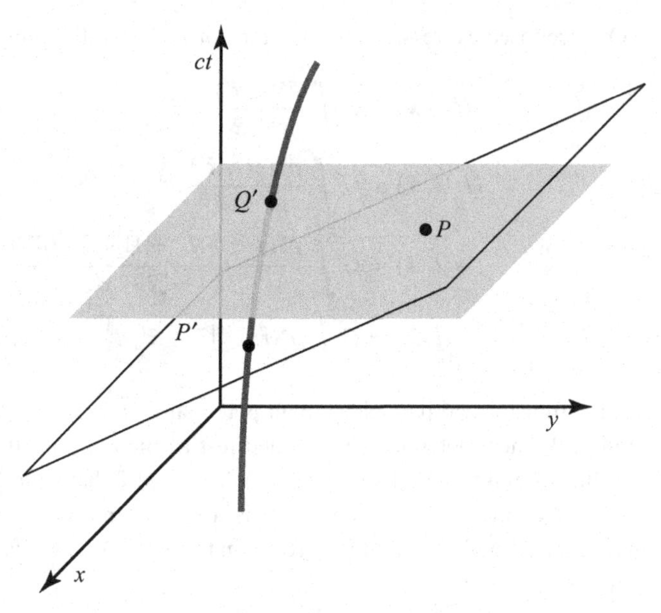

Fig. 8.1 World line of a selected fluid element viewed in the frame S. The grey plane is a hypersurface $t =$ constant, and the white plane is a hypersurface $\bar{t} =$ constant. The field point P is simultaneous with Q' in S, but it is simultaneous with P' in the frame $\bar{S}$.

We want to express $\boldsymbol{x}'$ in terms of $\bar{t}$, $\bar{\boldsymbol{x}}$, and $\bar{\boldsymbol{x}}'$, and this requires elimination of $\bar{\tau}$ and $\bar{\boldsymbol{\xi}}$ from Eqs. (8.87b). To achieve this we equate the t of Eqs. (8.85) with the t of Eqs. (8.87) and deduce that $\bar{\tau} = \bar{t} + c^{-2}V_j(\bar{x} - \bar{\xi})^j + O(c^{-4})$. We substitute this into the world-line equation $\bar{\boldsymbol{\xi}} = \bar{\boldsymbol{r}}(\bar{\tau})$ and get $\bar{\xi}^j = \bar{x}'^j + c^{-2}\bar{v}'^j V_k(\bar{x} - \bar{\xi})^k + O(c^{-4})$. Collecting results, we have obtained

$$\bar{\tau} = \bar{t} + \frac{V_k}{c^2}(\bar{x} - \bar{x}')^k + O(c^{-4}), \tag{8.88a}$$

$$\bar{\xi}^j = \bar{x}'^j + \frac{v'^j V_k}{c^2}(\bar{x} - \bar{x}')^k + O(c^{-4}). \tag{8.88b}$$

These are the coordinates of Q' in the frame $\bar{S}$, expressed in terms of the coordinates of both P and P'. Making the substitution in Eq. (8.87b), we arrive at

$$x'^j = \bar{x}'^j + V^j\bar{t} + \frac{1}{c^2}\left(\bar{v}'^j + V^j\right)V_k(\bar{x} - \bar{x}')^k + \frac{V^j}{2c^2}\left(V^2\bar{t} + V_k\bar{x}'^k\right) + O(c^{-4}), \tag{8.89}$$

the desired relation between $\boldsymbol{x}'$ and the coordinates of P and P' in the frame $\bar{S}$. To this equation we can adjoin

$$v'^j = \bar{v}'^j + V^j + O(c^{-2}), \tag{8.90}$$

the law of composition of velocities (truncated to the leading, Newtonian order). It follows from Eqs. (8.85) and (8.89) that

$$\frac{1}{|\boldsymbol{x}-\boldsymbol{x}'|} = \frac{1}{|\bar{\boldsymbol{x}}-\bar{\boldsymbol{x}}'|} + \frac{1}{c^2}\left(\bar{v}'_j V_k + \frac{1}{2}V_j V_k\right)\frac{(\bar{x}-\bar{x}')^j(\bar{x}-\bar{x}')^k}{|\bar{\boldsymbol{x}}-\bar{\boldsymbol{x}}'|^3} + O(c^{-4}), \tag{8.91}$$

and this is an important ingredient that enters the transformation of the Newtonian potential.

Another important ingredient is the statement that $\rho^* d^3x$ is invariant under the post-Galilean transformation. We express this as

$$\rho^*(t,\boldsymbol{x}')\,d^3x' = \bar{\rho}^*(\bar{t},\bar{\boldsymbol{x}}')\,d^3\bar{x}'. \tag{8.92}$$

The invariance of $dm := \rho^* d^3x$ reflects the simple fact that dm is the conserved rest-mass of a fluid element. Because this cannot be altered by a coordinate transformation, we have that $dm(Q') = d\bar{m}(Q')$, or $dm(t,\boldsymbol{x}') = d\bar{m}(\bar{\tau},\bar{\boldsymbol{\xi}})$. And because $d\bar{m}$ does not change as we follow the motion of the fluid element, we also have that $d\bar{m}(Q') = d\bar{m}(P')$, or $d\bar{m}(\bar{\tau},\bar{\boldsymbol{\xi}}) = d\bar{m}(\bar{t},\bar{\boldsymbol{x}}')$. We therefore arrive at Eq. (8.92), which is just the combined statement that $dm(Q') = d\bar{m}(P')$.

More formally, $dm(Q') = d\bar{m}(Q')$ is a consequence of the facts that (i) the proper mass density ρ is a scalar quantity; (ii) the spacetime volume element $\sqrt{-g}\,dt\,d^3x$ is invariant under a coordinate transformation; and (iii) the element of proper time $d\lambda$ along the world line also is an invariant. From all this it follows that $\rho\sqrt{-g}(dt/d\lambda)d^3x =: \rho^*\,d^3x$ is invariant under the transformation. On the other hand, that $d\bar{m}(Q') = d\bar{m}(P')$ follows formally from an application of the continuity equation, Eq. (7.3), to a single fluid element. The formal route also gives rise to the statement of Eq. (8.92).

When we substitute Eqs. (8.91) and (8.92) into the integral definition of the old Newtonian potential $U(t,\boldsymbol{x})$, we find that it can be expressed in terms of the new potentials as

$$U = \left(1 + \frac{V^2}{2c^2}\right)\bar{U} + \frac{V^j}{c^2}\bar{\Phi}_j - \frac{V^j V^k}{2c^2}\partial_{\bar{j}\bar{k}}\bar{X} + O(c^{-4}). \tag{8.93}$$

Here $\bar{\Phi}_j$ is the transformed version of the auxiliary potential defined by Eq. (8.8), and we make use of the identity of Eq. (8.11).

The remaining potentials transform in an analogous way. In fact, their transformation properties are much easier to identify, because here we do not need to calculate the correction terms of order c^{-2}; these impact the metric beyond the first post-Newtonian order. Taking into account Eq. (8.90) and the fact that p and Π transform as scalar quantities, we quickly obtain

$$U^j = \bar{U}^j + V^j\bar{U} + O(c^{-2}), \tag{8.94}$$

as well as

$$\Phi_1 = \bar{\Phi}_1 + 2V^j\bar{U}_j + V^2\bar{U} + O(c^{-2}), \tag{8.95a}$$
$$\Phi_2 = \bar{\Phi}_2 + O(c^{-2}), \tag{8.95b}$$
$$\Phi_3 = \bar{\Phi}_3 + O(c^{-2}), \tag{8.95c}$$
$$\Phi_4 = \bar{\Phi}_4 + O(c^{-2}), \tag{8.95d}$$
$$\Phi_5 = \bar{\Phi}_5 + O(c^{-2}), \tag{8.95e}$$
$$\Phi_6 = \bar{\Phi}_6 + 2V^j\bar{\Phi}_j + V^2\bar{U} - V^jV^k\partial_{\bar{j}\bar{k}}\bar{X} + O(c^{-2}). \tag{8.95f}$$

These equations imply that the post-Newtonian potential transforms as

$$\Psi = \bar{\Psi} + V^j\big(4\bar{U}_j - \bar{\Phi}_j\big) + \frac{3}{2}V^2\bar{U} + \frac{1}{2}V^jV^k\partial_{\bar{j}\bar{k}}\bar{X} + O(c^{-2}) \tag{8.96}$$

under a post-Galilean transformation.

Our final task is to verify that the transformed metric $\bar{g}_{\alpha\beta}$ takes the standard post-Newtonian form of Eq. (8.20) when it is expressed in terms of the transformed potentials $\bar{U}$, $\bar{U}^j$, and $\bar{\Psi}$. This is straightforward. We first specialize the general transformation equations (8.34) to the post-Galilean case of Eqs. (8.85) and get the components $\bar{g}_{00} = -1 + 2c^{-2}U + 2c^{-4}(\Psi - U^2 - 4V^jU_j + 2V^2U) + O(c^{-6})$, $\bar{g}_{0j} = -4c^{-3}(U_j - V_jU) + O(c^{-5})$, and $\bar{g}_{jk} = \delta_{jk}(1 + 2c^{-2}U) + O(c^{-4})$ for the new metric tensor. This is still expressed in terms of the old potentials, and we complete the calculation by involving Eqs. (8.93), (8.94), and (8.96). Our final result is

$$\bar{g}_{00} = -1 + \frac{2}{c^2}\bar{U} + \frac{2}{c^4}\big(\bar{\Psi} - \bar{U}^2\big) + O(c^{-6}), \tag{8.97a}$$

$$\bar{g}_{0j} = -\frac{4}{c^3}\bar{U}_j + O(c^{-5}), \tag{8.97b}$$

$$\bar{g}_{jk} = \left(1 + \frac{2}{c^2}\bar{U}\right)\delta_{jk} + O(c^{-4}), \tag{8.97c}$$

the statement that the transformed metric does indeed take the standard post-Newtonian form in terms of the proposed new potentials of Eqs. (8.86).

This completes our discussion of post-Galilean transformations. To sum up, we have established that a post-Galilean transformation describes a boost from a frame S to a new frame $\bar{S}$ that moves relative to S with a constant velocity V. In this new frame the metric keeps its standard post-Newtonian form, but the potentials are now defined by Eqs. (8.86); they refer to the fluid variables $\bar{\rho}^*$, $\bar{p}$, $\bar{\Pi}$, and $\bar{\boldsymbol{v}}$ that are measured in the new frame.

8.3.7 Pure-gauge transformations

Another interesting subclass of post-Newtonian transformations is obtained by setting

$$A = r^j = H^j = R^j = 0 \tag{8.98}$$

and retaining only the freedom contained in β. This class of transformations is described by

$$t = \bar{t} + \frac{1}{c^4}\beta(\bar{t}, \bar{x}^j) + O(c^{-6}), \tag{8.99a}$$

$$x^j = \bar{x}^j + O(c^{-4}), \tag{8.99b}$$

and the potentials change according to

$$\bar{U} = U, \tag{8.100a}$$

$$\bar{U}^j = U^j + \frac{1}{4}\partial_{\bar{j}}\beta, \tag{8.100b}$$

$$\bar{\Psi} = \Psi - \partial_{\bar{t}}\beta. \tag{8.100c}$$

In this case we no longer need to distinguish between the hatted potentials and their original expressions. Equations (8.100) take the appearance of an electromagnetic-type gauge transformation that links the potentials U^j and Ψ. We refer to this subclass of transformations as the *pure-gauge transformations* of post-Newtonian theory. When β is a harmonic function, the gauge transformation converts a set of harmonic coordinates to another set of harmonic coordinates.

The transformation may be exploited to remove the superpotential term from g_{00} and put it instead in g_{0j}. We refer to the decomposition of Eq. (8.2), and to eliminate the term $\frac{1}{2}\partial_{tt}X$ from g_{00} we choose

$$\beta = \frac{1}{2}\partial_{\bar{t}}X. \tag{8.101}$$

Note that this is not a harmonic function (because $\nabla^2 X = 2U$), so that the transformation does not preserve the harmonic coordinate condition. With this expression for β we find that the new metric is given by

$$\bar{g}_{00} = -1 + \frac{2}{c^2}U + \frac{2}{c^4}(\psi - U^2) + O(c^{-6}), \tag{8.102a}$$

$$\bar{g}_{0j} = -\frac{4}{c^3}U_j - \frac{1}{2c^3}\partial_{\bar{t}\bar{j}}X + O(c^{-5}), \tag{8.102b}$$

$$\bar{g}_{jk} = \left(1 + \frac{2}{c^2}U\right)\delta_{jk} + O(c^{-4}). \tag{8.102c}$$

This choice of coordinate system defines the so-called *standard gauge* of post-Newtonian theory. As we have pointed out in the introductory section, this choice of gauge was popularized by Chandrasekhar, and it was once widely utilized by researchers in the post-Newtonian community. Like most current workers in the field, however, we prefer to use the harmonic gauge, and we have made this choice consistently throughout the book, except in Chapter 13 where we examine alternative theories of gravity. To be sure, the choice of gauge is mostly a matter of taste and convenience. But there are, nevertheless, important advantages in using the harmonic coordinates: it is in this gauge that post-Newtonian theory can naturally be embedded within the wider context of post-Minkowskian theory. And as we explained back in Sec. 8.2, it is by doing this that the foundations of post-Newtonian theory can be made secure.

8.4 Post-Newtonian hydrodynamics

8.4.1 Introduction

The dynamics of perfect fluids has been a recurring topic in this book. We examined this first in the context of Newtonian physics in Sec. 1.4, we gave the theory a special-relativistic formulation in Sec. 4.2, and we promoted this to curved spacetime in Sec. 5.3. In Sec. 7.1.1 we introduced the variables $\{\rho^*, p, \Pi, \boldsymbol{v}\}$ and incorporated slowly-moving fluids within the post-Minkowskian approximation.

In Sec. 7.1.1 we saw that the behavior of a perfect fluid is governed by the continuity equation

$$\partial_t \rho^* + \partial_j(\rho^* v^j) = 0, \tag{8.103}$$

and in Sec. 7.3.2 we got re-acquainted with the Euler equation of Chapter 1,

$$\rho^* \frac{dv^j}{dt} = \rho^* \partial_j U - \partial_j p + O(c^{-2}). \tag{8.104}$$

We recall that $\rho^* := \sqrt{-g}\gamma\rho$, with $\gamma := u^0/c$, is the conserved mass density, and at Newtonian order there is no distinction between this and the proper density ρ; v^j is the fluid's velocity field, defined with respect to the time coordinate t, p is the pressure, and $d/dt = \partial_t + v^k\partial_k$ is the Lagrangian time derivative. We recall also that the exact statement of the first law of thermodynamics for perfect fluids is $d\Pi = (p/\rho^2)\,d\rho$, which we write as

$$\frac{d\Pi}{dt} = \frac{p}{\rho^{*2}} \frac{d\rho^*}{dt} + O(c^{-2}). \tag{8.105}$$

Here Π is the internal energy of a fluid element divided by its mass.

In this section we calculate the post-Newtonian corrections to Euler's equation (8.104). In addition, we derive expressions for the fluid's conserved mass-energy M, its total momentum P^j, and its center-of-mass position R^j. We shall not alter Eq. (8.103), which is exact, and we shall not need the $O(c^{-2})$ corrections to Eq. (8.105). In Chapter 9 we apply these results to situations in which the fluid breaks up into a number of separated components; this defines the post-Newtonian N-body problem.

8.4.2 Energy-momentum conservation

The components of the energy-momentum tensor of a perfect fluid were listed back in Eq. (8.7). The equation of energy-momentum conservation is

$$0 = \nabla_\beta T^{\alpha\beta} = \partial_\beta T^{\alpha\beta} + \Gamma^\alpha_{\mu\beta} T^{\mu\beta} + \Gamma^\beta_{\mu\beta} T^{\alpha\mu}, \tag{8.106}$$

and this can be simplified if we recall from Sec. 5.2 that $\Gamma^{\beta}_{\mu\beta} = (-g)^{-1/2}\partial_\beta(-g)^{1/2}$. We therefore have

$$0 = \partial_\beta\big(\sqrt{-g}T^{\alpha\beta}\big) + \Gamma^{\alpha}_{\beta\gamma}\big(\sqrt{-g}T^{\beta\gamma}\big), \tag{8.107}$$

and this form of the conservation equation is optimal for the following computations. We recall that the square root of the metric determinant is given by $\sqrt{-g} = 1 + 2c^{-2}U + O(c^{-4})$.

The zeroth component of Eq. (8.107) gives rise to a statement of energy conservation. When fully expanded the equation is

$$\begin{aligned} 0 &= \frac{1}{c}\partial_t\big(\sqrt{-g}T^{00}\big) + \partial_j\big(\sqrt{-g}T^{0j}\big) \\ &\quad + \Gamma^0_{00}\big(\sqrt{-g}T^{00}\big) + 2\Gamma^0_{0j}\big(\sqrt{-g}T^{0j}\big) + \Gamma^0_{jk}\big(\sqrt{-g}T^{jk}\big), \end{aligned} \tag{8.108}$$

and this becomes

$$\begin{aligned} 0 &= c\Big[\partial_t\rho^* + \partial_j(\rho^* v^j)\Big] \\ &\quad + \frac{1}{c}\bigg\{\partial_t\Big[\rho^*\Big(\frac{1}{2}v^2 + U + \Pi\Big)\Big] + \partial_j\Big[\rho^* v^j\Big(\frac{1}{2}v^2 + U + \Pi\Big)\Big] \\ &\quad + \partial_j(pv^j) - \rho^*\partial_t U - 2\rho^* v^j \partial_j U\bigg\} + O(c^{-3}), \end{aligned} \tag{8.109}$$

after inserting the components of $T^{\alpha\beta}$ and the Christoffel symbols of Eq. (8.15). At order c we recover the continuity equation (8.103), and at order c^{-1} we get

$$0 = \rho^*\partial_t\Big(\frac{1}{2}v^2 + \Pi\Big) + \rho^* v^j\partial_j\Big(\frac{1}{2}v^2 + \Pi\Big) + \partial_j(pv^j) - \rho^* v^j\partial_j U \tag{8.110}$$

after simplification. This is the equation that expresses the local conservation of energy within the fluid.

The spatial components of Eq. (8.107) provide a statement of momentum conservation. We write the equation in fully expanded form as

$$\begin{aligned} 0 &= \frac{1}{c}\partial_t\big(\sqrt{-g}T^{0j}\big) + \partial_k\big(\sqrt{-g}T^{jk}\big) \\ &\quad + \Gamma^j_{00}\big(\sqrt{-g}T^{00}\big) + 2\Gamma^j_{0k}\big(\sqrt{-g}T^{0k}\big) + \Gamma^j_{kn}\big(\sqrt{-g}T^{kn}\big), \end{aligned} \tag{8.111}$$

and we eventually arrive at

$$\begin{aligned} 0 &= \partial_t(\mu\rho^* v^j) + \partial_k(\mu\rho^* v^j v^k) + \partial_j p - \rho^*\partial_j U - \frac{\rho^*}{c^2}\Big(\frac{3}{2}v^2 - 3U + \Pi + p/\rho^*\Big)\partial_j U \\ &\quad + \frac{\rho^*}{c^2}\Big[2v_j(\partial_t U + v^k\partial_k U) - 4\partial_t U_j - 4v^k(\partial_k U_j - \partial_j U_k) - \partial_j\Psi\Big] + O(c^{-4}) \end{aligned} \tag{8.112}$$

after some algebra and simplification. We have introduced

$$\mu := 1 + \frac{1}{c^2}\left(\frac{1}{2}v^2 + U + \Pi + p/\rho^*\right) + O(c^{-4}), \tag{8.113}$$

and Eq. (8.112) expresses the local conservation of momentum within the fluid.

8.4.3 Post-Newtonian Euler equation

We next work on Eq. (8.112) and bring it to the form of a relativistic generalization of Eq. (8.104). We begin with the observation that the first two terms on the right-hand side of Eq. (8.112) can be expressed as

$$\partial_t(\mu\rho^* v^j) + \partial_k(\mu\rho^* v^j v^k) = \mu\rho^*\frac{dv^j}{dt} + \rho^* v^j\frac{d\mu}{dt} \tag{8.114}$$

after making use of the continuity equation (8.103). If we make the substitution in Eq. (8.112) and truncate the result at Newtonian order, we recover

$$\rho^*\frac{dv^j}{dt} = \rho^*\partial_j U - \partial_j p + O(c^{-2}), \tag{8.115}$$

the correct expression of Euler's equation. We are therefore on the right track, and we may now retrieve the neglected terms of order c^{-2}.

Differentiation of Eq. (8.113) yields

$$\frac{d\mu}{dt} = \frac{1}{c^2}\left(v_j\frac{dv^j}{dt} + \frac{dU}{dt} + \frac{d\Pi}{dt} + \frac{1}{\rho^*}\frac{dp}{dt} - \frac{p}{\rho^{*2}}\frac{d\rho^*}{dt}\right) + O(c^{-4}), \tag{8.116}$$

and this becomes

$$\frac{d\mu}{dt} = \frac{1}{c^2}\left(\partial_t U + 2v^k\partial_k U + \frac{1}{\rho^*}\partial_t p\right) + O(c^{-4}) \tag{8.117}$$

after insertion of Euler's equation and Eq. (8.105).

Substitution of Eqs. (8.114) and (8.117) into Eq. (8.112) produces

$$\begin{aligned}\mu\rho^*\frac{dv^j}{dt} &= -\partial_j p + \rho^*\partial_j U - \frac{1}{c^2}v^j\partial_t p + \frac{1}{c^2}\rho^*\left(\frac{3}{2}v^2 - 3U + \Pi + \frac{p}{\rho^*}\right)\partial_j U \\ &\quad -\frac{1}{c^2}\rho^*\Big[v^j\left(3\partial_t U + 4v^k\partial_k U\right) - 4\partial_t U_j - 4v^k\left(\partial_k U_j - \partial_j U_k\right) - \partial_j\Psi\Big] + O(c^{-4}),\end{aligned} \tag{8.118}$$

and this becomes

$$\begin{aligned}\rho^* \frac{dv^j}{dt} &= -\partial_j p + \rho^* \partial_j U \\ &+ \frac{1}{c^2}\Big[\Big(\frac{1}{2}v^2 + U + \Pi + \frac{p}{\rho^*}\Big)\partial_j p - v^j \partial_t p\Big] \\ &+ \frac{1}{c^2}\rho^*\Big[(v^2 - 4U)\partial_j U - v^j\big(3\partial_t U + 4v^k \partial_k U\big) \\ &\qquad + 4\partial_t U_j + 4v^k\big(\partial_k U_j - \partial_j U_k\big) + \partial_j \Psi\Big] \\ &+ O(c^{-4}) \end{aligned} \tag{8.119}$$

after multiplication of each side by μ^{-1}. Equation (8.119) is the post-Newtonian version of Euler's equation. This equation, together with the continuity equation (8.103) and an equation of state relating the pressure, density, and internal energy, completely determines the behavior of a slowly-moving fluid in a weak gravitational field.

8.4.4 Interlude: Integral identities

We now interrupt the main development and establish a number of identities that will be required in the following discussion. We first derive the results displayed back in Eqs. (8.10), and next we prove the integral identities

$$\int \rho^* \partial_j U \, d^3x = 0 \,, \tag{8.120a}$$

$$\int \rho^* U^j \, d^3x = \int \rho^* U v^j \, d^3x \,, \tag{8.120b}$$

$$\int \rho^* \partial_j \psi \, d^3x = -\int \rho^* \big(\tfrac{3}{2}v^2 - U + \Pi + 3p/\rho^*\big)\partial_j U \, d^3x \,, \tag{8.120c}$$

$$\int \rho^* v^j \partial_j U \, d^3x = \frac{1}{2}\frac{d}{dt}\int \rho^* U \, d^3x \,, \tag{8.120d}$$

$$\int \rho^* v^k \partial_j U_k \, d^3x = 0 \,, \tag{8.120e}$$

$$\int \rho^* v^k \partial_j \Phi_k \, d^3x = 0 \,, \tag{8.120f}$$

$$\int \rho^* x^j \big(\partial_t U - v^k \partial_k U\big) \, d^3x = \int \rho^* \Phi^j \, d^3x \,. \tag{8.120g}$$

The potentials U, U^j, ψ, and X are expressed in terms of the fluid variables in Eqs. (8.4); the auxiliary potential Φ^j was introduced in Eq. (8.8).

As a consequence of Eqs. (8.10) and (8.120b) we also find that

$$\int \rho^* \big(U v_j + \partial_{tj} X\big) \, d^3x = \int \rho^* \Phi_j \, d^3x \,. \tag{8.121}$$

And combining Eqs. (8.10), (8.120e), and (8.120f) yields

$$\int \rho^* v^k \partial_{tjk} X \, d^3x = 0 \,, \tag{8.122}$$

another useful integral identity.

Box 8.4 **Integration and time differentiation**

The integral tricks reviewed here were first introduced back in Sec. 1.4.3. We recall that

$$\frac{d}{dt} \int \rho^* f(t, \boldsymbol{x}) \, d^3x = \int \rho^* \frac{df}{dt} \, d^3x \tag{1}$$

comes as an immediate consequence of the continuity equation (8.103); here ρ^* is a function of t and $\boldsymbol{x}$, f is an arbitrary function of its arguments, and $df/dt = \partial_t f + v^k \partial_k f$ is its total time derivative. We generalize this identity by allowing f to be a function of $\boldsymbol{x}'$ also, and we define $F(t, \boldsymbol{x}) := \int \rho^{*\prime} f(t, \boldsymbol{x}, \boldsymbol{x}') \, d^3x'$, with $\rho^{*\prime}$ standing for $\rho^*(t, \boldsymbol{x}')$. If we keep $\boldsymbol{x}$ fixed in this equation, Eq. (1) tells us that

$$\partial_t F = \int \rho^{*\prime} \left(\partial_t f + v'^k \partial_{k'} f \right) d^3x' \,, \tag{2}$$

where v'^k is the velocity field expressed as a function of t and $\boldsymbol{x}'$, and $\partial_{k'}$ indicates partial differentiation with respect to the primed coordinates. The total time derivative of F is $\partial_t F + v^k \partial_k F$, and this can be expressed as $dF/dt = \int \rho^{*\prime} (\partial_t f + v^k \partial_k f + v'^k \partial_{k'} f) \, d^3x'$. The quantity within brackets is recognized as the total time derivative of the function $f(t, \boldsymbol{x}, \boldsymbol{x}')$, and we write our identity as

$$\frac{dF}{dt} = \int \rho^{*\prime} \frac{df}{dt} \, d^3x' \,, \tag{3}$$

with $df/dt = \partial_t f + v^k \partial_k f + v'^k \partial_{k'} f$. Finally, we define the function $\mathcal{F}(t) := \int \rho^* F(t, \boldsymbol{x}) \, d^3x = \int \rho^* \rho^{*\prime} f(t, \boldsymbol{x}, \boldsymbol{x}') \, d^3x' d^3x$ and insert F in place of f within Eq. (1). After also using Eq. (3) we obtain

$$\frac{d\mathcal{F}}{dt} = \int \rho^* \rho^{*\prime} \frac{df}{dt} \, d^3x' d^3x \,, \tag{4}$$

with df/dt defined as in Eq. (3).

To establish these results we rely on the integral tricks reviewed in Box 8.4. To obtain Eqs. (8.10) we first differentiate $X = G \int \rho^{*\prime} |\boldsymbol{x} - \boldsymbol{x}'| \, d^3x'$ with respect to time. Using Eq. (2) of Box 8.4, we find $\partial_t X = G \int \rho^{*\prime} v'^k \partial_{k'} |\boldsymbol{x} - \boldsymbol{x}'| \, d^3x'$, or

$$\partial_t X = -G \int \rho^{*\prime} v'_k \frac{(x - x')^k}{|\boldsymbol{x} - \boldsymbol{x}'|} \, d^3x' \,. \tag{8.123}$$

We next differentiate this with respect to x^j and get

$$\partial_{tj} X = -G \int \rho^{*\prime} v'_k \partial_j \frac{(x - x')^k}{|\boldsymbol{x} - \boldsymbol{x}'|} \, d^3x' \,, \tag{8.124}$$

which becomes

$$\partial_{tj} X = -G \int \frac{\rho^{*\prime} v'_j}{|\boldsymbol{x} - \boldsymbol{x}'|} d^3x' + G \int \rho^{*\prime} v'_k \frac{(x - x')^j (x - x')^k}{|\boldsymbol{x} - \boldsymbol{x}'|^3} \tag{8.125}$$

after evaluation of the partial derivative. In view of the definitions for U_j and Φ_j, this is just the first of Eqs. (8.10). If we differentiate instead with respect to time, we get

$$\partial_{tt} X = -G \int \rho^{*\prime} \left[\frac{dv'_k}{dt} \frac{(x - x')^k}{|\boldsymbol{x} - \boldsymbol{x}'|} + v'_k v'^j \partial_{j'} \frac{(x - x')^k}{|\boldsymbol{x} - \boldsymbol{x}'|} \right] d^3x' .$$

The second term within the integral is handled as before, but to evaluate the first we need an expression for $\rho^{*\prime} dv'_k/dt$. This was obtained back in Eq. (8.119), in the form of the post-Newtonian Euler equation, and we may substitute this here. But since X is a post-Newtonian potential that always appears with an accompanying factor of c^{-2}, it is appropriate to truncate the Euler equation to its Newtonian form, and we therefore use $\rho^{*\prime} dv'_k/dt = -\partial_{k'} p' + \rho^{*\prime} \partial_{k'} U' + O(c^{-2})$. Our expression for $\partial_{tt} X$ becomes

$$\begin{aligned} \partial_{tt} X = {} & G \int \partial_{k'} p' \frac{(x - x')^k}{|\boldsymbol{x} - \boldsymbol{x}'|} d^3x' - G \int \rho^{*\prime} \partial_{k'} U' \frac{(x - x')^k}{|\boldsymbol{x} - \boldsymbol{x}'|} d^3x' \\ & + G \int \frac{\rho^{*\prime} v'^2}{|\boldsymbol{x} - \boldsymbol{x}'|} d^3x' - G \int \rho^{*\prime} v'_j v'_k \frac{(x - x')^j (x - x')^k}{|\boldsymbol{x} - \boldsymbol{x}'|^3} d^3x' , \end{aligned} \tag{8.126}$$

and the first term can be changed to $2G \int p' |\boldsymbol{x} - \boldsymbol{x}'|^{-1} d^3x'$ by integration by parts. Taking into account the definitions of Eqs. (8.8), we see that this is just the second of Eqs. (8.10).

Moving on to the integral identities of Eqs. (8.120), we first differentiate $U = \int \rho^{*\prime} |\boldsymbol{x} - \boldsymbol{x}'|^{-1} d^3x'$ with respect to x^j, multiply by ρ^*, and integrate. This gives

$$\int \rho^* \partial_j U \, d^3x = G \int \rho^* \rho^{*\prime} \frac{(x - x')^j}{|\boldsymbol{x} - \boldsymbol{x}'|^3} d^3x' d^3x,$$

and by switching the identities of the integration variables ($\boldsymbol{x} \leftrightarrow \boldsymbol{x}'$), we may also express the right-hand side as $G \int \rho^{*\prime} \rho^* (x' - x)^j |\boldsymbol{x}' - \boldsymbol{x}|^{-3} d^3x d^3x'$. This is equal and opposite to the original integral, and we conclude that the integral vanishes; Eq. (8.120a) is thus established. (We frequently exploit the "switch trick" in the following manipulations.)

With $U^j = G \int \rho^{*\prime} v'^j |\boldsymbol{x} - \boldsymbol{x}'|^{-1} d^3x'$ we find that the integral of $\rho^* U^j$ is given by

$$\int \rho^* U^j \, d^3x = G \int \frac{\rho^* \rho^{*\prime} v'^j}{|\boldsymbol{x} - \boldsymbol{x}'|} d^3x' d^3x. \tag{8.127}$$

Applying the switch trick, we write the right-hand side as $G \int \rho^{*\prime} \rho^* v^j |\boldsymbol{x} - \boldsymbol{x}'|^{-1} d^3x d^3x'$, which we recognize as the integral of $\rho^* U v^j$ with respect to d^3x. Equation (8.120b) is thus established, and Eq. (8.120c) is obtained with very similar manipulations.

Moving on to Eq. (8.120d), we write

$$\int \rho^* v^j \partial_j U \, d^3x = G \int \rho^* \rho^{*\prime} v^j \partial_j |\boldsymbol{x} - \boldsymbol{x}'|^{-1} d^3x' d^3x, \tag{8.128}$$

and note that we can re-express the right-hand side as $G\int \rho^{*\prime}\rho^* v'^j \partial_{j'}|\boldsymbol{x}-\boldsymbol{x}'|^{-1}\,d^3x d^3x'$. Adding the expressions and multiplying by $\frac{1}{2}$, we have

$$\frac{1}{2}G\int \rho^*\rho^{*\prime}\big(v^j\partial_j + v'^j\partial_{j'}\big)|\boldsymbol{x}-\boldsymbol{x}'|^{-1}\,d^3x'd^3x,$$

and according to Eq. (4) of Box 8.4, this is

$$\frac{1}{2}G\frac{d}{dt}\int \rho^*\rho^{*\prime}|\boldsymbol{x}-\boldsymbol{x}'|^{-1}\,d^3x'd^3x.$$

We recognize this as the time derivative of $\frac{1}{2}\int \rho^* U\,d^3x$, and we have established Eq. (8.120d).

The identities of Eqs. (8.120e) and (8.120f) are obtained almost immediately by writing out the integrals and exploiting the switch trick. For Eq. (8.120g) we differentiate the Newtonian potential and construct the integrals that appear on the left-hand side. We have

$$\int \rho^* x^j\partial_t U\,d^3x = G\int \rho^*\rho^{*\prime}v'_k x^j\frac{(x-x')^k}{|\boldsymbol{x}-\boldsymbol{x}'|^3}\,d^3x'd^3x \tag{8.129}$$

and

$$\begin{aligned}\int \rho^* x^j v^k\partial_k U\,d^3x &= -G\int \rho^*\rho^{*\prime}v_k x^j\frac{(x-x')^k}{|\boldsymbol{x}-\boldsymbol{x}'|^3}\,d^3x'd^3x \\ &= G\int \rho^{*\prime}\rho^* v'_k x'^j\frac{(x-x')^k}{|\boldsymbol{x}-\boldsymbol{x}'|^3}\,d^3xd^3x',\end{aligned} \tag{8.130}$$

and combining the results produces

$$\int \rho^* x^j(\partial_t U - v^k\partial_k U)\,d^3x = G\int \rho^*\rho^{*\prime}v'_k\frac{(x-x')^j(x-x')^k}{|\boldsymbol{x}-\boldsymbol{x}'|^3}\,d^3x'd^3x. \tag{8.131}$$

In view of the definition of Φ^j provided by Eq. (8.8), this is just Eq. (8.120g).

8.4.5 Conservation of mass-energy

We now resume our development of post-Newtonian hydrodynamics. In this and the following sections we obtain expressions for the total mass-energy M and momentum P^j of a fluid system, as well as an expression for the position R^j of the center-of-mass, defined in such a way that $M\dot{R}^j = P^j$. Our strategy is to manipulate the local conservation statements of Eqs. (8.110) and (8.112) to obtain integral statements; the conserved integrals are then identified with M and P^j.

The total material mass of the fluid system is

$$m := \int \rho^*\,d^3x, \tag{8.132}$$

and this is conserved by virtue of Eq. (1) of Box 8.4: substituting $f=1$ gives $dm/dt=0$ immediately.

To derive an expression for the total energy E we write Eq. (8.110) as

$$0 = \rho^*\frac{d}{dt}\Big(\frac{1}{2}v^2 + \Pi\Big) - \rho^* v^j\partial_j U + \partial_j(pv^j) + O(c^{-2}). \tag{8.133}$$

Integrating this over the volume occupied by the fluid, we find that the first term gives rise to

$$\frac{d}{dt}\int \rho^*\Big(\frac{1}{2}v^2 + \Pi\Big)\,d^3x$$

after taking the total time derivative outside the integral. By virtue of Eq. (8.120d) we find that the second term contributes

$$\frac{d}{dt}\int \rho^*\Big(-\frac{1}{2}U\Big)\,d^3x,$$

and the third term vanishes (by Gauss's theorem) after integration. Writing

$$E = \mathcal{T} + \Omega + E_{\rm int} + O(c^{-2}) \tag{8.134}$$

with

$$\mathcal{T} := \frac{1}{2}\int \rho^* v^2\,d^3x, \tag{8.135a}$$

$$\Omega := -\frac{1}{2}\int \rho^* U\,d^3x, \tag{8.135b}$$

$$E_{\rm int} := \int \rho^* \Pi\,d^3x, \tag{8.135c}$$

we have shown that $dE/dt = 0$. We recognize in $\mathcal{T}$ the total kinetic energy of the fluid, Ω is the total gravitational potential energy, and $E_{\rm int}$ is the total internal energy; these add up to the conserved energy E.

The total mass-energy M of the fluid system is defined by $M := m + E/c^2$. Combining Eqs. (8.132) and (8.134), this is

$$M := \int \rho^*\Big[1 + \frac{1}{c^2}\Big(\frac{1}{2}v^2 - \frac{1}{2}U + \Pi\Big)\Big]\,d^3x + O(c^{-4}), \tag{8.136}$$

and we have that $dM/dt = 0$. We have encountered the expression of Eq. (8.136) before, back in Sec. 7.3.2, in the context of the post-Minkowskian approximation. In Chapter 7 the total mass-energy was defined by $M := c^{-2}\int(-g)(T^{00} + t^{00}_{\rm LL})\,d^3x$, in terms of the fluid's energy-momentum tensor and the Landau–Lifshitz pseudotensor. In the present context, our expression for M was obtained by manipulating the fluid equations, and it is reassuring that we have complete consistency between the two approaches.

8.4.6 Conservation of momentum

More work is required to identify the total momentum P^j and show that $dP^j/dt = 0$. We return to Eq. (8.112) and examine the second group of post-Newtonian terms. We see that the terms involving the Newtonian potential can be grouped into

$$2\rho^* v_j\frac{dU}{dt} = 2\rho^*\frac{d}{dt}(Uv_j) - 2\rho^* U\frac{dv_j}{dt}, \tag{8.137}$$

or

$$2\rho^* v_j\frac{dU}{dt} = 2\rho^*\frac{d}{dt}(Uv_j) + 2U\partial_j p - 2\rho^* U\partial_j U + O(c^{-2}) \tag{8.138}$$

after inserting the Newtonian version of Euler's equation. The terms involving the vector potential can similarly be expressed as $-4\rho^*(dU_j/dt - v^k\partial_j U_k)$, and we find that the local statement of momentum conservation becomes

$$\begin{aligned}0 &= \partial_t(\mu\rho^* v^j) + \partial_k(\mu\rho^* v^j v^k) + \partial_j p - \rho^*\partial_j U + \frac{2}{c^2}U\partial_j p \\ &\quad - \frac{1}{c^2}\rho^*\Big(\frac{3}{2}v^2 - U + \Pi + p/\rho^*\Big)\partial_j U + \frac{2}{c^2}\rho^*\frac{d}{dt}\Big(Uv_j - 2U_j\Big) \\ &\quad + \frac{4}{c^2}\rho^* v^k\partial_j U_k - \frac{1}{c^2}\rho^*\partial_j\Psi + O(c^{-4}),\end{aligned} \tag{8.139}$$

with μ defined by Eq. (8.113).

We next integrate this equation over the volume occupied by the fluid. Examining each term in turn, we find that the integral of the first term contributes

$$\frac{d}{dt}\int \mu\rho^* v^j\, d^3x,$$

but that the integrals of the second, third, and fourth terms vanish by virtue of Gauss's theorem and the identity of Eq. (8.120a). For the fifth term we get $-2c^{-2}\int p\partial_j U\, d^3x$ after integration by parts. Leaving the sixth term alone for the time being, we note that the seventh term becomes

$$\frac{2}{c^2}\frac{d}{dt}\int \rho^*(Uv_j - 2U_j)\, d^3x = -\frac{2}{c^2}\frac{d}{dt}\int \rho^* U v_j\, d^3x \tag{8.140}$$

after involvement of Eq. (8.120b). And finally, we obtain

$$\begin{aligned}0 &= \frac{d}{dt}\int \mu\rho^* v^j\, d^3x - \frac{2}{c^2}\frac{d}{dt}\int \rho^* U v_j\, d^3x \\ &\quad - \frac{1}{c^2}\int \rho^*\Big(\frac{3}{2}v^2 - U + \Pi + 3p/\rho^*\Big)\partial_j U\, d^3x + \frac{4}{c^2}\int \rho^* v^k\partial_j U_k\, d^3x \\ &\quad - \frac{1}{c^2}\int \rho^*\partial_j\psi\, d^3x - \frac{1}{2c^2}\int \rho^*\partial_{ttj}X\, d^3x + O(c^{-4})\end{aligned} \tag{8.141}$$

after inserting $\Psi = \psi + \frac{1}{2}\partial_{tt}X$ within the last term. This simplifies to

$$0 = \frac{d}{dt}\int \mu\rho^* v^j\, d^3x - \frac{2}{c^2}\frac{d}{dt}\int \rho^* U v_j\, d^3x - \frac{1}{2c^2}\int \rho^*\partial_{ttj}X\, d^3x + O(c^{-4}) \tag{8.142}$$

when we invoke the identities of Eqs. (8.120c) and (8.120e).

We must now work on the term involving the superpotential. We write

$$\rho^*\partial_{ttj}X = \rho^*\frac{d}{dt}\big(\partial_{tj}X\big) - \rho^* v^k\partial_{tjk}X \tag{8.143}$$

and integrate. We obtain

$$\int \rho^*\partial_{ttj}X\, d^3x = -\frac{d}{dt}\int \rho^* U_j\, d^3x + \frac{d}{dt}\int \rho^*\Phi_j\, d^3x \tag{8.144}$$

after using Eqs. (8.10) and (8.122), and we note that by virtue of Eq. (8.120b), the first integral on the right-hand side can also be expressed as $\int \rho^* U v_j\, d^3x$.

Collecting results, we have obtained the conservation statement $dP^j/dt = 0$, with the total momentum P^j identified as

$$P^j := \int \rho^* v^j \biggl[1 + \frac{1}{c^2}\Bigl(\frac{1}{2}v^2 - \frac{1}{2}U + \Pi + p/\rho^* \Bigr) \biggr] d^3x - \frac{1}{2c^2} \int \rho^* \Phi^j \, d^3x + O(c^{-4}). \tag{8.145}$$

We recall that the potential Φ^j was defined by Eq. (8.8). In Chapter 7 the total momentum was defined as $P^j := c^{-1} \int (-g)(T^{0j} + t^{0j}_{\rm LL})\, d^3x$, and this led to the expression displayed in Eq. (7.65). A glance at Eq. (8.120g) confirms that the two expressions are equivalent.

The total momentum of a post-Newtonian spacetime can always be made to vanish by performing a post-Galilean transformation of the type described in Sec. 8.3.6. The transformation is characterized by the velocity vector $V^j = P^j/M$.

8.4.7 Center-of-mass

Inspection of Eq. (8.132) suggests that a plausible expression for the position of the center-of-mass might be

$$R^j := \frac{1}{M} \int \rho^* x^j \biggl[1 + \frac{1}{c^2}\Bigl(\frac{1}{2}v^2 - \frac{1}{2}U + \Pi \Bigr) \biggr] d^3x + O(c^{-4}). \tag{8.146}$$

This matches the result obtained back in Sec. 7.3.2 on the basis of the post-Minkowskian definition $R^j := (Mc^2)^{-1} \int (-g)(T^{00} + t^{00}_{\rm LL}) x^j \, d^3x$. We confirm this result by proving that with Eq. (8.146), we can produce the expected center-of-mass relation

$$M\dot{R}^j = P^j + O(c^{-4}). \tag{8.147}$$

Once P^j has been set equal to zero by performing a post-Galilean transformation, the position of the center-of-mass is fixed in space, and a constant translation of the spatial coordinates allows us to set $R^j = 0$. The conditions $P^j = 0$ and $R^j = 0$ define the *center-of-mass frame* of the fluid system.

We express Eq. (8.146) as $MR^j = \int \nu\rho^* x^j \, d^3x + O(c^{-4})$, with $\nu = 1 + c^{-2}(\frac{1}{2}v^2 - \frac{1}{2}U + \Pi)$. Differentiation with respect to time produces

$$M\dot{R}^j = \int \nu\rho^* v^j \, d^3x + \int \rho^* x^j \frac{d\nu}{dt} \, d^3x, \tag{8.148}$$

and this can be written in the form

$$M\dot{R}^j = P^j - \frac{1}{c^2} \int p v^j \, d^3x + \frac{1}{2c^2} \int \rho^* \Phi^j \, d^3x + \int \rho^* x^j \frac{d\nu}{dt} \, d^3x \tag{8.149}$$

after incorporating Eq. (8.145). The derivative of ν can be evaluated with the help of Euler's equation (8.104) and the first law of thermodynamics, Eq. (8.105). We obtain

$$\rho^* x^j \frac{d\nu}{dt} = -\frac{1}{c^2} x^j \partial_k (p v^k) - \frac{1}{2c^2} \rho^* x^j \bigl(\partial_t U - v^k \partial_k U \bigr) + O(c^{-4}), \tag{8.150}$$

and integration yields

$$\int \rho^* x^j \frac{dv}{dt}\, d^3x = \frac{1}{c^2}\int p v^j\, d^3x - \frac{1}{2c^2}\int \rho^* \Phi^j\, d^3x + O(c^{-4}); \tag{8.151}$$

we integrated the first term by parts and made use of Eq. (8.120g) for the second term. Making the substitution into Eq. (8.149), we see that Eq. (8.147) is indeed satisfied.

8.5 Bibliographical notes

The Maxwell-like formulation of the equations of post-Newtonian theory, as reviewed in Box 8.1, has received a number of presentations in the literature. One of the earliest incarnations was provided by Braginskii, Caves, and Thorne (1977).

The classic approach to post-Newtonian theory can be traced to the earliest days of general relativity. Representative works are Lorentz and Droste (1917), Eddington and Clark (1938), Einstein, Infeld, and Hoffmann (1938), and the treatise by Fock (1959). The work was invigorated by Chandrasekhar in the nineteen sixties, through a series of papers written with students and collaborators: Chandrasekhar and Contopoulos (1967), Chandrasekhar (1965 and 1969), Chandrasekhar and Nutku (1969), and Chandrasekhar and Esposito (1970).

The theory of post-Newtonian coordinate transformations developed in Sec. 8.3 was first initiated by Damour, Soffel, and Xu (1991); our treatment follows Racine and Flanagan (2005). The post-Galilean subclass of transformations was first investigated in Chandrasekhar and Contopoulos (1967). The rotating coordinates of Box 8.3 and the darned disk are described in some detail in Robertson and Noonan (1968).

The post-Newtonian theory of fluid dynamics was first developed in Chandrasekhar (1965 and 1969). Our treatment in Sec. 8.4 follows the master's work quite closely.

8.6 Exercises

8.1 Show that the inverse to the metric of Eqs. (8.20) is given by

$$g^{00} = -1 - \frac{2}{c^2}U - \frac{2}{c^4}\left(\Psi + U^2\right) + O(c^{-6}),$$

$$g^{0j} = -\frac{4}{c^3}U^j + O(c^{-5}),$$

$$g^{jk} = \left(1 - \frac{2}{c^2}U\right)\delta^{jk} + O(c^{-4}),$$

where $U^j := \delta^{jk}U_k$. Show that the metric determinant is $\sqrt{-g} = 1 + 2U/c^2 + O(c^{-4})$. Verify Eqs. (8.15) for the Christoffel symbols.

8.2 Show that the post-Newtonian version of the geodesic equation $Du^\alpha/d\tau = 0$ can be presented in the form

$$\frac{d(g_s \gamma \boldsymbol{v})}{dt} = \gamma\Big(\boldsymbol{E}_g + \boldsymbol{v} \times \boldsymbol{B}_g + v^2 \boldsymbol{\nabla} g_s\Big) + O(c^{-4}),$$

where $\boldsymbol{E}_g$ and $\boldsymbol{B}_g$ are the gravitational fields defined in Box 8.1, $\gamma := dt/d\tau$, and $g_s = 1 + 2c^{-2}U$ is the coefficient of the spatial part of the metric.

8.3 (a) Show that the coordinate transformation

$$t = \bar{t}, \qquad x^j = \bar{x}^j + \frac{\lambda}{c^2}\partial^{\bar{j}}\bar{X},$$

in which λ is a constant, produces a new post-Newtonian metric given by

$$\bar{g}_{00} = g_{00} - \frac{2\lambda}{c^4}(\bar{U}^2 - \bar{\Phi}_2 + \bar{\Phi}_W),$$

$$\bar{g}_{0j} = g_{0j} + \frac{\lambda}{c^3}\partial_{\bar{t}\bar{j}}\bar{X},$$

$$\bar{g}_{jk} = g_{jk} + \frac{2\lambda}{c^2}\partial_{\bar{j}\bar{k}}\bar{X},$$

where $g_{\alpha\beta}$ denotes the original post-Newtonian metric, with all potentials defined in terms of $\bar{x}^\alpha$, and where $\bar{\Phi}_W$ is an auxiliary potential (known as the Whitehead potential) defined by

$$\bar{\Phi}_W := G^2 \int \bar{\rho}^{*\prime}\bar{\rho}^{*\prime\prime}\frac{(\bar{x}-\bar{x}')_j}{|\bar{\boldsymbol{x}}-\bar{\boldsymbol{x}}'|^3}\left[\frac{(\bar{x}'-\bar{x}'')^j}{|\bar{\boldsymbol{x}}-\bar{\boldsymbol{x}}''|} - \frac{(\bar{x}-\bar{x}'')^j}{|\bar{\boldsymbol{x}}'-\bar{\boldsymbol{x}}''|}\right] d^3\bar{x}' d^3\bar{x}''.$$

(b) Consider a static system with a "point mass" at the origin. This assumption allows us to ignore the potentials $\bar{\psi}$, $\bar{\Phi}_2$, and $\bar{\Phi}_W$, and to set all time derivatives to zero. Find the value of λ for which the metric is linear in $\bar{U}$ to post-Newtonian order. Show that the line element in spherical polar coordinates can be expressed to post-Newtonian order as

$$ds^2 = -(1 - R/\bar{r})(cd\bar{t})^2 + (1 - R/\bar{r})^{-1}d\bar{r}^2 + \bar{r}^2(d\bar{\theta}^2 + \sin^2\bar{\theta}d\bar{\phi}^2),$$

where $R = (2G/c^2)\int \rho^* d^3x = 2Gm/c^2$. What is this metric?

8.4 (a) Using the expressions for the Landau–Lifshitz pseudotensor given in Eqs. (7.48) and (7.49), together with the post-Newtonian expression for the potentials $h_2^{\alpha\beta}$ from Box 7.5, show that the components of the effective energy-momentum pseudotensor τ^{0j} and τ^{jk} are given to post-Newtonian order by

$$c^{-1}\tau^{0j} = \rho^* v^j\left[1 + \frac{1}{c^2}\left(\frac{1}{2}v^2 + 3U + \Pi + p/\rho^*\right)\right] + \frac{1}{4\pi G c^2}\Big[3\partial_t U \partial^j U + 4(\partial^j U^k - \partial^k U^j)\partial_k U\Big] + O(c^{-4}),$$

and

$$\tau^{jk} = \rho^* v^j v^k \left[1 + \frac{1}{c^2}\left(\frac{1}{2}v^2 + 3U + \Pi + p/\rho^*\right)\right] + p\left(1 + \frac{2}{c^2}U\right)\delta^{jk}$$
$$+ \frac{1}{4\pi G}\left(\partial^j U \partial^k U - \frac{1}{2}\delta^{jk}\partial_n U \partial^n U\right)$$
$$\frac{1}{4\pi G c^2}\Big\{2\partial^{(j} U \partial^{k)}\Psi - 16\partial^{[j} U^{n]}\partial^{[k} U^{n]} + 8\partial^{(j} U \partial_t U^{k)}$$
$$- \delta^{jk}\Big[\partial^n U \partial^n \Psi - 4\partial^{[m} U^{n]}\partial^{[m} U^{n]} + 4\partial^n U \partial_t U^n + \frac{3}{2}(\partial_t U)^2\Big]\Big\}$$
$$+ O(c^{-4}),$$

where $\Psi = \psi + \frac{1}{2}\partial_{tt} X$.

(b) Show that the conservation statement $\partial_\beta \tau^{j\beta} = 0$ yields the post-Newtonian version of Euler's equation, as displayed in Eq. (8.119). You may make use of the continuity equation for ρ^*, the first law of thermodynamics, and the Poisson equations satisfied by the various potentials.

8.5 In this problem we consider the equilibrium structure of a spherical body in post-Newtonian theory, thereby generalizing the Newtonian discussion of Sec. 2.2. We assume that the matter distribution is static and spherically symmetric, so that all variables depend on r only. Show that under these conditions, the gravitational potentials are determined by the set of equations

$$\frac{dU}{dr} = -\frac{Gm}{r^2}, \qquad \frac{dm}{dr} = 4\pi r^2 \rho^*,$$

$$\frac{d\Psi}{dr} = -\frac{Gn}{r^2}, \qquad \frac{dn}{dr} = 4\pi r^2 \rho^*(-U + \Pi + 3p/\rho^*),$$

where n is a post-Newtonian auxiliary variable analogous to the Newtonian mass function m. Show also that the equation of hydrostatic equilibrium becomes

$$\frac{dp}{dr} = -\frac{G\rho^*}{r^2}\left\{m + \frac{1}{c^2}\Big[(-3U + \Pi + p/\rho^*)m + n\Big]\right\} + O(c^{-4}).$$

These equations are to be supplemented by an equation of state $p = p(\rho^*)$ and the first law of thermodynamics, $d\Pi = (p/\rho^{*2})\,d\rho^* + O(c^{-2})$.

8.6 The equations derived in the preceding problem should agree with the exact formulation of the structure equations, as presented in Sec. 5.6.5, when these are expressed as a post-Newtonian expansion. The comparison, however, is not entirely straightforward, because the formulations use different variables and coordinates.

(a) By comparing the metric of Eq. (5.156) with the post-Newtonian metric of Eq. (8.2), relate the radial coordinate $\bar{r}$ of Sec. 5.6.5 (perversely denoted r there) to the harmonic radial coordinate r employed in this chapter.

(b) Show that the metric functions are related by $\Phi = U + \Psi/c^2 + O(c^{-4})$ and $\bar{m} = m + O(c^{-2})$, where $\bar{m}$ is the mass function defined by Eq. (5.158) (and denoted m there).

(c) Prove that when Eqs. (5.215), (5.216), and (5.218) are expanded through order c^{-2} in a post-Newtonian expansion, they agree with the equations derived in the preceding problem.

(d) Show that the comparison relies on the identification

$$\bar{m} = m + \frac{1}{c^2}\big(n + mU - Gm^2/r - 4\pi r^3 p\big) + O(c^{-4})$$

for the relativistic mass function.

8.7 Using the Newtonian Euler equation and the first law of thermodynamics, verify the energy conservation equation (8.110).

8.8 Show that the conserved angular momentum tensor for an isolated system is given to post-Newtonian order by

$$J^{jk} = 2\int \rho^* x^{[j}\bigg\{v^{k]} + \frac{v^{k]}}{c^2}\bigg(\frac{1}{2}v^2 + 3U + \Pi + p/\rho^*\bigg) - \frac{1}{c^2}\bigg(4U^{k]} - \frac{1}{2}\partial_t^{k]}X\bigg)\bigg\}d^3x\,.$$

8.9 Verify the integral identities of Eqs. (8.120e) and (8.120f).

9 Post-Newtonian theory: System of isolated bodies

In this chapter we apply the results of Chapter 8 to situations in which a fluid distribution breaks up into a collection of separated bodies. Our aim is to go from a fine-grained description involving the fluid variables $\{\rho^*, p, \Pi, \boldsymbol{v}\}$ to a coarse-grained description involving a small number of center-of-mass variables for each body. We accomplish this reduction in Sec. 9.1, and in Sec. 9.2 we apply it to a calculation of the spacetime metric in the empty region between bodies; the metric is thus expressed in terms of the mass-energy M_A, position $\boldsymbol{r}_A(t)$, and velocity $\boldsymbol{v}_A(t)$ of each body. In Sec. 9.3 we derive post-Newtonian equations of motion for the center-of-mass positions, and in Sec. 9.4 we show that the same equations apply to compact bodies with strong internal gravity. In Sec. 9.5 we allow the bodies to rotate, and we calculate the influence of the spins on the metric and equations of motion; we also derive evolution equations for the spin vectors. We conclude in Sec. 9.6 with a discussion of how point particles can be usefully incorporated within post-Newtonian theory, in spite of their infinite densities and diverging gravitational potentials.

9.1 From fluid configurations to isolated bodies

We consider a situation in which a distribution of perfect fluid breaks up into a number N of separated components. We call each component a "body," and label each body with the index $A = 1, 2, \ldots, N$. Mathematically, this means that we can express the fluid density as

$$\rho^* = \sum_A \rho^*_A; \tag{9.1}$$

the sum extends over each body, and ρ^*_A is zero everywhere except within the volume occupied by body A. In this situation we forbid the presence of matter between the bodies; for example, there is no mass transfer between any members of the N-body system.

9.1.1 Center-of-mass variables

The material mass of body A is

$$m_A := \int_A \rho^* \, d^3x. \tag{9.2}$$

The domain of integration V_A is a time-independent region of three-dimensional space that extends beyond the volume occupied by the body. It is sufficiently large that in a time interval dt, the body will not cross the domain's boundary S_A; but it is sufficiently small that V_A does not include nor intersect another body within the system. We could have inserted ρ_A^* in place of ρ^* inside the integral, but since $\rho^* = \rho_A^*$ within V_A, this distinction is not necessary. By virtue of Eq. (1) in Box 8.4, we have that $dm_A/dt = 0$.

We *define* the position of the center-of-mass of body A by

$$\boldsymbol{r}_A(t) := \frac{1}{m_A}\int_A \rho^* \boldsymbol{x}\, d^3x. \tag{9.3}$$

This definition is largely arbitrary (as we have observed before in Box 1.7), but it proves convenient for our purposes: It produces simple expressions for the gravitational potentials, and the equations of motion for each body take a simple form when expressed in terms of $\boldsymbol{r}_A$. We next introduce

$$\boldsymbol{v}_A(t) := \frac{d\boldsymbol{r}_A}{dt} = \frac{1}{m_A}\int_A \rho^* \boldsymbol{v}\, d^3x \tag{9.4}$$

as the velocity of the body taken as a whole, and the body's acceleration is

$$\boldsymbol{a}_A(t) := \frac{d\boldsymbol{v}_A}{dt} = \frac{1}{m_A}\int_A \rho^* \frac{d\boldsymbol{v}}{dt}\, d^3x. \tag{9.5}$$

We evaluate this in Sec. 9.3 by inserting the post-Newtonian Euler equation within the integral.

9.1.2 Relative variables; reflection symmetry

To carry out the integrations over the domain V_A it is convenient to introduce the relative variables

$$\bar{\boldsymbol{x}} := \boldsymbol{x} - \boldsymbol{r}_A(t), \qquad \bar{\boldsymbol{v}} := \boldsymbol{v} - \boldsymbol{v}_A(t); \tag{9.6}$$

the vector $\bar{\boldsymbol{x}}$ gives the position of a fluid element relative to the center-of-mass $\boldsymbol{r}_A(t)$, while $\bar{\boldsymbol{v}}$ measures the velocity of this fluid element relative to the body velocity $\boldsymbol{v}_A(t)$. Under this transformation the domain V_A is translated by $-\boldsymbol{r}_A(t)$ and becomes a neighborhood of the origin, and the volume element becomes $d^3\bar{x} = d^3x$.

For technical reasons it will be useful to assume that each body is *reflection-symmetric* about its center-of-mass. Mathematically, this means that we shall impose the property

$$\rho^*(t, \boldsymbol{r}_A - \bar{\boldsymbol{x}}) = \rho^*(t, \boldsymbol{r}_A + \bar{\boldsymbol{x}}), \tag{9.7}$$

and subject the fluid pressure p and the specific internal energy density Π to the same symmetry. There is nothing particularly deep about this assumption, which is far less restrictive, for example, than demanding that each body be spherically symmetric. A generic body will not satisfy this symmetry requirement, but computations carried out for this generic body will involve a lot of additional details that can be avoided if we impose Eq. (9.7). In particular, the symmetry implies that *any body integral that involves the product of an odd number of internal vectors*, such as $\bar{\boldsymbol{x}}$ or $\bar{\boldsymbol{v}}$, *will vanish identically*.

For example, the integral $\int_A \rho^*(\boldsymbol{r}_A + \bar{\boldsymbol{x}})\bar{x}^j\bar{x}^k\bar{v}^n\, d^3\bar{x}$ becomes $-\int_A \rho^*(\boldsymbol{r}_A - \bar{\boldsymbol{x}})\bar{x}^j\bar{x}^k\bar{v}^n\, d^3\bar{x}$ under a reflection across the center-of-mass, and Eq. (9.7) implies that the integral must vanish. In the following we frequently exploit this observation and eliminate many such odd integrals; and this is, ultimately, the sole reason for introducing the symmetry requirement of Eq. (9.7). In the course of our development we will accumulate a lot of evidence to support the claim that the gravitational field outside the bodies, and the motion of these bodies, are largely insensitive to details of internal structure. We might as well, therefore, impose the reflection symmetry and benefit from its great convenience, feeling sure that the longer computations required for a generic body would lead to the same answers.

9.1.3 Structure integrals; equilibrium conditions

As we shall see, the internal structure of each body is characterized by a number of structure integrals, which we introduce here. We first have the scalar quantities

$$\mathcal{T}_A := \frac{1}{2}\int_A \rho^* \bar{v}^2\, d^3\bar{x}, \tag{9.8a}$$

$$\Omega_A := -\frac{1}{2}G\int_A \frac{\rho^*\rho^{*\prime}}{|\bar{\boldsymbol{x}} - \bar{\boldsymbol{x}}'|}\, d^3\bar{x}'d^3\bar{x}, \tag{9.8b}$$

$$P_A := \int_A p\, d^3\bar{x}, \tag{9.8c}$$

$$E_A^{\rm int} := \int_A \rho^*\Pi\, d^3\bar{x}, \tag{9.8d}$$

$$H_A := G\int_A \rho^*\rho^{*\prime}\frac{\bar{v}'_j(\bar{x} - \bar{x}')^j}{|\bar{\boldsymbol{x}} - \bar{\boldsymbol{x}}'|^3}\, d^3\bar{x}'d^3\bar{x}, \tag{9.8e}$$

which are all functions of time t. Here ρ^*, p, Π, and $\bar{\boldsymbol{v}}$ are all expressed as functions of t and position $\boldsymbol{r}_A(t) + \bar{\boldsymbol{x}}$, and $\rho^{*\prime}$ stands for $\rho^*(t, \boldsymbol{r}_A + \bar{\boldsymbol{x}}')$. We have encountered some of these quantities before: $\mathcal{T}_A$ is recognized as the total kinetic energy of body A (as measured in A's comoving frame), Ω_A is the total gravitational potential energy, P_A is the integrated pressure, and $E_A^{\rm int}$ is the total internal energy; H_A is a new quantity.

We also have the tensorial quantities

$$I_A^{jk} := \int_A \rho^*\bar{x}^j\bar{x}^k\, d^3\bar{x}, \tag{9.9a}$$

$$S_A^{jk} := \int_A \rho^*\left(\bar{x}^j\bar{v}^k - \bar{x}^k\bar{v}^j\right) d^3\bar{x}, \tag{9.9b}$$

$$\mathcal{T}_A^{jk} := \frac{1}{2}\int_A \rho^*\bar{v}^j\bar{v}^k\, d^3\bar{x}, \tag{9.9c}$$

$$L_A^{jk} := \int_A \bar{v}^j\partial_k p\, d^3\bar{x}, \tag{9.9d}$$

$$\Omega_A^{jk} := -\frac{1}{2}G\int_A \rho^*\rho^{*\prime}\frac{(\bar{x} - \bar{x}')^j(\bar{x} - \bar{x}')^k}{|\bar{\boldsymbol{x}} - \bar{\boldsymbol{x}}'|^3}\, d^3\bar{x}'d^3\bar{x}, \tag{9.9e}$$

(continued overleaf)

$$H_A^{jk} := G \int_A \rho^* \rho^{*\prime} \frac{\bar{v}'^j (\bar{x} - \bar{x}')^k}{|\bar{x} - \bar{x}'|^3} \, d^3\bar{x}' d^3\bar{x}, \tag{9.9f}$$

$$K_A^{jk} := G \int_A \rho^* \rho^{*\prime} \frac{\bar{v}'_n (\bar{x} - \bar{x}')^n (\bar{x} - \bar{x}')^j (\bar{x} - \bar{x}')^k}{|\bar{x} - \bar{x}'|^5} \, d^3\bar{x}' d^3\bar{x}, \tag{9.9g}$$

where $\bar{v}'^j$ is the relative velocity field at position $\boldsymbol{r}_A + \bar{\boldsymbol{x}}'$. Some of these structure integrals, like the quadrupole moment I_A^{jk}, the angular-momentum tensor S_A^{jk}, the kinetic-energy tensor $\mathcal{T}_A^{jk}$, and the potential-energy tensor Ω_A^{jk}, were encountered before; note that $\mathcal{T}_A = \delta_{jk} \mathcal{T}_A^{jk}$ and $\Omega_A = \delta_{jk} \Omega_A^{jk}$. The others, H_A^{jk}, K_A^{jk}, and L_A^{jk}, are new; note that $H_A = \delta_{jk} H_A^{jk}$.

As an additional assumption concerning the bodies, we shall take them to be in *dynamical equilibrium*. By this we mean that each body has had time, under its own internal dynamics, to relax to a steady state in which its structure properties do not depend on time. This means, in particular, that the structure integrals listed in Eqs. (9.8) and (9.9) can all be taken to be time-independent. And as we show below in Sec. 9.1.7, the assumption also implies the validity of the *equilibrium conditions*

$$2\mathcal{T}_A^{jk} + \Omega_A^{jk} + \delta^{jk} P_A = O(c^{-2}) \tag{9.10}$$

and

$$4H_A^{(jk)} - 3K_A^{jk} + \delta^{jk} \dot{P}_A - 2L_A^{(jk)} = O(c^{-2}). \tag{9.11}$$

We also record the trace of Eq. (9.10),

$$2\mathcal{T}_A + \Omega_A + 3P_A = O(c^{-2}). \tag{9.12}$$

It is important to understand that the equilibrium conditions are valid only approximately. We have insisted that each body should reach an equilibrium state under its own internal dynamics, which involves hydrodynamical processes and the body's own gravitational field. Each body, however, is also subjected to the gravitational influence of the remaining bodies, and this external dynamics comes in two different guises. The first effect, and by far the most important one, is the motion of the body's center-of-mass in the field of the external bodies; this will be considered in Sec. 9.3, and the key point here is that the motion of the body as a whole does not prevent it from reaching an internal equilibrium state. The second effect, however, produces a small deviation from equilibrium; this is the tidal interaction between the body and its companions, produced by inhomogeneities in the external gravitational field across the volume occupied by the body. This effect, however, is small when the bodies are widely separated (we shall quantify this in the following subsection), and we shall ignore it in this and the following sections. Our conclusion is that while the equilibrium conditions are approximate, they hold to a large degree of accuracy, and this degree is quite sufficient for our purposes.

9.1.4 Multipole structure

In principle the mass distribution of each body is characterized by an infinite number of mass multipole moments $I_A^L(t)$, and the fluid motions within each body are characterized by an infinite number of current multipole moments $J_A^{jL}(t)$; here $L := k_1 k_2 \cdots k_\ell$ is a multi-index that contains a number l of individual indices. The most important moments have already been introduced: we have the mass monopole moment m_A, the quadrupole moment I_A^{jk}, and the angular-momentum tensor S_A^{jk}. We shall first simplify this description by demanding that the bodies be well separated.

This condition can be formulated as follows. Let R_A be a length scale associated with the volume occupied by body A, and let $s_A := |\boldsymbol{x} - \boldsymbol{r}_A|$ be the distance from A's center-of-mass. We assume that in most applications that will interest us,

$$R_A \ll s_A. \tag{9.13}$$

This implies that when, for example, we evaluate the gravitational potentials outside each body, we are allowed to ignore all terms generated by the quadrupole moment I_A^{jk} and its higher-order analogues. To see this, recall that relative to the monopole term in the Newtonian potential, the quadrupole term scales as $I_A^{jk}/(m_A s_A^2)$, or $(R_A/s_A)^2$, which is small by virtue of Eq. (9.13). The suppression is even more dramatic when the body is intrinsically spherical, and deformed ever so slightly by its tidal coupling with the other bodies; in this case I_A^{jk} scales as $m_A R_A^5/r_{AB}^3$ instead of $m_A R_A^2$, where $r_{AB} := |\boldsymbol{r}_A - \boldsymbol{r}_B|$ is the inter-body distance, and the quadrupole term is smaller than the monopole term by a factor of order $R_A^5/(r_{AB}^3 s_A^2) \ll 1$. On the other hand, the condition of Eq. (9.13) is obviously invalid when we examine the internal structure of each body.

In our initial treatment of the N-body system, we shall also simplify the multipole description by taking each body to be non-rotating. We shall therefore set

$$S_A^{jk} = 0, \tag{9.14}$$

and maintain this assumption until Sec. 9.5, in which we finally incorporate the spinning motion of each body into the metric and equations of motion.

9.1.5 Internal and external potentials

In the following computations we shall have to distinguish between the gravitational potentials produced by body A and those produced by the remaining bodies in the system. To accomplish this we proceed as in Sec. 1.6.3 and introduce a decomposition of each potential into internal and external pieces. For example, the Newtonian potential U is decomposed as

$$U = U_A + U_{\neg A}, \tag{9.15}$$

with

$$U_A(t, \boldsymbol{x}) = G \int_A \frac{\rho^*(t, \boldsymbol{x}')}{|\boldsymbol{x} - \boldsymbol{x}'|} d^3x' \tag{9.16}$$

denoting the internal piece, and

$$U_{\neg A}(t, \boldsymbol{x}) = \sum_{B \neq A} G \int_B \frac{\rho^*(t, \boldsymbol{x}')}{|\boldsymbol{x} - \boldsymbol{x}'|} d^3x' \tag{9.17}$$

denoting the external piece.

As another example we examine the auxiliary potential Φ_2 introduced in Eq. (8.8). Its decomposition is

$$\Phi_2 = \Phi_{2,A} + \Phi_{2,\neg A}, \tag{9.18}$$

with

$$\Phi_{2,A} = G\int_A \frac{\rho^{*\prime}U'}{|\boldsymbol{x}-\boldsymbol{x}'|}\,d^3x' \tag{9.19}$$

and

$$\Phi_{2,\neg A} = \sum_{B\neq A} G\int_B \frac{\rho^{*\prime}U'}{|\boldsymbol{x}-\boldsymbol{x}'|}\,d^3x'. \tag{9.20}$$

But these expressions involve the Newtonian potential, and this should also be decomposed into internal and external pieces. For $\Phi_{2,A}$ we express U' as in Eq. (9.15) and get

$$\Phi_{2,A} = G\int_A \frac{\rho^{*\prime}U'_A}{|\boldsymbol{x}-\boldsymbol{x}'|}\,d^3x' + G\int_A \frac{\rho^{*\prime}U'_{\neg A}}{|\boldsymbol{x}-\boldsymbol{x}'|}\,d^3x'. \tag{9.21}$$

For $\Phi_{2,\neg A}$ we must be more careful, because the internal-external decomposition should now refer to each body B instead of body A. In fact, to be fully clear we should refine our decomposition and write $U' = U'_A + U'_B + U'_{\neg AB}$, with $U'_{\neg AB} := \sum_{C\neq A,B}\int_C \rho^{*\prime\prime}|\boldsymbol{x}'-\boldsymbol{x}''|^{-1}\,d^3x''$ now excluding bodies A and B. Then Eq. (9.20) becomes

$$\Phi_{2,\neg A} = \sum_{B\neq A}\left\{G\int_B \frac{\rho^{*\prime}U'_A}{|\boldsymbol{x}-\boldsymbol{x}'|}\,d^3x' + G\int_B \frac{\rho^{*\prime}U'_B}{|\boldsymbol{x}-\boldsymbol{x}'|}\,d^3x' + G\int_B \frac{\rho^{*\prime}U'_{\neg AB}}{|\boldsymbol{x}-\boldsymbol{x}'|}\,d^3x'\right\}. \tag{9.22}$$

Subtleties like these arise also in the decomposition of the potential Φ_5; they are, of course, a consequence of the non-linear nature of the field equations.

9.1.6 Total mass-energy

The total mass-energy of body A is $M_A := m_A + (\mathcal{T}_A + \Omega_A + E_A^{\rm int})/c^2 + O(c^{-4})$, or

$$M_A := \int_A \rho^*\left[1 + \frac{1}{c^2}\left(\frac{1}{2}\bar{v}^2 - \frac{1}{2}U_A + \Pi\right)\right]d^3x + O(c^{-4}) \tag{9.23}$$

in view of Eqs. (9.8) and (9.16). In Sec. 8.4.5 we learned that the total mass-energy M of the entire fluid system is a conserved quantity, $dM/dt = 0$. The manipulations that led to this conclusion involved the integration of the fluid equations over the volume occupied by the entire fluid. The integrations, however, can be limited to the volume occupied by body A, and it is easy to see that the same manipulations would now reveal that each M_A is conserved when the time dependence of external multipole moments can be ignored:

$$\frac{dM_A}{dt} = 0. \tag{9.24}$$

The relation between M and the sum $\sum_A M_A$ is elucidated below in Sec. 9.3.6; the gravitational interaction between bodies prevents these quantities from being equal to each other.

We shall see below that it is only the combination M_A that appears in the metric of an N-body system; the components m_A, $\mathcal{T}_A$, Ω_A, or $E_A^{\rm int}$ do not make individual appearances. We shall see also that only M_A appears in the equations of motion. The spacetime of an N-body system therefore depends only on the total mass-energy of each body, and the decomposition of this mass-energy into material mass, kinetic energy, gravitational potential energy, and internal energy is of no consequence; two bodies with vastly different internal compositions but the same mass-energy will produce the same spacetime and move in identical manners. This is a statement of the *strong equivalence principle* in post-Newtonian theory. (Refer to Sec. 5.1 for a discussion of the weak, strong, and Einstein versions of the principle of equivalence.)

The important physical role of the total mass-energy M_A suggests that we might refine our notion of center-of-mass and adopt

$$\boldsymbol{R}_A := \frac{1}{M_A}\int_A \rho^* \boldsymbol{x}\biggl[1 + \frac{1}{c^2}\biggl(\frac{1}{2}\bar{v}^2 - \frac{1}{2}U_A + \Pi\biggr)\biggr]\, d^3x + O(c^{-4}) \tag{9.25}$$

instead of Eq. (9.3) as a proper definition of center-of-mass position. Fortunately, the definitions are equivalent under the symmetry assumption of Eq. (9.7):

$$\boldsymbol{R}_A = \boldsymbol{r}_A + O(c^{-4}). \tag{9.26}$$

To prove this we insert $\boldsymbol{x} = \boldsymbol{r}_A + \bar{\boldsymbol{x}}$ inside the integral, and see immediately that $\boldsymbol{R}_A$ differs from $\boldsymbol{r}_A$ by the integral

$$\frac{1}{M_A c^2}\int_A \rho^* \bar{\boldsymbol{x}}\biggl(\frac{1}{2}\bar{v}^2 - \frac{1}{2}U_A + \Pi\biggr)\, d^3\bar{x}.$$

The first and third terms vanish after integration, because the integrals involve an odd number of internal vectors. The second term also leads to a vanishing integral, because

$$\int_A \rho^* \bar{\boldsymbol{x}} U_A\, d^3\bar{x} = G\int_A \rho^* \rho^{*\prime} \frac{\bar{\boldsymbol{x}}}{|\bar{\boldsymbol{x}} - \bar{\boldsymbol{x}}'|}\, d^3\bar{x}' d^3\bar{x} \tag{9.27}$$

is odd under a reflection across the center-of-mass. The integral vanishes, and we have established the equality of $\boldsymbol{R}_A$ and $\boldsymbol{r}_A$. There is no need to modify our definition for the center-of-mass position.

9.1.7 Virial identities

In this subsection we derive the virial identities that give rise to the equilibrium conditions of Eqs. (9.10) and (9.11). They are

$$\frac{1}{2}\dot{I}_A^{jk} = \frac{1}{2}S_A^{jk} + \int_A \rho^* \bar{v}^j \bar{x}^k\, d^3\bar{x}, \tag{9.28a}$$

$$\frac{1}{2}\ddot{I}_A^{jk} = 2\mathcal{T}_A^{jk} + \Omega_A^{jk} + \delta^{jk}P_A + \int_A \rho^* \bar{x}^{(j}\partial^{k)}U_{\neg A}\, d^3\bar{x} + O(c^{-2}), \tag{9.28b}$$

$$\begin{aligned}\frac{1}{2}\dddot{I}_A^{jk} &= 4H_A^{(jk)} - 3K_A^{jk} + \delta^{jk}\dot{P}_A - 2L_A^{(jk)} + \int_A \rho^* \bar{x}^{(j}\frac{d}{dt}\partial^{k)}U_{\neg A}\, d^3\bar{x}\\ &\quad + 3\int_A \rho^* \bar{v}^{(j}\partial^{k)}U_{\neg A}\, d^3\bar{x} + O(c^{-2}).\end{aligned} \tag{9.28c}$$

The identities are generated by repeated differentiation of $I_A^{jk} = \int_A \rho^* \bar{x}^j \bar{x}^k \, d^3\bar{x}$ with respect to time. Equation (9.28a) follows easily from a first differentiation and the definition of the angular-momentum tensor; when the body is non-spinning and in dynamical equilibrium, the identity reveals that

$$\int_A \rho^* \bar{v}^j \bar{x}^k \, d^3\bar{x} = 0. \tag{9.29}$$

Equation (9.28b) features an integral involving the external Newtonian potential in addition to the structure integrals introduced in Eq. (9.9); under the condition of Eq. (9.13) the external term is suppressed by a factor of order $(R_A/r_{AB})^2 \ll 1$ with respect to Ω_A^{jk}, and it can be neglected. At equilibrium the identity gives rise to Eq. (9.10). And similarly, the external terms can be neglected in Eq. (9.28c), and we recover Eq. (9.11) when the body is in dynamical equilibrium.

We now proceed with the derivation of Eq. (9.28b). Taking two derivatives of the quadrupole-moment tensor produces

$$\frac{1}{2}\ddot{I}_A^{jk} = \int_A \rho^* \bar{x}^{(j} \bar{a}^{k)} \, d^3\bar{x} + \int_A \rho^* \bar{v}^j \bar{v}^k \, d^3\bar{x}, \tag{9.30}$$

where $\bar{a}^k = dv^k/dt - a_A^j$ is the acceleration of a fluid element relative to $a_A^j := dv_A^j/dt$, the acceleration of the center-of-mass. The second integral is $2\mathcal{T}_A^{jk}$, and to evaluate the first we substitute the Newtonian version of Euler's equation, $\rho^* dv^k/dt = -\partial_k p + \rho^* \partial_k U + O(c^{-2})$. We obtain

$$\frac{1}{2}\ddot{I}_A^{jk} = 2\mathcal{T}_A^{jk} - \int_A \bar{x}^{(j} \partial^{k)} p \, d^3\bar{x} + \int_A \rho^* \bar{x}^{(j} \partial^{k)} U \, d^3\bar{x} + O(c^{-2}) \tag{9.31}$$

after noting that the terms involving $\boldsymbol{a}_A$ vanish identically. We next integrate the pressure term by parts and decompose the potential term into internal and external pieces; this yields

$$\frac{1}{2}\ddot{I}_A^{jk} = 2\mathcal{T}_A^{jk} + \delta^{jk} P_A + \int_A \rho^* \bar{x}^{(j} \partial^{k)} U_A \, d^3\bar{x} + \int_A \rho^* \bar{x}^{(j} \partial^{k)} U_{\neg A} \, d^3\bar{x} + O(c^{-2}). \tag{9.32}$$

The integral involving U_A is

$$-G \int_A \rho^* \rho^{*\prime} \frac{\bar{x}^{(j} (\bar{x} - \bar{x}')^{k)}}{|\bar{\boldsymbol{x}} - \bar{\boldsymbol{x}}'|^3} \, d^3\bar{x}' d^3\bar{x}, \tag{9.33}$$

and by switching the identity of the integration variables we can also write it as

$$+G \int_A \rho^{*\prime} \rho^* \frac{\bar{x}'^{(j} (\bar{x} - \bar{x}')^{k)}}{|\bar{\boldsymbol{x}} - \bar{\boldsymbol{x}}'|^3} \, d^3\bar{x} d^3\bar{x}', \tag{9.34}$$

or as

$$-\frac{1}{2} G \int_A \rho^* \rho^{*\prime} \frac{(\bar{x} - \bar{x}')^j (\bar{x} - \bar{x}')^k}{|\bar{\boldsymbol{x}} - \bar{\boldsymbol{x}}'|^3} \, d^3\bar{x}' d^3\bar{x}, \tag{9.35}$$

which we recognize as Ω_A^{jk}. Making the substitution in Eq. (9.32) gives us Eq. (9.28b).

Moving on to Eq. (9.28c), we return to Eq. (9.32) and differentiate it with respect to time. We get

$$\frac{1}{2}\dddot{I}_A^{jk} = 2\dot{\mathcal{T}}_A^{jk} + \delta^{jk}\dot{P}_A + \int_A \rho^* \bar{x}^{(j} \frac{d}{dt}\partial^{k)} U_A \, d^3\bar{x} + \int_A \rho^* \bar{v}^{(j}\partial^{k)} U_A \, d^3\bar{x}$$
$$+ \int_A \rho^* \bar{x}^{(j} \frac{d}{dt}\partial^{k)} U_{\neg A} \, d^3\bar{x} + \int_A \rho^* \bar{v}^{(j}\partial^{k)} U_{\neg A} \, d^3\bar{x} + O(c^{-2}). \tag{9.36}$$

The derivative of the kinetic-energy tensor is $\dot{\mathcal{T}}_A^{jk} = \int_A \rho^* \bar{v}^{(j}\bar{a}^{k)} \, d^3\bar{x}$, and this becomes

$$\dot{\mathcal{T}}_A^{jk} = -\int_A \bar{v}^{(j}\partial^{k)} p \, d^3\bar{x} + \int_A \rho^* \bar{v}^{(j}\partial^{k)} U_A \, d^3\bar{x} + \int_A \rho^* \bar{v}^{(j}\partial^{k)} U_{\neg A} \, d^3\bar{x} + O(c^{-2}) \tag{9.37}$$

after invoking Euler's equation and decomposing U into internal and external pieces. The pressure integral is $L_A^{(jk)}$, and we note that the integral involving U_A also appears in Eq. (9.36); it is given by

$$-G\int_A \rho^* \rho^{*\prime} \frac{\bar{v}^{(j}(\bar{x}-\bar{x}')^{k)}}{|\bar{\boldsymbol{x}}-\bar{\boldsymbol{x}}'|^3} \, d^3\bar{x}' d^3\bar{x} = +G\int_A \rho^{*\prime} \rho^* \frac{\bar{v}'^{(j}(\bar{x}-\bar{x}')^{k)}}{|\bar{\boldsymbol{x}}-\bar{\boldsymbol{x}}'|^3} \, d^3\bar{x} d^3\bar{x}', \tag{9.38}$$

which is recognized as $H_A^{(jk)}$. We therefore have

$$\dot{\mathcal{T}}_A^{jk} = H_A^{(jk)} - L_A^{(jk)} + \int_A \rho^* \bar{v}^{(j}\partial^{k)} U_{\neg A} \, d^3\bar{x} + O(c^{-2}), \tag{9.39}$$

and this can be inserted within Eq. (9.36).

We next work on the integral involving $d(\partial^k U_A)/dt$ in Eq. (9.36). We have

$$\frac{d}{dt}\partial^k U_A = \partial_t \partial^k U_A + v^n \partial_n \partial^k U_A$$
$$= -G\int_A \rho^{*\prime} v'^n \partial_{n'} \frac{(x-x')^k}{|\boldsymbol{x}-\boldsymbol{x}'|^3} \, d^3x' - G\int_A \rho^{*\prime} v^n \partial_n \frac{(x-x')^k}{|\boldsymbol{x}-\boldsymbol{x}'|^3} \, d^3x', \tag{9.40}$$

and evaluation of the partial derivatives yields

$$\frac{d}{dt}\partial^k U_A = -G\int_A \rho^{*\prime} \frac{(v-v')^k}{|\boldsymbol{x}-\boldsymbol{x}'|^3} \, d^3x' + 3G\int_A \rho^{*\prime} \frac{(v-v')_n (x-x')^n (x-x')^k}{|\boldsymbol{x}-\boldsymbol{x}'|^5} \, d^3x'. \tag{9.41}$$

Now

$$\int_A \rho^* \bar{x}^j \frac{d}{dt}\partial^k U_A \, d^3\bar{x} = -G\int_A \rho^* \rho^{*\prime} \frac{\bar{x}^j(\bar{v}-\bar{v}')^k}{|\bar{\boldsymbol{x}}-\bar{\boldsymbol{x}}'|^3} \, d^3\bar{x}' d^3\bar{x}$$
$$+ 3G\int_A \rho^* \rho^{*\prime} \frac{(\bar{v}-\bar{v}')_n (\bar{x}-\bar{x}')^n \bar{x}^j (\bar{x}-\bar{x}')^k}{|\bar{\boldsymbol{x}}-\bar{\boldsymbol{x}}'|^5} \, d^3\bar{x}' d^3\bar{x}, \tag{9.42}$$

and we see that by playing with the identities of the integration variables – our old "switch trick" $\bar{\boldsymbol{x}} \leftrightarrow \bar{\boldsymbol{x}}'$ – we can bring the first integral to the form of H_A^{kj}, while the second integral is recognized as $-K_A^{jk}$. Our final expression is therefore

$$\int_A \rho^* \bar{x}^{(j} \frac{d}{dt}\partial^{k)} U_A \, d^3\bar{x} = H_A^{(jk)} - 3K_A^{jk}. \tag{9.43}$$

Collecting the results displayed in Eqs. (9.36), (9.39), and (9.43), we finally obtain the virial identity of Eq. (9.28c).

9.2 Inter-body metric

9.2.1 Introduction

The post-Newtonian metric was first written down in Eq. (8.2); it is

$$g_{00} = -1 + \frac{2}{c^2}U + \frac{2}{c^4}(\Psi - U^2) + O(c^{-6}), \tag{9.44a}$$

$$g_{0j} = -\frac{4}{c^3}U_j + O(c^{-5}), \tag{9.44b}$$

$$g_{jk} = \left(1 + \frac{2}{c^2}U\right)\delta_{jk} + O(c^{-4}), \tag{9.44c}$$

and the gravitational potentials were expressed in terms of fluid variables in Eqs. (8.4). The Newtonian and vector potentials are

$$U = G\int \frac{\rho^{*\prime}}{|\boldsymbol{x} - \boldsymbol{x}'|}\, d^3x', \tag{9.45a}$$

$$U^j = G\int \frac{\rho^{*\prime} v^{\prime j}}{|\boldsymbol{x} - \boldsymbol{x}'|}\, d^3x', \tag{9.45b}$$

in which $\rho^{*\prime}$ and $v^{\prime j}$ are respectively the density and velocity field expressed as functions of time t and position $\boldsymbol{x}'$. The post-Newtonian potential can be written as

$$\Psi = 2\Phi_1 - \Phi_2 + \Phi_3 + 4\Phi_4 - \frac{1}{2}\Phi_5 - \frac{1}{2}\Phi_6, \tag{9.46}$$

in terms of the auxiliary potentials introduced in Eq. (8.8):

$$\Phi_1 = G\int \frac{\rho^{*\prime} v^{\prime 2}}{|\boldsymbol{x} - \boldsymbol{x}'|}\, d^3x', \tag{9.47a}$$

$$\Phi_2 = G\int \frac{\rho^{*\prime} U'}{|\boldsymbol{x} - \boldsymbol{x}'|}\, d^3x', \tag{9.47b}$$

$$\Phi_3 = G\int \frac{\rho^{*\prime} \Pi'}{|\boldsymbol{x} - \boldsymbol{x}'|}\, d^3x', \tag{9.47c}$$

$$\Phi_4 = G\int \frac{p'}{|\boldsymbol{x} - \boldsymbol{x}'|}\, d^3x', \tag{9.47d}$$

$$\Phi_5 = G\int \rho^{*\prime} \partial_{j'} U' \frac{(x - x')^j}{|\boldsymbol{x} - \boldsymbol{x}'|}\, d^3x', \tag{9.47e}$$

$$\Phi_6 = G\int \rho^{*\prime} v'_j v'_k \frac{(x - x')^j (x - x')^k}{|\boldsymbol{x} - \boldsymbol{x}'|^3}\, d^3x'. \tag{9.47f}$$

Our task in this section is to coarse-grain this fluid description and evaluate the potentials for a system of N separated bodies. We shall do this at a fair distance from each body, and exploit the condition $s_A \gg R_A$ of Eq. (9.13); here s_A is the distance to A's center-of-mass, and R_A is the body radius, a length scale associated with the volume occupied by body A.

In the course of our calculations we will use all the tricks reviewed in Sec. 9.1, and our results will be expressed in terms of the structure integrals of Eqs. (9.8) and (9.9). To simplify the notation it is convenient to introduce

$$\boldsymbol{s}_A := \boldsymbol{x} - \boldsymbol{r}_A(t), \qquad s_A := |\boldsymbol{x} - \boldsymbol{r}_A(t)|, \qquad \boldsymbol{n}_A := \frac{\boldsymbol{s}_A}{s_A}, \tag{9.48}$$

as well as

$$\boldsymbol{r}_{AB} := \boldsymbol{r}_A(t) - \boldsymbol{r}_B(t), \qquad r_{AB} := |\boldsymbol{r}_A(t) - \boldsymbol{r}_B(t)|, \qquad \boldsymbol{n}_{AB} := \frac{\boldsymbol{r}_{AB}}{r_{AB}}. \tag{9.49}$$

We also need the identities

$$\partial_j s_A = n_A^j, \tag{9.50a}$$

$$\partial_{jk} s_A = \frac{1}{s_A}\left(\delta^{jk} - n_A^j n_A^k\right), \tag{9.50b}$$

$$\partial_{jkn} s_A = -\frac{1}{s_A^2}\left(\delta^{jk} n_A^n + \delta^{jn} n_A^k + \delta^{kn} n_A^j - 3 n_A^j n_A^k n_A^n\right), \tag{9.50c}$$

which are established by straightforward computations.

9.2.2 Potentials

Newtonian potential U

To evaluate the Newtonian potential we begin with Eq. (9.45a), introduce the sum over bodies, and change the integration variables to the relative position $\bar{\boldsymbol{x}}' := \boldsymbol{x}' - \boldsymbol{r}_A$. This gives

$$U(t, \boldsymbol{x}) = \sum_A G \int_A \frac{\rho^*(t, \boldsymbol{r}_A + \bar{\boldsymbol{x}}')}{|\boldsymbol{s}_A - \bar{\boldsymbol{x}}'|}\, d^3\bar{x}', \tag{9.51}$$

where $\boldsymbol{s}_A := \boldsymbol{x} - \boldsymbol{r}_A$ was introduced in Eq. (9.48). In the next step we invoke the condition $R_A \ll s_A$ and express $|\boldsymbol{s}_A - \bar{\boldsymbol{x}}'|^{-1}$ as a Taylor expansion in powers of $\bar{x}'^j$,

$$\frac{1}{|\boldsymbol{s}_A - \bar{\boldsymbol{x}}'|} = \frac{1}{s_A} - \bar{x}'^j \partial_j \frac{1}{s_A} + \frac{1}{2}\bar{x}'^j \bar{x}'^k \partial_{jk} \frac{1}{s_A} + \cdots. \tag{9.52}$$

Substitution within each integral produces

$$\begin{aligned}\int_A \frac{\rho^{*\prime}}{|\boldsymbol{s}_A - \bar{\boldsymbol{x}}'|}\, d^3\bar{x}' &= \frac{1}{s_A}\int_A \rho^{*\prime}\, d^3\bar{x}' - \left(\partial_j \frac{1}{s_A}\right)\int_A \rho^{*\prime}\bar{x}'^j\, d^3\bar{x}' \\ &\quad + \frac{1}{2}\left(\partial_{jk}\frac{1}{s_A}\right)\int_A \rho^{*\prime}\bar{x}'^j \bar{x}'^j\, d^3\bar{x}' + \cdots. \end{aligned} \tag{9.53}$$

The first integral gives m_A, the material mass of body A. The second integral vanishes by virtue of the definition of $\boldsymbol{r}_A$ given by Eq. (9.3). The third integral gives I_A^{jk}, the quadrupole moment of the mass distribution. Our expression for U is therefore

$$U = \sum_A \left(\frac{Gm_A}{s_A} + \frac{1}{2} I_A^{jk} \partial_{jk} \frac{1}{s_A} + \cdots \right), \tag{9.54}$$

and we shall truncate this to

$$U = \sum_A \frac{Gm_A}{s_A} + \cdots, \tag{9.55}$$

noting that the quadrupole term is smaller than the monopole term by a factor of order $(R_A/s_A)^2 \ll 1$. This degree of accuracy is maintained in all following calculations.

Vector potential U^j

The steps involved in evaluation of the vector potential are very similar. We begin with Eq. (9.45b), introduce the sum over bodies, and make the substitutions $\boldsymbol{x}' = \boldsymbol{r}_A + \bar{\boldsymbol{x}}'$ and $\boldsymbol{v}' = \boldsymbol{v}_A + \bar{\boldsymbol{v}}'$ within the integrals. We get

$$U^j = \sum_A \left(v_A^j G \int_A \frac{\rho^{*\prime}}{|\boldsymbol{s}_A - \bar{\boldsymbol{x}}'|} d^3\bar{x}' + G \int_A \frac{\rho^{*\prime} \bar{v}'^j}{|\boldsymbol{s}_A - \bar{\boldsymbol{x}}'|} d^3\bar{x}' \right). \tag{9.56}$$

The Taylor expansion of $|\boldsymbol{s}_A - \bar{\boldsymbol{x}}'|^{-1}$ gives rise to $Gm_A v_A^j/s_A + \cdots$ for the first integral, where we again neglect the quadrupole term. For the second integral we have

$$\begin{aligned} \int_A \frac{\rho^{*\prime} \bar{v}'^j}{|\boldsymbol{s}_A - \bar{\boldsymbol{x}}'|} d^3\bar{x}' = {} & \frac{1}{s_A} \int_A \rho^{*\prime} \bar{v}'^j \, d^3\bar{x}' - \left(\partial_k \frac{1}{s_A} \right) \int_A \rho^{*\prime} \bar{v}'^j \bar{x}'^k \, d^3\bar{x}' \\ & + \frac{1}{2} \left(\partial_{kn} \frac{1}{s_A} \right) \int_A \rho^{*\prime} \bar{v}'^j \bar{x}'^k \bar{x}'^n \, d^3\bar{x}' + \cdots. \end{aligned} \tag{9.57}$$

The first integral vanishes by virtue of the definition of $\boldsymbol{v}_A$ given by Eq. (9.4). The second integral is the same one that appears in Eq. (9.29), and it vanishes for non-spinning bodies in equilibrium. And the third integral vanishes for bodies that are reflection-symmetric about the center-of-mass, as was discussed in Sec. 9.1.2. The neglected terms are suppressed by a factor of order $(R_A/s_A)^3$ relative to the leading term, and we conclude that

$$U^j = \sum_A \frac{Gm_A v_A^j}{s_A} + \cdots, \tag{9.58}$$

up to neglected terms of fractional order $(R_A/s_A)^2$.

Auxiliary potential Φ_1

Following the same steps we obtain

$$\begin{aligned} \Phi_1 = \sum_A \Bigg(& v_A^2 G \int_A \frac{\rho^{*\prime}}{|\boldsymbol{s}_A - \bar{\boldsymbol{x}}'|} d^3\bar{x}' + 2 v_A^j G \int_A \frac{\rho^{*\prime} \bar{v}'_j}{|\boldsymbol{s}_A - \bar{\boldsymbol{x}}'|} d^3\bar{x}' \\ & + G \int_A \frac{\rho^{*\prime} \bar{v}'^2}{|\boldsymbol{s}_A - \bar{\boldsymbol{x}}'|} d^3\bar{x}' \Bigg) \end{aligned} \tag{9.59}$$

from Eq. (9.47a). The first two terms are handled just as before, and the third integral is

$$\int_A \frac{\rho^{*\prime}\bar{v}^{\prime 2}}{|\boldsymbol{s}_A - \bar{\boldsymbol{x}}'|} d^3\bar{x}' = \frac{1}{s_A}\int_A \rho^{*\prime}\bar{v}^{\prime 2}\, d^3\bar{x}' - \left(\partial_j \frac{1}{s_A}\right)\int_A \rho^{*\prime}\bar{v}^{\prime 2}\bar{x}^{\prime j}\, d^3\bar{x}' + \frac{1}{2}\left(\partial_{jk}\frac{1}{s_A}\right)\int_A \rho^{*\prime}\bar{v}^{\prime 2}\bar{x}^{\prime j}\bar{x}^{\prime k}\, d^3\bar{x}' + \cdots . \tag{9.60}$$

The first integral is $2\mathcal{T}_A$, the second vanishes for reflection-symmetric bodies, and the third term is suppressed by a factor of order $(R_A/s_A)^2$ relative to the first. Collecting results, we have obtained

$$\Phi_1 = 2\sum_A \frac{G\mathcal{T}_A}{s_A} + \sum_A \frac{Gm_A v_A^2}{s_A} + \cdots . \tag{9.61}$$

Auxiliary potential Φ_2

This computation requires a little more care, because the Newtonian potential U' appears inside the Poisson integral that defines Φ_2. Decomposing this as $U'_A + U'_{\neg A}$, Eq. (9.47b) becomes

$$\Phi_2 = \sum_A G\int_A \frac{\rho^{*\prime}U'_A}{|\boldsymbol{x}-\boldsymbol{x}'|}\, d^3x' + \sum_A G\int_A \frac{\rho^{*\prime}U'_{\neg A}}{|\boldsymbol{x}-\boldsymbol{x}'|}\, d^3x' . \tag{9.62}$$

In the first group of terms in Eq. (9.62) we insert $U'_A = G\int_A \rho^{*\prime\prime}|\boldsymbol{x}'-\boldsymbol{x}''|^{-1}\, d^3x''$, as well as $\boldsymbol{x}' = \boldsymbol{r}_A + \bar{\boldsymbol{x}}'$ and $\boldsymbol{x}'' = \boldsymbol{r}_A + \bar{\boldsymbol{x}}''$. We obtain a sum of terms of the form

$$G^2 \int_A \frac{\rho^{*\prime}\rho^{*\prime\prime}}{|\bar{\boldsymbol{x}}'-\bar{\boldsymbol{x}}''||\boldsymbol{s}_A - \bar{\boldsymbol{x}}'|}\, d^3\bar{x}'' d^3\bar{x}' . \tag{9.63}$$

We next express $|\boldsymbol{s}_A - \bar{\boldsymbol{x}}'|^{-1}$ as a Taylor expansion in powers of $\bar{x}^{\prime j}$. The leading term gives rise to the contribution

$$\frac{G^2}{s_A}\int_A\int_A \frac{\rho^{*\prime}\rho^{*\prime\prime}}{|\bar{\boldsymbol{x}}'-\bar{\boldsymbol{x}}''|}\, d^3\bar{x}'' d^3\bar{x}' , \tag{9.64}$$

and we recognize this as $-2G\Omega_A/s_A$. The linear term produces an odd integral that vanishes, and the quadratic term gives rise to a negligible contribution of fractional order $(R_A/s_A)^2$. We therefore obtain $-2\sum_A G\Omega_A/s_A$ for the first group of terms in Eq. (9.62).

In the second group of terms we insert $U'_{\neg A} = \sum_{B\neq A} G\int_B \rho^{*\prime\prime}|\boldsymbol{x}'-\boldsymbol{x}''|^{-1}\, d^3x''$, as well as $\boldsymbol{x}' = \boldsymbol{r}_A + \bar{\boldsymbol{x}}'$ and $\boldsymbol{x}'' = \boldsymbol{r}_B + \bar{\boldsymbol{x}}''$. We obtain a double sum over bodies A and B of terms of the form

$$G^2 \int_A\int_B \frac{\rho^{*\prime}\rho^{*\prime\prime}}{|\boldsymbol{r}_{AB} + \bar{\boldsymbol{x}}' - \bar{\boldsymbol{x}}''||\boldsymbol{s}_A - \bar{\boldsymbol{x}}'|}\, d^3\bar{x}'' d^3\bar{x}' . \tag{9.65}$$

To evaluate this we express $|\boldsymbol{s}_A - \bar{\boldsymbol{x}}'|^{-1}$ as a Taylor expansion in powers of $\bar{x}^{\prime j}$, and we simultaneously expand $|\boldsymbol{r}_{AB} + \bar{\boldsymbol{x}}' - \bar{\boldsymbol{x}}''|^{-1}$ in powers of $(\bar{x}' - \bar{x}'')^j$. Making the substitution, we find that as before, the integral is dominated by the leading term in the expansion, that the linear terms do not contribute at all, and that the quadratic terms can be neglected. We arrive at the expression $\sum_A \sum_{B\neq A} G^2 m_A m_B/(r_{AB}s_A)$ for the second group of terms in Eq. (9.62).

Collecting results, we find that our final expression for Φ_2 is

$$\Phi_2 = -2\sum_A \frac{G\Omega_A}{s_A} + \sum_A \sum_{B\neq A} \frac{G^2 m_A m_B}{r_{AB}s_A} + \cdots . \tag{9.66}$$

Auxiliary potentials Φ_3, Φ_4, and Φ_5

The potentials Φ_3 and Φ_4 are computed exactly as the Newtonian potential U. We immediately obtain

$$\Phi_3 = \sum_A \frac{GE_A^{\text{int}}}{s_A} + \cdots \tag{9.67}$$

and

$$\Phi_4 = \sum_A \frac{GP_A}{s_A} + \cdots \tag{9.68}$$

The computation of Φ_5 resembles that of Φ_2. We write Eq. (9.47e) in the form

$$\Phi_5 = \sum_A G \int_A \rho^{*\prime} \partial_{j'} U'_A \frac{(x-x')^j}{|\boldsymbol{x}-\boldsymbol{x}'|} d^3x' + \sum_A G \int_A \rho^{*\prime} \partial_{j'} U'_{\neg A} \frac{(x-x')^j}{|\boldsymbol{x}-\boldsymbol{x}'|} d^3x', \tag{9.69}$$

and work on each group of terms separately.

The first group is a sum of terms of the form

$$-G^2 \int_A \rho^{*\prime}\rho^{*\prime\prime} \frac{(\bar{x}'-\bar{x}'')^j (s_A - \bar{x}')_j}{|\bar{\boldsymbol{x}}'-\bar{\boldsymbol{x}}''|^3 |\boldsymbol{s}_A - \bar{\boldsymbol{x}}'|} d^3\bar{x}'' d^3\bar{x}', \tag{9.70}$$

and to evaluate this we expand $(s_A - \bar{x}')_j/|\boldsymbol{s}_A - \bar{\boldsymbol{x}}'| = \partial_j |\boldsymbol{s}_A - \boldsymbol{x}'|$ in powers of $\bar{x}'^k$. We obtain

$$\begin{aligned}\int_A \rho^{*\prime} \partial_{j'} U'_A \frac{(x-x')^j}{|\boldsymbol{x}-\boldsymbol{x}'|} d^3x' = &-G(\partial_j s_A) \int_A \rho^{*\prime}\rho^{*\prime\prime} \frac{(\bar{x}'-\bar{x}'')^j}{|\bar{\boldsymbol{x}}'-\bar{\boldsymbol{x}}''|^3} d^3\bar{x}'' d^3\bar{x}' \\ &+ G(\partial_{jk} s_A) \int_A \rho^{*\prime}\rho^{*\prime\prime} \frac{(\bar{x}'-\bar{x}'')^j \bar{x}'^k}{|\bar{\boldsymbol{x}}'-\bar{\boldsymbol{x}}''|^3} d^3\bar{x}'' d^3\bar{x}' \\ &- G(\partial_{jkn} s_A) \int_A \rho^{*\prime}\rho^{*\prime\prime} \frac{(\bar{x}'-\bar{x}'')^j \bar{x}'^k \bar{x}'^n}{|\bar{\boldsymbol{x}}'-\bar{\boldsymbol{x}}''|^3} d^3\bar{x}'' d^3\bar{x}' \\ &+ \cdots \end{aligned} \tag{9.71}$$

The first and third integrals are odd in the number of internal vectors and therefore vanish. The second integral can be written as

$$\frac{1}{2}\int_A \rho^{*\prime}\rho^{*\prime\prime} \frac{(\bar{x}'-\bar{x}'')^j (\bar{x}'-\bar{x}'')^k}{|\bar{\boldsymbol{x}}'-\bar{\boldsymbol{x}}''|^3} d^3\bar{x}'' d^3\bar{x}', \tag{9.72}$$

which we recognize as $-\Omega_A^{jk}/G$. The neglected terms are of fractional order $(R_A/s_A)^2$, and we conclude that the first group of terms in Eq. (9.69) equals $-\sum_A G\Omega_A^{jk}\partial_{jk}s_A$.

The second group is a double sum of terms

$$-G^2 \int_A \int_B \rho^{*\prime}\rho^{*\prime\prime} \frac{(r_{AB}+\bar{x}'-\bar{x}'')^j(s_A-\bar{x}')_j}{|\boldsymbol{r}_{AB}+\bar{\boldsymbol{x}}'-\bar{\boldsymbol{x}}''|^3|\boldsymbol{s}_A-\bar{\boldsymbol{x}}'|}\, d^3\bar{x}''d^3\bar{x}', \tag{9.73}$$

and the integrals can be evaluated by expanding the factor multiplying $\rho^{*\prime}\rho^{*\prime\prime}$ in a double Taylor series in powers of $(\bar{x}'-\bar{x}'')^k$ and $\bar{x}'^n$. Once more the dominant contribution comes from the zeroth-order term, and we find that the second group of terms in Eq. (9.69) can be approximated by $-\sum_A\sum_{B\neq A} G^2 m_A m_B(\boldsymbol{r}_{AB}\cdot\boldsymbol{s}_A)/(r_{AB}^3 s_A)$.

Collecting results, we have that Φ_5 is given by

$$\Phi_5 = -\sum_A G\Omega_A^{jk}\partial_{jk}s_A - \sum_A\sum_{B\neq A} G^2 m_A m_B \frac{\boldsymbol{r}_{AB}\cdot\boldsymbol{s}_A}{r_{AB}^3 s_A} + \cdots, \tag{9.74}$$

or

$$\Phi_5 = -\sum_A \frac{G\Omega_A}{s_A} + \sum_A G\Omega_A^{jk}\frac{n_{Aj}n_{Ak}}{s_A} - \sum_A\sum_{B\neq A} G^2 m_A m_B \frac{\boldsymbol{n}_{AB}\cdot\boldsymbol{n}_A}{r_{AB}^2} + \cdots \tag{9.75}$$

after using Eq. (9.50) to evaluate $\partial_{jk}s_A$.

Auxiliary potential Φ_6

Equation (9.47f) becomes

$$\begin{aligned}\Phi_6 = \sum_A G\bigg[& v_A^j v_A^k \int_A \rho^{*\prime}\frac{(s_A-\bar{x}')_j(s_A-\bar{x}')_k}{|\boldsymbol{s}_A-\bar{\boldsymbol{x}}'|^3}\, d^3\bar{x}' \\ & + 2v_A^j \int_A \rho^{*\prime}\bar{v}'^k \frac{(s_A-\bar{x}')_j(s_A-\bar{x}')_k}{|\boldsymbol{s}_A-\bar{\boldsymbol{x}}'|^3}\, d^3\bar{x}' \\ & + \int_A \rho^{*\prime}\bar{v}'^j\bar{v}'^k \frac{(s_A-\bar{x}')_j(s_A-\bar{x}')_k}{|\boldsymbol{s}_A-\bar{\boldsymbol{x}}'|^3}\, d^3\bar{x}'\bigg]\end{aligned} \tag{9.76}$$

after making the substitutions $\boldsymbol{x}' = \boldsymbol{r}_A + \bar{\boldsymbol{x}}'$ and $\boldsymbol{v}' = \boldsymbol{v}_A + \bar{\boldsymbol{v}}'$. In each integral we perform the usual trick of expanding $(s_A-\bar{x}')_j(s_A-\bar{x}')_k|\boldsymbol{s}_A-\bar{\boldsymbol{x}}'|^{-3}$ in powers of $\bar{x}'^n$. For the first integral we find that the zeroth-order term gives the dominant contribution, that the linear terms give rise to a vanishing integral, and that the quadratic terms can be neglected. For the second integral we find that the zeroth-order term vanishes (because the integral is odd in the number of internal vectors), that the first-order term vanishes by virtue of the non-spinning condition of Eq. (9.29), and that the second-order term also vanishes (another odd integral). And finally, the third integral is evaluated just as the first, and we retain only the zeroth-order contribution.

The end result is

$$\Phi_6 = 2\sum_A GT_A^{jk}\frac{n_{Aj}n_{Ak}}{s_A} + \sum_A Gm_A\frac{(\boldsymbol{n}_A\cdot\boldsymbol{v}_A)^2}{s_A} + \cdots. \tag{9.77}$$

Post-Newtonian potential Ψ

Combining Eqs. (9.46), (9.61), (9.66), (9.67), (9.68), (9.75), and (9.77) we arrive at the following expression for the post-Newtonian potential:

$$\begin{aligned}\Psi = &\sum_A \frac{G}{s_A}\left(4\mathcal{T}_A + \frac{5}{2}\Omega_A + E_A^{\text{int}} + \frac{9}{2}P_A\right) \\ &- \frac{1}{2}\sum_A \frac{G}{s_A}\left(2\mathcal{T}_A^{jk} + \Omega_A^{jk} + \delta^{jk}P_A\right)n_{Aj}n_{Ak} \\ &+ \sum_A \frac{Gm_A}{s_A}\left[2v_A^2 - \frac{1}{2}(\boldsymbol{n}_A \cdot \boldsymbol{v}_A)^2\right] \\ &- \sum_A \sum_{B\neq A} \frac{G^2 m_A m_B}{r_{AB}s_A}\left(1 - \frac{\boldsymbol{n}_{AB}\cdot \boldsymbol{s}_A}{2r_{AB}}\right).\end{aligned} \tag{9.78}$$

This can be simplified. We note first that the second group of terms, involving the structure tensors $\mathcal{T}_A^{jk}$, Ω_A^{jk}, and $\delta^{jk}P_A$, vanishes by virtue of the equilibrium condition of Eq. (9.10). On the other hand, according to Eq. (9.12) the first group of terms can be altered at will by the insertion of any multiple of $2\mathcal{T}_A + \Omega_A + 3P_A = 0$; we exploit this freedom to eliminate the P_A term in the first sum. And finally, we use the identity

$$\boldsymbol{n}_{AB}\cdot \boldsymbol{s}_A = \frac{s_B^2 - s_A^2 - r_{AB}^2}{2r_{AB}} \tag{9.79}$$

to alter the appearance of the double sum.

After implementing these changes, our final expression for Ψ is

$$\begin{aligned}\Psi = &\sum_A \frac{G}{s_A}\left(\mathcal{T}_A + \Omega_A + E_A^{\text{int}}\right) + \sum_A \frac{Gm_A}{s_A}\left[2v_A^2 - \frac{1}{2}(\boldsymbol{n}_A\cdot \boldsymbol{v}_A)^2\right] \\ &- \sum_A \sum_{B\neq A} \frac{G^2 m_A m_B}{r_{AB}s_A}\frac{5r_{AB}^2 + s_A^2 - s_B^2}{4r_{AB}^2}.\end{aligned} \tag{9.80}$$

Note that the first sum, which features the only remaining terms that involve the structure integrals, depends on the combination $E_A := \mathcal{T}_A + \Omega_A + E_A^{\text{int}}$; this is the *total energy* of body A, the sum of kinetic, gravitational, and internal energies.

9.2.3 At long last, the metric

We insert Eq. (9.55), (9.58), and (9.80) within the post-Newtonian metric of Eqs. (9.44). The first group of terms in the post-Newtonian potential Ψ combines with the Newtonian potential U, and the combination gives rise to a contribution

$$\frac{2}{c^2}\sum_A \frac{G(m_A + E_A/c^2)}{s_A}$$

to g_{00}. This result implies that the Newtonian piece of the metric is naturally expressed in terms of $M_A := m_A + E_A/c^2$, the total mass-energy of each body, as defined by Eq. (9.23). Furthermore, the post-Newtonian terms also can be expressed in terms of M_A, because the difference between this and m_A is of order c^{-2} and can be transferred to the (neglected) 2PN terms. The conclusion is that the metric involves M_A only, and is completely insensitive to its decomposition in terms of material mass m_A, kinetic energy $\mathcal{T}_A$, gravitational potential energy Ω_A, and internal energy $E_A^{\rm int}$. This conclusion was anticipated in Sec. 9.1.6: the 1PN metric does indeed satisfy the strong formulation of the principle of equivalence.

Our final expression for the metric is

$$g_{00} = -1 + \frac{2}{c^2}\sum_A \frac{GM_A}{s_A} + \frac{1}{c^4}\sum_A \frac{GM_A}{s_A}\left[4v_A^2 - (\boldsymbol{n}_A\cdot\boldsymbol{v}_A)^2 - 2\frac{GM_A}{s_A}\right] - \frac{1}{c^4}\sum_A\sum_{B\neq A}\frac{G^2M_AM_B}{s_A}\left(\frac{2}{s_B} + \frac{5}{2r_{AB}} + \frac{s_A^2 - s_B^2}{2r_{AB}^3}\right) + O(c^{-6}), \tag{9.81a}$$

$$g_{0j} = -\frac{4}{c^3}\sum_A \frac{GM_Av_A^j}{s_A} + O(c^{-5}), \tag{9.81b}$$

$$g_{jk} = \left(1 + \frac{2}{c^2}\sum_A\frac{GM_A}{s_A}\right)\delta_{jk} + O(c^{-4}). \tag{9.81c}$$

We recall the notation

$$\boldsymbol{s}_A := \boldsymbol{x} - \boldsymbol{r}_A(t), \qquad s_A := |\boldsymbol{x} - \boldsymbol{r}_A(t)|, \qquad \boldsymbol{n}_A := \frac{\boldsymbol{s}_A}{s_A}, \tag{9.82}$$

as well as

$$r_{AB} := |\boldsymbol{r}_A(t) - \boldsymbol{r}_B(t)|. \tag{9.83}$$

And we recall that the mass-energy parameter M_A was introduced in Eq. (9.23):

$$M_A := \int_A \rho^*\left[1 + \frac{1}{c^2}\left(\frac{1}{2}\bar{v}^2 - \frac{1}{2}U_A + \Pi\right)\right]d^3x + O(c^{-4}), \tag{9.84}$$

where $\bar{\boldsymbol{v}} := \boldsymbol{v} - \boldsymbol{v}_A$ is the velocity field in the comoving frame of body A, and U_A is the internal gravitational potential. The mass-energy M_A is a constant of the post-Newtonian motion.

The metric of Eqs. (9.81) is valid outside each body, at a distance s_A that is much larger than each body radius R_A. Only the leading terms are displayed, and our expressions leave out terms that are suppressed by factors of order $(R_A/s_A)^2 \ll 1$. They also leave out terms that are contributed by the spin of each body, which was assumed to vanish; these will be incorporated in Sec. 9.5. The metric is expressed in terms of the center-of-mass position $\boldsymbol{r}_A(t)$ and velocity $\boldsymbol{v}_A(t) = d\boldsymbol{r}_A/dt$ of each body. These have not yet been determined, and this shall be our task in the following section.

9.3 Motion of isolated bodies

9.3.1 Strategy

To find equations of motion for the center-of-mass positions $\boldsymbol{r}_A(t)$, we return to Eq. (9.5), which we copy as

$$m_A \boldsymbol{a}_A = \int_A \rho^* \frac{d\boldsymbol{v}}{dt}\, d^3x. \tag{9.85}$$

Here $m_A := \int_A \rho^*\, d^3x$ is the material mass of body A, $\boldsymbol{a}_A := d^2\boldsymbol{r}_A/dt^2$ is the coordinate acceleration of its center-of-mass, and $\boldsymbol{v}$ is the fluid's velocity field. The domain of integration V_A is a time-independent region of three-dimensional space that extends beyond the volume occupied by the body. As stated previously, it is sufficiently large that in a time interval dt, the body will not cross the domain's boundary S_A; but it is sufficiently small that V_A does not include nor intersect another body within the system.

In this equation we insert the post-Newtonian Euler equation (8.119), which is derived in Sec. 8.4.3. Taking into account Eq. (8.12), this gives rise to

$$m_A a_A^j = F_0^j + \sum_{n=1}^{18} F_n^j + O(c^{-4}), \tag{9.86}$$

where

$$F_0^j := \int_A \big(-\partial_j p + \rho^* \partial_j U\big)\, d^3x \tag{9.87}$$

is the Newtonian contribution to the force acting on body A, while the eighteen terms that make up the post-Newtonian contribution are

$$F_1^j := \frac{1}{2c^2} \int_A v^2 \partial_j p\, d^3x, \tag{9.88a}$$

$$F_2^j := \frac{1}{c^2} \int_A U \partial_j p\, d^3x, \tag{9.88b}$$

$$F_3^j := \frac{1}{c^2} \int_A \Pi \partial_j p\, d^3x, \tag{9.88c}$$

$$F_4^j := \frac{1}{c^2} \int_A \frac{p}{\rho^*} \partial_j p\, d^3x, \tag{9.88d}$$

$$F_5^j := -\frac{1}{c^2} \int_A v^j \partial_t p\, d^3x, \tag{9.88e}$$

$$F_6^j := \frac{1}{c^2} \int_A \rho^* v^2 \partial_j U\, d^3x, \tag{9.88f}$$

$$F_7^j := -\frac{4}{c^2} \int_A \rho^* U \partial_j U\, d^3x, \tag{9.88g}$$

$$F_8^j := -\frac{3}{c^2} \int_A \rho^* v^j \partial_t U\, d^3x, \tag{9.88h}$$

(continued overleaf)

$$F_9^j := -\frac{4}{c^2}\int_A \rho^* v^j v^k \partial_k U \, d^3x, \tag{9.88i}$$

$$F_{10}^j := \frac{4}{c^2}\int_A \rho^* \partial_t U^j \, d^3x, \tag{9.88j}$$

$$F_{11}^j := \frac{4}{c^2}\int_A \rho^* v^k \partial_k U^j \, d^3x, \tag{9.88k}$$

$$F_{12}^j := -\frac{4}{c^2}\int_A \rho^* v^k \partial_j U_k \, d^3x, \tag{9.88l}$$

$$F_{13}^j := \frac{2}{c^2}\int_A \rho^* \partial_j \Phi_1 \, d^3x, \tag{9.88m}$$

$$F_{14}^j := -\frac{1}{c^2}\int_A \rho^* \partial_j \Phi_2 \, d^3x, \tag{9.88n}$$

$$F_{15}^j := \frac{1}{c^2}\int_A \rho^* \partial_j \Phi_3 \, d^3x, \tag{9.88o}$$

$$F_{16}^j := \frac{4}{c^2}\int_A \rho^* \partial_j \Phi_4 \, d^3x, \tag{9.88p}$$

$$F_{17}^j := -\frac{1}{2c^2}\int_A \rho^* \partial_j \Phi_5 \, d^3x, \tag{9.88q}$$

$$F_{18}^j := -\frac{1}{2c^2}\int_A \rho^* \partial_j \Phi_6 \, d^3x. \tag{9.88r}$$

The auxiliary potentials Φ_n were introduced in Eqs. (8.8).

To evaluate the force integrals we rely on the techniques introduced in Sec. 9.1. We assume that each body is reflection-symmetric about the center-of-mass (Sec. 9.1.2), so that each variable (such as ρ^*, p, and Π) that specifies its internal structure can be taken to be invariant under the reflection $\bar{\boldsymbol{x}} \to -\bar{\boldsymbol{x}}$, where $\bar{\boldsymbol{x}} := \boldsymbol{x} - \boldsymbol{r}_A(t)$ is the position of a fluid element relative to the center-of-mass; this property allows us to eliminate all body integrals that contain an odd number of internal vectors (such as $\bar{\boldsymbol{x}}$, $\bar{\boldsymbol{v}} := \boldsymbol{v} - \boldsymbol{v}_A$, or $\boldsymbol{\nabla} p$). We express our results in terms of the structure integrals introduced in Sec. 9.1.3, and assume that the bodies are in dynamical equilibrium, so that Eqs. (9.10), (9.11), and (9.12) are satisfied. We continue to assume that each body is non-spinning (Sec. 9.1.4), so that its angular-momentum tensor S_A^{jk} vanishes. We rely on the decomposition of all gravitational potentials into internal and external pieces, as explained in Sec. 9.1.5. And finally, we assume that the bodies are well separated (Sec. 9.1.4) and allow ourselves to neglect terms that are suppressed by a factor of order $(R_A/r_{AB})^2$ relative to the leading-order contribution to each force integral; here R_A is the typical body radius and $r_{AB} := |\boldsymbol{r}_A - \boldsymbol{r}_B|$ the typical inter-body distance.

We exploit the condition $R_A \ll r_{AB}$ to simplify expressions involving the external potentials. The wide separation between bodies implies that when (say) the external potential $U_{\neg A}$ is evaluated within body A, it can be usefully expressed as the Taylor expansion

$$U_{\neg A}(t, \boldsymbol{x}) = U_{\neg A}(t, \boldsymbol{r}_A) + \bar{x}^j \partial_j U_{\neg A}(t, \boldsymbol{r}_A) + \frac{1}{2}\bar{x}^j \bar{x}^k \partial_{jk} U_{\neg A}(t, \boldsymbol{r}_A) + \cdots, \tag{9.89}$$

where $\bar{\boldsymbol{x}} := \boldsymbol{x} - \boldsymbol{r}_A$. With respect to the leading term $U_{\neg A}(t, \boldsymbol{r}_A)$, the linear term in Eq. (9.89) is suppressed by a factor of order R_A/r_{AB}, while the quadratic term is smaller by a factor of order $(R_A/r_{AB})^2$; according to the rules spelled out previously, this last term can be discarded.

9.3.2 Results and sample computations

It would be exhausting (both for the reader and the authors) to present a detailed calculation of each one of the nineteen force integrals F_n^j. We shall instead state the results, and present a small (but representative) sample of the calculations that are required to obtain these results.

The individual contributions to the gravitational force are

$$F_0^j = m_A \partial_j U_{\neg A}(t, \boldsymbol{r}_A) \tag{9.90}$$

and

$$c^2 F_1^j = L_A^{kj} v_A^k, \tag{9.91a}$$

$$c^2 F_2^j = -P_A \partial_j U_{\neg A}, \tag{9.91b}$$

$$c^2 F_3^j = 0, \tag{9.91c}$$

$$c^2 F_4^j = 0, \tag{9.91d}$$

$$c^2 F_5^j = -\dot{P}_A v_A^j + L_A^{jk} v_A^k, \tag{9.91e}$$

$$c^2 F_6^j = 2H_A^{kj} v_A^k + 2\mathcal{T}_A \partial_j U_{\neg A} + m_A v_A^2 \partial_j U_{\neg A}, \tag{9.91f}$$

$$c^2 F_7^j = -4\Omega_A^{jk} \partial_k U_{\neg A} + 8\Omega_A \partial_j U_{\neg A} - 4m_A U_{\neg A} \partial_j U_{\neg A}, \tag{9.91g}$$

$$c^2 F_8^j = 3H_A^{jk} v_A^k - 3H_A v_A^j - 3m_A v_A^j \partial_t U_{\neg A}, \tag{9.91h}$$

$$c^2 F_9^j = -4H_A^{jk} v_A^k - 4H_A v_A^j - 8\mathcal{T}_A^{jk} \partial_k U_{\neg A} - 4m_A v_A^j v_A^k \partial_k U_{\neg A}, \tag{9.91i}$$

$$c^2 F_{10}^j = 4H_A^{jk} v_A^k + 4H_A v_A^j - 8\Omega_A \partial_j U_{\neg A} + 4m_A \partial_t U_{\neg A}^j, \tag{9.91j}$$

$$c^2 F_{11}^j = -4H_A^{jk} v_A^k + 4H_A v_A^j + 4m_A v_A^k \partial_k U_{\neg A}^j, \tag{9.91k}$$

$$c^2 F_{12}^j = -4m_A v_A^k \partial_j U_{\neg A}^k, \tag{9.91l}$$

$$c^2 F_{13}^j = -4H_A^{kj} v_A^k + 2m_A \partial_j \Phi_{1,\neg A}, \tag{9.91m}$$

$$c^2 F_{14}^j = \Omega_A^{jk} \partial_k U_{\neg A} - m_A \partial_j \Phi_{2,\neg A}, \tag{9.91n}$$

$$c^2 F_{15}^j = m_A \partial_j \Phi_{3,\neg A}, \tag{9.91o}$$

$$c^2 F_{16}^j = 4m_A \partial_j \Phi_{4,\neg A}, \tag{9.91p}$$

$$c^2 F_{17}^j = -\Omega_A^{jk} \partial_k U_{\neg A} + \Omega_A \partial_j U_{\neg A} - \frac{1}{2} m_A \partial_j \Phi_{5,\neg A}, \tag{9.91q}$$

$$c^2 F_{18}^j = -H_A^{jk} v_A^k - H_A v_A^j + 3K_A^{jk} v_A^k - \frac{1}{2} m_A \partial_j \Phi_{6,\neg A}. \tag{9.91r}$$

Here $U_{\neg A}$ is a shorthand notation for $U_{\neg A}(t, \boldsymbol{x} = \boldsymbol{r}_A)$, with the rule extending to all other external potentials. These are differentiated with respect to t and x^j *before* being evaluated at $\boldsymbol{x} = \boldsymbol{r}_A$.

To illustrate the method of derivation we begin with the simplest case, the result of Eq. (9.90), which is obtained from Eq. (9.87); these calculations were also presented back in Sec. 1.6. The integral of $\partial_j p$ over the three-dimensional domain V_A can be expressed as the surface integral $\oint_{S_A} p\, dS_j$, and this vanishes because the boundary surface S_A lies outside of body A. The integral of $\partial_j U$ is handled by decomposing the Newtonian potential as $U = U_A + U_{\neg A}$. The first contribution to F_0^j is the self-interaction term $\int_A \rho^* \partial_j U_A\, d^3x$, which vanishes; this is the statement of Eq. (8.120a), adapted to the current situation in which the integration extends over the volume occupied by body A. The sole contribution to the Newtonian force is therefore $F_0^j = \int_A \rho^* \partial_j U_{\neg A}\, d^3x$. In this we insert the Taylor expansion of Eq. (9.89). The first term gives rise to Eq. (9.90). The second term vanishes, because $\int_A \rho^* \bar{x}^j\, d^3x = 0$ by virtue of the definition of the center-of-mass position $\boldsymbol{r}_A$. The third term gives rise to a contribution $\frac{1}{2} I_A^{kn} \partial_{jkn} U_{\neg A}(t, \boldsymbol{r}_A)$ to the Newtonian force, and this is smaller than the leading term by a factor of order $(R_A/r_{AB})^2$; we discard this as well as all higher-order terms. Our final result is Eq. (9.90).

We next examine a more complicated example, the computation of F_6^j. Here again we write $U = U_A + U_{\neg A}$, and in addition we express the velocity field $\boldsymbol{v}$ as $\boldsymbol{v}_A + \bar{\boldsymbol{v}}$. This gives rise to a sum of six terms,

$$
\begin{aligned}
c^2 F_6^j &= v_A^2 \int_A \rho^* \partial_j U_A\, d^3x + 2v_A^k \int_A \rho^* \bar{v}_k \partial_j U_A\, d^3x + \int_A \rho^* \bar{v}^2 \partial_j U_A\, d^3x \\
&\quad + v_A^2 \int_A \rho^* \partial_j U_{\neg A}\, d^3x + 2v_A^k \int_A \rho^* \bar{v}_k \partial_j U_{\neg A}\, d^3x + \int_A \rho^* \bar{v}^2 \partial_j U_{\neg A}\, d^3x .
\end{aligned}
\tag{9.92}
$$

The first integral vanishes, as was observed previously. In the second integral we insert $U_A = G \int_A \rho^{*\prime} |\boldsymbol{x} - \boldsymbol{x}'|^{-1}\, d^3x'$ which we differentiate with respect to x^j. The result is

$$
-G \int_A \rho^* \rho^{*\prime} \bar{v}_k \frac{(\bar{x} - \bar{x}')^j}{|\bar{\boldsymbol{x}} - \bar{\boldsymbol{x}}'|^3}\, d^3\bar{x}' d^3\bar{x}
$$

after changing the variables of integration from $\boldsymbol{x}$ and $\boldsymbol{x}'$ to $\bar{\boldsymbol{x}} := \boldsymbol{x} - \boldsymbol{r}_A$ and $\bar{\boldsymbol{x}}' := \boldsymbol{x}' - \boldsymbol{r}_A$. This can be written as

$$
+G \int_A \rho^{*\prime} \rho^* \bar{v}_k' \frac{(\bar{x} - \bar{x}')^j}{|\bar{\boldsymbol{x}} - \bar{\boldsymbol{x}}'|^3}\, d^3\bar{x} d^3\bar{x}'
$$

by switching the identities of the integration variables (the old "switch trick"). This is recognized as the structure integral H_A^{kj}, and the second contribution to $c^2 F_A^j$ is $2H_A^{kj} v_A^k$. The third integral vanishes, because it contains an odd number of internal vectors $\bar{\boldsymbol{x}}$, $\bar{\boldsymbol{x}}'$, and $\bar{\boldsymbol{v}}$. In the fourth integral we insert the Taylor expansion of the external potential and retain the leading term only; the fourth contribution to $c^2 F_6^j$ is $m_A v_A^2 \partial_j U_{\neg A}(t, \boldsymbol{r}_A)$, plus terms that are smaller than this by a factor of order $(R_A/r_{AB})^2$. In the fifth integral the Taylor expansion gives rise to a leading term proportional to $\int_A \rho^* \bar{v}_k\, d^3\bar{x}$, which vanishes by virtue of Eq. (9.4). The second term is proportional to $\int_A \rho^* \bar{v}_k \bar{x}^n \, d^3\bar{x}$, which vanishes by virtue of Eq. (9.29). We neglect the third term and conclude that the fifth contribution to $c^2 F_6^j$ is negligible. And finally, the sixth integral produces the final contribution $2\mathcal{T}_A \partial_j U_{\neg A}(t, \boldsymbol{r}_A)$ to $c^2 F_6^j$. Collecting results, we obtain the expression displayed in Eq. (9.91f).

As our final example we go through the computations that lead to Eq. (9.91n). After decomposing Φ_2 into internal and external pieces, we have that

$$c^2 F_{14}^j = -\int_A \rho^* \partial_j \Phi_{2,A} \, d^3x - \int_A \rho^* \partial_j \Phi_{2,\neg A} \, d^3x. \tag{9.93}$$

In the second integral we substitute the Taylor expansion for $\Phi_{2,\neg A}$, and this gives rise to a contribution $-m_A \partial_j \Phi_{2,\neg A}(t, \boldsymbol{r}_A)$ to $c^2 F_{14}^j$. Working now on the first integral, we invoke the definition of Φ_2 from Eqs. (8.8) on page 374, decompose U' into internal and external pieces, and write

$$\partial_j \Phi_{2,A} = -G \int_A \rho^{*\prime} U_A' \frac{(x - x')^j}{|\boldsymbol{x} - \boldsymbol{x}'|^3} \, d^3x' - G \int_A \rho^{*\prime} U_{\neg A}' \frac{(x - x')^j}{|\boldsymbol{x} - \boldsymbol{x}'|^3} \, d^3x'. \tag{9.94}$$

Making the substitution, we find that the first term produces a contribution

$$G \int_A \rho^* \rho^{*\prime} U_A' \frac{(\bar{x} - \bar{x}')^j}{|\bar{\boldsymbol{x}} - \bar{\boldsymbol{x}}'|^3} \, d^3\bar{x}' d^3\bar{x} \tag{9.95}$$

to $c^2 F_{14}^j$; this integral vanishes because it contains an odd number of internal vectors. The second term produces

$$G \int_A \rho^* \rho^{*\prime} U_{\neg A}' \frac{(\bar{x} - \bar{x}')^j}{|\bar{\boldsymbol{x}} - \bar{\boldsymbol{x}}'|^3} \, d^3\bar{x}' d^3\bar{x}, \tag{9.96}$$

and in this we substitute the Taylor expansion for the external Newtonian potential. The leading term gives rise to an odd integral, and the next term produces a contribution

$$\partial_k U_{\neg A}(t, \boldsymbol{r}_A) \, G \int_A \rho^* \rho^{*\prime} \frac{\bar{x}'^k (\bar{x} - \bar{x}')^j}{|\bar{\boldsymbol{x}} - \bar{\boldsymbol{x}}'|^3} \, d^3\bar{x}' d^3\bar{x} \tag{9.97}$$

to $c^2 F_{14}^j$. By making use of the switch trick we re-express this as

$$-\partial_k U_{\neg A}(t, \boldsymbol{r}_A) \, G \int_A \rho^{*\prime} \rho^* \frac{\bar{x}^k (\bar{x} - \bar{x}')^j}{|\bar{\boldsymbol{x}} - \bar{\boldsymbol{x}}'|^3} \, d^3\bar{x} d^3\bar{x}', \tag{9.98}$$

and averaging the results produces the final expression

$$-\frac{1}{2} \partial_k U_{\neg A}(t, \boldsymbol{r}_A) \, G \int_A \rho^* \rho^{*\prime} \frac{(\bar{x} - \bar{x}')^j (\bar{x} - \bar{x}')^k}{|\bar{\boldsymbol{x}} - \bar{\boldsymbol{x}}'|^3} \, d^3\bar{x}' d^3\bar{x}; \tag{9.99}$$

this is equal to $\Omega_A^{jk} \partial_k U_{\neg A}(t, \boldsymbol{r}_A)$. The higher-order terms in the Taylor expansion can be neglected, and collecting results, we have established Eq. (9.91n).

9.3.3 Equations of motion (in terms of external potentials)

Substitution of Eqs. (9.90) and (9.91) into Eq. (9.86) returns

$$\begin{aligned} a_A^j = \frac{1}{m_A c^2}\Big[& \big(2L_A^{(jk)} - 4H_A^{(jk)} + 3K_A^{jk} - \delta^{jk}\dot{P}_A\big)v_A^k \\ & - 4\big(2\mathcal{T}_A^{jk} + \Omega_A^{jk} + \delta^{jk}P_A\big)\partial_k U_{\neg A} + \big(2\mathcal{T}_A + \Omega_A + 3P_A\big)\partial_j U_{\neg A}\Big] \\ + \partial_j U_{\neg A} + \frac{1}{c^2}\Big[& \big(v_A^2 - 4U_{\neg A}\big)\partial_j U_{\neg A} - v_A^j\big(4v_A^k\partial_k U_{\neg A} + 3\partial_t U_{\neg A}\big) \\ & - 4v_A^k\big(\partial_j U_{\neg A}^k - \partial_k U_{\neg A}^j\big) + 4\partial_t U_{\neg A}^j + \partial_j \Psi_{\neg A}\Big] + O(c^{-4}) \end{aligned} \tag{9.100}$$

after some algebra and simplification. The self-interaction terms can all be eliminated by taking into account the equilibrium conditions of Eqs. (9.10), (9.11), and (9.12), as well as setting $\dot{P}_A = 0$. The equations of motion for body A reduce to

$$\begin{aligned} a_A^j = \partial_j U_{\neg A} + \frac{1}{c^2}\Big[& \big(v_A^2 - 4U_{\neg A}\big)\partial_j U_{\neg A} - v_A^j\big(4v_A^k\partial_k U_{\neg A} + 3\partial_t U_{\neg A}\big) \\ & - 4v_A^k\big(\partial_j U_{\neg A}^k - \partial_k U_{\neg A}^j\big) + 4\partial_t U_{\neg A}^j + \partial_j \Psi_{\neg A}\Big] + O(c^{-4}). \end{aligned} \tag{9.101}$$

The acceleration vector is currently expressed in terms of the external piece of the Newtonian potential U, the vector potential U^j, and the post-Newtonian potential Ψ; these are evaluated at $\boldsymbol{x} = \boldsymbol{r}_A(t)$ after differentiation. The external post-Newtonian potential is given by

$$\Psi_{\neg A} = 2\Phi_{1,\neg A} - \Phi_{2,\neg A} + \Phi_{3,\neg A} + 4\Phi_{4,\neg A} - \frac{1}{2}\Phi_{5,\neg A} - \frac{1}{2}\Phi_{6,\neg A} \tag{9.102}$$

in terms of the auxiliary external potentials $\Phi_{n,\neg A}$. Our next task is to evaluate the external potentials, and find their expressions as explicit functions of the positions $\boldsymbol{r}_A$ and velocities $\boldsymbol{v}_A$.

Before we proceed it is interesting to compare Eq. (9.101), which governs the motion of body A among a system of N gravitating bodies, with the geodesic equation (8.16), which determines the motion of a test mass in a pre-determined spacetime with gravitational potentials U, U^j, and Ψ. The equations are formally identical, and this allows us to conclude that body A moves as if it were on a geodesic in a spacetime with gravitational potentials $U_{\neg A}$, $U_{\neg A}^j$, and $\Psi_{\neg A}$. This conclusion rests on our assumptions that each body is non-spinning and well separated from any other body, so that the effects of the higher-order multipole moments (such as the quadrupole moment I_A^{jk} and the angular-momentum tensor S_A^{jk}) can be neglected in the equations of motion; inclusion of these effects would produce deviations from geodesic motion. The conclusion must also be formulated with care, because the external potentials are not truly independent of body A; as we shall see, the non-linear nature of the field equations implies that $\partial_t U^j_{\neg A}$ and $\partial_j \Phi_{2,\neg A}$ depend on m_A in addition to all other, external masses.

9.3.4 Evaluation of the external potentials

Once again we shall state our results and go through the details of only a small subset of computations. After evaluation at $\boldsymbol{x} = \boldsymbol{r}_A(t)$ we find that the derivatives of the external potentials are given by

$$\partial_j U_{\neg A} = -\sum_{B\neq A} \frac{Gm_B n^j_{AB}}{r^2_{AB}}, \tag{9.103a}$$

$$\partial_t U_{\neg A} = \sum_{B\neq A} \frac{Gm_B(\boldsymbol{n}_{AB}\cdot\boldsymbol{v}_B)}{r^2_{AB}}, \tag{9.103b}$$

$$\partial_k U^j_{\neg A} = -\sum_{B\neq A} \frac{Gm_B v^j_B n^k_{AB}}{r^2_{AB}}, \tag{9.103c}$$

$$\begin{aligned}\partial_t U^j_{\neg A} &= \sum_{B\neq A} G\big(2\mathcal{T}^{jk}_B + \Omega^{jk}_B + \delta^{jk}P_A\big)\frac{n^k_{AB}}{r^2_{AB}} + \sum_{B\neq A}\frac{Gm_B(\boldsymbol{n}_{AB}\cdot\boldsymbol{v}_B)v^j_B}{r^2_{AB}} \\ &\quad + \sum_{B\neq A}\frac{G^2 m_A m_B n^j_{AB}}{r^3_{AB}} - \sum_{B\neq A}\sum_{C\neq A,B}\frac{G^2 m_B m_C n^j_{BC}}{r_{AB}r^2_{BC}},\end{aligned} \tag{9.103d}$$

$$\partial_j \Phi_{1,\neg A} = -\sum_{B\neq A}\frac{2G\mathcal{T}_B n^j_{AB}}{r^2_{AB}} - \sum_{B\neq A}\frac{Gm_B v^2_B n^j_{AB}}{r^2_{AB}}, \tag{9.103e}$$

$$\begin{aligned}\partial_j \Phi_{2,\neg A} &= \sum_{B\neq A}\frac{2G\Omega_B n^j_{AB}}{r^2_{AB}} - \sum_{B\neq A}\frac{G^2 m_A m_B n^j_{AB}}{r^3_{AB}} \\ &\quad - \sum_{B\neq A}\sum_{C\neq A,B}\frac{G^2 m_B m_C n^j_{AB}}{r^2_{AB} r_{BC}},\end{aligned} \tag{9.103f}$$

$$\partial_j \Phi_{3,\neg A} = -\sum_{B\neq A}\frac{GE^{\text{int}}_B n^j_{AB}}{r^2_{AB}}, \tag{9.103g}$$

$$\partial_j \Phi_{4,\neg A} = -\sum_{B\neq A}\frac{GP_B n^j_{AB}}{r^2_{AB}}, \tag{9.103h}$$

$$\begin{aligned}\partial_j \Phi_{5,\neg A} &= -\sum_{B\neq A} G\Omega^{kn}_B \partial_{jkn} r_{AB} \\ &\quad - \sum_{B\neq A}\sum_{C\neq A,B}\frac{G^2 m_B m_C}{r_{AB}r^2_{BC}}\Big[n^j_{BC} - (\boldsymbol{n}_{AB}\cdot\boldsymbol{n}_{BC})n^j_{AB}\Big],\end{aligned} \tag{9.103i}$$

(continued overleaf)

$$\partial_j \Phi_{6,\neg A} = -\sum_{B\neq A} 2G\mathcal{T}_B^{kn}\partial_{jkn}r_{AB} - \sum_{B\neq A} \frac{2G\mathcal{T}_B n^j_{AB}}{r^2_{AB}} + \sum_{B\neq A} \frac{Gm_B(\boldsymbol{n}_{AB}\cdot\boldsymbol{v}_B)}{r^2_{AB}}\Big[2v^j_B - 3(\boldsymbol{n}_{AB}\cdot\boldsymbol{v}_B)n^j_{AB}\Big]. \tag{9.103j}$$

Once more we have assumed that the bodies are well separated, and terms that are suppressed by factors of order $(R_A/r_{AB})^2$ relative to the leading contributions have been freely discarded. We make use of the notation introduced in Eq. (9.49),

$$\boldsymbol{r}_{AB} := \boldsymbol{r}_A - \boldsymbol{r}_B, \qquad r_{AB} := |\boldsymbol{r}_A - \boldsymbol{r}_B|, \qquad \boldsymbol{n}_{AB} := \frac{\boldsymbol{r}_{AB}}{r_{AB}}, \tag{9.104}$$

and $\partial_{jkn}r_{AB}$ stands for the third derivative of the inter-body distance r_{AB} with respect to the vector $\boldsymbol{r}_{AB}$.

To illustrate how these results are obtained we begin with the simplest case, the evaluation of $\partial_j U_{\neg A}$. Introducing the notation $s := |\boldsymbol{x} - \boldsymbol{x}'|$, we recall first that the Newtonian potential is given by $U = G\int \rho^{*\prime} s^{-1}\, d^3x'$, so that $\partial_j U = G\int \rho^{*\prime}\partial_j s^{-1}\, d^3x'$; the external piece of this is

$$\partial_j U_{\neg A} = \sum_{B\neq A} G\int_B \rho^{*\prime}\partial_j s^{-1}\, d^3x'. \tag{9.105}$$

In this we substitute $\boldsymbol{x}' = \boldsymbol{r}_B(t) + \bar{\boldsymbol{x}}'$, so that s becomes $s = |\boldsymbol{x} - \boldsymbol{r}_B - \bar{\boldsymbol{x}}'|$. We next expand $\partial_j s^{-1}$ in powers of $\bar{x}'^k$:

$$\partial_j s^{-1} = \partial_j s_B^{-1} - \bar{x}'^k\partial_{jk}s_B^{-1} + \frac{1}{2}\bar{x}'^k\bar{x}'^n\partial_{jkn}s_B^{-1} + \cdots, \tag{9.106}$$

where $s_B := |\boldsymbol{x} - \boldsymbol{r}_B|$. Making the substitution in $\partial_j U_{\neg A}$, we find that it becomes

$$\partial_j U_{\neg A} = \sum_{B\neq A}\Big(Gm_B\partial_j s_B^{-1} + \frac{1}{2}GI_B^{kn}\partial_{jkn}s_B^{-1} + \cdots\Big). \tag{9.107}$$

The term involving $\partial_{jk}s_B^{-1}$ vanishes, because $\int_B \rho^{*\prime}\bar{x}'^k\, d^3x' = 0$ by virtue of the definition of the center-of-mass position r^k_B. The term involving the quadrupole moment tensor is smaller than the leading term by a factor of order $(R_B/s_B)^2$, and we discard it. After evaluation of $\partial_j s_B^{-1}$ using Eqs. (9.50), we set $\boldsymbol{x} = \boldsymbol{r}_A(t)$ and arrive at Eq. (9.103a).

We next tackle a more complicated case, the evaluation of $\partial_t U^j_{\neg A}$. The vector potential is $U^j = G\int \rho^{*\prime}v'^j s^{-1}\, d^3x'$, and using the rules spelled out in Box 8.4, we find that its time derivative is given by $\partial_t U^j = G\int \rho^{*\prime}(dv'^j/dt + v'^j v'^k\partial_{k'})s^{-1}\, d^3x'$. In this we substitute the Newtonian version of Euler's equation, Eq. (8.104), and obtain

$$\partial_t U^j = -G\int(\partial_{j'}p')s^{-1}\, d^3x' + G\int \rho^{*\prime}(\partial_{j'}U')s^{-1}\, d^3x' - G\int \rho^{*\prime}v'^j v'^k\partial_k s^{-1}\, d^3x'; \tag{9.108}$$

in the last term we have made use of the identity $\partial_{k'} s^{-1} = -\partial_k s^{-1}$, and our expression for $\partial_t U^j$ is accurate up to terms of order c^{-2}. The external piece of this is

$$\partial_t U^j_{\neg A} = -\sum_{B\neq A} G\int_B (\partial_{j'} p')s^{-1}\, d^3x' + \sum_{B\neq A} G\int_B \rho^{*\prime}(\partial_{j'} U')s^{-1}\, d^3x' - \sum_{B\neq A} G\int_B \rho^{*\prime} v'^j v'^k \partial_k s^{-1}\, d^3x'. \tag{9.109}$$

We initially examine the first group of terms. We substitute $\boldsymbol{x}' = \boldsymbol{r}_B(t) + \bar{\boldsymbol{x}}'$ inside the integral and expand s^{-1} in powers of $\bar{x}'^k$; it becomes

$$s_B^{-1}\int_B \partial_{j'} p'\, d^3\bar{x}' - \partial_k s_B^{-1}\int_B \bar{x}'^k \partial_{j'} p'\, d^3\bar{x}' + \frac{1}{2}\partial_{kn} s_B^{-1}\int_B \bar{x}'^k \bar{x}'^n \partial_{j'} p'\, d^3\bar{x}' + \cdots \tag{9.110}$$

The first integral vanishes automatically, the third vanishes because it contains an odd number of internal vectors, and after integration by parts the second integral returns $-\delta^{jk} P_B$. The first group of terms in Eq. (9.109) is therefore

$$-\sum_{B\neq A} G\int_B (\partial_{j'} p')s^{-1}\, d^3x' = \sum_{B\neq A} G P_B \partial_j s_B^{-1}, \tag{9.111}$$

after discarding contributions that are smaller by a factor of order $(R_B/s_B)^2$.

We turn next to the second group of terms in Eq. (9.109), in which we insert $\partial_{j'} U' = \partial_{j'} U'_A + \partial_{j'} U'_B + \sum_{C\neq A,B} \partial_{j'} U'_C$. The integral that involves $\partial_{j'} U'_A = G\int_A \rho^{*\prime\prime} \partial_{j'} s'^{-1}\, d^3x''$ is

$$G\int_B\int_A \rho^{*\prime}\rho^{*\prime\prime} s^{-1} \partial_{j'} s'^{-1}\, d^3x''\, d^3x', \tag{9.112}$$

where $s' := |\boldsymbol{x}' - \boldsymbol{x}''|$. In this we substitute $\boldsymbol{x}' = \boldsymbol{r}_B + \bar{\boldsymbol{x}}'$, $\boldsymbol{x}'' = \boldsymbol{r}_A + \bar{\boldsymbol{x}}''$ and express $s^{-1}\partial_{j'} s'^{-1}$ as a double Taylor expansion in powers of $\bar{x}'^k$ and $\bar{x}''^n$. Only the leading term is required, and we arrive at

$$G\int_B \rho^{*\prime}(\partial_{j'} U'_A)s^{-1}\, d^3x' = G^2 m_A m_B s_B^{-1} \partial_{j'} r_{AB}^{-1}, \tag{9.113}$$

in which $\partial_{j'}$ is interpreted as a partial derivative with respect to r_B^j. The integral that involves $\partial_{j'} U'_C$ is evaluated in the same way, and we get

$$G\int_B \rho^{*\prime}(\partial_{j'} U'_C)s^{-1}\, d^3x' = G^2 m_B m_C s_B^{-1} \partial_{j'} r_{BC}^{-1}. \tag{9.114}$$

The integral that involves $\partial_{j'} U'_B$ is

$$-G\int_B \rho^{*\prime}\rho^{*\prime\prime} s^{-1} \frac{(x'-x'')^j}{|\boldsymbol{x}'-\boldsymbol{x}''|^3}\, d^3x''\, d^3x', \tag{9.115}$$

and in this we substitute $\boldsymbol{x}' = \boldsymbol{r}_B + \bar{\boldsymbol{x}}'$ and $\boldsymbol{x}'' = \boldsymbol{r}_B + \bar{\boldsymbol{x}}''$. We expand $s = |\boldsymbol{x} - \boldsymbol{r}_B - \bar{\boldsymbol{x}}'|$ in powers of $\bar{x}'^k$, eliminate all odd integrals, and get

$$G\partial_k s_B^{-1}\int_B \rho^{*\prime}\rho^{*\prime\prime} \frac{\bar{x}'^k(\bar{x}'-\bar{x}'')^j}{|\bar{\boldsymbol{x}}'-\bar{\boldsymbol{x}}''|^3}\, d^3\bar{x}''\, d^3\bar{x}'. \tag{9.116}$$

After symmetrization in the primed and double-primed variables, we arrive at

$$G\int_B \rho^{*\prime}(\partial_{j'}U'_B)s^{-1}\,d^3x' = -G\Omega_B^{jk}\partial_k s_B^{-1}. \tag{9.117}$$

Collecting results, we find that the second group of terms in Eq. (9.109) is

$$\sum_{B\neq A} G\int_B \rho^{*\prime}(\partial_{j'}U')s^{-1}\,d^3x' = -\sum_{B\neq A} G\Omega_B^{jk}\partial_k s_B^{-1} + \sum_{B\neq A} G^2 m_A m_B s_B^{-1}\partial_{j'}r_{AB}^{-1}$$
$$+\sum_{B\neq A}\sum_{C\neq A,B} G^2 m_B m_C s_B^{-1}\partial_{j'}r_{BC}^{-1}. \tag{9.118}$$

We recall that in the second and third sums, $\partial_{j'}$ is interpreted as a partial derivative with respect to r_B^j.

We finally examine the third group of terms in Eq. (9.109). Here the manipulations are simple: we insert $\boldsymbol{x}' = \boldsymbol{r}_B + \bar{\boldsymbol{x}}'$, $\boldsymbol{v}' = \boldsymbol{v}_B + \bar{\boldsymbol{v}}'$ and expand $\partial_k s^{-1}$ in powers of $\bar{x}'^n$. We quickly arrive at

$$-\sum_{B\neq A} G\int_B \rho^{*\prime} v'^j v'^k \partial_k s^{-1}\,d^3x' = -\sum_{B\neq A} 2GT_B^{jk}\partial_k s_B^{-1} - \sum_{B\neq A} Gm_B v_B^j v_B^k \partial_k s_B^{-1}. \tag{9.119}$$

Substitution of Eqs. (9.111), (9.118), and (9.119) into Eq. (9.109) returns Eq. (9.103d), after computation of the derivatives of s_B, r_{AB}, and r_{BC} and evaluation of all expressions at $\boldsymbol{x} = \boldsymbol{r}_A(t)$.

9.3.5 Equations of motion (final form)

When Eqs. (9.103) are inserted within Eqs. (9.101) and (9.102), we finally obtain an explicit expression for a_A^j, the coordinate acceleration of body A. It can be decomposed as

$$\boldsymbol{a}_A = \boldsymbol{a}_A[\text{0PN}] + \boldsymbol{a}_A[\text{1PN}] + \boldsymbol{a}_A[\text{STR}] + O(c^{-4}), \tag{9.120}$$

where

$$\boldsymbol{a}_A[\text{0PN}] = -\sum_{B\neq A}\frac{Gm_B}{r_{AB}^2}\boldsymbol{n}_{AB} \tag{9.121}$$

is the Newtonian acceleration, while

$$\begin{aligned}
c^2\boldsymbol{a}_A[\text{1PN}] = &-\sum_{B\neq A}\frac{Gm_B}{r_{AB}^2}\biggl[v_A^2 - 4(\boldsymbol{v}_A\cdot\boldsymbol{v}_B) + 2v_B^2 - \frac{3}{2}(\boldsymbol{n}_{AB}\cdot\boldsymbol{v}_B)^2 \\
&\qquad - \frac{5Gm_A}{r_{AB}} - \frac{4Gm_B}{r_{AB}}\biggr]\boldsymbol{n}_{AB} \\
&+\sum_{B\neq A}\frac{Gm_B}{r_{AB}^2}\Bigl[\boldsymbol{n}_{AB}\cdot(4\boldsymbol{v}_A - 3\boldsymbol{v}_B)\Bigr](\boldsymbol{v}_A - \boldsymbol{v}_B) \\
&+\sum_{B\neq A}\sum_{C\neq A,B}\frac{G^2 m_B m_C}{r_{AB}^2}\biggl[\frac{4}{r_{AC}} + \frac{1}{r_{BC}} - \frac{r_{AB}}{2r_{BC}^2}(\boldsymbol{n}_{AB}\cdot\boldsymbol{n}_{BC})\biggr]\boldsymbol{n}_{AB} \\
&-\frac{7}{2}\sum_{B\neq A}\sum_{C\neq A,B}\frac{G^2 m_B m_C}{r_{AB}r_{BC}^2}\boldsymbol{n}_{BC}
\end{aligned} \tag{9.122}$$

is the post-Newtonian piece of the acceleration vector.

The third contribution $\boldsymbol{a}_A[\text{STR}]$ is generated by the structure-integral terms that are scattered throughout Eqs. (9.103); it is given by

$$\begin{aligned} c^2 a_A^j[\text{STR}] = \sum_{B\neq A} \biggl[& 4G\bigl(2\mathcal{T}_B^{jk} + \Omega_B^{jk} + \delta^{jk} P_B\bigr)\frac{n_{AB}^k}{r_{AB}^2} \\ & + \frac{1}{2}G\bigl(2\mathcal{T}_B^{kn} + \Omega_B^{kn} + \delta^{kn} P_B\bigr)\partial_{jkn} r_{AB} \\ & - G\bigl(2\mathcal{T}_B + \Omega_B + 3P_B\bigr)\frac{n_{AB}^j}{r_{AB}^2} - \frac{GE_B}{r_{AB}^2} n_{AB}^j \biggr], \end{aligned} \tag{9.123}$$

where $E_B := \mathcal{T}_B + \Omega_B + E_B^{\text{int}}$ is the total energy contained in body B, the sum of kinetic, gravitational, and internal energies. To obtain this expression we added $\frac{1}{2}G\delta^{kn} P_B \partial_{jkn} r_{AB}$ to the second group of terms and subtracted the same thing, $\frac{1}{2}GP_B\partial_{jkk} r_{AB} = -P_B n_{AB}^j / r_{AB}^2$, from the third group. Most of $a_A^j[\text{STR}]$ vanishes after imposing the equilibrium conditions of Eqs. (9.10) and (9.12); what survives is

$$\boldsymbol{a}_A[\text{STR}] = -\sum_{B\neq A} \frac{G(E_B/c^2)}{r_{AB}^2}\boldsymbol{n}_{AB}, \tag{9.124}$$

and this makes a contribution to the acceleration at 1PN order.

We observe that $\boldsymbol{a}[\text{0PN}]$ and $\boldsymbol{a}[\text{STR}]$ combine nicely to give

$$\boldsymbol{a}[\text{0PN}] + \boldsymbol{a}[\text{STR}] = -\sum_{B\neq A}\frac{GM_B}{r_{AB}^2}\boldsymbol{n}_{AB}, \tag{9.125}$$

where

$$M_B := m_B + E_B/c^2 + O(c^{-4}) \tag{9.126}$$

is the total mass-energy of body B. And we observe that the substitution $m_B = M_B + O(c^{-2})$ can be made in $\boldsymbol{a}_A[\text{1PN}]$ without altering the form of the equations of motion at 1PN order. Our conclusion is that the equations of motion, like the inter-body metric of Eqs. (9.81), depend on the total mass-energy parameters M_B only, and not on their decomposition in terms of material mass m_B, kinetic energy $\mathcal{T}_B$, gravitational potential energy Ω_B, and internal energy E_B^{int}. The equations of motion, like the inter-body metric, are compatible with the strong formulation of the principle of equivalence.

Our final expression for the equations of motion is

$$\begin{aligned} \boldsymbol{a}_A = & -\sum_{B\neq A}\frac{GM_B}{r_{AB}^2}\boldsymbol{n}_{AB} \\ & + \frac{1}{c^2}\biggl\{ -\sum_{B\neq A}\frac{GM_B}{r_{AB}^2}\biggl[v_A^2 - 4(\boldsymbol{v}_A\cdot\boldsymbol{v}_B) + 2v_B^2 - \frac{3}{2}(\boldsymbol{n}_{AB}\cdot\boldsymbol{v}_B)^2 \\ & \qquad - \frac{5GM_A}{r_{AB}} - \frac{4GM_B}{r_{AB}}\biggr]\boldsymbol{n}_{AB} \end{aligned}$$

(continued overleaf)

$$
\begin{aligned}
&+\sum_{B\neq A}\frac{GM_B}{r_{AB}^2}\Big[\boldsymbol{n}_{AB}\cdot(4\boldsymbol{v}_A-3\boldsymbol{v}_B)\Big](\boldsymbol{v}_A-\boldsymbol{v}_B)\\
&+\sum_{B\neq A}\sum_{C\neq A,B}\frac{G^2M_BM_C}{r_{AB}^2}\left[\frac{4}{r_{AC}}+\frac{1}{r_{BC}}-\frac{r_{AB}}{2r_{BC}^2}(\boldsymbol{n}_{AB}\cdot\boldsymbol{n}_{BC})\right]\boldsymbol{n}_{AB}\\
&-\frac{7}{2}\sum_{B\neq A}\sum_{C\neq A,B}\frac{G^2M_BM_C}{r_{AB}r_{BC}^2}\boldsymbol{n}_{BC}\Bigg]+O(c^{-4}).
\end{aligned}
\tag{9.127}
$$

We recall our notation: $\boldsymbol{r}_A(t)$ is the position of body A in harmonic coordinates, $\boldsymbol{v}_A(t) := d\boldsymbol{r}_A/dt$ is its velocity, and $\boldsymbol{a}_A := d\boldsymbol{v}_A/dt$ is the coordinate acceleration. The vector $\boldsymbol{r}_{AB} := \boldsymbol{r}_A - \boldsymbol{r}_B$ points from body B to body A; its length $r_{AB} := |\boldsymbol{r}_A - \boldsymbol{r}_B|$ is the inter-body distance, and $\boldsymbol{n}_{AB} := \boldsymbol{r}_{AB}/r_{AB}$.

The equations of motion (9.127) apply to each body A within the N-body system. The bodies are assumed to be non-spinning and sufficiently well separated that the effects of higher-order multipole moments can be ignored. These equations have a rich history that was well summarized by Peter Havas in a 1989 essay. The equations were first formulated in 1917 by Lorentz and Droste, who published their results in Dutch in a communication to the Dutch Academy; their breakthrough remained unnoticed by the few researchers involved in the early development of general relativity. The equations of motion were also obtained at about the same time by de Sitter, who made use of the post-Newtonian metric previously derived by Droste, and postulated that the bodies should move on geodesics of the external metric; because of a calculational error, de Sitter's equations differed from the correct post-Newtonian equations by one term, and led to the wrong prediction that the system's barycenter should undergo a secular acceleration. The error was discovered and corrected in 1938 by Eddington and Clark (the twenty-year delay indicating the low level of activity in general relativity at the time), and in the same year a new derivation of the equations of motion was produced by Einstein, Infeld, and Hoffmann. In spite of the much earlier work of Lorentz and Droste, which eventually came to light thanks to an English translation published in 1937, the equations became known as the *EIH equations of motion*.

9.3.6 Conserved quantities

In Secs. 8.4.5, 8.4.6, and 8.4.7 we established the existence of conserved quantities associated with the dynamics of a perfect fluid in a post-Newtonian spacetime. We identified

$$
M := \int \rho^*\left[1+\frac{1}{c^2}\left(\frac{1}{2}v^2-\frac{1}{2}U+\Pi\right)\right]d^3x + O(c^{-4}) \tag{9.128}
$$

as the total mass-energy of the fluid system,

$$
P^j := \int \rho^* v^j\left[1+\frac{1}{c^2}\left(\frac{1}{2}v^2-\frac{1}{2}U+\Pi+p/\rho^*\right)\right]d^3x - \frac{1}{2c^2}\int \rho^*\Phi^j\,d^3x + O(c^{-4}) \tag{9.129}
$$

as the total momentum, with Φ^j defined by

$$\Phi^j := G \int \rho^{*\prime} v_k' \frac{(x-x')^j (x-x')^k}{|\boldsymbol{x}-\boldsymbol{x}'|^3} \, d^3x', \tag{9.130}$$

and

$$R^j := \frac{1}{M} \int \rho^* x^j \left[1 + \frac{1}{c^2}\left(\frac{1}{2}v^2 - \frac{1}{2}U + \Pi \right) \right] d^3x + O(c^{-4}) \tag{9.131}$$

as the position of the center-of-mass for the entire fluid system. The total mass-energy and momentum are constants of the fluid's motion; the position of the center-of-mass satisfies $M\dot{R}^j = P^j$.

The conserved quantities keep their usefulness when the fluid distribution is broken up into a collection of N separated bodies. In this case the integrals of Eqs. (9.128), (9.129), and (9.131) become a sum of N individual integrals, and the conserved quantities become

$$M = \sum_A M_A + \frac{1}{c^2} \sum_A \frac{1}{2} M_A v_A^2 - \frac{1}{c^2} \sum_A \sum_{B \neq A} \frac{G M_A M_B}{2 r_{AB}} + O(c^{-4}), \tag{9.132a}$$

$$\begin{aligned} \boldsymbol{P} = {} & \sum_A M_A \boldsymbol{v}_A + \frac{1}{c^2} \sum_A \frac{1}{2} M_A v_A^2 \boldsymbol{v}_A \\ & - \frac{1}{c^2} \sum_A \sum_{B \neq A} \frac{G M_A M_B}{2 r_{AB}} \Big[\boldsymbol{v}_A + (\boldsymbol{n}_{AB} \cdot \boldsymbol{v}_A) \boldsymbol{n}_{AB} \Big] + O(c^{-4}), \end{aligned} \tag{9.132b}$$

$$M\boldsymbol{R} = \sum_A M_A \boldsymbol{r}_A + \frac{1}{c^2} \sum_A \frac{1}{2} M_A v_A^2 \boldsymbol{r}_A - \frac{1}{c^2} \sum_A \sum_{B \neq A} \frac{G M_A M_B}{2 r_{AB}} \boldsymbol{r}_A + O(c^{-4}). \tag{9.132c}$$

As usual the expressions of Eqs. (9.132) apply to bodies that are well separated; terms of fractional order $(R_A/r_{AB})^2$ have been neglected. It is straightforward (though tedious) to show directly that $dM/dt = 0$, $d\boldsymbol{P}/dt = 0$, and $M d\boldsymbol{R}/dt = \boldsymbol{P}$ by virtue of the post-Newtonian equations of motion. The expression for M reveals that the total mass-energy of the N-body system consists of a sum of mass-energies from each body, plus the total kinetic energy of the system (divided by c^2), plus the total gravitational potential energy of the system (also divided by c^2).

To derive these results we rely on the techniques introduced in the preceding subsections. Starting from Eq. (9.128) we find that the total mass-energy is given by

$$\begin{aligned} M = {} & \sum_A \left(\int_A \rho^* \, d^3x + \frac{1}{2c^2} \int_A \rho^* v^2 \, d^3x - \frac{1}{2c^2} \int_A \rho^* U \, d^3x + \frac{1}{c^2} \int_A \rho^* \Pi \, d^3x \right) \\ & + O(c^{-4}). \end{aligned} \tag{9.133}$$

The ρ^* integral gives m_A, the $\rho^* v^2$ integral produces $m_A v_A^2 + 2\mathcal{T}_A$ after expressing $\boldsymbol{v}$ as $\boldsymbol{v}_A + \bar{\boldsymbol{v}}$, and the $\rho^*\Pi$ integral gives E_A^{int}. In the $\rho^* U$ integral we write $U = U_A + U_{\neg A}$ and observe that the first term produces $-2\Omega_A$, while the second term gives $m_A U_{\neg A}(t, \boldsymbol{r}_A)$ after substitution of the Taylor expansion for the external potential; it is here that terms of fractional order $(R_A/r_{AB})^2$ are discarded. Collecting results, and noting that $m_A + (\mathcal{T}_A + \Omega_A + E_A^{\text{int}})/c^2 = M_A$, we arrive at Eq. (9.132a) after making the substitution

$m_A = M_A + O(c^{-2})$ in the post-Newtonian terms. The derivation of Eq. (9.132c) proceeds in exactly the same way, and there is no need to go through the details here.

Equation (9.129) implies that the total momentum of the N-body system is given by

$$P^j = \sum_A \biggl(\int_A \rho^* v^j \, d^3x + \frac{1}{2c^2} \int_A \rho^* v^2 v^j \, d^3x - \frac{1}{2c^2} \int_A \rho^* U v^j \, d^3x + \frac{1}{c^2} \int_A \rho^* \Pi v^j \, d^3x + \frac{1}{c^2} \int_A p v^j \, d^3x - \frac{1}{2c^2} \int_A \rho^* \Phi^j \, d^3x \biggr) + O(c^{-4}). \tag{9.134}$$

The first four integrals are evaluated as we did previously, and the fifth integral produces $P_A v_A^j$. In the sixth and final integral we decompose Φ^j as $\Phi_A^j + \Phi_{\neg A}^j$. The internal piece produces $-2\Omega_A^{jk} v_A^k$, and the external piece gives $m_A \Phi_{\neg A}^j(t, \boldsymbol{r}_A)$ after Taylor expansion of the external potential. From Eq. (9.130) we get

$$\Phi_{\neg A}^j := \sum_{B \neq A} G \int_B \rho^{*\prime} v_k^{\prime} \frac{(x - x')^j (x - x')^k}{|\boldsymbol{x} - \boldsymbol{x}'|^3} \, d^3x', \tag{9.135}$$

and in this we substitute $\boldsymbol{x}' = \boldsymbol{r}_B + \bar{\boldsymbol{x}}'$ and $\boldsymbol{v}' = \boldsymbol{v}_B + \bar{\boldsymbol{v}}'$. To leading order in an expansion in powers of $\bar{x}'^j$, we find that $\Phi_{\neg A}^j(t, \boldsymbol{r}_A) = \sum_{B \neq A} G m_B (\boldsymbol{n}_{AB} \cdot \boldsymbol{v}_B) n_{AB}^j / r_{AB}$. Collecting results, we have that

$$P^j = \sum_A \biggl[m_A + \frac{1}{c^2} \bigl(\mathcal{T}_A + \Omega_A + E_A^{\rm int} \bigr) \biggr] v_A^j + \frac{1}{c^2} \sum_A \bigl(2\mathcal{T}_A^{jk} + \Omega_A^{jk} + \delta^{jk} P_A \bigr) v_A^k + \frac{1}{c^2} \sum_A \frac{1}{2} m_A v_A^2 v_A^j - \frac{1}{c^2} \sum_A \sum_{B \neq A} \frac{G m_A m_B}{2 r_{AB}} \Bigl[v_A^j + (\boldsymbol{n}_{AB} \cdot \boldsymbol{v}_B) n_{AB}^j \Bigr] + O(c^{-4}). \tag{9.136}$$

In the first group of terms we recognize $m_A + E_A/c^2 = M_A$, and we eliminate the second group by invoking the equilibrium condition of Eq. (9.10). In the remaining groups we insert $m_A = M_A + O(c^{-2})$, and in the last step we rearrange the double sum that gives rise to the last group: We switch the identities of bodies A and B and re-express the sums as

$$\sum_B \sum_{A \neq B} \frac{G M_B M_A}{2 r_{BA}} (\boldsymbol{n}_{BA} \cdot \boldsymbol{v}_A) n_{BA}^j; \tag{9.137}$$

because $\boldsymbol{n}_{BA} = -\boldsymbol{n}_{AB}$ and $r_{BA} = r_{AB}$, this is

$$\sum_A \sum_{B \neq A} \frac{G M_A M_B}{2 r_{AB}} (\boldsymbol{n}_{AB} \cdot \boldsymbol{v}_A) n_{AB}^j, \tag{9.138}$$

and we have arrived at Eq. (9.132b).

9.3.7 Binary systems

The equations obtained in the preceding subsections apply to any number of well-separated bodies. To conclude this section we examine the special case $N = 2$, that is, the case of a binary system. In the Newtonian context reviewed in Sec. 1.6.7, we saw that the description of the motion simplified when the origin of the coordinate system was attached to the barycenter $\boldsymbol{R}$, and that the position of each body could be determined in terms of the separation vector. The same simplification occurs in the post-Newtonian context.

The binary system consists of a first body of mass-energy M_1, position $\boldsymbol{r}_1$, and velocity $\boldsymbol{v}_1$, and a second body of mass-energy M_2, position $\boldsymbol{r}_2$, and velocity $\boldsymbol{v}_2$. To simplify the notation we introduce the mass parameters

$$m := M_1 + M_2, \qquad \eta := \frac{M_1 M_2}{(M_1 + M_2)^2}, \qquad \Delta := \frac{M_1 - M_2}{M_1 + M_2}, \tag{9.139}$$

so that m is a kind of total mass, η a symmetric mass ratio, and Δ a dimensionless measure of the mass difference; it should be noted that m differs from the total mass-energy M introduced in Eq. (9.132) by terms of order c^{-2}. We introduce also the separation $\boldsymbol{r} := \boldsymbol{r}_1 - \boldsymbol{r}_2$, the relative velocity $\boldsymbol{v} := \boldsymbol{v}_1 - \boldsymbol{v}_2$, and we shall set $r := |\boldsymbol{r}| = r_{12}$, $\boldsymbol{n} := \boldsymbol{r}/r = \boldsymbol{n}_{12}$, and $v := |\boldsymbol{v}|$.

According to Eq. (9.132), the position of the system's barycenter is given by

$$M\boldsymbol{R} = M_1\left[1 + \frac{1}{2c^2}\left(v_1^2 - \frac{GM_2}{r}\right)\right]\boldsymbol{r}_1 + M_2\left[1 + \frac{1}{2c^2}\left(v_2^2 - \frac{GM_1}{r}\right)\right]\boldsymbol{r}_2, \tag{9.140}$$

and we wish to impose the condition $\boldsymbol{R} = \boldsymbol{0}$. This allows us to solve for $\boldsymbol{r}_1$ and $\boldsymbol{r}_2$ in terms of $\boldsymbol{r}$, and the result is

$$\boldsymbol{r}_1 = \frac{M_2}{m}\boldsymbol{r} + \frac{\eta\Delta}{2c^2}\left(v^2 - \frac{Gm}{r}\right)\boldsymbol{r}, \tag{9.141a}$$

$$\boldsymbol{r}_2 = -\frac{M_1}{m}\boldsymbol{r} + \frac{\eta\Delta}{2c^2}\left(v^2 - \frac{Gm}{r}\right)\boldsymbol{r}. \tag{9.141b}$$

These equations imply that $\boldsymbol{v}_1 = (M_2/m)\boldsymbol{v} + O(c^{-2})$ and $\boldsymbol{v}_2 = -(M_1/m)\boldsymbol{v} + O(c^{-2})$.

An equation of motion for $\boldsymbol{r}$ can be obtained by computing the relative acceleration $\boldsymbol{a} := \boldsymbol{a}_1 - \boldsymbol{a}_2$ from Eq. (9.127). Taking into account that $\boldsymbol{n}_{21} = -\boldsymbol{n}_{12} = -\boldsymbol{n}$, the final result after simplification is

$$\boldsymbol{a} = -\frac{Gm}{r^2}\boldsymbol{n} - \frac{Gm}{c^2 r^2}\left\{\left[(1+3\eta)v^2 - \frac{3}{2}\eta(\boldsymbol{n}\cdot\boldsymbol{v})^2 - 2(2+\eta)\frac{Gm}{r}\right]\boldsymbol{n} - 2(2-\eta)(\boldsymbol{n}\cdot\boldsymbol{v})\boldsymbol{v}\right\} + O(c^{-4}). \tag{9.142}$$

This is a second-order differential equation for $\boldsymbol{r}(t)$, and its solution provides, through $\boldsymbol{r}_1$ and $\boldsymbol{r}_2$, complete information regarding the motion of the binary system.

9.4 Motion of compact bodies

The post-Newtonian equations of motion (9.127) apply to fluid bodies that are well separated from one another, so that their mutual gravitational interaction is weak. The method of derivation relied on the post-Newtonian fluid equations, and these rest on an assumption that the self-gravity of each body is also weak. In this context, therefore, the gravitational field is assumed to be weak *everywhere*. In this section we examine a different context in which we retain the *weakness of the mutual gravity* between bodies, but allow each body to be *strongly self-gravitating*. We demonstrate that Eq. (9.127) continues to apply in these situations.

In the new context the bodies can be arbitrarily compact, and can possess an arbitrarily strong internal gravitational field. The bodies are not necessarily built from a perfect fluid, and indeed, we shall have no interest in their internal constitution. The only assumption concerning them shall be that they are spherically symmetric – a restrictive assumption that was not made in the fluid case. Each body may thus be a neutron star, a black hole, or any other object with strong internal gravity; it may still be, of course, a diffuse perfect-fluid body with weak internal gravity. We maintain, however, the requirement that the bodies be well separated, so that gravity is allowed to be weak between the bodies; it is in these inter-body regions that the post-Newtonian metric provides a good approximation to the true gravitational field.

We shall focus our attention on the vacuum region external to one of the compact bodies, and our new derivation of its equations of motion will be based entirely on solving the Einstein field equations in this region. Matter variables never enter this discussion. Our strategy is based instead on the transformation between the inertial frame of the post-Newtonian spacetime and the moving frame of the compact body. To describe this we rely on the theory of post-Newtonian coordinate transformations that was developed back in Sec. 8.3; please refer to the summary provided in Box 8.2. We follow the treatment provided in the 2008 article by Taylor and Poisson; the paper offers additional details that are not covered in this briefer treatment.

9.4.1 Zones and matching strategy

We select one of the compact bodies as our reference body, and we henceforth refer to it as "the body." We introduce three distinct zones in spacetime (see Fig. 9.1). The first is the *body zone*, the body's immediate neighborhood; in the body zone the gravitational field is dominated by the body's own field, and the contribution from external bodies is small. If $\bar{r}$ is the distance from the body's center-of-mass, then the body zone is defined by $\bar{r} < r_{\rm max}$, with $r_{\rm max} \ll r_{AB}$ marking the zone's boundary; r_{AB} is the inter-body distance, and the condition $\bar{r} \ll r_{AB}$ ensures that the external gravity is indeed small. The second zone is the *post-Newtonian zone*, in which gravity is weak everywhere. The outer boundary of the post-Newtonian zone coincides with the boundary of the near zone, as was discussed in Sec. 8.1. The inner boundary is a sphere of radius $\bar{r} = r_{\rm min}$, within which the body's gravity becomes strong; we demand that $r_{\rm min} \gg GM/c^2$, where M is the body's mass. When the bodies are well separated we have that $GM/c^2 \ll r_{AB}$, and we can ensure that $r_{\rm min} < r_{\rm max}$. The region $r_{\rm min} < \bar{r} < r_{\rm max}$ is the *overlap zone*, the intersection between the body and post-Newtonian zones.

Our strategy behind the new derivation of the equations of motion is based on the following key idea: We construct independently two solutions to the vacuum field equations

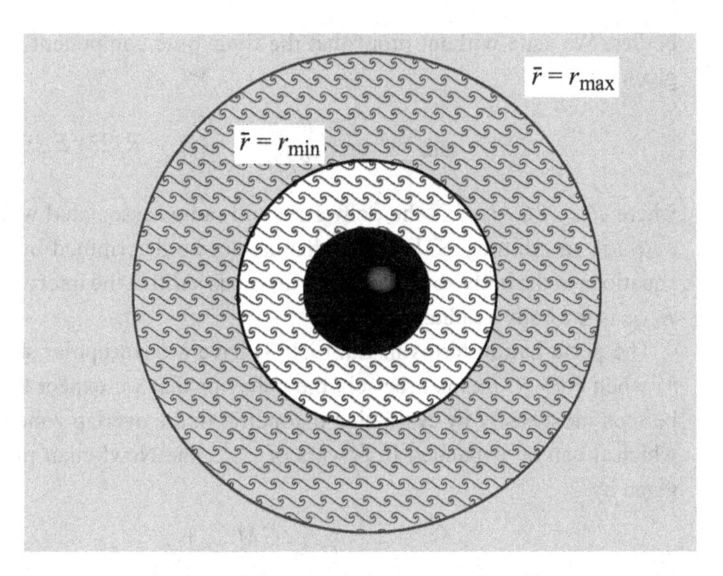

Fig. 9.1 Body, post-Newtonian, and overlap zones. The compact body is shown in black. The body zone is shown in wavy texture, and is restricted by $\bar{r} < r_{\max} \ll r_{AB}$, where $\bar{r}$ is the distance to the body. The post-Newtonian zone is tinted, and is restricted by $\bar{r} > r_{\min} \gg GM/c^2$. The overlap zone is shown in both wavy texture and tint, and is restricted by $r_{\min} < \bar{r} < r_{\max}$.

in two overlapping regions of spacetime, and we match these solutions in the overlap; the equations of motion follow as an outcome of the matching procedure. The first solution is constructed in the body zone, in the body's comoving frame, and the metric is presented in harmonic coordinates $(c\bar{t}, \bar{x}^j)$ that are attached to the body's center-of-mass. The second solution is constructed in the post-Newtonian zone, in the global inertial frame, and the metric is presented in a different set of harmonic coordinates (ct, x^j) that are attached to the center-of-mass of the entire N-body system. To match the solutions in the overlap zone we must first reconcile the coordinate systems, and we therefore transform the post-Newtonian metric from the global coordinates (ct, x^j) to the body coordinates $(c\bar{t}, \bar{x}^j)$. We next compare the post-Newtonian metric to the body metric, and demand that they agree. The matching procedure determines (i) unknown functions within each metric, (ii) unknown functions within the coordinate transformation, and finally, (iii) the body's equations of motion.

9.4.2 Body metric

We wish to find the metric of a spherical body placed in the presence of remote external bodies. The metric will be presented in harmonic coordinates $(c\bar{t}, \bar{x}^j)$, in the body's own moving frame, and it will be valid in the body zone, in which $\bar{r} \ll r_{AB}$. We do not need the metric inside the body, and indeed, the details of internal structure are entirely irrelevant for our purposes. If the body were in complete isolation, its external metric would be given by the Schwarzschild solution of Eq. (5.171) – refer to Sec. 5.6. What we need here, however, is a perturbed version of this metric that accounts for the weak gravity created by the external bodies. We state without proof that the time–time component of this perturbed metric is given by

$$g_{\bar{0}\bar{0}} = -\frac{1 - R/2\bar{r}}{1 + R/2\bar{r}} - \frac{1}{c^2}(1 - R/2\bar{r})^2 \mathcal{E}_{jk}\bar{x}^j\bar{x}^k, \tag{9.143}$$

where $R := 2GM/c^2$ is the Schwarzschild radius associated with the body mass M, and $\mathcal{E}_{jk}(\bar{t})$ is an arbitrary STF tensor that cannot be determined by solving the vacuum field equations in the body zone only. This metric is valid in the interval $R_{\text{body}} < \bar{r} < r_{\text{max}}$, where R_{body} is the body's radius.

The perturbation terms in Eq. (9.143) have a quadrupolar structure, and they grow as $\bar{r}^2$ when $\bar{r} \gg R$; this is precisely the behavior that we expect from a tidal field. This can be seen most easily by evaluating the metric in the overlap zone ($GM/c^2 \ll \bar{r} \ll r_{AB}$), in which it can be expanded in powers of c^{-2}. The Newtonian piece of the body metric is given by

$$\bar{U} = \frac{GM}{\bar{r}} - \frac{1}{2}\mathcal{E}_{jk}\bar{x}^j\bar{x}^k, \tag{9.144}$$

and we may compare this with Eq. (2.261) or with Eq. (8.73). The comparison reveals that $\mathcal{E}_{jk}(\bar{t}) = -\partial_{jk}U_{\text{ext}}(\bar{t}, \boldsymbol{r})$, with $\boldsymbol{r}(\bar{t})$ denoting the body's position, and that the second term in Eq. (9.144) is the leading term in an expansion of the external potential in powers of $\bar{x}^j$. The tensor $\mathcal{E}_{jk}(\bar{t})$ therefore characterizes the body's tidal environment, and Eq. (9.143) neglects higher-order terms in the expansion of the tidal potential.

The metric neglects other terms as well. A spherical body subjected to tidal forces normally suffers a deformation, but this effect was not included in Eq. (9.143). As we learned back in Sec. 2.5.1, the deformation is measured by the quadrupole-moment tensor $I_{\langle jk\rangle}$, and dimensional analysis requires a relationship of the form $GI_{\langle jk\rangle} = -\frac{2}{3}k_2 R^5_{\text{body}}\mathcal{E}_{jk}$, in which k_2 is the body's gravitational Love number, which depends on the details of its internal structure. The quadrupole-moment term in $\bar{U}$ would decay as $\bar{r}^{-3}$, and would be of fractional order $(R_{\text{body}}/\bar{r})^5$ relative to the tidal term; we choose to neglect it in $\bar{U}$, and choose to neglect it in $g_{\bar{0}\bar{0}}$ also, recalling that our accuracy requirements were more modest in the preceding sections.

We shall not be interested in tidal effects in the following considerations, and for our purposes it is sufficient to know that in the overlap zone, the Newtonian piece of the body metric is given by $\bar{U} = GM/\bar{r} + O(\bar{r}^2)$. Similarly, we need to know that the post-Newtonian piece obtained from Eq. (9.143) is given by $\bar{\Psi} - \bar{U}^2 = -G^2M^2/\bar{r}^2 + O(\bar{r}^2)$, so that

$$\bar{\Psi} = O(\bar{r}^2). \tag{9.145}$$

We shall not require the time–space components of the metric, apart from the knowledge that they contain tidal terms only; from this we infer that

$$\bar{U}^j = O(\bar{r}^2). \tag{9.146}$$

And we shall not need the space–space components of the metric; for our purposes they provide only redundant information.

Our derivation of the equations of motion relies on the expression of Eq. (9.143) for the body metric, and we present it here without a derivation. The result is based on a straightforward application of gravitational perturbation theory applied to spherical bodies, but to go through the details of its construction would take us too far afield. Some key points, however, are worth a mention. The unperturbed metric is given by the Schwarzschild solution, and its spherical symmetry guarantees that once a (linear) perturbation field is decomposed into spherical harmonics, each mode decouples from any other mode. The metric perturbation of Eq. (9.143) is a pure quadrupole field ($\ell = 2$), as can be seen from the fact that it is generated by a second-rank STF tensor $\mathcal{E}_{jk}(\bar{t})$. It can be shown that there is no non-trivial monopole field ($\ell = 0$) in vacuum (a trivial field would correspond to a meaningless shift in the mass parameter), and the statement that the dipole field ($\ell = 1$) vanishes amounts to making a specific choice of center-of-mass position for the body. It is therefore the quadrupole field that describes the leading-order tidal effects, and higher-order multipole fields correspond to higher-order terms in the Taylor expansion of the external Newtonian potential; we neglect these additional terms.

We make a final remark before moving on: The Newtonian potential of Eq. (9.144) is expressed in terms of M, the mass parameter of the Schwarzschild metric. This is the body's *total mass-energy*, which in principle could be expressed in exact relativistic form in terms of the material mass and all relevant forms of energy. Our current convention for the Newtonian potential is therefore different from our previous usage, in which U accounted for the material mass m only, while Ψ accounted for the body's energy E. Here we let the Newtonian potential account for the body's total mass-energy, and as a result, the expected energy terms do not appear within the post-Newtonian potential of Eq. (9.145).

9.4.3 Post-Newtonian metric

Gravity is weak everywhere in the post-Newtonian zone, and here we may express the metric as the post-Newtonian expansion

$$g_{00} = -1 + \frac{2}{c^2}U + \frac{2}{c^4}(\Psi - U^2) + O(c^{-4}). \tag{9.147}$$

The metric is presented in the global inertial frame, in harmonic coordinates (ct, x^j) attached to the center-of-mass of the entire N-body system. The post-Newtonian zone coincides with the near zone, but it excludes a sphere of radius $\bar{r} = r_{\text{min}}$ centered on the body; r_{min} is chosen to be much larger than GM/c^2 to ensure that gravity does not get too strong as we approach the body.

In the overlap zone ($GM/c^2 \ll r_{\text{min}} < \bar{r} < r_{\text{max}} \ll r_{AB}$) the post-Newtonian metric must satisfy the vacuum field equations. According to Eqs. (8.24), the gravitational potentials must be solutions to $\nabla^2 U = 0$, $\nabla^2 U^j = 0$, $\nabla^2 \psi = 0$, and $\nabla^2 X = 2U$; the post-Newtonian potential is $\Psi = \psi + \frac{1}{2}\partial_{tt}X$. The solutions must contain the information that a compact body is situated at $\boldsymbol{x} = \boldsymbol{r}(t)$, and that external bodies are to be found outside the overlap zone. We model the body as a post-Newtonian monopole (an assumption that is justified by the matching procedure to be carried out later), and we write

$$U = \frac{GM}{s} + U_{\text{ext}}, \tag{9.148a}$$

$$U^j = \frac{GMv^j}{s} + U^j_{\text{ext}}, \tag{9.148b}$$

$$\psi = \frac{GM\mu}{s} + \psi_{\text{ext}}, \tag{9.148c}$$

$$X = GMs + X_{\text{ext}}, \tag{9.148d}$$

where $s := |\boldsymbol{x} - \boldsymbol{r}(t)|$ is the length of the vector $\boldsymbol{s} := \boldsymbol{x} - \boldsymbol{r}(t)$. We have introduced $\boldsymbol{v}(t) := d\boldsymbol{r}/dt$ as the body's velocity, and its presence in the vector potential is required by the harmonic gauge condition, $\partial_t U + \partial_j U^j = 0$. The quantity $\mu(t)$ is a post-Newtonian correction to the mass parameter M, and this cannot be determined by solving the field equations in the post-Newtonian zone only. The external potentials separately satisfy the vacuum field equations. From Eq. (9.148) we get

$$\Psi = \frac{GM}{s}\left[\mu + \frac{1}{2}v^2 - \frac{1}{2}(\boldsymbol{n}\cdot\boldsymbol{v})^2\right] - \frac{1}{2}GM\boldsymbol{n}\cdot\boldsymbol{a} + \Psi_{\text{ext}}, \tag{9.149}$$

where $\boldsymbol{n} := \boldsymbol{s}/s$ and $\Psi_{\text{ext}} = \psi_{\text{ext}} - \frac{1}{2}\partial_{tt}X_{\text{ext}}$. We have also introduced $\boldsymbol{a}(t) := d\boldsymbol{v}/dt$ as the body's acceleration, and this is the quantity that we wish to determine.

Notice that the body potentials are all singular at $\boldsymbol{x} = \boldsymbol{r}(t)$; this is to be expected from a monopole field. The singularity, however, is not physical – the body is not a point mass, but a fully extended object. The singularity is also only apparent, because the point $\boldsymbol{x} = \boldsymbol{r}(t)$ *lies outside the post-Newtonian zone*; the potentials are valid for $s \gg GM/c^2$ only.

9.4.4 Transformation to the comoving frame

The coordinate transformation from the inertial system (ct, x^j) to the comoving system $(c\bar{t}, \bar{x}^j)$ is described in Box 8.2. It is characterized by a number of free functions that must be determined in the course of the matching procedure. The most important piece of information is the vector $\boldsymbol{r}(\bar{t})$, which determines the position of the body. The transformation also involves $A(\bar{t})$, $H^j(\bar{t})$, $R^j(\bar{t})$, and the harmonic function $\gamma(\bar{t}, \bar{x}^j)$.

The first step is to construct the "hatted potentials," the original gravitational potentials evaluated at time $t = \bar{t}$ and position $\boldsymbol{x} = \bar{\boldsymbol{x}} + \boldsymbol{r}(\bar{t})$. From Eqs. (9.148) and (9.149) we get

$$\hat{U} = \frac{GM}{\bar{r}} + \hat{U}_{\text{ext}}, \tag{9.150a}$$

$$\hat{U}^j = \frac{GMv^j}{\bar{r}} + \hat{U}^j_{\text{ext}}, \tag{9.150b}$$

$$\hat{\Psi} = \frac{GM}{\bar{r}}\left[\mu + \frac{1}{2}v^2 - \frac{1}{2}(\bar{\boldsymbol{n}}\cdot\boldsymbol{v})^2\right] - \frac{1}{2}GM\bar{\boldsymbol{n}}\cdot\boldsymbol{a} + \hat{\Psi}_{\text{ext}}, \tag{9.150c}$$

where $\boldsymbol{v}$, μ, and $\boldsymbol{a}$ are now functions of $\bar{t}$, and $\bar{\boldsymbol{n}} := \bar{\boldsymbol{x}}/\bar{r}$.

The transformed potentials $\bar{U}$, $\bar{U}^j$, and $\bar{\Psi}$ will contain terms that are singular in the formal limit $\bar{r} \to 0$, and terms that are well-behaved in this limit. The external potentials contribute non-singular terms, and as usual it is convenient to express them as Taylor expansions in powers of $\bar{x}^j$. We write, for example,

$$\hat{U}_{\rm ext}(\bar{t}, \bar{\boldsymbol{x}}) = \hat{U}_{\rm ext}(\bar{t}, \boldsymbol{0}) + \bar{x}^j \partial_j \hat{U}_{\rm ext}(\bar{t}, \boldsymbol{0}) + \frac{1}{2}\bar{x}^j \bar{x}^k \partial_{jk} \hat{U}_{\rm ext}(\bar{t}, \boldsymbol{0}) + \cdots . \tag{9.151}$$

The harmonic function γ also contributes non-singular terms, and we also express it as a Taylor expansion:

$$\gamma(\bar{t}, \bar{\boldsymbol{x}}) = C(\bar{t}) + \gamma_j(\bar{t})\bar{x}^j + \frac{1}{2}\gamma_{jk}(\bar{t})\bar{x}^j \bar{x}^k + \cdots ; \tag{9.152}$$

here $C(\bar{t})$ and $\gamma_j(\bar{t})$ are arbitrary functions of time $\bar{t}$, and $\gamma_{jk}(\bar{t})$ is an arbitrary STF tensor (so that γ can be a solution to Laplace's equation).

For our purposes here it is useful to refine the notation of Box 8.2 and decompose the acceleration vector $\boldsymbol{a}(\bar{t})$ into Newtonian and post-Newtonian pieces. We write

$$\boldsymbol{a} = \boldsymbol{a}[0\text{PN}] + \boldsymbol{a}[1\text{PN}] + O(c^{-4}), \qquad \boldsymbol{a}[1\text{PN}] =: c^{-2}\boldsymbol{\alpha}, \tag{9.153}$$

and re-express the transformation of the Newtonian potential as

$$\bar{U} = \hat{U} - \dot{A} + \frac{1}{2}v^2 - a_j[0\text{PN}]\bar{x}^j . \tag{9.154}$$

The missing piece involving $\boldsymbol{a}[1\text{PN}]$ is transferred to the post-Newtonian potential $\bar{\Psi}$. This has the effect of altering the expression of G_j by an additional term $-\alpha_j$.

A lengthy computation reveals that the transformed potentials are given by

$$\bar{U} = \frac{GM}{\bar{r}} + {}_0U + {}_1U_j \bar{x}^j + \cdots , \tag{9.155a}$$

$$\bar{U}^j = {}_0U^j + {}_1U^j{}_k \bar{x}^k + \cdots , \tag{9.155b}$$

$$\bar{\Psi} = -\frac{GM}{\bar{r}^3}(H_j - Av_j)\bar{x}^j + \frac{GM}{\bar{r}}(\mu + \dot{A} - 2v^2) + {}_0\Psi + {}_1\Psi_j \bar{x}^j + \cdots , \tag{9.155c}$$

with

$${}_0U = \frac{1}{2}v^2 - \dot{A} + \hat{U}_{\rm ext}, \tag{9.156a}$$

$${}_1U_j = -a_j[0\text{PN}] + \partial_{\bar{j}}\hat{U}_{\rm ext}, \tag{9.156b}$$

$${}_0U^j = \hat{U}^j_{\rm ext} - v^j \hat{U}_{\rm ext} + \frac{1}{4}(2\dot{A} - v^2)v^j - \frac{1}{4}\dot{H}^j + \frac{1}{4}\epsilon^j{}_{pq}v^p R^q + \frac{1}{4}\gamma^j, \tag{9.156c}$$

$$\begin{aligned} {}_1U^j{}_k &= \partial_{\bar{k}}\hat{U}^j_{\rm ext} - v^j \partial_{\bar{k}}\hat{U}_{\rm ext} + \frac{3}{8}v^j a_k[0\text{PN}] + \frac{1}{8}a^j[0\text{PN}]v_k \\ &\quad + \frac{1}{4}\delta^j{}_k\left(\frac{4}{3}\ddot{A} - 2v^n a_n[0\text{PN}]\right) - \frac{1}{4}\epsilon^j{}_{kp}\dot{R}^p + \frac{1}{4}\gamma^j{}_k, \end{aligned} \tag{9.156d}$$

$$\begin{aligned} {}_0\Psi &= \hat{\Psi}_{\rm ext} - 4v_j\hat{U}^j_{\rm ext} + 2v^2\hat{U}_{\rm ext} + A\partial_{\bar{t}}\hat{U}_{\rm ext} + (H^j - Av^j)\partial_{\bar{j}}\hat{U}_{\rm ext} \\ &\quad + \frac{1}{2}\dot{A}^2 - \dot{A}v^2 + \frac{1}{4}v^4 + \dot{H}^j v_j - \dot{C}, \end{aligned} \tag{9.156e}$$

(continued overleaf)

$$
\begin{aligned}
{}_1\Psi_j &= \partial_{\bar{j}}\hat{\Psi}_{\rm ext} - 4v_k\partial_{\bar{j}}\hat{U}^k_{\rm ext} + \left(\frac{5}{2}v^2 - \dot{A}\right)\partial_{\bar{j}}\hat{U}_{\rm ext} - \frac{1}{2}v_j v^k\partial_{\bar{k}}\hat{U}_{\rm ext} + v_j\partial_{\bar{t}}\hat{U}_{\rm ext} \\
&\quad + \dot{A}\partial_{\bar{t}\bar{j}}\hat{U}_{\rm ext} + (H^k - Av^k)\partial_{\bar{j}\bar{k}}\hat{U}_{\rm ext} - \alpha_j + \left(\dot{A} - \frac{1}{2}v^2\right)a_j[0\text{PN}] \\
&\quad - \left(\ddot{A} - \frac{3}{2}v^n a_n[0\text{PN}]\right)v_j - \epsilon_{jpq}\left(\partial^{\bar{p}}\hat{U}_{\rm ext}R^q + v^p\dot{R}^q\right) - \dot{\gamma}_j.
\end{aligned}
\tag{9.156f}
$$

It is understood that in these expressions, the external potentials are all evaluated at $\bar{\boldsymbol{x}} = \boldsymbol{0}$ after differentiation.

For later convenience it is useful to decompose ${}_1U_{jk}$ into its trace, symmetric-tracefree, and antisymmetric parts. We have

$$
{}_1U_{jk} = \frac{1}{3}\delta_{jk}\,{}_1U + {}_1U_{\langle jk\rangle} + {}_1U_{[jk]} \tag{9.157}
$$

with

$$
{}_1U = \ddot{A} - \partial_{\bar{t}}\hat{U}_{\rm ext} - v^j a_j[0\text{PN}], \tag{9.158a}
$$

$$
{}_1U_{\langle jk\rangle} = \partial_{\langle\bar{j}}\hat{U}^{\rm ext}_{k\rangle} - v_{\langle j}\partial_{\bar{k}\rangle}\hat{U}_{\rm ext} + \frac{1}{2}v_{\langle j}a_{k\rangle}[0\text{PN}] + \frac{1}{4}\gamma_{jk}, \tag{9.158b}
$$

$$
{}_1U_{[jk]} = -\partial_{[\bar{j}}\hat{U}^{\rm ext}_{k]} - v_{[j}\partial_{\bar{k}]}\hat{U}_{\rm ext} + \frac{1}{4}v_{[j}a_{k]}[0\text{PN}] - \frac{1}{4}\epsilon_{jkp}\dot{R}^p. \tag{9.158c}
$$

9.4.5 Matching

Comparison of Eqs. (9.144), (9.145), and (9.146) with Eqs. (9.155) reveals that agreement is achieved if and only if the matching conditions

$$
{}_0U = {}_1U_j = {}_0U^j = {}_1U_{jk} = {}_0\Psi = {}_1\Psi_j = 0, \tag{9.159}
$$

and

$$
H^j = Av^j, \qquad \mu = 2v^2 - \dot{A}, \tag{9.160}
$$

are satisfied. By virtue of Eq. (9.157), the condition ${}_1U_{jk} = 0$ implies the independent conditions ${}_1U = {}_1U_{\langle jk\rangle} = {}_1U_{[jk]} = 0$. We now proceed to extract the information contained in these equations.

The condition ${}_0U = 0$ implies that $\dot{A} = \frac{1}{2}v^2 + \hat{U}_{\rm ext}$. This, together with Eq. (9.160), tells us that the metric function μ is given by

$$
\mu = \frac{3}{2}v^2 - \hat{U}_{\rm ext}(\bar{t}, \boldsymbol{0}). \tag{9.161}
$$

The condition ${}_1U_j = 0$ implies that

$$
a_j[0\text{PN}] = \partial_{\bar{j}}\hat{U}_{\rm ext}(\bar{t}, \boldsymbol{0}), \tag{9.162}
$$

and we have established the fact that to leading order in a post-Newtonian expansion of the mutual gravitational interaction, the compact body moves just as any Newtonian body.

The condition ${}_0U_j = 0$ reveals that $\gamma_j = -4\hat{U}^{\rm ext}_j + (\frac{1}{2}v^2 + 3\hat{U}_{\rm ext})v_j + A\partial_{\bar{j}}\hat{U}_{\rm ext} - \epsilon_{jpq}v^pR^q$. The condition ${}_1U = 0$ is automatically satisfied, and ${}_1U_{\langle jk\rangle} = 0$ determines γ_{jk}, but this quantity is not required in the derivation of the equations of motion. The condition ${}_1U_{[jk]} = 0$ reveals that $\epsilon_{jkp}\dot{R}^p = -4\partial_{[\bar{j}}\hat{U}^{\rm ext}_{k]} - 3v_{[j}\partial_{\bar{k}]}\hat{U}_{\rm ext}$, and this equation determines the vector R^j. The condition ${}_0\Psi = 0$ determines $\dot{C}$, but this quantity also is not required. And finally, ${}_1\Psi_j = 0$ determines the post-Newtonian piece of the acceleration vector; after some algebra and simplification, we arrive at

$$\alpha_j = \big(v^2 - 4\hat{U}_{\rm ext}\big)\partial_{\bar{j}}\hat{U}_{\rm ext} - v_j\big(v^k\partial_{\bar{k}}\hat{U}_{\rm ext} + 3\partial_{\bar{t}}\hat{U}_{\rm ext}\big) - 4v_k\partial_{\bar{j}}\hat{U}^k_{\rm ext} + 4\partial_{\bar{t}}\hat{U}^j_{\rm ext} + \partial_{\bar{j}}\hat{\Psi}_{\rm ext}. \tag{9.163}$$

As before it is understood that the external potentials are evaluated at $\bar{\boldsymbol{x}} = \boldsymbol{0}$ after differentiation.

The matching conditions have determined the unknown pieces of the coordinate transformation, the functions A, H^j, R^j, C, γ_j, γ_{jk}, and the all-important vector $\boldsymbol{r}(\bar{t})$, which is recovered by solving the equations of motion $\ddot{\boldsymbol{r}} = \boldsymbol{a}[0\text{PN}] + c^{-2}\boldsymbol{\alpha} + O(c^{-4})$. In addition, the matching conditions have determined the unknown metric function μ, and further analysis would also produce the tidal moments $\mathcal{E}_{jk}$. The problem, therefore, is solved completely: we have the metric, the coordinate transformation, and the equations of motion.

9.4.6 Equations of motion

The final form of the equations of motion is obtained by inserting Eqs. (9.162) and (9.163) within Eq. (9.153). We evaluate each quantity at time $\bar{t} = t$ and replace the hatted potentials by their original version in Eqs. (9.148) and (9.149). After paying careful attention to the rules of partial differentiation, which are spelled out at the end of Box 8.2, we arrive at

$$a^j = \partial_j U_{\rm ext} + \frac{1}{c^2}\Big[\big(v^2 - 4U_{\rm ext}\big)\partial_j U_{\rm ext} - v^j\big(4v^k\partial_k U_{\rm ext} + 3\partial_t U_{\rm ext}\big) - 4v_k\big(\partial_j U^k_{\rm ext} - \partial_k U^j_{\rm ext}\big) + 4\partial_t U^j_{\rm ext} + \partial_j\Psi_{\rm ext}\Big] + O(c^{-4}). \tag{9.164}$$

The acceleration vector $\boldsymbol{a}$ now stands for $d^2\boldsymbol{r}/dt^2$, and the equations of motion are expressed in the global inertial frame. These should be compared with Eq. (9.101), which determines the motion of a fluid body. The equations are identical, and we have established the important statement that *the compact body moves in exactly the same way as a weakly self-gravitating body.* This is a nice confirmation of the fact that general relativity is compatible with the strong formulation of the principle of equivalence.

In principle the calculation should be completed with the determination of the external potentials and the conversion of Eq. (9.164) to an explicit system of differential equations for the position vectors $\boldsymbol{r}_A(t)$. There is no need to go through these details here; it is clear that the computation would lead back to Eq. (9.127).

A few comments, however, may be helpful. The decompositions of Eqs. (9.148) and (9.149) distinguish between "the body" and the external objects. To recover the expressions of Eqs. (9.55), (9.58), and (9.80), we must adapt our notation and assign the label $A = 1$ (say) to our reference body. The internal term GM/s in U then becomes GM_1/s_1. Because the Newtonian potential satisfies a linear field equation ($\nabla^2 U = 0$), the external piece can be written as a sum of similar terms, so that $U_{\rm ext} = \sum_{A\neq 1} GM_A/s_A$. This reproduces the expression of Eq. (9.55), and the same procedure also gives rise to Eq. (9.58).

The procedure works also for Ψ, which also satisfies a linear field equation. In the new notation, and with μ_1 given by Eq. (9.161), the internal piece of Ψ is

$$\Psi_1 = \frac{GM_1}{s_1}\Big[2v_1^2 - \frac{1}{2}(\boldsymbol{n}_1\cdot\boldsymbol{v}_1)^2 - U_{\rm ext}\Big] - \frac{1}{2}GM_1\boldsymbol{n}_1\cdot\boldsymbol{a}_1. \tag{9.165}$$

With $U_{\rm ext} = \sum_{B\neq 1} GM_B/r_{1B}$ and $\boldsymbol{a}_1 = \nabla U_{\rm ext} + O(c^{-2})$, this is

$$\Psi_1 = \frac{GM_1}{s_1}\Big[2v_1^2 - \frac{1}{2}(\boldsymbol{n}_1\cdot\boldsymbol{v}_1)^2\Big] - \sum_{B\neq 1}\frac{G^2M_1M_B}{r_{1B}s_1}\bigg(1 - \frac{\boldsymbol{n}_{1B}\cdot\boldsymbol{s}_1}{2r_{1B}}\bigg). \tag{9.166}$$

Taking $\Psi_{\rm ext}$ to be a sum of similar terms, we have reproduced Eq. (9.80), except for a noticeable difference: the sum $\sum_A GE_A/s_A$ is present in the original expression, but it is absent here. The reason for this discrepancy was explained at the end of Sec. 9.4.2: here the energy terms have been incorporated within the Newtonian potential, which is expressed in terms of M_A instead of m_A; they must therefore be removed from the post-Newtonian potential.

9.5 Motion of spinning bodies

Back in Sec. 9.1.4 we imposed the important restriction that each body should have a vanishing angular-momentum tensor, and subsequent developments relied heavily on this assumption. But rotation is everywhere, and it is crucially important to incorporate it in a description of the motion of an N-body system. In fact, non-rotating bodies are about as rare as a relativist at a biophysics convention, and failure to account for rotation would be a significant shortcoming. In atomic, nuclear, and particle physics, the effects of quantum-mechanical spin are known to be of central importance. In gravitational physics, it is becoming increasingly clear that spin effects play a similarly central role in such phenomena as binary black-hole inspirals, gravitational collapse, accretion onto compact objects, and the emission of gravitational radiation. In addition, several key experimental tests of general relativity have involved the effects of spin. By spin, of course, we mean the macroscopic rotation of an extended body, and not the quantum-mechanical spin of an elementary particle. We will use the term "spin" to describe the intrinsic (as opposed to orbital) angular momentum of a rotating body, and will refer to spin–orbit and spin–spin effects in the orbital motion of an N-body system; these effects are purely classical, but they have direct analogues in quantum physics.

In this section we compute the inter-body metric of a system of spinning bodies, obtain the equations of motion for the center-of-mass positions of each body, derive evolution equations for the intrinsic angular momentum of each body, and examine issues associated with the choice of center-of-mass.

9.5.1 Definitions of spin

The intrinsic angular-momentum tensor (or spin tensor) of a rotating body was defined back in Eq. (9.9b),

$$S_A^{jk} := \int_A \rho^* \left(\bar{x}^j \bar{v}^k - \bar{x}^k \bar{v}^j\right) d^3\bar{x}, \tag{9.167}$$

in which $\bar{\boldsymbol{x}} := \boldsymbol{x} - \boldsymbol{r}_A$ is the position of a fluid element relative to the body's center-of-mass, and $\bar{\boldsymbol{v}} := \boldsymbol{v} - \boldsymbol{v}_A$ is its relative velocity. We may also introduce a vectorial version of the spin angular momentum, defined by

$$\boldsymbol{S}_A := \int_A \rho^* \bar{\boldsymbol{x}} \times \bar{\boldsymbol{v}}\, d^3\bar{x}. \tag{9.168}$$

It is easy to show that the tensor and vector are related by

$$S_A^j = \frac{1}{2}\epsilon^{jpq} S_A^{pq}, \qquad S_A^{jk} = \epsilon^{jkp} S_A^p. \tag{9.169}$$

In the rest of the section we frequently go back and forth between the vectorial and tensorial notions of intrinsic angular momentum.

We continue to assume that our bodies are in dynamical equilibrium, and the virial identity of Eq. (9.28a) implies that the spin tensor can also be expressed as

$$S_A^{jk} = 2\int_A \rho^* \bar{x}^j \bar{v}^k\, d^3\bar{x}. \tag{9.170}$$

It is important to understand that this relation holds in dynamical equilibrium only; the definition of Eq. (9.167) is completely general.

We shall have occasion, below, to refine the definition of the spin vector by the inclusion of post-Newtonian terms at order c^{-2}. Our final definition of the spin vector is

$$\bar{S}_A^j = S_A^j + \Delta_{\rm int} S_A^j + \Delta_{\rm ext} S_A^j, \tag{9.171}$$

in which

$$\begin{aligned}\Delta_{\rm int} S_A^j := \frac{1}{c^2}\epsilon^{jpq}\biggl[&\int_A \rho^* \bar{x}^p \bar{v}^q \left(\frac{1}{2}\bar{v}^2 + 3U_A + \Pi + \frac{p}{\rho^*}\right) d^3\bar{x} \\ &- \int_A \rho^* \bar{x}^p \left(4U_A^q + \frac{1}{2}\partial_{tq} X_A\right) d^3\bar{x}\biggr]\end{aligned} \tag{9.172}$$

is a post-Newtonian correction that originates from the body's internal motion and gravitational potentials, and

$$\Delta_{\rm ext} S_A^j := \frac{1}{c^2}\left[\left(v_A^2 + 3U_{\neg A}\right) S_A^j - \frac{1}{2}(\boldsymbol{v}_A \cdot \boldsymbol{S}_A) v_A^j\right] \tag{9.173}$$

is another correction that originates from the orbital motion and external Newtonian potential. We shall motivate these post-Newtonian additions below, but for the time being we proceed with the original definition of Eqs. (9.167) and (9.168).

9.5.2 Equilibrium conditions

To prepare the way for our subsequent computations, we revisit the equilibrium conditions of Sec. 9.1.3 to see how they must be amended to account for spin. It is easy to see that the only affected condition is Eq. (9.11), which becomes

$$4H_A^{(jk)} - 3K_A^{jk} + \delta^{jk}\dot{P}_A - 2L_A^{(jk)} + S_A^{p(j}\partial^{k)}_{\ p}U_{\neg A}(\boldsymbol{r}_A) = O(c^{-2}) \tag{9.174}$$

in the presence of spin.

To establish this generalized form of the equilibrium condition, we return to the virial identity of Eq. (9.28c), which is exact and requires no modification to account for spin. In the no-spin context of Sec. 9.1.3, the terms involving the external Newtonian potential $U_{\neg A}$ could be neglected, as they give rise to contributions of fractional order $(R_A/r_{AB})^2 \ll 1$ to the equilibrium condition. In our current context, however, these terms contain spin contributions that must be identified and included in the equilibrium condition.

The first term to examine is

$$A^{jk} := \int_A \rho^* \bar{x}^j \frac{d}{dt}\partial_k U_{\neg A}\, d^3\bar{x}. \tag{9.175}$$

We follow the familiar method of expressing the external potential as a Taylor expansion about $\boldsymbol{x} = \boldsymbol{r}_A$, inserting the expansion inside the integral, evaluating the result, and discarding terms that are suppressed by a factor of order $(R_A/r_{AB})^2$. In the current case we have

$$\frac{d}{dt}\partial_k U_{\neg A}(\boldsymbol{x}) = \frac{d}{dt}\partial_k U_{\neg A}(\boldsymbol{r}_A) + \bar{v}^p \partial_{pk} U_{\neg A}(\boldsymbol{r}_A) + \bar{x}^p \frac{d}{dt}\partial_{pk} U_{\neg A}(\boldsymbol{r}_A) + \cdots, \tag{9.176}$$

in which d/dt on the right-hand side is to be interpreted as $\partial_t + v_A^q \partial_q$, and the spatial derivatives act on the variables $\boldsymbol{r}_A$. Only the second term contributes to A^{jk}, and making use of Eq. (9.170), we find that

$$A^{jk} = -\frac{1}{2} S_A^{pj}\, \partial_{pk} U_{\neg A}(\boldsymbol{r}_A). \tag{9.177}$$

The second term to examine is

$$B^{jk} := 3\int_A \bar{v}^j \partial_k U_{\neg A}\, d^3\bar{x}, \tag{9.178}$$

and its evaluation proceeds along similar lines. In this case we find

$$B^{jk} = \frac{3}{2} S_A^{pj}\, \partial_{pk} U_{\neg A}(\boldsymbol{r}_A), \tag{9.179}$$

and inclusion of these terms in Eq. (9.28c) gives rise to the modified equilibrium condition of Eq. (9.174).

9.5.3 Inter-body metric of spinning bodies

Our first task is to re-calculate the inter-body metric of an N-body system to account for the spin of the bodies. We follow the general methods introduced in Sec. 9.2, but now retain the terms that used to vanish under the no-spin condition. The idea is to expand the gravitational potentials about the center-of-mass of each body, and to find the terms that combine an $\bar{x}^j$ with a $\bar{v}^k$ so as to give rise to a spin tensor S_A^{jk} under Eq. (9.170). Terms of higher order in $\bar{x}^j$, such as $\bar{v}^j\bar{x}^k\bar{x}^m\bar{x}^n$ or $\bar{v}^j\bar{v}^k\bar{x}^m\bar{x}^n$, are discarded, because they give rise to negligible contributions of fractional order $(R_A/s_A)^2 \ll 1$ to the potentials. In this spirit, we also neglect terms in the potentials that are quadratic in the body spins.

Going over the computations of Sec. 9.2, we see that spin terms will appear in the post-Newtonian potentials U^j, Φ_1, and Φ_6, but that the remaining potentials are not affected. Straightforward computations reveal that the additional spin terms are given by

$$\Delta U^j = -\frac{1}{2}\sum_A \frac{G S_A^{jk} n_A^k}{s_A^2}, \tag{9.180a}$$

$$\Delta \Phi_1 = -\sum_A \frac{G v_A^j S_A^{jk} n_A^k}{s_A^2}, \tag{9.180b}$$

$$\Delta \Phi_6 = -\sum_A \frac{G v_A^j S_A^{jk} n_A^k}{s_A^2}, \tag{9.180c}$$

in which $\boldsymbol{s}_A := \boldsymbol{x} - \boldsymbol{r}_A$, $s_A := |\boldsymbol{s}_A|$, and $\boldsymbol{n}_A := \boldsymbol{s}_A/s_A$. These results imply that the main post-Newtonian potential Ψ is changed by

$$\Delta \Psi = -\frac{3}{2}\sum_A \frac{G v_A^j S_A^{jk} n_A^k}{s_A^2}. \tag{9.181}$$

These results are expressed in terms of the spin tensor. Transforming to the spin vector, we have that $S_A^{jk} n_A^k = (\boldsymbol{n}_A \times \boldsymbol{S}_A)^j$ and $v_A^j S_A^{jk} n_A^k = -(\boldsymbol{n}_A \times \boldsymbol{v}_A)\cdot \boldsymbol{S}_A$.

Inserting the potentials within the metric of Eq. (9.44), we find that the changes to the inter-body metric are given by

$$\Delta g_{00} = \frac{3}{c^4}\sum_A \frac{G(\boldsymbol{n}_A \times \boldsymbol{v}_A)\cdot \boldsymbol{S}_A}{s_A^2} + O(c^{-6}), \tag{9.182a}$$

$$\Delta g_{0j} = \frac{2}{c^3}\sum_A \frac{G(\boldsymbol{n}_A \times \boldsymbol{S}_A)^j}{s_A^2} + O(c^{-5}), \tag{9.182b}$$

$$\Delta g_{jk} = O(c^{-4}). \tag{9.182c}$$

These additional terms affect the motion of test masses around each body, and we shall see that spin effects also modify the center-of-mass motion of each body within the system.

To illustrate the gravitational influence of spin on a test mass, we specialize the metric to a single body. We place the body at $\boldsymbol{r}_A = \boldsymbol{0}$, set $\boldsymbol{v}_A = \boldsymbol{0}$, and import the additional terms from Eq. (9.81). The metric becomes

$$g_{00} = -1 + \frac{2}{c^2}\frac{GM}{r} - \frac{2}{c^4}\left(\frac{GM}{r}\right)^2 + O(c^{-6}), \tag{9.183a}$$

$$g_{0j} = \frac{2}{c^3}\frac{G(\boldsymbol{x}\times\boldsymbol{S})^j}{r^3} + O(c^{-5}), \tag{9.183b}$$

$$g_{jk} = \left(1 + \frac{2}{c^2}\frac{GM}{r}\right)\delta_{jk} + O(c^{-4}), \tag{9.183c}$$

in which M is the body's mass, $\boldsymbol{S}$ its spin vector, and $r := |\boldsymbol{x}|$. We examine the geodesic motion of a test particle in this spacetime. We describe the motion in terms of the position vector $\boldsymbol{r}(t)$, and we recall from Sec. 5.2.3 (see Eq. (5.52)) that the motion follows from the Lagrangian $L = -mc\sqrt{-g_{\alpha\beta}v^\alpha v^\beta}$, in which m is the particle's mass and $v^\alpha = (c, \boldsymbol{v})$ with $\boldsymbol{v} = d\boldsymbol{r}/dt$. Making the substitutions, and keeping only the Newtonian and spin terms in the Lagrangian, we find that it is given approximately by

$$L = -mc^2 + \frac{1}{2}mv^2 + \frac{GmM}{r} - \frac{2Gm(\boldsymbol{x}\times\boldsymbol{v})\cdot\boldsymbol{S}}{c^2r^3}. \tag{9.184}$$

Aligning the spin direction with the z-axis, and expressing the Lagrangian in spherical polar coordinates, we obtain our final expression

$$L = -mc^2 + \frac{1}{2}m\left(\dot{r}^2 + r^2\dot{\theta}^2 + r^2\sin^2\theta\,\dot{\phi}^2\right) + \frac{GmM}{r} - \frac{2GmS\sin^2\theta\,\dot{\phi}}{c^2r}, \tag{9.185}$$

in which an overdot indicates differentiation with respect to t.

The Lagrangian implies the existence of a conserved angular momentum,

$$h := \frac{1}{m}\frac{\partial L}{\partial\dot{\phi}} = r^2\sin^2\theta\left(\dot{\phi} - \frac{2GS}{c^2r^3}\right), \tag{9.186}$$

and this equation captures the essential features of motion around a spinning body. Consider a particle released from rest at infinity. The particle has no angular momentum, $h = 0$, but when it reaches a position r, it has nevertheless acquired an angular velocity $2GS/(c^2r^3)$. A particle with no angular momentum is therefore compelled to rotate in the same direction as the spinning body, as if it were dragged along by the rotational motion of the central body. Conversely, a test particle with a vanishing angular velocity at a position r has a negative angular momentum, $h = -2GS\sin^2\theta/(c^2r)$, as if it were counter-rotating relative to the local spacetime. In the spacetime of a rotating body, therefore, zero angular momentum does not imply zero angular velocity, and zero angular velocity does not imply zero angular momentum. The phrase *dragging of inertial frames* is often attached to these phenomena. We return to the observable consequences of frame dragging in Chapter 10.

9.5.4 Spin–orbit and spin–spin accelerations

We next turn to the task of calculating the center-of-mass acceleration $\boldsymbol{a}_A$ of a spinning body. Our strategy here is identical to the one adopted in Sec. 9.3. First, we insert the post-Newtonian Euler equation within the definition

$$m_A \boldsymbol{a}_A = \int_A \rho^* \frac{d\boldsymbol{v}}{dt} d^3x \tag{9.187}$$

and evaluate the resulting integrals, as listed in Eqs. (9.88). Second, we invoke the equilibrium conditions of Secs. 9.1.3 and 9.5.2 to eliminate all terms that depend on the body's internal structure. And third, we evaluate the external potentials and their derivatives at the position of each body. As we shall show below, the final result of this computation is the expression

$$\boldsymbol{a}_A = \boldsymbol{a}_A[\text{0PN}] + \boldsymbol{a}_A[\text{1PN}] + \boldsymbol{a}_A[\text{SO}] + \boldsymbol{a}_A[\text{SS}] + O(c^{-4}), \tag{9.188}$$

in which the Newtonian and post-Newtonian terms are given by Eq. (9.127),

$$a_A^j[\text{SO}] = \frac{3}{2c^2} \sum_{B \neq A} \frac{GM_B}{r_{AB}^3} \Big\{ n_{AB}^{\langle jk \rangle} \Big[v_A^p (3\hat{S}_A^{kp} + 4\hat{S}_B^{kp}) - v_B^p (3\hat{S}_B^{kp} + 4\hat{S}_A^{kp}) \Big] + n_{AB}^{\langle kp \rangle} (v_A - v_B)^p (3\hat{S}_A^{jk} + 4\hat{S}_B^{jk}) \Big\} \tag{9.189}$$

is the spin–orbit acceleration, which is linear in each spin tensor, and

$$a_A^j[\text{SS}] = -\frac{15}{c^2} \sum_{B \neq A} \frac{GM_B}{r_{AB}^4} \hat{S}_A^{kp} \hat{S}_B^{kq} n_{AB}^{\langle jpq \rangle} \tag{9.190}$$

is the spin–spin acceleration, bilinear in the spins. We recall the symbols $\boldsymbol{r}_{AB} := \boldsymbol{r}_A - \boldsymbol{r}_B$, $r_{AB} := |\boldsymbol{r}_{AB}|$, $\boldsymbol{n}_{AB} := \boldsymbol{r}_{AB}/r_{AB}$, that angular brackets such as $\langle jpq \rangle$ indicate a symmetric-tracefree combination, and we have introduced

$$\hat{S}_A^{jk} := \frac{1}{M_A} S_A^{jk} \tag{9.191}$$

to denote the spin tensor divided by the body's total mass-energy M_A.

To arrive at Eq. (9.188) we return to the listing of partial forces in Eqs. (9.88) and identify the ones that produce a dependence upon spin. It is easy to see that F_6^j, F_8^j, F_9^j, F_{11}^j, and F_{12}^j all contain terms proportional to S_A^{jk}, and that F_{10}^j, F_{11}^j, F_{12}^j, F_{13}^j, and F_{18}^j contain terms proportional to S_B^{jk} contributed by the external potentials $U_{\neg A}^j$, $\Phi_{1,\neg A}$, and $\Phi_{6,\neg A}$.

Making use of the calculational tools developed in Sec. 9.3.2, we find that the first group of terms evaluates to

$$c^2 \Delta F_6^j = -v_A^k S_A^{kp} \partial_{pj} U_{\neg A}, \tag{9.192a}$$

$$c^2 \Delta F_8^j = \frac{3}{2} S_A^{jp} \partial_{tp} U_{\neg A}, \tag{9.192b}$$

$$c^2 \Delta F_9^j = 2 v_A^k S_A^{jp} \partial_{kp} U_{\neg A}, \tag{9.192c}$$

$$c^2 \Delta F_{11}^j = 0, \tag{9.192d}$$

$$c^2 \Delta F_{12}^j = 2 S_A^{kp} \partial_{jp} U_{\neg A}^k, \tag{9.192e}$$

in which the external potentials are evaluated at $\boldsymbol{r}_A$ after differentiation. The second group of terms is

$$c^2 \Delta F^j_{10} = 4m_A \Delta \partial_t U^j_{\neg A}, \tag{9.193a}$$

$$c^2 \Delta F^j_{11} = 4m_A v^k_A \Delta \partial_k U^j_{\neg A}, \tag{9.193b}$$

$$c^2 \Delta F^j_{12} = -4m_A v^k_A \Delta \partial_j U^k_{\neg A}, \tag{9.193c}$$

$$c^2 \Delta F^j_{13} = 2m_A \Delta \partial_j \Phi_{1,\neg A}, \tag{9.193d}$$

$$c^2 \Delta F^j_{18} = -\frac{1}{2} m_A \Delta \partial_j \Phi_{6,\neg A}, \tag{9.193e}$$

in which, for example, $\Delta \partial_k U^j_{\neg A}$ denotes the spin-dependent terms in the gradient of the external vector potential. In addition to these contributions we must also account for the spin-dependent term in the equilibrium condition of Eq. (9.174), which, according to Eq. (9.100), gives rise to the shift

$$c^2 \Delta F^j = \Delta\big(2L^{(jk)}_A - 4H^{(jk)}_A + 3K^{jk}_A - \delta^{jk} \dot{P}_A\big) v^k_A = v^k_A S^{p(k}_A \partial^{j)}_p U_{\neg A}. \tag{9.194}$$

In the next step we employ the methods described in Sec. 9.3.4 to evaluate the derivatives of external potentials that occur in Eqs. (9.192), (9.193), and (9.194). We have

$$\partial_{jk} U_{\neg A} = 3 \sum_{B \neq A} \frac{G m_B n^{\langle jk \rangle}_{AB}}{r^3_{AB}}, \tag{9.195a}$$

$$\partial_{tk} U_{\neg A} = -3 \sum_{B \neq A} \frac{G m_B n^{\langle kp \rangle}_{AB} v^p_B}{r^3_{AB}}, \tag{9.195b}$$

$$\partial_{jp} U^k_{\neg A} = \sum_{B \neq A} \left(3 \frac{G m_B n^{\langle jp \rangle}_{AB} v^k_B}{r^3_{AB}} - \frac{15}{2} \frac{G S^{kq}_B n^{\langle jpq \rangle}_{AB}}{r^4_{AB}} \right), \tag{9.195c}$$

$$\Delta \partial_j U^k_{\neg A} = \frac{3}{2} \sum_{B \neq A} \frac{G S^{kp}_B n^{\langle jp \rangle}_{AB}}{r^3_{AB}}, \tag{9.195d}$$

$$\Delta \partial_t U^j_{\neg A} = -\frac{3}{2} \sum_{B \neq A} \frac{G S^{jp}_B v^k_B n^{\langle kp \rangle}_{AB}}{r^3_{AB}}, \tag{9.195e}$$

$$\Delta \partial_j \Phi_{1,\neg A} = 3 \sum_{B \neq A} \frac{G S^{kp}_B v^k_B n^{\langle jp \rangle}_{AB}}{r^3_{AB}}, \tag{9.195f}$$

$$\Delta \partial_j \Phi_{6,\neg A} = 3 \sum_{B \neq A} \frac{G S^{kp}_B v^k_B n^{\langle jp \rangle}_{AB}}{r^3_{AB}}. \tag{9.195g}$$

Making the substitutions and adding up the partial forces, we finally obtain the spin–orbit and spin–spin accelerations of Eqs. (9.189) and (9.190). In the last step we make the replacement $m_A \to M_A + O(c^{-2})$ in the accelerations, so as to express them in terms of the total mass-energy M_A instead of the material mass m_A.

9.5.5 Conserved quantities

The conserved quantities of Sec. 9.3.6 also acquire spin-dependent terms. While it is easy to see that the total mass-energy M is spin-independent, a close examination of Eqs. (9.129) and (9.131) reveals that the center-of-mass position $\boldsymbol{R}$ and total momentum $\boldsymbol{P}$ contain spin terms. In the case of the center-of-mass position, these contributions originate from the term $\frac{1}{2}\rho^* x^j v^2$ inside the integral. In the case of the total momentum, they arise from $\rho^* v^j U$ and $\rho^* \Phi^j$.

We calculate the spin-dependent terms by following the approach detailed in Sec. 9.3.6, and obtain

$$\Delta(MR^j) = \frac{1}{2c^2}\sum_A S_A^{jk} v_A^k = \frac{1}{2c^2}\sum_A \big(\boldsymbol{v}_A \times \boldsymbol{S}_A\big)^j \tag{9.196}$$

for the change in the barycenter position, and

$$\Delta P^j = -\frac{1}{2c^2}\sum_A \sum_{B\neq A} \frac{GM_B}{r_{AB}^2} S_A^{jk} n_{AB}^k = -\frac{1}{2c^2}\sum_A \sum_{B\neq A} \frac{GM_B}{r_{AB}^2}\big(\boldsymbol{n}_{AB} \times \boldsymbol{S}_A\big)^j \tag{9.197}$$

for the change in the total momentum. It is easy to show that $d\Delta(M\boldsymbol{R})/dt = \Delta\boldsymbol{P} + O(c^{-4})$ and $d\Delta\boldsymbol{P}/dt = O(c^{-4})$ provided that $dS_A^{jk}/dt = O(c^{-2})$.

9.5.6 Spin precession

Our next task in this survey of spinning bodies is to derive an evolution equation for each spin vector $\boldsymbol{S}_A$. This question was first considered back in Sec. 1.6.8 in the context of Newtonian gravity, where we showed that $\boldsymbol{S}_A$ changes as a result of a coupling between the body's multipole moments I_A^L and inhomogeneities in the external Newtonian potential $U_{\neg A}$. We have so far neglected all effects associated with multipole moments, and we shall continue to do so in our treatment of spin evolution. As a result, we shall find that the changes in $\boldsymbol{S}_A$ occur at post-Newtonian order, and result from a coupling between the body's spin and gradients of the external potentials.

Our final expression of the spin evolution equation involves the refined spin vector of Eq. (9.171). It takes the form of the precession equation

$$\frac{d\bar{\boldsymbol{S}}_A}{dt} = \boldsymbol{\Omega}_A \times \bar{\boldsymbol{S}}_A + O(c^{-4}), \tag{9.198}$$

in which the precessional angular velocity is given by

$$\boldsymbol{\Omega}_A = \boldsymbol{\Omega}_A[\mathrm{so}] + \boldsymbol{\Omega}_A[\mathrm{ss}], \tag{9.199}$$

with

$$\boldsymbol{\Omega}_A[\mathrm{so}] = \frac{1}{2c^2}\sum_{B\neq A}\frac{GM_B}{r_{AB}^2}\boldsymbol{n}_{AB} \times \big(3\boldsymbol{v}_A - 4\boldsymbol{v}_B\big), \tag{9.200a}$$

$$\boldsymbol{\Omega}_A[\mathrm{ss}] = \frac{1}{c^2}\sum_{B\neq A}\frac{G}{r_{AB}^3}\Big[3\big(\bar{\boldsymbol{S}}_B \cdot \boldsymbol{n}_{AB}\big)\boldsymbol{n}_{AB} - \bar{\boldsymbol{S}}_B\Big]. \tag{9.200b}$$

The first term describes a spin–orbit interaction, and the second a spin–spin interaction. Equation (9.198) implies that the magnitude $|\bar{S}_A|$ of the spin vector is a constant of the motion; the equation describes a precession of the spin around the time-dependent angular-velocity vector $\boldsymbol{\Omega}_A$.

The spin-precession equation involves the refined spin vector $\bar{\boldsymbol{S}}_A$ instead of the original vector $\boldsymbol{S}_A$, but the inter-body metric and equations of motion were previously written in terms of the original spin. There is no obstacle, however, in expressing all previous results in terms of the refined spin. Because $\boldsymbol{S}_A = \bar{\boldsymbol{S}}_A + O(c^{-2})$, and because all spin terms occur at post-Newtonian order in the metric and equations of motion, the substitution affects only the neglected terms at 2PN order. The final expression of our results, therefore, involves only the refined spin vector.

Partial torques

The derivation of Eq. (9.198) begins with

$$\frac{dS_A^j}{dt} = \epsilon^{jpq} \int_A \rho^* \bar{x}^p \frac{dv^q}{dt}\, d^3\bar{x}, \tag{9.201}$$

which is obtained by straightforward differentiation of Eq. (9.168); the integral initially features $d\bar{v}^p/dt = dv^p/dt - a_A^j$, but the center-of-mass condition $\int_A \rho^* \bar{x}^p\, d^3\bar{x} = 0$ implies that the second term leads to a vanishing contribution to the torque. In this we insert the post-Newtonian version of Euler's equation and end up with a sum of terms that bears a close resemblance to Eq. (9.86). For our purposes here it is helpful to write the Euler equation – refer to Eq. (8.119) – in the alternative form

$$\begin{aligned}\rho^* \frac{dv^j}{dt} &= -\partial_j p + \rho^* \partial_j U + \frac{1}{c^2}\biggl\{\biggl(\frac{1}{2}v^2 + U + \Pi + \frac{p}{\rho^*}\biggr)\partial_j p - v^j \partial_t p \\ &\quad + \rho^*\biggl[(v^2 - 4U)\partial_j U - 3v^j \frac{dU}{dt} - v^j v^k \partial_k U + 4\frac{dU^j}{dt} - 4v^k \partial_j U^k + \partial_j \Psi\biggr]\biggr\} \\ &\quad + O(c^{-4}),\end{aligned} \tag{9.202}$$

in which $\Psi = \psi + \frac{1}{2}\partial_{tt} X$ and $\psi = \frac{3}{2}\Phi_1 - \Phi_2 + \Phi_3 + 3\Phi_4$ – refer to Eq. (8.3) and Eq. (8.9). Inserting this within Eq. (9.201), we have that

$$\frac{dS_A^j}{dt} = \frac{1}{c^2}\epsilon^{jpq} \sum_{n=1}^{9} G_n^{pq} + O(c^{-4}), \tag{9.203}$$

where

$$G_1^{jk} := \int_A \bar{x}^j \biggl(\frac{1}{2}v^2 + U + \Pi + \frac{p}{\rho^*}\biggr)\partial_k p\, d^3\bar{x}, \tag{9.204a}$$

$$G_2^{jk} := -\int_A \bar{x}^j v^k \partial_t p\, d^3\bar{x}, \tag{9.204b}$$

$$G_3^{jk} := \int_A \rho^* \bar{x}^j (v^2 - 4U)\partial_k U\, d^3\bar{x}, \tag{9.204c}$$

(continued overleaf)

$$G_4^{jk} := -3 \int_A \rho^* \bar{x}^j v^k \frac{dU}{dt}\, d^3\bar{x}, \tag{9.204d}$$

$$G_5^{jk} := -\int_A \rho^* \bar{x}^j v^k v^p \partial_p U\, d^3\bar{x}, \tag{9.204e}$$

$$G_6^{jk} := 4 \int_A \rho^* \bar{x}^j \frac{dU^k}{dt}\, d^3\bar{x}, \tag{9.204f}$$

$$G_7^{jk} := -4 \int_A \rho^* \bar{x}^j v^p \partial_k U^p\, d^3\bar{x}, \tag{9.204g}$$

$$G_8^{jk} := \int_A \rho^* \bar{x}^j \partial_k \psi\, d^3\bar{x}, \tag{9.204h}$$

$$G_9^{jk} := \frac{1}{2} \int_A \rho^* \bar{x}^j \partial_{ttk} X\, d^3\bar{x}, \tag{9.204i}$$

are the partial torques. As motivated previously, we have discarded the Newtonian terms in Eq. (9.203), and factorized the overall factor of c^{-2}.

The remaining computations are lengthy, but they proceed along the same lines as the evaluation of the partial forces in Sec. 9.3.2. For example, the velocity vector is decomposed as $\boldsymbol{v} = \boldsymbol{v}_A + \bar{\boldsymbol{v}}$, the potentials are decomposed into internal and external pieces, the external potentials are expressed as Taylor expansions about $\boldsymbol{r}_A$, integrals featuring an odd number of internal vectors are set to zero, and terms of fractional order $(R_A/r_{AB})^2 \ll 1$ are neglected. In addition, double volume integrals are symmetrized with respect to the integration variables $\bar{\boldsymbol{x}}$ and $\bar{\boldsymbol{x}}'$, and all terms in G_n^{jk} that are symmetric in jk are discarded, because they do not survive the antisymmetrization operation contained in Eq. (9.203). Skipping over these computational details, we obtain

$$G_1^{jk} = \int_A \bar{x}^j \left(\frac{1}{2}\bar{v}^2 + U_A + \Pi + \frac{p}{\rho^*} \right) \partial_k p\, d^3\bar{x}, \tag{9.205a}$$

$$G_2^{jk} = -\int_A \bar{x}^j \bar{v}^k \partial_t p\, d^3\bar{x}, \tag{9.205b}$$

$$G_3^{jk} = \int_A \rho^* \bar{x}^j (\bar{v}^2 - 4U_A) \partial_k U_A\, d^3\bar{x} + S_A^{jp} v_A^p \partial_k U_{\neg A}, \tag{9.205c}$$

$$G_4^{jk} = -3 \int_A \rho^* \bar{x}^j \bar{v}^k \frac{dU_A}{dt}\, d^3\bar{x} - \frac{3}{2} v_A^k S_A^{jp} \partial_p U_{\neg A} - \frac{3}{2} S_A^{jk} \frac{dU_{\neg A}}{dt}, \tag{9.205d}$$

$$G_5^{jk} = -v_A^k \Omega_A^{jp} v_A^p - \int_A \rho^* \bar{x}^j \bar{v}^k \bar{v}^p \partial_p U_A\, d^3\bar{x} - \frac{1}{2}\left(v_A^k S_A^{jp} + v_A^p S_A^{jk} \right) \partial_p U_{\neg A}, \tag{9.205e}$$

$$G_6^{jk} = 4 \frac{d}{dt} \int_A \rho^* \bar{x}^j U_A^k\, d^3\bar{x} + 2 S_A^{jp} \partial_p U_{\neg A}^k, \tag{9.205f}$$

$$G_7^{jk} = -2 S_A^{jp} \partial_k U_{\neg A}^p, \tag{9.205g}$$

$$G_8^{jk} = G \int_A \rho^* \rho^{*\prime} \bar{x}^j \left(-\frac{3}{2}\bar{v}'^2 + U_A' - \Pi' - 3\frac{p'}{\rho^{*\prime}} \right) \frac{(\bar{x} - \bar{x}')^k}{|\bar{\boldsymbol{x}} - \bar{\boldsymbol{x}}'|^3}\, d^3\bar{x}' d^3\bar{x}, \tag{9.205h}$$

$$G_9^{jk} = \frac{1}{2} \frac{d}{dt} \int_A \rho^* \bar{x}^j \partial_{tk} X_A\, d^3\bar{x} + v_A^k \Omega_A^{jk} v_A^p, \tag{9.205i}$$

in which the external potentials are evaluated at $\boldsymbol{r}_A$ after differentiation. The partial torques involve some external pieces that depend on the spin tensor S_A^{jk}, and some internal integrals that are not directly featured in the final result displayed in Eq. (9.198). Many of these integrals actually cancel out, as can be seen for the terms involving the potential energy tensor Ω_A^{jk}.

Other cancellations are produced once we express G_1^{jk} in a different form. We eliminate $\partial_k p$ from the integral by making use of the Newtonian version of Euler's equation, which we write in the approximate form $\rho^* d\bar{v}^k/dt = -\partial_k p + \rho^* \partial_k U_A$, having equated dv_A^k/dt with $\partial_k U_{\neg A}(\boldsymbol{r}_A)$. The partial torque becomes

$$
\begin{aligned}
G_1^{jk} = &- \int_A \rho^* \bar{x}^j \left(\frac{1}{2}\bar{v}^2 + 3U_A + \Pi + \frac{p}{\rho^*} \right) \frac{d\bar{v}^k}{dt}\, d^3\bar{x} + 2 \int_A \rho^* \bar{x}^j U_A \frac{d\bar{v}^k}{dt}\, d^3\bar{x} \\
&+ \int_A \rho^* \bar{x}^j \left(\frac{1}{2}\bar{v}^2 + U_A + \Pi + \frac{p}{\rho^*} \right) \partial_k U_A\, d^3\bar{x}.
\end{aligned}
\tag{9.206}
$$

The integrand of the first term is $\rho^* \bar{x}^j A d\bar{v}^k/dt$ with $A := \frac{1}{2}\bar{v}^2 + 3U_A + \Pi + p/\rho^*$, and we express it as

$$
\rho^* \frac{d}{dt}(\bar{x}^j \bar{v}^k A) - \rho^* \bar{v}^j \bar{v}^k A - \rho^* \bar{x}^j \bar{v}^k \frac{dA}{dt},
$$

with $dA/dt = (\partial_t p + v_A^n \partial_n p)/\rho^* + \bar{v}^n \partial_n U_A + 3dU_A/dt$; to arrive at this result we have invoked Euler's equation once more, and made use of $d\Pi/dt = (p/\rho^{*2})d\rho^*/dt$, the statement of the first law of thermodynamics. We make the substitutions within the first integral in G_A^{jk}, and for the second integral we make use of Euler's equation to express $\bar{x}^j U_A \partial_k p$ as

$$
\partial_k(\bar{x}^j p\, U_A) - \delta^{jk} p\, U_A - \bar{x}^j p\, \partial_k U_A,
$$

with the first term producing a vanishing surface integral. Collecting results, we obtain

$$
\begin{aligned}
G_1^{jk} = &- \frac{d}{dt} \int_A \rho^* \bar{x}^j \bar{v}^k \left(\frac{1}{2}\bar{v}^2 + 3U_A + \Pi + \frac{p}{\rho^*} \right) d^3\bar{x} + \int_A \bar{x}^j \bar{v}^k \partial_t p\, d^3\bar{x} \\
&+ \int_A \rho^* \bar{x}^j \bar{v}^k \bar{v}^p \partial_p U_A\, d^3\bar{x} + 3 \int_A \rho^* \bar{x}^j \bar{v}^k \frac{dU_A}{dt}\, d^3\bar{x} \\
&+ \int_A \rho^* \bar{x}^j \left(\frac{1}{2}\bar{v}^2 + 3U_A + \Pi + 3\frac{p}{\rho^*} \right) \partial_k U_A\, d^3\bar{x},
\end{aligned}
\tag{9.207}
$$

and recognize some of the internal integrals that were encountered previously.

Internal and external torques

Summing over the partial torques, we find that the internal pieces collect themselves into

$$
\begin{aligned}
G_{\text{int}}^{jk} = -\frac{d}{dt} \bigg[&\int_A \rho^* \bar{x}^j \bar{v}^k \left(\frac{1}{2}\bar{v}^2 + 3U_A + \Pi + \frac{p}{\rho^*} \right) d^3\bar{x} \\
&- \int_A \rho^* \bar{x}^j \left(4U_A^k + \frac{1}{2}\partial_{tk} X_A \right) d^3\bar{x} \bigg]
\end{aligned}
$$

(continued overleaf)

$$- G\int_A \rho^*\rho^{*\prime}\bar{x}^j\left(\frac{3}{2}\bar{v}^2 - U_A + \Pi + 3\frac{p}{\rho^*}\right)\frac{(\bar{x}-\bar{x}')^k}{|\bar{\boldsymbol{x}}-\bar{\boldsymbol{x}}'|^3}\,d^3\bar{x}'d^3\bar{x}$$
$$- G\int_A \rho^*\rho^{*\prime}\bar{x}^j\left(\frac{3}{2}\bar{v}'^2 - U_A' + \Pi' + 3\frac{p'}{\rho^{*\prime}}\right)\frac{(\bar{x}-\bar{x}')^k}{|\bar{\boldsymbol{x}}-\bar{\boldsymbol{x}}'|^3}\,d^3\bar{x}'d^3\bar{x}. \tag{9.208}$$

The last term can be re-expressed as

$$+G\int_A \rho^*\rho^{*\prime}\bar{x}'^j\left(\frac{3}{2}\bar{v}^2 - U_A + \Pi + 3\frac{p}{\rho^*}\right)\frac{(\bar{x}-\bar{x}')^k}{|\bar{\boldsymbol{x}}-\bar{\boldsymbol{x}}'|^3}\,d^3\bar{x}'d^3\bar{x}$$

and combined with the previous integral to give something symmetric in jk. Discarding this, we find that the internal pieces give rise to a total time derivative that can be moved to the left-hand side of Eq. (9.203) and absorbed into a re-definition of the spin vector. This is the origin of the term $\Delta_{\rm int}S_A^j$ in Eq. (9.171).

With the redefinition $\bar{S}_A^j := S_A^j + \Delta_{\rm int}S_A^j$, which for the moment excludes the external shift also present in Eq. (9.171), we find that the spin-evolution equation becomes

$$\frac{d\bar{S}_A^j}{dt} = \frac{1}{c^2}\epsilon^{jpq}G^{pq}_{\rm ext} + O(c^{-4}) \tag{9.209}$$

with

$$G^{jk}_{\rm ext} := \bar{S}_A^{jp}v_A^p\partial_k U_{\neg A} - 2\bar{S}_A^{jp}v_A^k\partial_p U_{\neg A} - \frac{1}{2}\bar{S}_A^{jk}v_A^p\partial_p U_{\neg A} - \frac{3}{2}\bar{S}_A^{jk}\frac{dU_{\neg A}}{dt}$$
$$+ 2\bar{S}_A^{jp}\left(\partial_p U^k_{\neg A} - \partial_k U^p_{\neg A}\right), \tag{9.210}$$

in which the original spin tensor S_A^{jk} has been replaced by its refinement $\bar{S}_A^{jk} = S_A^{jk} + O(c^{-2})$. Our expression for $G^{jk}_{\rm ext}$ is not unique, because we may again collect some of its terms in a total time derivative that can be moved to the left-hand side of the spin-evolution equation. We can use this freedom judiciously to ensure that the resulting equation takes the form of a precession equation, as we have it in Eq. (9.198). In these manipulations we use the fact that $d\bar{S}_A^{jk}/dt = O(c^{-2})$, as well as the Newtonian equations of motion $dv_A^j/dt = \partial_j U_{\neg A}(\boldsymbol{r}_A) + O(c^{-2})$. For example, the $v_A^p\partial_p U_{\neg A}$ factor in the third term of Eq. (9.210) can be written as

$$v_A^p\partial_p U_{\neg A} = v_A^p\frac{dv_A^p}{dt} = \frac{1}{2}\frac{d}{dt}\left(v_A^2\right), \tag{9.211}$$

and this term can indeed be expressed as a total time derivative. As another example, we have

$$v_A^k\partial_p U_{\neg A} = \frac{d}{dt}\left(v_A^k v_A^p\right) - v_A^p\partial_k U_{\neg A}, \tag{9.212}$$

and the substitution can be made in the second term of Eq. (9.210). After simplification we find that our new expression for $G^{jk}_{\rm ext}$ becomes

$$G^{jk}_{\rm ext} = -\frac{d}{dt}\left(\frac{1}{2}\bar{S}_A^{jp}v_A^k v_A^p + \frac{1}{4}\bar{S}_A^{jk}v_A^2 + \frac{3}{2}\bar{S}_A^{jk}U_{\neg A}\right)$$
$$+ \frac{3}{2}\bar{S}_A^{jp}\left(v_A^p\partial_k U_{\neg A} - v_A^k\partial_p U_{\neg A}\right) - 2\bar{S}_A^{jp}\left(\partial_k U^p_{\neg A} - \partial_p U^k_{\neg A}\right). \tag{9.213}$$

Moving the time derivative to the left-hand side of Eq. (9.203), we see that the extra terms give rise to the additional shift $\Delta_{\rm ext} S^j_A$ in Eq. (9.171); we have arrived at our final version of the spin vector, $\bar{S}^j_A = S^j_A + \Delta_{\rm int} S^j_A + \Delta_{\rm ext} S^j_A = S^j_A + O(c^{-2})$.

In the final step we evaluate the derivatives of the external potentials, taking into account the spin terms in the vector potential – refer to Eq. (9.195d) – and make the substitutions in Eq. (9.213). Excluding now the total time derivative, we have that

$$G^{jk}_{\rm ext} = G^{jk}[\mathrm{so}] + G^{jk}[\mathrm{ss}], \tag{9.214}$$

with

$$G^{jk}[\mathrm{so}] = -\frac{1}{2}\bar{S}^{jp}_A \sum_{B\neq A} \frac{Gm_B}{r^2_{AB}} \Big[(3v_A - 4v_B)^p n^k_{AB} - (3v_A - 4v_B)^k n^p_{AB}\Big] \tag{9.215}$$

and

$$G^{jk}[\mathrm{ss}] = \bar{S}^{jp}_A \sum_{B\neq A} \frac{G}{r^3_{AB}} \Big[3n^p_{AB}\bar{S}^{kq}_B n^q_{AB} - 3n^k_{AB}\bar{S}^{pq}_B n^q_{AB} + 2\bar{S}^{pk}_B\Big]. \tag{9.216}$$

Inserting this within Eq. (9.203), expressing the spin tensor in terms of the spin vector $\bar{S}^j_A$, and simplifying, we finally arrive at the spin-precession equation of Eq. (9.198).

9.5.7 Comoving frame and proper spin

The partial redefinition $\boldsymbol{S}_A \to \boldsymbol{S}_A + \Delta_{\rm int}\boldsymbol{S}_A$ of the spin vector was performed to eliminate the total time derivative from the right-hand side of Eq. (9.208), and this operation was unique. The further shift of the spin vector by $\Delta_{\rm ext}\boldsymbol{S}_A$ was introduced specifically to cast the spin evolution equation in the form of a precession equation, and this operation reflected a choice on our part; the equation could have been left in its original form, or manipulated into yet a different form, in spite of the fact that these alternative versions would have been less compelling than Eq. (9.198). In this section we provide additional motivation for the external shift. We shall show that

$$\bar{\boldsymbol{S}}_A := \boldsymbol{S}_A + \Delta_{\rm ext}\boldsymbol{S}_A = \boldsymbol{S}_A + \frac{1}{c^2}\Big[\big(v^2_A + 3U_{\neg A}\big)\boldsymbol{S}_A - \frac{1}{2}(\boldsymbol{v}_A\cdot\boldsymbol{S}_A)\boldsymbol{v}_A\Big] \tag{9.217}$$

is the spin vector as measured in a non-inertial frame that is at all times moving with the body; we call it the *proper spin*, or *comoving spin*. Our considerations here exclude the internal shift contributed by the internal motions and potentials.

The transformation from the original, inertial frame (t, x^j) to the non-inertial, comoving frame $(\bar{t}, \bar{x}^j)$ is described by the class of post-Newtonian transformations presented in Sec. 8.3 – refer to Box 8.2. The transformation is given by

$$t = \bar{t} + c^{-2}\alpha(\bar{t}, \bar{x}^j) + O(c^{-4}), \tag{9.218a}$$

$$x^j = \bar{x}^j + r^j(t) + c^{-2}h^j(\bar{t}, \bar{x}^j) + O(c^{-4}), \tag{9.218b}$$

in which

$$\alpha = A(\bar{t}) + \dot{r}_p \bar{x}^p, \tag{9.219a}$$

$$h^j = H^j(\bar{t}) + H^j{}_p(\bar{t})\bar{x}^p + \frac{1}{2} H^j{}_{pq}(\bar{t})\bar{x}^p \bar{x}^q, \tag{9.219b}$$

with

$$H_{jp} = \frac{1}{2}\dot{r}_j \dot{r}_p - \delta_{jp}\left(\dot{A} - \frac{1}{2}\dot{r}^2\right), \tag{9.220a}$$

$$H_{jpq} = -\delta_{jp}\ddot{r}_q - \delta_{jq}\ddot{r}_p + \delta_{pq}\ddot{r}_j; \tag{9.220b}$$

the functions $r^j(\bar{t})$, $A(\bar{t})$, and $H^j(\bar{t})$ are for now arbitrary, overdots indicate differentiation with respect to $\bar{t}$, and $\dot{r}^2 := \delta_{pq}\dot{r}^p\dot{r}^q$. The terms of order c^{-4} in the time transformation are not required here, and the functions $R^j(\bar{t})$ that would normally appear in H_{jp} were set equal to zero.

The starting point of our discussion of proper spin is the selection of a representative world line $z_A(t)$ within body A, which is meant to track the motion of its "center-of-mass." An issue that we shall have to address is the precise meaning of this phrase. In our previous considerations, the center-of-mass was always defined in the inertial frame, but here we have the option of defining the center-of-mass in the body's comoving frame; these definitions are *not equivalent* when the body is spinning, and we shall have to decide which center-of-mass the representative world line is supposed to track. We leave it arbitrary for the time being, and think of it as the world line of an arbitrarily-selected fluid element within the body. The frame $(\bar{t}, \bar{x}^j)$ is attached to the representative world line, so that $\bar{x}^j = 0$ on the representative world line.

Our spin vector shall now be defined with respect to the representative world line. We have

$$\boldsymbol{S}_A := \int_A \rho^*(\boldsymbol{x} - \boldsymbol{z}_A) \times (\boldsymbol{v} - \boldsymbol{w}_A)\, d^3x \tag{9.221}$$

in the inertial frame, in which $\boldsymbol{w}_A := d\boldsymbol{z}_A/dt$ and the integral is evaluated on a surface $t =$ constant in spacetime (so that all the sampled points $\boldsymbol{x}$ are simultaneous with $\boldsymbol{z}_A$ in the inertial frame), and

$$\bar{\boldsymbol{S}}_A := \int_A \bar{\rho}^* \bar{\boldsymbol{x}} \times \bar{\boldsymbol{v}}\, d^3\bar{x} \tag{9.222}$$

in the comoving frame, in which the integral is evaluated on a surface $\bar{t} =$ constant (all the sampled points $\bar{\boldsymbol{x}}$ are simultaneous in the comoving frame). We wish to express $\boldsymbol{S}_A$ in terms of $\bar{\boldsymbol{S}}_A$, and for this we must relate $\boldsymbol{x} - \boldsymbol{z}_A$ to $\bar{\boldsymbol{x}}$, and $\boldsymbol{v} - \boldsymbol{w}_A$ to $\bar{\boldsymbol{v}}$, and take into account the fact that $dm := \rho^*\, d^3x = \bar{\rho}^*\, d^3\bar{x}$ is an invariant – recall the discussion around Eq. (8.92).

To achieve this we examine the world line of an arbitrary fluid element that passes through a spacetime point P within the body, to which we assign the coordinates (t, x^j) in the inertial frame, and the coordinates $(\bar{t}, \bar{x}^j)$ in the comoving frame; the transformation is given by Eq. (9.218). Together with this world line we consider the body's representative world line, and on it we select the spacetime point Q that is simultaneous with P in the inertial frame; its coordinates are (t, z^j_A) in the inertial frame, $(\bar{t}_A, 0)$ in the comoving frame, and the transformation is given by

$$t = \bar{t}_A + c^{-2}A(\bar{t}_A) + O(c^{-4}), \qquad z_A^j = r^j(\bar{t}_A) + c^{-2}H^j(\bar{t}_A) + O(c^{-4}). \tag{9.223}$$

From Eqs. (9.218) and (9.223) we find that

$$\bar{t} = \bar{t}_A - c^{-2}\dot{r}_p(\bar{t}_A)\bar{x}^p + O(c^{-4}) \tag{9.224}$$

and $(x - z_A)^j = \bar{x}^j + r^j(\bar{t}) - r^j(\bar{t}_A) + c^{-2}[h^j(\bar{t}, \bar{x}^j) - H^j(\bar{t}_A)] + O(c^{-4})$. The last equation appears to give us what we need, but it requires additional work because the various terms on the right-hand side refer to different times; what we want instead is for all the terms to be simultaneous in the comoving frame. In particular, $\bar{x}^j$ refers to P, which is not simultaneous, and we wish to express this in terms of the coordinates of the point P' on the fluid element's world line that is simultaneous with Q. If the functions $\bar{x}^j(\bar{t})$ describe the world line in the comoving frame, then $\bar{x}^j(\bar{t})$ refers to P, while $\bar{x}^j(\bar{t}_A)$ refers to P', and the relation $\bar{x}^j = \bar{x}^j(\bar{t}_A) - c^{-2}\bar{v}^j\dot{r}_p\bar{x}^p + O(c^{-4})$ follows by simple Taylor expansion; $\bar{v}^j$ is the fluid's velocity field in the comoving frame. Proceeding similarly with $r^j(\bar{t})$, we find that

$$(x - z_A)^j = \bar{x}^j + c^{-2}\biggl[\bigl(H^j_{\ p} - \dot{r}^j\dot{r}_p - \bar{v}^j\dot{r}_p\bigr)\bar{x}^p + \frac{1}{2}H^j_{\ pq}\bar{x}^p\bar{x}^q\biggr] + O(c^{-4}), \tag{9.225}$$

in which all terms on the right-hand side refer to P' and are evaluated at comoving time $\bar{t}_A$.

The left-hand side of Eq. (9.225) can be differentiated with respect to t, the right-hand side can be differentiated with respect to $\bar{t}_A$, and Eq. (9.223) can be used to relate the time differentials. The end result is

$$\begin{aligned}(v - w_A)^j = \bar{v}^j + c^{-2}\biggl[&\bigl(\dot{H}^j_{\ p} - \dot{r}^j\ddot{r}_p - \bar{v}^j\ddot{r}_p - \ddot{r}^j\dot{r}_p - \bar{a}^j\dot{r}_p\bigr)\bar{x}^p - \dot{A}\bar{v}^j \\ &+ \bigl(H^j_{\ p} - \dot{r}^j\dot{r}_p - \bar{v}^j\dot{r}_p\bigr)\bar{v}^p + \frac{1}{2}\dot{H}^j_{\ pq}\bar{x}^p\bar{x}^q + H^j_{\ pq}\bar{x}^p\bar{v}^q\biggr],\end{aligned} \tag{9.226}$$

in which all terms on the right-hand side continue to be evaluated at comoving time $\bar{t}_A$.

We next insert Eqs. (9.225) and (9.226) within Eq. (9.221) and evaluate the integral. We implement our usual simplification rules by discarding all terms that involve an odd number of internal vectors, and neglecting all terms that scale as R_A^2. We insert the previously displayed expression for H_{jk}, simplify the result, and arrive at

$$\boldsymbol{S}_A = \bar{\boldsymbol{S}}_A + \frac{1}{c^2}\biggl[\biggl(\frac{1}{2}\dot{r}^2 - 3\dot{A}\biggr)\bar{\boldsymbol{S}}_A + \frac{1}{2}(\dot{\boldsymbol{r}}\cdot\bar{\boldsymbol{S}}_A)\dot{\boldsymbol{r}}\biggr] + O(c^{-4}), \tag{9.227}$$

which can be inverted to give

$$\bar{\boldsymbol{S}}_A = \boldsymbol{S}_A + \frac{1}{c^2}\biggl[\biggl(3\dot{A} - \frac{1}{2}\dot{r}^2\biggr)\boldsymbol{S}_A - \frac{1}{2}(\dot{\boldsymbol{r}}\cdot\boldsymbol{S}_A)\dot{\boldsymbol{r}}\biggr] + O(c^{-4}). \tag{9.228}$$

This equation looks vaguely like Eq. (9.217), but we must justify the identification $\dot{\boldsymbol{r}} = \boldsymbol{v}_A$ and determine the arbitrary function $A(\bar{t})$.

In fact, our discussion thus far has left many quantities undetermined. We have yet to specify the functions $r^j(\bar{t})$, $A(\bar{t})$, and $H^j(\bar{t})$, and we have yet to make a choice of representative world line. We first tackle the determination of r^j and H^j, and to achieve this we return to Eq. (9.223), which we write as $\bar{t}_A = t - c^{-2}A(\bar{t}_A) + O(c^{-4})$ and

$$z_A^j(t) = r^j(\bar{t}_A) + c^{-2}H^j(\bar{t}_A) + O(c^{-4}). \tag{9.229}$$

If we express the functions on the right-hand side in terms of t, we have that

$$z_A^j(t) = r^j(t) + c^{-2}\big[H^j(t) - A(t)\dot{r}^j(t)\big] + O(c^{-4}). \tag{9.230}$$

This equation becomes $z_A^j = r^j + O(c^{-4})$ when we set $H^j = A\dot{r}^j$, and we may then identify $r^j(\bar{t})$ with the position z_A^j of the representative world line evaluated at the time $t = \bar{t}$. In this manner we identity $\boldsymbol{r}$ with $\boldsymbol{z}_A$, and $\dot{\boldsymbol{r}}$ with $\boldsymbol{w}_A$.

We next tackle determination of the representative world line. As a guide, we insert Eq. (9.225) within $\int_A \rho^*(x - z_A)^j\, d^3x$, evaluate the integral, and get

$$\begin{aligned}(r_A - z_A)^j &= \left[1 - \frac{1}{c^2}\left(\dot{A} - \frac{1}{2}w_A^2\right)\right]\bar{r}_A^j - \frac{1}{2c^2}(\boldsymbol{w}_A \cdot \bar{\boldsymbol{r}}_A)w_A^j \\ &\quad + \frac{1}{2m_A c^2}\bar{S}_A^{jk} w_A^k + O(c^{-4}),\end{aligned} \tag{9.231}$$

in which

$$\boldsymbol{r}_A := \frac{1}{m_A}\int_A \rho^* \boldsymbol{x}\, d^3x \tag{9.232}$$

is the center-of-mass position in the inertial frame,

$$\bar{\boldsymbol{r}}_A := \frac{1}{m_A}\int_A \bar{\rho}^* \bar{\boldsymbol{x}}\, d^3\bar{x} \tag{9.233}$$

is the center-of-mass position in the comoving frame, and

$$\bar{S}_A^{jk} := \int_A \bar{\rho}^*\left(\bar{x}^j \bar{v}^k - \bar{x}^k \bar{v}^j\right) d^3\bar{x} \tag{9.234}$$

is the comoving spin tensor.

There are two obvious ways of selecting a representative world line. The first is to declare that it will track the body's center-of-mass as defined in the inertial frame. To effect this choice we set $\boldsymbol{z}_A = \boldsymbol{r}_A$, and Eq. (9.231) informs us that the position of the comoving center-of-mass is given by

$$\bar{r}_A^j = -\frac{1}{2m_A c^2}\bar{S}_A^{jk} v_A^k + O(c^{-4}). \tag{9.235}$$

With this choice of representative world line, we find that the spatial origin of the comoving frame $(\bar{t}, \bar{x}^j)$ does not coincide with the comoving center-of-mass, which is offset by the vector $\bar{\boldsymbol{r}}_A$. This situation invites us to make a distinction between the comoving spin $\bar{\boldsymbol{S}}_A$, defined relative to the origin of the comoving frame, and a proper spin defined relative to the comoving center-of-mass. The proper spin would be defined by

$$\int_A \bar{\rho}^*(\bar{\boldsymbol{x}} - \bar{\boldsymbol{r}}_A) \times (\bar{\boldsymbol{v}} - \bar{\boldsymbol{v}}_A)\, d^3\bar{x},$$

but it is easy to show that since $\bar{\boldsymbol{r}}_A = O(c^{-2})$ and $\bar{\boldsymbol{v}}_A = O(c^{-2})$, this is equal to the comoving spin $\bar{\boldsymbol{S}}_A$ up to terms of order c^{-4}. So while the distinction is important as a matter of principle, it has no practical implications at the level of accuracy maintained in our discussion of spinning bodies.

The second way of selecting a representative world line is to declare that it will track the body's center-of-mass as defined in the comoving frame. In this case we set $\bar{\boldsymbol{r}}_A = \boldsymbol{0}$, and Eq. (9.231) implies that

$$z_A^j = r_A^j - \frac{1}{2m_A c^2} S_A^{jk} v_A^k + O(c^{-4}). \tag{9.236}$$

This equation states that the representative world line is offset from the inertial-frame center-of-mass by the vector $S_A^{jk} v_A^k/(2m_A c^2)$. In this case there is no distinction to be made between comoving spin and proper spin, and the distinction between the inertial spin of Eq. (9.221) and the usual spin vector $\int_A \rho^*(\boldsymbol{x} - \boldsymbol{r}_A) \times (\boldsymbol{v} - \boldsymbol{v}_A)\, d^3x$ occurs only at order c^{-4} and is therefore unimportant.

Either way of selecting the representative world line leads to the identifications $\boldsymbol{r} = \boldsymbol{z}_A = \boldsymbol{r}_A + O(c^{-2})$ and $\dot{\boldsymbol{r}} = \boldsymbol{w}_A = \boldsymbol{v}_A + O(c^{-2})$, and Eq. (9.228) becomes

$$\bar{\boldsymbol{S}}_A = \boldsymbol{S}_A + \frac{1}{c^2}\biggl[\Bigl(3\dot{A} - \frac{1}{2}v_A^2\Bigr)\boldsymbol{S}_A - \frac{1}{2}(\boldsymbol{v}_A \cdot \boldsymbol{S}_A)\boldsymbol{v}_A\biggr] + O(c^{-4}). \tag{9.237}$$

Our only remaining task is to determine the function $A(\bar{t})$ that appears in the transformation between the internal and comoving frames. Here we take our guidance from Sec. 8.3.5, in which we first examined the post-Newtonian transformation between a global, inertial frame to a local, comoving frame. There it was shown that in order to account for the special-relativistic and gravitational effects of time dilation, the function $A(\bar{t})$ must be a solution to

$$\dot{A} = \frac{1}{2}v_A^2 + U_{\neg A}, \tag{9.238}$$

in which the external potential is evaluated on the representative world line. Making the substitution in Eq. (9.237), we arrive at Eq. (9.217), and our justification of the external shift $\Delta_{\text{ext}}\boldsymbol{S}_A$ is complete.

9.5.8 Choice of representative world line

The discussion of the previous subsection revealed an ambiguity in the choice of representative world line when the body is spinning. We considered two canonical choices, one in which the representative world line tracks the body's center-of-mass as defined in the global, inertial frame, and another in which it tracks the center-of-mass as defined in the local, comoving frame. While these choices lead to the same notion of proper spin, they lead to different representations of the center-of-mass motion.

To explore the impact of this ambiguity on the equations of motion, we enlarge our freedom of choice and examine a one-parameter family of representative world lines described by

$$\tilde{r}_A^j = r_A^j - \frac{\lambda}{2m_A c^2} S_A^{jk} v_A^k + O(c^{-4}), \tag{9.239}$$

in which $\boldsymbol{r}_A$ is the center-of-mass position in the inertial frame, S_A^{jk} is the spin tensor of Eq. (9.167), and λ is a dimensionless parameter. (In this subsection we prefer to use the notation $\tilde{\boldsymbol{r}}_A$ for the representative world line, instead of $\boldsymbol{z}_A$ as we did in the preceding subsection.) The assignment $\lambda = 0$ makes the representative world line track the inertial center-of-mass, $\lambda = 1$ makes it track the comoving center-of-mass, and it is easy to show that Eq. (9.239) gives rise to the comoving spin vector of Eq. (9.217) for any choice of λ.

Because the difference between $\tilde{\boldsymbol{r}}_A$ and $\boldsymbol{r}_A$ is of order c^{-2}, the transformation has no impact on the spin-precession equations (9.198). It also has no direct impact on the post-Newtonian terms in the center-of-mass accelerations of Eq. (9.188). There is, however, an indirect impact, because the Newtonian acceleration will undergo a change of order c^{-2}, and this change can be transferred to the post-Newtonian terms; because the transformation is linear in the spin, the transfer affects the form of the spin–orbit acceleration. In addition, we can see that the transformation will have no direct impact on the post-Newtonian potentials within the metric, but that the Newtonian potential will be changed by a term of order c^{-2} that can be transferred to the post-Newtonian potentials.

We begin with a computation of the transformed Newtonian potential, which is given by $U = \sum_A Gm_A/s_A$ with $s_A := |\boldsymbol{x} - \boldsymbol{r}_A|$ when it is evaluated far away from each body. This expression refers to the inertial center-of-mass of each body, and our goal here is to shift the reference to the representative world line. We write $\boldsymbol{r}_A = \tilde{\boldsymbol{r}}_A + \delta\boldsymbol{r}_A$, with $\delta\boldsymbol{r}_A$ given by (minus) the second term in Eq. (9.239), and we express s_A as $|\tilde{\boldsymbol{s}}_A - \delta\boldsymbol{r}_A|$, with $\tilde{\boldsymbol{s}}_A := \boldsymbol{x} - \tilde{\boldsymbol{r}}_A$ denoting the separation between the field point $\boldsymbol{x}$ and the representative world line. Performing a Taylor expansion in powers of $\delta\boldsymbol{r}_A$, we quickly arrive at

$$U = \tilde{U} + \frac{\lambda}{2c^2}\sum_A \frac{G(\tilde{\boldsymbol{n}}_A \times \tilde{\boldsymbol{v}}_A)\cdot\boldsymbol{S}_A}{\tilde{s}_A^2} + O(c^{-4}), \tag{9.240}$$

in which $\tilde{U} := \sum_A Gm_A/\tilde{s}_A$ is the shifted potential, $\tilde{s}_A := |\tilde{\boldsymbol{s}}_A|$, $\tilde{\boldsymbol{n}}_A := \tilde{\boldsymbol{s}}_A/\tilde{s}_A$, and $\tilde{\boldsymbol{v}}_A := d\tilde{\boldsymbol{r}}_A/dt$ is the velocity of the representative world line. The transformation of the Newtonian potential creates a spin contribution to g_{00} that must be added to the one already listed in Eq. (9.182). We find that

$$\Delta g_{00} = \frac{3+\lambda}{c^4}\sum_A \frac{G(\tilde{\boldsymbol{n}}_A \times \tilde{\boldsymbol{v}}_A)\cdot\boldsymbol{S}_A}{\tilde{s}_A^2} + O(c^{-6}) \tag{9.241}$$

is the spin term under the new description of the body's motion. There are no additional changes to g_{0j} and g_{jk}.

We next consider the changes in the equations of motion. The first source of change is the difference implied by Eq. (9.239) between the acceleration $\tilde{\boldsymbol{a}}_A$ of the representative world line and the acceleration $\boldsymbol{a}_A$ of the inertial center-of-mass. To calculate this difference we first differentiate Eq. (9.239) with respect to time and obtain $\tilde{\boldsymbol{v}}_A$ in terms of $\boldsymbol{v}_A$. The term involving $dS_A^{jk}/dt = O(c^{-2})$ can be neglected, and the term involving a_A^k can be written in terms of the Newtonian acceleration. The end result is

$$\tilde{v}_A^j = v_A^j + \frac{\lambda}{2m_Ac^2}\sum_{B\neq A}\frac{GM_B}{r_{AB}^2}S_A^{jk}n_{AB}^k. \tag{9.242}$$

Another differentiation with respect to t produces

$$\tilde{a}_A^j = a_A^j - \frac{3\lambda}{2c^2} \sum_{B \neq A} \frac{GM_B}{r_{AB}^3} n_{AB}^{\langle kp \rangle} (v_A - v_B)^p \hat{S}_A^{jk}, \tag{9.243}$$

in which $\hat{S}_A^{jk} := S_A^{jk}/m_A = S_A^{jk}/M_A + O(c^{-2})$.

The second source of change is the shift of the Newtonian acceleration $\boldsymbol{a}_A[0\text{PN}]$ that occurs when the acceleration is made to refer to the representative world line instead of the inertial center-of-mass. This shift is analogous to the one calculated previously for the Newtonian potential, and indeed, it can be computed in the same way, by expanding the acceleration in powers of $\delta \boldsymbol{r}_A$. The outcome of this computation is

$$a_A^j[0\text{PN}] = -\sum_{B \neq A} \frac{GM_B}{\tilde{r}_{AB}^2} \tilde{n}_{AB}^j + \frac{3\lambda}{2c^2} \sum_{B \neq A} \frac{GM_B}{\tilde{r}_{AB}^3} \tilde{n}_{AB}^{\langle jk \rangle} \left(\hat{S}_A^{kp} \tilde{v}_A^p - \hat{S}_B^{kp} \tilde{v}_B^p \right), \tag{9.244}$$

in which $\tilde{\boldsymbol{r}}_{AB} := \tilde{\boldsymbol{r}}_A - \tilde{\boldsymbol{r}}_B$, $\tilde{r}_{AB} := |\tilde{\boldsymbol{r}}_{AB}|$, and $\tilde{\boldsymbol{n}}_{AB} := \tilde{\boldsymbol{r}}_{AB}/\tilde{r}_{AB}$.

Collecting results, we find that the changes to the acceleration are all linear in the spin tensors, and that they contribute to a shift in the spin–orbit acceleration of Eq. (9.189). The shifted expression is

$$\begin{aligned} \tilde{a}_A^j[\text{so}] := \frac{3}{2c^2} \sum_{B \neq A} \frac{GM_B}{\tilde{r}_{AB}^3} \biggl\{ & \tilde{n}_{AB}^{\langle jk \rangle} \Bigl[\tilde{v}_A^p \Bigl((3+\lambda) \hat{S}_A^{kp} + 4\hat{S}_B^{kp} \Bigr) - \tilde{v}_B^p \Bigl((3+\lambda) \hat{S}_B^{kp} + 4\hat{S}_A^{kp} \Bigr) \Bigr] \\ & + \tilde{n}_{AB}^{\langle kp \rangle} (\tilde{v}_A - \tilde{v}_B)^p \Bigl((3-\lambda) \hat{S}_A^{jk} + 4\hat{S}_B^{jk} \Bigr) \biggr\}, \end{aligned} \tag{9.245}$$

and there are no additional changes to the post-Newtonian and spin–spin accelerations.

Ambiguities tend to make one feel uncomfortable. Things ought to be well defined, one feels, and there ought to be a "correct" value of λ. There is no such thing, however, and one must learn to accept the freedom associated with the choice of representative world line. In a way, the freedom to shift the world line is analogous to the inherent freedom in general relativity to shift the coordinate system. The coordinate freedom is complete, and one knows that a coordinate transformation will produce a change in the metric, and a change in the equations of motion. One learns to live with this freedom, and to eliminate the coordinate ambiguity by formulating well-posed questions that have coordinate-independent answers. The situation is similar in the case of the representative world line. Here also the freedom is complete (even though we have restricted it to a one-parameter family in this discussion), and here also a shift in world line produces a change in the metric and the equations of motion. One must learn to live with the freedom, and to eliminate the ambiguity by formulating meaningful questions with precise answers.

In the literature the spin–orbit acceleration has traditionally been presented in two canonical forms corresponding to $\lambda = 0$ and $\lambda = 1$, respectively, and the choice of λ is often described as "imposing a spin supplementary condition;" for example, the choice $\lambda = 1$ is described as the "covariant spin supplementary condition," for reasons that we won't care to go into here. Each form is acceptable, but one must be sure to implement the choice of λ consistently in the equations of motion, the metric, and any other quantity computed from these ingredients.

9.5.9 Binary systems

To conclude our discussion of spinning bodies, we specialize our results to the case of a binary system involving a first body of mass-energy M_1, position $\boldsymbol{r}_1$, velocity $\boldsymbol{v}_1$, and spin $\boldsymbol{S}_1$, and a second body of mass-energy M_2, position $\boldsymbol{r}_2$, velocity $\boldsymbol{v}_2$, and spin $\boldsymbol{S}_2$. We adopt the one-parameter family of representative world lines introduced in the preceding subsection, but omit the tildes on the position and velocity vectors to keep the notation uncluttered.

Returning to the discussion of Sec. 9.3.7 but incorporating the changes coming from the spins, we find that the system's barycenter is now situated at

$$\begin{aligned} M\boldsymbol{R} = {} & M_1\biggl[1+\frac{1}{2c^2}\biggl(v_1^2-\frac{GM_2}{r}\biggr)\biggr]\boldsymbol{r}_1 + M_2\biggl[1+\frac{1}{2c^2}\biggl(v_2^2-\frac{GM_1}{r}\biggr)\biggr]\boldsymbol{r}_2 \\ & + \frac{1+\lambda}{2c^2}\Bigl(\boldsymbol{v}_1\times\boldsymbol{S}_1+\boldsymbol{v}_2\times\boldsymbol{S}_2\Bigr) + O(c^{-4}). \end{aligned} \tag{9.246}$$

Imposing the barycentric condition $\boldsymbol{R} = \boldsymbol{0}$ allows us to express $\boldsymbol{r}_1$ and $\boldsymbol{r}_2$ in terms of the separation $\boldsymbol{r} := \boldsymbol{r}_1 - \boldsymbol{r}_2$ and relative velocity $\boldsymbol{v} := \boldsymbol{v}_1 - \boldsymbol{v}_2$. We find that $\boldsymbol{r}_1 = (M_2/m)\boldsymbol{r} + \boldsymbol{P}$ and $\boldsymbol{r}_2 = -(M_1/m)\boldsymbol{r} + \boldsymbol{P}$, with

$$\boldsymbol{P} := \frac{\eta\Delta}{2c^2}\biggl(v^2-\frac{Gm}{r}\biggr)\boldsymbol{r} - \frac{1+\lambda}{2mc^2}\boldsymbol{v}\times\bigl(M_2\boldsymbol{S}_1 - M_1\boldsymbol{S}_2\bigr) + O(c^{-4}), \tag{9.247}$$

in which $m := M_1 + M_2$, $\eta := M_1M_2/(M_1+M_2)^2$, and $\Delta := (M_1 - M_2)/(M_1+M_2)$. These equations imply that $\boldsymbol{v}_1 = (M_2/m)\boldsymbol{v} + O(c^{-2})$ and $\boldsymbol{v}_2 = -(M_1/m)\boldsymbol{v} + O(c^{-2})$.

The relative acceleration $\boldsymbol{a} := \boldsymbol{a}_1 - \boldsymbol{a}_2$ can be expressed as

$$\boldsymbol{a} = \boldsymbol{a}[\text{0PN}] + \boldsymbol{a}[\text{1PN}] + \boldsymbol{a}[\text{SO}] + \boldsymbol{a}[\text{SS}] + O(c^{-4}), \tag{9.248}$$

with $\boldsymbol{a}[\text{0PN}]$ and $\boldsymbol{a}[\text{1PN}]$ given by Eq. (9.142), and the spin–orbit and spin–spin accelerations given by

$$a^j[\text{SO}] = \frac{3G}{2c^2r^3}\Bigl\{n^{\langle jk\rangle}v^p\bigl[(3+\lambda)\sigma^{kp}+4S^{kp}\bigr] + n^{\langle kp\rangle}v^p\bigl[(3-\lambda)\sigma^{jk}+4S^{jk}\bigr]\Bigr\} \tag{9.249}$$

and

$$a^j[\text{SS}] = -\frac{15Gm}{c^2r^4}\hat{S}_1^{kp}\hat{S}_2^{kq}n^{\langle jqp\rangle}, \tag{9.250}$$

respectively, in which $r := |\boldsymbol{r}|$, $\boldsymbol{n} := \boldsymbol{r}/r$, and

$$\sigma^{jk} := \frac{M_2}{M_1}S_1^{jk} + \frac{M_1}{M_2}S_2^{jk}, \qquad S^{jk} := S_1^{jk} + S_2^{jk}. \tag{9.251}$$

These become

$$\begin{aligned} \boldsymbol{a}[\text{SO}] = \frac{G}{c^2r^3}\biggl\{ & \frac{3}{2}(\boldsymbol{n}\times\boldsymbol{v})\cdot\bigl[(3+\lambda)\boldsymbol{\sigma}+4\boldsymbol{S}\bigr]\boldsymbol{n} + \frac{3}{2}(\boldsymbol{n}\cdot\boldsymbol{v})\boldsymbol{n}\times\bigl[(3-\lambda)\boldsymbol{\sigma}+4\boldsymbol{S}\bigr] \\ & - \boldsymbol{v}\times(3\boldsymbol{\sigma}+4\boldsymbol{S})\biggr\} \end{aligned} \tag{9.252}$$

and

$$\boldsymbol{a}[\mathrm{ss}] = -\frac{3Gm}{c^2 r^4}\Big[(\hat{\boldsymbol{S}}_1 \cdot \hat{\boldsymbol{S}}_2)\boldsymbol{n} - 5(\hat{\boldsymbol{S}}_1 \cdot \boldsymbol{n})(\hat{\boldsymbol{S}}_2 \cdot \boldsymbol{n})\boldsymbol{n} + (\hat{\boldsymbol{S}}_1 \cdot \boldsymbol{n})\hat{\boldsymbol{S}}_2 + (\hat{\boldsymbol{S}}_2 \cdot \boldsymbol{n})\hat{\boldsymbol{S}}_1\Big], \tag{9.253}$$

when we express the accelerations in terms of the spin vectors; here we have that $\boldsymbol{\sigma} := (M_2/M_1)\boldsymbol{S}_1 + (M_1/M_2)\boldsymbol{S}_2$, $\boldsymbol{S} := \boldsymbol{S}_1 + \boldsymbol{S}_2$, $\hat{\boldsymbol{S}}_1 := \boldsymbol{S}_1/M_1$, and $\hat{\boldsymbol{S}}_2 := \boldsymbol{S}_2/M_2$.

The spin-precession equations become

$$\frac{d\bar{\boldsymbol{S}}_1}{dt} = \boldsymbol{\Omega}_1 \times \bar{\boldsymbol{S}}_1, \tag{9.254}$$

in which $\bar{\boldsymbol{S}}_1$ is the proper spin of the first body, and

$$\boldsymbol{\Omega}_1 = \boldsymbol{\Omega}_1[\mathrm{so}] + \boldsymbol{\Omega}_1[\mathrm{ss}], \tag{9.255}$$

with

$$\boldsymbol{\Omega}_1[\mathrm{so}] = \frac{2\eta Gm}{c^2 r^2}\left(1 + \frac{3M_2}{4M_1}\right)\boldsymbol{n} \times \boldsymbol{v}, \tag{9.256a}$$

$$\boldsymbol{\Omega}_1[\mathrm{ss}] = \frac{G}{c^2 r^3}\big[3(\bar{\boldsymbol{S}}_2 \cdot \boldsymbol{n})\boldsymbol{n} - \bar{\boldsymbol{S}}_2\big], \tag{9.256b}$$

the precessional angular velocity. The equations for the second body are obtained by a simple exchange of labels $1 \leftrightarrow 2$.

9.6 Point particles

The calculations that led to the inter-body metric of Eq. (9.81) were laborious, and most of this labor was spent on the computation of terms that depend on the internal structure of each body. These terms, however, all cancel out after invoking the equilibrium conditions, and they do not appear in the final expression for the metric. This effort is not entirely wasted, because it generates considerable evidence that general relativity satisfies the strong formulation of the principle of equivalence, but one wonders whether a shortcut to the final result might not exist.

In this final section of Chapter 9 we examine the shortcut that results when the bodies are modeled as point particles. We show that the road to the metric is made much shorter indeed, but that this efficiency comes at a high price: the need to regularize divergent integrals. The origin of the problem is easy to identify: a point particle possesses an infinite mass density ρ^* and produces a Newtonian potential U that diverges at the particle's position; because the product $\rho^* U$ acts as a source for the post-Newtonian potential Ψ, the mathematical existence of this object becomes questionable. We shall show, however, that a simple and well-motivated regularization prescription allows us to make sense of the divergent integrals, and the method reproduces the results displayed in Eq. (9.81).

We use the shortcut extensively in Chapter 11, when we calculate the gravitational waves produced by the motion of an N-body system. Throughout this section we take the bodies to have no spin, so that $S_A^{jk} = 0$.

9.6.1 Energy-momentum tensor

The description of a point particle moving in a curved spacetime was developed in Sec. 5.3.4. The particle has a mass m, it moves on a world line described by the parametric relations $r^\alpha(\tau)$, and its velocity four-vector is $u^\alpha = dr^\alpha/d\tau$; τ is proper time on the particle's world line. Its energy-momentum tensor was displayed in Eq. (5.108); it is

$$T^{\alpha\beta} = mc \int u^\alpha u^\beta \frac{\delta\big(x^\mu - r^\mu(\tau)\big)}{\sqrt{-g}}\, d\tau. \tag{9.257}$$

This expression can be simplified if we change the variable of integration from τ to $r^0(\tau)$. This permits an integration over the delta function $\delta(x^0 - r^0)$, and we obtain

$$T^{\alpha\beta} = \frac{mu^\alpha u^\beta}{\gamma\sqrt{-g}}\, \delta\big(\boldsymbol{x} - \boldsymbol{r}(t)\big), \tag{9.258}$$

where $\gamma := u^0/c$. The particle's world line is now described by the relations $\boldsymbol{r}(t)$ with $t := x^0/c$, and the velocity four-vector is decomposed as $u^\alpha = \gamma(c, \boldsymbol{v})$ with $\boldsymbol{v} := d\boldsymbol{r}/dt$.

Equation (9.258) can be compared with Eq. (5.91), which gives the energy-momentum tensor of a perfect fluid. The comparison reveals that the point particle is a limiting case of a perfect fluid, with a proper energy density given by $\mu = mc^2\gamma^{-1}(-g)^{-1/2}\delta(\boldsymbol{x} - \boldsymbol{r})$, and a vanishing pressure p. Since $\mu = \rho c^2 + \epsilon$, where ρ is the proper mass density and ϵ is the proper density of internal energy, we see that we can effectively set $\epsilon = 0$, or equivalently set the internal energy per unit mass Π to zero. The conserved mass density is

$$\rho^* = m\delta\big(\boldsymbol{x} - \boldsymbol{r}(t)\big). \tag{9.259}$$

The fluid's velocity field, in this case, reduces to the particle's velocity $\boldsymbol{v}(t)$.

The preceding description applies to a single particle. For a system of N particles we add the contributions from each particle, and the conserved mass density becomes

$$\rho^* = \sum_A m_A \delta\big(\boldsymbol{x} - \boldsymbol{r}_A(t)\big), \tag{9.260}$$

with m_A denoting the mass of each particle, and $\boldsymbol{r}_A(t)$ the individual trajectories. For the system we still have

$$\Pi = p = 0, \tag{9.261}$$

and the velocity field reduces to the individual velocities $\boldsymbol{v}_A(t)$.

9.6.2 Regularization

Equation (9.261) implies that each "body" can be assigned a zero integrated pressure P_A and a zero internal energy $E_A^{\rm int}$. The internal kinetic energy $\mathcal{T}_A$ vanishes also, and the equilibrium condition of Eq. (9.12) indicates that the gravitational potential energy Ω_A must also be assigned a value of zero. This sensible conclusion, however, creates a mathematical inconsistency.

Going back to the definition of Eq. (9.8), the potential-energy integral is

$$\Omega_A = -\frac{1}{2} G \int_A \frac{\rho^* \rho^{*\prime}}{|\boldsymbol{x} - \boldsymbol{x}'|} \, d^3x' d^3x, \tag{9.262}$$

where ρ^* stands for $\rho^*(t, \boldsymbol{x})$, while $\rho^{*\prime}$ stands for $\rho^*(t, \boldsymbol{x}')$. The integral involves a product of delta functions, and its value is mathematically ill-defined. It is not clear, therefore, that Ω_A can be set equal to zero. To explore this we substitute $m_A \delta(\boldsymbol{x} - \boldsymbol{r}_A)$ for ρ^*, $m_A \delta(\boldsymbol{x}' - \boldsymbol{r}_A)$ for $\rho^{*\prime}$, and we integrate with respect to d^3x'; the result is

$$\Omega_A = -\frac{1}{2} G m_A^2 \int_A \frac{\delta(\boldsymbol{x} - \boldsymbol{r}_A)}{|\boldsymbol{x} - \boldsymbol{r}_A|} \, d^3x, \tag{9.263}$$

and we see why the integral is ill-defined: the quantity $\delta(\boldsymbol{x} - \boldsymbol{r}_A)/|\boldsymbol{x} - \boldsymbol{r}_A|$ is not defined as a distribution, and a blind evaluation would return $1/0$. This mathematical difficulty illustrates rather well the spectacular failure of the point particle to provide a sensible model for an extended body in general relativity; the non-linearity of the field equations simply won't allow it.

All is not lost, however. We can reconcile the diverging values for Ω_A if we introduce the seemingly nonsensical *regularization prescription*

$$\frac{\delta(\boldsymbol{x} - \boldsymbol{r}_A)}{|\boldsymbol{x} - \boldsymbol{r}_A|} \equiv 0. \tag{9.264}$$

With this rule the integral becomes well-defined, and we arrive at the desired result, $\Omega_A = 0$. As we shall see, the regularization prescription is the only additional rule that is required to make sense of all the ill-defined integrals that we shall encounter; and with this prescription we shall be able to recover the metric of Eq. (9.81) on the basis of the point-particle model.

The regularization prescription of Eq. (9.264) is a special case of a more powerful method known as *Hadamard regularization*, which was used to great benefit by our friends Luc Blanchet and Thibault Damour in their work (along with their collaborators) on high-order post-Newtonian theory. The method works as follows.

Let $F(\boldsymbol{x}; \boldsymbol{r})$ be a function of $\boldsymbol{x}$ that diverges when $\boldsymbol{x}$ approaches the point $\boldsymbol{r}$. Specifically, assume that its behavior near $\boldsymbol{x} = \boldsymbol{r}$ is given by the Laurent series

$$F(\boldsymbol{x}; \boldsymbol{r}) = \sum_{n=0}^{n_{\max}} s^{-n} f_n(\boldsymbol{n}; \boldsymbol{r}) + O(s), \tag{9.265}$$

where $\boldsymbol{s} := \boldsymbol{x} - \boldsymbol{r}$, $s := |\boldsymbol{x} - \boldsymbol{r}|$, and $\boldsymbol{n} := \boldsymbol{s}/s$. The function therefore diverges as $s^{-n_{\max}}$ when $\boldsymbol{x} \to \boldsymbol{r}$, and it clearly does not have a well-defined value at $\boldsymbol{x} = \boldsymbol{r}$. We regularize it by extracting its *partie finie* at the singular point $\boldsymbol{x} = \boldsymbol{r}$. This is defined by

$$\lfloor F \rfloor(\boldsymbol{r}) := \frac{1}{4\pi} \int f_0(\boldsymbol{n}; \boldsymbol{r}) \, d\Omega(\boldsymbol{n}), \tag{9.266}$$

in which $d\Omega(\boldsymbol{n})$ is an element of solid angle in the direction of the unit vector $\boldsymbol{n}$. Thus, the *partie finie* of F is the angular average of the zeroth term $f_0(\boldsymbol{n}; \boldsymbol{r})$ in its Laurent series. The *partie finie* can be used to make sense of the product of F with the delta function $\delta(\boldsymbol{x} - \boldsymbol{r})$: We declare that

$$F(\boldsymbol{x}; \boldsymbol{r}) \delta(\boldsymbol{x} - \boldsymbol{r}) \equiv \lfloor F \rfloor(\boldsymbol{r}) \, \delta(\boldsymbol{x} - \boldsymbol{r}). \tag{9.267}$$

It follows immediately from this rule that $\int F(\boldsymbol{x}; \boldsymbol{r}) \delta(\boldsymbol{x} - \boldsymbol{r}) \, d^3x = \lfloor F \rfloor(\boldsymbol{r})$.

We now see that Eq. (9.264) is indeed a special case of Hadamard regularization. In this case $F = |\boldsymbol{x} - \boldsymbol{r}|^{-1}$, and its *partie finie* vanishes; Eq. (9.267) then implies $\delta(\boldsymbol{x} - \boldsymbol{r})/|\boldsymbol{x} - \boldsymbol{r}| \equiv 0$. Hadamard regularization even allows us to generalize the rule to $\delta(\boldsymbol{x} - \boldsymbol{r})/|\boldsymbol{x} - \boldsymbol{r}|^n \equiv 0$ for any positive integer n.

9.6.3 Potentials

The gravitational potentials are computed by substituting Eqs. (9.260) and (9.261) into the Poisson integrals of Eqs. (8.4). We immediately obtain

$$U = \sum_A \frac{GM_A}{s_A}, \tag{9.268a}$$

$$U^j = \sum_A \frac{GM_A v_A^j}{s_A}, \tag{9.268b}$$

$$X = \sum_A GM_A s_A, \tag{9.268c}$$

where s_A is the length of the vector $\boldsymbol{s}_A := \boldsymbol{x} - \boldsymbol{r}_A(t)$, and we have used the fact that since $\mathcal{T}_A = \Omega_A = E_A^{\rm int} = 0$, the total mass-energy M_A of a point particle is equal to its material mass m_A.

The potential ψ requires more work. The starting point is

$$\psi = G \int \frac{\rho^{*\prime}\left(\frac{3}{2}v'^2 - U'\right)}{|\boldsymbol{x} - \boldsymbol{x}'|} d^3x', \tag{9.269}$$

and we decompose the Newtonian potential as $U' = U'_A + U'_{\neg A}$, where $U'_A = GM_A/s'_A$ and $U'_{\neg A} = \sum_{B \neq A} GM_B/s'_B$, with $s'_A := |\boldsymbol{x}' - \boldsymbol{r}_A(t)|$. The integral involving U'_A is

$$-\sum_A G^2 M_A^2 \int \frac{\delta(\boldsymbol{x}' - \boldsymbol{r}_A)}{|\boldsymbol{x}' - \boldsymbol{r}_A|} \frac{1}{|\boldsymbol{x} - \boldsymbol{x}'|} d^3x',$$

and this is ill-defined. The regularization prescription of Eq. (9.264), however, dictates that the integral vanishes. The remaining piece of ψ is

$$\sum_A \frac{GM_A}{s_A}\left[\frac{3}{2}v_A^2 - U_{\neg A}(\boldsymbol{r}_A)\right],$$

and we arrive at

$$\psi = \frac{3}{2}\sum_A \frac{GM_A v_A^2}{s_A} - \sum_A \sum_{B \neq A} \frac{GM_A M_B}{s_A r_{AB}}, \tag{9.270}$$

where r_{AB} is the length of the vector $\boldsymbol{r}_{AB} := \boldsymbol{r}_A(t) - \boldsymbol{r}_B(t)$.

According to Eq. (8.3), the post-Newtonian potential is $\Psi = \psi + \frac{1}{2}\partial_{tt}X$. Differentiation of the superpotential yields

$$\partial_{tt}X = \sum_A \frac{GM_A}{s_A}\left[v_A^2 - (\boldsymbol{n}_A \cdot \boldsymbol{v}_A)^2\right] - \sum_A GM_A \boldsymbol{n}_A \cdot \boldsymbol{a}_A, \tag{9.271}$$

where $\boldsymbol{n}_A := \boldsymbol{s}_A/s_A$. To evaluate this fully we need an expression for $\boldsymbol{a}_A$, the acceleration of particle A. For our purposes here it suffices to use the Newtonian expression $-\sum_{B\neq A} GM_B\boldsymbol{n}_{AB}/r_{AB}^2$, where $\boldsymbol{n}_{AB} := \boldsymbol{r}_{AB}/r_{AB}$. This yields

$$\partial_{tt}X = \sum_A \frac{GM_A}{s_A}\Big[v_A^2 - (\boldsymbol{n}_A\cdot\boldsymbol{v}_A)^2\Big] + \sum_A\sum_{B\neq A}\frac{G^2M_AM_B(\boldsymbol{n}_{AB}\cdot\boldsymbol{n}_A)}{r_{AB}^2}. \tag{9.272}$$

Collecting results, the post-Newtonian potential is

$$\Psi = \sum_A \frac{GM_A}{s_A}\Big[2v_A^2 - \frac{1}{2}(\boldsymbol{n}_A\cdot\boldsymbol{v}_A)^2\Big] - \sum_A\sum_{B\neq A}\frac{G^2M_AM_B}{r_{AB}s_A}\bigg(1 - \frac{\boldsymbol{n}_{AB}\cdot\boldsymbol{s}_A}{2r_{AB}}\bigg). \tag{9.273}$$

Our expressions for U, U^j, and Ψ agree with Eqs. (9.55), (9.58), and (9.80), respectively, and we recover the inter-body metric of Eqs. (9.81). It should be evident that the computations carried out here were far less tedious than those presented in Sec. 9.2. The point-particle model, in spite of its mathematical difficulties and the need to regularize divergent integrals, has clear merits.

9.7 Bibliographical notes

The post-Newtonian equations of motion derived in Sec. 9.3 were first obtained by Lorentz and Droste (1917). A version of the equations, containing an error, was obtained independently by de Sitter (1916); the mistake was eventually corrected by Eddington and Clark (1938). A definitive treatment of the problem of motion was provided by Einstein, Infeld, and Hoffmann (1938), and the history of this fascinating episode in the development of general relativity is related in Havas (1989). Another fine survey of the "problem of motion" in Einstein's theory is Damour (1987). The method of derivation adopted in Sec. 9.3 is adapted from Will (1993).

The alternative method employed in Sec. 9.4 to derive equations of motion for compact bodies originates in the work of Demianski and Grishchuk (1974), D'Eath (1975), and Damour (1983). The method was developed systematically in Damour, Soffel, and Xu (1992), and generalized to bodies of arbitrary shape and composition by Racine and Flanagan (2005). The particular approach adopted in Sec. 9.4 is adapted from Taylor and Poisson (2008).

The motion of spinning bodies in curved spacetime has a long history, which is also summarized in Havas (1989). The equations of motion were first derived on the basis of point-particle models by Mathisson (1937) and Papapetrou (1951), and the importance of imposing a "spin supplementary condition" was stressed by Barker and O'Connell (1974). Derivations making use of extended bodies were provided later, and our methods in Sec. 9.5 are based on Kidder (1995) and Will (2005). An elegant alternative method was devised by Damour, Soffel, and Xu (1993).

The mathematical theory of Hadamard regularization, touched upon very briefly in Sec. 9.6, is developed fully in Sellier (1994) and Blanchet and Faye (2000).

9.8 Exercises

9.1 Verify all the results listed in Eqs. (9.91).

9.2 Verify all the results listed in Eqs. (9.103).

9.3 Show that the post-Newtonian equations of motion for a system of N bodies can be derived from the Lagrangian

$$L = -\sum_A M_A c^2 \left[1 - \frac{1}{2}(v_A/c)^2 - \frac{1}{8}(v_A/c)^4\right] + \frac{1}{2}\sum_{A,B\neq A} \frac{GM_A M_B}{r_{AB}} \times \left\{1 + \frac{1}{c^2}\left[3v_A^2 - \frac{7}{2}\boldsymbol{v}_A\cdot\boldsymbol{v}_B - \frac{1}{2}(\boldsymbol{n}_{AB}\cdot\boldsymbol{v}_A)(\boldsymbol{n}_{AB}\cdot\boldsymbol{v}_B) - \sum_{C\neq A}\frac{GM_C}{r_{AC}}\right]\right\}.$$

Find the canonical momentum $\boldsymbol{P}_A$ for this Lagrangian, and show that $\sum_A \boldsymbol{P}_A$ equals the conserved total momentum of Eq. (9.132b).

9.4 Verify Eq. (9.141).

9.5 Verify all the results listed in Eqs. (9.205).

9.6 To conclude the exploration of the one-parameter family of representative world lines in Sec. 9.5.8, calculate the changes in the barycenter position $\boldsymbol{R}$ and total momentum $\boldsymbol{P}$ induced by the transformation of Eq. (9.239).

9.7 We saw in Sec. 9.6 that the gravitational potentials of a spinless body can be computed efficiently by modeling the body as a point mass with density $m_A\delta(\boldsymbol{x} - \boldsymbol{r}_A)$. Here we wish to show that the body's spin can be accommodated by making the substitution

$$v^j \to v_A^j + \frac{1}{2m_A} S_A^{jk}\partial_k$$

in the potentials. Here the derivative operator attached to the spin tensor is meant to act on the delta function supplied by the density. The prescription is valid to first order in the spin tensor, and terms quadratic in the spin must be neglected. Use the prescription to compute the potentials U^j, Φ_1, and Φ_6, and compare your results to those displayed in Sec. 9.5.3.

10 Post-Newtonian celestial mechanics, astrometry and navigation

In November 1915, Einstein completed a calculation whose result so agitated him that he worried that he might be having a heart attack. He later wrote to a friend that "for several days I was beside myself in joyous excitement." What Einstein calculated was the contribution to the advance of the perihelion of Mercury from the first post-Newtonian corrections to Newtonian gravity provided by his newly completed theory of general relativity. This had been a notorious and unsolved problem in astronomy, ever since Le Verrier pointed out in 1859 that there was a discrepancy of approximately 43 arcseconds per century in the rate of advance between what was observed and what could be accounted for in Newtonian theory from planetary perturbations (refer to Secs. 3.1 and 3.4). Many earlier attempts to devise relativistic theories of gravity, including Einstein's own "Entwurf" (outline) theory of 1913 with Marcel Grossmann, had failed to give the correct answer. Now armed with the correct field equations, Einstein found an approximate vacuum solution that could be applied to the geodesic motion of Mercury around the Sun. He found that the orbit was almost Keplerian, but with a perihelion that advances at a rate that matched Le Verrier's observations.

For Einstein, this success with Mercury was the first concrete evidence that his theory, over which he had struggled so mightily for the past four years, might actually be correct. His prediction for the deflection of light by the Sun, completed that same month, and which doubled the value that he had derived in 1907 using just the principle of equivalence, would not be confirmed until 1919.

For the next 60 years, until the discovery of the first binary pulsar by Hulse and Taylor, the main testing-ground for general relativity would be the solar system, where gravitational fields are weak and motions are slow, so that the conditions for the post-Newtonian approximation are valid. In this chapter we take the formal development of post-Newtonian theory, presented in Chapters 8 and 9, and apply it to the real world of solar-system dynamics, solar-system experiments, and the dynamics of binary star (and binary black-hole) systems. We will encounter some of the famous experimental tests of general relativity, and will see how the precision of modern tools such as atomic clocks, satellite navigation systems, radio interferometry, and laser ranging make it necessary to take relativistic effects into account in a number of practical applications.

We begin in Sec. 10.1 with a description of the post-Newtonian motion of two self-gravitating bodies; it is in this section that we emulate Einstein with his calculation of the perihelion advance of Mercury. In Sec. 10.2 we examine the motion of light in weak gravitational fields, and describe such phenomena as the deflection of a light ray by a massive body, gravitational lenses, and the Shapiro time delay. The practically important issue of clock synchronization in the presence of gravitation is the topic of Sec. 10.3. We conclude the chapter in Sec. 10.4 with a discussion of the motion of a binary system of spinning bodies.

10.1 Post-Newtonian two-body problem

We begin our exploration of post-Newtonian dynamics with an examination of the two-body problem, the relativistic generalization of the Kepler problem reviewed in Sec. 3.2. The foundations of the post-Newtonian problem were presented in Chapter 9, and the equations that govern the motion of binary systems were derived in Sec. 9.3.7. While the method of derivation employed in most of Chapter 9 was restricted to bodies with weak internal gravity, we saw back in Sec. 9.4 that in fact the equations of motion apply just as well to compact bodies such as neutron stars and black holes; the internal gravity of each body can be arbitrarily strong, but if the mutual gravity is weak, the equations apply.

The equations of motion can be applied to any binary-star system, but they can also be applied to the solar system, in spite of the fact that the number of bodies exceeds two. The reason is that the Sun dominates the mass of the solar system by a factor of 1000, so that from the point of view of relativistic effects, which scale as $GM/(c^2r)$, the Sun is the elephant in the room, making the other planets largely irrelevant. Consider, for example, the perihelion advance of Mercury, which is produced in part by relativistic effects, and in part by perturbations generated by other planets (mostly Jupiter). As we shall see below, the relativistic part of the advance per orbit is of order $GM_\odot/(c^2a) \sim 3 \times 10^{-8}$ radians, where a is Mercury's semi-major axis. On the other hand, we recall from Sec. 3.4.1 (refer to Eq. (3.83)), that the advance produced by a perturbing planet scales as $(m_p/M_\odot)(a/R)^3$, where m_p and R are the mass and semi-major axis of the perturbing planet; this is of order 4×10^{-7} radians per orbit when Jupiter is the perturbing planet, and is therefore comparable to the relativistic contribution. We can expect the post-Newtonian corrections to Newtonian third-body perturbations, as well as third-body contributions to the post-Newtonian perturbation, to be of the same order of magnitude as the product of these two small factors, and thus to be unmeasurably small. It follows that for most solar-system applications, we can safely calculate post-Newtonian two-body effects and Newtonian N-body effects separately, and simply add them together. The conclusion of this discussion is that the relativistic two-body problem provides an adequate foundation to describe most relativistic effects in the solar system.

An important exception to this rule occurs in the Earth–Moon system, for which there is a measurable post-Newtonian three-body effect produced by the Sun, a contribution to the precession of the lunar orbital plane. This effect is often called *de Sitter precession*, because it was predicted in 1916 by the Dutch astronomer Willem de Sitter, on the basis of his post-Newtonian equations of motion. We shall conclude this section with a description of the de Sitter precession.

10.1.1 Equations of motion

The equations that govern the motion of a two-body system were obtained back in Sec. 9.3.7, where they were cast in the form of an effective one-body problem. We work in terms of the separation $\boldsymbol{r} := \boldsymbol{r}_1 - \boldsymbol{r}_2$, relative velocity $\boldsymbol{v} := \boldsymbol{v}_1 - \boldsymbol{v}_2$, relative acceleration $\boldsymbol{a} := \boldsymbol{a}_1 - \boldsymbol{a}_2$, and involve the mass parameters $m := M_1 + M_2$, $\eta := M_1 M_2/(M_1 + M_2)^2$, and $\Delta := (M_1 - M_2)/(M_1 + M_2)$. The relative acceleration is given by Eq. (9.142),

$$\boldsymbol{a} = -\frac{Gm}{r^2}\boldsymbol{n} - \frac{Gm}{c^2 r^2}\left\{\left[(1+3\eta)v^2 - \frac{3}{2}\eta\dot{r}^2 - 2(2+\eta)\frac{Gm}{r}\right]\boldsymbol{n} - 2(2-\eta)\dot{r}\boldsymbol{v}\right\} + O(c^{-4}), \tag{10.1}$$

where $r := |\boldsymbol{r}|$ is the inter-body distance, $\boldsymbol{n} := \boldsymbol{r}/r$ a unit vector that points from body 2 to body 1, and $\dot{r} := \boldsymbol{v}\cdot\boldsymbol{n}$ the radial component of the velocity vector. Equation (10.1) is a second-order differential equation for $\boldsymbol{r}(t)$, and its solution determines the position of each body. This information is provided by Eq. (9.141),

$$\boldsymbol{r}_1 = \frac{M_2}{m}\boldsymbol{r} + \frac{\eta\Delta}{2c^2}\left(v^2 - \frac{Gm}{r}\right)\boldsymbol{r} + O(c^{-4}), \tag{10.2a}$$

$$\boldsymbol{r}_2 = -\frac{M_1}{m}\boldsymbol{r} + \frac{\eta\Delta}{2c^2}\left(v^2 - \frac{Gm}{r}\right)\boldsymbol{r} + O(c^{-4}). \tag{10.2b}$$

These equations imply that $\boldsymbol{v}_1 = (M_2/m)\boldsymbol{v} + O(c^{-2})$ and $\boldsymbol{v}_2 = -(M_1/m)\boldsymbol{v} + O(c^{-2})$.

We saw back in Sec. 3.2 that the Newtonian two-body problem admits constants of the motion, the orbital energy $E = \mu\varepsilon$ and the orbital angular momentum $\boldsymbol{L} = \mu\boldsymbol{h}$, in which $\mu := M_1 M_2/(M_1 + M_2) = \eta m$ is the system's reduced mass. The post-Newtonian problem also admits a conserved energy and a conserved angular-momentum vector, and it is not difficult to deduce their expressions. We know that at the Newtonian level, the conserved energy is given by $\varepsilon = \frac{1}{2}v^2 - Gm/r$, and we also know that post-Newtonian corrections come with a multiplicative factor of c^{-2}. Possible contributions must then be proportional to $v^4, \dot{r}^2 v^2, \dot{r}^4, v^2 Gm/r, \dot{r}^2 Gm/r$, and $(Gm/r)^2$, and the correct combination of such terms can be identified by including them all (with unknown coefficients) in a trial expression for ε, and demanding that $d\varepsilon/dt = 0$ by virtue of the post-Newtonian dynamics of Eq. (10.1). The end result of this simple exercise is the expression

$$\varepsilon := \frac{1}{2}v^2 - \frac{Gm}{r} + \frac{1}{c^2}\left\{\frac{3}{8}(1-3\eta)v^4 + \frac{Gm}{2r}\left[(3+\eta)v^2 + \eta\dot{r}^2 + \frac{Gm}{r}\right]\right\} + O(c^{-4}) \tag{10.3}$$

for the post-Newtonian energy of the two-body system. In a similar way we can show that

$$\boldsymbol{h} := \left\{1 + \frac{1}{c^2}\left[\frac{1}{2}(1-3\eta)v^2 + (3+\eta)\frac{Gm}{r}\right]\right\}(\boldsymbol{r}\times\boldsymbol{v}) + O(c^{-4}) \tag{10.4}$$

is the post-Newtonian angular momentum.

Box 10.1 Ambiguities in energy and angular momentum

The expressions of Eqs. (10.3) and (10.4) are actually not unique. For example, six arbitrary coefficients are needed when constructing a post-Newtonian trial expression for ε, and six conditions are found when $d\varepsilon/dt$ is required to vanish after involving the post-Newtonian equations of motion. Two of the conditions, however, turn out to be redundant, and as a result, one coefficient cannot be determined. It is easy to show that the free coefficient represents the freedom to add to the expression for ε an arbitrary amount of a post-Newtonian contribution $c^{-2}(v^2 - 2Gm/r)^2$. This is constant by virtue of the Newtonian equations of motion, and so the constancy of ε through $O(c^{-2})$ still holds. This freedom reflects the fact that the zero of energy is not fixed in classical mechanics. To arrive at Eq. (10.3) we fixed the free coefficient so that the v^4 term in ε has the factor shown. With this choice, the limit $Gm/r \to 0$ of ε matches the Newtonian and post-Newtonian terms in the expansion of the special relativistic energy of two bodies, given by $\gamma_1 m_1 c^2 + \gamma_2 m_2 c^2$, where $\gamma_A = (1 - v_A^2/c^2)^{-1/2}$. Similar considerations apply to the conserved angular momentum $\boldsymbol{h}$.

It is interesting to note that although $\boldsymbol{h}$ is conserved in the post-Newtonian dynamics, the vector $\boldsymbol{r} \times \boldsymbol{v}$ is not; this is a point of departure from the Newtonian situation. But while $\boldsymbol{r} \times \boldsymbol{v}$ is no longer constant in magnitude, it is still constant *in direction*, and this is sufficient to establish that *the orbital motion proceeds within a fixed orbital plane*, just as in the Newtonian situation. As in Sec. 3.2, we may simplify the mathematical description of the post-Newtonian motion by taking the orbital plane to coincide with the x-y plane of the coordinate system, and by introducing the vectorial basis $\boldsymbol{n} := [\cos\phi, \sin\phi, 0]$, $\boldsymbol{\lambda} := [-\sin\phi, \cos\phi, 0]$, and $\boldsymbol{e}_z := [0, 0, 1]$, in which ϕ is the orbital angle. In terms of the orbital basis we have $\boldsymbol{r} = r\,\boldsymbol{n}$, $\boldsymbol{v} = \dot{r}\,\boldsymbol{n} + r\dot{\phi}\,\boldsymbol{\lambda}$,

$$\boldsymbol{a} = \left(\ddot{r} - r\dot{\phi}^2\right)\boldsymbol{n} + \frac{1}{r}\frac{d}{dt}\left(r^2\dot{\phi}\right)\boldsymbol{\lambda}, \tag{10.5}$$

and $\boldsymbol{r} \times \boldsymbol{v} = (r^2\dot{\phi})\,\boldsymbol{e}_z$. The equations of motion become

$$\ddot{r} = r\dot{\phi}^2 - \frac{Gm}{r^2} + \frac{Gm}{c^2r^2}\left[\frac{1}{2}(6 - 7\eta)\dot{r}^2 - (1 + 3\eta)(r\dot{\phi})^2 + 2(2 + \eta)\frac{Gm}{r}\right] + O(c^{-4}), \tag{10.6a}$$

$$\frac{d}{dt}\left(r^2\dot{\phi}\right) = 2(2 - \eta)\frac{Gm}{c^2}\dot{r}\dot{\phi} + O(c^{-4}), \tag{10.6b}$$

when they are expressed in terms of the dynamical variables $r(t)$ and $\phi(t)$.

The integration of Eqs. (10.6) is greatly facilitated by the existence of ε and $h := |\boldsymbol{h}|$ as constants of the motion. In fact, the conserved quantities can be used to express $\dot{r}^2$ and $\dot{\phi}$ as simple polynomials in $1/r$. A simple computation reveals that

$$\dot{r}^2 = 2\varepsilon\left[1 - \frac{3}{2}(1 - 3\eta)\frac{\varepsilon}{c^2}\right] + 2\frac{Gm}{r}\left[1 - (6 - 7\eta)\frac{\varepsilon}{c^2}\right] - \frac{h^2}{r^2}\left[1 - 2(1 - 3\eta)\frac{\varepsilon}{c^2}\right] - 5(2 - \eta)\frac{(Gm)^2}{c^2r^2} + (8 - 3\eta)\frac{Gmh^2}{c^2r^3} + O(c^{-4}) \tag{10.7}$$

and

$$\dot{\phi} = \frac{h}{r^2}\left[1 - (1 - 3\eta)\frac{\varepsilon}{c^2}\right] - 2(2 - \eta)\frac{Gmh}{c^2 r^3} + O(c^{-2}). \tag{10.8}$$

These equations can be compared with their Keplerian version displayed in Eqs. (3.10) and (3.13).

10.1.2 Circular orbits

Our next task is the integration of the post-Newtonian equations of motion. We begin with the simple case of a circular orbit of radius r. Setting $\dot{r} = 0$ in Eq. (10.6b) reveals that the orbital angular velocity $\dot{\phi}$ is a constant that we denote Ω. Setting $\ddot{r} = 0$ in Eq. (10.6a) allows us to relate Ω to r, and we obtain

$$\Omega^2 = \frac{Gm}{r^3}\left[1 - (3 - \eta)\frac{Gm}{c^2 r} + O(c^{-4})\right], \tag{10.9}$$

the relativistic version of the familiar Keplerian relation $\Omega^2 = Gm/r^3$. The orbital velocity is then given by

$$v^2 = (r\Omega)^2 = \frac{Gm}{r}\left[1 - (3 - \eta)\frac{Gm}{c^2 r} + O(c^{-4})\right], \tag{10.10}$$

and the constants of the motion reduce to

$$\varepsilon = -\frac{Gm}{2r}\left[1 - \frac{1}{4}(7 - \eta)\frac{Gm}{c^2 r} + O(c^{-4})\right] \tag{10.11}$$

and

$$h = \sqrt{Gmr}\left[1 + 2\frac{Gm}{c^2 r} + O(c^{-4})\right]. \tag{10.12}$$

10.1.3 Perturbed Keplerian orbits

A possible way of integrating the equations of motion is to treat the post-Newtonian terms in Eq. (10.1) as a perturbing force $\boldsymbol{f}$ in the perturbed Kepler problem described in Sec. 3.3. (Recall our admonition from Chapter 3, that even though $\boldsymbol{f}$ is an acceleration, we nevertheless use the conventional term "force" to describe it.) We exploit the method of osculating orbital elements, in which the perturbed motion is represented by a Keplerian orbit with time-dependent orbital elements, which vary in response to the perturbing force. The components of the perturbing force in the orbital basis $(\boldsymbol{n}, \boldsymbol{\lambda}, \boldsymbol{e}_z)$ are

$$\mathcal{R} = \frac{Gm}{c^2 r^2}\left[-(1 + 3\eta)v^2 + \frac{1}{2}(8 - \eta)\dot{r}^2 + 2(2 + \eta)\frac{Gm}{r}\right], \tag{10.13a}$$

$$\mathcal{S} = \frac{Gm}{c^2 r^2}\left[2(2 - \eta)\dot{r}(r\dot{\phi})\right], \tag{10.13b}$$

$$\mathcal{W} = 0, \tag{10.13c}$$

in which we insert the Keplerian relations $r = p/(1+e\cos f)$, $\dot{r} = \sqrt{Gm/p}\, e \sin f$, $r\dot{\phi} = \sqrt{Gm/p}\,(1+e\cos f)$, where p is the Keplerian semi-latus rectum, e is the Keplerian eccentricity, and $f := \phi - \omega$ is the true anomaly, with ω denoting the Keplerian longitude of pericenter. Substituting the force components within Eqs. (3.69) produces

$$\frac{dp}{df} = 4(2-\eta)\frac{Gm}{c^2} e \sin f, \tag{10.14a}$$

$$\frac{de}{df} = \frac{Gm}{c^2 p}\left\{\left[3 - \eta + \frac{1}{8}(56 - 47\eta)e^2\right]\sin f + (5 - 4\eta)e \sin 2f - \frac{3}{8}\eta e^2 \sin 3f\right\}, \tag{10.14b}$$

$$\frac{d\omega}{df} = \frac{1}{e}\frac{Gm}{c^2 p}\left\{3e - \left[3 - \eta - \frac{1}{8}(8 + 21\eta)e^2\right]\cos f - (5 - 4\eta)e\cos 2f + \frac{3}{8}\eta e^2 \cos 3f\right\}. \tag{10.14c}$$

The vanishing of $\mathcal{W} := \boldsymbol{f} \cdot \boldsymbol{e}_z$ implies that the inclination angle ι and longitude of ascending node Ω are not affected by the perturbing force; this is a consequence of the fact that the post-Newtonian motion proceeds in a fixed orbital plane.

The perturbing force $\boldsymbol{f}$ is of first post-Newtonian order, and working consistently at this order, it is appropriate to integrate Eqs. (10.14) while keeping the orbital elements constant on the right-hand side of the equations. In this way we obtain

$$p(f) = p_0\left[1 + 4(2-\eta)\frac{Gme_0}{c^2 p_0}(1 - \cos f)\right] \tag{10.15}$$

and more complicated expressions for $e(f)$ and $\omega(f)$; here $p_0 := p(f = 0)$. These expressions can then be inserted within the original Keplerian relations to obtain the complete solutions for $\boldsymbol{r}(f)$ and $\boldsymbol{v}(f)$. The post-Newtonian motion is thus parameterized by the true anomaly f, and its description in terms of time t can be obtained by integrating Eq. (3.70).

10.1.4 Pericenter advance

The description of the post-Newtonian motion in terms of osculating Keplerian orbits is perfectly adequate from a mathematical point of view, but it is fairly awkward to use and not well suited to practical implementations. An illustration of this is the curious fact that the circular orbit of radius r examined in Sec. 10.1.2 is one with Keplerian orbital parameters $e = (3-\eta)Gm/(c^2 r)$ and $p = r[1 - (3-\eta)Gm/(c^2 r)]$; this orbit has a constant true anomaly given by $f = \pi$, so that the pericenter advances at the same rate as the body itself. We shall give the motion a much better description in the next subsection, but let's not give up just yet on the osculating formulation.

In the applications of the method of osculating orbital elements examined in Sec. 3.4, we saw that the method's most powerful insights are delivered when it is asked to reveal the *secular changes* in the orbital elements, and allowed to discard any information about the periodic changes that average out after each orbital cycle. We shall adopt this wisdom here, and calculate the secular changes in p, e, and ω produced by the post-Newtonian perturbation of Eqs. (10.13). They are obtained by integrating Eqs. (10.14) over a complete orbital period (from $f = 0$ to $f = 2\pi$), and the equations produce $\Delta p = 0$, $\Delta e = 0$, as well as

$$\Delta\omega = 6\pi \frac{Gm}{c^2 p}. \tag{10.16}$$

The fact that p undergoes no secular change is a consequence of angular-momentum conservation, and the absence of a secular change in e (and therefore in the Keplerian semi-major axis a) is a consequence of energy conservation. The only parameter that undergoes a secular evolution is the longitude of periastron ω, and Eq. (10.16) describes the pericenter advance that was so famously calculated by Einstein.

Einstein's method of derivation was very different. He did not have access to the post-Newtonian metric of an N-body system, and he did not have access to the N-body equations of motion. What he did was to obtain the post-Newtonian metric of a single body (the Sun), and to calculate the motion of a second body (Mercury) under the assumption that it is a test mass that moves on a geodesic of the spacetime. This is a sensible assumption, given that the Mercury–Sun mass ratio is approximately one to 6 million. Einstein obtained the result of Eq. (10.16), but with the total mass $m := M_1 + M_2$ well approximated by M_1, the mass of the Sun. Our result is more general, and it applies to a much broader range of situations.

The advance per orbit can be converted to a rate of advance by dividing by the orbital period. We can also eliminate the semi-major axis appearing in $p = a(1-e^2)$ by using Kepler's third law, $a = (Gm)^{1/3}(P/2\pi)^{2/3}$, where P is the orbital period; in principle the relation should be modified by a post-Newtonian correction, but the modification is irrelevant because $\Delta\omega$ is already of first post-Newtonian order. The result is

$$\begin{aligned}\left(\frac{d\omega}{dt}\right)_{\text{sec}} &= \frac{3}{1-e^2}\frac{(Gm/c^3)^{2/3}}{(P/2\pi)^{5/3}} \\ &= 716.25\frac{1}{1-e^2}\left(\frac{m}{M_\odot}\right)^{2/3}\left(\frac{P}{1\text{ day}}\right)^{-5/3} \text{ as/yr.}\end{aligned} \tag{10.17}$$

Substituting the values for Mercury, $e = 0.2056$ and $P = 87.97$ days, we obtain 42.98 arcseconds per century. As we saw back in Table 3.1, the modern difference between the measured advance and the one predicted by Newtonian N-body perturbations is 42.98 ± 0.04 arcseconds per century, in 0.1 percent agreement with the relativistic prediction.

The discovery of binary-pulsar systems with total masses of 2 to 3 solar masses, and with orbital periods as small as fractions of a day, resulted in the observation of periastron advances of several degrees per year. The relativistic periastron advance plays an interesting role in these systems. In the solar system, Gm for the Sun is known to high precision from the measured orbital period and orbital radius of the Earth, combined with Kepler's third law. By contrast, the masses of the neutron stars are not known, apart from the general expectation that they should be around the Chandrasekhar limit of $1.4\,M_\odot$, based on models of how such systems might have formed. In the famous Hulse–Taylor binary pulsar, the first such system to be discovered, it was possible to measure the orbital eccentricity, the orbital period, and the pericenter advance very accurately; the current values are $e = 0.6171338(4)$, $P = 0.322997448930(4)$ day, and $\dot{\omega} = 4.226595(5)$ deg/yr, in which the number in parentheses denotes the error in the final digit. Assuming that there is no other source of periastron advance, we can turn Eq. (10.17) around and use it to measure the total mass of the system. The result is $m = 2.828296(5)\,M_\odot$. The recently discovered double pulsar J0737-3039A/B, in which both stars are observed as pulsars, has $e = 0.0877775(9)$, $P = 0.10225156248(5)$ day, and $\dot{\omega} = 16.8995(7)$ deg/yr, giving a total mass of $2.5871(2)\,M_\odot$. These are remarkably accurate measurements of an astrophysical quantity, and as we can see, general relativity plays a central role in the analysis.

10.1.5 Integration of the equations of motion

To integrate the post-Newtonian equations of motion listed in Sec. 10.1.1, we adopt the approach followed by Damour and Deruelle in their seminal 1985 paper. Our starting point is the observation that the transformation

$$r = \bar{r} - \frac{1}{2}(8 - 3\eta)\frac{Gm}{c^2} + O(c^{-4}) \tag{10.18}$$

turns Eq. (10.7) into the simpler polynomial

$$\dot{\bar{r}}^2 = 2\varepsilon_{\rm K} + 2\frac{Gm_{\rm K}}{\bar{r}} - \frac{h_{\rm K}^2}{\bar{r}^2}, \tag{10.19}$$

in which

$$\varepsilon_{\rm K} := \varepsilon\biggl[1 - \frac{3}{2}(1 - 3\eta)\frac{\varepsilon}{c^2} + O(c^{-4})\biggr], \tag{10.20a}$$

$$m_{\rm K} := m\biggl[1 - (6 - 7\eta)\frac{\varepsilon}{c^2} + O(c^{-4})\biggr], \tag{10.20b}$$

$$h_{\rm K}^2 := h^2\biggl[1 - 2(1 - 3\eta)\frac{\varepsilon}{c^2} + 2(1 - \eta)\frac{(Gm)^2}{c^2h^2} + O(c^{-4})\biggr]. \tag{10.20c}$$

The radial equation is an exact replica of the Keplerian equation displayed in Eq. (3.13), and it therefore admits the same solution. We adopt a representation in terms of an eccentric anomaly u, according to which $\bar{r} = \bar{a}(1 - \bar{e}\cos u)$ and $t - T = \sqrt{\bar{a}^3/(Gm_{\rm K})}(u - \bar{e}\sin u)$, in which $\bar{a}$ is a post-Newtonian semi-major axis and $\bar{e}$ a post-Newtonian eccentricity, defined in the same way as in the Keplerian problem: $\varepsilon_{\rm K} = -Gm_{\rm K}/(2\bar{a})$ and $h_{\rm K}^2 = Gm_{\rm K}\bar{a}(1 - \bar{e}^2)$.

With the solution for $\bar{r}$ thus obtained, we apply the transformation of Eq. (10.18) to express the motion directly in terms of r. Simple manipulations produce

$$r = a(1 - e\cos u), \tag{10.21}$$

in which $a := \bar{a}[1 - \frac{1}{2}(8 - 3\eta)Gm/(c^2\bar{a})]$ and $e = \bar{e}[1 + \frac{1}{2}(8 - 3\eta)Gm/(c^2\bar{a})]$ are new post-Newtonian orbital elements. Adopting these as the primary elements, it is a simple matter to express the conserved energy ε and conserved angular momentum h in terms of a and e; we obtain

$$\varepsilon = -\frac{Gm}{2a}\biggl[1 - \frac{1}{4}(7 - \eta)\frac{Gm}{c^2a} + O(c^{-4})\biggr] \tag{10.22}$$

and

$$h^2 = Gma(1-e^2)\biggl[1 + \frac{4+(2-\eta)e^2}{1-e^2}\frac{Gm}{c^2a} + O(c^{-4})\biggr]. \tag{10.23}$$

The time function can then be expressed as

$$t - T = \frac{P}{2\pi}\bigl(u - e_t \sin u\bigr), \tag{10.24}$$

in which

$$e_t := e\biggl[1 - \frac{1}{2}(8-3\eta)\frac{Gm}{c^2a} + O(c^{-4})\biggr] \tag{10.25}$$

is a second eccentricity parameter (which differs from e by a post-Newtonian correction), and

$$P := 2\pi\sqrt{\frac{a^3}{Gm}}\biggl[1 + \frac{1}{2}(9-\eta)\frac{Gm}{c^2a} + O(c^{-4})\biggr] \tag{10.26}$$

is the post-Newtonian period.

To obtain ϕ as a function of u we begin with the transformation

$$r = \tilde{r} - (2-\eta)\frac{Gm}{c^2} + O(c^{-4}), \tag{10.27}$$

which turns Eq. (10.8) into

$$\dot{\phi} = \frac{\tilde{h}}{\tilde{r}^2}, \tag{10.28}$$

where

$$\tilde{h} := \sqrt{Gma(1-e^2)}\biggl[1 + \frac{5-3\eta+(1+2\eta)e^2}{2(1-e^2)}\frac{Gm}{c^2a} + O(c^{-4})\biggr]. \tag{10.29}$$

We next use Eq. (10.21) to write $\tilde{r} = \tilde{a}(1-\tilde{e}\cos u)$, in which $\tilde{a} = a[1+(2-\eta)Gm/(c^2a)]$ and $\tilde{e} = e[1-(2-\eta)Gm/(c^2a)]$, and Eq. (10.24) to obtain

$$\frac{d\phi}{du} = \frac{P}{2\pi}\frac{\tilde{h}}{\tilde{a}^2}\frac{1-e_t\cos u}{(1-\tilde{e}\cos u)^2} + O(c^{-4}), \tag{10.30}$$

which can be integrated to yield the orbital angle ϕ in terms of eccentric anomaly u. This equation can be compared with the Keplerian version of Eq. (3.33), and here we see a substantial difference in the form of the equations. In addition to the multiplicative factor that appears on the right-hand side, we see that the post-Newtonian expression involves $(1-e_t\cos u)/(1-\tilde{e}\cos u)^2$ while the Keplerian version features the simpler factor of $(1-e\cos u)^{-1}$. The forms, however, can be reconciled with a simple trick. We first observe that e_t differs from $\tilde{e}$ by a post-Newtonian correction that we denote ϵ. We next define a third eccentricity parameter by $e_\phi := \tilde{e} - \epsilon$, and factorize $(1-\tilde{e}\cos u)^2$ as

$$\begin{aligned}(1-\tilde{e}\cos u)^2 &= \bigl[1-(\tilde{e}+\epsilon)\cos u\bigr]\bigl[1-(\tilde{e}-\epsilon)\cos u\bigr] + O(c^{-4}) \\ &= (1-e_t\cos u)(1-e_\phi\cos u) + O(c^{-4}).\end{aligned} \tag{10.31}$$

This allows us to re-express $d\phi/du$ as

$$\frac{d\phi}{du} = (1+k)\frac{(1-e_\phi^2)^{1/2}}{1-e_\phi\cos u} + O(c^{-4}), \tag{10.32}$$

which now looks identical to Eq. (3.33), except for the factor $1+k := (P/2\pi)(\tilde{h}/\tilde{a}^2)(1-e_\phi^2)^{-1/2}$.

The solution to this equation can be expressed as in Eqs. (3.30) and (3.32). We have that

$$\cos\left(\frac{\phi}{1+k}\right) = \frac{\cos u - e_\phi}{1-e_\phi\cos u}, \qquad \sin\left(\frac{\phi}{1+k}\right) = \frac{(1-e_\phi^2)^{1/2}\sin u}{1-e_\phi\cos u}, \tag{10.33}$$

or that

$$\tan\left[\frac{\phi}{2(1+k)}\right] = \sqrt{\frac{1+e_\phi}{1-e_\phi}}\tan\frac{u}{2}. \tag{10.34}$$

Simple algebra confirms that

$$e_\phi = e\left[1 + \frac{1}{2}\eta\frac{Gm}{c^2a} + O(c^{-4})\right] \tag{10.35}$$

and

$$k = \frac{3}{1-e^2}\frac{Gm}{c^2a} + O(c^{-4}). \tag{10.36}$$

The meaning of this last quantity is easy to extract from Eqs. (10.33) and (10.34). These relations inform us that in the course of a complete radial cycle, as u runs from 0 to 2π, the orbital angle ϕ runs from 0 to $2\pi(1+k)$. The excess angle,

$$2\pi k = 6\pi\frac{Gm}{c^2a(1-e^2)} + O(c^{-4}), \tag{10.37}$$

is the pericenter advance of Eq. (10.16).

It is a remarkable fact that the two-body equations of motion of post-Newtonian theory can be integrated in the same manner as the Keplerian equations, with only the small cost of introducing two additional eccentricity parameters (e_t and e_ϕ, which differ from e by post-Newtonian corrections), and a pericenter-advance parameter $k := \Delta\omega/(2\pi)$. This implies that the post-Newtonian motion can be computed with great ease, by exploiting the tried and true methods of celestial mechanics.

The individual motion of each body can be obtained from Eq. (10.2). It is easy to show that

$$r_1 := |\boldsymbol{r}_1| = a_1(1-e_1\cos u), \tag{10.38}$$

in which $a_1 := (M_2/m)a$ is the semi-major axis of the first body, while

$$e_1 = e\left[1 - \frac{M_1(M_1-M_2)}{2m^2}\frac{Gm}{c^2a}\right] \tag{10.39}$$

is its eccentricity. The corresponding results for the second body are obtained by a suitable exchange of labels.

10.1.6 de Sitter precession

The de Sitter precession of the lunar orbit is a relativistic three-body effect that involves the Moon, the Earth, and the Sun; it is a secular advance of the line of nodes of the lunar orbital plane. You will recall from our discussion in Chapter 3 that the motion of the Moon is a notoriously difficult problem in Newtonian gravity, mainly because the strong perturbations caused by the Sun lead to a poorly convergent sequence of corrections. Special formulations of the perturbation theory, such as the Hill–Brown theory, were required for high precision, and today the equations of motion are solved directly using computers. With the motion of the Moon in Newtonian gravity now under control, discrepancies between the predicted and observed motions can be attributed to relativistic effects.

Relativistic three-body effects can be investigated on the basis of the N-body equations listed in Sec. 9.3.5 – refer to Eq. (9.127). As in Sec. 3.4.1, in which we considered third-body effects in Newtonian theory, we focus our attention on a two-body system (the Earth and the Moon) and examine the perturbations produced by a third body (the Sun). We let the Moon be the first body (mass M_1, position $\boldsymbol{r}_1$), the Earth be the second body (mass M_2, position $\boldsymbol{r}_2$), and the Sun be the third body (mass M_3, position $\boldsymbol{r}_3$). We let $\boldsymbol{r} := \boldsymbol{r}_1 - \boldsymbol{r}_2$ be the separation between the Moon and the Earth, and $\boldsymbol{R} := \boldsymbol{r}_2 - \boldsymbol{r}_3$ be the separation between the Earth and the Sun. Similarly, we let $\boldsymbol{v} := \boldsymbol{v}_1 - \boldsymbol{v}_2$ be the Moon's velocity relative to Earth's, and $\boldsymbol{V} := \boldsymbol{v}_2 - \boldsymbol{v}_3$ be the Earth's velocity relative to the Sun. For simplicity we imagine that the Sun is at rest at the spatial origin of the coordinate system, so that $\boldsymbol{r}_3 = \boldsymbol{0} = \boldsymbol{v}_3$.

The relativistic three-body equations lead to effects of various sizes on the two-body system. We wish to focus our attention on the dominant effects, and to allow ourselves to neglect the smaller ones, or to ignore relativistic effects that would be masked by much larger Newtonian effects. To seek guidance in the identification of which terms must be kept in the equations of motion, and which terms can be neglected, we first examine the various scales of the problem.

Our first observation is that M_1, the mass of the Moon, is much smaller than both M_2 and M_3, and we shall therefore neglect all terms in the equations of motion that involve M_1. Our second observation is that the acceleration of the Earth–Moon system toward the Sun is approximately twice the acceleration of the Moon toward the Earth, $GM_3/R^2 \sim 2GM_2/r^2$; this comes about because $r/R \sim 2 \times 10^{-3}$, which counteracts the large value of M_3/M_2. Our third observation is that the post-Newtonian corrections scale as $GM_3/(c^2R) \sim 10^{-8}$ for the Earth–Moon system moving around the Sun, so that $V/c \sim 10^{-4}$, and as $GM_2/(c^2r) \sim 10^{-11}$ for the Moon moving around the Earth, so that $v/c \sim 3 \times 10^{-6}$; the motion around the Sun is more relativistic than the motion around the Earth. We rely on this observation to neglect $GM_2/(c^2r)$ compared to $GM_3/(c^2R)$ in the equations of motion, and to neglect $(v/c)^2$ compared to $(V/c)^2$. Another source of simplification is the previously noted smallness of r/R, which allows us to ignore the inhomogeneity of the Sun's gravitational field across the Earth–Moon orbit; this means that $\boldsymbol{r}_{13} = \boldsymbol{R} + \boldsymbol{r}$ can be safely approximated by $\boldsymbol{R}$ in the equations of motion.

With these simplifications, it is a straightforward exercise to apply Eq. (9.127) to the Earth–Moon system perturbed by the Sun. We find that the post-Newtonian term in the

Moon's acceleration reduces to

$$\begin{aligned}
\boldsymbol{a}_1[\mathrm{PN}] &= \frac{GM_2}{c^2r^2}\left[V^2 + 2\boldsymbol{V}\cdot\boldsymbol{v} + \frac{3}{2}(\boldsymbol{n}\cdot\boldsymbol{V})^2 + \frac{5GM_3}{R}\right]\boldsymbol{n} + \frac{GM_2}{c^2r^2}(\boldsymbol{n}\cdot\boldsymbol{V})\boldsymbol{v} \\
&\quad - \frac{GM_3}{c^2R^2}\left[V^2 + 2\boldsymbol{V}\cdot\boldsymbol{v} - \frac{4GM_3}{R}\right]\boldsymbol{N} \\
&\quad + \frac{4GM_3}{c^2R^2}\left[(\boldsymbol{N}\cdot\boldsymbol{V})\boldsymbol{V} + (\boldsymbol{N}\cdot\boldsymbol{V})\boldsymbol{v} + (\boldsymbol{N}\cdot\boldsymbol{v})\boldsymbol{V}\right],
\end{aligned} \tag{10.40}$$

in which $\boldsymbol{n} := \boldsymbol{r}/r$ and $\boldsymbol{N} := \boldsymbol{R}/R$, and that the Earth's post-Newtonian acceleration becomes

$$\boldsymbol{a}_2[\mathrm{PN}] = -\frac{GM_3}{c^2R^2}\left(V^2 - \frac{4GM_3}{R}\right)\boldsymbol{N} + \frac{4GM_3}{c^2R^2}(\boldsymbol{N}\cdot\boldsymbol{V})\boldsymbol{V}. \tag{10.41}$$

The relative acceleration $\boldsymbol{a}[\mathrm{PN}] := \boldsymbol{a}_1[\mathrm{PN}] - \boldsymbol{a}_2[\mathrm{PN}]$ is then

$$\begin{aligned}
\boldsymbol{a}[\mathrm{PN}] &= \frac{GM_2}{c^2r^2}\left[V^2 + \frac{3}{2}(\boldsymbol{n}\cdot\boldsymbol{V})^2 + \frac{5GM_3}{R}\right]\boldsymbol{n} \\
&\quad + \frac{2GM_3}{c^2R^2}\left[2(\boldsymbol{N}\cdot\boldsymbol{V})\boldsymbol{v} + 2(\boldsymbol{N}\cdot\boldsymbol{v})\boldsymbol{V} - (\boldsymbol{V}\cdot\boldsymbol{v})\boldsymbol{N}\right],
\end{aligned} \tag{10.42}$$

in which we have neglected $2\boldsymbol{V}\cdot\boldsymbol{v}$ compared to V^2 in the first group of terms. This can be treated as a perturbing force $\boldsymbol{f}$ on the Keplerian system formed by the Earth and the Moon, and it can be used in a perturbative evolution of the orbital elements along the lines developed in Sec. 3.4.

The motion of the Moon around the Earth is described by an osculating Keplerian orbit of elements $(p, e, \omega, \iota, \Omega)$ perturbed by the external force of Eq. (10.42). To simplify the description of the evolution of the orbital elements, we follow the approach of Sec. 3.4.1 and place the Earth on a circular orbit of radius R, angular velocity Ω_{orb}, and orbital phase $F = \Omega_{\mathrm{orb}}t$ around the Sun; the orbit is situated in the fundamental X-Y plane. It is then a straightforward task to calculate the components $(\mathcal{R}, \mathcal{S}, \mathcal{W})$ of the perturbing force, to insert these into the osculating equations (3.69), to obtain the total changes $(\Delta p, \Delta e, \Delta\omega, \Delta\iota, \Delta\Omega)$ over a complete lunar orbit, and finally, to average these over the motion of the Earth around the Sun. Our final results are that $\langle\Delta p\rangle = \langle\Delta e\rangle = \langle\Delta\iota\rangle = \langle\Delta\omega\rangle = 0$, but that

$$\langle\Delta\Omega\rangle = \frac{3\pi G}{c^2}\frac{M_\odot^{3/2}p^{3/2}}{M_\oplus^{1/2}R^{5/2}(1-e^2)^{3/2}}, \tag{10.43}$$

in which we have substituted the standard symbols $M_\odot$ and $M_\oplus$ for the mass of the Sun and the Earth, respectively. Thus, the only long-term impact of relativistic three-body effects on the lunar orbit is a precession of the line of nodes, whose longitude Ω advances at the averaged rate

$$\left(\frac{d\Omega}{dt}\right)_{\mathrm{sec}} = \frac{3}{2}\sqrt{\frac{GM_\odot}{R^3}}\frac{GM_\odot}{c^2R}, \tag{10.44}$$

obtained by dividing $\langle\Delta\Omega\rangle$ by the Moon's orbital period.

According to Eq. (10.44), the line of nodes of the lunar orbit advances at an averaged rate of 19.1 arcseconds per century. The effect was first predicted in 1916 by de Sitter, but at the time it was far too small to be detected on top of the 19.3 *degrees* of advance per year produced by the Sun's Newtonian perturbations. It is a remarkable feat of modern precision instrumentation that thanks to lunar laser ranging (a technique described in some detail in Box 13.2), the de Sitter precession has been measured to a precision of better than one percent.

10.2 Motion of light in post-Newtonian gravity

Most of the information about astrophysical objects comes in the form of electromagnetic signals, and confrontation between theory and observations must account for the curved path of a light ray in a gravitational field. Indeed, electromagnetic waves are deflected and delayed by a massive body, and these measurable effects must be taken into account in high-accuracy astronomical observations.

10.2.1 Motion of a photon

In the geometric-optics approximation of electromagnetism (refer to Box 5.6), light rays behave as massless particles – photons – that move on null geodesics of a curved spacetime; the geodesic equation for a photon in a post-Newtonian spacetime was obtained in Sec. 8.1.4. The particle moves on a trajectory $\boldsymbol{r}(t)$ with a velocity $\boldsymbol{v} = d\boldsymbol{r}/dt$ that can be expressed as

$$\boldsymbol{v} = c\left(1 - \frac{2}{c^2}U\right)\boldsymbol{n} + O(c^{-3}), \tag{10.45}$$

in terms of a unit vector $\boldsymbol{n}$ that specifies the direction of propagation. This satisfies the differential equation

$$\frac{dn^j}{dt} = \frac{2}{c}(\delta^{jk} - n^j n^k)\partial_k U + O(c^{-2}), \tag{10.46}$$

in which U is the Newtonian potential evaluated at $\boldsymbol{x} = \boldsymbol{r}(t)$. It should be noted that in this section, $\boldsymbol{r}(t)$ stands for the photon's trajectory and not the inter-body separation of a two-body system, and $\boldsymbol{n}$ denotes the direction of propagation and not the vector $\boldsymbol{r}/r$.

The leading-order solution to Eq. (10.46) is $\boldsymbol{n} = \boldsymbol{k} + O(c^{-2})$, in which $\boldsymbol{k}$ is a constant vector. At this order, the photon's trajectory is described by the straight path

$$\boldsymbol{r}(t) = \boldsymbol{r}_e + c\boldsymbol{k}(t - t_e) + O(c^{-2}), \tag{10.47}$$

in which $\boldsymbol{r}_e = \boldsymbol{r}(t = t_e)$ is the position of the source, and t_e is the emission time. At the next order we have

$$\boldsymbol{n} = \boldsymbol{k} + \boldsymbol{\alpha} + O(c^{-4}), \tag{10.48}$$

in which the deflection vector $\boldsymbol{\alpha}$ satisfies

$$\frac{d\alpha^j}{dt} = \frac{2}{c}(\delta^{jk} - k^j k^k)\partial_k U, \tag{10.49}$$

with $\partial_k U$ evaluated at $\boldsymbol{x} = \boldsymbol{r}(t)$, as given by Eq. (10.47). We integrate Eq. (10.49) with the initial conditions

$$\boldsymbol{\alpha}(t = t_e) = \boldsymbol{0}, \tag{10.50}$$

so that $\boldsymbol{n}(t = t_e)$, the initial direction of propagation, coincides with the vector $\boldsymbol{k}$.

Inserting Eq. (10.48) within Eq. (10.45), we find that

$$\boldsymbol{v} = c\left(1 - \frac{2}{c^2}U\right)\boldsymbol{k} + c\boldsymbol{\alpha} + O(c^{-3}). \tag{10.51}$$

Note that $\boldsymbol{\alpha}$ is necessarily orthogonal to $\boldsymbol{k}$, so that the second term in Eq. (10.51) describes a transverse deflection of the photon; the longitudinal aspects of the correction are captured by the first term.

To integrate Eq. (10.49) we substitute $\partial_k U = -G\int \rho' s^{-3} s^k \, d^3x'$ on the right-hand side, where $\rho' := \rho(t, \boldsymbol{x}')$ is the mass density of the matter distribution, $\boldsymbol{s} := \boldsymbol{r}(t) - \boldsymbol{x}'$, and $s := |\boldsymbol{s}|$. This gives

$$\frac{d\boldsymbol{\alpha}}{dt} = -\frac{2G}{c}\int \rho' \frac{\boldsymbol{b}}{s^3} \, d^3x', \tag{10.52}$$

in which

$$\boldsymbol{b} := \boldsymbol{s}_e - (\boldsymbol{s}_e \cdot \boldsymbol{k})\,\boldsymbol{k}, \qquad \boldsymbol{s}_e := \boldsymbol{r}_e - \boldsymbol{x}'; \tag{10.53}$$

the vector $\boldsymbol{b}$ points from $\boldsymbol{x}'$ to the point of closest approach reached by a photon emitted from a position $\boldsymbol{r}_e$ in a direction $\boldsymbol{k}$. We next invoke the easily established identity

$$\frac{1}{c}\frac{d}{dt}\left(\frac{\boldsymbol{s} \cdot \boldsymbol{k}}{s}\right) = \frac{b^2}{s^3} + O(c^{-2}) \tag{10.54}$$

to express $d\boldsymbol{\alpha}/dt$ as

$$\frac{d\boldsymbol{\alpha}}{dt} = -\frac{2G}{c^2}\int \rho' \frac{\boldsymbol{b}}{b^2}\frac{d}{dt}\left(\frac{\boldsymbol{s} \cdot \boldsymbol{k}}{s}\right) d^3x' = -\frac{2G}{c^2}\frac{d}{dt}\int \rho' \frac{\boldsymbol{b}}{b^2}\frac{\boldsymbol{s} \cdot \boldsymbol{k}}{s} \, d^3x'. \tag{10.55}$$

Integration of this equation is immediate.

Taking into account Eq. (10.50), we find that the deflection vector is given by

$$\boldsymbol{\alpha}(t) = -\frac{2G}{c^2}\int \rho(t, \boldsymbol{x}')\frac{\boldsymbol{b}}{b^2}\left(\frac{\boldsymbol{s} \cdot \boldsymbol{k}}{s} - \frac{\boldsymbol{s}_e \cdot \boldsymbol{k}}{s_e}\right) d^3x' \tag{10.56}$$

for any matter distribution. Further progress on evaluating the deflection vector requires specification of the mass density.

With the photon's velocity $\boldsymbol{v}$ given by Eqs. (10.51) and (10.56), we may now determine the photon's trajectory $\boldsymbol{r}(t)$. We express the solution to $d\boldsymbol{r}/dt = \boldsymbol{v}$ as

$$\boldsymbol{r}(t) = \boldsymbol{r}_e + c\boldsymbol{k}(t - t_e) + \boldsymbol{k}\,\delta r_{\parallel}(t) + \delta\boldsymbol{r}_{\perp}(t) + O(c^{-4}), \tag{10.57}$$

in which $\delta r_{\parallel}(t)$ is the longitudinal displacement determined by $d(\delta r_{\parallel})/dt = -2U/c$, and $\delta \boldsymbol{r}_{\perp}(t)$ is the transverse displacement determined by $d(\delta \boldsymbol{r}_{\perp})/dt = c\boldsymbol{\alpha}$. The longitudinal term in Eq. (10.57) does not alter the path of the photon with respect to its unperturbed description, but it affects the relationship between position and time. The transverse term represents a deviation from the path, in response to the deflection vector $\boldsymbol{\alpha}$.

To calculate the longitudinal displacement we insert the usual expression for the Newtonian potential and get $d(\delta r_{\parallel})/dt = -(2G/c)\int \rho' s^{-1}\, d^3x'$. The identity

$$\frac{1}{c}\frac{d}{dt}\ln(s + \boldsymbol{s}\cdot\boldsymbol{k}) = \frac{1}{s} + O(c^{-2}) \tag{10.58}$$

permits an immediate integration, and we obtain

$$\delta r_{\parallel}(t) = -\frac{2G}{c^2}\int \rho(t,\boldsymbol{x}')\ln\left(\frac{s + \boldsymbol{s}\cdot\boldsymbol{k}}{s_e + \boldsymbol{s}_e\cdot\boldsymbol{k}}\right) d^3x', \tag{10.59}$$

which reflects the initial condition $\delta r_{\parallel}(t=0) = 0$. The factorization $b^2 = (s_e - \boldsymbol{s}_e\cdot\boldsymbol{k})(s_e + \boldsymbol{s}_e\cdot\boldsymbol{k})$ allows us to write this in the alternative form

$$\delta r_{\parallel}(t) = -\frac{2G}{c^2}\int \rho(t,\boldsymbol{x}')\ln\left[\frac{(s + \boldsymbol{s}\cdot\boldsymbol{k})(s_e - \boldsymbol{s}_e\cdot\boldsymbol{k})}{b^2}\right] d^3x' \tag{10.60}$$

involving the vector $\boldsymbol{b}$.

To obtain the transverse displacement we invoke Eq. (10.56) and get

$$\frac{d}{dt}\delta\boldsymbol{r}_{\perp} = -\frac{2G}{c}\int \rho(t,\boldsymbol{x}')\frac{\boldsymbol{b}}{b^2}\left(\frac{\boldsymbol{s}\cdot\boldsymbol{k}}{s} - \frac{\boldsymbol{s}_e\cdot\boldsymbol{k}}{s_e}\right) d^3x'. \tag{10.61}$$

Once more integration is immediate thanks to the identity $c^{-1}ds/dt = \boldsymbol{s}\cdot\boldsymbol{k}/s + O(c^{-2})$ and the fact that the second term within brackets is constant in time. We arrive at

$$\delta\boldsymbol{r}_{\perp}(t) = -\frac{2G}{c^2}\int \rho(t,\boldsymbol{x}')\frac{\boldsymbol{b}}{b^2}\left(s - \frac{\boldsymbol{s}\cdot\boldsymbol{s}_e}{s_e}\right) d^3x', \tag{10.62}$$

which also reflects the initial condition $\delta\boldsymbol{r}_{\perp}(t=0) = \boldsymbol{0}$. The motion of the photon in the post-Newtonian spacetime is now completely determined.

10.2.2 Deflection by a spherical body

The simplest application of light deflection involves a single, spherically-symmetric body of mass M, which we place at the spatial origin of the coordinate system. Because the photon must travel outside the body to be observable, the gravitational potential U can be equated to its external expression $GM/|\boldsymbol{x}|$, and Eq. (10.49) can be integrated for this special case. Alternatively, and more simply, we can insert $\rho(\boldsymbol{x}') = M\delta(\boldsymbol{x}')$ within Eq. (10.56) and get

$$\boldsymbol{\alpha}(t) = -\frac{2GM}{c^2}\frac{\boldsymbol{b}}{b^2}\left[\frac{\boldsymbol{r}(t)\cdot\boldsymbol{k}}{r(t)} - \frac{\boldsymbol{r}_e\cdot\boldsymbol{k}}{r_e}\right], \tag{10.63}$$

in which $\boldsymbol{r}(t) := \boldsymbol{r}_e + c\boldsymbol{k}(t - t_e) + O(c^{-2})$, and

$$\boldsymbol{b} := \boldsymbol{r}_e - (\boldsymbol{r}_e\cdot\boldsymbol{k})\boldsymbol{k} \tag{10.64}$$

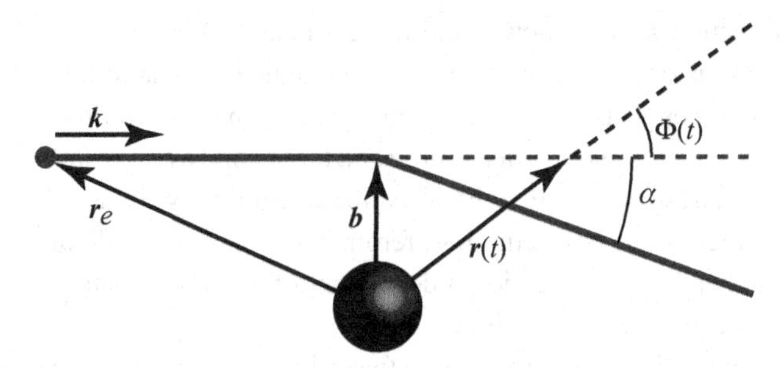

Fig. 10.1 Deflection of light by a spherical body.

is now a constant vector that points from the body's center-of-mass to the photon's point of closest approach (see Fig. 10.1); $b := |\boldsymbol{b}|$ is the impact parameter. Note that the deflection vector always points in the direction of $-\boldsymbol{b}$, corresponding to a deflection toward the massive body, and varies in magnitude in response to the change in the photon's position.

The deflection vector can also be expressed as

$$\boldsymbol{\alpha}(t) = -\frac{2GM}{c^2}\frac{\boldsymbol{b}}{b^2}\left[\cos\Phi(t) + \sqrt{1-(b/r_e)^2}\right], \tag{10.65}$$

in which $\cos\Phi(t) := (\boldsymbol{r}\cdot\boldsymbol{k})/r$, so that $\Phi(t)$ is the angle between the photon's current position $\boldsymbol{r}(t)$ and its initial direction $\boldsymbol{k}$ (see Fig. 10.1), and the square root is an alternative expression for $-(\boldsymbol{r}_e\cdot\boldsymbol{k})/r_e$. If we specialize to a situation in which the distance of closest approach b is very much smaller than the distance to the source r_e, then the deflection vector simplifies to

$$\boldsymbol{\alpha}(t) = -\frac{4GM}{c^2}\frac{\boldsymbol{b}}{b^2}\frac{1+\cos\Phi(t)}{2}. \tag{10.66}$$

Evaluating this as $t \to \infty$, long after the photon has passed the point of closest approach, we find that $\boldsymbol{r}(t)$ becomes increasingly aligned with $\boldsymbol{k}$, so that $\cos\Phi(t\to\infty) \to 1$ and

$$\boldsymbol{\alpha}(t\to\infty) = -\frac{4GM}{c^2}\frac{\boldsymbol{b}}{b^2}. \tag{10.67}$$

From Fig. 10.1 it is easy to see that the photon's total angle of deflection α is given by $\alpha \simeq \tan\alpha = |\boldsymbol{n}\cdot\boldsymbol{b}/b|/(\boldsymbol{n}\cdot\boldsymbol{k})$, in which $\boldsymbol{n} = \boldsymbol{k} + \boldsymbol{\alpha} + O(c^{-4})$ is evaluated at late times. This gives $\alpha = |\boldsymbol{\alpha}\cdot\boldsymbol{b}|/b = |\boldsymbol{\alpha}|$, and we find that

$$\alpha = \frac{4GM}{c^2 b}. \tag{10.68}$$

This is the famous *deflection angle* of a light ray passing near a spherical body, as first calculated by Einstein in 1915.

10.2.3 Measurement of light deflection

The preceding results indicate that light is indeed deflected by a massive body, but they do not yet offer a means to put the prediction to a test. The reason is that while the final direction of propagation $\boldsymbol{k} + \boldsymbol{\alpha}$ of a photon can easily be measured by a telescope, the initial direction $\boldsymbol{k}$ is unknown, and the extraction of $\boldsymbol{\alpha}$ from the observations is impossible. To perform a test it is necessary to manipulate the equations to produce a relationship between quantities that can be measured directly. The way to proceed is to involve two sources of light, the first a *reference source* and the second a *target source*, which represents the emitter of interest. The angle θ between the two light rays, when they are received simultaneously at a telescope or a radio interferometer, is a measurable quantity. It can be given a precise mathematical expression independent of any coordinate system, and it is this relationship that can be tested by observations.

In the following we assume that all measurements are carried out by an observer at rest in the post-Newtonian spacetime. In reality, of course, observations would be performed by an astronomer on the moving Earth, and her motion would have to be incorporated in the analysis of the measurement; we ignore such effects here. The observer has a spacetime velocity vector u^α, and her reference frame is spanned by the spatial unit vectors $e^\alpha_{(j)}$ first introduced in Sec. 4.1.5; the label (j) runs from (1) to (3). The vectors are mutually orthogonal, they are also orthogonal to u^α, and they satisfy the identity

$$P^{\alpha\beta} := g^{\alpha\beta} + u^\alpha u^\beta/c^2 = e^\alpha_{(1)}e^\beta_{(1)} + e^\alpha_{(2)}e^\beta_{(2)} + e^\alpha_{(3)}e^\beta_{(3)}. \tag{10.69}$$

The tensor $P^\alpha_{\ \beta}$ projects any vector A^β in the directions orthogonal to u^α, which define the observer's reference frame; it was first introduced in Sec. 4.1.5 in the context of flat spacetime, and this is its incarnation in a curved spacetime with metric $g_{\alpha\beta}$.

Let the target source emit a photon with velocity vector $v^\alpha = (c, \boldsymbol{v})$, as given by Eq. (10.51), and let the reference source emit another photon with velocity vector $v'^\alpha = (c, \boldsymbol{v}')$. With no loss of generality we orient the vectorial basis $e^\alpha_{(j)}$ in such a way that both photons move in the x-y plane of the observer's reference frame. As shown back in Sec. 4.1.6 (see Eq. (4.28)), the angle ϕ made by the target photon with respect to the observer's x-axis is determined by

$$\cos\phi = c\frac{v_\alpha e^\alpha_{(1)}}{-v_\beta u^\beta}, \qquad \sin\phi = c\frac{v_\alpha e^\alpha_{(2)}}{-v_\beta u^\beta}. \tag{10.70}$$

Similarly, the angle ϕ' made by the reference photon is given by

$$\cos\phi' = c\frac{v'_\alpha e^\alpha_{(1)}}{-v'_\beta u^\beta}, \qquad \sin\phi' = c\frac{v'_\alpha e^\alpha_{(2)}}{-v'_\beta u^\beta}. \tag{10.71}$$

The angle between the two photons is $\theta := \phi - \phi'$, and this is given by

$$\cos\theta = \cos\phi\cos\phi' + \sin\phi\sin\phi' = \frac{v_\alpha v'_\beta\big(e^\alpha_{(1)}e^\beta_{(1)} + e^\alpha_{(2)}e^\beta_{(2)}\big)}{v_\mu v'_\nu(u^\mu u^\nu/c^2)}. \tag{10.72}$$

Because $v_\alpha e^\alpha_{(3)} = 0 = v'_\alpha e^\alpha_{(3)}$, the quantity within brackets can be replaced by the projector of Eq. (10.69), and we arrive at

$$\cos\theta = 1 + \frac{g_{\alpha\beta}v^\alpha v'^\beta}{v^\mu v'^\nu(u_\mu u_\nu/c^2)}, \tag{10.73}$$

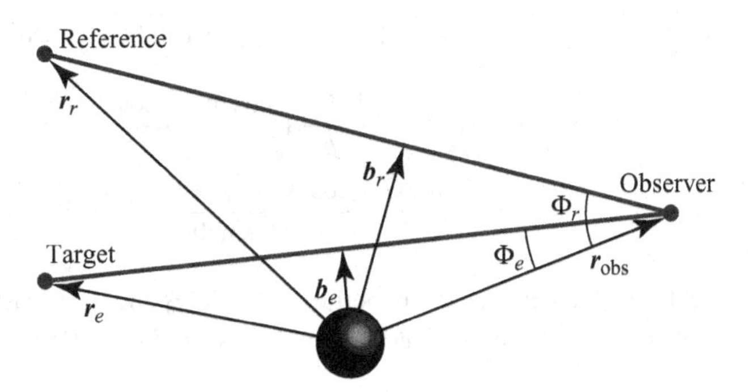

Fig. 10.2 Geometry of light deflection measurements.

our final expression for the relative angle. Note that the right-hand side of Eq. (10.73) is a spacetime invariant that can be evaluated in any coordinate system; this guarantees that θ is a precisely defined, observable quantity.

Our next task is to evaluate Eq. (10.73) for our situation, which involves the post-Newtonian spacetime, the observer at rest, and the target and reference photons. The relevant components of the metric tensor are $g_{00} = -(1 - 2U/c^2) + O(c^{-4})$, $g_{jk} = (1 + 2U/c^2)\delta_{jk} + O(c^{-4})$; the only non-vanishing component of the observer's velocity vector is $u^0 = c(1 + U/c^2) + O(c^{-3})$, and the photon's velocities are given by $\boldsymbol{v} = c(1 - 2U/c^2)\boldsymbol{k} + c\boldsymbol{\alpha} + O(c^{-3})$ and $\boldsymbol{v}' = c(1 - 2U/c^2)\boldsymbol{k}' + c\boldsymbol{\alpha}' + O(c^{-3})$. Inserting all this within Eq. (10.73) gives

$$\cos\theta = \boldsymbol{k} \cdot \boldsymbol{k}' + \boldsymbol{k}' \cdot \boldsymbol{\alpha} + \boldsymbol{k} \cdot \boldsymbol{\alpha}' + O(c^{-4}). \tag{10.74}$$

The relative angle is thus given a simple expression in terms of $\boldsymbol{k}$ and $\boldsymbol{\alpha}$, the initial direction and deflection of the target photon, as well as $\boldsymbol{k}'$ and $\boldsymbol{\alpha}'$, the initial direction and deflection of the reference photon.

We next insert Eq. (10.66) within Eq. (10.74) and obtain

$$\cos\theta = \cos\theta_0 - \frac{4MG}{c^2}\frac{\boldsymbol{k}' \cdot \boldsymbol{b}_e}{b_e^2}\frac{1 + \cos\Phi_e}{2} - \frac{4MG}{c^2}\frac{\boldsymbol{k} \cdot \boldsymbol{b}_r}{b_r^2}\frac{1 + \cos\Phi_r}{2}, \tag{10.75}$$

in which $\cos\theta_0 := \boldsymbol{k} \cdot \boldsymbol{k}'$, so that θ_0 is the relative angle in the absence of deflection, $\cos\Phi_e := \boldsymbol{r}_{\rm obs} \cdot \boldsymbol{k}/r_{\rm obs}$, so that Φ_e is the angle between the observer's position $\boldsymbol{r}_{\rm obs}$ and the direction $\boldsymbol{k}$ of the target photon, and $\cos\Phi_r := \boldsymbol{r}_{\rm obs} \cdot \boldsymbol{k}'/r_{\rm obs}$, so that Φ_r is the angle between the observer's position and the direction of the reference photon (see Fig. 10.2); we also have $\boldsymbol{b}_e := \boldsymbol{r}_e - (\boldsymbol{r}_e \cdot \boldsymbol{k})\boldsymbol{k}$, where $\boldsymbol{r}_e$ is the position of the target source (the emitter), and $\boldsymbol{b}_r := \boldsymbol{r}_r - (\boldsymbol{r}_r \cdot \boldsymbol{k}')\boldsymbol{k}'$, where $\boldsymbol{r}_r$ is the position of the reference source.

Our expression for $\cos\theta$ can be cleaned up if we express $\boldsymbol{k}' \cdot \boldsymbol{b}_e$ and $\boldsymbol{k} \cdot \boldsymbol{b}_r$ in terms of the angles Φ_e, Φ_r, and θ_0. To accomplish this we note that since the vector $\boldsymbol{r}_{\rm obs} - \boldsymbol{r}_e$ is directed along $\boldsymbol{k}$ (apart from the small correction from α), we can write $\boldsymbol{b}_e = \boldsymbol{r}_{\rm obs} - (\boldsymbol{r}_{\rm obs} \cdot \boldsymbol{k})\boldsymbol{k}$; similarly we have that $\boldsymbol{b}_r = \boldsymbol{r}_{\rm obs} - (\boldsymbol{r}_{\rm obs} \cdot \boldsymbol{k}')\boldsymbol{k}'$. These expressions imply $\boldsymbol{k}' \cdot \boldsymbol{b}_e = r_{\rm obs}(\cos\Phi_r - \cos\Phi_e\cos\theta_0)$ and $\boldsymbol{k} \cdot \boldsymbol{b}_r = r_{\rm obs}(\cos\Phi_e - \cos\Phi_r\cos\theta_0)$. Noting in addition that $\sin\Phi_e = b_e/r_{\rm obs}$ and $\sin\Phi_r = b_r/r_{\rm obs}$, as can be gleaned from Fig. 10.2, we arrive at

$$\cos\theta = \cos\theta_0 - \frac{4MG}{c^2 b_e}\left(\frac{\cos\Phi_r - \cos\Phi_e\cos\theta_0}{\sin\Phi_e}\right)\frac{1+\cos\Phi_e}{2} - \frac{4MG}{c^2 b_r}\left(\frac{\cos\Phi_e - \cos\Phi_r\cos\theta_0}{\sin\Phi_r}\right)\frac{1+\cos\Phi_r}{2}. \tag{10.76}$$

The difference between the unperturbed angle θ_0 and the measured angle θ is denoted $\delta\theta$, and since $\cos\theta = \cos(\theta_0 + \delta\theta) \simeq \cos\theta_0 - \sin\theta_0\delta\theta$, we find that

$$\delta\theta = \frac{4MG}{c^2 b_e}\left(\frac{\cos\Phi_r - \cos\Phi_e\cos\theta_0}{\sin\Phi_e\sin\theta_0}\right)\frac{1+\cos\Phi_e}{2} + \frac{4MG}{c^2 b_r}\left(\frac{\cos\Phi_e - \cos\Phi_r\cos\theta_0}{\sin\Phi_r\sin\theta_0}\right)\frac{1+\cos\Phi_r}{2}. \tag{10.77}$$

We recall that θ_0 is the angle subtended by the target and reference sources, that Φ_e is the angle subtended by the target source and the deflecting body, and that Φ_r is the angle subtended by the reference source and the deflecting body. In our discussion so far, these angles have been defined by the unperturbed, straight motion of the light rays in flat spacetime, as depicted in Fig. 10.2. But since the angles occur within a post-Newtonian expression, and the difference between the perturbed and unperturbed angles is itself of post-Newtonian order, we can actually interpret θ_0, Φ_e, and Φ_r in Eq. (10.77) as if they were the *observed angles* on the sky; the difference manifests itself at second post-Newtonian order only, and is negligible.

Box 10.2 **Spherical trigonometry**

Angular measurements in astronomy are performed on the celestial sphere, a fictitious sphere of large radius centered at the position of the observer, on which all astronomical bodies are imagined to be situated. Relationships between angles are then clarified with the rules of spherical trigonometry.

For the angles shown in Fig. 10.3, we have the relations

$$\cos\theta_0 = \cos\Phi_r\cos\Phi_e + \sin\Phi_r\sin\Phi_e\cos\chi,$$
$$\cos\Phi_r = \cos\theta_0\cos\Phi_e + \sin\theta_0\sin\Phi_e\cos B,$$
$$\cos\Phi_e = \cos\theta_0\cos\Phi_r + \sin\theta_0\sin\Phi_r\cos A,$$

and

$$\frac{\sin\Phi_e}{\sin A} = \frac{\sin\Phi_r}{\sin B} = \frac{\sin\theta_0}{\sin\chi}.$$

With these we find that Eq. (10.77) simplifies to

$$\delta\theta = \frac{4MG}{c^2 b_e}\cos B\,\frac{1+\cos\Phi_e}{2} + \frac{4MG}{c^2 b_r}\cos A\,\frac{1+\cos\Phi_r}{2}.$$

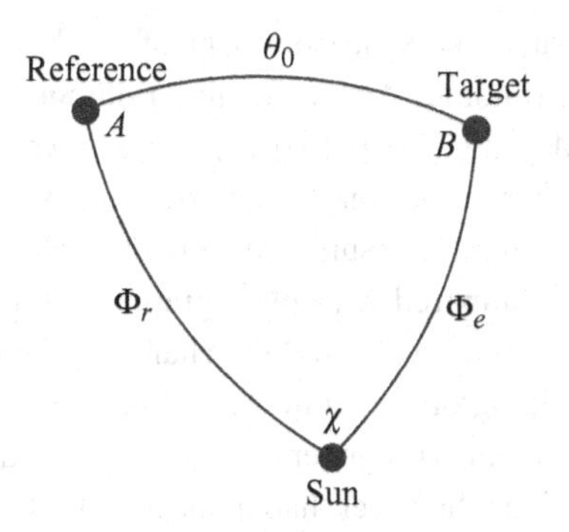

Fig. 10.3 Angles of sources on the celestial sphere as seen from the Earth. When the Sun is behind the Earth, the point labelled "Sun" becomes the extension of the Sun–Earth line into the sky, and the angles Φ_r and Φ_e become $\pi - \Phi_r$ and $\pi - \Phi_e$, respectively.

This equation states that the change in angular separation between the target and reference sources is just the sum of the apparent displacement of each image away from the deflecting body, projected along the line joining the reference and target sources.

Equation (10.77) is what we need to put the prediction of light deflection to a test, except for the fact that while we can always measure θ for the target and reference sources, we still do not have access to θ_0, its unperturbed value. To get around this, all measurements of the light deflection are carried out *differentially*. In this method, the angle between the target and reference stars is measured at different times, first when the Sun is nowhere in the vicinity of the stars, and again when the light from the target star passes near the Sun. The first measurement yields the unperturbed angle, which does not change with time unless the stars have significant proper motion. The second measurement yields the deflected angle θ, and a measurement of $\delta\theta = \theta - \theta_0$ can finally be compared with Eq. (10.77).

Equation (10.77) simplifies a little when the target source is very close to the Sun as seen from Earth, so that $\Phi_e \ll 1$. An expansion of $\cos\theta_0 = \cos\Phi_r \cos\Phi_e + \sin\Phi_r \sin\Phi_e \cos\chi$ (see Box 10.2) in powers of Φ_e reveals that

$$\theta_0 = \Phi_r - \cos\chi\, \Phi_e + \frac{\cos\Phi_r \sin^2\chi}{2\sin\Phi_r}\, \Phi_e^2 + O(\Phi_e^3), \tag{10.78}$$

and substitution within Eq. (10.77) produces

$$\delta\theta = \frac{4GM}{c^3 b_e}\left[-\cos\chi + \frac{1 + (1 + 2\sin^2\chi)\cos\theta_0}{2\sin\theta_0}\, \Phi_e + O(\Phi_e^2) \right], \tag{10.79}$$

in which χ is the angle subtended by the reference and target sources as seen from the Sun (as shown in Fig. 10.3). This expression reveals clearly how the angular separation between the target and reference sources changes with time as the Sun moves across the sky, causing χ and Φ_e to vary with time. Inserting $b_e = r_{\rm obs} \sin\Phi_e = r_{\rm obs}\Phi_e + O(\Phi_e^3)$ reveals also that the theoretical prediction can be expressed entirely in terms of observable quantities.

The first successful measurement of the bending of light by the Sun was carried out by British astronomer Arthur Stanley Eddington and his colleagues during the total solar eclipse of May 29, 1919. Two expeditions were sent out to measure the eclipse, one to Brazil, the other to the small island of Principe, off the west coast of Africa. In both cases it was a differential measurement: Photographs of the stars near the Sun taken during the eclipse were compared with photographs of the same stars taken at night from the same locations later in the year, and the changes in angles between pairs of stars were carefully measured. Eddington's announcement in November 1919 that the bending measurements were in agreement with general relativity helped make Einstein an international celebrity. The observations, however, had an accuracy of approximately 30 percent, and succeeding eclipse measurements were not much better. The results were scattered between one half and twice the Einstein prediction, but in spite of such limited success, Einstein was declared victorious.

The subsequent development of long-baseline radio interferometry greatly improved the measurement of the light deflection. These techniques now have the capability of measuring angular separations and changes in angles to accuracies of tens of micro-arcseconds. Early measurements took advantage of the fact that a few quasistellar radio sources – quasars – pass very close to the Sun as seen from the Earth. As the Earth moves in its orbit, changing the lines of sight of the quasars relative to the Sun, the angular separation $\delta\theta$ between pairs of quasars varies. The time evolution of χ and Φ_e in Eq. (10.79) is determined using an accurate ephemeris for the Earth and initial directions for the quasars, and the resulting prediction for $\delta\theta$ as a function of time is compared with the measured values. A number of measurements of this kind over the period 1969–1975 yielded results in agreement with general relativity to a few parts in 10^3. In recent years, transcontinental and intercontinental VLBI observations of quasars and radio galaxies have been made primarily to monitor the Earth's rotation and to establish a highly accurate reference frame for astronomy and navigation. These measurements are sensitive to the deflection of light over almost the entire celestial sphere. For example, the deflection of a ray approaching the Earth from a direction 90° away from the Sun is 4 milli-arcseconds, easily measurable by modern VLBI techniques. A 2004 analysis of nearly 2 million VLBI observations of 541 radio sources, made by 87 VLBI sites over a 20-year period, verified Einstein's prediction to a few parts in 10^4. Analysis of observations made by the Hipparcos optical astrometry satellite yielded a test at the level of 0.3 percent, and a future orbiting observatory named GAIA will have the capability of testing the deflection to parts per million.

Box 10.3 The "Newtonian" deflection of light

At the time of Eddington's attempt to measure the deflection of light, people envisioned three possible outcomes for the experiment: no deflection, the Einsteinian deflection, or one half of Einstein's prediction, commonly called the "Newtonian" deflection. The Newtonian deflection can be derived in a variety of ways. One way is to assume that light behaves as a particle, to recall that the trajectory of a particle is independent of its mass (weak equivalence principle), and to calculate the deflection of its trajectory in the limit in which the particle's speed approaches the speed of light. Such an approach would have made sense in Newton's day, when light was really viewed as a "corpuscle," and indeed, Newton himself speculated on the possible effect of gravity on light.

The English physicist Henry Cavendish may have been the first person to calculate the bending explicitly, possibly as early as 1784, although evidence for this was not discovered until around 1914, during an effort to compile and publish his entire body of work – publication never being high on Cavendish's list of priorities. In fact, all that was found was a scrap of paper in Cavendish's handwriting stating that he had done the calculation, and giving the answer.

Independently of Cavendish, the Bavarian astronomer Johann von Soldner did publish in 1803 a detailed calculation of the Newtonian bending in a German astronomical journal. Strangely, von Soldner's calculation was largely forgotten until it was resurrected in 1921 by Phillip Lenard as part of a campaign to discredit the "Jewish" relativity of Einstein by publicizing the earlier work of the "Aryan" von Soldner. Apparently, Lenard was not deterred by the fact that the 1919 observations actually favored general relativity over the Newtonian deflection.

Unaware of the earlier work, Einstein himself derived the "Newtonian" deflection in 1911. He argued, as we have back in Sec. 5.1.3, that gravity requires replacement of the Minkowski metric of flat spacetime by the Newtonian metric of Eq. (5.12),

$$ds^2 = -(1 - 2U/c^2)\,d(ct)^2 + dx^2 + dy^2 + dz^2.$$

Geodesic motion for a test particle in this spacetime reproduces Newtonian gravity, and geodesic motion for a photon gives the Newtonian deflection. Another derivation using only the equivalence principle imagines a sequence of freely falling frames through which a light ray passes as it travels near a gravitating body. Each frame is momentarily at rest at the moment the light ray enters it. Although the path of the ray is a straight line within each frame, the frame picks up a downward velocity during the ray's traversal, because of the body's gravitational attraction. When the adjacent frame receives the light ray, it is deflected toward the body because the downward motion of the previous frame induces aberration on the received ray. By adding up all the tiny aberrations over a sequence of frames, one arrives at the Newtonian deflection.

But the full theory of general relativity doubles the deflection, because the spatial part of the metric now comes with the multiplying factor $(1 + 2U/c^2)$. This represents spatial curvature, which could not be taken into account either by Newtonian gravity or by the principle of equivalence. So the total deflection can be viewed as a sum of a Newtonian deflection relative to locally straight lines, plus the bending of locally straight lines relative to straight lines at infinity; each effect contributes exactly half the total deflection.

10.2.4 Gravitational lenses

The deflection of light has become a cornerstone of the empirical edifice that supports general relativity. But in 1979 the phenomenon became much more than that. That year, astronomers Dennis Walsh, Robert Carswell and Ray Weymann discovered the "double quasar" Q0957+561, which consisted of two quasar images about 6 arcseconds apart, with almost the same redshift ($z = 1.41$) and very similar spectra. Given that quasars are thought to be among the most distant objects in the universe, the probability of finding two so close together was low. It was soon realized that there was in fact just one quasar, but that intervening matter in the form of a galaxy or a cluster of galaxies was bending the light from the quasar and producing two separate images.

Since then, over 60 lensed quasars have been discovered. But more importantly, gravitational lensing has become a major tool in efforts to map the distribution of mass around galaxies and clusters, and in searches for dark matter, dark energy, compact objects, and extrasolar planets. Many subtopics of gravitational lensing have been developed to cover different astronomical realms: microlensing for the search for dim compact objects and extrasolar planets, the use of luminous arcs to map the distribution of mass and dark matter, and weak lensing to measure the properties of dark energy. Lensing has to be taken into account in interpreting certain aspects of the cosmic microwave background radiation, and in extracting information from gravitational waves emitted by sources at cosmological distances. These topics are beyond the scope of this book, but we can extend our discussion of the deflection of light to provide many of the basic concepts and results.

Deflection vector

We consider light rays emitted by a remote source like a quasar to be observed by an astronomer on Earth. The rays pass through a distribution of matter – the gravitational lens – on their way to the observer, and undergo a deflection described by the vector $\boldsymbol{\alpha}$ of Eq. (10.56). In this situation the source is almost directly behind the lens, so that $\boldsymbol{s}_e \cdot \boldsymbol{k}/s_e \simeq -1$, the observer is almost directly in front of the lens, so that $\boldsymbol{s}_{\rm obs} \cdot \boldsymbol{k}/s_{\rm obs} \simeq 1$, and the deflection vector can be simplified to

$$\boldsymbol{\alpha} = -\frac{4G}{c^2} \int \rho(\boldsymbol{x}') \frac{\boldsymbol{b}}{b^2}\, d^3x', \tag{10.80}$$

in which $\boldsymbol{b} := \boldsymbol{s}_e - (\boldsymbol{s}_e \cdot \boldsymbol{k})\boldsymbol{k}$, where $\boldsymbol{s}_e := \boldsymbol{r}_e - \boldsymbol{x}'$; here $\boldsymbol{r}_e$ is the position of the source relative to the center-of-mass of the matter distribution, and $\boldsymbol{k}$ is the ray's initial direction; its final direction as measured by the observer is $\boldsymbol{k} + \boldsymbol{\alpha}$. We have assumed that the distribution of matter is time independent, so that ρ depends on the spatial variables $\boldsymbol{x}'$ only.

To express the deflection vector in a more convenient form we introduce the vectors

$$\boldsymbol{\xi} := \boldsymbol{r}_e - (\boldsymbol{r}_e \cdot \boldsymbol{k})\boldsymbol{k}, \qquad \boldsymbol{\xi}' := \boldsymbol{x}' - (\boldsymbol{x}' \cdot \boldsymbol{k})\boldsymbol{k}, \tag{10.81}$$

so that $\boldsymbol{\xi}$ is the projection of the source's position $\boldsymbol{r}_e$ in the *lens plane*, defined as the plane perpendicular to the direction of propagation $\boldsymbol{k}$ ($\boldsymbol{\xi}$ is also the photon's point of closest approach to the center-of-mass of the matter distribution), and similarly, $\boldsymbol{\xi}'$ is the projection of $\boldsymbol{x}'$ in the lens plane. In terms of these vectors we have that $\boldsymbol{b} = \boldsymbol{\xi} - \boldsymbol{\xi}'$, and the deflection vector becomes

$$\boldsymbol{\alpha} = -\frac{4G}{c^2} \int \rho(\boldsymbol{x}') \frac{\boldsymbol{\xi} - \boldsymbol{\xi}'}{|\boldsymbol{\xi} - \boldsymbol{\xi}'|^2}\, d^3x'. \tag{10.82}$$

To simplify this further we implement a coordinate transformation from the old system $\boldsymbol{x}'$ to a new system $(\boldsymbol{\xi}', \ell')$, in which $\ell' := \boldsymbol{x}' \cdot \boldsymbol{k}$ is the distance along the line of sight, perpendicular to the lens plane. Integration over $d\ell'$ involves $\rho(\boldsymbol{\xi}', \ell')$ only, and it gives rise to

$$\Sigma(\boldsymbol{\xi}') := \int \rho(\boldsymbol{\xi}', \ell')\, d\ell', \tag{10.83}$$

the projected, two-dimensional mass density (per unit area) of the matter distribution. Our final expression for the deflection vector is

$$\boldsymbol{\alpha}(\boldsymbol{\xi}) = -\frac{4G}{c^2} \int \Sigma(\boldsymbol{\xi}') \frac{\boldsymbol{\xi} - \boldsymbol{\xi}'}{|\boldsymbol{\xi} - \boldsymbol{\xi}'|^2}\, d^2\xi'; \tag{10.84}$$

it reveals that $\boldsymbol{\alpha}$ is a function of $\boldsymbol{\xi}$ only, and that it lies within the lens plane. Alternative expressions for the deflection vector are formulated in Exercise 10.5.

The deflection vector acquires a particularly simple form when the matter distribution is axially symmetric, so that the surface density Σ depends only on the magnitude ξ' of the vector $\boldsymbol{\xi}'$. In this case we can adopt a system of polar coordinates (ξ', ϕ') in the lens plane, and carry out the integration over the angle ϕ'. The result is

$$\boldsymbol{\alpha}(\boldsymbol{\xi}) = -\frac{4G}{c^2} \frac{m(\xi)}{\xi^2} \boldsymbol{\xi}, \tag{10.85}$$

in which

$$m(\xi) := 2\pi \int_0^\xi \Sigma(\xi')\xi'\, d\xi' \tag{10.86}$$

is the mass inside a circle of radius ξ in the lens plane. In this case the deflection vector necessarily points in the direction opposite to $\boldsymbol{\xi}$.

Lens equation

We now consider the situation depicted in Fig. 10.4. We have a source at a distance D_S from the observer, and a lens between the source and observer, at a distance D_L from the observer; $D_{LS} := D_S - D_L$ is the distance between the lens and the source. The figure displays the lens plane, which is perpendicular to the initial direction of propagation of the light rays, and a source plane that contains the source, which is parallel to the lens plane. The figure also shows the optical axis, which passes through the observer and the center-of-mass of the lens; the optical axis is not necessarily perpendicular to the planes. The vector $\boldsymbol{\eta}$ gives the position of the source in the source plane, relative to the optical axis, and $\boldsymbol{\zeta}$ is the position of the image relative to the source; as usual $\boldsymbol{\xi}$ is the point of closest approach in the lens plane. We introduce the vector $\boldsymbol{\beta} := \boldsymbol{\eta}/D_S$, whose magnitude β is the angle between the source and the optical axis. We introduce also the vector $\boldsymbol{\theta} := (\boldsymbol{\eta} + \boldsymbol{\zeta})/D_S$, whose magnitude θ is the angle between the image and the optical axis.

Our main goal is to determine $\boldsymbol{\theta}$ for a given $\boldsymbol{\beta}$, and the key step is to recognize from Fig. 10.4 that the deflection vector $\boldsymbol{\zeta}$ can be expressed as $-\boldsymbol{\alpha} D_{LS}$, where the negative sign accounts for the fact that $\boldsymbol{\alpha}$ points in the direction opposite to $\boldsymbol{\xi}$. The definition of $\boldsymbol{\theta}$ then

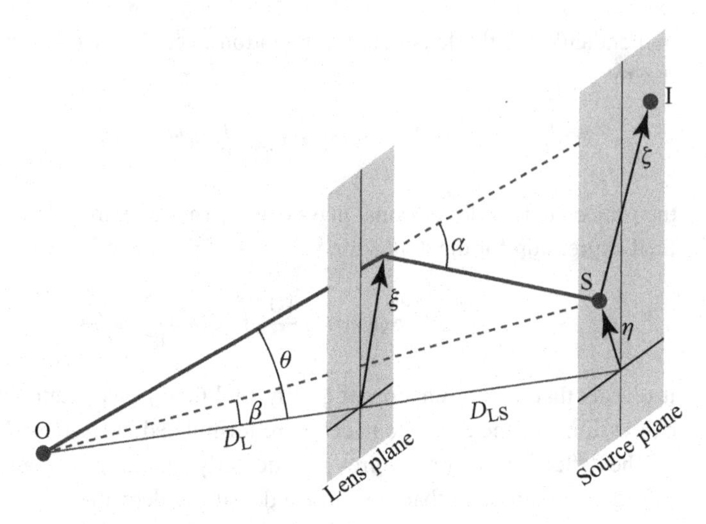

Fig. 10.4 Geometry of a gravitational lens. The observer is labeled O, the source S, and the image I.

implies $\boldsymbol{\theta} = (D_{\rm S}\boldsymbol{\beta} - D_{\rm LS}\boldsymbol{\alpha})/D_{\rm S}$, or

$$\boldsymbol{\theta} + \frac{D_{\rm LS}}{D_{\rm S}}\boldsymbol{\alpha} = \boldsymbol{\beta}. \tag{10.87}$$

This *lens equation* can be solved for $\boldsymbol{\theta}$ once we insert the deflection vector of Eq. (10.84), in which we substitute $\boldsymbol{\xi} = D_{\rm L}\boldsymbol{\theta}$, another relation that can be inferred from Fig. 10.4. Because the deflection vector is a non-linear function of its argument, the lens equation typically admits more than one solution for a given source position $\boldsymbol{\beta}$. To obtain these solutions it is necessary to know $\boldsymbol{\alpha}(\boldsymbol{\xi})$, and this requires an evaluation of the integral in Eq. (10.84). This can be accomplished when the density profile of the lens is known, but this information is usually not available in an astronomical context. The power of gravitational lensing in astronomy resides in the fact that the lens equation can be turned around: the measured positions of the multiple images of a given source can be used, through the deflection vector $\boldsymbol{\alpha}$, to deduce some features of the mass distribution.

Schwarzschild lens

The simplest instance of a gravitational lens is one created by a single, spherically-symmetric body of mass M. In this case the deflection vector is given by Eq. (10.67), or is obtained by letting $m(\xi) = M$ in Eq. (10.85). We have

$$\boldsymbol{\alpha}(\boldsymbol{\xi}) = -\frac{4GM}{c^2}\frac{\boldsymbol{\xi}}{\xi^2}, \tag{10.88}$$

and substitution within Eq. (10.87) gives rise to

$$\theta - \frac{\theta_{\rm E}^2}{\theta} = \beta, \tag{10.89}$$

in which

$$\theta_E^2 := \frac{4GM}{c^2}\frac{D_{LS}}{D_S D_L}. \tag{10.90}$$

In this case the lens equation reduces to a scalar equation, because by virtue of the axial symmetry of the lens (which follows from its spherical symmetry), all vectors point in the same direction. The parameter θ_E is known as the *Einstein angle*, and the corresponding length scale

$$\xi_E := D_L\theta_E = \sqrt{\frac{4GM}{c^2}\frac{D_{LS}D_L}{D_S}} \tag{10.91}$$

is the *Einstein radius*. The Einstein angle gives the characteristic scale of the deflection. For lenses of galactic scales with $M \sim 10^{12} M_\odot$, $\theta_E \simeq 1.8$ as, while for solar-mass lenses, $\theta_E \simeq 0.5$ mas. The lens equation is consistent whenever $\xi \gg 4GM/c^2$, so that higher post-Newtonian corrections are not required, and this condition implies that θ must be much larger than $(D_S/D_{LS})\theta_E^2$.

The solutions to the lens equation are

$$\theta_\pm = \frac{1}{2}\left(\beta \pm \sqrt{\beta^2 + 4\theta_E^2}\right), \tag{10.92}$$

and we see that the lens produces two images of the source. For large angles $\beta \gg \theta_E$ the solutions reduce to $\theta_+ = \beta + \theta_E^2/\beta + \cdots$ and $\theta_- = -\theta_E^2/\beta + \cdots$, which indicates that one image is displaced beyond the source, while the second image occurs below the optical axis and is closer to the lens than is the source itself. When β increases beyond unity, the second solution becomes smaller than θ_E^2 and enters a regime in which the lens equation is no longer consistent, because for this solution $\xi = D_L\theta = -D_L\theta_E^2/\beta \sim (4GM/c^2)(D_{LS}/D_S)/\beta$, which violates the condition $\xi \gg 4GM/c^2$. For small angles $\beta \ll \theta_E$ the solutions reduce to $\theta_+ = \theta_E + \frac{1}{2}\beta + \cdots$ and $\theta_- = -\theta_E + \frac{1}{2}\beta + \cdots$, which indicates that the images are at an angle approximately equal to θ_E above and below the optical axis. When $\beta = 0$, that is, when the source is directly behind the lens, the deflection angle is precisely equal to $\pm\theta_E$, and the axial symmetry of the situation forces the image to take the shape of a ring – known as an *Einstein ring* – around the optical axis.

When the source has a non-zero angular size, the lens continues to displace its images, but there is also a distortion of its shape. Points on opposite sides of the source perpendicular to the optical axis are displaced by an angle θ, and are therefore stretched by a corresponding factor, while points on either side parallel to the optical axis are stretched only by the difference in θ. When the source is circular, and in the limit where its angular size $\delta\beta$ is small compared to β, the image is an ellipse with an axis ratio $a/b = \beta/\sqrt{\beta^2 + 4\theta_E^2}$, with a denoting the semi-axis in the direction of the lens, and b the semi-axis in the perpendicular direction. For sources of larger angular size, the image can actually be distorted into an arc with a convex side; this occurs when

$$\delta\beta < \beta \quad \text{and} \quad \theta_E \geq \frac{\beta - \delta\beta}{2\delta\beta}\sqrt{\beta^2 - \delta\beta^2}. \tag{10.93}$$

The orientation and shapes of such luminous arcs have been used to deduce the mass distribution of the galaxies or clusters that act as lenses, a procedure sometimes dubbed *gravitational tomography*. Even when the lens produces elliptical distortions that are too small to be measured individually, there is a systematic effect, averaged over large collections of images, that is sensitive to the evolution of the universe over an epoch when dark energy began to be important; this is the realm of weak gravitational lensing.

Because the number of photons emitted per unit area and unit time is constant for a steady source, the brightness of the image is proportional to its observed area, which for small angles is proportional to $\int \theta \, d\theta d\phi$. And because the intrinsic area of the source is proportional to $\int \beta \, d\beta d\phi$, we find that each image is magnified by a factor

$$\mu_\pm = \frac{\theta_\pm d\theta_\pm}{\beta d\beta} = \pm\frac{1}{4}\left(\frac{\beta}{\sqrt{\beta^2 + 4\theta_E^2}} + \frac{\sqrt{\beta^2 + 4\theta_E^2}}{\beta} \pm 2\right). \tag{10.94}$$

The term in parentheses is always greater than zero, and we see that μ_+ is positive, while μ_- is negative, indicating that the image is inverted relative to the source. In addition, we see that μ_+ is always greater than unity, but that $|\mu_-|$ is smaller than unity, indicating that the second image is actually demagnified by the lens.

In microlensing situations the images are too close together to be resolved individually by the observer, and in such cases the total magnification is measured by

$$|\mu_+| + |\mu_-| = \frac{1}{2}\left(\frac{\beta}{\sqrt{\beta^2 + 4\theta_E^2}} + \frac{\sqrt{\beta^2 + 4\theta_E^2}}{\beta}\right), \tag{10.95}$$

which is always greater than unity. The technique of monitoring the variable brightness of lensing images was used in a series of experiments to search for "massive compact halo objects" (MACHOs) in our galaxy. If the galaxy contained a population of dark objects (black holes, neutron stars, brown dwarfs, or other exotic objects) with masses comparable to $M_\odot$, then the brightness of a star transiting behind such an object should behave in a way consistent with Eq. (10.95). This effect can be distinguished from the star's own variability, or from the absorption of starlight by intervening matter, because these tend to depend on wavelength, while the lensing is independent of wavelength. Searches for dark objects passing in front of the dense field of stars in the Large Magellanic Cloud and in the galactic center were carried out between 1993 and 2007, placing a stringent upper limit on the amount of halo mass that could be made up of such objects. This strengthened the conclusion that the vast majority of the halo mass must be made of non-baryonic dark matter.

In 2003, an extrasolar planetary system was discovered by microlensing. The combined lensing of a distant source by a Jupiter-scale companion and its host star was measured and could be deconvolved to determine the mass ratio and the approximate distance between the planet and the star. Additional systems were discovered subsequently, and gravitational lensing is proving to be a key tool in the search for exoplanets.

10.2.5 Shapiro time delay

Our discussion of the motion of light in a gravitational field has so far emphasized the transverse aspects of the motion – the deflection of a light ray with respect to its unperturbed, straight path. The longitudinal aspects – the changed relationship between position and time – are also important, and we conclude this section with an examination of the delay suffered by a photon as it travels across a gravitational potential well. We consider a light source situated at $\boldsymbol{r}_e$ relative to the center-of-mass of a spherical body of mass M, and an observer situated at $\boldsymbol{r}_{\rm obs}$. We imagine that the body is situated between the source and the observer, near the line of sight, and that the photon is at a distance b from the body when it reaches its point of closest approach. We wish to calculate the light-travel time between the source and observer.

The information is contained in Eq. (10.57),

$$\boldsymbol{r}(t) = \boldsymbol{r}_e + c\boldsymbol{k}(t - t_e) + \boldsymbol{k}\,\delta r_\parallel(t) + \delta\boldsymbol{r}_\perp(t) + O(c^{-4}), \tag{10.96}$$

in which we may insert Eq. (10.60) for the longitudinal displacement $\delta r_\parallel$, and Eq. (10.62) for the transverse displacement $\delta\boldsymbol{r}_\perp$. The equation is evaluated at $t = t_{\rm obs}$ so that $\boldsymbol{r}(t = t_{\rm obs}) = \boldsymbol{r}_{\rm obs}$, and the displacements must be specialized to the case at hand, for which the mass density can be expressed as $\rho(t, \boldsymbol{x}') = M\delta(\boldsymbol{x}')$. We do not need an explicit expression for $\delta\boldsymbol{r}_\perp$, but we find that the longitudinal displacement becomes

$$\delta r_\parallel = -\frac{2GM}{c^2}\ln\left[\frac{(r_{\rm obs} + \boldsymbol{r}_{\rm obs}\cdot\boldsymbol{k})(r_e - \boldsymbol{r}_e\cdot\boldsymbol{k})}{b^2}\right]. \tag{10.97}$$

The travel time $t_{\rm obs} - t_e$ is computed by squaring Eq. (10.96), which yields

$$|\boldsymbol{r}_{\rm obs} - \boldsymbol{r}_e|^2 = c^2(t_{\rm obs} - t_e)^2 + 2c(t_{\rm obs} - t_e)\delta r_\parallel + O(c^{-4}), \tag{10.98}$$

or

$$c(t_{\rm obs} - t_e) = |\boldsymbol{r}_{\rm obs} - \boldsymbol{r}_e| - \delta r_\parallel + O(c^{-4}). \tag{10.99}$$

Inserting our expression for $\delta r_\parallel$, we arrive at

$$t_{\rm obs} - t_e = \frac{1}{c}|\boldsymbol{r}_{\rm obs} - \boldsymbol{r}_e| + \frac{2GM}{c^3}\ln\left[\frac{(r_{\rm obs} + \boldsymbol{r}_{\rm obs}\cdot\boldsymbol{k})(r_e - \boldsymbol{r}_e\cdot\boldsymbol{k})}{b^2}\right] + O(c^{-5}). \tag{10.100}$$

The first term is obviously the time required to travel a distance $|\boldsymbol{r}_{\rm obs} - \boldsymbol{r}_e|$ in the absence of gravity, while the second term is the delay produced by the massive body. With the source assumed to be at a large distance behind the body, and the observer at a large distance in front of the body, we have that $\boldsymbol{r}_e\cdot\boldsymbol{k}/r_e \simeq -1$ and $\boldsymbol{r}_{\rm obs}\cdot\boldsymbol{k}/r_{\rm obs} \simeq 1$, and Eq. (10.100) simplifies to

$$t_{\rm obs} - t_e = \frac{1}{c}|\boldsymbol{r}_{\rm obs} - \boldsymbol{r}_e| + \frac{2GM}{c^3}\ln\frac{4r_{\rm obs}r_e}{b^2} + O(c^{-5}). \tag{10.101}$$

This will be our final expression for the light-travel time.

The situation examined thus far does not yet give rise to a means to measure the time delay, because the time of emission t_e is typically not known. A slight variation on the theme, however, gives us what we need. Imagine that the light source at $\boldsymbol{r}_e$ is replaced by a reflector, and that the observer at $\boldsymbol{r}_{\rm obs}$ sends a pulse of light to the reflector at a time t_0, to receive it back at a later time t_1. The light-travel time during the round trip is twice what was calculated previously, and we find that

$$t_1 - t_0 = \frac{2}{c}|\boldsymbol{r}_{\rm obs} - \boldsymbol{r}_e| + \frac{4GM}{c^3}\ln\frac{4r_{\rm obs}r_e}{b^2} + O(c^{-5}). \tag{10.102}$$

The last term is the famous *Shapiro time delay*. For solar-system situations it can be expressed as

$$\Delta t_{\rm Shapiro} = \frac{M}{M_\odot}\left\{240 - 20\ln\left[\left(\frac{b}{R_\odot}\right)^2\left(\frac{\rm AU^2}{r_{\rm obs}r_e}\right)\right]\right\}\ \mu{\rm s}, \tag{10.103}$$

in which $r_{\rm obs}$ and r_e are measured in astronomical units, the distance between the Earth and the Sun; the Shapiro time delay is measured in hundreds of microseconds.

While we are now closer to a measurement protocol, we are not there yet. For one thing, our expression for the time delay, Eq. (10.102), is given in terms of coordinate time and distances, and it should be converted to an expression that involves observable quantities only. For another, we do not have access to the unperturbed, Euclidean distance $|\boldsymbol{r}_{\rm obs} - \boldsymbol{r}_e|$ in order to separate out the relativistic effect. Just as in the case of the deflection of light reviewed in Sec. 10.2.3, it is essential to do a differential measurement of the variations in round-trip travel times during many repetitions of the experiment. Particularly important are "superior conjunction" configurations in which the reflector transits behind the Sun as viewed from Earth, leading to a strong modulation of the travel times by the logarithmic term in Eq. (10.102) as b changes with time. In order to do this accurately, however, one must take into account the variations in travel times that are due to the orbital motion of the reflector. This is done by accumulating radar-ranging data on the reflector when it is far from superior conjunction (so that the time-delay term is negligible) to determine its orbit, using the computed orbit to predict its coordinate trajectory $\boldsymbol{r}_e(t)$ near superior conjunction, and then combining this with the trajectory of the Earth $\boldsymbol{r}_\oplus(t)$ to determine the Newtonian round-trip time and the logarithmic term in Eq. (10.102). The prediction made on this basis can then be compared with the actual measurements obtained during superior conjunction.

It was radio astronomer Irwin I. Shapiro who first discovered, in 1964, that the time delay was a prediction of general relativity. (It was independently discovered by Jet Propulsion Laboratory scientists Duane Muhleman and Paul Reichley.) He called it the "fourth" test of general relativity, after the three classical tests – the periastron advance of Mercury, the light deflection, and the redshift of light as it climbs up a gravitational field. Shapiro and his colleagues carried out the first measurement of the time delay in 1967, by bouncing radar signals off the surface of Mercury. Later experiments involved radar echos from Venus and the tracking of the Mars exploration spacecraft, Mariners 6, 7 and 9, as well as the 1976 Viking landers and orbiters.

The most recent measurement of the Shapiro time delay exploited the Cassini spacecraft while it was on its way to Saturn. Several circumstances made this mission particularly favorable. One was the ability to do tracking measurements using both X-band (7175 MHz) and Ka-band (34316 MHz) radar, thereby significantly reducing the dispersive effects of the solar corona. Another was that the distance of closest approach of the radar signals to the

Sun was only 1.6 $R_\odot$ during Cassini's 2002 superior conjunction, when the spacecraft was at 8.43 AU from the Sun. Because the tracking involved Doppler measurements only, and not time delays, it was the rate of change of the Shapiro delay that was actually measured. The result was in agreement with general relativity to two parts in 10^5.

The Shapiro delay now figures in a wide range of astrophysical phenomena. In its one-way incarnation, it has been measured in a number of binary-pulsar systems, most notably in the double pulsar J0737-3039A/B, where the orbit is seen almost edge-on, and the pulsed radio signals from each pulsar pass close to the other pulsar once per orbit. It is also relevant in analyses of the spectra and time variations of X-ray emissions from accretion disks around black holes; in this case the modeling must be performed in full general relativity, because the post-Newtonian approximation breaks down near black holes.

10.3 Post-Newtonian gravity in timekeeping and navigation

In our development of the foundations of special and general relativity, we have emphasized that only physically measurable quantities are relevant, and that the selection of coordinates is completely arbitrary, devoid of physical meaning. On the other hand, to calculate such things as equations of motion or the Shapiro time delay, it is essential to have a suitable coordinate system. No one should consider doing such calculations using a proper time variable or using physically-measured lengths, because this would introduce unnecessary complications into the calculations. In this section we explore the relationship between the coordinates employed to chart a spacetime and the physical measurements carried out by observers moving in the spacetime.

10.3.1 "A brief history of time"

[The title of this section is obviously borrowed from Stephen Hawking's bestseller. We could not resist.]

Most of the calculations presented in this book employ a coordinate system x^α defined so that, far from any gravitating system, the coordinate $x^0/c := t$ represents proper time as measured by an inertial observer at rest with respect to the coordinate system, and x^j represent spatial coordinates whose scale is measured by rigid rulers also at rest. But we rarely have access to those ideal observers, and we make our measurements using instruments that reside in the gravitational field of the Earth, which is itself moving in the gravitational environment of the solar system.

Most measurements involve time. We determine the orbit of a planet by measuring the time at which it passes in front of a star or transits across the Sun. We determine the orbit of a spacecraft by measuring either the round-trip travel time of a radar signal (radar ranging) or the change in frequency of the return signal compared to the emitted signal (Doppler tracking). We determine the position of a quasar by measuring the difference in arrival time of a given radio wave front at two radio telescopes separated by a known baseline. We navigate around the countryside by measuring the times of arrival of coded signals from four or more GPS satellites orbiting the Earth. So a question is: What is the relationship between time as measured on Earth and the *coordinate time* t that we use to calculate the motion of planets and satellites or the trajectories of light rays?

Before about 1950, this question did not have much practical importance. Time was determined initially by the regular rotation of the Earth, so that the second was defined to be 1/86 400 of a day. But as measurements of the orbits of the Earth and planets improved, it was realized that the length of the day fluctuates seasonally because of large-scale atmospheric motions, and it increases with time because of tidal dissipation, making it ill-suited as a standard of time for precision measurements. The standard of time was then replaced by ephemeris time, for which the second is defined to be a fraction of the period of the Earth's orbit around the Sun. But even the Earth's orbital period is not precisely constant, because of planetary perturbations, and a better time standard had to be identified.

The development of atomic clocks during the period 1949–1955 made it possible to envision a standard of time based on fundamental physics instead of astronomy, with a potential for unprecedented accuracy. Today the second is defined to be the time elapsed during 9 192 631 770 cycles of the hyperfine transition in the ground state of cesium-133. In addition, the speed of light has now been *defined* to be exactly equal to 299 792 458 m/s, so that the meter is now defined in terms of the second, and is no longer tied to the length of a certain platinum rod in Paris.

During the 1960s and early 1970s, two developments signaled that general relativity would have to be incorporated into the practical definition of time. One was the development of numerical methods to integrate the equations of motion for bodies in the solar system, which replaced the largely analytic perturbative methods reviewed in Chapter 3. These techniques made it straightforward to include post-Newtonian corrections in the equations of motion, so that ephemeris time could be regarded as akin to the coordinate time t that appears in the post-Newtonian metric. The second development was the ongoing improvement in the accuracy and stability of atomic clocks, in particular those based on cesium-133, rubidium-87, and the 21-cm hyperfine transition of hydrogen. With stabilities better than a few parts in 10^{13}, these new standards were sensitive to the effects of gravity on time. For example, two clocks on the surface of the Earth differ in their rates by one part in 10^{13} for every kilometer in height difference, and this effect is easily measured by the latest clocks. The rate difference between the surface of the Earth and interplanetary space is a few parts in 10^{10}, so that the effect of gravity on time is very important in the determination of planetary orbits.

As a result, in 1976 the International Astronomical Union, one of whose functions is to establish the world-wide agreements and conventions for the establishment and dissemination of precise time, resolved to establish atomic time as the primary international standard, known as *Temps Atomique International* (TAI), for which the second is defined by a cesium-133 clock at rest on the Earth's geoid (see below). They also established the relationship between TAI and the time coordinate t involved in the post-Newtonian metric and in calculations of orbital motion. In 1991 the IAU endeavored to make these definitions as rigorous as possible from a relativistic point of view, and established working groups to monitor the definitions and their accuracy in the face of improving technology and more exacting requirements by users. This is a great example of the practical importance of general relativity.

The resolutions of the IAU are framed in the context of the post-Newtonian metric and equations of motion, and it has not yet proved necessary to extend these considerations to second post-Newtonian order. This is because the maximum size of the 2PN corrections is of order $(U/c^2)^2 \sim 10^{-16}$, which is just below the accuracy of the best current clocks (here the gravitational potential is dominated by the Sun, and it is evaluated at the Earth's orbit). One can well imagine, however, that the next generation of atomic clocks, based on phenomena like cold atom fountains, Bose–Einstein condensates, or atom interferometers, will be so accurate that 2PN corrections will be needed.

10.3.2 Reference frames

We begin our exploration of the measurement of time on, or near, a moving and rotating Earth with the specification of three reference frames. The first is the *barycentric frame* $(t_{\rm bary}, \boldsymbol{x}_{\rm bary})$, which is attached to the barycenter of the entire solar system; this is the frame in which the motion of planets would be calculated. The second is the *non-rotating geocentric frame* $(\bar{t}, \bar{\boldsymbol{x}})$, which is attached to the moving Earth, but is non-rotating with respect to the barycentric frame; this is the frame in which the motion of satellites around the Earth would be calculated, and the behavior of their clocks analyzed. And the third is the *rotating geocentric frame* $(t, \boldsymbol{x})$, which is also attached to the moving Earth, but rotating rigidly with Earth's angular velocity; this is the frame in which the behavior of clocks on the surface of the Earth would be analyzed.

The transformation between the barycentric and non-rotating geocentric frames is a special case of the class of post-Newtonian transformations developed in Sec. 8.3. This case was examined in Sec. 8.3.5, where it was shown that

$$t_{\rm bary} = \bar{t} + \frac{1}{c^2}(A + \boldsymbol{v}_\oplus \cdot \bar{\boldsymbol{x}}) + O(c^{-4}), \tag{10.104a}$$

$$\boldsymbol{x}_{\rm bary} = \bar{\boldsymbol{x}} + \boldsymbol{r}_\oplus + O(c^{-2}), \tag{10.104b}$$

in which $\boldsymbol{r}_\oplus(\bar{t})$ is the Earth's position in the barycentric frame, expressed as a function of geocentric time $\bar{t}$, $\boldsymbol{v}_\oplus(\bar{t})$ its velocity, and $A(\bar{t})$ is determined by

$$\frac{dA}{d\bar{t}} = \frac{1}{2}v_\oplus^2 + U_{\rm ext}, \tag{10.105}$$

with $U_{\rm ext}(\bar{t}, \boldsymbol{r}_\oplus)$ denoting the piece of the Newtonian potential produced by the bodies external to the Earth, including the Moon, Sun, and other solar-system bodies, evaluated at the Earth's position.

The discussion of Sec. 8.3.5 implies that in the non-rotating geocentric frame, the metric is given by

$$ds^2 = -\left[1 - \frac{2}{c^2}U + O(c^{-4})\right]d(c\bar{t})^2 + \left[1 + O(c^{-2})\right]\left(d\bar{x}^2 + d\bar{y}^2 + d\bar{z}^2\right), \tag{10.106}$$

in which U (which should be properly denoted $\bar{U}$) is the Earth's gravitational potential augmented by the tidal potential produced by the external bodies. For reasons that will be clarified shortly, we have neglected the post-Newtonian terms in the metric, and retained only the Newtonian term in $\bar{g}_{00}$.

We wish to use the metric of Eq. (10.106) to calculate the interval of proper time $d\tau$ measured by a clock moving with a velocity $\bar{\boldsymbol{v}}$ in the non-rotating geocentric frame. Because $d\tau$ is related to ds^2 by $c^2 d\tau^2 = -ds^2$, and because $d\bar{\boldsymbol{x}} = \bar{\boldsymbol{v}}\, d\bar{t}$ as we follow the clock's world line, we find that

$$d\tau = \left[1 - \frac{1}{c^2}\left(\frac{1}{2}\bar{v}^2 + U\right) + O(c^{-4})\right] d\bar{t}. \tag{10.107}$$

This is a key equation that allows us to relate $d\tau$, a quantity measured by an actual clock, to $d\bar{t}$, an interval of coordinate time that is measured by no one, but could be employed in many calculations. The relation involves the clock's speed $\bar{v}$ – a special relativistic effect – and the gravitational potential U – a general relativistic effect. The neglect of post-Newtonian terms in the metric can now be justified: they would lead to corrections of order c^{-4} in $d\tau$, which are negligible in the present context. We shall also neglect the tidal terms in U, because $U_{\text{tidal}}/c^2 \sim 10^{-17}$ is too small to be of practical significance, at least for today's technological state of the art.

The transformation between the rotating and non-rotating geocentric frames involves a rigid rotation of the coordinates around the Earth's rotation axis (which we choose to align with the z-direction), at the Earth's angular velocity ω. The transformation is described in great detail in Sec. 2.3.1, where we show that the velocities transform as $v^{\bar{a}} = \Lambda^{\bar{a}}{}_j(v^j + \epsilon^j{}_{kn}\omega^k x^n)$, in which $\bar{\boldsymbol{v}}$ is the clock's velocity in the non-rotating frame, $\boldsymbol{v}$ its velocity in the rotating frame, $\boldsymbol{\omega} = [0, 0, \omega]$ is the angular-velocity vector, and Λ is the transformation matrix. The relation implies

$$\bar{v}^2 = v^2 + 2\boldsymbol{v} \cdot (\boldsymbol{\omega} \times \boldsymbol{x}) + \omega^2(x^2 + y^2), \tag{10.108}$$

and substitution within Eq. (10.107) returns

$$d\tau = \left[1 - \frac{1}{c^2}\left(\frac{1}{2}v^2 + \boldsymbol{v} \cdot (\boldsymbol{\omega} \times \boldsymbol{x}) + \Phi\right) + O(c^{-4})\right] dt, \tag{10.109}$$

in which

$$\Phi = U + \Phi_C \tag{10.110}$$

is the Newtonian potential augmented by the centrifugal potential $\Phi_C := \frac{1}{2}\omega^2(x^2 + y^2)$. It should be noted that the time coordinate employed in the rotating frame is the same as in the non-rotating frame: $t := \bar{t}$. We can rely on Eq. (10.109) to relate a clock time interval $d\tau$ to the coordinate time interval dt.

10.3.3 Geoid

We now focus our attention on a clock situated on Earth's surface, or at a low elevation above the surface; the clock may be at rest or moving with a velocity $\boldsymbol{v}$. We shall perform our calculations in the rotating geocentric frame $(t, \boldsymbol{x})$, and therefore work with Eq. (10.109).

Our first item of business is to specify a reference surface from which all elevations will be measured. Because of the important role that the generalized potential Φ plays in Eq. (10.109), it is wise to choose a surface $\Phi =$ constant as our reference, which is then known as the *geoid*. From Sec. 2.3.1 we know that surfaces of constant ρ, p, and Φ all coincide for a fluid body, and since the body's boundary is situated at $p = 0$, the reference surface can be made to coincide with the boundary. The Earth is not quite a fluid body, but most of its surface is covered by oceans which do behave as perfect fluids. For the Earth, therefore, the geoid is chosen to coincide with the mean ocean surface extended through the continents.

To model the gravitational field of the Earth we take into account its rotational deformation, but neglect all other deformations produced by inhomogeneities in its mass density; this is a gross oversimplification of the actual situation. We assume that the deformation is axially symmetric, and express the gravitational potential as in Eq. (1.144),

$$U(r,\theta) = \frac{GM}{r}\left[1 - J_2\left(\frac{R}{r}\right)^2 P_2(\cos\theta)\right], \tag{10.111}$$

in which the multipole expansion was truncated to $\ell = 2$; here M is the mass of the Earth, R its equatorial radius, and J_2 its dimensionless quadrupole moment. The centrifugal potential is

$$\Phi_C = \frac{1}{2}\omega^2 r^2 \sin^2\theta, \tag{10.112}$$

and according to our discussion in Sec. 2.4.5, $J_2 = k_2\zeta$, in which k_2 is the Earth's gravitational Love number, and

$$\zeta := \frac{2\omega^2 R^3}{3GM} \tag{10.113}$$

provides the fractional scale of the rotational deformation. We shall assume that $\zeta \ll 1$.

Constructing $\Phi = U + \Phi_C$, it is easy to show that the geoid $\Phi(r,\theta) = \Phi_{\rm geoid} =$ constant is situated on the surface $r = R_{\rm geoid}(\theta)$, where

$$R_{\rm geoid}(\theta) = R(1 - f\cos^2\theta) \tag{10.114}$$

with

$$f = \frac{3}{2}J_2 + \frac{\omega^2 R^3}{2GM} = \frac{3}{4}(1 + 2k_2)\zeta. \tag{10.115}$$

The expression implies that $f = 1 - R(0)/R(\frac{\pi}{2})$, and the condition $\zeta \ll 1$ implies that f is small.

Another description of the geoid is given by the Cartesian form $x^2 + y^2 + (1 + 2f)z^2 = R^2$, and the unit normal $\boldsymbol{n}$ to the geoid is easily shown to have the components $n_x = (1 - 2f\cos^2\theta)\sin\theta\cos\phi$, $n_y = (1 - 2f\cos^2\theta)\sin\theta\sin\phi$, and $n_z = (1 + 2f\sin^2\theta)\cos\theta$. The *geographical latitude* ψ of a point on the geoid is the angle between $\boldsymbol{n}$ and the equatorial plane, so that $\sin\psi = n_z$. The relation implies

$$\cos\theta = (1 - 2f\cos^2\psi)\sin\psi. \tag{10.116}$$

In terms of the geographical latitude we have that $R_{\rm geoid} = R(1 - f\sin^2\psi)$.

For future reference we wish to find an approximate expression for the potential Φ for a point $\boldsymbol{x}$ at a height $h \ll R$ just above (or below) the geoid; the displacement is in the direction of the normal vector $\boldsymbol{n}$, at a constant geographical latitude ψ. Expressing Φ in terms of r and ψ, setting $r = R_{\rm geoid} + h$, and expanding in powers of h, we find that

$$\Phi = \Phi_{\rm geoid} - hg(\psi) + O(h^2), \tag{10.117}$$

where $g(\psi) := -\partial\Phi/\partial r$, given by

$$\begin{aligned} g(\psi) &= \frac{GM}{R^2}\left[1 + \frac{3}{2}J_2 - \frac{\omega^2 R^3}{GM} + \left(\frac{2\omega^2 R^3}{GM} - \frac{3}{2}J_2\right)\sin^2\psi\right] \\ &= \frac{GM}{R^2}\left[1 - \frac{3}{2}(1-k_2)\zeta + \frac{3}{2}(2-k_2)\zeta\,\sin^2\psi\right], \end{aligned} \tag{10.118}$$

is the local acceleration of gravity. For the Earth this evaluates to $g = (9.7803 + 0.0519\sin^2\psi)\ \mathrm{m/s^2}$.

10.3.4 Temps Atomique International

By definition, TAI is the time measured by an atomic clock at rest on the geoid. By virtue of Eq. (10.109), in which we insert $\boldsymbol{v} = \boldsymbol{0}$, we have that the relationship between TAI and coordinate time t is given by

$$d(\text{TAI}) = (1 - \Phi_{\rm geoid}/c^2)\,dt, \tag{10.119}$$

and constancy of Φ on the geoid guarantees that clocks at rest on the geoid mark time at exactly the same rate.

Actual atomic clocks are typically not situated on the geoid. Instead they tend to be housed in more convenient places at a range of altitudes above sea level, such as Boulder, Colorado (NIST), Washington, DC (USNO), Paris, France (BIPM), and so on. These clocks, therefore, do not keep time at the rate specified by TAI, and a correction must be introduced to account for their different placement in the Earth's potential. To calculate this we return to Eq. (10.109), insert Eq. (10.117), and get

$$d\tau = \left[1 - \Phi_{\rm geoid}/c^2 + hg(\psi)/c^2\right]dt \tag{10.120}$$

for the time τ measured by a clock at rest at a geographical latitude ψ and elevation h above the geoid. Combining this with Eq. (10.119) produces

$$d(\text{TAI}) = \left[1 - hg(\psi)/c^2\right]d\tau, \tag{10.121}$$

an equation that relates the measured time τ to the time standard TAI. The correction evaluates to $g/c^2 = (1.0882 + 0.00577\sin^2\psi) \times 10^{-16}\ \mathrm{m^{-1}}$. Thus, any atomic clock at rest on Earth can have its rate linked directly to TAI by a simple correction factor. In practice, TAI is defined by an average measurement over an ensemble of the best atomic clocks at different locations around the world, suitably corrected via Eq. (10.121).

The establishment of TAI as a precise time standard relies on measurements made by atomic clocks at multiple locations. How are these clocks synchronized? We shall not delve into the practical aspects of this matter, but wish to raise an important question of principle: How can clocks be synchronized in a rotating reference frame?

Suppose that we have a clock at position B that we wish to synchronize with another clock at position A. For simplicity we ignore effects associated with elevation, and assume that both locations are on the geoid. Each clock marks time at the rate indicated by Eq. (10.119), and integration yields $\tau[A] = (1 - \Phi_{\rm geoid}/c^2)t$ and $\tau[B] = (1 - \Phi_{\rm geoid}/c^2)t + \Delta\tau$, in which $\Delta\tau$ is an integration constant. The synchronization procedure is meant to achieve $\Delta\tau = 0$, so that $\tau[B] = \tau[A]$.

A possible way of establishing synchronization would be to rely on a portable clock that is transported slowly from A to B. This clock is initially synchronized with the A-clock, and during its trip to B it marks time at a rate $d\tau^*$ given by Eq. (10.109),

$$d\tau^* = \left[1 - \frac{1}{c^2}\Big(\boldsymbol{v}\cdot(\boldsymbol{\omega}\times\boldsymbol{r}) + \Phi_{\rm geoid}\Big)\right] dt, \tag{10.122}$$

in which $\boldsymbol{r}(t)$ is the path of the portable clock; we have neglected the $\frac{1}{2}v^2$ term under the assumption that the transport is slow. Integrating this for the whole trip, which begins at time t_1 and ends at time t_2, we find that

$$\tau_2^* - \tau_1^* = \left[1 - \Phi_{\rm geoid}/c^2\right](t_2 - t_1) - \frac{1}{c^2}\int_A^B (\boldsymbol{\omega}\times\boldsymbol{r})\cdot d\boldsymbol{r}. \tag{10.123}$$

Because the clock is synchronized with the A-clock at the beginning of the trip, we have that $\tau_1^* = \tau_1[A] = (1 - \Phi_{\rm geoid}/c^2)t_1$, in which $\tau_1[A]$ is the time recorded by the A-clock at departure. Expressing t_2 in terms of $\tau_2[B]$, the time recorded by the B-clock at the arrival of the portable clock, and solving for τ_2^*, we find that

$$\tau_2^* = \tau_2[B] - \Delta\tau - \frac{1}{c^2}\int_A^B (\boldsymbol{\omega}\times\boldsymbol{r})\cdot d\boldsymbol{r}. \tag{10.124}$$

We see that the path integral prevents us from achieving synchronization ($\Delta\tau = 0$) by doing the obvious, setting the B-clock so that $\tau_2[B] = \tau_2^*$. Instead, we must synchronize the B-clock by setting

$$\tau_2[B] = \tau_2^* + \frac{1}{c^2}\int_A^B (\boldsymbol{\omega}\times\boldsymbol{r})\cdot d\boldsymbol{r}; \tag{10.125}$$

a correction must be applied to reflect the transport of the clock in the rotating frame of the Earth. Note that the path integral is independent of $\boldsymbol{v}$; the effect survives even in the limit of a quasi-static transport.

This failure of simple-minded synchronization, known as the *Sagnac effect* after the French physicist Georges Sagnac (1869–1926), is a consequence of the Earth's rotation. For a path $\boldsymbol{r}$ confined to the geoid, is it is easy to show that

$$\frac{1}{c^2}\int_A^B (\boldsymbol{\omega}\times\boldsymbol{r})\cdot d\boldsymbol{r} = \frac{\omega R^2}{c^2}\int_A^B \sin^2\theta\, d\phi, \tag{10.126}$$

and the integral (multiplied by R^2) is twice the area swept out by a line joining the center of the Earth to a projection of the path onto the equatorial plane. For a closed path around the equator, the Sagnac term amounts to 207 ns, and the effect is very noticeable. Note that the Sagnac effect disappears when the clock is transported along a meridian, so that $d\phi = 0$.

Clock synchronization on a rotating Earth can nevertheless be achieved provided that one applies the Sagnac correction, which requires one to keep careful track of the path taken by the portable clock. As a result, all clocks on the geoid can be consistently synchronized to a single master clock, which could be situated in Washington DC, Greenwich UK, Beijing China, or at the North Pole (the latter being the most politically neutral). This method now forms the basis for all terrestrial timekeeping.

10.3.5 Orbiting clocks

Atomic clocks orbiting the Earth have become an important part of our lives thanks to the Global Positioning System (GPS; see Box 10.4), which would fail utterly if it did not account for relativistic effects in timekeeping. To analyze the behavior of orbiting clocks we put ourselves in the non-rotating geocentric frame $(\bar{t}, \bar{\boldsymbol{x}})$ and use Eq. (10.107),

$$d\tau[\text{orbit}] = \left[1 - \frac{1}{c^2}\left(\frac{1}{2}\bar{v}^2 + U\right)\right] d\bar{t}, \tag{10.127}$$

to relate the clock's proper time $\tau[\text{orbit}]$ to the coordinate time $\bar{t}$; here $\bar{\boldsymbol{v}}$ is the satellite's orbital velocity and U is the Earth's gravitational potential. In this case we may ignore the multipolar deformations, which are small at the high altitudes considered here ($r \simeq 4R_\oplus$ for GPS satellites), and set $U = GM/r$.

Each satellite moves on a Keplerian orbit around the Earth, and we may rely on the orbital equations derived in Sec. 3.2.4 to integrate Eq. (10.127). It is convenient to employ a representation of the motion in terms of the eccentric anomaly u, as displayed in Eqs. (3.34) and (3.35). We have that $\bar{r} = a(1 - e\cos u)$ and $\bar{v}^2 = (GM/a)(1 + e\cos u)/(1 - e\cos u)$, where a and e are the semi-major axis and eccentricity, respectively, and the eccentric anomaly is related to $\bar{t}$ by Kepler's equation

$$\bar{t} - T = \sqrt{\frac{a^3}{GM}}(u - e\sin u), \tag{10.128}$$

in which T is the time of perigee passage. It follows that

$$\frac{1}{2}\bar{v}^2 + \frac{GM}{r} = \frac{3}{2}\frac{GM}{a} + \frac{2GM}{a}\frac{e\cos u}{1 - e\cos u}, \tag{10.129}$$

and substitution within Eq. (10.127) yields

$$\tau[\text{orbit}] = \left(1 - \frac{3}{2}\frac{GM}{c^2 a}\right)\bar{t} - \frac{2GM}{c^2 a}\sqrt{\frac{a^3}{GM}} e\sin u + \tau_0, \tag{10.130}$$

where τ_0 is a constant of integration set by synchronization of the orbiting clock.

The time marked by the orbiting clock is expressed in terms of the coordinate time $\bar{t}$. To complete the description we should relate it to the time measured by a clock at rest (or moving slowly inside a car) at a geographical latitude ψ and elevation h on the surface of the Earth. The translation is provided by Eq. (10.120), which integrates to

$$\tau[\text{ground}] = \big[1 - \Phi_{\text{geoid}}/c^2 + hg(\psi)/c^2\big]\bar{t} + \tau_1, \tag{10.131}$$

with τ_1 another constant of integration. Making the substitution in Eq. (10.130), we find that

$$\begin{aligned}\tau[\text{orbit}] = {} & \left\{1 - \frac{1}{c^2}\left[\frac{3}{2}\frac{GM}{a} - \Phi_{\text{geoid}} + hg(\psi)\right]\right\}\tau[\text{ground}] \\ & - \frac{2GM}{c^2 a}\sqrt{\frac{a^3}{GM}}\, e \sin u + \tau_2, \end{aligned} \tag{10.132}$$

in which u is now expressed implicitly as a function of $\tau[\text{ground}]$.

Equation (10.132) relates the time measured by a clock on board a satellite to the time measured by a clock on the ground. The first group of terms describes a constant difference between the clock rates. The first correction term $-3GM/(2c^2a)$ comes with a negative sign, and represents a time-dilation effect produced by the orbital motion. The second term $+\Phi_{\text{geoid}}/c^2$ comes with a positive sign, and represents a gravitational blueshift associated with the transfer of the time signal from high to low altitude. The third term $-hg(\psi)/c^2$ is an elevation-dependent correction, and its sign depends on the sign of h. The second group of terms is a time-dependent modulation of $\tau[\text{orbit}]$ associated with the orbital eccentricity; this effect is usually very small, because the GPS satellites are in highly circular orbits ($e \sim 0.01$). A typical orbit has a period of 12 hours, corresponding to $a = 26\,630$ km, or about $4.2R_\oplus$, and this implies that the geoid term dominates the orbital term in the constant rate factor. This means that the gravitational blueshift dominates time dilation, so that GPS clocks tick faster than clocks on the geoid by a factor of 4.467×10^{-10}, or by about 38.6 microseconds per day. Note that the geoid and satellite terms cancel each other at an orbital radius of about $1.5R_\oplus$, and that time dilation would dominate for a lower orbit.

Box 10.4 **Global Positioning System**

The Global Positioning System (GPS) has become a ubiquitous fixture of our daily lives, being found in many smartphones and helping us navigate our cars in unfamiliar neighborhoods. The GPS consists of 24 satellites orbiting the Earth, with groups of four satellites on six distinct orbital planes oriented at 55 degrees relative to Earth's equatorial plane. The system operates in such a way that at least four satellites are visible above the horizon at any given time. Each satellite harbours a very accurate atomic clock; the accumulated time lapse of a given clock over a full day is measured in nanoseconds. The system is operated by the United States Department of Defense; it was deployed for military purposes before it was made available to civilians.

Each GPS satellite emits a sequence of radio pulses that contain information about the satellite's position and the time of emission. This information is used by a receiving unit on Earth to determine its own position. The receiver acquires the position $\boldsymbol{r}_A$ and time of emission t_A from each of the four visible satellites. Ignoring the relativistic effects described in the main text, the receiver determines its own position $\boldsymbol{r}$ and time t by solving the system of equations

$$c^2(t - t_A)^2 = |\boldsymbol{r} - \boldsymbol{r}_A|^2;$$

solving for t is necessary because the clock inside the receiver is not nearly as accurate as the atomic clocks, and because this clock has not been synchronized with sufficient precision. In this way the receiving unit can determine its position on Earth with an accuracy of a few meters and local time with an accuracy of 30 nanoseconds.

As explained in the main text, relativistic effects such as time dilation and gravitational blueshift must be taken into account for the successful operation of the GPS. For example, according to Eq. (10.132), the size of the constant rate difference between $\tau[\text{orbit}]$ and $\tau[\text{ground}]$ is given by

$$\frac{3}{2}\frac{GM}{c^2 a} - \frac{\Phi_{\text{geoid}}}{c^2} = -4.4647 \times 10^{-10}.$$

This may seem like a small number, but in the course of a full day it builds up to a time difference of $38.6\ \mu\text{s}$. This is *much, much larger* than the intrinsic accuracy of the atomic clocks, and translates (after multiplication by c) to a distance error of $12\ \text{km}$. Think of Einstein the next time your GPS saves you from an unfortunate detour in downtown Los Angeles.

10.3.6 Timing of binary pulsars

As a final application of timekeeping in general relativity, we examine binary pulsars, those important astrophysical systems that have been so fruitful in delivering tests of our favorite theory. A binary pulsar is a two-body system, of which at least one is an active pulsar that emits radio pulses at very regular intervals. The pulses are received on Earth by a radio telescope, and a careful timing allows the astronomer to extract detailed information about the orbital motion of the binary pulsar, including its post-Newtonian aspects. The companion star is generally assumed to be a compact object as well, either a neutron star or a white dwarf (both types of system have been observed) or a black hole (no evidence so far).

We wish to relate the proper time of emission τ_e of a radio pulse, as measured by a clock attached to the pulsar, to its arrival time t_a, as measured by a remote observer. Our timing model will include all the relevant relativistic effects, but for simplicity we shall neglect a number of effects that are also important to the radio astronomer. These include the motion of the observer relative to the barycenter of the binary system, the aberration associated with the rotational motion of the pulsar, and the dispersion of the radio pulses in the interstellar medium.

We consider a binary system consisting of a pulsar of mass M_1 and position $\boldsymbol{r}_1$, and a companion of mass M_2 and position $\boldsymbol{r}_2$. The motion of the binary is described in the system's barycentric frame, and the observer is situated at a remote position $\boldsymbol{r}_{\text{obs}}$ relative to the barycenter; the observer is imagined to be at rest at that position, and her clock measures coordinate time t. As usual we let $\boldsymbol{r} := \boldsymbol{r}_1 - \boldsymbol{r}_2$ denote the orbital separation, and $r := |\boldsymbol{r}|$.

Our first task is to relate the pulsar's proper time τ to the coordinate time t. Working to the leading, Newtonian order, the relation is provided by Eq. (10.107), which we evaluate at the position of the pulsar. Removing the overbars, which are no longer required, we have that

$$d\tau = \left[1 - \frac{1}{c^2}\left(\frac{1}{2}v_1^2 + U\right) + O(c^{-4})\right] dt, \tag{10.133}$$

in which $\boldsymbol{v}_1$ is the pulsar's velocity and U is the binary's gravitational potential, consisting of the pulsar's own constant potential U_1 and the companion's potential GM_2/r. To integrate this we once again make use of the Keplerian relations obtained in Sec. 3.2.4. After some algebra we find that

$$\frac{1}{2}v_1^2 + \frac{GM_2}{r} = \frac{GM_2}{a}\left(\frac{2M_1 + 3M_2}{2m} + \frac{M_1 + 2M_2}{m}\frac{e\cos u}{1 - e\cos u}\right), \tag{10.134}$$

in which $m := M_1 + M_2$ is the total mass, a is the binary's semi-major axis, e its eccentricity, and u is the eccentric anomaly, which is related to t by Kepler's equation. Making the substitution and integrating, we find that

$$\tau = \left(1 - \frac{2M_1 + 3M_2}{2m}\frac{GM_2}{c^2 a} - \frac{U_1}{c^2}\right)t - \frac{M_1 + 2M_2}{m}\sqrt{\frac{a^3}{Gm}}\frac{GM_2}{c^2 a}\, e\sin u + \tau_0, \tag{10.135}$$

in which τ_0 is a constant of integration. In practical applications τ_0 is unobservable, and so is the constant factor in front of t, which can be absorbed into a re-definition of the pulsar's intrinsic frequency, which is unknown; the key information is contained in the time-dependent term, which is proportional to the binary's eccentricity.

Equation (10.135) allows us to relate the proper time of emission τ_e to the coordinate time t_e. We must now obtain the relation between t_e and the time of arrival t_a, by calculating the time required by an electromagnetic signal to travel from the pulsar at $\boldsymbol{r}_1$ to the observer at $\boldsymbol{r}_{\rm obs}$. The work was already carried out in Sec. 10.2.5, and Eq. (10.100) informs us that

$$c(t_a - t_e) = |\boldsymbol{r}_{\rm obs} - \boldsymbol{r}_1| + \frac{2GM_2}{c^2}\ln\left[\frac{|\boldsymbol{r}_{\rm obs} - \boldsymbol{r}_2| + (\boldsymbol{r}_{\rm obs} - \boldsymbol{r}_2)\cdot\boldsymbol{k}}{r + \boldsymbol{r}\cdot\boldsymbol{k}}\right] + O(c^{-4}), \tag{10.136}$$

where

$$\boldsymbol{k} := \frac{\boldsymbol{r}_{\rm obs} - \boldsymbol{r}_1}{|\boldsymbol{r}_{\rm obs} - \boldsymbol{r}_1|} \tag{10.137}$$

is the direction of propagation. Some work was required to go from Eq. (10.100) to Eq. (10.136). First, the version here accounts for the fact that the delay is produced by the companion at $\boldsymbol{r}_2$ instead of a body situated at the spatial origin of the coordinate system. And second, the factor b^2 in Eq. (10.100) was factorized as $b^2 = (r - \boldsymbol{r}\cdot\boldsymbol{k})(r + \boldsymbol{r}\cdot\boldsymbol{k})$, a relation that follows directly from the definition $\boldsymbol{b} := \boldsymbol{r} - (\boldsymbol{r}\cdot\boldsymbol{k})\boldsymbol{k}$. Our result can be simplified if we recognize that the observer is situated at a large distance from the binary system. With $r_{\rm obs}/r_1 \gg 1$ and $r_{\rm obs}/r_2 \gg 1$, we find that Eq. (10.136) reduces to

$$c(t_a - t_e) = r_{\rm obs} - \boldsymbol{n}_{\rm obs}\cdot\boldsymbol{r}_1 - \frac{2GM_2}{c^2}\ln\left[\frac{(1 + \boldsymbol{n}_{\rm obs}\cdot\boldsymbol{n})r}{2r_{\rm obs}}\right] + O(c^{-4}), \tag{10.138}$$

in which $\boldsymbol{n} := \boldsymbol{r}/r$ and $\boldsymbol{n}_{\rm obs} := \boldsymbol{r}_{\rm obs}/r_{\rm obs}$. It is understood that $\boldsymbol{r}$ and $\boldsymbol{r}_1$ are evaluated at time $t = t_e$.

We now wish to put Eq. (10.138) in a more concrete form. We adopt the point of view of the astronomer, who looks up at the binary system and assigns to it the orbital parameters

$(a, e, \iota, \Omega, \omega)$, in which ι is the orbital inclination, Ω the longitude of the ascending node, and ω the longitude of pericenter. She erects a fundamental (X, Y, Z) frame and aligns it in such a way that the Z-axis points toward the binary. In this description we have that $\boldsymbol{n}_{\rm obs} = -\boldsymbol{e}_Z$, and $\boldsymbol{n}$ is given by Eq. (3.42); with this it follows that $\boldsymbol{n}_{\rm obs} \cdot \boldsymbol{n} = -\sin\iota \sin(\omega + f)$, in which f is the orbit's true anomaly.

In the term $\boldsymbol{n}_{\rm obs} \cdot \boldsymbol{r}_1$ on the right-hand side of Eq. (10.138) we insert the post-Newtonian description of the orbital motion given by Eq. (10.21). In the logarithmic term we may insert the usual Keplerian relations. Making the substitutions, we arrive at

$$
\begin{aligned}
c(t_a - t_e) &= r_{\rm obs} + \sin\iota \sin(\omega + f_e) a_1 (1 - e_1 \cos u_e) \\
&\quad - \frac{2GM_2}{c^2} \ln\Big\{\big[1 - \sin\iota \sin(\omega + f_e)\big](1 - e \cos u_e)\Big\} + O(c^{-4}),
\end{aligned} \tag{10.139}
$$

in which a_1 and e_1 are the pulsar's post-Newtonian orbital parameters, as defined in Eq. (10.38), f_e is the true anomaly at the time of emission, and u_e is the eccentric anomaly. The relation between f_e and u_e is given by Eq. (10.34), and the relation between u_e and t_e is expressed by Eq. (10.24).

Our task is completed. With Eq. (10.135) we relate the proper time of emission τ_e of a radio pulse to the coordinate time t_e, and with Eq. (10.139) we relate t_e to the arrival time t_a. Combining these expressions, we obtain the desired relation between t_a and τ_e. The relation is provided in a rather implicit form, but the building blocks are all sufficiently simple that it can easily be turned into a practical timing tool.

In Sec. 10.1.4 we pointed out that the measurement of the pericenter advance in a binary-pulsar system yields a very accurate determination of the system's total mass. The timing formulae of Eqs. (10.135) and (10.139) provide additional information. The time-dependent term in Eq. (10.135) involves a different combination of M_1 and M_2, and combining this measurement with a pericenter-advance measurement returns values for each mass. When the Shapiro delay is significant (which requires a nearly edge-on orbit), the amplitude of the delay reveals the companion mass M_2 straight away, while its detailed shape depends on $\sin\iota$. Such independent measurements of the masses in many binary pulsars turn out to give consistent results to high precision, affirming the correctness of the general relativistic description of the physics. In Chapter 12 we uncover yet another way to measure the masses, via the effect of gravitational radiation damping.

10.4 Spinning bodies

As relativistic effects go, those associated with a body's spin tend to be suppressed compared to the effects reviewed in Sec. 10.1. For example, in the spin–orbit contributions to the equations of motion, Eqs. (9.189) or Eq. (9.245), a relativistic factor v_A^2/c^2 has been replaced by $S_A v_A/(M_A c^2 r_{AB})$, which is smaller by a factor of order $(V_A/v_A)(R_A/r_{AB})$, where V_A is the equatorial rotation velocity of the spinning body and R_A is its radius. The rotation velocity for a body is limited by the Keplerian velocity $(GM_A/R_A)^{1/2}$, at which matter on the equator is no longer bound to the body, and the orbital velocity is itself comparable to $(GM_A/r_{AB})^{1/2}$, so that one would expect a spin–orbit effect to be smaller than an orbital post-Newtonian effect by a factor of order $(R_A/r_{AB})^{1/2}$. In the context of the theory of motion developed in Chapter 9, in which each body was assumed to be small compared with the inter-body separation, this is necessarily a small number. Furthermore, most bodies do not rotate at anything close to the Keplerian limit. For example, the Earth's equatorial rotational velocity is approximately 0.5 km/s, which is 16 times smaller than the Keplerian limit. Even the most rapidly spinning pulsar rotates at less than half its Keplerian limit. The sole exception to this rule is a rapidly rotating black hole, for which the effective V_A can be comparable to the speed of light.

So it would seem that to detect the effects of spin in the solar system would be hopeless, unless we could relax the constraint that $R_A \ll r_{AB}$, which means putting one body close to the surface of the rotating body. We can do this and retain the validity of the equations of motion obtained in Chapter 9 by demanding that the first body have a negligible mass. This is the context of Gravity Probe B, in which gyroscopes are placed on a low Earth orbit with $r_{AB} = R_A + 650$ km, and of the LAGEOS experiment, which tracked satellites on near-Earth orbits with $r_{AB} \approx 2R_A$. On the other hand, by virtue of the remarkable accuracy afforded by pulsar timing, relativistic spin–orbit effects have been measured in a number of binary pulsar systems.

10.4.1 Frame dragging and Gravity Probe B

Gravity Probe B will very likely go down in the history of science as one of the most ambitious, difficult, expensive, and controversial relativity experiments ever performed.[1] In fact, the list of superlatives associated with the project is formidable: it fabricated the world's most perfect spheres, it achieved the lowest magnetic fields, it built the best drag-free control, and so on. All these achievements and more were essential to the successful measurement of the minute precession of a spinning body as it orbits the Earth.

Precession of a gyroscope

We consider a binary system that consists of the Earth, with mass $M_\oplus$ and spin $\boldsymbol{S}_\oplus$, and a gyroscope of negligible mass and spin $\boldsymbol{S}$. The gyroscope orbits the Earth, and its spin precesses. The relevant equation was listed in Sec. 9.5.6. According to Eq. (9.198), we have that $d\boldsymbol{S}/dt = \boldsymbol{\Omega} \times \boldsymbol{S}$, where $\boldsymbol{\Omega} = \boldsymbol{\Omega}[\text{so}] + \boldsymbol{\Omega}[\text{ss}]$ is the precessional angular velocity decomposed in spin–orbit and spin–spin contributions, with

$$\boldsymbol{\Omega}[\text{so}] = \frac{3}{2}\frac{GM_\oplus}{c^2r^2}\boldsymbol{n} \times \boldsymbol{v}, \tag{10.140a}$$

$$\boldsymbol{\Omega}[\text{ss}] = \frac{G}{c^2r^3}\Big[3(\boldsymbol{S}_\oplus \cdot \boldsymbol{n})\boldsymbol{n} - \boldsymbol{S}_\oplus\Big], \tag{10.140b}$$

in which $\boldsymbol{r} = r\boldsymbol{n}$ is the gyroscope's orbital position, and $\boldsymbol{v}$ its orbital velocity.

[1] Full disclosure: one of us (CMW) served as Chair of an external NASA Science Advisory Committee for Gravity Probe B from 1998 to 2010.

We describe the orbital motion in the fundamental (X, Y, Z) frame introduced in Sec. 3.2.5. The frame is oriented in such a way that the X-Y plane coincides with the Earth's equatorial plane, and $\boldsymbol{S}_\oplus$ points in the Z-direction. We place the gyroscope on a circular orbit of radius a, inclination ι, and longitude of ascending node Ω. The direction to the gyroscope is given by Eq. (3.42),

$$\begin{aligned}\boldsymbol{n} = &\big[\cos\Omega\cos(\omega+f) - \cos\iota\sin\Omega\sin(\omega+f)\big]\,\boldsymbol{e}_X\\ &+\big[\sin\Omega\cos(\omega+f) + \cos\iota\cos\Omega\sin(\omega+f)\big]\,\boldsymbol{e}_Y\\ &+\sin\iota\sin(\omega+f)\,\boldsymbol{e}_Z,\end{aligned} \tag{10.141}$$

in which $f = 2\pi t/P$ is the orbital phase; $P = 2\pi a^{3/2}(GM_\oplus)^{-1/2}$ is the orbital period.

It is convenient to introduce a vectorial basis $(\boldsymbol{e}_p, \boldsymbol{e}_q, \boldsymbol{e}_z)$ adapted to the gyroscope's orbital plane. The normal to the plane is given by Eq. (3.45),

$$\boldsymbol{e}_z = \sin\iota\sin\Omega\,\boldsymbol{e}_X - \sin\iota\cos\Omega\,\boldsymbol{e}_Y + \cos\iota\,\boldsymbol{e}_Z. \tag{10.142}$$

As a second basis vector we select

$$\boldsymbol{e}_p = \cos\Omega\,\boldsymbol{e}_X + \sin\Omega\,\boldsymbol{e}_Y, \tag{10.143}$$

which lies in the plane and points toward the ascending node. And as a third vector, orthogonal to the first two, we get

$$\boldsymbol{e}_q = -\cos\iota\sin\Omega\,\boldsymbol{e}_X + \cos\iota\cos\Omega\,\boldsymbol{e}_Y + \sin\iota\,\boldsymbol{e}_Z. \tag{10.144}$$

In terms of the orbital basis we have that $\boldsymbol{n} = \cos f\,\boldsymbol{e}_p + \sin f\,\boldsymbol{e}_q$, $\boldsymbol{e}_Z = \sin\iota\,\boldsymbol{e}_q + \cos\iota\,\boldsymbol{e}_z$, and $\boldsymbol{n}\times\boldsymbol{v} = \sqrt{GM_\oplus/a}\,\boldsymbol{e}_z$.

We insert the foregoing relations within the precessional angular velocities, obtaining

$$\boldsymbol{\Omega}[\mathrm{so}] = \frac{3}{2c^2a}\left(\frac{GM_\oplus}{a}\right)^{3/2}\boldsymbol{e}_z, \tag{10.145a}$$

$$\boldsymbol{\Omega}[\mathrm{ss}] = \frac{GS_\oplus}{c^2a^3}\left[\frac{3}{2}\sin\iota\sin 2f\,\boldsymbol{e}_p + \frac{1}{2}\sin\iota(1-3\cos 2f)\,\boldsymbol{e}_q - \cos\iota\,\boldsymbol{e}_z\right], \tag{10.145b}$$

and substitute these into the precession equation. Because the changes in the gyroscope spin $\boldsymbol{S}$ are small, we may insert its initial value $\boldsymbol{S}_0$ on the right-hand side of the equation. And because we are interested only in the secular evolution of the spin, we average the precession equation over a complete cycle of the orbital motion. We arrive at

$$\left\langle\frac{d\boldsymbol{S}}{dt}\right\rangle_{\mathrm{SO}} = \frac{3}{2c^2a}\left(\frac{GM_\oplus}{a}\right)^{3/2}\boldsymbol{e}_z\times\boldsymbol{S}_0, \tag{10.146a}$$

$$\left\langle\frac{d\boldsymbol{S}}{dt}\right\rangle_{\mathrm{SS}} = \frac{GS_\oplus}{c^2a^3}\left(\frac{1}{2}\sin\iota\,\boldsymbol{e}_q - \cos\iota\,\boldsymbol{e}_z\right)\times\boldsymbol{S}_0. \tag{10.146b}$$

Inserting the appropriate numerical values for the Earth, the equations reveal that the precessional angular velocities are given by

$$\langle \Omega[\mathrm{so}] \rangle = \frac{3}{2c^2 a}\left(\frac{GM_\oplus}{a}\right)^{3/2} = 8.43\left(\frac{R_\oplus}{a}\right)^{5/2} \text{ as/yr}, \tag{10.147a}$$

$$\langle \Omega[\mathrm{ss}] \rangle = \frac{GS_\oplus}{c^2 a^3} = 0.109\left(\frac{R_\oplus}{a}\right)^{3} \text{ as/yr}; \tag{10.147b}$$

we see that the spin–orbit precession amounts to a few arcseconds per year, while the spin–spin precession is smaller by almost two orders of magnitude.

Geodetic precession; Schiff precession

The spin–orbit precession is frequently called the "geodetic precession" of a spin vector. Its underlying origin is the fact that spacetime is curved in the vicinity of the Earth, and that parallel transport of a vector around a closed path returns a vector that points in a direction that differs from the initial direction. The change is given by the product of components of the curvature tensor with the vector itself, multiplied by the area enclosed by the orbit. In order of magnitude we have that $\delta S \sim$ (Riemann)(area)S_0, and with Riemann $\sim GM_\oplus/(c^2 a^3)$, area $\sim a^2$, $P \sim \sqrt{a^3/(GM_\oplus)}$, we find that

$$\left\langle \frac{d\boldsymbol{S}}{dt} \right\rangle_{\mathrm{SO}} \sim \frac{\delta S}{P} \sim \frac{1}{c^2 a}\left(\frac{GM_\oplus}{a}\right)^{3/2} S_0, \tag{10.148}$$

in agreement with the more detailed calculation. The geodetic precession is independent of the orientation of the orbit, but is maximized when the spin $\boldsymbol{S}_0$ lies within the orbital plane; it vanishes when the spin is parallel to $\boldsymbol{e}_z$.

The spin–spin precession is also called the "Schiff precession" of a spin vector. In 1960, Leonard Schiff calculated the spin–orbit and spin–spin precessions of a gyroscope and suggested the possibility of measuring them. Unbeknownst to Schiff, a researcher at the US Pentagon named George Pugh had performed the same calculations a few months earlier. Pugh was working for the Weapons Systems Evaluation Group, assessing the use of high-performance gyroscopes in missile and aircraft guidance. He wondered how large the relativistic effects would be, and what it would take of a space experiment to measure them. Pugh's classified work could not be published in the open literature, and so Schiff was initially given credit for the idea; only later was Pugh's work discovered and recognized. It is worth noting that all this occurred in the very early days of space exploration, only two years after the launch of *Sputnik*.

The phrases "frame dragging" and "Lense–Thirring effects" are sometimes attached to the spin–spin precession of a spin vector. Lense and Thirring were the first, in 1918, to examine the metric of a rotating body in the weak-field limit, and to work out the effects of rotation on the orbits of bodies, a topic to which we return in the next subsection. They did not consider the motion of gyroscopes explicitly, but the basic phenomenon of frame dragging applies to them just as well: an inertial frame with axes defined by an array of perpendicularly oriented gyroscopes will precess about the axis defined by $\boldsymbol{\Omega}[\mathrm{ss}]$ relative to an inertial frame at infinity. This phenomenon is directly linked to the dragging of a particle moving on a geodesic near a rotating body, as discussed back in Sec. 9.5.3.

The Maxwell-like formulation of post-Newtonian theory summarized in Box 8.1 provides additional insight into the phenomenon of spin–spin precession. We recall that a rotating body produces a gravitomagnetic field $\boldsymbol{B}_g = \nabla \times \boldsymbol{A}_g$, in which $\boldsymbol{A}_g = -4\boldsymbol{U}/c^2$ is a rescaled version of the post-Newtonian vector potential. For a spinning Earth at the spatial origin of the coordinate system, we have that

$$\boldsymbol{A}_g = \frac{2G}{c^2} \frac{\boldsymbol{x} \times \boldsymbol{S}_\oplus}{|\boldsymbol{x}|^3}, \tag{10.149}$$

and this produces a dipolar gravitomagnetic field given by

$$\boldsymbol{B}_g = -\frac{2G}{c^2|\boldsymbol{x}|^3}\left[\frac{3(\boldsymbol{S}_\oplus \cdot \boldsymbol{x})\boldsymbol{x}}{|\boldsymbol{x}|^2} - \boldsymbol{S}_\oplus\right]. \tag{10.150}$$

Comparing this with Eq. (10.140), we see that $\boldsymbol{\Omega}[\mathrm{ss}] = -\frac{1}{2}\boldsymbol{B}_g(\boldsymbol{x} = \boldsymbol{r})$. In electrodynamics, the torque exerted by a magnetic field on a magnetic dipole moment $\boldsymbol{m}$ is $\boldsymbol{\tau} = \boldsymbol{m} \times \boldsymbol{B}$. In gravity we would expect that the torque exerted by a gravitomagnetic field on a gravitomagnetic dipole moment $\boldsymbol{m}_g := \frac{1}{2}\boldsymbol{S}$ is given by

$$\boldsymbol{\tau} = \boldsymbol{m}_g \times \boldsymbol{B}_g = \left(\frac{1}{2}\boldsymbol{S}\right) \times \left(-2\boldsymbol{\Omega}[\mathrm{ss}]\right) = \boldsymbol{\Omega}[\mathrm{ss}] \times \boldsymbol{S}, \tag{10.151}$$

in agreement with the precession equation.

Gravity Probe B

We now return to the description of the Gravity Probe B experiment. In order to maximize the spin–orbit precession, we align the gyroscope so that it points in the orbital plane. We write

$$\boldsymbol{S}_0 = S_0\left(\cos\psi\, \boldsymbol{e}_p + \sin\psi\, \boldsymbol{e}_q\right), \tag{10.152}$$

with ψ denoting the angle between $\boldsymbol{S}_0$ and the line of nodes. It is then easy to show that

$$\begin{aligned} \left\langle \frac{d\boldsymbol{S}}{dt} \right\rangle_{\mathrm{SS}} &\propto \left(\frac{1}{2}\sin\iota\, \boldsymbol{e}_q - \cos\iota\, \boldsymbol{e}_z\right) \times \boldsymbol{S}_0 \\ &\propto \cos\iota \sin\psi\, \boldsymbol{e}_p - \cos\iota\cos\psi\, \boldsymbol{e}_q - \frac{1}{2}\sin\iota\cos\psi\, \boldsymbol{e}_z, \end{aligned} \tag{10.153}$$

and that the length of this vector is proportional to $[1 - \sin^2\iota(1 - \frac{1}{4}\cos^2\psi)]^{1/2}$. It becomes clear that the effect is maximized by setting $\iota = 0$ and therefore placing the orbit in the Earth's equatorial plane. This, however, is not the optimal choice for the operations of Gravity Probe B, because in this case $\langle d\boldsymbol{S}/dt\rangle_{\mathrm{SO}}$ and $\langle d\boldsymbol{S}/dt\rangle_{\mathrm{SS}}$ are both aligned in the direction of $\boldsymbol{e}_z \times \boldsymbol{S}_0$. In this orientation the spin–spin precession would be very difficult to distinguish from the dominant spin–orbit precession, and the configuration does not permit a clean measurement of each effect separately. A much better strategy is to select an orbit that produces spin–orbit and spin–spin precessions that are perpendicular to each other. It is easy to show that the inner product between $\langle d\boldsymbol{S}/dt\rangle_{\mathrm{SO}}$ and $\langle d\boldsymbol{S}/dt\rangle_{\mathrm{SS}}$ is proportional to

$\cos\iota$, and setting $\iota = \frac{\pi}{2}$ forces this to be zero. The optimal orbit, therefore, is a *polar orbit*. With these specifications the spin-precession equations become

$$\left\langle \frac{d\boldsymbol{S}}{dt} \right\rangle_{\rm SO} = \frac{3}{2c^2 a}\left(\frac{GM_\oplus}{a}\right)^{3/2}\left(-\sin\psi\,\boldsymbol{e}_p + \cos\psi\,\boldsymbol{e}_q\right), \tag{10.154a}$$

$$\left\langle \frac{d\boldsymbol{S}}{dt} \right\rangle_{\rm SS} = -\frac{GS_\oplus}{2c^2 a^3}\cos\psi\,\boldsymbol{e}_z. \tag{10.154b}$$

In this geometry the spin–orbit precession is in the orbital plane, and the spin–spin precession is normal to the plane.

The Gravity Probe B spacecraft was launched into an almost perfectly circular polar orbit at an altitude of 642 km, with the orbital plane parallel to the direction of a guide star known as *IM Pegasi*. The spacecraft contained four spheres made of fused quartz, all spinning about the same axis (two were spun in the opposite direction), which was oriented to be in the orbital plane, pointing toward the guide star. An onboard telescope pointed continuously at the guide star, and the direction of each spin was compared with the direction to the star, which was at a declination of 16° relative to the Earth's equatorial plane. With ψ chosen such that the angle between $\boldsymbol{e}_q$ (now parallel to the Earth's rotation axis) and the spins is equal to 74°, we obtain a value of 6630 mas/yr for the spin–orbit precession, and of 38 mas/yr for the spin–spin precession, the first in the orbital plane (in the north–south direction) and the second perpendicular to it (in the east–west direction).

In order to reduce the non-relativistic torques on the rotors to an acceptable level, the rotors were fabricated to be both spherical and homogenous to better than a few parts in 10 million. Each rotor was coated with a thin film of niobium, and the experiment was conducted at cryogenic temperatures inside a dewar containing 2200 litres of superfluid liquid helium. As the niobium film becomes a superconductor, each rotor develops a magnetic moment parallel to its spin axis. Variations in the direction of the magnetic moment relative to the spacecraft were then measured using current loops surrounding each rotor. As the spacecraft orbits the Earth, the aberration of light from the guide star causes an artificial but predictable change in direction between the rotors and the on-board telescope; this was an essential tool for calibrating the conversion between the voltages read by the current loops and the actual angle between the rotors and the guide star.

The spacecraft was launched on April 20, 2004, and the mission ended in September 2005, as scheduled, when the remaining liquid helium boiled off. Although all subsystems of the spacecraft and the apparatus performed extremely well, they were not perfect. Calibration measurements carried out during the mission, both before and after the science phase, revealed unexpectedly large torques on the rotors, believed to be caused by electrostatic interactions between surface imperfections on the niobium films and the spherical housings surrounding each rotor. These effects and other anomalies greatly contaminated the data and complicated its analysis, but finally, in October 2010, the Gravity Probe B team announced that the experiment had successfully measured both the geodetic and frame-dragging precessions. The outcome was in agreement with general relativity, with a precision of 0.3 percent for the spin–orbit precession, and 20 percent for the spin–spin precession.

10.4.2 Frame dragging and LAGEOS satellites

We next consider the influence of the Earth's spin on the orbital motion of a satellite, which we take to have a negligible mass and spin. The spin–orbit contribution to the relative acceleration between the satellite and the Earth is given by Eq. (9.252), and with our assumptions we find that it becomes

$$\boldsymbol{a}[\mathrm{so}] = \frac{2G}{c^2 r^3}\Big[3(\boldsymbol{n}\times\boldsymbol{v})\cdot\boldsymbol{S}_\oplus\,\boldsymbol{n} + 3(\boldsymbol{n}\cdot\boldsymbol{v})\boldsymbol{n}\times\boldsymbol{S}_\oplus - 2\boldsymbol{v}\times\boldsymbol{S}_\oplus\Big], \tag{10.155}$$

in which $\boldsymbol{r}$ is the satellite's position relative to Earth, $\boldsymbol{v}$ its velocity, and as usual $\boldsymbol{n} := \boldsymbol{r}/r$. Recalling that $\boldsymbol{h} := \boldsymbol{r}\times\boldsymbol{v} = h\boldsymbol{e}_z$ is the conserved angular momentum (per unit reduced mass) of the Keplerian motion, and writing $\boldsymbol{S}_\oplus = S_\oplus \boldsymbol{e}_Z$, the spin–orbit acceleration becomes

$$\boldsymbol{a}[\mathrm{so}] = -\frac{2GS_\oplus}{c^2 r^3}\Big[2\boldsymbol{v}\times\boldsymbol{e}_Z - 3\dot{r}\,\boldsymbol{n}\times\boldsymbol{e}_Z - \frac{3h}{r}(\boldsymbol{e}_z\cdot\boldsymbol{e}_Z)\boldsymbol{n}\Big], \tag{10.156}$$

with $\dot{r} := \boldsymbol{n}\cdot\boldsymbol{v}$ denoting the radial component of the velocity vector.

We take the spin–orbit acceleration as a perturbing force $\boldsymbol{f}$ to be inserted in the formalism of osculating Keplerian orbits reviewed in Sec. 3.3.2. A short computation using $\boldsymbol{v} = \dot{r}\,\boldsymbol{n} + (h/r)\,\boldsymbol{\lambda}$ reveals that the components of $\boldsymbol{f}$ in the orbital basis $(\boldsymbol{n}, \boldsymbol{\lambda}, \boldsymbol{e}_z)$ introduced in Sec. 3.2.2 are

$$\mathcal{R} = \frac{2GS_\oplus}{c^2 r^3}\frac{h}{r}\boldsymbol{e}_z\cdot\boldsymbol{e}_Z, \tag{10.157a}$$

$$\mathcal{S} = -\frac{2GS_\oplus}{c^2 r^3}\dot{r}\,\boldsymbol{e}_z\cdot\boldsymbol{e}_Z, \tag{10.157b}$$

$$\mathcal{W} = \frac{2GS_\oplus}{c^2 r^3}\Big(\dot{r}\,\boldsymbol{\lambda}\cdot\boldsymbol{e}_Z + \frac{2h}{r}\boldsymbol{n}\cdot\boldsymbol{e}_Z\Big). \tag{10.157c}$$

In this we insert the Keplerian relations $h = \sqrt{Gmp}$, $r = p/(1+e\cos f)$, $\dot{r} = \sqrt{Gm/p}\,e\sin f$, in which $m \simeq M_\oplus$ is the total mass, p the semi-latus rectum, e the eccentricity, and f is the true anomaly. We also insert the vectorial relations listed in Sec. 3.2.5, according to which $\boldsymbol{e}_z\cdot\boldsymbol{e}_Z = \cos\iota$, $\boldsymbol{\lambda}\cdot\boldsymbol{e}_Z = \sin\iota\cos(\omega+f)$, and $\boldsymbol{n}\cdot\boldsymbol{e}_Z = \sin\iota\sin(\omega+f)$, in which ι is the orbit's inclination relative to Earth's equatorial plane, and ω is the longitude of perigee. Finally, we insert the components of the perturbing force in the osculating equations listed in Sec. 3.3.2 (refer to Eqs. (3.69)), and integrate over a complete orbital cycle to obtain the secular changes in the Keplerian orbital elements.

The final outcome of this sequence of steps is the statement that $\Delta p = \Delta e = \Delta\iota = 0$, but that

$$\Delta\omega = -12\pi\frac{GS_\oplus}{\sqrt{GM_\oplus p^3 c^2}}\cos\iota, \tag{10.158a}$$

$$\Delta\Omega = 4\pi\frac{GS_\oplus}{\sqrt{GM_\oplus p^3 c^2}}. \tag{10.158b}$$

These results can be combined to give

$$\Delta\omega + \cos\iota\,\Delta\Omega = -8\pi\frac{GS_\oplus}{\sqrt{GM_\oplus p^3 c^2}}\cos\iota, \tag{10.159}$$

Table 10.1 Orbital elements of laser-ranged satellites.

Satellite	Semi-major axis (km)	Orbital period (min)	Eccentricity	Inclination to equator (°)	Mass (kg)
LAGEOS I	12 257	225	0.0045	109.84	407
LAGEOS II	12 168	223	0.0135	52.64	405
LARES	7 821	115	0.0007	69.5	387

which describes the secular advance of the perigee relative to a fixed reference direction – the X-axis – in the fundamental plane. The nodal advance $\Delta\Omega$ over a complete orbit can be converted to an averaged rate of advance by dividing by the orbital period, and inserting numerical values appropriate for an Earth-orbiting satellite, we obtain

$$\left(\frac{d\Omega}{dt}\right)_{\rm sec} = 0.2188\left(\frac{R_\oplus}{a}\right)^3 (1-e^2)^{-3/2}\ \text{as/yr}. \tag{10.160}$$

This very small effect on the satellite's motion, measured in fractions of arcseconds per year, is not beyond the reach of modern spacecraft tracking using lasers. In particular, a pair of "Laser Geodynamics Satellites" (LAGEOS) are ideal for this purpose. Launched in 1976 and 1992, the satellites are massive spheres, 60 cm in diameter and weighing about 400 kg, placed in nearly circular orbits with semi-major axes approximately equal to $2R_\oplus$ (the precise orbital elements are listed in Table 10.1). The spheres are covered with laser retroreflectors, conical mirrors designed to reflect a laser beam back to the same direction from which it came. Because of their large mass-to-area ratio, the spheres are less affected by atmospheric drag than other satellites at similar altitudes. This, combined with the high precision of laser ranging (which routinely achieves millimeter-level precision), means that their orbits can be determined extremely precisely. The LAGEOS satellites were launched primarily to carry out studies in geodesy and geodynamics, but it was soon recognized that they were potentially capable of measuring the relativistic nodal advance, which under the stated conditions amounts to approximately 32 mas/yr.

A significant challenge in performing the experiment comes from the fact that the deformed figure of the Earth produces a huge contribution to the nodal advance. The effect of oblateness on orbital motion was examined back in Sec. 3.4.3, and an application of these results to the LAGEOS orbits reveals that the Newtonian perturbation produces an advance of approximately 120 degrees per year, more than 10 million times larger than the relativistic advance. To measure the relativistic contribution to 10 percent accuracy requires a very precise subtraction of the Newtonian effect, which requires knowledge of the Earth's quadrupole moment ($J_2 \sim 10^{-3}$) to better than a part in 10^5. The higher multipole moments J_4, J_6, and so on, also contribute substantially to the nodal advance. The Earth's multipole moments were not known sufficiently well at the time of the launch of the first LAGEOS satellite to permit a clean measurement of the relativistic contribution.

After the launch of LAGEOS I in 1976, it was recognized by Italian physicist Ignazio Ciufolini and others, building on an earlier idea by Stanford physicists Richard van Patten and Francis Everitt, that the Newtonian and relativistic contributions to $\Delta\Omega$ could actually be distinguished from one another. The key observation is that the Newtonian contribution is proportional to $\cos\iota$, while the relativistic contribution is independent of inclination. If a second LAGEOS satellite were to have an inclination supplementary to that of LAGEOS I, so that $\iota_2 = 180^\circ - \iota_1$, then the Newtonian advance would be equal and opposite for the two satellites. (This turns out to be true for the contributions of *all* even-ℓ multipole moments; see Exercise 3.5.) This means that if one measured the nodal advance of both satellites simultaneously and added them together, the large Newtonian effect would cancel out exactly, leaving the relativistic effect in full view.

Ciufolini and other relativists campaigned vigorously to have LAGEOS II launched with $\iota_2 = 70.16^\circ$, but other considerations prevailed in the end. LAGEOS II was launched with $\iota_2 = 52.64^\circ$, mainly to optimize coverage by the world's network of laser tracking stations, which was important for geophysics and geodynamics research. The fall-back option was then to combine the data from the two satellites as they were. One could still eliminate the largest Newtonian contribution coming from J_2 with a suitable linear combination of the two measured nodal advances, thereby revealing the relativistic contribution and those coming from higher-order multipole moments, whose uncertainties would contribute to the error made in measuring the relativistic effect. For a time, Ciufolini and collaborators tried to include a third piece of data, the perigee advance of LAGEOS II (which has a small eccentricity), as a way to also eliminate J_4, but this turned out to be plagued with systematic errors that were large and hard to control.

But then along came CHAMP and GRACE. Europe's CHAMP (Challenging Minisatellite Payload) and NASA's GRACE (Gravity Recovery and Climate Experiment) missions, launched in 2000 and 2002, respectively, use precision tracking of spacecraft to measure variations in the Earth's gravity on scales as small as several hundred kilometers, with unprecedented accuracies. GRACE consists of a pair of satellites flying in close formation (200 kilometers apart) on polar orbits. Each satellite carries an on-board accelerometer to measure non-gravitational perturbations, a satellite-to-satellite K-band radar to measure variations in the Earth's gravity gradient on short scales, and a GPS tracking unit to measure larger scale variations. With the dramatic improvements on J_ℓ obtained by CHAMP and GRACE, Ciufolini and his colleagues could now treat J_4 and higher multipole moments as known, and use the two LAGEOS nodal advances to determine J_2 and the relativistic contribution. The final outcome was a successful test of the relativistic prediction at the level of 10 percent accuracy, which was later confirmed by independent analyses of the LAGEOS/GRACE/CHAMP data.

On February 13, 2012, a third laser-ranged satellite, known as LARES (Laser Relativity Satellite) was launched by the Italian Space Agency. Its inclination was 69.5°, very close to the required supplementary angle relative to LAGEOS I, and its eccentricity was very nearly zero (see Table 10.1). Combining data from all three satellites with continually improving Earth data from GRACE, the LARES team led by Ciufolini hopes to achieve a test of frame-dragging at the 1 percent level.

10.4.3 Binary systems of spinning bodies

In our final exploration of the influence of spin in post-Newtonian gravity, we examine the motion of a binary system of spinning bodies.

Equations of motion and conserved quantities

The equations of motion for the relative orbit and spin vectors of a binary system were derived back in Sec. 9.5.9. The relative acceleration $\boldsymbol{a} := \boldsymbol{a}_1 - \boldsymbol{a}_2$ is expressed as $\boldsymbol{a} = \boldsymbol{a}[0\text{PN}] + \boldsymbol{a}[1\text{PN}] + \boldsymbol{a}[\text{SO}] + \boldsymbol{a}[\text{SS}] + O(c^{-4})$, with the two first terms given by Eq. (10.1), the spin–orbit contribution given by

$$\boldsymbol{a}[\text{SO}] = \frac{G}{c^2r^3}\biggl\{\frac{3}{2}(\boldsymbol{n}\times\boldsymbol{v})\cdot(3\boldsymbol{\sigma}+4\boldsymbol{S})\,\boldsymbol{n} + \frac{3}{2}(\boldsymbol{n}\cdot\boldsymbol{v})\,\boldsymbol{n}\times(3\boldsymbol{\sigma}+4\boldsymbol{S}) - \boldsymbol{v}\times(3\boldsymbol{\sigma}+4\boldsymbol{S}) + \frac{3\lambda}{2}\Bigl[\boldsymbol{n}(\boldsymbol{n}\times\boldsymbol{v})\cdot\boldsymbol{\sigma} - (\boldsymbol{n}\cdot\boldsymbol{v})\,\boldsymbol{n}\times\boldsymbol{\sigma}\Bigr]\biggr\}, \tag{10.161}$$

and the spin–spin contribution given by

$$\boldsymbol{a}[\text{SS}] = -\frac{3G}{\mu c^2 r^4}\Bigl[(\boldsymbol{S}_1\cdot\boldsymbol{S}_2)\boldsymbol{n} - 5(\boldsymbol{S}_1\cdot\boldsymbol{n})(\boldsymbol{S}_2\cdot\boldsymbol{n})\boldsymbol{n} + (\boldsymbol{S}_1\cdot\boldsymbol{n})\boldsymbol{S}_2 + (\boldsymbol{S}_2\cdot\boldsymbol{n})\boldsymbol{S}_1\Bigr]. \tag{10.162}$$

The accelerations are expressed in terms of the combinations $\boldsymbol{\sigma} := (M_2/M_1)\boldsymbol{S}_1 + (M_1/M_2)\boldsymbol{S}_2$ and $\boldsymbol{S} := \boldsymbol{S}_1 + \boldsymbol{S}_2$ of the spin vectors, and the arbitrary parameter λ was introduced back in Sec. 9.5.8 to characterize the choice of representative world line within each body; $m := M_1 + M_2$ and $\mu := M_1M_2/(M_1+M_2)$ are respectively the total and reduced masses of the binary system.

The spin-precession equation for the first body is $d\boldsymbol{S}_1/dt = \boldsymbol{\Omega}_1 \times \boldsymbol{S}_1$ with $\boldsymbol{\Omega}_1 = \boldsymbol{\Omega}_1[\text{SO}] + \boldsymbol{\Omega}_1[\text{SS}]$, where

$$\boldsymbol{\Omega}_1[\text{SO}] = \frac{2G\mu}{c^2r^2}\left(1 + \frac{3M_2}{4M_1}\right)\boldsymbol{n}\times\boldsymbol{v}, \tag{10.163a}$$

$$\boldsymbol{\Omega}_1[\text{SS}] = \frac{G}{c^2r^3}\bigl[3(\boldsymbol{S}_2\cdot\boldsymbol{n})\boldsymbol{n} - \boldsymbol{S}_2\bigr], \tag{10.163b}$$

are the precessional angular velocities. The equations for the second body are obtained by a simple exchange of labels $1 \leftrightarrow 2$.

Back in Sec. 10.1.1 we established that the equations of motion $\boldsymbol{a} = \boldsymbol{a}[0\text{PN}] + \boldsymbol{a}[1\text{PN}] + O(c^{-4})$ admit a conserved energy $E = \mu\varepsilon$ given by Eq. (10.3), and a conserved angular momentum $\boldsymbol{L} = \mu\boldsymbol{h}$ given by Eq. (10.4). It is possible to extend these results and show that the equations of motion with spin–orbit and spin–spin terms included also admit a conserved energy and a conserved angular momentum. The modified energy (per unit reduced mass) takes the form of $\varepsilon = \varepsilon[0\text{PN}] + \varepsilon[1\text{PN}] + \varepsilon[\text{SO}] + \varepsilon[\text{SS}] + O(c^{-4})$, with the first two terms as listed previously, the spin–orbit term given by

$$\varepsilon[\text{SO}] = \lambda\frac{G}{c^2r^2}(\boldsymbol{n}\times\boldsymbol{v})\cdot\boldsymbol{\sigma}, \tag{10.164}$$

and the spin–spin term given by

$$\varepsilon[\text{SS}] = -\frac{G}{\mu c^2 r^3}\Bigl[\boldsymbol{S}_1\cdot\boldsymbol{S}_2 - 3(\boldsymbol{n}\cdot\boldsymbol{S}_1)(\boldsymbol{n}\cdot\boldsymbol{S}_2)\Bigr]. \tag{10.165}$$

Note that the spin–orbit contribution to the conserved energy vanishes when $\lambda = 0$, that is, when the representative world line is chosen to track the body's center-of-mass in the barycentric frame (as opposed to the moving frame of each body). The modified total angular momentum takes the form

$$J = L + S + O(c^{-4}) \tag{10.166}$$

with $L = L[\text{0PN}] + L[\text{1PN}] + L[\text{SO}]$, where the Newtonian and post-Newtonian terms were given by Eq. (10.4), and the spin–orbit term is given by

$$\begin{aligned} L[\text{SO}] &= \frac{G\mu}{2c^2 r} n \times \big[n \times (3\sigma + 4S)\big] \\ &\quad - \frac{\lambda}{2c^2}\left[\frac{G\mu}{r} n \times (n \times \sigma) + \frac{\mu}{m} v \times (v \times \sigma)\right]; \end{aligned} \tag{10.167}$$

there is no spin–spin contribution to the total angular momentum. The facts that $d\varepsilon/dt = 0$ and $dJ/dt = 0$ by virtue of the post-Newtonian dynamics can be verified with a straightforward computation involving the complete set of equations of motion; for this purpose it is convenient to also make use of the evolution equation

$$\frac{dS}{dt} = \frac{G\mu}{2c^2 r^2}(n \times v) \times (3\sigma + 4S) + \frac{3G}{c^2 r^3}\Big[(S_1 \cdot n) n \times S_2 + (S_2 \cdot n) n \times S_1\Big] \tag{10.168}$$

for the total spin vector.

Simple precession

To explore the effects produced by the spin–orbit coupling in the equations of motion of a binary system, we examine first the case in which only one of the bodies has spin; this is sometimes called the "simple precession" case. Taking the first body to have spin, and setting $S_2 = 0$, the precession equation for $S = S_1$ can be expressed as

$$\frac{dS}{dt} = \frac{2G}{c^2 r^3}\left(1 + \frac{3M_2}{4M_1}\right) L[\text{0PN}] \times S, \tag{10.169}$$

where $L[\text{0PN}] = \mu r \times v$ is the Newtonian expression for the orbital angular momentum. Expressing $1 + 3M_2/(4M_1)$ as $\frac{1}{4}(1 + 3m/M_1)$, and $L[\text{0PN}]$ as $J - S + O(c^{-2})$, we find that the equation becomes

$$\frac{dS}{dt} = \frac{G}{2c^2 r^3}\left(1 + \frac{3m}{M_1}\right) J \times S, \tag{10.170}$$

in which corrections of order c^{-4} are neglected. Because $J = L + S$ is conserved, this equation implies that $dL/dt = -dS/dt$ is also proportional to $J \times S$. Replacing S by $J - L$ in this expression, we arrive at

$$\frac{dL}{dt} = \frac{G}{2c^2 r^3}\left(1 + \frac{3m}{M_1}\right) J \times L \tag{10.171}$$

for the rate of change of the orbital angular momentum. These equations reveal that both S and L precess about a fixed J at a common angular frequency

$$\Omega_{\text{prec}} = \frac{GJ}{2c^2 r^3}\left(1 + \frac{3m}{M_1}\right); \tag{10.172}$$

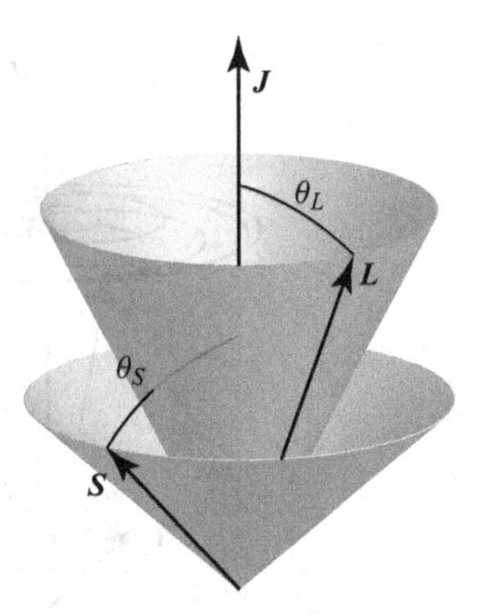

Fig. 10.5 Simple precession of $\boldsymbol{L}$ and $\boldsymbol{S}$ about a fixed total angular momentum $\boldsymbol{J}$.

the precession preserves the angle θ_L between $\boldsymbol{L}$ and $\boldsymbol{J}$, and the angle θ_S between $\boldsymbol{S}$ and $\boldsymbol{J}$ (see Fig. 10.5). The precession of $\boldsymbol{L}$ about $\boldsymbol{J}$ implies that the binary's orbital plane no longer has a fixed orientation in space, as it does in the Newtonian and spinless post-Newtonian dynamics. The normal to the orbital plane is aligned with $\boldsymbol{L}[0\text{PN}] = \mu \boldsymbol{r} \times \boldsymbol{v}$, and because this deviates from $\boldsymbol{L}$ by terms of order c^{-2}, we see that the orbital plane does not quite follow the precessional motion; there is an extra wobble produced by the spin–orbit contribution to the orbital angular momentum. On average, however, the motion of the orbital plane is described by a steady precession, and this phenomenon has already been encountered in our discussion of the LAGEOS satellites. It is easy to show that the precession angle integrated over a complete orbit corresponds precisely to the nodal advance given by Eq. (10.158).

Simple precession can also occur when both bodies are spinning, but in this case the orbital angular momentum is required to be *much larger* than the spin angular momenta, and the masses are required to be equal. The demand for a large $\boldsymbol{L}$ is to ensure that the spin–spin coupling continues to be negligible compared to the spin–orbit coupling, and the demand for equal masses is to ensure equality of the precessional angular velocities, which are proportional to $(1 + 3m/M_1)$ and $(1 + 3m/M_2)$, respectively. In such circumstances the precession equations become

$$\frac{d\boldsymbol{S}_1}{dt} \simeq \frac{7G}{2c^2r^3}\boldsymbol{L}[0\text{PN}] \times \boldsymbol{S}_1, \qquad \frac{d\boldsymbol{S}_2}{dt} \simeq \frac{7G}{2c^2r^3}\boldsymbol{L}[0\text{PN}] \times \boldsymbol{S}_2. \tag{10.173}$$

These equations can be combined to give a precession equation for the total spin $\boldsymbol{S} = \boldsymbol{S}_1 + \boldsymbol{S}_2$, and the $\boldsymbol{L}[0\text{PN}]$ factor on the right-hand side can again be expressed as $\boldsymbol{J} - \boldsymbol{S} + O(c^{-2})$. The result can then be used to derive a precession equation for $\boldsymbol{L}$, and we arrive at

$$\frac{d\boldsymbol{S}}{dt} \simeq \frac{7G}{2c^2r^3}\boldsymbol{J} \times \boldsymbol{S}, \qquad \frac{d\boldsymbol{L}}{dt} \simeq \frac{7G}{2c^2r^3}\boldsymbol{J} \times \boldsymbol{L}. \tag{10.174}$$

These equations imply that $\boldsymbol{S}$ and $\boldsymbol{L}$ both precess around $\boldsymbol{J}$ with a common angular velocity $\Omega_{\text{prec}} = 7GJ/(2c^2r^3)$.

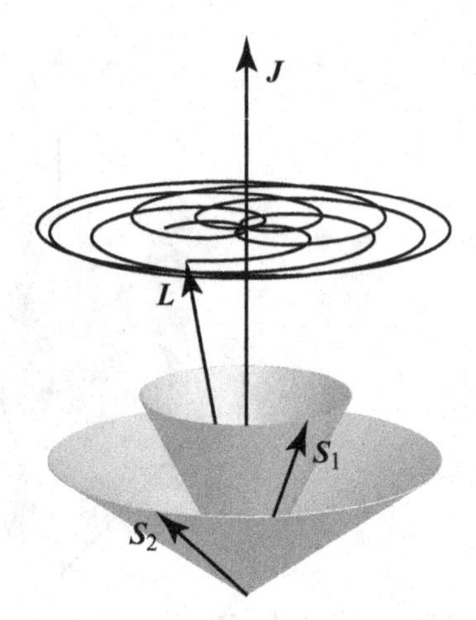

Fig. 10.6 Precession of $\boldsymbol{L}$ about a fixed $\boldsymbol{J}$ for two bodies with $M_1 = 2M_2$, with spins of equal magnitudes whose projections perpendicular to $\boldsymbol{J}$ are initially at right angles.

The precessions remain reasonably simple even when the masses are unequal, provided that we maintain the condition that $\boldsymbol{L}$ is much larger than both $\boldsymbol{S}_1$ and $\boldsymbol{S}_2$. In this case the factor $\boldsymbol{L}[0\text{PN}] \times \boldsymbol{S}_1$ in $d\boldsymbol{S}_1/dt$ can be expressed as $[\boldsymbol{J} - \boldsymbol{S}_1 - \boldsymbol{S}_2 + O(c^{-2})] \times \boldsymbol{S}_1$, and since $\boldsymbol{J}$ is dominated by orbital angular momentum, this can be approximated by $\boldsymbol{J} \times \boldsymbol{S}_1$. The resulting equations are

$$\frac{d\boldsymbol{S}_1}{dt} \simeq \frac{G}{2c^2r^3}\left(1 + \frac{3m}{M_1}\right)\boldsymbol{J} \times \boldsymbol{S}_1, \tag{10.175a}$$

$$\frac{d\boldsymbol{S}_2}{dt} \simeq \frac{G}{2c^2r^3}\left(1 + \frac{3m}{M_2}\right)\boldsymbol{J} \times \boldsymbol{S}_2, \tag{10.175b}$$

and the spins precess with different angular velocities. The motion of $\boldsymbol{L} = \boldsymbol{J} - \boldsymbol{S}_1 - \boldsymbol{S}_2$ incorporates both precessions, and this leads to a complicated trajectory of the type displayed in Fig. 10.6.

Spin–spin coupling

Thus far we have considered situations in which the spin–spin aspects of the orbital dynamics could be neglected. To conclude our survey of the dynamics of a binary system of spinning bodies, we examine the conditions under which the spin–spin aspects of the dynamics can become important.

The importance of the spin–spin coupling on body 1 (say) can be measured by the ratio

$$\frac{\text{ss}}{\text{so}} := \frac{\Omega_1[\text{ss}]}{\Omega_1[\text{so}]}, \tag{10.176}$$

in which $\Omega_1[\text{ss}]$ is the magnitude of the spin–spin precessional angular velocity vector, and $\Omega_1[\text{so}]$ the magnitude of the spin–orbit angular velocity. In order of magnitude we have that $\Omega_1[\mathrm{so}] = GL(1+3m/M_1)/(2c^2r^3)$ and $\Omega_1[\mathrm{ss}] = 2GS_2/(c^2r^3)$, so that

$$\frac{\mathrm{ss}}{\mathrm{so}} = \frac{4}{1+3m/M_1}\frac{S_2}{L}. \tag{10.177}$$

To estimate this we take body 2 to have a radius R_2 and a rotational frequency ω_2, so that $S_2 \sim M_2R_2^2\,\omega_2$. We also let $L \sim \mu r^2\Omega_K$, in which $\Omega_K = \sqrt{Gm/r^3}$ is the Keplerian angular velocity of the orbital motion. Making the substitutions, we obtain

$$\frac{\mathrm{ss}}{\mathrm{so}} = \frac{4m}{3m+M_1}\left(\frac{R_2}{r}\right)^2\frac{\omega_2}{\Omega_K}. \tag{10.178}$$

A useful bound on this is obtained by recalling that ω_2 is limited by the Keplerian angular velocity at the body's surface, so that $\omega_2 < \sqrt{GM_2/R_2^3}$. This gives

$$\frac{\mathrm{ss}}{\mathrm{so}} < \frac{4m}{3m+M_1}\left(\frac{M_2}{m}\right)^{1/2}\left(\frac{R_2}{r}\right)^{1/2}. \tag{10.179}$$

This bound, together with our usual requirement that $R_2 \ll r$, implies that $\mathrm{ss}/\mathrm{so} \ll 1$, and that the spin–spin aspects of the dynamics are typically small.

An exception to this rule arises when body 1 is very small (in both size and mass), so that it can approach body 2 without suffering from tidal disruption. In such circumstances Eq. (10.178) simplifies to $\mathrm{ss}/\mathrm{so} \simeq (4/3)(\omega_2/\Omega_K)$, which can be large whenever ω_2 is comparable to Ω_K. This is precisely the situation for Gravity Probe B: plugging in the angular velocities for the Earth and for a low Earth orbit, we find that ss/so is comparable to the ratio identified in Sec. 10.4.1 between the 43 mas frame-dragging precession and the 6600 mas geodetic precession.

Another exception to the general rule arises in the case of a compact object, a neutron star or a black hole, for which the rotational velocity can approach the speed of light, producing a very large spin. The spin of a compact object scales as

$$S_2 = \chi\frac{GM_2^2}{c}, \tag{10.180}$$

in which χ is a dimensionless parameter that ranges between 0 and approximately 0.6 for a neutron star, and between 0 and 1 for a Kerr black hole. Inserting this and $L \sim \mu\sqrt{Gmr}$ within Eq. (10.177), we find that

$$\frac{\mathrm{ss}}{\mathrm{so}} = \frac{4M_2}{4M_1+3M_2}\chi\sqrt{\frac{Gm}{c^2r}} \tag{10.181}$$

for a binary system involving compact objects. For a very tight binary at the natural endpoint of its orbital evolution under gravitational radiation reaction (refer to Chapter 12 below), the spin–spin aspects of the dynamics become important when r drops below approximately $10Gm/c^2$. This is especially true when one of the spins becomes aligned with $\boldsymbol{L}$ so that the spin–orbit term passes through zero. The system then loses its gyroscopic bearing, and $\boldsymbol{L}$, $\boldsymbol{S}_1$, and $\boldsymbol{S}_2$ can all precess wildly. The orbital angular momentum can even change sign, so that an initially clockwise orbit flips over and becomes counter-clockwise. These strange gyrations, however, occur in a strong-field regime where post-Newtonian theory is gradually breaking down, and it is only with the advent of numerical relativity that the rich dynamical behavior of spinning binaries could be explored to the fullest extent.

10.5 Bibliographical notes

Post-Newtonian celestial mechanics provides a vast field of study that we have sampled only sparsely in this chapter. For a more comprehensive survey the reader is invited to consult the treatises by Soffel (1989) and Brumberg (1991).

The methods introduced in Sec. 10.1.5 to integrate the post-Newtonian equations of motion of a two-body system originate from Damour and Deruelle (1985). The relativistic precession of the line of nodes of the lunar orbit, reviewed in Sec. 10.1.6, was first predicted by de Sitter (1916), and the effect was first measured by Shapiro *et al.* (1988).

The story of the 1919 eclipse expedition to measure the relativistic deflection of light, mentioned in Sec. 10.2.3, is engagingly told in the book *Einstein's Jury* by Jeffrey Crelinsten (2006), and in Daniel Kennefick's *Physics Today* article from 2009. The official report on the expedition is published as Dyson, Eddington, and Davidson (1920). The results of light-deflection measurements carried out by the VLBI between 1979 and 1999 are compiled in Shapiro *et al.* (2004), and the Hipparcos measurements are reported in Froeschlé, Mignard, and Arenou (1997). The "Newtonian" calculation described in Box 10.3, which originates in the independent work of Cavendish and von Soldner, is described in more detail in Will (1988).

The theory of gravitational lenses, touched upon in Sec. 10.2.4, is developed systematically in the excellent monograph by Schneider, Ehlers, and Falco (1992). The discovery of the first lens, in the form of the double quasar Q0957+561, was reported by Walsh, Carswell, and Weymann (1979). Results of the MACHO experiment are recorded in Alcock *et al.* (2000), and the first discovery of an exoplanet via gravitational lensing is described in Bond *et al.* (2004).

The calculation of the Shapiro time delay (Sec. 10.2.5) was first published in Shapiro (1964), and the results of the first measurement were reported in Shapiro *et al.* (1968). The improved experiment involving the Cassini spacecraft is described in Bertotti, Iess, and Tortora (2003).

The relativistic aspects of the Global Positioning System, described in Sec. 10.3.5 and Box 10.4, are treated systematically in Neil Ashby's excellent *Living Reviews* article from 2003. Our discussion of the timing of binary pulsars in Sec. 10.3.6 borrows heavily from Blandford and Teukolsky (1976) and Damour and Deruelle (1986).

The final results of the Gravity Probe B experiment, described in Sec. 10.4.1, were reported in Everitt *et al.* (2011). The dragging of inertial frames was first studied by Thirring and Lense (1918), and the calculation of the spin–spin precession was first presented in the open literature in Schiff (1960). The independent classified calculation by Pugh (1959) was later reprinted in 2003. The successful measurement of the dragging of inertial frames on the orbital motion of the LAGEOS satellites, as described in Sec. 10.4.2, was reported in Ciufolini and Pavlis (2004).

10.6 Exercises

10.1 Show that the post-Newtonian energy of Eq. (10.3) and the post-Newtonian angular momentum of Eq. (10.4) are conserved under the two-body dynamics reviewed in Sec. 10.1.1.

10.2 In the formulation of the post-Newtonian two-body problem in Sec. 10.1.3, express the post-Newtonian energy ε and post-Newtonian angular momentum $\boldsymbol{h}$ explicitly in terms of the osculating orbital elements. Using Eq. (10.14), show that ε and $\boldsymbol{h}$ are constant at 1PN order.

10.3 Consider the osculating equations (10.14) in the limit of small eccentricity. Show that a circular orbit (an orbit with r constant) does *not* correspond to $e = 0$. Find a solution for p, e, and ω that corresponds to a circular post-Newtonian orbit, and give an interpretation of the orbit in the language of osculating Keplerian orbits.

10.4 Calculate the "Newtonian deflection of light" according to the prescription of Box 10.3. Show that the result is half the deflection predicted by general relativity.

10.5 In this problem we formulate alternative expressions for the deflection vector of a gravitational lens.

(a) Show that $\boldsymbol{\alpha}(\boldsymbol{\xi}) = \boldsymbol{\nabla}_{\xi}\psi$, in which ∇_{ξ} is the gradient operator associated with the vector $\boldsymbol{\xi}$, and

$$\psi(\boldsymbol{\xi}) = -\frac{4G}{c^2}\int \Sigma(\boldsymbol{\xi}')\ln|\boldsymbol{\xi} - \boldsymbol{\xi}'|\,d^2\xi'$$

is a deflection potential.

(b) Show that the deflection angle can also be expressed as

$$\boldsymbol{\alpha}(\boldsymbol{\xi}) = \frac{2}{c^2}\int \boldsymbol{\nabla}_{\xi}U\,d\ell,$$

in which U is the Newtonian potential expressed in terms of the coordinates $(\boldsymbol{\xi}, \ell)$ introduced in Sec. 10.2.4.

10.6 Show that the magnification of the images of a Schwarzschild gravitational lens can be written in the form

$$\mu_{\pm} = \frac{1}{1 - (\theta_{\rm E}/\theta_{\pm})^4},$$

in which $\theta_{\rm E}$ is the Einstein angle and $\theta_{\pm}$ are the two solutions to the lens equation.

10.7 Consider a Schwarzschild gravitational lens, and a circularly symmetric source whose center is at an undeflected angle β_0 from the lens in the x-direction. Assume that the source has an angular diameter 2χ, with $\chi < \beta_0$, and model any point on the edge of the source as being on a circle described by $\beta(\phi) = (\beta_0 + \chi\cos\phi)\boldsymbol{e}_x + \chi\sin\phi\,\boldsymbol{e}_y$, with ϕ ranging from 0 to 2π.

(a) In the limit $\chi \ll \beta_0$, show that the image is distorted into an ellipse, with a minor axis parallel to the direction of the image displacement, and with the ratio of minor to major axes given by $\beta_0/\sqrt{\beta_0^2 + 4\theta_E^2}$.

(b) As χ increases for fixed β_0, show that the ellipse becomes concave, i.e. becomes an arc, when

$$\frac{\chi}{(\beta_0 - \chi)\sqrt{\beta_0^2 - \chi^2}} \geq \frac{1}{2\theta_E}.$$

10.8 The Schwarzschild lens is a rather poor model of a gravitational lens produced by a galaxy, and a better approximation of a galactic mass distribution is provided by the Plummer expression

$$m(\xi) = M\frac{(\xi/\xi_c)^2}{1 + (\xi/\xi_c)^2},$$

in which M is the total mass of the galaxy, and ξ_c is a "core radius" inside which most of the mass density resides.

(a) Calculate the mass density $\Sigma(\xi)$ and plot it as a function of ξ/ξ_c.

(b) Show that the lens equation for a Plummer lens is given by

$$\theta - \left(\frac{\theta_E}{\theta_c}\right)^2 \frac{\theta}{1 + (\theta/\theta_c)^2} = \beta,$$

in which θ_E is the Einstein angle of Eq. (10.90), and $\theta_c := \xi_c/D_L$.

(c) The lens equation is now a cubic equation for θ, and the number of solutions will depend on the sign of the discriminant Δ: the lens produces three images when $\Delta > 0$, and a single image when $\Delta < 0$. Show that the discriminant is given by

$$\Delta = -4\beta^4 - \left(8 + 20\theta_E^2 - \theta_E^4\right)\beta^2 + 4\left(\theta_E^2 - 1\right)^3,$$

in which β and θ_E are now expressed in units of θ_c.

(d) Show that the sign of Δ is dictated by the parameter β_* defined by

$$\beta_*^2 = -1 - \frac{5}{2}\theta_E^2 + \frac{1}{8}\theta_E^4 + \frac{1}{8}\theta_E\left(\theta_E^2 + 8\right)^{3/2},$$

so that $\Delta > 0$ when $\beta < \beta_*$, while $\Delta < 0$ when $\beta > \beta_*$.

The Plummer lens therefore produces three images when β is smaller than the critical value β_*. It is known that the production of an *odd number of images* is typical in axially symmetric lenses with an extended distribution of mass.

10.9 A rocket is launched from the surface of a body of mass M and radius R on a radial orbit, reaching a maximum distance $r_{\rm max}$ before returning to the planet. The rocket carries an atomic clock, while an identical clock remains on the body's surface. During the flight, the clock on the ground emits a signal with a frequency f_0, and this signal is received by the rocket with a frequency f' as measured onboard the rocket. A signal of that same frequency is then generated and returned to the ground (the device that achieves this is called a transponder); there it is measured to have a frequency f_1. This is the method used for Doppler tracking. At the same moment as the tracking signal is transponded back, the rocket emits its own signal of frequency f_0 as measured onboard. This signal is received on the ground, and the frequency is measured to be f_2.

(a) Show that

$$\frac{f_1}{f_0} = \frac{1 - \dot{r}/c}{1 + \dot{r}/c} + O(c^{-3}),$$

$$\frac{f_2}{f_0} = 1 - \frac{\dot{r}}{c} + \frac{GM}{c^2}\left(\frac{1}{R} - \frac{1}{r_{\max}}\right) + O(c^{-3}),$$

where $\dot{r}$ is the radial velocity of the rocket.

(b) Define the combination of frequency ratios $\Delta := (f_2/f_0) - (f_1/2f_0) - 1/2$. Show that this combination eliminates the first-order Doppler effect and is given by

$$\Delta = \frac{GM}{c^2}\left(\frac{1}{R} - \frac{2}{r} + \frac{1}{r_{\max}}\right) + O(c^{-3}). \tag{10.182}$$

Discuss the qualitative behavior of the frequency shift during the experiment.

(c) For the Earth, and for a rocket reaching a maximum altitude of 10 000 km, determine the maximum value of Δ. This was the basis for the 1976 Gravity Probe A experiment of Robert Vessot and colleagues, using hydrogen maser clocks and a Scout rocket launched from Wallops Island, Virginia. The experiment verified the prediction for Δ to about 70 parts per million.

10.10 Using the recipe concocted in Sec. 10.3.6, write a computer code that calculates t_a, the time of arrival of a radio pulse emitted by a binary pulsar, as a function of τ_e, the proper time of emission. The code should have the pulsar's intrinsic frequency ω, the masses M_1 and M_2, the post-Newtonian orbital elements a and e, and the positional elements ι and ω as input parameters. You have complete latitude over the design and implementation of the algorithm. Be clever!

10.11 This exercise approaches the de Sitter precession of the lunar orbit from a different viewpoint.

(a) Treat the Earth–Moon system as a giant gyroscope with spin $\boldsymbol{S} = \mu \boldsymbol{r} \times \boldsymbol{v}$, where μ is the reduced mass. Imagine that this gyroscope is on a circular orbit around the Sun. Show that the spin–orbit precession of $\boldsymbol{S}$ produces the de Sitter precession as given by Eq. (10.44).

(b) Using Eq. (10.42), show that the rate of change of the Moon's angular-momentum vector $\boldsymbol{h} = \boldsymbol{r} \times \boldsymbol{v}$ can be expressed as

$$\frac{d\boldsymbol{h}}{dt} = \frac{3}{2}\frac{GM_3}{c^2 R^2}(\boldsymbol{N} \times \boldsymbol{V}) \times \boldsymbol{h}, \tag{10.183}$$

where M_3 is the mass of the Sun, R the Earth–Sun distance, $\boldsymbol{N}$ a unit vector pointing from the Sun to the Earth, and $\boldsymbol{V}$ the Earth's velocity. Assume that the orbit of the Earth around the Sun is circular. Compare this result with your simple model in the first part of this exercise. *Hint:* Any post-Newtonian expression that can be converted to total time derivatives can be absorbed into the definition of $\boldsymbol{h}$; be careful to stay within the approximations used to construct Eq. (10.42), which amount to keeping only terms linear in $\boldsymbol{v}$, the Moon's velocity relative to Earth.

10.12 The gravitomagnetic viewpoint elaborated in Box 8.1 provides useful insights into a number of post-Newtonian effects, most notably the behavior of a gyroscope around a rotating body. Review this material and then provide qualitative answers to the following questions, using only "right-hand-rule" techniques and lessons learned from electromagnetism. To keep everything simple, model both the Earth and a gyroscope as loops of matter rotating about an axis perpendicular to the plane of the loop.

(a) Show that the gravitomagnetic field of the Earth has the same dipole form as the magnetic field of a loop of positive charge rotating in the same direction, but with the field lines emanating from the south pole and returning to the north pole.

(b) A gyroscope (loop) sits in a gravitomagnetic field that is perpendicular to the plane of the loop. Show that the torque on the gyroscope vanishes.

(c) A gyroscope sits in a gravitomagnetic field that lies in the plane of the loop. Show that the torque $d\boldsymbol{S}/dt$ on the gyroscope is in the direction of $\boldsymbol{S} \times \boldsymbol{B}_g$, where $\boldsymbol{S}$ is the spin vector of the gyroscope.

(d) Show that a gyroscope at the Earth's equator, with its spin axis parallel to the equator, precesses in a direction opposite to the Earth's rotation. Show that a similar gyroscope at the pole precesses in the same direction as the Earth's rotation. How do these qualitative conclusions compare with the corresponding precessions derived from Eq. (10.140b)?

10.13 In the formulation of the dynamics of spinning binaries provided in Sec. 10.4.3, show that the energy per unit reduced mass ε and the total angular momentum $\boldsymbol{J}$ are conserved quantities.

11 Gravitational waves

In the preceding three chapters we stayed safely in the near zone and ignored all radiative aspects of the motion of bodies subjected to a mutual gravitational interaction. In this chapter we move to the wave zone and determine the gravitational waves produced by the moving bodies. To achieve this goal we must return to the post-Minkowskian approximation developed in Chapters 6 and 7, because the post-Newtonian techniques of Chapter 8 are necessarily restricted to the near zone.

We begin in Sec. 11.1 by reviewing the notion of *far-away wave zone*, in which the gravitational-wave field can be extracted from the (larger set of) gravitational potentials $h^{\alpha\beta}$; we explain how to perform this extraction and obtain the gravitational-wave polarizations h_+ and $h_\times$. In Sec. 11.2 we derive the famous *quadrupole formula*, the leading term in an expansion of the gravitational-wave field in powers of v_c/c (with v_c denoting a characteristic velocity of the moving bodies); we flesh out this discussion by examining a number of applications of the formula. Section 11.3 is a very long excursion into a computation of the gravitational-wave field beyond the quadrupole formula, in which we add corrections of fractional order (v_c/c), $(v_c/c)^2$, and $(v_c/c)^3$ to the leading-order expression. The calculations are carried out for a system of N bodies, and they reveal a very interesting physical phenomenon: the fact that the waves propagate not in the fictitious flat spacetime of post-Minkowskian theory, but in a physical spacetime which is curved by the total mass-energy contained in the N-body system. The true waves are delayed with respect to the fictitious waves because they climb out of a gravitational potential well as they travel from the near zone to the wave zone. In Sec. 11.4 we convert the general formalism of the preceding section into concrete expressions for h_+ and $h_\times$ by restricting the number of bodies to two; we first derive general expressions for arbitrary (eccentric) motion, and next specialize our results to circular orbits. We conclude the chapter with Sec. 11.5, where we show how to relate the polarizations h_+ and $h_\times$ to the output channel of a laser interferometric gravitational-wave detector.

The radiative themes explored in this chapter are developed further in Chapter 12, in which we determine the effects of radiative losses on the motion of an N-body system. This is the phenomenon of *radiation reaction*, which reveals a direct link between the near zone and the wave zone.

11.1 Gravitational-wave field and polarizations

11.1.1 Far-away wave zone

The notion of wave zone was first introduced back in Sec. 6.3.2; this is the region of three-dimensional space in which $R := |\boldsymbol{x}|$ is larger than λ_c, the characteristic wavelength of the gravitational-wave field. (In previous chapters the length of the position vector was denoted r instead of R; for our purposes in this chapter we adapt the notation and keep r available for a later assignment.) By *far-away wave zone* we mean the farthest reaches of the wave zone, a neighborhood of spatial infinity in which the R^{-1} part of the gravitational potentials $h^{\alpha\beta}$ dominates over the parts that fall off as R^{-2} and faster.

To illustrate these notions, and the distinctions between behavior in the near zone and behavior in the wave zone, we return to the scalar dipole field of Box 6.6,

$$\psi(t, \boldsymbol{x}) = (\boldsymbol{p} \cdot \boldsymbol{N})\left(\frac{\cos \omega\tau}{R^2} - \frac{\omega}{c}\frac{\sin \omega\tau}{R}\right), \tag{11.1}$$

which is an exact solution to the wave equation $\Box\psi = 0$ outside a region of radius r_c that contains the source. The potential has the dimension of an inverse length, and $\boldsymbol{p}$ is a constant vector of order r_c; ω is the frequency of oscillation of the dipole, and the characteristic wavelength is $\lambda_c = 2\pi c/\omega$. We have re-introduced the retarded-time variable

$$\tau := t - R/c \tag{11.2}$$

and the unit vector

$$\boldsymbol{N} := \boldsymbol{x}/R\,, \tag{11.3}$$

which points in the direction of the field point $\boldsymbol{x}$. (In previous chapters the unit vector was denoted $\boldsymbol{n}$ instead of $\boldsymbol{N}$; here we change the notation and keep $\boldsymbol{n}$ available for a later assignment.) We assume that the dipole is subjected to a slow-motion condition (refer to Sec. 6.3.2), so that $r_c \ll \lambda_c$. With $t_c = \omega^{-1}$ denoting a characteristic time scale and $v_c = r_c/t_c$ a characteristic velocity, we have that $r_c/\lambda_c \sim v_c/c \ll 1$.

The *near zone* is defined to be the region of space where $R < \lambda_c = 2\pi c/\omega$. In the near zone the potential behaves as

$$\psi = (\boldsymbol{p} \cdot \boldsymbol{N})\frac{\cos \omega t}{R^2}\left[1 + \frac{1}{2}\left(\frac{\omega R}{c}\right)^2 + \cdots\right], \tag{11.4}$$

and we see that when $R \sim r_c$, the leading term is corrected by an expression of fractional order $(r_c/\lambda_c)^2 \sim (v_c/c)^2$. We may say that the correction is of 1PN order, and this is the same near-zone behavior that was identified for the gravitational potentials of post-Newtonian theory.

The *wave zone* is defined to be the region of space where $R > \lambda_c = 2\pi c/\omega$. In the wave zone the potential behaves as

$$\psi = -\frac{\omega}{cR}(\boldsymbol{p} \cdot \boldsymbol{N})\left(\sin \omega\tau - \frac{c}{\omega R}\cos \omega\tau\right), \tag{11.5}$$

and we see that the leading term is corrected by an expression of fractional order λ_c/R. This correction becomes increasingly irrelevant as R increases beyond λ_c, and it is completely negligible in the *far-away wave zone*, our adopted vantage point in this chapter.

The gravitational-wave field is obtained by evaluating the gravitational potentials $h^{\alpha\beta}$ in the far-away wave zone, where we neglect corrections of order $\lambda_c/R \ll 1$. How good is this approximation? For a source of gravitational waves with a frequency of 100 Hz ($\lambda_c \sim$ 3000 km) at a distance of 100 Mpc (3×10^{21} km), the neglected terms are of order 10^{-18}, making this an excellent approximation indeed. Only a subset of the potentials is actually involved: as we shall see below, the gravitational-wave field is the transverse-tracefree (TT) piece of the complete set of potentials, and this is what we aim to calculate in this chapter. We assume that the source of the gravitational field is a bounded distribution of matter subjected to a slow-motion condition, so that it lies well within the near zone.

11.1.2 Gravitational potentials in the far-away wave zone

In Chapter 7 we obtained expressions for the gravitational potentials $h^{\alpha\beta}$ that are applicable in the wave zone. These expressions were accurate in the second post-Minkowskian approximation of general relativity. From the summary provided in Box 7.7, we gather that their behavior in the far-away wave zone is given by

$$h^{00} = \frac{4GM}{c^2R} + \frac{G}{c^4R}C(\tau, \boldsymbol{N}), \tag{11.6a}$$

$$h^{0j} = \frac{G}{c^4R}D^j(\tau, \boldsymbol{N}), \tag{11.6b}$$

$$h^{jk} = \frac{G}{c^4R}A^{jk}(\tau, \boldsymbol{N}). \tag{11.6c}$$

Here, M is the total gravitational mass, as defined by Eq. (7.63) in the case of a fluid system, or by Eq. (9.132) in the case of an N-body system. The functions C, D^j, and A^{jk} depend on the retarded-time variable $\tau := t - R/c$ and the unit vector $\boldsymbol{N} := \boldsymbol{x}/R$. We shall not need their precise forms just yet. In fact, the validity of Eqs. (11.6) extends beyond the post-Minkowskian domain of Sec. 7.1.4. It is easy to show that these equations provide solutions to the wave equations $\Box h^{\alpha\beta} = -16\pi G\tau^{\alpha\beta}/c^4$ provided only that $\tau^{\alpha\beta}$, the effective energy-momentum pseudotensor, falls off at least as fast as R^{-2}. The impact of the harmonic gauge conditions $\partial_\beta h^{\alpha\beta} = 0$ on the solutions is examined below.

Before we proceed we introduce a useful differentiation rule that applies in the far-away far zone:

$$\partial_j h^{\alpha\beta} = -\frac{1}{c}N_j \partial_\tau h^{\alpha\beta}. \tag{11.7}$$

The rule follows from the fact that the potentials depend on the spatial coordinates x^j through the overall factor of R^{-1}, and through the dependence of the functions C, D^j, and A^{jk} on τ and $\boldsymbol{N}$. Because $\partial_j R^{-1} = O(R^{-2})$ and $\partial_j N_k = O(R^{-1})$, the only dependence that matters is in τ, and we have that $\partial_j \tau = -c^{-1}\partial_j R = -c^{-1}N_j$. This, finally, leads to Eq. (11.7), in which a correction term of order R^{-2} is omitted.

11.1.3 Decomposition into irreducible components

Before we examine the impact of the gauge conditions $\partial_\beta h^{\alpha\beta} = 0$ on the gravitational potentials, it is useful to decompose the vector D^j and the tensor A^{jk} into longitudinal and transverse components. The longitudinal direction is identified with $\boldsymbol{N}$, and the machinery to achieve the decomposition was developed back in Sec. 5.5; it is summarized in Box 5.7. The decomposition is simplified by the differential rule of Eq. (11.7).

We write

$$D^j = DN^j + D^j_{\mathrm{T}}, \tag{11.8}$$

with DN^j representing the longitudinal part of D^j, and D^j_{T} its transverse part; the latter is required to satisfy

$$N_j D^j_{\mathrm{T}} = 0. \tag{11.9}$$

The three components of D^j are therefore partitioned into one longitudinal component D, and two transverse components contained in D^j_{T}; these are functions of τ and $\boldsymbol{N}$. Similarly, we write

$$A^{jk} = \frac{1}{3}\delta^{jk}A + \left(N^jN^k - \frac{1}{3}\delta^{jk}\right)B + N^jA^k_{\mathrm{T}} + N^kA^j_{\mathrm{T}} + A^{jk}_{\mathrm{TT}}, \tag{11.10}$$

which is a decomposition of A^{jk} into a trace part $\frac{1}{3}\delta^{jk}A$, a longitudinal-tracefree part $(N^jN^k - \frac{1}{3}\delta^{jk})B$, a longitudinal-transverse part $N^jA^k_{\mathrm{T}} + N^kA^j_{\mathrm{T}}$, and a transverse-tracefree part A^{jk}_{TT}. We impose the constraints

$$N_jA^j_{\mathrm{T}} = 0 \tag{11.11}$$

and

$$N_jA^{jk}_{\mathrm{TT}} = 0 = \delta_{jk}A^{jk}_{\mathrm{TT}}, \tag{11.12}$$

so that the six independent components of A^{jk} are contained in two scalars A and B, two components of a transverse vector A^j_{T}, and two components of a transverse-tracefree tensor A^{jk}_{TT}. The last term in Eq. (11.10) is called the *transverse-tracefree part*, or TT part, of A^{jk}. As we shall see, the radiative parts of the gravitational potentials are contained entirely within A^{jk}_{TT}.

11.1.4 Harmonic gauge conditions

The harmonic gauge conditions are $c^{-1}\partial_\tau h^{00} + \partial_k h^{0k} = 0$ and $c^{-1}\partial_\tau h^{0j} + \partial_k h^{jk} = 0$. In the far-away wave zone they simplify to

$$\partial_\tau\left(h^{00} - h^{0k}N_k\right) = 0, \qquad \partial_\tau\left(h^{0j} - h^{jk}N_k\right) = 0, \tag{11.13}$$

thanks to the differentiation rule of Eq. (11.7). After substituting Eqs. (11.8) and (11.10) into Eqs. (11.6), and these into Eqs. (11.13), we find that the harmonic gauge conditions imply

$$C = D\,, \tag{11.14a}$$

$$D = \frac{1}{3}A + \frac{2}{3}B\,, \tag{11.14b}$$

$$D^j_{\rm T} = A^j_{\rm T}\,. \tag{11.14c}$$

We have set the constants of integration to zero, because an eventual τ-independent term in C would correspond to an unphysical shift in the total gravitational mass M, while a τ-independent term in D^j would be incompatible with the fact that the time-independent part of h^{0j} is associated with the total angular momentum, and must fall off as R^{-2} instead of R^{-1}.

Incorporating these constraints, the gravitational potentials become

$$h^{00} = \frac{4GM}{c^2R} + \frac{G}{c^4R}\frac{1}{3}\bigl(A+2B\bigr)\,, \tag{11.15a}$$

$$h^{0j} = \frac{G}{c^4R}\biggl[\frac{1}{3}(A+2B)N^j + A^j_{\rm T}\biggr]\,, \tag{11.15b}$$

$$h^{jk} = \frac{G}{c^4R}\biggl[\frac{1}{3}\delta^{jk}A + \Bigl(N^jN^k - \frac{1}{3}\delta^{jk}\Bigr)B + N^jA^k_{\rm T} + N^kA^j_{\rm T} + A^{jk}_{\rm TT}\biggr]\,, \tag{11.15c}$$

in which A, B, $A^j_{\rm T}$, and $A^{jk}_{\rm TT}$ are functions of τ and $\boldsymbol{N}$. We now have a total of six independent quantities: one in A, another in B, two in $A^j_{\rm T}$, and two more in $A^{jk}_{\rm TT}$. The gauge conditions have eliminated four redundant quantities.

11.1.5 Transformation to the TT gauge

It is possible, in the far-away wave zone, to specialize the harmonic gauge even further, and to eliminate four additional redundant quantities. We wish to implement a gauge transformation generated by a four-vector field $\zeta^\alpha(x^\beta)$ chosen so as to preserve the harmonic gauge conditions, $\partial_\beta h^{\alpha\beta} = 0$. We first figure out how such a transformation affects the gravitational potentials $h^{\alpha\beta}$.

We saw back in Sec. 5.5 that when the spacetime metric is expressed as $g_{\alpha\beta} = \eta_{\alpha\beta} + p_{\alpha\beta}$, where $\eta_{\alpha\beta}$ is the Minkowski metric and $p_{\alpha\beta}$ is a perturbation, a gauge transformation produces the change

$$p_{\alpha\beta} \rightarrow p_{\alpha\beta} - \partial_\alpha\zeta_\beta - \partial_\beta\zeta_\alpha \tag{11.16}$$

to first order in the small quantities $p_{\alpha\beta}$ and $\partial_\alpha\zeta_\beta$, where $\zeta_\alpha := \eta_{\alpha\beta}\zeta^\beta$; this is Eq. (5.122). To relate $p_{\alpha\beta}$ to the gravitational potentials we appeal to Eqs. (7.20), which states that

$$g_{\alpha\beta} = \eta_{\alpha\beta} + h_{\alpha\beta} - \frac{1}{2}h\,\eta_{\alpha\beta} + O(h^2)\,, \tag{11.17}$$

where $h_{\alpha\beta} = \eta_{\alpha\mu}\eta_{\beta\nu}h^{\mu\nu}$ and $h = \eta_{\mu\nu}h^{\mu\nu}$; in the far-away wave zone we can neglect the terms quadratic in $h^{\alpha\beta}$, because they fall off as R^{-2}. We find that

$$h_{\alpha\beta} = p_{\alpha\beta} - \frac{1}{2}p\,\eta_{\alpha\beta}\,, \tag{11.18}$$

where $p = \eta^{\mu\nu}p_{\mu\nu}$. It follows that the gauge transformation produces the change

$$h^{\alpha\beta} \to h^{\alpha\beta} - \partial^\alpha\zeta^\beta - \partial^\beta\zeta^\alpha + \left(\partial_\mu\zeta^\mu\right)\eta^{\alpha\beta} \tag{11.19}$$

in the gravitational potentials. We next find that $\partial_\beta h^{\alpha\beta} \to \partial_\beta h^{\alpha\beta} - \Box\zeta^\alpha$ and conclude that the harmonic gauge conditions will be preserved whenever the vector field satisfies the wave equation

$$\Box\zeta^\alpha = 0\,. \tag{11.20}$$

We wish to preserve the harmonic gauge, and we construct a solution to the wave equation by writing

$$\xi^0 = \frac{G}{c^3R}\alpha(\tau, \boldsymbol{N}) + O(R^{-2})\,, \tag{11.21a}$$

$$\xi^j = \frac{G}{c^3R}\beta^j(\tau, \boldsymbol{N}) + O(R^{-2})\,, \tag{11.21b}$$

where α and β^j are arbitrary functions of their arguments, and the factors of G/c^3 were inserted for convenience. As before we decompose the vector in terms of its irreducible components,

$$\beta^j = \beta N^j + \beta^j_{\mathrm{T}}\,, \qquad N_j\beta^j_{\mathrm{T}} = 0\,. \tag{11.22}$$

We differentiate ξ^0 and ξ^j using the differentiation rule of Eq. (11.7), and insert the results within Eq. (11.19). After also involving Eqs. (11.15), we eventually deduce that the gauge transformation produces the changes

$$A \to A + 3\partial_\tau\alpha - \partial_\tau\beta\,, \tag{11.23a}$$

$$B \to B + 2\partial_\tau\beta\,, \tag{11.23b}$$

$$A^j_{\mathrm{T}} \to A^j_{\mathrm{T}} + \partial_\tau\beta^j_{\mathrm{T}}\,, \tag{11.23c}$$

$$A^{jk}_{\mathrm{TT}} \to A^{jk}_{\mathrm{TT}} \tag{11.23d}$$

in the irreducible pieces of the gravitational potentials.

We see that the transverse-tracefree part of A^{jk} is invariant under the gauge transformation. We see also that α, β, and β^j_{T} can be chosen so as to set A, B, and A^j_{T} all equal to zero. Implementing this gauge transformation, we arrive at the simplest expression for the gravitational potentials in the far-away wave zone:

$$h^{00} = \frac{4GM}{c^2R}\,, \tag{11.24a}$$

$$h^{0j} = 0\,, \tag{11.24b}$$

$$h^{jk} = \frac{G}{c^4R}A^{jk}_{\mathrm{TT}}(\tau, \boldsymbol{N})\,. \tag{11.24c}$$

By virtue of the conditions imposed in Eq. (11.12), $N_j A_{\rm TT}^{jk} = 0 = \delta_{jk} A_{\rm TT}^{jk}$, the number of time-dependent quantities has been reduced to two. The gravitational potentials of Eqs. (11.24) are said to be in the *transverse-tracefree gauge*, or *TT gauge*, a specialization of the harmonic gauge that can be achieved in the far-away wave zone. It is clear that the radiative degrees of freedom of the gravitational field must be contained in the two independent components of $A_{\rm TT}^{jk}$.

11.1.6 Geodesic deviation

This conclusion, that $A_{\rm TT}^{jk}$ contains the radiative degrees of freedom, is reinforced by the following argument. Suppose that a gravitational-wave detector consists of two test masses that are moving freely in the far-away wave zone. The masses are separated by a spacetime vector ξ^α, and they move with a four-velocity u^α. Assuming that the distance between the masses is small compared with the radiation's characteristic wavelength (this defines a short gravitational-wave detector such as the LIGO instrument), the behavior of the separation vector is governed by the equation of geodesic deviation,

$$\frac{D^2\xi^\alpha}{ds^2} = -R^\alpha_{\ \beta\gamma\delta} u^\beta \xi^\gamma u^\delta , \tag{11.25}$$

in which D/ds indicates covariant differentiation in the direction of u^α, and where $R^\alpha_{\ \beta\gamma\delta}$ is the Riemann tensor. This equation was first encountered back in Sec. 5.2; see Eq. (5.67). Assuming in addition that the test masses are moving slowly, this equation reduces to the approximate form of Eq. (5.68),

$$\frac{d^2\xi^j}{dt^2} = -c^2 R_{0j0k}\xi^k ; \tag{11.26}$$

this involves ordinary differentiation with respect to t, as well as the spatial components of the separation vector.

It is a straightforward exercise to compute the Riemann tensor associated with the metric $g_{\alpha\beta} = \eta_{\alpha\beta} + h_{\alpha\beta} - \frac{1}{2}h\eta_{\alpha\beta}$, even when the gravitational potentials are expressed in their general form of Eqs. (11.15). Alternatively, one can proceed from Eqs. (11.24) and appeal to the fact that the Riemann tensor is invariant under a gauge transformation (as was established back in Sec. 5.5). In any event, the computation reveals that

$$c^2 R_{0j0k} = -\frac{G}{2c^4 R}\partial_{\tau\tau} A_{\rm TT}^{jk} , \tag{11.27}$$

and the equation of geodesic deviation becomes

$$\frac{d^2\xi^j}{dt^2} = \frac{G}{2c^4 R}(\partial_{\tau\tau} A_{\rm TT}^{jk})\xi_k = \frac{1}{2}(\partial_{\tau\tau} h_{\rm TT}^{jk})\xi_k . \tag{11.28}$$

This equation can be integrated immediately if we assume that $h_{\rm TT}^{jk}$ is small. We have that

$$\xi^j(t) = \xi^j(0) + \frac{1}{2}h_{\rm TT}^{jk}(t - R/c)\xi_k(0) , \tag{11.29}$$

and we see that changes in the displacement vector are driven by $h_{\rm TT}^{jk}$ and proportional to the initial separation $\xi^k(0)$.

We conclude that our gravitational-wave detector is driven by the TT piece of the gravitational potentials, which therefore captures the radiative degrees of freedom. The remaining pieces contain no radiative information; the fact that they can be eliminated by a coordinate transformation makes it clear that they contain only information about the coordinate system. Henceforth we shall refer to $h^{jk}_{\rm TT}$ specifically as the *gravitational-wave field*; we shall continue to refer to h^{jk} as the (spatial components of) the gravitational potentials.

11.1.7 Extraction of the TT part

Given gravitational potentials presented in the general form of Eq. (11.6),

$$h^{jk} = \frac{G}{c^4 R} A^{jk}(\tau, \boldsymbol{N}), \tag{11.30}$$

the radiative pieces can be extracted by isolating the transverse-tracefree part of A^{jk}. This can be done efficiently by introducing the TT projector $(\mathrm{TT})^{jk}{}_{pq}$, and by writing

$$A^{jk}_{\rm TT} = (\mathrm{TT})^{jk}{}_{pq} A^{pq}. \tag{11.31}$$

The TT projector is constructed as follows. We first introduce the transverse projector

$$P^j{}_k := \delta^j{}_k - N^j N_k, \tag{11.32}$$

which removes the longitudinal components of vectors and tensors. For example, for a vector $A^j = AN^j + A^j_{\rm T}$ with $N_j A^j_{\rm T} = 0$, we have that $P^j{}_k A^k = A^j_{\rm T}$. The transverse projector satisfies

$$P^j{}_k N^k = 0, \qquad P^j{}_j = 2, \qquad P^j{}_p P^p{}_k = P^j{}_k. \tag{11.33}$$

The TT projector is obtained by acting with the transverse projector twice and removing the trace:

$$(\mathrm{TT})^{jk}{}_{pq} := P^j{}_p P^k{}_q - \frac{1}{2} P^{jk} P_{pq}. \tag{11.34}$$

It is easy to see that this possesses the required properties. First, $(\mathrm{TT})^{jk}{}_{pq} N^q = 0$; second, $(\mathrm{TT})^{jk}{}_{pq} \delta^{pq} = 0$; and third, $(\mathrm{TT})^{jk}{}_{pq} A^{pq}_{\rm TT} = A^{jk}_{\rm TT}$ if the tensor $A^{jk}_{\rm TT}$ is already transverse and tracefree. For a general symmetric tensor A^{jk} decomposed as in Eq. (11.10), it is easy to verify that

$$(\mathrm{TT})^{jk}{}_{pq} A^{pq} = A^{jk}_{\rm TT}. \tag{11.35}$$

This equation informs us that the TT part of any symmetric tensor A^{jk} can be extracted by acting with the TT projector defined by Eq. (11.34).

To carry out these manipulations it is convenient to introduce a vectorial basis in the transverse subspace. Having previously selected $\boldsymbol{N}$ as the longitudinal direction, we parameterize it with the polar angles ϑ and φ by writing

$$\boldsymbol{N} := [\sin\vartheta\cos\varphi, \sin\vartheta\sin\varphi, \cos\vartheta]. \tag{11.36}$$

We next introduce the unit vectors

$$\boldsymbol{\vartheta} := [\cos\vartheta\cos\varphi, \cos\vartheta\sin\varphi, -\sin\vartheta] \tag{11.37}$$

and

$$\boldsymbol{\varphi} := [-\sin\varphi, \cos\varphi, 0]\,, \tag{11.38}$$

which are orthogonal to $\boldsymbol{N}$ and to each other. The vector $\boldsymbol{\vartheta}$ points in the direction of increasing colatitude on the surface of a sphere, while $\boldsymbol{\varphi}$ points in the direction of increasing longitude; they span the transverse subspace orthogonal to $\boldsymbol{N}$, which is normal to the sphere. The basis gives us the completeness relations

$$\delta^{jk} = N^j N^k + \vartheta^j \vartheta^k + \varphi^j \varphi^k\,, \tag{11.39}$$

and it follows from Eq. (11.32) that the transverse projector is given by

$$P^{jk} = \vartheta^j \vartheta^k + \varphi^j \varphi^k\,. \tag{11.40}$$

This can be inserted within Eq. (11.34) to form the TT projector.

The transverse basis formed by $\boldsymbol{\vartheta}$ and $\boldsymbol{\varphi}$ is not unique. For any longitudinal direction $\boldsymbol{N}$ we may rotate the unit vectors by an angle ψ around $\boldsymbol{N}$ and thereby obtain a new basis $(\boldsymbol{\vartheta}', \boldsymbol{\varphi}')$. The operation is described by

$$\boldsymbol{\vartheta}' = \cos\psi\, \boldsymbol{\vartheta} + \sin\psi\, \boldsymbol{\varphi}, \qquad \boldsymbol{\varphi}' = -\sin\psi\, \boldsymbol{\vartheta} + \cos\psi\, \boldsymbol{\varphi}\,. \tag{11.41}$$

The equations (11.39) and (11.40) are invariant under such a rotation.

Any symmetric, transverse, and tracefree tensor $A^{jk}_{\rm TT}$ can be decomposed in a tensorial basis that is built entirely from the vectors $\boldsymbol{\vartheta}$ and $\boldsymbol{\varphi}$. Such a tensor has two independent components, which we denote A_+ and $A_\times$ and call the *polarizations* of the tensor $A^{jk}_{\rm TT}$. We write

$$A^{jk}_{\rm TT} = A_+ \big(\vartheta^j\vartheta^k - \varphi^j\varphi^k\big) + A_\times \big(\vartheta^j\varphi^k + \varphi^j\vartheta^k\big)\,, \tag{11.42}$$

so that A_+ represents the ϑ-ϑ component of the tensor (and also minus the φ-φ component, in order to satisfy the tracefree condition), while $A_\times$ represents its ϑ-φ component. It is easy to check that Eq. (11.42) implies

$$A_+ = \frac{1}{2}\big(\vartheta_j\vartheta_k - \varphi_j\varphi_k\big) A^{jk}_{\rm TT}\,, \tag{11.43a}$$

$$A_\times = \frac{1}{2}\big(\vartheta_j\varphi_k + \varphi_j\vartheta_k\big) A^{jk}_{\rm TT}\,. \tag{11.43b}$$

Because the tensorial operators acting on $A^{jk}_{\rm TT}$ are already transverse and tracefree, this can also be written as

$$A_+ = \frac{1}{2}\big(\vartheta_j\vartheta_k - \varphi_j\varphi_k\big) A^{jk}\,, \tag{11.44a}$$

$$A_\times = \frac{1}{2}\big(\vartheta_j\varphi_k + \varphi_j\vartheta_k\big) A^{jk}\,, \tag{11.44b}$$

in which the projection operators are acting on the original tensor A^{jk} instead of its TT part $A^{jk}_{\rm TT}$.

Under the rotation of Eq. (11.41) the polarizations of $A^{jk}_{\rm TT}$ transform according to

$$A'_+ = \cos 2\psi\, A_+ + \sin 2\psi\, A_\times\,, \qquad A'_\times = -\sin 2\psi\, A_+ + \cos 2\psi\, A_\times\,. \tag{11.45}$$

It is easy to verify that Eqs. (11.41) and (11.45) ensure that Eqs. (11.42), (11.43), and (11.44) stay invariant under a rotation of the transverse basis.

Equations (11.44), together with the definitions of Eqs. (11.37) and (11.38), provide an efficient way of extracting the polarizations A_+ and $A_\times$ from a general tensor A^{jk}. The end results are

$$\begin{aligned} A_+ = &-\frac{1}{4}\sin^2\vartheta(A^{xx}+A^{yy})+\frac{1}{4}(1+\cos^2\vartheta)\cos 2\varphi\,(A^{xx}-A^{yy}) \\ &+\frac{1}{2}(1+\cos^2\vartheta)\sin 2\varphi\,A^{xy}-\sin\vartheta\cos\vartheta\cos\varphi\,A^{xz}-\sin\vartheta\cos\vartheta\sin\varphi\,A^{yz} \\ &+\frac{1}{2}\sin^2\vartheta\,A^{zz}, \end{aligned} \tag{11.46a}$$

$$\begin{aligned} A_\times = &-\frac{1}{2}\cos\vartheta\sin 2\varphi\,(A^{xx}-A^{yy})+\cos\vartheta\cos 2\varphi\,A^{xy}+\sin\vartheta\sin\varphi\,A^{xz} \\ &-\sin\vartheta\cos\varphi\,A^{yz}. \end{aligned} \tag{11.46b}$$

With A_+ and $A_\times$ known, $A^{jk}_{\rm TT}$ can be constructed with the help of Eq. (11.42); the complete listing of components is

$$A^{xx}_{\rm TT} = -\frac{1}{2}\big[\sin^2\vartheta-(1+\cos^2\vartheta)\cos 2\varphi\big]A_+ - \cos\vartheta\sin 2\varphi\,A_\times, \tag{11.47a}$$

$$A^{xy}_{\rm TT} = \frac{1}{2}(1+\cos^2\vartheta)\sin 2\varphi\,A_+ + \cos\vartheta\cos 2\varphi\,A_\times, \tag{11.47b}$$

$$A^{xz}_{\rm TT} = -\sin\vartheta\cos\vartheta\cos\varphi\,A_+ + \sin\vartheta\sin\varphi\,A_\times, \tag{11.47c}$$

$$A^{yy}_{\rm TT} = -\frac{1}{2}\big[\sin^2\vartheta+(1+\cos^2\vartheta)\cos 2\varphi\big]A_+ + \cos\vartheta\sin 2\varphi\,A_\times, \tag{11.47d}$$

$$A^{yz}_{\rm TT} = -\sin\vartheta\cos\vartheta\sin\varphi\,A_+ - \sin\vartheta\cos\varphi\,A_\times, \tag{11.47e}$$

$$A^{zz}_{\rm TT} = \sin^2\vartheta\,A_+. \tag{11.47f}$$

For example, when the wave travels in the y-direction, so that $\vartheta=\varphi=\frac{\pi}{2}$, we have that $A_+=\frac{1}{2}(A^{zz}-A^{xx})$ and $A_\times = A^{xz}$. We also have $A^{zz}_{\rm TT}=-A^{xx}_{\rm TT}=A_+$ and $A^{xz}_{\rm TT}=A_\times$ as the only non-vanishing components of the transverse-tracefree tensor.

11.1.8 Distortion of a ring of particles by a gravitational wave

A useful way to visualize the gravitational-wave polarizations is to examine the geodesic deviations that they generate. Consider an initially circular ring of freely moving particles in an inertial frame. A gravitational wave travels in the z-direction past the ring, which lies in the x-y plane. In this case $\vartheta=0$, and we can choose $\varphi=0$. Equations (11.47) reveal that $A^{xx}_{\rm TT}=-A^{yy}_{\rm TT}=A_+$ and $A^{xy}_{\rm TT}=A^{yx}_{\rm TT}=A_\times$, and the other components vanish. The components are conveniently displayed as a matrix,

$$A^{jk}_{\rm TT} = \begin{pmatrix} A_+ & A_\times & 0 \\ A_\times & -A_+ & 0 \\ 0 & 0 & 0 \end{pmatrix}. \tag{11.48}$$

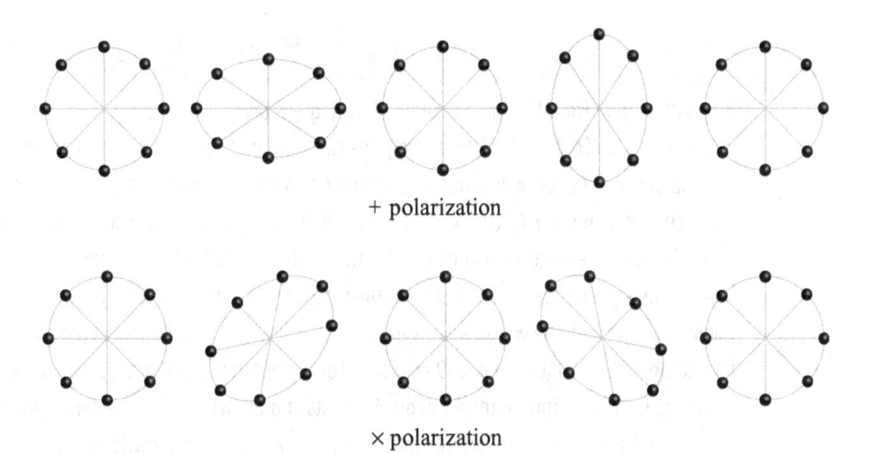

Fig. 11.1 Effect of the + and × gravitational-wave polarizations on a circular ring of freely falling particles. The wave propagates out of the page, and a complete wave cycle is shown from left to right.

The displacement of a given particle from the center of the ring is given by the solution (11.29) to the geodesic deviation equation. We have

$$\xi^j(t) = \xi^j(0) + \frac{G}{2c^4 R} A^{jk}_{\mathrm{TT}}\, \xi_k(0)\,, \tag{11.49}$$

or

$$x(t) = x_0 + \frac{G}{2c^4 R}\Big(A_+ x_0 + A_\times y_0\Big)\,, \tag{11.50a}$$

$$y(t) = y_0 + \frac{G}{2c^4 R}\Big(A_\times x_0 - A_+ y_0\Big)\,, \tag{11.50b}$$

$$z(t) = z_0\,, \tag{11.50c}$$

in terms of the (x, y, z) components of the deviation vector $\boldsymbol{\xi}$.

Consider now a pure + mode. It is simple to show that a circle of particles of unit radius will be distorted into an ellipse described by

$$\left(\frac{x}{1+\eta_+}\right)^2 + \left(\frac{y}{1-\eta_+}\right)^2 = 1\,, \tag{11.51}$$

where $\eta_+(t) = \frac{1}{2}(G/c^4 R)A_+(t)$ is assumed to be small. As $\eta_+(t)$ varies between its maximum and minimum value, the ellipse transforms between the shapes shown in the upper panel of Fig. 11.1, passing through a circular shape when $\eta_+(t) = 0$. Similarly, for a pure $\times$ mode the circle will be distorted into an ellipse described by

$$\frac{1}{2}\left(\frac{x+y}{1+\eta_\times}\right)^2 + \frac{1}{2}\left(\frac{x-y}{1-\eta_\times}\right)^2 = 1\,, \tag{11.52}$$

where $\eta_\times(t) = \frac{1}{2}(G/c^4 R)A_\times(t)$. This is the same as the first ellipse, except that it is rotated by 45 degrees. In both cases the area of the ellipse is constant to first order in η. The ring is unaffected in the z-direction, a reminder that the waves are transverse.

Box 11.1 **Why 45 degrees?**

It is evident from Fig. 11.1 that a rotation of 45 degrees takes a $+$ mode into a $\times$ mode and vice versa. This is also clear from Eq. (11.45). Alternatively, we can state that the two modes of polarization are related by a rotation of $\pi/4$ about the direction of propagation. In electromagnetism, the two modes of polarization are related by a rotation of $\pi/2$ (think of the electric field pointing along the x-axis versus the y-axis). Fundamentally this is because electrodynamics is associated with a vector field, while gravity is associated with a tensor field. There is a close connection between the rotation angle and the *helicity* or *spin* of the particle that one might associate with the waves: it is given by $\pi/(2S)$, where S is the spin of the particle in units of $\hbar$. Thus for photons ($S = 1$), the angle is 90 degrees. For the putative graviton that is often associated with gravity (although no fully quantized theory of gravity exists at present), $S = 2$, leading to the 45 degree angle. For a spin-$\frac{1}{2}$ particle like an electron, the rotation angle is π, as is well known from the Dirac equation.

11.2 The quadrupole formula

In the preceding section we saw that the gravitational-wave field is described by the transverse-tracefree piece of the potentials $h^{jk} = GA^{jk}/(c^4R)$, and we developed methods to extract these radiative pieces from a known tensor A^{jk}. In this section we provide an expression for A^{jk} and examine some applications of the resulting formalism.

11.2.1 Formulation

The tensor A^{jk} was, in fact, calculated back in Chapter 7, in the context of a post-Minkowskian approximation to general relativity. The gravitational potentials were computed for the specific case of a matter distribution consisting of a perfect fluid subjected to a slow-motion condition. The results are summarized in Box 7.7. To leading order in a post-Newtonian expansion in powers of v_c/c we have that $A^{jk} = 2\ddot{\mathcal{I}}^{jk}$, where

$$\begin{aligned}\mathcal{I}^{jk}(\tau) &:= \int_{\mathscr{M}} c^{-2}\tau^{00}(\tau, \boldsymbol{x}')x'^j x'^k \, d^3x' \\ &= \int \rho^*(\tau, \boldsymbol{x}')x'^j x'^k \, d^3x' + O(c^{-2}) \end{aligned} \tag{11.53}$$

is the mass quadrupole moment of the matter distribution. The *quadrupole formula* for the gravitational-wave field is therefore $h^{jk}_{\rm TT} = (2G/c^4R)\ddot{\mathcal{I}}^{jk}_{\rm TT}$, in which an overdot indicates differentiation with respect to τ.

We remark that this result was derived after *two iterations* of the relaxed Einstein equations. Two iterations were required to ensure that the fluid's equations of motion incorporate gravity at the Newtonian level. But the quadrupole formula appears to be linear in G, and one might be tempted to think that it could have been derived more simply using linearized theory, as presented in Sec. 5.5. One would be wrong, because in linearized theory the fluid does not respond to gravity, and the domain of validity of the result would be severely limited. The fact that the gravitational-wave field has such a "linearized" look is to some degree coincidental; it has been the source of endless confusion in the literature.

Equation (11.53) displays the leading contribution to $\mathcal{I}^{jk}$ in a post-Newtonian expansion. In Sec. 11.3 we obtain higher-order corrections to this expression, but in this section we stick to the lowest-order terms. With this in mind, we choose to reserve the notation $\mathcal{I}^{jk}$ for the formal, iterated expression for the quadrupole-moment tensor, and define a Newtonian quadrupole moment by

$$I^{jk}(\tau) := \int \rho^*(\tau, \boldsymbol{x}')x'^j x'^k \, d^3x' \,. \tag{11.54}$$

The lowest-order gravitational-wave field can then be written as

$$h^{jk}_{\mathrm{TT}} = \frac{2G}{c^4 R} \ddot{I}^{jk}_{\mathrm{TT}} \,, \tag{11.55}$$

in terms of the Newtonian moment. Note that we use ρ^* as our main density variable instead of the proper density ρ; since these differ by corrections of order c^{-2}, we may ignore the difference, but ρ^* is a more convenient density to use in a relativistic context.

Equation (11.55) is easily turned into a robust order-of-magnitude estimate for the gravitational-wave amplitude h_0; this is defined in such a way that a typical component of h^{jk}_{TT} is of the order of h_0. We imagine that the waves are produced by a matter distribution of mass M confined to a volume of radius r_c, and that changes in the matter distribution occur over a time scale t_c; the source's characteristic velocity is then $v_c \sim r_c/t_c$. The quadrupole-moment tensor scales as Mr_c^2, and $\ddot{I}^{jk}$ is of order $M(r_c/t_c)^2 \sim Mv_c^2$. Then Eq. (11.55) gives

$$h_0 \sim \frac{GM}{c^2 R}(v_c/c)^2 \,, \tag{11.56}$$

and we see that strong waves are produced when a large mass M is involved in a rapid process with $v_c \sim c$. It is important to understand that v_c characterizes only the part of the motion that deviates from spherical symmetry; a spherical matter distribution would have $I^{jk} \propto \delta^{jk}$, $I^{jk}_{\mathrm{TT}} = 0$, and would not emit gravitational waves. (This conclusion is not limited to the quadrupole approximation. It is an exact consequence of general relativity that a spherically-symmetric matter distribution does not emit gravitational waves. This is the statement of Birkhoff's theorem, first encountered in Sec. 5.6.2.) To estimate h_0 numerically we imagine an astrophysical process that involves a mass $M = 10\, M_\odot$ situated at a distance $R = 1$ Mpc, which corresponds to the approximate size of the local group of galaxies. Under these conditions Eq. (11.56) gives rise to the estimate

$$h_0 \sim 4.8 \times 10^{-19} \left(\frac{M}{10\, M_\odot}\right)\left(\frac{1\,\mathrm{Mpc}}{R}\right)(v_c/c)^2 . \tag{11.57}$$

This exercise reveals that even the most violent events in the universe produce tiny gravitational waves.

To obtain a more precise expression for h^{jk}_{TT} we must evaluate the time derivatives of the quadrupole-moment tensor and carry out the transverse-tracefree projection. The second operation is simple, and relies on the results displayed in Eqs. (11.44) and (11.46). The first operation relies on a knowledge of the fluid dynamics, which is governed by the Newtonian limit of the equations of hydrodynamics, $\rho^* dv^j/dt = \rho^* \partial^j U - \partial^j p$, where v^j is the fluid's velocity field, U the Newtonian gravitational potential, and p the pressure. This is the Euler equation, first encountered in its Newtonian form back in Sec. 1.4, and then in relativistic form in Sec. 7.3.2; refer to Eq. (7.58).

The second derivative of the quadrupole-moment tensor is provided by the *virial theorem*,

$$\frac{1}{2}\ddot{I}^{jk} = 2\mathcal{T}^{jk} + \Omega^{jk} + P\delta^{jk}, \tag{11.58}$$

where

$$\mathcal{T}^{jk} = \frac{1}{2}\int \rho^* v^j v^k \, d^3x, \tag{11.59a}$$

$$\Omega^{jk} = -\frac{1}{2}G\int \rho^* \rho^{*\prime} \frac{(x-x')^j(x-x')^k}{|\boldsymbol{x}-\boldsymbol{x}'|^3} \, d^3x' d^3x, \tag{11.59b}$$

$$P = \int p \, d^3x. \tag{11.59c}$$

The virial theorem is a direct consequence of the Euler equation; it was first derived in the context of Newtonian mechanics back in Sec. 1.4.3; see Eq. (1.88). It is understood that here, $\mathcal{T}^{jk}$, Ω^{jk}, and P are all functions of retarded time τ; inside the integrals ρ^*, v^j, and p are functions of τ and $\boldsymbol{x}$, while $\rho^{*\prime}$ is a function of τ and $\boldsymbol{x}'$.

When the virial theorem is inserted in Eq. (11.55), h^{jk} is seen to contain terms that are both linear and quadratic in G; the linear terms come from $2\mathcal{T}^{jk}$ and $P\delta^{jk}$, while the quadratic terms come from Ω^{jk}. The virial theorem implies that in general, the contributions from $2\mathcal{T}^{jk}$, Ω^{jk}, and $P\delta^{jk}$ are all comparable to each other, because the sum of terms must vanish on the average. This indicates that the terms of order G^2 in h^{jk} are comparable to the terms of order G, and that a proper derivation of the quadrupole formula *must* be based on a second post-Minkowskian approximation to general relativity. A derivation based on the linearized theory (first post-Minkowskian approximation) would omit the G^2 terms and give rise to the wrong answer for the gravitational-wave field. As we observed previously, the additional factor of G does not show up when the field is expressed in terms of $\ddot{I}^{jk}$, but in fact it is hidden within the time derivatives, which demand the use of the Newtonian equations of motion.

An exception to this rule occurs when the fluid dynamics is dominated by pressure gradients and gravity is relatively unimportant. In this case $2\mathcal{T}^{jk}$ and $P\delta^{jk}$ are both much larger than Ω^{jk}, and the terms of order G^2 can be neglected in h^{jk}. In this restricted context the quadrupole formula can be reliably derived on the basis of the linearized theory or a single iteration of the relaxed Einstein equations. One obtains Eq. (11.55), but with the important restriction that the source dynamics cannot involve the gravitational field. Such a derivation would be valid for gravitational waves emitted by a source with negligible self-gravity, such as a rotating dumbbell.

In the formulation of the quadrupole formula given here, the fluid system can be of one continuous piece, or it can be broken up into N separated components; this would represent an N-body system of fluid bodies. When the internal structure of each body can be ignored, we may adopt the point-mass description developed in Sec. 9.6 and set

$$\rho^* = \sum_A M_A \delta(\boldsymbol{x} - \boldsymbol{r}_A), \tag{11.60}$$

where M_A is the total mass-energy of the body identified by the label A, and $\boldsymbol{r}_A$ is its position vector evaluated at time τ. In this case the quadrupole moment tensor reduces to

$$I^{jk} = \sum_A M_A r_A^j r_A^k, \tag{11.61}$$

and the dynamics of the system is governed by Newton's equations of motion. These are $\boldsymbol{a}_A = -\sum_{B \neq A} GM_B\, \boldsymbol{n}_{AB}/r_{AB}^2$, where $\boldsymbol{r}_{AB} := \boldsymbol{r}_A - \boldsymbol{r}_B, r_{AB} := |\boldsymbol{r}_A - \boldsymbol{r}_B|$, and $\boldsymbol{n}_{AB} := \boldsymbol{r}_{AB}/r_{AB}$. In this case the virial theorem becomes

$$\frac{1}{2}\ddot{I}^{jk} = -\frac{1}{2}\sum_A \sum_{B \neq A} \frac{GM_A M_B}{r_{AB}} n_{AB}^j n_{AB}^k + \sum_A M_A v_A^j v_A^k, \tag{11.62}$$

where $\boldsymbol{v}_A := d\boldsymbol{r}_A/d\tau$ is the velocity vector of body A. This can be obtained by differentiating Eq. (11.61) and involving the equations of motion, or directly from Eq. (11.58) by exploiting the regularization techniques developed in Sec. 9.6.

Box 11.2 The quadrupole-formula controversy

In a remarkable pair of papers published in 1916 and 1918, Einstein calculated the gravitational-wave field and radiated energy of a time-dependent source, such as a rotating dumbbell, for which self-gravity is unimportant. He performed this computation in a slow-motion approximation, using the linearized Einstein equations, and obtained the quadrupole formula (11.55). It is perhaps a slight exaggeration to say that it was all downhill from there, at least until 1979.

It didn't help that Einstein made a calculational error in his 1918 paper, leading to a wrong factor of 2, discovered later by Eddington. Nor did it help that Eddington, concerned about the gauge freedom available in the description of gravitational waves, wondered in 1922 whether aspects of gravitational waves were physically real or purely coordinate artifacts; as he put it, perhaps they "propagate with the speed of thought." Although Eddington understood that the gauge-invariant modes were physical and believed that gravitational waves did exist, his remark, taken out of context, had the effect of making the entire subject seem dubious.

To make matters worse, in 1936 Einstein and his young colleague Nathan Rosen (of Einstein–Podolsky–Rosen paradox fame) submitted a paper to *The Physical Review* with the provocative title "Do gravitational waves exist?". They thought they had found an exact solution of the field equations describing a plane gravitational wave, but because the solution had a singularity, it could not be physically valid, and they concluded that gravitational waves could not exist. *The Physical Review* sent the paper for review, and the report that came back pointed out that the Einstein–Rosen solution in fact described a cylindrical wave, and that the singularity was merely a harmless coordinate singularity associated with the axis. So the solution was perfectly valid, and in fact it supported the existence of gravitational waves. Einstein was so angry that the journal had sent his paper out to be refereed, a practice that was unfamiliar to him, that he withdrew the paper and never published again in that journal. Shortly thereafter, however, Einstein was convinced by another of his assistants, Leopold Infeld (who had been approached by the anonymous referee), that the referee had been perfectly correct. Einstein rewrote the paper with the opposite conclusion and published it under the title "On gravitational radiation" (but not in *The Physical Review*). While there has been plenty of speculation as to the identity of the anonymous referee, it wasn't until 2005 that our friend Daniel Kennefick was allowed access to the records of the journal and revealed conclusively that the referee was the well-known Princeton and Caltech relativist H.P. Robertson (the co-discoverer of the Robertson–Walker metric for cosmology).

This episode did not end the debate over the existence of gravitational waves. Even if one accepts the validity of Einstein's prediction that a rotating dumbbell will radiate gravitational waves, the argument was made that a binary-star system would *not* radiate. After all, each body is moving on a geodesic, and is therefore unaccelerated relative to a local freely falling frame. Without acceleration, the argument went, there should be no radiation. Peter Havas was one of the proponents of this possibility.

Beginning in the late 1940s, numerous attempts were made to calculate the "back reaction" forces that would alter the motion of a binary system in response to the radiation of energy and angular momentum (this is the primary subject of Chapter 12). Yet different workers got different answers.

By 1974, while most researchers in the field accepted the reality of gravitational waves and the validity of the quadrupole formula for slowly moving binary systems, a vocal minority remained skeptical. This "quadrupole-formula controversy" came to a head with the September 1974 discovery of the first binary pulsar by Russell Hulse and Joseph Taylor. It was immediately clear that it would be possible to test the quadrupole formula by exploiting the high-precision timing of the pulsar's radio signals to measure the slow variation in the orbit induced by the loss of orbital energy to radiation.

But in a letter published in the *Astrophysical Journal* in 1976, Jürgen Ehlers, Arnold Rosenblum, Joshua Goldberg, and Peter Havas argued that the quadrupole formula could not be justified as a theoretical prediction of general relativity. They presented a laundry list of theoretical problems that they claimed had been swept under the rug by proponents of the quadrupole formula. Among them were these: people assumed energy balance to infer the reaction of the source to the flux of radiation, but there was no proof that this was a valid assumption; no reliable calculation of the equations of motion that included radiation reaction had (in their opinion) ever been carried out; many "derivations" of the quadrupole formula relied on the linearized theory, which was clearly wrong for binary systems; since higher-order corrections had not been calculated, it was impossible to know if the quadrupole formula was even a good approximation; even worse, higher-order terms were known to be rife with divergent integrals.

There was considerable annoyance among holders of the "establishment" viewpoint when this paper appeared, mainly because it was realized that its criticisms had considerable merit. As a result many research groups embarked on a program to return to the fundamentals and to develop approximation schemes for equations of motion and gravitational radiation that would not be subject to the flaws that so disturbed Ehlers *et al*. Among the noteworthy outcomes of this major effort was the fully developed post-Minkowskian formalism that forms the heart of this book. Toward the end of his life, Jürgen Ehlers, one of the great relativists of his time, admitted to one of us (after some prodding, to be sure, and only up to a point!) that the justification of the quadrupole formula was in much better shape than it was in 1976.

Experimentally, the situation was not at all controversial. By 1979, Taylor and his colleagues had measured the damping of the binary pulsar's orbit, in agreement with the quadrupole formula to about 10 percent; by 2005, the agreement was at the 0.2 percent level. The formula has also been beautifully confirmed in a number of other binary-pulsar systems.

11.2.2 Application: Binary system

As a first application of the quadrupole formula, we examine the gravitational waves emitted by a binary system of orbiting bodies. We rely on the Newtonian description of the orbital motion reviewed in Sec. 3.2.

Orbital motion and gravitational-wave field

The position of the first body, of mass m_1, is $\boldsymbol{r}_1(t)$ relative to the system's barycenter, and its velocity is $\boldsymbol{v}_1(t)$; similarly, the position of the second body, of mass m_2, is $\boldsymbol{r}_2(t)$ and its velocity is $\boldsymbol{v}_2(t)$. In terms of the separation vector $\boldsymbol{r} := \boldsymbol{r}_{12} = \boldsymbol{r}_1 - \boldsymbol{r}_2$ we have that

$$\boldsymbol{r}_1 = \frac{m_2}{m}\boldsymbol{r}, \qquad \boldsymbol{r}_2 = -\frac{m_1}{m}\boldsymbol{r}, \tag{11.63}$$

where $m := m_1 + m_2$ is the total mass of the system. We also have

$$\boldsymbol{v}_1 = \frac{m_2}{m}\boldsymbol{v}, \qquad \boldsymbol{v}_2 = -\frac{m_1}{m}\boldsymbol{v}, \tag{11.64}$$

where $\boldsymbol{v} := \boldsymbol{v}_1 - \boldsymbol{v}_2$ is the relative velocity vector. For later purposes we introduce the notations

$$r := |\boldsymbol{r}|, \qquad \boldsymbol{n} := \boldsymbol{r}/r, \tag{11.65}$$

together with

$$\eta := \frac{m_1 m_2}{(m_1 + m_2)^2}; \tag{11.66}$$

this quantity is known as the *symmetric mass ratio* of the binary system.

Making the substitutions in the quadrupole-moment tensor of Eq. (11.61) reveals that $I^{jk} = \eta m r^j r^k$, and Eq. (11.62) becomes $\frac{1}{2}\ddot{I}^{jk} = \eta m[v^j v^k - (Gm/r)n^j n^k]$. We then obtain

$$h^{jk} = \frac{4G\eta m}{c^4 R}\left(v^j v^k - \frac{Gm}{r}n^j n^k\right) \tag{11.67}$$

for the gravitational potentials created by a binary system. To proceed further we need expressions for $\boldsymbol{r}$ and $\boldsymbol{v}$.

To describe the orbital motion we introduce first an "orbit-adapted" coordinate system (x, y, z) that possesses the following properties. First, the origin of the coordinates coincides with the system's barycenter. Second, the orbital plane coincides with the x-y plane, and the z-axis points in the direction of the angular-momentum vector. And third, the x-axis is aligned with the orbit's major axis, while the y-axis is aligned with the minor axis. The relative orbit is described by the Keplerian equations

$$r = \frac{p}{1 + e\cos\phi}, \qquad \dot{\phi} = \sqrt{\frac{Gm}{p^3}}(1 + e\cos\phi)^2, \tag{11.68}$$

in which ϕ is the angle from the x-axis (also known as the *true anomaly*). In addition, p is the orbit's semi-latus rectum, and e is the eccentricity; these orbital elements are constants of the motion that can be related to the system's total energy and angular momentum. In the orbit-adapted coordinates (x, y, z) we have that the unit vectors

$$\boldsymbol{n} = [\cos\phi, \sin\phi, 0], \qquad \boldsymbol{\lambda} = [-\sin\phi, \cos\phi, 0], \tag{11.69}$$

form a basis in the orbital plane. In terms of these

$$\boldsymbol{r} = r\,\boldsymbol{n}, \qquad \boldsymbol{v} = \dot{r}\,\boldsymbol{n} + r\dot{\phi}\,\boldsymbol{\lambda}, \tag{11.70}$$

and the description of the motion is complete. Taking into account Eqs. (11.68) and (11.70), Eq. (11.67) becomes

$$\begin{aligned} h^{jk} = \frac{4\eta}{c^4 R}\frac{(Gm)^2}{p}\Big[&-(1 + e\cos\phi - e^2\sin^2\phi)n^j n^k \\ &+ e\sin\phi(1 + e\cos\phi)\big(n^j\lambda^k + \lambda^j n^k\big) + (1 + e\cos\phi)^2\lambda^j\lambda^k\Big]. \end{aligned} \tag{11.71}$$

The components of h^{jk} in the orbit-adapted frame can then be obtained with the help of Eq. (11.69).

Polarizations

In order to construct the gravitational-wave polarizations h_+ and $h_\times$, it is helpful to introduce, in addition to the original system (x, y, z), a "detector-adapted" coordinate system (X, Y, Z) that possesses the following properties. First, the origin of the coordinates coincides with the origin of the system (x, y, z). Second, the Z-axis points in the direction of the gravitational-wave detector, at which the polarizations are being measured. And third, the X-Y plane is orthogonal to the Z-axis and coincides with the plane of the sky from the detector's point of view, and the X-axis is aligned with the *line of nodes*, the line at which the orbital plane cuts the reference plane; by convention the X-axis points toward the *ascending node*, the point at which the orbit cuts the plane from below. The construction was detailed in Sec. 3.2, and we recall that in the original (x, y, z) coordinates, the new coordinate directions are described by

$$\boldsymbol{e}_X = [\cos\omega, \ -\sin\omega, \ 0], \tag{11.72a}$$

$$\boldsymbol{e}_Y = [\cos\iota\sin\omega, \ \cos\iota\cos\omega, \ -\sin\iota], \tag{11.72b}$$

$$\boldsymbol{e}_Z = [\sin\iota\sin\omega, \ \sin\iota\cos\omega, \ \cos\iota] = \boldsymbol{N}. \tag{11.72c}$$

When viewed in the detector-adapted frame (X, Y, Z), the *inclination angle* ι measures the inclination of the orbital plane with respect to the X-Y plane, while the *longitude of pericenter* ω is the angle between the pericenter and the line of nodes, as measured in the orbital plane. A third angle, the *longitude of ascending node* Ω, was also introduced back in Sec. 3.2, but it is not needed here; we have set $\Omega = 0$ by convention. The vectors $\boldsymbol{n}$ and $\boldsymbol{\lambda}$ are given by

$$\boldsymbol{n} = \big[\cos(\omega + \phi), \cos\iota\sin(\omega + \phi), \sin\iota\sin(\omega + \phi)\big] \tag{11.73}$$

and

$$\boldsymbol{\lambda} = \big[-\sin(\omega+\phi), \cos\iota\cos(\omega+\phi), \sin\iota\cos(\omega+\phi)\big] \tag{11.74}$$

when expressed in the detector-adapted coordinates (X, Y, Z).

Because the gravitational waves propagate from the binary system to the detector along the Z-axis, we may adopt $\boldsymbol{e}_X$ and $\boldsymbol{e}_Y$ as a vectorial basis in the transverse subspace. And having made this choice, the polarizations h_+ and $h_\times$ may be computed according to Eq. (11.44); we have that

$$h_+ = \frac{1}{2}\big(e^j_X e^k_X - e^j_Y e^k_Y\big)h_{jk}, \tag{11.75a}$$

$$h_\times = \frac{1}{2}\big(e^j_X e^k_Y + e^j_Y e^k_X\big)h_{jk}. \tag{11.75b}$$

Note that this choice of transverse basis differs only in notation from the description given back in Eqs. (11.36), (11.37), and (11.38). In the old notation we have that $\boldsymbol{N} = \boldsymbol{e}_Z, \boldsymbol{\vartheta} = \boldsymbol{e}_Y$, and $\boldsymbol{\varphi} = -\boldsymbol{e}_X$; the old angles are related to the new ones by $\vartheta = \iota$ and $\varphi = \frac{\pi}{2} - \omega$.

Inserting Eqs. (11.71), (11.73), (11.74) within Eq. (11.75) reveals that in the selected transverse basis, the gravitational-wave polarizations are given by

$$h_+ = h_0\, H_+, \qquad h_\times = h_0\, H_\times, \tag{11.76}$$

where

$$h_0 = \frac{2\eta}{c^4 R}\frac{(Gm)^2}{p} \tag{11.77}$$

is the gravitational-wave amplitude, and

$$\begin{aligned} H_+ = -(1+\cos^2\iota)\bigg[&\cos(2\phi+2\omega) + \frac{5}{4}e\cos(\phi+2\omega) + \frac{1}{4}e\cos(3\phi+2\omega) \\ &+ \frac{1}{2}e^2\cos 2\omega\bigg] + \frac{1}{2}e\sin^2\iota\big(\cos\phi + e\big), \end{aligned} \tag{11.78a}$$

$$\begin{aligned} H_\times = -2\cos\iota\bigg[&\sin(2\phi+2\omega) + \frac{5}{4}e\sin(\phi+2\omega) + \frac{1}{4}e\sin(3\phi+2\omega) \\ &+ \frac{1}{2}e^2\sin 2\omega\bigg] \end{aligned} \tag{11.78b}$$

are scale-free polarizations. Plots of H_+ and $H_\times$ are displayed in Fig. 11.2.

Circular motion

When $e = 0$ the orbit is circular, and ϕ increases linearly with time, at a uniform rate equal to $\Omega := \sqrt{Gm/p^3}$. In this case the polarizations simplify to

$$H_+ = -(1+\cos^2\iota)\cos 2(\Omega\tau + \omega), \qquad H_\times = -2\cos\iota\sin 2(\Omega\tau+\omega), \tag{11.79}$$

where $\tau := t - R/c$ is retarded time. We see that the waves oscillate at *twice* the orbital frequency; this doubling of frequency is a consequence of the quadrupolar nature of the wave.

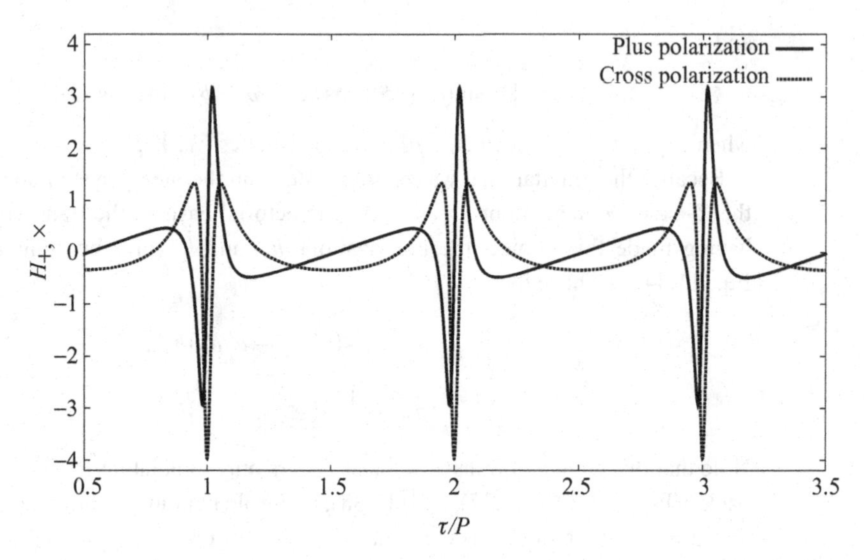

Fig. 11.2 The polarizations H_+ and $H_\times$ as functions of retarded time τ, in units of the orbital period P. The curves were constructed from Eqs. (11.78), and ϕ is related to τ by integrating Eq. (11.68), which was done numerically. The curves are displayed for an eccentricity $e = 0.7$, an inclination angle $\iota = 30°$, and a longitude of pericenter $\omega = 45°$. We see that most of the emission takes place near the pericenter, where the orbit is smallest and the motion fastest.

Numerical estimate

The gravitational-wave amplitude of Eq. (11.77) can also be expressed in terms of the so-called *chirp mass*

$$\mathcal{M} := \eta^{3/5} m = \left(\frac{m_1^3 m_2^3}{m}\right)^{1/5} \tag{11.80}$$

and the orbital period

$$P := 2\pi\sqrt{\frac{a^3}{Gm}}, \tag{11.81}$$

where $a := p/(1-e^2)$ is the orbit's semi-major axis. The expression is

$$h_0 = \frac{2}{c^4 R}(G\mathcal{M})^{5/3}\left(\frac{2\pi}{P}\right)^{2/3}\frac{1}{1-e^2}. \tag{11.82}$$

We evaluate this for a binary system of black holes on a very tight orbit, moments before they are about to plunge toward each other and merge into a single, final black hole. We take $m_1 = 25\, M_\odot$ and $m_2 = 22\, M_\odot$, so that $\mathcal{M}$ is approximately equal to $20\, M_\odot$. We imagine that the orbital period is of the order of 10 ms, and that the binary is situated at a distance $R = 100$ Mpc, sufficiently far that the probability of occurrence of such an event is reasonable. These numbers give rise to the estimate

$$h_0 = \frac{3.0\times 10^{-21}}{1-e^2}\left(\frac{\mathcal{M}}{20\, M_\odot}\right)^{5/3}\left(\frac{10\,\text{ms}}{P}\right)^{2/3}\left(\frac{100\,\text{Mpc}}{R}\right), \tag{11.83}$$

and this indicates that gravitational waves from realistic astrophysical events are exceedingly weak. They are not, however, impossible to detect, and the search for such signals is on.

11.2.3 Application: Rotating neutron star

As a second application of the quadrupole formula we calculate the gravitational waves emitted by a deformed body that rotates around one of its principal axes. We think of this body as a rotating neutron star, but the calculation applies to any type of rigid body, irrespective of its composition and internal structure.

General description

The simplest description of the body is given in a "body-adapted" frame (x', y', z') that co-rotates with the body, and in which the quadrupole moment tensor $I^{a'b'}$ is diagonal. The coordinates are directed along the body's principal axes, and we assume that $I^{a'b'}$ does not depend upon time t. We assume also that the body is rotating uniformly around the z'-axis, with an angular velocity Ω. The transformation to the non-rotating frame (x, y, z) is given by

$$x = x' \cos\Omega t - y' \sin\Omega t, \tag{11.84a}$$

$$y = x' \sin\Omega t + y' \cos\Omega t, \tag{11.84b}$$

$$z = z'. \tag{11.84c}$$

The components of the quadrupole-moment tensor in the non-rotating frame are given by

$$I^{jk} = \frac{\partial x^j}{\partial x^{a'}} \frac{\partial x^k}{\partial x^{b'}} I^{a'b'}, \tag{11.85}$$

and the transformation implies that I^{jk}, unlike $I^{a'b'}$, depends on time.

It is customary to encode the three independent components of $I^{a'b'}$ into the *principal moments of inertia*

$$I_1 := \int \rho(\boldsymbol{x}')(y'^2 + z'^2)\, d^3x' = I^{y'y'} + I^{z'z'}, \tag{11.86a}$$

$$I_2 := \int \rho(\boldsymbol{x}')(x'^2 + z'^2)\, d^3x' = I^{x'x'} + I^{z'z'}, \tag{11.86b}$$

$$I_3 := \int \rho(\boldsymbol{x}')(x'^2 + y'^2)\, d^3x' = I^{x'x'} + I^{y'y'}. \tag{11.86c}$$

A body with $I_1 = I_2 = I_3$ is spherically symmetric, and such a body would not emit gravitational waves. A body with $I_1 = I_2 \neq I_3$ is symmetric about the axis of rotation, and such a body also would not emit gravitational waves. To produce waves the body must be sufficiently deformed, and a convenient measure of the deformation is the *ellipticity parameter*

$$\varepsilon := \frac{I_1 - I_2}{I_3}. \tag{11.87}$$

As we shall see, the gravitational-wave field is proportional to $(I_1 - I_2) = \varepsilon I_3$.

The components of the quadrupole-moment tensor in the co-rotating frame are $I^{x'x'} = \frac{1}{2}(I_2 + I_3 - I_1)$, $I^{y'y'} = \frac{1}{2}(I_3 + I_1 - I_2)$, and $I^{z'z'} = \frac{1}{2}(I_1 + I_2 - I_3)$. As a consequence of Eq. (11.85), we find that they are given by

$$I^{xx} = \frac{1}{2}I_3 - \frac{1}{2}(I_1 - I_2)\cos 2\Omega t, \tag{11.88a}$$

$$I^{xy} = -\frac{1}{2}(I_1 - I_2)\sin 2\Omega t, \tag{11.88b}$$

$$I^{yy} = \frac{1}{2}I_3 + \frac{1}{2}(I_1 - I_2)\cos 2\Omega t, \tag{11.88c}$$

$$I^{zz} = \frac{1}{2}(I_1 + I_2 - I_3), \tag{11.88d}$$

in the non-rotating frame; the components I^{xz} and I^{yz} vanish. The non-vanishing components of $\ddot{I}^{jk}$ are

$$\ddot{I}^{xx} = 2\varepsilon I_3 \Omega^2 \cos 2\Omega t, \tag{11.89a}$$

$$\ddot{I}^{xy} = 2\varepsilon I_3 \Omega^2 \sin 2\Omega t, \tag{11.89b}$$

$$\ddot{I}^{yy} = -2\varepsilon I_3 \Omega^2 \cos 2\Omega t. \tag{11.89c}$$

These expressions are ready to be inserted within Eq. (11.55) to obtain the gravitational-wave field $h^{jk}_{\rm TT}$.

To extract the polarizations h_+ and $h_\times$ we adopt the same conventions as in Sec. 11.2.2. We specify the direction of the gravitational-wave detector in the non-rotating frame (x, y, z) by the polar angles (ι, ω), and use the vectors $\boldsymbol{e}_X$ and $\boldsymbol{e}_Y$ of Eqs. (11.72) as a basis in the transverse subspace. In this case, ι is the angle between the body's rotation axis and the direction to the detector, and ω is the angle, at $t = 0$, between the intersection of the body's equatorial plane with the plane of the sky and the direction of the body's long axis. The polarizations are defined as in Eqs. (11.75), and a quick calculation returns the expressions

$$h_+ = \frac{1}{2}(1 + \cos^2 \iota) h_0 \cos 2(\Omega\tau + \omega), \qquad h_\times = \cos\iota\, h_0 \sin 2(\Omega\tau + \omega), \tag{11.90}$$

where $\tau := t - R/c$ is retarded time, and

$$h_0 = \frac{4G}{c^4 R}\varepsilon I_3 \Omega^2 \tag{11.91}$$

is the gravitational-wave amplitude.

Mountain on a spherical star

A simple model of a deformed neutron star features a mountain on the surface of an otherwise spherical body. The body has a mass M and radius a, and for simplicity we take its density to be uniform. The mountain has a mass $m \ll M$ and is situated on the surface at a position determined by the polar angles (θ, ϕ) in the body-adapted frame (x', y', z'); we model it as a point mass with a mass density $\rho = m\delta(\boldsymbol{x}' - \boldsymbol{\xi})$, with $\boldsymbol{\xi} := [a\sin\theta\cos\phi, a\sin\theta\sin\phi, a\cos\theta]$ giving the position of the mountain in the co-rotating frame.

It is easy to calculate that the body's contribution to the moments of inertia is given by

$$I_1^{\text{body}} = I_2^{\text{body}} = I_3^{\text{body}} = \frac{2}{5}Ma^2. \tag{11.92}$$

We see that as expected, the body makes no contribution to the ellipticity ε. The mountain, on the other hand, gives rise to

$$I_1^{\text{mountain}} - I_2^{\text{mountain}} = -ma^2 \sin^2\theta \cos 2\phi, \tag{11.93}$$

as well as a contribution to I_3 that is smaller than I_3^{body} by a factor of order $m/M \ll 1$. Neglecting this, we find that the model produces

$$\varepsilon = -\frac{5}{2}\frac{m}{M}\sin^2\theta \cos 2\phi, \qquad I_3 = \frac{2}{5}Ma^2. \tag{11.94}$$

These expressions can then be inserted within Eq. (11.91) to calculate the gravitational-wave amplitude.

Ellipsoid of uniform density

Another model of a deformed neutron star puts it in the shape of an ellipsoid of principal axes a, b, and c. The surface is thus described by the equation

$$\frac{x'^2}{a^2} + \frac{y'^2}{b^2} + \frac{z'^2}{c^2} = 1, \tag{11.95}$$

and we take the body to have a uniform mass density ρ. To carry out the integrations over the star's interior, it is useful to adopt the ellipsoidal coordinates (r, θ, ϕ) which are related to the original Cartesian coordinates by $x' = ar\sin\theta\cos\phi$, $y' = br\sin\theta\sin\phi$, and $z' = cr\cos\theta$. The radial coordinate r is dimensionless and ranges from 0 to 1; the polar angles (θ, ϕ) have their usual ranges. The volume element is $d^3x' = abc\,r^2\sin\theta\, dr d\theta d\phi$ in the ellipsoidal coordinates.

The mass of the body is given by $M = (4\pi/3)\rho abc$, and a straightforward calculation reveals that the moments of inertia are

$$I_1 = \frac{1}{5}M(b^2 + c^2), \tag{11.96a}$$

$$I_2 = \frac{1}{5}M(a^2 + c^2), \tag{11.96b}$$

$$I_3 = \frac{1}{5}M(a^2 + b^2). \tag{11.96c}$$

This produces an ellipticity given by

$$\varepsilon = \frac{b^2 - a^2}{b^2 + a^2}. \tag{11.97}$$

These expressions can be inserted within Eq. (11.91) to obtain the gravitational-wave amplitude.

Realistic neutron stars

The degree of deformation of a realistic neutron star is largely unknown, as are the exact mechanisms that would be involved in supporting a long-lived ellipticity ε. The most popular models feature either a genuine mountain that might reach deep below the crust, a deformation driven and sustained by accretion of matter from a companion, or a deformation created by a large toroidal magnetic field. These models suggest that $\varepsilon < 10^{-6}$ for conventional models of neutron stars (which involve a solid crust resting on a liquid core), but larger values might be possible for more exotic objects such as quark stars.

A typical neutron star has a mass $M = 1.4M_{\odot}$ and radius $a = 12$ km, and this gives rise to a moment of inertia of the order of $I_3 = \frac{2}{5}Ma^2 = 1.6 \times 10^{38}$ kg m^2. A fast pulsar rotates with a period $P = 10$ ms and might be situated at a distance $R = 1$ kpc. Using $\varepsilon = 10^{-6}$ as a typical value for the ellipticity, Eq. (11.91) gives rise to the estimate

$$h_0 \simeq 6.8 \times 10^{-25} \left(\frac{\varepsilon}{10^{-6}}\right) \left(\frac{I_3}{1.6 \times 10^{38}\ \mathrm{kg\,m^2}}\right) \left(\frac{10\ \mathrm{ms}}{P}\right)^2 \left(\frac{1\ \mathrm{kpc}}{R}\right). \tag{11.98}$$

Gravitational waves produced by rotating neutron stars are exceedingly small, but coherent integration of a signal of known frequency over a very long time builds up a signal-to-noise ratio that may exceed the detection threshold of a gravitational-wave detector. (The frequency can be measured in radio waves if the rotating neutron star is a known pulsar.) The search is on!

11.2.4 Application: Tidally deformed star

As a third and final application of the quadrupole formula we calculate the gravitational waves emitted during a tidal interaction between a fluid body and a nearby object. For concreteness and simplicity we take the body to be non-rotating and to have a uniform density, and we place the external object on a parabolic trajectory. We work in the moving frame of the body, and ignore the gravitational waves produced by the center-of-mass motion (these were considered previously, in the case of elliptical and circular motion); as we shall see, the tidal gravitational waves are typically much weaker than the waves produced by the orbital motion. The body's tidal dynamics was studied in some detail back in Sec. 2.5.3, and we begin our discussion with a recollection of the main results.

Tidal dynamics

The body is assumed to be spherical and in hydrostatic equilibrium in the absence of a tidal interaction; in its unperturbed state it has a mass M, a radius a, and a uniform density ρ_0. The body is perturbed by an external object of mass M' at a position $x^j = rn^j$ relative to the body's center-of-mass. This object produces a tidal potential $U_{\mathrm{tidal}} = -\frac{1}{2}\mathcal{E}_{jk}(t)\,x^j x^k$ inside the body, with

$$\mathcal{E}_{jk} = \frac{GM'}{r^3}\left(\delta_{jk} - 3n_j n_k\right), \tag{11.99}$$

denoting the tidal quadrupole moment. As we saw back in Sec. 2.5.3, the body's deformation in response to the perturbation is measured by its mass quadrupole moment $I^{\langle jk\rangle}(t)$, which, according to Eq. (2.289), is given by

$$I^{\langle jk\rangle} = -\frac{2}{5}Ma^2\mathcal{F}^{jk}, \tag{11.100}$$

where

$$\mathcal{F}^{jk}(t) := \frac{1}{\omega_2}\int_{-\infty}^{t} \mathcal{E}^{jk}(t')\sin\omega_2(t-t')\,dt' \tag{11.101}$$

is the body's response function, with

$$\omega_2 := \sqrt{\frac{4}{5}\frac{GM}{a^3}} \tag{11.102}$$

denoting the body's f-mode frequency for a quadrupole deformation.

Gravitational waves

Differentiation of Eq. (11.100) gives

$$\ddot{I}^{\langle jk\rangle} = -\frac{2}{5}Ma^2\mathcal{G}^{jk}, \tag{11.103}$$

in which

$$\mathcal{G}^{jk}(t) := \frac{1}{\omega_2}\int_{-\infty}^{t} \ddot{\mathcal{E}}^{jk}(t')\sin\omega_2(t-t')\,dt' \tag{11.104}$$

is the response function associated with $\ddot{\mathcal{E}}^{jk}$ instead of $\mathcal{E}^{jk}$; two integrations by parts were required to arrive at this result. Substituting this into the quadrupole formula of Eq. (11.55), we find that the gravitational-wave field is given by

$$h^{jk}_{\rm TT}(t,\boldsymbol{x}) = -\frac{4}{5}\frac{GMa^2}{c^4R}\mathcal{G}^{jk}_{\rm TT}(\tau), \tag{11.105}$$

in which $R := |\boldsymbol{x}|$ is the distance to the detector and $\tau := t - R/c$ is retarded time.

We can use Eq. (11.105) to estimate the magnitude of the gravitational waves produced by a tidal interaction. For an external object of mass M' at a distance r, the tidal moment scales as $\mathcal{E}_{jk} \sim GM'/r^3$, and it changes over a time scale comparable to Ω_c^{-1}, in which $\Omega_c := \sqrt{G(M+M')/r^3}$ is a characteristic frequency of the orbital motion. This yields $\ddot{\mathcal{E}}_{jk} \sim \Omega_c^2\mathcal{E}_{jk}$, and substitution within Eq. (11.104) returns the estimate $\mathcal{G}_{jk} \sim \omega_2^{-2}\ddot{\mathcal{E}}_{jk} \sim (\Omega_c/\omega_2)^2\mathcal{E}_{jk}$. Inserting this within Eq. (11.105), we arrive at

$$h \sim \frac{G^2(M+M')M'}{c^4R}\frac{a^5}{r^6}. \tag{11.106}$$

It is useful to compare this with Eq. (11.77), which provides an estimate for the gravitational waves produced by the orbital motion. According to this, we find that the ratio of wave amplitudes is estimated as

$$\frac{h_{\text{tidal}}}{h_{\text{orbital}}} \sim \frac{M'}{M+M'}\left(\frac{a}{r}\right)^5. \tag{11.107}$$

The five powers of the ratio of length scales imply that when r is even modestly larger than a, the waves emitted during the tidal interaction are very weak compared with the waves produced by the orbital motion. The waves can be comparable only for very close encounters with $r \sim a$.

Parabolic encounter

To give concreteness to these considerations, we examine the tidal interaction that results when the external object is placed on a parabolic trajectory described by setting $e = 1$ in Eq. (11.68). The motion of the external object is parameterized by $p := 2r_{\text{min}}$, in which r_{min} is the distance of closest approach. It can also be parameterized by the frequency

$$\Omega := \sqrt{\frac{G(M+M')}{p^3}}, \tag{11.108}$$

which is such that the angular velocity $\dot{\phi}_{\text{max}}$ at closest approach is equal to $\dot{\phi}_{\text{max}} = 4\Omega$. It is useful to note that

$$\left(\frac{\omega_2}{\Omega}\right)^2 = \frac{32}{5}\frac{M}{M+M'}\left(\frac{r_{\text{min}}}{a}\right)^3. \tag{11.109}$$

It is straightforward to differentiate Eq. (11.99) twice with respect to time and to insert the result within Eq. (11.104), which must then be evaluated numerically. We extract the gravitational-wave polarizations h_+ and $h_\times$ from Eq. (11.105) by adopting the same conventions as in Sec. 11.2.2. We obtain the expressions

$$h_{+,\times} = -h_0\, H_{+,\times}, \tag{11.110}$$

where

$$h_0 = 3\frac{G^2(M+M')M'}{c^4 R}\frac{a^5}{p^6} \tag{11.111}$$

is the gravitational-wave amplitude, and the scale-free polarizations $H_{+,\times}$ are extracted from $H^{jk} := \frac{1}{3}[(M+M')/M'](\omega_2/\Omega^2)^2\mathcal{G}^{jk}$. These are plotted in Fig. 11.3 for selected values of ω_2/Ω, and the caption describes their main properties.

11.3 Beyond the quadrupole formula: Waves at 1.5PN order

We now embark on a long journey to improve our description of gravitational waves by going beyond the quadrupole formula of Eq. (11.55). This, we recall, is the leading term in an expansion of the gravitational-wave field in powers of v_c/c, where v_c is a characteristic

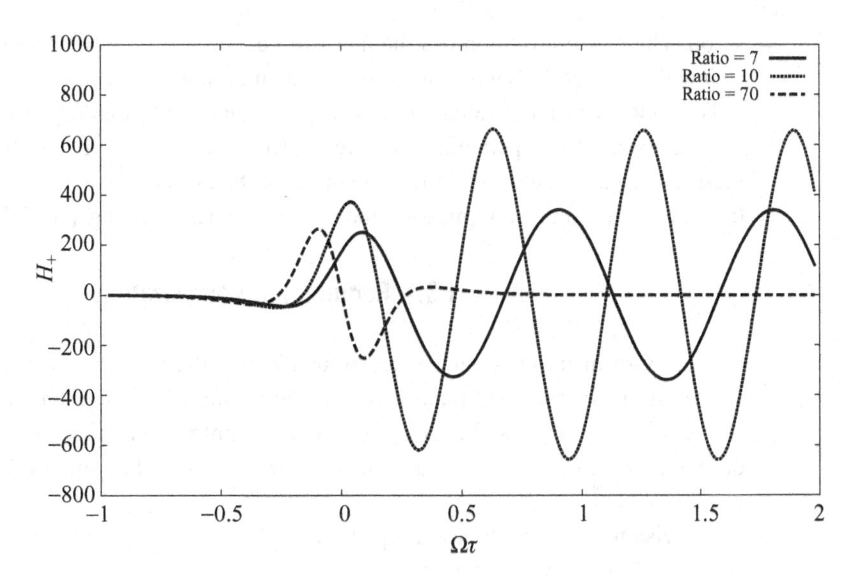

Fig. 11.3 The polarization H_+ as a function of the dimensionless retarded time $\Omega\tau$, for selected values of the ratio ω_2/Ω. The curves are displayed for an inclination angle $\iota = 30°$, and a longitude of pericenter $\omega = 45°$. We see that most of the emission takes place at and after the moment of closest approach ($\tau = 0$). When ω_2 is comparable to Ω, the parabolic encounter ignites a fluid mode of frequency ω_2 which produces a gravitational wave of frequency ω_2. The effect is maximized when the resonant condition $2\dot{\phi}_{\rm max} = 8\Omega \simeq \omega_2$ is met, and the wave is heavily suppressed when $\omega_2 \gg \Omega$. In a more realistic model of the tidal interaction, the wave would eventually be damped by dissipative processes taking place inside the body; in our simplified model the fluid mode goes on forever.

velocity of the source. We shall call this leading term the *Newtonian contribution* to the gravitational-wave field, and in this section we will compute corrections of order (v_c/c), $(v_c/c)^2$, and $(v_c/c)^3$ to the quadrupole formula; in other words, we shall calculate $h^{jk}_{\rm TT}$ through 1.5PN order in a post-Newtonian expansion.

We note that the post-Newtonian counting described here differs from the convention adopted back in Sec. 7.4; see Box 7.7. In the original convention the quadrupole terms in the gravitational potentials were given a 1PN label instead of the 0PN label assigned here. The reason for this can be gathered from the following expression for h^{00},

$$h^{00} = \frac{4G}{c^2 R}\left[M + \frac{1}{2c^2}\ddot{I}^{jk}N_j N_k + \cdots\right], \tag{11.112}$$

which holds in the far-away wave zone. The leading term in this expression is the mass term, and in the old convention this was given a sensible 0PN label. The quadrupole term is smaller than this by a factor of order $(v_c/c)^2$, and this was given a 1PN label. Our new convention differs from this because our focus is now different: We are interested only in the spatial components of the gravitational potentials, and these do not contain a mass term. And since the leading term involves the *Newtonian* quadrupole moment I^{jk}, it is convenient to reset the post-Newtonian counter and call the right-hand side of Eq. (11.55) the Newtonian contribution to $h^{jk}_{\rm TT}$. Additional terms are labeled 0.5PN, 1PN, and 1.5PN, and so on. The new convention has the merit of keeping the post-Newtonian orders of h^{jk} in step with those of the source, and those of the multipole moments.

For concreteness the calculations will be specialized to a system of N bodies that we assume to be well separated; these are identified by a label A, and body A has a total mass-energy M_A, a position $\boldsymbol{r}_A(t)$, and moves with a velocity $\boldsymbol{v}_A(t)$. For simplicity we shall take the bodies to be point masses, and rely on the description given in Sec. 9.6.

11.3.1 Requirements and strategy

Our purpose in this first subsection is to identify the tasks that lie ahead: we map the terrain of our journey and plan the calculational strategy. The computations will be long and tedious, and they will occupy us in the remaining nine subsections. The reader who does not wish to follow the details can skip ahead to Box 11.4 and find a summary of our results.

We wish to integrate the wave equation

$$\Box h^{jk} = -\frac{16\pi G}{c^4}\tau^{jk} \tag{11.113}$$

for the spatial components of the gravitational potentials, and evaluate the solution in the far-away wave zone. Here, $\tau^{jk} = (-g)(T^{jk} + t^{jk}_{\rm LL} + t^{jk}_{\rm H})$ are the spatial components of the effective energy-momentum pseudotensor first introduced in Sec. 6.2.1, decomposed into a material contribution T^{jk}, the Landau–Lifshitz pseudotensor $t^{jk}_{\rm LL}$, and the harmonic-gauge contribution $t^{jk}_{\rm H}$. We wish to integrate the wave equation to a degree of accuracy that surpasses what was achieved in Chapter 7; this amounts to constructing a *third post-Minkowskian approximation* to the exact gravitational potentials. And we wish to extract from h^{jk} the transverse-tracefree pieces that truly represent the gravitational-wave field.

Techniques to integrate Eq. (11.113) were developed in Chapter 6 and summarized in Box 6.7. In Sec. 6.3 we learned to express the solution as an integral over the past light cone of the field point $(t, \boldsymbol{x})$, which is decomposed as

$$h^{jk} = h^{jk}_{\mathscr{N}} + h^{jk}_{\mathscr{W}}. \tag{11.114}$$

The near-zone piece $h^{jk}_{\mathscr{N}}$ comes from the portion of the light cone that lies within the near zone (where $|\boldsymbol{x}'| < \mathcal{R}$), and the wave-zone piece $h^{jk}_{\mathscr{W}}$ comes from the portion that lies in the wave zone (where $|\boldsymbol{x}'| > \mathcal{R}$); the boundary between the zones is arbitrarily positioned at the radius $|\boldsymbol{x}'| = \mathcal{R} \sim \lambda_c$. In Sec. 6.3.4 we derived an expression for $h^{jk}_{\mathscr{N}}$ that is valid in the far-away wave zone; this is given by Eq. (6.91), which we copy as

$$h^{jk}_{\mathscr{N}} = \frac{4G}{c^4 R}\sum_{\ell=0}^{\infty}\frac{1}{\ell! c^\ell} N_L \left(\frac{d}{d\tau}\right)^\ell \int_{\mathscr{M}} \tau^{jk}(\tau, \boldsymbol{x}')x'^L\, d^3x', \tag{11.115}$$

where $\boldsymbol{N} := \boldsymbol{x}/R$ is a unit radial vector, L a multi-index that contains a number ℓ of individual indices, $N_L := N_{j_1}N_{j_2}\cdots N_{j_\ell}$, $x'^L := x'^{j_1}x'^{j_2}\cdots x'^{j_\ell}$, and where the domain of integration $\mathscr{M}$ is defined by $|\boldsymbol{x}'| < \mathcal{R}$.

We consider $h^{jk}_{\mathscr{W}}$ at a later stage. For the time being we focus our attention on $h^{jk}_{\mathscr{N}}$, and write Eq. (11.115) in a form that showcases the first few terms:

$$h^{jk}_{\mathscr{N}} = \frac{4G}{c^4 R}\biggl\{\int_{\mathscr{M}} \tau^{jk}\, d^3x' + \frac{1}{c} N_a \frac{d}{d\tau}\int_{\mathscr{M}} \tau^{jk} x'^a\, d^3x' + \frac{1}{2c^2} N_a N_b \frac{d^2}{d\tau^2}\int_{\mathscr{M}} \tau^{jk} x'^a x'^b\, d^3x' + \frac{1}{6c^3} N_a N_b N_c \frac{d^3}{d\tau^3}\int_{\mathscr{M}} \tau^{jk} x'^a x'^b x'^c\, d^3x' + [\ell \geq 4]\biggr\}, \tag{11.116}$$

where $[\ell \geq 4]$ stands for the remaining terms in the sum over ℓ. In keeping with our previous discussion, we say that the first term on the right-hand side of Eq. (11.116) makes a 0PN contribution to h^{jk} (together with a correction of 1PN order), the second term makes a 0.5PN contribution (together with a correction of 1.5PN order), the third term makes a 1PN contribution, and the fourth term a 1.5PN contribution; the $[\ell \geq 4]$ terms contribute at 2PN and higher orders, and we shall not keep them in the calculation.

To help with the first two integrals we invoke the conservation identities of Eqs. (7.14), which we copy here as

$$\tau^{jk} = \frac{1}{2c^2}\frac{\partial^2}{\partial\tau^2}\bigl(\tau^{00}x^j x^k\bigr) + \frac{1}{2}\partial_p\bigl(\tau^{jp}x^k + \tau^{kp}x^j - \partial_q \tau^{pq} x^j x^k\bigr), \tag{11.117a}$$

$$\tau^{jk}x^a = \frac{1}{2c}\frac{\partial}{\partial\tau}\bigl(\tau^{0j}x^k x^a + \tau^{0k}x^j x^a - \tau^{0a}x^j x^k\bigr) + \frac{1}{2}\partial_p\bigl(\tau^{jp}x^k x^a + \tau^{kp}x^j x^a - \tau^{ap}x^j x^k\bigr). \tag{11.117b}$$

Making the substitutions and introducing some notation to simplify the writing, we find that $h^{jk}_{\mathscr{N}}$ can be expressed as

$$h^{jk}_{\mathscr{N}} = \frac{2G}{c^4 R}\frac{\partial^2}{\partial\tau^2}\biggl\{Q^{jk} + Q^{jka}N_a + Q^{jkab}N_a N_b + \frac{1}{3}Q^{jkabc}N_a N_b N_c + [\ell \geq 4]\biggr\} + \frac{2G}{c^4 R}\biggl\{P^{jk} + P^{jka}N_a\biggr\}, \tag{11.118}$$

in which the radiative multipole moments are defined by

$$Q^{jk} := \int_{\mathscr{M}} c^{-2}\tau^{00} x'^j x'^k\, d^3x', \tag{11.119a}$$

$$Q^{jka} := \frac{1}{c}\int_{\mathscr{M}} \bigl(c^{-1}\tau^{0j}x'^k x'^a + c^{-1}\tau^{0k}x'^j x'^a - c^{-1}\tau^{0a}x'^j x'^k\bigr)\, d^3x', \tag{11.119b}$$

$$Q^{jkab} := \frac{1}{c^2}\int_{\mathscr{M}} \tau^{jk} x'^a x'^b\, d^3x', \tag{11.119c}$$

$$Q^{jkabc} := \frac{1}{c^3}\frac{d}{d\tau}\int_{\mathscr{M}} \tau^{jk} x'^a x'^b x'^c\, d^3x', \tag{11.119d}$$

and where

$$P^{jk} := \oint_{\partial\mathcal{M}} \left(\tau^{jp}x'^k + \tau^{kp}x'^j - \partial'_q \tau^{pq}x'^j x'^k\right) dS_p, \tag{11.120a}$$

$$P^{jka} := \frac{1}{c}\frac{d}{d\tau}\oint_{\partial\mathcal{M}} \left(\tau^{jp}x'^k x'^a + \tau^{kp}x'^j x'^a - \tau^{ap}x'^j x'^k\right) dS_p, \tag{11.120b}$$

are surface integrals that also contribute to $h^{jk}_{\mathcal{N}}$. In the radiative moments, τ^{jk} is expressed as a function of τ and $\boldsymbol{x}'$. The same is true within the surface integrals, except for the fact that x'^j is now equal to $\mathcal{R}N^j$; the surface element on $\partial\mathcal{M}$ is $dS_j := \mathcal{R}^2 N_j\, d\Omega$. The multipole moments and surface integrals are functions of τ only.

In the following subsections we will endeavor to calculate the quantities that appear within Eq. (11.118). As was stated previously, we wish to compute h^{jk} accurately through 1.5PN order. In a schematic notation, what we want is

$$h^{jk} = \frac{G}{c^4 R}\left(c^0 + c^{-1} + c^{-2} + c^{-3} + \cdots\right), \tag{11.121}$$

in which the leading, c^0 term is the 0PN contribution, the correction of order c^{-1} a 0.5PN term, and so on. To achieve this we need to calculate $c^{-2}\tau^{00} = c^0 + c^{-2} + \cdots$ to obtain $Q^{jk} = c^0 + c^{-2} + \cdots$, $c^{-1}\tau^{0j} = c^0 + c^{-2} + \cdots$ to obtain $Q^{jka} = c^{-1} + c^{-3} + \cdots$, and $\tau^{jk} = c^0 + \cdots$ to obtain $Q^{jkab} = c^{-2} + \cdots$ and $Q^{jkabc} = c^{-3} + \cdots$. And on $\partial\mathcal{M}$ we need to calculate $\tau^{jk} = c^0 + c^{-2} + \cdots$ to obtain $P^{jk} = c^0 + c^{-2} + \cdots$ and $P^{jka} = c^{-1} + c^{-3} + \cdots$ All in all, this will give us the 1.5PN accuracy that we demand for h^{jk}.

Our considerations have so far excluded $h^{jk}_{\mathcal{W}}$. We postpone a detailed discussion until Sec. 11.3.7, where we compute this contribution to the gravitational potentials. For the time being it suffices to say that $h^{jk}_{\mathcal{W}}$ contributes at 1.5PN order. It is therefore needed to achieve the required level of accuracy for h^{jk}.

The calculations that follow are lengthy. They are simplified considerably by the observation that ultimately we wish to extract the transverse-tracefree part of h^{jk}. It is therefore superfluous to calculate any term that will not survive the TT projection introduced in Sec. 11.1.7. For example, a term in h^{jk} that is known to be proportional to δ^{jk}, or to N^j, will not survive the projection, and does not need to be computed. There are many such terms, and ignoring them is a substantial time saver. As another example, terms in Q^{jkab} that are proportional to δ^{ja}, or δ^{ka}, or δ^{jb}, or δ^{kb} (but not δ^{ab}!), can all be ignored because they produce contributions to h^{jk} that are proportional to N^j or N^k, and these will not survive the TT projection. To indicate equality modulo terms that do not survive the transverse-tracefree projection, we introduce the notation $\stackrel{\text{TT}}{=}$, so that

$$A^{jk} \stackrel{\text{TT}}{=} B^{jk} \tag{11.122}$$

whenever

$$(\text{TT})^{jk}{}_{pq} A^{pq} = (\text{TT})^{jk}{}_{pq} B^{pq}. \tag{11.123}$$

In other words, A^{jk} and B^{jk} differ by a tensor C^{jk} that contains no TT part: $(\text{TT})^{jk}{}_{pq} C^{pq} = 0$.

An additional source of simplification – an important one – was exploited previously in Sec. 7.4, with a justification provided in Sec. 6.3.3: we are free to ignore all $\mathcal{R}$-dependent terms in $h^{jk}_{\mathcal{N}}$, and all $\mathcal{R}$-dependent terms in $h^{jk}_{\mathcal{W}}$, because any dependence on the arbitrary cutoff parameter $\mathcal{R}$ (the radius of the artificial boundary between the near zone and the wave zone) is guaranteed to cancel out when $h^{jk}_{\mathcal{N}}$ and $h^{jk}_{\mathcal{W}}$ are added together to form the complete potentials h^{jk}. The freedom to discard all $\mathcal{R}$-dependent terms is another substantial time saver.

11.3.2 Integration techniques for field integrals

In the course of our calculations we shall encounter a number of field integrals, a representative example of which is

$$E^{jk} := \frac{1}{4\pi} \int_{\mathscr{M}} U \partial^j U x^k \, d^3x, \tag{11.124}$$

where $\mathscr{M}$ is the domain of integration $|\boldsymbol{x}| < \mathcal{R}$, and where

$$U := \sum_A \frac{GM_A}{|\boldsymbol{x} - \boldsymbol{r}_A|} \tag{11.125}$$

is the Newtonian potential for a system of point masses. In this subsection we introduce techniques to evaluate such integrals. We will examine the specific case of Eq. (11.124), but the techniques are quite general, and they apply just as well to many similar field integrals.

Explicit form of E^{jk}; change of integration variables

After evaluating $\partial^j U$ we find that the field integral can be expressed in the more explicit form

$$E^{jk} = -\sum_A G^2 M_A^2 E_A^{jk} - \sum_A \sum_{B \neq A} G^2 M_A M_B E_{AB}^{jk}, \tag{11.126}$$

where

$$E_A^{jk} := \frac{1}{4\pi} \int_{\mathscr{M}} \frac{(x - r_A)^j x^k}{|\boldsymbol{x} - \boldsymbol{r}_A|^4} \, d^3x, \tag{11.127a}$$

$$E_{AB}^{jk} := \frac{1}{4\pi} \int_{\mathscr{M}} \frac{(x - r_B)^j x^k}{|\boldsymbol{x} - \boldsymbol{r}_A||\boldsymbol{x} - \boldsymbol{r}_B|^3} \, d^3x. \tag{11.127b}$$

To evaluate the first integral we make the substitution $\boldsymbol{x} = \boldsymbol{r}_A + \boldsymbol{y}$ and integrate with respect to the new variables $\boldsymbol{y}$. This leads to

$$E_A^{jk} = \frac{1}{4\pi} \int_{\mathscr{M}} \frac{y^j y^k}{y^4} \, d^3y + \frac{r_A^k}{4\pi} \int_{\mathscr{M}} \frac{y^j}{y^4} \, d^3y, \tag{11.128}$$

where $y := |\boldsymbol{y}|$. To evaluate the second integral we use instead $\boldsymbol{x} = \boldsymbol{r}_B + \boldsymbol{y}$ and integrate with respect to $\boldsymbol{y}$. This leads to

$$E_{AB}^{jk} = \frac{1}{4\pi} \int_{\mathscr{M}} \frac{1}{|\boldsymbol{y} - \boldsymbol{r}_{AB}|} \frac{y^j y^k}{y^3} \, d^3y + \frac{r_B^k}{4\pi} \int_{\mathscr{M}} \frac{1}{|\boldsymbol{y} - \boldsymbol{r}_{AB}|} \frac{y^j}{y^3} \, d^3y, \tag{11.129}$$

where $\boldsymbol{r}_{AB} := \boldsymbol{r}_A - \boldsymbol{r}_B$.

Translation of the domain of integration

Each one of the integrals that appears in Eqs. (11.128) and (11.129) is of the form

$$\int_{\mathscr{M}} f(\boldsymbol{y})\, d^3y,$$

where f is a function of the vector $\boldsymbol{y}$, which is related to the original variables $\boldsymbol{x}$ by a translation $\boldsymbol{x} = \boldsymbol{y} + \boldsymbol{r}$, with $\boldsymbol{r}$ independent of $\boldsymbol{x}$. The domain of integration $\mathscr{M}$ is defined by $|\boldsymbol{x}| < \mathcal{R}$, or $|\boldsymbol{y} + \boldsymbol{r}| < \mathcal{R}$, and it will be convenient to replace it by the simpler domain $\mathscr{M}_y$ defined by $y := |\boldsymbol{y}| < \mathcal{R}$.

To effect this replacement we note that the cutoff radius $\mathcal{R}$ can be assumed to be large compared with $r := |\boldsymbol{r}|$. (Recall the discussion of Sec. 6.3.3, in which $\mathcal{R}$ is chosen to be comparable to λ_c, the characteristic wavelength of the gravitational radiation. Recall also the discussion of Sec. 6.3.2, in which λ_c is shown to be large compared with both $|\boldsymbol{r}_A|$ and $|\boldsymbol{r}_{AB}|$, because in a slow-motion situation the matter distribution is always situated deep within the near zone. Conclude from these observations that $r/\mathcal{R} \ll 1$, as claimed.) The condition that defines $\mathscr{M}$ is $y^2 + 2\boldsymbol{r}\cdot\boldsymbol{y} + r^2 < \mathcal{R}^2$, and this can be expressed more simply as

$$y < \mathcal{R} - r\cos\gamma + O(r^2/\mathcal{R}), \tag{11.130}$$

when $r/\mathcal{R} \ll 1$; here γ is the angle between the vectors $\boldsymbol{y}$ and $\boldsymbol{r}$.

Switching to spherical polar coordinates (y, θ, ϕ) associated with the vector $\boldsymbol{y}$, the integral is

$$\begin{aligned}\int_{\mathscr{M}} f(\boldsymbol{y})\, d^3y &= \int d\Omega \int_0^{\mathcal{R} - r\cos\gamma + \cdots} f(y,\theta,\phi)\, y^2 dy \\ &= \int d\Omega \int_0^{\mathcal{R}} f(y,\theta,\phi)\, y^2 dy + \int d\Omega \int_{\mathcal{R}}^{\mathcal{R} - r\cos\gamma + \cdots} f(y,\theta,\phi)\, y^2 dy,\end{aligned} \tag{11.131}$$

where $d\Omega = \sin\theta\, d\theta d\phi$ is an element of solid angle. In the second line, the first integral is over the domain $\mathscr{M}_y$, while the second integral is

$$\int (-r\cos\gamma)\mathcal{R}^2 f(\mathcal{R},\theta,\phi)\, d\Omega = -\oint_{\partial\mathscr{M}_y} f(\boldsymbol{y})\, \boldsymbol{r}\cdot d\boldsymbol{S} \tag{11.132}$$

to first order in $r/\mathcal{R}$; here, $dS^j := \mathcal{R}^2 N^j\, d\Omega$, with $\boldsymbol{N} := \boldsymbol{y}/y$, is the surface element on $\partial\mathscr{M}_y$, the boundary of $\mathscr{M}_y$ described by the equation $y = \mathcal{R}$.

We have obtained the useful approximation

$$\int_{\mathscr{M}} f(\boldsymbol{y})\, d^3y = \int_{\mathscr{M}_y} f(\boldsymbol{y})\, d^3y - \oint_{\partial\mathscr{M}_y} f(\boldsymbol{y})\, \boldsymbol{r}\cdot d\boldsymbol{S} + \cdots, \tag{11.133}$$

in which the domain of integration $\mathscr{M}_y$ is defined by $y := |\boldsymbol{y}| < \mathcal{R}$, and $\partial\mathscr{M}_y$ is its boundary at $y = \mathcal{R}$. It is clear that the surface integral in Eq. (11.133) is smaller than the volume integral by a factor of order $r/\mathcal{R} \ll 1$; the neglected terms are smaller still.

Evaluation of E_A^{jk}

We now return to the field integral of Eq. (11.128). We begin by working on the first term, which we copy as

$$\frac{1}{4\pi}\int_{\mathcal{M}} \frac{y^j y^k}{y^4}\, d^3y.$$

Inserting this within Eq. (11.133), we find that the volume integral is

$$\frac{1}{4\pi}\int_{\mathcal{M}_y} \frac{y^j y^k}{y^4}\, d^3y = \frac{1}{4\pi}\int_{\mathcal{M}_y} N^j N^k\, dy d\Omega = \langle\langle N^j N^k \rangle\rangle \int_0^{\mathcal{R}} dy = \frac{1}{3}\delta^{jk}\mathcal{R}, \tag{11.134}$$

in which $\langle\langle \cdots \rangle\rangle := (4\pi)^{-1}\int (\cdots)\, d\Omega$ denotes an angular average; the identity $\langle\langle N^j N^k \rangle\rangle = \frac{1}{3}\delta^{jk}$ was established back in Sec. 1.5.3, along with other similar results. This contribution to E_A^{jk} can be discarded because it is proportional to $\mathcal{R}$, and it was agreed near the end of Sec. 11.3.1 that all $\mathcal{R}$-dependent terms can indeed be ignored. With the understanding that $\boldsymbol{r}$ stands for $\boldsymbol{r}_A$, the surface integral is

$$-\frac{1}{4\pi}\oint_{\partial\mathcal{M}_y} \frac{y^j y^k}{y^4}\, \boldsymbol{r}\cdot d\boldsymbol{S} = -\frac{1}{4\pi}\int N^j N^k r_p N^p\, d\Omega = -r_p \langle\langle N^j N^k N^p \rangle\rangle = 0. \tag{11.135}$$

The neglected terms in Eq. (11.133) are of order $\mathcal{R}^{-1}$ and smaller, and because they depend on $\mathcal{R}$, they can be freely discarded. We conclude that the first term in Eq. (11.128) evaluates to zero.

We next set to work on the second term, which involves the integral

$$\frac{1}{4\pi}\int_{\mathcal{M}} \frac{y^j}{y^4}\, d^3y.$$

Inserting this within Eq. (11.133), we find that the volume integral is

$$\frac{1}{4\pi}\int_{\mathcal{M}_y} \frac{y^j}{y^4}\, d^3y = \langle\langle N^j \rangle\rangle \int_0^{\mathcal{R}} \frac{dy}{y} = 0. \tag{11.136}$$

It is a fortunate outcome that the logarithmic divergence at $y = 0$ (which occurs because the matter distribution is modeled as a collection of point masses) requires no explicit regularization, because the angular integration vanishes identically. The surface integral is

$$-\frac{1}{4\pi}\oint_{\mathcal{M}_y} \frac{y^j}{y^4}\boldsymbol{r}\cdot d\boldsymbol{S} = -\frac{r_p}{\mathcal{R}}\langle\langle N^j N^p \rangle\rangle = -\frac{1}{3}\frac{r^j}{\mathcal{R}}, \tag{11.137}$$

in which $\boldsymbol{r}$ stands for $\boldsymbol{r}_A$. The additional terms in Eq. (11.133) are smaller by additional powers of $r/\mathcal{R} \ll 1$, and because they all depend on $\mathcal{R}$, they can be freely discarded. We conclude that the second term in Eq. (11.128) evaluates to zero.

We have arrived at

$$E_A^{jk} = 0, \tag{11.138}$$

modulo $\mathcal{R}$-dependent terms that can be freely discarded.

Evaluation of E_{AB}^{jk}

To evaluate the right-hand side of Eq. (11.129) we continue to use Eq. (11.133) to express an integral over the domain $\mathscr{M}$ in terms of a volume integral over $\mathscr{M}_y$ and a surface integral over $\partial\mathscr{M}_y$. We also make use of the addition theorem for spherical harmonics,

$$\frac{1}{|\boldsymbol{y}-\boldsymbol{r}_{AB}|} = \sum_{\ell=0}^{\infty}\sum_{m=-\ell}^{\ell} \frac{4\pi}{2\ell+1}\frac{r_<^\ell}{r_>^{\ell+1}} Y^*_{\ell m}(\boldsymbol{n}_{AB}) Y^{\ell m}(\boldsymbol{N}), \tag{11.139}$$

in which $r_< := \min(y, r_{AB})$, $r_> = \max(y, r_{AB})$, $\boldsymbol{N} := \boldsymbol{y}/y$, and $\boldsymbol{n}_{AB} := \boldsymbol{r}_{AB}/r_{AB}$. We insert Eq. (11.139) within the first integral on the right-hand side of Eq. (11.129). Recalling Eq. (11.133), we approximate this by

$$\begin{aligned}
&\frac{1}{4\pi}\int_{\mathscr{M}_y} \frac{1}{|\boldsymbol{y}-\boldsymbol{r}_{AB}|}\frac{y^j y^k}{y^3}\, d^3y \\
&= \frac{1}{4\pi}\int_{\mathscr{M}_y} \frac{1}{|\boldsymbol{y}-\boldsymbol{r}_{AB}|} N^j N^k y\, dy d\Omega \\
&= \sum_\ell \frac{1}{2\ell+1}\int_0^{\mathcal{R}} dy\, y \frac{r_<^\ell}{r_>^{\ell+1}} \sum_m Y^*_{\ell m}(\boldsymbol{n}_{AB}) \int Y^{\ell m}(\boldsymbol{N}) N^j N^k\, d\Omega.
\end{aligned} \tag{11.140}$$

To evaluate the angular integral we express $N^j N^k$ as

$$N^j N^k = N^{\langle jk\rangle} + \frac{1}{3}\delta^{jk}, \tag{11.141}$$

where $N^{\langle jk\rangle}$ is an STF tensor of the sort introduced back in Sec. 1.5.3, and we invoke the identity of Eq. (1.171),

$$\sum_{m=-\ell}^{\ell} Y^*_{\ell m}(\boldsymbol{n}_{AB}) \int Y_{\ell m}(\boldsymbol{N}) N^{\langle L'\rangle}\, d\Omega = \delta_{\ell\ell'}\, n_{AB}^{\langle L\rangle}. \tag{11.142}$$

This produces

$$\frac{1}{4\pi}\int_{\mathscr{M}_y} \frac{1}{|\boldsymbol{y}-\boldsymbol{r}_{AB}|}\frac{y^j y^k}{y^3}\, d^3y = \frac{1}{5}K(2,1)\, n_{AB}^{\langle jk\rangle} + \frac{1}{3}K(0,1)\,\delta^{jk}, \tag{11.143}$$

where the radial integrals

$$K(\ell, n) := \int_0^{\mathcal{R}} y^n \frac{r_<^\ell}{r_>^{\ell+1}}\, dy \tag{11.144}$$

are evaluated below. This expression must be corrected by the surface integral of Eq. (11.133). We have

$$\frac{1}{4\pi}\oint_{\partial\mathscr{M}_y} \frac{1}{|\boldsymbol{y}-\boldsymbol{r}_{AB}|}\frac{y^j y^k}{y^3}\, \boldsymbol{r}\cdot d\boldsymbol{S} = \frac{\mathcal{R} r_p}{4\pi}\int \frac{1}{|\boldsymbol{y}-\boldsymbol{r}_{AB}|} N^j N^k N^p\, d\Omega, \tag{11.145}$$

in which $\boldsymbol{r}$ stands for $\boldsymbol{r}_B$. Because the leading term of $|\boldsymbol{y}-\boldsymbol{r}_{AB}|^{-1}$ in an expansion in powers of $r_{AB}/\mathcal{R} \ll 1$ is equal to $\mathcal{R}^{-1}$, the surface integral potentially gives rise to an $\mathcal{R}$-independent contribution to E^{jk}_{AB}. But this leading term is proportional to $\langle\langle N^j N^k N^p\rangle\rangle = 0$, and we find that the surface integral does not actually contribute. At this stage we have obtained

$$\frac{1}{4\pi}\int_{\mathscr{M}} \frac{1}{|\boldsymbol{y}-\boldsymbol{r}_{AB}|}\frac{y^j y^k}{y^3}\,d^3y = \frac{1}{5}K(2,1)\,n_{AB}^{\langle jk\rangle} + \frac{1}{3}K(0,1)\,\delta^{jk} \tag{11.146}$$

for the first integral on the right-hand side of Eq. (11.129).

We next set to work on the second integral, and we begin by evaluating

$$\begin{aligned}
&\frac{1}{4\pi}\int_{\mathscr{M}_y} \frac{1}{|\boldsymbol{y}-\boldsymbol{r}_{AB}|}\frac{y^j}{y^3}\,d^3y \\
&= \frac{1}{4\pi}\int_{\mathscr{M}_y} \frac{1}{|\boldsymbol{y}-\boldsymbol{r}_{AB}|} N^j\,dyd\Omega \\
&= \sum_\ell \frac{1}{2\ell+1}\int_0^{\mathcal{R}} dy\,\frac{r_<^\ell}{r_>^{\ell+1}} \sum_m Y^*_{\ell m}(\boldsymbol{n}_{AB}) \int Y^{\ell m}(\boldsymbol{N}) N^j\,d\Omega.
\end{aligned} \tag{11.147}$$

Using Eqs. (11.142) and (11.144), this is

$$\frac{1}{4\pi}\int_{\mathscr{M}_y} \frac{1}{|\boldsymbol{y}-\boldsymbol{r}_{AB}|}\frac{y^j}{y^3}\,d^3y = \frac{1}{3}K(1,0)\,n_{AB}^j\,. \tag{11.148}$$

This must be corrected by the surface integral of Eq. (11.133), and it is easy to show that in this case also, the result scales as $\mathcal{R}^{-1}$ and does not contribute. We have therefore obtained $\frac{1}{3}K(1,0)n_{AB}^j$ for the second integral on the right-hand side of Eq. (11.129).

Collecting results, we find that

$$E_{AB}^{jk} = \frac{1}{5}K(2,1)\,n_{AB}^{\langle jk\rangle} + \frac{1}{3}K(0,1)\,\delta^{jk} + \frac{1}{3}K(1,0)\,n_{AB}^j r_B^k\,. \tag{11.149}$$

Radial integrals

To complete the computation we must now evaluate the radial integrals defined by Eq. (11.144),

$$K(\ell,n) := \int_0^{\mathcal{R}} y^n \frac{r_<^\ell}{r_>^{\ell+1}}\,dy, \tag{11.150}$$

in which $r_< := \min(y,r)$ and $r_> = \max(y,r)$, with r standing for r_{AB}.

Excluding the case $n=\ell$, which never occurs in practice, we have

$$\begin{aligned}
K(\ell,n) &= \frac{1}{r^{\ell+1}}\int_0^r y^{\ell+n}\,dy + r^\ell \int_r^{\mathcal{R}} y^{n-\ell-1}\,dy \\
&= \frac{r^n}{\ell+n+1} - \frac{r^n}{n-\ell}\Big[1-(r/\mathcal{R})^{\ell-n}\Big].
\end{aligned} \tag{11.151}$$

We discard the last term because it depends on the cutoff radius $\mathcal{R}$, and we conclude that

$$K(\ell,n) = \frac{2\ell+1}{(\ell-n)(\ell+n+1)}|\boldsymbol{r}_{AB}|^n \qquad (\ell\neq n). \tag{11.152}$$

In particular, $K(2,1) = \frac{5}{4}r_{AB}$, $K(0,1) = -\frac{1}{2}r_{AB}$, and $K(1,0) = \frac{3}{2}$.

Final answer

Substituting Eq. (11.152) into Eq. (11.149), we find that E^{jk}_{AB} becomes

$$E^{jk}_{AB} = \frac{1}{4} r_{AB} n^{\langle jk \rangle}_{AB} - \frac{1}{6} r_{AB} \delta^{jk} + \frac{1}{2} n^j_{AB} r^k_B. \tag{11.153}$$

This, together with Eq. (11.138) for E^{jk}_A, can now be inserted within Eq. (11.126). We arrive at

$$E^{jk} = -\sum_A \sum_{B \neq A} G^2 M_A M_B \left(\frac{1}{4} r_{AB} n^{\langle jk \rangle}_{AB} - \frac{1}{6} r_{AB} \delta^{jk} + \frac{1}{2} n^j_{AB} r^k_B \right),$$

and this can also be expressed as

$$E^{jk} = -\sum_A \sum_{B \neq A} G^2 M_A M_B \left(\frac{1}{4} r_{AB} n^{\langle jk \rangle}_{AB} - \frac{1}{6} r_{AB} \delta^{jk} - \frac{1}{2} n^j_{AB} r^k_A \right)$$

if we interchange the identities of bodies A and B and recall that $\boldsymbol{n}_{BA} = -\boldsymbol{n}_{AB}$. When we add these expressions and divide by 2, we obtain the symmetrized form

$$E^{jk} = -\sum_A \sum_{B \neq A} G^2 M_A M_B \left(\frac{1}{4} r_{AB} n^{\langle jk \rangle}_{AB} - \frac{1}{6} r_{AB} \delta^{jk} - \frac{1}{4} r_{AB} n^j_{AB} n^k_{AB} \right).$$

This becomes

$$E^{jk} = \frac{1}{4} \delta^{jk} \sum_A \sum_{B \neq A} G^2 M_A M_B |\boldsymbol{r}_A - \boldsymbol{r}_B| \tag{11.154}$$

after simplification, and this is our final answer.

Box 11.3 — Field integrals

Let us retrace the main steps that led us from the definition

$$E^{jk} = \frac{1}{4\pi} \int_{\mathscr{M}} U \partial^j U x^k \, d^3x,$$

to its evaluation

$$E^{ab} = \frac{1}{4} \delta^{jk} \sum_A \sum_{B \neq A} G^2 M_A M_B |\boldsymbol{r}_A - \boldsymbol{r}_B|.$$

These steps will allow us to evaluate many similar field integrals.

After inserting the Newtonian potential and its derivative within the integral, we change the variables of integration from $\boldsymbol{x}$ to $\boldsymbol{y} = \boldsymbol{x} - \boldsymbol{r}$, in which $\boldsymbol{r}$ stands for either $\boldsymbol{r}_A$ or $\boldsymbol{r}_B$, depending on the context. We also translate the domain of integration from $\mathscr{M}$ (defined by $|\boldsymbol{x}| < \mathcal{R}$) to $\mathscr{M}_y$ (defined by $|\boldsymbol{y}| < \mathcal{R}$), and we make use of the identity

$$\int_{\mathscr{M}} f(y) \, d^3y = \int_{\mathscr{M}_y} f(y) \, d^3y - \oint_{\partial \mathscr{M}_y} f(y) \boldsymbol{r} \cdot d\boldsymbol{S} + \cdots, \tag{1}$$

in which the surface integral is smaller than the volume integral by a factor of order $|\boldsymbol{r}|/\mathcal{R} \ll 1$ (and the dotted terms are smaller still).

Next we invoke the addition theorem for spherical harmonics,

$$\frac{1}{|\boldsymbol{y} - \boldsymbol{r}_{AB}|} = \sum_{\ell=0}^{\infty} \sum_{m=-\ell}^{\ell} \frac{4\pi}{2\ell+1} \frac{r_<^\ell}{r_>^{\ell+1}} Y_{\ell m}^*(\boldsymbol{n}_{AB}) Y^{\ell m}(\boldsymbol{N}), \tag{2}$$

in which $\boldsymbol{r}_{AB} = \boldsymbol{r}_A - \boldsymbol{r}_B, r_< := \min(y, r_{AB}), r_> = \max(y, r_{AB}), \boldsymbol{N} := \boldsymbol{y}/y$, and $\boldsymbol{n}_{AB} := \boldsymbol{r}_{AB}/r_{AB}$. After expressing all factors such as N^L in terms of STF tensors, the angular integrations are carried out with the help of the identity

$$\sum_{m=-\ell}^{\ell} Y_{\ell m}^*(\boldsymbol{n}_{AB}) \int Y_{\ell m}(\boldsymbol{N}) N^{\langle L' \rangle} \, d\Omega = \delta_{\ell \ell'} \, n_{AB}^{\langle L \rangle}. \tag{3}$$

We rely also on the following listing of angular averages:

$$\langle\langle N^j \rangle\rangle = 0, \tag{4a}$$

$$\langle\langle N^j N^k \rangle\rangle = \frac{1}{3}\delta^{jk}, \tag{4b}$$

$$\langle\langle N^j N^k N^p \rangle\rangle = 0, \tag{4c}$$

$$\langle\langle N^j N^k N^p N^q \rangle\rangle = \frac{1}{15}\left(\delta^{jk}\delta^{pq} + \delta^{jp}\delta^{kq} + \delta^{jq}\delta^{kp}\right), \tag{4d}$$

where $\langle\langle \cdots \rangle\rangle := (4\pi)^{-1} \int (\cdots) \, d\Omega$; these results were obtained back in Sec. 1.5.3.

This leaves us with a number of radial integrations to work out, and these are given by

$$K(\ell, n) := \int_0^{\mathcal{R}} y^n \frac{r_<^\ell}{r_>^{\ell+1}} \, dy = \frac{2\ell+1}{(\ell-n)(\ell+n+1)} |\boldsymbol{r}_{AB}|^n, \tag{5}$$

provided that $\ell \neq n$.

And at last, after simplification, we obtain our final expression for the field integral. All the while we are justified to throw away any term that contains an explicit dependence on the arbitrary cutoff radius $\mathcal{R}$.

11.3.3 Radiative quadrupole moment

We launch our calculation of the gravitational-wave field with a computation of Q^{jk}, the radiative quadrupole moment. According to Eq. (11.119), this is defined by

$$Q^{jk}(\tau) := \frac{1}{c^2} \int_{\mathcal{M}} \tau^{00}(\tau, \boldsymbol{x}) x^j x^k \, d^3x, \tag{11.155}$$

in which $\tau := t - R/c$ is retarded time, and where we suppress the primes on the integration variables to simplify the notation. (It should be kept in mind that R is the distance to the field point, which is distinct from the source point now identified by the vector $\boldsymbol{x}$.) We show below that the radiative quadrupole moment is given by

$$Q^{jk} \stackrel{\text{TT}}{=} \sum_A M_A \left(1 + \frac{1}{2}\frac{v_A^2}{c^2} - \frac{1}{2}\frac{\lfloor U \rfloor_A}{c^2}\right) r_A^j r_A^k + O(c^{-4}), \tag{11.156}$$

where

$$\lfloor U \rfloor_A := \sum_{B \neq A} \frac{GM_B}{r_{AB}} \tag{11.157}$$

is the *partie finie* of the Newtonian potential $U(\boldsymbol{x})$ evaluated at $\boldsymbol{x} = \boldsymbol{r}_A$. The expression of Eq. (11.156) leaves out terms proportional to δ^{jk} that would not survive the action of the transverse-tracefree projector $(\text{TT})^{jk}{}_{pq}$, as well as $\mathcal{R}$-dependent terms that can be freely discarded. It is understood that the position vectors $\boldsymbol{r}_A$, and the velocity vectors $\boldsymbol{v}_A$, are evaluated at the retarded time τ.

According to the discussion of Sec. 11.3.1, to calculate Q^{jk} to the required degree of accuracy, we need an expression for $c^{-2}\tau^{00}$ that includes terms of order c^0 (Newtonian, or 0PN) and terms of order c^{-2} (1PN). Such an expression was obtained back in Sec. 7.3.1 in the case of a matter distribution that consists of a perfect fluid. According to Eq. (7.54a), we have that

$$c^{-2}\tau^{00} = \rho^* \left[1 + \frac{1}{c^2}\left(\frac{1}{2}v^2 + 3U + \Pi\right)\right] - \frac{7}{8\pi Gc^2}\partial_p U \partial^p U + O(c^{-4}). \tag{11.158}$$

For a system of point particles we have that $\rho^* = \sum_A M_A \delta(\boldsymbol{x} - \boldsymbol{r}_A)$ and $\Pi = 0$, and the Newtonian potential reduces to $U = \sum_B GM_B |\boldsymbol{x} - \boldsymbol{r}_B|^{-1}$.

This expression for $c^{-2}\tau^{00}$ is ill-defined for point particles, because the term $B = A$ in U gives rise to a term $\sum_A GM_A^2 |\boldsymbol{x} - \boldsymbol{r}_A|^{-1} \delta(\boldsymbol{x} - \boldsymbol{r}_A)$ in $\rho^* U$. This is not defined as a distribution, and such a term gives rise to an ambiguity in the evaluation of the radiative quadrupole moment. We have, however, encountered a similar situation before, and learned how to deal with it. Indeed, suitable regularization methods were developed back in Sec. 9.6, where it was shown that ambiguous integrals can be made well-defined by adopting the regularization prescription

$$\frac{\delta(\boldsymbol{x} - \boldsymbol{r}_A)}{|\boldsymbol{x} - \boldsymbol{r}_A|} \equiv 0. \tag{11.159}$$

The rule removes the offending term in $\rho^* U$, and the piece of the Newtonian potential that survives multiplication by $\delta(\boldsymbol{x} - \boldsymbol{r}_A)$ is the *partie finie* displayed in Eq. (11.157). With this prescription, our expression for the effective mass density becomes

$$c^{-2}\tau^{00} = \sum_A M_A \left(1 + \frac{v_A^2}{2c^2} + \frac{3\lfloor U \rfloor_A}{c^2}\right) \delta(\boldsymbol{x} - \boldsymbol{r}_A) - \frac{14}{16\pi Gc^2}\partial_p U \partial^p U + O(c^{-4}). \tag{11.160}$$

The radiative quadrupole moment can be decomposed as

$$Q^{jk} = Q^{jk}[\text{M}] + Q^{jk}[\text{F}] + O(c^{-4}). \tag{11.161}$$

It contains a matter contribution that comes from the delta functions in τ^{00}, and a field contribution that comes from the term involving $\partial_p U \partial^p U$. The matter contribution can be calculated at once:

$$Q^{jk}[\mathrm{M}] = \sum_A M_A\left(1 + \frac{v_A^2}{2c^2} + \frac{3\lfloor U\rfloor_A}{c^2}\right) r_A^j r_A^k. \tag{11.162}$$

The field contribution is

$$Q^{jk}[\mathrm{F}] = -\frac{14}{16\pi G c^2}\int_{\mathscr{M}} \partial_p U \partial^p U x^j x^k \, d^3x, \tag{11.163}$$

and its computation requires a lot more work.

To evaluate the field integral of Eq. (11.163) we first express the integrand in the equivalent form

$$\begin{aligned}\partial_p U \partial^p U x^j x^k = {}& \partial_p\left(U\partial^p U x^j x^k\right) - \frac{1}{2}\partial^j\left(U^2 x^k\right) - \frac{1}{2}\partial^k\left(U^2 x^j\right) \\ & + U^2\delta^{jk} - U(\nabla^2 U) x^j x^k,\end{aligned} \tag{11.164}$$

which allows us to integrate by parts. We may discard the term $U^2\delta^{jk}$ on the grounds that it will not survive the TT projection introduced in Sec. 11.1.7. We may also replace $\nabla^2 U$ by $-4\pi G\sum_A M_A\delta(\boldsymbol{x} - \boldsymbol{r}_A)$, and write

$$\begin{aligned}\int_{\mathscr{M}} \partial_p U \partial^p U x^j x^k \, d^3x \overset{\text{TT}}{=} {}& \oint_{\partial\mathscr{M}} U\partial^p U x^j x^k \, dS_p - \oint_{\partial\mathscr{M}} U^2 x^{(j} dS^{k)} \\ & + 4\pi G\sum_A M_A\lfloor U\rfloor_A r_A^j r_A^k,\end{aligned} \tag{11.165}$$

where the notation $\overset{\text{TT}}{=}$ was introduced near the end of Sec. 11.3.1, and where $dS^j = \mathcal{R}^2 N^j \, d\Omega$ is the surface element on $\partial\mathscr{M}$. Note that we have once more made use of the regularization prescription of Eq. (11.159). Making the substitution, we obtain

$$\begin{aligned}Q^{jk}[\mathrm{F}] \overset{\text{TT}}{=} {}& -\frac{7}{2Gc^2}\left(\mathcal{R}^4\langle\!\langle U\partial_p U N^j N^k N^p\rangle\!\rangle - \mathcal{R}^3\langle\!\langle U^2 N^j N^k\rangle\!\rangle\right) \\ & - \frac{7}{2c^2}\sum_A M_A\lfloor U\rfloor_A r_A^j r_A^k,\end{aligned} \tag{11.166}$$

in which the angular brackets denote an average over the unit two-sphere.

We must now evaluate the surface integrals, on which $\boldsymbol{x}$ is set equal to $\mathcal{R}\boldsymbol{N}$. Recalling that $\mathcal{R}$ is large compared with $\boldsymbol{r}_A$ (refer to Sec. 11.3.2), it is appropriate to expand U in inverse powers of $r := |\boldsymbol{x}|$ before we insert it within the integrals. We have

$$U = \frac{Gm}{r} + \frac{1}{2}GI^{jk}\partial_{jk}r^{-1} + O(r^{-3}), \tag{11.167}$$

where $m := \sum_A M_A$ is the total mass, and $I^{jk} := \sum_A M_A r_A^j r_A^k$ is the Newtonian quadrupole moment of the mass distribution. It is important to note that the Newtonian dipole moment, $I^j := \sum_A M_A r_A^j$, has been set equal to zero. This is allowed, because $\boldsymbol{I} = m\boldsymbol{R} + O(c^{-2})$, where $\boldsymbol{R}$ is the post-Newtonian barycenter (refer to Sec. 9.3.6), and we work in a coordinate system for which $\boldsymbol{R} = \boldsymbol{0}$. From the expansion of the Newtonian potential we also get

$$\partial_j U = Gm\partial_j r^{-1} + \frac{1}{2}GI^{kp}\partial_{jkp}r^{-1} + O(r^{-4}). \tag{11.168}$$

These results indicate that on $\partial\mathcal{M}$, the potential and its gradient are given schematically by $U = \mathcal{R}^{-1} + \mathcal{R}^{-3} + \cdots$ and $\partial_j U = \mathcal{R}^{-2} + \mathcal{R}^{-4} + \cdots$. This implies, for example, that $\mathcal{R}^4 U \partial_p U = \mathcal{R} + \mathcal{R}^{-1} + \cdots$ and $\mathcal{R}^3 U^2 = \mathcal{R} + \mathcal{R}^{-1} + \cdots$. This reveals, finally, that the surface integrals produce no $\mathcal{R}$-independent contributions to $Q^{jk}[\mathrm{F}]$.

We have obtained

$$Q^{jk}[\mathrm{F}] \stackrel{\mathrm{TT}}{=} -\frac{7}{2c^2} \sum_A M_A \lfloor U \rfloor_A r_A^j r_A^k, \tag{11.169}$$

and combining this with Eq. (11.162), we conclude that the radiative quadrupole moment of Eq. (11.161) is indeed given by Eq. (11.156).

11.3.4 Radiative octupole moment

We turn next to computation of Q^{jka}, the radiative octupole moment. According to Eq. (11.119), this is defined by

$$Q^{jka} := A^{jka} + A^{kja} - A^{ajk}, \tag{11.170}$$

where

$$A^{jka}(\tau) := \frac{1}{c^2} \int_{\mathcal{M}} \tau^{0j}(\tau, \boldsymbol{x}) x^k x^a \, d^3x. \tag{11.171}$$

We show below that this is given by

$$\begin{aligned}
A^{jka} \stackrel{\mathrm{TT}}{=} & \frac{1}{c} \sum_A M_A \left(1 + \frac{v_A^2}{2c^2}\right) v_A^j r_A^k r_A^a \\
& - \frac{1}{2c^3} \sum_A \sum_{B \neq A} \frac{G M_A M_B}{r_{AB}} \Big[(\boldsymbol{n}_{AB} \cdot \boldsymbol{v}_A) n_{AB}^j r_A^k r_A^a + v_A^j r_A^k r_A^a \Big] \\
& + \frac{1}{2c^3} \sum_A \sum_{B \neq A} G M_A M_B \Big[(\boldsymbol{n}_{AB} \cdot \boldsymbol{v}_A) n_{AB}^j n_{AB}^{(k} r_A^{a)} - 7 n_{AB}^j v_A^{(k} r_A^{a)} \\
& \qquad + 7 v_A^j n_{AB}^{(k} r_A^{a)} \Big] \\
& - \frac{1}{6c^3} \sum_A \sum_{B \neq A} G M_A M_B r_{AB} \Big[(\boldsymbol{n}_{AB} \cdot \boldsymbol{v}_A) n_{AB}^j n_{AB}^k n_{AB}^a - 11 n_{AB}^j n_{AB}^{(k} v_A^{a)} \\
& \qquad + 11 v_A^j n_{AB}^k n_{AB}^a \Big] + O(c^{-5}).
\end{aligned} \tag{11.172}$$

This expression leaves out terms that would not survive a transverse-tracefree projection, as well as $\mathcal{R}$-dependent terms that can be freely discarded. It is understood that the position vectors $\boldsymbol{r}_A$ and the velocity vectors $\boldsymbol{v}_A$ are evaluated at retarded time τ.

Matter and field contributions

According to the discussion of Sec. 11.3.1, to calculate Q^{jka} to the required degree of accuracy we need an expression for $c^{-2}\tau^{0j}$ that includes terms of order c^{-1} (0.5PN) and terms of order c^{-3} (1.5PN). Such an expression was worked out in Sec. 7.3.1 in the case of a matter distribution that consists of a perfect fluid. According to Eq. (7.54b), we have that

$$c^{-2}\tau^{0j} = \frac{1}{c}\rho^* v^j \left[1 + \frac{1}{c^2}\left(\frac{1}{2}v^2 + 3U + \Pi + p/\rho^* \right) \right] + \frac{1}{4\pi G c^3}\left[3\partial_t U \partial^j U + 4\left(\partial^j U^k - \partial^k U^j\right)\partial_k U \right] + O(c^{-5}). \tag{11.173}$$

For a system of point particles, $U^j = \sum_B GM_B v_B^j/|\boldsymbol{x} - \boldsymbol{r}_B|$. Our expression for $c^{-2}\tau^{0j}$ must be regularized with the help of Eq. (11.159), and the end result is

$$c^{-2}\tau^{0j} = \frac{1}{c}\sum_A M_A v_A^j \left(1 + \frac{v_A^2}{2c^2} + \frac{3\lfloor U \rfloor_A}{c^2} \right) \delta(\boldsymbol{x} - \boldsymbol{r}_A) + \frac{1}{16\pi G c^3}\left[12\partial_t U \partial^j U + 16\left(\partial^j U^k - \partial^k U^j\right)\partial_k U \right] + O(c^{-5}), \tag{11.174}$$

where $\lfloor U \rfloor_A$ is the *partie finie* of the Newtonian potential evaluated at $\boldsymbol{x} = \boldsymbol{r}_A$, as given by Eq. (11.157).

The octupole moment contains a contribution $Q^{jka}[\mathrm{M}]$ that comes directly from the matter distribution, and another contribution $Q^{jka}[\mathrm{F}]$ that comes from the gravitational field. They are obtained from $A^{jka} = A^{jka}[\mathrm{M}] + A^{jka}[\mathrm{F}] + O(c^{-5})$, which is then substituted into Eq. (11.170). We have introduced

$$A^{jka}[\mathrm{M}] := \frac{1}{c}\sum_A M_A v_A^j \left(1 + \frac{v_A^2}{2c^2} + \frac{3\lfloor U \rfloor_A}{c^2} \right) r_A^k r_A^a \tag{11.175}$$

and

$$A^{jka}[\mathrm{F}] := \frac{1}{4\pi G c^3}\int_{\mathscr{M}} \left[3\partial_t U \partial^j U + 4\left(\partial^j U^p - \partial^p U^j\right)\partial_p U \right] x^k x^a \, d^3x, \tag{11.176}$$

and the remainder of this subsection is devoted to computation of $A^{jka}[\mathrm{F}]$.

Computation of the field integral: Organization

To simplify our computations we invoke the identity of Eq. (7.40),

$$\partial_t U + \partial_j U^j = 0. \tag{11.177}$$

We recall that this is a direct consequence of the statement of mass conservation, $\partial_t \rho^* + \partial_j(\rho^* v^j) = 0$. We use the identity to eliminate $\partial_t U$ from Eq. (11.176), which becomes

$$A^{jka}[\mathrm{F}] = \frac{1}{Gc^3}\left(-3B_1^{jka} + 4B_2^{jka} - 4B_3^{jka} \right), \tag{11.178}$$

where

$$B_1^{jka} := \frac{1}{4\pi} \int_{\mathscr{M}} \partial^j U \partial_p U^p x^k x^a \, d^3x, \tag{11.179a}$$

$$B_2^{jka} := \frac{1}{4\pi} \int_{\mathscr{M}} \partial_p U \partial^j U^p x^k x^a \, d^3x, \tag{11.179b}$$

$$B_3^{jka} := \frac{1}{4\pi} \int_{\mathscr{M}} \partial_p U \partial^p U^j x^k x^a \, d^3x. \tag{11.179c}$$

After integration by parts, which is designed to leave one factor of U undifferentiated, we find that each field integral B^{jka} breaks up into a volume integral $B^{jka}[\mathscr{M}]$ and a surface integral $B^{jka}[\partial\mathscr{M}]$. A number of terms are found to be proportional to δ^{jk}, or δ^{ja}, or δ^{ka}. All such terms will not survive a transverse-tracefree projection, and according to our discussion near the end of Sec. 11.3.1, they can all be discarded. If, for example, B^{jka} contains a term $\delta^{jk} B^a$, then its contribution to Q^{jka} will be of the form $2\delta^{jk} B^a - \delta^{ja} B^k$. The first term is a pure trace, and the second term is longitudinal, because it becomes proportional to N^j after Q^{jka} is multiplied by N_a; in each case the contribution does not survive the TT projection.

After eliminating all such terms, we find that

$$B_1^{jka} \overset{\text{TT}}{=} B_1^{jka}[\mathscr{M}] + B_1^{jka}[\partial\mathscr{M}], \tag{11.180a}$$

$$B_1^{jka}[\mathscr{M}] := -\frac{1}{4\pi} \int_{\mathscr{M}} U \partial^j{}_p U^p \, x^k x^a \, d^3x \,, \tag{11.180b}$$

$$B_1^{jka}[\partial\mathscr{M}] := \frac{1}{4\pi} \oint_{\partial\mathscr{M}} U \partial_p U^p \, x^k x^a \, dS^j \,, \tag{11.180c}$$

that

$$B_2^{jka} \overset{\text{TT}}{=} B_2^{jka}[\mathscr{M}] + B_2^{jka}[\partial\mathscr{M}], \tag{11.181a}$$

$$B_2^{jka}[\mathscr{M}] := -\frac{1}{4\pi} \int_{\mathscr{M}} U \big(\partial^j{}_p U^p \, x^k x^a + \partial^j U^k \, x^a + \partial^j U^a \, x^k\big) \, d^3x \,, \tag{11.181b}$$

$$B_2^{jka}[\partial\mathscr{M}] := \frac{1}{4\pi} \oint_{\partial\mathscr{M}} U \partial^j U^p \, x^k x^a \, dS_p \,, \tag{11.181c}$$

and that

$$B_3^{jka} \overset{\text{TT}}{=} B_3^{jka}[\mathscr{M}] + B_3^{jka}[\partial\mathscr{M}], \tag{11.182a}$$

$$B_3^{jka}[\mathscr{M}] := -\frac{1}{4\pi} \int_{\mathscr{M}} U \big(\nabla^2 U^j \, x^k x^a + \partial^k U^j \, x^a + \partial^a U^j \, x^k\big) \, d^3x \,, \tag{11.182b}$$

$$B_3^{jka}[\partial\mathscr{M}] := \frac{1}{4\pi} \oint_{\partial\mathscr{M}} U \partial^p U^j \, x^k x^a \, dS_p \,. \tag{11.182c}$$

There are many volume integrals to evaluate, but they are all of the form

$$C^{mnpab} := -\frac{1}{4\pi} \int_{\mathscr{M}} U \partial^{mn} U^p \, x^a x^b \, d^3x \tag{11.183}$$

and

$$D^{mna} := -\frac{1}{4\pi}\int_{\mathscr{M}} U\partial^m U^n\, x^a\, d^3x\,. \tag{11.184}$$

Specifically,

$$B_1^{jka}[\mathscr{M}] = C^{j\ \ pka}_{\ p}\,, \tag{11.185a}$$

$$B_2^{jka}[\mathscr{M}] = C^{j\ \ pka}_{\ p} + D^{jka} + D^{jak}\,, \tag{11.185b}$$

$$B_3^{jka}[\mathscr{M}] = C^{p\ \ jka}_{\ p} + D^{kja} + D^{ajk}\,. \tag{11.185c}$$

Similarly, the surface integrals are of the form

$$E^{mnabp} := \frac{1}{4\pi}\oint_{\partial\mathscr{M}} U\partial^m U^n\, x^a x^b\, dS^p\,, \tag{11.186}$$

with

$$B_1^{jka}[\partial\mathscr{M}] = E_p^{\ pkaj}\,, \quad B_2^{jka}[\partial\mathscr{M}] = E^{jpka}_{\quad\ \ p}\,, \quad B_3^{jka}[\partial\mathscr{M}] = E^{pjka}_{\quad\ \ p}\,. \tag{11.187}$$

The key is therefore the evaluation of the generic volume integrals of Eqs. (11.183) and (11.184), as well as the surface integral of Eq. (11.186). Once these are in hand, the computation of B_1^{jka}, B_2^{jka}, and B_3^{jka} is soon completed, and Eq. (11.178) gives us $A^{jka}[\mathrm{F}]$. Adding the $A^{jka}[\mathrm{M}]$ of Eq. (11.175) produces A^{jka}, and from Eq. (11.170) we get our final answer for Q^{jka}.

Computation of C^{mnpab}

We follow the general methods described in Sec. 11.3.2. We begin with the differentiation of the vector potential $U^p = \sum_B GM_B v_B^p/|\boldsymbol{x} - \boldsymbol{r}_B|$, which returns

$$\partial^{mn} U^p = -\sum_B GM_B v_B^p\biggl[-3\frac{(x-r_B)^m(x-r_B)^n}{|\boldsymbol{x}-\boldsymbol{r}_B|^5} + \frac{\delta^{mn}}{|\boldsymbol{x}-\boldsymbol{r}_B|^3} + \frac{4\pi}{3}\delta^{mn}\delta(\boldsymbol{x}-\boldsymbol{r}_B)\biggr]. \tag{11.188}$$

The last term, involving the delta function, does not appear in a straightforward computation of $\partial^{mn}U^p$, in which one implicitly assumes that $\boldsymbol{x} \neq \boldsymbol{r}_B$. Without it, however, our expression would be wrong, because it would give rise to $\nabla^2 U^p = 0$ instead of the correct $\nabla^2 U^p = -4\pi G\sum_B M_B v_B^p \delta(\boldsymbol{x} - \boldsymbol{r}_B)$. The distributional term is therefore inserted to produce the correct answer when $\boldsymbol{x} = \boldsymbol{r}_B$, and to ensure that U^p satisfies the appropriate Poisson equation. After insertion of $U = \sum_A GM_A/|\boldsymbol{x} - \boldsymbol{r}_A|$ and some algebra, Eq. (11.183) becomes

$$C^{mnpab} = \sum_A G^2 M_A^2 v_A^p\Bigl(\delta^{mn}F_A^{ab} - 3F_A^{mnab}\Bigr) + \sum_A\sum_{B\neq A} G^2 M_A M_B v_B^p\biggl(\delta^{mn}F_{AB}^{ab} - 3F_{AB}^{mnab} + \frac{1}{3}\delta^{mn}\frac{r_B^a r_B^b}{r_{AB}}\biggr), \tag{11.189}$$

where

$$F_A^{mnab} := \frac{1}{4\pi}\int_{\mathcal{M}} \frac{(x-r_A)^m(x-r_A)^n}{|\boldsymbol{x}-\boldsymbol{r}_A|^6} x^a x^b\, d^3x\,, \tag{11.190a}$$

$$F_{AB}^{mnab} := \frac{1}{4\pi}\int_{\mathcal{M}} \frac{1}{|\boldsymbol{x}-\boldsymbol{r}_A|}\frac{(x-r_B)^m(x-r_B)^n}{|\boldsymbol{x}-\boldsymbol{r}_B|^5} x^a x^b\, d^3x\,, \tag{11.190b}$$

and where

$$F_A^{ab} := \delta_{mn} F_A^{mnab}\,, \qquad F_{AB}^{ab} := \delta_{mn} F_{AB}^{mnab}\,. \tag{11.191}$$

The term involving $r_B^a r_B^b / r_{AB}$ in Eq. (11.189) originates from the distributional term in $\partial^{mn} U^p$; a similar term involving $r_A^a r_A^b / r_{AA}$ was set equal to zero by invoking the regularization prescription of Eq. (11.159).

We first set to work on F_A^{mnab}. Following the general strategy summarized in Box 11.3, we substitute $\boldsymbol{x} = \boldsymbol{y} + \boldsymbol{r}_A$ inside the integral, and get

$$\begin{aligned} F_A^{mnbc} &= \frac{1}{4\pi}\int_{\mathcal{M}} \frac{y^m y^n y^a y^b}{y^6}\, d^3y + \frac{r_B^a}{4\pi}\int_{\mathcal{M}} \frac{y^m y^n y^b}{y^6}\, d^3y \\ &\quad + \frac{r_B^b}{4\pi}\int_{\mathcal{M}} \frac{y^m y^n y^a}{y^6}\, d^3y + \frac{r_B^a r_B^b}{4\pi}\int_{\mathcal{M}} \frac{y^m y^n}{y^6}\, d^3y. \end{aligned} \tag{11.192}$$

According to Eq. (1) of Box 11.3, each integral over $\mathcal{M}$ can be expressed as a volume integral over the simpler domain $\mathcal{M}_y$ defined by $y := |\boldsymbol{y}| < \mathcal{R}$, plus a correction of fractional order $|\boldsymbol{r}_B|/\mathcal{R}$ given by a surface integral over $\partial\mathcal{M}_y$.

The first integral produces

$$\begin{aligned} \frac{1}{4\pi}\int_{\mathcal{M}_y} \frac{y^m y^n y^a y^b}{y^6}\, d^3y &= \langle\!\langle N^m N^n N^a N^b \rangle\!\rangle \int_0^{\mathcal{R}} dy \\ &= \frac{1}{15}\mathcal{R}\big(\delta^{mn}\delta^{ab} + \delta^{ma}\delta^{nb} + \delta^{mb}\delta^{na}\big), \end{aligned} \tag{11.193}$$

where we involve Eq. (4d) of Box 11.3. Because it is proportional to $\mathcal{R}$, this contribution to F_A^{mnab} can be discarded. The surface integral that corrects this will potentially give rise to an $\mathcal{R}$-independent contribution, and it should be evaluated carefully. It turns out, however, that it is proportional to $r_B^p \langle\!\langle N^m N^n N^a N^b N_p \rangle\!\rangle$, and it vanishes because the angular average of a product of an odd number of vectors $\boldsymbol{N}$ is necessarily zero. The neglected terms in Eq. (1) are of order $\mathcal{R}^{-1}$ and higher, and we conclude that the first integral in F_A^{mnab} makes no contribution to C^{mnpab}.

The second and third integrals produce terms such as

$$\frac{1}{4\pi}\int_{\mathcal{M}_y} \frac{y^m y^n y^b}{y^6}\, d^3y = \langle\!\langle N^m N^n N^b \rangle\!\rangle \int_0^{\mathcal{R}} \frac{dy}{y}, \tag{11.194}$$

and this vanishes by virtue of Eq. (4c) of Box 11.3; the logarithmic divergence of the radial integration requires no explicit regularization. The surface integral that corrects this is easily shown to be of order $\mathcal{R}^{-1}$, and we conclude that the second and third integrals do not contribute to C^{mnpab}.

The fourth integral produces

$$\frac{1}{4\pi}\int_{\mathcal{M}_y} \frac{y^m y^n}{y^6}\, d^3y = \langle\langle N^m N^n \rangle\rangle \int_0^{\mathcal{R}} \frac{dy}{y^2} = \frac{1}{3}\delta^{mn} \int_0^{\mathcal{R}} \frac{dy}{y^2}, \tag{11.195}$$

and this involves a radial integration that is formally divergent. Once more the surface integral does not contribute, and we have obtained

$$F_A^{mnab} = \frac{1}{3}\delta^{mn} r_A^a r_A^b \int_0^{\mathcal{R}} \frac{dy}{y^2} \tag{11.196}$$

for the field integral of Eq. (11.190), modulo $\mathcal{R}$-dependent terms that can be freely discarded. It is disturbing to see that F_A^{mnab} is proportional to a diverging integral, but it is a fortunate outcome that the combination $\delta^{mn} F_A^{ab} - 3F_A^{mnab}$ that appears in C^{mnpab} happens to vanish identically. The divergence does not require explicit regularization, and all in all we find that F_A^{mnab} makes no contribution to C^{mnpab}.

We next set to work on F_{AB}^{mnab}. Once more we follow the general strategy summarized in Box 11.3, and we substitute $\boldsymbol{x} = \boldsymbol{y} + \boldsymbol{r}_B$ inside the integral. We get

$$\begin{aligned} F_{AB}^{mnbc} &= \frac{1}{4\pi}\int_{\mathcal{M}} \frac{1}{|\boldsymbol{y} - \boldsymbol{r}_{AB}|} \frac{y^m y^n y^a y^b}{y^5}\, d^3y + \frac{r_B^a}{4\pi}\int_{\mathcal{M}} \frac{1}{|\boldsymbol{y} - \boldsymbol{r}_{AB}|} \frac{y^m y^n y^b}{y^5}\, d^3y \\ &\quad + \frac{r_B^b}{4\pi}\int_{\mathcal{M}} \frac{1}{|\boldsymbol{y} - \boldsymbol{r}_{AB}|} \frac{y^m y^n y^a}{y^5}\, d^3y + \frac{r_B^a r_B^b}{4\pi}\int_{\mathcal{M}} \frac{1}{|\boldsymbol{y} - \boldsymbol{r}_{AB}|} \frac{y^m y^n}{y^5}\, d^3y. \end{aligned} \tag{11.197}$$

We begin with the first integral, which produces

$$\frac{1}{4\pi}\int_{\mathcal{M}_y} \frac{1}{|\boldsymbol{y} - \boldsymbol{r}_{AB}|} \frac{y^m y^n y^a y^b}{y^5}\, d^3y.$$

To evaluate this we involve Eq. (2) of Box 11.3, and we express $N^m N^n N^a N^b$ as

$$\begin{aligned} N^m N^n N^a N^b &= N^{\langle mnab\rangle} + \frac{1}{7}\Big(\delta^{mn} N^{\langle ab\rangle} + \delta^{ma} N^{\langle nb\rangle} + \delta^{mb} N^{\langle na\rangle} + \delta^{na} N^{\langle mb\rangle} \\ &\quad + \delta^{nb} N^{\langle ma\rangle} + \delta^{ab} N^{\langle mn\rangle}\Big) + \frac{1}{15}\Big(\delta^{mn}\delta^{ab} + \delta^{ma}\delta^{nb} + \delta^{mb}\delta^{na}\Big), \end{aligned} \tag{11.198}$$

in terms of the angular STF tensors $N^{\langle mnab\rangle}$ and $N^{\langle mn\rangle}$. We perform the angular integrations with the help of Eq. (3) of Box 11.3, and the remaining radial integrals are in the form of Eq. (5). After some algebra, we obtain the expression

$$\begin{aligned} &\frac{1}{9}K(4,1) n_{AB}^{\langle mnab\rangle} + \frac{1}{35}K(2,1)\Big(\delta^{mn} n_{AB}^{\langle ab\rangle} + \text{permutations}\Big) \\ &\quad + \frac{1}{15}K(0,1)\Big(\delta^{mn}\delta^{ab} + \delta^{ma}\delta^{nb} + \delta^{mb}\delta^{na}\Big) \end{aligned}$$

for the volume integral. The corresponding surface integral is easily seen to be of order $\mathcal{R}^{-1}$, and we arrive at

$$\frac{1}{4\pi}\int_{\mathcal{M}} \frac{1}{|\boldsymbol{y}-\boldsymbol{r}_{AB}|}\frac{y^m y^n y^a y^b}{y^5}\,d^3y = \frac{1}{18} r_{AB} n_{AB}^{\langle mnab\rangle} + \frac{1}{28} r_{AB}\Big(\delta^{mn} n_{AB}^{\langle ab\rangle} + \delta^{ma} n_{AB}^{\langle nb\rangle} + \delta^{mb} n_{AB}^{\langle na\rangle} + \delta^{na} n_{AB}^{\langle mb\rangle} + \delta^{nb} n_{AB}^{\langle ma\rangle} + \delta^{ab} n_{AB}^{\langle mn\rangle}\Big) - \frac{1}{30} r_{AB}\Big(\delta^{mn}\delta^{ab} + \delta^{ma}\delta^{nb} + \delta^{mb}\delta^{na}\Big) \tag{11.199}$$

after using Eq. (5) of Box 11.3 to evaluate the radial integrals.

We next turn to the second and third integrals, which are both approximated by

$$\frac{1}{4\pi}\int_{\mathcal{M}_y} \frac{1}{|\boldsymbol{y}-\boldsymbol{r}_{AB}|}\frac{y^m y^n y^a}{y^5}\,d^3y.$$

To evaluate this we involve Eq. (2), and we express $N^m N^n N^a$ as

$$N^m N^n N^a = N^{\langle mna\rangle} + \frac{1}{5}\Big(\delta^{mn} N^a + \delta^{ma} N^n + \delta^{na} N^m\Big), \tag{11.200}$$

in terms of the angular STF tensor $N^{\langle mna\rangle}$. We carry out the angular integrations with the help of Eq. (3), and the remaining radial integrals are in the form of Eq. (5). After some algebra, we obtain the expression

$$\frac{1}{7} K(3,0) n_{AB}^{\langle mna\rangle} + \frac{1}{15} K(1,0)\Big(\delta^{mn} n_{AB}^a + \delta^{ma} n_{AB}^n + \delta^{na} n_{AB}^m\Big)$$

for the volume integral. The corresponding surface integral is once more of order $\mathcal{R}^{-1}$, and we arrive at

$$\frac{1}{4\pi}\int_{\mathcal{M}} \frac{1}{|\boldsymbol{y}-\boldsymbol{r}_{AB}|}\frac{y^m y^n y^a}{y^5}\,d^3y = \frac{1}{12} n_{AB}^{\langle mna\rangle} + \frac{1}{10}\Big(\delta^{mn} n_{AB}^a + \delta^{ma} n_{AB}^n + \delta^{na} n_{AB}^m\Big) \tag{11.201}$$

after using Eq. (5) to evaluate the radial integrals.

The final step in the computation of F_{AB}^{mnab} is the evaluation of the fourth integral, which is approximated by

$$\frac{1}{4\pi}\int_{\mathcal{M}_y} \frac{1}{|\boldsymbol{y}-\boldsymbol{r}_{AB}|}\frac{y^m y^n}{y^5}\,d^3y.$$

After following the same familiar steps, this becomes $\frac{1}{5}K(2,-1) n_{AB}^{\langle mn\rangle} + \frac{1}{3}K(0,-1)\delta^{mn}$, and the corresponding surface integral is of order $\mathcal{R}^{-2}$. We arrive at

$$\frac{1}{4\pi}\int_{\mathcal{M}} \frac{1}{|\boldsymbol{y}-\boldsymbol{r}_{AB}|}\frac{y^m y^n}{y^5}\,d^3y = \frac{1}{6 r_{AB}} n_{AB}^{\langle mn\rangle} + \frac{1}{3}K(0,-1)\delta^{mn}, \tag{11.202}$$

and we note that $K(0,-1)$ is formally a divergent integral of the sort encountered in Eq. (11.196). We shall see that this divergence requires no explicit regularization, because (as happened before) it eventually drops out of the calculation.

Collecting results, we have obtained

$$\begin{aligned} F_{AB}^{mnab} &= \frac{1}{18} r_{AB} n_{AB}^{\langle mnab \rangle} \\ &+ \frac{1}{28} r_{AB} \Big(\delta^{mn} n_{AB}^{\langle ab \rangle} + \delta^{ma} n_{AB}^{\langle nb \rangle} + \delta^{mb} n_{AB}^{\langle na \rangle} + \delta^{na} n_{AB}^{\langle mb \rangle} + \delta^{nb} n_{AB}^{\langle ma \rangle} + \delta^{ab} n_{AB}^{\langle mn \rangle} \Big) \\ &- \frac{1}{30} r_{AB} \Big(\delta^{mn} \delta^{ab} + \delta^{ma} \delta^{nb} + \delta^{mb} \delta^{na} \Big) \\ &+ \frac{1}{12} n_{AB}^{\langle mna \rangle} r_B^b + \frac{1}{10} \Big(\delta^{mn} n_{AB}^a + \delta^{ma} n_{AB}^n + \delta^{na} n_{AB}^m \Big) r_B^b \\ &+ \frac{1}{12} n_{AB}^{\langle mnb \rangle} r_B^a + \frac{1}{10} \Big(\delta^{mn} n_{AB}^b + \delta^{mb} n_{AB}^n + \delta^{nb} n_{AB}^m \Big) r_B^a \\ &+ \frac{1}{6 r_{AB}} r_{AB}^{\langle mn \rangle} r_B^a r_B^b + \frac{1}{3} K(0,-1) \delta^{mn} r_B^a r_B^b \end{aligned} \tag{11.203}$$

for the field integral of Eq. (11.190), modulo $\mathcal{R}$-dependent terms that can be freely discarded. The trace of this is

$$F_{AB}^{ab} = \frac{1}{4} r_{AB} n_{AB}^{\langle ab \rangle} - \frac{1}{6} r_{AB} \delta^{ab} + \frac{1}{2} n_{AB}^a r_B^b + \frac{1}{2} n_{AB}^b r_B^a + K(0,-1) r_B^a r_B^b, \tag{11.204}$$

and we see that, as claimed, the terms involving $K(0,-1)$ cancel out in the combination $\delta^{mn} F_{AB}^{ab} - 3 F_{AB}^{mnab}$ that appears in Eq. (11.189); these terms make no contribution to C^{mnpab}.

We may now substitute Eqs. (11.196), (11.203), and (11.204) into Eq. (11.189). After simplification, our final result is

$$\begin{aligned} C^{mnpab} &= \sum_A \sum_{B \neq A} G^2 M_A M_B v_B^p \bigg[-\frac{1}{6} r_{AB} n_{AB}^{\langle mnab \rangle} \\ &- \frac{3}{28} r_{AB} \Big(\delta^{ma} n_{AB}^{\langle nb \rangle} + \delta^{mb} n_{AB}^{\langle na \rangle} + \delta^{na} n_{AB}^{\langle mb \rangle} + \delta^{nb} n_{AB}^{\langle ma \rangle} + \delta^{ab} n_{AB}^{\langle mn \rangle} \Big) \\ &+ r_{AB} \delta^{mn} \Big(\frac{1}{7} n_{AB}^{\langle ab \rangle} - \frac{1}{15} \delta^{ab} \Big) + \frac{1}{10} r_{AB} \Big(\delta^{ma} \delta^{nb} + \delta^{mb} \delta^{na} \Big) \\ &- \frac{1}{4} n_{AB}^{\langle mna \rangle} r_B^b - \frac{1}{4} n_{AB}^{\langle mnb \rangle} r_B^a - \frac{3}{10} \Big(\delta^{ma} n_{AB}^n + \delta^{na} n_{AB}^m \Big) r_B^b \\ &- \frac{3}{10} \Big(\delta^{mb} n_{AB}^n + \delta^{nb} n_{AB}^m \Big) r_B^a + \frac{1}{5} \delta^{mn} \Big(n_{AB}^a r_B^b + n_{AB}^b r_B^a \Big) \\ &+ \frac{1}{r_{AB}} \Big(-\frac{1}{2} n_{AB}^{\langle mn \rangle} + \frac{1}{3} \delta^{mn} \Big) r_B^a r_B^b \bigg]. \end{aligned} \tag{11.205}$$

Computation of D^{mna}

After inserting the expressions for U and U^n within Eq. (11.184), we obtain

$$D^{mna} = \sum_A G^2 M_A^2 v_A^n E_A^{ma} + \sum_A \sum_{B \neq A} G^2 M_A M_B v_B^n E_{AB}^{ma}, \tag{11.206}$$

where

$$E_A^{ma} := \frac{1}{4\pi}\int_{\mathcal{M}} \frac{(x-r_A)^m x^a}{|\boldsymbol{x}-\boldsymbol{r}_A|^4}\, d^3x, \tag{11.207a}$$

$$E_{AB}^{ma} := \frac{1}{4\pi}\int_{\mathcal{M}} \frac{(x-r_B)^m x^a}{|\boldsymbol{x}-\boldsymbol{r}_A||\boldsymbol{x}-\boldsymbol{r}_B|^3}\, d^3x, \tag{11.207b}$$

were introduced back in Eqs. (11.127). These integrals were evaluated in Sec. 11.3.2, and we obtained

$$E_A^{ma} = 0, \tag{11.208a}$$

$$E_{AB}^{ma} = \frac{1}{4}r_{AB}n_{AB}^{\langle ma\rangle} - \frac{1}{6}r_{AB}\delta^{ma} + \frac{1}{2}n_{AB}^m r_B^a; \tag{11.208b}$$

these are Eqs. (11.138) and (11.153), respectively. Making the substitutions, we arrive at

$$D^{mna} = \sum_A\sum_{B\neq A} G^2 M_A M_B v_B^n\biggl(\frac{1}{4}r_{AB}n_{AB}^{\langle ma\rangle} - \frac{1}{6}r_{AB}\delta^{ma} + \frac{1}{2}n_{AB}^m r_B^a\biggr). \tag{11.209}$$

Computation of E^{mnabp}

The surface integrals

$$E^{mnabp} = \frac{1}{4\pi}\oint_{\partial\mathcal{M}} U\partial^m U^n\, x^a x^b\, dS^p \tag{11.210}$$

are evaluated at $|\boldsymbol{x}| = \mathcal{R}$. On $\partial\mathcal{M}$ the Newtonian potential has the schematic form $U = \mathcal{R}^{-1} + \mathcal{R}^{-3} + \cdots$, and the vector potential can be similarly expressed as $U^n = \mathcal{R}^{-2} + \mathcal{R}^{-3} + \cdots$. This implies that $\partial^m U^n = \mathcal{R}^{-3} + \mathcal{R}^{-4} + \cdots$. We recall that U does not include an $\mathcal{R}^{-2}$ term because the Newtonian dipole moment $\boldsymbol{I} := \sum_A M_A \boldsymbol{r}_A$ can be set equal to zero, and similarly, U^n does not contain an $\mathcal{R}^{-1}$ term because $\dot{I}^j = \sum_A M_A v_A^j = 0$. With $x^j = \mathcal{R}N^j$ and $dS^j = \mathcal{R}^2 N^j\, d\Omega$, we find that the leading term in the surface integral is of order $\mathcal{R}^0$, and that it must be evaluated carefully. Further investigation reveals that at this order, $\partial^m U^n$ involves an even number of angular vectors $\boldsymbol{N}$, which implies that the surface integral involves an odd number of such vectors. This guarantees that

$$E^{mnabp} = 0, \tag{11.211}$$

modulo $\mathcal{R}$-dependent terms that can be freely discarded.

Computation of $A^{jka}[\mathrm{F}]$

It is now a straightforward task to substitute Eq. (11.205) for C^{mnpab}, Eq. (11.209) for D^{mna}, and Eq. (11.211) for E^{mnabp} into Eqs. (11.185) and (11.187). These results, in turn, can be inserted within Eq. (11.180) for B_1^{abc}, Eq. (11.181) for B_2^{abc}, and Eq. (11.182) for B_3^{abc}. The final step is to substitute these expressions into the right-hand side of Eq. (11.178). The end result, after much simplification, and after discarding terms that will not survive the TT projection, is

$$A^{jka}[\mathrm{F}] \overset{\mathrm{TT}}{=} \frac{1}{c^3} \sum_A \sum_{B \neq A} G M_A M_B \Biggl\{ r_{AB} \biggl[-\frac{1}{6} (\boldsymbol{n}_{AB} \cdot \boldsymbol{v}_B) n_{AB}^j n_{AB}^k n_{AB}^a + \frac{11}{12} n_{AB}^j \bigl(n_{AB}^k v_B^a + v_B^k n_{AB}^a \bigr) - \frac{11}{6} v_B^j n_{AB}^k n_{AB}^a \biggr] - \frac{1}{4} (\boldsymbol{n}_{AB} \cdot \boldsymbol{v}_B) n_{AB}^j \bigl(n_{AB}^k r_B^a + r_B^k n_{AB}^a \bigr) + \frac{7}{4} n_{AB}^j \bigl(v_B^k r_B^a + r_B^k v_B^a \bigr) - \frac{7}{4} v_B^j \bigl(n_{AB}^k r_B^a + r_B^k n_{AB}^a \bigr) - \frac{1}{r_{AB}} \biggl[\frac{1}{2} (\boldsymbol{n}_{AB} \cdot \boldsymbol{v}_B) n_{AB}^j r_B^k r_B^a + \frac{7}{2} v_B^j r_B^k r_B^a \biggr] \Biggr\}. \tag{11.212}$$

Final answer

Equation (11.212) for $A^{jka}[\mathrm{F}]$ and Eq. (11.175) for $A^{jka}[\mathrm{M}]$ can finally be combined to form A^{jka}, as defined by Eq. (11.171). After inserting $\sum_{B \neq A} GM_B/r_{AB}$ for $\lfloor U \rfloor_A$ and additional simplification, we obtain Eq. (11.172). To arrive at this result we rearrange some of the sums in Eq. (11.212) and switch the identities of bodies A and B; this permutation affects the signs of some terms, because $\boldsymbol{n}_{BA} = -\boldsymbol{n}_{AB}$.

11.3.5 Radiative 4-pole and 5-pole moments

Our next step is computation of Q^{jkab}, the radiative 4-pole moment, and Q^{jkabc}, the radiative 5-pole moment. These are defined by Eq. (11.119),

$$Q^{jkab}(\tau) := \frac{1}{c^2} \int_{\mathscr{M}} \tau^{jk}(\tau, \boldsymbol{x}) x^a x^b \, d^3x \tag{11.213}$$

and

$$Q^{jkabc}(\tau) := \frac{1}{c^3} \frac{\partial}{\partial \tau} \int_{\mathscr{M}} \tau^{jk}(\tau, \boldsymbol{x}) x^a x^b x^c \, d^3x. \tag{11.214}$$

We show below that these are given by

$$Q^{jkab} \overset{\mathrm{TT}}{=} \frac{1}{c^2} \sum_A M_A v_A^j v_A^k r_A^a r_A^b - \frac{1}{2c^2} \sum_A \sum_{B \neq A} \frac{G M_A M_B}{r_{AB}} n_{AB}^j n_{AB}^k r_A^a r_A^b + \frac{1}{12c^2} \sum_A \sum_{B \neq A} G M_A M_B r_{AB} n_{AB}^j n_{AB}^k \bigl(n_{AB}^a n_{AB}^b - \delta^{ab} \bigr) + O(c^{-4}) \tag{11.215}$$

and

$$Q^{jkabc} \overset{\text{TT}}{=} \frac{1}{c^3}\frac{\partial}{\partial\tau}\biggl[\sum_A M_A v_A^j v_A^k r_A^a r_A^b r_A^c - \frac{1}{2}\sum_A\sum_{B\neq A}\frac{GM_AM_B}{r_{AB}} n_{AB}^j n_{AB}^k r_A^a r_A^b r_A^c$$
$$+ \frac{1}{4}\sum_A\sum_{B\neq A} GM_AM_B r_{AB} n_{AB}^j n_{AB}^k r_A^{(a}\Bigl(n_{AB}^b n_{AB}^{c)} - \delta^{bc)}\Bigr)\biggr]$$
$$+ O(c^{-5}). \tag{11.216}$$

The index symmetrization in the last sum of Eq. (11.216) is over the trio of indices abc. We leave the differentiation with respect to τ unevaluated for the time being; it is advantageous to take care of this at a later stage.

According to the discussion of Sec. 11.3.1, to calculate Q^{jkab} to the required degree of accuracy we need an expression for τ^{jk} that includes terms of order c^0 only. According to Eq. (7.54c), we have that

$$\tau^{jk} = \rho^* v^j v^k + p\,\delta^{jk} + \frac{1}{4\pi G}\Bigl(\partial^j U\partial^k U - \frac{1}{2}\delta^{jk}\partial_p U\partial^p U\Bigr) + O(c^{-2}), \tag{11.217}$$

and this becomes

$$\tau^{jk} = \sum_A M_A v_A^j v_A^k \delta(\boldsymbol{x} - \boldsymbol{r}_A) + \frac{1}{4\pi G}\Bigl(\partial^j U\partial^k U - \frac{1}{2}\delta^{jk}\partial_p U\partial^p U\Bigr) + O(c^{-2}) \tag{11.218}$$

for a system of point particles. The matter contribution can be calculated at once:

$$Q^{jkab}[\mathrm{M}] = \frac{1}{c^2}\sum_A M_A v_A^j v_A^k r_A^a r_A^b. \tag{11.219}$$

The field contribution is

$$Q^{jkab}[\mathrm{F}] = \frac{1}{4\pi Gc^2}\int_{\mathscr{M}} \partial^j U\partial^k U x^a x^b\, d^3x - \frac{1}{8\pi Gc^2}\delta^{jk}\int_{\mathscr{M}} \partial_p U\partial^p U x^a x^b\, d^3x, \tag{11.220}$$

and the second term, because it comes with a factor δ^{jk} in front of the integral, will not survive a TT projection. The complete 4-pole moment is $Q^{jkab} = Q^{jkab}[\mathrm{M}] + Q^{jkab}[\mathrm{F}] + O(c^{-4})$.

To evaluate the first integral we employ our usual strategy of integrating by parts so as to leave one factor of U undifferentiated. We find that the integral splits into a volume integral over the domain $\mathscr{M}$ and a surface integral over $\partial\mathscr{M}$, and that Eq. (11.220) becomes

$$Q^{jkab}[\mathrm{F}] \overset{\text{TT}}{=} Q^{jkab}[\mathrm{F}, \mathscr{M}] + Q^{jkab}[\mathrm{F}, \partial\mathscr{M}], \tag{11.221}$$

where

$$Q^{jkab}[\mathrm{F}, \mathscr{M}] := -\frac{1}{4\pi Gc^2}\int_{\mathscr{M}} U\partial^{jk}U\, x^a x^b\, d^3x, \tag{11.222a}$$

$$Q^{jkab}[\mathrm{F}, \partial\mathscr{M}] := \frac{1}{4\pi Gc^2}\oint_{\partial\mathscr{M}} U\partial^k U\, x^a x^b\, dS^j. \tag{11.222b}$$

To arrive at Eq. (11.221) we have discarded additional terms that will not survive a TT projection. For example, a contribution to Q^{jkab} of the form $\delta^{ja}A^{kb}$ would become $N^j A^{kb} N_b$

after contraction with $N_a N_b$, and this would make an irrelevant, longitudinal contribution to h^{jk}.

To evaluate the volume integral in Eqs. (11.222) we insert the familiar expression for U, as well as

$$\partial^{jk} U = -\sum_A GM_A \left[-3\frac{(x-r_A)^j(x-r_A)^k}{|\boldsymbol{x}-\boldsymbol{r}_A|^5} + \frac{\delta^{jk}}{|\boldsymbol{x}-\boldsymbol{r}_A|^3} + \frac{4\pi}{3}\delta^{jk}\delta(\boldsymbol{x}-\boldsymbol{r}_A) \right]. \tag{11.223}$$

Once more we ignore the terms in δ^{jk}, and find that

$$Q^{jkcd}[\mathrm{F}, \mathscr{M}] \stackrel{\text{TT}}{=} -\frac{3}{c^2}\sum_A GM_A^2 F_A^{jkab} - \frac{3}{c^2}\sum_A\sum_{B\neq A} GM_A M_B F_{BA}^{jkab}, \tag{11.224}$$

where the field integrals F_A^{jkab} and F_{BA}^{jkab} were introduced in Sec. 11.3.4; they are defined by Eqs. (11.190), and evaluated in Eqs. (11.196) and (11.203). From these results we learn that F_A^{jkab} is proportional to δ^{jk} and will not survive a TT projection, and that F_{BA}^{jkab} can be expressed as

$$F_{BA}^{jkab} \stackrel{\text{TT}}{=} \frac{1}{36} r_{AB} n_{AB}^j n_{AB}^k \left(2n_{AB}^a n_{AB}^b + \delta^{ab}\right) - \frac{1}{6} n_{AB}^j n_{AB}^k n_{AB}^{(a} r_A^{b)} + \frac{1}{6r_{AB}} n_{AB}^j n_{AB}^k r_A^a r_A^b, \tag{11.225}$$

after discarding terms that will be projected out and further simplification.

Inserting these expressions within $Q^{jkab}[\mathrm{F}, \mathscr{M}]$, we arrive at

$$\begin{aligned} Q^{jkab}[\mathrm{F}, \mathscr{M}] \stackrel{\text{TT}}{=} &-\frac{1}{12c^2}\sum_A\sum_{B\neq A} GM_A M_B r_{AB} n_{AB}^j n_{AB}^k \left(2n_{AB}^a n_{AB}^b + \delta^{ab}\right) \\ &+\frac{1}{2c^2}\sum_A\sum_{B\neq A} GM_A M_B n_{AB}^j n_{AB}^k n_{AB}^{(a} r_A^{b)} \\ &-\frac{1}{2c^2}\sum_A\sum_{B\neq A} \frac{GM_A M_B}{r_{AB}} n_{AB}^j n_{AB}^k r_A^a r_A^b. \end{aligned} \tag{11.226}$$

This expression can be simplified. We examine the second line, which we write as

$$\frac{1}{4c^2}\sum_A\sum_{B\neq A} GM_A M_B n_{AB}^j n_{AB}^k n_{AB}^a r_A^b + (a \leftrightarrow b).$$

By rearranging the sums, we see that this is also

$$\frac{1}{4c^2}\sum_A\sum_{B>A} GM_A M_B n_{AB}^j n_{AB}^k \left(n_{AB}^a r_A^b + n_{BA}^a r_B^b\right) + (a \leftrightarrow b),$$

or

$$\frac{1}{4c^2}\sum_A\sum_{B>A} GM_A M_B n_{AB}^j n_{AB}^k n_{AB}^a \left(r_A^b - r_B^b\right) + (a \leftrightarrow b).$$

The term within brackets is $r_{AB}n^b_{AB}$, and we see that the second line in $Q^{jkab}[\mathrm{F}, \mathscr{M}]$ can be joined with the first. Our final expression is

$$Q^{jkab}[\mathrm{F}, \mathscr{M}] \overset{\text{TT}}{=} \frac{1}{12c^2}\sum_A\sum_{B\neq A} GM_AM_Br_{AB}n^j_{AB}n^k_{AB}\big(n^a_{AB}n^b_{AB} - \delta^{ab}\big) - \frac{1}{2c^2}\sum_A\sum_{B\neq A}\frac{GM_AM_B}{r_{AB}}n^j_{AB}n^k_{AB}r^a_Ar^b_A + O(c^{-4}). \tag{11.227}$$

Moving on to the surface integral of Eq. (11.222), we recall our previous work in Sec. 11.3.3, in which U had the schematic structure $U = \mathcal{R}^{-1} + \mathcal{R}^{-3} + \cdots$ when evaluated on $\partial\mathscr{M}$, while its gradient is given by $\partial^k U = \mathcal{R}^{-2} + \mathcal{R}^{-4} + \cdots$ With $x^a = \mathcal{R}N^a$ and $dS^j = \mathcal{R}^2N^j\,d\Omega$, these statements imply that $Q^{jkab}[\mathrm{F}, \partial\mathscr{M}]$ contains terms at orders $\mathcal{R}$, $\mathcal{R}^{-1}$, and so on, but that there is no $\mathcal{R}$-independent contribution. For this reason, we may set

$$Q^{jkab}[\mathrm{F}, \partial\mathscr{M}] = 0, \tag{11.228}$$

modulo $\mathcal{R}$-dependent terms that can be freely discarded.

Collecting results, we find that the radiative 4-pole moment is given by the expression displayed back in Eq. (11.215). The computation of the radiative 5-pole moment is accomplished by following the same familiar steps. We shall not labor through the details here, but simply state that the final answer is the expression displayed back in Eq. (11.216).

11.3.6 Surface integrals

At this stage we have computed all the radiative multipole moments that contribute to $h^{jk}_{\mathcal{N}}$ through 1.5PN order. The multipole expansion of Eq. (11.118), however, also involves a pair of surface integrals, P^{jk} and P^{jka}, which are defined by Eqs. (11.120). Our task in this subsection is to evaluate them. We shall find that they make no contributions to $h^{jk}_{\mathcal{N}}$.

We begin with

$$P^{jk} := \oint_{\partial\mathscr{M}}\big(\tau^{jp}x^k + \tau^{kp}x^j - \partial_q\tau^{pq}x^jx^k\big)\,dS_p, \tag{11.229}$$

in which τ^{jk} is expressed as a function of τ and $\boldsymbol{x}$, and where we suppress the primes on the integration variables to simplify the notation. The effective stress tensor τ^{jk} is given to leading order by Eq. (11.218), and this reduces to

$$\tau^{jk} \overset{\text{TT}}{=} \frac{1}{4\pi G}\partial^jU\partial^kU + O(c^{-2}) \tag{11.230}$$

when it is evaluated on $\partial\mathscr{M}$, where the matter contribution vanishes. This expression, however, is not sufficient to achieve the required degree of accuracy for the surface integrals (as specified back in Sec. 11.3.1); for this we must also incorporate terms of order c^{-2}. An improved expression can be obtained from Eqs. (7.49) and (7.52). In this we substitute the near-zone gravitational potentials of Box 7.5, and we obtain

$$\tau^{jk} \stackrel{\mathrm{TT}}{=} \frac{1}{4\pi G}\partial^j U \partial^k U + \frac{1}{4\pi G c^2}\Big[2\partial^{(j} U \partial^{k)}\psi + \partial^{(j} U \partial^{k)}\partial_{tt} X + 8\partial^{(j} U \partial_t U^{k)} - 4\big(\partial^j U_p - \partial_p U^j\big)\big(\partial^k U^p - \partial^p U^k\big)\Big] + O(c^{-4}), \tag{11.231}$$

after discarding all terms proportional to δ^{jk}, for the usual reason that they will not survive a TT projection. This expression involves our old friends the Newtonian potential U and the vector potential U^j, but it involves also the post-Newtonian potentials ψ and X that were first introduced in Sec. 7.3. These were evaluated for a system of point masses in Sec. 9.6; refer to Eqs. (9.268), (9.270), and (9.272).

To calculate P^{jk} we also need $\partial_q \tau^{pq}$, which we express as $-c^{-1}\partial_t \tau^{0p}$ by involving the conservation identities $\partial_\beta \tau^{\alpha\beta} = 0$. With Eq. (11.174), this is

$$\partial_q \tau^{pq} = \frac{1}{4\pi G c^2}\frac{\partial}{\partial \tau}\Big[3\partial_q U^q \partial^p U - 4\big(\partial^p U^q - \partial^q U^p\big)\partial_q U\Big], \tag{11.232}$$

in which we have inserted the identity $\partial_t U + \partial_q U^q = 0$. The derivative operator can be taken outside of the surface integral.

From the explicit expressions obtained in Sec. 9.6 for U, U^j, ψ, and $\partial_{tt} X$, we observe that each one of these quantities has the schematic form $|\boldsymbol{x}|^{-1} + |\boldsymbol{x}|^{-2} + \cdots$ when expanded in inverse powers of $|\boldsymbol{x}|$. It follows that when $\partial^j U$, $\partial^j \psi$, $\partial^j_{tt} X$, $\partial^k U^j$, and $\partial_t U^j$ are evaluated on $\partial\mathcal{M}$, they each have the schematic form $\mathcal{R}^{-2} + \mathcal{R}^{-3} + \cdots$ This means that $\tau^{jk} = O(\mathcal{R}^{-4})$, and it follows that a quantity such as $\tau^{jp} x^k dS_p$ must scale as $\mathcal{R}^{-1}$; this does not give rise to an $\mathcal{R}$-independent contribution to the surface integral. A similar argument reveals that $\partial_q \tau^{pq} = O(\mathcal{R}^{-5})$, so that $\partial_q \tau^{pq} x^j x^k dS_p$ scales as $\mathcal{R}^{-1}$; this also makes no contribution. We conclude that

$$P^{jk} = 0, \tag{11.233}$$

modulo $\mathcal{R}$-dependent terms that can be freely discarded.

We next evaluate

$$P^{jka} := \frac{1}{c}\frac{\partial}{\partial \tau}\oint_{\partial\mathcal{M}} \big(\tau^{jp} x^k x^a + \tau^{kp} x^j x^a - \tau^{ap} x^j x^k\big)\, dS_p, \tag{11.234}$$

using the effective stress tensor displayed in Eq. (11.231). Relative to P^{jk}, this surface integral involves an additional power of $\boldsymbol{x}$, and therefore an additional power of $\mathcal{R}$; because P^{jk} was seen to be of order $\mathcal{R}^{-1}$, there is a chance that the surface integral might contain an $\mathcal{R}$-independent contribution. As we shall see presently, however, this does not happen, and as a matter of fact,

$$P^{jka} = 0, \tag{11.235}$$

modulo $\mathcal{R}$-dependent terms that can be freely discarded. This conclusion emerges as a result of a closer examination of the terms that make up τ^{jk}. It was stated previously that at leading order, $\partial^j U$, $\partial^j \psi$, $\partial^j_{tt} X$, $\partial^k U^j$, and $\partial_t U^j$ all scale as $\mathcal{R}^{-2}$ when they are evaluated on

$\partial\mathscr{M}$, so that $\tau^{jk} = O(\mathcal{R}^{-4})$. With the four powers of $\mathcal{R}$ that are contained in the position vectors and the surface element, we find that the integral does indeed scale as $\mathcal{R}^0$. It can be verified, however, that $\partial^j U$, $\partial^j \psi$, $\partial^j_{tt} X$, $\partial^k U^j$, and $\partial_t U^j$ are all proportional to a product of an *odd number* of angular vectors $\boldsymbol{N}$. This implies that τ^{jk} involves an *even number* of such vectors, and this, in turn, implies that the integrand in Eq. (11.234) contains an *odd number* of angular vectors. Integration gives zero, and we have established the statement of Eq. (11.235).

11.3.7 Tails: Wave-zone contribution to the gravitational waves

We departed on our long journey to calculate the gravitational-wave field back in Sec. 11.3.1, and all the while we have focused our attention on the near-zone piece $h^{jk}_{\mathcal{N}}$. We have ignored the wave-zone piece $h^{jk}_{\mathcal{W}}$, except to announce that it makes a relevant contribution at 1.5PN order. In this subsection we make amends and calculate this final contribution to the gravitational-wave field, which is generated entirely by field energy situated in the wave zone. We shall show that it is given by the so-called *tail integral*

$$h^{jk}_{\mathcal{W}} \overset{\text{TT}}{=} \frac{4G}{c^4 R}\frac{GM}{c^3}\int_0^\infty \overset{(4)}{I}{}^{\langle jk\rangle}(\tau - \zeta)\left(\ln\frac{\zeta}{\zeta + 2R/c} + \frac{11}{12}\right)d\zeta, \tag{11.236}$$

which involves the entire past history of the system, from the infinite past at $\zeta = \infty$ to the current (retarded) time at $\zeta = 0$. The wave-zone contribution depends on M, the total gravitational mass of the system, as well as I^{jk}, the Newtonian quadrupole moment of the matter distribution; in Eq. (11.236) the quadrupole moment is made tracefree and differentiated four times with respect to its argument. Recalling our discussion near the end of Sec. 11.3.1, we see that $h^{jk}_{\mathcal{W}}$ is a correction of order c^{-3} relative to the leading, quadrupole term in h^{jk}; the wave-zone contribution to the gravitational-wave field is therefore a term of 1.5PN order. In the course of our calculations we shall discover that $h^{jk}_{\mathcal{W}}$ comes about because the gravitational waves propagate not in the fictitious flat spacetime of post-Minkowskian theory, but in a physical spacetime that is curved by the presence of a mass M.

Wave-zone integrals

We begin our derivation of Eq. (11.236) by recalling that back in Sec. 6.3.5, we devised a method to calculate $h^{jk}_{\mathcal{W}}$ when τ^{jk} can be expressed as a sum of terms of the form

$$\tau^{jk}[\ell, n] = \frac{1}{4\pi}\frac{f(\tau)}{R^n}N^{\langle L\rangle}, \tag{11.237}$$

in which f is an arbitrary function of τ, n is an arbitrary integer, and $N^{\langle L\rangle}$ is an angular STF tensor of degree ℓ of the sort introduced back in Sec. 1.5.3. According to Eq. (6.105), $h^{jk}_{\mathcal{W}}$ is then a sum of terms of the form

$$h^{jk}_{\mathcal{W}}[\ell, n] = \frac{4G}{c^4 R}N^{\langle L\rangle}\left\{\int_0^{\mathcal{R}} ds f(\tau - 2s/c)A(s, R) + \int_{\mathcal{R}}^\infty ds f(\tau - 2s/c)B(s, R)\right\}, \tag{11.238}$$

where

$$A(s, R) = \int_{\mathcal{R}}^{R+s} \frac{P_\ell(\xi)}{p^{n-1}}\, dp, \qquad B(s, R) = \int_{s}^{R+s} \frac{P_\ell(\xi)}{p^{n-1}}\, dp, \tag{11.239}$$

in which P_ℓ is a Legendre polynomial of argument $\xi = (R+2s)/R - 2s(R+s)/(Rp)$. We shall rely on these results in the remainder of this subsection.

Construction of the source term

The wave-zone contribution to h^{jk} is obtained by evaluating the integrals displayed in Eq. (11.238), and this relies on a decomposition of τ^{jk} into irreducible pieces of the form of Eq. (11.237). Our first order of business, therefore, is to obtain an appropriate expression for the effective stress tensor; this expression must be valid everywhere in the wave zone.

The source term is constructed from the gravitational potentials, and wave-zone expressions for these were obtained in Sec. 7.4. According to the summary presented in Box 7.7, we have

$$h^{00} = \frac{4G}{c^2}\left[\frac{M}{R} + \frac{1}{2}\partial_{jk}\left(\frac{\mathcal{I}^{jk}}{R}\right) + \cdots\right], \tag{11.240a}$$

$$h^{0j} = \frac{4G}{c^2}\left[-\frac{1}{2c}\mathcal{J}^{jk}\frac{N_k}{R^2} - \frac{1}{2c}\partial_k\left(\frac{\dot{\mathcal{I}}^{jk}}{R}\right) + \cdots\right], \tag{11.240b}$$

$$h^{jk} = \frac{4G}{c^2}\left[\frac{1}{2c^2}\frac{\ddot{\mathcal{I}}^{jk}}{R} + \cdots\right]. \tag{11.240c}$$

The potentials are expressed in terms of $R := |\boldsymbol{x}|$, $\boldsymbol{N} := \boldsymbol{x}/R$, and the multipole moments that were introduced back in Sec. 7.1.2. For a system of N bodies, and to lowest PN order, we have the total gravitational mass $M = \sum_A M_A + O(c^{-2})$, the angular-momentum tensor $\mathcal{J}^{jk} = \sum_A M_A(v_A^j r_A^k - r_A^j v_A^k) + O(c^{-2})$, and the quadrupole moment

$$\mathcal{I}^{jk}(\tau) = \sum_A M_A r_A^j r_A^k + O(c^{-2}). \tag{11.241}$$

These expressions are obtained from the equations listed in Box 7.7 by specializing them to a system of point masses. The mass and angular momentum are conserved quantities, while $\mathcal{I}^{jk}$ depends on retarded time $\tau := t - R/c$. For the rest of this discussion we replace the formal post-Minkowskian moment $\mathcal{I}^{jk}$ with its Newtonian expression $I^{jk} = \sum_A M_A r_A^j r_A^k$.

The post-Newtonian order of each term in Eqs. (11.240) was identified in Box 7.7: relative to $GM/(c^2 R)$, each term involving I^{jk} is of 1PN order, and the term involving the angular-momentum tensor is also of 1PN order; the expressions are therefore truncated at 1PN order, and the neglected terms are of 1.5PN order. The rules to count post-Newtonian orders in wave-zone potentials were derived back in Sec. 7.2.3. It is useful to recall that in the wave zone, R is larger than λ_c, the characteristic wavelength of the gravitational radiation; it follows that if r_c is a characteristic length scale of the source, then $r_c/R \sim r_c/\lambda_c \sim v_c/c$, where v_c is the source's characteristic velocity.

In the wave zone, away from the matter distribution, the effective stress tensor τ^{jk} is made up of the Landau–Lifshitz pseudotensor $(-g)t_{\rm LL}^{jk}$ and the harmonic-gauge contribution

$(-g)t_{\rm H}^{jk}$. Sufficiently accurate expressions for these quantities were obtained in Sec. 7.3.1. The leading term comes from the Landau–Lifshitz pseudotensor of Eq. (7.48); this is

$$\frac{c^4}{64\pi G}\partial^j h^{00}\partial^k h^{00},$$

where we ignore the term proportional to δ^{jk} because, as we observed many times before, it will not survive a TT projection. Using Eq. (11.240), we find that this is equal to

$$\frac{G}{4\pi}\left[\frac{M^2}{R^4}N^j N^k - \frac{M}{R^2}N^{(j}\partial^{k)}_{\ \ pq}\left(\frac{I^{pq}}{R}\right) + \cdots\right].$$

It is easy to show that relative to GM^2/R^4, the second term is of order $(v_c/c)^2$, and the neglected terms are smaller by an additional power of v_c/c. Take, for example, the term that arises when R^{-1} is differentiated three times. This is of the schematic form $(M/R^2)(I/R^4)$, and relative to M^2/R^4 this is of order $(r_c/R)^2 \sim (r_c/\lambda_c)^2 \sim (v_c/c)^2$. As another example, take the term that arises when I^{pq} is differentiated three times. This is of the schematic form $(M/R^2)(\dddot{I}/c^3R)$, and relative to M^2/R^4 this is of order $r_c^2 R/(ct_c)^3 \sim (r_c/\lambda_c)^2 \sim (v_c/c)^2$.

We wish our expression for τ^{jk} to be as accurate as what was displayed previously. In particular, we want to be sure that our expression contains all occurrences of terms involving a product of M with I^{jk} or its derivatives; all such terms contribute at order $(v_c/c)^2$ relative to GM^2/R^4, and they must all be included. A careful examination of Eq. (7.49) reveals that the relevant terms are contained in

$$\begin{aligned}(-g)t_{\rm LL}^{jk} = \frac{c^4}{16\pi G}\bigg[&\frac{1}{4}\partial^j h^{00}\partial^k h^{00} + \partial^j h^{00}\partial_0 h^{0k} + \partial^k h^{00}\partial_0 h^{0j} \\ &+ \frac{1}{4}\partial^j h^{00}\partial^k h^p_{\ p} + \frac{1}{4}\partial^k h^{00}\partial^j h^p_{\ p} + \cdots\bigg],\end{aligned} \tag{11.242}$$

and that the additional terms are smaller by additional powers of v_c/c.

A careful examination of Eq. (7.53) reveals that

$$(-g)t_{\rm H}^{jk} = \frac{c^4}{16\pi G}\left[-h^{00}\partial_{00}h^{jk} + \cdots\right] \tag{11.243}$$

is also a relevant term. It is easy to see why: after writing $\partial_{00} = c^{-2}\partial_{\tau\tau}$, we find that this contribution to τ^{jk} is schematically

$$\frac{c^2}{G}h^{00}\partial_{\tau\tau}h^{jk} \sim \frac{G}{c^4}\frac{M\overset{(4)}{I}{}^{jk}}{R^2}, \tag{11.244}$$

in which the label (4) indicates that the quadrupole moment tensor is differentiated four times with respect to proper time τ. We have that $d^4I^{jk}/d\tau^4 \sim Mr_c^2/t_c^4$, $R > \lambda_c = ct_c$, and all this implies that this term is of order $(v_c/c)^2$ relative to GM^2/R^4.

This is the first time that $(-g)t_{\rm H}^{\alpha\beta}$ explicitly enters a computation. As we saw back in Sec. 6.2.1, this contribution to $\tau^{\alpha\beta}$ comes from the difference between $\partial_{\mu\nu}H^{\alpha\mu\beta\nu}$ and $-\Box h^{\alpha\beta}$ on the left-hand side of the Landau–Lifshitz formulation of the Einstein field equations. It is this term that informs us that the gravitational waves are propagating not in flat spacetime, but in a curved spacetime whose metric $g_{\alpha\beta}$ must be obtained self-consistently from the gravitational potentials (refer to Box 6.3). It is this contribution to $\tau^{\alpha\beta}$, therefore, that reveals the differences between the light cones of the mathematical flat spacetime and those of the physical curved spacetime. And as we shall see, it is this term that generates the tail integral of Eq. (11.236).

Collecting results, we find that the appropriate starting expression for the source term is

$$\tau^{jk} = \frac{c^4}{16\pi G}\biggl[\frac{1}{4}\partial^j h^{00}\partial^k h^{00} + \frac{1}{c}\partial^j h^{00}\partial_\tau h^{0k} + \frac{1}{c}\partial^k h^{00}\partial_\tau h^{0j} + \frac{1}{4}\partial^j h^{00}\partial^k h^p{}_p + \frac{1}{4}\partial^k h^{00}\partial^j h^p{}_p - \frac{1}{c^2}h^{00}\partial_{\tau\tau}h^{jk} + \cdots\biggr]. \tag{11.245}$$

We must now turn this into something more concrete, a set of expressions that are ready for insertion within Eq. (11.238).

Evaluation of the source term

The first step is to insert Eqs. (11.240) within Eq. (11.245). We need

$$\partial^j h^{00} = \frac{4G}{c^2}\biggl[-\frac{M}{R^2}N^j + \frac{1}{2}\partial^j{}_{pq}\biggl(\frac{I^{pq}}{R}\biggr) + \cdots\biggr], \tag{11.246a}$$

$$\partial_\tau h^{0j} = \frac{4G}{c^2}\biggl[-\frac{1}{2c}\partial_p\biggl(\frac{\ddot{I}^{jp}}{R}\biggr) + \cdots\biggr], \tag{11.246b}$$

$$\partial^j h^p{}_p = \frac{4G}{c^2}\biggl[-\frac{1}{2c^2}\frac{\ddot{I}}{R^2}N^j + \cdots\biggr], \tag{11.246c}$$

$$\partial_{\tau\tau}h^{jk} = \frac{4G}{c^2}\biggl[\frac{1}{2c^2 R}\overset{(4)}{I}{}^{jk} + \cdots\biggr], \tag{11.246d}$$

in which $\ddot{I} := \ddot{I}^{pp}$. After some algebra, we obtain

$$\tau^{jk} = \frac{GM}{4\pi R^2}\biggl[\frac{M}{R^2}N^j N^k - N^{(j}\partial^{k)}{}_{pq}\biggl(\frac{I^{pq}}{R}\biggr) + \frac{4}{c^2}N^{(j}\partial_p\biggl(\frac{\ddot{I}^{k)p}}{R}\biggr) + \frac{1}{c^2}\biggl(\frac{\ddot{I}}{R^2} + \frac{1}{c}\frac{\dddot{I}}{R}\biggr)N^j N^k - \frac{2}{c^4}\overset{(4)}{I}{}^{jk} + \cdots\biggr]. \tag{11.247}$$

The next step is to evaluate the derivatives. We recall that $\partial_j R = N_j$ and $\partial_j N_k = R^{-1}(\delta_{jk} - N_j N_k)$. We recall also that I^{jk} depends on the spatial coordinates through $\tau = t - R/c$, so that $\partial_p I^{jk} = -c^{-1}\dot{I}^{jk}N_p$. Using these rules, we find that

$$\partial_p\biggl(\frac{\ddot{I}^{jk}}{R}\biggr) = -\biggl(\frac{\ddot{I}^{jk}}{R^2} + \frac{1}{c}\frac{\dddot{I}^{jk}}{R}\biggr)N_p \tag{11.248}$$

and

$$\partial^j{}_{pq}\biggl(\frac{I^{pq}}{R}\biggr) = -\biggl(15\frac{I^{pq}}{R^4} + \frac{15}{c}\frac{\dot{I}^{pq}}{R^3} + \frac{6}{c^2}\frac{\ddot{I}^{pq}}{R^2} + \frac{1}{c^3}\frac{\dddot{I}^{pq}}{R}\biggr)N^j N_p N_q + \biggl(3\frac{I^{pq}}{R^4} + \frac{3}{c}\frac{\dot{I}^{pq}}{R^3} + \frac{1}{c^2}\frac{\ddot{I}^{pq}}{R^2}\biggr)(N^j\delta_{pq} + \delta^j{}_p N_q + \delta^j{}_q N_p). \tag{11.249}$$

With these results, Eq. (11.247) becomes

$$\begin{aligned}\tau^{jk} = \frac{GM^2}{4\pi R^4} N^j N^k + \frac{GM}{4\pi R^2} \biggl[& \biggl(15 \frac{I^{pq}}{R^4} + \frac{15}{c} \frac{\dot{I}^{pq}}{R^3} + \frac{6}{c^2} \frac{\ddot{I}^{pq}}{R^2} + \frac{1}{c^3} \frac{\dddot{I}^{pq}}{R} \biggr) N^j N^k N_p N_q \\ & - \biggl(3 \frac{I}{R^4} + \frac{3}{c} \frac{\dot{I}}{R^3} - \frac{1}{c^3} \frac{\dddot{I}}{R} \biggr) N^j N^k \\ & - \biggl(3 \frac{I^{jp}}{R^4} + \frac{3}{c} \frac{\dot{I}^{jp}}{R^3} + \frac{3}{c^2} \frac{\ddot{I}^{jp}}{R^2} + \frac{2}{c^3} \frac{\dddot{I}^{jp}}{r} \biggr) N^k N_p \\ & - \biggl(3 \frac{I^{kp}}{R^4} + \frac{3}{c} \frac{\dot{I}^{kp}}{R^3} + \frac{3}{c^2} \frac{\ddot{I}^{kp}}{R^2} + \frac{2}{c^3} \frac{\dddot{I}^{kp}}{R} \biggr) N^j N_p - \frac{2}{c^4} \overset{(4)}{I}{}^{jk} + \cdots \biggr]. \end{aligned} \tag{11.250}$$

The final step is to express the angular dependence of τ^{jk} in terms of STF tensors $N^{\langle L \rangle}$. We involve the definition of Eq. (1.155), and write $N^j N^k N^p N^q$ in terms of $N^{\langle jkpq \rangle}$, $N^j N^k N^p$ in terms of $N^{\langle jkp \rangle}$, and $N^j N^k$ in terms of $N^{\langle jk \rangle}$. After discarding all terms proportional to δ^{jk}, our final expression for the effective stress tensor is

$$\begin{aligned}\tau^{jk} = \frac{GM^2}{4\pi R^4} N^{\langle jk \rangle} + \frac{GM}{4\pi R^2} \biggl[& \biggl(15 \frac{I_{pq}}{R^4} + \frac{15}{c} \frac{\dot{I}_{pq}}{R^3} + \frac{6}{c^2} \frac{\ddot{I}_{pq}}{R^2} + \frac{1}{c^3} \frac{\dddot{I}_{pq}}{R} \biggr) N^{\langle jkpq \rangle} \\ & + \biggl(-\frac{6}{7} \frac{I}{R^4} - \frac{6}{7c} \frac{\dot{I}}{R^3} + \frac{6}{7c^2} \frac{\ddot{I}}{R^2} + \frac{8}{7c^3} \frac{\dddot{I}}{R} \biggr) N^{\langle jk \rangle} \\ & + 2 \biggl(\frac{9}{7} \frac{I^{\langle j}{}_p}{R^4} + \frac{9}{7c} \frac{\dot{I}^{\langle j}{}_p}{R^3} - \frac{9}{7c^2} \frac{\ddot{I}^{\langle j}{}_p}{R^2} - \frac{12}{7c^3} \frac{\dddot{I}^{\langle j}{}_p}{R} \biggr) N^{\langle k \rangle p \rangle} \\ & - \frac{6}{5c^2} \frac{\ddot{I}^{\langle jk \rangle}}{R^2} - \frac{6}{5c^3} \frac{\dddot{I}^{\langle jk \rangle}}{R} - \frac{2}{c^4} \overset{(4)}{I}{}^{\langle jk \rangle} + \cdots \biggr]. \end{aligned} \tag{11.251}$$

This expression is a sum of terms that have the structure of Eq. (11.237). For example, the first group of terms inside the square brackets has $\ell = 4$, and it consists of four terms with $n = 6$, $n = 5$, $n = 4$, and $n = 3$; for each one of these contributions we can easily read off the appropriate function f.

We shall keep in mind that it is the last term of Eq. (11.251), the one involving four derivatives of $I^{\langle jk \rangle}(\tau)$, that originated from $(-g)t^{jk}_{\rm H}$. It is this term that will reveal the differences between the light cones of the mathematical flat spacetime and those of the physical curved spacetime.

Evaluation of the wave-zone integrals

Each term $\tau^{jk}[\ell, n]$ in Eq. (11.251) makes a contribution to the gravitational-wave field h^{jk} given by Eq. (11.238). To see how these integrals are evaluated, we shall work through the representative case of $\ell = 0$ and $n = 3$.

We begin by extracting the relevant piece of τ^{jk} from Eq. (11.251). Comparing with Eq. (11.237), we find that in this case the function f is given by

$$f(\tau) = -\frac{6}{5} \frac{GM}{c^3} \dddot{I}^{\langle jk \rangle}. \tag{11.252}$$

We next evaluate the functions A and B. With $\ell = 0$ and $n = 3$, the computations are elementary, and the results are

$$A(s, R) = \frac{1}{\mathcal{R}} - \frac{1}{R+s}, \qquad B(s, R) = \frac{1}{s} - \frac{1}{R+s}. \tag{11.253}$$

We now set to work on the integrals that appear in Eq. (11.238). The first is

$$F_A := \int_0^{\mathcal{R}} ds f(\tau - 2s/c) A(s, R) = \int_0^{\mathcal{R}} ds f(\tau - 2s/c)\left(\frac{1}{\mathcal{R}} - \frac{1}{R+s}\right), \tag{11.254}$$

and we rewrite it as

$$F_A = \frac{1}{\mathcal{R}} \int_0^{\mathcal{R}} f(\tau - 2s/c)\, ds - \int_0^{\mathcal{R}} f(\tau - 2s/c)\, d\ln(R+s). \tag{11.255}$$

After integrating the second term by parts, our final expression is

$$\begin{aligned} F_A = &-f(\tau - 2\mathcal{R}/c)\ln(R + \mathcal{R}) + f(\tau)\ln R + \frac{1}{\mathcal{R}}\int_0^{\mathcal{R}} f(\tau - 2s/c)\, ds \\ &- \frac{2}{c}\int_0^{\mathcal{R}} \dot{f}(\tau - 2s/c)\ln\frac{R+s}{s}\, ds - \frac{2}{c}\int_0^{\mathcal{R}} \dot{f}(\tau - 2s/c)\ln s\, ds. \end{aligned} \tag{11.256}$$

The second integral is

$$F_B := \int_{\mathcal{R}}^{\infty} ds f(\tau - 2s/c) B(s, R) = \int_{\mathcal{R}}^{\infty} ds f(\tau - 2s/c)\left(\frac{1}{s} - \frac{1}{R+s}\right), \tag{11.257}$$

and we rewrite it as

$$F_B = -\int_{\mathcal{R}}^{\infty} f(\tau - 2s/c)\, d\ln\frac{R+s}{s}. \tag{11.258}$$

Integration by parts yields

$$F_B = f(\tau - 2\mathcal{R}/c)\ln\frac{R+\mathcal{R}}{\mathcal{R}} - \frac{2}{c}\int_{\mathcal{R}}^{\infty} \dot{f}(\tau - 2s/c)\ln\frac{R+s}{s}\, ds, \tag{11.259}$$

assuming that $f(\tau - 2s/c)$ goes to zero sufficiently rapidly as $s \to \infty$ to ensure that there is no boundary term at $s = \infty$. (Physically, this condition implies that the system is only weakly dynamical in the infinite past.)

The sum of F_A and F_B is

$$\begin{aligned} F = &-f(\tau - 2\mathcal{R}/c)\ln\mathcal{R} + f(\tau)\ln R + \frac{1}{\mathcal{R}}\int_0^{\mathcal{R}} f(\tau - 2s/c)\, ds \\ &- \frac{2}{c}\int_0^{\mathcal{R}} \dot{f}(\tau - 2s/c)\ln s\, ds - \frac{2}{c}\int_0^{\infty} \dot{f}(\tau - 2s/c)\ln\frac{R+s}{s}\, ds. \end{aligned} \tag{11.260}$$

This result is exact, but to simplify it we exploit the fact that we may remove from this all $\mathcal{R}$-dependent pieces. As a formal tool to achieve this, we express $f(\tau - 2s/c)$ and its derivative as an infinite Taylor series in powers of s, and we evaluate the two integrals from $s = 0$ to $s = \mathcal{R}$. We find that they combine to give $f(\tau)$, plus terms that can be discarded because they come with explicit factors of $\mathcal{R}$. After also expanding $f(\tau - 2\mathcal{R}/c)$ in powers of $\mathcal{R}$, we find that

$$F = f(\tau)\Big[1 + \ln(R/\mathcal{R})\Big] - \frac{2}{c}\int_0^\infty \dot{f}(\tau - 2s/c)\ln\frac{R+s}{s}\,ds, \tag{11.261}$$

modulo $\mathcal{R}$-dependent terms that can be freely discarded. This still contains a logarithmic dependence on $\mathcal{R}$, but it could be removed by writing $\ln(R/\mathcal{R}) = \ln(R/r_0) + \ln(r_0/\mathcal{R})$ and discarding the second term. This alternative expression would then contain a dependence on an arbitrary constant r_0, and it is perhaps preferable to stick with the original form, in spite of the residual $\mathcal{R}$-dependence.

The final answer is obtained by inserting our expressions for $f(\tau)$ and F within Eq. (11.238). We get

$$h^{jk}_{\mathcal{W}}[0,3] = \frac{4GM}{c^4R}\left\{-\frac{6G}{5c^3}\Big[1 + \ln(R/\mathcal{R})\Big]\dddot{I}^{\langle jk\rangle} + \frac{12}{5}K^{jk}\right\}, \tag{11.262}$$

in which the *tail integral*

$$K^{jk}(\tau, R) := \frac{G}{c^4}\int_0^\infty \overset{(4)}{I}{}^{\langle jk\rangle}(\tau - 2s/c)\ln\frac{R+s}{s}\,ds \tag{11.263}$$

must be left unevaluated. Note that the tail integral involves the entire past history of the system, from the infinite past (at $s = \infty$) to the current retarded time (at $s = 0$). We shall see what fate awaits the logarithmic term $\ln(R/\mathcal{R})$ in $h^{jk}_{\mathcal{W}}[0,3]$, when this contribution to $h^{jk}_{\mathcal{W}}$ is combined with others.

The same techniques are employed to calculate all other contributions to $h^{jk}_{\mathcal{W}}$. We shall not labor through the details here, but simply list the final results:

$$h^{jk}_{\mathcal{W}}[0,2] = \frac{4GM}{c^4R}\left\{-2K^{jk}\right\}, \tag{11.264a}$$

$$h^{jk}_{\mathcal{W}}[0,3] = \frac{4GM}{c^4R}\left\{-\frac{6G}{5c^3}\Big[1 + \ln(R/\mathcal{R})\Big]\dddot{I}^{\langle jk\rangle} + \frac{12}{5}K^{jk}\right\}, \tag{11.264b}$$

$$h^{jk}_{\mathcal{W}}[0,4] = \frac{4GM}{c^4R}\left\{\frac{6G}{5c^3}\Big[\frac{3}{2} + \ln(R/\mathcal{R})\Big]\dddot{I}^{\langle jk\rangle} - \frac{12}{5}K^{jk}\right\}, \tag{11.264c}$$

$$h^{jk}_{\mathcal{W}}[2,3] = \frac{4GM}{c^4R}\left\{-\frac{2G}{7c^3}\dddot{I}^{j}{}_{p}\right\}N^{\langle pk\rangle} + (j \leftrightarrow k), \tag{11.264d}$$

$$h^{jk}_{\mathcal{W}}[2,4] = \frac{4GM}{c^4R}\left\{-\frac{3G}{28c^3}\dddot{I}^{j}{}_{p}\right\}N^{\langle pk\rangle} + (j \leftrightarrow k), \tag{11.264e}$$

$$h^{jk}_{\mathcal{W}}[2,5] = \frac{4GM}{c^4R}\left\{\frac{G}{c^3}\Big[\frac{47}{700} + \frac{3}{35}\ln(R/\mathcal{R})\Big]\dddot{I}^{j}{}_{p} - \frac{6}{35}K^{j}{}_{p}\right\}N^{\langle pk\rangle} + (j \leftrightarrow k), \tag{11.264f}$$

$$h^{jk}_{\mathcal{W}}[2,6] = \frac{4GM}{c^4R}\left\{\frac{G}{c^3}\Big[-\frac{97}{700} - \frac{3}{35}\ln(R/\mathcal{R})\Big]\dddot{I}^{j}{}_{p} + \frac{6}{35}K^{j}{}_{p}\right\}N^{\langle pk\rangle} + (j \leftrightarrow k), \tag{11.264g}$$

$$h^{jk}_{\mathcal{W}}[4,3] = \frac{4GM}{c^4R}\left\{\frac{G}{20c^3}\dddot{I}_{pq}\right\}N^{\langle jkpq\rangle}, \tag{11.264h}$$

(continued overleaf)

$$h^{jk}_{\mathcal{W}}[4,4] = \frac{4GM}{c^4R}\left\{\frac{G}{30c^3}\dddot{I}_{pq}\right\}N^{\langle jkpq\rangle}, \tag{11.264i}$$

$$h^{jk}_{\mathcal{W}}[4,5] = \frac{4GM}{c^4R}\left\{\frac{G}{42c^3}\dddot{I}_{pq}\right\}N^{\langle jkpq\rangle}, \tag{11.264j}$$

$$h^{jk}_{\mathcal{W}}[4,6] = \frac{4GM}{c^4R}\left\{\frac{G}{56c^3}\dddot{I}_{pq}\right\}N^{\langle jkpq\rangle}. \tag{11.264k}$$

To arrive at these results we have freely discarded all $\mathcal{R}$-dependent terms, except when the dependence is logarithmic. In some cases we have also removed terms that fall off as R^{-2}, R^{-3}, or faster, because these are negligible in the far-away wave zone.

From the preceding listing of results we find that the sums of contributions for $\ell = 0$, $\ell = 2$, and $\ell = 4$ are

$$h^{jk}_{\mathcal{W}}[\ell=0] = \frac{4GM}{c^4R}\left\{\frac{3G}{5c^3}\dddot{I}^{\langle jk\rangle} - 2K^{jk}\right\}, \tag{11.265a}$$

$$h^{jk}_{\mathcal{W}}[\ell=2] = \frac{4GM}{c^4R}\left\{-\frac{13G}{28c^3}\dddot{I}^{j}{}_{p}N^{\langle pk\rangle} + (j \leftrightarrow k)\right\}, \tag{11.265b}$$

$$h^{jk}_{\mathcal{W}}[\ell=4] = \frac{4GM}{c^4R}\left\{\frac{G}{8c^3}\dddot{I}_{pq}N^{\langle jkpq\rangle}\right\}. \tag{11.265c}$$

Note that the logarithmic terms have all canceled out, and that the tail integral K^{jk} appears only within the $\ell = 0$ contribution. Tracing the origin of the tail integral, we see that it comes from $\tau^{jk}[0,2]$, the term in τ^{jk} that involves four derivatives of the Newtonian quadrupole moment. This term, the last one in Eq. (11.250), originates from $(-g)t^{jk}_{\mathrm{H}}$, and as we have observed previously, it reveals the differences between the light cones of the mathematical flat spacetime and those of the physical curved spacetime. The tail integral, therefore, informs us that the gravitational waves are propagating in a curved spacetime instead of the fictitious flat spacetime of post-Minkowskian theory.

Final answer

Adding the contributions from $\ell = 0$, $\ell = 2$, and $\ell = 4$, we find that the wave-zone piece of the gravitational-wave field is given by

$$h^{jk}_{\mathcal{W}} = \frac{4GM}{c^4R}\left\{\frac{3G}{5c^3}\dddot{I}^{\langle jk\rangle} - 2K^{jk} - \frac{13G}{28c^3}\left(\dddot{I}^{j}{}_{p}N^{\langle pk\rangle} + \dddot{I}^{k}{}_{p}N^{\langle pj\rangle}\right) + \frac{G}{8c^3}\dddot{I}_{pq}N^{\langle jkpq\rangle}\right\}. \tag{11.266}$$

From this we may remove any term that will not survive a TT projection. For example, we may use

$$\dddot{I}^{j}{}_{p}N^{\langle pk\rangle} = \dddot{I}^{j}{}_{p}\left(N^pN^k - \frac{1}{3}\delta^{pk}\right) \overset{\mathrm{TT}}{=} -\frac{1}{3}\dddot{I}^{\langle jk\rangle} \tag{11.267}$$

and

$$\dddot{I}_{pq}N^{\langle jkpq\rangle} \overset{\mathrm{TT}}{=} \frac{2}{35}\dddot{I}^{\langle jk\rangle} \tag{11.268}$$

to simplify the expression, which becomes

$$h^{jk}_{\mathcal{W}} \overset{\text{TT}}{=} \frac{4GM}{c^4R}\left\{\frac{11G}{12c^3}\overset{\cdots}{I}{}^{\langle jk\rangle} - 2K^{jk}\right\}. \tag{11.269}$$

To arrive at the final form of Eq. (11.236), we substitute Eq. (11.263) for the tail integral and clean things up by setting $s = \frac{1}{2}c\,\zeta$, thereby adopting ζ as a new integration variable. This gives us

$$h^{jk}_{\mathcal{W}} \overset{\text{TT}}{=} \frac{4G^2M}{c^7R}\left\{\frac{11}{12}\overset{\cdots}{I}{}^{\langle jk\rangle}(\tau) + \int_0^\infty \overset{(4)}{I}{}^{\langle jk\rangle}(\tau-\zeta)\ln\frac{\zeta}{\zeta+2R/c}\,d\zeta\right\}, \tag{11.270}$$

and it is easy to show that this is equivalent to Eq. (11.236).

11.3.8 Summary: Gravitational-wave field

Our computation of the gravitational-wave field generated by an N-body system is now essentially complete. For easy reference we copy in Box 11.4 the main results obtained in the preceding five subsections.

Box 11.4 Gravitational-wave field to 1.5PN order

The gravitational potentials h^{jk} are decomposed according to

$$h^{jk} = h^{jk}_{\mathcal{N}} + h^{jk}_{\mathcal{W}}, \tag{1}$$

and the near-zone piece is expressed as the multipole expansion

$$h^{jk}_{\mathcal{N}}(t,\boldsymbol{x}) = \frac{2G}{c^4R}\frac{\partial^2}{\partial\tau^2}\left\{Q^{jk} + Q^{jka}N_a + Q^{jkab}N_aN_b + \frac{1}{3}Q^{jkabc}N_aN_bN_c\right\}, \tag{2}$$

in which $R := |\boldsymbol{x}|$, $\boldsymbol{N} := \boldsymbol{x}/R$, and $\tau := t - R/c$ is retarded time. The radiative multipole moments are given by

$$Q^{jk} \overset{\text{TT}}{=} \sum_A M_A\left(1 + \frac{1}{2}\frac{v_A^2}{c^2}\right)r_A^jr_A^k - \frac{1}{2c^2}\sum_A\sum_{B\neq A}\frac{GM_AM_B}{r_{AB}}r_A^jr_A^k + O(c^{-4}),$$

$$Q^{jka} = A^{jka} + A^{kja} - A^{ajk},$$

$$\begin{aligned}A^{jka} \overset{\text{TT}}{=} &\frac{1}{c}\sum_A M_A\left(1+\frac{v_A^2}{2c^2}\right)v_A^jr_A^kr_A^a\\ &-\frac{1}{2c^3}\sum_A\sum_{B\neq A}\frac{GM_AM_B}{r_{AB}}\left[(\boldsymbol{n}_{AB}\cdot\boldsymbol{v}_A)n_{AB}^jr_A^kr_A^a + v_A^jr_A^kr_A^a\right]\\ &+\frac{1}{2c^3}\sum_A\sum_{B\neq A}GM_AM_B\left[(\boldsymbol{n}_{AB}\cdot\boldsymbol{v}_A)n_{AB}^jn_{AB}^{(k}r_A^{a)} - 7n_{AB}^jv_A^{(k}r_A^{a)} + 7v_A^jn_{AB}^{(k}r_A^{a)}\right]\end{aligned}$$

(continued overleaf)

$$-\frac{1}{6c^3}\sum_A\sum_{B\neq A} GM_AM_Br_{AB}\Big[(\boldsymbol{n}_{AB}\cdot\boldsymbol{v}_A)n^j_{AB}n^k_{AB}n^a_{AB} - 11n^j_{AB}n^{(k}_{AB}v^{a)}_A + 11v^j_An^k_{AB}n^a_{AB}\Big] + O(c^{-5}),$$

$$Q^{jkab} \overset{\text{TT}}{=} \frac{1}{c^2}\sum_A M_Av^j_Av^k_Ar^a_Ar^b_A - \frac{1}{2c^2}\sum_A\sum_{B\neq A}\frac{GM_AM_B}{r_{AB}}n^j_{AB}n^k_{AB}r^a_Ar^b_A + \frac{1}{12c^2}\sum_A\sum_{B\neq A} GM_AM_Br_{AB}n^j_{AB}n^k_{AB}\left(n^a_{AB}n^b_{AB} - \delta^{ab}\right) + O(c^{-4}),$$

$$Q^{jkabc} \overset{\text{TT}}{=} \frac{1}{c^3}\frac{\partial}{\partial\tau}\Big[\sum_A M_Av^j_Av^k_Ar^a_Ar^b_Ar^c_A - \frac{1}{2}\sum_A\sum_{B\neq A}\frac{GM_AM_B}{r_{AB}}n^j_{AB}n^k_{AB}r^a_Ar^b_Ar^c_A + \frac{1}{4}\sum_A\sum_{B\neq A} GM_AM_Br_{AB}n^j_{AB}n^k_{AB}r^{(a}_A\left(n^b_{AB}n^{c)}_{AB} - \delta^{bc)}\right)\Big] + O(c^{-5}).$$

They are expressed in terms of the mass-energy M_A of each body, its position $\boldsymbol{r}_A$, and velocity $\boldsymbol{v}_A$; all position and velocity vectors are evaluated at the retarded time τ, and the radiative moments are functions of τ only. We use $r_{AB} := |\boldsymbol{r}_A - \boldsymbol{r}_B|$ to denote the distance between bodies A and B, and $\boldsymbol{n}_{AB} = (\boldsymbol{r}_A - \boldsymbol{r}_B)/r_{AB}$ is a unit vector that points from body B to body A. With this listing of radiative multipole moments, the multipole expansion is accurate through 1.5PN order.

The wave-zone piece is given by the tail integral

$$h^{jk}_{\mathscr{W}}(t,\boldsymbol{x}) \overset{\text{TT}}{=} \frac{4G}{c^4R}\frac{GM}{c^3}\int_0^\infty \overset{(4)}{I}{}^{\langle jk\rangle}(\tau-\zeta)\left(\ln\frac{\zeta}{\zeta+2R/c} + \frac{11}{12}\right)d\zeta, \tag{3}$$

which involves the entire past history of the system. It depends on the total gravitational mass of the system, $M = \sum_A M_A + O(c^{-2})$, as well as the Newtonian quadrupole moment of the matter distribution, $I^{jk} = \sum_A M_Ar^j_Ar^k_A$, which is made tracefree and differentiated four times with respect to its argument. The wave-zone piece makes a contribution at 1.5PN order to the gravitational-wave field.

The computation is essentially complete, but much work remains to be done to turn these expressions into something more concrete. First, the derivatives with respect to τ must be evaluated, and this will require a large effort. Second, the projection to the transverse subspace must be fully carried out, because our multipole moments still contain pieces that can be removed by acting with $(\text{TT})^{jk}{}_{pq}$. The ultimate goal is to obtain the polarizations h_+ and $h_\times$ expressed entirely in terms of the positions $\boldsymbol{r}_A$ and velocities $\boldsymbol{v}_A$. We shall proceed toward this goal in the following section.

11.4 Gravitational waves emitted by a two-body system

To simplify the task of producing concrete expressions for h_+ and $h_\times$, we choose at this stage to specialize our discussion to a binary system of orbiting bodies. The system will therefore involve the masses M_1 and M_2, the positions $\boldsymbol{r}_1$ and $\boldsymbol{r}_2$, and the velocities $\boldsymbol{v}_1$ and $\boldsymbol{v}_2$. The dynamics of the binary system is described by the post-Newtonian equations of motion obtained back in Sec. 9.3.7.

11.4.1 Motion in the barycentric frame

We work in the post-Newtonian barycentric frame ($\boldsymbol{R} = \boldsymbol{0}$), and according to Eqs. (9.141), the position vector of each body is given by

$$\boldsymbol{r}_1 = \frac{M_2}{m}\boldsymbol{r} + \frac{\eta\Delta}{2c^2}\left(v^2 - \frac{Gm}{r}\right)\boldsymbol{r} + O(c^{-4}), \tag{11.271a}$$

$$\boldsymbol{r}_2 = -\frac{M_1}{m}\boldsymbol{r} + \frac{\eta\Delta}{2c^2}\left(v^2 - \frac{Gm}{r}\right)\boldsymbol{r} + O(c^{-4}). \tag{11.271b}$$

They are expressed in terms of the separation vector $\boldsymbol{r} := \boldsymbol{r}_{12} := \boldsymbol{r}_1 - \boldsymbol{r}_2$ and the relative velocity $\boldsymbol{v} := \boldsymbol{v}_{12} := \boldsymbol{v}_1 - \boldsymbol{v}_2$; these have magnitudes $r = |\boldsymbol{r}|$ and $v = |\boldsymbol{v}|$, respectively. We have re-introduced the mass parameters

$$m := M_1 + M_2, \tag{11.272a}$$

$$\eta := \frac{M_1 M_2}{(M_1 + M_2)^2}, \tag{11.272b}$$

$$\Delta := \frac{M_1 - M_2}{M_1 + M_2}. \tag{11.272c}$$

Differentiation of Eqs. (11.271) returns the velocity vector of each body:

$$\boldsymbol{v}_1 = \frac{M_2}{m}\boldsymbol{v} + \frac{\eta\Delta}{2c^2}\left[\left(v^2 - \frac{Gm}{r}\right)\boldsymbol{v} - \frac{Gm}{r}\dot{r}\boldsymbol{n}\right] + O(c^{-4}), \tag{11.273a}$$

$$\boldsymbol{v}_2 = -\frac{M_1}{m}\boldsymbol{v} + \frac{\eta\Delta}{2c^2}\left[\left(v^2 - \frac{Gm}{r}\right)\boldsymbol{v} - \frac{Gm}{r}\dot{r}\boldsymbol{n}\right] + O(c^{-4}), \tag{11.273b}$$

where $\dot{r} := \boldsymbol{n} \cdot \boldsymbol{v}$ is the radial component of the velocity vector, and $\boldsymbol{n} := \boldsymbol{r}/r$ is a unit vector that points from body 2 to body 1. To arrive at these expressions we have involved the relative acceleration $\boldsymbol{a} := \boldsymbol{a}_1 - \boldsymbol{a}_2$, which according to Eq. (9.142) is given by

$$\boldsymbol{a} = -\frac{Gm}{r^2}\boldsymbol{n} - \frac{Gm}{c^2r^2}\left\{\left[(1+3\eta)v^2 - \frac{3}{2}\eta\dot{r}^2 - 2(2+\eta)\frac{Gm}{r}\right]\boldsymbol{n} - 2(2-\eta)\dot{r}\,\boldsymbol{v}\right\} + O(c^{-4}). \tag{11.274}$$

11.4.2 Radiative multipole moments

We make the substitutions within the radiative multipole moments of Box 11.4 and simplify the resulting expressions. The sums that appear in these equations must be specialized to two bodies, and in these we set $r_{12} = r_{21} = r$ and $\boldsymbol{n}_{12} = -\boldsymbol{n}_{21} = \boldsymbol{n}$. In the course of these (lengthy, but straightforward) computations we encounter various functions of M_1 and M_2 that can be rewritten in terms of the mass parameters of Eqs. (11.272). For example, it is easy to show that

$$\frac{M_1^2 + M_2^2}{(M_1+M_2)^2} = 1 - 2\eta, \tag{11.275a}$$

$$\frac{M_1^3 + M_2^3}{(M_1+M_2)^3} = 1 - 3\eta, \tag{11.275b}$$

$$\frac{M_1^4 - M_2^4}{(M_1+M_2)^4} = \Delta(1 - 2\eta), \tag{11.275c}$$

and we make many such substitutions while simplifying our expressions.

We obtain

$$Q^{jk} = \eta m\left[1 + \frac{1}{2}(1-3\eta)\frac{v^2}{c^2} - \frac{1}{2}(1-2\eta)\frac{Gm}{c^2 r} + O(c^{-4})\right] r^j r^k, \tag{11.276a}$$

$$\begin{aligned} Q^{jka} = \frac{\eta m \Delta}{c}\biggl\{ & r^j r^k v^a - (v^j r^k + r^j v^k) r^a \\ & - \left[\frac{1}{2}(1-5\eta)\frac{v^2}{c^2} + \frac{1}{6}(7+12\eta)\frac{Gm}{c^2 r}\right](v^j r^k + r^j v^k) r^a \\ & + \left[\frac{1}{2}(1-5\eta)\frac{v^2}{c^2} + \frac{1}{6}(17+12\eta)\frac{Gm}{c^2 r}\right] r^j r^k v^a \\ & + \frac{1}{6}(1-6\eta)\frac{Gm}{c^2 r}\dot{r}\, n^j r^k r^a + O(c^{-4})\biggr\}, \end{aligned} \tag{11.276b}$$

$$\begin{aligned} Q^{jkab} = \frac{\eta m}{c^2}\biggl\{ & (1-3\eta) v^j v^k r^a r^b - \frac{1}{3}(1-3\eta)\frac{Gm}{r} n^j n^k r^a r^b \\ & - \frac{1}{6}\frac{Gm}{r} r^j r^k \delta^{ab} + O(c^{-2})\biggr\}, \end{aligned} \tag{11.276c}$$

$$\begin{aligned} Q^{jkabc} = \frac{\eta m \Delta}{c^3}\frac{\partial}{\partial \tau}\biggl\{ & -(1-2\eta) v^j v^k r^a r^b r^c + \frac{1}{4}(1-2\eta)\frac{Gm}{r} n^j n^k r^a r^b r^c \\ & + \frac{1}{4}\frac{Gm}{r} r^j r^k r^{(a}\delta^{bc)} + O(c^{-2})\biggr\}. \end{aligned} \tag{11.276d}$$

We observe that in order to simplify the writing, we have replaced the qualified equality sign $\stackrel{\text{TT}}{=}$ ("equal after a TT projection") by the usual equality sign.

11.4.3 Computation of retarded-time derivatives

The near-zone contribution to h^{jk} is given by Eq. (2) of Box 11.4, and in this we must insert the radiative multipole moments displayed in the preceding subsection; the computation involves taking two retarded-time derivatives of these moments. Similarly, the wave-zone contribution to h^{jk} is given by Eq. (3) of Box 11.4, and this involves taking four retarded-time derivatives of $I^{jk} = \eta m r^j r^k$. Our immediate task in this subsection is to compute these derivatives.

The general strategy is clear. The radiative multipole moments of Eqs. (11.276) are expressed explicitly in terms of the position and velocity vectors, and these are functions of the retarded time τ. Differentiating with respect to τ therefore involves taking derivatives of the position and velocity vectors. Differentiating $\boldsymbol{r}$ gives $\boldsymbol{v}$, and differentiating $\boldsymbol{v}$ gives $\boldsymbol{a}$, the post-Newtonian acceleration vector of Eq. (11.274). After making this substitution, the result is once more expressed in terms of $\boldsymbol{r}$ and $\boldsymbol{v}$, and is ready for further differentiation.

More concretely, consider the task of computing $\ddot{Q}^{jk}$. The quadrupole moment is a function of $\boldsymbol{r}$ at order c^0, and a function of $\boldsymbol{r}$ and $\boldsymbol{v}$ at order c^{-2}. Taking a first derivative with respect to τ produces terms in $\boldsymbol{r}$ and $\boldsymbol{v}$ at order c^0, and terms in $\boldsymbol{r}$, $\boldsymbol{v}$, and $\boldsymbol{a}$ at order c^{-2}. In the post-Newtonian term we can substitute the Newtonian expression for the acceleration vector, $\boldsymbol{a} = -Gm\boldsymbol{r}/r^3 + O(c^{-2})$, because the error incurred occurs at order c^{-4} in $\dot{Q}^{jk}$. The end result is a function of $\boldsymbol{r}$ and $\boldsymbol{v}$ at order c^0, another function of $\boldsymbol{r}$ and $\boldsymbol{v}$ at order c^{-2}, and neglected terms at order c^{-4}. Taking a second derivative introduces the acceleration vector at orders c^0 and c^{-2}. In the Newtonian term we must now substitute the post-Newtonian expression for the acceleration vector, because its PN term will influence the c^{-2} piece of $\ddot{Q}^{jk}$; but we are still allowed to insert the Newtonian acceleration within the c^{-2} piece of the second derivative. The end result for $\ddot{Q}^{jk}$ is a function of $\boldsymbol{r}$ and $\boldsymbol{v}$ at order c^0, and another function of $\boldsymbol{r}$ and $\boldsymbol{v}$ at order c^{-2}.

Derivatives of other multipole moments are computed in a similar way. These computations are tedious and lengthy, but they are completely straightforward. They are aided by the identities

$$v\dot{v} = -\frac{Gm}{r^2}\dot{r} + O(c^{-2}), \qquad r\ddot{r} = v^2 - \dot{r}^2 - \frac{Gm}{r} + O(c^{-2}), \tag{11.277}$$

which are consequences of the Newtonian expression for the acceleration vector.

We display the final results:

$$\begin{aligned}
\ddot{Q}^{jk} &= 2\eta m\left(v^j v^k - \frac{Gm}{r}n^j n^k\right) \\
&\quad + \frac{\eta m}{c^2}\Biggl\{\left[-\frac{1}{2}(7+2\eta)v^2 + \frac{3}{2}(1-2\eta)\dot{r}^2 + \frac{19}{2}\frac{Gm}{r}\right]\frac{Gm}{r}n^j n^k \\
&\qquad + \left[(1-3\eta)v^2 - (1-2\eta)\frac{Gm}{r}\right]v^j v^k + (3+2\eta)\frac{Gm}{r}\dot{r}\left(v^j n^k + n^j v^k\right)\Biggr\} \\
&\quad + O(c^{-4}),
\end{aligned} \tag{11.278a}$$

(continued overleaf)

$$
\begin{aligned}
\ddot{Q}^{jka} &= \frac{\eta m \Delta}{c}\biggl\{-3\frac{Gm}{r}\dot{r}\,n^j n^k n^a + 3\frac{Gm}{r}\bigl(v^j n^k + n^j v^k\bigr)n^a + \frac{Gm}{r}n^j n^k v^a - 2v^j v^k v^a\biggr\} \\
&+ \frac{\eta m \Delta}{c^3}\biggl\{\biggl[\frac{3}{2}(2-\eta)v^2 + \frac{9}{2}(1+\eta)\dot{r}^2 - \frac{1}{3}(31-9\eta)\frac{Gm}{r}\biggr]\frac{Gm}{r}\bigl(v^j n^k + n^j v^k\bigr)n^a \\
&\quad - (15+2\eta)\frac{Gm}{r}\dot{r}\,v^j v^k n^a \\
&\quad + \biggl[-\frac{3}{2}(4-3\eta)v^2 + \frac{5}{2}(1-3\eta)\dot{r}^2 + \frac{2}{3}(29-3\eta)\frac{Gm}{r}\biggr]\frac{Gm}{r}\dot{r}\,n^j n^k n^a \\
&\quad + \biggl[\frac{1}{2}(4-\eta)v^2 - \frac{3}{2}(1-\eta)\dot{r}^2 - \frac{1}{3}(25-3\eta)\frac{Gm}{r}\biggr]\frac{Gm}{r}n^j n^k v^a \\
&\quad - (3+2\eta)\frac{Gm}{r}\dot{r}\bigl(v^j n^k + n^j v^k\bigr)v^a \\
&\quad + \biggl[-(1-5\eta)v^2 + (1-4\eta)\frac{Gm}{r}\biggr]v^j v^k v^a\biggr\} + O(c^{-5}), \qquad (11.278b)
\end{aligned}
$$

$$
\begin{aligned}
\ddot{Q}^{jkab} &= \frac{\eta m}{c^2}\biggl\{5(1-3\eta)\frac{Gm}{r}\dot{r}\bigl(v^j n^k + n^j v^k\bigr)n^a n^b \\
&\quad + (1-3\eta)\biggl(v^2 - 5\dot{r}^2 + \frac{7}{3}\frac{Gm}{r}\biggr)\frac{Gm}{r}n^j n^k n^a n^b \\
&\quad - \frac{14}{3}(1-3\eta)\frac{Gm}{r}v^j v^k n^a n^b \\
&\quad - \frac{8}{3}(1-3\eta)\frac{Gm}{r}\bigl(v^j n^k + n^j v^k\bigr)\bigl(v^a n^b + n^a v^b\bigr) \\
&\quad + 2(1-3\eta)v^j v^k v^a v^b + 2(1-3\eta)\frac{Gm}{r}\dot{r}\,n^j n^k\bigl(v^a n^b + n^a v^b\bigr) \\
&\quad - \frac{2}{3}(1-3\eta)\frac{Gm}{r}n^j n^k v^a v^b + \frac{1}{6}\frac{Gm}{r}\biggl(v^2 - 3\dot{r}^2 + \frac{Gm}{r}\biggr)n^j n^k \delta^{ab} \\
&\quad + \frac{1}{3}\frac{Gm}{r}\dot{r}\bigl(v^j n^k + n^j v^k\bigr)\delta^{ab} - \frac{1}{3}\frac{Gm}{r}v^j v^k \delta^{ab}\biggr\} + O(c^{-4}), \qquad (11.278c)
\end{aligned}
$$

$$
\begin{aligned}
\ddot{Q}^{jkabc} &= \frac{\eta m \Delta}{c^3}\biggl\{-\frac{1}{4}(1-2\eta)\biggl(21v^2 - 105\dot{r}^2 + 44\frac{Gm}{r}\biggr)\frac{Gm}{r}\bigl(v^j n^k + n^j v^k\bigr)n^a n^b n^c \\
&\quad + \frac{1}{4}(1-2\eta)\biggl(45v^2 - 105\dot{r}^2 + 90\frac{Gm}{r}\biggr)\frac{Gm}{r}\dot{r}\,n^j n^k n^a n^b n^c \\
&\quad - \frac{51}{2}(1-2\eta)\frac{Gm}{r}\dot{r}\,v^j v^k n^a n^b n^c \\
&\quad - \frac{27}{2}(1-2\eta)\frac{Gm}{r}\dot{r}\bigl(v^j n^k + n^j v^k\bigr)\bigl(v^a n^b n^c + n^a v^b n^c + n^a n^b v^c\bigr) \\
&\quad - \frac{1}{4}(1-2\eta)\biggl(9v^2 - 45\dot{r}^2 + 28\frac{Gm}{r}\biggr)\frac{Gm}{r}n^j n^k\bigl(v^a n^b n^c + n^a v^b n^c + n^a n^b v^c\bigr)
\end{aligned}
$$

(continued overleaf)

$$+\frac{29}{2}(1-2\eta)\frac{Gm}{r}v^j v^k\big(v^a n^b n^c + n^a v^b n^c + n^a n^b v^c\big)$$
$$+\frac{15}{2}(1-2\eta)\frac{Gm}{r}\big(v^j n^k + n^j v^k\big)\big(v^a v^b n^c + v^a n^b v^c + n^a v^b v^c\big)$$
$$-6(1-2\eta)v^j v^k v^a v^b v^c - \frac{9}{2}(1-2\eta)\frac{Gm}{r}\dot{r}\, n^j n^k\big(v^a v^b n^c + v^a n^b v^c + n^a v^b v^c\big)$$
$$+\frac{3}{2}(1-2\eta)\frac{Gm}{r}n^j n^k v^a v^b v^c + \frac{1}{4}\left(9v^2 - 15\dot{r}^2 + 10\frac{Gm}{r}\right)\frac{Gm}{r}\dot{r}\, n^j n^k n^{(a}\delta^{bc)}$$
$$-\frac{1}{4}\left(3v^2 - 9\dot{r}^2 + 4\frac{Gm}{r}\right)\frac{Gm}{r}\big(v^j n^k + n^j v^k\big)n^{(a}\delta^{bc)}$$
$$-\frac{1}{4}\left(3v^2 - 9\dot{r}^2 + 4\frac{Gm}{r}\right)\frac{Gm}{r}n^j n^k v^{(a}\delta^{bc)}$$
$$-\frac{3}{2}\frac{Gm}{r}\dot{r}\, v^j v^k n^{(a}\delta^{bc)} - \frac{3}{2}\frac{Gm}{r}\dot{r}\big(v^j n^k + n^j v^k\big)v^{(a}\delta^{cd)}$$
$$\left.+\frac{3}{2}\frac{Gm}{r}\dot{r}\, v^j v^k v^{(a}\delta^{bc)}\right] + O(c^{-5}). \tag{11.278d}$$

In addition, we have that

$$\overset{(4)}{I}{}^{jk} = 2\eta m\frac{Gm}{r^3}\left[\left(3v^2 - 15\dot{r}^2 + \frac{Gm}{r}\right)n^j n^k + 9\dot{r}\big(v^j n^k + n^j v^k\big) - 4v^j v^k\right]$$
$$+ O(c^{-2}). \tag{11.279}$$

11.4.4 Gravitational-wave field

We may now substitute Eqs. (11.278) and (11.279) into Eqs. (2) and (3) of Box 11.4 and obtain the gravitational-wave field. These computations are straightforward, and we express the result as

$$h^{jk}(t,\boldsymbol{x}) = \frac{2\eta Gm}{c^4 R}\Big[A^{jk}[0\text{PN}] + A^{jk}[0.5\text{PN}] + A^{jk}[1\text{PN}]$$
$$+ A^{jk}[1.5\text{PN}] + A^{jk}[\text{tail}] + O(c^{-4})\Big], \tag{11.280}$$

in which we group terms according to their post-Newtonian order (the last term, with the label "tail," is also of 1.5PN order). We have

$$A^{jk}[0\text{PN}] = 2\left\{v^j v^k - \frac{Gm}{r}n^j n^k\right\}, \tag{11.281a}$$
$$A^{jk}[0.5\text{PN}] = \frac{\Delta}{c}\left\{3\frac{Gm}{r}(\boldsymbol{n}\cdot\boldsymbol{N})\big(v^j n^k + n^j v^k - \dot{r}\, n^j n^k\big)\right.$$
$$\left.+(\boldsymbol{v}\cdot\boldsymbol{N})\left(-2v^j v^k + \frac{Gm}{r}n^j n^k\right)\right\}, \tag{11.281b}$$

(continued overleaf)

$$
\begin{aligned}
A^{jk}[1\mathrm{PN}] = \frac{1}{c^2}\Bigg\{ & \frac{1}{3}\left[3(1-3\eta)v^2 - 2(2-3\eta)\frac{Gm}{r}\right]v^j v^k \\
& + \frac{2}{3}(5+3\eta)\frac{Gm}{r}\dot{r}\left(v^j n^k + n^j v^k\right) \\
& + \frac{1}{3}\frac{Gm}{r}\left[-(10+3\eta)v^2 + 3(1-3\eta)\dot{r}^2 + 29\frac{Gm}{r}\right]n^j n^k \\
& + \frac{2}{3}(1-3\eta)(\boldsymbol{v}\cdot\boldsymbol{N})^2\left(3v^j v^k - \frac{Gm}{r}n^j n^k\right) \\
& + \frac{4}{3}(1-3\eta)(\boldsymbol{v}\cdot\boldsymbol{N})(\boldsymbol{n}\cdot\boldsymbol{N})\frac{Gm}{r}\left[-4\left(v^j n^k + n^j v^k\right) + 3\dot{r}\, n^j n^k\right] \\
& + \frac{1}{3}(1-3\eta)(\boldsymbol{n}\cdot\boldsymbol{N})^2\frac{Gm}{r}\Bigg[-14 v^j v^k + 15\dot{r}\left(v^j n^k + n^j v^k\right) \\
& + \left(3v^2 - 15\dot{r}^2 + 7\frac{Gm}{r}\right)n^j n^k\Bigg]\Bigg\}\,,
\end{aligned}
\tag{11.281c}
$$

$$
\begin{aligned}
A^{jk}[1.5\mathrm{PN}] = \frac{\Delta}{c^3}\Bigg\{ & \frac{1}{12}(\boldsymbol{v}\cdot\boldsymbol{N})\Bigg\{-6\left[2(1-5\eta)v^2 - (3-8\eta)\frac{Gm}{r}\right]v^j v^k \\
& - 6(7+4\eta)\frac{Gm}{r}\dot{r}\left(v^j n^k + n^j v^k\right) \\
& + \frac{Gm}{r}\left[3(7-2\eta)v^2 - 9(1-2\eta)\dot{r}^2 - 4(26-3\eta)\frac{Gm}{r}\right]n^j n^k\Bigg\} \\
& + \frac{1}{12}(\boldsymbol{n}\cdot\boldsymbol{N})\frac{Gm}{r}\Bigg\{-6(31+4\eta)\dot{r}\, v^j v^k \\
& + \left[3(11-6\eta)v^2 + 9(7+6\eta)\dot{r}^2 - 4(32-9\eta)\frac{Gm}{r}\right]\left(v^j n^k + n^j v^k\right) \\
& - \dot{r}\left[9(7-6\eta)v^2 - 15(1-6\eta)\dot{r}^2 - 2(121-12\eta)\frac{Gm}{r}\right]n^j n^k\Bigg\} \\
& + \frac{1}{2}(1-2\eta)(\boldsymbol{v}\cdot\boldsymbol{N})^3\left\{-4v^j v^k + \frac{Gm}{r}n^j n^k\right\} \\
& + \frac{3}{2}(1-2\eta)(\boldsymbol{v}\cdot\boldsymbol{N})^2(\boldsymbol{n}\cdot\boldsymbol{N})\frac{Gm}{r}\left\{5\left(v^j n^k + n^j v^k\right) - 3\dot{r}\, n^j n^k\right\} \\
& + \frac{1}{4}(1-2\eta)(\boldsymbol{v}\cdot\boldsymbol{N})(\boldsymbol{n}\cdot\boldsymbol{N})^2\frac{Gm}{r}\Bigg\{58 v^j v^k - 54\dot{r}\left(v^j n^k + n^j v^k\right) \\
& - \left[9v^2 - 45\dot{r}^2 + 28\frac{Gm}{r}\right]n^j n^k\Bigg\} \\
& + \frac{1}{12}(1-2\eta)(\boldsymbol{n}\cdot\boldsymbol{N})^3\frac{Gm}{r}\Bigg\{-102\dot{r}\, v^j v^k \\
& - \left[21v^2 - 105\dot{r}^2 + 44\frac{Gm}{r}\right]\left(v^j n^k + n^j v^k\right) \\
& + 15\dot{r}\left[3v^2 - 7\dot{r}^2 + 6\frac{Gm}{r}\right]n^j n^k\Bigg\}\Bigg\}\,,
\end{aligned}
\tag{11.281d}
$$

(continued overleaf)

$$A^{jk}[\text{tail}] = \frac{4Gm}{c^3}\int_0^\infty \left\{\frac{Gm}{r^3}\left[\left(3v^2 - 15\dot{r}^2 + \frac{Gm}{r}\right)n^jn^k + 9\dot{r}\left(v^jn^k + n^jv^k\right) - 4v^jv^k\right]\right\}_{\tau-\zeta}\left[\ln\left(\frac{\zeta}{\zeta + 2R/c}\right) + \frac{11}{12}\right]d\zeta\,. \tag{11.281e}$$

The gravitational-wave field is expressed in terms of the separation vector $\boldsymbol{r} = \boldsymbol{r}_1 - \boldsymbol{r}_2$, the relative velocity $\boldsymbol{v} = \boldsymbol{v}_1 - \boldsymbol{v}_2$, the radial velocity $\dot{r} = \boldsymbol{n}\cdot\boldsymbol{v}$, and the mass parameters $m = M_1 + M_2$, $\eta = M_1M_2/m^2$, and $\Delta = (M_1 - M_2)/m$. In addition, h^{jk} depends on distance $R := |\boldsymbol{x}|$, retarded time $\tau = t - R/c$, and on the angular vector $\boldsymbol{N} := \boldsymbol{x}/R$ that specifies the direction from the barycenter to the field point $\boldsymbol{x}$. In the tail integral the terms within the large round brackets are evaluated at $\tau - \zeta$ instead of τ, and the integration from $\zeta = 0$ to $\zeta = -\infty$ involves the entire past history of the two-body system.

11.4.5 Polarizations

Our final task is to carry out the projection to the transverse subspace, and extract the gravitational-wave polarizations

$$h_{+,\times} = \frac{2\eta Gm}{c^4R}\Big[A_{+,\times}[0\text{PN}] + A_{+,\times}[0.5\text{PN}] + A_{+,\times}[1\text{PN}] + A_{+,\times}[1.5\text{PN}] + A_{+,\times}[\text{tail}] + O(c^{-4})\Big] \tag{11.282}$$

from Eq. (11.280). We adopt the same conventions as in Sec. 11.2.2. We re-introduce the "orbit-adapted" coordinate system (x, y, z) and express $\boldsymbol{n}$ in terms of $\phi(\tau)$, the (retarded) angular position of the relative orbit. The expression is $\boldsymbol{n} = [\cos\phi, \sin\phi, 0]$, and to this we adjoin another basis vector $\boldsymbol{\lambda} = [-\sin\phi, \cos\phi, 0]$, which also lies in the fixed orbital plane. We express the relative position and velocity vectors as

$$\boldsymbol{r} = r\,\boldsymbol{n}, \qquad \boldsymbol{v} = \dot{r}\,\boldsymbol{n} + r\dot{\phi}\,\boldsymbol{\lambda}, \tag{11.283}$$

where $r(\tau)$ is the (retarded) distance between the two bodies. And we re-introduce the directions

$$\boldsymbol{e}_X = [\cos\omega, -\sin\omega, 0], \tag{11.284a}$$

$$\boldsymbol{e}_Y = [\cos\iota\sin\omega, \cos\iota\cos\omega, -\sin\iota], \tag{11.284b}$$

$$\boldsymbol{e}_Z = [\sin\iota\sin\omega, \sin\iota\cos\omega, \cos\iota] = \boldsymbol{N}, \tag{11.284c}$$

which depend on the polar angles (ι, ω) that specify the direction of the detector relative to the (x, y, z) system. We use $\boldsymbol{e}_X$ and $\boldsymbol{e}_Y$ as a vectorial basis in the subspace transverse to the direction of propagation, and let

$$h_+ = \frac{1}{2}\left(e_X^je_X^k - e_Y^je_Y^k\right)h_{jk}, \tag{11.285a}$$

$$h_\times = \frac{1}{2}\left(e_X^je_Y^k + e_Y^je_X^k\right)h_{jk}, \tag{11.285b}$$

denote the gravitational-wave polarizations. The construction extends to each A_+ and $A_\times$ defined by Eq. (11.282).

The manipulations that return h_+ and $h_\times$ from Eqs. (11.280) and (11.281) are straightforward, but as usual they are long and tedious. The expressions that result from $A^{jk}[0\text{PN}]$ are simple, and they can easily be displayed here:

$$\begin{aligned} A_+[0\text{PN}] &= \frac{1}{2}\big[\dot{r}^2 + (r\dot{\phi})^2 - (GM/r)\big]S^2 \\ &\quad + \frac{1}{2}\big[\dot{r}^2 - (r\dot{\phi})^2 - (GM/r)\big](1+C^2)\cos 2\psi \\ &\quad - \dot{r}(r\dot{\phi})(1+C^2)\sin 2\psi, \end{aligned} \tag{11.286a}$$

$$A_\times[0\text{PN}] = \big[\dot{r}^2 - (r\dot{\phi})^2 - (GM/r)\big]C\sin 2\psi + 2\dot{r}(r\dot{\phi})C\cos 2\psi, \tag{11.286b}$$

where $S := \sin\iota$, $C := \cos\iota$, and $\psi := \phi(\tau) + \omega$. The polarizations associated with $A^{jk}[\text{tail}]$ are also relatively simple:

$$\begin{aligned} A_+[\text{tail}] &= \frac{Gm}{c^3}S^2\int_0^\infty \left[\frac{Gm}{r^3}\big[2\dot{r}^2 - (r\dot{\phi})^2 + (GM/r)\big]\right]_{\tau-\zeta} \Xi\, d\zeta \\ &\quad + \frac{Gm}{c^3}(1+C^2)\int_0^\infty \left[\frac{Gm}{r^3}\big[2\dot{r}^2 + 7(r\dot{\phi})^2 + (GM/r)\big]\cos 2\psi\right]_{\tau-\zeta} \Xi\, d\zeta \\ &\quad - 10\frac{Gm}{c^3}(1+C^2)\int_0^\infty \left[\frac{Gm}{r^3}\dot{r}(r\dot{\phi})\sin 2\psi\right]_{\tau-\zeta} \Xi\, d\zeta, \end{aligned} \tag{11.287a}$$

$$\begin{aligned} A_\times[\text{tail}] &= 2\frac{Gm}{c^3}C\int_0^\infty \left[\frac{Gm}{r^3}\big[2\dot{r}^2 + 7(r\dot{\phi})^2 + (GM/r)\big]\sin 2\psi\right]_{\tau-\zeta} \Xi\, d\zeta \\ &\quad + 20\frac{Gm}{c^3}C\int_0^\infty \left[\frac{Gm}{r^3}\dot{r}(r\dot{\phi})\cos 2\psi\right]_{\tau-\zeta} \Xi\, d\zeta, \end{aligned} \tag{11.287b}$$

where

$$\Xi := \ln\left(\frac{\zeta}{\zeta + 2R/c}\right) + \frac{11}{12}. \tag{11.288}$$

The expressions for the remaining polarizations are much, much larger, and we shall not display them here.

11.4.6 Specialization to circular orbits

Circular motion

In this subsection we make a further specialization to circular orbital motion. This is defined by the condition

$$\dot{r} = 0, \tag{11.289}$$

so that the two bodies move while maintaining a constant relative separation. This is undoubtedly a restriction on all possible motions, but more than this, Eq. (11.289) is also an approximation, because as the system loses energy to gravitational radiation (an effect examined in Chapter 12), the orbital separation slowly decreases, and $\dot{r}$ should actually be negative even for orbits that are otherwise circular. But because this radiation-reaction effect appears at 2.5PN order in the equations of motion, we are justified to neglect it here.

The post-Newtonian motion of a binary system in circular orbit was described back in Sec. 10.1.2. There we showed that the angular velocity $\Omega := \dot{\phi}$ of an orbit of radius r is a constant given by

$$\Omega^2 = \frac{Gm}{r^3}\left[1 - (3-\eta)\frac{Gm}{c^2 r} + O(c^{-4})\right]. \tag{11.290}$$

This is a post-Newtonian generalization of the usual Keplerian relation $\Omega^2 = Gm/r^3$. (When radiation-reaction effects are included, r slowly decreases, and this causes Ω to slowly increase.) It follows that the orbital velocity $v = r\Omega$ is given by

$$v^2 = \frac{Gm}{r}\left[1 - (3-\eta)\frac{Gm}{c^2 r} + O(c^{-4})\right]. \tag{11.291}$$

Post-Newtonian expansion parameter

The post-Newtonian expansion of the gravitational-wave field is formally an expansion in powers of c^{-1}, but physically it is an expansion in powers of a dimensionless quantity such as v/c. There are many such quantities that could be adopted as an expansion parameter. Equations (11.290) and (11.291) suggest, for example, that $\sqrt{Gm/(c^2 r)}$ could be selected, and this would indeed be a valid substitute for v/c. Another choice is

$$\beta := \left(\frac{Gm\Omega}{c^3}\right)^{1/3}, \tag{11.292}$$

which has the important advantage of being defined in terms of the orbital frequency Ω. As we shall see below, Ω is intimately related to the frequency of the gravitational waves, and it can therefore be measured directly. This is unlike v or r, which are coordinate-dependent variables that cannot be measured. It is easy to show, using Eqs. (11.290) and (11.291), that

$$v/c = \beta\left[1 - \frac{1}{3}(3-\eta)\beta^2 + O(\beta^4)\right] \tag{11.293}$$

and

$$\frac{Gm}{c^2 r} = \beta^2\left[1 + \frac{1}{3}(3-\eta)\beta^2 + O(\beta^4)\right]. \tag{11.294}$$

We shall henceforth adopt β as a meaningful post-Newtonian parameter, and re-express the gravitational-wave polarizations of Eq. (11.282) as expansions in powers of β.

Gravitational-wave polarizations

The polarizations produced by a binary system in circular orbit are obtained by following the general recipe described in Sec. 11.4.5, making use of Eqs. (11.289), (11.293), and

(11.294). After expanding in powers of β and evaluating the tail integrals (as detailed below), we arrive at

$$h_{+,\times} = \frac{2\eta Gm}{c^2 R}\left(\frac{Gm\Omega}{c^3}\right)^{2/3} H_{+,\times}, \tag{11.295a}$$

$$H_{+,\times} := H^{[0]}_{+,\times} + \Delta\beta H^{[1/2]}_{+,\times} + \beta^2 H^{[1]}_{+,\times} + \Delta\beta^3 H^{[3/2]}_{+,\times} + \beta^3 H^{\rm tail}_{+,\times} + O(\beta^4), \tag{11.295b}$$

where

$$H^{[0]}_{+} = -(1+C^2)\cos 2\Psi, \tag{11.296a}$$

$$H^{[1/2]}_{+} = -\frac{1}{8}S(5+C^2)\cos\Psi + \frac{9}{8}S(1+C^2)\cos 3\Psi, \tag{11.296b}$$

$$\begin{aligned} H^{[1]}_{+} &= \frac{1}{6}\Big[(19+9C^2-2C^4) - (19-11C^2-6C^4)\eta\Big]\cos 2\Psi \\ &\quad - \frac{4}{3}(1-3\eta)S^2(1+C^2)\cos 4\Psi, \end{aligned} \tag{11.296c}$$

$$\begin{aligned} H^{[3/2]}_{+} &= \frac{1}{192}S\Big[(57+60C^2-C^4) - 2(49-12C^2-C^4)\eta\Big]\cos\Psi \\ &\quad - \frac{9}{128}S\Big[(73+40C^2-9C^4) - 2(25-8C^2-9C^4)\eta\Big]\cos 3\Psi \\ &\quad + \frac{625}{384}(1-2\eta)S^3(1+C^2)\cos 5\Psi, \end{aligned} \tag{11.296d}$$

$$H^{\rm tail}_{+} = -4(1+C^2)\left\{\frac{\pi}{2}\cos 2\Psi + \big[\gamma + \ln(4\Omega R/c)\big]\sin 2\Psi\right\}, \tag{11.296e}$$

and

$$H^{[0]}_{\times} = -2C\sin 2\Psi, \tag{11.297a}$$

$$H^{[1/2]}_{\times} = -\frac{3}{4}SC\sin\Psi + \frac{9}{4}SC\sin 3\Psi, \tag{11.297b}$$

$$\begin{aligned} H^{[1]}_{\times} &= \frac{1}{3}C\Big[(17-4C^2) - (13-12C^2)\eta\Big]\sin 2\Psi \\ &\quad - \frac{8}{3}(1-3\eta)S^2 C\sin 4\Psi, \end{aligned} \tag{11.297c}$$

$$\begin{aligned} H^{[3/2]}_{\times} &= \frac{1}{96}SC\Big[(63-5C^2) - 2(23-5C^2)\eta\Big]\sin\Psi \\ &\quad - \frac{9}{64}SC\Big[(67-15C^2) - 2(19-15C^2)\eta\Big]\sin 3\Psi \\ &\quad + \frac{625}{192}(1-2\eta)S^3 C\sin 5\Psi, \end{aligned} \tag{11.297d}$$

$$H^{\rm tail}_{\times} = -8C\left\{\frac{\pi}{2}\sin 2\Psi - \big[\gamma + \ln(4\Omega R/c)\big]\cos 2\Psi\right\}, \tag{11.297e}$$

where

$$\Psi := \phi + \omega = \Omega(t - R/c) + \omega. \tag{11.298}$$

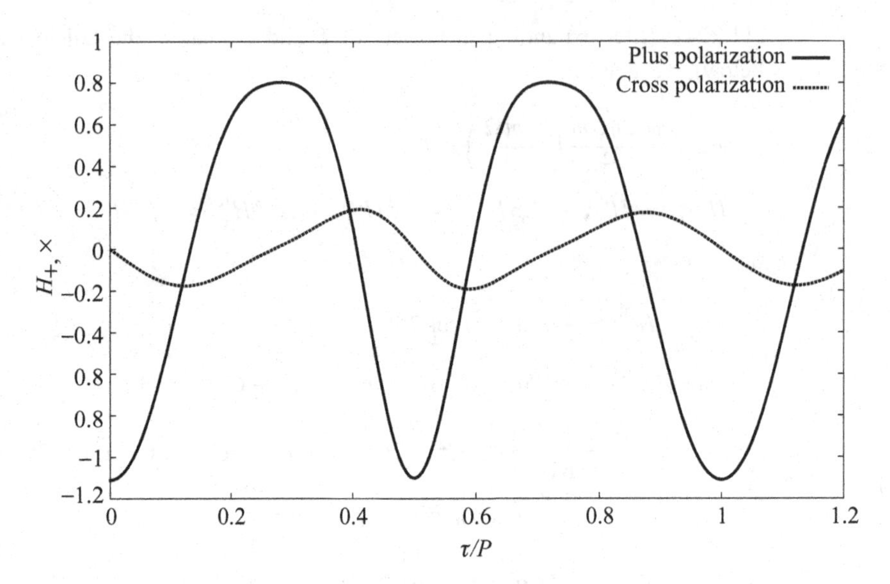

Fig. 11.4 The polarizations H_+ and $H_\times$ as functions of retarded time τ, in units of the orbital period P. The curves are displayed for a mass ratio $M_1/M_2 = 10$, a post-Newtonian parameter $\beta = 0.4$, an inclination angle $\iota = 85°$, and a longitude of pericenter $\omega = 0°$.

We recall that $m = M_1 + M_2$, $\eta = M_1 M_2/m^2$, $\Delta = (M_1 - M_2)/m$, $C := \cos\iota$, and $S := \sin\iota$. Equations (11.296) and (11.297) imply that at leading order in a post-Newtonian expansion, the gravitational wave oscillates at twice the orbital frequency; the post-Newtonian corrections contribute additional frequencies and the signal is therefore modulated. See Fig. 11.4 for an illustration.

The tail terms listed in Eqs. (11.296) and (11.297) are interesting. They involve the mathematical constants π and $\gamma \simeq 0.5772$ (Euler's number), and they also involve a logarithmic term that depends on $\Omega R/c$. The tail terms are best interpreted as giving rise to a correction to Ψ, the quantity that determines the phase of the gravitational wave. Indeed, it is a simple matter to show that the Newtonian and tail contributions to h_+ and $h_\times$ can be combined and expressed as

$$H_+^{[0]} + \beta^3 H_+^{\rm tail} = -(1 + C^2)\big(1 + 2\pi\beta^3\big)\cos 2\Psi^*, \tag{11.299a}$$

$$H_\times^{[0]} + \beta^3 H_\times^{\rm tail} = -2C\big(1 + 2\pi\beta^3\big)\sin 2\Psi^*. \tag{11.299b}$$

These expressions involve an amplitude correction equal to $2\pi\beta^3$, and a new phase function given by

$$\Psi^* = \Psi - 2\beta^3\big[\gamma + \ln(4\Omega R/c)\big] = \Omega\left(t - R/c - \frac{2Gm}{c^3}\ln\frac{4\Omega R}{c} + \text{constant}\right). \tag{11.300}$$

It is this shifted phase function that informs us, at long last, that the radiation propagates not along the mathematical light cones of Minkowski spacetime, but along the true, physical light cones of a curved spacetime. Indeed, the logarithmic term in Eq. (11.300) represents the familiar Shapiro time delay, the extra time required by a light wave, or a gravitational wave, to climb up a gravitational potential well created by a distribution of matter with total mass m. This effect was studied back in Sec. 10.2.5, and apart from irrelevant constant factors, the term $(2Gm/c^3)\ln R$ can be seen to originate from Eq. (10.100), in the special case in which the wave travels in the radial direction, so that $r_{\rm obs} + \boldsymbol{r}_{\rm obs}\cdot\boldsymbol{k} = 2r_{\rm obs}$.

Evaluation of the tail integrals

We must still evaluate the tail integrals, and show that they lead to the results displayed in Eqs. (11.296) and (11.297). We start with Eqs. (11.287), which we specialize to circular orbits by involving Eqs. (11.289), (11.293), and (11.294). After conversion to the notation of Eq. (11.295), we find that

$$H_+^{\rm tail} = 8(1+C^2)\Omega \int_0^\infty \cos(2\Psi - 2\Omega\zeta)\biggl[\ln\frac{\zeta}{\zeta + 2R/c} + \frac{11}{12}\biggr]\,d\zeta, \tag{11.301a}$$

$$H_\times^{\rm tail} = 16C\Omega \int_0^\infty \sin(2\Psi - 2\Omega\zeta)\biggl[\ln\frac{\zeta}{\zeta + 2R/c} + \frac{11}{12}\biggr]\,d\zeta. \tag{11.301b}$$

To evaluate this we change the variable of integration to $y := 2\Omega\zeta$ and introduce $k := 4\Omega R/c$ to simplify the notation. The tail integrals become

$$H_+^{\rm tail} = 4(1+C^2) \int_0^\infty \cos(2\Psi - y)\biggl[\ln\frac{y}{y+k} + \frac{11}{12}\biggr]\,dy, \tag{11.302a}$$

$$H_\times^{\rm tail} = 8C \int_0^\infty \sin(2\Psi - y)\biggl[\ln\frac{y}{y+k} + \frac{11}{12}\biggr]\,dy. \tag{11.302b}$$

Expanding the trigonometric functions, this is

$$H_+^{\rm tail} = 4(1+C^2)\bigl(J_{\rm c}\cos 2\Psi + J_{\rm s}\sin 2\Psi\bigr), \tag{11.303a}$$

$$H_\times^{\rm tail} = 8C\bigl(J_{\rm c}\sin 2\Psi - J_{\rm s}\cos 2\Psi\bigr), \tag{11.303b}$$

where

$$J_{\rm c} := \int_0^\infty \cos(y)\biggl[\ln\frac{y}{y+k} + \frac{11}{12}\biggr]\,dy, \tag{11.304a}$$

$$J_{\rm s} := \int_0^\infty \sin(y)\biggl[\ln\frac{y}{y+k} + \frac{11}{12}\biggr]\,dy. \tag{11.304b}$$

These integrals are ill-defined, because the function within square brackets behaves as $\frac{11}{12} - k/y$ for large y, and the constant term prevents the convergence of each integral. This, however, is an artificial problem that comes as a consequence of our (unphysical) approximation $\Omega =$ constant. In reality, the two-body system undergoes radiation reaction, and Ω slowly decreases as ζ increases toward ∞. (Recall that r decreases as time increases, which causes Ω to increase as time increases; but recall also that the tail term integrates towards the past, so that Ω decreases as ζ increases.) This effect does not alter substantially the logarithmic portion of the integral, but it is sufficient to ensure the convergence of the constant term.

The integrals can be defined properly by inserting a convergence factor within the integrand. Alternatively, and this practice is consistent with what was done back in Sec. 11.3.7, we can integrate by parts and simply discard an ambiguous (and unphysical) boundary term at $y = \infty$. Proceeding along those lines, we find that our integrals are equivalent to

$$J_c = -\int_0^\infty \frac{k \sin y}{y(y+k)}\, dy, \tag{11.305a}$$

$$J_s = \int_0^\infty \frac{k(\cos y - 1)}{y(y+k)}\, dy, \tag{11.305b}$$

which are well defined. They can even be evaluated in closed form:

$$\begin{aligned} J_c &= -\frac{\pi}{2} + \frac{\pi}{2}\cos k + \mathrm{Ci}(k)\sin k - \mathrm{Si}(k)\cos k \\ &= -\frac{\pi}{2} + O(k^{-1}), \end{aligned} \tag{11.306a}$$

$$\begin{aligned} J_s &= -\gamma - \ln k - \frac{\pi}{2}\sin k + \mathrm{Ci}(k)\cos k + \mathrm{Si}(k)\sin k \\ &= -\gamma - \ln k + O(k^{-2}), \end{aligned} \tag{11.306b}$$

where γ is Euler's constant, $\mathrm{Ci}(k)$ is the cosine integral, and $\mathrm{Si}(k)$ is the sine integral (defined, for example, in Sec. 5.2 of Abramowitz and Stegun's *Handbook of Mathematical Functions* (1975)). The approximate forms neglect terms of order $k^{-1} = (4\Omega R/c)^{-1} \sim (\lambda_c/R)$ and smaller, and these are small by virtue of the fact that the gravitational-wave field is evaluated in the far-away wave zone, where $R \gg \lambda_c$.

Collecting results, we find that

$$H_+^{\rm tail} = -4(1+C^2)\left\{\frac{\pi}{2}\cos 2\Psi + \left[\gamma + \ln(4\Omega R/c)\right]\sin 2\Psi\right\}, \tag{11.307a}$$

$$H_\times^{\rm tail} = -8C\left\{\frac{\pi}{2}\sin 2\Psi - \left[\gamma + \ln(4\Omega R/c)\right]\cos 2\Psi\right\}, \tag{11.307b}$$

and these expressions have already been presented in Eqs. (11.296) and (11.297).

11.4.7 Beyond 1.5PN order

The calculations of this section were long and arduous, but as it turns out, they were merely child's play. At the time of writing, the gravitational waves for binary systems in circular motion have been calculated all the way out to 3.5PN order, and this is a much, much larger challenge. At 2PN order, for example, one finds not only the expected "standard" corrections of order β^4, but also tail contributions generated by the 0.5PN order terms. At 2.5PN order one finds tails generated by the 1PN terms, 1PN corrections to the 1.5PN tail terms, as well as standard 2.5PN terms. At 3PN order there are, in addition to the standard terms, tails generated by the normal 1.5PN terms, 1.5PN corrections to the 1.5PN tail terms, and completely new "tails of tails" terms: tails generated by the 1.5PN tails. These formidable calculations have been carried out by a number of groups around the world, at an enormous cost of labor and sweat (perhaps even blood). There was a strong motivation behind this large effort: the measurement by laser interferometric detectors of gravitational waves emitted by compact binary systems involving neutron stars or black holes relies in an essential way on these very accurate theoretical predictions, which allow the extremely weak signals to be distinguished from noise. Data analysis relies on a bank of templates constructed from these waveforms, and cross-correlation of the detector output with the templates can reveal a signal that would otherwise be lost in the noisy data stream. In this way theorists, who build the templates, work hand in hand with experimentalists, who build the detectors, toward the successful measurement of gravitational waves.

11.5 Gravitational waves and laser interferometers

Thus far, in this long chapter on gravitational waves, we have said very little about the actual detection of these waves, and indeed we intend to leave it that way. The attempts to detect gravitational waves, from the pioneering experiments carried out by Joseph Weber in the 1960s and 1970s using suspended cylinders of aluminum, to the present international effort involving laser interferometry, pulsar timing, and cosmic microwave background observations, is a story rich in sociology, history, technological development, and big-science politics. But it is not the main focus of this book. We refer readers who wish to learn more about the detection aspects to a number of excellent resources, listed at the end of this chapter.

Having come this far, however, and having produced the waveforms h_+ and $h_\times$ in a ready-to-use form for various sources and in various approximations, our coverage would seem incomplete if we did not make some attempt to connect them with the output of a gravitational-wave detector. It therefore seems appropriate to conclude this chapter by showing how h_+ and $h_\times$ can be measured in one of the leading approaches to gravitational-wave detection, laser interferometry.

In its most schematic realization, a laser interferometric gravitational-wave detector works just like the interferometer used by Michelson in the late 1800s to measure the speed of light and search for evidence of an "aether." The real-life interferometers at the Earth-based LIGO, Virgo, Geo600, and KAGRA observatories, and the one envisioned for a space-based detector (known in 2013 as eLISA), are much more sophisticated than this, but this simple model is adequate and captures the essential physics.

A laser interferometer consists of a laser source, a beam splitter, and two end mirrors mounted on test masses imagined to be freely moving in spacetime (although in reality they can be suspended by thin wires). The arms of the interferometers are taken to be perpendicular to each other, although as we shall see, this is not an essential feature of the design. The laser beam is divided in two at the beam splitter, and each beam travels along one arm of the interferometer, reflects off the test mass, and returns to the beam splitter to be recombined with the other beam. The relative phase of the beams determines whether they produce a bright or dark spot when the recombined beam is measured by a photon detector. Since the initial phases at the beam splitter are identical, the phase difference $\Delta\Phi$

depends on the difference in travel time along the two arms. We can write

$$\Delta\Phi = 2\pi\nu(2L_1/c - 2L_2/c), \tag{11.308}$$

where ν is the frequency of the laser light, L_1 and L_2 are the armlengths, and $2L_1/c$ and $2L_2/c$ are the travel times along each arm (forward and back).

With the origin of the coordinate system placed at the beam splitter, the test mass at the end of the first arm is at a position $\boldsymbol{\xi}_1(t)$, and the test mass at the end of the second arm is at $\boldsymbol{\xi}_2(t)$. In the absence of a gravitational wave, the arms would have equal lengths, and we would have $\boldsymbol{\xi}_1 = L_0\boldsymbol{e}_1$ and $\boldsymbol{\xi}_2 = L_0\boldsymbol{e}_2$, in which L_0 is the unperturbed length of each arm, and $\boldsymbol{e}_1$, $\boldsymbol{e}_2$ are unit vectors pointing in the direction of each arm. In the presence of a gravitational wave, the position of each test mass varies with time. Assuming that the armlength L_0 is much shorter than the wavelength λ of the gravitational wave, the displacement is described by Eq. (11.29), and we have

$$\xi_1^j = L_0\left(e_1^j + \frac{1}{2}h_{\mathrm{TT}}^{jk}e_1^k\right), \tag{11.309a}$$

$$\xi_2^j = L_0\left(e_2^j + \frac{1}{2}h_{\mathrm{TT}}^{jk}e_2^k\right). \tag{11.309b}$$

The length of each arm is then given by

$$L_1 = L_0\left(1 + \frac{1}{2}h_{\mathrm{TT}}^{jk}e_1^j e_1^k\right), \tag{11.310a}$$

$$L_2 = L_0\left(1 + \frac{1}{2}h_{\mathrm{TT}}^{jk}e_2^j e_2^k\right), \tag{11.310b}$$

to first order in h_{TT}^{jk}, and the phase difference at beam recombination is

$$\Delta\Phi = \frac{4\pi\nu L_0}{c}\frac{1}{2}\left(e_1^j e_1^k - e_2^j e_2^k\right)h_{\mathrm{TT}}^{jk}. \tag{11.311}$$

If we express the gravitational-wave field as in Eq. (11.6), $h^{jk} = (G/c^4R)A^{jk}$, this is

$$\Delta\Phi = \frac{4\pi\nu G L_0}{c^5 R}S(t), \tag{11.312}$$

in which

$$S(t) = \frac{1}{2}\left(e_1^j e_1^k - e_2^j e_2^k\right)A_{\mathrm{TT}}^{jk}(\tau, \boldsymbol{N}) \tag{11.313}$$

is the detector's response function. We recall that R is the distance to the source, that $\boldsymbol{N}$ is a unit vector pointing from the source to the detector, and that $\tau := t - R/c$ is retarded time.

To calculate $S(t)$ we decompose A_{TT}^{jk} in a transverse basis formed by the unit vectors $\boldsymbol{e}_X$ and $\boldsymbol{e}_Y$, which are perpendicular to the direction of propagation $\boldsymbol{e}_Z = \boldsymbol{N}$. This decomposition was detailed back in Sec. 11.1.7, and we use the notation introduced in Sec. 11.2.2. We have that

$$A_{\mathrm{TT}}^{jk} = \left(e_X^j e_X^k - e_Y^j e_Y^k\right)A_+ + \left(e_X^j e_Y^k + e_Y^j e_X^k\right)A_\times, \tag{11.314}$$

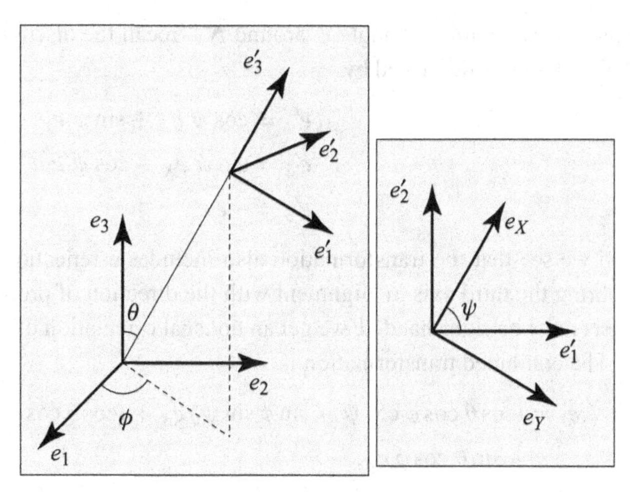

Fig. 11.5 Relation between the detector basis ($\boldsymbol{e}_1$, $\boldsymbol{e}_2$) and the transverse basis ($\boldsymbol{e}_X$, $\boldsymbol{e}_Y$).

and substitution within Eq. (11.313) produces

$$S(t) = F_+ A_+(t - R/c) + F_\times A_\times(t - R/c), \tag{11.315}$$

in which

$$F_+ := \frac{1}{2}\left(e_1^j e_1^k - e_2^j e_2^k\right)\left(e_X^j e_X^k - e_Y^j e_Y^k\right), \tag{11.316a}$$

$$F_\times := \frac{1}{2}\left(e_1^j e_1^k - e_2^j e_2^k\right)\left(e_X^j e_Y^k + e_Y^j e_X^k\right), \tag{11.316b}$$

are the *detector pattern functions* of the laser interferometer, which describe the angular response of the detector to each gravitational-wave polarization. Note that in general, the detector measures a linear superposition of the gravitational-wave polarizations.

To calculate F_+ and $F_\times$ we must relate the detector basis $\boldsymbol{e}_1$ and $\boldsymbol{e}_2$ to the transverse basis $\boldsymbol{e}_X$ and $\boldsymbol{e}_Y$. We imagine that when viewed from the detector's vantage point, the source of gravitational waves is situated in a direction $-\boldsymbol{N} = [\sin\theta\cos\phi, \sin\theta\sin\phi, \cos\theta]$, described by polar angles (θ, ϕ) defined relative to the detector basis. The vectors $(\boldsymbol{e}_1, \boldsymbol{e}_2, \boldsymbol{e}_3)$ can then be related to $(\boldsymbol{e}_X, \boldsymbol{e}_Y, \boldsymbol{e}_Z)$ by a sequence of simple operations illustrated in Fig. 11.5. From the detector basis we first form an intermediate basis $(\boldsymbol{e}'_1, \boldsymbol{e}'_2, \boldsymbol{e}'_3)$ by performing two elementary rotations. The first is a rotation by an angle ϕ around the $\boldsymbol{e}_3$-axis, to align the rotated $\boldsymbol{e}_1$-axis in the direction of $-\boldsymbol{N}$ projected down to the 1-2 plane. The second is a rotation by an angle θ around the new $\boldsymbol{e}_2$-axis, to align the rotated $\boldsymbol{e}_3$-axis in the direction of $-\boldsymbol{N}$. It is easy to show that the detector basis is related to the intermediate basis by

$$\boldsymbol{e}_1 = \cos\theta\cos\phi\,\boldsymbol{e}'_1 - \sin\phi\,\boldsymbol{e}'_2 + \sin\theta\cos\phi\,\boldsymbol{e}'_3, \tag{11.317a}$$

$$\boldsymbol{e}_2 = \cos\theta\sin\phi\,\boldsymbol{e}'_1 + \cos\phi\,\boldsymbol{e}'_2 + \sin\theta\sin\phi\,\boldsymbol{e}'_3, \tag{11.317b}$$

$$\boldsymbol{e}_3 = -\sin\theta\,\boldsymbol{e}'_1 + \cos\theta\,\boldsymbol{e}'_3. \tag{11.317c}$$

The vectors $\boldsymbol{e}'_1$ and $\boldsymbol{e}'_2$ are transverse to the direction of propagation $\boldsymbol{N} = -\boldsymbol{e}'_3$, but they are not equal to the transverse vectors $\boldsymbol{e}_X$ and $\boldsymbol{e}_Y$. Indeed, these vectors will in general be related by a rotation of angle ψ around $\boldsymbol{N}$ – recall the discussion surrounding Eq. (11.41). This rotation is described by

$$e'_1 = \cos\psi\, \boldsymbol{e}_X + \sin\psi\, \boldsymbol{e}_Y, \tag{11.318a}$$

$$e'_2 = \sin\psi\, \boldsymbol{e}_X - \cos\psi\, \boldsymbol{e}_Y, \tag{11.318b}$$

$$e'_3 = -\boldsymbol{e}_Z, \tag{11.318c}$$

and we see that the transformation also includes a reflection across the transverse plane, to bring the third axis in alignment with the direction of propagation. Since each vectorial basis must be right-handed, we get an unusual orientation of the $\boldsymbol{e}_Y$ vector relative to $\boldsymbol{e}'_2$.

The combined transformation is

$$\boldsymbol{e}_1 = (\cos\theta\cos\phi\cos\psi - \sin\phi\sin\psi)\,\boldsymbol{e}_X + (\cos\theta\cos\phi\sin\psi + \sin\phi\cos\psi)\,\boldsymbol{e}_Y - \sin\theta\cos\phi\,\boldsymbol{e}_Z, \tag{11.319a}$$

$$\boldsymbol{e}_2 = (\cos\theta\sin\phi\cos\psi + \cos\phi\sin\psi)\,\boldsymbol{e}_X + (\cos\theta\sin\phi\sin\psi - \cos\phi\cos\psi)\,\boldsymbol{e}_Y - \sin\theta\sin\phi\,\boldsymbol{e}_Z, \tag{11.319b}$$

$$\boldsymbol{e}_3 = -\sin\theta\cos\psi\,\boldsymbol{e}_X - \sin\theta\sin\psi\,\boldsymbol{e}_Y - \cos\theta\,\boldsymbol{e}_Z, \tag{11.319c}$$

and making the substitutions in Eq. (11.316) returns

$$F_+ = \frac{1}{2}(1+\cos^2\theta)\cos 2\phi\cos 2\psi - \cos\theta\sin 2\phi\sin 2\psi, \tag{11.320a}$$

$$F_\times = \frac{1}{2}(1+\cos^2\theta)\cos 2\phi\sin 2\psi + \cos\theta\sin 2\phi\cos 2\psi, \tag{11.320b}$$

after simplification. We see that there are directions in the sky, for example $\theta = \frac{\pi}{2}$ and $\phi = \frac{\pi}{4}$, for which the laser interferometer is unable to detect any wave.

Measurement of gravitational waves with laser interferometry does not require the arms to be perpendicular to each other, as we have taken them to be in this discussion. For the proposed space-based eLISA interferometer, for example, the angle between the arms will be 60°. In fact, it can be shown (see Exercise 11.8) that for an interferometer whose arms make an angle χ, the response function $S(t)$ is the same as in Eqs. (11.315) and (11.320), but with the overall response reduced by a factor of $\sin\chi$. This simple result follows when the arms are oriented symmetrically in the laboratory basis, so that each arm makes the same angle $\frac{\pi}{4} - \frac{1}{2}\chi$ with respect to the $\boldsymbol{e}_1$ and $\boldsymbol{e}_2$ axes.

11.6 Bibliographical notes

The physics of gravitational waves features a wealth of aspects that could not all be covered in this book, even in such a long chapter. For a more comprehensive treatment the reader is invited to consult the few books devoted entirely to this rich subject, including Saulson (1994), Maggiore (2007), and Creighton and Anderson (2011).

The quadrupole-formula controversy described in Box 11.2 is related in much more detail in Daniel Kennefick's wonderful book *Traveling at the Speed of Thought*, published in 2007; the phrase is due to Eddington (1922), who was referring specifically to the unphysical gauge modes contained in the theory. The Einstein–Rosen paper eventually appeared in 1937 in the *Journal of the Franklin Institute*, a publication that specializes in engineering and applied mathematics; the identity of the mysterious *Physical Review* referee was finally revealed by Kennefick in his 2005 *Physics Today* article. Another reference cited in Box 11.2 is Ehlers *et al.* (1976), and a useful summary of the controversy is presented in Walker and Will (1980).

The gravitational-wave polarizations corresponding to a binary system in eccentric motion (Sec. 11.2.2) were first calculated by Wahlquist (1987). The elastic deformation of a neutron star and its potential for gravitational-wave emissions (Sec. 11.2.3) has been investigated by a number of researchers; representative papers are Ushomirsky, Cutler and Bildsten (2000) and Owen (2005). Gravitational waves produced during tidal encounters (Sec. 11.2.4) were first studied by Turner (1977) and Will (1983).

Wagoner and Will (1976) were the first to calculate the post-Newtonian corrections to the gravitational-wave signal of a binary system, though with considerably less rigor than displayed in Secs. 11.3 and 11.4. Our calculations are patterned after Will and Wiseman (1996), who actually carry them out through second post-Newtonian order; the final results for the polarizations h_+ and $h_\times$ (through 2PN order) are neatly presented in Blanchet *et al.* (1996). Higher-order post-Newtonian calculations are reviewed in Blanchet (2006), and at the time of writing, the most recent results on 3.5PN waveforms were obtained by Faye *et al.* (2012).

The physics of gravitational-wave detectors, touched upon ever so briefly in Sec. 11.5, is described much more thoroughly in Saulson (1994), Maggiore (2007), and Creighton and Anderson (2011). A nice introduction to the workings of a laser interferometric detector is provided by Black and Gutenkunst (2003).

11.7 Exercises

11.1 Consider a gravitational-wave field $h^{\alpha\beta}$ in the far-away wave zone, satisfying the harmonic gauge condition. Prove by direct calculation that

$$R_{0j0k} = -\frac{1}{2c^2}(\mathrm{TT})^{jk}{}_{pq}\,\partial_{\tau\tau}h^{pq}\,.$$

11.2 We know that in the far-away wave zone, the effective energy-momentum pseudotensor falls off at least as fast as R^{-2}. Thus we can write the relaxed Einstein equation in harmonic gauge in the form

$$\Box h^{\alpha\beta} = O(R^{-2})\,.$$

Show that the general solution of this equation is given by

$$h^{\alpha\beta} = \frac{A^{\alpha\beta}(\tau, \boldsymbol{N})}{R} + O(R^{-2})\,,$$

where $A^{\alpha\beta}$ is an arbitrary function of $\tau = t - R/c$ and the unit vector $\boldsymbol{N}$. Show that the harmonic gauge condition $\partial_\beta h^{\alpha\beta} = 0$ gives rise to the constraint

$$\partial_\tau A^{\alpha 0} - N_j \partial_\tau A^{\alpha j} = O(R^{-1}).$$

11.3 An alternative way to study the polarizations of gravitational waves in the far-away wave zone is to focus on the Riemann tensor, and to exploit the fact that the waves, to lowest order in post-Minkowskian theory, propagate along null directions with respect to the background Minkowski spacetime. The idea, following Ted Newman and Roger Penrose, is to express the components of $R_{\alpha\beta\gamma\delta}$ on a basis of complex null vectors, defined by

$$\ell^\alpha := (1, \boldsymbol{N}), \qquad n^\alpha := \tfrac{1}{2}(1, -\boldsymbol{N}),$$
$$m^\alpha := \frac{1}{\sqrt{2}}(0, \boldsymbol{\vartheta} + i\boldsymbol{\varphi}), \qquad \bar{m}^\alpha := \frac{1}{\sqrt{2}}(0, \boldsymbol{\vartheta} - i\boldsymbol{\varphi}).$$

Here ℓ^α is an outgoing null vector tangent to the gravitational waves, n^α is an ingoing null vector, and $\boldsymbol{\vartheta}$ and $\boldsymbol{\varphi}$ are defined as in Eqs. (11.37) and (11.38). Complex conjugation converts m^α to $\bar{m}^\alpha$ and vice versa.

(a) Prove the following properties of the basis vectors:

$$\ell_\alpha = -c\partial_\alpha(t - R/c), \qquad n_\alpha = -\tfrac{1}{2}c\partial_\alpha(t + R/c),$$
$$\ell_\alpha \ell^\alpha = n_\alpha n^\alpha = m_\alpha m^\alpha = \bar{m}_\alpha \bar{m}^\alpha = 0,$$
$$\ell_\alpha n^\alpha = -1, \qquad m_\alpha \bar{m}^\alpha = 1,$$
$$\eta^{\alpha\beta} = -2\ell^{(\alpha} n^{\beta)} + 2m^{(\alpha}\bar{m}^{\beta)}.$$

(b) Assume that the Riemann tensor in the far-away wave zone can be expressed as $R_{\alpha\beta\gamma\delta} = A_{\alpha\beta\gamma\delta}/R + O(R^{-2})$, in which $A_{\alpha\beta\gamma\delta}$ is an arbitrary function of retarded time $\tau := t - R/c$ and the unit vector $\boldsymbol{N}$. Show that

$$\partial_\mu R_{\alpha\beta\gamma\delta} = -\frac{1}{c}\ell_\mu \partial_\tau R_{\alpha\beta\gamma\delta} + O(R^{-2}).$$

(c) Making use of this differentiation rule, use the linearized Bianchi identities

$$\partial_\epsilon R_{\alpha\beta\gamma\delta} + \partial_\delta R_{\alpha\beta\epsilon\gamma} + \partial_\gamma R_{\alpha\beta\delta\epsilon} = 0$$

to show that only the six components R_{npnq} can be non-zero, where the indices (p, q) run over the values ℓ, m, and $\bar{m}$. In this notation, for example, $R_{n\ell n\ell}$ stands for $R_{\alpha\beta\gamma\delta} n^\alpha \ell^\beta n^\gamma \ell^\delta$. You may ignore any constant of integration that arises when integrating with respect to retarded time.

(d) Calculate the Ricci tensor, and show that the vacuum Einstein field equations give rise to the four additional constraints

$$R_{n\ell n\ell} = R_{n\ell nm} = R_{n\ell n\bar{m}} = R_{nmn\bar{m}} = 0.$$

Show that there are only two unconstrained components, represented by R_{nmnm} and its complex conjugate (or equivalently, by its real and imaginary parts). These are the gravitational-wave modes, as represented by the Riemann tensor.

(e) Show that the link between the remaining components of the Riemann tensor and the gravitational-wave polarizations is provided by

$$R_{nmnm} = -\frac{1}{2c^2}\partial_{\tau\tau}h_{mm} = -\frac{1}{2c^2}\partial_{\tau\tau}(h_+ + ih_\times).$$

11.4 Consider the Landau–Lifshitz formulation of the Einstein field equations, as reviewed in Sec. 6.1.1. Assuming that the Landau–Lifshitz pseudotensor falls off as R^{-2} in the far-away wave zone, the vacuum field equations can be expressed as

$$\partial_{\mu\nu}H^{\alpha\mu\beta\nu} = O(R^{-2}),$$

where $H^{\alpha\mu\beta\nu} := \mathfrak{g}^{\alpha\beta}\mathfrak{g}^{\mu\nu} - \mathfrak{g}^{\alpha\mu}\mathfrak{g}^{\beta\nu}$. We wish to formulate these equations in a gauge that is not harmonic. Instead we choose to impose the gauge conditions

$$h^{0j} = 0, \qquad \eta_{\mu\nu}h^{\mu\nu} = 0,$$

in which $h^{\alpha\beta} := \eta^{\alpha\beta} - \mathfrak{g}^{\alpha\beta}$.

(a) Write out the (00), $(0j)$ and (jk) field equations explicitly, using the gauge conditions to simplify the expressions. Keep the equations linear in $h^{\alpha\beta}$, and set the right-hand sides equal to zero; the $O(R^{-2})$ residuals are not important for this problem.

(b) Show that the field equations are invariant under a further gauge transformation described by

$$h^{\alpha\beta} \to h^{\alpha\beta} - \partial^\alpha\zeta^\beta - \partial^\beta\zeta^\alpha + (\partial_\mu\zeta^\mu)\eta^{\alpha\beta},$$

in which ζ^α is chosen to preserve the four conditions already adopted.

(c) Show that the residual gauge freedom can be exploited to express the gravitational potentials in the TT gauge.

(d) Show that the field equations then reduce to wave equations for $h^{jk}_{\rm TT}$.

11.5 In this problem we examine the response of a free particle to a gravitational wave propagating in the z-direction, in the reference frame of an observer. In the absence of other forces, the particle's motion relative to the observer is governed by Eq. (11.28). Let the particle undergo a small displacement relative to an unperturbed location at a distance L from the observer. This is described by $\xi^j(t) = (L + \delta\xi)e^j$, in which $\delta\xi$ depends on time, and the direction e^j is expressed in terms of polar angles θ and ϕ.

(a) Show that

$$\frac{d}{dt}\delta\xi = \frac{1}{2}L\sin^2\theta\big(\cos 2\phi\,\partial_\tau h_+ + \sin 2\phi\,\partial_\tau h_\times\big),$$

where h_+ and $h_\times$ are the gravitational-wave polarizations as measured in the observer's reference frame.

(b) Letting $h_+ = A_+\cos\omega t$ and $h_\times = A_\times\sin\omega t$, calculate $\frac{1}{2}m\langle(d\delta\xi/dt)^2\rangle$, the time-averaged kinetic energy of the particle, as a function of A_+, $A_\times$, ω, and direction.

(c) The "antenna pattern" of this gravitational-wave detector is defined to be the averaged kinetic energy acquired by the particle for a given orientation relative to the incident wave, divided by the maximum kinetic energy. Plot the antenna pattern as a function of θ and ϕ for each polarization. Present them as parametric plots, either as 3D plots or as 2D plots in various planes. Discuss the properties of the patterns.

11.6 Consider an array of particles that are able to move freely in the x-y plane. A gravitational wave impinges on the plane in the z direction. It is described by polarizations h_+ and $h_\times$ defined relative to the x-y-z basis.

(a) Calculate the acceleration field $\ddot{\boldsymbol{\xi}}$ experienced by the particles. Draw the lines of force in the x-y plane when the wave is a pure $+$ polarization, and when it is a pure $\times$ polarization. How does the pattern change when the wave is a linear superposition of each polarization?

(b) Show that the local surface density of the particles is not affected by the gravitational wave, to first order in h_+ and $h_\times$. *Hint:* Evaluate the divergence of the displacement velocity field, $\boldsymbol{\nabla} \cdot \dot{\boldsymbol{\xi}}$.

(c) Show that the integral of the acceleration field around a closed path in the x-y plane always vanishes. Conclude that the acceleration field can be expressed as the gradient of a potential $\Phi_{\rm GW}$,

$$\ddot{\boldsymbol{\xi}} = \boldsymbol{\nabla} \Phi_{\rm GW} .$$

Determine $\Phi_{\rm GW}$ in terms of h_+ and $h_\times$.

11.7 The gravitational analogue of electromagnetic bremsstrahlung is a process in which a body of mass m_1 passes by a body of mass m_2 and is scattered by a small angle. This is the limit in which $v^2 \gg Gm/b$, where m is the total mass and b is the distance of closest approach. We still assume that $v \ll c$, and in this problem we employ the quadrupole formula to calculate the gravitational waves produced by the encounter.

The process corresponds to a Newtonian hyperbolic orbit with a very large eccentricity $e \gg 1$. (For $e > 1$ the semi-major axis a is not defined, but the semilatus rectum p is related as always to h, the angular momentum per unit reduced mass, by $h^2 = Gmp$.) We introduce the velocity at infinity defined by $v_\infty^2 := 2\varepsilon$, where ε is the conserved energy per unit reduced mass, and we define the impact parameter $b := p/e$.

(a) Using the Keplerian orbit formulae derived in Chapter 3, establish the following relations, assuming that the orbit is confined to the x-y plane, and that the orbit's pericenter is aligned with the x-direction (so that $\omega = 0$):

$$v_\infty = \sqrt{\frac{Gm}{p}}\, e \left[1 - \frac{1}{2} e^{-2} + O(e^{-4}) \right],$$

$$r = \frac{b}{\cos\phi} \left[1 - \frac{1}{e \cos\phi} + O(e^{-2}) \right],$$

$$\boldsymbol{v} = v_\infty \big[-e^{-1} \sin\phi, 1 + e^{-1} \cos\phi, 0 \big] + O(e^{-2}) .$$

(b) Integrate the orbital equation for ϕ to leading order in e^{-1}, and show that

$$\sin\phi = \frac{v_\infty t}{(b^2 + v_\infty^2 t^2)^{1/2}} + O(e^{-1}), \qquad \cos\phi = \frac{b}{(b^2 + v_\infty^2 t^2)^{1/2}} + O(e^{-1}) .$$

(c) Using the quadrupole formula, and taking the waves to be propagating in the direction of the vector $\boldsymbol{N} = [\sin\vartheta\cos\varphi, \sin\vartheta\sin\varphi, \cos\vartheta]$, show that the gravitational-wave polarizations are given by

$$h_{+,\times} = \frac{2\eta(Gm)^2}{c^4 bR} A_{+,\times},$$

in which $\eta := m_1 m_2/m^2$ and

$$\begin{aligned} A_+ &= -\frac{1}{2}(1+\cos^2\vartheta)\big[\cos 2\varphi\,(C_1+2C_3) + 2\sin 2\varphi\,(S_1+S_3)\big] \\ &\quad - \frac{1}{2}\sin^2\vartheta\, C_1, \\ A_\times &= -\cos\vartheta\big[2\cos 2\varphi\,(S_1+S_3) - \sin 2\varphi\,(C_1+2C_3)\big], \end{aligned}$$

where $C_n := \cos^n\phi$ and $S_n := \sin\phi\cos^{n-1}\phi$. An unobservable constant contribution to $h_{+,\times}$ has been dropped.

(d) Plot A_+ and $A_\times$ as a function of time in units of $t_0 = b/v_\infty$ for the following sets of directions (in degrees): $(\vartheta, \varphi) = (0, 0), (45, 0), (90, 0), (90, 45), (90, 90), (45, 90)$, and $(60, 54.7)$, the last point corresponding to a direction in a plane tilted 45 degrees relative to the orbital plane, and 45 degrees from the y-direction in this plane. Running the plots from $t = -10t_0$ to $t = +10t_0$ will reveal the salient features.

(e) Some of the waveforms have an unusual feature. What is it? Discuss whether it might be observable to any practical gravitational wave detector.

11.8 Show that the angular pattern functions for an interferometer whose arms make an angle χ with each other are the same as in Eqs. (11.320), but multiplied by $\sin\chi$. *Hint*: Orient the arms in the 1-2 plane so that each one makes an angle $\frac{\pi}{4} - \frac{1}{2}\chi$ with respect to the $\boldsymbol{e}_1$ and $\boldsymbol{e}_2$ axes.

12 Radiative losses and radiation reaction

In Chapters 8, 9, and 10 we examined gravitational phenomena that take place in the near zone, the region of space which contains the source of the gravitational field, and which is confined to a radius R that is much smaller than λ_c, the characteristic wavelength of the emitted radiation. This near-zone physics excluded radiative phenomena, and the dynamics of the system was entirely conservative. In Chapter 11 we moved to the wave zone, situated at a distance R that is much larger than λ_c, and studied the gravitational waves produced by processes taking place in the near zone; this wave-zone physics is all about radiative phenomena. In the first part of this chapter we continue our exploration of wave-zone physics by describing how gravitational waves transport energy, momentum, and angular momentum away from their source. These radiative losses imply that the near-zone physics cannot be strictly conservative, and in the second part of the chapter we identify the radiation-reaction forces which produce the required dissipation within the system. This chapter, therefore, is all about the linkage between the near and wave zones.

Radiative losses and radiation reaction are subtle topics in general relativity, and the mathematical description of these phenomena is technically demanding. To ease our entry into this subject, in Sec. 12.1 we first review the situation in the simpler context of flat-spacetime electromagnetism. We return to gravity in Sec. 12.2, in which we develop a general description of radiative losses in general relativity. This relies on two major pillars: the Landau–Lifshitz formulation of the Einstein field equations, as reviewed in Chapter 6, and a shortwave approximation, in which the gravitational potentials are expanded in powers of $\lambda_c/R \ll 1$; this approximation is quite independent of the post-Newtonian and post-Minkowskian expansions introduced in previous chapters. In Sec. 12.3 we apply the general formalism to slowly-moving systems, thereby incorporating a post-Newtonian expansion within the shortwave approximation. In this context the radiative losses can be computed in terms of the mass and current multipole moments of the matter distribution; at the leading order we obtain the famous quadrupole formula for the rate at which gravitational waves remove energy from the system. In Sec. 12.4 we explore a number of astrophysical implications of these radiative losses.

In Sec. 12.5 we return to the near zone and identify the gravitational potentials that are involved in the dissipative dynamics of the system; these radiation-reaction terms are seen to occur at 2.5PN order in a post-Newtonian expansion of the potentials. We apply them to fluid dynamics in Sec. 12.6, and to the motion of an N-body system in Sec. 12.7. In Sec. 12.8 we simplify the description of the radiation-reaction potentials and forces by constructing transformations from the original (harmonic) spacetime coordinates to alternative coordinates; as special cases we shall encounter the Burke–Thorne and Schäfer radiation-reaction gauges. And finally, in Sec. 12.9 we analyze the equations that govern the dissipative dynamics of a binary system of gravitating bodies, focusing on the secular changes that occur in the orbital motion.

12.1 Radiation reaction in electromagnetism

Before we mount our direct attack on radiative losses and radiation reaction in gravitating systems, we take a moment to explore these themes in the simpler context of electromagnetism. To keep the discussion as simple as possible we ignore gravity entirely (so that the spacetime is flat), and we assume that the motion of the charged bodies is slow. This restricted situation is sufficient for our purposes: it features all of the essential physics that we shall encounter in the gravitational case, and this physics can be explored without the conceptual and technical difficulties that appear in the gravitational case. We shall describe the major differences between the two cases at the end of this section.

12.1.1 System of charged bodies

As in Chapter 9 we examine a system of well-separated bodies moving under the influence of their mutual interactions; each body consists of a perfect fluid described by its mass density ρ, pressure p, and velocity field $\boldsymbol{v}$. What is new here is that each body also possesses a charge distribution described by a charge density ρ_e and a current density $\boldsymbol{j}_e = \rho_e \boldsymbol{v}$, and that the interaction between bodies is mediated by electromagnetism instead of gravity.

Once more we work in terms of the center-of-mass variables that were first introduced in Sec. 1.6.5. The mass of body A is $m_A := \int_A \rho\, d^3x$, and its center-of-mass is situated at $\boldsymbol{r}_A := m_A^{-1} \int_A \rho \boldsymbol{x}\, d^3x$. The velocity vector of body A is $\boldsymbol{v}_A := m_A^{-1} \int_A \rho \boldsymbol{v}\, d^3x$, and its acceleration is given by

$$m_A \boldsymbol{a}_A := \int_A \rho \frac{d\boldsymbol{v}}{dt}\, d^3x. \tag{12.1}$$

In addition to these quantities we introduce the body's total charge

$$q_A := \int_A \rho_e\, d^3x, \tag{12.2}$$

which we take to be non-vanishing. Each body is assumed to be reflection-symmetric about its center-of-mass, in the sense specified back in Sec. 9.1.2. And each body is assumed to be moving slowly, in the sense that $v_A \ll c$. This implies that the bodies are situated deep within the near zone, as was first explained back in Sec. 6.3.2.

The *total dipole moment* of the charge distribution plays an important role in our discussion. This vector is defined by

$$\boldsymbol{I}_e(t) := \int \rho_e \boldsymbol{x}\, d^3x, \tag{12.3}$$

and the continuity equation $\partial_t \rho_e + \nabla \cdot \boldsymbol{j}_e = 0$ ensures that its time derivative is given by

$$\dot{\boldsymbol{I}}_e = \int \boldsymbol{j}_e \, d^3x. \tag{12.4}$$

The total dipole moment can also be expressed as $\boldsymbol{I}_e = \sum_A \int_A \rho_e(\boldsymbol{r}_A + \bar{\boldsymbol{x}})\, d^3\bar{x}$, with $\bar{\boldsymbol{x}}$ denoting the position of a fluid element relative to the center-of-mass of each body. Integration of $\rho_e \bar{\boldsymbol{x}}$ produces zero when the charge distribution is reflection-symmetric about the center-of-mass, and under these circumstances we obtain the simple expression

$$\boldsymbol{I}_e = \sum_A q_A \boldsymbol{r}_A \tag{12.5}$$

for the total dipole moment.

The equations that govern the coupled dynamics of the bodies and their electromagnetic field consist of Maxwell's equations

$$\nabla \cdot \boldsymbol{E} = \frac{1}{\epsilon_0} \rho_e, \tag{12.6a}$$

$$\nabla \cdot \boldsymbol{B} = 0, \tag{12.6b}$$

$$\nabla \times \boldsymbol{E} = -\frac{\partial \boldsymbol{B}}{\partial t}, \tag{12.6c}$$

$$\nabla \times \boldsymbol{B} = \mu_0 \boldsymbol{j}_e + \epsilon_0 \mu_0 \frac{\partial \boldsymbol{E}}{\partial t}, \tag{12.6d}$$

in which $\boldsymbol{E}$ is the electric field and $\boldsymbol{B}$ the magnetic field, and the generalization of Euler's equation given by

$$\rho \frac{d\boldsymbol{v}}{dt} = -\nabla p + \rho_e(\boldsymbol{E} + \boldsymbol{v} \times \boldsymbol{B}). \tag{12.7}$$

The coupling constants ϵ_0 and μ_0 are related to the speed of light by $\epsilon_0 \mu_0 = c^{-2}$.

Substitution of Eq. (12.7) into Eq. (12.1) yields

$$m_A \boldsymbol{a}_A = \int_A \rho_e \boldsymbol{E} \, d^3x + \int_A \rho_e \boldsymbol{v} \times \boldsymbol{B} \, d^3x. \tag{12.8}$$

Our first task in this section is to evaluate the right-hand side of Eq. (12.8) to leading order in an expansion in powers of v_A/c, which we shall refer to as a "post-Coulombian expansion," in an obvious analogy with the post-Newtonian expansions of preceding chapters. The answer, of course, is the well-known Coulomb law, given by Eq. (12.18) below. We shall, nevertheless, provide a fairly detailed derivation of this equation, because this allows us to establish a number of results used in the derivation of the radiation-reaction force in Sec. 12.1.4.

12.1.2 Motion of charged bodies

To obtain $\boldsymbol{a}_A$ we must first compute the fields $\boldsymbol{E}$ and $\boldsymbol{B}$, and this requires integration of Maxwell's equations. The homogeneous equations are solved automatically when we express the fields as

$$E = -\frac{\partial A}{\partial t} - \nabla\Phi, \qquad B = \nabla \times A, \tag{12.9}$$

in terms of a vector potential A and a scalar potential Φ. The potentials can be freely altered by a *gauge transformation* of the form $\Phi \to \Phi - \partial_t \chi$, $A \to A + \nabla\chi$, in which χ is an arbitrary scalar function; the transformation leaves the fields unchanged. The gauge freedom can be exploited to enforce the *Lorenz gauge condition*

$$\nabla \cdot A + \frac{1}{c^2}\frac{\partial \Phi}{\partial t} = 0, \tag{12.10}$$

and in this gauge the inhomogeneous Maxwell equations reduce to the wave equations

$$\Box\Phi = -4\pi\kappa\,\rho_e, \qquad \Box A = -\frac{4\pi\kappa}{c^2} j_e, \tag{12.11}$$

for the potentials; here $\kappa := (4\pi\epsilon_0)^{-1}$.

The solutions to the wave equations are

$$\Phi(t, x) = \kappa \int \frac{\rho_e(t - |x - x'|/c, x')}{|x - x'|}\, d^3x', \tag{12.12a}$$

$$A(t, x) = \frac{\kappa}{c^2} \int \frac{j_e(t - |x - x'|/c, x')}{|x - x'|}\, d^3x', \tag{12.12b}$$

in which the integration domains are limited to the volume occupied by the charge distribution. Techniques to evaluate such retarded integrals in the near zone were developed in Sec. 6.3.4. The main strategy is to express the delayed time dependence of each source function as a Taylor expansion about the current time t. For example, we express the charge density ρ_e as

$$\rho_e(t - |x - x'|/c) = \rho_e - \frac{1}{c}|x - x'|\frac{\partial \rho_e}{\partial t} + \frac{1}{2c^2}|x - x'|^2\frac{\partial^2 \rho_e}{\partial t^2} - \frac{1}{6c^3}|x - x'|^3\frac{\partial^3 \rho_e}{\partial t^3} + O(c^{-4}), \tag{12.13}$$

in which each term on the right-hand side is evaluated at time t; the slow-motion condition ensures that each term is smaller than the preceding one by a factor of order v_A/c.

To leading order in a post-Coulombian expansion (0PC order), the scalar potential is given by $\Phi = U_e + O(c^{-2})$, where

$$U_e(t, x) = \kappa \int \frac{\rho_e(t, x')}{|x - x'|}\, d^3x' \tag{12.14}$$

is an instantaneous potential of a sort encountered before in the context of Newtonian and post-Newtonian gravity. It is easy to show that the term of order c^{-1} in Φ vanishes by virtue of charge conservation. The corresponding expression for the vector potential is of order c^{-2} and makes no leading-order contribution to the equations of motion; it participates at 1PC order, along with the neglected terms of order c^{-2} in Φ.

Substitution of the potentials into Eq. (12.9), and these fields into Eq. (12.8), produces

$$m_A a_A^j = -\int_A \rho_e \partial_j U_e \, d^3x' + O(c^{-2}), \tag{12.15}$$

an expression that is familiar from Newtonian gravity. As in Chapters 1 and 9 the potential can be decomposed as $U_e = U_{e,A} + U_{e,\neg A}$, in terms of a piece $U_{e,A}$ sourced by body A and an external piece $U_{e,\neg A}$ sourced by the remaining bodies. As before the body's self-field does not contribute to the force, and the external potential can be expressed as a Taylor expansion about the center-of-mass $\boldsymbol{r}_A$ – refer to Eq. (9.89). Our assumption that the bodies are well separated allows us to retain only the leading term in this expansion, and we arrive at

$$m_A a_A^j = -q_A \partial_j U_{e,\neg A}(t, \boldsymbol{r}_A) + O(c^{-2}). \tag{12.16}$$

In addition to the post-Coulombian corrections of order c^{-2}, this expression for the force neglects terms that are smaller by a factor of order $(R_A/r_{AB})^2 \ll 1$, in which R_A is the size of body A while r_{AB} is the typical inter-body distance.

The remaining steps are also familiar from Chapters 1 and 9. We differentiate the external potential, evaluate the result at $\boldsymbol{x} = \boldsymbol{r}_A$, and take the limit of well-separated bodies. We arrive at

$$\partial_j U_{e,\neg A}(t, \boldsymbol{r}_A) = -\sum_{B \neq A} \kappa q_B \frac{n_{AB}^j}{r_{AB}^2}, \tag{12.17}$$

where $\boldsymbol{n}_{AB} := (\boldsymbol{r}_A - \boldsymbol{r}_B)/|\boldsymbol{r}_A - \boldsymbol{r}_B|$ is a unit vector that points from body B to body A, and $r_{AB} := |\boldsymbol{r}_A - \boldsymbol{r}_B|$. Our final expression for the force acting on body A is then

$$m_A \boldsymbol{a}_A = \sum_{B \neq A} \frac{\kappa q_A q_B}{r_{AB}^2} \boldsymbol{n}_{AB} + O(c^{-2}), \tag{12.18}$$

a result that can be obtained directly from Newtonian gravity by making the replacements $G \to \kappa$, $m_A \to -q_A$ (on the right-hand side of the equation only), and $m_B \to q_B$. Slow-motion electromagnetism, like slow-motion gravity, is characterized by a force that is inversely proportional to the squared distance. The main difference concerns the product of charges: while $m_A m_B$ is necessarily positive in gravity, $q_A q_B$ can be of either sign in electromagnetism; the force can be repulsive as well as attractive.

The (mechanical plus field) energy of the system of charged bodies is given by

$$E = \sum_A \frac{1}{2} m_A v_A^2 + \sum_A \sum_{B \neq A} \frac{\kappa q_A q_B}{2 r_{AB}} + O(c^{-2}), \tag{12.19}$$

and it is easy to show that the motion keeps the energy constant: $dE/dt = 0$. The slow-motion dynamics of charged bodies is conservative.

12.1.3 Radiative losses

The system, however, is known to emit electromagnetic waves, and the radiation takes energy away from the system. The rate at which it does so, to leading order in a post-Coulombian expansion of the radiation, is given by the *electric dipole formula*,

$$\mathcal{P} = \frac{2\kappa}{3c^3}\left|\ddot{\boldsymbol{I}}_e\right|^2 + O(c^{-5}), \tag{12.20}$$

in which $\mathcal{P}$ is the total energy radiated per unit time, and $\ddot{\boldsymbol{I}}_e$ is the electric dipole moment of Eq. (12.5) differentiated twice with respect to time. The terms of order c^{-5} in Eq. (12.20) are produced by changes in the magnetic dipole moment and electric quadrupole moment of the charge distribution; higher-order terms are produced by changes in higher-order multipole moments. For a system consisting of a single charged body moving on an accelerated trajectory, we have $\ddot{\boldsymbol{I}}_e = q\boldsymbol{a}$, where $\boldsymbol{a}$ is the acceleration, and this yields the well-known Larmor formula $\mathcal{P} = (2\kappa q^2/3c^2)|\boldsymbol{a}|^2$.

The radiative losses described by the electric dipole formula are not accounted for in the conservative dynamics described in the preceding subsection. We should expect, however, that a more accurate treatment of the equations of motion would convert the statement $dE/dt = 0$ into something like $dE/dt = -\mathcal{P}$, which properly expresses conservation of total (mechanical plus field plus radiation) energy in the presence of dissipation. This is our second main task in this section: To identify the radiation-reaction terms in the equations of motion, and to show that these produce (something like) the expected energy-balance equation. The change in the system's energy will correspond to the rate at which the radiation-reaction forces do work on the bodies. This is expressed mathematically by

$$\frac{dE}{dt} = \sum_A \boldsymbol{F}_A[\mathrm{rr}] \cdot \boldsymbol{v}_A, \tag{12.21}$$

in which $\boldsymbol{F}_A[\mathrm{rr}]$ is the radiation-reaction force acting on body A.

12.1.4 Radiation reaction

Because the radiated power $\mathcal{P}$ scales as c^{-3} to leading order in a post-Coulombian expansion, our improved treatment of the equations of motion must include terms of order c^{-3} in order to account for radiation-reaction effects. A systematic post-Coulombian expansion of the equations of motion would therefore commence with Eq. (12.18) at leading order, then incorporate 1PC terms at order c^{-2}, 1.5PC terms at order c^{-3}, 2PC terms at order c^{-4}, and so on. Here we shall jump directly from 0PC order to 1.5PC order, and bypass the additional conservative terms that appear at 1PC order; we shall not pursue the post-Coulombian expansion beyond this point.

That radiation-reaction terms should appear in the equations of motion with an *odd power* of c^{-1} reflects a deep fact about radiating systems, that they necessarily break the *time-reversal invariance* of the underlying theory. Here the radiation is electromagnetic in nature, and the theory is Maxwell's electrodynamics. Later in the chapter the radiation is gravitational in nature, and the theory is Einstein's general relativity. In each case the fundamental theory is time-reversal invariant, but the selected solutions specify a direction for the arrow of time.

In the present context the choice of direction was made in Eqs. (12.12), when we selected the *retarded solutions* to the wave equations satisfied by the potentials. This choice is physically well motivated: it produces fields that are causally related to the source (the field now depends on the past behavior of the source), it produces waves that are propagating outward, and it produces a transfer of energy away from the source. But to select the retarded solutions is a choice nevertheless, and this choice is not dictated by the fundamental equations of Maxwell's theory. We could easily make a different choice and adopt instead the *advanced solutions* to the wave equations. This would produce fields that are in an anti-causal relation with the source (the field now would depend on the future behavior of the source), radiation that is propagating inward, and a transfer of energy towards the source. In terms of radiation reaction, the advanced solutions would give rise to an equation of the form $dE/dt = +\mathcal{P}$ instead of the expected $dE/dt = -\mathcal{P}$, and this would express the (unphysical) fact that the incoming radiation brings energy to the system. And because dE/dt is related to the radiation-reaction force by Eq. (12.21), we see that the switch from retarded to advanced solutions *changes the sign* of $\boldsymbol{F}_A^{\rm rr}$.

These considerations lead directly to the expectation that the radiation-reaction force should involve odd powers of c^{-1} only, because an advanced solution to the wave equation can be obtained from a retarded solution by making the substitution $c \to -c$. Such a change in the sign of c turns outgoing waves into incoming waves, turns the flow of energy from outward to inward, and changes the sign of the radiation-reaction force. All terms that come with odd powers of c^{-1} in the potentials Φ and $\boldsymbol{A}$ therefore contribute to the radiation-reaction force; terms that come with even powers of c^{-1} do not change sign under the substitution $c \to -c$, and these give rise to the purely conservative part of the force. To leading order in a post-Coulombian expansion, the radiation-reaction force comes from terms that scale as c^{-3} in Φ and $\boldsymbol{A}$.

To obtain the radiation-reaction piece of Φ we return to Eq. (12.13), pluck out the term of order c^{-3}, and insert it within Eq. (12.12). Proceeding similarly with $\boldsymbol{A}$, we get

$$\Phi_{\rm rr} = -\frac{\kappa}{6c^3}\frac{\partial^3}{\partial t^3}\int \rho_e(t,\boldsymbol{x}')|\boldsymbol{x}-\boldsymbol{x}'|^2\,d^3x' + O(c^{-5}), \tag{12.22a}$$

$$\boldsymbol{A}_{\rm rr} = -\frac{\kappa}{c^3}\frac{d}{dt}\int \boldsymbol{j}_e(t,\boldsymbol{x}')\,d^3x' + O(c^{-5}). \tag{12.22b}$$

Further manipulations using Eqs. (12.3) and (12.4) reveal that the radiation-reaction potentials are given by

$$\Phi_{\rm rr} = \frac{\kappa}{6c^3}\left(2\boldsymbol{x}\cdot\dddot{\boldsymbol{I}}_e - \dddot{I}_e^{kk}\right) + O(c^{-5}), \tag{12.23a}$$

$$\boldsymbol{A}_{\rm rr} = -\frac{\kappa}{c^3}\ddot{\boldsymbol{I}}_e + O(c^{-5}), \tag{12.23b}$$

where $I_e^{kk} := \int \rho_e r^2\,d^3x$ is the trace of the quadrupole-moment tensor of the charge distribution. In the scalar potential the electric dipole moment $\boldsymbol{I}_e$ is differentiated three times with respect to t; in the vector potential it is differentiated twice.

The radiation-reaction fields are obtained by inserting the potentials within Eq. (12.9). This yields

$$\boldsymbol{E}_{\rm rr} = \frac{2\kappa}{3c^3}\dddot{\boldsymbol{I}}_e + O(c^{-5}), \qquad \boldsymbol{B}_{\rm rr} = O(c^{-5}). \tag{12.24}$$

We see that the magnetic field vanishes at leading order, and that the electric field depends on t only; it is uniform within the distribution of charge. Substitution of the fields into Eq. (12.8) produces

$$\boldsymbol{F}_A[\mathrm{rr}] = \frac{2\kappa}{3c^3} q_A \dddot{\boldsymbol{I}}_e + O(c^{-5}). \tag{12.25}$$

This is the *radiation-reaction force* acting on body A.

12.1.5 Energy balance

The total energy of the system is given by Eq. (12.19), and differentiation of the right-hand side with respect to t gives rise to Eq. (12.21), noting that the work done by the conservative piece of the force accounts for the changes in the system's potential energy. Substitution of Eq. (12.25) produces

$$\frac{dE}{dt} = \frac{2\kappa}{3c^3} \dddot{\boldsymbol{I}}_e \cdot \sum_A q_A \boldsymbol{v}_A = \frac{2\kappa}{3c^3} \dddot{\boldsymbol{I}}_e \cdot \dot{\boldsymbol{I}}_e, \tag{12.26}$$

where we have made use of Eq. (12.5). This can also be expressed as

$$\frac{dE}{dt} = \frac{d}{dt}\left(\frac{2\kappa}{3c^3} \ddot{\boldsymbol{I}}_e \cdot \dot{\boldsymbol{I}}_e\right) - \frac{2\kappa}{3c^3} \left|\ddot{\boldsymbol{I}}_e\right|^2 \tag{12.27}$$

by transferring a time derivative from one dipole moment to the other. This, finally, we write in the form

$$\frac{dE}{dt} = -\frac{dE'}{dt} - \mathcal{P}, \tag{12.28}$$

where

$$E' := -\frac{2\kappa}{3c^3} \ddot{\boldsymbol{I}}_e \cdot \dot{\boldsymbol{I}}_e \tag{12.29}$$

has the dimension of an energy, and $\mathcal{P}$ is the radiated power of Eq. (12.20).

Equation (12.28) is our statement of energy balance, the expression of conservation of total (mechanical plus field plus radiation) energy. We see, however, that the statement is not quite the expected $dE/dt = -\mathcal{P}$; the additional term $-dE'/dt$ requires an interpretation.

The statement of Eq. (12.28) forces us to recognize that the exchange of energy between the charged bodies and the electromagnetic field may take different forms. First, the kinetic energy of the bodies may change, and so can the system's potential energy, as defined by the second term on the right-hand side of Eq. (12.19); these changes are accounted for by the term dE/dt in Eq. (12.28). Second, part of the system's energy is converted into radiation, which propagates outward and irreversibly leaves the system; this is described by the term $-\mathcal{P}$ in Eq. (12.28). But potential energy and radiation are not the only relevant forms of field energy; there is also a form that stays bound to the system and is *reversibly* exchanged between the electromagnetic field and the charged bodies. This is what is described by E', as defined by Eq. (12.29). Equation (12.28) therefore describes an exchange between all relevant forms of energy: kinetic, potential, radiation, and bound field energy.

There are ways by which the statement of Eq. (12.28) can be simplified and turned into the expected $dE/dt = -\mathcal{P}$. One way is to assume that the conservative motion is periodic, or that it begins and ends in a static state, and to average Eq. (12.28) over time. This produces a coarse-grained statement of energy balance,

$$\langle dE/dt\rangle = -\langle \mathcal{P}\rangle. \tag{12.30}$$

The contribution from dE'/dt no longer appears, because its average is proportional to the total change in E' over the time interval, and this vanishes at order c^{-3} under the stated conditions. Another way is to modify the definition of the system's total energy so that it now includes the bound field energy E'. The redefinition $E \to E + E'$ obviously turns Eq. (12.28) into the simpler statement $dE/dt = -\mathcal{P}$.

Box 12.1 Redefining the energy

The reader might feel that it is inappropriate to redefine the energy E for the sole purpose of obtaining a desired statement of energy balance. We wish to reassure the reader: there was nothing sacred about the initial definition of the system's energy, and a redefinition can be entirely appropriate. It is helpful to recall that the energy is not some rigid quantity defined a priori, but that its definition is provided by the dynamics of the mechanical system. For a conservative system, it is whatever function of the positions and velocities that happens to be conserved. Because the zero point of energy is not fixed (at least at the non-quantum level), the definition is not even unique.

Energy is not even conserved for radiative systems, and its definition is even more ambiguous. A good definition of energy would be one that is approximately constant when the radiation can be ignored, and one that accurately reflects the long-term behavior of the system under the damping effects of the radiation leaving the system. The short-term behavior reveals fluctuations caused by a reversible transfer of energy between the system and the field, and an ambiguity arises because it is impossible to decompose the field energy into a piece that is unambiguously associated with the reversible transfer, and another piece that is unambiguously associated with the radiation. The ambiguity, however, is largely irrelevant when the fluctuations are small compared with the dominant contribution to the energy, which is approximately conserved in the short term.

In the electromagnetic case examined here, the Coulombian energy E is of order $mv^2 \sim \kappa q^2/r$, and the field energy E' can be seen from Eq. (12.29) to be of order $\kappa^2 q^4 v/(mc^3r^2) \sim E(v/c)^3$. This is a small correction that leads to a small ambiguity.

These remarks apply to any dissipative system. They keep their relevance when we turn to gravity, and establish an equivalence between the work done by the radiation-reaction forces and the energy carried away by gravitational waves.

12.1.6 Looking ahead: gravity

The remainder of this chapter is devoted to a discussion of radiative losses and radiation reaction in the context of gravity. The conservative dynamics of fluids and bodies moving under a mutual gravitational attraction was explored in Chapters 8, 9, and 10, where the equations of motion were obtained accurately through the first post-Newtonian order. The gravitational waves emitted by these bodies were examined in Chapter 11, and our first order of business in this chapter (Sec. 12.2) is to calculate the rates at which the waves transport energy, linear momentum, and angular momentum away from the system. In the case of radiated energy, we shall derive (Sec. 12.3) the famous *quadrupole formula*, the gravitational analogue of Eq. (12.20); because there is no dipole radiation in general relativity, the radiated power $\mathcal{P}$ must involve the quadrupole-moment tensor of the mass distribution. Astrophysical consequences of radiative losses are explored in Sec. 12.4. We then (Sec. 12.5) obtain the gravitational potentials – analogous to those of Eqs. (12.23) – that give rise to the gravitational radiation-reaction force, which is computed and applied in Secs. 12.6 through 12.9.

The treatment of the gravitational case involves computations that are much more challenging than those reviewed in this section. The reason for this is two-fold. First, general relativity, even in its weak-field and slow-motion formulation, is intrinsically more complicated than Maxwell's electrodynamics. And second, the absence of dipole radiation implies that the radiated power scales as c^{-5} instead of c^{-3}, which was the scaling observed in electromagnetism. This implies that the radiation-reaction terms in the equations of motion will appear at order c^{-5} in a post-Newtonian expansion, an order that is far removed from the c^{-2} order that was achieved back in Chapter 9. In terms of the usual post-Newtonian counting, we are looking for terms of 2.5PN order in the equations of motion, while the results obtained in Chapter 9 are accurate only through 1PN order. A systematic treatment of the equations of motion to this order of accuracy would have to provide conservative terms at 2PN order in addition to the desired terms at 2.5PN order. It is a fortunate circumstance, however, that the derivation of the radiation-reaction terms at 2.5PN order is insensitive to the details that appear at 2PN order. We may, therefore, bypass the 2PN terms entirely, and jump directly to the radiation-reaction terms that actually interest us. This is analogous to what was observed in the electromagnetic case: the radiation-reaction terms at order c^{-3} could be derived independently of the conservative corrections at order c^{-2}.

The difficulties associated with radiative losses and radiation reaction in gravity are not just technical in nature. There are also conceptual issues that arise as a consequence of the principle of equivalence. Consider a freely-moving observer in spacetime. As was discussed at length in Chapter 5, such an observer moves without ever noticing a local gravitational field, and would therefore assign a zero value to any proposed measure of gravitational energy flux. This is quite unlike the value that he would assign to the Poynting vector $\mu_0^{-1}\boldsymbol{E} \times \boldsymbol{B}$, which measures the flux of electromagnetic energy. Given, then, that the principle of equivalence forbids the very existence of a "gravitational Poynting vector," how is one to calculate the total radiated power $\mathcal{P}$? In electromagnetism one simply integrates the normal component of the Poynting vector over a spherical surface situated in the wave zone, and the result is the electric-dipole formula of Eq. (12.20). A corresponding recipe is harder to identify in general relativity, and any proposed recipe must be handled with care in light of the principle of equivalence. That systems of moving bodies governed by gravitational interactions *must* lose energy to gravitational radiation is, however, very clear on physical grounds.

12.2 Radiative losses in gravitating systems

Our main task in this section is to establish the validity of balance equations for energy, momentum, and angular momentum, and to find expressions for the rates at which gravitational waves carry energy, momentum, and angular momentum away from their source. We pursue a dual goal: to provide a sound foundation for a discussion of radiative losses, and to develop a practical formalism to calculate these losses.

12.2.1 Balance equations

Our description of radiative losses in gravitating systems relies on the Landau–Lifshitz formulation of the Einstein field equations reviewed in Sec. 6.1; in particular, it relies on the conservation identities derived in Sec. 6.1.3. These consist of balance equations involving the energy, linear momentum, and angular momentum of the gravitating system. The energy-balance equation was first displayed in Eq. (6.30), which we copy here as

$$\frac{dE}{dt} = -\mathcal{P}, \tag{12.31}$$

where $E = Mc^2$ is the total energy contained in the system, as defined by Eq. (6.36), while

$$\mathcal{P} := c \oint_\infty (-g)t^{0k}_{\rm LL}\, dS_k \tag{12.32}$$

is the rate at which gravitational waves remove energy from the system. The momentum-balance equation was first displayed in Eq. (6.31), which we copy as

$$\frac{dP^j}{dt} = -\mathcal{F}^j, \tag{12.33}$$

where P^j is the total momentum of the system, as defined by Eq. (6.37), while

$$\mathcal{F}^j := \oint_\infty (-g)t^{jk}_{\rm LL}\, dS_k \tag{12.34}$$

is the rate at which gravitational waves remove (linear) momentum from the system. And finally, the statement of conservation of angular momentum is

$$\frac{dJ^{jk}}{dt} = -\mathcal{T}^{jk}, \tag{12.35}$$

where J^{jk} is the total angular-momentum tensor, as defined by Eq. (6.38), while

$$\mathcal{T}^{jk} := \oint_\infty \left[x^j(-g)t^{kn}_{\rm LL} - x^k(-g)t^{jn}_{\rm LL}\right] dS_n \tag{12.36}$$

is the rate at which gravitational waves remove angular momentum from the system; Eq. (12.35) is a re-statement of Eq. (6.34). The angular-momentum vector is given by $J^j = \frac{1}{2}\epsilon^j{}_{pq} J^{pq}$, and we may define a "gravitational-wave torque vector" by $\mathcal{T}^j := \frac{1}{2}\epsilon^j{}_{pq}\mathcal{T}^{pq}$, so that $dJ^j/dt = -\mathcal{T}^j$. In these equations, $t^{\alpha\beta}_{\rm LL}$ is the Landau–Lifshitz pseudotensor of Eq. (6.5). The surface integrals are evaluated in the limit $R \to \infty$, and $dS_j = R^2 N_j\, d\Omega$

is an outward-directed surface element; we use the notation of Chapter 11, with $R := |\boldsymbol{x}|$, $\boldsymbol{N} := \boldsymbol{x}/R$, and $d\Omega$ is an element of solid angle in the direction of the unit vector $\boldsymbol{N}$.

The Landau–Lifshitz formulation of the Einstein field equations supplies a sound foundation for a description of radiative losses because it achieves three essential goals: first, it provides expressions for the radiative fluxes $\mathcal{P}$, $\mathcal{F}^j$, and $\mathcal{T}^{jk}$; second, it provides definitions for the total energy E, total momentum P^j, and total angular momentum J^{jk}; and third, it establishes precise balance equations that describe how E, P^j, and J^{jk} change as a result of gravitational-wave emissions. It should be pointed out that this formulation is not unique: a different formulation of the Einstein field equations, based on other variables, other coordinate systems, and a different choice of pseudotensors, may well produce different definitions for E, P^j, and J^{jk}, and different expressions for $\mathcal{P}$, $\mathcal{F}^j$, and $\mathcal{T}^{jk}$. These differences are of no concern, however, so long as the proposed definitions for E, P^j, and J^{jk} are physically sensible. As we have seen previously in Secs. 7.3.2, 8.4.5, and 8.4.6, the Landau–Lifshitz definitions, when implemented within post-Minkowskian theory or post-Newtonian theory, are physically sensible.

12.2.2 The shortwave approximation

We begin by evaluating the gravitational-wave fluxes $\mathcal{P}$, $\mathcal{F}^j$, and $\mathcal{T}^{jk}$. We wish to do so in a framework that is broader than, but includes, the post-Minkowskian approximation formulated in Chapters 6 and 7. The post-Minkowskian approximation is based on an expansion of the gravitational potentials $h^{\alpha\beta}$ in powers of the gravitational constant G. Our approach here is distinct, and our expressions will be accurate to all orders in G. Our approach is based on a different type of approximation, which, following Misner, Thorne, and Wheeler (1973), we term *shortwave approximation*. It relies on the fact that we are interested in the gravitational potentials in the *far-away wave zone*, as introduced in Sec. 11.1.1. In this region of spacetime R is much larger than λ_c, the characteristic wavelength of the gravitational radiation, and we exploit this fact by using $\lambda_c/R \ll 1$ as an expansion parameter. The shortwave approximation has already been invoked within the context of post-Minkowskian theory in Chapter 7 (refer to Box 7.7), and it has also been invoked independently of the post-Minkowskian approximation back in Sec. 11.1. In this subsection we develop it more systematically.

The shortwave approximation is based on an expansion of the gravitational potentials in powers of λ_c/R, where $R := |\boldsymbol{x}|$ is the distance to the system's center-of-mass. We write

$$h^{\alpha\beta} = (\lambda_c/R) f_1^{\alpha\beta} + (\lambda_c/R)^2 f_2^{\alpha\beta} + \cdots, \tag{12.37}$$

in which $f_n^{\alpha\beta}$ is assumed to be a function of retarded time $\tau := t - R/c$ and the angles contained in the unit vector $\boldsymbol{N} := \boldsymbol{x}/R$. In Sec. 11.1 the shortwave expansion was truncated at leading order; here we keep track of higher-order terms, because they are needed in our computation of the angular-momentum flux $\mathcal{T}^{jk}$. As was pointed out previously, each $f_n^{\alpha\beta}$ can be considered to be accurate to all orders in the gravitational constant G. A refined formulation of the approximation would involve the insertion of $\ln R$ terms in Eq. (12.37) to accommodate the presence of wave tails in the gravitational potentials; refer to Sec. 11.3.7.

Our considerations below are not affected by these logarithmic terms, and for simplicity we prefer to omit them from the expressions.

The assumed dependence of $f_n^{\alpha\beta}$ on retarded time $\tau := t - R/c$ is motivated by the fact that according to Eq. (6.51), the gravitational potentials must satisfy a wave equation of the form $\Box h^{\alpha\beta} = -(16\pi G/c^4)\tau^{\alpha\beta}$, where $\tau^{\alpha\beta}$ is the effective energy-momentum pseudotensor of Eq. (6.52). By virtue of the scaling of $h^{\alpha\beta}$ with λ_c/R, $\tau^{\alpha\beta}$ is guaranteed to be of order $(\lambda_c/R)^2$, and this implies that the time dependence of $f_1^{\alpha\beta}$ must be contained in the combination $t - R/c$. The same dependence is then fed into the remaining terms in the expansion. These considerations rely on the fact that the gravitational potentials are taken to satisfy the harmonic gauge condition of Eq. (6.49), $\partial_\beta h^{\alpha\beta} = 0$.

Differentiation of the potentials is facilitated by employing the identities $\partial_j R = N_j$ and $\partial_j N_k = P_{jk}/R$, where $P_{jk} := \delta_{jk} - N_j N_k$. Recalling that $x^0 = ct$ and using these differentiation rules, we obtain

$$c\partial_0 h^{\alpha\beta} = (\lambda_c/R)\partial_\tau f_1^{\alpha\beta} + (\lambda_c/R)^2\partial_\tau f_2^{\alpha\beta} + \cdots, \tag{12.38a}$$

$$c\partial_j h^{\alpha\beta} = -(\lambda_c/R)N_j\partial_\tau f_1^{\alpha\beta} - (\lambda_c/R)^2\left(N_j\partial_\tau f_2^{\alpha\beta} + \frac{c}{\lambda_c}N_j f_1^{\alpha\beta} - \frac{c}{\lambda_c}P_j{}^k\frac{\partial f_1^{\alpha\beta}}{\partial N^k}\right) + \cdots \tag{12.38b}$$

The expansion of $c\partial_j h^{\alpha\beta}$ contains a number of contributions at second order. It is easy to see that the term involving $f_2^{\alpha\beta}$ is smaller than the leading term by a factor of order $\lambda_c/R \ll 1$. The terms involving $f_1^{\alpha\beta}$ are multiplied by $c/\lambda_c = 1/t_c$, where t_c is the characteristic time scale associated with changes in the potentials, and $(c/\lambda_c)f_1^{\alpha\beta}$ is comparable in size to $\partial_\tau f_1^{\alpha\beta}$; these terms also are smaller than the leading term by a factor of order $\lambda_c/R \ll 1$.

To leading order in the shortwave approximation we can write that

$$\partial_\mu h^{\alpha\beta} = -\frac{1}{c}k_\mu\partial_\tau h^{\alpha\beta} + O(\lambda_c^2/R^2), \tag{12.39}$$

in which $k_\mu := (-1, \boldsymbol{N})$ is a spacetime vector that satisfies the null condition $\eta^{\mu\nu}k_\mu k_\nu = 0$. At this order the only spatial dependence that matters is contained in the overall factor of R^{-1} and in the proper time τ; the dependence contained in the vector $\boldsymbol{N}$ is subdominant, and appears at the next order.

12.2.3 Energy and momentum fluxes

To compute $\mathcal{P}$ and $\mathcal{F}^j$ we insert Eq. (12.39) within the Landau–Lifshitz pseudotensor of Eq. (6.5) and get, after much simplification,

$$(-g)t_{\rm LL}^{\alpha\beta} = \frac{c^2}{32\pi G}\left(\partial_\tau h_{\rm TT}^{jk}\,\partial_\tau h^{\rm TT}_{jk}\right)k^\alpha k^\beta, \tag{12.40}$$

where $k^\alpha = (1, \boldsymbol{N})$ and

$$h_{\rm TT}^{jk} = ({\rm TT})^{jk}{}_{pq}h^{pq}, \qquad ({\rm TT})^{jk}{}_{pq} := P^j{}_p P^k{}_q - \frac{1}{2}P^{jk}P_{pq}, \tag{12.41}$$

is the transverse-tracefree piece of the gravitational potentials, as defined back in Sec. 11.1.7. This result is invariant under any gauge transformation (described in Sec. 11.1.5) that preserves the harmonic gauge conditions $\partial_\beta h^{\alpha\beta} = 0$. The easiest way to establish Eq. (12.40) is therefore to refine the harmonic gauge to the TT gauge of Sec. 11.1.5, in which $h^{00} = O(\lambda_c^2/R^2)$, $h^{0j} = O(\lambda_c^2/R^2)$, and $h^{jk} = h^{jk}_{\rm TT} + O(\lambda_c^2/R^2)$; the properties $N_j h^{jk}_{\rm TT} = 0 = \delta_{jk} h^{jk}_{\rm TT}$ simplify the calculations significantly.

The expression of Eq. (12.40) is valid to leading order in an expansion in powers of λ_c/R; it is of second order, and it neglects terms of order $(\lambda_c/R)^3$. This expression is sufficiently accurate for substitution within Eqs. (12.32) and (12.34), and we immediately obtain

$$\mathcal{P} = \frac{c^3}{32\pi G} \int R^2 \Big(\partial_\tau h^{pq}_{\rm TT}\, \partial_\tau h^{\rm TT}_{pq} \Big)\, d\Omega \tag{12.42}$$

and

$$\mathcal{F}^j = \frac{c^2}{32\pi G} \int R^2 N^j \Big(\partial_\tau h^{pq}_{\rm TT}\, \partial_\tau h^{\rm TT}_{pq} \Big)\, d\Omega. \tag{12.43}$$

Because the integral is evaluated in the limit $R \to \infty$, higher-order terms in the expansion of $(-g)t^{\alpha\beta}_{\rm LL}$ do not contribute to $\mathcal{P}$ and $\mathcal{F}^j$. The expressions of Eqs. (12.42) and (12.43) are exact.

The energy and momentum fluxes can alternatively be expressed in terms of the gravitational-wave polarizations h_+ and $h_\times$ introduced in Sec. 11.1.7. These are defined by the decomposition

$$h^{jk}_{\rm TT} = h_+ \big(\vartheta^j \vartheta^k - \varphi^j \varphi^k \big) + h_\times \big(\vartheta^j \varphi^k + \varphi^j \vartheta^k \big) \tag{12.44}$$

of the TT field in terms of the transverse basis provided by the unit vectors $\boldsymbol{\vartheta} := [\cos\vartheta \cos\varphi, \cos\vartheta \sin\varphi, -\sin\vartheta]$ and $\boldsymbol{\varphi} := [-\sin\varphi, \cos\varphi, 0]$; these were introduced in Eqs. (11.37) and (11.38) and they are orthogonal to $\boldsymbol{N} := [\sin\vartheta\cos\varphi, \sin\vartheta\sin\varphi, \cos\vartheta]$. After inserting this decomposition within Eqs. (12.42) and (12.43) we obtain

$$\mathcal{P} = \frac{c^3}{16\pi G} \int R^2 \Big[\big(\partial_\tau h_+\big)^2 + \big(\partial_\tau h_\times\big)^2 \Big]\, d\Omega \tag{12.45}$$

and

$$\mathcal{F}^j = \frac{c^3}{16\pi G} \int R^2 N^j \Big[\big(\partial_\tau h_+\big)^2 + \big(\partial_\tau h_\times\big)^2 \Big]\, d\Omega. \tag{12.46}$$

These expressions, like Eqs. (12.42) and (12.43), are exact within the shortwave approximation. They are also accurate to all orders in a post-Minkowskian expansion of the potentials in powers of G.

12.2.4 Angular-momentum flux

The leading-order expression of Eq. (12.40), unfortunately, is not sufficiently accurate for insertion within Eq. (12.36). This can be seen from that fact that since $(-g)t^{jk}_{\rm LL}$ is proportional to $N^j N^k$, the integrand in Eq. (12.36) is proportional to $(N^j N^k - N^k N^j)N^n$ and therefore vanishes. This is actually a fortunate circumstance, because the presence of $x^j = RN^j$ within the integral implies that $\mathcal{T}^{jk}$ would otherwise diverge in the limit

$R \to \infty$. The computation of the angular-momentum flux, therefore, requires terms of order $(\lambda_c/R)^3$ in the expansion of the energy-momentum pseudotensor.

The result

It is wise to state the final answer before we present this long calculation:

$$\mathcal{T}^{jk} = \frac{c^3}{16\pi G}\int R^2\biggl[h^{jp}_{\rm TT}\,\partial_\tau h^{kp}_{\rm TT} - h^{kp}_{\rm TT}\,\partial_\tau h^{jp}_{\rm TT} - \frac{1}{2}\partial_\tau h^{pq}_{\rm TT}\bigl(x^j\partial^k - x^k\partial^j\bigr)h^{\rm TT}_{pq}\biggr]\,d\Omega\,. \tag{12.47}$$

Notice that when $x^j\partial^k - x^k\partial^j$ acts on the spatial dependence contained in $\tau := t - R/c$ within $h^{\rm TT}_{pq}$, or on the overall $1/R$ dependence, it produces $x^jN^k - x^kN^j$, which vanishes identically; the operation is sensitive only to the angular dependence contained in $\boldsymbol{N}$ within $h^{\rm TT}_{pq}$. In terms of the gravitational-wave polarizations we have

$$\mathcal{T}^{jk} = -\frac{c^3}{16\pi G}\int R^2 q^{jk}\,d\Omega \tag{12.48}$$

with

$$\begin{aligned} q^{jk} := \bigl(\partial_\tau h_+\bigr)&\Bigl[\bigl(x^j\partial^k - x^k\partial^j\bigr)h_+ - 2\,\mathrm{cosec}\,\vartheta\,\bigl(e^j_z\varphi^k - \varphi^je^k_z\bigr)h_\times\Bigr] \\ &+ \bigl(\partial_\tau h_\times\bigr)\Bigl[\bigl(x^j\partial^k - x^k\partial^j\bigr)h_\times + 2\,\mathrm{cosec}\,\vartheta\,\bigl(e^j_z\varphi^k - \varphi^je^k_z\bigr)h_+\Bigr], \end{aligned} \tag{12.49}$$

where $\boldsymbol{e}_z := [0, 0, 1]$ is a constant unit vector that points in the direction of the z-axis, relative to which the polar angles (ϑ, φ) are defined. To go from Eq. (12.47) to Eq. (12.48) we had to account for the spatial dependence of the transverse basis vectors. If the polarizations are expressed as functions of τ, ϑ, and φ, then the non-vanishing components of q^{jk} are given explicitly by

$$q^{xy} = \bigl(\partial_\tau h_+\bigr)\bigl(\partial_\varphi h_+\bigr) + \bigl(\partial_\tau h_\times\bigr)\bigl(\partial_\varphi h_\times\bigr), \tag{12.50a}$$

$$\begin{aligned} q^{yz} &= \bigl(\partial_\tau h_+\bigr)\bigl(-\sin\varphi\,\partial_\vartheta h_+ - \cot\vartheta\,\cos\varphi\,\partial_\varphi h_+ + 2\,\mathrm{cosec}\,\vartheta\,\cos\varphi\,h_\times\bigr) \\ &\quad + \bigl(\partial_\tau h_\times\bigr)\bigl(-\sin\varphi\,\partial_\vartheta h_\times - \cot\vartheta\,\cos\varphi\,\partial_\varphi h_\times - 2\,\mathrm{cosec}\,\vartheta\,\cos\varphi\,h_+\bigr), \end{aligned} \tag{12.50b}$$

$$\begin{aligned} q^{zx} &= \bigl(\partial_\tau h_+\bigr)\bigl(\cos\varphi\,\partial_\vartheta h_+ - \cot\vartheta\,\sin\varphi\,\partial_\varphi h_+ + 2\,\mathrm{cosec}\,\vartheta\,\sin\varphi\,h_\times\bigr) \\ &\quad + \bigl(\partial_\tau h_\times\bigr)\bigl(\cos\varphi\,\partial_\vartheta h_\times - \cot\vartheta\,\sin\varphi\,\partial_\varphi h_\times - 2\,\mathrm{cosec}\,\vartheta\,\sin\varphi\,h_+\bigr). \end{aligned} \tag{12.50c}$$

The expression of Eq. (12.47) was first obtained by Bryce DeWitt in 1971 while teaching a course on general relativity at Stanford University; his lecture notes, published only recently in 2011, contain very few details. The calculation was later repeated by Kip Thorne, who simply stated the final result in his 1980 review article on multipole expansions. Pending evidence to the contrary, the derivation presented below may well be the only one available in the literature.

Derivation

We now proceed with the derivation of Eq. (12.47). Terms that give rise to the subdominant, $(\lambda_c/R)^3$ piece of $(-g)t^{jk}_{\rm LL}$ are displayed in Eq. (12.38); these are the second-order terms in $\partial_\mu h^{\alpha\beta}$ generated by $\partial_\tau f_2^{\alpha\beta}$, $N_j f_1^{\alpha\beta}$, and derivatives of $f_1^{\alpha\beta}$ with respect to the vector $\boldsymbol{N}$. In addition to these we have first-order terms in $h^{\alpha\beta}$ that enter in the metric $g_{\alpha\beta}$, which also appears in the Landau–Lifshitz pseudotensor. A computation of $(-g)t^{jk}_{\rm LL}$ that includes all these contributions would be exceedingly tedious. A careful examination, however, reveals that most do not appear in the final result for $\mathcal{T}^{jk}$, because they produce terms proportional to N^j that cancel out in the end. This analysis allows us to state that $f_2^{\alpha\beta}$ makes no appearance in the final result, and that corrections of order λ_c/R to the Minkowski metric $\eta_{\alpha\beta}$ are similarly irrelevant. The computation, therefore, can be carried out safely by setting $h_{\alpha\beta} = (\lambda_c/R)f_1^{\alpha\beta}$ and $g_{\alpha\beta} = \eta_{\alpha\beta}$; derivatives of $h_{\alpha\beta}$, however, must be evaluated accurately to second order in (λ_c/R), beyond the expression displayed in Eq. (12.39). As we saw previously, the computation is simplified by adopting the TT gauge for the gravitational potentials.

A suitable starting point for the computation is

$$(-g)t^{jk}_{\rm LL} = \frac{c^4}{16\pi G}\biggl(-\partial^j h^{pq}\,\partial_p h^{kq} - \partial^k h^{pq}\,\partial_p h^{jq} - \partial_0 h^{jp}\,\partial_0 h^{kp} + \partial_p h^{jq}\,\partial^p h^{kq} + \frac{1}{2}\partial^j h^{pq}\,\partial^k h_{pq}\biggr), \tag{12.51}$$

which is obtained from Eq. (6.5) and simplified by incorporating the observations made in the preceding paragraph. In particular, h^{jk} is taken to be in the TT gauge, so that $N_j h^{jk} = 0$ and $\delta_{jk}h^{jk} = 0$. In addition, all terms proportional to δ^{jk} were discarded, because they cancel out when inserted within Eq. (12.36).

In Eq. (12.51), the operator ∂_j is understood in the usual way as holding t fixed while differentiating with respect to x^j. To proceed with the calculation, we express ∂_j acting on a retarded function like h^{pq} in the form

$$\partial_j h^{pq} = -(N_j/c)\partial_\tau h^{pq} + \eth_j h^{pq}, \tag{12.52}$$

where the operator $\eth_j$ bypasses the retarded-time dependence and operates only on the remaining spatial dependence. When $\eth_j$ acts on h^{pq}, it generates a term of second order in λ_c/R, while acting with ∂_τ produces a first-order term. Keeping our expressions accurate through order $(\lambda_c/R)^3$, we find that Eq. (12.51) becomes

$$(-g)t^{jk}_{\rm LL} = \frac{c^3}{16\pi G}\biggl\{N^j(\partial_\tau h^{pq})\Bigl(\eth_p h^{kq} - \frac{1}{2}\eth^k h_{pq}\Bigr) + N^k(\partial_\tau h^{pq})\Bigl(\eth_p h^{jq} - \frac{1}{2}\eth^j h_{pq}\Bigr) - N^p\Bigl[(\partial_\tau h^{jq})(\eth_p h^{kq}) + (\partial_\tau h^{kq})(\eth_p h^{jq})\Bigr] - \frac{1}{R}(h^{jq}\partial_\tau h^{kq} + h^{kq}\partial_\tau h^{jq})\biggr\} \tag{12.53}$$

when we discard terms proportional to N^jN^k that eventually cancel out, and exploit the following properties of the TT gauge: $N_j\partial_\tau h^{jk} = 0$ and $N_j\eth_p h^{jk} = -h^{pk}/R$. Substitution of Eq. (12.53) into Eq. (12.36) produces

$$\mathcal{T}^{jk} = \frac{c^3}{16\pi G}\int R^3 \big(\partial_\tau h^{pq}\big)\bigg[N^j \eth_p h^{kq} - N^k \eth_p h^{jq} - \frac{1}{2}\big(N^j \eth^k - N^k \eth^j\big) h_{pq}\bigg]\, d\Omega. \tag{12.54}$$

To bring Eq. (12.54) to the standard form of Eq. (12.47), it is helpful to decompose the differential operator $\eth_j$ into longitudinal and transverse pieces according to $\eth_j = N_j \eth_R + \eth^{\rm T}_j$, where $\eth_R := N^j \eth_j$ is a partial derivative with respect to R, while $\eth^{\rm T}_j := P_j{}^k \eth_k$ is the transverse derivative (related to partial derivatives with respect to ϑ and φ). Because $N_p h^{pq} = 0$, it is easy to see that $\eth_R$ plays no role in the first two terms of Eq. (12.54). The next step is to integrate these terms by parts. We write

$$\big(\partial_\tau h^{pq}\big) N^j \eth^{\rm T}_p h^{kq} = \eth^{\rm T}_p \big(\partial_\tau h^{pq} N^j h^{kq}\big) - \partial_\tau \big(\eth^{\rm T}_p h^{pq}\big) N^j h^{kq} - \big(\partial_\tau h^{pq}\big)\big(\eth^{\rm T}_p N^j\big) h^{kq}, \tag{12.55}$$

and observe that in the TT gauge, the harmonic condition $\partial_\alpha h^{\alpha q} = 0$ yields

$$\begin{aligned} \partial_p h^{pq} &= -N_p \partial_\tau h^{pq} + N_p \eth_R h^{pq} + \eth^{\rm T}_p h^{pq} \\ &= -N_p R^{-1} h^{pq} + \eth^{\rm T}_p h^{pq} \\ &= \eth^{\rm T}_p h^{pq} \\ &= 0. \end{aligned} \tag{12.56}$$

With $\eth^{\rm T}_p N^j = P^j{}_p / R$ we find that

$$\big(\partial_\tau h^{pq}\big) N^j \eth^{\rm T}_p h^{kq} = \eth^{\rm T}_p \big(\partial_\tau h^{pq} N^j h^{kq}\big) - \frac{1}{R}\big(\partial_\tau h^{jp}\big) h^{kp}, \tag{12.57}$$

which we insert within Eq. (12.54). The total transverse derivative gives no contribution to the integral (because the angular domain of integration has no boundary), and we finally arrive at Eq. (12.47). At this stage the distinction between $\eth_j$ and ∂_j is no longer important, because of the antisymmetry of the second group of terms in Eq. (12.47).

To go from Eq. (12.47) to Eq. (12.48) we substitute Eq. (12.44) within the integral and take into account the fact that the basis vectors $\boldsymbol{\vartheta}$ and $\boldsymbol{\varphi}$ depend on position. A simple calculation shows that their derivatives are given by

$$R\, \partial_k \vartheta^j = -N^j \vartheta_k + \cot\vartheta\, \varphi^j \varphi_k, \tag{12.58a}$$

$$R\, \partial_k \varphi^j = -N^j \varphi_k - \cot\vartheta\, \vartheta^j \varphi_k, \tag{12.58b}$$

and $\boldsymbol{e}_z$ appears in Eq. (12.49) by virtue of the identity $\boldsymbol{e}_z = \cos\vartheta\, \boldsymbol{N} - \sin\vartheta\, \boldsymbol{\vartheta}$.

The manipulations that lead to Eq. (12.50) involve relations $x\partial_y - y\partial_x = \partial_\varphi$, $y\partial_z - z\partial_y = -\sin\varphi\, \partial_\vartheta - \cot\vartheta \cos\varphi\, \partial_\varphi$, and $z\partial_x - x\partial_z = \cos\varphi\, \partial_\vartheta - \cot\vartheta\, \sin\varphi\, \partial_\varphi$, which are well known from the theory of orbital angular momentum in quantum mechanics.

12.2.5 Isaacson's effective energy-momentum tensor

In a series of two papers published in 1968, Richard Isaacson formulated and developed the shortwave approximation of Sec. 12.2.2. His formulation was more general and ambitious than what was presented there. Instead of limiting his attention to the far-away wave zone of an asymptotically-flat spacetime, Isaacson considered a high-frequency gravitational wave in an arbitrary vacuum region of spacetime in which the radius of curvature $\mathcal{R}$ is large compared to λ_c, and developed a systematic expansion of the field equations in powers of $\lambda_c/\mathcal{R}$. Writing the exact Einstein tensor as $G^{\alpha\beta} = G_0^{\alpha\beta} + G_1^{\alpha\beta} + G_2^{\alpha\beta} + \cdots$, where $G_0^{\alpha\beta}$ is the Einstein tensor of the background spacetime, $G_1^{\alpha\beta}$ the correction of order $\lambda_c/\mathcal{R}$, and $G_2^{\alpha\beta}$ the correction of order $(\lambda_c/\mathcal{R})^2$, he showed that $G_1^{\alpha\beta} = 0$ produces a linearized propagation equation for the gravitational waves, while

$$G_0^{\alpha\beta} = \frac{8\pi G}{c^4} T_{\text{eff}}^{\alpha\beta} := -G_2^{\alpha\beta} + \cdots \tag{12.59}$$

determines how the energy-momentum in the waves affects the background spacetime. After a suitable averaging procedure to eliminate terms that oscillate rapidly, the Isaacson effective energy-momentum tensor $T_{\text{eff}}^{\alpha\beta}$ can be involved in equations such as Eqs. (12.32), (12.34), and (12.36) to compute fluxes of energy, linear momentum, and angular momentum. Computation of the effective energy-momentum tensor in the far-away wave zone of an asymptotically-flat spacetime actually returns the right-hand side of Eq. (12.40), and therefore leads to the same results as Eqs. (12.42) and (12.43).

There are two reasons why the Isaacson approach fails to provide a complete foundation for the description of radiative losses in gravitating systems. First, while the Isaacson approach correctly reproduces the expressions for $\mathcal{P}$ and $\mathcal{F}^j$ that we obtained in Sec. 12.2.3, it does not provide definitions for the system's total energy E and total momentum P^j. In other words, Isaacson manages to supply the right-hand sides of the balance equations (12.31) and (12.33), but fails to reveal the identity of the left-hand sides. Since the main purpose of obtaining expressions for $\mathcal{P}$ and $\mathcal{F}^j$ is to involve them in statements of energy and momentum balance, the Isaacson approach falls short of providing a complete package. The second failure of the Isaacson approach has to do with the angular-momentum flux $\mathcal{T}^{jk}$. Isaacson's effective energy-momentum tensor agrees with Eq. (12.40) in the far-away wave zone of an asymptotically-flat spacetime, but we know from Sec. 12.2.4 that this approximation is too crude to deliver the expression of Eq. (12.47); the flux of angular momentum simply cannot be calculated in the Isaacson approach.

12.3 Radiative losses in slowly-moving systems

As was indicated in Sec. 12.2.2, the formalism developed in Sec. 12.2 rests on a shortwave approximation which permits an expansion of the gravitational potentials in powers of $\lambda_c/R \ll 1$. This approximation scheme is quite independent of the post-Minkowskian and post-Newtonian approximations developed in preceding chapters, and the expressions for $\mathcal{P}$, $\mathcal{F}^j$, and $\mathcal{T}^{jk}$ obtained in Secs. 12.2.3 and 12.2.4 have a wide domain of validity.

In this section we impose a slow-motion condition and calculate the gravitational-wave fluxes for situations in which the gravitational-wave field can be expressed as a post-Newtonian expansion. We shall derive ready-to-use expressions for $\mathcal{P}$, $\mathcal{F}^j$, and $\mathcal{T}^{jk}$ that involve multipole moments of the mass and current distribution, and recover the famous *quadrupole formula* for the rate at which gravitational waves carry energy away from their source. As an application of this slow-motion formalism, we shall examine the radiative losses suffered by a two-body system in bound orbital motion.

12.3.1 Leading-order multipole radiation

The gravitational potentials that result from a post-Minkowskian approximation of the field equations were calculated in Chapter 7, and expressions appropriate for the far-away wave zone were listed in Box 7.7. According to these expressions, the potentials are expressed as a multipole expansion,

$$h^{jk} = \frac{2G}{c^4 R}\left[\ddot{\mathcal{I}}^{jk} + \frac{1}{3c}\left(\dddot{\mathcal{I}}^{jkn} + 2\epsilon^{mnj}\ddot{\mathcal{J}}^{mk} + 2\epsilon^{mnk}\ddot{\mathcal{J}}^{mj}\right)N_n + O(c^{-2})\right], \tag{12.60}$$

automatically incorporating a post-Newtonian expansion in powers of $v_c/c \ll 1$, with v_c denoting a characteristic velocity of the matter distribution. At the leading post-Newtonian order, the formal moments $\mathcal{I}^{jk}$, $\mathcal{I}^{jkn}$, and $\mathcal{J}^{mj}$ can be replaced by their Newtonian expressions defined in terms of the conserved density ρ^*; we have the mass quadrupole moment

$$I^{jk}(\tau) := \int \rho^* x^j x^k \, d^3x\,, \tag{12.61}$$

the mass octupole moment

$$I^{jkn}(\tau) := \int \rho^* x^j x^k x^n \, d^3x\,, \tag{12.62}$$

and the current quadrupole moment

$$J^{jk}(\tau) := \epsilon^{jab} \int \rho^* x^a v^b x^k \, d^3x\,. \tag{12.63}$$

In these equations, the mass density ρ^* and the velocity field $\boldsymbol{v}$ are given as functions of retarded time $\tau := t - R/c$ and the spatial coordinates $\boldsymbol{x}$. In Eq. (12.60) the term involving the mass quadrupole moment is the leading-order, Newtonian contribution to the gravitational-wave field, while the remaining terms are 0.5PN corrections that are smaller by a factor of order $v_c/c \ll 1$. Our expression for h^{jk} neglects corrections of order $(v_c/c)^2$ and higher.

To properly represent the gravitational-wave field, the potentials of Eq. (12.60) must be subjected to the transverse-tracefree projection introduced in Sec. 11.1.7. According to Eqs. (11.31) and (11.34), we have that

$$h^{jk}_{\mathrm{TT}} = (\mathrm{TT})^{jk}{}_{pq} h^{pq}\,, \tag{12.64}$$

where

$$(\mathrm{TT})^{jk}{}_{pq} := P^j{}_p P^k{}_q - \frac{1}{2} P^{jk} P_{pq}\,, \qquad P^j{}_k := \delta^j{}_k - N^j N_k\,. \tag{12.65}$$

To facilitate this projection it is helpful to decompose the multipole moments into their irreducible components, according to

$$I^{jk} = I^{\langle jk\rangle} + \frac{1}{3}\delta^{jk} I^{pp}, \tag{12.66a}$$

$$I^{jkn} = I^{\langle jkn\rangle} + \frac{1}{5}\big(\delta^{jk} I^{npp} + \delta^{jn} I^{kpp} + \delta^{kn} I^{jpp}\big), \tag{12.66b}$$

$$J^{jk} = J^{\langle jk\rangle} + J^{[jk]}, \tag{12.66c}$$

in which the angular brackets indicate the symmetric-tracefree operation first introduced in Sec. 1.5.3, and the square brackets indicate antisymmetrization.

In terms of the irreducible moments we have that the gravitational potentials of Eq. (12.60) can be expressed as

$$h^{jk} \overset{\text{TT}}{=} \frac{2G}{c^4 R}\bigg[\ddot{I}^{\langle jk\rangle} + \frac{1}{3c}\Big(\dddot{I}^{\langle jkn\rangle} + 2\epsilon^{mnj}\ddot{J}^{\langle mk\rangle} + 2\epsilon^{mnk}\ddot{J}^{\langle mj\rangle}\Big)N_n + O(c^{-2})\bigg], \tag{12.67}$$

where the equality sign $\overset{\text{TT}}{=}$, first introduced near the end of Sec. 11.3.1, indicates that terms proportional to δ^{jk}, N^j, or N^k have been discarded; these will not survive the TT projection of Eq. (12.64). Note that the gravitational-wave field involves only the STF pieces of the radiative multipole moments.

12.3.2 Leading-order fluxes

It is straightforward to insert Eq. (12.67) within the flux formulae of Eqs. (12.42), (12.43), and (12.47), evaluate the angular integrals, and arrive at

$$\mathcal{P} = \frac{G}{5c^5}\dddot{I}^{\langle pq\rangle}\dddot{I}^{\langle pq\rangle}, \tag{12.68a}$$

$$\mathcal{F}^j = \frac{G}{c^7}\bigg(\frac{2}{63}\overset{(4)}{I}{}^{\langle jpq\rangle}\dddot{I}^{\langle pq\rangle} - \frac{16}{45}\epsilon^j{}_{pq}\dddot{J}^{\langle pr\rangle}\dddot{I}^{\langle qr\rangle}\bigg), \tag{12.68b}$$

$$\mathcal{T}^{jk} = \frac{2G}{5c^5}\Big(\ddot{I}^{\langle jp\rangle}\dddot{I}^{\langle kp\rangle} - \ddot{I}^{\langle kp\rangle}\dddot{I}^{\langle jp\rangle}\Big). \tag{12.68c}$$

We shall sketch the steps leading to these results in a moment, but we first observe that our result for $\mathcal{P}$ is the famous *quadrupole formula* which relates the energy radiated in gravitational waves to the quadrupole-moment tensor of the matter distribution. (Refer to Box 11.2 for a discussion of the historical controversy that has surrounded this formula.) This scales as G/c^5, and additional contributions from higher-order multipole moments would appear at order G/c^7 and higher. Our result for $\mathcal{F}^j$ involves a coupling between the mass quadrupole and octupole moments, as well as one between the mass and current quadrupole moments; we see that $\mathcal{F}^j$ scales as G/c^7 instead of G/c^5. Our result for the angular-momentum flux can also be expressed in terms of the torque vector $\mathcal{T}^j := \frac{1}{2}\epsilon^j{}_{pq}\mathcal{T}^{pq}$,

$$\mathcal{T}^j = \frac{2G}{5c^5}\epsilon^j{}_{pq}\ddot{I}^{\langle pr\rangle}\dddot{I}^{\langle qr\rangle}; \tag{12.69}$$

this involves the mass quadrupole moment only, and here also contributions from higher-order moments would appear at order G/c^7 and higher.

To obtain Eq. (12.68a) for $\mathcal{P}$ we truncate Eq. (12.67) to its leading-order term, carry out the TT projection of Eq. (12.64), and insert the result within Eq. (12.42). After simplification we obtain

$$\mathcal{P} = \frac{G}{2c^5}\Big(\delta_{pr}\delta_{qs} - 2\delta_{pr}\langle\!\langle N_q N_s\rangle\!\rangle + \frac{1}{2}\langle\!\langle N_p N_q N_r N_s\rangle\!\rangle\Big) f^{pq} f^{rs}, \tag{12.70}$$

with $f^{jk} := \dddot{I}^{\langle jk\rangle}$ and $\langle\!\langle \cdots \rangle\!\rangle := (4\pi)^{-1}\int(\cdots)\,d\Omega$ indicating an angular average. These are easily evaluated with the help of the identities

$$\langle\!\langle N_j N_k\rangle\!\rangle = \frac{1}{3}\delta_{jk}, \tag{12.71a}$$

$$\langle\!\langle N_j N_k N_p N_q\rangle\!\rangle = \frac{1}{15}\big(\delta_{jk}\delta_{pq} + \delta_{jp}\delta_{kq} + \delta_{jq}\delta_{kp}\big), \tag{12.71b}$$

$$\langle\!\langle N_j N_k N_p N_q N_r N_s\rangle\!\rangle = \frac{1}{105}\big(\delta_{jk}\delta_{pq}\delta_{rs} + \text{distinct permutations}\big), \tag{12.71c}$$

which were established back in Sec. 1.5.3. Making use of the STF properties of f^{jk}, we arrive at Eq. (12.68a).

To obtain Eq. (12.68b) from Eq. (12.43) we must include the subleading terms in Eq. (12.67), because a leading-order calculation involves angular integrations of odd products of the radial vector $\boldsymbol{N}$, which all vanish. After the TT projection we obtain

$$\mathcal{F}^j = \frac{G}{3c^7}\Big(\delta_{pr}\delta_{qs}\langle\!\langle N^j N_n\rangle\!\rangle - 2\delta_{pr}\langle\!\langle N^j N_q N_s N_n\rangle\!\rangle + \frac{1}{2}\langle\!\langle N^j N_p N_q N_r N_s N_n\rangle\!\rangle\Big) f^{pq} f^{rsn}, \tag{12.72}$$

with $f^{jk} := \dddot{I}^{\langle jk\rangle}$ and $f^{jkn} := \ddddot{I}^{\langle jkn\rangle} + 2\epsilon^{mnj}\dddot{J}^{\langle mk\rangle} + 2\epsilon^{mnk}\dddot{J}^{\langle mj\rangle}$. After evaluation of the angular integrals and simplification (taking advantage of the STF properties of f^{jk} and f^{jkn}), we obtain

$$\mathcal{F}^j = \frac{G}{3c^7}\left(\frac{22}{105} f_{pq} f^{pqj} - \frac{4}{35} f_{pq} f^{jpq}\right), \tag{12.73}$$

which leads directly to Eq. (12.68b).

The computations that lead to Eq. (12.68c) from Eq. (12.47) are similar. Here also we truncate Eq. (12.67) to its leading-order term, and the gravitational-wave field must be differentiated only after the TT projection has been accomplished. (This step is facilitated by invoking the identity $R\partial_j P_{pq} = -P_{jp}N_q - P_{jq}N_p$.) The remaining manipulations are straightforward, and we arrive at Eq. (12.68c).

12.3.3 Application: Newtonian binary system

As an application of the formalism developed in this section, we calculate the gravitational-wave fluxes $\mathcal{P}$, $\mathcal{F}^j$, and $\mathcal{T}^{jk}$ for a binary system consisting of a body of mass m_1 at position $\boldsymbol{r}_1(t)$ and a body of mass m_2 at position $\boldsymbol{r}_2(t)$; the positions are given in relation to the system's barycenter, chosen to be at the origin of the spatial coordinates. We describe the mutual gravitational attraction at the leading-order, Newtonian level, and we employ the formulae listed in Eqs. (12.68) and (12.69) to calculate the fluxes. The Keplerian orbital motion was first examined back in Sec. 3.2, and considered again in the context of gravitational-wave emissions in Sec. 11.2.2; the required equations can be found at the beginning of Sec. 11.2.2. The main orbital parameters are the semi-latus rectum p and the eccentricity e. The orbital period is

$$P = 2\pi\sqrt{\frac{a^3}{Gm}}, \tag{12.74}$$

where $a := p/(1-e^2)$ is the semi-major axis and $m := m_1 + m_2$. The orbital energy and angular momentum are

$$E = -\frac{\eta G m^2}{2a}, \qquad \boldsymbol{L} = \eta m h\, \boldsymbol{e}_z, \tag{12.75}$$

where $\eta := m_1 m_2/m^2$, $h := |\boldsymbol{r}\times\boldsymbol{v}| = \sqrt{Gmp}$ with $\boldsymbol{r} := \boldsymbol{r}_1 - \boldsymbol{r}_2$ and $\boldsymbol{v} := \boldsymbol{v}_1 - \boldsymbol{v}_2$, and $\boldsymbol{e}_z := [0, 0, 1]$ is a unit vector normal to the orbital plane. It is useful to note that the equations of motion imply that $v\dot{v} = -Gm\dot{r}/r^2$ and $r\ddot{r} = (r\dot{\phi})^2 - Gm/r$, in which $r := |\boldsymbol{r}|$, $v := |\boldsymbol{v}|$, $\dot{\phi} = h/r^2$, and an overdot indicates differentiation with respect to time.

The multipole moments that appear in the flux formulae are

$$I^{jk} = \eta m\, r^j r^k, \tag{12.76a}$$

$$I^{jkn} = -\Delta \eta m\, r^j r^k r^n, \tag{12.76b}$$

$$J^{jk} = -\Delta \eta m h\, e_z^j r^k, \tag{12.76c}$$

where $\Delta := (m_1 - m_2)/m$. These expressions follow directly from the definitions of Eqs. (12.61), (12.62), and (12.63) when the mass density is given by $\rho^* = m_1\delta(\boldsymbol{x}-\boldsymbol{r}_1) + m_2\delta(\boldsymbol{x}-\boldsymbol{r}_2)$ and the individual positions are expressed in terms of the separation $\boldsymbol{r}$. The result for J^{jk} takes into account the fact that $h\boldsymbol{e}_z = \boldsymbol{r}\times\boldsymbol{v}$.

Differentiation of the multipole moments and involvement of the Keplerian equations yields

$$\begin{aligned}
\ddot{I}^{jk} &= 2\eta m\left(v^j v^k - \frac{Gm}{r^3} r^j r^k\right) \\
&= 2\eta m\Big[(\dot{r}^2 - Gm/r)n^j n^k + \dot{r}(r\dot{\phi})(n^j\lambda^k + \lambda^j n^k) + (r\dot{\phi})^2\lambda^j\lambda^k\Big] \\
&= 2\eta m\frac{Gm}{p}\Big[-(1+e\cos\phi - e^2\sin^2\phi)n^j n^k \\
&\qquad + e\sin\phi\,(1+e\cos\phi)(n^j\lambda^k + \lambda^j n^k) + (1+e\cos\phi)^2\lambda^j\lambda^k\Big],
\end{aligned} \tag{12.77a}$$

$$\begin{aligned}
\dddot{I}^{jk} &= -2\eta m\frac{Gm}{r^2}\left[\frac{2}{r}(r^j v^k + v^j r^k) - \frac{3\dot{r}}{r^2} r^j r^k\right] \\
&= -2\eta m\frac{Gm}{r^2}\Big[\dot{r}\, n^j n^k + 2(r\dot{\phi})(n^j\lambda^k + \lambda^j n^k)\Big] \\
&= -2\eta m\frac{(Gm)^{3/2}}{p^{5/2}}(1+e\cos\phi)^2\Big[e\sin\phi\, n^j n^k + 2(1+e\cos\phi)(n^j\lambda^k + \lambda^j n^k)\Big],
\end{aligned} \tag{12.77b}$$

(continued overleaf)

$$\overset{(4)}{I}{}^{jkn} = \Delta\eta m \frac{Gm}{r^2}\biggl[\frac{20}{r}\bigl(v^j v^k r^n + v^j r^k v^n + r^j v^k v^n\bigr) - \frac{30\dot{r}}{r^2}\bigl(v^j r^k r^n + r^j v^k r^n + r^j r^k v^n\bigr) - \frac{3}{r^3}(3v^2 - 15\dot{r}^2 + 4Gm/r) r^j r^k r^n\biggr]$$
$$= \Delta\eta m \frac{Gm}{r^2}\Bigl\{3\bigl[2\dot{r}^2 - 3(r\dot{\phi})^2 - 4Gm/r\bigr]n^j n^k n^n + 10\dot{r}(r\dot{\phi})\bigl(n^j n^k \lambda^n + n^j \lambda^k n^n + \lambda^j n^k n^n\bigr) + 20(r\dot{\phi})^2\bigl(n^j \lambda^k \lambda^n + \lambda^j n^k \lambda^n + \lambda^j \lambda^k n^n\bigr)\Bigr\}$$
$$= \Delta\eta m \frac{(Gm)^2}{p^3}(1 + e\cos\phi)^2\Bigl[-3(7 + 10e\cos\phi + 5e^2\cos^2\phi - 2e^2)n^j n^k n^n + 10e\sin\phi\,(1 + e\cos\phi)\bigl(n^j n^k \lambda^n + n^j \lambda^k n^n + \lambda^j n^k n^n\bigr) + 20(1 + e\cos\phi)^2\bigl(n^j \lambda^k \lambda^n + \lambda^j n^k \lambda^n + \lambda^j \lambda^k n^n\bigr)\Bigr], \tag{12.77c}$$

$$\dddot{J}{}^{jk} = \Delta\eta m h \frac{Gm}{r^3} e_z^j \left(v^k - \frac{3\dot{r}}{r} r^k\right)$$
$$= -\Delta\eta m h \frac{Gm}{r^3} e_z^j \bigl[2\dot{r}\, n^k - (r\dot{\phi})\lambda^k\bigr]$$
$$= -\Delta\eta m \frac{(Gm)^2}{p^3}(1 + e\cos\phi)^3 e_z^j \bigl[2e\sin\phi\, n^k - (1 + e\cos\phi)\lambda^k\bigr], \tag{12.77d}$$

where $\boldsymbol{n} = [\cos\phi, \sin\phi, 0]$ and $\boldsymbol{\lambda} = [-\sin\phi, \cos\phi, 0]$ are basis vectors in the orbital plane. After symmetrizing $\dddot{J}{}^{jk}$, making all the moments tracefree, and inserting the results within the flux formulae of Eqs. (12.68) and (12.69), we obtain

$$\mathcal{P} = \frac{32}{5}\eta^2 \frac{c^5}{G}\left(\frac{Gm}{c^2 p}\right)^5 (1 + e\cos\phi)^4 \left[1 + 2e\cos\phi + \frac{1}{12}e^2(1 + 11\cos^2\phi)\right], \tag{12.78a}$$

$$\mathcal{F}^x = \frac{464}{105}\Delta\eta^2 \frac{c^4}{G}\left(\frac{Gm}{c^2 p}\right)^{11/2} \sin\phi\,(1 + e\cos\phi)^4 \times \left[1 + \frac{175}{58}e\cos\phi + \frac{2}{29}e^2(3 + 40\cos^2\phi) + \frac{5}{58}e^3\cos\phi\,(2 + 9\cos^2\phi)\right], \tag{12.78b}$$

$$\mathcal{F}^y = -\frac{464}{105}\Delta\eta^2 \frac{c^4}{G}\left(\frac{Gm}{c^2 p}\right)^{11/2} (1 + e\cos\phi)^4 \left[\cos\phi - \frac{1}{58}e(9 - 175\cos^2\phi) - \frac{1}{29}e^2\cos\phi\,(1 - 80\cos^2\phi) + \frac{1}{58}e^3(2 + 3\cos^2\phi + 45\cos^4\phi)\right], \tag{12.78c}$$

$$\mathcal{T}^{xy} = \frac{32}{5}\eta^2 m c^2 \left(\frac{Gm}{c^2 p}\right)^{7/2} (1 + e\cos\phi)^3 \left[1 + \frac{3}{2}e\cos\phi - \frac{1}{4}e^2(1 - 3\cos^2\phi)\right], \tag{12.78d}$$

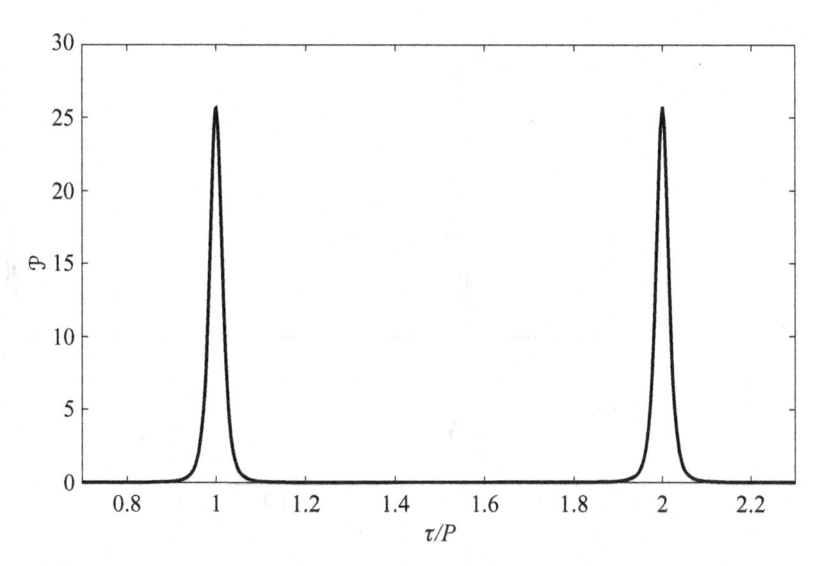

Fig. 12.1 The energy flux $\mathcal{P}$, in units of $(32/5)\eta^2(c^5/G)(Gm/c^2p)^5$, as a function of retarded time τ, in units of the orbital period P. The relation between τ and the orbital phase ϕ is obtained by numerical integration of the Newtonian equations of motion. The orbital eccentricity is set equal to $e = 0.7$, and the gravitational-wave polarizations corresponding to this orbital configuration are displayed in Fig. 11.2. We see that most of the energy is emitted when the orbit is near pericenter, where the motion is fastest.

while $\mathcal{F}^z = 0$ and $\mathcal{T}^{yz} = \mathcal{T}^{zx} = 0$. Plots of $\mathcal{P}$, $\mathcal{F}^x$, and $\mathcal{F}^y$ as functions of time are presented in Figs. 12.1 and 12.2; a plot of $\mathcal{T}^{xy}$ would look very similar to $\mathcal{P}$.

When the orbit is circular (so that $e = 0$) the fluxes simplify to

$$\mathcal{P} = \frac{32}{5}\eta^2\frac{c^5}{G}(v/c)^{10}, \tag{12.79a}$$

$$\mathcal{F}^x = \frac{464}{105}\Delta\eta^2\frac{c^4}{G}(v/c)^{11}\sin\phi, \tag{12.79b}$$

$$\mathcal{F}^y = -\frac{464}{105}\Delta\eta^2\frac{c^4}{G}(v/c)^{11}\cos\phi, \tag{12.79c}$$

$$\mathcal{T}^{xy} = \frac{32}{5}\eta^2 mc^2(v/c)^7, \tag{12.79d}$$

where $v := \sqrt{Gm/p}$ is the orbital velocity. In this case the orbital phase is given by $\phi = \Omega\tau$, with $\Omega := \sqrt{Gm/p^3}$ denoting the orbital angular velocity. We note that in this case of circular motion, the energy and angular-momentum fluxes are related by $\mathcal{P} = \Omega\,\mathcal{T}^{xy}$. We also see that the momentum flux is aligned with $\boldsymbol{v}$ when $m_1 < m_2$ (so that $\Delta < 0$), and that it is anti-aligned when $m_1 > m_2$; the momentum flux is therefore always in the direction of motion of the lighter body.

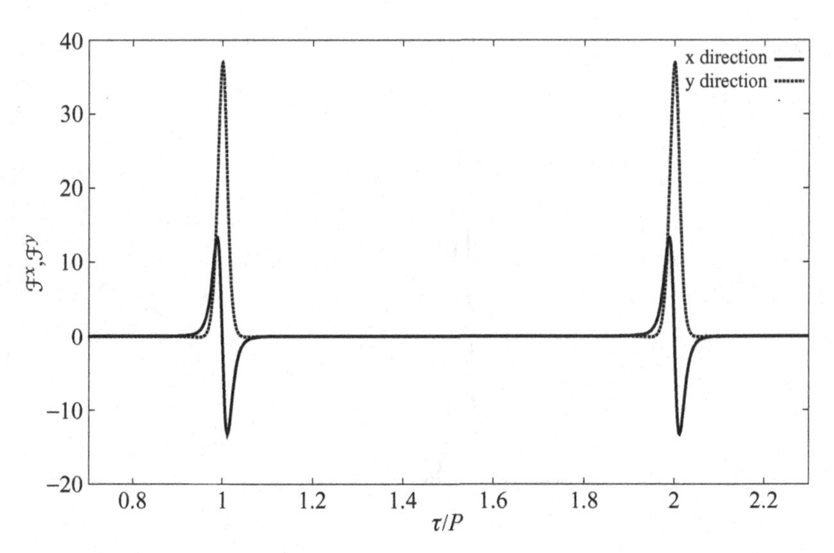

Fig. 12.2 The momentum fluxes $\mathcal{F}^x$ and $\mathcal{F}^y$, in units of $-\Delta\eta^2(c^4/G)(Gm/c^2p)^{11/2}$, as functions of retarded time τ, in units of the orbital period P. A minus sign is inserted in the unit of momentum flux to reflect the fact that $\Delta < 0$ when $m_1 < m_2$. As in Fig. 12.1 the orbital eccentricity is set equal to $e = 0.7$. We see that most of the momentum is emitted when the orbit is near pericenter. The emission in the x-direction cancels out over a complete orbital cycle, but the emission in the y-direction builds up coherently.

Box 12.2 **Momentum flux and gravitational-wave beaming**

Our friend Alan Wiseman has proposed a useful electromagnetic analogy to understand why the momentum flux is always in the direction of motion of the lighter body. Because the lighter body has a higher velocity, the energy flux associated with its motion is beamed in the forward direction, just as the electromagnetic radiation from a rapidly moving body is beamed by relativistic effects. Since energy carries momentum, there is a preferential emission of momentum in the direction of motion of the lighter body. This is admittedly an imperfect analogy, because the energy flux is predominantly quadrupolar, not dipolar, and thus it cannot have a unique direction associated with it. On the other hand, the momentum flux arises from an interference between the quadrupole moment I^{jk} and the 0.5PN corrections in Eq. (12.67), which depend on the octupole and current quadrupole moments; although the latter are not vectors, they have an odd parity, and they can therefore single out a preferred direction. Scaling the energy flux by mc^2, the momentum flux by mc, and taking the ratio gives $(\mathcal{F}^j/mc)/(\mathcal{P}/mc^2) \sim v/c$, precisely the kind of v/c effect that gives rise to relativistic beaming.

Returning to non-circular orbits, it is instructive to average the fluxes over a complete orbital cycle. We define the orbital average of a quantity f by

$$\langle f \rangle := \frac{1}{P}\int_0^P f(t)\,dt, \tag{12.80}$$

where P is the orbital period of Eq. (12.74). This can also be expressed as

$$\langle f \rangle = (1-e^2)^{3/2} \frac{1}{2\pi} \int_0^{2\pi} (1 + e\cos\phi)^{-2} f(\phi)\, d\phi \tag{12.81}$$

by switching integration variables from t to ϕ with the help of the Keplerian equations of motion. Substitution of Eqs. (12.78) into Eq. (12.81) yields

$$\langle \mathcal{P} \rangle = \frac{32}{5}\eta^2 \frac{c^5}{G}\left(\frac{Gm}{c^2 p}\right)^5 (1-e^2)^{3/2}\left(1 + \frac{73}{24}e^2 + \frac{37}{96}e^4\right), \tag{12.82a}$$

$$\langle \mathcal{F}^x \rangle = 0, \tag{12.82b}$$

$$\langle \mathcal{F}^y \rangle = -\frac{52}{5}\Delta\eta^2 \frac{c^4}{G}\left(\frac{Gm}{c^2 p}\right)^{11/2} e(1-e^2)^{3/2}\left(1 + \frac{19}{13}e^2 + \frac{37}{312}e^4\right), \tag{12.82c}$$

$$\langle \mathcal{T}^{xy} \rangle = \frac{32}{5}\eta^2 mc^2 \left(\frac{Gm}{c^2 p}\right)^{7/2} (1-e^2)^{3/2}\left(1 + \frac{7}{8}e^2\right). \tag{12.82d}$$

That $\langle \mathcal{F}^x \rangle = 0$ and $\langle \mathcal{F}^y \rangle \neq 0$ reflects the fact that while the orbital motion is reflection-symmetric across the x-axis, it is asymmetrical with respect to the y-axis: the motion is faster when the orbit crosses the x-axis at pericenter, and slower when it crosses it at apocenter, moving in the opposite direction. Most of the gravitational waves are emitted near pericenter, where the motion is fastest, and most of the momentum flux is therefore generated near $\phi = 0$. When $\Delta < 0$, that is, when $m_1 < m_2$, we have that $\mathcal{F}^y > 0$ at pericenter, and the averaged flux also is positive; this is the situation represented in Fig. 12.2. On the other hand, when $\Delta > 0$ and $m_1 > m_2$, we have that $\mathcal{F}^y < 0$ at pericenter, and the averaged flux also is negative. The momentum flux vanishes when the masses are equal, and the averaged flux vanishes when $e = 0$.

Bodies in unbound orbits, with $e \geq 1$, also radiate energy, momentum, and angular momentum. In this case there is no orbital period and no useful notion of averaged fluxes, but one can calculate the total energy, momentum, and angular momentum radiated during the encounter. These losses raise the interesting possibility of a "gravitational-wave capture," whereby two bodies in an unbound orbit lose enough energy to end up in a bound orbit. This possibility is explored in Exercise 12.8.

It is tempting to exploit the balance equations in averaged form,

$$\frac{dE}{dt} = -\langle \mathcal{P} \rangle, \qquad \frac{dL}{dt} = -\langle \mathcal{T}^{xy} \rangle, \tag{12.83}$$

to infer the rates at which the orbital elements a, p, and e must change in order to reflect the loss of energy and angular momentum to gravitational waves. This crude and tentative analysis is refined in Sec. 12.9, where we calculate how the orbital elements change instantaneously as a result of the action of the gravitational radiation-reaction force.

The rate of change of a is obtained by inserting Eq. (12.75) into the first balance equation. With Eq. (12.82) we obtain

$$\frac{da}{dt} = -\frac{64}{5}\eta c \left(\frac{Gm}{c^2 a}\right)^3 \frac{1 + \frac{73}{24}e^2 + \frac{37}{96}e^4}{(1-e^2)^{7/2}}. \tag{12.84}$$

Kepler's third law, $P \propto a^{3/2}$, allows us to compute how the orbital period changes as a result of the orbital evolution; we find that it decreases according to

$$\frac{dP}{dt} = -\frac{192\pi}{5}\left(\frac{G\mathcal{M}}{c^3}\frac{2\pi}{P}\right)^{5/3}\frac{1+\frac{73}{24}e^2+\frac{37}{96}e^4}{(1-e^2)^{7/2}}, \tag{12.85}$$

where $\mathcal{M} := \eta^{3/5}m$ is the chirp mass, as defined in Eq. (11.80). The rate of change of p is obtained by inserting Eq. (12.75) into the second balance equation. Here we obtain

$$\frac{dp}{dt} = -\frac{64}{5}\eta c\left(\frac{Gm}{c^2p}\right)^3(1-e^2)^{3/2}\left(1+\frac{7}{8}e^2\right). \tag{12.86}$$

The rate of change of e can then be inferred from the relation $p/a = 1-e^2$; we get

$$\frac{de}{dt} = -\frac{304}{15}\eta c\frac{e}{a}\left(\frac{Gm}{c^2a}\right)^3\frac{1+\frac{121}{304}e^2}{(1-e^2)^{5/2}} \tag{12.87a}$$

$$= -\frac{304}{15}\eta c\frac{e}{p}\left(\frac{Gm}{c^2p}\right)^3(1-e^2)^{3/2}\left(1+\frac{121}{304}e^2\right). \tag{12.87b}$$

We see that radiative losses tend to decrease both the size of the orbit and its eccentricity; in the course of time the orbit shrinks and becomes increasingly circular. The orbital evolution proceeds on a *radiation-reaction time scale* of the order of

$$T_{\rm rr} = \frac{1}{\eta}\left(\frac{Gm}{c^2a}\right)^{-5/2}\frac{P}{2\pi}; \tag{12.88}$$

this is longer than the orbital period by a factor of order $(v_c/c)^{-5} \gg 1$, where $v_c \sim \sqrt{Gm/a}$ is the characteristic orbital velocity.

The evolution equations for a, p, and e can be integrated to give

$$a = a_0(e/e_0)^{12/19}\left(\frac{1-e_0^2}{1-e^2}\right)\left(\frac{1+\frac{121}{304}e^2}{1+\frac{121}{304}e_0^2}\right)^{870/2299} \tag{12.89}$$

and

$$p = p_0(e/e_0)^{12/19}\left(\frac{1+\frac{121}{304}e^2}{1+\frac{121}{304}e_0^2}\right)^{870/2299}, \tag{12.90}$$

which express a and p as functions of the eccentricity; here a_0 and p_0 are respectively the values of a and p when $e = e_0$. When the eccentricity is always small, so that both e and e_0 are much smaller than unity, these results can be approximated as $a \simeq a_0(e/e_0)^{12/19}$ or $e \simeq e_0(a/a_0)^{19/12}$; in this approximation $p \simeq a$.

12.4 Astrophysical implications of radiative losses

12.4.1 Binary pulsars

For many years, any discussion of losses due to gravitational-wave emissions, such as the one presented in the previous section, was purely academic, because the effects are so

Table 12.1 Orbital parameters of binary pulsars. The parameter $a_1 \sin \iota$ is the projection of the pulsar's semi-major axis along the line of sight. The parameter $\dot{\omega}$ is defined by Eq. (10.17). The parameter γ' is the coefficient $[(M_1 + 2M_2)/m](a^3/Gm)^{1/2}(GM_2/c^2 a)e$ that appears in front of $\sin u$ in Eq. (10.135). The period derivative $\dot{P}$ is given by Eq. (12.85). The parameter r is the coefficient GM_2/c^2 that appears in front of the logarithmic term in Eq. (10.139), and $s = \sin \iota$ appears within the logarithm. The numbers in parentheses indicate the error in the last digit.

Parameter	PSR 1913+16	J0737-3039A
Keplerian parameters		
$a_1 \sin \iota/c$ (s)	2.341782(3)	1.415032(1)
Eccentricity, e	0.6171334(5)	0.0877775(9)
Orbital period, P (day)	0.322997448911(4)	0.10225156248(5)
Relativistic parameters		
Pericenter advance, $\dot{\omega}$ ($^\circ$ yr^{-1})	4.226598(5)	16.8995(7)
Redshift/time dilation, γ' (ms)	4.2992(8)	0.386(3)
Period derivative, $\dot{P}$ (10^{-12})	−2.423(1)	−1.25(2)
Shapiro delay, r (μs)		6.2(3)
Shapiro delay, s		0.9997(4)

extremely small. The effects of radiative losses on the Earth–Sun or Earth–Moon orbits, for example, are utterly undetectable, as you will discover in Exercise 12.4. The complete absence of experimental evidence bearing on radiative losses was a key ingredient in the persistence of the quadrupole-formula controversy, as reviewed in Box 11.2. All this changed in the fall of 1974.

During the summer of 1974, radio astronomer Joseph H. Taylor and his graduate student Russell Hulse were conducting a systematic search for new pulsars using the Arecibo radio telescope in Puerto Rico. Pulsars are rotating neutron stars that emit a beam of radio waves detectable as a pulse every time the beam crosses a radio receiver. Since the discovery of the first pulsar in 1967, almost 100 had been found by 1974 (nearly 2000 are known today). But this new pulsar, labeled PSR 1913+16 according to its position on the sky, was different from all the others because its pulse period, nominally about 59 milliseconds, varied by plus or minus 40 microseconds on a roughly eight-hour period. Hulse and Taylor quickly concluded that the variation was a result of the Doppler shift of the period caused by the pulsar's orbit about an unseen companion. By December they had measured the key orbital elements, the orbital period, the eccentricity, the semi-major axis projected along the line of sight, as well as the relativistic pericenter advance (the current values are listed in Table 12.1).

It was already clear from the first discovery announcement in late September that the binary pulsar would provide an opportunity to test the quadrupole formula. Inserting the measured period P and eccentricity e into Eq. (12.85), and scaling the chirp mass by the solar mass, we find that the orbital period should decrease at a rate of $53.2\,(\mathcal{M}/M_\odot)^{5/3}$ microseconds per year. Out of an orbital period of almost eight hours, this is a dauntingly small effect, but one that could nevertheless be measured. This required a change of data-analysis strategy, from measuring pulse periods to measuring pulse arrival times (as described in Sec. 10.3.6), and improvements of the instrumentation at the Arecibo telescope, to better handle the effects of interstellar dispersion on the radio signals. After collecting data for a few short years, Taylor and colleagues were able to report the first measurement of the period decrease in December 1979.

The measurement of dP/dt can only be compared with the relativistic prediction when the chirp mass $\mathcal{M}$ is known. Fortunately, the measured pericenter advance of 4.2 degrees per year gives the total mass – refer to Eq. (10.17) – and the arrival-time analysis also yields the combination of masses $(M_1 + 2M_2)M_2$ via the periodic term in Eq. (10.135). These data produced precise values for the pulsar and companion masses, $(1.4398 \pm 0.0002)M_\odot$ and $(1.3886 \pm 0.0002)M_\odot$ respectively, and thus a precise prediction of $dP/dt = -(2.402531 \pm 0.000014) \times 10^{-12}$, in beautiful agreement with the measurement.

While the value of dP/dt reported by Taylor in 1979 had measurement errors of 10 percent, the errors have steadily decreased over time, and they are now at the level of about 0.05 percent. At this level of precision it is necessary to take into account the relative acceleration between the binary pulsar and the solar system caused by the differential rotation of the galaxy. This causes all observed periods in the binary-pulsar system to vary slowly with time. The best estimates of the galactic effect yield a small correction $(dP/dt)_{\rm gal} = -(0.027 \pm 0.005) \times 10^{-12}$, so that the agreement between the observed radiative loss and the prediction of the quadrupole formula can be expressed as

$$\frac{(dP/dt)_{\rm obs}}{(dP/dt)_{\rm GR}} = 0.997 \pm 0.002\,. \tag{12.91}$$

This has been widely hailed as definitive, albeit indirect, evidence for the existence of gravitational radiation, and Taylor and Hulse were rewarded for this discovery with the 1993 Nobel Prize in Physics.

Interestingly, while continued observation of the system will provide more precise measurements of dP/dt (the statistical errors on such drifting parameters tend to decrease as $T^{-3/2}$, where T is the observation time), they will *not* produce an improved test of the quadrupole formula. This is because the measurement of dP/dt is now limited by our poor understanding of the galaxy's differential rotation, and the uncertainties associated with it are unlikely to ever be reduced. In fact, the tables can be turned: the assumed validity of the quadrupole formula combined with the observed dP/dt provides a better measure of the relative acceleration than is likely to be obtained by standard astronomical techniques.

Many other binary pulsars have been found since Hulse and Taylor's discovery, almost 100 to date and counting. One of the most remarkable is the "double pulsar" J0737-3039A&B, discovered in 2003. Whereas a single active pulsar was detected in the Hulse–Taylor system (the companion is believed to be a very old, essentially "dead" pulsar), here *both objects*, denoted A and B, were detected as pulsars. The orbit is very relativistic, with an orbital period of only a tenth of a day, and a pericenter advance of almost 17 degrees per year (see Table 12.1). The orbit is observed almost edge on, so that in addition to the three relativistic effects measured in the Hulse–Taylor binary pulsar (pericenter advance, redshift-time dilation effect, and period decrease), it is also possible to measure the Shapiro time delay of the signal from the stronger pulsar as it passes by the companion once per orbit. This delivers a precise determination of the masses – $M_A = (1.3381 \pm 0.0007)M_\odot$

and $M_B = (1.2489 \pm 0.0007)M_\odot$ – and the period derivative has already been measured to two-percent accuracy. Ultimately this system will yield a more precise test of the quadrupole formula than did PSR 1913+16, partly because the main pulsar has a narrower pulse profile, leading to a more precise timing of the pulse arrivals, but mainly because the system is closer to the Earth and in a favorable location within the galaxy, so that the galactic acceleration is essentially negligible (the acceleration is predominantly transverse to the line of sight). Another remarkable feature of the double pulsar is that it is no longer double: since its discovery the spin axis of the secondary pulsar has shifted in direction, thanks to the relativistic geodetic precession reviewed in Sec. 10.4.3, and as a consequence the pulsar beam is no longer intersecting the Earth. This is another example of a post-Newtonian effect with practical consequences, at least for the radio astronomers who keep observing the double pulsar.

12.4.2 Inspiralling compact binaries

In due course, radiative losses bring the two neutron stars of a binary-pulsar system into such close proximity that their orbital periods are measured in fractions of a second rather than hours. At this stage the gravitational waves emitted by the binary system enter the frequency band of a ground-based laser interferometric detector such as LIGO and Virgo, namely the band which lies between approximately 10 Hz and 1000 Hz. It is totally pointless to wait around for this to happen, however, because the required time is extremely long. In the limit of small eccentricities, it is straightforward to estimate the inspiral time by integrating Eq. (12.84) from a's present value to a final value of effectively zero. The present value and the total mass m can be conveniently expressed in terms of the orbital period P and the observed rate of pericenter advance $\dot{\omega}$. For the special case of almost equal masses ($\eta = 1/4$), this calculation yields

$$T = \frac{5}{128}\left(\frac{6\pi}{\dot{\omega}}\right)^{5/2}\left(\frac{1}{P}\right)^{3/2} \approx 1.9 \times 10^8 \left(\frac{5^\circ/\text{yr}}{\dot{\omega}}\right)^{5/2}\left(\frac{1\ \text{day}}{P}\right)^{3/2}\ \text{yr}, \tag{12.92}$$

and we see that for a typical binary pulsar, the inspiral time is of the order of 200 million years. A more eccentric orbit starting with the same a would give rise to a shorter time scale, because of the excess radiative loss induced by the rapidly changing velocities near pericenter (refer to Fig. 12.1).

The important point about this time scale, long as it may seem, is that it is still short compared to the age of typical galaxies like the Milky Way. There is therefore a chance that a binary pulsar born 200 million years ago in a nearby galaxy would be in its final death spiral today, emitting a detectable burst of gravitational waves. Such systems, called "inspiralling compact binaries" (they could also involve black holes), are a highly anticipated source of gravitational waves for detection by the LIGO–Virgo class of interferometers. Based on the small number of relativistic binary pulsars that have been observed to date, it is possible to estimate that one such inspiral could occur once every million years (give or take an order or magnitude) in a galaxy like ours. With detectors capable of measuring gravitational waves in a volume that contains millions of galaxies, as will be the case with advanced versions of the LIGO and Virgo instruments, the number of detectable events becomes interesting.

Another important feature of radiative losses is the circularization of orbits. When the gravitational waves from an inspiralling compact binary enter the frequency band of the LIGO and Virgo instruments, the system's semi-major axis a is of the order of 1000 km or less, and is therefore much smaller than its initial value a_0. Equation (12.89) then implies that the orbital eccentricity will be vanishingly small, irrespective of its initial value. This circumstance greatly simplifies the nature of the expected gravitational-wave signal, and reduces the challenge of devising data-analysis strategies to measure the waves.

Because the interferometers are broad-band detectors, they effectively detect the gravitational wavetrain cycle by cycle, measuring not only the varying strength within each cycle, but also the varying frequency as the inspiral proceeds. Studies of optimal detection strategies have shown that these measurements, particularly of the varying wave frequency or phase, can be very accurate, depending of course on the strength of the signal compared to instrumental noise. To exploit these optimal strategies, one must be able to predict the evolution of the phase to high accuracy, and this requires going beyond the quadrupole formula by incorporating higher-order post-Newtonian corrections in calculations of the gravitational-wave signal.

With this in mind, we now calculate the first post-Newtonian corrections to the fluxes of energy, momentum, and angular momentum emitted by a binary system in circular orbit, and infer the rate at which the orbital frequency changes with time. The gravitational waves emitted by a post-Newtonian binary system in circular motion were calculated in Sec. 11.4.6. There we introduced

$$\beta := \left(\frac{Gm\Omega}{c^3} \right)^{1/3} \tag{12.93}$$

as a meaningful post-Newtonian expansion parameter, defined directly in terms of the total mass m and orbital frequency Ω. The orbital velocity is then given by $v/c = \beta[1 - \frac{1}{3}(3-\eta)\beta^2 + O(\beta^4)]$, and the orbital radius is determined from $Gm/(c^2 r) = \beta^2[1 + \frac{1}{3}(3 - \eta)\beta^2 + O(\beta^4)]$. The gravitational-wave polarizations are expressed as

$$h_{+,\times} = \frac{2\eta Gm}{c^2 R} \beta^2 H_{+,\times}, \tag{12.94}$$

with $H_{+,\times}$ expanded as

$$H_{+,\times} = H^{[0]}_{+,\times} + \Delta\beta H^{[1/2]}_{+,\times} + \beta^2 H^{[1]}_{+,\times} + \Delta\beta^3 H^{[3/2]}_{+,\times} + \beta^3 H^{\rm tail}_{+,\times} + O(\beta^4), \tag{12.95}$$

where $\Delta := (M_1 - M_2)/m$ is the dimensionless mass difference. Each contribution $H^{[n]}_{+,\times}$ is a function of the inclination angle ι and of the orbital phase $\Psi := \Omega\tau + \omega$, where ω is a second orientation angle. Explicit expressions appear in Eqs. (11.296) and (11.297).

The most efficient way to compute $\mathcal{P}$, $\mathcal{F}^j$, and $\mathcal{T}^{jk}$ is to proceed via Eqs. (12.45), (12.46), and (12.48), which express the fluxes directly in terms of the gravitational-wave polarizations. To reflect the choice of transverse basis made here (and inherited from Sec. 11.2.2), we make the identifications $\vartheta = \iota$ and $\varphi = \frac{\pi}{2} - \omega$. Because the time dependence of each polarization is contained in the orbital phase Ψ, derivatives with respect to τ can be evaluated as $\partial_\tau = \Omega\partial_\Psi$, and derivatives with respect to φ can similarly be expressed as $-\partial_\Psi$.

Simple computations return

$$\mathcal{P} = \frac{32}{5}\eta^2 \frac{c^5}{G}\beta^{10}\biggl[1 - \Bigl(\frac{1247}{336} + \frac{35}{12}\eta\Bigr)\beta^2 + 4\pi\beta^3 + O(\beta^4)\biggr], \tag{12.96a}$$

$$\mathcal{F}^x = \frac{464}{105}\Delta\eta^2 \frac{c^4}{G}\beta^{11}\biggl[1 - \Bigl(\frac{452}{87} + \frac{1139}{522}\eta\Bigr)\beta^2 + O(\beta^3)\biggr]\sin\Omega\tau, \tag{12.96b}$$

$$\mathcal{F}^y = -\frac{464}{105}\Delta\eta^2 \frac{c^4}{G}\beta^{11}\biggl[1 - \Bigl(\frac{452}{87} + \frac{1139}{522}\eta\Bigr)\beta^2 + O(\beta^3)\biggr]\cos\Omega\tau. \tag{12.96c}$$

We also have that $\mathcal{F}^z = 0 = \mathcal{T}^{yz} = \mathcal{T}^{zx}$, and $\mathcal{T}^{xy}$ is related to $\mathcal{P}$ by $\mathcal{P} = \Omega\,\mathcal{T}^{xy}$. We see that these expressions reduce to Eqs. (12.79) when $\beta \ll 1$.

We observe that $\mathcal{P}$ contains no correction term at order β, in spite of the fact that the gravitational-wave polarizations do possess such terms; a 0.5PN correction to the energy flux would have to come from an interaction between the $H^{[0]}$ and $H^{[1/2]}$ terms within H, but because these signals are out of phase, the interaction produces no flux. We observe also that $\mathcal{P}$ contains a term at order β^3, and such a term has three possible origins. First, it might have originated from an interaction between $H^{[0]}$ and $H^{[3/2]}$, but this produces no flux because these signals also are out of phase. Second, it might have originated from an interaction between $H^{[1]}$ and $H^{[1/2]}$, but this does not contribute for the same reason. The only remaining possibility is an interaction between $H^{[0]}$ and $H^{\rm tail}$; these signals are in phase, and their interaction does indeed contribute to the energy flux. The $4\pi\beta^3$ term within Eq. (12.96a) has its origin in the tail effect; it is a *wave-propagation correction* to the leading-order expression that appears outside the large square brackets.

Taking inspiration from Sec. 12.3.3, we insert the energy flux of Eq. (12.96) within the energy-balance equation $dE/dt = -\mathcal{P}$ to determine the orbital evolution of a post-Newtonian binary system. From Eqs. (10.9) and (10.11) we know that the orbital energy of the two-body system is given by

$$E = -\frac{1}{2}\eta m c^2\beta^2\biggl[1 - \frac{1}{12}(9+\eta)\beta^2 + O(\beta^4)\biggr], \tag{12.97}$$

and dE/dt is therefore related to $d\beta/dt$, the rate of change of the velocity parameter of Eq. (12.93). Solving for this yields

$$\frac{d\beta}{dt} = \frac{32}{5}\eta\frac{c^3}{Gm}\beta^9\biggl[1 - \Bigl(\frac{743}{336} + \frac{11}{4}\eta\Bigr)\beta^2 + 4\pi\beta^3 + O(\beta^4)\biggr], \tag{12.98}$$

and integration produces Ω as a function of time.

We have computed the orbital evolution of a circular binary system through 1.5PN order beyond the leading order, quadrupole-formula expression. This degree of accuracy is not sufficient for the purpose of measuring gravitational waves with the advanced LIGO and Virgo detectors. For this it is proving necessary to calculate the energy flux and the resulting expression for $d\beta/dt$ through 3.5PN order beyond the quadrupole formula.

12.4.3 "How black holes get their kicks"

[The title of this section is borrowed from a seminal paper on black-hole recoils by Marc Favata, Scott Hughes, and Daniel Holz (2004). It inspired others to title their papers with equally bad puns. Among the notable cases we find "Getting a kick out of numerical relativity" by Baker *et al.* (2006), and "Total recoil" by Gonzalez *et al.* (2007).]

Another radiative loss with potentially important astrophysical consequences is the radiation of linear momentum and the resulting recoil of the system. This is particularly important for binary inspirals. Because momentum is radiated in the direction of motion of the lighter body, the center-of-mass of the system must recoil in the opposite direction. But this direction is continually changing as the bodies move on their orbits, and the center-of-mass spirals outward from its initial position, with a growing radius and an ever increasing speed as the radiation gains in intensity. Finally, after the bodies have merged and settled down to a stationary state (frequently a rotating black hole), the center-of-mass moves off with whatever velocity it had when the radiation terminated.

This so-called "kick velocity" can easily be estimated for an inspiralling circular orbit. We appeal to Eqs. (12.79b) and (12.79c), which inform us that the center-of-mass must recoil according to

$$\frac{d\boldsymbol{P}}{dt} = -\mathcal{F} = \frac{464}{105}\Delta\eta^2\frac{c^4}{G}(v/c)^{11}\boldsymbol{\lambda}, \tag{12.99}$$

where $\boldsymbol{\lambda} = [-\sin\phi, \cos\phi, 0]$ is tangent to the orbit. As we observed back in Eqs. (12.82b) and (12.82c), the averaged momentum flux vanishes when the orbital speed and frequency are constant. During an inspiral, however, the speed and frequency increase with time (over a very long radiation-reaction time scale), and the long-term average does not vanish. Integrating Eq. (12.99) from $t = -\infty$ to the present time using an adiabatic approximation, we find that the final kick velocity is given by (see Exercise 12.7)

$$\boldsymbol{V}_{\text{kick}} = \frac{464}{105}\Delta\eta^2\left(\frac{Gm}{c^2a}\right)^4 c\,\boldsymbol{n}, \tag{12.100}$$

modulo corrections of higher post-Newtonian order. Note that for a given final orbital velocity $v \sim \sqrt{Gm/a}$, the kick velocity is independent of the total mass of the system. It is directed along $\boldsymbol{n} = [\cos\phi, \sin\phi, 0]$, toward the more massive body. It vanishes in the limit of equal masses ($\Delta = 0$) by symmetry.

For a final separation of $a = 10Gm/c^2$, the recoil speed is equal to $133\Delta\eta^2$ km/s. Because Eq. (12.100) is valid only for $Gm/c^2a \ll 1$, it cannot be trusted for smaller separations, but it should continue to give a reliable order-of-magnitude estimate of the recoil. Numerical simulations of the final inspiral of two non-rotating black holes have produced maximum kick velocities of around 175 km/s, for the mass ratio that maximizes the coefficient $\Delta\eta^2$. Such velocities are probably too small to be of astrophysical importance. When the black holes are spinning rapidly, however, with spin axes misaligned relative to each other and to the orbital angular momentum, the simulations have shown that the combination of momentum flux and strong precessions of the orbital plane can result in "superkicks" as high as several thousand kilometers per second. Such velocities are astrophysically relevant for supermassive black-hole binaries, leading possibly to the ejection of the final black hole from the galaxy that hosted the inspiralling progenitor, or at least to a significant displacement of the final hole from the central core of the galaxy. There is now preliminary evidence for situations in which an accreting massive black hole is apparently displaced and moving away from the center of its host galaxy, with a velocity characteristic of a superkick.

12.5 Radiation-reaction potentials

In this section and the following ones we return to the near zone and work on obtaining expressions for the radiation-reaction force acting within a matter distribution. The analogous situation in electromagnetism was examined in Sec. 12.1.4, where we explained that the radiation-reaction force arises from terms in the potentials that come with an *odd power* of the expansion parameter c^{-1}. In electromagnetism the leading-order term in the radiation-reaction force scales as c^{-3}, and represents a 1.5PC correction to the equations of motion; the scaling reflects the fact that radiative losses are dominated by electric dipole radiation. We recall that we were able to obtain the radiation-reaction force at 1.5PC order without having to calculate conservative corrections to the equations of motion at 1PC order.

The situation is similar in gravitation. Here also the radiation-reaction force arises from terms in the potentials that come with odd powers of c^{-1} in a post-Newtonian expansion. In this case, however, we shall find that the force scales as c^{-5} to reflect the fact that radiative losses are dominated by mass quadrupole radiation; the radiation-reaction force represents a 2.5PN correction to the equations of motion. Conservative corrections at 1PN order were obtained in Secs. 8.4 and 9.3, and we shall see that the computations at 2.5PN order are insensitive to corrections at 2PN order; these are not required for our discussion, and they will not be computed.

In this section we calculate the gravitational potentials that give rise to the radiation-reaction force. Our final results are summarized in Box 12.3. In Sec. 12.6 we involve these potentials in a calculation of the radiation-reaction force acting within a perfect fluid. In Sec. 12.7 we specialize these results to a system of well-separated bodies. In Sec. 12.8 we examine alternative gauges that simplify the description of radiation reaction, and in Sec. 12.9 we calculate the orbital evolution of a two-body system under radiation reaction.

12.5.1 Near-zone potentials

A large part of the work required to identify the potentials responsible for radiation reaction has already been done back in Sec. 7.1.2. There we saw that the potentials are decomposed as $h^{\alpha\beta} = h^{\alpha\beta}_{\mathscr{N}} + h^{\alpha\beta}_{\mathscr{W}}$, into contributions from a near-zone domain $\mathscr{N}$ and a complementary wave-zone domain $\mathscr{W}$. The potentials $h^{\alpha\beta}_{\mathscr{N}}$ were expanded in powers of c^{-1}, up to and including $O(c^{-7})$ for $h^{00}_{\mathscr{N}}$, $O(c^{-6})$ for $h^{0j}_{\mathscr{N}}$, and $O(c^{-5})$ for $h^{jk}_{\mathscr{N}}$; judicious use was made of the conservation equations $\partial_\beta \tau^{\alpha\beta} = 0$ to convert some terms into surface integrals at the boundary of the near zone. In Eqs. (7.15) we obtained the expressions

$$
\begin{aligned}
h^{00}_{\mathcal{N}} = \frac{4G}{c^2}\biggl\{ & \int_{\mathcal{M}} \frac{c^{-2}\tau^{00}}{|\boldsymbol{x}-\boldsymbol{x}'|}\, d^3x' + \frac{1}{2c^2}\frac{\partial^2}{\partial t^2}\int_{\mathcal{M}} c^{-2}\tau^{00}|\boldsymbol{x}-\boldsymbol{x}'|\, d^3x' \\
& - \frac{1}{6c^3}\overset{(3)}{\mathcal{I}}{}^{kk}(t) + \frac{1}{24c^4}\frac{\partial^4}{\partial t^4}\int_{\mathcal{M}} c^{-2}\tau^{00}|\boldsymbol{x}-\boldsymbol{x}'|^3\, d^3x' \\
& - \frac{1}{120c^5}\Bigl[(4x^kx^l + 2r^2\delta^{kl})\overset{(5)}{\mathcal{I}}{}^{kl}(t) - 4x^k\overset{(5)}{\mathcal{I}}{}^{kll}(t) + \overset{(5)}{\mathcal{I}}{}^{kkll}(t)\Bigr] \\
& + O(c^{-6})\biggr\} + h^{00}[\partial\mathcal{M}],
\end{aligned}
\tag{12.101a}
$$

$$
\begin{aligned}
h^{0j}_{\mathcal{N}} = \frac{4G}{c^3}\biggl\{ & \int_{\mathcal{M}} \frac{c^{-1}\tau^{0j}}{|\boldsymbol{x}-\boldsymbol{x}'|}\, d^3x' + \frac{1}{2c^2}\frac{\partial^2}{\partial t^2}\int_{\mathcal{M}} c^{-1}\tau^{0j}|\boldsymbol{x}-\boldsymbol{x}'|\, d^3x' \\
& + \frac{1}{18c^3}\Bigl[3x^k\overset{(4)}{\mathcal{I}}{}^{jk}(t) - \overset{(4)}{\mathcal{I}}{}^{jkk}(t) + 2\epsilon^{mjk}\overset{(3)}{\mathcal{J}}{}^{mk}(t)\Bigr] \\
& + O(c^{-4})\biggr\} + h^{0j}[\partial\mathcal{M}],
\end{aligned}
\tag{12.101b}
$$

$$
\begin{aligned}
h^{jk}_{\mathcal{N}} = \frac{4G}{c^4}\biggl\{ & \int_{\mathcal{M}} \frac{\tau^{jk}}{|\boldsymbol{x}-\boldsymbol{x}'|}\, d^3x' - \frac{1}{2c}\overset{(3)}{\mathcal{I}}{}^{jk}(t) + \frac{1}{2c^2}\frac{\partial^2}{\partial t^2}\int_{\mathcal{M}} \tau^{jk}|\boldsymbol{x}-\boldsymbol{x}'|\, d^3x' \\
& - \frac{1}{36c^3}\Bigl[3r^2\overset{(5)}{\mathcal{I}}{}^{jk}(t) - 2x^m\overset{(5)}{\mathcal{I}}{}^{jkm}(t) - 8x^n\epsilon^{mn(j}\overset{(4)}{\mathcal{J}}{}^{m|k)}(t) + 6\overset{(3)}{\mathcal{M}}{}^{jkmm}(t)\Bigr] \\
& + O(c^{-4})\biggr\} + h^{jk}[\partial\mathcal{M}].
\end{aligned}
\tag{12.101c}
$$

In each Poisson integral the components of the effective energy-momentum pseudotensor $\tau^{\alpha\beta}$ are expressed as functions of t and $\boldsymbol{x}'$, and the integration is over the near-zone domain $\mathcal{M}$ defined by $r' := |\boldsymbol{x}'| < \mathcal{R}$, where $\mathcal{R}$ is the arbitrary cutoff radius between the near and wave zones. As we explained back in Sec. 6.3.3, we are interested in the $\mathcal{R}$-independent pieces of each integral.

The various multipole moments that appear in the potentials were defined by Eqs. (7.16); for convenience we list them explicitly here:

$$
\mathcal{I}^{kl}(t) := \int_{\mathcal{M}} c^{-2}\tau^{00}(t,\boldsymbol{x})x^kx^l\, d^3x\,, \tag{12.102a}
$$

$$
\mathcal{I}^{jkm}(t) := \int_{\mathcal{M}} c^{-2}\tau^{00}(t,\boldsymbol{x})x^jx^kx^l\, d^3x\,, \tag{12.102b}
$$

$$
\mathcal{I}^{kll}(t) := \int_{\mathcal{M}} c^{-2}\tau^{00}(t,\boldsymbol{x})x^kr^2\, d^3x\,, \tag{12.102c}
$$

$$
\mathcal{I}^{kkll}(t) := \int_{\mathcal{M}} c^{-2}\tau^{00}(t,\boldsymbol{x})r^4\, d^3x\,, \tag{12.102d}
$$

$$
\mathcal{J}^{mk}(t) := \epsilon^{mab}\int_{\mathcal{M}} c^{-1}\tau^{0b}(t,\boldsymbol{x})x^ax^k\, d^3x\,, \tag{12.102e}
$$

$$
\mathcal{M}^{jkmm} := \int_{\mathcal{M}} \tau^{jk}(t,\boldsymbol{x})r^2\, d^3x\,. \tag{12.102f}
$$

We explained in Box 7.1 how the multipole moments end up appearing in the expansions of $h^{\alpha\beta}_{\mathcal{N}}$ in powers of c^{-1}, and how these manipulations give rise to additional boundary terms $h^{\alpha\beta}[\partial\mathcal{M}]$. All such boundary terms must be carefully examined, and we assert that *at all post-Newtonian orders to be considered in this section, the boundary terms contain no $\mathcal{R}$-independent pieces, and they can therefore be safely discarded.* We shall not prove this assertion here, but you may be comforted with the assurance that each boundary term in Eq. (12.101) was examined by us and shown to make no $\mathcal{R}$-independent contribution to the potentials.

It was established back in Sec. 7.3.4 that $h^{\alpha\beta}_{\mathcal{W}}$ first makes a contribution at 3PN order in the near zone, and because our considerations in this section are limited to 2.5PN order, this can be ignored. The complete near-zone potentials $h^{\alpha\beta}$ can therefore be identified with those of Eqs. (12.101).

12.5.2 Odd terms in the potentials

The terms that come with an odd power of c^{-1} within the curly brackets in Eqs. (12.101) will be the focus of our attention; it is these that are responsible for the radiation reaction. We have to keep in mind that while some factors of c^{-1} appear explicitly in these equations, some are contained implicitly in the source functions $\tau^{\alpha\beta}$. To identify the explicit and implicit odd terms, it is useful to introduce the notation

$$h^{00} := \frac{4}{c^2}V, \qquad h^{0j} := \frac{4}{c^3}V^j, \qquad h^{jk} := \frac{4}{c^4}W^{jk}, \tag{12.103}$$

for the potentials, and the notation

$$\tau^{00} := \varrho c^2, \qquad \tau^{0j} := s^j c \qquad \tau^{jk} := \tau^{jk}, \tag{12.104}$$

for the source functions. We next express the post-Newtonian expansion of the potentials as

$$V = V[0] + c^{-2}V[2] + c^{-4}V[4] + O(c^{-6}) + c^{-3}V[3] + c^{-5}V[5] + O(c^{-7}), \tag{12.105a}$$

$$V^j = V^j[0] + c^{-2}V^j[2] + O(c^{-4}) + c^{-3}V^j[3] + O(c^{-5}), \tag{12.105b}$$

$$W^{jk} = W^{jk}[0] + c^{-2}W^{jk}[2] + O(c^{-4}) + c^{-1}W^{jk}[1] + c^{-3}W^{jk}[3] + O(c^{-5}), \tag{12.105c}$$

where we have separated the odd-order terms from the even-order terms, and we express the post-Newtonian expansion of the source functions as

$$\varrho = \varrho[0] + c^{-2}\varrho[2] + c^{-4}\varrho[4] + O(c^{-6}) + c^{-5}\varrho[5] + O(c^{-7}), \tag{12.106a}$$

$$s^j = s^j[0] + c^{-2}s^j[2] + O(c^{-4}) + c^{-5}s^j[5] + O(c^{-7}), \tag{12.106b}$$

$$\tau^{jk} = \tau^{jk}[0] + c^{-2}\tau^{jk}[2] + O(c^{-4}) + c^{-5}\tau^{jk}[5] + O(c^{-7}). \tag{12.106c}$$

The expansions of Eqs. (12.105) are directly imported from Eqs. (12.101). The expansions of Eqs. (12.106), however, incorporate an assumption that the odd terms in the source functions begin at order c^{-5}; this assumption will be justified below.

When we substitute Eqs. (12.106) into Eqs. (12.101) and read off the various odd potentials from Eqs. (12.105), we find that

$$V[3] = -\frac{1}{6} G \overset{(3)}{\mathcal{I}}{}^{kk}[0], \tag{12.107a}$$

$$V[5] = G \int_{\mathscr{M}} \frac{\varrho[5]}{|\boldsymbol{x} - \boldsymbol{x}'|} d^3x' - \frac{1}{6} G \overset{(3)}{\mathcal{I}}{}^{kk}[2] \\ - \frac{1}{120} G \Big\{ (4x^k x^l + 2r^2 \delta^{kl}) \overset{(5)}{\mathcal{I}}{}^{kl}[0] - 4x^k \overset{(5)}{\mathcal{I}}{}^{kll}[0] + \overset{(5)}{\mathcal{I}}{}^{kkll}[0] \Big\}, \tag{12.107b}$$

$$V^j[3] = \frac{1}{18} G \Big\{ 3x^k \overset{(4)}{\mathcal{I}}{}^{jk}[0] - \overset{(4)}{\mathcal{I}}{}^{jkk}[0] + 2\epsilon^{mjk} \overset{(3)}{\mathcal{J}}{}^{mk}[0] \Big\}, \tag{12.107c}$$

$$W^{jk}[3] = -\frac{1}{2} G \overset{(3)}{\mathcal{I}}{}^{jk}[0], \tag{12.107d}$$

$$W^{jk}[5] = -\frac{1}{2} G \overset{(3)}{\mathcal{I}}{}^{jk}[2] - \frac{1}{36} G \Big\{ 3r^2 \overset{(5)}{\mathcal{I}}{}^{jk}[0] - 2x^m \overset{(5)}{\mathcal{I}}{}^{jkm}[0] \\ - 8x^n \epsilon^{mn(j} \overset{(4)}{\mathcal{J}}{}^{m|k)}[0] + 6 \overset{(3)}{\mathcal{M}}{}^{jkmm}[0] \Big\}. \tag{12.107e}$$

The bracket notation was extended to the multipole moments; for example $\mathcal{I}^{jk}[0]$ is the quadrupole moment constructed from $\varrho[0]$, the 0PN piece of the effective mass density. We can now appreciate what was meant by implicit and explicit odd terms: Examining the expression for $V[5]$, we see that the terms involving the multipole moments were listed explicitly in Eqs. (12.101), but that the Poisson integral involving $\varrho[5]$ was present only implicitly; it has now become fully explicit, thanks to the expansions of Eqs. (12.106). For future reference we note that

$$W[1] := \delta_{jk} W^{jk}[1] = 3V[3] = -\frac{1}{2} G \overset{(3)}{\mathcal{I}}{}^{kk}[0] \tag{12.108}$$

is an immediate consequence of Eqs. (12.107).

12.5.3 Odd terms in the effective energy-momentum tensor

To proceed we must justify the assumption made in Eqs. (12.106), that the odd terms in ϱ, s^j, and τ^{jk} begin at order c^{-5}, and we must obtain an expression for $\varrho[5]$. We recall that the effective energy-momentum pseudotensor is defined by Eq. (6.52); it is given by

$$\tau^{\alpha\beta} = (-g)\big(T^{\alpha\beta} + t_{\rm LL}^{\alpha\beta} + t_{\rm H}^{\alpha\beta}\big), \tag{12.109}$$

in terms of a material contribution $T^{\alpha\beta}$, the Landau–Lifshitz pseudotensor $t_{\rm LL}^{\alpha\beta}$ of Eq. (6.5), and the harmonic pseudotensor $t_{\rm H}^{\alpha\beta}$ of Eq. (6.53).

We begin with an examination of the material contribution. As usual we take the matter distribution to consist of a perfect fluid, and we adopt the set of variables $\mathsf{m} := \{\rho^*, p, \Pi, \boldsymbol{v}\}$ that was first introduced in Sec. 7.1.1. Here $\rho^* := \sqrt{-g}\gamma\rho$ is the conserved mass density, p is the pressure, Π is the specific internal energy, and $\boldsymbol{v}$ is the velocity field; we also have that ρ is the fluid's proper mass density, g is the metric determinant, and $\gamma := u^0/c$ is determined by the normalization condition $g_{\alpha\beta}u^\alpha u^\beta = -c^2$ for the spacetime velocity field $u^\alpha := \gamma(c, \boldsymbol{v})$. We recall that ρ^* satisfies the continuity equation $\partial_t \rho^* + \partial_j(\rho^* v^j) = 0$, which expresses conservation of rest mass, and that the fluid's energy-momentum tensor is given by

$$T^{\alpha\beta} = (\rho + \epsilon/c^2 + p/c^2)u^\alpha u^\beta + p g^{\alpha\beta}, \tag{12.110}$$

where $\epsilon = \rho\Pi$ is the proper density of internal energy.

Because the metric enters explicitly in Eq. (12.110), the energy-momentum tensor possesses a dependence upon the gravitational potentials, and it therefore contains terms that are odd in c^{-1}. To find these we first calculate the metric, which can be obtained from Eqs. (7.23); a short computation reveals that

$$g_{00} = -1 + (\text{even}) + \frac{2}{c^5}\big(V[3] + W[1]\big) + O(c^{-7}), \tag{12.111a}$$

$$g_{jk} = \delta_{jk} + (\text{even}) + O(c^{-5}), \tag{12.111b}$$

$$g^{00} = -1 + (\text{even}) - \frac{2}{c^5}\big(V[3] + W[1]\big) + O(c^{-7}), \tag{12.111c}$$

$$g^{jk} = \delta^{jk} + (\text{even}) + O(c^{-5}), \tag{12.111d}$$

$$\sqrt{-g} = 1 + (\text{even}) + \frac{2}{c^5}\big(V[3] - W[1]\big) + O(c^{-7}), \tag{12.111e}$$

where (even) designates terms of order c^{-2}, c^{-4}, and so on. As a consequence of Eq. (12.101) we find that the odd terms in g_{0j} first appear at order c^{-6} and can be neglected. (We are aware that 6 is an even number; we stubbornly call it odd because g_{0j} is generated mostly by $h^{0j} = 4V^j/c^3$, whose leading radiation-reaction piece comes from $V^j[3]$. Recall that g_{0j} couples to v^j/c in the energy-momentum tensor, converting the c^{-6} into a genuinely odd term of order c^{-7}.) From these expressions for the metric we obtain

$$\gamma = 1 + (\text{even}) + \frac{1}{c^5}\big(V[3] + W[1]\big) + O(c^{-7}), \tag{12.112}$$

and conclude that odd terms first appear at order c^{-5} in the energy-momentum tensor. For example, the material contribution to ϱ, given by $c^{-2}(-g)T^{00}$, is

$$\varrho_{\text{matter}} = \rho^*\left[1 + (\text{even}) + \frac{1}{c^5}\big(3V[3] - W[1]\big) + O(c^{-7})\right]. \tag{12.113}$$

The relation of Eq. (12.108) implies that the $O(c^{-5})$ term actually vanishes, and this means that there is no matter contribution to $\varrho[5]$.

We next examine the Landau–Lifshitz pseudotensor. Its definition is provided by Eq. (6.5), and we express it in the schematic form

$$(-g)t^{\alpha\beta}_{\rm LL} = \frac{c^4}{16\pi G}(\partial h \partial h)^{\alpha\beta}. \tag{12.114}$$

The overall factor of c^4 on the right-hand side implies that a term of order c^{-7} in $(\partial h \partial h)^{00}$ makes a contribution to $\varrho[5]$, a term of order c^{-8} in $(\partial h \partial h)^{0j}$ makes a contribution to $s^j[5]$, and a term of order c^{-9} in $(\partial h \partial h)^{jk}$ makes a contribution to $\tau^{jk}[5]$. The various derivatives of the potentials are given by

$$\partial_n h^{00} = \frac{4}{c^2}\{\partial_n V[0] + O(c^{-2}) + c^{-5}\partial_n V[5] + O(c^{-7})\}, \tag{12.115a}$$

$$\partial_n h^{0j} = \frac{4}{c^3}\{\partial_n V^j[0] + O(c^{-2}) + c^{-3}\partial_n V^j[3] + O(c^{-5})\}, \tag{12.115b}$$

$$\partial_n h^{jk} = \frac{4}{c^4}\{\partial_n W^{jk}[0] + O(c^{-2}) + c^{-3}\partial_n W^{jk}[3] + O(c^{-5})\}, \tag{12.115c}$$

$$\partial_0 h^{00} = \frac{4}{c^3}\{\partial_t V[0] + O(c^{-2}) + c^{-3}\partial_t V[3] + O(c^{-5})\}, \tag{12.115d}$$

$$\partial_0 h^{0j} = \frac{4}{c^4}\{\partial_t V^j[0] + O(c^{-2}) + c^{-3}\partial_t V^j[3] + O(c^{-5})\}, \tag{12.115e}$$

$$\partial_0 h^{jk} = \frac{4}{c^5}\{\partial_t W^{jk}[0] + O(c^{-2}) + c^{-1}\partial_t W^{jk}[1] + O(c^{-3})\}, \tag{12.115f}$$

in which we have included only the required terms; note that $V[3]$ and $W^{jk}[1]$ do not appear in $\partial_n h^{00}$ and $\partial_n h^{jk}$, because they do not depend on the spatial coordinates.

Referring to Eq. (6.5), we observe that a typical term in $(\partial h \partial h)^{\alpha\beta}$ actually has the form of $gg\partial h \partial h$. (There are also terms of the form $gggg\partial h \partial h$, but they need not be distinguished for the purpose of this argument.) There are two ways of generating terms that contain an odd power of c^{-1}. The first is to let $\partial h \partial h$ be odd in c^{-1}, and to keep the factor gg even; the second is to let gg be odd, and to keep $\partial h \partial h$ even.

In the first scenario we need to multiply an even term in one of the factors ∂h by an odd term in the remaining ∂h. Using the expansions of Eqs. (12.115), we find that the dominant scaling of such products is c^{-8}, and that it is produced by the set

$$\mathscr{S}_1 = \left\{\partial_p h^{00}\partial_q h^{0j},\ \partial_p h^{00}\partial_0 h^{00},\ \partial_p h^{00}\partial_q h^{jk}\right\}. \tag{12.116}$$

We also find that the set of products

$$\mathscr{S}_2 = \Big\{\partial_p h^{00}\partial_q h^{00},\ \partial_p h^{00}\partial_q h^{jk},\ \partial_p h^{00}\partial_0 h^{0j},\ \partial_p h^{0j}\partial_q h^{0k},$$
$$\partial_p h^{0j}\partial_0 h^{00},\ \partial_p h^{0j}\partial_0 h^{kn},\ \partial_0 h^{00}\partial_0 h^{00},\ \partial_0 h^{00}\partial_0 h^{jk}\Big\} \tag{12.117}$$

participates at order c^{-9}.

In the second scenario we let the factors of g supply the odd terms, and we keep $\partial h \partial h$ even. The leading odd terms in g come at order c^{-5} in g_{00}, order c^{-6} in g_{0j}, and order c^{-5} in g_{jk}. The leading even term in $\partial h \partial h$ comes from $\partial_p h^{00}\partial_q h^{00}$ at order c^{-4}. After multiplication we find that the set

$$\mathscr{S}_3 = \left\{ h^{00}\partial_p h^{00}\partial_q h^{00},\ h^{jk}\partial_p h^{00}\partial_q h^{00} \right\} \tag{12.118}$$

also participates at order c^{-9}.

The next step is to decide how the various terms listed in $\mathscr{S}_1$, $\mathscr{S}_2$, and $\mathscr{S}_3$ enter in the components of the Landau–Lifshitz pseudotensor. A careful examination of Eq. (6.5) reveals that $\mathscr{S}_1$ appears only in $(\partial h \partial h)^{0j}$, whose dominant odd term therefore scales as c^{-8}; this produces a contribution to $s^j[5]$. It reveals also that $\mathscr{S}_2$ and $\mathscr{S}_3$ appear in $(\partial h \partial h)^{00}$ and $(\partial h \partial h)^{jk}$, whose dominant odd terms scale as c^{-9}; this produces a contribution to $\varrho[7]$ and $\tau^{jk}[5]$. We conclude from all this that the Landau–Lifshitz pseudotensor makes contributions at the expected odd order to the source functions, but that there is no contribution to $\varrho[5]$.

Finally we examine the harmonic pseudotensor $(-g)t^{\alpha\beta}_{\rm H}$. A similar sequence of steps allows us to conclude that the leading odd terms make contributions to $\varrho[5]$, $s^j[5]$, and $\tau^{jk}[5]$; the assumption contained in Eq. (12.106) is now fully justified. The contribution to $\varrho[5]c^2$ is produced by $-(c^4/16\pi G)h^{jk}\partial_{jk}h^{00}$ on the right-hand side of Eq. (6.53), and from Eqs. (12.105) and (12.107) we find that this is given by

$$\varrho[5] = -\frac{1}{\pi G} W^{jk}[1]\partial_{jk}V[0] = \frac{1}{2\pi}\dddot{\mathcal{I}}^{jk}[0]\partial_{jk}V[0]. \tag{12.119}$$

To put this in its final form we recall that the [0] label refers to the Newtonian limit. The quadrupole moment is therefore the Newtonian moment I^{jk}, and $V[0]$ is the Newtonian potential, which was denoted U in previous chapters. What we have obtained, therefore, is

$$\varrho[5] = \frac{1}{2\pi}\dddot{I}^{jk}\partial_{jk}U. \tag{12.120}$$

We have also established that no other implicit, odd-order contribution to the effective energy-momentum pseudotensor arises at 2.5PN order.

12.5.4 Radiation-reaction potentials: Final expressions

With $\varrho[5]$ in hand, we may now return to Eq. (12.107) and complete the computation of $V[5]$. We need to evaluate the Poisson integral

$$\int_{\mathscr{M}} \frac{\varrho[5]}{|\boldsymbol{x}-\boldsymbol{x}'|}\,d^3x',$$

in which $\varrho[5]$ is expressed as a function of t and $\boldsymbol{x}'$. Making the substitution from Eq. (12.120) gives

$$\int_{\mathscr{M}} \frac{\varrho[5]}{|\boldsymbol{x}-\boldsymbol{x}'|}\,d^3x' = \frac{1}{2\pi}\dddot{I}^{jk}\int_{\mathscr{M}} \frac{\partial_{j'k'}U(t,\boldsymbol{x}')}{|\boldsymbol{x}-\boldsymbol{x}'|}\,d^3x', \tag{12.121}$$

and to deal with this we appeal to integration by parts. We begin by writing

$$\frac{\partial_{j'k'}U}{|\boldsymbol{x}-\boldsymbol{x}'|} = \partial_{j'}\left(\frac{\partial_{k'}U}{|\boldsymbol{x}-\boldsymbol{x}'|}\right) - \partial_{k'}U\partial_{j'}\frac{1}{|\boldsymbol{x}-\boldsymbol{x}'|}. \tag{12.122}$$

Noting that $\partial_{j'}|\boldsymbol{x}-\boldsymbol{x}'| = -\partial_j|\boldsymbol{x}-\boldsymbol{x}'|$, this is

$$\frac{\partial_{j'k'}U}{|\boldsymbol{x}-\boldsymbol{x}'|} = \partial_{j'}\left(\frac{\partial_{k'}U}{|\boldsymbol{x}-\boldsymbol{x}'|}\right) + \partial_j\left(\frac{\partial_{k'}U}{|\boldsymbol{x}-\boldsymbol{x}'|}\right). \tag{12.123}$$

Applying this trick once more, we obtain

$$\frac{\partial_{j'k'}U}{|\boldsymbol{x}-\boldsymbol{x}'|} = \partial_{j'}\left(\frac{\partial_{k'}U}{|\boldsymbol{x}-\boldsymbol{x}'|}\right) + \partial_{jk'}\left(\frac{U}{|\boldsymbol{x}-\boldsymbol{x}'|}\right) + \partial_{jk}\left(\frac{U}{|\boldsymbol{x}-\boldsymbol{x}'|}\right). \tag{12.124}$$

Integration yields

$$\int_{\mathscr{M}} \frac{\partial_{j'k'}U}{|\boldsymbol{x}-\boldsymbol{x}'|}\, d^3x' = \oint_{\partial\mathscr{M}} \frac{\partial_{k'}U}{|\boldsymbol{x}-\boldsymbol{x}'|}\, dS_j + \partial_j \oint_{\partial\mathscr{M}} \frac{U}{|\boldsymbol{x}-\boldsymbol{x}'|}\, dS_k + \partial_{jk}\int_{\mathscr{M}} \frac{U}{|\boldsymbol{x}-\boldsymbol{x}'|}\, d^3x'. \tag{12.125}$$

Inspection of the surface integrals reveals that they scale as $\mathcal{R}^{-1}$ and do not give rise to $\mathcal{R}$-independent contributions to $V[5]$. The remaining volume integral is nothing but $-2\pi X$, up to an $\mathcal{R}$-dependent constant, where X is the post-Newtonian superpotential defined in Box 7.3. Our final expression for the integral is therefore $-2\pi\,\partial_{jk}X$.

We have arrived at

$$\int_{\mathscr{M}} \frac{\rho[5]}{|\boldsymbol{x}-\boldsymbol{x}'|}\, d^3x' = -\overset{\cdots}{I}{}^{jk}\partial_{jk}X. \tag{12.126}$$

This result might have been anticipated on the grounds that the left-hand side is a solution to $\nabla^2(\text{LHS}) = -4\pi\rho[5] = -2\overset{\cdots}{I}{}^{jk}\partial_{jk}U = \nabla^2(-\overset{\cdots}{I}{}^{jk}\partial_{jk}X)$, since the superpotential is itself a solution to $\nabla^2 X = 2U$. The discussion of Box 7.3, however, warns us that integration of each side over a truncated domain $\mathscr{M}$ gives rise to boundary integrals that must be examined closely. In this case we have the happy circumstance that all such integrals give rise to $\mathcal{R}$-dependent terms that can be ignored. In other circumstances the surface integrals might have made important contributions, and the detailed calculation presented here would have revealed them.

This computation completes the determination of the radiation-reaction potentials. Our results are summarized in Box 12.3.

Box 12.3 Radiation-reaction potentials

The gravitational potentials in the near zone can be expanded as

$$h^{00} = \frac{4}{c^2}\Big\{U + O(c^{-2}) + c^{-3}V[3] + c^{-5}V[5] + O(c^{-7})\Big\},$$

$$h^{0j} = \frac{4}{c^3}\Big\{U^j + O(c^{-2}) + c^{-3}V^j[3] + O(c^{-5})\Big\},$$

$$h^{jk} = \frac{4}{c^4}\Big\{P^{jk} + O(c^{-2}) + c^{-1}W^{jk}[1] + c^{-3}W^{jk}[3] + O(c^{-5})\Big\},$$

in which $U := V[0]$, $U^j := V^j[0]$, and $P^{jk} := W^{jk}[0]$ are the leading-order, near-zone potentials listed in Box 7.5. The terms that come with an odd power of c^{-1} are the radiation-reaction potentials, and they are given by

$$V[3] = -\frac{1}{6}G\dddot{I}_{kk},$$

$$V[5] = G\Big[-\ddot{I}^{jk}\partial_{jk}X - \frac{1}{6}\dddot{\mathcal{I}}_{kk}[2] - \frac{1}{60}(r^2\delta^{jk} + 2x^jx^k)\overset{(5)}{I}{}^{jk} + \frac{1}{30}x^j\overset{(5)}{I}{}^{jkk} - \frac{1}{120}\overset{(5)}{I}{}^{jjkk}\Big],$$

$$V^j[3] = G\Big[\frac{1}{6}x^k\overset{(4)}{I}{}^{jk} - \frac{1}{18}\overset{(4)}{I}{}^{jkk} - \frac{1}{9}\dddot{J}^{jkk}\Big],$$

$$W^{jk}[1] = -\frac{1}{2}G\dddot{I}^{jk},$$

$$W^{jk}[3] = G\Big[-\frac{1}{2}\dddot{\mathcal{I}}^{jk}[2] - \frac{1}{12}r^2\overset{(5)}{I}{}^{jk} + \frac{1}{18}x^p\overset{(5)}{I}{}^{jkp} + \frac{1}{9}x^p(\overset{(4)}{J}{}^{jpk} + \overset{(4)}{J}{}^{kpj}) - \frac{1}{9}\dddot{M}^{jkpp}\Big].$$

They are expressed in terms of the Newtonian multipole moments

$$I^{jk} := \int \rho^* x^j x^k\, d^3x,$$

$$I^{jkn} := \int \rho^* x^j x^k x^n\, d^3x,$$

$$I^{jknp} := \int \rho^* x^j x^k x^n x^p\, d^3x,$$

$$J^{jkn} := \int \rho^*(v^j x^k - v^k x^j)x^n\, d^3x,$$

$$M^{jknp} := \int (\rho^* v^j v^k + p\,\delta^{jk})x^n x^p\, d^3x.$$

These are functions of t, and the number of overdots, or the number within an overlaid bracket, indicates the number of differentiations with respect to t. The potentials also depend on $c^{-2}\mathcal{I}^{jk}[2]$, the post-Newtonian correction to the quadrupole moment; an expression for this is not required in subsequent calculations. In addition, $V[5]$ depends on the post-Newtonian superpotential X, defined by

$$X = G\int \rho^*|\boldsymbol{x} - \boldsymbol{x}'|\, d^3x',$$

in which the mass density ρ^* is expressed as a function of t and $\boldsymbol{x}'$.

12.6 Radiation reaction of fluid systems

Our main goal in this section is to calculate the radiation-reaction force density acting within a fluid distribution of conserved mass density ρ^*, pressure p, and velocity field $\boldsymbol{v}$. We rely on the formulation of post-Newtonian fluid dynamics initiated in Sec. 8.4, and work to obtain the terms of order c^{-5} that must be inserted within the post-Newtonian generalization of Euler's equation,

$$\rho^* \frac{dv^j}{dt} = \rho^* \partial_j U - \partial_j p + (\text{even}) + f^j[\text{rr}]. \tag{12.127}$$

The first two terms on the right-hand side govern the Newtonian dynamics of the fluid, (even) designates the conservative corrections at 1PN and 2PN orders, and $f^j[\text{rr}]$ is the radiation-reaction force density that we wish to obtain.

12.6.1 Metric, Christoffel symbols, and matter variables

We first determine the pieces of the metric that are involved in the derivation of the radiation-reaction force. These are obtained by inserting the potentials of Box 12.3 within Eqs. (7.20). For our purposes here the non-linear terms are important, and after a straightforward computation we obtain

$$\begin{aligned} g_{00} &= -1 + 2c^{-2}U + O(c^{-4}) + 8c^{-5}V[3] + O(c^{-6}) \\ &\quad + 2c^{-7}\big(V[5] + W[3]\big) - 24c^{-7}V[3]U + O(c^{-8}), \end{aligned} \tag{12.128a}$$

$$g_{0j} = O(c^{-3}) + O(c^{-5}) - 4c^{-6}V_j[3] + O(c^{-7}), \tag{12.128b}$$

$$g_{jk} = \delta_{jk} + O(c^{-2}) + O(c^{-4}) + 4c^{-5}\big(W_{jk}[1] - \delta_{jk}V[3]\big) + O(c^{-6}), \tag{12.128c}$$

$$\begin{aligned} g^{00} &= -1 - 2c^{-2}U + O(c^{-4}) - 8c^{-5}V[3] + O(c^{-6}) \\ &\quad - 2c^{-7}\big(V[5] + W[3]\big) - 8c^{-7}V[3]U + O(c^{-8}), \end{aligned} \tag{12.128d}$$

$$g^{0j} = O(c^{-3}) + O(c^{-5}) - 4c^{-6}V^j[3] + O(c^{-7}), \tag{12.128e}$$

$$g^{jk} = \delta^{jk} + O(c^{-2}) + O(c^{-4}) - 4c^{-5}\big(W^{jk}[1] - \delta^{jk}V[3]\big) + O(c^{-6}), \tag{12.128f}$$

$$\sqrt{-g} = 1 + O(c^{-2}) + O(c^{-4}) - 4c^{-5}V[3] + O(c^{-6}). \tag{12.128g}$$

To arrive at these results we have taken into account the fact that $W[1] = 3V[3]$.

The metric allows us to obtain the radiation-reaction terms in the relevant Christoffel symbols, and another straightforward computation returns

$$\begin{aligned} c^2\Gamma^j_{00} &= -\partial_j U + O(c^{-2}) + O(c^{-4}) - c^{-5} \\ &\quad \times \Big(\partial^j V[5] + \partial^j W[3] + 4\partial_t V^j[3] - 4W^{jk}[1]\partial_k U - 8V[3]\partial^j U\Big) + O(c^{-6}), \end{aligned} \tag{12.129a}$$

$$\begin{aligned} c\Gamma^j_{0k} &= O(c^{-2}) + O(c^{-4}) + 2c^{-5}\Big(\partial_j V_k[3] - \partial_k V_j[3] + \partial_t W_{jk}[1] \\ &\quad - \delta_{jk}\partial_t V[3]\Big) + O(c^{-6}), \end{aligned} \tag{12.129b}$$

$$\Gamma^j_{kn} = O(c^{-2}) + O(c^{-4}) + O(c^{-6}). \tag{12.129c}$$

We observe that since the c^{-5} term in g_{00} depends on time only, there is no contribution at order c^{-3} in Γ^j_{00}. This is connected to the fact, discussed back in Sec. 7.3.5, that the c^{-5} term in g_{00} can be removed by the coordinate transformation

$$t = t' - \frac{2G}{3c^5}\ddot{I}^{kk}(t') + O(c^{-7}), \tag{12.130}$$

so that it has no physical effect on the motion of matter. This removal works only at order c^{-5}; the transformation generates corrections at order c^{-7} that depend on $\dddot{I}^{kk}$, as can be expected from the presence of $-8V[3]\partial^j U$ in Γ^j_{00}. Similarly, because the c^{-5} term in g_{jk} depends on time only, there is no contribution at order c^{-5} in the purely spatial components of the Christoffel symbols.

We next turn to the matter variables, which were introduced in Sec. 12.5.3. We re-express Eq. (12.112) as

$$\gamma = 1 + O(c^{-2}) + O(c^{-4}) + 4c^{-5}V[3] + O(c^{-6}) \tag{12.131}$$

and observe that the terms of order c^{-5} cancel out in

$$\gamma\sqrt{-g} = 1 + O(c^{-2}) + O(c^{-4}) + O(c^{-6}). \tag{12.132}$$

From Eq. (12.110) we obtain

$$c^{-2}\sqrt{-g}T^{00} = \gamma\rho^* + O(c^{-2}) + O(c^{-4}) + O(c^{-6}), \tag{12.133a}$$

$$c^{-1}\sqrt{-g}T^{0j} = \gamma\rho^* v^j + O(c^{-2}) + O(c^{-4}) + O(c^{-6}), \tag{12.133b}$$

$$\begin{aligned}\sqrt{-g}T^{jk} &= \gamma\rho^* v^j v^k + p\,\delta^{jk} + O(c^{-2}) + O(c^{-4}) \\ &\quad - 4c^{-5}W^{jk}[1]p + O(c^{-6}).\end{aligned} \tag{12.133c}$$

12.6.2 Radiation-reaction force density

The post-Newtonian generalization of Euler's equation is obtained by invoking the local statement of momentum conservation, as expressed by Eq. (8.111):

$$\begin{aligned}0 &= c^{-1}\partial_t\big(\sqrt{-g}T^{0j}\big) + \partial_k\big(\sqrt{-g}T^{jk}\big) \\ &\quad + \Gamma^j_{00}\big(\sqrt{-g}T^{00}\big) + 2\Gamma^j_{0k}\big(\sqrt{-g}T^{0k}\big) + \Gamma^j_{kn}\big(\sqrt{-g}T^{kn}\big).\end{aligned} \tag{12.134}$$

After substitution of Eqs. (12.129) and (12.133) and simplification, we obtain

$$\begin{aligned}0 &= \partial_t\big(\rho^* v^j\big) + \partial_k\big(\rho^* v^j v^k\big) - \rho^*\partial_j U + \partial_j p + O(c^{-2}) + O(c^{-4}) \\ &\quad - c^{-5} f^j[5] + O(c^{-6}),\end{aligned} \tag{12.135}$$

where

$$\begin{aligned}f^j[5] &:= \rho^*\partial_j\big(V[5] + W[3]\big) + 4\rho^*\partial_t V^j[3] - 4\rho^* v^k\big(\partial_j V_k[3] - \partial_k V_j[3] + \partial_t W_{jk}[1]\big) \\ &\quad - 4\rho^* W^{jk}[1]\partial_k U - 8\rho^* V[3]\partial_j U + 4W^{jk}[1]\partial_k p + 4V[3]\partial_j p.\end{aligned} \tag{12.136}$$

We next involve the continuity equation $\partial_t\rho^* + \partial_k(\rho^* v^k) = 0$ and insert the radiation-reaction potentials listed in Box 12.3. The end result is the Euler equation of Eq. (12.127), with $f^j[\mathrm{rr}] = c^{-5} f^j[5] + O(c^{-7})$.

Our final expression for the radiation-reaction force density is

$$f^j[\mathrm{rr}] = \frac{G}{c^5}\biggl(-\rho^* \dddot{I}^{pq} \partial_{jpq} X + 2\rho^* \dddot{I}^{jk} \partial_k U + \frac{4}{3}\rho^* \dddot{I}_{pp} \partial_j U - 2\dddot{I}^{jk} \partial_k p - \frac{2}{3}\dddot{I}^{pp} \partial_j p$$
$$+ 2\rho^* \overset{(4)}{I}{}^{jk} v_k + \frac{3}{5}\rho^* \overset{(5)}{I}{}^{jk} x^k - \frac{1}{5}\rho^* \overset{(5)}{I}{}^{pp} x^j - \frac{2}{15}\rho^* \overset{(5)}{I}{}^{jpp} - \frac{2}{3}\rho^* \overset{(4)}{J}{}^{jpp} \biggr)$$
$$+ O(c^{-7}). \tag{12.137}$$

The various multipole moments that appear in this expression are defined in Box 12.3.

12.6.3 Energy balance

Before we apply the radiation-reaction force of Eq. (12.137) to a system of well-separated bodies, we verify that it leads to the expressions of energy, momentum, and angular-momentum balance that were stated back in Sec. 12.2. We begin with energy balance. According to Eq. (12.31), the total energy E of the fluid distribution should vary in time according to

$$\frac{dE}{dt} = -\mathcal{P} = -\frac{G}{5c^5} \dddot{I}^{\langle jk \rangle} \dddot{I}^{\langle jk \rangle} + O(c^{-7}), \tag{12.138}$$

where the expression for the energy flux $\mathcal{P}$ has been copied from Sec. 12.3.2.

For our purposes here it is sufficient to import an expression for the total energy E that is accurate to the leading, Newtonian order, but it is still necessary to recognize that E also contains terms at 1PN, 2PN, and 2.5PN orders. The Newtonian term was computed in Sec. 8.4.5, and according to Eq. (8.134), we have that

$$E = \mathcal{T} + \Omega + E_{\mathrm{int}} + O(c^{-2}) + O(c^{-4}) + c^{-5} E[5] + O(c^{-6}), \tag{12.139}$$

where $\mathcal{T} := \frac{1}{2}\int \rho^* v^2\, d^3x$ is the fluid's total kinetic energy, $\Omega := -\frac{1}{2}\int \rho^* U\, d^3x$ its gravitational potential energy, and $E_{\mathrm{int}} := \int \rho^* \Pi\, d^3x$ its total internal energy. The 1PN and 2PN terms are not required in this discussion, but the 2.5PN term is important and its precise identity will be revealed in due course.

We differentiate each term in Eq. (12.139) with respect to time. For the kinetic energy we get $d\mathcal{T}/dt = \int \rho^* v_j (dv^j/dt)\, d^3x$, in which we substitute the Euler equation of Eq. (12.127); the contribution at order c^{-5} is $\int v_j f^j[\mathrm{rr}]\, d^3x$, the rate at which the radiation-reaction force does work on the fluid. For the potential energy we get that $d\Omega/dt$ is given exactly by Eq. (12.144d) below and gives no contribution at order c^{-5}. For the internal energy we have that $dE_{\mathrm{int}}/dt = \int \rho^* (d\Pi/dt)\, d^3x$, in which we insert the exact statement of the first law of thermodynamics, as stated back in Sec. 8.4; we have that $d\Pi/dt = (p/\rho^2) d\rho/dt$, where ρ is the fluid's proper-mass density. Because ρ differs from ρ^* by a factor $\gamma\sqrt{-g}$ that contains no term of order c^{-5} – refer to Eq. (12.132) – we find that dE_{int}/dt makes no contribution at 2.5PN order. Finally, observing that all c^0, c^{-2}, and c^{-4} terms in E are conserved by virtue of the conservative dynamics at 2PN order, we are left with

$$\int v_j f^j[\mathrm{rr}]\, d^3x + \frac{d}{dt}\Bigl(c^{-5} E[5]\Bigr) = -\frac{G}{5c^5} \dddot{I}^{\langle jk \rangle} \dddot{I}^{\langle jk \rangle} + O(c^{-7}) \tag{12.140}$$

as a precise statement of energy balance. This equation states that up to an overall change in total energy at 2.5PN order, the work done by the radiation-reaction force is equal to the energy radiated by gravitational waves.

In situations in which the Newtonian dynamics of the system is periodic, or when the evolution begins and ends in a static state, Eq. (12.140) can be averaged over time, and the resulting coarse-grained statement of energy balance no longer involves $E[5]$. In this formulation we would write

$$\left\langle \int v_j f^j[\text{rr}]\, d^3x \right\rangle = -\frac{G}{5c^5}\left\langle \dddot{I}^{\langle jk\rangle} \dddot{I}_{\langle jk\rangle} \right\rangle + O(c^{-7}), \tag{12.141}$$

with the angular brackets indicating the averaging procedure. In this coarse-grained statement we recover the expected equality between work done and radiated energy. This statement is obviously less general than the fine-grained statement of Eq. (12.140), and is subjected to the assumption that the initial and final values of $E[5]$ must be equal.

Before we proceed with the derivation of Eq. (12.140) and the determination of $E[5]$, we re-introduce the tensorial generalizations of $\mathcal{T}$ and Ω,

$$\mathcal{T}^{jk} := \frac{1}{2}\int \rho^* v^j v^k \, d^3x, \tag{12.142a}$$

$$\Omega^{jk} := -\frac{1}{2}G\int \rho^* \rho^{*\prime} \frac{(x-x')^j (x-x')^k}{|\boldsymbol{x}-\boldsymbol{x}'|^3}\, d^3x' d^3x, \tag{12.142b}$$

which were first encountered back in Sec. 1.4.3. In terms of these we have that $\mathcal{T} = \delta_{jk}\mathcal{T}^{jk}$ and $\Omega = \delta_{jk}\Omega^{jk}$. These quantities are involved in a number of helpful identities, including the familiar virial theorem

$$\frac{1}{2}\ddot{I}^{jk} = 2\mathcal{T}^{jk} + \Omega^{jk} + P\delta^{jk}, \tag{12.143}$$

with $P := \int p\, d^3x$ denoting the integrated pressure. Other identities include

$$\dot{\mathcal{T}}_{jk} = \int \rho^* v_{(j}\partial_{k)} U \, d^3x - \int v_{(j}\partial_{k)} p \, d^3x + O(c^{-2}), \tag{12.144a}$$

$$\dot{\Omega}_{jk} - \dot{\Omega}\delta_{jk} = \int \rho^* v^p \partial_{pjk} X \, d^3x, \tag{12.144b}$$

$$\dot{\mathcal{T}} = \int \rho^* v^j \partial_j U \, d^3x - \int v^j \partial_j p \, d^3x + O(c^{-2}), \tag{12.144c}$$

$$\dot{\Omega} = -\int \rho^* v^j \partial_j U \, d^3x. \tag{12.144d}$$

These identities are similar to those encountered previously in Sec. 8.4.4, and it is a simple matter to establish them. For example, Eq. (12.144a) follows directly from the definition of $\mathcal{T}^{jk}$ and involvement of Euler's equation. Equation (12.144c) is obtained by taking the trace of Eq. (12.144a). The derivation of Eq. (12.144b) is more involved, but Eq. (12.144d) follows by taking its trace.

To establish Eq. (12.144b) we begin with the definition of the superpotential, $X = G \int \rho^{*\prime} |\boldsymbol{x} - \boldsymbol{x}'|\, d^3x'$, from which we obtain

$$\int \rho^* v^p \partial_{pjk} X\, d^3x = G \int \rho^* \rho^{*\prime} v^p \partial_{pjk} |\boldsymbol{x} - \boldsymbol{x}'|\, d^3x' d^3x \tag{12.145a}$$

$$= G \int \rho^{*\prime} \rho^* v'^p \partial_{p'j'k'} |\boldsymbol{x}' - \boldsymbol{x}|\, d^3x\, d^3x' \tag{12.145b}$$

$$= -G \int \rho^* \rho^{*\prime} v'^p \partial_{pjk} |\boldsymbol{x} - \boldsymbol{x}'|\, d^3x' d^3x \tag{12.145c}$$

$$= \frac{1}{2} G \int \rho^* \rho^{*\prime} (v - v')^p \partial_{pjk} |\boldsymbol{x} - \boldsymbol{x}'|\, d^3x' d^3x \tag{12.145d}$$

$$= \frac{1}{2} G \frac{d}{dt} \int \rho^* \rho^{*\prime} \partial_{jk} |\boldsymbol{x} - \boldsymbol{x}'|\, d^3x' d^3x. \tag{12.145e}$$

We swap the identities of the integration variables to go from the first to the second line, and to go to the third line we observe that $\partial_{j'} |\boldsymbol{x} - \boldsymbol{x}'| = -\partial_j |\boldsymbol{x} - \boldsymbol{x}'|$. We symmetrize the expressions in the fourth line, and in the fifth line we invoke an integral identity derived in Box 8.4. The final result of Eq. (12.144b) follows after evaluating the remaining derivatives.

Returning to Eq. (12.140), we insert the force density of Eq. (12.137) and involve the identities of Eq. (12.144) to evaluate the integrals. We simplify our expressions by setting $\int \rho^* v^j\, d^3x + O(c^{-2}) = P^j = 0$, and obtain

$$\begin{aligned} \frac{c^5}{G} \int v_j f^j[\mathrm{rr}]\, d^3x = {} & -\dddot{I}_{jk} \big(2\dot{\mathcal{T}}^{jk} + \dot{\Omega}^{jk}\big) + \frac{1}{3} \dddot{I}_{pp} \big(2\dot{\mathcal{T}} + \dot{\Omega}\big) \\ & + \frac{3}{10} \overset{(5)}{I}{}^{jk} \dot{I}^{jk} - \frac{1}{10} \overset{(5)}{I}{}^{pp} \dot{I}^{pp} + 4 \frac{d}{dt} \Big(\dddot{I}^{jk} \mathcal{T}^{jk} \Big) \end{aligned} \tag{12.146}$$

for the rate of work done. With the virial theorem of Eq. (12.143) we can express this as

$$\begin{aligned} \frac{c^5}{G} \int v_j f^j[\mathrm{rr}]\, d^3x = {} & -\frac{1}{2} \dddot{I}^{jk} \dddot{I}^{jk} + \frac{1}{6} \dddot{I}^{pp} \dddot{I}^{pp} + \frac{3}{10} \overset{(5)}{I}{}^{jk} \dot{I}^{jk} - \frac{1}{10} \overset{(5)}{I}{}^{pp} \dot{I}^{pp} \\ & + 4 \frac{d}{dt} \Big(\dddot{I}^{jk} \mathcal{T}^{jk} \Big), \end{aligned} \tag{12.147}$$

and we simplify it further by distributing the time derivatives. We write, for example,

$$\overset{(5)}{I}{}^{jk} \dot{I}^{jk} = \frac{d}{dt} \Big(\overset{(4)}{I}{}^{jk} \dot{I}^{jk} - \dddot{I}^{jk} \ddot{I}^{jk} \Big) + \dddot{I}^{jk} \dddot{I}^{jk}, \tag{12.148}$$

and obtain the final expression

$$\begin{aligned} \frac{c^5}{G} \int v_j f^j[\mathrm{rr}]\, d^3x = {} & -\frac{1}{5} \dddot{I}^{jk} \dddot{I}^{jk} + \frac{1}{15} \dddot{I}^{pp} \dddot{I}^{pp} + \frac{d}{dt} \bigg(4 \dddot{I}^{jk} \mathcal{T}^{jk} + \frac{3}{10} \overset{(4)}{I}{}^{jk} \dot{I}^{jk} - \frac{1}{10} \overset{(4)}{I}{}^{pp} \dot{I}^{pp} \\ & - \frac{3}{10} \dddot{I}^{jk} \ddot{I}^{jk} + \frac{1}{10} \dddot{I}^{pp} \ddot{I}^{pp} \bigg) \end{aligned} \tag{12.149}$$

for the rate at which the radiation-reaction force does work on the fluid.

After simplification achieved by expressing I^{jk} in terms of its trace and tracefree pieces, we find that we do indeed recover the energy-balance statement of Eq. (12.140). The total-derivative term in Eq. (12.149) implies that

$$c^{-5}E[5] = -\frac{G}{c^5}\left(4\dddot{I}^{jk}\mathcal{T}^{jk} + \frac{3}{10}\overset{(4)}{I}{}^{\langle jk\rangle}\dot{I}^{\langle jk\rangle} - \frac{3}{10}\dddot{I}^{\langle jk\rangle}\ddot{I}^{\langle jk\rangle}\right) \tag{12.150}$$

is the 2.5PN contribution to the total energy.

12.6.4 Momentum balance

We next turn to momentum balance, and verify that the radiation-reaction force density of Eq. (12.137) is compatible with the statement of Eq. (12.33),

$$\frac{dP^j}{dt} = -\mathcal{F}^j = O(c^{-7}); \tag{12.151}$$

the scaling of the momentum flux $\mathcal{F}^j$ with c^{-7} was obtained in Sec. 12.3.2. The most important observation here is that P^j is conserved at 2.5PN order; radiation reaction acts on the total momentum at 3.5PN order.

According to Eq. (8.145), the total momentum of the fluid distribution is given by

$$P^j = \int \rho^* v^j\, d^3x + O(c^{-2}) + O(c^{-4}) + c^{-5}P^j[5] + O(c^{-6}), \tag{12.152}$$

with $c^{-5}P^j[5]$ denoting the contribution at 2.5PN order. Differentiation with respect to t and substitution of Eq. (12.127) gives rise to

$$\int f^j[\mathrm{rr}]\, d^3x + \frac{d}{dt}\left(c^{-5}P^j[5]\right) = O(c^{-7}), \tag{12.153}$$

which is our precise statement of momentum balance. We must verify that the integrated radiation-reaction force is equal to a total time derivative, and the computation will reveal the identity of $P^j[5]$.

To evaluate the integral we require the identities

$$\int \rho^*\partial_j U\, d^3x = 0, \tag{12.154a}$$

$$\int \rho^*\partial_{jpq} X\, d^3x = 0, \tag{12.154b}$$

$$\int \partial_j p\, d^3x = 0. \tag{12.154c}$$

The first is familiar; it was encountered back in Sec. 1.4.3 and then again in Sec. 8.4.4, and its demonstration proceeds by inserting the Newtonian potential and noting that the integrand is odd in the integration variables $\boldsymbol{x}$ and $\boldsymbol{x}'$. The second identity is new, but it follows after a similar sequence of steps. The third identity also is familiar, and it follows by applying Gauss's theorem and noting that $p = 0$ on any closed surface that bounds the distribution of matter.

Integration of Eq. (12.137) yields

$$\int f^j[\mathrm{rr}]\, d^3x = -\frac{2}{15}\frac{Gm}{c^5}\left(\overset{(5)}{I}{}^{jpp} + 5\overset{(4)}{J}{}^{jpp}\right) + O(c^{-7}), \tag{12.155}$$

where $m := \int \rho^* \, d^3x$ is the total material mass, and where the multipole moments I_{jkn} and J_{jkn} are defined in Box 12.3. To arrive at this expression we have used the fact that since P^j is conserved at 2.5PN order, we may work in the center-of-mass frame and set $\int \rho^* x^j \, d^3x = O(c^{-5})$ and $\int \rho^* v^j \, d^3x = O(c^{-5})$. The integrated radiation-reaction force is indeed a total time derivative, and from its expression we deduce that the 2.5PN contribution to the total momentum is

$$c^{-5} P^j[5] = \frac{2}{15} \frac{Gm}{c^5} \Big(\overset{(4)}{I}{}^{jpp} + 5 \dddot{J}^{jpp} \Big). \tag{12.156}$$

We conclude that Eq. (12.137) is indeed compatible with the precise statement of momentum balance.

12.6.5 Angular-momentum balance

Finally we examine the statement of angular-momentum balance, which is provided by Eq. (12.35),

$$\frac{dJ^{jk}}{dt} = -\mathcal{T}^{jk} = -\frac{2G}{5c^5} \Big(\ddot{I}^{jp} \dddot{I}^{kp} - \ddot{I}^{kp} \dddot{I}^{jp} \Big) + O(c^{-7}), \tag{12.157}$$

where the expression for the angular-momentum flux $\mathcal{T}^{jk}$ has been copied from Sec. 12.3.2; it is easy to show that the result displayed here is equivalent to that of Eq. (12.68c), in spite of the fact that our expression omits the STF projection of the quadrupole-moment tensor.

The total angular-momentum tensor of the fluid distribution is given by

$$J^{jk} = \int \rho^* \big(x^j v^k - v^j x^k \big) \, d^3x + O(c^{-2}) + O(c^{-4}) + c^{-5} J^{jk}[5] + O(c^{-6}), \tag{12.158}$$

with $c^{-5} J^{jk}[5]$ denoting the contribution at 2.5PN order. The precise statement of angular-momentum balance is

$$\int \big(x^j f^k[\mathrm{rr}] - x^k f^j[\mathrm{rr}] \big) \, d^3x + \frac{d}{dt} \Big(c^{-5} J^{jk}[5] \Big) = -\frac{2G}{5c^5} \Big(\ddot{I}^{jp} \dddot{I}^{kp} - \ddot{I}^{kp} \dddot{I}^{jp} \Big) + O(c^{-7}), \tag{12.159}$$

and in this we must substitute Eq. (12.137).

To evaluate the integral on the left-hand side we require the identities

$$\int \rho^* x^j \partial_k U \, d^3x = \Omega_{jk}, \tag{12.160a}$$

$$\int \rho^* x^j \partial_{kpq} X \, d^3x = \Omega_{jk} \delta_{pq} + \Omega_{jp} \delta_{kq} + \Omega_{jq} \delta_{kp} - 3\Omega_{jkpq}, \tag{12.160b}$$

$$\int x^j \partial_k p \, d^3x = -\delta_{jk} P, \tag{12.160c}$$

$$\int \rho^* x^j v^k \, d^3x = \frac{1}{2} \big(\dot{I}^{jk} + J^{jk} \big) + O(c^{-2}), \tag{12.160d}$$

in which Ω_{jk} is the tensorial quantity defined by Eq. (12.142),

$$\Omega_{jkpq} := -\frac{1}{2} G \int \rho^* \rho^{*\prime} \frac{(x - x')^j (x - x')^k (x - x')^p (x - x')^q}{|\boldsymbol{x} - \boldsymbol{x}'|^5} \, d^3x' d^3x, \tag{12.161}$$

and $P := \int p\, d^3x$ is the integrated pressure. The first identity is familiar from the proof of the virial theorem presented back in Sec. 1.4.3, and the second follows after a similar sequence of steps. The third identity is an immediate consequence of integration by parts, and the fourth follows directly from the definitions of I^{jk} and J^{jk}.

After substitution of Eq. (12.137) and involvement of the identities we obtain

$$\int x^j f^k[\mathrm{rr}]\, d^3x = \frac{G}{c^5}\left(\frac{3}{5} I^{jp} \overset{(5)}{I}{}^{kp} + \dot{I}^{jp} \overset{(4)}{I}{}^{kp} + J^{jp} \overset{(4)}{I}{}^{kp}\right), \tag{12.162}$$

up to a number of terms that are symmetric in the pair of indices jk. This becomes

$$\int x^j f^k[\mathrm{rr}]\, d^3x = -\frac{2G}{5c^5} \ddot{I}^{jk} \dddot{I}^{kp} + \frac{G}{5c^5}\frac{d}{dt}\left(3 I^{jp} \overset{(4)}{I}{}^{kp} + 2\dot{I}^{jp} \dddot{I}^{kp} + 5 J^{jp} \dddot{I}^{kp}\right) \tag{12.163}$$

after distributing the time derivatives. Comparing with Eq. (12.159), we conclude that the statement of angular-momentum balance is indeed satisfied, and deduce that the 2.5PN contribution to the total angular-momentum tensor is

$$\begin{aligned} c^{-5} J^{jk}[5] = -\frac{G}{5c^5}\Big[& 3\big(I^{jp} \overset{(4)}{I}{}^{kp} - I^{kp} \overset{(4)}{I}{}^{jp}\big) + 2\big(\dot{I}^{jp} \dddot{I}^{kp} - \dot{I}^{kp} \dddot{I}^{jp}\big) \\ & + 5\big(J^{jp} \dddot{I}^{kp} - J^{kp} \dddot{I}^{jp}\big)\Big]. \end{aligned} \tag{12.164}$$

In this expression J^{jk} stands for the Newtonian piece of the angular-momentum tensor, obtained from Eq. (12.158) by ignoring all contributions at higher post-Newtonian orders.

12.7 Radiation reaction of N-body systems

We now specialize the results of the preceding section to a case in which the fluid distribution consists of N well-separated bodies. We wish to calculate the radiation-reaction force acting on the center-of-mass of each body, and in order to achieve this we shall exploit the techniques developed back in Chapter 9.

12.7.1 N bodies

As in Sec. 9.1 we assign to each body a label $A = 1, 2, \ldots, N$, and for each body we define the center-of-mass variables

$$m_A := \int_A \rho^*\, d^3x, \tag{12.165a}$$

$$\boldsymbol{r}_A := \frac{1}{m_A}\int_A \rho^* \boldsymbol{x}\, d^3x, \tag{12.165b}$$

$$\boldsymbol{v}_A := \frac{1}{m_A}\int_A \rho^* \boldsymbol{v}\, d^3x, \tag{12.165c}$$

$$\boldsymbol{a}_A := \frac{1}{m_A}\int_A \rho^* \frac{d\boldsymbol{v}}{dt}\, d^3x; \tag{12.165d}$$

the domain of integration is the region of space occupied by body A. We calculate the radiation-reaction force $\boldsymbol{F}_A[\mathrm{rr}]$ acting on body A by inserting Eq. (12.127) within $\boldsymbol{a}_A$; we find

$$\boldsymbol{F}_A[\mathrm{rr}] = m_A \boldsymbol{a}_A[\mathrm{rr}] = \int_A \boldsymbol{f}[\mathrm{rr}]\, d^3x, \tag{12.166}$$

in which $\boldsymbol{f}[\mathrm{rr}]$ is the radiation-reaction force density of Eq. (12.137).

To evaluate this we proceed as in Sec. 9.1.5 and decompose the Newtonian potential U and the post-Newtonian superpotential X into internal and external pieces. We have, for example, $U = U_A + U_{\neg A}$, with

$$U_A = G \int_A \frac{\rho^*(t, \boldsymbol{x}')}{|\boldsymbol{x} - \boldsymbol{x}'|}\, d^3x' \tag{12.167}$$

denoting the internal piece of the Newtonian potential, and

$$U_{\neg A} = \sum_{B \neq A} G \int_B \frac{\rho^*(t, \boldsymbol{x}')}{|\boldsymbol{x} - \boldsymbol{x}'|}\, d^3x' \tag{12.168}$$

denoting its external piece. As in Sec. 9.3 we exploit the assumption that the bodies are well separated to express each external potential as a Taylor expansion about the center-of-mass $\boldsymbol{r}_A$; we ignore the multipole structure of the body and retain only the leading term in the expansion. The internal potentials require no approximation, and for these we invoke the identities of Eq. (12.154),

$$\int_A \rho^* \partial_j U_A\, d^3x = 0, \tag{12.169a}$$

$$\int_A \rho^* \partial_{jpq} X_A\, d^3x = 0, \tag{12.169b}$$

$$\int_A \partial_j p\, d^3x = 0, \tag{12.169c}$$

which imply that the internal potentials and the pressure make no contribution to the radiation-reaction force.

Following through with the computations, we quickly obtain

$$\begin{aligned} a_A^j[\mathrm{rr}] = \frac{G}{c^5}\biggl(&-\dddot{I}^{pq} \partial_{jpq} X_{\neg A} + 2\dddot{I}^{jk} \partial_k U_{\neg A} + \frac{4}{3}\dddot{I}^{pp} \partial_j U_{\neg A} + 2\overset{(4)}{I}{}^{jk} v_A^k \\ &+ \frac{3}{5}\overset{(5)}{I}{}^{jk} r_A^k - \frac{1}{5}\overset{(5)}{I}{}^{pp} r_A^j - \frac{2}{15}\overset{(5)}{I}{}^{jpp} - \frac{2}{3}\overset{(4)}{J}{}^{jpp}\biggr) + O(c^{-7}), \end{aligned} \tag{12.170}$$

in which it is understood that the external potentials are evaluated at $\boldsymbol{x} = \boldsymbol{r}_A$ after differentiation. For a more concrete expression we must evaluate the external potentials. These manipulations are familiar from Sec. 9.3.4, and we arrive at

$$\partial_j U_{\neg A} = -\sum_{B \neq A} \frac{GM_B}{r_{AB}^2} n_{AB}^j, \tag{12.171a}$$

$$\partial_{jpq} X_{\neg A} = -\sum_{B \neq A} \frac{GM_B}{r_{AB}^2} \left(n_{AB}^j \delta_{pq} + n_{AB}^p \delta_{jq} + n_{AB}^q \delta_{jp} - 3 n_{AB}^j n_{AB}^p n_{AB}^q \right), \tag{12.171b}$$

in which $\boldsymbol{r}_{AB} := \boldsymbol{r}_A - \boldsymbol{r}_B$, $r_{AB} := |\boldsymbol{r}_{AB}|$, and $\boldsymbol{n}_{AB} := \boldsymbol{r}_{AB}/r_{AB}$; we take the liberty of expressing the material mass m_A in terms of the total mass-energy M_A of Eq. (9.23), noting that the difference is of order c^{-2} and therefore affects the radiation-reaction force at order c^{-7} only.

Making the substitutions returns our final expression for the radiation-reaction force (per unit mass) acting on body A:

$$a_A^j[\text{rr}] = \frac{G}{c^5}\biggl(-3\dddot{I}^{pq}\sum_{B\neq A}\frac{GM_B}{r_{AB}^2}n_{AB}^j n_{AB}^p n_{AB}^q - \frac{1}{3}\dddot{I}^{pp}\sum_{B\neq A}\frac{GM_B}{r_{AB}^2}n_{AB}^j + 2\overset{(4)}{I}{}^{jk}v_A^k + \frac{3}{5}\overset{(5)}{I}{}^{\langle jk\rangle}r_A^k - \frac{2}{15}\overset{(5)}{I}{}^{jpp} - \frac{2}{3}\overset{(4)}{J}{}^{jpp}\biggr) + O(c^{-7}). \qquad (12.172)$$

To complete the evaluation of $\boldsymbol{a}_A[\text{rr}]$ we should calculate the time derivatives of all the multipole moments that occur in Eq. (12.172). Because the end result is long and unwieldy, we postpone the pursuit of these details until we specialize our discussion to a two-body system.

12.7.2 Two bodies

We next apply the general results of the preceding subsection to the case $N = 2$. Because the total momentum $\boldsymbol{P}$ of the system is conserved at order c^{-5}, we may set it equal to zero and work in the barycentric frame in which $\boldsymbol{r}_1 = (M_2/m)\boldsymbol{r} + O(c^{-2})$ and $\boldsymbol{r}_2 = -(M_1/m)\boldsymbol{r} + O(c^{-2})$, where $m := M_1 + M_2$ and $\boldsymbol{r} := \boldsymbol{r}_1 - \boldsymbol{r}_2$. Equation (12.172) then provides the radiation-reaction acceleration of the relative orbit, $\boldsymbol{a}[\text{rr}] := \boldsymbol{a}_1[\text{rr}] - \boldsymbol{a}_2[\text{rr}]$; we get

$$a^j[\text{rr}] = \frac{G}{c^5}\biggl[-\frac{Gm}{r^2}\Bigl(3\dddot{I}^{pq}n^p n^q + \frac{1}{3}\dddot{I}^{pp}\Bigr)n^j + 2\overset{(4)}{I}{}^{jk}v^k + \frac{3}{5}\overset{(5)}{I}{}^{\langle jk\rangle}r^k\biggr] + O(c^{-7}), \quad (12.173)$$

where $r := |\boldsymbol{r}|$ and $\boldsymbol{n} := \boldsymbol{r}/r$. We observe that the multipole moments I_{jpp} and J_{jpp} no longer appear in this expression; as we saw back in Sec. 12.6.4, their role is limited to providing a contribution to the total momentum at 2.5PN order.

The mass quadrupole moment is $I^{jk} = \eta m r^j r^k$, where $\eta := M_1 M_2/m^2$ is the symmetric mass ratio, and to evaluate its derivatives we rely on the Keplerian dynamics reviewed in Sec. 3.2. From the Newtonian equation of motion $\boldsymbol{a} = -Gm\boldsymbol{n}/r^2$ we deduce that $v\dot{v} = -Gm\dot{r}/r^2$ and $r\ddot{r} = v^2 - \dot{r}^2 - Gm/r$, up to corrections of order c^{-2}, and a straightforward computation reveals that

$$\dddot{I}^{jk} = -2\eta m\frac{Gm}{r^2}\Bigl[2\bigl(n^j v^k + v^j n^k\bigr) - 3\dot{r}\,n^j n^k\Bigr], \qquad (12.174a)$$

$$\overset{(4)}{I}{}^{jk} = -2\eta m\frac{Gm}{r^3}\Bigl[-9\dot{r}\bigl(n^j v^k + v^j n^k\bigr) + 4v^j v^k + \bigl(15\dot{r}^2 - 3v^2 - Gm/r\bigr)n^j n^k\Bigr], \qquad (12.174b)$$

$$\overset{(5)}{I}{}^{jk} = -2\eta m\frac{Gm}{r^4}\Bigl[4\bigl(15\dot{r}^2 - 3v^2 + Gm/r\bigr)\bigl(n^j v^k + v^j n^k\bigr) - 30\dot{r}\,v^j v^k + 15\dot{r}\bigl(3v^2 - 7\dot{r}^2\bigr)n^j n^k\Bigr]. \qquad (12.174c)$$

Making the substitutions in Eq. (12.173), we eventually arrive at our final expression for the radiation-reaction acceleration:

$$\boldsymbol{a}[\mathrm{rr}] = \frac{8}{5}\eta \frac{(Gm)^2}{c^5 r^3}\left[\left(3v^2 + \frac{17}{3}\frac{Gm}{r}\right)\dot{r}\,\boldsymbol{n} - \left(v^2 + 3\frac{Gm}{r}\right)\boldsymbol{v}\right]. \tag{12.175}$$

We return to this expression in Sec. 12.9, when we examine the orbital motion of a binary system under the action of the radiation-reaction force.

12.8 Radiation reaction in alternative gauges

The radiation-reaction force density of Eq. (12.137), and the body forces of Eqs. (12.172) and (12.175), are expressed in a harmonic coordinate system in which the spacetime metric takes the form of Eq. (12.128). The metric and the resulting expressions for the radiation-reaction force appear to be more complicated than they need to be. For example, the metric involves the multipole moments $\mathcal{I}^{jk}[2]$, I^{jknp}, and M^{jknp}, but these make no appearance in the radiation-reaction force. As another example, $f^j[\mathrm{rr}]$ involves the multipole moments I_{jpp} and J_{jpp}, in spite of the fact that they produce no physical consequences; as we saw, their role is limited to providing a contribution to the total momentum $\boldsymbol{P}$ at 2.5PN order. In view of this unnecessary complexity, it is worthwhile to seek coordinate transformations that could simplify the form of the metric and of the radiation-reaction force. To explore this freedom is our purpose in this section. We shall identify two radiation-reaction gauges that offer an optimum of simplicity: the famous *Burke–Thorne gauge* of Eq. (12.198) below, and the *Schäfer gauge* of Eq. (12.199).

12.8.1 Coordinate transformation

We follow the general framework of Sec. 8.3 and consider a class of coordinate transformations described by

$$t = \bar{t} + c^{-5}\alpha(\bar{t}, \bar{x}^j) + c^{-7}\beta(\bar{t}, \bar{x}^j) + O(c^{-9}), \tag{12.176a}$$

$$x^j = \bar{x}^j + c^{-5}h^j(\bar{t}, \bar{x}^k) + O(c^{-7}), \tag{12.176b}$$

in which $(\bar{t}, \bar{x}^j)$ are the new coordinates, (t, x^j) are the old harmonic coordinates, and α, β, and h^j are functions that will be determined in the course of our investigation. The transformation impacts the metric in the way described by Eq. (8.34). When we insert the metric of Eq. (12.128) and read off the terms of relevant orders in the post-Newtonian expansion, we find that the radiation-reaction terms become

$$\bar{g}_{00}[5] = 2\big(-\partial_{\bar{t}}\alpha + 4V[3]\big), \tag{12.177a}$$

$$\bar{g}_{00}[7] = 2\big(-\partial_{\bar{t}}\beta + V[5] + W[3] - \Delta U[5]\big) + 4U\big(\partial_{\bar{t}}\alpha - 6V[3]\big), \tag{12.177b}$$

$$\bar{g}_{0j}[4] = -\partial_{\bar{j}}\alpha, \tag{12.177c}$$

$$\bar{g}_{0j}[6] = 2U\partial_{\bar{j}}\alpha - \partial_{\bar{j}}\beta + \partial_{\bar{t}}h_j - 4V_j[3], \tag{12.177d}$$

$$\bar{g}_{jk}[5] = \partial_{\bar{j}}h_k + \partial_{\bar{k}}h_j + 4\big(W_{jk}[1] - \delta_{jk}V[3]\big). \tag{12.177e}$$

The original radiation-reaction potentials are listed in Box 12.3, and

$$c^{-5}\Delta U[5] := \bar{U} - U \tag{12.178}$$

is the difference between the Newtonian potential $\bar{U}$ expressed in terms of the new coordinates $(\bar{t}, \bar{x}^j)$ and the original Newtonian potential U.

We see from Eqs. (12.177) that the transformation produces a new radiation-reaction term in $\bar{g}_{0j}$, at order c^{-4}. This term is undesirable, and we eliminate it by demanding that α be independent of the spatial coordinates $\bar{x}^j$. To complete the determination of α we also choose to eliminate $\bar{g}_{00}[5]$, another undesirable term at 1.5PN order. This can be achieved by setting $\partial_{\bar{t}}\alpha = 4V[3]$, or

$$\alpha(\bar{t}) = -\frac{2}{3}G\ddot{I}^{pp}. \tag{12.179}$$

In this expression the mass quadrupole moment is expressed as a function of the new time variable $\bar{t}$. The transformation $t = \bar{t} - \frac{2}{3}(G/c^5)\ddot{I}^{pp}(\bar{t}) + O(c^{-7})$ is the one displayed in Eq. (12.130).

To calculate $\Delta U[5]$ we rely on the discussion of Sec. 8.3.6, in which a "boosted" Newtonian potential $\bar{U}$ was obtained as a result of a post-Galilean transformation of the coordinate system. The original Newtonian potential is

$$U(t, \boldsymbol{x}) = G\int \frac{\rho^*(t, \boldsymbol{x}')}{|\boldsymbol{x} - \boldsymbol{x}'|}\, d^3x', \tag{12.180}$$

where it is understood that the source point $\boldsymbol{x}'$ is simultaneous with the field point $\boldsymbol{x}$ in the harmonic reference frame. The new potential is defined to be

$$\bar{U}(\bar{t}, \bar{\boldsymbol{x}}) = G\int \frac{\bar{\rho}^*(\bar{t}, \bar{\boldsymbol{x}}')}{|\bar{\boldsymbol{x}} - \bar{\boldsymbol{x}}'|}\, d^3\bar{x}', \tag{12.181}$$

with $\bar{\boldsymbol{x}}'$ and $\bar{\boldsymbol{x}}$ simultaneous in the new reference frame. Because the relation between t and $\bar{t}$ is independent of the spatial coordinates at order c^{-5}, simultaneity in one frame implies simultaneity in the other frame, and we find that $(x - x')^j = (\bar{x} - \bar{x}')^j + c^{-5}\Delta h^j$, where

$$\Delta h^j := h^j(\bar{t}, \bar{\boldsymbol{x}}) - h^j(\bar{t}, \bar{\boldsymbol{x}}'). \tag{12.182}$$

Noting also that $\rho^*\, d^3x' = \bar{\rho}^*\, d^3\bar{x}'$ – refer to Eq. (8.92) – we find that the original potential can be expressed as

$$U = G\int \frac{\bar{\rho}^*}{|\bar{\boldsymbol{x}} - \bar{\boldsymbol{x}}'|}\left(1 - \frac{1}{c^5}\frac{(\bar{x} - \bar{x}')^j\Delta h_j}{|\bar{\boldsymbol{x}} - \bar{\boldsymbol{x}}'|^2}\right) d^3\bar{x}'. \tag{12.183}$$

In view of Eq. (12.178), this means that

$$\Delta U[5] = G\int \bar{\rho}^*(\bar{t}, \bar{\boldsymbol{x}}')\frac{(\bar{x} - \bar{x}')^j\Delta h_j}{|\bar{\boldsymbol{x}} - \bar{\boldsymbol{x}}'|^3}\, d^3\bar{x}' \tag{12.184}$$

is our final expression for the change in the Newtonian potential at order c^{-5}.

12.8.2 Two-parameter family of radiation-reaction gauges

To proceed with this exploration of coordinate transformations, we restrict our considerations to radiation-reaction gauges in which

$$\bar{g}_{0j}[6] = 0, \tag{12.185a}$$

$$\bar{g}_{jk}[5] = 2a\, G\dddot{I}^{\langle jk\rangle} + 2b\,\delta_{jk}\, G\dddot{I}^{pp}, \tag{12.185b}$$

where a and b are dimensionless parameters that can be chosen freely. As we shall see, the choice of Eq. (12.185) does not fully exhaust the coordinate freedom, and additional choices will be made to specify the form of $\bar{g}_{00}[7]$.

According to Eq. (12.177) and the potentials of Box 12.3, to achieve the required form for $\bar{g}_{jk}[5]$ we need h^j to be a solution to

$$\partial_{\bar{j}} h_k + \partial_{\bar{k}} h_j = 2(1+a)G\dddot{I}^{\langle jk\rangle} + 2b\,\delta_{jk}\, G\dddot{I}^{pp}. \tag{12.186}$$

A sufficiently general solution is

$$G^{-1} h_j = f_j + \big[(1+a)\ddot{I}^{\langle jk\rangle} + b\,\delta_{jk}\,\ddot{I}^{pp}\big]\bar{x}^k, \tag{12.187}$$

in which f_j is an arbitrary function of $\bar{t}$ that will be specified at a later stage. To achieve the required form for $\bar{g}_{0j}[6]$ we need β to be a solution to

$$G^{-1}\partial_{\bar{j}}\beta = \dot{f}_j + \frac{2}{9}\overset{(4)}{I}{}^{jpp} + \frac{4}{9}\dddot{J}^{jpp} + \left[\left(\frac{1}{3}+a\right)\overset{(4)}{I}{}^{\langle jk\rangle} + \left(b-\frac{2}{9}\right)\delta_{jk}\overset{(4)}{I}{}^{pp}\right]\bar{x}^k, \tag{12.188}$$

and in this case we find that

$$\begin{aligned} G^{-1}\beta &= \gamma + \left(\dot{f}_j + \frac{2}{9}\overset{(4)}{I}{}^{jpp} + \frac{4}{9}\dddot{J}^{jpp}\right)\bar{x}^j \\ &\quad + \frac{1}{2}\left[\left(\frac{1}{3}+a\right)\overset{(4)}{I}{}^{\langle jk\rangle} + \left(b-\frac{2}{9}\right)\delta_{jk}\overset{(4)}{I}{}^{pp}\right]\bar{x}^j\bar{x}^k, \end{aligned} \tag{12.189}$$

in which γ is an arbitrary function of $\bar{t}$.

We next incorporate these results in a computation of $\bar{g}_{00}[7]$. The first step is to insert Eq. (12.187) within Eq. (12.184), which yields

$$\Delta U[5] = (1+a)G\dddot{I}^{\langle jk\rangle}\bar{U}_{jk} + b\, G\dddot{I}^{pp}\bar{U}, \tag{12.190}$$

where

$$\bar{U}^{jk} := G\int \bar{\rho}^* \frac{(\bar{x}-\bar{x}')^j(\bar{x}-\bar{x}')^k}{|\bar{\boldsymbol{x}}-\bar{\boldsymbol{x}}'|^3}\, d^3\bar{x}'. \tag{12.191}$$

We bring this to its final form of

$$\Delta U[5] = -(1+a)G\dddot{I}^{\langle jk\rangle}\partial_{\bar{j}\bar{k}}\bar{X} + b\, G\dddot{I}^{pp}\bar{U} \tag{12.192}$$

by invoking the identity $\partial_{jk}X = \delta_{jk}U - U_{jk}$; here $\bar{X}$ is the superpotential expressed in terms of the new coordinates.

Making the substitutions in Eq. (12.177), we find that $\bar{g}_{00}[7]$ contains a term independent of the spatial coordinates that can be eliminated by choosing

$$\gamma = -\frac{2}{3}\ddot{\mathcal{I}}^{pp}[2] - \frac{1}{120}\overset{(4)}{I}{}^{ppqq} - \frac{1}{9}\ddot{M}^{ppqq}. \tag{12.193}$$

It also contains a term linear in $\bar{x}^j$ that can be eliminated by choosing

$$f_j = -\frac{2}{15}\dddot{I}^{jpp} - \frac{2}{3}\ddot{J}^{jpp}. \tag{12.194}$$

At this stage the coordinate freedom is exhausted, and the resulting expression for $\bar{g}_{00}[7]$ is

$$\bar{g}_{00}[7] = 2a\, G\dddot{I}^{\langle jk\rangle}\partial_{\bar{j}\bar{k}}\bar{X} - 2b\, G\dddot{I}^{pp}\bar{U} - \biggl[\biggl(\frac{2}{5}+a\biggr)G\overset{(5)}{I}{}^{\langle jk\rangle} + b\,\delta_{jk}\,G\overset{(5)}{I}{}^{pp}\biggr]\bar{x}^j\bar{x}^k. \tag{12.195}$$

The calculation of the metric in the new coordinate system is now complete.

To summarize, the two-parameter family of radiation-reaction gauges is described by the metric

$$g_{00} = -1 + \frac{2}{c^2}(U + \mathcal{U}) + O(c^{-9}), \tag{12.196a}$$

$$g_{0j} = O(c^{-8}), \tag{12.196b}$$

$$g_{jk} = \delta_{jk} + 2\mathcal{V}_{jk} + O(c^{-7}), \tag{12.196c}$$

with the radiation-reaction potentials

$$\mathcal{U} = \frac{G}{c^5}\biggl\{a\,\dddot{I}^{\langle jk\rangle}\partial_{jk}X - b\,\dddot{I}^{pp}U - \frac{1}{2}\biggl[\biggl(\frac{2}{5}+a\biggr)\overset{(5)}{I}{}^{\langle jk\rangle} + b\,\delta_{jk}\overset{(5)}{I}{}^{pp}\biggr]x^jx^k\biggr\}, \tag{12.197a}$$

$$\mathcal{V}_{jk} = \frac{G}{c^5}\Bigl(a\,\dddot{I}^{\langle jk\rangle} + b\,\delta_{jk}\dddot{I}^{pp}\Bigr). \tag{12.197b}$$

This expression for the metric omits all terms of 1PN and 2PN orders that have nothing to do with radiation reaction. To simplify the notation we have removed the (now redundant) overbars on the coordinates and potentials.

Two special cases are especially interesting and simple. When we set $a = b = 0$ we find that the radiation-reaction potentials become

$$\mathcal{U} = -\frac{G}{5c^5}\overset{(5)}{I}{}^{\langle jk\rangle}x^jx^k,$$

$$\mathcal{V}_{jk} = 0; \tag{12.198}$$

this choice defines the *Burke–Thorne radiation-reaction gauge*. This gauge is especially attractive, because it encapsulates all radiation-reaction effects in a $O(c^{-5})$ correction to the Newtonian potential. On the other hand, when we set $a = -\frac{2}{5}$ and $b = 0$ we find that all terms involving the mass quadrupole moment differentiated five times disappear, and we are left with

$$\mathcal{U} = -\frac{2G}{5c^5}\dddot{I}^{\langle jk\rangle}\partial_{jk}X,$$

$$\mathcal{V}_{jk} = -\frac{2G}{5c^5}\dddot{I}^{\langle jk\rangle}; \tag{12.199}$$

this choice defines the *Schäfer radiation-reaction gauge*.

12.8.3 Radiation-reaction force

The computation of the radiation-reaction force density $f^j[\mathrm{rr}]$ in the two-parameter family of radiation-reaction gauges proceeds just as in Sec. 12.6. In this case we have that $\sqrt{-g} = 1 + \mathcal{V} + O(c^{-7})$, $\gamma = 1 + O(c^{-7})$, and the components of the energy-momentum tensor are

$$c^{-2}\sqrt{-g}T^{00} = \rho^* + O(c^{-7}), \tag{12.200a}$$

$$c^{-1}\sqrt{-g}T^{0j} = \rho^* v^j + O(c^{-7}), \tag{12.200b}$$

$$\sqrt{-g}T^{jk} = \rho^* v^j v^k + p\,\delta^{jk} - \left(2\mathcal{V}^{jk} - \delta^{jk}\mathcal{V}\right)p + O(c^{-7}). \tag{12.200c}$$

We use the notation $\mathcal{V} := \delta^{jk}\mathcal{V}_{jk}$ and continue to omit all terms of 1PN and 2PN orders that have nothing to do with radiation reaction. The relevant Christoffel symbols are

$$c^2\Gamma^j_{00} = -\partial_j(U + \mathcal{U}) + 2\mathcal{V}^{jk}\partial_k U + O(c^{-7}), \tag{12.201a}$$

$$c\Gamma^j_{0k} = \partial_t \mathcal{V}_{jk} + O(c^{-7}), \tag{12.201b}$$

$$\Gamma^j_{kn} = O(c^{-7}). \tag{12.201c}$$

Making the substitutions in the momentum-conservation equation of Eq. (12.134), we arrive at the post-Newtonian Euler equation of Eq. (12.127), with

$$f^j[\mathrm{rr}] = \rho^*\partial_j\mathcal{U} - 2\rho^* v^k \partial_t \mathcal{V}_{jk} - 2\mathcal{V}^{jk}\left(\rho^*\partial_k U - \partial_k p\right) - \mathcal{V}\,\partial_j p. \tag{12.202}$$

When we next insert the radiation-reaction potentials of Eqs. (12.197), we obtain

$$f^j[\mathrm{rr}] = \frac{G}{c^5}\biggl\{a\,\dddot{I}^{\langle pq\rangle}\rho^*\partial_{jpq}X - 2a\,\dddot{I}^{\langle jk\rangle}\left(\rho^*\partial_k U - \partial_k p\right) - b\,\dddot{I}^{pp}\left(3\rho^*\partial_j U + \partial_j p\right)$$
$$- 2\rho^*\left(a\overset{(4)}{I}{}^{\langle jk\rangle} + b\,\delta_{jk}\overset{(4)}{I}{}^{pp}\right)v^k - \rho^*\left[\left(\frac{2}{5} + a\right)\overset{(5)}{I}{}^{\langle jk\rangle} + b\,\delta_{jk}\overset{(5)}{I}{}^{pp}\right]x^k\biggr\} \tag{12.203}$$

as our final expression for the radiation-reaction force density.

The force simplifies considerably in the Burke–Thorne gauge:

$$f^j[\mathrm{rr}] = -\frac{2G}{5c^5}\rho^*\overset{(5)}{I}{}^{\langle jk\rangle}x^k. \tag{12.204}$$

While its expression is more complicated in the Shäfer gauge, it is nevertheless useful because the number of time derivatives acting on the quadrupole-moment tensor goes down from five to four. This can be a great advantage in numerical work, because the estimation of derivatives on a finite grid generates numerical noise that can be minimized with a smaller number of derivatives. In some applications the term involving four derivatives can be small compared with terms containing three derivatives; this occurs, for example, when the system undergoes small oscillations and v^2 is small compared with U (contrary to what might be expected from the virial theorem).

We next calculate the radiation-reaction force acting on the center-of-mass of each body A within an N-body system. The steps involved are virtually identical to those encountered previously in Sec. 12.7, and we arrive at

$$a^j_A[\text{rr}] = \frac{G}{c^5}\biggl\{a\,\dddot{I}^{\langle pq\rangle}\partial_{jpq}X_{\neg A} - 2a\,\dddot{I}^{\langle jk\rangle}\partial_k U_{\neg A} - 3b\,\dddot{I}^{pp}\partial_j U_{\neg A} - 2\Bigl(a\,\overset{(4)}{I}{}^{\langle jk\rangle} + b\,\delta_{jk}\overset{(4)}{I}{}^{pp}\Bigr)v^k_A - \biggl[\Bigl(\frac{2}{5}+a\Bigr)\overset{(5)}{I}{}^{\langle jk\rangle} + b\,\delta_{jk}\overset{(5)}{I}{}^{pp}\biggr]r^k_A\biggr\}, \tag{12.205}$$

where $U_{\neg A}$ and $X_{\neg A}$ are the potentials produced by the bodies external to A, evaluated at $\boldsymbol{x} = \boldsymbol{r}_A$ after differentiation. Taking care of these manipulations, we obtain our final expression

$$a^j_A[\text{rr}] = \frac{G}{c^5}\biggl\{3\sum_{B\neq A}\frac{GM_B}{r^2_{AB}}\Bigl(a\,\dddot{I}^{\langle pq\rangle}n^p_{AB}n^q_{AB} + b\,\dddot{I}^{pp}\Bigr)n^j_{AB} - 2\Bigl(a\,\overset{(4)}{I}{}^{\langle jk\rangle} + b\,\delta_{jk}\overset{(4)}{I}{}^{pp}\Bigr)v^k_A - \biggl[\Bigl(\frac{2}{5}+a\Bigr)\overset{(5)}{I}{}^{\langle jk\rangle} + b\,\delta_{jk}\overset{(5)}{I}{}^{pp}\biggr]r^k_A\biggr\}, \tag{12.206}$$

in which $\boldsymbol{r}_{AB} := \boldsymbol{r}_A - \boldsymbol{r}_B$, $r_{AB} := |\boldsymbol{r}_{AB}|$, and $\boldsymbol{n}_{AB} := \boldsymbol{r}_{AB}/r_{AB}$.

The reduction to a two-body system produces

$$a^j[\text{rr}] = \frac{G}{c^5}\biggl\{3\frac{Gm}{r^2}\Bigl(a\,\dddot{I}^{\langle pq\rangle}n^p n^q + b\,\dddot{I}^{pp}\Bigr)n^j - 2\Bigl(a\,\overset{(4)}{I}{}^{\langle jk\rangle} + b\,\delta_{jk}\overset{(4)}{I}{}^{pp}\Bigr)v^k - \biggl[\Bigl(\frac{2}{5}+a\Bigr)\overset{(5)}{I}{}^{\langle jk\rangle} + b\,\delta_{jk}\overset{(5)}{I}{}^{pp}\biggr]r^k\biggr\} \tag{12.207}$$

for the relative acceleration $\boldsymbol{a}[\text{rr}] := \boldsymbol{a}_1[\text{rr}] - \boldsymbol{a}_2[\text{rr}]$. Completing the calculation with the help of Eqs. (12.174), we finally arrive at

$$\boldsymbol{a}[\text{rr}] = \frac{8}{5}\eta\frac{(Gm)^2}{c^5r^3}\biggl\{\biggl[\Bigl(18+15a-\frac{45}{4}b\Bigr)v^2 + \Bigl(\frac{2}{3}-\frac{10}{3}a+\frac{25}{4}b\Bigr)\frac{Gm}{r} - \Bigl(25+25a-\frac{75}{4}b\Bigr)\dot{r}^2\biggr]\dot{r}\,\boldsymbol{n} - \biggl[\Bigl(6+\frac{35}{6}a-\frac{5}{2}b\Bigr)v^2 - \Bigl(2+\frac{35}{6}a-\frac{5}{2}b\Bigr)\frac{Gm}{r} - \Bigl(15+\frac{35}{2}a-\frac{15}{2}b\Bigr)\dot{r}^2\biggr]\boldsymbol{v}\biggr\}. \tag{12.208}$$

In the Burke–Thorne gauge ($a = b = 0$) the relative acceleration reduces to

$$\boldsymbol{a}[\text{rr}] = \frac{8}{5}\eta\frac{(GM)^2}{c^5r^3}\biggl[\Bigl(18v^2 + \frac{2}{3}\frac{GM}{r} - 25\dot{r}^2\Bigr)\dot{r}\,\boldsymbol{n} - \Bigl(6v^2 - 2\frac{GM}{r} - 15\dot{r}^2\Bigr)\boldsymbol{v}\biggr], \tag{12.209}$$

while in the Schäfer gauge ($a = -2/5$, $b = 0$) we have

$$a[\text{rr}] = \frac{8}{5}\eta\frac{(GM)^2}{c^5r^3}\left[\left(12v^2 + 2\frac{GM}{r} - 15\dot{r}^2\right)\dot{r}\,\boldsymbol{n} - \left(\frac{11}{3}v^2 + \frac{1}{3}\frac{GM}{r} - 8\dot{r}^2\right)\boldsymbol{v}\right]. \tag{12.210}$$

In addition to these choices, it is interesting to note that the settings $a = -2/3$ and $b = 4/9$ produce

$$a[\text{rr}] = \frac{8}{5}\eta\frac{(GM)^2}{c^5r^3}\left[\left(3v^2 + \frac{17}{3}\frac{GM}{r}\right)\dot{r}\,\boldsymbol{n} - \left(v^2 + 3\frac{GM}{r}\right)\boldsymbol{v}\right], \tag{12.211}$$

the same expression as in Eq. (12.175); the two-parameter family of radiation-reaction gauges can therefore reproduce the original expression for the relative acceleration, which was obtained in the harmonic gauge. It should be kept in mind that while the relative accelerations do correspond when $a = -2/3$ and $b = 4/9$, the coordinate systems do not coincide, and the radiation-reaction potentials are quite different in the two gauges. This selection of parameters is often called the Damour–Deruelle gauge.

12.8.4 Balance equations

To finish our discussion of the two-parameter family of radiation-reaction gauges, we verify that the radiation-reaction force density of Eq. (12.203) is compatible with the statements of energy, momentum, and angular-momentum balance.

We begin with energy balance, and follow the developments of Sec. 12.6.3. Here the precise statement of energy balance is more complicated than in Eq. (12.140), because $\rho \neq \rho^*$ in the two-parameter family of radiation-reaction gauges. Instead we have that

$$\rho = (1 - \mathcal{V})\rho^* + O(c^{-7}), \tag{12.212}$$

up to corrections of 1PN and 2PN orders, and the presence of $\mathcal{V}$ affects the statement of the first law of thermodynamics. The exact formulation, we recall, is $d\Pi/dt = (p/\rho^2)d\rho/dt$, and this implies that

$$\rho^*\frac{d\Pi}{dt} = (1 + \mathcal{V})\frac{p}{\rho^*}\frac{d\rho^*}{dt} - p\,\partial_t\mathcal{V} + O(c^{-7}), \tag{12.213}$$

where we use the fact that $\mathcal{V}$ depends on time only. The continuity equation satisfied by ρ^* allows us to replace $d\rho^*/dt$ with $-\rho^*\partial_j v^j$, and integrating over the volume occupied by the fluid, we obtain

$$\frac{dE_{\text{int}}}{dt} = (1 + \mathcal{V})\int v^j\partial_j p\,d^3x - P\,\partial_t\mathcal{V} + O(c^{-7}) \tag{12.214}$$

for the rate of change of the fluid's internal energy; here $P := \int p\,d^3x$ is the integrated pressure. The terms in $\mathcal{V}$ contribute to the change in total energy at 2.5PN order, and instead of Eq. (12.140) we find that

$$\int v_j f^j[\text{rr}]\,d^3x + \mathcal{V}\int v^j\partial_j p\,d^3x - P\,\partial_t\mathcal{V} + \frac{d}{dt}\left(c^{-5}E[5]\right) = -\frac{G}{5c^5}\dddot{I}^{\langle jk\rangle}\dddot{I}^{\langle jk\rangle} + O(c^{-7}) \tag{12.215}$$

is the precise statement of energy balance in the two-parameter family of radiation-reaction gauges. The integrals can be evaluated in the same way as in Sec. 12.6.3, and after some calculations we confirm that the radiation-reaction force density of Eq. (12.203) is indeed compatible with energy balance. In the course of this computation we also find that

$$\begin{aligned} c^{-5}E[5] = \frac{G}{c^5}\biggl[& 4a\,\dddot{I}^{\langle jk\rangle}\mathcal{T}^{jk} + \frac{1}{2}\left(\frac{2}{5}+a\right)\left(\overset{(4)}{I}{}^{\langle jk\rangle}\dot{I}^{\langle jk\rangle} - \dddot{I}^{\langle jk\rangle}\ddot{I}^{\langle jk\rangle}\right) \\ & + b\,\dddot{I}^{pp}(4\mathcal{T}+3P) + \frac{1}{2}b\left(\overset{(4)}{I}{}^{pp}\dot{I}^{qq} - \dddot{I}^{pp}\ddot{I}^{qq}\right)\biggr] \end{aligned} \tag{12.216}$$

is the appropriate expression for the 2.5PN contribution to the total energy.

The statement of momentum balance requires no modification from Eq. (12.153), and the computations carried out in Sec. 12.6.4 can easily be adapted to the two-parameter family of radiation-reaction gauges. We confirm that the radiation-reaction force density of Eq. (12.203) is indeed compatible with momentum balance, and that

$$c^{-5}P^j[5] = 0 \tag{12.217}$$

in these gauges; there is no contribution to the total momentum at 2.5PN order.

The same statements apply to the expression of angular-momentum balance. Here also the precise statement of Eq. (12.159) requires no modification, and here also we find that it is compatible with Eq. (12.203). The 2.5PN contribution to the total angular-momentum tensor is given by

$$\begin{aligned} c^{-5}J^{jk}[5] = \frac{G}{c^5}\biggl[& \left(\frac{2}{5}+a\right)\left(I^{jp}\overset{(4)}{I}{}^{kp} - I^{kp}\overset{(4)}{I}{}^{jp}\right) + a\left(J^{jp}\dddot{I}^{kp} - J^{kp}\dddot{I}^{jp}\right) \\ & - \frac{2}{5}\left(\dot{I}^{jp}\dddot{I}^{kp} - \dot{I}^{kp}\dddot{I}^{jp}\right)\biggr] \end{aligned} \tag{12.218}$$

in the two-parameter family of radiation-reaction gauges.

12.9 Orbital evolution under radiation reaction

In this last section of Chapter 12 we describe how the radiation-reaction force of Eq. (12.208) affects the orbital motion of a two-body system. The system's dynamics is dominated by the Newtonian gravitational attraction between the two bodies, and the radiation-reaction force creates a perturbation. We wish to determine the effect of this perturbing force over a time scale that is much longer than the orbital period; we are primarily interested in the *secular effects* of the radiation-reaction force. The Newtonian dynamics was investigated in detail back in Sec. 3.2, and a formalism to describe perturbed Keplerian orbits, based on osculating elliptical orbits and evolving orbital elements, was introduced in Sec. 3.3; our analysis in this section employs this formalism as an essential foundation.

12.9.1 Evolution of orbital elements

We return to Eq. (12.208) and substitute Keplerian expressions for $\boldsymbol{r} := \boldsymbol{r}_1 - \boldsymbol{r}_2$ and $\boldsymbol{v} := \boldsymbol{v}_1 - \boldsymbol{v}_2$, which we decompose as $\boldsymbol{r} = r\,\boldsymbol{n}$ and $\boldsymbol{v} = \dot{r}\,\boldsymbol{n} + (r\dot{\phi})\boldsymbol{\lambda}$, with $\dot{\phi}$ denoting the orbital angular velocity. The unit vectors $\boldsymbol{n}$ and $\boldsymbol{\lambda}$ are tangent to the orbital plane and mutually orthogonal; in Cartesian coordinates (x, y, z) oriented in such a way that the orbital plane coincides with the surface $z = 0$, we have that $\boldsymbol{n} = [\cos\phi, \sin\phi, 0]$ and $\boldsymbol{\lambda} = [-\sin\phi, \cos\phi, 0]$, where $\phi(t)$ is the angle between $\boldsymbol{r}$ and the x-axis. The Keplerian relations are

$$r = \frac{p}{1 + e\cos f}, \qquad \dot{r} = \sqrt{\frac{Gm}{p}}\,e\sin f, \qquad r\dot{\phi} = \sqrt{\frac{Gm}{p}}(1 + e\cos f), \tag{12.219}$$

where p is the semi-latus rectum, e the eccentricity, $f := \phi - \omega$ the true anomaly, and ω the longitude of pericenter. The orbital period is

$$P = \frac{2\pi}{\sqrt{Gm}}\left(\frac{p}{1 - e^2}\right)^{3/2}, \tag{12.220}$$

and $p/(1 - e^2)$ is the semi-major axis.

Making the substitutions in Eq. (12.208) returns

$$\boldsymbol{a}[\mathrm{rr}] = \frac{8}{5}\eta\frac{(Gm)^{7/2}}{c^5 p^{9/2}}(1 + e\cos f)^3\big[(e\sin f)A\,\boldsymbol{n} - (1 + e\cos f)B\,\boldsymbol{\lambda}\big], \tag{12.221}$$

with

$$\begin{aligned} A :=& \frac{44}{3} + \frac{35}{3}a - 5b + e\left(\frac{80}{3} + \frac{125}{6}a - \frac{55}{4}b\right)\cos f \\ &+ e^2\left[2 + \frac{5}{3}a + \frac{5}{2}b + \left(10 + \frac{15}{2}a - \frac{45}{4}b\right)\cos^2 f\right], \end{aligned} \tag{12.222a}$$

$$\begin{aligned} B :=& 4 + e\left(10 + \frac{35}{6}a - \frac{5}{2}b\right)\cos f \\ &- e^2\left[9 + \frac{35}{3}a - 5b - \left(15 + \frac{35}{2}a - \frac{15}{2}b\right)\cos^2 f\right]. \end{aligned} \tag{12.222b}$$

We recall that a and b are parameters that specify the choice of radiation-reaction gauge. In the Burke–Thorne gauge we set $a = 0$ and $b = 0$, in the Schäfer gauge we set $a = -2/5$ and $b = 0$, and the assignments $a = -2/3$ and $b = 4/9$ produce the harmonic-gauge expression for the radiation-reaction force.

The effect of a perturbing force on the Keplerian orbital elements p, e, and ω was described in detail back in Sec. 3.3. In this description, the orbital motion is described at all times by the Keplerian relations of Eq. (12.219), but the orbital elements acquire a time dependence from the perturbing force. This description is exact, and the method of osculating orbits is a powerful starting point for an approximate treatment of the orbital evolution. The relevant equations are displayed in Eqs. (3.69); we have

$$\frac{dp}{df} = \frac{2p^3}{Gm}\frac{1}{(1+ec)^3}\mathcal{S}, \tag{12.223a}$$

$$\frac{de}{df} = \frac{p^2}{Gm}\left[\frac{s}{(1+ec)^2}\mathcal{R} + \frac{e+2c+ec^2}{(1+ec)^3}\mathcal{S}\right], \tag{12.223b}$$

$$e\frac{d\omega}{df} = \frac{p^2}{Gm}\left[-\frac{c}{(1+ec)^2}\mathcal{R} + \frac{s(2+ec)}{(1+ec)^3}\mathcal{S}\right], \tag{12.223c}$$

where $c := \cos f$, $s := \sin f$, $\mathcal{R} := \boldsymbol{n}\cdot\boldsymbol{a}[\mathrm{rr}]$ is the radial component of the perturbing acceleration, and $\mathcal{S} := \boldsymbol{\lambda}\cdot\boldsymbol{a}[\mathrm{rr}]$ is the tangential component. These equations must be supplemented with

$$\frac{dt}{df} = \sqrt{\frac{p^3}{Gm}}\frac{1}{(1+ec)^2}\left\{1 - \frac{1}{e}\frac{p^2}{Gm}\left[\frac{c}{(1+ec)^2}\mathcal{R} - \frac{2+ec}{(1+ec)^3}\sin f\,\mathcal{S}\right]\right\}, \tag{12.224}$$

which describes how the orbit evolves in time; this equation can be integrated once the system of Eqs. (12.223) has been solved.

To analyze the equations it is helpful to turn p into a dimensionless variable $\mathsf{p} := p/p^*$ by dividing it by a representative scale p^*; this may, for example, be chosen as the initial value $p(f=0)$. It is also helpful to introduce

$$\epsilon := \frac{8}{5}\eta\left(\frac{Gm}{c^2 p^*}\right)^{5/2} \tag{12.225}$$

as a dimensionless measure of the strength of the radiation-reaction force. Because the orbital velocity v is comparable to $\sqrt{Gm/p^*}$ during the evolution, ϵ is of the same order of magnitude as $(v/c)^5$ and therefore quite small. Finally, it is helpful to make time dimensionless by defining $\mathsf{t} := t/t^*$, with $t^* := \sqrt{p^{*3}/(Gm)}$.

In terms of these new variables the evolution equations are

$$\frac{d\mathsf{p}}{df} = -2\epsilon\mathsf{p}^{-3/2}(1+ec)B, \tag{12.226a}$$

$$\frac{de}{df} = \epsilon\mathsf{p}^{-5/2}(1+ec)\big[es^2 A - (e+2c+ec^2)B\big], \tag{12.226b}$$

$$e\frac{d\omega}{df} = -\epsilon\mathsf{p}^{-5/2}s(1+ec)\big[ec\,A + (2+ec)B\big], \tag{12.226c}$$

$$\frac{d\mathsf{t}}{df} = \frac{\mathsf{p}^{3/2}}{(1+ec)^2}\left\{1 - \epsilon\frac{(1+ec)s}{e\mathsf{p}^{5/2}}\big[ec\,A + (2+ec)B\big]\right\}. \tag{12.226d}$$

Substitution of A and B returns expressions that are quite large. For our purposes below it is sufficient to present their schematic structure, which is as follows:

$$\frac{d\mathsf{p}}{df} = -\frac{\epsilon}{\mathsf{p}^{3/2}}\left[8\left(1+\frac{7}{8}e^2\right) + k_1^p \cos f + k_2^p \cos 2f + k_3^p \cos 3f\right], \tag{12.227a}$$

$$\frac{de}{df} = -\frac{\epsilon}{\mathsf{p}^{5/2}}\left[\frac{38}{3}e\left(1+\frac{121}{304}e^2\right) + k_1^e \cos f + k_2^e \cos 2f + k_3^e \cos 3f + k_4^e \cos 4f + k_5^e \cos 5f\right], \tag{12.227b}$$

$$\frac{d\omega}{df} = -\frac{\epsilon}{e\mathsf{p}^{5/2}}\left[k_1^\omega \sin f + k_2^\omega \sin 2f + k_3^\omega \sin 3f + k_4^\omega \sin 4f + k_5^\omega \sin 5f\right]. \tag{12.227c}$$

The various coefficients k_n that come with the trigonometric functions depend on e as well as the gauge parameters a and b. The schematic forms of Eq. (12.227) reveal that the driving forces for p and e contain steady pieces that produce secular changes, in addition to oscillatory pieces that average out after each orbital cycle. And as we can see, the oscillatory pieces are gauge-dependent, while the secular pieces are independent of the choice of gauge. The driving force for ω is entirely oscillatory.

12.9.2 Multi-scale analysis of orbital evolution

Apart from the post-Newtonian expansion that gave rise to the radiation-reaction force, no approximations have entered the formulation of the orbital equations as displayed in Eqs. (12.226). We now wish to integrate these equations, and we shall take advantage of the fact that $\epsilon \ll 1$ to find approximate solutions. We must, however, be cognizant of the fact that changes in the orbital elements occur over two very distinct time scales. The first is the orbital time scale P, which is short, and the second is the radiation-reaction time scale $T_{\rm rr}$ – refer to Eq. (12.88) – which is longer than P by a factor of order $\epsilon^{-1} \gg 1$. In terms of the true anomaly f, we have changes over the short angular scale of 2π, and changes over the much longer scale ϵ^{-1}. Our perturbative approach must allow us to probe these widely separated scales; the method of choice is a *multi-scale analysis*, as presented in the excellent book by our friend Carl Bender and his late colleague Steven Orzag (1978), and summarized in Box 12.4.

Box 12.4 Multi-scale analysis

To introduce the method we work through a simple example involving a damped harmonic oscillator. We wish to integrate the differential equation

$$\ddot{x} + 2\epsilon\dot{x} + x = 0$$

with boundary conditions $x(0) = 1, \dot{x}(0) = 0$, in a regime in which $\epsilon \ll 1$. Here all variables are dimensionless, and an overdot indicates differentiation with respect to t. The equation can be integrated exactly, and the solution is

$$x = e^{-\epsilon t}\Big(\cos\omega t + \frac{\epsilon}{\omega}\sin\omega t\Big),$$

with $\omega := \sqrt{1-\epsilon^2}$. The solution reveals features that occur over widely separated time scales: there are rapid oscillations over a time scale of order unity, and a slow damping of the envelope over a time scale of order $\epsilon^{-1} \gg 1$. We would like to capture these features with a perturbative analysis of the differential equation.

A straightforward expansion of the solution in powers of ϵ would fail in this endeavor. Suppose that we write $x = x_0(t) + \epsilon x_1(t) + O(\epsilon^2)$, insert this within the differential equation, and equate terms of like powers of ϵ to zero. We would obtain the differential equations $\ddot{x}_0 + x_0 = 0$ and $\ddot{x}_1 + x_1 = -2\dot{x}_0$, and these can be integrated in turn for $x_0(t)$ and $x_1(t)$. Keeping the boundary conditions in mind, we get $x_0 = \cos t$ and observe that the driving force for x_1, equal to $2\sin t$, oscillates at the same frequency as the oscillator's own natural frequency. This creates a resonant behavior that leads to unbounded growth, as can be seen in the solution $x_1 = \sin t - t\cos t$. The end result of this standard perturbative analysis is an approximate solution

$$x = (1-\epsilon t)\cos t + \epsilon\sin t + O(\epsilon^2)$$

that becomes wholly inaccurate over times $t \simeq \epsilon^{-1}$. This approximation, in particular, does not capture the damping that occurs over the long time scale.

In a multi-scale treatment of this problem one introduces a slow-time variable $\tilde{t} := \epsilon t$ in addition to the fast time t, and postulates a solution of the form $x = x_0(\tilde{t}, t) + \epsilon x_1(\tilde{t}, t) + O(\epsilon^2)$, in which x_0 and x_1 are assumed to be bounded functions. The original differential equation is generalized in such a way that $\tilde{t}$ and t are treated as independent variables, and the total time derivative is interpreted as

$$\frac{d}{dt} = \frac{\partial}{\partial t} + \epsilon\frac{\partial}{\partial\tilde{t}}.$$

In our case we have that $\dot{x} = \partial x_0/\partial t + \epsilon\partial x_0/\partial\tilde{t} + \epsilon\partial x_1/\partial t + O(\epsilon^2)$ and $\ddot{x} = \partial^2 x_0/\partial t^2 + 2\epsilon\partial^2 x_0/\partial\tilde{t}\partial t + \epsilon\partial^2 x_1/\partial t^2 + O(\epsilon^2)$, and the differential equation becomes the set of equations

$$\frac{\partial^2 x_0}{\partial t^2} + x_0 = 0,$$
$$\frac{\partial^2 x_1}{\partial t^2} + x_1 = -2\Big(\frac{\partial x_0}{\partial t} + \frac{\partial^2 x_0}{\partial\tilde{t}\partial t}\Big).$$

The solution to the first equation is $x_0 = A(\tilde{t})\cos t$, in which the function $A(\tilde{t})$ plays the role of constant of integration; this function cannot be determined until we proceed to the next order. We observe that a second solution to the differential equation, $B(\tilde{t})\sin t$, should in principle be added to the first; it will, however, eventually be rejected by the boundary conditions.

At the next order we find that the driving force for x_1 is now equal to $2(A + A')\sin t$, in which a prime indicates differentiation with respect to $\tilde{t}$. We recognize that such a resonant force would produce a growth in t, and that this would violate our assumption that x_1 should stay bounded. We eliminate this eventuality by demanding that $A(\tilde{t})$ be a solution to $A + A' = 0$, which implies that $A = A_0\exp(-\tilde{t})$. Keeping the boundary conditions in mind, we find that the zeroth-order solution to the differential equation is

$$x = \exp(-\tilde{t}) \cos t + O(\epsilon).$$

As we can see, this captures the essential aspects of the exact solution, including the damping that occurs over the long time scale.

The zeroth-order solution can be refined by integrating what has become of the first-order differential equation. We now have $\partial^2 x_1/\partial t^2 + x_1 = 0$, whose solution is $x_1 = C(\tilde{t}) \sin t$ (up to the addition of a second solution, which we discard by virtue of the boundary conditions). Once more we find that the unknown function $C(\tilde{t})$ cannot be determined until we proceed to the next order. It would, however, be determined in the same way, by ensuring that the driving force for x_2 is free of a resonant term. Such an analysis would reveal that $C = \exp(-\tilde{t})$, and the first-order solution to the differential equation is

$$x = \exp(-\tilde{t})\big(\cos t + \epsilon \sin t\big) + O(\epsilon^2).$$

This captures even more of the exact solution, but the central message of this exercise is that x_0 by itself does a very fine job of reproducing the essential behavior of the exact solution.

General theory of multi-scale orbital evolution

We now perform a multi-scale analysis of the evolution equations for the orbital elements p, e, and ω. Before we return to the specific situation that concerns us in this section (orbital evolution under radiation reaction), we develop a fairly general formulation that can be applied to many different situations. This formulation was devised by our friend Adam Pound in his PhD dissertation (2010). Similar (but less general) formulations can be found in a 1990 article by Lincoln and Will, and a 2004 paper by Mora and Will.

We collect the orbital elements into the vector μ^a, and express their evolution equations as

$$\frac{d\mu^a}{df} = \epsilon F^a(f, \mu^b), \tag{12.228}$$

in which ϵ is a small parameter, and each driving force F^a is assumed to be periodic in f, so that $F^a(f + 2\pi, \mu^b) = F^a(f, \mu^b)$. These equations are supplemented with

$$\frac{dt}{df} = T_0(f, \mu^a) + \epsilon T_1(f, \mu^a), \tag{12.229}$$

which governs the behavior of the time function; T_0 and T_1 are also assumed to be periodic in f. The exact specifications of F^a, T_0, and T_1 for our particular problem can be obtained from Eqs. (12.223) and (12.224), but our considerations here are more general.

In a multi-scale analysis of these differential equations we postulate the existence of solutions of the form

$$\mu^a = \mu_0^a(\tilde{f}, f) + \epsilon \mu_1^a(\tilde{f}, f) + \epsilon^2 \mu_2^a(\tilde{f}, f) + O(\epsilon^3) \tag{12.230}$$

and

$$t = \epsilon^{-1}t_{-1}(\tilde{f}, f) + t_0(\tilde{f}, f) + \epsilon t_1(\tilde{f}, f) + O(\epsilon^2), \tag{12.231}$$

in which $\tilde{f} := \epsilon f$ is the slow variable. Each function μ_n^a and t_n is assumed to be periodic in f. The presence of $\epsilon^{-1}t_{-1}$ in t is required because the time variable grows secularly even in the absence of a perturbation; when $\tilde{f}$ is of order unity and f of order ϵ^{-1}, t also is of order ϵ^{-1}.

To proceed with Eqs. (12.228) we treat $\tilde{f}$ and f as independent variables and interpret a total derivative with respect to f as

$$\frac{d}{df} = \frac{\partial}{\partial f} + \epsilon \frac{\partial}{\partial \tilde{f}}. \tag{12.232}$$

Inserting Eq. (12.230) within the differential equations and equating like powers of ϵ, we obtain

$$\frac{\partial \mu_0^a}{\partial f} = 0, \tag{12.233a}$$

$$\frac{\partial \mu_0^a}{\partial \tilde{f}} + \frac{\partial \mu_1^a}{\partial f} = F^a(f, \mu_0^b), \tag{12.233b}$$

$$\frac{\partial \mu_1^a}{\partial \tilde{f}} + \frac{\partial \mu_2^a}{\partial f} = \mu_1^c \partial_c F^a(f, \mu_0^b); \tag{12.233c}$$

in the last equation the driving force F^a is differentiated with respect to μ^c, and the repeated index indicates a summation over all orbital elements.

The equations of the system (12.233) can be integrated in turn. Equation (12.233a) implies that μ_0^a is a function of the slow variable $\tilde{f}$ only. To integrate Eq. (12.233b) we first average the equation over a complete cycle of the fast variable. We use the notation

$$\langle q \rangle(\tilde{f}) := \frac{1}{2\pi} \int_0^{2\pi} q(\tilde{f}, f)\, df \tag{12.234}$$

to indicate such an average; since $\tilde{f}$ and f are independent variables the integral is evaluated while keeping $\tilde{f}$ fixed. Because the first term of Eq. (12.233b) is independent of f, its average returns $d\mu_0^a/d\tilde{f}$. The average of the second term vanishes, because of the assumed periodicity of μ_1^a. We are left with

$$\frac{d\mu_0^a}{d\tilde{f}} = \langle F^a \rangle(\mu_0^b), \tag{12.235}$$

a system of equations that determine $\mu_0^a(\tilde{f})$. Subtracting this from Eq. (12.233b), we next obtain $\partial \mu_1^a/\partial f = F^a - \langle F^a \rangle$, which integrates to

$$\mu_1^a = \mu_{1,\mathrm{osc}}^a(\tilde{f}, f) + \mu_{1,\mathrm{sec}}^a(\tilde{f}), \tag{12.236}$$

where

$$\mu_{1,\mathrm{osc}}^a = \int \left[F^a(f, \mu_0^b) - \langle F^a \rangle(\mu_0^b) \right] df \tag{12.237}$$

is a periodic function of f, and $\mu^a_{1,\rm sec}$ acts as a constant of integration. This is determined at the next order, by averaging Eq. (12.233c). The first term gives rise to $d\mu^a_{1,\rm sec}/d\tilde{f}$, the second term contributes nothing, and we arrive at

$$\frac{d\mu^a_{1,\rm sec}}{d\tilde{f}} = \big\langle \mu^c_{1,\rm osc}\partial_c F^a(f, \mu^b_0)\big\rangle. \tag{12.238}$$

We note that only $\mu^c_{1,\rm osc}$ appears on the right-hand side, instead of the complete expression μ^c_1, because the average of the term involving $\mu^a_{1,\rm sec}$ vanishes. At this stage we have all the ingredients required in the construction of μ^a to first order in ϵ; to obtain the terms of order ϵ^2 in Eq. (12.230) we would proceed to the next order.

We follow the same recipe to integrate Eq. (12.229). Skipping over the details, we find that t_{-1} is a function of $\tilde{f}$ only determined by

$$\frac{dt_{-1}}{d\tilde{f}} = \langle T_0\rangle(\mu^a_0), \tag{12.239}$$

and that t_0 can be expressed as

$$t_0 = t_{0,\rm osc}(\tilde{f}, f) + t_{0,\rm sec}(\tilde{f}) \tag{12.240}$$

with

$$t_{0,\rm osc} = \int \big[T_0(f, \mu^a_0) - \langle T_0\rangle(\mu^a_0)\big]\, df \tag{12.241}$$

and

$$\frac{dt_{0,\rm sec}}{d\tilde{f}} = \big\langle \mu^b_{1,\rm osc}\partial_b T_0(f, \mu^a_0) + T_1(f, \mu^a_0)\big\rangle. \tag{12.242}$$

These equations allow us to construct t to zeroth order in ϵ; to obtain the terms of order ϵ in Eq. (12.231) we would proceed to the next order.

The most important pieces of the orbital elements μ^a are those that grow secularly and depend on the slow variable $\tilde{f}$. These are contained in μ^a_0 and $\mu^a_{1,\rm sec}$, and a useful *secular approximation* to the orbital elements is given by

$$\mu^a_{\rm sec} = \mu_0(\tilde{f}) + \epsilon\mu^a_{1,\rm sec}(\tilde{f}) + O(\epsilon^2); \tag{12.243}$$

this ignores the unimportant periodic oscillations contained in $\mu^a_{1,\rm osc}$, which average out after each orbital cycle. Similarly, a useful secular approximation to the time function is given by

$$t_{\rm sec} = \epsilon^{-1}t_{-1}(\tilde{f}) + t_{0,\rm sec}(\tilde{f}) + O(\epsilon). \tag{12.244}$$

It is important to understand that while the oscillatory terms $\mu^a_{1,\rm osc}$ do not appear in these equations, they are nevertheless required in the construction of the secular terms $\mu^a_{1,\rm sec}$ and $t_{0,\rm sec}$. This can be seen from Eqs. (12.238) and (12.242), which reveal that oscillations in μ^a_1 can combine with oscillations in $\partial_c F^a(f, \mu^b_0)$ and $\partial_b T_0(f, \mu^a_0)$ to produce secular terms.

Radiation reaction

We now have all the required tools at our disposal, and we may return to the original problem, the determination of the orbital evolution under radiation reaction. The orbital elements are $\mu^a := (\mathsf{p}, e, \omega)$, and the driving forces F^a can be extracted from Eq. (12.227). Their averages are easily computed, and Eq. (12.235) returns

$$\frac{d\mathsf{p}_0}{d\tilde{f}} = -8\mathsf{p}_0^{-3/2}\left(1 + \frac{7}{8}e_0^2\right), \tag{12.245a}$$

$$\frac{de_0}{d\tilde{f}} = -\frac{38}{3}e_0\mathsf{p}_0^{-5/2}\left(1 + \frac{121}{304}e_0^2\right), \tag{12.245b}$$

$$\frac{d\omega_0}{d\tilde{f}} = 0, \tag{12.245c}$$

where $\mu_0^a := (\mathsf{p}_0, e_0, \omega_0)$ are the zeroth-order approximations to the orbital elements. From Eq. (12.227) and (12.237) we also obtain the oscillatory first-order corrections,

$$\mathsf{p}_{1,\mathrm{osc}} = -\frac{1}{\mathsf{p}_0^{3/2}}\left(k_1^p \sin f + \frac{1}{2}k_2^p \sin 2f + \frac{1}{3}k_3^p \sin 3f\right), \tag{12.246a}$$

$$e_{1,\mathrm{osc}} = -\frac{1}{\mathsf{p}_0^{5/2}}\left(k_1^e \sin f + \frac{1}{2}k_2^e \sin 2f + \frac{1}{3}k_3^e \sin 3f + \frac{1}{4}k_4^e \sin 4f + \frac{1}{5}k_5^e \sin 5f\right), \tag{12.246b}$$

$$\omega_{1,\mathrm{osc}} = \frac{1}{e_0\mathsf{p}_0^{5/2}}\left(k_1^\omega \cos f + \frac{1}{2}k_2^\omega \cos 2f + \frac{1}{3}k_3^\omega \cos 3f + \frac{1}{4}k_4^\omega \cos 4f + \frac{1}{5}k_5^\omega \cos 5f\right). \tag{12.246c}$$

These are combined with derivatives of the driving forces to compute the right-hand side of Eq. (12.238). We obtain

$$\frac{d\mathsf{p}_{1,\mathrm{sec}}}{d\tilde{f}} = \frac{de_{1,\mathrm{sec}}}{d\tilde{f}} = 0, \tag{12.247}$$

and conclude that there is no secular growth in p and e at first order in ϵ. The calculation reveals also that $d\omega_{1,\mathrm{sec}}/d\tilde{f} \neq 0$, but we shall not be concerned with this small amount of secular growth in ω.

Proceeding with the time function, we insert Eq. (12.226) within Eq. (12.239) and get

$$\frac{d\mathsf{t}_{-1}}{d\tilde{f}} = \left(\frac{\mathsf{p}_0}{1 - e_0^2}\right)^{3/2}. \tag{12.248}$$

From Eq. (12.242) we next obtain

$$\frac{d\mathsf{t}_{0,\mathrm{sec}}}{d\tilde{f}} = 0, \tag{12.249}$$

and observe the absence of secular growth at zeroth order in ϵ.

Conclusion

The multi-scale analysis of the orbital evolution equations is now completed. We have found that the secular changes in the orbital elements are governed by the system of differential equations

$$\left.\frac{d\mathsf{p}}{df}\right|_{\rm sec} = -\epsilon\frac{8}{\mathsf{p}^{3/2}}\left(1+\frac{7}{8}e^2\right) + O(\epsilon^3), \tag{12.250a}$$

$$\left.\frac{de}{df}\right|_{\rm sec} = -\epsilon\frac{38e}{3\mathsf{p}^{5/2}}\left(1+\frac{121}{304}e^2\right) + O(\epsilon^3), \tag{12.250b}$$

$$\left.\frac{d\mathsf{t}}{df}\right|_{\rm sec} = \left(\frac{\mathsf{p}}{1-e^2}\right)^{3/2} + O(\epsilon^2), \tag{12.250c}$$

in which we have removed the now-redundant suffixes on p, e, and t.

To display the equations in their final form we re-introduce the scales p^* and t^*, substitute ϵ from Eq. (12.225), and eliminate f. This yields

$$\left.\frac{dp}{dt}\right|_{\rm sec} = -\frac{64}{5}\eta c\left(\frac{GM}{c^2p}\right)^3(1-e^2)^{3/2}\left(1+\frac{7}{8}e^2\right), \tag{12.251a}$$

$$\left.\frac{de}{dt}\right|_{\rm sec} = -\frac{304}{15}\eta c\frac{e}{p}\left(\frac{GM}{c^2p}\right)^3(1-e^2)^{3/2}\left(1+\frac{121}{304}e^2\right). \tag{12.251b}$$

These are the same equations that were obtained in Sec. 12.3.3 on the basis of averaged statements of energy and angular-momentum balance – refer to Eqs. (12.86) and (12.87). The treatment given here, based on a multi-scale analysis of the evolution equations, is much more satisfactory than the earlier work: We were able to prove that Eqs. (12.251) do indeed capture the secular behavior of the orbital elements, and that the fractional error incurred is of order ϵ^2.

In Figs. 12.3 and 12.4 we compare the approximate evolution obtained on the basis of Eqs. (12.250) to an exact numerical integration of Eqs. (12.226). We can see that the agreement is excellent.

12.10 Bibliographical notes

Radiation-reaction effects in electromagnetism are discussed in many textbooks, including Jackson's *Classical Electrodynamics* (1998). The treatment is usually restricted to point charges, and much space is devoted to the curious fact that the radiation-reaction force depends on the rate of change of the particle's acceleration. The literature has debated this issue endlessly, and many misconceptions have taken hold — don't get us started. For a well-balanced overview, refer to Gralla, Harte, and Wald (2009).

The shortwave approximation of Sec. 12.2 was first formulated by Isaacson (1968a and 1968b). The angular-momentum flux of gravitational waves was first calculated by Bryce DeWitt in the early nineteen seventies, and his results were eventually published in DeWitt

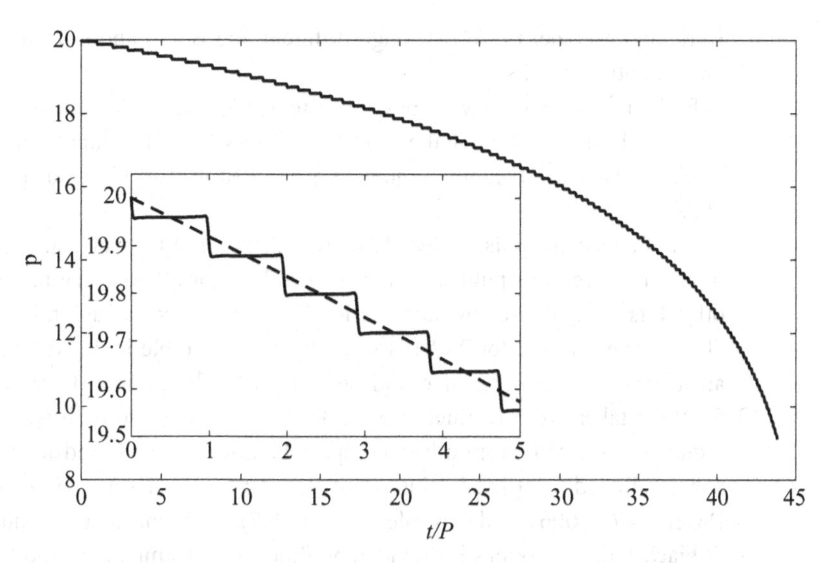

Fig. 12.3 Evolution of the dimensionless semi-latus rectum p under radiation reaction. The evolution begins with $p = 20$ and $e = 0.7$, and it proceeds for 100 orbital cycles with $\epsilon = 0.1$. Time is expressed in units of the initial orbital period P. The exact curve (solid) displays oscillations around the mean curve (dashed), and it was obtained by numerical integration of Eqs. (12.226). These computations were carried out in the Burke–Thorne gauge, with $a = b = 0$. The mean curve is obtained on the basis of the secular approximation of Eqs. (12.250).

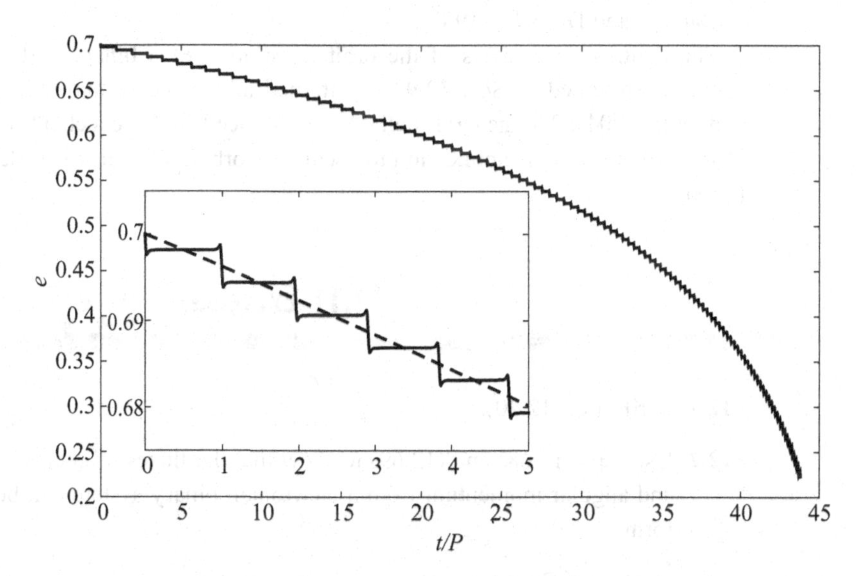

Fig. 12.4 Evolution of the eccentricity e under radiation reaction. The details are as in Fig. 12.3.

(2011); the final expression also appeared in Thorne (1980). An alternative approach to the description of radiative losses in general relativity, widely considered to be more rigorous and convincing than the Landau–Lifshitz approach adopted here, was formulated by Bondi and his colleagues. Representative papers are Sachs (1961 and 1962) and Bondi, van der

Burg, and Metzner (1962). Though different, the Bondi and Landau–Lifshitz approaches yield identical results.

Radiative losses of slowly-moving systems (Sec. 12.3), including binary stars, were first explored by Peters and Mathews (1963), Peters (1964), Bekenstein (1973), and Fitchett (1983). Wiseman's beaming argument, presented in Box 12.2, was published in Wiseman (1992).

The first binary pulsar (Sec. 12.4) was discovered by Hulse and Taylor in 1974; their discovery paper was published as Hulse and Taylor (1975). The first measurement of the orbital damping caused by radiative losses was reported in Taylor, Fowler, and McCulloch (1979). The numbers for PSR 1913+16 displayed in Table 12.1 and the following discussion are taken from Weisberg, Nice, and Taylor (2010). The numbers for the double pulsar J0737-3039 are taken from Kramer *et al.* (2006). Calculations to high post-Newtonian order of radiative losses for inspiralling compact binaries are reviewed in Blanchet (2006). The papers alluded to at the beginning of Sec. 12.4.3 are Favata, Hughes, and Holz (2004), Baker *et al.* (2006), and Gonzalez *et al.* (2007). A recent review of numerical simulations of black-hole superkicks is provided by Zlochower, Campanelli, and Lousto (2011).

The approach adopted in Secs. 12.5–12.7 to calculate radiation-reaction potentials and forces is patterned after Pati and Will (2000 and 2001). The Burke–Thorne gauge featured in Sec. 12.8 was formulated in Thorne (1969), Burke (1971), and Misner, Thorne, and Wheeler (1973); see, however, the criticisms expressed by Walker and Will (1980). The Schäfer gauge was formulated in Schäfer (1983), and the Damour–Deruelle gauge refers to Damour and Deruelle (1981).

The multi-scale analysis of the orbital evolution of a binary system under radiation reaction, presented in Sec. 12.9, is patterned after Lincoln and Will (1990) and Mora and Will (2004). The general method is introduced in the excellent text by Bender and Orzag (1978), and the application to osculating orbital elements was developed in Pound (2010).

12.11 Exercises

12.1 Verify Eq. (12.40).

12.2 Use Eqs. (12.68) and (12.69) to show that the fluxes of energy, linear momentum, and angular momentum from a Newtonian binary system can be expressed in the form

$$\mathcal{P} = \frac{8}{15}\eta^2 \frac{c^3}{G}\left(\frac{Gm}{c^2 r}\right)^4 \left(12v^2 - 11\dot{r}^2\right),$$

$$\mathcal{F}^j = -\frac{8}{105}\Delta\eta^2 \frac{c}{G}\left(\frac{Gm}{c^2 r}\right)^4 \left[v^j\left(50v^2 - 38\dot{r}^2 + 8\frac{Gm}{r}\right) - \dot{r}n^j\left(55v^2 - 45\dot{r}^2 + 12\frac{Gm}{r}\right)\right],$$

$$\mathcal{T}^j = \frac{8}{5}\eta^2 \frac{c}{G}\left(\frac{Gm}{c^2 r}\right)^3 h^j\left(2v^2 - 3\dot{r}^2 + 2\frac{Gm}{r}\right).$$

By incorporating a total time derivative into $P^j[5]$ and $J^j[5]$, show that the momentum and angular-momentum fluxes can be written in the simpler forms

$$\mathcal{F}^j = -\frac{4}{15}\Delta\eta^2\frac{c}{G}\left(\frac{Gm}{c^2r}\right)^4 v^j\left(13v^2 - 15\dot{r}^2\right),$$

$$\mathcal{T}^j = \frac{8}{5}\eta^2\frac{c}{G}\left(\frac{Gm}{c^2r}\right)^3 h^j\left(4v^2 - 9\dot{r}^2\right),$$

and verify that the momentum flux is in the direction of motion of the lighter body.

12.3 Consider a Keplerian, circular orbit of radius a in the x-y plane. Show that when averaged over a complete orbit, the power emitted in gravitational waves per unit solid angle is given by

$$\frac{d\mathcal{P}}{d\Omega} = \frac{1}{2\pi}\eta^2\frac{c^5}{G}\left(\frac{Gm}{c^2a}\right)^5\left(1 + 6\cos^2\theta + \cos^4\theta\right),$$

where θ is the angle between the z-axis and $\boldsymbol{N}$, the direction of emission. Verify that the total power integrated over all solid angles agrees with Eq. (12.82a) when $e = 0$.

12.4 Consider a Keplerian orbit that is circular apart from the slow decrease in radius a caused by the energy lost to gravitational radiation. As a function of η, m, and the initial radius a_0, calculate the lifetime of the binary system and the number of completed orbits before the radiation reaction brings the radius to zero. Give alternative expressions for the lifetime and number of orbits in terms of η, m, and the initial orbital period P. Using these results, carry out the following estimates:

(a) the remaining lifetime of the Hulse–Taylor binary pulsar PSR 1913+16, with $M_1 \approx M_2 \approx 1.4M_\odot$ and $P = 7.75$ hours (assume that the orbit is circular);

(b) the total time and number of cycles in the gravitational-wave signal from an inspiralling binary system of two $1.4M_\odot$ compact objects, from the time it enters the LIGO–Virgo frequency band with a gravitational-wave frequency of 10 Hz to the end of the inspiral (when $a = 0$);

(c) the remaining lifetime of the Earth–Sun system.

12.5 The current eccentricity of the Hulse–Taylor binary pulsar orbit is $e_0 \approx 0.6$, and its orbital period is 7.75 hours. Estimate the orbital eccentricity when gravitational waves from the system first enter the LIGO–Virgo band at 10 Hz. You may treat the eccentricity as if it were much smaller than unity when making your estimate.

12.6 Consider a binary pulsar of equal masses, for which the only orbital variables measured initially are the orbital period P and the pericenter advance $\dot{\omega}$. Using the approximation of small eccentricity, show that the lifetime of the system is given by

$$T = \frac{5}{128}\left(\frac{6\pi}{\dot{\omega}}\right)^{5/2}\left(\frac{1}{P}\right)^{3/2}.$$

On the $\dot{\omega}$-P plane, plot the curve corresponding to $T = 1$ billion years. Plot the curve from approximately 0.1 to 30 degrees per year for $\dot{\omega}$, and from approximately 1 to

30 hours for P. How is the curve shifted when the initial orbit has a high eccentricity? How does the curve change if the masses are not equal?

12.7 For an inspiralling circular orbit of two bodies, show that the accumulated velocity of the center-of-mass as a result of radiative recoil is given by

$$V_{\text{kick}} = \frac{464}{105}\Delta\eta^2 \left(\frac{Gm}{c^2 a}\right)^4 c\,\boldsymbol{n}\,.$$

Assume that the time scale for radiative losses is much longer than the orbital period, so that $d\Omega/dt \sim O(c^{-5/2})\Omega^2$. Find the mass ratio that maximizes the recoil velocity.

12.8 Consider a two-body system on a Newtonian hyperbolic orbit described by

$$r = \frac{p}{1+e\cos\phi}\,, \qquad \frac{d\phi}{dt} = \sqrt{\frac{Gm}{p^3}}(1+e\cos\phi)^2\,,$$

where m is the total mass, and p is the semi-latus rectum, related to the angular momentum per unit reduced mass h by $h^2 = Gmp$. The system's energy is given by

$$E = \eta\frac{Gm^2}{2p}(e^2-1)\,.$$

Note that $e > 1$, $E > 0$, and the orbit comes in from infinity at $\phi = -\arccos(-1/e)$, reaching pericenter at $\phi = 0$.

(a) Show that the total energy emitted in gravitational waves by the time the orbit reaches pericenter is given by

$$\Delta E = \frac{32}{5}\eta^2\frac{Gm^2}{p}\left(\frac{Gm}{c^2 p}\right)^{5/2} f(e)\,,$$

$$f(e) = \left(1+\frac{73}{24}e^2+\frac{37}{96}e^4\right)\arccos\left(-\frac{1}{e}\right) + \frac{301}{144}\left(1+\frac{673}{602}e^2\right)\sqrt{e^2-1}\,.$$

(b) In the limit where e is close to unity, that is, when $e = 1+\epsilon$ with $\epsilon \ll 1$, show that the energy loss will convert the hyperbolic orbit into a bound orbit before pericenter when the following inequality is satisfied:

$$\frac{85\pi\eta}{3\epsilon}\left(\frac{Gm}{c^2 p}\right)^{5/2} > 1\,.$$

(c) Instead of the parameterization (p, ϵ) for the orbit, adopt (b, v_∞), in which b is an impact parameter and v_∞ is the orbital velocity at infinite separation. These quantities are defined by $v_\infty^2 = (Gm/p)(e^2-1) \simeq 2(Gm/p)\epsilon$ and $h = bv_\infty$. Express the criterion of part (b) as an inequality for b.

(d) Show that the cross-section $\sigma_{\text{GW}} := \pi b^2_{\text{capture}}$ for gravitational-wave capture is given by

$$\sigma_{\text{GW}} = \pi\left(\frac{170\pi}{3}\eta\right)^{2/7}\left(\frac{Gm}{c^2}\right)^2\left(\frac{c}{v_\infty}\right)^{18/7}\,.$$

Such a process could play an important role in the evolution of dense star clusters.

12.9 Consider the radiation-reaction acceleration for a binary system. Given that it must be a 2.5PN correction of the Newtonian acceleration, that it must vanish in the test-body limit, and that it must be related to the mutual accelerations of the bodies, it is possible to show, without doing any work, that it must be of the general form

$$\boldsymbol{a}[\mathrm{rr}] = \frac{8\eta}{5c^3}\frac{Gm}{r^2}\frac{Gm}{c^2r}\left[\left(a_1v^2 + a_2\frac{Gm}{r} + a_3\dot{r}^2\right)\dot{r}\,\boldsymbol{n} + \left(b_1v^2 + b_2\frac{Gm}{r} + b_3\dot{r}^2\right)\boldsymbol{v}\right],$$

where $m = M_1 + M_2$, $\eta = M_1M_2/m$, and a_n and b_n are arbitrary parameters.

(a) Justify the form shown above.

(b) Calculate the energy and angular momentum losses, dE/dt and $d\boldsymbol{J}/dt$, implied by the proposed radiation-reaction acceleration.

(c) Using Newtonian theory, prove the following useful identity:

$$\frac{d}{dt}\left(\frac{v^{2s}\dot{r}^p}{r^q}\right) = \frac{v^{2s-2}\dot{r}^{p-1}}{r^{q+1}}\left[pv^4 - pv^2\frac{Gm}{r} - (p+q)v^2\dot{r}^2 - 2s\frac{Gm}{r}\dot{r}^2\right],$$

where p, q and s are integers.

(d) By considering the three cases $(p, q, s) = (1, 2, 1), (3, 2, 0)$ and $(1, 3, 0)$, use the identity to show that the various numerical coefficients in the expressions for dE/dt and $d\boldsymbol{J}/dt$ can be altered by absorbing total time derivatives into $E[5]$ and $\boldsymbol{J}[5]$. Show that the freedom contained in these redefinitions is described by a three-parameter family.

(e) Assume now that dE/dt and $d\boldsymbol{J}/dt$, as calculated previously, and including the three-parameter family of redefinitions, match the gravitational-wave fluxes of energy and angular momentum, given by

$$\mathcal{P} = \frac{8}{15}\eta^2\frac{c^3}{G}\left(\frac{Gm}{c^2r}\right)^4\left(12v^2 - 11\dot{r}^2\right),$$

$$\mathcal{T}^j = \frac{8}{5}\eta^2\frac{c}{G}\left(\frac{Gm}{c^2r}\right)^3 h^j\left(2v^2 - 3\dot{r}^2 + 2\frac{Gm}{r}\right),$$

where $\boldsymbol{h} = \boldsymbol{r}\times\boldsymbol{v}$. Obtain constraints on the coefficients a_n and b_n in $\boldsymbol{a}[\mathrm{rr}]$. Show that a_n and b_n can be determined up to two unknown parameters. Show that this freedom corresponds precisely to the two-parameter gauge freedom described in Eq. (12.185), and that one can recover the Burke–Thorne, Schäfer, and Damour–Deruelle expressions for the radiation-reaction acceleration. This approach to finding the radiation-reaction acceleration was taken by Iyer and Will (1995).

12.10 The radiative losses of energy and angular momentum cause the eccentricity of a binary system to decrease. This implies that in the past, the eccentricity must always have been larger. From Eqs. (12.251), which describe the secular evolution of the orbital elements, show that as $t \to -\infty$, the eccentricity tends to unity and the semi-latus rectum p tends to a constant p_∞. Show that they evolve according to

$$1 - e = \frac{85}{72} \ln(p_\infty/p)\big[1 + O(1 - e)\big] .$$

Using Eq. (12.224), show that $f \to -\pi$ in the limit $t \to -\infty$, and thus that the two bodies were at an infinite separation on a parabolic orbit. A rigorous analysis taking into account the periodic variations in the orbital elements reveals that a generic orbit in the infinite past is actually hyperbolic ($e > 1$), and that the parabolic orbit is a special case.

13 Alternative theories of gravity

From Chapter 5 until now we have confined our attention to Einstein's general theory of relativity. But general relativity is not the only possible relativistic theory of gravity. Even in the late 1800s, well before Einstein began his epochal work on special and general relativity, there were attempts to devise theories of gravity that went beyond Newtonian theory. Some attempts were modeled on Maxwell's electrodynamics. Some replaced ∇^2 with a wave operator in Poisson's equation of Newtonian gravity, in an attempt to formulate a theory that was invariant under Lorentz transformations. None of these attempts was very successful; for example, most theories could not account for the anomalous perihelion advance of Mercury. In 1913, before Einstein completed the general theory of relativity, Nordström proposed a theory involving a curved spacetime; the metric was expressed as $g_{\alpha\beta} = \Phi\eta_{\alpha\beta}$, with the scalar field Φ satisfying a Lorentz-invariant wave equation. But the theory automatically predicts a zero deflection of light, and ultimately it failed the test of experiment.

Alternative proposals appeared even after the publication of general relativity and the empirical successes with Mercury and the deflection of light. The eminent mathematician and philosopher Alfred North Whitehead formulated such an alternative theory in 1922. Troubled by the fact that in general relativity the causal relationships in spacetime are not known a priori, but only after the metric has been determined for a given distribution of matter, he devised a theory with a background Minkowski metric in order to put causality on a "firmer" ground. Initially, Whitehead's theory was found to agree with the deflection of light and Mercury's perihelion advance, and so for many years it was considered a viable, if not particularly attractive, alternative to general relativity. Indeed, it was not shown to be in serious violation of experimental results until 1971.

The continuing weakness of experimental confirmations of general relativity left plenty of openings for alternative proposals. Two influential papers from the early 1960s illustrate the situation: A 1962 review of experiments in gravitation by Bruno Bertotti, Dieter Brill, and Ronald Krotkov demonstrated just how thin was the evidence supporting general relativity, and a review by Gerald Whitrow and Georg Morduch (1965) listed scores of nominally viable alternative theories of gravity.

It was no coincidence that Bertotti, Brill, and Krotkov were working in Robert Dicke's research group at Princeton University at the time, for Dicke was seriously interested in his own alternative theory of gravity, and he wanted to know exactly what the experimental constraints were. He had published the theory in 1961 with his student Carl Brans. Although it was based in part on earlier theories by Markus Fierz and Pascal Jordan, the theory nevertheless became known as the Brans–Dicke theory, and it had a major impact on the development of theoretical and experimental gravitational physics. It was the simplest modification of general relativity, retaining the concepts of curved spacetime and the Einstein equivalence principle, but modifying the way matter generates curvature by the added effect of a scalar field. As a mathematically consistent and well-motivated theory, it made calculable predictions for experiments, and those were only slightly different from the predictions of general relativity, in all cases within the experimental bounds that were known in the early 1960s. The Brans–Dicke theory played a primary role in a flowering of experimental gravity during the late 1960s and the 1970s. Ironically, this led to the demise of the theory, as experiments continued to support general relativity, and the constraints became too tight to leave much room for the alternative theory.

But recent years have witnessed a resurgence of alternative theories of gravity, or theories that go "beyond" Einstein. Some of this interest comes from the direction of elementary particle physics, especially the development of superstring theory, whose low-energy limit is *not* general relativity but a variant of Brans–Dicke theory, with a scalar field related to the dilaton and moduli fields that are central ingredients of string theory. Other alternative theories have emerged from attempts to formulate laws of physics in spacetimes of higher dimensionality. The observational evidence that as much as 25 percent of the universe is made up of dark matter has spawned alternative theories that attempt to account for the anomalous rotation curves of galaxies by modifying gravity instead of introducing a distribution of dark matter. The 1998 discovery that the expansion of the universe is accelerating has motivated the development of theories that modify general relativity on the largest scales, while leaving intact its predictions for solar-system and stellar-scale phenomena. Finally, the ongoing search for a quantum theory of gravity leaves open the possibility that the classical limit may be a theory radically different from general relativity.

The subject of alternative theories is too large and too active at the time of writing to do it justice in a single chapter. We merely touch on aspects of alternative theories and their experimental tests for which the post-Newtonian methods developed in this book are most useful. We begin in Sec. 13.1 with a general discussion of metric theories of gravity and their relation to the strong equivalence principle. In Sec. 13.2 we introduce a very general and powerful formalism, the *parameterized post-Newtonian framework*, to extract the experimental consequences of a broad class of alternative theories and subject them to empirical tests. We describe such tests in Sec. 13.3, and in Sec. 13.4 we explore the physics of gravitational waves in a broad class of alternative theories. Finally, in Sec. 13.5 we examine the predictions of a class of theories known as *scalar–tensor gravity*, in which the description of the gravitational field involves a scalar field in addition to the usual metric tensor; the Brans–Dicke theory is a particular member of this class of theories.

13.1 Metric theories and the strong equivalence principle

We wish to consider alternatives to general relativity, but we shall limit the scope of this generalization by insisting that each theory should satisfy the Einstein equivalence principle and be subjected to the metric-theory principles listed in Sec. 5.1. This defines the class of *metric theories of gravitation*, to which general relativity belongs.

The central aspect of the Einstein equivalence principle is that matter should couple in a universal manner to a single tensorial gravitational field, the metric $g_{\alpha\beta}$. There could be other gravitational fields present in spacetime, but these are prevented from interacting with the matter, and only serve to mediate the manner in which matter generates the metric. The additional fields could be scalars (as in the Brans–Dicke theory), vectors, tensors, or more exotic mathematical objects. They could be dynamical agents, governed by their own field equations, or they could be non-dynamical, fixed in some manner independently of the behavior of matter and fields (as for the Minkowski metric of Whitehead's theory). Some theories have only the metric as the basic ingredient, as in general relativity, but propose alternative field equations. One class of such theories postulates a gravitational Lagrangian density that is a general function of the Ricci scalar, rather than the Ricci scalar itself; these are called "$f(R)$ theories," devised to alter the behavior of gravity on cosmological scales. Another class of theories adds quadratic and higher-order curvature terms to the general relativistic Lagrangian density; this alters the behavior of the metric on short scales, and the higher-order terms are sometimes viewed as representing quantum corrections to classical general relativity.

Thinking about alternative theories of gravity from this broad but circumscribed point of view, we can draw some general conclusions about the nature of gravity in different metric theories. Consider a local reference frame, moving freely in a spacetime described by a given metric theory of gravity. Let this frame be sufficiently small that inhomogeneities in the external gravitational fields are suitably small throughout its volume. On the other hand, let the frame be sufficiently large that it can contain a system of gravitating matter and its associated gravitational fields; the system could be a star, a black hole, the solar system, or a Cavendish experiment set up to measure Newton's constant G. This is the kind of frame that was described at length back in Sec. 9.4; call it a "quasi-local Lorentz frame."

To determine the behavior of the system we must calculate the metric generated by the material bodies and other fields. The computation proceeds in two stages. First, we determine the external behavior of the metric and other gravitational fields, thereby establishing boundary values that must be imposed upon the fields generated by the local system, at the boundary of the quasi-local frame far from the local system. Second, we solve for the fields generated by the local system. Because the metric is coupled directly or indirectly to the other fields of the theory, its structure and evolution inside the quasi-local frame are influenced by the boundary values taken on by these fields. This is true even when we work in a coordinate system in which the asymptotic form of $g_{\alpha\beta}$ in the boundary region between the local system and the external world is that of the Minkowski metric, modulo tidal potentials from the external bodies. The gravitational environment in which the local system resides can therefore influence the metric generated by the local system via the boundary values of the auxiliary fields. Consequently, the results of local gravitational experiments may depend on the position and velocity of the quasi-local frame relative to the external environment. (A non-gravitational experiment is unaffected, because the gravitational fields it generates are negligible, and because the apparatus couples only to the metric, whose form can always be made locally Minkowskian at a given event in spacetime.) A local gravitational experiment might consist of a Cavendish-type experiment, a measurement of the acceleration of massive self-gravitating bodies, a study of the structure of stars and planets, or an analysis of the periods of planetary orbits.

We can now make a number of statements regarding different kinds of metric theories. First, a theory containing only the metric $g_{\alpha\beta}$ and no other fields yields a local gravitational physics that is independent of the position and velocity of the local system. This follows from the fact that the only field coupling the local system to the environment is $g_{\alpha\beta}$, and it is always possible to find a coordinate system in which $g_{\alpha\beta}$ takes the Minkowski form at the boundary between the local system and the external environment, modulo tidal potentials. The asymptotic values of $g_{\alpha\beta}$ are therefore constants independent of position, and they are asymptotically Lorentz invariant, independent of the velocity of the quasi-local Lorentz frame. General relativity is an example of such a theory.

Second, a theory containing the metric $g_{\alpha\beta}$ and a number of dynamical scalar fields ϕ_A yields a local gravitational physics that may depend on the position of the frame but is independent of its velocity. This follows from the asymptotic Lorentz invariance of the Minkowski metric and the scalar fields, but now the asymptotic values of the scalar fields may depend on the position of the frame. An example is Brans–Dicke theory, in which the asymptotic scalar field determines the effective value of the gravitational constant, which therefore varies as ϕ varies. The scalar field can vary in time because of cosmological evolution, or it can vary in space because of the proximity of matter outside the quasi-local frame.

Third, a theory containing the metric $g_{\alpha\beta}$ and additional dynamical or non-dynamical vector or tensor fields yields a local gravitational physics that may have both position- and velocity-dependent effects. For example, a timelike vector field K^α, whose value is determined by the distribution of matter in the universe, will have only a time component K^0 in a reference frame in which the large-scale distribution of matter is isotropic, presumably the rest-frame of the cosmic background radiation. But in a quasi-local Lorentz frame that is moving relative to this frame with a velocity $\boldsymbol{v}$, the asymptotic form of K^α will have spatial components $K^j \propto K^0 v^j$, and these velocity-dependent components can then feed into the local form of the metric.

These ideas can be framed in the context of the strong equivalence principle:

> If a test body is placed at an initial event in spacetime and given an initial velocity there, and if the body subsequently moves freely, then its world line will be independent of its mass, internal structure, and composition, *whether it is self-gravitating or not.*
>
> The outcome of any local non-gravitational or gravitational test experiment performed by a freely-moving apparatus is independent of the velocity of the apparatus and independent of when and where it is carried out.

Compare this with the statement of the Einstein equivalence principle provided in Sec. 5.1.1. The distinction between the strong and Einstein versions is the inclusion of bodies with self-gravitational interactions (planets, stars) and of experiments involving gravitational forces (Cavendish-type experiments, gravimeter measurements). Note that the strong principle is indeed stronger than the Einstein principle, and contains it in the limit in which local gravitational forces can be ignored. In fact, this principle is so strong that general relativity is one of the very few metric theories that actually satisfy it. Another example is Nordström's theory, which was described previously, and revealed not to be compatible with experimental tests.

Our previous discussion of the coupling of auxiliary fields to local gravitating systems suggests that alternative theories involving additional fields will tend to violate the strong equivalence principle, because the results of gravitational experiments are expected to depend on position and velocity. The discussion also suggests that when the strong equivalence principle is strictly respected, the description of gravity should involve a single universal field, the metric $g_{\alpha\beta}$. The argument is only a suggestion, however, and no rigorous proofs of these statements are available at present. Empirically it has been found that every metric theory that introduces auxiliary fields, either dynamical or non-dynamical, predicts violations of the strong equivalence principle at some level.

13.2 Parameterized post-Newtonian framework

13.2.1 A class of post-Newtonian theories

Metric theories of gravity demand that matter and non-gravitational fields respond only to the metric, and that they be completely oblivious to other fields that the theory might contain. The only gravitational field that can enter the matter's equations of motion, derived from the conservation equation $\nabla_\beta T^{\alpha\beta} = 0$, is therefore the metric $g_{\alpha\beta}$; the other fields are present only to help generate the metric. In this context the metric tensor and the equations of motion for matter become the primary entities for calculating observable effects, and all that distinguishes one metric theory from another is the particular form of the metric generated by a given distribution of matter.

The situation becomes particularly simple when we consider systems involving slow motions and weak gravitational fields. In this post-Newtonian limit, the spacetime metric $g_{\alpha\beta}$ produced by nearly every metric theory of gravity has the same formal structure. It can be written as an expansion in powers of c^{-2} about the Minkowski metric $\eta_{\alpha\beta}$, in terms of the Newtonian potential U and many of the same post-Newtonian potentials that we encountered back in Chapter 8. (New potentials are required for some theories.) The only aspect that changes from one theory to the next is the numerical value of the various coefficients that appear in front of the potentials.

This is a very fortunate circumstance, because the post-Newtonian limit is sufficient to describe the gravitational physics of the solar system and the experimental tests one can perform there. And to a limited degree, it can also describe the gravity of binary-pulsar systems. If we therefore replace the numerical coefficients in front of all potentials in the post-Newtonian metric of general relativity with arbitrary parameters, and add a few new potentials with their own parameters, we obtain a framework that encompasses a broad spectrum of alternative theories, and that can be used to calculate a wide range of testable phenomena.

This framework is called the parameterized post-Newtonian (PPN) framework, and the main idea originated with Eddington. In his classic 1922 textbook, *The Mathematical Theory of Relativity*, he parameterized the post-Newtonian limit of the Schwarzschild metric (in isotropic coordinates) by expressing it in the form

$$ds^2 = -\left[1 - 2\alpha\frac{GM}{c^2r} + 2\beta\left(\frac{GM}{c^2r}\right)^2\right]d(ct)^2 + \left[1 + 2\gamma\frac{GM}{c^2r}\right](dx^2 + dy^2 + dz^2), \tag{13.1}$$

with α, β, and γ denoting free parameters. Eddington calculated the motion of planets and the trajectories of light rays in this metric, and obtained predictions for various measurable quantities, such as the perihelion advance of Mercury and the deflection of light. He then interpreted the measurements as empirical constraints on the free parameters, with $\alpha = \beta = \gamma = 1$ confirming the predictions of general relativity. Eddington did not realize at the time that α is actually a redundant parameter, because it can always be absorbed into GM, which determines the Newtonian dynamics of the system and can be measured by observing its Keplerian orbital motion.

The modern version of the PPN framework, involving a gravitating system of point masses, was first developed by Kenneth Nordtvedt, Jr. in 1968. The framework was extended by one of us (CMW) to incorporate self-gravitating fluid systems. In 1972 the two approaches were unified by Nordtvedt and Will, and this gave rise to the modern version of the PPN framework, as it is used today.

13.2.2 Parameterized post-Newtonian metric

The metric and its potentials are displayed in Box 13.1. As we discussed previously, there is no need to introduce a parameter α in front of the Newtonian potential in g_{00}. Setting aside $\Phi^{\rm PF}$ and $\Phi^{\rm PF}_j$ for the moment, we see that there are ten independent potentials in the post-Newtonian part of the metric. In g_{00} we recognize U^2 and the six potentials in ψ, in g_{0j} we find U_j and $\partial_{tj}X$, and in g_{jk} there is U. There is a one-to-one correspondence with the ten PPN parameters, although the assignment of parameters may appear a little strange at first.

Box 13.1 Parameterized post-Newtonian metric

Parameters: $\gamma,\ \beta,\ \xi,\ \alpha_1,\ \alpha_2,\ \alpha_3,\ \zeta_1,\ \zeta_2,\ \zeta_3,\ \zeta_4$.

Metric:

$$g_{00} = -1 + \frac{2}{c^2}U + \frac{2}{c^4}(\psi - \beta U^2) + \frac{1}{c^4}\Phi^{\rm PF} + O(c^{-6}),$$

$$g_{0j} = -\frac{1}{c^3}\left[2(1+\gamma) + \frac{1}{2}\alpha_1\right]U_j - \frac{1}{2c^3}\left[1 + \alpha_2 - \zeta_1 + 2\xi\right]\partial_{tj}X + \frac{1}{c^3}\Phi^{\rm PF}_j + O(c^{-5}),$$

$$g_{jk} = \left(1 + \frac{2}{c^2}\gamma U\right)\delta_{jk} + O(c^{-4});$$

(*continued overleaf*)

$$\psi := \frac{1}{2}(2\gamma + 1 + \alpha_3 + \zeta_1 - 2\xi)\Phi_1 - (2\beta - 1 - \zeta_2 - \xi)\Phi_2 + (1 + \zeta_3)\Phi_3 + (3\gamma + 3\zeta_4 - 2\xi)\Phi_4 - \frac{1}{2}(\zeta_1 - 2\xi)\Phi_6 - \xi\Phi_W .$$

Potentials:

$$U := G\int \frac{\rho^{*\prime}}{|\boldsymbol{x} - \boldsymbol{x}'|} d^3x' ,$$

$$\Phi_1 := G\int \frac{\rho^{*\prime} v'^2}{|\boldsymbol{x} - \boldsymbol{x}'|} d^3x' ,$$

$$\Phi_2 := G\int \frac{\rho^{*\prime} U'}{|\boldsymbol{x} - \boldsymbol{x}'|} d^3x' ,$$

$$\Phi_3 := G\int \frac{\rho^{*\prime} \Pi'}{|\boldsymbol{x} - \boldsymbol{x}'|} d^3x' ,$$

$$\Phi_4 := G\int \frac{p'}{|\boldsymbol{x} - \boldsymbol{x}'|} d^3x' ,$$

$$\Phi_6 := G\int \rho^{*\prime} v'_j v'_k \frac{(x - x')^j (x - x')^k}{|\boldsymbol{x} - \boldsymbol{x}'|^3} d^3x' ,$$

$$\Phi_W := G^2 \int \rho^{*\prime} \rho^{*\prime\prime} \frac{(x - x')_j}{|\boldsymbol{x} - \boldsymbol{x}'|^3} \left[\frac{(x' - x'')^j}{|\boldsymbol{x} - \boldsymbol{x}''|} - \frac{(x - x'')^j}{|\boldsymbol{x}' - \boldsymbol{x}''|} \right] d^3x' d^3x''$$

$$= -U^2 - \Phi_2 - \nabla U \cdot \nabla X + G\nabla \cdot \int \frac{\rho^{*\prime}}{|\boldsymbol{x} - \boldsymbol{x}'|} \nabla' X' d^3x' ,$$

$$U^j := G\int \frac{\rho^* v^j(t, \boldsymbol{x}')}{|\boldsymbol{x} - \boldsymbol{x}'|} d^3x' ,$$

$$X := G\int \rho^*(t, \boldsymbol{x}')|\boldsymbol{x} - \boldsymbol{x}'|\, d^3x' .$$

Preferred-frame potentials:

$$\Phi^{\rm PF} := (\alpha_3 - \alpha_1) w^2 U + \alpha_2 w^j w^k \partial_{jk} X + (2\alpha_3 - \alpha_1) w^j U_j ,$$

$$\Phi^{\rm PF}_j := -\frac{1}{2}\alpha_1 w_j U + \alpha_2 w^k \partial_{jk} X .$$

The quantity w^j is the velocity of the PPN coordinate frame relative to a universal preferred frame.

When $\beta = \gamma = 1$ and all other parameters vanish, the metric reduces to the familiar post-Newtonian metric of general relativity, as constructed back in Sec. 8.1. The metric, however, does not quite reduce to the form displayed in Eq. (8.2) – see also Eq. (8.9). The reason is that the coordinate system (ct, x^j) is not harmonic. Indeed, the PPN metric is cast in a generalization of the *standard post-Newtonian gauge* described back in Sec. 8.3.7. There is no deep reason behind this choice of coordinates; it merely reflects the historical development of the PPN framework.

Table 13.1 PPN parameters and their physical significance. Note that α_3 is listed twice to indicate that it is a measure of two separate effects.

Parameter	What it measures relative to GR	GR value	Semi-conservative theories	Fully-conservative theories
γ	How much spatial curvature produced by unit rest mass?	1	γ	γ
β	How much "nonlinearity" in the superposition of gravity?	1	β	β
ξ	Preferred-location effects?	0	ξ	ξ
α_1	Preferred-frame effects?	0	α_1	0
α_2		0	α_2	0
α_3		0	0	0
α_3	Is total momentum conserved?	0	0	0
ζ_1		0	0	0
ζ_2		0	0	0
ζ_3		0	0	0
ζ_4		0	0	0

The choice of parameters is made (with considerable hindsight, to be sure) to ensure that their values are particularly simple in general relativity, but more importantly, to ensure also that specific physical meanings can be attached to them. These heuristic meanings are summarized in Table 13.1. The parameters β and γ are directly related to Eddington's original parameters. Roughly speaking, β measures how non-linear gravity is, in that it multiplies the U^2 term in g_{00}, and γ measures the spatial curvature generated by a body relative to what general relativity would predict; specifically, a calculation of the Riemann curvature tensor for the three-dimensional subspace defined by $dt = 0$ gives a result proportional to $\gamma GM/(c^2 r^3)$. These meanings are not to be taken literally, because they are very much tied to the choice of coordinates. In general relativity, for example, the Schwarzschild metric expressed in the standard Schwarzschild coordinates has no term quadratic in $GM/(c^2 r)$ in g_{00}; the post-Newtonian limit of the metric takes the Eddington form of Eq. (13.1) with $\beta = \gamma = 1$ *only* in isotropic coordinates. The lesson is that the interpretation of the PPN parameters must only be viewed as a rough heuristic.

The parameters α_n are linked to violations of local Lorentz invariance in gravitational physics, which is predicted by some theories. Suppose that all three parameters vanish. If we perform a post-Galilean transformation of the PPN metric to a coordinate system moving with respect to the first, as we did back in Sec. 8.3.6 for general relativity, we discover that the metric is of the same form in the moving frame as it was in the original frame, with all variables (such as positions, velocities, and densities) referring to the new frame. In this case, the physics of an isolated gravitating system does not depend on its velocity; we can analyze it in a reference frame at rest with respect to the system's center-of-mass, or in a frame moving relative to it with an arbitrary velocity, and the results are the same.

Now suppose that one or more of the α_n parameters is non-zero. In this case, a post-Galilean transformation of the metric generates new terms, and these are precisely the PF terms displayed in Box 13.1. This implies that the physics of a gravitating system depends explicitly on the system's velocity relative to our coordinate system. But what can this possibly mean, given that our coordinate system was chosen arbitrarily? It must mean that there is a preferred reference frame in the universe, a frame whose velocity $\boldsymbol{w}$ can be taken to vanish. What is this frame? The answer must be provided by the underlying theory of gravity, and it will differ from theory to theory. An example is a theory with a dynamical timelike vector field K^α in addition to the metric. If K^α is determined by some field equations with a source related to the distribution of matter, then in a homogeneous, isotropic universe there exists a reference frame in which the vector has only a time component. By symmetry, this must be the mean rest frame of the large-scale distribution of matter and radiation, or equivalently, the rest frame of the cosmic microwave background radiation. The specifics of how this circumstance arises may vary from theory to theory, but it is reasonable to suppose that in general, the frame in which $K^j = 0$ can be identified with the preferred reference frame. Because $\boldsymbol{w} = \boldsymbol{0}$ in the preferred frame, we would find that the potentials $\Phi^{\rm PF}$ and $\Phi_j^{\rm PF}$ vanish in this frame. But performing a post-Galilean transformation to any other frame automatically introduces the PF terms into the PPN metric, with a velocity $\boldsymbol{w}$ given by the new frame's velocity with respect to the cosmic background radiation. The PF potentials have observable consequences, and in Sec. 13.3 we see how experiments have placed very tight constraints on the α_n parameters, all consistent with the validity of post-Galilean invariance.

One might worry that the existence of a preferred frame violates the general covariance that is presumably built into all such alternative theories. In fact, general covariance is not broken: as long as the PF potentials are included in the metric, we can perform calculations in any reference frame we like. For example, instead of adopting the solar system's rest frame to do planetary calculations, we could adopt the rest frame of the galaxy. In this case, the $\boldsymbol{w}$ that appears in the metric would be the velocity of the galaxy relative to the cosmic background radiation, and the velocities of all solar-system bodies would be defined with respect to the galactic center. It turns out in the end that any measurable effect depends only on the velocity of the relevant body relative to the preferred frame; the velocity of the coordinate frame always drops out. The preferred frame comes about as a result of a *physical interaction*, say between a vector field and the metric; it does not represent a violation of general covariance. Fundamentally, this is no different from our ability to determine the rest frame of the cosmic background radiation by interacting with its photons, except that in alternative theories with preferred-frame effects, the physical mechanism behind the existence of a preferred frame is more subtle and indirect.

The four parameters ζ_n, together with the parameter α_3 (which plays a dual role), indicate whether the theory admits a conservation law for the total momentum of an isolated system. Energy is not included in this discussion, because in the present context of 1PN gravity, the relevant energy is rest-mass energy plus Newtonian total energy (the sum of kinetic, gravitational, and internal energies), which is conserved by virtue of the Newtonian dynamics.

Recall the Landau–Lifshitz formulation of the Einstein field equations, as presented back in Sec. 6.1, and the conservation laws that followed from them. The basic idea was that the equations of energy-momentum conservation could be expressed in the form $\partial_\beta \tau^{\alpha\beta} = 0$, where $\tau^{\alpha\beta}$ is a pseudotensor made up of the matter's energy-momentum tensor $T^{\alpha\beta}$ and a contribution from the gravitational field (the Landau–Lifshitz pseudotensor). It follows from this that the quantity $P^j := c^{-1} \int \tau^{0j}\, d^3x$ is constant in time, apart from the small effects associated with gravitational radiation. Total momentum is therefore conserved in general relativity, and we wish to see if the conclusion extends to other theories.

In any metric theory the equation of energy-momentum conservation is given by $\nabla_\beta T^{\alpha\beta} = 0$, and in the post-Newtonian limit the covariant derivatives can be computed with the PPN metric displayed in Box 13.1; the metric also appears in appropriate places within the energy-momentum tensor. It may then be asked whether one can define an object $\Theta^{\alpha\beta}$ such that the conservation equation can be expressed as $\partial_\beta \Theta^{\alpha\beta} = 0$, with partial derivatives replacing the covariant derivatives. Presumably, such an object would take a form such as $(1 - aU/c^2)(T^{\alpha\beta} + t^{\alpha\beta})$, with a a constant to be determined and $t^{\alpha\beta}$ a pseudotensor constructed from the various gravitational potentials. If this object exists, then it can be exploited to define a conserved total momentum $P^j := c^{-1} \int \Theta^{0j}\, d^3x$, and the theory would indeed be compatible with momentum conservation. It turns out that $\Theta^{\alpha\beta}$ exists provided that $\zeta_1 = \zeta_2 = \zeta_3 = \zeta_4 = \alpha_3 = 0$. Theories of gravity with this property are called "semi-conservative." It can be shown that any metric theory whose field equations can be derived from an invariant action principle automatically comes with a conserved momentum.

The equation $\partial_\beta \Theta^{\alpha\beta} = 0$ guarantees only that total momentum is conserved. If we further demand that angular momentum should be conserved, and that the center-of-mass should possess a uniform motion, we must also impose that $\Theta^{\alpha\beta}$ be a symmetric pseudotensor. It turns out that this requires $\alpha_1 = \alpha_2 = 0$ in addition to the other constraints. Theories with this property are called "fully conservative," and we note that it is possible for a theory formulated in terms of an action principle (especially ones with vector or tensor fields in addition to the metric) to fail to conserve angular momentum. A fully conservative theory is therefore one with no preferred-frame effects (at least at post-Newtonian order). It may seem odd that conservation of total angular momentum $J^{\alpha\beta}$, including the center-of-mass quantity J^{0j}, should require post-Galilean invariance, but there is nothing surprising about this: since J^{0j} mixes the time and space components of a tensor, an expectation of its conservation implicitly assumes invariance under boosts.

The final parameter ξ is tied to the potential Φ_W, called the Whitehead potential because it was first spotted in Whitehead's theory of gravity, although it was later found to occur in a broad class of similar theories. The potential looks particularly nasty, especially in its integral form, but it can be understood as having a simple origin. We have assumed that the spatial part of the PPN metric is diagonal, given by $g_{jk} = (1 + 2\gamma U/c^2)\delta_{jk}$. We could, however, have included in g_{jk} a perfectly acceptable term of the form $2\lambda \partial_{jk} X/c^2$, where λ is an arbitrary parameter. But this term can be removed by a coordinate transformation

$x^j = \bar{x}^j - \lambda \partial_j X/c^2$, which changes the metric according to

$$\bar{g}_{00}(\bar{x}^\mu) = g_{00}(\bar{x}^\mu) + 2\lambda \Gamma^j_{00} \partial_j X/c^2 \,, \tag{13.2a}$$
$$\bar{g}_{0j}(\bar{x}^\mu) = g_{0j}(\bar{x}^\mu) - \lambda \partial_{tj} X/c^3 \,, \tag{13.2b}$$
$$\bar{g}_{jk}(\bar{x}^\mu) = g_{jk}(\bar{x}^\mu) - 2\lambda \partial_{jk} X/c^2 \,. \tag{13.2c}$$

The change in g_{0j} merely modifies the coefficient of the pre-existing term involving $\partial_{tj} X/c^3$, but the change in g_{00} is more complicated. First, there is the term involving Γ^j_{00}. Second, while the Newtonian potential in g_{00} is to be evaluated at $\bar{x}^\mu$, it is still expressed as an integral over the old coordinate $\boldsymbol{x}'$, and we must transform the integration variables. The quantity $\rho^* d^3x'$ is invariant, so $\rho^* d^3x' = \bar{\rho}^* d^3\bar{x}'$, but we must write

$$\frac{1}{|\bar{\boldsymbol{x}} - \boldsymbol{x}'|} = \frac{1}{|\bar{\boldsymbol{x}} - \bar{\boldsymbol{x}}'|} + \frac{\lambda}{c^2} \bar{\nabla} \left(\frac{1}{|\bar{\boldsymbol{x}} - \bar{\boldsymbol{x}}'|} \right) \cdot \bar{\nabla}' \bar{X}' \,. \tag{13.3}$$

Inserting this within the Newtonian potential, and $\Gamma^j_{00} = -\partial_j U/c^2$ within $\bar{g}_{00}$, we obtain

$$\begin{aligned}
\bar{g}_{00}(\bar{x}^\mu) &= g_{00}(\bar{x}^\mu) - \frac{2}{c^4} \lambda \bar{\nabla} \bar{U} \cdot \bar{\nabla} \bar{X} + \frac{2G}{c^4} \lambda \bar{\nabla} \cdot \int \frac{\bar{\rho}^{*\prime}}{|\bar{\boldsymbol{x}} - \bar{\boldsymbol{x}}'|} \bar{\nabla}' \bar{X}' \, d^3\bar{x}' \\
&= g_{00}(\bar{x}^\mu) + \frac{2}{c^4} \lambda (\bar{\Phi}_W + \bar{U}^2 + \bar{\Phi}_2) \,,
\end{aligned} \tag{13.4}$$

where g_{00} is now expressed entirely in terms of the barred coordinates. So we see that far from being an anomaly, the Whitehead potential can be linked to gauges in which the spatial metric is not diagonal. This occurs even in general relativity: we would pick up a Whitehead term in g_{00} if we were to transform the metric from the standard post-Newtonian gauge to one in which g_{jk} is not diagonal. In fact, there is a close analogy between this coordinate transformation in post-Newtonian theory and the transformation between isotropic and Schwarzschild coordinates for the Schwarzschild metric.

13.2.3 Equations of hydrodynamics

We develop the equations of hydrodynamics in the PPN framework just as we did in Chapters 7 and 8 in the case of general relativity. As in Sec 7.1, we introduce a conserved density [refer to Eq. (7.4)]

$$\rho^* := \sqrt{-g} \rho u^0 / c \,, \tag{13.5}$$

where ρ is the proper mass density. Using the normalization condition $g_{\alpha\beta} u^\alpha u^\beta = -c^2$ and the PPN metric, we find that the densities are related by

$$\rho = \left[1 - \frac{1}{c^2} \left(\frac{1}{2} v^2 + 3\gamma U \right) + O(c^{-4}) \right] \rho^* \,, \tag{13.6}$$

which may be compared with Eq. (7.46). The components of the energy-momentum tensor are given by

$$c^{-2}T^{00} = \rho^*\left\{1 + \frac{1}{c^2}\left[\frac{1}{2}v^2 - (3\gamma - 2)U + \Pi\right]\right\} + O(c^{-2}), \tag{13.7a}$$

$$c^{-1}T^{0j} = \rho^* v^j\left\{1 + \frac{1}{c^2}\left[\frac{1}{2}v^2 - (3\gamma - 2)U + \Pi + p/\rho^*\right]\right\} + O(c^{-3}), \tag{13.7b}$$

$$\begin{aligned} T^{jk} &= \rho^* v^j v^k\left\{1 + \frac{1}{c^2}\left[\frac{1}{2}v^2 - (3\gamma - 2)U + \Pi + p/\rho^*\right]\right\} \\ &\quad + p\left(1 - \frac{2}{c^2}\gamma U\right)\delta^{jk} + O(c^{-4}), \end{aligned} \tag{13.7c}$$

to the required post-Newtonian order; compare these with Eq. (8.7).

Calculating the Christoffel symbols $\Gamma^\alpha_{\beta\gamma}$ from the PPN metric and inserting them into the conservation equation $\nabla_\beta T^{\alpha\beta} = 0$, we eventually obtain the PPN version of Euler's equation,

$$\begin{aligned} \rho^*\frac{dv^j}{dt} &= -\partial_j p + \rho^*\partial_j U \\ &\quad + \frac{1}{c^2}\left\{\left[\frac{1}{2}v^2 + (2-\gamma)U + \Pi + \frac{p}{\rho^*}\right]\partial_j p - v^j\partial_t p\right\} \\ &\quad + \frac{1}{c^2}\rho^*\left\{\left[\gamma v^2 - 2(\gamma + \beta)U\right]\partial_j U\right. \\ &\qquad - v^j\left[(2\gamma + 1)\partial_t U + 2(\gamma + 1)v^k\partial_k U\right] \\ &\qquad \left. + \frac{1}{2}(4\gamma + 4 + \alpha_1)\left[\partial_t U_j + v^k\left(\partial_k U_j - \partial_j U_k\right)\right] + \partial_j\Psi\right\} \\ &\quad + \frac{1}{c^2}\rho^*\left[\frac{1}{2}\partial_j\Phi^{\rm PF} - \partial_t\Phi^{\rm PF}_j - v^k\left(\partial_k\Phi^{\rm PF}_j - \partial_j\Phi^{\rm PF}_k\right)\right] \\ &\quad + O(c^{-4}), \end{aligned} \tag{13.8}$$

where

$$\Psi := \psi + \frac{1}{2}(1 + \alpha_2 - \zeta_1 + 2\xi)\partial_{tt}X, \tag{13.9}$$

and where ψ, $\Phi^{\rm PF}$, and $\Phi^{\rm PF}_j$ are displayed in Box 13.1. Recall that $\partial_{tt}X = \Phi_1 + 2\Phi_4 - \Phi_5 - \Phi_6$. Note that the PPN equation of hydrodynamics parallels closely the general relativistic equation, displayed in Eq. (8.119). The main differences are in the numerical coefficients in front of the post-Newtonian potentials, the inclusion of the Whitehead potential in ψ, and the addition of the "preferred-frame" terms.

13.2.4 Motion of isolated bodies

We now wish to obtain equations of motion for a system of isolated, non-rotating bodies, as we did back in Chapter 9 for the post-Newtonian limit of general relativity. The foundation for this work is Eq. (13.8), which closely resembles its general-relativistic counterpart. This implies that much of the machinery developed in Chapter 9 can be directly imported here. Additional work is required to handle the Whitehead potential and the preferred-frame terms.

The idea is to integrate Eq. (13.8) over body A to find the net force acting on its center-of-mass. The force can be decomposed into "self-force" terms involving the variables of body A only, and external forces produced by the other bodies. The 18 integrals of Eqs. (9.88), as evaluated in Eqs. (9.91), make contributions to these forces, and the contribution from the Whitehead potential is

$$\begin{aligned} F^j_{19} &:= -\frac{\xi}{c^2}\int_A \rho^* \partial_j \Phi_W \, d^3x \\ &= -\frac{G^2\xi}{c^2}\sum_{B,C}\int_A \rho^* \partial_j \int_B \rho^{*\prime} \int_C \rho^{*\prime\prime} \frac{(x-x')_k}{|\boldsymbol{x}-\boldsymbol{x}'|^3}\left[\frac{(x-x'')^k}{|\boldsymbol{x}'-\boldsymbol{x}''|} - \frac{(x'-x'')^k}{|\boldsymbol{x}-\boldsymbol{x}''|}\right] \\ &\quad \times d^3x d^3x' d^3x'' , \end{aligned} \tag{13.10}$$

where the double sum over bodies B and C includes body A. When B and C both refer to body A, it is straightforward to see that the integral vanishes because of the assumed reflection symmetry of the density distribution (refer to Sec. 9.1.2). When B refers to body A, while C refers to an external body, we expand $\boldsymbol{x}-\boldsymbol{x}''$ and $\boldsymbol{x}'-\boldsymbol{x}''$ about the center-of-mass of body A. The method is very similar to, though slightly more tedious than, the method used to evaluate F^j_{14} in Sec. 9.3.2. We find that some of the double integrals over $\boldsymbol{x}$ and $\boldsymbol{x}'$ in body A vanish by symmetry, while others lead to terms involving Ω^{jk}_A and Ω_A. Higher-order terms in the expansions can be neglected because of our assumption that the bodies are well separated. Similar considerations apply when C refers to body A and B refers to an external body, leading to terms involving Ω^{jk}_A and Ω_A, as well as a term involving $U_{\neg A}$ evaluated at body B. The final case, with both B and C referring to an external body, involves external potentials only. We denote the collection of terms involving external potentials by $\Phi_{W,\neg A}$, and the final result for F^j_{19} is

$$c^2 F^j_{19} = 6\Omega_A \partial_j U_{\neg A} - 3\Omega^{jk}_A \partial_k U_{\neg A} - \Omega^{kl}_A \partial_{jkl} X_{\neg A} + m_A \partial_j \Phi_{W,\neg A} . \tag{13.11}$$

The contribution to the force that comes from the preferred-frame potentials is

$$F^j_{\rm PF} = \frac{1}{c^2}\int_A \rho^* \left(\frac{1}{2}\partial_j \Phi^{\rm PF} - \partial_t \Phi^{\rm PF}_j - 2v^k \partial_{[j}\Phi^{\rm PF}_{k]}\right) d^3x . \tag{13.12}$$

This can be handled in much the same way, with the help of some results from Sec. 9.3. We merely quote the final answer:

$$\begin{aligned} c^2 F^j_{\rm PF} &= \left[\alpha_1 H_A \delta^{jk} - \alpha_2\left(2H^{(jk)}_A - 3K^{jk}_A + H_A\delta^{jk}\right) - \alpha_3 H^{kj}_A\right] w^k \\ &\quad + \frac{1}{2}(\alpha_3-\alpha_1) m_A w^2 \partial_j U_{\neg A} + \frac{1}{2}\alpha_2 m_A w^k w^l \partial_{jkl} X_{\neg A} \\ &\quad + \frac{1}{2}(2\alpha_3-\alpha_1) m_A w^k \partial_j U_{k,\neg A} + \frac{1}{2}\alpha_1 m_A w^j \partial_t U_{\neg A} \\ &\quad - \alpha_2 m_A w^k \partial_{tjk} X_{\neg A} + \alpha_1 m_A v^k_A w_{[j}\partial_{k]} U_{\neg A} . \end{aligned} \tag{13.13}$$

Collecting results, we find that the acceleration of body A is given by

$$a_A^j = \partial_j U_{\neg A} + (a_A^j)_{\rm self} + (a_A^j)_{\rm ext} + (a_A^j)_{\rm PF} + O(c^{-4}). \qquad (13.14)$$

The first term is of course the Newtonian acceleration created by the external bodies. The second term contains all self-interaction effects, including those coming from the preferred-frame potentials:

$$\begin{aligned} m_A c^2 (a_A^j)_{\rm self} = &\Big[2L_A^{(jk)} - (2+\alpha_2)H_A^{jk} - (2+\alpha_2+\alpha_3)H_A^{kj} + 3(1+\alpha_2)K_A^{jk} \\ &+ (\alpha_1-\alpha_2)\delta^{jk}H_A - \delta^{jk}\dot{P}_A\Big]v_A^k \\ &+ \Big[\alpha_1 H_A\delta^{jk} - \alpha_2(2H_A^{(jk)} - 3K_A^{jk} + H_A\delta^{jk}) - \alpha_3 H_A^{kj}\Big]w^k \\ &+ \Big[2\gamma\mathcal{T}_A + (4\beta - 3 - \alpha_1 + \alpha_2 - \zeta_1 - 4\xi)\Omega_A + 3\gamma P_A\Big]\partial_j U_{\neg A} \\ &- \Big[4(\gamma+1)\mathcal{T}_A^{jk} + (2\gamma + 2 + \alpha_2 - \zeta_1 + \zeta_2)\Omega_A^{jk} \\ &+ 2(\gamma+1)\delta^{jk}P_A\Big]\partial_k U_{\neg A} + \xi\Omega_A^{kl}\partial_{jkl}X_{\neg A}. \end{aligned} \qquad (13.15)$$

As we did back in Chapter 9, we invoke the conditions (9.10)–(9.12) satisfied by a body in internal dynamical equilibrium,

$$2\mathcal{T}_A^{jk} + \Omega_A^{jk} + \delta^{jk}P_A = O(c^{-2}), \qquad (13.16a)$$

$$4H_A^{(jk)} - 3K_A^{jk} - 2L_A^{(jk)} + \delta^{jk}\dot{P}_A = O(c^{-2}), \qquad (13.16b)$$

$$2\mathcal{T}_A + \Omega_A + 3P_A = O(c^{-2}). \qquad (13.16c)$$

We also invoke the additional steady-state conditions

$$\dot{\Omega}_A^{jk} = 2H_A^{(jk)} - 3K_A^{jk} = 0, \qquad (13.17a)$$

$$\dot{\Omega}_A = -H_A = 0, \qquad (13.17b)$$

and obtain the final expression

$$\begin{aligned} (a_A^j)_{\rm self} = &-\frac{\alpha_3}{m_A c^2}H_A^{kj}(w_k + v_{Ak}) \\ &+ (4\beta - \gamma - 3 - 4\xi - \alpha_1 + \alpha_2 - \zeta_1)\frac{\Omega_A}{m_A c^2}\partial_j U_{\neg A} \\ &- (\alpha_2 - \zeta_1 + \zeta_2)\frac{\Omega_A^{jk}}{m_A c^2}\partial_k U_{\neg A} + \xi\frac{\Omega_A^{kl}}{m_A c^2}\partial_{jkl}X_{\neg A} \end{aligned} \qquad (13.18)$$

for the self-acceleration. The acceleration produced by the external post-Newtonian potentials is given by

$$\begin{aligned} c^2(a_A^j)_{\rm ext} = &\big[\gamma v_A^2 - 2(\gamma+\beta)U_{\neg A}\big]\partial_j U_{\neg A} \\ &- v_A^j\big[2(\gamma+1)v_A^k\partial_k U_{\neg A} + (2\gamma+1)\partial_t U_{\neg A}\big] \\ &+ \frac{1}{2}(4\gamma + 4 + \alpha_1)\big[\partial_t U_{\neg A}^j + v_A^k(\partial_k U_{j\neg A} - \partial_j U_{k\neg A})\big] \\ &+ \partial_j \Psi_{\neg A}, \end{aligned} \qquad (13.19)$$

where

$$\Psi_{\neg A} = \frac{1}{2}(2\gamma + 2 + \alpha_2 + \alpha_3)\Phi_{1,\neg A} - (2\beta - 1 - \zeta_2 - \xi)\Phi_{2,\neg A}$$
$$+ (1 + \zeta_3)\Phi_{3,\neg A} + (3\gamma + 1 + \alpha_2 + 3\zeta_4 - \zeta_1)\Phi_{4,\neg A}$$
$$- \frac{1}{2}(1 + \alpha_2 - \zeta_1 + 2\xi)\Phi_{5,\neg A} - \frac{1}{2}(1 + \alpha_2)\Phi_{6,\neg A} - \xi\Phi_{W,\neg A}, \qquad (13.20)$$

and the preferred-frame acceleration is

$$c^2(a^j_A)_{\rm PF} = +\frac{1}{2}(\alpha_3 - \alpha_1)w^2\partial_j U_{\neg A} + \frac{1}{2}\alpha_2 w^k w^l \partial_{jkl} X_{\neg A}$$
$$+ \frac{1}{2}(2\alpha_3 - \alpha_1)w^k \partial_j U_{k,\neg A} + \frac{1}{2}\alpha_1 w^j \partial_t U_{\neg A}$$
$$- \alpha_2 w^k \partial_{tjk} X_{\neg A} + \alpha_1 v^k_A w_{[j}\partial_{k]} U_{\neg A}. \qquad (13.21)$$

We must now evaluate the derivatives of the external potentials. Again we find that most of the hard work has already been carried out in Chapter 9, and the results of Eqs. (9.103) can all be imported here. The only additional work required is to evaluate $\partial_j \Phi_{W,\neg A}$. When B and C refer to different external bodies, we can treat each body as a point mass and calculate the gradient of the potential directly in terms of the masses and separations. But when B and C both refer to the same external body B, we must expand $\boldsymbol{x} - \boldsymbol{x}'$ and $\boldsymbol{x} - \boldsymbol{x}''$ about the center-of-mass of body B. This generates a structure-dependent term proportional to $\Omega_B \boldsymbol{n}_{AB}/r^2_{AB}$. A similar term arises when B is in an external body and C refers to body A. The final result is

$$\partial_j \Phi_{W,\neg A} = -\sum_{B\neq A} \frac{2G\Omega_B n^j_{AB}}{r^2_{AB}} - \sum_{B\neq A} \frac{G^2 m_A m_B n^j_{AB}}{r^3_{AB}}$$
$$+ \sum_{B\neq A}\sum_{C\neq A,B} \frac{G^2 m_B m_C}{r^2_{AB}} \left\{ \frac{\delta^{jk} - 3n^j_{AB}n^k_{AB}}{r_{AB}} \left[\frac{r_{BC}}{r_{AC}} n^k_{BC} - \frac{r_{AC}}{r_{BC}} n^k_{AC} \right] \right.$$
$$\left. - \frac{n^k_{AB}}{r_{AC}} \left[\frac{r_{AC}}{r_{BC}} \delta^{jk} + \frac{r_{BC}}{r_{AC}} n^j_{AC} n^k_{BC} \right] \right\}. \qquad (13.22)$$

Collecting results, we obtain an explicit expression for the acceleration of body A. As we did back in Sec. 9.3.5, we decompose it as

$$\boldsymbol{a}_A = \boldsymbol{a}_A[\text{EXT}] + \boldsymbol{a}_A[\text{STR}] + O(c^{-4}), \qquad (13.23a)$$
$$\boldsymbol{a}_A[\text{EXT}] = \boldsymbol{a}_A[0\text{PN}] + \boldsymbol{a}_A[1\text{PN}], \qquad (13.23b)$$

where $\boldsymbol{a}_A[0\text{PN}]$ is the Newtonian acceleration, $\boldsymbol{a}_A[1\text{PN}]$ groups all the post-Newtonian corrections that do not depend on the internal structures of the bodies, and $\boldsymbol{a}_A[\text{STR}]$ contains all the structure-dependent terms. As we also did back in Sec. 9.3.5, we pull out a term

$$\boldsymbol{a}_A[\text{STR}]' = -\sum_{B\neq A} \frac{G(E_B/c^2)}{r^2_{AB}} \boldsymbol{n}_{AB}, \qquad (13.24)$$

from $\boldsymbol{a}_A[\text{STR}]$, where $E_B = \mathcal{T}_B + \Omega_B + E_B^{\text{int}}$ is the total energy of body B, and combine it with the 0PN term to obtain

$$\boldsymbol{a}_A[0\text{PN}] + \boldsymbol{a}_A[\text{STR}]' = -\sum_{B \neq A} \frac{GM_B}{r_{AB}^2} \boldsymbol{n}_{AB} , \tag{13.25}$$

where $M_B := m_B + E_B/c^2 + O(c^{-4})$ is the total gravitational mass of body B. We can then insert $m_B = M_B + O(c^{-2})$ in all post-Newtonian terms in the acceleration without changing the equations of motion at 1PN order. In Chapter 9 this procedure eliminated all structure-dependent terms from the equations of motion. Here, in contrast, the remaining terms are given by

$$\begin{aligned} c^2 a_A^j[\text{STR}] = & -\frac{\alpha_3}{M_A} H_A^{kj}(w_k + v_{Ak}) \\ & + \sum_{B \neq A} GM_B \biggl\{ (4\beta - \gamma - 3 - 4\xi - \alpha_1 + \alpha_2 - \zeta_1) \frac{\Omega_A}{M_A} \partial_j (r_{AB})^{-1} \\ & \qquad - (\alpha_2 - \zeta_1 + \zeta_2) \frac{\Omega_A^{jk}}{M_A} \partial_k (r_{AB})^{-1} + \xi \frac{\Omega_A^{kl}}{M_A} \partial_{jkl} r_{AB} \\ & \qquad + \left(4\beta - \gamma - 3 - 4\xi - \frac{1}{2}\alpha_3 - 2\zeta_2 \right) \frac{\Omega_B}{M_B} \partial_j (r_{AB})^{-1} \\ & \qquad + \left(\xi - \frac{1}{2}\zeta_1 \right) \frac{\Omega_B^{kl}}{M_B} \partial_{jkl} r_{AB} - \zeta_3 \frac{E_B^{\text{int}}}{M_B} \partial_j (r_{AB})^{-1} \\ & \qquad + \left(\frac{3}{2}\alpha_3 - 3\zeta_4 + \zeta_1 \right) \frac{P_B}{M_B} \partial_j (r_{AB})^{-1} \biggr\} . \end{aligned} \tag{13.26}$$

These structure-dependent terms all vanish when the PPN parameters take their general-relativistic values. In general, however, the acceleration of a body in the field of other bodies can depend on its internal structure (dependence on Ω_A), and can even contain a component in a direction perpendicular to the direction of the other body (dependence on Ω_A^{jk}). Furthermore, the gravitational attraction produced by an external body can depend on *its own* internal structure (dependence on Ω_B, E_B^{int}, P_B). These violations of the strong equivalence principle can persist even in a fully conservative theory ($\alpha_n = \zeta_n = 0$), depending on the values of γ, β, and ξ. We notice the remarkable fact that the term proportional to α_3 depends on the body's net velocity $\boldsymbol{w} + \boldsymbol{v}_A$ relative to the preferred frame; as we have observed before, general covariance implies that the equations of motion cannot depend on the velocity of the PPN frame relative to the preferred frame.

The remaining terms in the post-Newtonian equations of motion are the N-body terms analogous to those displayed in Eq. (9.127):

$$\begin{aligned} \boldsymbol{a}_A[\text{EXT}] = & -\sum_{B \neq A} \frac{GM_B}{r_{AB}^2} \boldsymbol{n}_{AB} \\ & + \frac{1}{c^2} \biggl[-\sum_{B \neq A} \frac{GM_B}{r_{AB}^2} \biggl\{ \gamma v_A^2 - (2\gamma + 2)(\boldsymbol{v}_A \cdot \boldsymbol{v}_B) + (\gamma + 1) v_B^2 \end{aligned}$$

(continued overleaf)

$$
\begin{aligned}
&\quad -\frac{3}{2}(\boldsymbol{n}_{AB}\cdot\boldsymbol{v}_B)^2+\frac{1}{2}(\alpha_2+\alpha_3)(w+v_B)^2 \\
&\quad -\frac{1}{2}\alpha_1(\boldsymbol{w}+\boldsymbol{v}_A)\cdot(\boldsymbol{w}+\boldsymbol{v}_B)-\frac{3}{2}\alpha_2[\boldsymbol{n}_{AB}\cdot(\boldsymbol{w}+\boldsymbol{v}_B)]^2 \\
&\quad -\left(2\gamma+2\beta+1+\frac{1}{2}\alpha_1-\zeta_2\right)\frac{GM_A}{r_{AB}}-(2\gamma+2\beta)\frac{GM_B}{r_{AB}}\Bigg\}\boldsymbol{n}_{AB} \\
&+\sum_{B\neq A}\frac{GM_B}{r_{AB}^2}\Big\{\boldsymbol{n}_{AB}\cdot\big[(2\gamma+2)\boldsymbol{v}_A-(2\gamma+1)\boldsymbol{v}_B\big]\Big\}(\boldsymbol{v}_A-\boldsymbol{v}_B) \\
&-\sum_{B\neq A}\frac{GM_B}{r_{AB}^2}\Big\{\boldsymbol{n}_{AB}\cdot\big[\alpha_2(\boldsymbol{w}+\boldsymbol{v}_B)+\frac{1}{2}\alpha_1(\boldsymbol{v}_A-\boldsymbol{v}_B)\big]\Big\}(\boldsymbol{w}+\boldsymbol{v}_B) \\
&+\sum_{B\neq A}\sum_{C\neq A,B}\frac{G^2M_BM_C}{r_{AB}^2}\Bigg[(2\gamma+2\beta-2\xi)\frac{1}{r_{AC}}+(2\beta-1-2\xi-\zeta_2)\frac{1}{r_{BC}} \\
&\quad -\frac{1}{2}(1+2\xi+\alpha_2-\zeta_1)\frac{r_{AB}}{r_{BC}^2}(\boldsymbol{n}_{AB}\cdot\boldsymbol{n}_{BC})-\xi\frac{r_{BC}}{r_{AC}^2}(\boldsymbol{n}_{AC}\cdot\boldsymbol{n}_{BC})\Bigg]\boldsymbol{n}_{AB} \\
&-\frac{1}{2}(4\gamma+3-2\xi+\alpha_1-\alpha_2+\zeta_1)\sum_{B\neq A}\sum_{C\neq A,B}\frac{G^2M_BM_C}{r_{AB}r_{BC}^2}\boldsymbol{n}_{BC} \\
&-\xi\sum_{B\neq A}\sum_{C\neq A,B}\frac{G^2M_BM_C}{r_{AB}^3}\Big[(\boldsymbol{n}_{AC}-\boldsymbol{n}_{BC})-3\boldsymbol{n}_{AB}\cdot(\boldsymbol{n}_{AC}-\boldsymbol{n}_{BC})\boldsymbol{n}_{AB}\Big]\Bigg] \\
&+O(c^{-4}).
\end{aligned}
\tag{13.27}
$$

Setting the PPN parameters equal to their general relativistic values allows us to recover Eq. (9.127).

The appearance of structure-dependent terms in the equations of motion is a radical departure from general relativity. We conclude this section with a discussion of this remarkable phenomenon. In discussions of the equivalence principle in Newtonian gravitation, it can be useful to identify three types of mass that characterize gravitating bodies: the *inertial mass* $M_{\mathcal{I}}$, which relates momentum to velocity, or force to acceleration; the *passive gravitational mass* $M_{\mathcal{P}}$, which relates gravitational force to the gradient of the potential; and the *active gravitational mass* $M_{\mathcal{A}}$, which determines the potential of the gravitating body. With these definitions, we write

$$
\boldsymbol{F}_A=(M_{\mathcal{I}})_A\frac{d\boldsymbol{v}_A}{dt}=(M_{\mathcal{P}})_A\nabla\sum_{B\neq A}\frac{(M_{\mathcal{A}})_B}{r_{AB}}, \tag{13.28}
$$

or

$$
\boldsymbol{a}_A=\left(\frac{M_{\mathcal{P}}}{M_{\mathcal{I}}}\right)_A\sum_{B\neq A}\nabla\frac{(M_{\mathcal{A}})_B}{r_{AB}}. \tag{13.29}
$$

So if the inertial and passive masses of body A are not equal, the motion of body A will depend on its internal structure. From Eq. (13.26) we see not only that the masses are not generally equal in PPN gravity, but that they are actually tensorial in nature. We also observe that the active mass of each body is tensorial and structure-dependent. When each body is spherically symmetric we can use the fact that $\Omega_A^{jk} = \frac{1}{3}\delta^{jk}\Omega_A$ to simplify the structure-dependent terms. In this case the active and passive masses reduce to

$$\frac{(M_{\mathcal{P}})_A}{M_A} = 1 + \left(4\beta - \gamma - 3 - \frac{10}{3}\xi - \alpha_1 + \frac{2}{3}\alpha_2 - \frac{2}{3}\zeta_1 - \frac{1}{3}\zeta_2\right)\frac{\Omega_A}{M_A c^2}, \tag{13.30a}$$

$$\frac{(M_A)_B}{M_B} = 1 + \left(4\beta - \gamma - 3 - \frac{10}{3}\xi - \frac{1}{2}\alpha_3 - \frac{1}{3}\zeta_1 - 2\zeta_2\right)\frac{\Omega_B}{M_B c^2}$$
$$+ \zeta_3 \frac{E_B^{\text{int}}}{M_B c^2} - \left(\frac{3}{2}\alpha_3 + \zeta_1 - 3\zeta_4\right)\frac{P_B}{M_B c^2}. \tag{13.30b}$$

We now observe that when we sum Eq. (13.28) over all bodies, we obtain

$$\sum_A (M_{\mathcal{I}})_A \frac{d\boldsymbol{v}_A}{dt} = \sum_A \sum_{B \neq A} (M_{\mathcal{P}})_A (M_A)_B \frac{\boldsymbol{n}_{AB}}{r_{AB}^2}. \tag{13.31}$$

The double sum vanishes when the product $(M_{\mathcal{P}})_A (M_A)_B$ is symmetric under an exchange of A and B; under such a circumstance the system's total momentum is conserved, and the center-of-mass moves uniformly with a constant speed. This occurs, for example, when the active mass of each body is equal to its passive mass, and this is an expression of Newton's third law. When do we expect this to occur for PPN gravity? If we examine Eq. (13.26) and impose the constraints of a fully conservative theory of gravity, $\alpha_n = \zeta_n = 0$, we see that the structure-dependent terms depend only on Ω_A, Ω_B and their associated tensors, and that the contributions are then *symmetric* in A and B. Under such a circumstance we have that the sum $\sum_A M_A a_A^j[\text{STR}]$ vanishes, and the same is true for $\sum_A M_A a_A^j[\text{EXT}]$. There are therefore theories that violate the strong equivalence principle ($M_{\mathcal{P}} \neq M_{\mathcal{I}}$) but still satisfy Newton's third law ($M_{\mathcal{P}} = M_A$).

13.2.5 Motion of light

To the post-Newtonian order required to describe the deflection of light and the Shapiro time delay, we can approximate the PPN metric to

$$ds^2 = -(1 - 2U/c^2)\,d(ct)^2 + (1 + 2\gamma U/c^2)(dx^2 + dy^2 + dz^2). \tag{13.32}$$

The null condition $g_{\alpha\beta} v^\alpha v^\beta = 0$ implies that $(v/c)^2 = 1 - 2(1+\gamma)U/c^2 + O(c^{-3})$, and $\boldsymbol{v}$ can therefore be expressed as

$$\boldsymbol{v} = c\left[1 - (1+\gamma)\frac{U}{c^2}\right]\boldsymbol{n} + O(c^{-3}), \tag{13.33}$$

in terms of a unit vector $\boldsymbol{n}$. Calculating the Christoffel symbols to the appropriate order and inserting them into the geodesic equation produces

$$\frac{dn^j}{dt} = \frac{1+\gamma}{c}(\delta^{jk} - n^j n^k)\partial_k U + O(c^{-2}) \tag{13.34}$$

after making use of Eq. (13.33). Comparing this with the general relativistic equations displayed in Eqs. (8.17) and (8.19), we see that the factor 2 in each equation has been replaced by $1+\gamma$. Otherwise, the equations that describe the propagation of light in PPN gravity are the same as in general relativity.

13.2.6 Metric near a moving body and local gravitational constant

Back in Sec. 9.4 we carried out a transformation of the general relativistic post-Newtonian metric to a non-inertial frame that moved with one body in a system of isolated bodies. With a suitable choice of transformation functions we could make the metric in the region outside the selected body look precisely like the Schwarzschild metric (expanded to 1PN order), plus correction terms proportional to $\bar{r}^2$ representing tidal potentials created by the other bodies in the system ($\bar{r}$ is the distance away from the selected body). To make this work we had to set the acceleration of the moving frame to be *precisely equal* to the 1PN acceleration of the selected body. Furthermore, the body's internal structure played no role in the analysis (although we did assume that it was spherically symmetric for simplicity); we only used the fact that its exterior geometry could be described by the Schwarzschild metric, apart from the tidal corrections. This exercise illustrated the notion that in general relativity, the equations of motion of an isolated body can be obtained by exploiting only the vacuum field equations, which apply between bodies, an idea that goes all the way back to the famous 1938 paper by Einstein, Infeld and Hoffmann. But it illustrated also the validity of the strong equivalence principle in general relativity, because the motion of the body was seen to be manifestly independent of its internal structure.

This exercise can be repeated using the PPN metric, but the results are very different, and they lead to additional tests of gravitational theories. The calculation is long and tedious, and we will not trouble the reader with it here. For simplicity we assume that the body does not have a significant gravitational binding energy. As before we find that we must require the non-inertial frame to move with an acceleration equal to the PPN acceleration, as given by Eq. (13.27). The final form of the transformed metric to 1PN order is

$$g_{00} = -1 + 2\frac{G_{\rm eff}M}{c^2\bar{r}} - 2\beta\left(\frac{G_{\rm eff}M}{c^2\bar{r}}\right)^2 + O(\bar{r}^2) + O(c^{-6}), \tag{13.35a}$$

$$g_{0j} = -\frac{1}{2}\alpha_1(w+v)_j\frac{G_{\rm eff}M}{c^2\bar{r}} + O(\bar{r}^2) + O(c^{-5}), \tag{13.35b}$$

$$g_{jk} = \left(1 + 2\gamma\frac{G_{\rm eff}M}{c^2\bar{r}}\right)\delta_{jk} + O(\bar{r}^2) + O(c^{-4}), \tag{13.35c}$$

where the effective gravitational "constant" $G_{\rm eff}$ is given by

$$\begin{aligned} G_{\rm eff} := G\biggl\{ & 1 - \frac{1}{c^2}(4\beta - \gamma - 3 - 4\xi - \zeta_2)\hat{U}_{\rm ext} - \frac{1}{c^2}\xi\,\bar{n}^j\bar{n}^k\partial_{\bar{j}\bar{k}}\hat{X}_{\rm ext} \\ & - \frac{1}{2c^2}(\alpha_1 - \alpha_2 - \alpha_3)(w+v)^2 - \frac{1}{2c^2}\alpha_2\bigl[(\boldsymbol{w}+\boldsymbol{v})\cdot\bar{\boldsymbol{n}}\bigr]^2 \\ & - \frac{1}{2c^2}(4\beta - \gamma - 3 - 4\xi + \alpha_2 - \zeta_1)\bar{x}^k\partial_{\bar{k}}\hat{U}_{\rm ext} \\ & - \frac{1}{2c^2}\xi\,\bar{r}\bar{n}^j\bar{n}^k\bar{n}^l\partial_{\bar{j}\bar{k}\bar{l}}\hat{X}_{\rm ext}\biggr\}, \end{aligned} \tag{13.36}$$

where $\bar{n}^j = \bar{x}^j/\bar{r}$ and the "hatted potentials" were introduced in Sec. 8.3; they denote the original external potentials (the ones that appear in the inertial-frame metric) evaluated at the center-of-mass position of the moving body.

There are several interesting observations to be made about this result. The first is that in general relativity ($\beta = \gamma = 1$, all other parameters vanishing), the coordinate transformation eliminates all reference to the external universe, up to the expected tidal terms at order $\bar{r}^2$, and returns the post-Newtonian limit of the Schwarzschild metric with $G_{\text{eff}} = G$. This reproduces what we found back in Sec. 9.4. But this is not so in a generic alternative theory: external bodies and the existence of a preferred frame *can* influence the geometry of our local frame. This is a clear example of a violation of the strong equivalence principle, as it was introduced in Sec. 13.1. The violations can be encapsulated in an effective gravitational parameter G_{eff}, which is no longer constant, and an additional gravitomagnetic term in g_{0j}.

The second observation is that the metric and G_{eff} depend on the combination $\boldsymbol{w} + \boldsymbol{v}$, the body's velocity relative to the preferred universal frame; the arbitrarily chosen velocity $\boldsymbol{w}$ of the initial PPN frame is by itself irrelevant. The third observation is that G_{eff} varies with time as $\hat{U}_{\text{ext}}$ or $\boldsymbol{v}$ vary, and that it also varies with direction, as represented by the unit vector $\bar{\boldsymbol{n}}$; we see this in the terms involving $\bar{n}^j \bar{n}^k \partial_{\bar{j}\bar{k}} \hat{X}_{\text{ext}}$ and $[(\boldsymbol{w} + \boldsymbol{v}) \cdot \bar{\boldsymbol{n}}]^2$. There is also a variation with $\bar{r}$ seen in the last two terms.

Our expression for G_{eff} is not an artifact of the coordinate transformation from the inertial frame to the body's comoving frame. It can be obtained in an invariant manner with an alternative calculation. The gravitational constant is a physically measurable quantity, determined with a Cavendish-type experiment, in which the force between two known masses at a determined separation is measured; Newton's constant is then the measured coefficient in the expression $F = Gm_1m_2/r^2$. To be more precise, we imagine a situation in which a body of mass M moves freely through spacetime, while a test body of negligible mass is held at a fixed proper distance s from the source mass by a four-acceleration a^μ; the test body is assumed to be non-rotating relative to the source body. An invariant radial unit vector e^μ points from the test body to the source. The gravitational constant is then defined by Newton's law, which states that $a^\mu e_\mu = G_L M/s^2$, where we use the notation G_L to indicate that this is the invariant, locally-measured gravitational constant. A detailed calculation of G_L using the PPN metric and equations of motion reveals that $G_L = G_{\text{eff}}$, as given by Eq. (13.36).

13.2.7 Spin dynamics

The methods that were exploited to derive the equations of motion of spinning bodies in general relativity can be carried over to the PPN framework with only a few modifications, which come from the new parameters and the addition of the Whitehead and preferred-frame potentials.

We again define the spin tensor and vector of each body by Eqs. (9.167) and (9.168), and expand all potentials about the center-of-mass of each body, as defined by Eq. (9.3). We keep terms that involve products of $\bar{v}^k$ with $\bar{x}^j$, along with the usual "self-terms" involving H_A^{ij}, Ω_A, and so on. But we discard terms that vanish in the limit in which the size of each body tends to zero. The integrals of Eqs. (9.192) and (9.193) carry over to this calculation, with the caveat that some coefficients in the expressions must be replaced with their PPN counterparts. We also have to calculate the spin contributions to the preferred-frame terms in Eq. (13.12); these are given by

$$c^2 \Delta F^j_{\rm PF} = -\frac{1}{4}\alpha_1 w^k S^{lk}_A \partial_{jl} U_{\neg A} + \frac{1}{2}(2\alpha_3 - \alpha_1) M_A w^k \partial_k U^j_{\neg A} \,. \tag{13.37}$$

The Whitehead potential makes no spin contribution to the equations of motion, because it is independent of velocity.

Collecting results, we find that the PPN spin–orbit and spin–spin accelerations are given by

$$\begin{aligned} a^j_A[\mathrm{so}] = \frac{3}{2c^2} \sum_{B \neq A} \frac{GM_B}{r^3_{AB}} \Big\{ & n^{\langle jk \rangle}_{AB} \Big[v^p_A \Big((2\gamma + 1)\hat{S}^{kp}_A + (2\gamma + 2)\hat{S}^{kp}_B \Big) \\ & - v^p_B \Big((2\gamma + 1)\hat{S}^{kp}_B + (2\gamma + 2)\hat{S}^{kp}_A \Big) \Big] \\ & + n^{\langle kp \rangle}_{AB} (v_A - v_B)^p \Big((2\gamma + 1)\hat{S}^{jk}_A + (2\gamma + 2)\hat{S}^{jk}_B \Big) \\ & + \frac{1}{2}\alpha_1 n^{\langle jk \rangle}_{AB} \Big[(w + v_A)^p \hat{S}^{kp}_B - (w + v_B)^p \hat{S}^{kp}_A \Big] \\ & + \frac{1}{2}\alpha_1 n^{\langle kp \rangle}_{AB} (v_A - v_B)^p \hat{S}^{jk}_B - \alpha_3 n^{\langle jl \rangle}_{AB} (w + v_B)^p \hat{S}^{lk}_B \Big\} \,, \end{aligned} \tag{13.38a}$$

$$a^j_A[\mathrm{ss}] = -\frac{15}{8c^2}(4\gamma + 4 + \alpha_1) \sum_{B \neq A} \frac{GM_B}{r^4_{AB}} \hat{S}^{kp}_A \hat{S}^{kq}_B n^{\langle jpq \rangle}_{AB} \,, \tag{13.38b}$$

where $\hat{S}^{jk}_A := S^{jk}_A / M_A$ and M_A is the total mass-energy of body A. Comparing with Eqs. (9.189) and (9.190), we see that the coefficients of 3 and 4 in the spin–orbit terms have been replaced by $(2\gamma + 1)$ and $(2\gamma + 2)$, respectively, and that the spin–spin term is now modulated by the factor $\frac{1}{8}(4\gamma + 4 + \alpha_1)$. The terms involving the preferred-frame parameters α_1 and α_3 depend either on the combinations $\boldsymbol{w} + \boldsymbol{v}_A$ or $\boldsymbol{w} + \boldsymbol{v}_B$, or on the difference $\boldsymbol{v}_A - \boldsymbol{v}_B$, showing that all velocities are measured relative to the preferred universal frame or to each other; once more we find that the arbitrary velocity of the PPN frame is irrelevant.

In the same manner we can derive the PPN equations for the evolution of each spin vector. We begin with the general form of Eq. (9.203),

$$\frac{dS^j_A}{dt} = \frac{1}{c^2}\epsilon^{jpq} \sum_{i=1}^{11} G^{pq}_n + O(c^{-4}) \,, \tag{13.39}$$

and for n ranging from 1 to 9 we can import G^{pq}_n directly from Eqs. (9.205), with the appropriate insertion of PPN parameters. A tenth contribution, from the Whitehead potential, is found most simply by adopting the second form displayed in Box 13.1,

$$\Phi_W = -U^2 - \Phi_2 - \boldsymbol{\nabla} U \cdot \boldsymbol{\nabla} X + G \boldsymbol{\nabla} \cdot \int \frac{\rho^{*\prime}}{|\boldsymbol{x} - \boldsymbol{x}'|} \boldsymbol{\nabla}' X' \, d^3x' \,. \tag{13.40}$$

After some tedious calculations we arrive at

$$c^2 G_{10}^{jk} = \xi \frac{d}{dt} \int_A \rho^* (\bar{x}^j \bar{v}^l \partial_{kl} \bar{X}_A + \bar{v}^j \partial_k \bar{X}_A)\, d^3\bar{x} - 2\xi \int_A p\, \bar{x}^j \partial_k \bar{U}_A\, d^3\bar{x}$$
$$- \xi \int_A \rho^* (\bar{v}^j \partial_{tk} \bar{X}_A + 2\bar{v}^j \bar{v}^l \partial_{kl} \bar{X}_A + \bar{x}^j \bar{v}^l \partial_{tkl} \bar{X}_A + \bar{x}^j \bar{v}^l \bar{v}^m \partial_{klm} \bar{X}_A)\, d^3\bar{x}$$
$$+ 2\xi \Omega_A^{jl} \partial_{kl} X_{\neg A}\,. \tag{13.41}$$

The eleventh and final contribution comes from the preferred-frame terms in Eq. (13.8). Collecting results, we find three classes of terms, those that involve integrals over the variables of body A only, those that involve Ω_A and external variables such as $\boldsymbol{v}_A$, and spin terms. Most, but not all, of the self-terms can be manipulated to obtain a total time derivative, with the result

$$\left(\frac{dS_A^j}{dt}\right)_{\text{self}} = -\epsilon^{jpq} \frac{1}{c^2} \frac{d}{dt} \int_A \rho^* \bar{x}^p \left[\bar{v}^q \left(\frac{1}{2}\bar{v}^2 + (2\gamma + 1)U_A + \Pi + \frac{p}{\rho^*} \right) \right.$$
$$\left. - \frac{1}{2}(4\gamma + 4 + \alpha_1)U_A^q - \frac{1}{2}(1 + \alpha_2)\partial_{tq} X_A \right] d^3\bar{x}$$
$$+ \epsilon^{jpq} \frac{1}{c^2} \int_A \rho^* \bar{x}^p \left[\frac{1}{2}\alpha_3 \partial_q \Phi_{1A} + \zeta_2 \partial_q \Phi_{2A} + \zeta_3 \partial_q \Phi_{3A} \right.$$
$$\left. + (3\zeta_4 - \zeta_1)\partial_q \Phi_{4A} + \frac{1}{2}\zeta_1 \partial_q \Phi_{5A} \right] d^3\bar{x}\,. \tag{13.42}$$

As before, the total time derivative can be moved to the left-hand side of the spin-evolution equation and absorbed into an internal correction $\Delta_{\text{int}} S_A^j$ of the spin vector. The remaining terms are present only in non-conservative theories. For a body that is spherically symmetric to a good approximation, the integrals will all be proportional to δ^{pq}, which is killed by the contraction with ϵ^{jpq}. For a stationary axisymmetric body, the only quantities available to construct a two-index tensor are δ^{pq} and $e^p e^q$, where $\boldsymbol{e}$ is a unit vector in the direction of the symmetry axis, and these also are killed by a contraction with ϵ^{jpq}. The strange non-conservative precessions are therefore relevant only for rather oddly shaped bodies.

Other spin-evolution terms involve the body's self-gravitational energy Ω_A; they are given by

$$\left(\frac{dS_A^j}{dt}\right)_{\Omega_A} = \epsilon^{jpq} \frac{1}{c^2} \Omega_A^{pn} \left[\alpha_2 (w + v_A)^q (w + v_A)^n - 2\xi \sum_{B \neq A} \frac{m_B}{r_{AB}} n_{AB}^{qn} \right]. \tag{13.43}$$

The first term depends on the velocity of body A relative to the preferred frame, and the second term is the contribution from the Whitehead potential. For a spherically symmetric body, both terms are killed by the contraction with ϵ^{jpq}.

Finally, we collect the spin-evolution terms that depend only on the spins. By defining the proper spin $\bar{\boldsymbol{S}}_A$ according to [compare with Eq. (9.217)]

$$\bar{\boldsymbol{S}}_A := \boldsymbol{S}_A + \Delta_{\text{ext}} \boldsymbol{S}_A = \boldsymbol{S}_A + \frac{1}{c^2} \left\{ \left[v_A^2 + (2\gamma + 1)U_{\neg A} \right] \boldsymbol{S}_A - \frac{1}{2}(\boldsymbol{v}_A \cdot \boldsymbol{S}_A)\boldsymbol{v}_A \right\}, \tag{13.44}$$

we obtain the PPN equation of spin precession,

$$\frac{d\bar{\boldsymbol{S}}_A}{dt} = \boldsymbol{\Omega}_A \times \bar{\boldsymbol{S}}_A + O(c^{-4}), \tag{13.45}$$

where

$$\boldsymbol{\Omega}_A = \boldsymbol{\Omega}_A[\text{SO}] + \boldsymbol{\Omega}_A[\text{SS}] + \boldsymbol{\Omega}_A[\text{PF}], \tag{13.46a}$$

$$\boldsymbol{\Omega}_A[\text{SO}] = \frac{1}{2c^2}\sum_{B\neq A}\frac{GM_B}{r_{AB}^2}\boldsymbol{n}_{AB}\times\big[(2\gamma+1)\boldsymbol{v}_A-(2\gamma+2)\boldsymbol{v}_B\big], \tag{13.46b}$$

$$\boldsymbol{\Omega}_A[\text{SS}] = \frac{4\gamma+4+\alpha_1}{8c^2}\sum_{B\neq A}\frac{G}{r_{AB}^3}\big[3\boldsymbol{n}_{AB}(\boldsymbol{n}_{AB}\cdot\bar{\boldsymbol{S}}_B)-\bar{\boldsymbol{S}}_B\big], \tag{13.46c}$$

$$\boldsymbol{\Omega}_A[\text{PF}] = -\frac{\alpha_1}{4c^2}\sum_{B\neq A}\frac{GM_B}{r_{AB}^2}\boldsymbol{n}_{AB}\times(\boldsymbol{w}+\boldsymbol{v}_B). \tag{13.46d}$$

The spin–orbit and spin–spin pieces of $\boldsymbol{\Omega}_A$ can be compared with Eqs. (9.199); the preferred-frame piece is new.

Applying these equations to the Gravity Probe B experiment, we may choose to work in a PPN frame at rest with respect to the Earth, and set $\boldsymbol{v}_B = 0$. Comparing Eqs. (13.46) with Eqs. (9.199), we can see that the spin–orbit precession is proportional to the PPN coefficient $\frac{1}{3}(2\gamma+1)$, while the spin–spin precession is proportional to $\frac{1}{8}(4\gamma+4+\alpha_1)$. The implications for an experiment like Gravity Probe B can be explored by reviewing the discussion of Sec. 10.4.1. The preferred-frame precession produces a purely periodic motion of the gyroscopes that is too small to be detected.

13.3 Experimental tests of gravitational theories

13.3.1 Two-body problem and pericenter advance

We now specialize the PPN equations of motion obtained in Sec. 13.2.4 to a system of two bodies, such as the Sun–Mercury system or a binary-star system. To keep things simple we drop the preferred-frame terms from the equations of motion; these can be studied separately, as we shall do in Sec. 13.3.4. We include the structure dependence on the passive and active gravitational masses, assuming the bodies to be spherically symmetric. With these simplifications, the equations of motion for the first body become

$$\boldsymbol{a}_1 = -\left(\frac{M_P}{M}\right)_1\frac{G(M_A)_2}{r^2}\boldsymbol{n}$$
$$+\frac{1}{c^2}\Bigg\{-\frac{GM_2}{r^2}\Big[\gamma v_1^2-\frac{1}{2}(4\gamma+4+\alpha_1)(\boldsymbol{v}_1\cdot\boldsymbol{v}_2)$$

(continued overleaf)

$$
\begin{aligned}
&\qquad + \frac{1}{2}(2\gamma + 2 + \alpha_2 + \alpha_3)v_2^2 - \frac{3}{2}(1+\alpha_2)(\boldsymbol{n}\cdot\boldsymbol{v}_2)^2 \\
&\qquad - \Big(2\gamma + 2\beta + 1 + \frac{1}{2}\alpha_1 - \zeta_2\Big)\frac{GM_1}{r} - (2\gamma+2\beta)\frac{GM_2}{r}\Bigg]\boldsymbol{n} \\
&\quad + \frac{GM_2}{r^2}\Big\{\boldsymbol{n}\cdot\big[(2\gamma+2)\boldsymbol{v}_1 - (2\gamma+1)\boldsymbol{v}_2\big]\Big\}\boldsymbol{v}_1 \\
&\quad - \frac{1}{2}\frac{GM_2}{r^2}\Big\{\boldsymbol{n}\cdot\big[(4\gamma+4+\alpha_1)\boldsymbol{v}_1 - (4\gamma+2+\alpha_1-2\alpha_2)\boldsymbol{v}_2\big]\Big\}\boldsymbol{v}_2\Bigg] \\
&+ O(c^{-4}),
\end{aligned}
\tag{13.47}
$$

where $\boldsymbol{r} := \boldsymbol{r}_1 - \boldsymbol{r}_2$, $r := |\boldsymbol{r}|$, $\boldsymbol{n} := \boldsymbol{r}/r$, and $M_{\mathcal{P}}$, $M_{\mathcal{A}}$ for each body are given by Eqs. (13.30). The equations of motion for the second body are obtained by interchanging the variables according to $M_1 \leftrightarrow M_2$, $\boldsymbol{v}_1 \leftrightarrow \boldsymbol{v}_2$, and $\boldsymbol{n} \to -\boldsymbol{n}$.

We wish to convert the equations of motion into an effective one-body problem, as we did back in Sec. 9.3.7. We have to be careful when we are not working with a conservative theory of gravity, because the system's barycenter will not necessarily be at rest or move uniformly. But this failure occurs at post-Newtonian order, and since we need to relate $\boldsymbol{v}_1$ and $\boldsymbol{v}_2$ to the relative velocity $\boldsymbol{v}$ in 1PN terms only, the non-conservative effects are beyond our order of approximation. We can therefore choose our PPN coordinate system so that the barycenter is approximately at rest – we set $M_1\boldsymbol{v}_1 + M_2\boldsymbol{v}_2 = O(c^{-2})$. Defining $\boldsymbol{v} := \boldsymbol{v}_1 - \boldsymbol{v}_2$ and $m := M_1 + M_2$, we can to sufficient accuracy replace $\boldsymbol{v}_1$ and $\boldsymbol{v}_2$ in the 1PN terms by the relations

$$
\boldsymbol{v}_1 = \frac{M_2}{m}\boldsymbol{v} + O(c^{-2}), \qquad \boldsymbol{v}_2 = -\frac{M_1}{m}\boldsymbol{v} + O(c^{-2}),
\tag{13.48}
$$

and obtain an expression for the relative acceleration $\boldsymbol{a} := \boldsymbol{a}_1 - \boldsymbol{a}_2$. We find

$$
\begin{aligned}
\boldsymbol{a} = -\frac{Gm^*}{r^2}\boldsymbol{n} - \frac{Gm}{r^2c^2}\Bigg\{\Bigg[&\Big(\gamma + \frac{1}{2}\eta[6+\alpha_1+\alpha_2+\alpha_3]\Big)v^2 \\
&- \frac{3}{2}\eta(1+\alpha_2)(\boldsymbol{n}\cdot\boldsymbol{v})^2 - \big(2\gamma+2\beta+\eta[2+\alpha_1-2\zeta_2]\big)\frac{Gm}{r}\Bigg]\boldsymbol{n} \\
&- \big(2\gamma+2-\eta[2-\alpha_1-\alpha_2]\big)(\boldsymbol{n}\cdot\boldsymbol{v})\boldsymbol{v}\Bigg\} + O(c^{-4}),
\end{aligned}
\tag{13.49}
$$

where $\eta := M_1M_2/m^2$ and

$$
\begin{aligned}
m^* &:= \left(\frac{M_{\mathcal{P}}}{M}\right)_1 (M_{\mathcal{A}})_2 + \left(\frac{M_{\mathcal{P}}}{M}\right)_2 (M_{\mathcal{A}})_1 \\
&= m + O(c^{-2}) \times [\text{structure-dependent terms}].
\end{aligned}
\tag{13.50}
$$

Note that it is m^* that now plays the role of the Kepler-measured mass of the two-body system; because it differs from m by 1PN corrections, we can replace m by m^* in all 1PN terms. The bottom line is that the structure-dependent effects are unmeasurable in a two-body system. We can therefore drop the distinction between m^* and m, and express

$\boldsymbol{a}$ entirely in terms of m. Equation (13.49) then has exactly the same form as Eq. (10.1), except for the numerical coefficients in front of each term.

We next involve the acceleration of Eq. (13.49) in a calculation of the orbital motion. We adopt the methods of Sec. 10.1.3: the post-Newtonian terms in the acceleration are collected into a perturbing force $\boldsymbol{f}$, and the motion is described using the formalism of osculating orbital elements. We shall not go into the details here, but simply state that over a complete orbital period, the only orbital element that undergoes a net change is the longitude of pericenter ω. The PPN equations of motion imply that it advances by

$$\Delta\omega = \frac{6\pi Gm}{pc^2}\left[\frac{1}{3}(2+2\gamma-\beta)+\frac{1}{6}(2\alpha_1-\alpha_2+\alpha_3+2\zeta_2)\eta\right] \tag{13.51}$$

in the course of each orbit. Because the perturbing force involves terms proportional to either $\boldsymbol{n}$ or $\boldsymbol{v}$, the orientation of the orbital plane is unaffected by the perturbation. The second term in Eq. (13.51) vanishes in any fully conservative theory of gravity ($\alpha_n = \zeta_2 = 0$). It is also negligible in the case of Mercury orbiting the Sun, because $\eta \simeq 2\times 10^{-7}$. The predicted PPN perihelion advance of Mercury is therefore

$$\left(\frac{d\omega}{dt}\right)_{\text{sec}} = 42.98\left[\frac{1}{3}(2+2\gamma-\beta)\right]\text{as/century}. \tag{13.52}$$

As we saw back in Sec. 10.1.4, the advance of Mercury's perihelion beyond what can be explained by planetary perturbations is 42.98 arcseconds per century, accurate to about one part in 10^3. Data from helioseismology have shown that the complicating effect of a solar quadrupole moment is an order of magnitude smaller than the observational error (refer to Sec. 3.4.3). Combining these data with the bound on γ obtained below in Sec. 13.3.2 – it must be equal to unity at the level of parts in 10^5 – we see that β must be equal to unity to about 3 parts in 10^3.

13.3.2 Light deflection and Shapiro time delay

As we saw back in Sec. 13.2.5, the propagation of light in the PPN framework is the same as in general relativity, except that the overall magnitude of any post-Newtonian effect is proportional to the factor $\frac{1}{2}(1+\gamma)$. The deflection of a photon by a body of mass M is therefore described by the vector

$$\boldsymbol{\alpha}(t) = -\left(\frac{1+\gamma}{2}\right)\frac{4GM}{c^2}\frac{\boldsymbol{b}}{b^2}\frac{1+\cos\Phi(t)}{2}, \tag{13.53}$$

which is imported directly from Eq. (10.66). Here $\boldsymbol{b}$ is a vector that points from the body to the photon's point of closest approach, $b := |\boldsymbol{b}|$ is the impact parameter, and $\Phi(t)$ is the angle between the photon's current position and its initial direction (see Fig. 10.1). The total deflection angle at $t\to\infty$ is then

$$\alpha = \left(\frac{1+\gamma}{2}\right)\frac{4GM}{c^2 b}, \tag{13.54}$$

and evaluating this for the Sun gives

$$\alpha = 1.7504\left(\frac{1+\gamma}{2}\right)\frac{R_\odot}{b}\ \text{as}\,. \tag{13.55}$$

These results can be compared with Eq. (10.68).

The PPN expressions for the light deflection complement our discussion of the "Newtonian deflection" in Box 10.3. In the factor $\frac{1}{2}(1+\gamma)$, the "1/2" piece comes from g_{00}, the Newtonian part of the metric, and this corresponds to the Newtonian deflection. The "$\gamma/2$" piece comes from g_{jk}, and this reflects the impact of spatial curvature, which bends locally straight lines (spatial geodesics) near the Sun relative to locally straight lines far from the Sun. The total deflection is the sum of these two effects. Only the part resulting from the spatial curvature can vary from one metric theory to another. It is entirely coincidental that the two parts happen to be equal in general relativity.

The expression for the Shapiro time delay is altered from Eq. (10.100) by the same overall factor. The propagation time of a light signal between an emitter at position $\boldsymbol{r}_e$ and an observer at position $\boldsymbol{r}_{\text{obs}}$ is therefore given by

$$t_{\text{obs}} - t_e = \frac{1}{c}|\boldsymbol{r}_{\text{obs}} - \boldsymbol{r}_e| + \left(\frac{1+\gamma}{2}\right)\frac{2GM}{c^3}\ln\left[\frac{(r_{\text{obs}} + \boldsymbol{r}_{\text{obs}}\cdot\boldsymbol{k})(r_e - \boldsymbol{r}_e\cdot\boldsymbol{k})}{b^2}\right], \tag{13.56}$$

where $\boldsymbol{k}$ is the initial direction of the light signal. We can use high-precision measurements of the deflection of light and Shapiro time delay to place bounds on the parameter γ. These experiments were described in Secs. 10.2.3 and 10.2.5. The best current limit, obtained from Doppler tracking measurements of the Cassini spacecraft, is $|\gamma - 1| < 4\times10^{-5}$, in excellent agreement with general relativity.

Gravitational lensing also scales by the $\frac{1}{2}(1+\gamma)$ factor. Because of the large uncertainties in the gravitational potential of the lensing galaxies and clusters, it is more useful to exploit the high-precision bounds on γ from solar-system measurements to turn lensing into a tool for mapping the gravitational potential, and thus for mapping the distribution of dark matter in and around galaxies and clusters. One remarkable test of γ on galactic scales was nevertheless reported in 2006. This interesting measurement used data on gravitational lensing by 15 elliptical galaxies, collected by the Sloan Digital Sky Survey. The Newtonian potential U of each lensing galaxy (including the contribution from dark matter) could be determined using a Newtonian model derived from the observed velocity dispersion of stars in the galaxy, essentially exploiting the virial theorem which relates v^2 to U. Comparing the observations with the lensing predicted by the models provided a 10 percent bound on γ, in agreement with general relativity. Unlike the much tighter bound described previously, which was obtained on the scale of the solar system, this bound was obtained on a galactic scale.

13.3.3 Tests of the strong equivalence principle: Nordtvedt effect

We have seen that the structure-dependent contributions to the active and passive masses of a body are unobservable in a two-body system. They are, however, observable in a three-body system, and the corresponding effect, known as the Nordtvedt effect, has given rise to an important test of general relativity.

Working at the level of Newtonian gravity, but allowing the inertial, passive, and active masses of bodies to be different from each other, we write down the equations of motion of a two-body system in the presence of a third body [refer to Eqs. (3.73) and (3.74) for a strictly Newtonian formulation],

$$\begin{aligned} \boldsymbol{a}_1 &= -\left(\frac{M_{\mathcal{P}}}{M}\right)_1 \left[G(M_{\mathcal{A}})_2 \frac{\boldsymbol{r}_{12}}{r_{12}^3} + G(M_{\mathcal{A}})_3 \frac{\boldsymbol{r}_{13}}{r_{13}^3}\right], \\ \boldsymbol{a}_2 &= \left(\frac{M_{\mathcal{P}}}{M}\right)_2 \left[G(M_{\mathcal{A}})_1 \frac{\boldsymbol{r}_{12}}{r_{12}^3} - G(M_{\mathcal{A}})_3 \frac{\boldsymbol{r}_{23}}{r_{23}^3}\right]. \end{aligned} \tag{13.57}$$

We denote the barycentric position of the two-body system by

$$\boldsymbol{r}_c := \frac{M_1}{m}\boldsymbol{r}_1 + \frac{M_2}{m}\boldsymbol{r}_2, \tag{13.58}$$

where $m := M_1 + M_2$, and assuming that $r_{12} \ll r_{23}$, we expand

$$\frac{r_{13}^j}{r_{13}^3} = \frac{r_{c3}^j}{r_{c3}^3} - \sum_{\ell=1}^{\infty} \frac{1}{\ell!} r_{1c}^L \partial^{\langle jL \rangle} \left(\frac{1}{r_{c3}}\right), \tag{13.59}$$

where $\boldsymbol{r}_{c3} := \boldsymbol{r}_c - \boldsymbol{r}_3$ and $\boldsymbol{r}_{1c} := \boldsymbol{r}_1 - \boldsymbol{r}_c$; we develop a similar expansion for r_{23}^j/r_{23}^3. We define $R := r_{c3} = |\boldsymbol{r}_{c3}|$, $\boldsymbol{N} := \boldsymbol{r}_{c3}/r_{c3}$, $r_{1c} := |\boldsymbol{r}_{1c}|$, $\boldsymbol{n}_{1c} := \boldsymbol{r}_{1c}/r_{1c}$, $r_{2c} := |\boldsymbol{r}_{2c}|$, $\boldsymbol{n}_{2c} := \boldsymbol{r}_{2c}/r_{2c}$, and note that

$$r_{1c} = (M_2/m)r, \qquad r_{2c} = -(M_1/m)r, \qquad \boldsymbol{n}_{1c} = -\boldsymbol{n}_{2c} = \boldsymbol{n}, \tag{13.60}$$

where $r := |\boldsymbol{r}_{12}|$ and $\boldsymbol{n} := \boldsymbol{r}_{12}/r_{12}$. With this notation we find that the relative acceleration $\boldsymbol{a} := \boldsymbol{a}_1 - \boldsymbol{a}_2$ is given by

$$\begin{aligned} a^j = &-\frac{Gm^*}{r^2}n^j - \frac{GM_3}{R^2}N^j \left[\left(\frac{M_{\mathcal{P}}}{M}\right)_1 - \left(\frac{M_{\mathcal{P}}}{M}\right)_2\right] \\ &- \frac{GM_3}{R^2} \sum_{\ell=1}^{\infty} \frac{(-1)^\ell (2\ell+1)!!}{\ell!} \left(\frac{r}{R}\right)^\ell n^L N^{\langle jL \rangle} \\ &\times \left[\left(\frac{M_{\mathcal{P}}}{M}\right)_1 \left(\frac{M_2}{M}\right)^\ell - (-1)^\ell \left(\frac{M_{\mathcal{P}}}{M}\right)_2 \left(\frac{M_1}{M}\right)^\ell\right], \end{aligned} \tag{13.61}$$

where m^* is defined by Eq. (13.50), and we have dropped the $\mathcal{A}$ subscript on M_3. Introducing now

$$\alpha := \frac{1}{2}\left[\left(\frac{M_{\mathcal{P}}}{M}\right)_1 + \left(\frac{M_{\mathcal{P}}}{M}\right)_2\right], \tag{13.62a}$$

$$\delta := \left[\left(\frac{M_{\mathcal{P}}}{M}\right)_1 - \left(\frac{M_{\mathcal{P}}}{M}\right)_2\right], \tag{13.62b}$$

and keeping terms only through $\ell = 2$ in Eq. (13.61), we obtain

$$a^j = -\frac{Gm^*}{r^2}n^j - \delta\frac{GM_3}{R^2}N^j + 3\alpha\frac{GM_3 r}{R^3}n^k N^{\langle jk\rangle} - \frac{15}{2}\alpha\Delta\frac{GM_3 r^2}{R^4}n^{pq}N^{\langle jpq\rangle}$$
$$+\frac{3}{2}\delta\Delta\frac{GM_3 r}{R^3}n^k N^{\langle jk\rangle} - \frac{15}{4}\delta(1-2\eta)\frac{GM_3 r^2}{R^4}n^{pq}N^{\langle jpq\rangle}, \tag{13.63}$$

where $\Delta := (M_1 - M_2)/(M_1 + M_2)$ and $\eta := M_1 M_2/(M_1 + M_2)$. The first term in Eq. (13.63) is the standard Newtonian acceleration; as we saw previously, the fact that m^* incorporates structure-dependent corrections is irrelevant since it is m^* that represents the Kepler-measured mass of the two-body system. The second term is a relative acceleration that stretches or shrinks the orbit along a line directed toward the third body. Whether the stretching occurs when $\boldsymbol{n}$ is parallel or antiparallel to $\boldsymbol{N}$ depends on the sign of δ. From Eq. (13.30a) we have that

$$\delta = \left(4\beta - \gamma - 3 - \frac{10}{3}\xi - \alpha_1 + \frac{2}{3}\alpha_2 - \frac{2}{3}\zeta_1 - \frac{1}{3}\zeta_2\right)\left(\frac{\Omega_1}{M_1c^2} - \frac{\Omega_2}{M_2c^2}\right), \tag{13.64}$$

from which we conclude that the sign of δ depends on the PPN parameters and the sign of the difference between gravitational binding energies. The third and fourth terms originate from the tidal interactions with the third body; note that $\alpha = 1 + O(\Omega_A/M_Ac^2)$. The final two terms are corrections to these tidal perturbations that result from a non-zero δ.

We apply the equations of motion to the Earth–Moon system, with the Sun making up the third body. We set $M_1 = M_\oplus$, $M_2 = M$, $M_3 = M_\odot$, and we first estimate the size of the various terms in the equations of motion. We recall that $Gm^*/r^3 \approx M_\oplus/r^3 \approx \omega_\oplus^2$ and $GM_\odot/R^3 \approx \omega_\odot^2$, where $\omega_\oplus$ is the angular velocity of the Moon around the Earth, and $\omega_\odot$ is the angular velocity of the Earth around the Sun, with $\omega_\oplus/\omega_\odot \approx 13.4$. We also have

$$\frac{\Omega_\oplus}{M_\oplus c^2} = -4.6 \times 10^{-10}, \tag{13.65a}$$

$$\frac{\Omega}{Mc^2} = -0.2 \times 10^{-10}, \tag{13.65b}$$

and $R/r \approx 395$. As a consequence of these numerical relations, we find that the second term in Eq. (13.63) is smaller than the first by a factor of $(\omega_\odot/\omega_\oplus)^2(R/r)\delta \sim 10^{-10}$. The next two terms are smaller than the first by factors $(\omega_\odot/\omega_\oplus)^2 \sim (1/13.4)^2 \sim 5\times10^{-3}$ and $(\omega_\odot/\omega_\oplus)^2(r/R) \sim 1.4 \times 10^{-5}$. And the final two terms are smaller than these by an additional factor of 10^{-10}. Dropping these final terms and setting $\alpha = 1$, the equations of motion for the Earth–Moon system reduce to

$$\boldsymbol{a} = -\frac{Gm^*}{r^2}\boldsymbol{n} - \delta\frac{GM_3}{R^2}\boldsymbol{N} + \frac{GM_3 r}{R^3}\big[3\boldsymbol{N}(\boldsymbol{N}\cdot\boldsymbol{n}) - \boldsymbol{n}\big]$$
$$-\frac{3}{2}\frac{GM_3 r^2}{R^4}\Delta\big[5\boldsymbol{N}(\boldsymbol{N}\cdot\boldsymbol{n})^2 - 2\boldsymbol{n}(\boldsymbol{N}\cdot\boldsymbol{n}) - \boldsymbol{N}\big]. \tag{13.66}$$

The second term, proportional to δ, is a structure-dependent contribution to the acceleration, and it is this term that gives rise to the Nordtvedt effect. The remaining terms originate from the tidal interactions with the Sun, and they give rise to orbital perturbations that are actually larger than the Nordtvedt effect.

To analyze the violation of the strong equivalence principle on the lunar orbit, we consider a simplified situation in which the Moon–Earth and Earth–Sun orbits lie in the same plane (their actual relative inclination is about 5 degrees), and are approximately circular (the eccentricity of the Moon–Earth orbit is 0.055, and the eccentricity of the Earth–Sun orbit is 0.017). We calculate the first-order perturbations of the lunar orbit created by the last three terms in Eq. (13.66). In the limit of small eccentricity, we can write $d\phi/dt = \omega_\oplus =$ constant, so that

$$r = a(1 - e\cos f) + O(e^2), \tag{13.67a}$$

$$f = \phi - \omega = \omega_\oplus t - \omega + O(e), \tag{13.67b}$$

$$\boldsymbol{n} = \boldsymbol{e}_X \cos(\omega_\oplus t) + \boldsymbol{e}_Y \sin(\omega_\oplus t) + O(e), \tag{13.67c}$$

$$\boldsymbol{\lambda} = -\boldsymbol{e}_X \sin(\omega_\oplus t) + \boldsymbol{e}_Y \cos(\omega_\oplus t) + O(e), \tag{13.67d}$$

$$\boldsymbol{N} = \boldsymbol{e}_X \cos(\omega_\odot t - \Phi) + \boldsymbol{e}_Y \sin(\omega_\odot t - \Phi), \tag{13.67e}$$

in which e is the eccentricity of the lunar orbit, and Φ is the initial phase of the Earth's orbit around the Sun, which we take to be perfectly circular. Note that

$$\boldsymbol{n}\cdot\boldsymbol{N} = \cos(\Lambda t + \Phi), \tag{13.68a}$$

$$\boldsymbol{\lambda}\cdot\boldsymbol{N} = -\sin(\Lambda t + \Phi), \tag{13.68b}$$

where $\Lambda := \omega_\oplus - \omega_\odot$ is the *synodic frequency*, the angular frequency of the lunar orbit relative to the Sun. This is to be distinguished from the *sidereal frequency* $\omega_\oplus$, which refers to the barycentric frame.

We wish to involve $\boldsymbol{f} := \boldsymbol{a} + Gm^*\boldsymbol{n}/r^2$ in a perturbative calculation of the orbital motion, making use of the formalism of osculating orbital elements reviewed in Sec. 3.3.2. The formalism must be applied with some care, because the orbital eccentricity is small, and ω loses its meaning when $e \to 0$. In such cases it is best to use the alternative variables $A := e\cos\omega$ and $B := e\sin\omega$, and to re-express the osculating equations in terms of A and B. We therefore write

$$r = a\big[1 - A\cos(\omega_\oplus t) - B\sin(\omega_\oplus t)\big] + O(e^2), \tag{13.69}$$

and convert Eqs. (3.64) to the form

$$\frac{da}{dt} = \frac{2}{\omega_\oplus}\mathcal{S} + O(e), \tag{13.70a}$$

$$\frac{dA}{dt} = \frac{1}{\omega_\oplus a}\Big[\mathcal{R}\sin(\omega_\oplus t) + 2\mathcal{S}\cos(\omega_\oplus t)\Big] + O(e), \tag{13.70b}$$

$$\frac{dB}{dt} = \frac{1}{\omega_\oplus a}\Big[-\mathcal{R}\cos(\omega_\oplus t) + 2\mathcal{S}\sin(\omega_\oplus t)\Big] + O(e), \tag{13.70c}$$

where $\mathcal{R}$ and $\mathcal{S}$ are the radial and tangential components of the perturbing acceleration. To lowest order in e they are given by

$$\mathcal{R} = \omega_\odot^2 a \left\{ \left[3\cos^2(\Lambda t + \Phi) - 1\right] - \frac{3}{2}\zeta \cos(\Lambda t + \Phi)\left[5\cos^2(\Lambda t + \Phi) - 3\right] - \delta(R/a)\cos(\Lambda t + \Phi) \right\}, \tag{13.71a}$$

$$\mathcal{S} = \omega_\odot^2 a \left\{ -3\sin(\Lambda t + \Phi)\cos(\Lambda t + \Phi) + \frac{3}{2}\zeta \sin(\Lambda t + \Phi)\left[5\cos^2(\Lambda t + \Phi) - 1\right] + \delta(R/a)\sin(\Lambda t + \Phi) \right\}, \tag{13.71b}$$

where $\zeta := \Delta(a/R)$. Because the orbital planes are taken to coincide, there are no perturbations to the inclination ι or line of nodes Ω.

Substituting Eqs. (13.71b) into Eqs. (13.70), integrating subject to the boundary conditions $a = a_0$, $A = 0$, $B = 0$ at $t = 0$, and inserting the results within Eq. (13.69), we eventually obtain

$$\begin{aligned} \delta r(t) = a_0 \left(\frac{\omega_\odot}{\omega_\oplus}\right)^2 \Bigg\{ & \frac{3}{2}\frac{\omega_\oplus^2(1 + \omega_\oplus/\Lambda)}{\omega_\oplus^2 - 4\Lambda^2}\cos 2(\Lambda t + \Phi) \\ & - \zeta\left[\frac{3}{8}\frac{\omega_\oplus^2(3 + 2\omega_\oplus/\Lambda)}{\omega_\oplus^2 - \Lambda^2}\cos(\Lambda t + \Phi) + \frac{5}{8}\frac{\omega_\oplus^2(3 + 2\omega_\oplus/\Lambda)}{\omega_\oplus^2 - 9\Lambda^2}\cos 3(\Lambda t + \Phi)\right] \\ & - \delta\frac{R}{a_0}\frac{\omega_\oplus^2(1 + 2\omega_\oplus/\Lambda)}{\omega_\oplus^2 - \Lambda^2}\cos(\Lambda t + \Phi)\Bigg\} \end{aligned} \tag{13.72}$$

for the perturbation in the Earth–Moon distance. Not surprisingly, because we have a sinusoidal driving force acting on a sinusoidal oscillator, the perturbation takes the form of a resonant response, with denominators of the form $1/(\omega_\oplus^2 - N^2\Lambda^2)$, with N representing the harmonic degree. Note that if we had taken the Sun to be fixed in space, with $\omega_\odot = 0$, then $\Lambda = \omega_\oplus$ and the response of the fundamental harmonic would have produced a linear growth in the semi-major axis. Here the response is strictly harmonic, but because $\omega_\odot = \omega_\oplus/13.4$, the response to the $\cos(\Lambda t + \Phi)$ harmonic is enhanced by the resonant factor $\omega_\oplus^2/(\omega_\oplus^2 - \Lambda^2) \approx 7.0$.

Inserting the relevant numbers for the Earth–Moon system, setting $\delta \simeq 4.2 \times 10^{-10}$ [refer to Eq. (13.65)], and taking into account the fact that $a_0 \approx 3.84 \times 10^5$ km, we see that the two tidal perturbations and the Nordtvedt effect have approximate amplitudes of 2700 km, 73 km, and 8 meters, respectively. The leading tidal perturbation occurs at twice the synodic frequency (2Λ), whereas the Nordtvedt effect occurs at the synodic frequency; the tidal perturbation is much larger, but it can be cleanly separated from the Nordtvedt effect by observing many lunar orbits. The subleading tidal perturbation, however, has contributions at frequencies Λ and 3Λ, albeit at a much smaller amplitude of 73 km. Because this perturbation depends on parameters ($\omega_\oplus$, $\omega_\odot$, R, a_0, Δ) that are very accurately measured by other means, it can be predicted in advance to an accuracy well below the amplitude of the Nordtvedt effect.

In fact, an accurate calculation of the relevant amplitudes would require us to go beyond the first-order perturbation analysis carried out here. The perturbations induced by each term affect the behavior of the other terms, and so it is necessary to go to higher order in the orbital perturbation equations, and to include higher values of the multipole index ℓ. The problem is that the solar perturbation of the lunar orbit is so large, as seen by the 2700 km leading tidal amplitude, that one must employ more sophisticated techniques, such as the Hill–Brown lunar theory, in order to find a sequence of perturbations that converges in a reasonable way. The final conclusion of such calculations is that the effective amplitude of the Nordtvedt term is increased from the amplitude shown in Eq. (13.72) by a factor approximately equal to $1 + 2\omega_\odot/\omega_\oplus \simeq 1.15$, leading to an amplitude of 9.2 meters. Similarly, the amplitude of the competing synodic term is increased from 75 km to 110 km, but it can still be predicted accurately enough to be subtracted from the data.

Inserting the values for $\Omega/(Mc^2)$ for the Earth and Moon, the resulting prediction for the Nordtvedt effect is

$$\delta r(t) \simeq 9.2\eta_N \cos(\Lambda t + \Phi)\,\mathrm{m}\,, \tag{13.73}$$

where η_N is the Nordtvedt parameter

$$\eta_N := 4\beta - \gamma - 3 - \frac{10}{3}\xi - \alpha_1 + \frac{2}{3}\alpha_2 - \frac{2}{3}\zeta_1 - \frac{1}{3}\zeta_2\,. \tag{13.74}$$

As we explain in Box 13.2, long-term monitoring of the lunar orbit has revealed no sign of a Nordtvedt effect, and η_N is currently constrained to be smaller than 4.4×10^{-4}. This implies that violations of the strong equivalence principle are very small in the Earth–Moon–Sun system. And this, of course, is compatible with general relativity, which predicts $\eta_N = 0$.

Box 13.2 Lunar laser ranging and the Nordtvedt effect

Lunar laser ranging (LLR) illustrates the importance of new technology and broad theoretical frameworks in the program to test general relativity.

Except for the de Sitter precession of the lunar orbit, described in Sec. 10.1.6, most general relativistic effects on the lunar orbit are so small as to be virtually undetectable. But in the late 1950s, Robert Dicke was thinking beyond Einstein, and he wondered whether one could measure a variation with time of Newton's constant G, a feature that he would soon incorporate into his own alternative theory of gravity. By the early 1960s, the development of pulsed ruby lasers and the rapid build-up of the lunar space program led him and others to propose making very accurate measurements of the Earth–Moon distance by bouncing laser pulses off specially designed reflectors, to be placed on the lunar surface by either unmanned or manned landers. Such measurements would provide tests of general relativity, but they would also have other important scientific benefits, such as improving our understanding of the Earth–Moon orbit, the librations of the Moon, and even the motions of the Earth-bound laser sources resulting from continental drift.

The theoretical discovery of the Nordtvedt effect added to the science case. The discovery required a broader theoretical framework than general relativity; if the focus had never deviated from general relativity, there would have been no case. The Nordtvedt effect and the possibility that G could vary with time provided specific experimental targets, even if the conventional expectation (though perhaps not Dicke's) was for the answers to be "zero" in both cases. (As Ken Nordtvedt was fond of saying, zero is as good as any other number.)

The first retroreflector was deployed on the Moon by US astronaut Neil Armstrong on July 21, 1969, and within a month, the first successful acquisition was made of a reflected laser signal. Two other US and two French-built reflectors were subsequently placed on the Moon by US astronauts and Soviet unmanned landers. Strangely, the French reflectors were never detected via laser bounces, until just recently.

Since that time a worldwide network of observatories has made regular measurements of the round-trip travel time to the three US lunar retroreflectors, with accuracies that are routinely at the 50 ps (1 cm) level, and that are approaching 5 ps (1 mm). These measurements are fitted using the method of least-squares to a theoretical model for the lunar motion that takes into account perturbations created by the Sun and the other planets, tidal interactions, and post-Newtonian gravitational effects. The predicted round-trip travel times between retroreflector and telescope also take into account the librations of the Moon, the orientation of the Earth, the location of the observatories, and atmospheric effects on the signal propagation. The Nordtvedt parameter is then estimated in the least-squares fit, along with several other important parameters of the model.

From the first published analyses of LLR data in 1976 to the present, there has been absolutely no evidence, within experimental uncertainty, for the Nordtvedt effect. The best current bound on the Nordtvedt parameter is

$$\eta_N = (4.4 \pm 4.5) \times 10^{-4} .$$

This is equivalent to an orbital perturbation $\delta r(t) = (2.8 \pm 4.1)\,\mathrm{mm}\,\cos(\Lambda t + \Phi)$, and represents a limit on a possible violation of the strong equivalence principle of about 2 parts in 10^{13}.

At this level of precision, however, we cannot regard the results of LLR as a completely "clean" test of the Nordtvedt effect until we consider the possibility of a compensating violation of the weak equivalence principle for the Earth and Moon. This is because the chemical compositions of the Earth and Moon differ: the Earth is richer in the iron group elements, while the Moon is richer in silicates. To address this issue, the Eöt-Wash group at the University of Washington in Seattle (refer to Box 1.1) carried out a novel torsion-balance test of the weak equivalence principle by fabricating laboratory bodies whose chemical compositions mimic that of the Earth and Moon. They found that the mini-Earth and mini-Moon fell with the same acceleration to 1.4 parts in 10^{13}. The uncertainty implied by this possible effect has been incorporated into the bound on η_N quoted above.

The Apache Point Observatory for Lunar Laser ranging Operation (APOLLO) project, a joint effort by researchers from the University of Washington, Seattle, and the University of California, San Diego, is using enhanced laser and telescope technology, together with a good, high-altitude site in New Mexico, to improve the LLR bound by as much as an order of magnitude.

Tests of the Nordtvedt effect for neutron stars have also been carried out using a class of systems known as wide-orbit binary millisecond pulsars (WBMSP), which are pulsar–white-dwarf binary systems with small orbital eccentricities. In the gravitational field of the galaxy, a non-zero Nordtvedt effect can induce an anomalous eccentricity pointed toward the galactic center. This can be bounded using statistical methods, given a sufficient number of WBMSPs. Using data from 21 WBMSPs, including recently discovered highly circular systems, Ingrid Stairs and her colleagues obtained the bound $|\eta_N(\Omega/Mc^2)_{\rm NS}| < 5.6 \times 10^{-3}$. Because $(\Omega/Mc^2)_{\rm NS} \sim 0.1$ for typical neutron stars, this bound on η_N does not compete with the bound from LLR; on the other hand, the presence of neutron stars implies that these systems test the strong equivalence principle in the strong-field regime.

Lunar laser ranging also demonstrated that G is not changing significantly; the bound is $|\dot{G}/G| < 9 \times 10^{-13}\,{\rm yr}^{-1}$, a limit 83 times smaller than the inverse age of the universe.

13.3.4 Tests of the strong equivalence principle: preferred-frame and preferred-location effects

Back in Sec. 13.2.6 we saw that the metric near a body moving in the field of external bodies can be expressed as in Eq. (13.35), with an effective gravitational parameter $G_{\rm eff}$ that varies with time and position. The various contributions to $G_{\rm eff}$, as displayed in Eq. (13.36), have observable consequences. The most dramatic come from the anisotropies contained in the terms involving $\bar{n}^j\bar{n}^k\partial_{\bar{j}\bar{k}}\hat{X}_{\rm ext}$ and $[(\boldsymbol{w}+\boldsymbol{v})\cdot\bar{\boldsymbol{n}}]^2$. For a body such as the Earth, these terms make the gravitational forces holding the body together weaker in one direction than in another, with the result that the Earth is distorted into an elliptical shape; this is quite analogous to the effects of the luni-solar tides. A gravimeter will then measure a varying acceleration g as the Earth rotates. Assuming that $\boldsymbol{w}$ is the velocity of the solar system relative to the cosmic background radiation, with magnitude 375 km/s, then $w^2/c^2 \simeq 10^{-6}$. From the point of view of the Earth, the dominant external body is actually the galaxy, with $\hat{U}_{\rm gal} \simeq 5 \times 10^{-7}$ (only the anisotropic part of the potential is relevant, so the much more massive dark matter halo that reportedly surrounds the galaxy can be ignored, as can the mass distribution on larger cosmic scales).

The local g reading of a gravimeter will actually be affected by three factors. The first is the raw variation in $G_{\rm eff}$. The second is the distortion of the Earth generated by the variation in $G_{\rm eff}$, which displaces the gravimeter toward or away from the Earth's center, thus causing the local g to vary. The third is the redistribution of matter caused by the distortion. The latter two effects are controlled by the Love numbers h and k – refer to Sec. 2.4 – whose values depend on the detailed structure of the Earth. The overall factor in gravimeter readings is approximately 1.16 times the raw variation in $G_{\rm eff}$. Of course, the Sun and Moon generate similar effects via their tidal gravitational fields, called "solid Earth tides" to distinguish them from the more complex oceanic tides. The effect of tides on gravimeters is governed by the same Love numbers. But whereas the PPN variations are oriented relative to the directions of $\boldsymbol{w}$ and the galactic center, which are fixed in inertial space, the luni-solar tides are oriented toward the Sun and Moon, which revolve around the Earth (as seen from the Earth's point of view). The PPN effects therefore vary dominantly at multiples of a sidereal day, while the luni-solar tides vary at multiples of the solar and lunar day. One can in principle separate the PPN effects from the luni-solar tides by exploiting their slightly different time dependences. In fact, because of the slight inclination of the lunar orbit relative to the ecliptic, and the small eccentricities of the lunar and terrestrial orbits, there are smaller sidebands of the luni-solar tides at the same sidereal frequencies as the PPN tides. But tidal theory is sufficiently advanced that these Newtonian tides can be predicted quite accurately.

The dominant luni-solar tides have amplitudes $\Delta g/g \sim 2 \times 10^{-8}$, and measurements using arrays of superconducting gravimeters in the western USA during the middle 1970s showed that there were no anomalous solid tides at the sidereal frequencies, down to the $\Delta g/g \sim 4 \times 10^{-10}$ level. These experiments place upper bounds on the PPN parameters, given by

$$|\alpha_2| < 4 \times 10^{-4}, \qquad |\xi| < 10^{-3}. \tag{13.75}$$

Another bound can be obtained by examining the cross-term $\boldsymbol{w} \cdot \boldsymbol{v}$ in G_{eff}. As the Earth orbits the Sun, this term varies annually with an amplitude of around 10^{-7}. This variation in the magnitude of G_{eff} causes the Earth to "breathe" in and out; the resulting variations in its moment of inertia cause its rotation rate to vary on an annual basis. Annual variations in the Earth's rotation rate are well measured, and are known to be related to seasonal changes in atmospheric winds. Again measurements show no evidence of an anomaly, and an upper limit on a combination of α_1, α_2, and α_3 could be obtained at the level of about 0.02.

The term involving $\hat{U}_{\text{ext}}$ in G_{eff} is intrinsically interesting, because it reveals that nearby matter can affect the locally measured constant of gravitation. But it is not so interesting observationally, because it is approximately constant in time, and of order 10^{-8} for the Sun; because the absolute value of G is known to about one part in 10^5, such constant factors are unmeasurable. Even the variation in $\hat{U}_{\text{ext}}$ at the level of 10^{-10}, because of the eccentricity of the Earth's orbit, is too small to have any measurable consequences. The final two terms in Eq. (13.36) are also too small, in the context of the Earth, to provide useful tests.

13.4 Gravitational radiation in alternative theories of gravity

Back in Sec. 11.1 we learned that in general relativity, the gravitational-wave field in the far-away wave zone is characterized by two polarization modes, $A_+(\tau)$ and $A_\times(\tau)$. For a wave traveling in the z-direction, this means that the non-vanishing components of the wave field are $h^{xx} = -h^{yy} = GA_+/(c^4 R)$ and $h^{xy} = h^{yx} = GA_\times/(c^4 R)$; the wave is transverse to the direction of propagation, and it satisfies a tracefree condition. Here R is the distance to the source, and $\tau := t - R/c$ is retarded time.

We wish to generalize this discussion to a class of alternative metric theories of gravity, and we will show that in this context, the most general gravitational wave is characterized by six modes of polarization. The only assumptions we make are that we are examining the spacetime metric in the far-away wave zone, that the field equations reduce to wave equations in the far-away wave zone, and that the propagation speed of the waves is the same as the speed of light. This last assumption is somewhat restrictive, because many alternative theories of gravity predict that gravitational waves have a speed that differs from c. Some predict a speed that varies with the wavelength λ, as if the hypothetical quantum particle associated with the waves – the graviton – possessed a non-zero mass; in such a case the speed of gravitational waves would be given by $v_g/c = \sqrt{1-(\lambda/\lambda_c)^2}$, where $\lambda_c = h/(mc)$ is the Compton wavelength of a particle of mass m. Some theories predict that gravitational waves propagate along the null cones of a second metric, distinct from the metric that couples to matter and light. Nevertheless, for simplicity we shall restrict our attention to theories for which gravitational waves propagate at the speed of light. As a matter of fact, our conclusions about the polarization modes do not depend much on the speed of the waves, and similar conclusions would be reached in a more general setting.

13.4.1 Gravitational potentials in the far-away wave zone

We choose a reference frame in which the center-of-mass of the system is at rest, and we first examine the metric of a stationary system in this frame. Inspecting Box 13.1, we see that the leading contribution to the 1PN metric when $R \to \infty$ is given by

$$g_{00} = -1 + \frac{2GM}{c^2 R} + \frac{GM}{c^4 R}\Big[(\alpha_2 + \alpha_3 - \alpha_1)w^2 - \alpha_2(\boldsymbol{w}\cdot\boldsymbol{N})^2\Big], \tag{13.76a}$$

$$g_{0j} = \frac{GM}{2c^3 R} w^k \Big[(2\alpha_2 - \alpha_1)\delta_{jk} - 2\alpha_2 N_j N_k\Big], \tag{13.76b}$$

$$g_{jk} = \left(1 + \frac{2\gamma GM}{c^2 R}\right)\delta_{jk}, \tag{13.76c}$$

where M is the (active) gravitational mass of the stationary source, $\boldsymbol{w}$ is the velocity of the chosen frame relative to the preferred universal frame, $R := |\boldsymbol{x}|$, and $\boldsymbol{N} := \boldsymbol{x}/R$. From this we can construct the gravitational potentials

$$h^{\alpha\beta} := \eta^{\alpha\beta} - \sqrt{-g}g^{\alpha\beta}, \tag{13.77}$$

which formed the basis of our development of post-Minkowskian theory in Chapter 6, and our discussion of gravitational radiation in Chapter 11. To order c^{-3} we find that the potentials are given by

$$h^{00} = (3\gamma + 1)\frac{GM}{c^2 R} + O(c^{-4}), \tag{13.78a}$$

$$h^{0j} = \frac{GM}{2c^3 R} w^k \Big[(2\alpha_2 - \alpha_1)\delta_{jk} - 2\alpha_2 N_j N_k\Big] + O(c^{-5}), \tag{13.78b}$$

$$h^{jk} = (1-\gamma)\frac{GM}{c^2 R}\delta^{jk} + O(c^{-4}). \tag{13.78c}$$

We see that there is a c^{-2} contribution to h^{jk} whenever $\gamma \neq 1$; in general relativity this term vanishes, and indeed, when $\gamma = 1$ and $\alpha_1 = \alpha_2 = 0$ we recover the stationary limit of the expressions listed in Eq. (11.6).

We next allow the source to be time-dependent, and to emit gravitational waves. We assume that in such a situation, the stationary potentials in the far-away wave zone are supplemented by

$$\Delta h^{\alpha\beta} = \frac{G}{c^4 R} A^{\alpha\beta}(\tau, \boldsymbol{N}). \tag{13.79}$$

We assume specifically that the field equations of the alternative theory reduce in a suitable way to wave equations in the far-away wave zone, and that the propagation speed is equal to c; this accounts for the assumed dependence of the amplitudes $A^{\alpha\beta}$ on retarded time τ. (We could allow for a different speed v_g and generalize the definition of retarded time to $t - R/v_g$, but this would have little impact on our discussion of polarization modes, as observed previously.)

We wish to identify the physical meaning of the amplitudes $A^{\alpha\beta}$. When we performed this exercise in Sec. 11.1 in the context of general relativity, we eliminated the redundant components of $A^{\alpha\beta}$ by appealing to the harmonic gauge condition $\partial_\beta h^{\alpha\beta} = 0$ and its further refinement to the transverse-tracefree gauge; we ended up with two physical polarization modes encoded in $A^{jk}_{\rm TT}$. The situation is more complicated in a generic metric theory of gravity. In this case we cannot rely on the harmonic gauge condition, and the number of polarization modes is larger. We can, nevertheless, simplify the description of these modes by appealing to the standard freedom to transform the coordinates.

To effect this simplification we follow the approach described back in Sec. 11.1.3. We decompose $\Delta h^{\alpha\beta}$ into irreducible pieces according to

$$\Delta h^{00} = \frac{G}{c^4 R} C(\tau, \boldsymbol{N}), \tag{13.80a}$$

$$\Delta h^{0j} = \frac{G}{c^4 R} D^j(\tau, \boldsymbol{N}), \tag{13.80b}$$

$$\Delta h^{jk} = \frac{G}{c^4 R} A^{jk}(\tau, \boldsymbol{N}), \tag{13.80c}$$

and

$$D^j = D N^j + D^j_{\rm T}, \tag{13.81a}$$

$$A^{jk} = \frac{1}{3}\delta^{jk} A + \left(N^j N^k - \frac{1}{3}\delta^{jk} \right) B + 2N^{(j} A^{k)}_{\rm T} + A^{jk}_{\rm TT}, \tag{13.81b}$$

in which $A^j_{\rm T}$ and $D^j_{\rm T}$ are transverse vector fields satisfying $N_j A^j_{\rm T} = N_j D^j_{\rm T} = 0$, and $A^{jk}_{\rm TT}$ is a transverse-tracefree tensor field satisfying $N_j A^{jk}_{\rm TT} = \delta_{jk} A^{jk}_{\rm TT} = 0$. The decomposition involves ten independent functions of τ and $\boldsymbol{N}$.

The freedom to transform the coordinates must be restricted by the requirement that in the far-away wave zone, $h^{\alpha\beta}$ must always be of the general form described by Eqs. (13.78) and (13.79). This implies that the allowed transformations are small deformations described by $x^\alpha \to x'^\alpha = x^\alpha + \zeta^\alpha(x^\beta)$, with a gauge vector ζ^α restricted to be of the general form

$$\zeta^0 = \frac{G}{c^3 R} \alpha(\tau, \boldsymbol{N}) + O(R^{-2}), \tag{13.82a}$$

$$\zeta^j = \frac{G}{c^3 R} \beta^j(\tau, \boldsymbol{N}) + O(R^{-2}), \tag{13.82b}$$

first introduced in Sec. 11.1.5. Note that the gauge vector field satisfies the wave equation $\Box \zeta^\alpha = O(R^{-2})$. This does not occur because of a requirement to keep x'^α within the class of harmonic coordinates, as was the case back in Sec. 11.1.5. Instead, this is an automatic property of the assumed form for ζ^α, which is designed to preserve the form of the gravitational potentials.

We decompose β^j into longitudinal and transverse pieces, $\beta^j = N^j\beta + \beta^j_{\rm T}$, and performing the gauge transformation on the potentials, we find that

$$\Delta h'^{00} = \frac{G}{c^4 R} C'(\tau, \boldsymbol{N}), \tag{13.83a}$$

$$\Delta h'^{0j} = \frac{G}{c^4 R} D'^j(\tau, \boldsymbol{N}), \tag{13.83b}$$

$$\Delta h'^{jk} = \frac{G}{c^4 R} A'^{jk}(\tau, \boldsymbol{N}), \tag{13.83c}$$

with

$$C' = C + \partial_\tau(\alpha + \beta), \tag{13.84a}$$

$$D' = D + \partial_\tau(\alpha + \beta), \tag{13.84b}$$

$$D'^j_{\rm T} = D^j_{\rm T} + \partial_\tau \beta^j_{\rm T}, \tag{13.84c}$$

$$A' = A + \partial_\tau(3\alpha - \beta), \tag{13.84d}$$

$$B' = B + 2\partial_\tau \beta, \tag{13.84e}$$

$$A'^j_{\rm T} = A^j_{\rm T} + \partial_\tau \beta^j_{\rm T}, \tag{13.84f}$$

$$A'^{jk}_{\rm TT} = A^{jk}_{\rm TT}. \tag{13.84g}$$

With the freedom to specify the four functions α, β, and $\beta^j_{\rm T}$, we could decide to make A'^{jk} purely transverse and tracefree, but in this case we would find that C' and D'^j do not vanish. Alternatively, we could decide to make D'^j vanish, but in this case C' cannot be set equal to zero, and while the remaining freedom could be used to eliminate either A' or B', we cannot achieve both, and we cannot eliminate $A'^j_{\rm T}$. The bottom line is that no matter what choices are made, six independent degrees of freedom remain in the specification of $A'^{\alpha\beta}$.

13.4.2 Polarizations

The physical meaning of these six degrees of freedom is best identified by examining how a gravitational wave interacts with an actual detector. As we reviewed back in Sec. 11.1.6, the interaction with a short detector is governed by the equation of geodesic deviation,

$$\frac{d^2\xi_j}{dt^2} = -c^2 R_{0j0k}\xi^k, \tag{13.85}$$

in which the components R_{0j0k} of the linearized Riemann tensor are given by

$$R_{0j0k} = -\frac{1}{2}\left(\partial_{00}h^{jk} - \frac{1}{2}\partial_{00}h\delta_{jk} + \partial_{jk}h^{00} + \frac{1}{2}\partial_{jk}h + \partial_{0j}h^{0k} + \partial_{0k}h^{0j}\right), \tag{13.86}$$

where $h := \eta_{\alpha\beta}h^{\alpha\beta} = -h^{00} + h^{kk}$. In general relativity R_{0j0k} can be expressed entirely in terms of $h^{jk}_{\rm TT}$, the transverse-tracefree piece of the gravitational potentials. In the current context we have instead

$$c^2 R_{0j0k} = -\frac{G}{2c^4 R}\frac{\partial^2}{\partial\tau^2} S^{jk}(\tau, \boldsymbol{N}), \tag{13.87}$$

where

$$S^{jk} := \left(\delta^{jk} - N^j N^k\right) A_{\mathrm{S}} + N^j N^k A_{\mathrm{L}} + 2N^{(j} A_{\mathrm{V}}^{k)} + A_{\mathrm{TT}}^{jk}, \tag{13.88}$$

with

$$A_{\mathrm{S}} := -\frac{1}{6}(A + 2B - 3C), \tag{13.89a}$$

$$A_{\mathrm{L}} := \frac{1}{3}(A + 2B + 3C - 6D), \tag{13.89b}$$

$$A_{\mathrm{V}}^k := A_{\mathrm{T}}^k - D_{\mathrm{T}}^k. \tag{13.89c}$$

To arrive at this result we have made use of the fact that for a function f of τ and $\boldsymbol{N}$, $\partial_j f = -c^{-1} N_j \partial_\tau f + O(R^{-2})$. It can be checked that A_{S}, A_{L}, A_{V}^j, and A_{TT}^{jk} are all invariant under a gauge transformation described by Eq. (13.82). This is as it should be, because the linearized Riemann tensor is necessarily gauge invariant, and the equation of geodesic deviation describes physically measurable motions. In general relativity, the conditions $C = D$, $A + 2B = 3D$, and $D_{\mathrm{T}}^k = A_{\mathrm{T}}^k$ that arise from the harmonic-gauge condition – refer to Eq. (11.14) – imply that A_{S}, A_{V}^j, and A_{L} all vanish.

Integrating the equation of geodesic deviation to first order in the displacement yields

$$\xi^j(t) = \xi^j(0) + \frac{G}{2c^4 R} S^{jk}(\tau, \boldsymbol{N}) \xi^k(0). \tag{13.90}$$

We see that the detector's response is governed by a number of gravitational-wave modes: a scalar mode A_{S} which is transverse to the direction of propagation (but not tracefree), a longitudinal mode A_{L}, two vector modes A_{V}^j which are partly longitudinal and partly transverse, and the familiar transverse-tracefree modes A_{TT}^{jk}. In general relativity the absence of scalar, longitudinal, and vector modes implies that the response is governed entirely by the transverse-tracefree modes.

To describe the gravitational-wave modes in more concrete terms, we make use of the vector basis $(\boldsymbol{N}, \boldsymbol{\vartheta}, \boldsymbol{\varphi})$ first introduced in Sec. 11.1.7. Employing the polar angles (ϑ, φ) to describe the direction of propagation, we write

$$\boldsymbol{N} := [\sin\vartheta \cos\varphi,\ \sin\vartheta \sin\varphi,\ \cos\vartheta], \tag{13.91a}$$

$$\boldsymbol{\vartheta} := [\cos\vartheta \cos\varphi,\ \cos\vartheta \sin\varphi,\ -\sin\vartheta], \tag{13.91b}$$

$$\boldsymbol{\varphi} := [-\sin\varphi,\ \cos\varphi,\ 0], \tag{13.91c}$$

and we define the gravitational-wave polarizations

$$A_{\mathrm{V1}} := \vartheta_k A_{\mathrm{V}}^k, \tag{13.92a}$$

$$A_{\mathrm{V2}} := \varphi_k A_{\mathrm{V}}^k, \tag{13.92b}$$

$$A_{+} := \frac{1}{2}\left(\vartheta_j \vartheta_k - \varphi_j \varphi_k\right) A_{\mathrm{TT}}^{jk}, \tag{13.92c}$$

$$A_{\times} := \frac{1}{2}\left(\vartheta_j \varphi_k + \varphi_j \vartheta_k\right) A_{\mathrm{TT}}^{jk}. \tag{13.92d}$$

Making the substitutions in S^{jk}, we obtain

$$\begin{aligned} S^{jk} &= A_{\rm S}(\vartheta^j\vartheta^k + \varphi^j\varphi^k) + A_{\rm L}N^jN^k \\ &\quad + 2A_{\rm V1}N^{(j}\vartheta^{k)} + 2A_{\rm V2}N^{(j}\varphi^{k)} \\ &\quad + A_+(\vartheta^j\vartheta^k - \varphi^j\varphi^k) + A_\times(\vartheta^j\varphi^k + \varphi^j\vartheta^k)\,. \end{aligned} \tag{13.93}$$

For a wave traveling in the z-direction, S^{jk} can be displayed as the matrix

$$S^{jk} = \begin{pmatrix} A_{\rm S}+A_+ & A_\times & A_{\rm V1} \\ A_\times & A_{\rm S}-A_+ & A_{\rm V2} \\ A_{\rm V1} & A_{\rm V2} & A_{\rm L} \end{pmatrix}. \tag{13.94}$$

Box 13.3 Distortion of a ring of particles by a gravitational wave

The discussion of Sec. 11.1.8 can be generalized to the case in which a gravitational wave includes all six polarization modes. For a gravitational wave traveling in the z-direction past an initially circular ring of particles, the displacement of a given particle from the center of the ring is given by Eq. (13.90). In terms of the (x, y, z) components of the vector $\boldsymbol{\xi}$, we have

$$\begin{aligned} x(t) &= x_0 + \frac{G}{2c^4R}\Big[(A_{\rm S}+A_+)x_0 + A_\times y_0 + A_{\rm V1}z_0\Big], \\ y(t) &= y_0 + \frac{G}{2c^4R}\Big[A_\times x_0 + (A_{\rm S}-A_+)y_0 + A_{\rm V2}z_0\Big], \\ z(t) &= z_0 + \frac{G}{2c^4R}\Big[A_{\rm V1}x_0 + A_{\rm V2}y_0 + A_{\rm L}z_0\Big], \end{aligned}$$

with $\boldsymbol{\xi}(0) = [x_0, y_0, z_0]$. For pure $+$ or $\times$ modes, the patterns are as previously displayed in Eqs. (11.50). A pure S mode is also transverse, and causes a circle in the x-y plane to shrink and grow while remaining a circle. A V1 mode distorts a circle in the x-z plane into an ellipse rotated by 45 degrees, while a V2 mode does the same to a circle in the y-z plane. Finally, an L mode takes a circle in any plane parallel to the z-axis and stretches and shrinks it in the z-direction. These deformations are illustrated in Fig. 13.1.

13.4.3 Interaction with a laser interferometer

The interaction of a gravitational wave with a laser interferometer was described back in Sec. 11.5 in the context of general relativity. This discussion can easily be generalized to incorporate all six polarization modes allowed by a generic alternative theory. In Sec. 11.5 we found that the detector's response, measured by the phase difference between the light signals traveling in both arms of the interferometer, is given by $\Delta\Phi = (4\pi\nu GL_0/c^5R)S(t)$, where ν is the frequency of the laser light, L_0 is the length of the unperturbed interferometer arm, R is the distance from the source, and $S(t)$ is the detector's response function. In general relativity this was given by Eq. (11.313). In a generic alternative theory the response function is given instead by

$$S(t) = \frac{1}{2}(e_1^je_1^k - e_2^je_2^k)S^{jk}(\tau, \boldsymbol{N}), \tag{13.95}$$

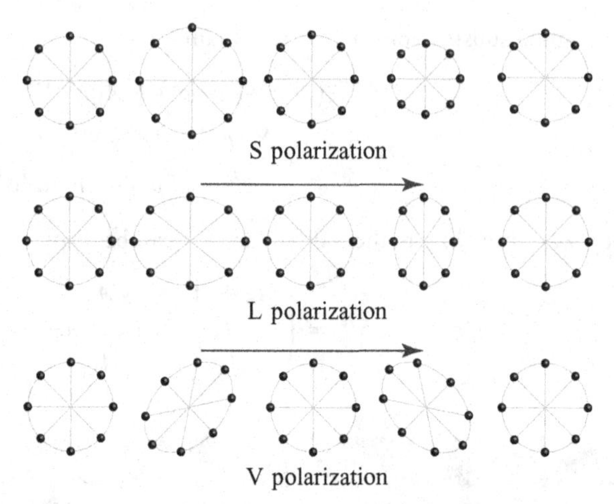

Fig. 13.1 Effect of the S, L, and V gravitational-wave polarizations on a circular ring of freely-moving particles. The wave travels in the z-direction. The effect of the $+$ and $\times$ polarizations was displayed in Fig. 11.1. For the S mode, the wave propagates into the page, and the ring is placed in the x-y plane. For the L mode, the wave propagates from left to right (in the direction of the arrow), and the ring is placed either in the x-z plane or in the y-z plane. For the V modes, the wave also propagates from left to right; in the case of the V1 mode, the ring is placed in the x-z plane, and in the case of the V2 mode, the ring is placed in the y-z plane.

where $\boldsymbol{e}_1$ and $\boldsymbol{e}_2$ are unit vectors pointing in the direction of the interferometer arms, and S^{jk} is given by Eqs. (13.88) or (13.93).

Adopting the same vectorial basis as in Sec. 11.5, we write

$$\begin{aligned} S^{jk} &= \left(e_X^j e_X^k + e_Y^j e_Y^k\right) A_{\rm S} + e_Z^j e_Z^k A_{\rm L} \\ &\quad + \left(e_X^j e_Z^k + e_Z^j e_X^k\right) A_{\rm V1} + \left(e_Y^j e_Z^k + e_Z^j e_Y^k\right) A_{\rm V2} \\ &\quad + \left(e_X^j e_X^k - e_Y^j e_Y^k\right) A_+ + \left(e_X^j e_Y^k + e_Y^j e_X^k\right) A_\times \,, \end{aligned} \tag{13.96}$$

and we insert the expressions for $\boldsymbol{e}_1$ and $\boldsymbol{e}_2$ displayed in Eq. (11.319). After some manipulations we obtain

$$S(t) = F_{\rm S} A_{\rm S} + F_{\rm L} A_{\rm L} + F_{\rm V1} A_{\rm V1} + F_{\rm V2} A_{\rm V2} + F_+ A_+ + F_\times A_\times \,, \tag{13.97}$$

where the angular pattern functions $F_A(\theta, \phi, \psi)$ are given by

$$F_{\rm S} = -\frac{1}{2} \sin^2\theta \cos 2\phi \,, \tag{13.98a}$$

$$F_{\rm L} = \frac{1}{2} \sin^2\theta \cos 2\phi \,, \tag{13.98b}$$

$$F_{\rm V1} = -\sin\theta (\cos\theta \cos 2\phi \cos\psi - \sin 2\phi \sin\psi) \,, \tag{13.98c}$$

$$F_{\rm V2} = -\sin\theta (\cos\theta \cos 2\phi \sin\psi + \sin 2\phi \cos\psi) \,, \tag{13.98d}$$

$$F_+ = \frac{1}{2}(1 + \cos^2\theta) \cos 2\phi \cos 2\psi - \cos\theta \sin 2\phi \sin 2\psi \,, \tag{13.98e}$$

$$F_\times = \frac{1}{2}(1 + \cos^2\theta) \cos 2\phi \sin 2\psi + \cos\theta \sin 2\phi \cos 2\psi \,. \tag{13.98f}$$

The clear detection of a scalar, longitudinal, or vector polarization in a gravitational-wave interferometer would provide serious evidence against general relativity.

13.4.4 Multipolar structure of gravitational waves

In Chapter 11 we learned that in general relativity, the gravitational waves emitted by a slowly-moving system are dominantly quadrupolar; the waves are generated mostly by variations in the quadrupole-moment tensor of the mass distribution, and there is no monopole or dipole radiation. As we discussed back in Sec. 7.1.4, this is because the field equations of general relativity insist that the monopole moment of the gravitational field in the wave zone be the total mass of the system, which is constant up to small changes resulting from radiative losses. The field equations also demand that the time derivative of the dipole moment be the total momentum of the system, which is also constant up to radiative losses, and can be set to zero by a suitable choice of reference frame. It is therefore the quadrupole moment that leads off in generating gravitational waves. As we have also seen, in general relativity the gravitational waves are described by two modes of polarizations, h_+ and $h_\times$, and the radiative fluxes of energy, momentum, and angular momentum can all be obtained from h_+ and $h_\times$.

No such rules apply in alternative metric theories of gravity. There is no reason to expect that a generic theory will predict the suppression of monopole and dipole radiation, and as we have seen, there is every reason to expect that the gravitational waves will display more than two polarization modes. In addition, the radiative fluxes can be a lot more complicated, involving all polarization modes and the additional fields that might be present in the theory. It is therefore difficult to make general statements about the generation of gravitational waves in alternative theories of gravity, and one is forced to explore the predictions of each theory separately. Studies carried out in the wake of the discovery of the Hulse–Taylor binary pulsar in 1974 revealed the unusual fact that a number of otherwise respectable theories actually predicted the emission of *negative energy*. Once the orbital period of the binary pulsar was shown to decrease (as opposed to increase) in response to the emission of gravitational waves, these theories found themselves in the gravitational dust-bin. But one class of alternative theories has stubbornly refused to die, in spite of all the strong empirical evidence in support of general relativity. This is the class of scalar–tensor theories, to which we turn next.

13.5 Scalar–tensor gravity

One of the simplest ways to formulate an alternative metric theory of gravity is to postulate that the gravitational field is represented by a scalar ϕ in addition to the metric $g_{\alpha\beta}$. As we have emphasized previously in Sec. 13.1, the matter fields are still assumed to respond only to the metric, and there is no direct interaction with the scalar field. But the field equations relating the matter distribution to the gravitational field will involve this additional degree of freedom. The first scalar–tensor theory of gravitation was formulated by Brans and Dicke, building on previous work by Fierz and Jordan. The theory introduced here is a generalization of the original Brans–Dicke theory; it is in fact the most general theory involving a scalar field in addition to the metric tensor. We shall study a few aspects of this theory, including its predictions regarding weak-field and slow-motion situations, and the generation of gravitational radiation.

13.5.1 Field equations

The field equations of scalar–tensor gravity are best obtained by formulating an action principle, much as we did back in Sec. 5.4 in the case of general relativity. The complete action functional for the gravitating system is

$$S = S_{\rm grav} + S_{\rm matter} , \tag{13.99}$$

with a gravitational action given by

$$S_{\rm grav} = \frac{c^3}{16\pi G_0} \int \biggl[\phi R - \frac{\omega(\phi)}{\phi} g^{\alpha\beta} \partial_\alpha \phi \, \partial_\beta \phi - U(\phi) \biggr] \sqrt{-g}\, d^4x , \tag{13.100}$$

where G_0 is a "bare" gravitational constant that will later be related to the locally measured constant G, and ω and U are arbitrary functions of the scalar field. Note that the coupling between the scalar field and the metric appears in two guises. First, there is a factor of ϕ in front of the Ricci scalar, and this can be thought of as a local redefinition of the gravitational constant, $G_0 \to G_0/\phi$, which may now depend on position in spacetime. Second, the metric appears in the kinetic-energy term for the scalar field, proportional to $\partial_\alpha \phi \partial_\beta \phi$. The matter action is given by

$$S_{\rm matter} = \int \mathscr{L}(\mathsf{m}, g_{\alpha\beta}) \sqrt{-g}\, d^4x , \tag{13.101}$$

in which $\mathscr{L}$ is a Lagrangian density involving the matter variables m and the metric $g_{\alpha\beta}$. Note that ϕ does not appear in the matter action; this is the statement that the scalar field does not couple directly to the matter variables.

It is a straightforward exercise to vary S with respect to the field variables ϕ and $g_{\alpha\beta}$ to obtain the field equations of scalar–tensor gravity. The resulting equations, however, are rather complicated and not the most useful for practical calculations. For an optimal formulation of the field equations we introduce an auxiliary metric $\tilde{g}_{\alpha\beta}$ related to the physical metric $g_{\alpha\beta}$ by the *conformal transformation*

$$g_{\alpha\beta} := (\phi_0/\phi)\, \tilde{g}_{\alpha\beta} , \tag{13.102}$$

in which ϕ_0 is an arbitrary constant that will be selected in later applications. The transformation to the new metric represents a rescaling of spatial and temporal intervals by a factor $(\phi/\phi_0)^{1/2}$ that depends on position in spacetime, but we attach no physical significance to $\tilde{g}_{\alpha\beta}$; it is merely a convenient auxiliary quantity that will be involved in computations, before all final results are expressed in terms of the physical metric $g_{\alpha\beta}$. The conformal transformation implies that $g^{\alpha\beta} = (\phi/\phi_0)\tilde{g}^{\alpha\beta}$ and $\sqrt{-g} = (\phi_0/\phi)^2\sqrt{-\tilde{g}}$, and it can be shown that the Ricci scalar transforms according to

$$R = (\phi/\phi_0)^2 \bigl(\tilde{R} + 6\tilde{\nabla}_\alpha \tilde{B}^\alpha - 6\tilde{B}_\alpha \tilde{B}^\alpha\bigr), \tag{13.103}$$

in which $\tilde{R}$ and $\tilde{\nabla}_\alpha$ are the Ricci scalar and covariant derivative defined in terms of $\tilde{g}_{\alpha\beta}$, respectively, $\tilde{B}_\alpha := \frac{1}{2}\partial_\alpha \ln\phi$, and $\tilde{B}^\alpha := \tilde{g}^{\alpha\beta}\tilde{B}_\beta$.

The gravitational action becomes

$$S_{\rm grav} = \frac{c^3}{16\pi\tilde{G}} \int \left[\tilde{R} - \frac{2\omega(\phi)+3}{2\phi^2}\,\tilde{g}^{\alpha\beta}\,\partial_\alpha\phi\,\partial_\beta\phi - V(\phi)\right]\sqrt{-\tilde{g}}\,d^4x\,, \tag{13.104}$$

in which $\tilde{G} := G_0/\phi_0$ and $V(\phi) := \phi_0 U(\phi)/\phi^2$. To arrive at this result we have eliminated the term $6\tilde{\nabla}_\alpha\tilde{B}^\alpha$ from the action, because by the four-dimensional version of Gauss's theorem, it can be expressed as an irrelevant surface integral. Note that ϕ no longer appears in front of the Ricci scalar; this has the virtue of simplifying the field equations, and the conformal transformation was introduced for this specific purpose. The matter action becomes

$$S_{\rm matter} = \int (\phi_0/\phi)^2 \mathscr{L}(\mathsf{m}, \phi, \tilde{g}_{\alpha\beta})\sqrt{-\tilde{g}}\,d^4x\,. \tag{13.105}$$

Note that the scalar field now makes an appearance in the matter action; but the coupling between m and $\tilde{g}_{\alpha\beta}$ in $\mathscr{L}$ is still required to occur through the physical metric $(\phi_0/\phi)\tilde{g}_{\alpha\beta}$.

Variation of S with respect to the auxiliary metric $\tilde{g}_{\alpha\beta}$ gives rise to the tensorial field equation

$$\tilde{G}_{\alpha\beta} - \frac{1}{2}\tilde{\Theta}_{\alpha\beta} = \frac{8\pi\tilde{G}}{c^4}\tilde{T}_{\alpha\beta}\,, \tag{13.106}$$

in which $\tilde{G}_{\alpha\beta}$ is the Einstein tensor associated with the auxiliary metric,

$$\tilde{\Theta}_{\alpha\beta} := \frac{2\omega+3}{\phi^2}\left(\partial_\alpha\phi\,\partial_\beta\phi - \frac{1}{2}\tilde{g}_{\alpha\beta}\tilde{g}^{\mu\nu}\,\partial_\mu\phi\,\partial_\nu\phi\right) - V(\phi)\tilde{g}_{\alpha\beta}\,, \tag{13.107}$$

and $\tilde{T}_{\alpha\beta} := (\phi_0/\phi)T_{\alpha\beta}$ is an auxiliary energy-momentum tensor obtained from the variation of $S_{\rm matter}$; $T_{\alpha\beta}$ is the physical energy-momentum tensor. Variation of S with respect to the scalar field ϕ yields the scalar field equation

$$\tilde{g}^{\alpha\beta}\tilde{\nabla}_\alpha\tilde{\nabla}_\beta\phi + \tilde{F} = \frac{8\pi\tilde{G}}{c^4}\frac{\phi}{2\omega+3}\tilde{T}\,, \tag{13.108}$$

in which

$$\tilde{F} := \frac{1}{2}\frac{d}{d\phi}\left[\ln\left(\frac{2\omega+3}{\phi^2}\right)\right]\tilde{g}^{\alpha\beta}\,\partial_\alpha\phi\,\partial_\beta\phi - \frac{\phi^2}{2\omega+3}\frac{dV}{d\phi}\,, \tag{13.109}$$

and $\tilde{T} := \tilde{g}^{\alpha\beta}\tilde{T}_{\alpha\beta}$. These are the field equations of scalar–tensor gravity.

13.5.2 Post-Minkowskian formulation

We wish to explore the consequences of scalar–tensor gravity in the post-Newtonian regime, both in the near zone, where they will be related to the PPN framework of Sec. 13.2, and in the far-away wave zone, where they will be related to the general discussion of gravitational waves in Sec. 13.4. To perform the required calculations it is advantageous to rely on techniques that proved so powerful in the context of general relativity. We shall therefore subject the field equations of scalar–tensor gravity to a post-Minkowskian reformulation, relying on lessons learned in Chapter 6. We shall next carry out a post-Newtonian expansion of these equations, relying on techniques introduced in Chapter 7.

We adapt the post-Minkowskian techniques of Chapter 6 to the current situation by working in terms of the auxiliary metric $\tilde{g}_{\alpha\beta}$ instead of the physical metric $g_{\alpha\beta}$. We introduce the gothic inverse metric

$$\tilde{\mathfrak{g}}^{\alpha\beta} := \sqrt{-\tilde{g}}\tilde{g}^{\alpha\beta}\,, \tag{13.110}$$

the tensor density

$$\tilde{H}^{\alpha\mu\beta\nu} := \tilde{\mathfrak{g}}^{\alpha\beta}\tilde{\mathfrak{g}}^{\mu\nu} - \tilde{\mathfrak{g}}^{\alpha\nu}\tilde{\mathfrak{g}}^{\beta\mu}\,, \tag{13.111}$$

and we rely on the identity of Eq. (6.4),

$$\partial_{\mu\nu}\tilde{H}^{\alpha\mu\beta\nu} = 2(-\tilde{g})\tilde{G}^{\alpha\beta} + \frac{16\pi\tilde{G}}{c^4}(-\tilde{g})\tilde{t}^{\alpha\beta}_{\rm LL}\,, \tag{13.112}$$

in which $\tilde{t}^{\alpha\beta}_{\rm LL}$ is a Landau–Lifshitz pseudotensor defined as in Eq. (6.5), but in terms of the auxiliary metric. Making the substitutions in Eq. (13.106) produces

$$\partial_{\mu\nu}\tilde{H}^{\alpha\mu\beta\nu} = \frac{16\pi\tilde{G}}{c^4}(-\tilde{g})\Big(\tilde{T}^{\alpha\beta} + \tilde{t}^{\alpha\beta}_{\phi} + \tilde{t}^{\alpha\beta}_{\rm LL}\Big)\,, \tag{13.113}$$

where

$$\tilde{t}^{\alpha\beta}_{\phi} := \frac{c^4}{16\pi\tilde{G}}\tilde{\Theta}^{\alpha\beta}\,. \tag{13.114}$$

At this stage we introduce the potentials

$$\tilde{h}^{\alpha\beta} := \eta^{\alpha\beta} - \tilde{\mathfrak{g}}^{\alpha\beta}\,, \tag{13.115}$$

impose the conformal harmonic gauge condition

$$\partial_{\beta}\tilde{h}^{\alpha\beta} = 0\,, \tag{13.116}$$

and make use of the identity

$$\partial_{\mu\nu}\tilde{H}^{\alpha\mu\beta\nu} = -\Box\tilde{h}^{\alpha\beta} - \frac{16\pi\tilde{G}}{c^4}(-\tilde{g})\tilde{t}^{\alpha\beta}_{\rm H}\,, \tag{13.117}$$

where $\tilde{t}^{\alpha\beta}_{\rm H}$ is a harmonic-gauge pseudotensor defined as in Eq. (6.53), but in terms of the potentials $\tilde{h}^{\alpha\beta}$. All this produces the wave equation

$$\Box\tilde{h}^{\alpha\beta} = -\frac{16\pi\tilde{G}}{c^4}\tilde{\tau}^{\alpha\beta}\,, \tag{13.118}$$

in which

$$\tilde{\tau}^{\alpha\beta} := (-\tilde{g})\Big(\tilde{T}^{\alpha\beta} + \tilde{t}^{\alpha\beta}_{\phi} + \tilde{t}^{\alpha\beta}_{\rm LL} + \tilde{t}^{\alpha\beta}_{\rm H}\Big) \tag{13.119}$$

plays the role of an effective energy-momentum pseudotensor. The wave equation can be compared with Eq. (6.51); the only differences concern the definition of the potentials, which involves the auxiliary metric instead of the physical metric, and the additional contribution to $\tilde{\tau}^{\alpha\beta}$ that comes from the scalar field. Note that as in Eq. (6.54), the effective energy-momentum pseudotensor necessarily satisfies the conservation identity

$$\partial_\beta \tilde{\tau}^{\alpha\beta} = 0 . \tag{13.120}$$

This is equivalent to the conservation equation $\nabla_\beta T^{\alpha\beta} = 0$ for the physical energy-momentum tensor.

The scalar field equation (13.108) can also be expressed in the form of a wave equation in flat spacetime. For this we rely on the identity

$$\tilde{g}^{\alpha\beta}\tilde{\nabla}_\alpha \tilde{\nabla}_\beta \phi = \frac{1}{\sqrt{-\tilde{g}}}\partial_\alpha\left(\tilde{\mathfrak{g}}^{\alpha\beta}\partial_\beta\phi\right) = \frac{1}{\sqrt{-\tilde{g}}}\left(\Box\phi - \tilde{h}^{\alpha\beta}\partial_{\alpha\beta}\phi\right), \tag{13.121}$$

where we make use of Eq. (13.116) in the second step. Making the substitution in Eq. (13.108), we arrive at

$$\Box\phi = -\frac{8\pi\tilde{G}}{c^4}\tau_{\rm s} , \tag{13.122}$$

in which

$$\begin{aligned}\tau_{\rm s} = {} & -\sqrt{-\tilde{g}}\frac{\phi}{2\omega+3}\tilde{T} + \frac{c^4}{16\pi\tilde{G}}\frac{d}{d\phi}\left[\ln\left(\frac{2\omega+3}{\phi^2}\right)\right]\left(\eta^{\alpha\beta} - \tilde{h}^{\alpha\beta}\right)\partial_\alpha\phi\,\partial_\beta\phi \\ & - \frac{c^4}{8\pi\tilde{G}}\left(\tilde{h}^{\alpha\beta}\partial_{\alpha\beta}\phi + \sqrt{-\tilde{g}}\frac{\phi^2}{2\omega+3}\frac{dV}{d\phi}\right)\end{aligned} \tag{13.123}$$

is an effective source for the scalar field.

13.5.3 Slow-motion condition

To proceed with the integration of Eqs. (13.118) and (13.123), we set $V(\phi) = 0$ for simplicity, and assume that the matter distribution is subjected to a slow-motion condition, so that $v_c \ll c$, with v_c denoting a characteristic velocity of the matter variables. As we discussed back in Sec. 7.1, this assumption implies the existence of a hierarchy between the components of the energy-momentum tensor, so that $T^{00} = O(c^2)$, $T^{0j} = O(c)$, and $T^{jk} = O(1)$. The hierarchy is inherited by the gravitational potentials, and to reflect this we introduce the notation

$$\tilde{h}^{00} := \frac{4}{c^2}\tilde{V} , \tag{13.124a}$$

$$\tilde{h}^{0j} := \frac{4}{c^3}\tilde{V}^j , \tag{13.124b}$$

$$\tilde{h}^{jk} := \frac{4}{c^4}\tilde{W}^{jk} , \tag{13.124c}$$

in which $\tilde{V}$, $\tilde{V}^j$, and $\tilde{W}^{jk}$ are assumed to scale as $O(1)$. We also introduce the notation

$$\phi = \phi_0\left(1 + \frac{2}{c^2}f\right) \tag{13.125}$$

for the scalar field, which reflects the expectation that variations in ϕ will scale as $O(c^{-2})$. Taking f to approach zero as $r \to \infty$, the expression also assigns a meaning to the constant ϕ_0 introduced in Eq. (13.102): it represents the asymptotic value $\phi(\infty)$ of the scalar field far away from the matter distribution. In principle this could depend on time, because $\phi(\infty)$ is determined by the conditions that prevail in the asymptotic regions of the spacetime. For example, the asymptotic behavior of the scalar field could be tied to the cosmological expansion, which would dictate that $\phi(\infty)$ should indeed depend on time. But this dependence can be ignored whenever its characteristic time scale is very long compared with the dynamical time scale associated with the system itself. Under such circumstances, we can safely take $\phi(\infty)$ to be a constant, and associate it with ϕ_0.

The relations displayed in Eq. (7.24) allow us to express the auxiliary metric in terms of the potentials $\tilde{V}$, $\tilde{V}^j$, and $\tilde{W}^{jk}$. We have

$$\tilde{g}_{00} = -1 + \frac{2}{c^2}\tilde{V} + \frac{2}{c^4}(\tilde{W} - 3\tilde{V}^2) + O(c^{-6}), \tag{13.126a}$$

$$\tilde{g}_{0j} = -\frac{4}{c^3}\tilde{V}^j + O(c^{-5}), \tag{13.126b}$$

$$\tilde{g}_{jk} = \delta_{jk}\left(1 + \frac{2}{c^2}\tilde{V}\right) + O(c^{-4}), \tag{13.126c}$$

$$(-\tilde{g}) = 1 + \frac{4}{c^2}\tilde{V} + O(c^{-4}), \tag{13.126d}$$

in which $\tilde{W} := \delta_{jk}\tilde{W}^{jk}$. From Eqs. (13.102) and (13.125) we can then obtain the components of the physical metric:

$$g_{00} = -1 + \frac{2}{c^2}(\tilde{V} + f) + \frac{2}{c^4}(\tilde{W} - 3\tilde{V}^2 - 2f\tilde{V} - 2f^2) + O(c^{-6}), \tag{13.127a}$$

$$g_{0j} = -\frac{4}{c^3}\tilde{V}^j + O(c^{-5}), \tag{13.127b}$$

$$g_{jk} = \delta_{jk}\left[1 + \frac{2}{c^2}(\tilde{V} - f)\right] + O(c^{-4}). \tag{13.127c}$$

The next order of business is to compute the effective energy-momentum pseudotensor $\tilde{\tau}^{\alpha\beta}$. We begin with an examination of the matter contribution, assuming that the matter consists of a perfect fluid. The physical energy-momentum tensor $T^{\alpha\beta}$ is expressed as in Sec. 7.1.1, in terms of the metric $g_{\alpha\beta}$ and the matter variables $\mathsf{m} = \{\rho^*, p, \Pi, \boldsymbol{v}\}$ – refer to Eq. (7.1) and the following equations. The field equations, however, involve the auxiliary energy-momentum tensor

$$\tilde{T}^{\alpha\beta} = (\phi_0/\phi)^3\, T^{\alpha\beta}. \tag{13.128}$$

Using the techniques developed in Secs. 7.2.1 and 7.3.1, and incorporating the metric of Eq. (13.127), we quickly obtain

$$c^{-2}(-\tilde{g})\tilde{T}^{00} = \rho^*\left[1 + \frac{1}{c^2}\left(\frac{1}{2}v^2 + 3\tilde{V} + \Pi - f\right) + O(c^{-4})\right], \tag{13.129a}$$

$$c^{-1}(-\tilde{g})\tilde{T}^{0j} = \rho^* v^j + O(c^{-2}), \tag{13.129b}$$

$$(-\tilde{g})\tilde{T}^{jk} = \rho^* v^j v^k + p\,\delta^{jk} + O(c^{-2}), \tag{13.129c}$$

for the components of the energy-momentum tensor, expressed to a sufficient degree of accuracy as post-Newtonian expansions.

The contribution from the scalar field can be obtained from Eqs. (13.107), (13.114), (13.125), and (13.126). The calculation relies on an expansion of the arbitrary function $\omega(\phi)$ in powers of c^{-2}, which can be expressed as

$$\omega(\phi) = \omega_0 + \frac{2}{c^2}\phi_0\omega_0' f + O(c^{-4}), \tag{13.130}$$

in which

$$\omega_0 := \omega(\phi_0), \qquad \omega_0' := \frac{d\omega}{d\phi}\bigg|_{\phi_0}. \tag{13.131}$$

A straightforward computation returns

$$c^{-2}(-\tilde{g})t_\phi^{00} = \frac{2\omega_0+3}{8\pi\tilde{G}c^2}\,\partial_n f\partial_n f + O(c^{-4}), \tag{13.132a}$$

$$c^{-1}(-\tilde{g})t_\phi^{0j} = O(c^{-2}), \tag{13.132b}$$

$$(-\tilde{g})t_\phi^{jk} = \frac{2\omega_0+3}{4\pi\tilde{G}c^2}\biggl(\partial_j f\partial_k f - \frac{1}{2}\delta_{jk}\partial_n f\partial_n f\biggr) + O(c^{-2}). \tag{13.132c}$$

The Landau–Lifshitz contribution to the effective energy-momentum pseudotensor can be obtained directly from Eq. (7.48b) and the listing of potentials provided in Eq. (13.124). We get

$$c^{-2}(-\tilde{g})t_{\rm LL}^{00} = -\frac{7}{8\pi\tilde{G}c^2}\partial_n\tilde{V}\partial_n\tilde{V} + O(c^{-4}), \tag{13.133a}$$

$$c^{-1}(-\tilde{g})t_{\rm LL}^{0j} = O(c^{-2}), \tag{13.133b}$$

$$(-\tilde{g})t_{\rm LL}^{jk} = \frac{1}{4\pi\tilde{G}}\biggl(\partial_j\tilde{V}\partial_k\tilde{V} - \frac{1}{2}\delta_{jk}\partial_n\tilde{V}\partial_n\tilde{V}\biggr) + O(c^{-2}). \tag{13.133c}$$

Similarly, the harmonic-gauge contribution can be obtained directly from Eq. (7.52), and we get

$$c^{-2}(-\tilde{g})t_{\rm H}^{00} = O(c^{-4}), \tag{13.134a}$$

$$c^{-1}(-\tilde{g})t_{\rm H}^{0j} = O(c^{-4}), \tag{13.134b}$$

$$(-\tilde{g})t_{\rm H}^{jk} = O(c^{-2}). \tag{13.134c}$$

Collecting results, we arrive at complete expressions for the components of the effective energy-momentum pseudotensor,

$$c^{-2}\tilde{\tau}^{00} = \rho^*\biggl[1 + \frac{1}{c^2}\biggl(\frac{1}{2}v^2 + 3\tilde{V} + \Pi - f\biggr)\biggr] + \frac{1}{4\pi\tilde{G}c^2}\biggl[-\frac{7}{2}\partial_n\tilde{V}\partial_n\tilde{V} + \frac{1}{2}(2\omega_0+3)\partial_n f\partial_n f\biggr] + O(c^{-4}), \tag{13.135a}$$

$$c^{-1}\tilde{\tau}^{0j} = \rho^* v^j + O(c^{-2}), \tag{13.135b}$$

$$\tilde{\tau}^{jk} = \rho^* v^j v^k + p\,\delta^{jk} + \frac{1}{4\pi\tilde{G}}\biggl[\partial_j\tilde{V}\partial_k\tilde{V} - \frac{1}{2}\delta_{jk}\partial_n\tilde{V}\partial_n\tilde{V} + (2\omega_0+3)\biggl(\partial_j f\partial_k f - \frac{1}{2}\delta_{jk}\partial_n f\partial_n f\biggr)\biggr] + O(c^{-2}). \tag{13.135c}$$

The same expansion techniques can be used to compute the scalar source of Eq. (13.123). Here we find that only the first two terms contribute at the required order, and we get

$$c^{-2}\tau_s = \frac{\phi_0 \rho^*}{2\omega_0 + 3}\left[1 - \frac{1}{c^2}\left(\frac{1}{2}v^2 + \tilde{V} - \Pi + 3p/\rho^*\right) + \frac{1}{c^2}\left(1 - \frac{4\phi_0\omega_0'}{2\omega_0 + 3}\right)f\right]$$
$$- \frac{\phi_0}{2\pi \tilde{G} c^2}\left(1 - \frac{\phi_0\omega_0'}{2\omega_0 + 3}\right)\partial_n f \partial_n f + O(c^{-4}). \tag{13.136}$$

The source terms for the wave equations have now been computed to the required degree of accuracy. The next task is the integration of these equations.

13.5.4 Near-zone solution: PPN metric

We employ the techniques developed in Sec. 6.3 to integrate the wave equations (13.118) and (13.122). Each equation is of the form $\Box\psi = -4\pi\mu$, and we recall that $\psi(t, \boldsymbol{x})$ can be expressed as an integral over the past light cone of the field point $(t, \boldsymbol{x})$. The domain of integration $\mathscr{C}$ is decomposed as $\mathscr{C} = \mathscr{N} + \mathscr{W}$, into a near-zone domain $\mathscr{N}$ and a wave-zone domain $\mathscr{W}$, and ψ is similarly decomposed as $\psi = \psi_{\mathscr{N}} + \psi_{\mathscr{W}}$. With $(t, \boldsymbol{x})$ situated in the near zone, an expression for $\psi_{\mathscr{N}}$ was displayed in Eq. (6.92), and the discussion of Sec. 7.3.4 indicates that $\psi_{\mathscr{W}}$ makes no contribution at 1PN order, the level of accuracy maintained in our computations. With all this in mind, the near-zone solution to the wave equation can be expressed as

$$\psi(t, \boldsymbol{x}) = \int_{\mathscr{M}} \frac{\mu(t, \boldsymbol{x}')}{|\boldsymbol{x} - \boldsymbol{x}'|}\, d^3x' - \frac{1}{c}\frac{d}{dt}\int_{\mathscr{M}} \mu(t, \boldsymbol{x}')\, d^3x'$$
$$+ \frac{1}{2c^2}\frac{\partial^2}{\partial t^2}\int_{\mathscr{M}} \mu(t, \boldsymbol{x}')|\boldsymbol{x} - \boldsymbol{x}'|\, d^3x' + O(c^{-3}), \tag{13.137}$$

where the domain of integration $\mathscr{M}$ is described by $|\boldsymbol{x}'| < \mathcal{R}$, with $\mathcal{R}$ denoting the arbitrary radius of the boundary between the near and wave zones. To evaluate the right-hand side of Eq. (13.137) we require an expression for $\mu(t, \boldsymbol{x}')$ in the near zone, and such expressions for $\tilde{\tau}^{\alpha\beta}$ and τ_s were obtained in the preceding subsection.

Before we proceed we introduce the notation

$$\zeta := \frac{1}{2\omega_0 + 4}, \qquad \lambda := \frac{\phi_0\omega_0'}{(2\omega_0 + 3)(2\omega_0 + 4)}. \tag{13.138}$$

These new quantities act as substitutes for ω_0 and ω_0', as defined by Eq. (13.131). In terms of ζ and λ we have that $2\omega_0 + 3 = (1 - \zeta)/\zeta$ and $\phi_0\omega_0' = \lambda(1 - \zeta)/\zeta^2$.

Inspection of Eqs. (13.135) and (13.136) reveals that the source functions $\tilde{\tau}^{\alpha\beta}$ and τ_s involve the variables $\tilde{V}$ and f that appear on the left-hand side of the wave equations. This situation is familiar from the post-Minkowskian formulation of general relativity reviewed in Chapter 6, and to handle it we rely on the iterative strategy described in Sec. 6.2.3. To obtain $\tilde{V}$ and f to leading order in a post-Newtonian expansion we set $c^{-2}\tilde{\tau}^{00} = \rho^* + O(c^{-2})$ and $c^{-2}\tau_s = \zeta\phi_0\rho^*/(1 - \zeta) + O(c^{-2})$, insert Eqs. (13.124) and (13.125) within Eqs. (13.118) and (13.122), and get the solutions from Eq. (13.137). We find

$$\tilde{V} = \tilde{G} \int \frac{\rho^*(t, \boldsymbol{x}')}{|\boldsymbol{x} - \boldsymbol{x}'|} d^3x' + O(c^{-2}), \tag{13.139a}$$

$$f = \frac{\zeta \tilde{G}}{1 - \zeta} \int \frac{\rho^*(t, \boldsymbol{x}')}{|\boldsymbol{x} - \boldsymbol{x}'|} d^3x' + O(c^{-2}), \tag{13.139b}$$

and we note that $\tilde{V}$ and f bear a striking resemblance to the Newtonian potential

$$U = G \int \frac{\rho^*(t, \boldsymbol{x}')}{|\boldsymbol{x} - \boldsymbol{x}'|} d^3x', \tag{13.140}$$

which is defined in terms of the physically measured gravitational constant G. The Newtonian potential is formally defined in terms of the physical metric by $g_{00} = -1 + 2U/c^2 + O(c^{-4})$, and the precise relationship with $\tilde{V}$ and f is given by Eq. (13.127), which reveals that

$$U = \tilde{V} + f = \frac{\tilde{G}}{1 - \zeta} \int \frac{\rho^*(t, \boldsymbol{x}')}{|\boldsymbol{x} - \boldsymbol{x}'|} d^3x'. \tag{13.141}$$

Comparison with Eq. (13.140) gives us

$$G = \frac{\tilde{G}}{1 - \zeta} = \frac{2\omega_0 + 4}{2\omega_0 + 3} \tilde{G}, \tag{13.142}$$

the relationship between the theoretical parameter $\tilde{G} = G_0/\phi_0$ and the physically measured gravitational constant.

We next proceed with the second iteration of the wave equations. We insert $\tilde{G} = (1 - \zeta)G$, $\tilde{V} = (1 - \zeta)U + O(c^{-2})$, and $f = \zeta U + O(c^{-2})$ within Eqs. (13.135) and (13.136), and obtain new expressions for the source functions:

$$c^{-2}\tilde{\tau}^{00} = \rho^* \left\{ 1 + \frac{1}{c^2} \left[\frac{1}{2} v^2 + (3 - 4\zeta)U + \Pi \right] \right\} - \frac{7 - 8\zeta}{8\pi G c^2} \partial_k U \partial_k U + O(c^{-4}), \tag{13.143a}$$

$$c^{-1}\tilde{\tau}^{0j} = \rho^* v^j + O(c^{-2}), \tag{13.143b}$$

$$\tilde{\tau}^{kk} = \rho^* v^2 + 3p - \frac{1}{8\pi G} \partial_k U \partial_k U + O(c^{-2}), \tag{13.143c}$$

$$c^{-2}\tau_s = \frac{\zeta \phi_0}{1 - \zeta} \left[\rho^* \left\{ 1 - \frac{1}{c^2} \left[\frac{1}{2} v^2 + (1 - 2\zeta + 4\lambda)U - \Pi + 3p/\rho^* \right] \right\} - \frac{\zeta - \lambda}{2\pi G c^2} \partial_k U \partial_k U \right] + O(c^{-4}); \tag{13.143d}$$

we shall not need a complete expression for $\tilde{\tau}^{jk}$, because our main goal is to obtain the near-zone metric, which involves only $\tilde{W} := \delta_{jk} \tilde{W}^{jk}$. The source functions can also be expressed as

$$c^{-2}\tilde{\tau}^{00} = \rho^*\left\{1 + \frac{1}{c^2}\left[\frac{1}{2}v^2 - \frac{1}{2}U + \Pi\right]\right\} - \frac{7-8\zeta}{16\pi G c^2}\nabla^2 U^2 + O(c^{-4}), \tag{13.144a}$$

$$c^{-1}\tilde{\tau}^{0j} = \rho^* v^j + O(c^{-2}), \tag{13.144b}$$

$$\tilde{\tau}^{kk} = \rho^*\left(v^2 - \frac{1}{2}U\right) + 3p - \frac{1}{16\pi G}\nabla^2 U^2 + O(c^{-2}), \tag{13.144c}$$

$$\begin{aligned} c^{-2}\tau_s = \frac{\zeta\phi_0}{1-\zeta}\Bigg[&\rho^*\left\{1 - \frac{1}{c^2}\left[\frac{1}{2}v^2 + (1+2\lambda)U - \Pi + 3p/\rho^*\right]\right\} \\ &- \frac{\zeta-\lambda}{4\pi G c^2}\nabla^2 U^2\Bigg] + O(c^{-4}), \end{aligned} \tag{13.144d}$$

by invoking the identity $\partial_k U \partial_k U = \frac{1}{2}\nabla^2 U^2 + 4\pi G \rho^* U$ first encountered in Eq. (7.70).

Integration of the wave equations is now straightforward. Once more we insert Eqs. (13.124) and (13.125) within Eqs. (13.118) and (13.122), and get the solutions from Eq. (13.137). Each Poisson integral involving the fluid variables gives rise to one of the potentials listed in Box 13.1, and the Poisson integral involving $\nabla^2 U^2$ is evaluated as in Eq. (7.75). We arrive at

$$\begin{aligned} \tilde{V} = (1-\zeta)U + \frac{1-\zeta}{c^2}&\left[\frac{1}{2}\Phi_1 - \frac{1}{2}\Phi_2 + \Phi_3 + \frac{1}{4}(7-8\zeta)U^2 + \frac{1}{2}\partial_{tt}X\right] \\ &+ O(c^{-3}), \end{aligned} \tag{13.145a}$$

$$\tilde{V}^j = (1-\zeta)U^j + O(c^{-2}), \tag{13.145b}$$

$$\tilde{W} = (1-\zeta)\left[\Phi_1 - \frac{1}{2}\Phi_2 + 3\Phi_4 + \frac{1}{4}U^2\right] + O(c^{-1}), \tag{13.145c}$$

$$\begin{aligned} f = \zeta U + \frac{\zeta}{c^2}&\left[-\frac{1}{2}\Phi_1 - (1+2\lambda)\Phi_2 + \Phi_3 - 3\Phi_4 + (\zeta-\lambda)U^2 + \frac{1}{2}\partial_{tt}X\right] \\ &+ O(c^{-3}). \end{aligned} \tag{13.145d}$$

Making the substitutions within Eq. (13.127) produces

$$g_{00} = -1 + \frac{2}{c^2}U + \frac{2}{c^4}\left[\psi - (1+\zeta\lambda)U^2 + \frac{1}{2}\partial_{tt}X\right] + O(c^{-5}), \tag{13.146a}$$

$$g_{0j} = -\frac{4}{c^3}(1-\zeta)U^j + O(c^{-5}), \tag{13.146b}$$

$$g_{jk} = \delta_{jk}\left[1 + \frac{2}{c^2}(1-2\zeta)U\right] + O(c^{-3}), \tag{13.146c}$$

in which

$$\psi := \frac{1}{2}(3-4\zeta)\Phi_1 - (1+2\zeta\lambda)\Phi_2 + \Phi_3 + 3(1-2\zeta)\Phi_4. \tag{13.147}$$

The physical metric is presented in the conformal harmonic gauge introduced in Eq. (13.116). To bring it to the standard post-Newtonian gauge of Box 13.1 we implement the coordinate transformation $t = \bar{t} + \frac{1}{2}c^{-4}\partial_{\bar{t}}X + O(c^{-6})$ and $x^j = \bar{x}^j$, which was first encountered in Sec. 8.3.7. When we transform the metric to the new coordinates $(\bar{t}, \bar{x}^j)$ and drop the overbars on the new variables, we obtain

$$g_{00} = -1 + \frac{2}{c^2}U + \frac{2}{c^4}\Big[\psi - (1+\zeta\lambda)U^2\Big] + O(c^{-5}), \tag{13.148a}$$

$$g_{0j} = -\frac{4}{c^3}(1-\zeta)U^j - \frac{1}{2c^3}\partial_{tj}X + O(c^{-5}), \tag{13.148b}$$

$$g_{jk} = \delta_{jk}\bigg[1 + \frac{2}{c^2}(1-2\zeta)U\bigg] + O(c^{-3}). \tag{13.148c}$$

This is the near-zone metric produced by a generic scalar–tensor theory of gravity. It is parameterized by two numbers, λ and ζ, which are related by Eqs. (13.131) and (13.138) to the theory's coupling function $\omega(\phi)$.

The metric can be compared with the PPN metric displayed in Box 13.1. This reveals that the PPN parameters of scalar–tensor gravity are given by

$$\beta = 1 + \zeta\lambda = 1 + \frac{\phi_0\omega_0'}{(2\omega_0+3)(2\omega_0+4)^2} \tag{13.149}$$

and

$$\gamma = 1 - 2\zeta = \frac{\omega_0+1}{\omega_0+2}, \tag{13.150}$$

together with $\xi = \alpha_n = \zeta_n = 0$. The vanishing of the preferred-frame parameters α_n should not come as a surprise, because a scalar field is invariant under Lorentz transformations, and there is no way for a preferred frame to be selected by the theory. The vanishing of ζ_n is a consequence of the fact that scalar–tensor gravity is based on the action principle of Eq. (13.100), which necessarily makes it a conservative theory. There is no fundamental reason why the Whitehead parameter ξ should vanish in scalar–tensor gravity; this is probably due to the simplicity of the theory. The best empirical constraint on the parameters of scalar–tensor gravity comes from the experimental bound on γ provided by the Cassini tracking measurement of the Shapiro time delay. From $|\gamma - 1| = |2\zeta| < 2.3 \times 10^{-5}$ we can conclude that $\omega_0 > 40\,000$.

Box 13.4 Nordtvedt effect and the variation of $G_{\rm eff}$

Two important features of the post-Newtonian limit of scalar–tensor gravity should be emphasized, both representing violations of the strong equivalence principle. The first is that the theory predicts a non-vanishing Nordtvedt effect; from Eq. (13.74) we see that the Nordtvedt parameter is given by

$$\eta_N = 2\zeta(1+2\lambda) = \frac{1}{\omega_0+2}\bigg[1 + \frac{\phi_0\omega_0'}{(\omega_0+2)(2\omega_0+3)}\bigg].$$

The second is that the effective, or locally measured, gravitational constant depends on the presence of nearby matter; from Eq. (13.36) we find that

$$G_{\rm eff} = G\bigg(1 - \eta_N\frac{\hat{U}_{\rm ext}}{c^2}\bigg),$$

where we have ignored the small term linear in $\bar{r}$.

The fact that the variation in $G_{\rm eff}$ depends on the same parameter as the Nordtvedt effect is no coincidence. Soon after Nordtvedt informed Dicke (around 1967) that his theory predicted a violation of the strong equivalence principle for self-gravitating bodies, Dicke overcame his initial skepticism by devising an elegant energy-conservation argument that showed that it must be so. The argument goes as follows.

Consider a collection of N particles, each of mass m, on the surface of the Earth. Extract from a reservoir a sufficient amount of energy, $Nmgh$, to raise the particles to a height h, where g is the acceleration of gravity. At this height, assemble the particles into a gravitationally bound body of mass $M = Nm - E_B(h)/c^2$, where $E_B(h)$ is the body's binding energy, and convert the released binding energy into more particles. Let these particles fall to the ground, convert their mass into energy, and add this together with the accumulated kinetic energy to the reservoir, which thereby acquires an energy $E_B(h)(1+gh/c^2)$. Now let the self-gravitating body fall to the ground, but to be open minded, let its acceleration be a instead of g. The kinetic energy it has when it reaches the surface is given by Mah. Add this energy to the reservoir. Now extract enough energy $E_B(0)$ from the reservoir to pull the body apart and return it to the N separate particles, each of mass m. We have returned the system to its initial state, and if energy is to be conserved (otherwise we would have a perpetual motion machine and would make zillions of dollars), the reservoir must now be empty. Doing the accounting, we see that

$$-Nmgh + E_B(h)(1+gh/c^2) + Mah - E_B(0) = 0\,,$$

and this implies that

$$a = g + \frac{E_B(h) - E_B(0)}{Mh}\,.$$

But the binding energy depends on the local value of the gravitational constant. If we postulate that $G_{\rm eff} = G(1 - \eta_N U_\oplus/c^2)$ in the vicinity of the Earth, where η_N is an arbitrary parameter, and if we use the properties that $E_B \propto G_{\rm eff}$ and $dU_\oplus/dh = g$, then we can show that

$$\begin{aligned} a &= g + \frac{1}{M}\frac{dE_B}{dh} = g + \frac{1}{M}\frac{dE_B}{dG_{\rm eff}}\frac{dG_{\rm eff}}{dh} = g + \frac{1}{M}\frac{E_B}{G}\left(-\frac{\eta_N g}{c^2}\right)G \\ &= g\left(1 - \eta_N\frac{E_B}{Mc^2}\right)\,. \end{aligned}$$

This is precisely a description of the Nordtvedt effect, and we see that it is indeed tied to the variation of $G_{\rm eff}$.

13.5.5 Wave-zone solution: gravitational waves

We next move to the far-away wave zone and examine the predictions of scalar–tensor gravity regarding the generation of gravitational waves. We rely on the discussion of Sec. 7.1.4, which reveals that when the field point $(t, \boldsymbol{x})$ is in the far-away wave zone, the potentials $\tilde{h}^{\alpha\beta}$ can be expressed as the multipole expansion

$$\tilde{h}^{00} = \frac{4\tilde{G}\tilde{M}}{c^2 R} + \frac{2\tilde{G}}{c^4 R}\ddot{\tilde{\mathcal{I}}}^{jk} N_j N_k + O(c^{-5}), \tag{13.151a}$$

$$\tilde{h}^{0j} = \frac{2\tilde{G}}{c^4 R}\ddot{\tilde{\mathcal{I}}}^{jk} N_k + O(c^{-5}), \tag{13.151b}$$

$$\tilde{h}^{jk} = \frac{2\tilde{G}}{c^4 R}\ddot{\tilde{\mathcal{I}}}^{jk} + O(c^{-5}), \tag{13.151c}$$

in which $R := |\boldsymbol{x}|$, $\boldsymbol{N} := \boldsymbol{x}/R$,

$$\tilde{M} := \int_{\mathcal{M}} c^{-2}\tilde{\tau}^{00}(\tau, \boldsymbol{x})\, d^3x, \tag{13.152}$$

and

$$\tilde{\mathcal{I}}^{jk}(\tau) := \int_{\mathcal{M}} c^{-2}\tilde{\tau}^{00}(\tau, \boldsymbol{x}) x^j x^k\, d^3x, \tag{13.153}$$

where $\tau := t - R/c$. The multipole expansion incorporates the conservation equation $\partial_\beta \tilde{\tau}^{\alpha\beta} = 0$, which implies that $\tilde{M}$ is constant up to small radiation-reaction effects, and that the dipole moment $\tilde{I}^j := \int_{\mathcal{M}} c^{-2}\tilde{\tau}^{00} x^j\, d^3x$ can be set equal to zero (also up to small radiation-reaction effects). The potentials $\tilde{h}^{\alpha\beta}$ are identified with the near-zone contributions $\tilde{h}^{\alpha\beta}_{\mathcal{N}}$ because, as was discovered back in Sec. 7.4, the wave-zone contributions $\tilde{h}^{\alpha\beta}_{\mathcal{W}}$ produce terms that fall off as R^{-2} at the relevant post-Newtonian order; these can be neglected in the far-away wave zone. The potentials of Eq. (13.151) can be compared with the general relativistic expressions displayed in Box 7.7.

The "tensor mass" $\tilde{M}$ can be evaluated by inserting $c^{-2}\tilde{\tau}^{00}$ from Eqs. (13.144) into Eq. (13.152). The term involving $\nabla^2 U^2$ produces a surface integral that scales as $\mathcal{R}^{-1}$ and therefore makes no $\mathcal{R}$-independent contribution to the potentials. Discarding this term, we find that

$$\tilde{M} = M + O(c^{-3}), \tag{13.154}$$

in which M is the total mass-energy of the system, as defined in general relativity by Eq. (7.63). We also find that to leading order in a post-Newtonian expansion,

$$\tilde{\mathcal{I}}^{jk} = I^{jk} + O(c^{-2}), \tag{13.155}$$

where $I^{jk} := \int \rho^* x^j x^k\, d^3x$ is the Newtonian quadrupole-moment tensor.

Similarly, the solution to the scalar wave equation can be expressed as

$$\phi = \phi_0 + \frac{2\tilde{G}}{c^2 R}\left[\mathcal{I}_s + \frac{1}{c}\dot{\mathcal{I}}_s^j N_j + \frac{1}{2c^2}\ddot{\mathcal{I}}_s^{jk} N_j N_k + O(c^{-3})\right], \tag{13.156}$$

where

$$\mathcal{I}_s(\tau) := \int_{\mathcal{M}} \tau_s(\tau, \boldsymbol{x})\, d^3x, \tag{13.157a}$$

$$\mathcal{I}_s^j(\tau) := \int_{\mathcal{M}} \tau_s(\tau, \boldsymbol{x}) x^j\, d^3x, \tag{13.157b}$$

$$\mathcal{I}_s^{jk}(\tau) := \int_{\mathcal{M}} \tau_s(\tau, \boldsymbol{x}) x^j x^k\, d^3x, \tag{13.157c}$$

are the multipole moments of the scalar source τ_s. Because this does not satisfy any conservation equation, the monopole moment $\mathcal{I}_s$ cannot be expected to be constant, and the dipole moment $\mathcal{I}_s^j$ cannot be set equal to zero.

To evaluate these moments it is useful to introduce an effective energy density μ defined by the relation

$$c^{-2}\tau_s = \frac{\zeta\phi_0}{1-\zeta}\left[c^{-2}\tilde{\tau}^{00} - \frac{1}{c^2}\mu + O(c^{-3})\right]. \tag{13.158}$$

This expresses the fact, apparent from Eqs. (13.144), that τ_s and $\zeta\phi_0\tilde{\tau}^{00}/(1-\zeta)$ are equal to each other to leading order in a post-Newtonian expansion, and differ by a post-Newtonian correction. We have

$$\mu = \rho^*\left[v^2 + \left(\frac{1}{2} + 2\lambda\right)U\right] + 3p - \frac{1}{4\pi Gc^2}\left(\frac{7}{4} - 3\zeta + \lambda\right)\nabla^2 U^2 . \tag{13.159}$$

We define multipole moments associated with μ,

$$\mathcal{E}(\tau) := \int_{\mathcal{M}} \mu(\tau, \boldsymbol{x})\, d^3x , \tag{13.160a}$$

$$\mathcal{E}^j(\tau) := \int_{\mathcal{M}} \mu(\tau, \boldsymbol{x})x^j\, d^3x , \tag{13.160b}$$

$$\mathcal{E}^{jk}(\tau) := \int_{\mathcal{M}} \mu(\tau, \boldsymbol{x})x^j x^k\, d^3x , \tag{13.160c}$$

and write

$$\mathcal{I}_s = \frac{\zeta\phi_0}{1-\zeta}\left[\tilde{M} - \frac{1}{c^2}\mathcal{E} + O(c^{-3})\right], \tag{13.161a}$$

$$\mathcal{I}_s^j = \frac{\zeta\phi_0}{1-\zeta}\left[-\frac{1}{c^2}\mathcal{E}^j + O(c^{-3})\right], \tag{13.161b}$$

$$\mathcal{I}_s^{jk} = \frac{\zeta\phi_0}{1-\zeta}\left[\tilde{\mathcal{I}}^{jk} - \frac{1}{c^2}\mathcal{E}^{jk} + O(c^{-3})\right]. \tag{13.161c}$$

From Eq. (13.159) we find that

$$\mathcal{E} = \int \rho^*\left[v^2 + \left(\frac{1}{2} + 2\lambda\right)U + 3p/\rho^*\right] d^3x , \tag{13.162a}$$

$$\mathcal{E}^j = \int \rho^*\left[v^2 + \left(\frac{1}{2} + 2\lambda\right)U + 3p/\rho^*\right] x^j\, d^3x , \tag{13.162b}$$

and we shall not need an explicit expression for $\mathcal{E}^{jk}$; the term involving $\nabla^2 U^2$ in μ makes no $\mathcal{R}$-independent contributions to the monopole and dipole moments.

Making the substitutions in Eq. (13.156) returns

$$\phi/\phi_0 = 1 + \frac{2\zeta G}{c^2 R}\left[M - \frac{1}{c^2}\mathcal{A}(\tau, \boldsymbol{N}) + O(c^{-3})\right], \tag{13.163}$$

where

$$\mathcal{A} := \mathcal{E}(\tau) + \frac{1}{c}\dot{\mathcal{E}}^j(\tau)N_j - \frac{1}{2}\ddot{I}^{jk}(\tau)N_j N_k \tag{13.164}$$

is the radiative piece of the scalar field. To arrive at this result we have incorporated the information displayed in Eqs. (13.154) and (13.155). Note that the term involving the dipole moment $\mathcal{E}^j$ has been retained in the expression for $\mathcal{A}$, in spite of the fact that it is formally of order c^{-3} and therefore of the same order as the neglected terms in Eq. (13.163). The reason for this is revealed below.

We are now ready to convert these results into a description of gravitational waves in scalar–tensor gravity. As was reviewed back in Secs. 13.4.1 and 13.4.2, the gravitational-wave polarizations can be extracted from potentials $h^{\alpha\beta}$ defined in terms of the physical metric $g_{\alpha\beta}$ by Eq. (13.77). The relation with $\tilde{h}^{\alpha\beta}$ is provided by Eqs. (13.102) and (13.115), and we get

$$h^{\alpha\beta} = (\phi_0/\phi)\tilde{h}^{\alpha\beta} + (1 - \phi_0/\phi)\eta^{\alpha\beta} . \tag{13.165}$$

Making the substitutions from Eqs. (13.151) and (13.163), we arrive at

$$h^{00} = 2(2 - 3\zeta)\frac{GM}{c^2 R} + \frac{G}{c^4 R}\Big[2(1-\zeta)\ddot{I}^{jk}N_j N_k + 2\zeta\mathcal{A} + O(c^{-1})\Big], \tag{13.166a}$$

$$h^{0j} = \frac{G}{c^4 R}\Big[2(1-\zeta)\ddot{I}^{jk}N_k + O(c^{-1})\Big], \tag{13.166b}$$

$$h^{jk} = 2\zeta\frac{GM}{c^2 R}\delta^{jk} + \frac{G}{c^4 R}\Big[2(1-\zeta)\ddot{I}^{jk} - 2\zeta\mathcal{A}\delta^{jk} + O(c^{-1})\Big]. \tag{13.166c}$$

The assignments of Eqs. (13.149) and (13.150) ensure that the stationary terms, involving the total mass-energy M, agree with the general expression of Eq. (13.78). Comparison with Eq. (13.80) then allows us to read off the radiative fields C, D^j, and A^{jk}, which are decomposed into irreducible pieces as in Eqs. (13.81). We get

$$A = 2(1-\zeta)\ddot{I}^{pp} - 6\zeta\mathcal{A}, \tag{13.167a}$$

$$B = 3(1-\zeta)\ddot{I}^{\langle jk\rangle}N_j N_k, \tag{13.167b}$$

$$C = 2(1-\zeta)\ddot{I}^{\langle jk\rangle}N_j N_k + \frac{2}{3}(1-\zeta)\ddot{I}^{pp} + 2\zeta\mathcal{A}, \tag{13.167c}$$

$$D = 2(1-\zeta)\ddot{I}^{\langle jk\rangle}N_j N_k + \frac{2}{3}(1-\zeta)\ddot{I}^{pp}, \tag{13.167d}$$

$$A^j_{\rm T} = 2(1-\zeta)P^j_{\ p}\ddot{I}^{\langle pk\rangle}N_k, \tag{13.167e}$$

$$D^j_{\rm T} = 2(1-\zeta)P^j_{\ p}\ddot{I}^{\langle pk\rangle}N_k, \tag{13.167f}$$

$$A^{jk}_{\rm TT} = 2(1-\zeta)\ddot{I}^{jk}_{\rm TT}, \tag{13.167g}$$

in which $P_{jk} := \delta_{jk} - N_j N_k$ is the projector to the subspace transverse to N^j. Finally, Eq. (13.89) allows us to obtain the gauge-invariant gravitational-wave amplitudes,

$$A_{\rm S} = 2\zeta\mathcal{A}, \tag{13.168a}$$

$$A_{\rm L} = 0, \tag{13.168b}$$

$$A^j_{\rm V} = 0, \tag{13.168c}$$

$$A^{jk}_{\rm TT} = 2(1-\zeta)\ddot{I}^{jk}_{\rm TT}. \tag{13.168d}$$

We see that in addition to the familiar transverse-tracefree polarizations involving the quadrupole-moment tensor, scalar–tensor gravity gives rise to a scalar mode involving the combination of monopole, dipole, and quadrupole moments contained in $\mathcal{A}$.

The generation of monopole and dipole waves in scalar–tensor gravity is a major departure from general relativity, and we conclude this section with a discussion of these new features. The monopole and dipole moments $\mathcal{E}$ and $\mathcal{E}^j$ are defined by Eq. (13.162), and for the purpose of this discussion we evaluate them for a system of well-separated bodies, using techniques developed in Chapter 9 – refer to Sec. 9.3.6. We find that the monopole moment evaluates to

$$\mathcal{E} = \sum_A \left[2\mathcal{T}_A - (1+4\lambda)\Omega_A + 3P_A + m_A v_A^2 + \frac{1}{2}(1+4\lambda) m_A U_{\neg A} \right], \tag{13.169}$$

in which $\mathcal{T}_A$ is the total kinetic energy of body A, Ω_A its gravitational potential energy, P_A the integrated pressure, and $U_{\neg A}$ the gravitational potential of the external bodies evaluated at the position of body A. This becomes

$$\mathcal{E} = -2(1+2\lambda)\sum_A \Omega_A + \sum_A M_A v_A^2 + (1+4\lambda)\sum_A \sum_{B\neq A} \frac{GM_A M_B}{2r_{AB}} + O(c^{-2}) \tag{13.170}$$

after we make use of the virial theorem of Eq. (13.16), insert the familiar expression for the external potential, and make the replacements $m_A = M_A + O(c^{-2})$. For a binary system the monopole moment reduces to

$$\mathcal{E} = -2(1+2\lambda)(\Omega_1 + \Omega_2) + \eta m \left[v^2 + (1+4\lambda)\frac{Gm}{r} \right] + O(c^{-2}), \tag{13.171}$$

in which $m := M_1 + M_2$, $\eta := M_1 M_2/m^2$, $r := |\boldsymbol{r}_1 - \boldsymbol{r}_2|$, and $v := |\boldsymbol{v}_1 - \boldsymbol{v}_2|$. The monopole moment contains a contribution from each body's gravitational potential energy Ω_A, and neglecting tidal interactions between bodies, this is constant for bodies in hydrodynamical equilibrium; these terms do not participate in the generation of gravitational waves. The remaining terms, however, coming from the orbital motion of each body, are time-dependent and do participate in the production of gravitational waves. It is easy to see that the monopole term in $\mathcal{A}$ is of the same order of magnitude as the quadrupole term $\frac{1}{2}\ddot{I}^{jk}N_j N_k$.

After making use of the virial theorem, we find that the dipole moment is given by

$$\begin{aligned} \mathcal{E}^j = &-2(1+2\lambda)\sum_A \Omega_A r_A^j + \sum_A M_A v_A^2 r^j \\ &+ (1+4\lambda)\sum_A \sum_{B\neq A} \frac{GM_A M_B}{2r_{AB}} r_A^j + O(c^{-2}) \end{aligned} \tag{13.172}$$

for a system of well-separated bodies. For a binary system this reduces to

$$\mathcal{E}^j = -2(1+2\lambda)\eta m c^2 \mathcal{S} r^j - \eta \Delta m \left[v^2 + (1+4\lambda)\frac{Gm}{2r} \right] r^j + O(c^{-2}), \tag{13.173}$$

where $\boldsymbol{r} := \boldsymbol{r}_1 - \boldsymbol{r}_2$, $\Delta := (M_1 - M_2)/m$, and

$$\mathcal{S} := \frac{\Omega_1}{M_1 c^2} - \frac{\Omega_2}{M_2 c^2} . \tag{13.174}$$

We find that the dipole moment also contains terms involving the gravitational potential energies in addition to orbital terms. These terms come now with a time-dependent separation vector $\boldsymbol{r}$, and they do participate in the production of gravitational waves. For weakly bound bodies, $\Omega_A/(M_A c^2)$ is of the same order of magnitude as $(v_A/c)^2$ and $GM_B/(c^2 r_{AB})$, and the dipole term in $\mathcal{A}$ can be seen to be a factor v_c/c smaller than the monopole and quadrupole terms (with v_c denoting a characteristic orbital velocity). In this case the dipole term should be lumped together with the error terms of order c^{-1} in Eq. (13.166).

The situation changes dramatically when the bodies are compact, as in the case of neutron stars. In this situation $\Omega_A/(M_A c^2) \sim 0.1$, and the internal terms strongly dominate over the orbital terms in the dipole moment. In this case we find that the dipole term in $\mathcal{A}$ is a factor c/v_c *larger* than the monopole and quadrupole terms. The orbital motion of compact bodies, therefore, can in principle give rise to gravitational waves that are dominantly dipolar, thanks to the dependence of the dipole moment on each body's gravitational potential energy. In practice, however, the effect can be strongly suppressed. For a binary system the dipolar radiation is controlled by the *difference* in $\Omega/(Mc^2)$ between bodies, and it is suppressed whenever the bodies are very similar. This occurs, for example, for most binary pulsars, which involve neutron stars with masses that are tightly clustered around $1.4\, M_\odot$, producing very small values of $\mathcal{S}$. Furthermore, $\mathcal{A}$ is multiplied by ζ in the gravitational-wave field, and ζ is already constrained to be small by solar-system measurements of the Shapiro time delay. As a result, binary neutron-star systems have yet to provide interesting tests of scalar–tensor gravity, except for theories that come with anomalously large values of the parameter λ. It is possible, however, to test scalar–tensor gravity with the few binary-pulsar systems that are known to involve a white-dwarf companion. With $\Omega_A/(M_A c^2) \sim 10^{-4}$ for a white dwarf, $\mathcal{S}$ is numerically large, and the emission of dipole radiation can be important. Data from one such system, known as J1738+0333, provides a bound on ζ that is beginning to compete with constraints from the solar system.

13.6 Bibliographical notes

The literature on alternative theories of gravity and their experimental tests is far too vast to attempt even a partial summary. At the risk of self-promotion, we refer the reader to the book *Theory and Experiment in Gravitational Physics* by one of us (CMW) (1993), which gives a lot more detail than we were able to provide in this chapter. In addition, Will's *Living Reviews* article (2006b) describes the experimental situation up to that time; it is scheduled to be updated by 2014. Another useful resource is Will (2010), an annotated compilation of almost 100 references on tests of gravitational theories.

Nordström's relativistic theory of gravity, mentioned in the introduction to the chapter, is published as Nordström (1913). Whitehead's theory, involving a non-dynamical Minkowski metric in addition to a dynamical metric, was proposed in his 1922 book; its multiple deaths are described in Gibbons and Will (2008). The experimental basis of general relativity and alternative theories in the early nineteen sixties is reviewed in Bertotti, Brill, and Krotkov (1962) and Whitrow and Morduch (1965). The Brans–Dicke theory originates in their 1961 article, which built upon previous work by Fierz (1956) and Jordan (1959).

The parameterized post-Newtonian framework, reviewed in Sec. 13.2, was first initiated in Eddington (1922). It was developed systematically by Nordtvedt (1968a and 1968b) for systems of point particles, and by Thorne and Will (1970) and Will (1971a, 1971b, and 1971c) for perfect-fluid systems. The two versions of the formalism were consolidated in Will and Nordtvedt (1972a and 1972b).

The measurement of the PPN parameter γ on galactic scales using gravitational lensing, described in Sec. 13.3.2, is reported in Bolton, Rappaport, and Burles (2006). The effect on the lunar orbit of a failure of the strong equivalence principle (Sec. 13.3.3) was first discovered by Nordtvedt (1968c), who also proposed a detection scheme based on laser ranging. The state of the art on lunar laser ranging and the measurement of the Nordtvedt effect (Box 13.2) is summarized in Nordtvedt (1999), Williams, Turyshev, and Boggs (2009), and Merkowitz (2010). The Eöt-Wash test of the weak equivalence principle on a mini-Moon and mini-Earth system was described in Baessler *et al.* (1999). Tests of the Nordtvedt effect on pulsar–white dwarf systems were reported by Stairs *et al.* (2005).

The discussion of gravitational-wave polarizations in alternative theories of gravity, provided in Sec. 13.4.2, is based on Eardley *et al.* (1973).

Our treatment of scalar–tensor gravity in Sec. 13.5 was inspired by the seminal 1992 paper by Damour and Esposito-Farèse. Dicke's argument of Box 13.4, connecting the variation of $G_{\rm eff}$ to the Nordtvedt effect, originated in Dicke (1970); an expanded version of the argument is presented in the Appendix of Will (1971a). The proof that the stationary black holes of scalar–tensor gravity are the same as those of general relativity (Exercise 13.5) was formulated by Hawking (1972) in the case of Brans–Dicke theory, and by Sotiriou and Faraoni (2012) in the general case.

13.7 Exercises

13.1 Consider a semi-conservative theory of gravity, with $\alpha_3 = \zeta_n = 0$. Using the PPN equation of hydrodynamics, Eq. (13.8), and making use of the method described in Sec. 8.4.6, show that the total momentum defined by

$$\begin{aligned} P^j := &\int \rho^* v^j \biggl[1 + \frac{1}{c^2}\Bigl(\frac{1}{2}v^2 - \frac{1}{2}U + \Pi + p/\rho^* \Bigr) \biggr] d^3x - \frac{1}{2c^2} \int \rho^* \Phi^j \, d^3x \\ &- \frac{1}{2c^2}\alpha_1 \int \rho^* (v^j + w^j) U \, d^3x - \frac{1}{2c^2}\alpha_2 \int \rho^* \bigl(\partial_{tj} X - w^k \partial_{jk} X \bigr) \, d^3x \\ &+ O(c^{-4}) \end{aligned}$$

is conserved.

13.2 Consider a semi-conservative theory of gravity, with $\alpha_3 = \zeta_n = 0$. A single, gravitationally-bound body A in isolation moves with a velocity $\boldsymbol{v}_A$ relative to the PPN coordinate system, which itself moves with a velocity $\boldsymbol{w}$ relative to the preferred universal frame. Using Eq. (13.26), show that to 1PN order, there exist a conserved momentum and energy, given by

$$P^j = (M_I^{jk})_A (w + v_A)^k \,, \qquad E = \frac{1}{2}(M_I^{jk})_A (w + v_A)^j (w + v_A)^k \,,$$

where $(M_I^{jk})_A$ is an "inertial mass tensor" given by

$$(M_I^{jk})_A = M_A \delta^{jk}\left[1 + (\alpha_1 - \alpha_2)\frac{\Omega_A}{M_A c^2}\right] + \alpha_2 \frac{\Omega_A^{jk}}{M_A c^2} \,.$$

13.3 Consider the structure-dependent term in Eq. (13.26),

$$c^2 a_A^j[\text{STR}] = -\frac{\alpha_3}{M_A} H_A^{kj}(w_k + v_{Ak}) \,,$$

where H_A^{kj} is given by Eq. (9.9f).

(a) Assuming that the body is rotating uniformly with an angular velocity $\boldsymbol{\omega}$, and that it is approximately spherically symmetric, show that this acceleration is given by

$$c^2 \boldsymbol{a}_A[\text{STR}] = -\frac{1}{3}\alpha_3 \frac{\Omega_A}{M_A} \boldsymbol{\omega} \times (\boldsymbol{w} + \boldsymbol{v}_A) \,,$$

where Ω_A is the body's gravitational potential energy.

(b) Show that the acceleration $\boldsymbol{a}$ of a pulsar leads to an observed rate of change of its pulse period given by $\dot{P} = -(\boldsymbol{a} \cdot \boldsymbol{N})P$, where N is a unit vector along the line of sight.

(c) The isolated pulsar PSR 1937+21 has a rotation period of 1.56 milliseconds, and an observed $\dot{P}$ of 10^{-19}. Assuming a value $|\Omega_A/M_A| \sim 0.1$ for the pulsar, a value of 300 km/s for the velocity of the pulsar relative to the preferred frame, and optimum alignment for the preferred-frame effect, place a bound on the parameter α_3.

13.4 In this problem we generalize the Newman–Penrose description of gravitational waves, introduced in Exercise 11.3, from general relativity to the more general framework of Sec. 13.4. The component notation is explained in the previous exercise, and it involves the vectors $\ell^\alpha := (1, \boldsymbol{N})$, $n^\alpha := \frac{1}{2}(1, -\boldsymbol{N})$, $m^\alpha := 2^{-1/2}(0, \boldsymbol{\vartheta} + i\boldsymbol{\varphi})$, and $\bar{m}^\alpha := 2^{-1/2}(0, \boldsymbol{\vartheta} - i\boldsymbol{\varphi})$, which are defined in the asymptotically flat spacetime of the far-away wave zone.

(a) Show that the six non-zero components of $R_{\alpha\beta\gamma\delta}$ in the far-away wave zone are related to the polarization amplitudes of Sec. 13.4.2 by

$$R_{n\ell n\ell} = -\frac{G}{2c^4 R}\frac{\partial^2}{\partial \tau^2} A_{\text{L}} \,,$$

$$R_{n\ell nm} = -\frac{G}{2c^4 R}\frac{\partial^2}{\partial \tau^2} A_{\text{V}}^m \,,$$

(continued overleaf)

$$R_{nmn\bar{m}} = -\frac{G}{2c^4 R}\frac{\partial^2}{\partial\tau^2}A_{\rm S}\,,$$

$$R_{nmnm} = -\frac{G}{2c^4 R}\frac{\partial^2}{\partial\tau^2}A^{mm}_{\rm TT}\,.$$

(b) By relating the Ricci scalar $R := g^{\alpha\beta}R_{\alpha\beta}$ to the non-zero components of the Riemann tensor, show that any theory of gravity for which the Ricci scalar vanishes or falls off faster than $1/R$ in the far-away wave zone has a vanishing longitudinal gravitational-wave mode $A_{\rm L}$.

13.5 In this problem we examine black-hole solutions in scalar–tensor gravity.

(a) Prove that when $V(\phi_0) = V'(\phi_0) = 0$, a configuration $(g_{\alpha\beta}, \phi_0)$ is an exact vacuum solution to the scalar–tensor field equations, provided that $g_{\alpha\beta}$ is a solution to the vacuum field equations in general relativity. This shows in particular that with the stated assumptions on the potential, the Schwarzschild metric is an exact solution of scalar–tensor gravity.

(b) We wish to see if other black-hole solutions might be possible. We consider a deformation

$$g_{\alpha\beta} = g^{\rm Schw}_{\alpha\beta} + \delta g_{\alpha\beta}\,, \qquad \phi = \phi_0 + \delta\phi\,,$$

of the previous configuration, in which the perturbations $\delta g_{\alpha\beta}$ and $\delta\phi$ are assumed to depend on the radial coordinate r only. To first order in perturbation theory, $\delta\phi$ satisfies a differential equation formulated in the background spacetime of the Schwarzschild spacetime. Derive this differential equation, and show that the solution is either singular at infinity, or singular at the event horizon. Conclude that scalar–tensor gravity does not admit static black-hole solutions that deviate slightly from the Schwarzschild solution.

A general proof that the stationary black holes of scalar–tensor gravity are the same as those of general relativity was provided by Thomas Sotiriou and Valerio Faraoni in 2012. Their work generalizes a 1972 paper by Stephen Hawking, which was restricted to the Brans–Dicke theory.

13.6 In many alternative theories of gravity, G becomes a function of time via the asymptotic boundary conditions on the auxiliary fields – scalar–tensor theory is a notable example. Because of the expansion of the universe, G may thus vary on a Hubble timescale. One way to model the effect of such a variation on solar-system dynamics is to write the effective Newtonian equation of motion as follows:

$$\boldsymbol{a} = -G(t)m\boldsymbol{x}/r^3\,,$$

where on solar-system time scales we approximate $G(t)$ by

$$G(t) = G_0 + \dot{G}_0 t + O(G_0 t^2/t_H^2)\,,$$

where t_H is the Hubble time, and t is chosen to be zero at the beginning of a fiducial orbit.

(a) Assuming that the osculating Keplerian orbit is governed by G_0, write down the components $\mathcal{R}$, $\mathcal{S}$, and $\mathcal{W}$ of the disturbing force.
(b) Find the change in the orbital elements a, e, ω, ι and Ω over one orbit. What is the change in orbital angular momentum $h = \sqrt{G_0 m p}$?
(c) By integrating the perturbed time equation (3.70) for dt/df over 2π, show that the change in orbital period over one orbit is given by $\Delta P/P = -\frac{1}{2}(\dot{G}_0/G_0)P$.
(d) The Earth–Moon distance (semi-major axis) is known to be increasing at a rate of about 3.8 cm/yr. Assuming that tidal dissipation accounts for this at a level of 10 percent accuracy, what bound can you place on $\dot{G}_0/G_0$? How does this compare with the inverse Hubble time?

References

Abramowitz, M. and Stegun, I.A. 1975. *Handbook of Mathematical Functions*. Dover.

Alcock, C., Allsman, R.A., Alves, D.R., *et al.* 2000. The MACHO project: Microlensing results from 5.7 years of Large Magellanic Cloud observations. *Astrophys. J.* **542**, 281–307.

Alväger, T., Farley, F.J.M., Kjellman, J., and Wallin, I. 1964. Test of the second postulate of special relativity in the GeV region. *Phys. Lett.* **12**, 260–262.

Antoci, S. and Loinger, A. 1999. On the gravitational field of a mass point according to Einstein's theory (English translation of Schwarzschild's 1916 paper). `arXiv.org/abs/physics/9905030`.

Arfken, G.B., Weber H.J., and Harris, F.E. 2012. *Mathematical Methods for Physicists*. Seventh Edition: *A Comprehensive Guide.* Academic Press.

Arnowitt, R., Deser, S., and Misner, C.W. 1962. The dynamics of general relativity, in *Gravitation: An Introduction to Current Research*, edited by Witten, L., 227–265. Wiley.

Ashby, N. 2003. Relativity in the Global Positioning System. *Living Rev. Relativity* **6**. `http://www.livingreviews.org/lrr-2003-1`.

Baessler, S., Heckel, B.R., Adelberger, E.G., *et al.* 1999. Improved test of the equivalence principle for gravitational self-energy. *Phys. Rev. Lett.* **83**, 3585–3588.

Baker, J.G., Centrella, J., Choi, D.I., *et al.* 2006. Getting a kick out of numerical relativity. *Astrophys. J.* **653**, L93–L96.

Barker, B.M. and O'Connell, R.F. 1974. Nongeodesic motion in general relativity. *Gen. Relativ. Gravit.* **5**, 539–554.

Bekenstein, J.D. 1973. Gravitational-radiation recoil and runaway black holes. *Astrophys. J.* **183**, 657–664.

Bender, C.M. and Orzag, S.A. 1978. *Advanced Mathematical Methods for Scientists and Engineers*. McGraw-Hill.

Bertotti, B., Brill, D.R., and Krotkov, R.D. 1962. Experiments on gravitation, in *Gravitation: An Introduction to Current Research*, edited by Witten, L., 1–48. Wiley.

Bertotti, B., Iess, L., and Tortora, P. 2003. A test of general relativity using radio links with the Cassini spacecraft. *Nature* **425**, 374–376.

Black, E.D. and Gutenkunst, R.N. 2003. An introduction to signal extraction in interferometric gravitational wave detectors. *Am. J. Phys.* **71**, 365–378.

Blanchet, L. 2006. Gravitational radiation from post-Newtonian sources and inspiralling compact binaries. *Living Rev. Relativity* **9**. `http://www.livingreviews.org/lrr-2006-4`.

Blanchet, L., and Faye, G. 2000. Hadamard regularization. *J. Math. Phys.* **41**, 7675–7714.

Blanchet, L., Iyer, B.R., Will, C.M., and Wiseman, A.G. 1996. Gravitational waveforms from inspiralling compact binaries to second-post-Newtonian order. *Class. Quantum Grav.* **13**, 575–584.

Blandford, R. and Teukolsky, S.A. 1976. Arrival-time analysis for a pulsar in a binary system. *Astrophys. J.* **205**, 580–591.

Bolton, A.S., Rappaport, S., and Burles, S. 2006. Constraint on the post-Newtonian parameter γ on galactic size scales. *Phys. Rev. D* **74**, 061501(R) (5 pages).

Bond, I.A., Udalski, A., Jaroszynski, M., *et al.*, and OGLE Collaboration. 2004. OGLE 2003-BLG-235/MOA 2003-BLG-53: A planetary microlensing event. *Astrophys. J.* **606**, L155–L158.

Bondi, H., van der Burg, M.G.J., and Metzner, A.W.K. 1962. Gravitational waves in general relativity. VII. Waves from axi-symmetric isolated systems. *Proc. Roy. Soc. London* **A269**, 21–52.

Braginskii, V.B., Caves, C.M., and Thorne, K.S. 1977. Laboratory experiments to test relativistic gravity. *Phys. Rev. D* **15**, 2047–2068.

Brans, C.H. and Dicke, R.H. 1961. Mach's principle and a relativistic theory of gravitation. *Phys. Rev.* **124**, 925–935.

Brecher, K. 1977. Is the speed of light independent of the velocity of the source? *Phys. Rev. Lett.* **39**, 1051–1054.

Brooker, R.A. and Olle, T.W. 1955. Apsidal-motion constants for polytropic models. *Mon. Not. R. Astr. Soc.* **115**, 101–106.

Brouwer, D. and Clemence, G.M. 1961. *Methods of Celestial Mechanics*. Academic Press.

Brown, E.W. 1960. *An Introductory Treatise on the Lunar Theory*. Dover.

Brumberg, V.A. 1991. *Essential Relativistic Celestial Mechanics.* IOP Publishing.

Burke, W.L. 1971. Gravitational radiation damping of slowly moving systems calculated using matched asymptotic expansions. *J. Math. Phys.* **12**, 401–418.

Carroll, S. 2003. *Spacetime and Geometry: An Introduction to General Relativity*. Addison-Wesley.

Chandrasekhar, S. 1931. The maximum mass of ideal white dwarfs. *Astrophys. J.* **74**, 81–82.

Chandrasekhar, S. 1958. *An Introduction to the Study of Stellar Structure*. Dover.

Chandrasekhar, S. 1965. The post-Newtonian equations of hydrodynamics in general relativity. *Astrophys. J.* **142**, 1488–1512.

Chandrasekhar, S. 1969. Conservation laws in general relativity and in the post-Newtonian approximation. *Astrophys. J.* **158**, 45–54.

Chandrasekhar, S. 1987. *Ellipsoidal Figures of Equilibrium*. Dover.

Chandrasekhar, S. and Contopoulos, G. 1967. On a post-Galilean transformation appropriate to the post-Newtonian theory of Einstein, Infeld, and Hoffmann. *Proc. Roy. Soc. London* **A298**, 123–141.

Chandrasekhar, S. and Esposito, F.P. 1970. The 5/2-post-Newtonian equations of hydrodynamics and radiation reaction in general relativity. *Astrophys. J.* **160**, 153–179.

Chandrasekhar, S. and Nutku, Y. 1969. The second post-Newtonian equations of hydrodynamics in general relativity. *Astrophys. J.* **158**, 55–79.

Ciufolini, I. and Pavlis, E.C. 2004. A confirmation of the general relativistic prediction of the Lense–Thirring effect. *Nature* **431**, 958–960.

Cowling, T.G. 1941. The non-radial oscillations of polytropic stars. *Mon. Not. R. Astr. Soc.* **101**, 367–375.

Cox., A.N. 2001. *Allen's Astrophysical Quantities*. Fourth Edition. Springer.

Cox, J.P. 1980. *Theory of Stellar Pulsation*. Princeton University Press.

Creighton, J.D.E. and Anderson, W.G. 2011. *Gravitational-wave Physics and Astronomy: An Introduction to Theory, Experiment and Data Analysis*. Wiley-VCH.

Crelinsten, J. 2006. *Einstein's Jury: The Race to Test Relativity.* Princeton University Press.

D'Eath, P.D. 1975. Interaction of two black holes in the slow-motion limit. *Phys. Rev. D* **12**, 2183– 2199.

Damour, T. 1983. Gravitational radiation and the motion of compact bodies. in *Rayonnement Gravitationnel*, edited by Deruelle, N. and Piran, T., 59–144. North-Holland.

Damour, T. 1987. The problem of motion in Newtonian and Einsteinian gravity, in *Three Hundred Years of Gravitation*, edited by Hawking, S.W. and Israel, W., 128–198. Cambridge University Press.

Damour, T. and Deruelle, N. 1981. Radiation reaction and angular momentum loss in small angle gravitational scattering. *Phys. Lett. A* **87**, 81–84.

Damour, T. and Deruelle, N. 1985. General relativistic celestial mechanics of binary systems. I. The post-Newtonian motion. *Ann. Inst. H. Poincaré*, **A43**, 107–132.

Damour, T. and Deruelle, N. 1986. General relativistic celestial mechanics of binary systems. II. The post-Newtonian timing formula. *Ann. Inst. H. Poincaré*, **A44**, 263–292.

Damour, T. and Esposito-Farèse, G. 1992. Tensor–multi-scalar theories of gravitation. *Class. Quantum Grav.* **9**, 2093–2176.

Damour, T. and Iyer, B.R. 1991. Multipole analysis for electromagnetism and linearized gravity with irreducible Cartesian tensors. *Phys. Rev. D* **43**, 3259–3272.

Damour, T., Soffel, M., and Xu, C. 1991. General-relativistic celestial mechanics. I. Method and definition of reference systems. *Phys. Rev. D* **43**, 3273–3307.

Damour, T., Soffel, M., and Xu, C. 1992. General-relativistic celestial mechanics. II. Translational equations of motion. *Phys. Rev. D* **45**, 1017–1044.

Damour, T., Soffel, M., and Xu, C. 1993. General-relativistic celestial mechanics. III. Rotational equations of motion. *Phys. Rev. D* **47**, 3124–3135.

Demianski, M. and Grishchuck, L.P. 1974. Note on the motion of black holes. *Gen. Relativ. Gravit.* **5**, 673–679.

Demorest, P.B., Pennucci, T., Ransom, S.M., Roberts, M.S.E., and Hessels, J.W.T. 2010. A two-solar-mass neutron star measured using Shapiro delay. *Nature* **467**, 1081–1083.

de Sitter, W. 1916. On Einstein's theory of gravitation, and its astronomical consequences. Second paper. *Mon. Not. R. Astr. Soc.* **27**, 155–184.

DeWitt, B. 2011. *Bryce DeWitt's Lectures on Gravitation*. Lecture Notes in Physics, Volume 826, edited by Christensen, S.M. Springer-Verlag.

Dicke, R.H. 1970. *Gravitation and the Universe — Jayne Lectures for 1969*. American Philosophical Society.

Dicke, R.H. and Goldenberg, H.M. 1967. Solar oblateness and general relativity. *Phys. Rev. Lett.* **18**, 313–316.

Dyson, F.W., Eddington, A.S., and Davidson, C., 1920. A determination of the deflection of light by the Sun's gravitational field, from observations made at the total eclipse of May 29, 1919. *Phil. Trans. Roy. Soc. London* **A220**, 291–333.

Eardley, D.M., Lee, D.L., Lightman, A.P., Wagoner, R.V., and Will, C.M. 1973. Gravitational-wave observations as a tool for testing relativistic gravity. *Phys. Rev. Lett.* **30**, 884–886.

Eddington, A.S. 1922. *The Mathematical Theory of Relativity*. Cambridge University Press.

Eddington, A.S. and Clark, G.L. 1938. The problem of *n* bodies in general relativity theory. *Proc. Roy. Soc. London* **A166**, 465–475.

Ehlers, J., Rosenblum, A., Goldberg, J.N., and Havas, P. 1976. Comments on gravitational radiation damping and energy loss in binary systems. *Astrophys. J.* **208**, L77–L81.

Einstein, A. and Rosen, N. 1937. On gravitational waves. *J. of the Franklin Institute* **223**, 143–154.

Einstein, A., Infeld, L., and Hoffmann, B. 1938. The gravitational equations and the problem of motion. *Annals of Mathematics* **39**, 65–100.

Everitt, C.W.F., Debra, D.B., Parkinson, B.W., *et al.* 2011. Gravity Probe B: Final results of a space experiment to test general relativity. *Phys. Rev. Lett.* **106**, 221101 (4 pages).

Farley, F.J.M., Bailey, J., Brown, R.C.A., *et al.* 1966. The anomalous magnetic moment of the negative muon. *Nuovo Cimento* **45**, 281–286.

Favata, M., Hughes, S.A., and Holz, D.E. 2004. How black holes get their kicks: Gravitational radiation recoil revisited. *Astrophys. J.* **607**, L5–L8.

Faye, G., Marsat, S., Blanchet, L., and Iyer, B.R. 2012. The third and a half-post-Newtonian gravitational wave quadrupole mode for quasi-circular inspiralling compact binaries. *Class. Quantum Grav.* **29**, 175004 (16 pages).

Fierz, M, 1956. Über die physikalische Deutung der erweiterten Gravitationstheorie P. Jordans. *Helv. Phys. Acta* **29**, 128–134.

Fitchett, M.J. 1983. The influence of gravitational momentum losses on the centre of mass motion of a Newtonian binary system. *Mon. Not. R. Astr. Soc.* **203**, 1049–1062.

Flanagan, E.E. and Hughes, S.A. 2005. The basics of gravitational wave theory. *New J. Phys.* **7**, 204 (52 pages).

Fock, V.A. 1959. *Theory of Space, Time and Gravitation*. Pergamon.

French, A.P. 1968. *Special Relativity*. W.W. Norton & Company.

French, A.P. 1971. *Newtonian Mechanics*. W.W. Norton & Company.

Friedman, J.L. and Stergioulas, N. 2013. *Rotating Relativistic Stars*. Cambridge University Press.

Froeschlé, M., Mignard, F., and Arenou, F. 1997. Determination of the PPN parameter γ with the Hipparcos data, in *Proceedings from the Hipparcos Venice 97 Symposium*, edited by Battrick, B., 49–52. European Space Agency.

Gibbons, G. and Will, C.M. 2008. On the multiple deaths of Whitehead's theory of gravity. *Stud. Hist. Philos. Mod. Phys.* **39**, 41–61.

Glendenning, N.K. 2000. *Compact Stars: Nuclear Physics, Particle Physics, and General Relativity*, Second Edition. Springer.

Goldstein, H., Poole Jr, C.P., and Safko, J.L. 2001. *Classical Mechanics*. Third Edition. Addison-Wesley.

Gonzalez, J.A., Sperhake, U., Bruegmann, B., Hannam, M., and Husa, S. 2007. Total recoil: the maximum kick from nonspinning black-hole binary inspiral. *Phys. Rev. Lett.* **98**, 091101 (4 pages).

Gralla, S.E., Harte, A.I., and Wald, R.M. 2009. A rigorous derivation of electromagnetic self-force. *Phys. Rev. D* **80**, 024031 (22 pages).

Gullstrand, A. 1922. Allegemeine Lösung des statischen Einkörper-problems in der Einsteinschen Gravitations Theorie. *Arkiv. Mat. Astron. Fys.* **16**(8), 1–15.

Hansen, C.J., Kawaler, S.D., and Trimble, V. 2004. *Stellar Interiors – Physical Principles, Structure, and Evolution*. Second Edition. Springer.

Hartle, J.B. 2003. *Gravity: An Introduction to Einstein's General Relativity*. Addison-Wesley.

Havas, P. 1989. The early history of the 'problem of motion' in general relativity, in *Einstein and the History of General Relativity*, edited by Howard, D. and Stachel, J. Birkhäuser.

Havas, P. and Goldberg, J.N. 1962. Lorentz-invariant equations of motion of point masses in the general theory of relativity. *Phys. Rev.* **128**, 398–414.

Hawking, S.W. 1972. Black holes in the Brans–Dicke theory of gravitation. *Commun. Math. Phys.* **25**, 167–171.

Hawking, S.W. 1998. *A Brief History of Time. 10th Anniversary Edition*. Bantam.

Hulse, R.A. and Taylor, J.H. 1975. Discovery of a pulsar in a binary system. *Astrophys. J.* **195**, L51–L53.

Isaacson, R.A. 1968a. Gravitational radiation in the limit of high frequency. I. The linear approximation and geometrical optics. *Phys. Rev.* **166**, 1263–1271.

Isaacson, R.A. 1968b. Gravitational radiation in the limit of high frequency. II. Nonlinear terms and the effective stress tensor. *Phys. Rev.* **166**, 1272–1279.

Ives, H.E., and Stilwell, G.R. 1938. An experimental study of the rate of a moving atomic clock. *J. Opt. Soc. Am.* **28**, 215–226.

Iyer, B.R. and Will, C.M. 1995. Post-Newtonian gravitational radiation reaction for two-body systems: Nonspinning bodies. *Phys. Rev. D* **52**, 6882–6893.

Jackson, J.D. 1998. *Classical Electrodynamics*. Third Edition. Wiley.

Jordan, P. 1959. Zum gegenwärtigen Stand der Diracschen kosmologischen Hypothesen. *Z. Phys.* **157**, 112–121.

Kennefick, D. 2005. Einstein versus the *Physical Review*. *Physics Today* **58**(9), 43–48.

Kennefick, D. 2007. *Traveling at the Speed of Thought: Einstein and the Quest for Gravitational Waves*. Princeton University Press.

Kennefick, D. 2009. Testing relativity from the 1919 eclipse – A question of bias. *Physics Today* **62**(3), 37–42.

Kidder, L.E. 1995. Coalescing binary systems of compact objects to 2.5 post-Newtonian order. V. Spin effects. *Phys. Rev. D* **52**, 821–847.

Kopal, Z. 1959. *Close Binary Systems*. Chapman and Hall.

Kopal, Z. 1978. *Dynamics of Close Binary Systems*. Reidel.

Kozai, Y. 1962. Secular perturbations of asteroids with high inclination and eccentricity. *Astron. J.* **67**, 591–598.

Kramer, M., Stairs I.H., Manchester, R.N., *et al.* 2006. Tests of general relativity from timing the double pulsar. *Science* **314**, 97–102.

Kundu, P.K., Cohen, I.M., and Dowling, D.R. 2011. *Fluid Mechanics*. Fifth Edition. Academic Press.

Landau, L.D. 1932. On the theory of stars. *Phys. Z. Sowjetunion* **1**, 285–288.

Landau, L.D. and Lifshitz, E.M. 1976. *Mechanics*. Third Edition. Butterworth-Heinemann.

Landau, L.D. and Lifshitz, E.M. 1987. *Fluid Mechanics*. Second Edition. Butterworth-Heinemann.

Landau, L.D. and Lifshitz, E.M. 2000. *The Classical Theory of Fields*. Fourth Edition. Butterworth-Heinemann.

Lattimer, J.M. and Prakash, M. 2001. Neutron star structure and the equation of state. *Astrophys. J.* **550**, 426–442.

Lattimer, J.M. and Prakash, M. 2007. Neutron star observations: Prognosis for equation of state constraints. *Phys. Report* **442**, 109–165.

Lidov, M.L. 1962. The evolution of orbits of artificial satellites of planets under the action of gravitational perturbations of external bodies. *Planetary and Space Science* **9**, 719–759.

Lincoln, C.W. and Will, C.M. 1990. Coalescing binary systems of compact objects to 2.5 post-Newtonian order: Late-time evolution and gravitational-radiation emission. *Phys. Rev. D* **42**, 1123–1144.

Lorentz, H.A. and Droste, J. 1917. The motion of a system of bodies under the influence of their mutual attraction, according to Einstein's theory. *Versl. K. Akad. Wetensch. Amsterdam* **26**, 392. English translation in Lorentz, H.A. 1937. *Collected papers*, Vol. 5, edited by Zeeman, P. and Fokker, A.D. Martinus Nijhoff.

Love, A.E.H. 1911. *Some Problems of Geodynamics*. Cambridge University Press.

Maggiore, M. 2007. *Gravitational Waves*. Volume 1: *Theory and Experiments*. Oxford University Press.

Martel, K. and Poisson, E. 2001. Regular coordinate systems for Schwarzschild and other spherical spacetimes. *Am. J. Phys.* **69**, 476–480.

Mathisson, M. 1937. Neue Mechanik materieller Systeme. *Acta Phys. Polon.* **6**, 163–200.

Mccully, J.G. 2006. *Beyond the Moon: A Conversational, Common Sense Guide to Understanding the Tides*. World Scientific.

Merkowitz, S. 2010. Tests of gravity using lunar laser ranging. *Living Rev. Relativity* **13**. `http://www.livingreviews.org/lrr-2010-7`.

Mikheev, S.P. and Smirnov, A.Yu. 1985. Resonant amplification of neutrino oscillations in matter and spectroscopy of solar neutrinos. *Yad. Fiz.* **42**, 1441–1448. [*Sov. J. Nucl. Phys.* **42**, 913–917.]

Mikheev, S.P. and Smirnov, A.Yu. 1986. Resonant amplification of neutrino oscillations in matter and solar neutrino spectroscopy. *Nuovo Cimento C* **9**, 17–26.

Misner, C.W., Thorne, K.S., and Wheeler, J.A. 1973. *Gravitation*. Freeman.

Mora, T. and Will, C.M. 2004. Post-Newtonian diagnostic of quasiequilibrium binary configurations of compact objects. *Phys. Rev. D* **69**, 104021 (25 pages).

Moulton, F.R. 1984. *An Introduction to Celestial Mechanics*. Second Revised Edition. Dover.

Murray, C.D. and Dermott, S.F. 2000. *Solar System Dynamics*. Cambridge University Press.

Narayanan, A.S. 2012. *An Introduction to Waves and Oscillations in the Sun*. Springer.

Newton, I. 1999. *The Principia: Mathematical Principles of Natural Philosophy*. Translated and edited by Cohen, I.B., Whitman, A., and Budenz, J. University of California Press.

Nordström, G. 1913. Zur Theorie des Gravitation vom Standpunkt des Relativitätsmechanik. *Ann. Physik* **42**, 533–554.

Nordtvedt Jr, K. 1968a. Equivalence principle for massive bodies. I. Phenomenology. *Phys. Rev.* **169**, 1014–1016.

Nordtvedt Jr, K. 1968b. Equivalence principle for massive bodies. II. Theory. *Phys. Rev.* **169**, 1017–1025.

Nordtvedt Jr, K. 1968c. Testing relativity with laser ranging to the Moon. *Phys. Rev.* **170**, 1186–1187.

Nordtvedt Jr, K. 1999. 30 years of lunar laser ranging and the gravitational interaction. *Class. Quantum Grav.* **16**, A101–A112.

Owen, B.J. 2005. Maximum elastic deformations of compact stars with exotic equations of state. *Phys. Rev. Lett.* **95**, 211101 (4 pages).

Painlevé, P. 1921. La mécanique classique et la théorie de la relativité. *C. R. Acad. Sci. (Paris)*, **173**, 677–680.

Papapetrou, A. 1951. Spinning test-particles in general relativity. I. *Proc. Roy. Soc. London* **A209**, 248–258.

Pati, M.E. and Will, C.M. 2000. Post-Newtonian gravitational radiation and equations of motion via direct integration of the relaxed Einstein equations: Foundations. *Phys. Rev. D* **62**, 124015 (28 pages).

Pati, M.E. and Will, C.M. 2001. Post-Newtonian gravitational radiation and equations of motion via direct integration of the relaxed Einstein equations. II. Two-body equations of motion to second post-Newtonian order, and radiation-reaction to 3.5 post-Newtonian order. *Phys. Rev. D* **65**, 104008 (21 pages).

Peters, P.C. 1964. Gravitational radiation and the motion of two point masses. *Phys. Rev.* **136**, B1224–B1232.

Peters, P.C. and Mathews, J. 1963. Gravitational radiation from point masses in a Keplerian orbit. *Phys. Rev.* **131**, 435–440.

Pirani, F.A.E. 1964. Introduction to gravitational radiation theory, in *Lectures on General Relativity*, edited by Trautman, A., Pirani, F.A.E., and Bondi, H., 249–273. Prentice-Hall.

Pound, A. 2010. Motion of small bodies in general relativity: Foundations and implementations of the self-force. PhD thesis, University of Guelph. Available online at `arXiv.org/abs/1006.3903`.

Pugh, G.E. 1959. Proposal for a satellite test of the Coriolis predictions of general relativity. Weapons System Evaluation Group, Research Memorandum No. 111, Department of Defense (unpublished). Reprinted (2003) in *Nonlinear Gravitodynamics. The Lense–Thirring Effect*, edited by Ruffini, R.J. and Sigismondi, C., 414–426. World Scientific.

Racine, E. and Flanagan, E.E. 2005. Post-1-Newtonian equations of motion for systems of arbitrarily structured bodies. *Phys. Rev. D* **71**, 044010 (44 pages).

Reif, F. 2008. *Fundamentals of Statistical and Thermal Physics*. Waveland Pr. Inc.

Rindler, W. 1991. *Introduction to Special Relativity*. Second Edition. Oxford University Press.

Robertson, H.P. and Noonan, T.W. 1968. *Relativity and Cosmology*. W.B. Saunders.

Rosenblum, A. 1978. Gravitational radiation energy loss in scattering problems and the Einstein quadrupole formula. *Phys. Rev. Lett.* **41**, 1003–1005.

Rossi, B. and Hall, D.B. 1941. Variation of the rate of decay of mesotrons with momentum. *Phys. Rev.* **59**, 223–228.

Sachs, R.K. 1961. Gravitational waves in general relativity. VI. The outgoing radiation condition. *Proc. Roy. Soc. London* **A264**, 309–338.

Sachs, R.K. 1962. Gravitational waves in general relativity. VIII. Waves in asymptotically flat space-time. *Proc. Roy. Soc. London* **A270**, 103–126.

Saulson, P.R. 1994. *Fundamentals of Interferometric Gravitational Wave Detectors*. World Scientific.

Schäfer, G. 1983. On often used gauge transformations in gravitational radiation-reaction calculations. *Lett. Nuovo Cimento* **36**, 105–108.

Schiff, L.I. 1960. Motion of a gyroscope according to Einstein's theory of gravitation. *Proc. Nat. Acad. Sci. U.S.* **46**, 871–882.

Schneider, P., Ehlers, J., and Falco, E.E. 1992. *Gravitational Lenses*. Springer.

Schutz, B.F. 2003. *Gravity from the Ground Up*. Cambridge University Press.

Schutz, B.F. 2009. *A First Course in General Relativity*. Second Edition. Cambridge University Press.

Schwarzschild, K. 1916. Uber das Gravitationsfeld eines Massenpunktes nach der Einsteinschen Theorie. *Sitzber. Deut. Akad. Wiss. Berlin, Kl. Math.-Phys. Tech.*, 189–196. For an English translation, see `arXiv.org/abs/physics/9905030`.

Sellier, A. 1994. Hadamard's finite part concept in dimension $n \geq 2$, distributional definition, regularization forms and distributional derivatives. *Proc. R. Soc. London*, **A445**, 69–98.

Shapiro, I.I. 1964. Fourth test of general relativity. *Phys. Rev. Lett.* **13**, 789–791.

Shapiro, I.I., Pettengill, G.H., Ash, M.E., *et al.* 1968. Fourth test of general relativity: Preliminary results. *Phys. Rev. Lett.* **20**, 1265–1269.

Shapiro, I.I., Reasenberg, R.D., Chandler, J.F., and Babcock, R.W. 1988. Measurement of the de Sitter precession of the Moon: A relativistic three-body effect. *Phys. Rev. Lett.* **61**, 2643–2646.

Shapiro, S.L. and Teukolsky, S.A. 1983. *Black Holes, White Dwarfs and Neutron Stars: The Physics of Compact Objects*. Wiley.

Shapiro, S.S., Davis, J.L., Lebach, D.E., and Gregory, J.S. 2004. Measurement of the solar gravitational deflection of radio waves using geodetic very-long-baseline interferometry data, 1979–1999. *Phys. Rev. Lett.* **92**, 121101 (4 pages).

Smith, S.F. and Havas, P. 1965. Effects of gravitational radiation reaction in the general relativistic two-body problem by a Lorentz-invariant approximation method. *Phys. Rev.* **138**, B495–B508.

Soffel, M.H. 1989. *Relativity in Astrometry, Celestial Mechanics and Geodesy*. Springer-Verlag.

Sotiriou, T.P. and Faraoni, V. 2012. Black holes in scalar–tensor gravity. *Phys. Rev. Lett.* **108**, 081103 (4 pages).

Stairs, I.H., Faulkner, A.J., Lyne, A.G., *et al.* 2005. Discovery of three wide-orbit binary pulsars: Implications for binary evolution and equivalence principles. *Astrophys. J.* **632**, 1060–1068.

Steiner, A.W., Lattimer, J.M., and Brown, E.F. 2010. The equation of state from observed masses and radii of neutron stars. *Astrophys. J.* **722**, 33–54.

Steves, B.A. and Maciejewski, A.J. 2001. *The Restless Universe: Applications of Gravitational N-Body Dynamics to Planetary, Stellar and Galactic Systems*. Institute of Physics.

Su, Y., Heckel, B.R., Adelberger, E.G., *et al.* 1994. New tests of the universality of free fall. *Phys. Rev. D* **50**, 3614–3636.

Tassoul, J.L. 1978. *Theory of Rotating Stars*. Princeton University Press.

Taylor, J.H., Fowler, L.A., and McCulloch, P.M. 1979. Measurements of general relativistic effects in the binary pulsar PSR 1913+16. *Nature* **277**, 437–440.

Taylor, S. and Poisson, E. 2008. Nonrotating black hole in a post-Newtonian tidal environment. *Phys. Rev. D* **78**, 084016 (26 pages).

Thirring, H. and Lense, J. 1918. Uber den Einfluss der Eigenrotation der Zentralkörper auf die Bewegung der Planeten und Monde nach des Einsteinschen Gravitationstheorie. *Phys. Z.* **19**, 156–163.

Thorne, K.S. 1969. Nonradial pulsation of general-relativistic stellar models. IV. The weak-field limit. *Astrophys. J.* **158**, 997–1019.

Thorne, K.S. 1980. Multipole expansions of gravitational radiation, *Rev. Mod. Phys.* **52**, 299–340.

Thorne, K.S. and Kovacs, S.J. 1975. The generation of gravitational waves. I. Weak-field sources. *Astrophys. J.* **200**, 245–262.

Thorne, K.S. and Will, C.M. 1970. Theoretical frameworks for testing relativistic gravity. I. Foundations. *Astrophys. J.* **163**, 595–610.

Tooper, R.F. 1965. Adiabatic fluid spheres in general relativity. *Astrophys. J.* **142**, 1541–1562.

Turner, M. 1977. Tidal generation of gravitational waves from orbiting Newtonian stars. I. General formalism. *Astrophys. J.* **216**, 914–929.

Ushomirsky, G., Cutler, C., and Bildsten, L. 2000. Deformations of accreting neutron star crusts and gravitational wave emission. *Mon. Not. R. Astr. Soc.* **319**, 902–932.

Wagoner, R.V. and Will, C.M. 1976. Post-Newtonian gravitational radiation from orbiting point masses, *Astrophys. J.* **210**, 764–775.

Wahlquist, H. 1987. The Doppler response to gravitational waves from a binary star source. *Gen. Relativ. Gravit.* **19**, 1101–1113.

Wald, R.M. 1984. *General Relativity*. University of Chicago Press.

Walker, M. and Will, C.M. 1980. The approximation of radiative effects in relativistic gravity: Gravitational radiation reaction and energy loss in nearly Newtonian systems. *Astrophys. J.* **242**, L129–L133.

Walsh, D., Carswell, R.F., and Weymann, R.J. 1979. 0957 + 561 A, B – Twin quasistellar objects or gravitational lens. *Nature* **279**, 381–384.

Weinberg, S. 1972. *Gravitation and Cosmology*. Wiley.

Weisberg, J.M., Nice, D.J., and Taylor, J.H. 2010. Timing measurements of the relativistic binary pulsar PSR B1913+16. *Astrophys. J.* **722**, 1030–1034.

Whitehead, A.N. 1922. *The Principle of Relativity, with Applications to Physical Science*. Cambridge University Press.

Whitrow, G.J. and Morduch, G.E. 1965. Relativistic theories of gravitation: A comparative analysis with particular reference to astronomical tests. *Vistas in Astronomy* **6**, 1–67.

Will, C.M. 1971a. Theoretical frameworks for testing relativistic gravity. II. Parameterized post-Newtonian hydrodynamics, and the Nordtvedt effect. *Astrophys. J.* **163**, 611–628.

Will, C.M. 1971b. Theoretical frameworks for testing relativistic gravity. III. Conservation laws, Lorentz invariance, and values of the PPN parameters. *Astrophys. J.* **169**, 125–140.

Will, C.M. 1971c. Relativistic gravity in the solar system. II. Anisotropy in the Newtonian gravitational constant. *Astrophys. J.* **169**, 141–155.

Will, C.M. 1983. Tidal gravitational radiation from homogeneous stars. *Astrophys. J.* **274**, 858–874.

Will, C.M. 1988. Henry Cavendish, Johann von Soldner, and the deflection of light. *Am. J. Phys.* **56**, 413–415.

Will, C.M. 1993. *Theory and Experiment in Gravitational Physics*. Revised Edition. Cambridge University Press.

Will, C.M. 2005. Post-Newtonian gravitational radiation and equations of motion via direct integration of the relaxed Einstein equations. III. Radiation reaction for binary systems with spinning bodies. *Phys. Rev. D* **71**, 084027 (15 pages).

Will, C.M. 2006a. Special relativity: A centenary perspective. *Einstein 1905–2005: Poincaré Seminar 2005*, edited by Damour, T., Darrigol, O., Duplantier, B. and Rivasseau, V., 33–58. Birkhäuser Publishing.

Will, C.M. 2006b. The confrontation between general relativity and experiment. *Living Rev. Relativity* **9**. `http://www.livingreviews.org/lrr-2006-3/`.

Will, C.M. 2010. Resource letter PTG-1: Precision tests of gravity. *Am. J. Phys.* **78**, 1240–1247.

Will, C.M. and Nordtvedt Jr, K. 1972a. Conservation laws and preferred frames in relativistic gravity. I. Preferred-frame theories and an extended PPN formalism. *Astrophys. J.* **177**, 757–774.

Will, C.M. and Nordtvedt Jr, K. 1972b. Conservation laws and preferred frames in relativistic gravity. II. Experimental evidence to rule out preferred-frame theories of gravity. *Astrophys. J.* **177**, 775–792.

Will, C.M. and Wiseman, A.G. 1996. Gravitational radiation from compact binary systems: Gravitational waveforms and energy loss to second post-Newtonian order. *Phys. Rev. D* **54**, 4813–4848.

Williams, J.G., Turyshev, S.G., and Boggs, D.H. 2009. Lunar laser ranging tests of the equivalence principle with the Earth and Moon. *Int. J. Mod. Phys. D* **18**, 1129–1175.

Wiseman, A.G. 1992. Coalescing binary systems of compact objects to (post)$^{5/2}$-Newtonian order. II. Higher-order wave forms and radiation recoil. *Phys. Rev. D* **46**, 1517–1539.

Wiseman, A.G. and Will, C.M. 1991. Christodoulou's nonlinear gravitational-wave memory: Evaluation in the quadrupole approximation. *Phys. Rev. D* **44**, R2945–R2949.

Wolfenstein, L. 1978. Neutrino oscillations in matter. *Phys. Rev. D* **17**, 2369–2374.

Zlochower, Y., Campanelli, M. and Lousto, C.O. 2011. Modeling gravitational recoil from black-hole binaries using numerical relativity. *Class. Quantum Grav.* **28**, 114015 (11 pages).

Index

Printed in the United States
By Bookmasters